Lecture Notes in Computer Science 2775
Edited by G. Goos, J. Hartmanis, and J. van Leeuwen

Springer
Berlin
Heidelberg
New York
Hong Kong
London
Milan
Paris
Tokyo

Marco Conti Silvia Giordano
Enrico Gregori Stephan Olariu (Eds.)

Personal Wireless Communications

IFIP-TC6 8th International Conference, PWC 2003
Venice, Italy, September 23-25, 2003
Proceedings

Springer

Series Editors

Gerhard Goos, Karlsruhe University, Germany
Juris Hartmanis, Cornell University, NY, USA
Jan van Leeuwen, Utrecht University, The Netherlands

Volume Editors

Marco Conti
Enrico Gregori
Consiglio Nazionale delle Ricerche (CNR)
Istituto di Informatica e Telematica (IIT)
Via G. Moruzzi, 1, 56124 Pisa, Italy
E-mail: {marco.conti;enrico.gregori}@iit.cnr.it

Silvia Giordano
Scuola Universitaria Professionale della Svizzera italiana (SUPSI)
Dipartimento di Informatica ed Elettronica
Galleria 2, 6928 Manno, Switzerland
E-mail: silvia.giordano@die.supsi.ch

Stephan Olariu
Old Dominion University
Department of Computer Science
Norfolk, VA 23529-0162, USA
E-mail: olariu@cs.odu.edu

Cataloging-in-Publication Data applied for

A catalog record for this book is available from the Library of Congress.

Bibliographic information published by Die Deutsche Bibliothek
Die Deutsche Bibliothek lists this publication in the Deutsche Nationalbibliografie; detailed bibliographic data is available in the Internet at <http://dnb.ddb.de>.

CR Subject Classification (1998): C.2, H.4, H.3, D.2

ISSN 0302-9743
ISBN 3-540-20123-8 Springer-Verlag Berlin Heidelberg New York

Springer-Verlag Berlin Heidelberg New York
a member of BertelsmannSpringer Science+Business Media GmbH

http://www.springer.de

Typesetting: Camera-ready by author, data conversion by PTP-Berlin GmbH
Printed on acid-free paper SPIN: 10931400 06/3142 5 4 3 2 1 0

Preface

This book constitutes the refereed proceedings of the IFIP-TC6 Eighth International Conference on Personal Wireless Communications, PWC 2003. PWC 2003 is the flagship conference of the IFIP Working Group 6.8, *Mobile and Wireless Communications*, and is the premier international forum for discussions between researchers, practitioners, and students interested in the symbiosis of mobile computing and wireless networks.

It is a great pleasure to present the PWC 2003 technical program. This year the conference received 115 submissions from 27 countries indicating that PWC is a reference conference for worldwide researchers from the wireless and mobile community. With so many papers to choose from, the Technical Program Committee's job, to provide a conference program of the highest technical quality, was challenging and time consuming. From the 115 submissions, we finally selected 34 full papers and 15 short papers for presentation in the conference technical sessions.

The conference technical program was split into three days, and included, in addition to the 49 refereed contributions, 4 invited papers from top-level researchers from the mobile and wireless community.

To give researchers the opportunity to present ongoing work, and the novel ideas they are starting to explore, we included in the technical program two work-in-progress sessions and two novel-ideas sessions. The technical program also included a poster session devoted to presenting ongoing research projects on wireless and mobile communications.

This event would not have been possible without the enthusiastic and hard work of a number of colleagues. First of all, I would like to thank the special tracks chairs – Silvia Giordano and Stephan Olariu – for their valuable contribution in organizing two special sessions on very hot topics in mobile and wireless research: *Sensor Networks* and *Security and Cooperation in Ad Hoc Networks*. A special thanks to the TPC members, and all the referees, for their invaluable help in reviewing the papers for PWC 2003. I would also like to thank all the authors that submitted their papers to this conference for their interest and time. Last but not least, I would like to thank Patrizia Andronico and Alessandro Urpi. Patrizia designed and managed the PWC 2003 web site, while Alessandro designed and implemented *Colibrì*, the efficient system for the handling of the electronic submissions.

July 2003 Marco Conti

Message from the General Chairs

Personal Wireless Communications (PWC) is the flagship conference of the IFIP working group WG 6.8 (Wireless and Mobile Communications).

PWC 2003 was organized by IIT (Istituto di Informatica e Telematica) of the Italian National Research Council and by the Learning Services division of Telecom Italia.

The organization of such a complex event required a major effort and we wish to express our sincere appreciation to all the executive committee members for their excellent work.

We would like to express our special appreciation to the main conference technical program chair Marco Conti and to the special tracks chairs: Silvia Giordano and Stefan Olariu. The overall high quality of the conference technical sessions is the result of a complex evaluation process that they handled in an excellent way.

We are also indebted to our organizations, IIT-CNR and Telecom Italia Learning Services. IIT-CNR not only allowed Enrico Gregori and Marco Conti to dedicate considerable time to the organization of this event, but also financially supported the event. Telecom Italia provided not only relevant expertise but also an excellent conference center; the Telecom Italia Future Centre in Venice.

July 2003

Enrico Gregori
Fabrizio Davide

Organizers

Organization

Conference Executive Committee

General Chair:
Enrico Gregori, IIT-CNR, Italy

General Vice-Chair:
Fabrizio Davide, Telecom Italia, Italy

Technical Program Chair:
Marco Conti, IIT-CNR, Italy

Special Tracks Program Chairs:
Silvia Giordano, SUPSI, Switzerland
Stephan Olariu, Old Dominion University, USA

Publicity Co-chairs:
Raffaele Bruno, IIT-CNR, Italy
Alexandros Kaloxylos, University of Athens, Greece

Steering Committee Members:

Imrich Chlamtac, University of Trento, Italy (Chair)
Jon Crowcroft, University of Cambridge, UK
Sajal K. Das, The University of Texas at Arlington, USA
Anthony Ephremides, University of Maryland, USA
K. Ben Letaief, Hong Kong University, Hong Kong
Guy Omidyar, National University of Singapore, Singapore
Adam Wolisz, Technical University of Berlin, Germany

Electronic Submission:
Alessandro Urpi, University of Pisa, Italy

Web Designer:
Patrizia Andronico, IIT-CNR, Italy

Program Committee

Arup Acharya, IBM T.J. Watson Research, USA
Eitan Altman, INRIA, France
Giuseppe Anastasi, University of Pisa, Italy
Andrea Baiocchi, University of Rome "La Sapienza"
Roberto Battiti, University of Trento, Italy

Elizabeth M. Belding-Royer, UC Santa Barbara, USA
Giuseppe Bianchi, University of Palermo, Italy
Luciano Bononi, University of Bologna, Italy
Azzedine Boukerche, University of North Texas, USA
Raffaele Bruno, IIT-CNR, Italy
Levente Buttyan, Budapest University of Technology and Economics, Hungary
Andrew T. Campbell, Columbia University, USA
Carla-Fabiana Chiasserini, Politecnico di Torino, Italy
Raimo Kantola, Helsinki University of Technology, Finland
Holger Karl, Technische Universität Berlin, Germany
Mohan Kumar, University of Texas at Arlington, USA
Laura M. Feeney, SICS, Sweden
Gabor Fodor, Ericsson, Sweden
Luigi Fratta, Politecnico di Milano, Italy
Per Gunningberg, Uppsala University, Sweden
K. Ben Letaief, Hong Kong University, Hong Kong
Andrey Lyakhov, Russian Academy of Sciences, Russia
Martin Mauve, University of Mannheim, Germany
Franco Mazzenga, University of Rome Tor Vergata, Italy
Enzo Mingozzi, University of Pisa, Italy
Refik Molva, Institut EURECOM, France
Hiroyuki Morikawa, University of Tokyo, Japan
Sergio Palazzo, University of Catania, Italy
Michael Paterakis, Technical University of Crete, Greece
Niovi Pavlidou, Aristotle University of Thessaloniki, Greece
Apostolis Salkintzis, Motorola, Canada
Ioannis Stavrakakis, University of Athens, Greece
Ivan Stojmenovic, University of Ottawa, Canada
Samir Tohme, Ecole Nationale Superieure des Telecommunications (ENST), France
Christian Tschudin, University of Basel, Switzerland
Sami Uskela, Nokia Networks, Finland
Anna Vaccarelli, IIT-CNR, Italy
Andras Valko, Ericsson AB, Sweden
Hidetoshi Yokota, KDDI R&D Laboratories, Japan
Bernhard Walke, Aachen University of Technology, Germany

Referees

Arup Acharya
Eitan Altman
Giuseppe Anastasi
Andrea Baiocchi
Albert Banchs
Sandro Bartolini
Roberto R. Battiti
E.M. Belding-Royer
Paolo Bellavista
Silvello Betti
Mauro Biagi
Giuseppe Bianchi

Chatschik Bisdikian
Stefano Bistarelli
Li Bo
Luciano Bononi
Azzedine Boukerche
Mauro Brunato
Raffaele Bruno
Levente Buttyan
Andrew T. Campbell
Llorenç Cerdà
Rajiv Chakravorty
Carla F. Chiasserini
Ernestina Cianca
Marco Conti
Carlos M. Cordeiro
Gabriele D'Angelo
Andrea De Vendictis
Franca Delmastro
Andrea Detti
Felicita Di Giandomenico
Gianluca Dini
Khalil El-Khatib
Romano Fantacci
Laura M. Feeney
Mauro Femminella
Gabor A. Fodor
Luigi Fratta
Holger Füßler
Stefano Galli
Laura Galluccio
Michele Garetto
Leonidas Georgiadis
Paolo Giaccone
Mauro Giaconi
Romeo Giuliano
Per Gunningberg
S. Hadjiefthymiades
Annika Jonsson
Athanasios G. Kanatas
Raimo A. Kantola
Holger Karl
Bjorn Knutsson
Polychronis Koutsakis
Martin Kubisch
Mohan J. Kumar
Nikolaos Laoutaris
Conny J. Larsson
Khaled B. Letaief
Renato Lo Cigno
Pierpaolo Loreti
Andrey Lyakhov
Stefano Mangione
Fabio Martinelli
Ibrahim Matta
Martin Mauve
Franco Mazzenga
Tommaso Melodia
Pietro Michiardi
Armin Mikler
Enzo Mingozzi
Jelena Misic
Refik Molva
Alberto Montresor
Giacomo Morabito
Hiroyuki Morikawa
András Méhes
Giovanni Neglia
Jan Nilsson
Pavan Nuggehalli
Sergio Palazzo
Andrea Passarella
Nikos Passas
Michael Paterakis
Niovi Pavlidou
Davide Pedroni
Marinella Petrocchi
Christos Politis
Spyros M. Psychis
Fabrizio Quaglia
Andras Racz
Ram Ramanathan
Franklin Reynolds
Christian Rohner
Davide Rossi
Yves Roudier
Apostolis Salkintzis
Theodoros Salonidis
Stefano Salsano
Giovanni Santella
Giovanni Schembra
Nikos D. Sidiropoulos
Vikram Srinivasan
Ioannis Stavrakakis
Ivan Stojmenovic
Ilenia Tinnirello
Alfredo Todini
Samir Tohme
Matthias Transier
Christian F. Tschudin
Yu-Chee Tseng
Giovanni Turi
Alessandro Urpi
Sami Uskela
Anna Vaccarelli
Francesco Vacirca
Athanasios Vaios
Andras Valko
Alessio Vecchio
Bernhard H. Walke
Nathalie Weiler
Jörg Widmer
Hidetoshi Yokota
Gianluigi Zavattaro

Table of Contents

Collaborative and Distributed Computation in Mesh-Like Wireless Sensor Arrays*

Mitali Singh[1], Viktor K. Prasanna[1], Jose Rolim[2], and Cauligi S. Raghavendra[1]

[1] University of Southern California
Department of Computer Science
Los Angeles, CA 90089, USA
{mitali,prasanna,raghu}@usc.edu
http://pacman.usc.edu

[2] University of Geneva
Department of Informatics
Geneva 4, Switzerland
rolim@cui.unige.ch

Abstract. Sensor networks are being used for implementation of a large number of applications involving distributed and collaborative computation. Extensive research has focused upon design of time optimal parallel and distributed algorithms for two dimensional mesh connected computers (MCC). In this paper, we discuss a simple scheme for emulating the above algorithms for mesh-like sensor arrays. We show that a large set of parallel algorithms (see Property 1), that take time $T(n)$ on MCC of size n, can be implemented on a wireless sensor mesh of size n in time $O(r^2 + T(n/r^2).r^2)$. Here r represents the transmission range of the sensor nodes. We discuss implementation of algorithms for ranking and sorting using our techniques and analyze them for time and energy efficiency.

1 Introduction

Sensor networks can be considered as large scale dynamically configurable, distributed systems, where autonomous nodes (sensor nodes) collaborate among themselves to achieve a larger objective. Such networks have revolutionized data gathering and processing, and enabled a large range of applications such as unattended environment monitoring, traffic control, automatic target recognition, building vigilance, and hazard detection. In recent years, sensor networks have gained tremendous popularity in both the research community and the industry. Efforts are being made to design sensor nodes that are small, low cost and yet have large functionality built into them. Sensor applications often require deployment of nodes in inaccessible, remote areas. This makes energy management critical in sensor networks, where the functionality of the network is limited by

* This work is supported by the DARPA Power Aware Computing and Communication Program under contract no. F33615-C-00-1633.

M. Conti et al. (Eds.): PWC 2003, LNCS 2775, pp. 1–11, 2003.

the battery life of the nodes. Several research projects [9] [14] [16] [18] are focusing on design of energy efficient hardware and software for sensor applications.

We approach sensor networking from a parallel and distributed system's perspective and focus on developing algorithms for sensor networks using formal analysis. We observe that most of the general models of computation must be redefined in the context of sensor networks. Wireless communication and energy constraints are the two major factors that are responsible for the above. Moreover, while some sensor networks have regular topologies (such as traffic monitoring sensor nodes on cross-streets), most involve an ad hoc deployment of sensor nodes. Design of distributed algorithms for such networks is very challenging. However, we observe that a mesh like topology is intrinsic to densely populated sensor nodes uniformly distributed in a two dimensional plane (see Figure 2). The overheads for maintaining the topology are minimal if the network has been localized [4], synchronized [5] and each sensor node is aware of its location in an absolute or relative scale. We assume sensor nodes to be organized in mesh-like wireless sensor arrays and investigate design of time and energy efficient distributed algorithms for these systems.

A large number of distributed algorithms have been designed and analyzed in the past for time optimal implementation over two dimensional Mesh Connected computers (MCC) [7] [8] [10] [13]. We discuss a simple scheme for adapting these algorithms for implementation in wireless sensor networks. We observe that the efficiency of the algorithms is largely influenced by the transmission range of the sensor nodes. Larger range reduces the diameter of the network but also results in higher interference. This reduces the number of sensor nodes that can transmit concurrently. Moreover, larger range results in higher energy dissipation. We demonstrate that any algorithm with execution time $T(n)$ on MCC of size n can be emulated on a wireless mesh in time $O(n + T(n/r^2).r^2)$, where r is the transmission range and n is the total number of sensor nodes.

As an illustration, we discuss time optimal implementations of algorithms for sorting and ranking, and also analyze them for overall energy dissipation. We demonstrate that ranking can be performed in time $O(r^2 + \sqrt{n}/r.)$ with overall energy dissipation $O(n.r^2)$. Our algorithm for sorting executes in time $O(r^2 + r.\sqrt{n})$ and energy $O(n.r^2 + r.n\sqrt{n})$. We observe the results are asymptotically the same as the sorting algorithm for wireless networks discussed in [3].

The rest of the paper is organized as follows. Our model is presented in Section 2. The algorithms for ranking and sorting are discussed in Section 3. Finally, we conclude in Section 4.

2 Our Model

For our analysis, we consider a uniform distribution of sensor nodes over a two dimensional square plane of size $\sqrt{n} \times \sqrt{n}$ as illustrated in Figure 1. The sensor nodes communicate with each other over a single wireless channel with fixed transmission range r. We divide the plane into unit area cells such that each cell contains at least one sensor node. To ensure connectivity in the network

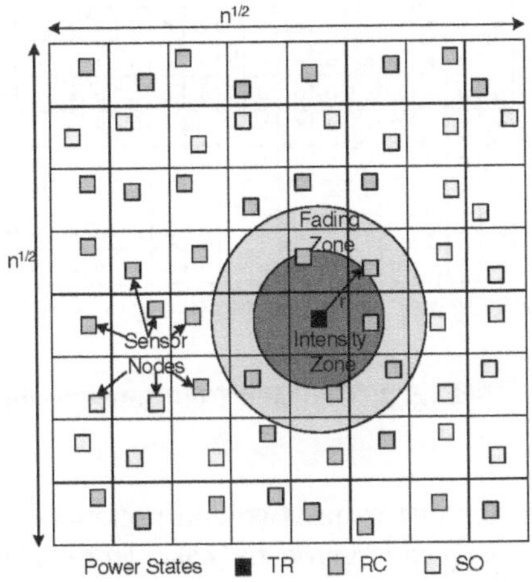

Fig. 1. Sensor network model

we assume $r \geq 1$. The sensor nodes, thus form a mesh like topology over the network. A similar topology is assumed by the Wireless Sensor Network (WSN) model described in [1]. Our model augments the WSN model with energy costs. Some of the key assumptions of our model are discussed below.

- **Network features:** We assume a homogeneous network comprising of n sensor nodes organized in a mesh like topology. Each sensor node has a unique id, and is aware of its relative position in the mesh. Energy efficient initialization schemes for wireless sensor networks are discussed in [11] [6]. The network is time synchronized. A sensor node has constant amount of memory denoted by $m > 1$, and **fixed** transmission radius $r \geq 1$.
- **Communication:** The sensor nodes communicate over a single wireless channel. The coverage area of a sensor with transmission range r, is defined as the distance till which it can be heard. All sensor nodes lying within distance r from the transmitting sensor are said to be in the *intensity zone* (see Figure 1) and are guaranteed to receive the transmission. The transmission signal strength reduces with distance as a function of r. We assume that it can be detected up to a distance $2r$ after which the signal power is below the reception level. The sensor nodes lying within distance $r < d \leq 2r$ lie in the *fading zone*. A COLLISION is said to occur if a sensor node lies within the coverage area of two or more transmitting sensor nodes. In any time step a sensor can receive or transmit one unit of data.
- **Power States:** A sensor node can be in three power states: transmit (TC), receive (RC), or switched off (SO) power state. A sensor can transmit in state TC and receive in state RC. No operation can be performed in state SO. Energy dissipation in state SO is considered to be negligible.
- **Communication Energy:** The communication energy of the network consists of the total transmission energy and the reception energy. Power dis-

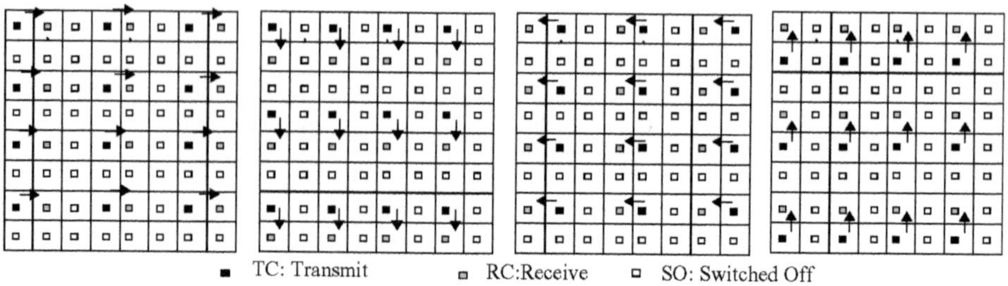

Fig. 2. Collision free transmission

sipation at the transmitter has two components, P_{Tx} and P_{Rad}. The power dissipation at a receiver is given by P_{Rx}. Here, P_{Tx} and P_{Rx} represent the power dissipated in processing a (transmitted or received) packet. This value depends on the electronics of the radio and the type of radio components integrated (such as frequency synthesizers, mixers, modulators) in the module and is equivalent. P_{Rad} denotes the radiation power at the transmitter, which is a function of the transmission range r, and the environment-dependent path loss exponent α. Typically α lies between $1.7 - 4$ [17]. We assume $\alpha = 2$. We define one energy unit as the energy dissipated in receiving one unit of data. Transmission of one unit of data dissipates energy r^2

- **Overall Time and Energy:** Energy and time costs for communication are much larger than computation in state-of-the-art sensor systems [19]. Thus **we consider only communication time and energy** for analysis of our algorithms. We do not consider energy dissipation or time taken for computation or sensing of data at any sensor node.

3 Mesh Emulation

Extensive literature exists for time and work optimal distributed algorithms designed for mesh connected computers (MCC). We discuss a simple scheme for emulating the above algorithms on mesh-like sensor arrays. Our goal is to minimize overall execution time and energy. Sensor networks use wireless technology for communication. Thus, in order to minimize interference and collisions in the network, the transmission schedules must be defined such that no receiving sensor node is in the coverage area of more than one transmitting sensor node. The coverage area of a sensor node is determined by the transmission range r. For $r = 1$, the behavior of the network is very similar to MCC. Lemma 1 demonstrates that time taken for emulation of an MCC algorithm is of the same order. For $r > 1$, the network diameter is reduced, but interference is increased, and the performance bounds for emulating the algorithm are discussed in Theorem 1.

Lemma 1. *Any parallel algorithm for MCC with time complexity $T(n)$ can be emulated on sensor mesh array with $r = 1$ in time $O(T(n))$.*

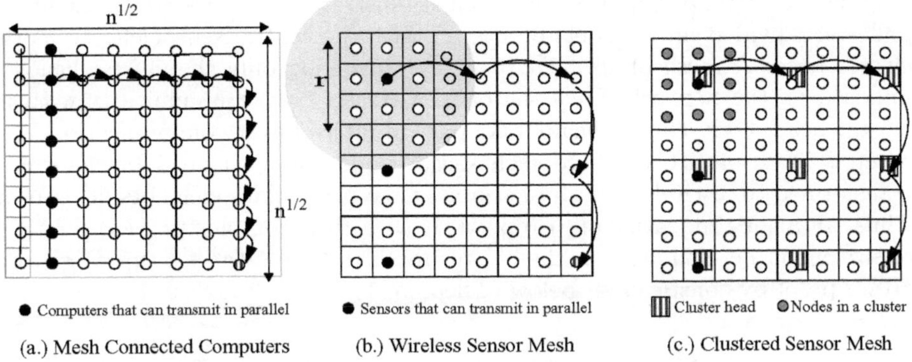

Fig. 3. Communication in MCC and sensor mesh

Proof: In the MCC model, a processing element can receive or transmit to each of its adjacent neighbors at any given time if the neighbors are idle. In a wireless mesh, the broadcast medium imposes some restrictions. To ensure collision free communication, no receiver should be in transmission range of more than one sensor node at any given time. This limits the number of sensor nodes that can broadcast concurrently reducing the bandwidth of the network. However, it can be shown that any operation of the MCC can be simulated on a wireless mesh in 24 time steps. In each time step, the mesh is assigned a specific power configuration, which limits the functionality of a sensor node in that time step. A sensor can transmit only in time steps, where it is scheduled to be in state TR. Figure 2 illustrates four time steps emulating four direction of communication. In any of the 4 time steps only $n/6$ sensor nodes can transmit. Thus, to emulate an operation of the MCC, 24 configurations are sufficient. The other 20 configurations can be obtained by changing the power states of the illustrated sensor nodes in a cyclic manner. Thus, time complexity of the algorithm when implemented on a wireless mesh is $24.T(n)$. □

Next we analyze the scenario where each sensor node has a transmission range $r > 1$. Each sensor can reach $O(r^2)$ adjacent sensor nodes in a single hop. In MCC, a processing element can communicate with only 4 adjacent processing elements in a single transmission. The network diameter of a $\sqrt{n} \times \sqrt{n}$ mesh is given by $\sqrt{n}$ in the MCC model and is reduced to $\sqrt{n}/r$ in a wireless mesh. Figure 3(a.) shows the number of hops required to traverse a mesh in MCC and Figure 3(b.) illustrates the number of hops required in a wireless mesh of same size. However, the broadcast medium of the wireless mesh imposes restrictions on the number of sensor nodes that can transmit concurrently. The coverage area of a sensor node is $O(r^2)$. Only one sensor in this area can transmit at a time. This reduces the mesh bandwidth by $O(r^2)$, as shown in Figure 3. All black processing elements in the MCC can transmit to their right neighbor in the same time slot. However, in the wireless mesh, the transmitters must be interleaved to ensure that the intended listener can only hear one transmission at a time.

This motivates a clustering approach as illustrated in Figure 3(c.). The sensor nodes are divided into clusters of size r^2, with the central sensor chosen as the cluster head. A mesh of size $\sqrt{n}/r \times \sqrt{n}/r$ involving only the cluster heads is imposed on the network. Each sensor in a cluster only communicates with its cluster head. Only cluster heads take part in inter-cluster communication. The wireless mesh of size n where each sensor node has a single data element is transformed to a wireless mesh of size n/r^2 having r^2 data elements per sensor. Thus, an algorithm satisfying Property 1, with time complexity $T(n)$ on MCC of size n can be emulated on the cluster mesh in time $O(r^2 + T(n/r^2).r^2)$. We give a proof by construction below (Theorem 1).

Property 1. Problem Size Linearity: An algorithm implemented on MCC is said to have problem size linearity if $T^p(n) = O(p.T(n))$. $T(n)$ represents the time complexity of the algorithm implemented on MCC of size n, where each processing element has a single unit of data. $T^p(n)$ denotes the time complexity of the algorithm implemented on a mesh of size n, where each processing element stores $p \geq 1$ data elements.

Remark 1. We observe that a very large set of algorithms satisfy problem size linearity. These include algorithms for problems such as sorting, ranking, matrix multiplication, sum, prefix sum, permutation routing, and matrix transposition.

Theorem 1. *A parallel algorithm implemented on a $\sqrt{n} \times \sqrt{n}$ MCC, that has problem size linearity and execution time $T(n)$ can be implemented on the mesh of sensor nodes of size $\sqrt{n} \times \sqrt{n}$ with transmission range r, in time $O(r^2 + T(n/r^2) \times r^2)$. Here, r represents the transmission range of a sensor.*

Proof: The proof is by construction. Consider an algorithm implemented on an $\sqrt{n} \times \sqrt{n}$ mesh with time complexity $T(n)$. We emulate the algorithm on an $\sqrt{n} \times \sqrt{n}$ sensor mesh, where each sensor node has a range r.

Step I: Divide the mesh into n/r^2 blocks of size $r \times r$. Let B_{ij} represent the block in the ith row and j^{th} column, where $1 \leq i, j \leq \sqrt{n}/r$. Node $N_{(i.r+\lceil r/2 \rceil),(j.r+\lceil r/2 \rceil)}$ is chosen to be the cluster head for sensor nodes in $B_{i,j}$ and is denoted by $S_{i,j}$. Initially all sensor nodes are in power state SO.

Step II: Data is aggregated from all sensor nodes in a block to the cluster head. Each block is scheduled to be either *active* or *inactive*. The power state of the sensor nodes in inactive blocks is SO. In each active block, the cluster head is in state RC for r^2 time slots. The $r^2 - 1$ sensor nodes transmit sequentially. To ensure collision free transmission, the active blocks must be interleaved by two inactive blocks as illustrated in Figure 4. This ensures that their is no interference from neighboring blocks at any time.

Step III: All sensor nodes except the cluster heads are put in state SO. Sensor nodes $S_{i,j}$ form a $\sqrt{n}/r \times \sqrt{n}/r$ mesh with r being the distance between adjacent cluster heads. Each cluster head represents r^2 data elements. The clusterhead mesh is similar to the sensor mesh with $r = 1$ as a sensor can only reach 4 adjacent blocks. A collision free emulation of an algorithm designed for MCC, can be ensured by the communication schedule defined earlier (in proof of Lemma 1) at the penalty of increase in time by a factor of 24.

The algorithm satisfies Property 1, thus on MCC of size n/r^2, where each processing element stores r^2, it takes time $O(T(n/r^2).r^2)$. Using Lemma 1, we can conclude that the time complexity of the simulation on the wireless mesh is given by $O(24.T(n/r^2).r^2)$.

Step IV: The cluster heads transmit the result to the sensor nodes using reverse scheduling of Step I.

Analysis: In step II, the aggregation completes in time $9r^2$. Step III takes time $O(T(n/r^2).r^2)$. Analysis of Step IV is analogous to step II. Thus, time complexity of the algorithm implemented on a wireless mesh of size n is given by $O(r^2 + T(n/r^2).r^2)$. □

Remark 2. Theorem 1 gives an upper bound on the time complexity of MCC algorithm implemented on a wireless mesh. It assumes that the cluster head represents a data set of size of $O(r^2)$ at all time steps. This is the worst case scenario. For several algorithms, the size of the data can be reduced. Consider the following two scenarios (1.) Only a fraction of sensor nodes per cluster contain useful data.(2.) Data aggregation can take place at the cluster head. Let us assume that only $1 \leq l \leq r^2$ sensor nodes per cluster collaborate. The data can be transmitted to the cluster head in time $O(l)$. Next the cluster head aggregates the data. For example, only retains the partial sum in an algorithm for summing. Then the data set represented by a cluster head is reduced to size $1 \leq k \leq l$. Thus, time taken by the algorithm is $O(l + T(n/r^2).k)$ where $1 \leq k \leq l \leq r^2$.

Remark 3. In our proof of Theorem 1, we considered data from all sensor nodes in a block to be collected at the cluster head. This implies, that the the cluster head must have a memory of size $O(r^2)$. The data aggregation at the cluster head was assumed to keep the proof simple. The memory constraint can be relaxed as follows. Each communication step at the cluster head is split into two steps. In the first step the cluster head transmits in its cluster, the id of the sensor whose data is required. All sensor nodes receive the message. In the next step, only the sensor that matches the id remains awake and transmits/receives the data from the cluster head. Every single communication step at the cluster head now requires three more communication steps instead of one. This increases the overall execution time of the algorithm only by a constant factor.

The above theorem is useful as it helps in evaluating the existing MCC algorithms for time performance when implemented on a wireless mesh and thus, in selection of the optimal candidate. We discuss implementations of algorithms for ranking and sorting and in a wireless, mesh-like sensor arrays.

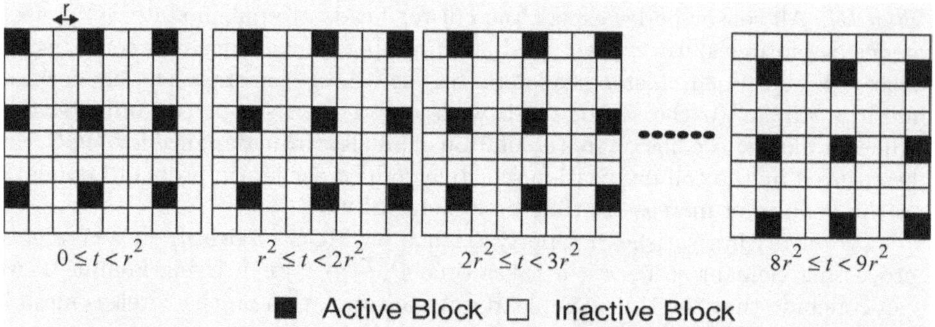


```
for t =0 to 9r^2-1 {
  for all 0 < i, j < (n^1/2/ r+1); 0 < k, l, < r^2 +1
     if (((i-1) mod 3= t / 3r^2) & ((j-1) mod 3= t/ r^2 mod 3) & (k= (t mod r^2/r + 1)& (l= t mod r +1))
         { State (Node i.r+k, j.r+1) = TR;
         State (S i, j) = RC ; }
     else{
         State (Node i.r+k, j.r+1) = SO;
         State (S i, j) = SO ; }
         }
```

Fig. 4. Power schedule for Step II

3.1 Ranking

The problem of ranking is described as follows. Given a set of n sensor node, determine the rank of a sensor node with id i, where $1 \leq i \leq n$. Ranking is a useful operation in several sensor applications. For example consider a scenario, where k highest power sensor nodes are required to be monitoring the field at any time. A sensor node may periodically poll the network to determine its rank to decide whether to monitor the field or go to sleep. Alternatively consider a scenario, where sensor nodes transmit in order of their rank. A sensor node must find its rank to determine its transmission slot. Ranking is also an important kernel for several filtering and image processing applications.

We consider the sensor field of size $\sqrt{n} \times \sqrt{n}$, where each sensor node has fixed transmission range r. Without loss of generality, let us assume sensor node $N_{1,1}$ requires to determine its rank.

Step I: Divide the mesh into blocks of size $r \times r$ and choose a cluster head $S_{i,j}$ for each block $B_{i,j}$. All sensor nodes except $N_{1,1}$ and $S_{1,1}$ are inactive. $N_{1,1}$ transmits its data to $S_{1,1}$.

Step II: In the $(\sqrt{n}/r)$ time steps, the value is transmitted to all the cluster heads in the row 1. In the next $\sqrt{n}/r$ time steps, the value is transmitted to all the cluster heads.

Step III: To avoid interference each block is scheduled to be active and inactive as illustrated in Figure 4 (discussed earlier in proof of Theorem 1). Each block is active for r^2 time steps. In the first time step the cluster head transmits the value to the cluster. All sensor nodes that have value larger than value transmitted transmit a response to the cluster head in their scheduled time slot. This can be accomplished in $r^2 - 1$ time steps without any collision. The cluster heads count the responses received. They compare the value to their own and increment the counter if required.

Step IV: The reverse schedule of step II is followed. Whenever, a cluster head receives a value, it adds its own counter to the value and transmits it to the next cluster head. This is accomplished in time $O(\sqrt{n}/r)$. $S_{1,1}$ transmits result to $N_{1,1}$.

Analysis: Each sensor node transmits and receives at most twice. Thus, energy of this algorithm is $O(n.r^2)$ and time taken is $O(r^2 + \sqrt{n}/r)$. The result is both time and energy optimal if the transmission range on the sensor nodes is fixed to r. The proof is trivial. Each sensor node must transmit its value at least once and each transmission costs energy r^2. Thus $E(n) = \Omega(n.r^2)$. The network diameter $\sqrt{n}$ must be traversed for data from $N_{1,1}$ to reach $N_{\sqrt{n},\sqrt{n}}$. Since the range of a sensor node is r, this requires at least $\sqrt{n}/r$ sequential operations. Thus, time complexity is $\Omega(\sqrt{n}/r)$ for $r < \sqrt{n}$. For $r \geq \sqrt{n}$, all sensor nodes can hear all transmissions. This implies only one sensor node may transmit at any time resulting in an overall time complexity of $\Omega(r^2)$.

3.2 Sorting

The problem of sorting n numbers is one of the most widely analyzed problem owing to its theoretical importance and use in a wide range of applications. Pixel sorting is required in several image compression and coding algorithms [15]. A sensor network can also be considered as a distributed data management system [2]. Sorting is an important kernel in a large number of data management and data mining applications. Sorting is useful when all sensors must determine their relative rank based on measured data or remaining battery power for filtering or network management applications. algorithms. The sorting problem is defined as follows. We are given a set of sensor nodes of n, where each sensor node contains a data element. Our goal is to redistribute the data among the sensors such that at the end of the sorting algorithm, for $1 \leq i \leq n$, sensor node with id i contains the data element with rank i.

Nassimi et al. [13] developed a parallel algorithm for bitonic sort that sorts the numbers on an $\sqrt{n} \times \sqrt{n}$ MCC in time $O(\sqrt{n})$. An implementation of the above algorithm on a wireless sensor mesh was discussed in [3]. The time complexity of the algorithm was demonstrated to be $O(r^2 + r\sqrt{n})$. Analysis using our model shows that the energy complexity of this algorithm is $O(n.r^2 + r.n\sqrt{n})$.

We observe that the algorithm bitonic sort satisfies the property of Problem size linearity (Property 1). Thus, a simple implementation of the algorithm bitonic sort can be achieved by using the clustering approach proposed in this paper. From Theorem 1, we conclude that using the clustering approach (discussed in proof of Theorem 1), bitonic sort can be implemented on a wireless mesh of size $\sqrt{n} \times \sqrt{n}$ in time $O(r^2 + \sqrt{n}/r.r^2) = O(r^2 + \sqrt{n}.r)$. Further analysis shows that the energy complexity of the algorithm is $O(n.r^2 + r.n\sqrt{n})$. Note the time performance and energy dissipation of our algorithm is same as the implementation discussed in [3].

The algorithm is time and energy optimal. Consider the scenario where $r < \sqrt{n}$. Let us assume that for all $1 \leq i, j, \sqrt{n}$, data element at position (i, j) in the mesh will be moved to position $((i + \sqrt{n}) mod \sqrt{n}, j)$ in the sorted distribution. Thus, each element must travel a distance of $\sqrt{n}$ to reach its final position (position after sorting is complete). This involves $\sqrt{n}/r$ hops. At any time only one sensor node out of adjacent r^2 (in a block of r^2) sensor nodes can transmit due to interference. Thus total time taken for data elements to reach their sorted position is $\Omega(r^2 + \sqrt{n}.r)$. Moreover, each data element makes $\sqrt{n}/r$ hops. This implies that the energy complexity of the algorithm is $\Omega((n.\sqrt{n}/r).r^2)$. Next we examine the scenario where $r \geq \sqrt{n}$. All sensor nodes can listen to each other. All sensors must transmit the value at least one. Thus, execution time is $\Omega(n)$ and energy dissipation is $\Omega(n.r^2)$.

4 Conclusions

In this paper, we illustrated a methodology for emulating parallel algorithms designed for mesh-connected computers onto wireless sensor meshes, and analyzed them as a function of the transmission range r of the sensor nodes. For a fixed transmission range r, the algorithms for ranking, and sorting are time and energy optimal.

The results obtained in this work are simple, and yet they demonstrate how analysis from prior research can be leveraged to design energy (or time) optimal algorithms for distributed sensor networks by exploiting the network configurability and density. The network model proposed in this work is an initial step towards understanding performance in these networks. The problem becomes more challenging when additional network parameters are considered to be variant such as number of communication channels, variable range control for transmission, fine tuned direction control at transmitter and receiver, among others. Moreover, a sensor network can be hierarchical in nature with some nodes being more powerful (more memory, faster computation, etc.) than others. Analysis of such networks require more sophisticated models.

In this paper, we assume that the sensor nodes organize themselves in a perfect mesh. It would be of interest to simulate a real scenario, and investigate the performance impact, when the sensor nodes are randomly distributed. The analysis will also provide some insight into the network design problem, as to with what density should the sensor nodes be deployed.

References

1. R. S. Bhuvaneswaran, L. J. Bordim, J. Cui, and K. Nakano, "Fundamental Protocols for Wireless Sensor Networks," International Parallel and Distributed Processing Symposium (IPDPS), Workshop on Advances in Parallel and Distributed Computatinal Models, April 2001
2. P. Bonnet, J. E. Gehrke, and P. Seshadri, "Towards Sensor Database Systems," International Conference on Mobile Data Management (MDM), January 2001
3. J. K. Bordim, K. Nakano, and H. Shen, "Sorting on Single-Channel Wireless Sensor Networks," International Symposium on Parallel Architectures, Algorithms, and Networks (ISPAN), May 2002
4. N. Bulusu, J. Heidemann, and D. Estrin, "GPS-less Low Cost Outdoor Localization For Very Small Devices," IEEE Personal Communications, October 2000
5. J. Elson and D. Estrin, "Time Synchronization in Wireless Sensor Networks," International Parallel and Distributed Processing Symposium (IPDPS), Workshop on Parallel and Distributed Computing Issues for Wireless and Mobile Systems, April 2001
6. T. Hayashi, K. Nakano, and S. Olariu, "Randomized Initialization Protocols for Packet Radio Networks," Discrete Mathematics and Theoretical Computer Science, SIAM Press (2000) 221–235
7. Joseph Ja Ja, "An Introduction to Parallel Algorithms," Addison Wesley Publishing Company (1992)
8. V. Kumar, A. Grama, A. Gupta, and G. Karypis, "Introduction to Parallel Computing: Design and Analysis of Algorithms," The Benjamin/Cummings Publishing Company (1994)
9. The μAMPS Project, http://www-mtl.mit.edu/research/icsystems/uamps/
10. R. Miller and Q. F. Stout, "Parallel Algorithms for Regular Architectures: Meshes and Pyramids," The MIT Press (1996)
11. K. Nakano and S. Olariu, "Energy-Efficient Initialization Protocols for Radio Networks with no Collision Detection," IEEE Transactions on Parallel and Distributed Systems, Vol. 11 (2000) 851–863
12. K. Nakano, S. Olariu, and J. L. Schwing, "Broadcast-Efficient Protocols for Mobile Radio Networks," IEEE Transactions on Parallel and Distributed Systems, Vol. 10 (1999) 1276–1289
13. D. Nassimi and S. Sahni, "Bitonic Sort on Mesh-Connected Computer," IEEE Transactions on Computers, Vol. c-27 January (1979)
14. The PACMAN Project, http://pacman.usc.edu
15. K. Peng and J. Kieffer, "Embedded Image Compression Based on Wavelet Pixel Classification and Sorting," International Conference on Acoustics, Speech, and Signal Processing (ICASSP), May 2002
16. The PicoRadio Project, http://bwrc.eecs.berkeley.edu/Research/Pico_Radio/Default.htm
17. T. S. Rappaport, "Wireless Communication," Prentice-Hall (1996)
18. The Smart Dust Project, http://robotics.eecs.berkeley.edu/~pister/SmartDust/
19. M. Singh and V. K. Prasanna, "System-Level Energy Tradeoffs for Collaborative Computation in Wireless Networks," International Conference on Communications (ICC), Workshop on Integrated Management of Power Aware Communications, Computing and NeTworking, May 2002

Prediction-Based Energy Map for Wireless Sensor Networks*

Raquel A.F. Mini[1], Antonio A.F. Loureiro[1], and Badri Nath[2]

[1] Department of Computer Science – Federal University of Minas Gerais
Belo Horizonte, MG, 31270-010, Brazil
{raquel,loureiro}@dcc.ufmg.br
[2] Department of Computer Science – Rutgers University
Piscataway, NJ, 08854, USA
badri@cs.rutgers.edu

Abstract. The key challenge in the design of wireless sensor networks is maximizing their lifetime. The information about the amount of available energy in each part of the network is called the energy map and can be useful to increase the lifetime of the network. In this paper, we address the problem of constructing the energy map of a wireless sensor network using prediction-based approaches. We also present an energy dissipation model that is used to simulate the behavior of a sensor node in terms of energy consumption. Simulation results compare the performance of the prediction-based approaches with a naive one in which no prediction is used. The results show that the prediction-based approaches outperform the naive in a variety of parameters.

Keywords: energy map, sensor networks, prediction.

1 Introduction

Wireless sensor networks are those in which nodes are low-cost sensors that can communicate with each other in a wireless manner, have limited computing capability and memory and operate with limited battery power. These sensors can produce a measurable response to changes in physical conditions, such as temperature or magnetic field. The main goal of such networks is to perform distributed sensing tasks, particularly for applications like environmental monitoring, smart spaces and medical systems. These networks form a new kind of ad hoc networks with a new set of characteristics and challenges.

Unlike conventional wireless ad hoc networks, a wireless sensor network potentially comprises of hundreds to thousands of nodes [14]. The sensors have to operate in noisy environments and, in order to achieve good sensing resolution, higher densities are required. Therefore, in a sensor network, scalability is a crucial factor. Different from nodes of a customary ad hoc network, sensor nodes

* This work has been partially supported by DARPA under contract number N-666001-00-1- 8953 and a grant from CISCO systems.

M. Conti et al. (Eds.): PWC 2003, LNCS 2775, pp. 12–26, 2003.

are generally stationary after deployment. Although the nodes are static, these networks still have dynamic network topology. During periods of low activity, the network may enter a dormant state in which many nodes go to sleep to conserve energy. Also, nodes go out of service when the energy of the battery runs out or when a destructive event takes place [9]. Another characteristic of these networks is that the sensors have limited resources, such as limited computing capability, memory and energy supplies, and they must balance these restricted resources in order to increase the lifetime of the network. In addition, the sensors will be battery powered and it is often very difficult to change or recharge batteries for these nodes. Therefore, in sensor networks, we are interested in prolonging the lifetime of the network and then the energy conservation is one of the most important aspects to be considered in the design of these networks.

The information about the remaining available energy in each part of the network is called the *energy map* and could aid in prolonging the lifetime of the network. We could represent the energy map of a sensor network as a gray level image, in which light shaded areas represent regions with more remaining energy, and regions short of energy are represented by dark shaded areas. Using the energy map, a user may be able to determine if any part of the network is about to suffer system failures in near future due to depleted energy [17]. The knowledge of low-energy areas can aid in incremental deployment of sensors because additional sensors can be placed selectively on those regions short of resources. The choice of the best location for the monitoring node can be made also based on the energy map. A monitoring node is a special node responsible for collecting information from the sensor nodes. Typically this node is named observer or end user and it is interested in obtaining information from the sensor nodes about the observed phenomenom. We know that nodes near the monitoring node probably will spend more energy because they are used more frequently to relay packets to the monitoring node. Consequently, if we move the monitoring node to areas with more remaining energy, we could prolong the lifetime of the network. Routing protocols can also take advantage of the available energy information in each part of the network. A routing algorithm can make a better use of the energy reserves if it selectively chooses routes that use nodes with more remaining energy, so that parts of the network with small reserves can be preserved. It can also form a virtual backbone based on connecting high energy islands. Other possible applications of the energy map are reconfiguration algorithms, query processing, data fusion, etc. In fact, it is difficult to think of an application and/or an algorithm that does not need to use an energy map. Therefore, the energy map is an important information for sensor networks. However, the naive approach, in which each node sends periodically only its available energy to the monitoring node, would spend so much energy due to communications that probably the utility of the energy information will not compensate the amount of energy spent in this process. For that reason, better energy-efficient techniques have to be devised to gather the information about the available energy in each part of a sensor network.

In this paper, we focus on proposing mechanisms to predict the energy consumption of a sensor node in order to construct the energy map of a wireless sensor network. There are situations in which the node can predict its energy consumption based on its own past history. If a sensor can predict efficiently the amount of energy it will dissipate in the future, it will not be necessary to transmit its available energy often. This node can just send one message with its available energy and the parameters of the model that describe its energy dissipation. With this information, the monitoring node can update often its local information about the available energy of this node. Clearly the effectiveness of this paradigm is dependent on the accuracy with which prediction models can be generated. We analyze the performance of probabilistic and statistical models, and compare them with a naive approach in which no prediction is used. In order to evaluate the approaches to construct the energy map, we have to have a clear idea of how is the energy drop in a sensor node. Thus, we also propose an energy dissipation model that is used to simulate the behavior of a sensor node in terms of energy consumption. Simulation results show that the use of prediction-based models decreases the amount of energy necessary to construct the energy map of wireless sensor networks.

The remainder of this article is organized in the following way. In Section 2, we briefly survey the related work. Section 3 describes the model that we propose to describe the behavior of a sensor node and, consequently, to simulate its energy drop. In Section 4, we describe two approaches to construct a prediction-based energy map for wireless sensor networks. We evaluate the performance of our approaches in Section 5 and conclude by giving directions for our future work in Section 6.

2 Related Work

In [1,7,10,12] the authors explore issues related to the design of sensors to be as energy-efficient as possible. In particular, the WINS [1,10] and PicoRadio [12] projects are seeking ways to integrate sensing, signal processing, and radio elements onto a single integrated circuit. The SmartDust project [7] aims to design millimeter-scale sensing and communicating nodes.

The energy efficiency is the primary concern in designing good media access control (MAC) protocols for the wireless sensor networks. Another important attribute is scalability with respect to network size, node density and topology. A good MAC protocol should easily accommodate such network changes [16]. In addition, a lot of energy-aware routing schemes have been proposed for wireless sensor networks. *Directed diffusion*, proposed in [6], is a new paradigm for communication between sensor nodes. In this paradigm, the data are named using attribute-value pairs and data aggregation techniques are used to dynamically select the best path for the packets. This enables diffusion to achieve energy savings.

The work proposed in [17] obtains the energy map of sensor networks by using an aggregation based approach. A sensor node only needs to report its local energy information when there is a significant energy level drop compared to the last time the node reported it. Energy information of neighbor nodes with similar available energy are aggregated in order to decrease the number of packets in the network. In [17], each node sends to the monitoring node only its available energy, whereas in our work each node sends also the parameters of a model that tries to predict the energy consumption in the near future. Thus, in our approach, each node sends to the monitoring node its available energy and also the parameters of the model chosen to represent its energy drop. With these parameters, the monitoring node can update locally its information about the current available energy at each node, decreasing the number of energy information packets in the network.

3 Energy Dissipation Model

In order to build the energy maps, we have to know how is the energy dissipation in the sensor nodes. To this end, we use an *energy dissipation model* that tries to describe the energy drop at each sensor node. To our knowledge, there is only one work that has addressed this problem [17]. In that work, two energy dissipation models are proposed. The first one is the *uniform dissipation* model. During a sensing event, each node n in the network has a probability p of initiating a local sensing activity, and every node within a circle of r centered at n consumes fixed amount of energy e. The other one is the *hotspot dissipation* model, where there are h fixed hotspots uniformly distributed randomly on the sensor field. Each node n has a probability of $p = f(d)$ to initiate a local sensing activity, and every node within a circle of r centered at n consumes fixed amount of energy e, where f is a density function and $d = min_{\forall i}\{|n - h_i|\}$ is the distance from n to the nearest hotspot. The main drawback of these models is that they do not take into account the fact that a lack of energy in these networks will influence their behaviors. For example, to conserve energy, some sensors have to sleep during some part of the time. Other problems include the assumption that all nodes working in a sensing event will consume the same amount of energy and that all events have the same radius of influence. In this work, we propose a model that tries to represent more realistically the behavior of a sensor network in terms of energy dissipation. In the following we describe our energy dissipation model.

The conservation of energy is the paramount issue to be considered in the design of sensor networks. The best way to save energy is to make unused components inactive whenever possible. This can be achieved in a framework in which the nodes have different modes of operation with different levels of activation and consequently different levels of energy consumption and, as soon as possible, they have to go to a mode that consumes less energy. In sensor networks, the nodes will have to change between different states of activation. Using this idea, we propose a model to describe the behavior of a sensor node and, consequently, to evaluate and simulate its energy dissipation. In this model, each node has four modes of operation: *state 1*: sensing off and radio off; *state 2*: sensing on and radio off; *state 3*: sensing on and radio receiving; *state 4*: sensing on and

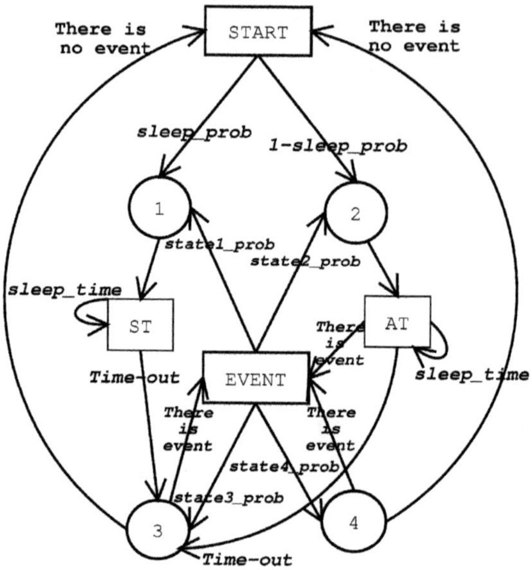

Fig. 1. Diagram of the state transition model: 1, 2, 3, and 4 represent the modes of operation of each node; ST and AT are synchronous and asynchronous timers respectively.

radio transmitting. These modes represent the simplicity of the hardware found in sensor nodes.

In this model, the following parameters are used: λ: arrival rate of the events; *sleep_time*: time in which the node will sleep; *sleep_prob*: when a node is not acting in a sensing event it will be in state 1 with probability *sleep_prob*, and in state 2 with probability (1− *sleep_prob*); *event_radius_min* and *event_radius_max*: the radius of each event will be a random variable uniformly distributed between *event_radius_min* and *event_radius_max*; *event_duration_min* and *event_duration_max*: the duration of each event will be a random variable uniformly distributed between *event_duration_min* and *event_duration_max*; *statei_prob*: probability of being in state *i* during an event; *dist_line*: distance of influence when an information is relayed to the monitoring node.

The behavior of the sensor node can be described by the diagram depicted in Figure 1. At the beginning of the simulation, each node goes to state 1 with probability *sleep_prob* or to state 2 with (1−*sleep_prob*).

When a node goes to state 1, it will be sleeping for *sleep_time* seconds. During this period, this node will be saving energy but it will not be able to communicate or to sense any event. After *sleep_time* seconds, the node wakes up and goes to state 3 to see if there is any event for it or if there is any node trying to communicate with it. If there is an event, the node will go to states 1, 2, 3 or 4 with probabilities *state1_prob*, *state2_prob*, *state3_prob* and *state4_prob*, respectively. If there is no event, the node will go to state 1 with probability *sleep_prob* and to state 2 with (1−*sleep_prob*).

If a node goes to state 2, it will be in this state for *sleep_time* seconds, but unlike in state 1, a node that is in state 2 can see the occurrence of an event because in this state the sensing is on. If an event occurs during the *sleep_time* seconds, the node will go to states 1, 2, 3 or 4 with probabilities *state1_prob*, *state2_prob*, *state3_prob* and *state4_prob*, respectively. If the time *sleep_time* ends and no event has happened, the node goes to state 3 to see if there is any node trying to communicate with it and again it will go to state 1 with probability *sleep_prob* and to state 2 with (1−*sleep_prob*).

In this model, the events are simulated by a Poisson process with parameter λ. Therefore, the number of events in each second of simulation is described by the random variable:

$$P(X = x) = \frac{\lambda^x e^{-\lambda}}{x!}. \tag{1}$$

When an event occurs, a position (X, Y) is randomly chosen for it. The radius of influence of each event is a random variable uniformly distributed between *event_radius_min* and *event_radius_max* and all nodes within the circle of influence of an event will be affected by it. This means that when these nodes realize that there is an event for them (the nodes have to be in states 2, 3 or 4), they will go to states 1, 2, 3 or 4 with probabilities *state1_prob*, *state2_prob*, *state3_prob* and *state4_prob*, respectively. The duration of each event is uniformly chosen between *event_duration_min* and *event_duration_max* seconds. After that time, the data have to be propagated to the monitoring node. We simulate this behavior making all nodes distant *dist_line* for the straight line between the point (X, Y) and the monitoring node go for a short time to state 3 and after to state 4.

The states transition described above tries to capture the behavior of a sensor node specially in terms of energy consumption. As there are no real large sensor networks implemented already, we have no information about the real energy dissipation of a sensor node. But, we believe that, for our purposes, this model can represent the energy drop in an acceptable way.

4 Prediction-Based Energy Map

As described earlier, the knowledge of the available energy reserves in each part of the network is an important information for sensor networks. The more natural way of thinking about the energy map construction is one in which periodically each node sends to the monitoring node its available energy. We call this the *naive* approach. As the sensor networks have lots of nodes with limited resources, the amount of energy spent in the naive approach will be prohibitive. For that reason, better energy-efficient techniques have to be designed to gather the information about the available energy at each part of a sensor network.

In this section, we discuss the possibilities of constructing the energy map using prediction-based approaches. Basically, each node sends to the monitoring node the parameters of the model that describes its energy drop and the monitoring node uses this information to update locally the information about the available energy at each node. The motivation that guided us to this work is that if a node is able to predict the amount of energy it will spend, it can send this information to the monitoring node and no more energy information will be sent during the period that the model can describe satisfactorily the energy dissipation. Then, if a node can efficiently predict the amount of energy it will dissipate in the future time, we can save energy in the process of constructing the energy map of a sensor network.

In order to predict the dissipated energy, we studied two models. In Section 4.1, we describe a probabilistic model based on the Markov chains, and, in Section 4.2, we present a statistical model in which the energy level is represented by a time series and the ARIMA (Autoregressive Integrated Moving Average) model is used to make the predictions.

4.1 Probabilistic Model

In this section, we claim that each sensor node can be modeled by a Markov chain. In this case, the node modes of operation are represented by the states of a Markov chain and the random variables represent the probability of staying at each state in a certain time. Then, if each sensor node has M modes of operations, each node will be modeled by a Markov chain with M states.

Using this model, at each node, we have a sequence of random variables $X_0, X_1, X_2, ...$ that represents its states during the time. Then, if $X_n = i$, we say that the sensor node is in mode of operation i at time-step[1] n. In addition, at each time the node is in state i, there is some fixed probability, P_{ij}, that the next state will be j. This probability can be represented by: $P_{ij} = P\{X_{m+1} = j | X_m = i\}$. We can also define the n-step transition probability, $P_{ij}^{(n)}$, that a node presently in state i will be in state j after n additional transitions [13]: $P_{ij}^{(n)} = \sum_{k=1}^{M} P_{ik}^{(r)} P_{kj}^{(n-r)}$, for any value of $0 < r < n$.

With the knowledge of the probabilities $P_{ij}^{(n)}$ for all nodes and the value of X_0 (initial state of each node), it is possible to estimate some information about the network that can be useful in many tasks. In this work, we will use these probabilities to predict the energy drop of a sensor node. The first step to make this prediction is to calculate for how many time-steps a node will be in a state s in the next T time-steps. If the node is in state i ($X_0 = i$), the number of time-steps a node will stay in the state s can be calculated by: $\sum_{t=1}^{T} P_{is}^{(t)}$. Also, if E_s is the amount of energy dissipated by a node that remains one time-step in state s, and the node is currently in state i, then the expected amount of energy spent in the next T times, E^T, is:

$$E^T = \sum_{s=1}^{M} (\sum_{t=1}^{T} P_{is}^{(t)}) \times E_s. \tag{2}$$

[1] A time-step is a small amount of time. We suppose that all state transitions occur at the beginning of any time-step.

Using the value E^T, each node can calculate its energy dissipation rate (ΔE) for the next T time-steps. Each node then sends its available energy and its ΔE to the monitoring node. The monitoring node can maintain an estimation for the dissipated energy at each node by decreasing the value ΔE periodically for the amount of remaining energy of each node. The better the estimation the node can do, the fewer the number of messages necessary to obtain the energy information and, consequently, the fewer the amount of energy spent in the process of getting the energy map.

4.2 Statistical Model

In this section, we present the statistical model used to forecast the energy level in the sensor nodes. In this model, we represent the energy drop of a sensor node as a time series. A time series is a set of observations x_t, each one being recorded at a specific time t [3]. A discrete- time time series is one in which the set T_0 of times at which observations are made is a discrete set. Continuous-time time series are obtained when observations are recorded continuously over some time interval. There are two main goals of time series analysis [15]: identifying the nature of the phenomenon represented by the sequence of observations, and forecasting (predicting future values of the time series variable). In this work, we are interested in using the time series analysis to forecast future values of the available energy in a sensor node. We will use the discrete-time time series in such a way that each node will verify its energy level in a discrete time interval.

We can observe that the time series which represents the energy drop of a sensor node has a clear decreasing trend[2] (we suppose that there is no replacement in the battery) and no seasonality[3]. The decreasing trend will also imply in a decreasing mean and then the energy level will also be a nonstationary time series[4].

In this work, we will use the ARIMA (Autoregressive Integrated Moving Average) model to predict future values of the time series. The ARIMA models were proposed by Box and Jenkins [2] and they consist of a systematic methodology for identifying and estimating models that could incorporate both autoregressive and moving average approaches. This makes ARIMA models a powerful and general class of models [8]. The "Integrated" part of the model is due to the differencing step necessary to make the series stationary.

The first step in developing an ARIMA model is to determine if the series is stationary. When the original series is not stationary, we need to difference it to achieve stationarity. Given the series Z_t, the differenced series is a new series $X_t = Z_t - Z_{t-1}$. The differenced data contain one less point than the original one. Although one can difference the data more than once, a small number of

[2] Trend refers to a gradual, long-term movement in the data.

[3] Seasonality refers to periodic fluctuations that are generally related to weather factors or to human-made factors such as holidays and vacations.

[4] A stationary time series is one whose statistical properties such as mean, variance, and autocorrelation, are all constant over time.

differences is usually sufficient to obtain a stationary time series [8]. The number of differencing applied in the original series is represented by the parameter d.

The next step in the construction of the ARIMA model is to identify the AR terms. An autoregressive model is simply a linear regression of the current value against one or more prior values of the series. The value of p is called the order of the AR model. Then, an autoregressive model of order p can be summarized by: $X_t = \phi_1 X_{t-1} + \phi_2 X_{t-2} + ... + \phi_p X_{t-p} + Z_t$, where X_t is the time series, $\phi_1, \phi_2, ..., \phi_p$ are the autoregressive model parameters, and Z_t represents normally distributed random errors.

After defining the differencing and the autoregressive parameters, we have to identify the MA terms. A moving average model is essentially a linear regression of the current value of the series against the random shocks of one or more prior values of the series [8]. The random shocks at each point are assumed to come from the same distribution, typically a normal distribution, with constant location and scale. The distinction in this model is that these random shocks are propagated to future values of the time series. A moving average model of order q is represented by: $X_t = Z_t + \theta_1 Z_{t-1} + \theta_2 Z_{t-2} + ... + \theta_q Z_{t-q}$, where X_t is the time series, $\theta_1, \theta_2, ..., \theta_q$ are the moving average model parameters and the Z_t are random shocks to the series.

Then, in order to use the ARIMA model we have to identify the values of p (order of the autoregressive model), d (number of differencing required to achieve stationarity), q (order of the moving average model) and the coefficients of the autoregressive and moving average models. Thus, a time series T_t can be represented by an ARIMA(p,d,q) model if, after differencing this series d times, we find a stationary time series X_t, such that for every t: $X_t = \phi_1 X_{t-1} + ... + \phi_p X_{t-p} + Z_t + \theta_1 Z_{t-1} + ... + \theta_q Z_{t-q}$.

When using equation above, we can predict the value of the time series in time t using the previous values and some random variables that represent the errors in the series. In general, the estimation of these parameters is not a trivial task. In [8,15], the authors describe some techniques to help in the process of parameters identification.

5 Simulation Results

In order to analyze the performance of the proposed schemes, we implemented the prediction-based energy maps in the ns-2 simulator. The approaches implemented were: the Markov, in which each node sends periodically to the monitoring node its available energy and its predicted energy consumption rate; and the ARIMA, in which each node sends to the monitoring node its available energy and the parameters of this model. These approaches are compared with the naive one in which each node sends periodically to the monitoring node only its available energy.

In our simulations, we use the energy dissipation model, presented in Section 3, to describe the behavior of sensor nodes and, consequently, to simulate their energy dissipation. Therefore, each node has four modes of operation: state

1 (sensing off, radio off), state 2 (sensing on, radio off), state 3 (sensing on, radio receiving) and state 4 (sensing on, radio transmitting). The values of power consumption for each state were calculated based on information presented in [5]: state 1: 25.5 μW, state 2: 38.72 mW, state 3: 52.2 mW and state 4: 74.7 mW. These values will be used throughout all simulations.

In the Markov model, each node sends its available energy and its energy dissipation rate to the monitoring node. To obtain its energy dissipation rate, each node locally calculates its own probabilities, $P_{ij}^{(n)}$. In this case, P_{ij} will be the number of times the node was in state i and went to state j divided by the total number of time-steps the node was in state i. With these probabilities, each node uses equation (2) to find its energy dissipation rate. If each node can predict efficiently its energy dissipation rate, this approach can save energy compared with the naive, because no more energy information packet has to be sent while the energy dissipation rate describes satisfactorily the energy drop in this node.

In the implementation of the ARIMA model, we have to identify the parameters p, d, q and to estimate the coefficients of the AR and MA models. The first step in fitting an ARIMA model is the determination of the order of differencing needed to stationarize the series (parameter d). Normally, the correct number of differencing is the lowest order of differencing that yields a time series which fluctuates around a well-defined mean value and whose autocorrelation function plot decays fairly rapidly to zero, either from above or below [4]. If the series still exhibits a long-term trend, i.e., a lack of tendency to return to its mean value, or if its autocorrelations are positive out to a high number of lags, it needs a higher order of differencing. In general, the optimal order of differencing is often the one at which the standard deviation is lowest [4]. In addition, if the lag 1 autocorrelation is -0.5 or more negative, the series may be over-differenced. In our simulation, we choose the smallest value of d that produces the lowest standard deviation in such a way that the lag 1 autocorrelation is not more negative than -0.5. The number of AR and MA terms was found using the autocorrelation and partial autocorrelation functions. The lag at which the partial autocorrelation function cuts off indicates the number of AR terms, and the number of MA terms is determined by the lag at which the autocorrelation function cuts off. The values of the coefficients of the AR and MA models were calculated based on a CSS-ML (minimize conditional sum-of- squares and maximum likelihood) method implemented in [11].

In all simulations we use the parameter *threshold* that determines the accuracy required or the maximum error acceptable in the energy map. If we define a threshold of 3%, a node will send another energy information to the monitoring node only when the error between the energy value predicted by the monitoring node and the correct value is greater than 3%. Each node can locally determine this error by just keeping the parameters of the last prediction sent to the monitoring node. Then, adjusting the value of the threshold, we can control the precision at which the energy maps are constructed.

The numerical values chosen for the base case of our simulations can be seen in Table 1. Unless specified otherwise, these values are used as the parameters

Table 1. Default values used in the simulations.

Parameter	Value	Parameter	Value
λ	0.5	*state1_prob*	0.01
sleep_time	10 sec	*state2_prob*	0.2
sleep_prob	0.7	*state3_prob*	0.45
event_radius_min	10 m	*state4_prob*	0.34
event_radius_max	30 m	Threshold	3%
event_duration_min	10 sec	Initial Energy	100 J
event_duration_max	50 sec	Communication Range	20 *m*
dist_line	20 m	Time-steps	1 sec

throughout the remainder of this work. Moreover, in all simulations, the monitoring node is positioned at the middle of the field at position $(50, 50)$, and all nodes are immobile and can communicate with other nodes within their communication range.

In order to analyze the performance of the approaches in situations where it is necessary an energy map with very low error (small threshold) and also when we can tolerate a greater error (big threshold), we changed the value of the parameter threshold. We ran the naive, Markov and ARIMA algorithms for 200 nodes in a $100 \times 100\text{m}^2$ field in which the average degree of each node is 22.7. Figure 2–a shows the average number of energy information packets that each node had to send to the monitoring node, during 1000 second simulation, in order to construct an energy map with an error no greater than the corresponding threshold. These results correspond to an average of these values and a 95% confidence interval. We can see that the Markov approach is better than the other two for all values of threshold. But its performance is very close to the ARIMA model, meaning that both approaches have similar power of prediction for all values of threshold. However, the graph of Figure 2–a is not a fair way of comparing the three approaches because when a node, running the naive algorithm, has to send an energy information packet, the size of the extra information required is only 4 bytes (its available energy). In the Markov algorithm, the overhead is of 8 bytes (its available energy and its current power consumption) and in the ARIMA model the overhead is about 40 bytes (with the parameters p, d, q and the coefficients of the AR and MA models). In order to perform a fair comparison between the three approaches, we have to analyze the average number of bytes that each node has to send when running the naive, Markov and ARIMA algorithms. Thus, the metric used to define energy efficiency will be the number of bytes transmitted. Figure 2–b compares the average number of bytes that each node had to send to the monitoring node if the normal packet size (or the per-packet header overhead) of a sensor network is 30 bytes. In this situation, each time a node has to send its energy information, it will have to send 34 bytes (30 bytes of the normal packet plus 4 bytes of the naive overhead) in the naive algorithm, 38 in the Markov and 70 bytes in the ARIMA. We can see that when we compare the number of bytes instead of the number of packets, the

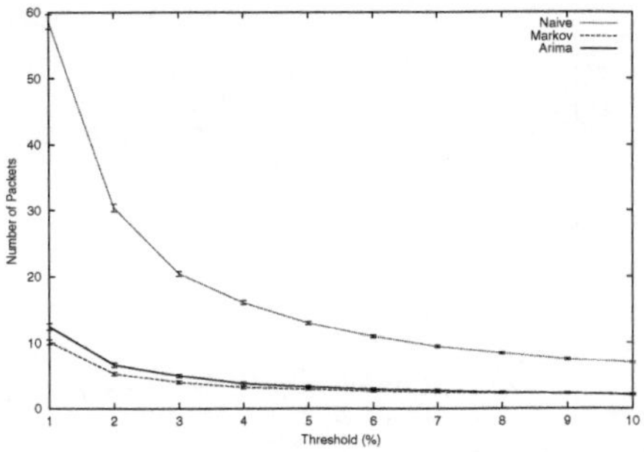

(a) Average number of packets.

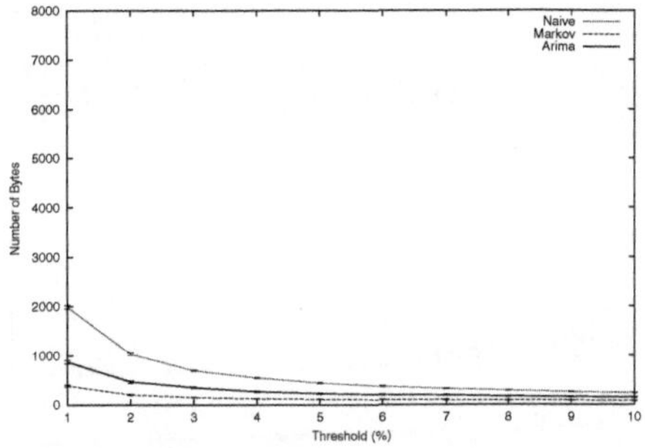

(b) Average number of bytes when the packet size is 30 bytes.

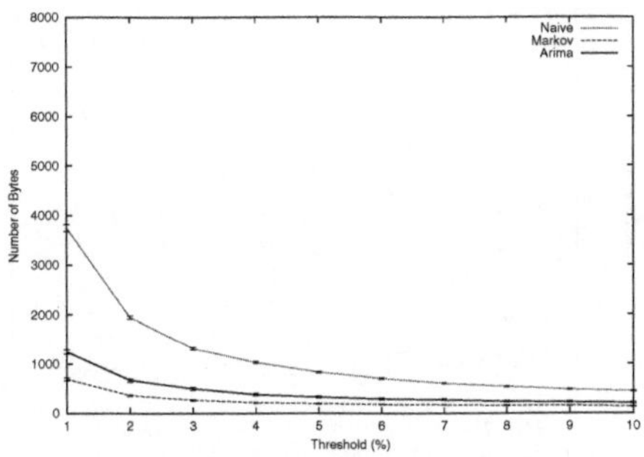

(c) Average number of bytes when the packet size is 60 bytes.

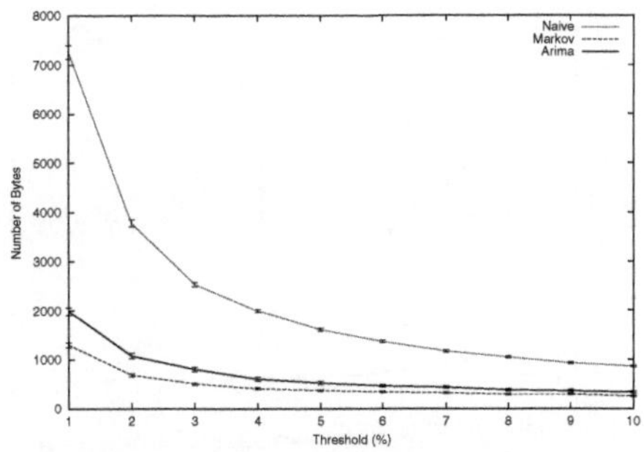

(d) Average number of bytes when the packet size is 120 bytes.

Fig. 2. Comparison between the three approaches when we change the value of the threshold.

performance of the ARIMA is closer to the naive, and the Markov is still the best of the three. Figures 2–c and 2–d show what happens when the normal size of a packet is 60 and 120 bytes, respectively. As the normal packet size increases, the naive becomes even worse because, in these situations, the overhead of the large amount of information required by the ARIMA has a smaller impact in the total number of bytes sent. Then, for all values of threshold analyzed, the Markov model was more energy-efficient than the other two models, and for sensor networks whose size of the packet is small, the performance of the ARIMA is very close to the naive approach.

Next we altered the value of the parameter λ in order to study the behavior of each approach when the number of events increases. We executed the three approaches using the same scenario described above, during 1000 seconds of simulation. Figure 3–a shows the average number of packets when we increase the number of events in the network. In these simulations, the threshold was

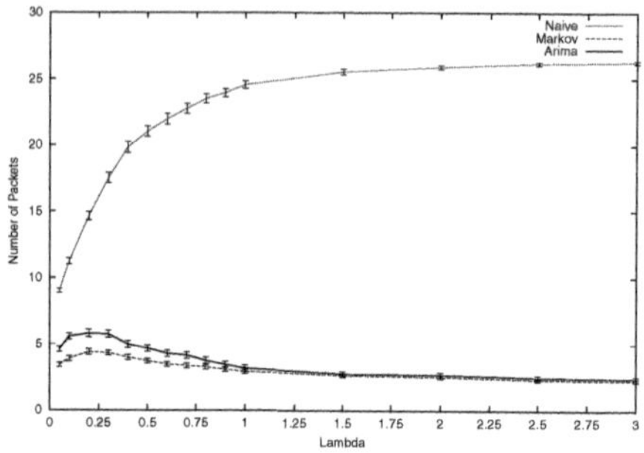

(a) Average number of packets.

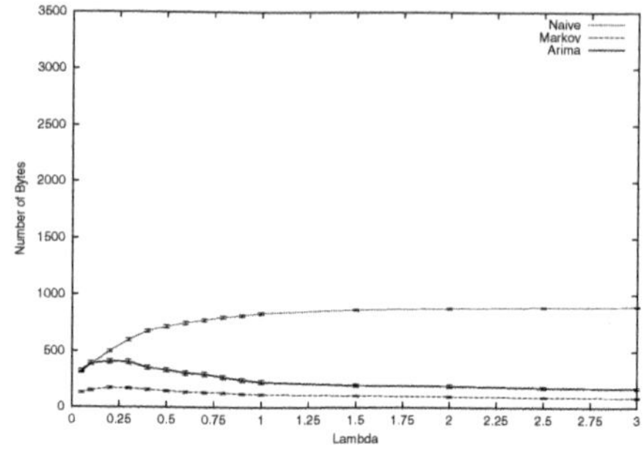

(b) Average number of bytes when the packet size is 30 bytes.

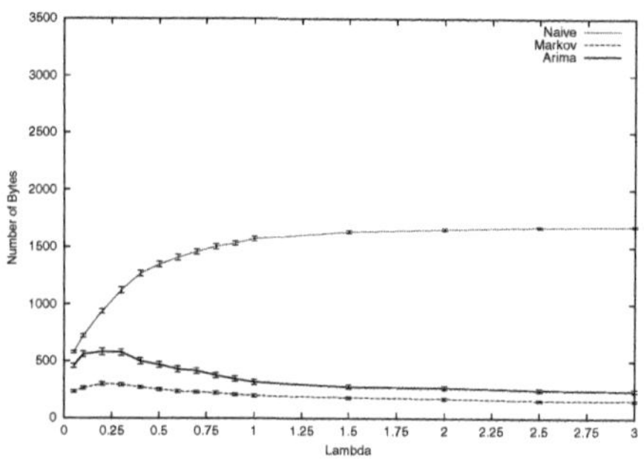

(c) Average number of bytes when the packet size is 60 bytes.

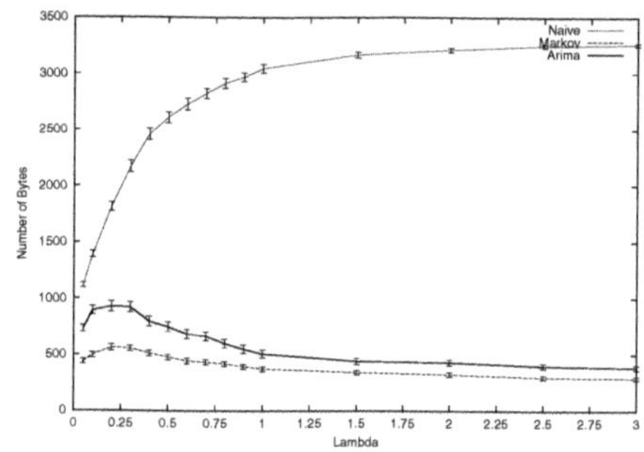

(d) Average number of bytes when the packet size is 120 bytes.

Fig. 3. Comparison between the three approaches when we change the value of the parameter λ.

fixed in 3%. We can see that the power of making prediction of the Markov model is very similar to the ARIMA, but still better for all values of λ. Also, as the network becomes more active, the difference between the number of packets required by the naive and by the prediction-based approaches is getting bigger. Nevertheless, as described above, to do a fair comparison, we have to analyze the number of bytes transmitted by each approach. These results are shown in Figures 3–b, 3–c and 3–d. We can see that the Markov approach is still better than the other two for all values of packet size, and also that when the packet size increases, the difference between the number of bytes transmitted by the prediction-based approaches and the naive increases. One interesting fact is that the prediction approaches have a better behavior when the number of events is very small or big. The worst case of these approaches happens for medium values of λ. This means that the fact of having more events does not make the problem of prediction more difficult. The more difficult situations for the prediction approaches are when there is a medium number of events. On the other hand, in the naive approach, as more events happen, more energy will be spent by a node and more often it will have to send energy information packets to the monitoring node. Then, the prediction approaches scale well when the number of events increases or, the power of making prediction does not decrease when the activity of the network increases.

Due to the nondeterministic characteristic of the sensor networks, it is better to perform predictions that are simple both in terms of the computation required to find the parameters of the prediction model and, mainly, in terms of the number of parameters that have to be sent to the monitoring node. This feature becomes clear when we compare the two prediction techniques. Even though both present similar capacity of making prediction, the Markov approach is better because, in this model, only one parameter describes the energy dissipation in a sensor node, and consequently only the available energy and the current dissipation rate have to be sent to the monitoring node. Thus, in the construction of prediction-based energy maps, it is better to use simple models instead of sophisticated predictions that demand a lot of communication between the sensors and the monitoring node.

6 Conclusions and Future Directions

In this work, we have studied the problem of constructing the energy map for wireless sensor networks. We analyzed two prediction-based energy maps based on probabilistic and statistical models. In the prediction-based energy maps, each node tries to estimate the amount of energy it will spend in the near future and it sends this information, along with its available energy, to the monitoring node. Using the energy dissipation model proposed in this paper, simulations were conducted in order to compare the performance of the two prediction-based approaches with a naive one, in which only the available energy is sent to the monitoring node. Simulation results indicate that the prediction-based approaches are more energy-efficient than the naive model, and also that these approaches are more scalable with respect to the number of sensing events.

As discussed here, prediction-based techniques are a good approach to construct the energy map for wireless sensor networks. We intend to extend this work by examining and evaluating other prediction models for obtaining the energy map.

Acknowledgment. The authors would like to thank the comments and suggestions of Prof. Narayan Mandayam and the members of the DATAMAN group of Rutgers University.

References

1. G. Asada, T. Dong, F. Lin, G. Pottie, W. Kaiser, and H. Marcy. Wireless integrated network sensors: Low power systems on a chip. In *European Solid State Circuits Conference*, The Hague, Netherlands, October 1998.
2. George E. P. Box and Gwilym M. Jenkins. *Time series analysis: forecasting and control*. San Francisco: Holden-Day, 1976.
3. Peter J. Brockwell and Richard A. Davis. *Introduction to time series and forecasting*. New York : Springer, 2nd edition, 2002.
4. Fuqua. School of business. Forecasting. http://www.duke.edu/ rnau/411home.htm, 2002.
5. J. Hill, R. Szewczyk, A. Woo, S. Hollar, D. Culler, and K. Pister. System architecture directions for networked sensors. In *Proceedings of the 9th International Conference on Architectural Support for Programming Languages and Operating Systems*, November 2000.
6. C. Intanagonwiwat, R. Govindan, and D. Estrin. Directed diffusion: a scalable and robust communication paradigm for sensor networks. In *Proceedings of the sixth annual international conference on Mobile computing and networking*, pages 56–67, Boston, MA USA, 2000.
7. J. M. Kahn, R. H. Katz, and K. S. J. Pister. Next century challenges: Mobile networking for smart dust. In *In Proceedings of MOBICOM*, pages 271–278, Seattle, 1999.
8. Nist. Nist/sematech – e-handbook of statistical methods. http://www.itl.nist.gov/div898/handbook, 2002.
9. S. Park, A. Savvides, and M. B. Srivastava. SensorSim: a simulation framework for sensor networks. In *Proceedings of the 3rd ACM international workshop on Modeling, analysis and simulation of wireless and mobile systems*, pages 104–111, Boston, MA USA, 2000.
10. G.J. Pottie and W.J. Kaiser. Wireless integrated network sensors. In *Communications of the ACM*, volume 43, pages 551–8, may 2000.
11. R-Project. The R project for statistical computing. http://www.r-project.org/, 2002.
12. Jan M. Rabaey, M. Josie Ammer, Julio L. da Silva Jr., Danny Patel., and Shad Roundy. Picoradio supports ad hoc ultra-low power wireless networking. *IEEE Computer*, 33(7), 2000 July.
13. Sheldon Ross. *A First Course in Probability*. Prentice Hall, fifth edition, 1998.
14. K. Sohrabi, J. Gao, V. Ailawadhi, and G.J. Pottie. Protocols for self-organization of a wireless sensor network. *IEEE Personal Communications*, 7:16–27, October 2000.
15. StatSoft. Inc. (2002). Electronic Statistics Textbook. Tulsa, OK: StatSoft. http://www.statsoft.com/textbook/stathome.html, 2002.
16. Alec Woo and David E. Culler. A transmission control scheme for media access in sensor networks. In *The seventh annual international conference on Mobile computing and networking 2001*, pages 221–235, Rome, Italy, July 2001.
17. Yonggang Jerry Zhao, Ramesh Govindan, and Deborah Estrin. Residual energy scans for monitoring wireless sensor networks. In *IEEE Wireless Communications and Networking Conference (WCNC'02)*, Orlando, FL, USA, March 2002.

A Formally Verified Decentralized Key Management Architecture for Wireless Sensor Networks

Yee Wei Law, Ricardo Corin, Sandro Etalle, and Pieter H. Hartel

Faculty of Electrical Engineering, Mathematics and Computer Science
University of Twente, PO Box 217, 7500 AE Enschede, The Netherlands*
{ywlaw,corin,etalle,pieter}@cs.utwente.nl
http://eyes.eu.org

Abstract. We present a decentralized key management architecture for wireless sensor networks, covering the aspects of key deployment, key refreshment and key establishment. Our architecture is based on a clear set of assumptions and guidelines. Balance between security and energy consumption is achieved by partitioning a system into two interoperable *security realms*: the *supervised realm* trades off simplicity and resources for higher security whereas in the *unsupervised realm* the vice versa is true. Key deployment uses minimal key storage while key refreshment is based on the well-studied scheme of Abdalla et al. The keying protocols involved use only symmetric cryptography and have all been verified with our constraint solving-based protocol verification tool CoProVe.

Keywords: Wireless sensor networks, key management, protocol verification.

1 Introduction

Wireless sensor networks (WSNs) are open architectures, where any potential intruder can easily intercept, eavesdrop and fake messages. Therefore, to guarantee any level of security (confidentiality, authentication etc.) one has to employ cryptographic protocols.

Key management is the process by which cryptographic keys are generated, stored, protected, transferred, loaded, used, and destroyed. Key management is a challenging problem for WSNs due to the hardware constraints of the sensors and the dynamic nature of WSNs themselves. The hardware constraints of sensors are in terms of:

1. **Cost:** Sensors are generally not tamper-resistant.
2. **Space:** They can only store as many keys as are usually allowed by the storage left over by the operating system and application code, which is not much.

* This work is partially sponsored by the European Union as part of the EYES project IST-2001-34734.

M. Conti et al. (Eds.): PWC 2003, LNCS 2775, pp. 27–39, 2003.

3. **Energy:** It is necessary to optimize the use of cryptography since cryptographic operations tend to be resource-intensive.
4. **Time:** Public-key cryptography should be avoided or at least limited to applications which are not time-constrained, because they are a few order of magnitude more resource *and* time consuming than symmetric-key cryptography [3].

The security and functional requirements of most WSNs are such that under the above constraints, the following guidelines need to be taken into account:

1. NO_SINGLE_KEY: The system should not operate on a system-wide key (or keys). Due to the sensor nodes' lack of tamper-resistance, having a vulnerable system-wide key is *not* better than having no key at all.
2. NO_SINGLE_POINT: The system should not have a single point (e.g. node) of failure.
3. SCALABILITY: The system should be scalable in the sense that the addition of new nodes should not cause excessive rise in computation, communication and administrative overhead in the network.

As of writing, we are not aware of any key management architecture that satisfies all the above hardware constraints as well as the guidelines. Take for example Basagni et al's proposal [2]. In this proposal, sensors are assumed to be tamper-resistant (contrary to our hardware constraints) and share a network-wide system key. Although repeated key refreshment (aka re-keying) would thwart cryptanalytic attampts on the traffic encryption key, the compromise of a single node compromises the system key and thus the whole network, leading to the violation of guideline NO_SINGLE_KEY and NO_SINGLE_POINT. In Perrig et al's proposal [12], the testbed architecture for the SPINS protocol suite possesses a single point of vulnerability that is the base station, violating guideline NO_SINGLE_POINT in addition to SCALABILITY.

The contributions of this paper are:

1. A decentralized key management architecture for WSNs, covering the aspects of key deployment, key refreshment and key establishment. Our decentralized key management architecture, which we call EYES Security Architecture 1 (ESA1), satisfies all the above hardware constraints and architecture guidelines. In addition, authenticated routing is directly implementable on top of the architecture, allowing an integrated security architecture to be specified.
2. Verification of all the protocols employed in our architecture using CoProVe [6]. The verification results imply that our protocols are secure in our standard set of test scenarios.

Overview. We start by giving the preliminaries in Section 2, which covers the notation that we use, and the overview and security objectives of our architecture. Section 3 proceeds to describe key deployment, i.e. the strategic distribution of keys on sensor nodes during the bootstrap phase; and key refreshment, i.e. the refreshment of keys for increasing the encryption threshold. Section 4 describes keying protocols, i.e. the establishment of common session keys to secure communication channels; and their verification. Section 5 discusses important related work. Section 6 concludes and wraps up with a list of future work.

2 Preliminaries

Notation. To simplify the discussion below, we are using the following standard notation:

- $E_K(M)$ represents the ciphertext as a result of encrypting plaintext M with key K.
- $MAC_K(M)$ represents the message authentication code (MAC) of message M using key K.
- If $A, B \in \{0,1\}^*$, then $A|B$ is their concatenation.

We also use the following non-standard notation:

- $A \rightsquigarrow B$: A's radio range reaches B, or equivalently A's transmission reaches B in a single hop; but not necessarily vice versa.
- $A \leftrightsquigarrow B$: $A \rightsquigarrow B$ and $B \rightsquigarrow A$.

Architecture. In ESA1, we employ a structure of supervised and unsupervised clusters. A *supervised* cluster consists of a supervisor node (e.g. S^α of cluster α, S^β of cluster β in Figure 1(a)) and a group of supervised nodes (e.g. N_1^α of cluster α, N_1^β and N_2^β of cluster β in Figure 1(a)). While the supervised nodes are normal sensor nodes, the supervisor is usually assumed to be tamper-resistant, to have higher energy and computational resources (like the Rich Uncles in [4]), and to have wider radio coverage compared with normal sensor nodes. An unsupervised cluster consists only of unsupervised nodes (e.g. N_1^γ of cluster γ in Figure 1(a)), which usually are just normal sensor nodes.

In our architecture, supervisors are needed to combine security (confidentiality and authentication) with energy efficiency and scalability. Nodes that are only sporadically involved in security-critical tasks requiring confidentiality and authentication can be left unsupervised.

Another motivation for having supervisors as specified is to support Di Pietro et al's secure selective exclusion [8] and intrusion detection, the details of which is however outside the scope of this paper. In other words, the above specification of the supervisors is *not* a limitation of our model. We allow as many or as few supervisors as desirable from a cost point of view.

Security Objectives. First, we define two *security realms*: the set of all supervised clusters constitute the supervised realm, while the set of all unsupervised clusters constitute the unsupervised realm.

- In the supervised realm, confidential and authenticated *node-to-node* communication within a single supervised cluster, and between nodes residing in different supervised clusters are realized through *intra-supervised cluster keying* (cf Section 4.2) and respectively *inter-supervised cluster keying*

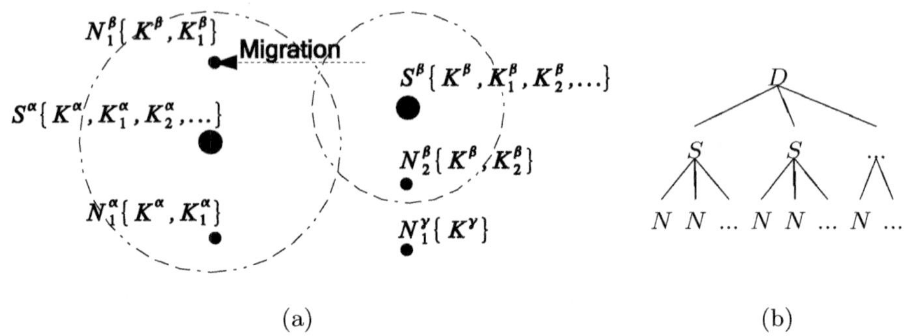

Fig. 1. (a) Key deployment in ESA1 (where $N_i^\alpha\{K^\alpha, K_i^\alpha\}$ means node N_i^α of cluster α with keys K^α, K_i^α; dashed circles represent supervisors' radio range). (b) Three-tier hierarchy for the supervised realm: D=Domain, S=Supervisor, N=Node.

(cf Section 4.3). In these modes of communication, the effect of any node's compromise is isolated, i.e. if a supervised node N_1^α is compromised, only sessions that involve N_1^α are compromised.

- In the unsupervised realm, confidential and authenticated *group* communication within a single unsupervised cluster is realized through *intra-unsupervised cluster keying* (cf Section 4.4). In this mode of communication, the compromise of a node compromises the security of the whole cluster. Between unsupervised clusters, communication is open and unauthenticated, i.e. no two unsupervised clusters trust each other.
- Between the supervised realm and the unsupervised realm, confidential and authenticated communication is realized through *unsupervised-to-supervised cluster keying* (cf Section 4.5). Since the supervised realm has a higher security level than the unsupervised realm, we do not think it makes sense to secure a communication that originates from a supervised node to an unsupervised node (drawing an analogy from the "no write-down" policy of the Bell LaPadula model). However the need for unsupervised-to-supervised node communication is believed to be quite possible, e.g. in the upload of sensitive data.

Independent of the confidentiality and authentication requirements, message integrity can and should always be ensured by using universal hash functions [5]. Universal hash functions are highly resistant to collisions and have also been recently shown to be resistant to algorithmic complexity-based denial of services attacks [7].

3 Key Deployment

We can now describe the main keys employed in ESA1 and their roles.

In ESA1, every supervised node N_i^α $(1 \leq i \leq n)$ of cluster α is equipped with a unique *node key* K_i^α and the supervisor S^α is equipped with the node keys of all the supervised nodes of its cluster. Additionally $S^\alpha, N_1^\alpha, \ldots, N_n^\alpha$ share the same *cluster key* K^α. The cluster key is a backup key that is only used when the supervisor is not available, in which case the supervised cluster downgrades to an unsupervised cluster. Therefore in contrast to what Figure 1(a) might suggest, the cluster membership of a node N in cluster α depends solely on (1) whether there is a unique shared key between N and S^α; and (2) whether N possesses the cluster key K^α. When a node, say N_1^β (cf Figure 1(a)), roams out of the radio range of its original supervisor S^β, to the radio range of another supervisor S^α, it will for efficiency initiate a process called *migration* that migrates its node key from S^β to S^α (cf Section 4.6). It is important to note that when we mention "migrating a key", we actually mean "migrating a derivation of the key", so that even if the migrated key is compromised backward secrecy is preserved.

In an unsupervised cluster γ, every node shares the same cluster key K^γ, and cluster membership depends solely on the possession of K^γ.

To establish trust between supervisors, we leave two options open. The first option involves a further domain tier higher in the hierarchy (cf Figure 1(b)), and it is to connect all the supervisors at bootstrap to a *domain controller*. The domain controller partitions the set of supervisors into *domains*. Only supervisors that belong to the same domain are given the same *domain key* (a symmetric key) and thus share the common trust. This scheme uses only symmetric cryptography but can only support a static and coarse notion of trust – the compromise of any supervisor compromises the whole domain. The second option is to resort to SSL/TLS (or similar public key-based mechanisms) to secure communication between supervisors. This scheme supports flexible and fine-grained access control at the expense of higher overhead and energy consumption. All in all, the suitability of each of these two options depends on the construction of the supervisors and the actual application scenario. In this paper, we focus on the node/cluster tiers of the hierarchy.

3.1 Key Refreshment

The effectiveness of a key is not only bounded by its length but also by the number of encryptions it has been used for, because with enough encryptions an intruder is able to launch birthday attacks (or cryptanalytic attacks which however typically require more encryptions [1]). Therefore, as part of any key management architecture, we need a scheme to refresh every now and then the deployed keys. For this, we refer to the well-studied scheme of Abdalla et al [1]. Given an initial shared key K_0 between two nodes A and B, the encryption and message authentication code (MAC) keys are derived *serially* at each i-th re-key as follows:

$$K_{enc,i} = MAC_{K_i}(1) \qquad \text{and} \qquad K_{MAC,i} = MAC_{K_i}(2) \tag{1}$$

where $K_{i+1} = MAC_{K_i}(0)$ and $i = 0, 1, 2, \ldots$. A convention in this paper is that a pair of encryption key and MAC key derived from key K is written as K' and K''.

In a supervised cluster, re-keying is message-count driven, i.e. when the encrypted message count approaches the encryption threshold (e.g. $2^{2k/3}$ for CBC mode, where k is the key length [1]), re-keying is initiated. In a supervised cluster, if the shared key is between a node and the superviser, or between a node and another node, re-keying can be initiated by either end of the secure channel. In an unsupervised cluster, re-keying of the cluster key can proceed in a similar fashion described by Basagni et al [2].

4 Verified Keying

In our end-to-end communication model, session establishment has two phases:

1. Keying: The initiator engages the node it wants to communicate with in a key establishment (keying for short) protocol, using an underlying, not necessarily secure routing protocol.
2. Communication: Then, by using the session key established, the two end-points start communicating with authenticated and optionally encrypted messages.

The following discusses the keying protocols and their formal verification.

4.1 Formal Protocol Verification

The keying protocols described in this section are verified using the protocol verifier CoProVe (available at `http://wwwes.cs.utwente.nl/24cqet/coprove.html` [6]). The verifier works by taking as input a specification of the protocol and a *system scenario* describing the roles involved in the protocol, e.g. the initiator, the responder or a server of the protocol. The system scenario is analyzed by the verifier in search of possible interleavings of the roles. Attacks manifest as interleavings not anticipated by the protocol designer.

In this paper, CoProVe is used to verify these properties:

– Confidentiality: A session key must only be known to the communicating nodes, and the supervisors involved in the keying protocol.
– Authentication: A keying protocol must end with every party properly authenticating the other parties it is communicating with. In other words, it must be impossible for any intruder I to impersonate another node A whose keys (used in the keying protocol) I does not have.
– Replay resistance: The meaning of replay attack on a role R is the possibility of unauthenticated parties to cause R to run, i.e. for R to process replayed messages. If R happens to maintain state of every run, then it would be maintaining incorrect states. In CoProVe, we verify replay resistance of R by making sure that two instances of R *cannot* complete regardless of the other roles.

Limitations. To ensure termination, system scenarios must be finite – a restriction shared by every model-checking-based approach. (Other approaches like theorem proving allow an unbounded number of sessions, but sacrifice termination guarantee, or need human intervention.) As such, a protocol is verified secure only in the finite tested scenarios. There is no assurance that an attack would not actually be found in a larger scenarios.

So far, we have tested all protocols in the context of a scenario containing a single session. Such tests turned out to be extremely useful, as they have allowed us for example to spot authentication failures resulting from the inappropriate use of unauthenticated nonces. Here, we also want to stress that these protocols are considerably more complex than many of the usual "toy" security protocols protocol verifiers are tested agains. In fact, considering that one of them involves nine message exchanges, we were ourselves pleasantly surprised to see that CoProVe could deal with them within a reasonable time-span (ranging from a fraction of a second to 2.5 days for the most complex protocol, on a standard Linux-i686 architecture). We are now carrying out extra tests involving larger scenarios.

4.2 Intra-supervised Cluster Keying

We start with the keying protocol that is used within a supervised cluster. Concerning the notation, N_A represents a nonce emitted by A, N_B represents a nonce emitted by B and so on; strings such as Ack and Success are constants. For the sake of notation simplicity, we leave out the superscripts we used in Figure 1. In addition, recall that the encryption key and the MAC key derived from key K are respectively denoted by K' and K''.

Suppose that in a cluster supervised by S, a node A (which shares key K_A with S) wants to initiate a session with B (which shares key K_B with S). The protocol is:

Protocol 1

1. $A \rightarrow S\colon N_A, B, \mathrm{MAC}_{K''_A}(N_A|B)$
2. $S \rightarrow A\colon E_{K'_A}(E_{K'_B}(N_S)|K_{AB}), \mathrm{MAC}_{K''_A}(N_A|B|E_{K'_B}(N_S)|K_{AB})$
3. $A \rightarrow B\colon A, E_{K'_B}(N_S)$
4. $B \rightarrow S\colon B, N_B, A, \mathrm{MAC}_{K''_B}(N_S|B|N_B|A)$
5. $S \rightarrow B\colon E_{K'_B}(K_{AB}), \mathrm{MAC}_{K''_B}(N_B|A|E_{K'_B}(K_{AB}))$
6. $B \rightarrow A\colon \mathit{Ack}, \mathrm{MAC}_{K''_{AB}}(\mathit{Ack})$

where K_{AB} is the final established session key. This protocol has been verified (1) secure in the confidentiality of K_{AB}, (2) secure in the mutual authentication between A and B, and (3) secure against replay attacks on S. Before we proceed to describe this protocol, we would like to state that an alternative keying protocol *could have been* derived from the node-to-node key agreement protocol of SPINS [12]:

Protocol 2

1. $A \to B$: N_A, A
2. $B \to S$: $N_A, N_B, A, B, \mathrm{MAC}_{K''_B}(N_A|N_B|A|B)$
3. $S \to A$: $E_{K'_A}(K_{AB}), \mathrm{MAC}_{K''_A}(N_A|B|E_{K'_A}(K_{AB}))$
4. $S \to B$: $E_{K'_B}(K_{AB}), \mathrm{MAC}_{K''_B}(N_B|A|E_{K'_B}(K_{AB}))$
5. $A \to B$: $Ack, \mathrm{MAC}_{K''_{AB}}(Ack)$

Here, the first four messages are from SPINS and we added the fifth message to prevent authentication failure. This protocol has been verified (1) secure in the confidentiality of K_{AB}, (2) secure in the mutual authentication between A and B, and (3) secure against replay attacks on S. The drawback of this protocol is that while there is no attack, it is far too easy for an intruder to generate random bogus requests (at step 1) and hence random bogus protocol instances, that would cause B to wait at step 5 for the bogus protocol instances to finish.

A design principle behind Protocol 1 is that A should do more work compared with B to initiate the protocol. Now it takes three messages for A to get B to respond, and S is able tell B whether A's request is genuine at step 5. Let's also have a look at the energy budget. We only consider the energy consumed in radio communication, since it generally dominates that used for computation. Assuming the following sizes: node names (IDs) and nonces are 64 bits long, MACs are 160 bits, keys are 128 bits, acknowledgements are 8 bits (to cater for other status codes), encryption is length-preserving; and assuming the energy consumption for receiving to be 40% that of sending [13], we found the energy consumption of Protocol 1 compared with Protocol 2 to be 52% higher for A, 6% for B and 21% for S. The fact that A has to consume more energy in Protocol 1 cannot be helped by our requirement that A authenticates itself with S first, but this is necessary to overcome the drawback of Protocol 2. Moreover, the energy cost can be amortized over A's period of operation, provided such keying is conducted infrequently.

Special attention is given to the MAC at step 6 of Protocol 1 because MAC'ed acknowledgements are susceptible to replays depending on the MAC mode used. By assuming that combined CTR and CBC mode (CTR + CBC-MAC) [10] is used, we should be able to minimize this risk. This also applies to other instances of MAC'ed acknowledgement in the other protocols in the rest of this paper.

4.3 Inter-supervised Cluster Keying

After solving the problem of keying two nodes in the same cluster, let us consider keying two nodes from two different clusters in the supervised realm. The protocol is:

Protocol 3

1. $A \rightarrow S_A$: $N_A, B, \mathrm{MAC}_{K''_A}(N_A|B)$
2. $S_A \rightarrow A$: $E_{K'_{SS}}(N_{S_A}), \mathrm{MAC}_{K''_A}(N_A|B|E_{K'_{SS}}(N_{S_A}))$
3. $A \rightarrow B$: $A, B, E_{K'_{SS}}(N_{S_A})$
4. $B \rightarrow S_B$: $B, N_B, A, E_{K'_{SS}}(N_{S_A}), \mathrm{MAC}_{K''_B}(B|N_B|A|E_{K'_{SS}}(N_{S_A}))$
5. $S_B \rightarrow S_A$: $N_{S_B}, A, B, E_{K'_{SS}}(K_{AB2}), \mathrm{MAC}_{K''_{SS}}(N_{S_A}|N_{S_B}|A|B|E_{K'_{SS}}(K_{AB2}))$
6. $S_A \rightarrow S_B$: $E_{K'_{SS}}(K_{AB1}), \mathrm{MAC}_{K''_{SS}}(N_{S_B}|A|B|E_{K''SS}(K_{AB1}))$
7. $S_A \rightarrow A$: $E_{K'_A}(K_{AB}), \mathrm{MAC}_{K''_A}(N_A|B|E_{K'_A}(K_{AB}))$
8. $S_B \rightarrow B$: $E_{K'_B}(K_{AB}), \mathrm{MAC}_{K''_B}(N_B|A|E_{K'_B}(K_{AB}))$
9. $B \rightarrow A$: $Ack, \mathrm{MAC}_{K_{AB}}(Ack)$

where K_{SS} is intuitively the shared key between S_A and S_B; and $K_{AB} = K_{AB1}|K_{AB2}$ is the final established session key. To divide the cost of generating the whole key, the key is generated as a concatenation of two sub-keys K_{AB1}, K_{AB2} respectively produced by S_A and S_B. Note that A does not need to know who B's supervisor is – that is S_A's (and B's) job. This protocol has been verified (1) secure in the confidentiality of K_{AB}, (2) secure in the mutual authentication between A and B, and (3) secure against replay attacks on S_A and S_B.

4.4 Intra-unsupervised Cluster Keying

When two nodes in an unsupervised cluster want to communicate with each other, the only key they can use is the cluster key. To state the obvious, the cluster key is never used directly, instead, for each session a new pair of encryption and MAC keys are derived from the cluster key using the mechanism of Section 3.1.

4.5 Unsupervised-to-Supervised Node Keying

As mentioned, there is by default no trust between any supervised cluster and unsupervised cluster, i.e. there is no *direct* shared key an unsupervised node can use to communicate with a supervised node. It is here that the idea of *visitor certificate* is employed. A visitor certificate takes the form $E_{K_g}(ID|K_v)$, where K_g is the *grant key*, K_v is the *visitor key* and ID is the visitor's name. K_g is held by a supervisor while K_v and the respective visitor certificate are held by an unsupervised node.

To start a session, the unsupervised node, say U (with visitor key K_v), submits its visitor certificate $E_{K_g}(U|K_v)$ to the supervisor, say S (with grant key K_g). Upon reception, S decrypts U's certificate with K_g to obtain K_v, and sends a new session key K_{US} to U, thereby establishing a secure session with U:

Protocol 4

1. $U \to S$: $N_U, S, E_{K_g}(U|K_v), \mathrm{MAC}_{K''_v}(N_U|S|E_{K_g}(U|K_v))$
2. $S \to U$: $N_S, E_{K'_v}(K_{US}), \mathrm{MAC}_{K''_v}(N_U|N_S|E_{K'_v}(K_{US}))$
3. $U \to S$: $Ack, \mathrm{MAC}_{K''_{US}}(N_S|Ack)$

Protocol 4 has been verified (1) secure in the secrecy of K_{US}, and (2) secure in the mutual authentication between U and S. From this point onwards, the keying protocol for the communication between U and any node supervised by S, let's say A will be similar to the intra-supervised cluster keying between U and A (cf Section 4.2).

The merit of this approach is that K_g and K_v can be pre-deployed, while the visitor certificate can be generated and distributed at runtime by a *visit granting agent* (VGA). A VGA keeps an application-determined set of visitor and grant keys. Under guideline **NO_SINGLE_POINT** (cf Section 1), no single VGA should possess all the visitor and grant keys of the whole network, therefore there should be several VGAs throughout the network, each safeguarding a preferrably non-intersecting set of visitor and grant keys. A VGA works with supervisors. Upon an unsupervised node U's request for a visitor certificate to visit supervisor S, a VGA

1. checks with S and U's neighbouring supervisors if U is a green node (i.e. a good node [8])
2. checks with itself if it has U's K_v and S's K_g

If all checks are positive, then the VGA will proceed to issue the requested visitor certificate to A. Otherwise the request would be denied. In this scheme, the only pre-deployed information is the visitor keys on the unsupervised nodes and grant keys on the supervisors, instead of (x = number of supervisors × number of unsupervised nodes) key pairs. Note that it is straightforward to extend the visitor certificate to include the expiry time of the certificate.

4.6 Node Migration

Intra-supervised cluster keying and inter-supervised cluster keying work most efficiently when $S_A \rightsquigarrow A$. If it happens that $S_A \not\rightsquigarrow A$ as a result of node mobility or other factors that break the radio link, A may 'employ' another supervisor, say S_B as its temporary supervisor if $S_B \rightsquigarrow A$, by migrating its node key from S_A to S_B, provided S_A and S_B have mutual trust. If A is not able to find such a supervisor, then no migration takes place. Note that A *never* becomes a member of S_B's cluster (which implies receiving the cluster key from S_B), because then a compromised node would be able to roam to every cluster and collect their cluster key.

Suppose A migrates from the cluster supervised by S_A to the cluster supervised by S_B, the security objective of this protocol is to prevent unauthorized migrations, i.e. if A did not authorize the migration, the migration should not occur. The protocol is:

Protocol 5

1. $A \rightarrow S_A \ : N_A, S_B, \text{MAC}_{K''_A}(N_A|S_B)$
2. $S_A \rightarrow A \ : E_{K'_A}(E_{K'_{SS}}(N_{S_A})), \text{MAC}_{K''_A}(N_A|S_B|E_{K'_{SS}}(N_{S_A}))$
3. $A \rightarrow S_B \ : A, S_A, E_{K'_{SS}}(N_{S_A})$
4. $S_B \rightarrow S_A$: $S_B, N_{S_B}, A, \text{MAC}_{K''_{SS}}(N_{S_A}|S_B|N_{S_B}|A)$
5. $S_A \rightarrow S_B$: $E_{K''_{SS}}(K_{A2}), \text{MAC}_{K''_{SS}}(N_{S_B}|A|E_{K''_{SS}}(K_{A2}))$
6. $S_B \rightarrow A \ : \textit{Sucess}, \text{MAC}_{K''_{A2}}(\textit{Sucess})$

where $K_{A2} = MAC_{K_A}(0)$ is the migrated key. Note that as mentioned, it is a derivation of A's node key that is migrated. This protocol has been verified (1) secure in the confidentiality of K_{A2}, (2) secure in the authentication of A to S_A and S_B, of S_A to A and S_B, of S_B to A and S_B, and (3) secure against replay attacks on S_A.

Suppose A migrates to S_C's cluster, if S_A, S_B and S_C trust each other, but $S_A \not\leadsto S_C$ and $S_B \leadsto S_C$, then S_B should for efficiency take over S_A's role of migrating A's key to S_C.

Supervisors like S_B and S_C are not intended to keep A's key when A leaves their cluster, however if A consistently returns, then it would not be efficient to purge A's key immediately once A leaves. It is best to have a mechanism for predicting A's movement so that A's key can be optimally cached, otherwise there should at least be a hysteresis period where A's key is cached. Note that S_A *never* purges A's key. This is to ensure that there is always one supervisor holding A's key.

5 Related Work

There is currently a lack of literature detailing key management architectures for WSNs. We mentioned Basagni et al's pebblenets and Perrig et al's SPINS in Section 1. We have also made a comparison between our intra-supervised cluster keying protocol with SPINS' node-to-node key agreement protocol in Section 4.2. Now we look at some other proposals.

Slijepcevic et al propose location-based keys [14]. All nodes start out sharing a set of master keys, and the network is divided into non-overlapping hexagonal cells. At runtime, a node selects and activates a key from the set of master keys based on the output of its pseudorandom generator *and* on which hexagonal cell the node is in, so that the compromise of a key in one cell does not affect the other cells. Seen another way, the problem of session establishment reduces to the problem of determining the location; and node mobility in this scheme does not necessitate the equivalent of migration in our proposal, because it is readily 'solved' by having all nodes sharing the same keys. However, this scheme does not fit our requirements, because of our assumption that sensor nodes are not tamper-resistant – the set of master keys, the corresponding pseudorandom number generator and the seed are easily exposed and hence the active key of any cell can be found out. The requirement of nodes being able to discover their coordinates is also too demanding for our case.

Luo et al's localized trust model [11] is not exactly a key management architecture proposal, but it can be regarded as such. This model is based on (k, n)-threshold secret sharing, wherein a system secret key (of a public-private key pair) is shared among every n nodes, and a node is only trusted if it acquires a certificate issued by k neighbours. Session establishment is presumed to proceed through Diffie-Hellman-derived protocols. This solution forgoes the need for imposing structure (e.g. clustering) on the network, but it also means that if a node cannot get certified by k neighbours, it would have to roam to another place where it can. More importantly, the extensive use of public-key cryptography is also too demanding within the resource constraints of WSNs.

6 Conclusion and Future Work

In this paper we have presented a decentralized key management architecture for WSNs, covering the aspects of key deployment, key refreshment and key establishment.

Motivated by the assumption that sensors are not tamper-resistant, we have devised an architecture in which two security realms namely the supervised realm and the unsupervised realm provide two levels of security. Communication in the supervised realm entails various protocols to implement intra- and inter-supervised cluster keying, which we have verified with our tool CoProVe. In the unsupervsied realm, security suffers from the fact that a cluster key is shared by all members of the same cluster. The trade-off provides WSN implementors with an option to partition their system according to the actual security requirements, and the network can be expanded by easily adding one node at a time, one cluster at a time, or even one domain at a time. To conclude, our architecture meets our hardware constraints, and the guidelines **NO_SINGLE_KEY**, **NO_SINGLE_POINT** and **SCALABILITY** of Section 1.

We have adopted a two-step design process by getting the protocols right first before proceeding to optimize them. Our next step is to validate our model through simulation and implementation. We are also motivated to extend our architecture to support directed diffusion (which implies multicast communication) [9]. As of now, our assumption is that a node knows exactly the name of the other node it wants to communicate with, which is not the case in directed diffusion.

Acknowledgement. The authors would like to thank Roberto di Pietro, Paul Havinga and the anonymous reviewers for their insightful comments.

References

1. M. Abdalla and M. Bellare. Increasing the lifetime of a key: A comparitive analysis of the security of rekeying techniques. In T. Okamoto, editor, *Advances in Cryptology – ASIACRYPT 2000*, volume 1976 of *LNCS*, pages 546–565. Springer-Verlag, 2000.
2. S. Basagni, K. Herrin, D. Bruschi, and E. Rosti. Secure pebblenets. In *Proceedings of the 2001 ACM Int. Symp. on Mobile Ad Hoc Networking and Computing*, pages 156–163. ACM Press, October 2001.
3. M. Brown, D. Cheung, D. Hankerson, J. L. Hernandez, M. Kirkup, and A. Menezes. PGP in Constrained Wireless Devices. In *9th USENIX Security Symposium*, pages 247–261. USENIX Association, Aug 2000.
4. D.W. Carman, P.S. Kruus, and B.J. Matt. Constraints and approaches for distributed sensor network security. Technical Report #00-010, NAI Labs, 2000.
5. J.L. Carter and M.N. Wegman. Universal classes of hash functions. *Journal of Computer and System Sciences*, 18:143–154, 1979.
6. R. Corin and S. Etalle. An improved constraint-based system for the verification of security protocols. In M. Hermenegildo and G. Puebla, editors, *9th Int. Static Analysis Symp. (SAS)*, volume 2477, pages 326–341, Madrid, Spain, Sep 2002. Springer-Verlag.
7. S.A. Crosby and D.S. Wallach. Denial of service via algorithmic complexity attacks. In *12th USENIX Security Symposium*. USENIX Association, 2003. To appear.
8. R. di Pietro, L.V. Mancini, and S. Jajodia. Secure selective exclusion in ad hoc wireless network. In M.A. Ghonaimy, M.T. El-Hadidi, and H.K. Aslan, editors, *Security in the Information Society: Visions and Perspectives*, pages 423–434. Kluwer Academic Publishers, 2002.
9. C. Intanagonwiwat, R. Govindan, and D. Estrin. Directed diffusion: A scalable and robust communication paradigm for sensor networks. In *6th Annual Int. Conf. on Mobile Computing and Networking (MobiCOM '00)*, pages 56–67, Boston, Massachusetts, United States, 2000. ACM Press.
10. J. Jonsson. On the security of CTR + CBC MAC. In K. Nyberg and H.Heys, editors, *Selected Areas in Cryptography, 9th Annual Workshop (SAC 2002)*, volume 2595 of *LNCS*, pages 76–93. Springer-Verlag, 2002.
11. H. Luo, P. Zerfos, J. Kong, S. Lu, and L. Zhang. Self-securing ad hoc wireless networks. In *7th IEEE Symp. on Computers and Communications*, pages 567–574, 2002.
12. A. Perrig, R. Szewczyk, V. Wen, D. Culler, and J.D. Tygar. SPINS: Security Protocols for Sensor Networks. In *Proceedings of the 7th Ann. Int. Conf. on Mobile Computing and Networking*, pages 189–199. ACM Press, 2001.
13. RF Monolithics, Inc. TR1001: 868.35 MHz Transceiver. Datasheet. `http://www.rfm.com/products/data/tr1001.pdf`.
14. S. Slijepcevic, V. Tsiatsis, S. Zimbeck, M.B. Srivastava, and M. Potkonjak. On communication security in wireless ad-hoc sensor networks. In *11th IEEE Int. Workshops on Enabling Technologies: Infrastructure for Collaborative Enterprises*, pages 139–144, Jun 2002.

JMS on Mobile Ad Hoc Networks*

Einar Vollset[1**], Dave Ingham[2], and Paul Ezhilchelvan[1]

[1] School of Computing Science,
University of Newcastle,
Newcastle upon Tyne, NE1 7RU,
United Kingdom
{einar.vollset,paul.ezhilchelvan}@ncl.ac.uk
[2] Arjuna Technologies Ltd,
Nanotechnology Centre,
Herschel Building,
Newcastle upon Tyne, NE1 7RU,
United Kingdom
dave.ingham@arjuna.com

Abstract. The Java Message Service (JMS) provides a standard asynchronous messaging API, which simplifies the construction of loosely coupled, distributed applications. This paper describes the design and implementation of a pure Java JMS solution for mobile adhoc networks (MANETs). The resulting JMS solution simplifies not only the construction of MANET applications but also the re-deployment of any existing JMS applications into a MANET context. The central contribution of this paper lies in comprehensively identifying and addressing the design challenges encountered. For example, common JMS implementations rely, for reasons of efficiency, on a central, reliable server for maintaining dynamic state information, e.g. routing information and group configuration. Construction of a JMS solution for MANETs, on the other hand, cannot rely on such a centralized server. Our server-less implementation involves building a new MANET transport module that is incorporated into an existing JMS product. This transport module implements a MANET multicast routing protocol that provides publish/subscribe semantics by mapping JMS topics to multicast addresses. To our knowledge, our implementation is the first ever Java middleware platform built for MANETs.

Keywords: MANETs, JMS, Mobile Ad-hoc Networks, Message-oriented Middleware, Multicast.

1 Introduction

A mobile ad-hoc network (MANET) is a network made up of a collection of mobile, autonomous nodes that communicate using wireless multi-hop links without any fixed

---

* This work has been supported in part by the EPSRC PACE project and Santosh Shrivastava at the Distributed Systems Research Group at the University of Newcastle upon Tyne, UK.

** Correspondence Author: Tel: +44 (0)191 2228546, Fax: +44 (0)191 2228232.

M. Conti et al. (Eds.): PWC 2003, LNCS 2775, pp. 40–52, 2003.

infrastructure such as base stations. Only pairs of nodes that lie within each other's transmission radius can communicate directly. However, each node in the network acts as a router and participates in forwarding packets for other nodes.

MANET technology is currently an active area of research. There is, for example, a large number of proposed routing protocols, such as Ad Hoc On-demand Distance Vector (AODV)[1] and Dynamic Source Routing (DSR)[2]. However, off-the-shelf wireless devices, such as PDAs with 802.11b networking, do not currently possess a standard built-in routing functionality necessary to form MANETs. This lack of standardized routing functionality can be argued to be holding back the development of applications for MANETs. In this paper we attempt to address the issue of simplifying the development of MANET applications. Our work is influenced by our experience in providing support for application development in the traditional wired networking world.

In the traditional wired networking world, distributed application development is greatly simplified through the use of middleware technologies, such as distributed objects systems, e.g., CORBA[3], and message-oriented middleware (MoM)[4]. Of these technologies, MoM appears to be the most appropriate model for the MANET environment, for two main reasons: a) The message exchange model of MoM is asynchronous (as opposed to RPC-style communication), which is the most appropriate model for MANETs. b) The most popular messaging model in MoM, namely publish/subscribe messaging, involves one-to-many communication which maps well to the broadcast nature of MANETs.

The goal of the research presented here was to develop MoM middleware for MANETs based on standard off-the-shelf hardware and operating systems. Specifically, this paper describes the design and implementation of a messageoriented middleware solution for MANETs that runs on top of a standard J2SE (Java 2 Standard Edition) virtual machine on a computer with 802.11b networking. The system is largely compliant with the Java Message Service (JMS) specification[5] which defines a standard asynchronous messaging API to message-oriented middleware. Designing such a system in the MANET context raises new challenges not present in the wired networking world. We formulate these challenges and describe how we addressed them in building a JMS system that supports publish/subscribe messaging. This system not only simplifies the development of new MANET applications, but also allows existing JMS applications to be deployed in a MANET environment. As a proof of concept we deployed an existing instant messaging application, originally developed for wired networks, on a real MANET.

The paper is structured as follows. To present the design challenges faced in designing a JMS system for MANETs into context, the next section provides a description of the JMS specification. This forms the background to section 3, which identifies the issues that need to be addressed when providing JMS on MANET infrastructure. In particular, we argue that the approaches adapted from wired networks are not appropriate. Sections 4 and 5 describe the design and implementation of our system. Section 6 describes testing and the proof of concept deployment of an instant messaging application, while section 7 describes how the system could be extended to provide more powerful message delivery semantics. Finally, we present our concluding remarks.

2 Java Message Service

The Java Message Service (JMS) is a specification developed by Sun Microsystems that provides a common way for Java applications to access each other via message-oriented middleware (MoM).

The JMS specification defines a set of interfaces and associated semantics that allow Java applications to create, send, receive, and read messages. The specification does not define how messages need to be transported with in a particular implementation, known as a *JMS Provider.* This clear separation of concerns was essential in order to allow vendors of existing messaging products, such as IBM MQSeries[6] and Tibco Rendezvous[7] to support the JMS specification. Furthermore, some JMS vendors provide multiple message-transportation implementations within the same product thereby providing the user with the ability to select the most appropriate transport technology for a particular deployment. An example is the Arjuna Message Service (Arjuna-MS)[8] (formerly the Hewlett Packard Message Service), which provides both server-based and server-less message transport technologies that can be mixed within the same application.

2.1 JMS Messaging Models

The JMS API includes support for the two most common enterprise-messaging models, namely, message queuing and publish/subscribe.

The message queuing model provides one-to-one delivery semantics. Clients send messages to, and receive messages from, *queues.* There may be multiple sender and receivers associated with a given queue, but each message sent by a sender is consumed by exactly one receiver. This means that if multiple receivers are associated with a queue, some sort of arbitration is required by the JMS Provider to decide which one will receive a given message.

The publish/subscribe (pub/sub) model complements the queuing model in that it provides one-to-many delivery semantics. Clients publish messages to and receive messages from *topics.* Each message may be consumed by zero, one, or more subscribers.

2.2 Reliability

JMS is able to support both best-effort and guaranteed delivery semantics called respectively *non-persistent* or *persistent* delivery modes. Non-persistent messages are delivered in a best-effort fashion by the JMS Provider. That is, they will generally be delivered but in the event of an abnormal condition, such as resource-exhaustion or a process failure, the message may be lost. Conversely, if a message producer specifies that persistent delivery mode should be used, then this is an indication to the JMS Provider that steps should be taken to ensure that the message is not lost in the presence of abnormal conditions.

Thus a full JMS implementation can provide 4 types of message delivery semantics: pub/sub non-persistent, pub/sub persistent, queuing non-persistent and queuing persistent.

3 Issues Arising When Providing JMS on MANETs

JMS products typically utilize a hub-and-spoke architecture in which clients connect to a central server (or cluster of servers) that manages the reliable message routing between them. However, a server-based JMS solution is not appropriate for mobile ad-hoc networks for a number of reasons. Mobile ad-hoc networks are aimed at environments where fixed infrastructure support is unavailable. The philosophy of MANET applications therefore is not to place reliance on centralized services, rather, such service functionality should be distributed across nodes in the MANET so as to minimize the dependency on any particular node and therefore improve overall system resilience. Furthermore, the nodes participating in MANETs are often resource-constrained devices such as PDAs or mobile phones. Because of this, it is advantageous for the processing and networking overheads to be *distributed* across the nodes rather than concentrated at a single server node.

Building a JMS solution for MANETs therefore requires a server-less architecture in which nodes collaborate in a peer-to-peer manner to provide the JMS functionality. We have identified the following issues (in increasing order of complexity), which need to be resolved in order to provide such a solution:

1. **Configuration:** Configuration is the means with which client discover the available queues and topics. A server-based JMS Provider can manage this configuration information on a central server and respond to client queries on configuration. In a server-less design, however, there is no such central server, and the information needs to be maintained in a distributed manner.
2. **Message transportation:** Server-based Providers typically use unicast communication for message transportation. Message producers send messages to the server, which stores and forwards them to the consumers, either a single receiver for the queuing model or several subscribers for the pub/sub model. This store-and-forward method is unsuitable for a MANET-based Provider as there can be no central message delivery mediator.
3. **Reliability:** Guaranteed message delivery in server-based JMS Provider is achieved by the server persisting the messages in a central database. Since the server manages the communication between all clients, it knows when a message has been consumed by all intended recipients and can be discarded from the database. Such a central database is not readily available in a MANET environment, as the availability of any one node at all times cannot be relied upon. This greatly increases the complexity involved in achieving guaranteed message delivery.
4. **Coordination:** As previously noted, if there are several receivers associated with a queue, then for each message, the provider has to select a single receiver to consume the message. With a server-based architecture it is straightforward for the server to make this decision. In a server-less architecture, distributed coordination mechanisms are required if fair, non-simplistic queue semantics are to be achieved (i.e. the messages are evenly distributed among the associated receivers). Such mechanisms require the use of an agreement (consensus) protocol. Note that the consensus problem cannot be deterministically solved in an asynchronous environment [23] and protocols are possible only when certain assumptions restricting asynchrony are possible to make.

In the following section, we describe how we addressed the first two of these issues, resulting in the construction of JMS system for MANETs with pub/sub non-persistent message delivery semantics. Section 6 addresses how the third issue can be solved and incorporated into the implemented system. We then indicate the ways of solving or circumventing the distributed coordination problem in MANETs. Our ongoing work involves the former.

4 Designing a JMS Provider for MANETs

In order to provide a complete JMS solution for MANETs, the design issues described in the previous section would all have to be dealt with.

The next two sections describe how we have designed a system based on an existing messaging product, which provides non-persistent JMS topic semantics by addressing the two primary issues (configuration and message transportation) required in order to develop a JMS solution for MANETs.

4.1 Configuration

The server-less nature of our JMS solution leads to the somewhat meta-physical question: when do queues and topics come into being, and how do the JMS clients find out about them? In traditional server-based JMS providers, an administrator creates a destination on the JMS server. It thus resides on the server as a tangible entity from creation until it is destroyed by an administrator. All clients then typically use JNDI to lookup by name, the queues and topics they require. With MANETs it is not so straightforward, as there is no central entity to lookup topics and queues from.

The approach taken in our system is to require all clients participating in a message exchange to have a local copy of an identical configuration file. This configuration file then contains all information the client needs to lookup and use the destination. This resolves both of the above issues, as a lookup of destinations only requires reading the configuration file, and a queue or a topic will have existed "forever" as far as the client is concerned. Although inflexible, it is worth considering that the creation of queues and topics on server-based JMS providers requires the intervention of an administrator on the JMS Server. Of course, there are other possible and more flexible solutions, including broadcasting destination information on a well-known multicast address, but these are outside the scope of this paper.

4.2 Message Transportation

In order to provide the required message transportation, an application-level implementation of a multi-hop routing protocol had to be developed. The reason for this is the lack of readily available MANET routing protocols integrated into the network layer in most commercially available mobile devices. As the main goal of this research was to provide JMS *topic* semantics to MANETs, it was decided to implement a *multicast* multi-hop routing protocol. In multicast, the key concept is that of a multicast group, which any node can send and receive packets to and from. In order for a packet to be delivered to all the members of a multicast group, a node only needs to send a packet once to a given multicast address. Clearly this functionality closely mirrors that of JMS topics, where a message producer only needs to send a message once to a given topic in order for all subscribers on that topic to receive the message.

Several multicast protocols specifically designed for MANETs have been proposed. These include ODMRP[9], AMRIS[10], CAMP[11] and MAODV[12]. ODMRP, or the On Demand Multicast Routing Protocol, was chosen because it has performed well with regards to throughput and control packet overhead in several simulation studies[13, 14].

A Java application-level implementation of ODMRP has been developed. This implementation has been named the *Java on-demand multicast routing protocol (Jomp).* Further details on Jomp can be found in the next section.

After developing Jomp, the required multicast functionality was now available for MANETs, and the Arjuna-MS client library has been extended to map between JMS publish-subscribe semantics and the multicast functionality provided by Jomp. Figure 1 shows a conceptual view of the Arjuna-MS client library and how it uses Jomp and its configuration file (as described above) to provide JMS messaging on MANETs.

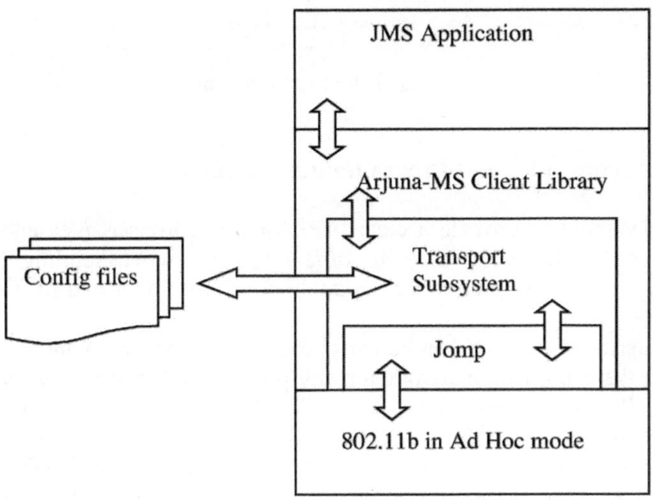

Fig. 1. Conceptual view of the Arjuna-MS client library for MANETs.

5 Implementing a JMS Provider for MANETs

This section describes in more detail the implementation aspects of extending Arjuna-MS to MANETs. As mentioned in section 4.2, a Java application-level version of the ODMRP protocol (Jomp) was implemented in order to provide the required message transportation for the system. Section 5.1 describes the architecture and programming interface to Jomp, while section 5.2 describes our experiences implementing an application-level routing protocol in Java. A complete description of the system is available in [24].

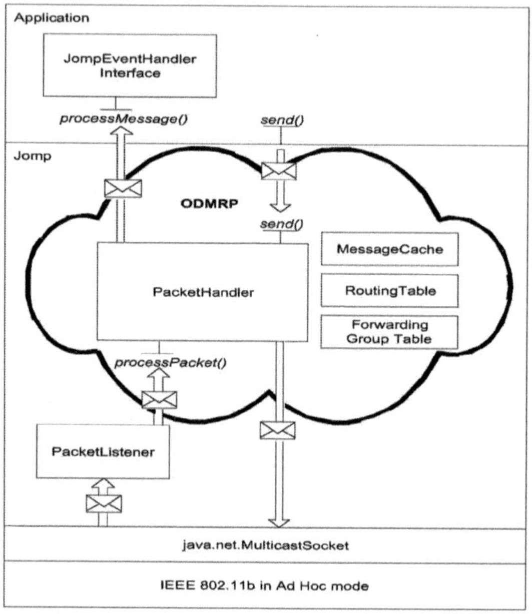

Fig. 2. Jomp architecture

5.1 Jomp Architecture and Programming Interface

Jomp was designed to provide a clean API for asynchronous message passing. This was achieved by designing a *Jomp object* for use when sending packets and a *JompEventHandler interface* which an application developer had to implement when using Jomp.

Jomp consists of 5 main components: the *PacketListener*, which is a thread that continuously listens on the network for any packets and passes any received packets up to the *PacketHandler*. The PacketHandler contains most of the ODMRP functionality and uses the *MessageCache, RoutingTable* and *ForwardingGroupTable* objects to make routing decisions (The last 3 are tables used by ODMRP, for further detail, see [9]). These routing decisions may result in the PacketHandler broadcasting a packet, passing the packet up to the application at the local host or discarding the packet.

Figure 2 shows a conceptual view of the architecture of Jomp, showing how these components interact and how Jomp interacts with the application using it and the network. The large "ODMRP cloud" indicates where the ODMRP protocol functionality is implemented.

5.2 Experiences Implementing a Java Application-Level Routing Protocol

Implementing anything in Java always raises the issues of performance, particularly when implementing something traditionally as low level as a routing protocol.

However, based on the fact that Jomp is an application-level protocol, and on the observation that previous application-level implementations of routing protocols in Java, such as LRMP[15], has adequate performance, we have come to the conclusion that performance in most cases are not adversely affected by the choice of an interpreted programming language.

A more pertinent issue when using Java is the lack of direct access to devices such as the 802.11b network card. Most simulations and implementations of MANET routing protocols imply that the network card has to be set to promiscuous mode, where it will receive packets destined to all addresses. This is not possible in Java, and could have been problematic, as without a work-around, only nodes interested in a given multicast address would be aiding in forwarding packets on that multicast group.

The implication of this with respect to using Jomp as the transport for Arjuna-MS is that a node will only aid in forwarding messages destined for topics that the node is subscribed to. Clearly this is not beneficial for the MANET as a whole. In fact the problem with "selfish" nodes is something that has been highlighted as a problem for a number of routing algorithms and applications in MANETs[16].

However, we worked around this limitation by sending all packets on a single, configurable multicast address. The packets that were sent on this address were all tagged with the relevant multicast addresses mapping to the topic to which they were destined. This setup means that all nodes running Jomp will participate in the routing and forwarding of all packets, not just packets destined for the multicast addresses they were interested in. It is thought that there is no performance hit due to this, as the sending and receiving of multicast packets using an 802.11b network card is the same no matter what multicast address the packets are sent to.

6 Simulation and Real World Deployment

Testing real world implementations of any type of application built for MANETs presents a number of difficulties due to the dynamic nature of the MANET environment. One of the main problems is the lack of control the tester has over the accuracy of test parameters, with issues such as interference from other radio sources and the difficulties in managing a (potentially high) number of mobile units making the testers task very difficult indeed. Because of this, a network simulator, the Java Network Simulator (JNS)[18], has been modified to enable it to use real world implementations as nodes in a simulation. This allows tests of real world MANET applications to be carried out in a controlled environment. The next section describes how we used JNS to verify the correct operation of Jomp and to run a simple JMS application.

In addition to this verification through simulation, we realize that any MANET middleware needs to be tested in the real world. As an indication of the potential of Arjuna-MS for MANETs we have deployed Arjuna Chat Demonstrator, the instant messaging demonstration application that ships with the standard, wired version of Arjuna-MS, in a real world scenario. Details of this deployment are described in section 6.2.

6.1 Testing

As ODMRP has already been shown in literature [13,14] to perform well compared to other MANET multicast protocols, the emphasis of the testing was to satisfy ourselves that Jomp adhered to the ODMRP specification.

In order to verify the correct operation of both Jomp itself and Arjuna-MS using Jomp, we setup a simulation environment with 10 nodes with partial connectivity, as could be found in a MANET at any instant in time. In this scenario, we made one of the nodes publish messages to 9 subscribers. The trace files describing the message flow between the nodes verify that Jomp act according to the ODMRP specification. Additionally, all the JMS subscribers received the published message, which shows that Arjuna-MS using Jomp also works. Figure 3 shows the initial broadcast from the source node. Space limitations prevent us from presenting the complete sequence of message exchanges, but a video clip showing the interaction and several screenshots is available at [19].

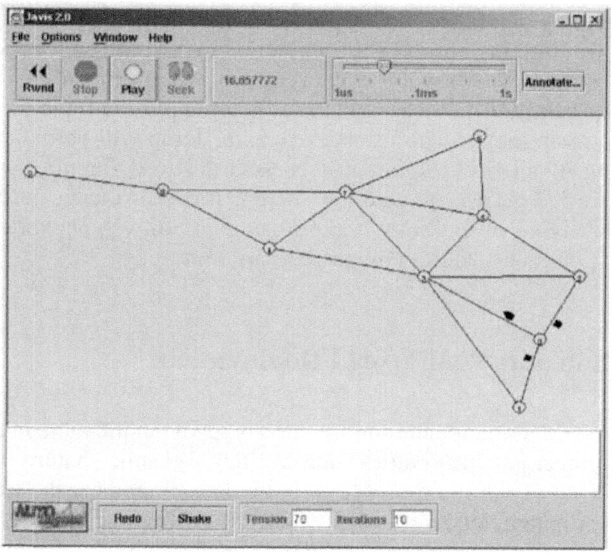

Fig. 3. A simulation run of Arjuna-MS running on JNS

6.2 Proof of Concept

In addition to the test runs on JNS, we have also deployed quite a large JMS application, the Arjuna Chat Demonstrator instant messaging demo application, in a real world scenario. This involved almost no change in the code base of the application, which shows the advantage of providing standard middleware for MANETs. The application, although not extensively tested, was successfully used in an office environment with several instances of the application running. Some anecdotal evidence of this successful deployment is available at [19].

7 Extensions and Future Work

7.1 Reliability

Currently, our JMS solution does not provide persistent messaging. This is because Jomp is an implementation of the basic operation of the ODMRP protocol, which only provides best effort message delivery guarantees.

Typical server-based JMS providers usually implement persistent messaging by putting the messages on stable storage on a central server. However, there are weaknesses even with this approach, as the assumption made when making a message persistent on the server is that the disk will not fail. This assumption is not necessarily 100% accurate in all situations, as disks might indeed fail. However, the underlying assumption is that the level of fault tolerance is dependant on the environment in which the JMS provider is operating. This implies that in an inherently unreliable environment such as MANETs, a JMS provider does not need to provide the same level of fault tolerance as it would do in for example a wired environment where the JMS provider could make use of for example RAID disks or off-site replication.

Based on this observation, if it is possible to increase the packet delivery ratio of Jomp to a high enough level, it could be argued that both of JMS' message delivery guarantees could be provided by Arjuna-MS over Jomp.

One method of extending Jomp to reach the required level of reliability is by adding *Anonymous Gossip (AG)* [20] to Jomp. In general, gossip is the name given to the technique where nodes share the data they possess with other nearby nodes, in order for all nodes to share a consistent view of what has been communicated.

Simulation has shown[20] that the addition of Anonymous Gossip to an unreliable multicast protocol will substantially increase the reliability of the protocol. Based on this observation, it is our opinion that adding Anonymous Gossip to Jomp would sufficiently increase the reliability of Jomp in order for us to be able to claim that Arjuna-MS on MANETs support persistent messaging.

7.2 Coordination

Extending Arjuna-MS on mobile ad-hoc networks to support queues is a different matter altogether, and the main problem is achieving the required semantics of a JMS queue without incurring too much of an overhead.

Broadly speaking there are two main approaches to achieving JMS queue semantics. The first is to use a multicast protocol implementation, such as Jomp, to disseminate a message sent to a queue to all nodes containing receivers on that queue. The nodes would then need to agree amongst themselves which one receiver is to be handed the message. This would be achieved through some sort of consensus algorithm, for example through the use of a distributed leader election algorithm, where the elected leader would pass the message up to its queue receiver. This approach has the benefit that we can use Jomp without modification and that fair, non-simplistic queue semantics is achieved. However, it has the drawback that comes with any currently available consensus algorithms for MANETs, in that it can be very costly both in terms processing power and network bandwidth.

The other approach is to weaken the fairness of JMS queue semantics and thus circumventing the distributed coordination problem. One way to do this is to handle all synchronization on the sender side. For example, JMS queue semantics could be achieved by having the node containing the queue sender know the identity of all the nodes with queue receivers. The node containing the queue sender would then simply pick one queue receiver and unicast the packet to that receiver. This approach has the benefit that no processing is required on the nodes with the queue receivers, and the required bandwidth is substantially less compared to the first approach. However, this approach does require the implementation of a unicast protocol, as well as requiring the nodes with queue senders to maintain some sort of membership information of the queue receivers. This in itself adds an overhead both in processing and network bandwidth. Another way to circumvent the coordination problem is through the use of an *anycast protocol.* Anycasting protocols are a fairly recent development in the mobile networking community, and essentially provide one-to-any delivery semantics. That is, a packet that is being anycasted will be delivered to only one of a group of nodes, typically the closest to the sender. An example of an anycast protocol is the extensions made to the Temporally Ordered Routing Algorithm (TORA)[21] when it was extended to support geocasting (GeoTORA)[22].

Clearly the two approaches circumventing the distributed coordination problem are fairly similar, diluting the message queuing semantics and fairness in order to reduce overhead, and choosing between these two and the use of a consensus protocol will have to be based on the desired properties of the system.

Our current research involves developing consensus algorithms specifically for MANETs, which aim to reduce the required network and processing bandwidth, thus making it more suitable for the MANET environment, and perhaps making the above choice a little easier.

8 Conclusion

Mobile ad-hoc networking is likely to become an important future communication technology; many interesting application areas have been suggested, including applications in vehicular and sensor networks. However, the development of such applications is currently held up by the unavailability of MANET-capable mobile devices. Current wireless devices, such as PDAs with 802.11b networking, do not support the necessary multi-hop routing protocols as standard.

Despite such disadvantages and due to potential application benefits, this paper has met the objective of designing and implementing a Java Message Service (JMS) solution for MANETs using an application-level multicast routing protocol. Since JMS is known to simplify the construction of loosely coupled distributed applications using familiar APIs and messaging models, the provision of a JMS solution for MANETs greatly eases the effort required to build MANET applications.

Our implementation was integrated into the Arjuna Message Service product as a pluggable transport module. Currently, our JMS solution for MANETs supports nonpersistent publish/subscribe messaging. Several existing JMS applications have been successfully deployed on top of it without the necessity for any code changes. Further verification of the correct operation of the underlying multicast routing protocol was obtained by extending an existing network simulator (JNS) to trace the behaviour of the executing JMS applications. Ideas for future work include support for the queuing messaging model through the development of distributed consensus algorithms for MANETs and reliable delivery through the use of message replication using the anonymous gossip protocol.

References

1. Charles E. Perkins, Elizabeth M. Royer, and S. Das,*Ad Hoc On Demand Distance Vector (AODV) Routing*. draft-ietf-manet-aodv-10.txt, 2002
2. Johnson, D.B. and D.A. Maltz, *Dynamic Source Routing in Ad Hoc Wireless Networks*, in *Mobile Computing*, H.F. Korth, Editor. 1996, KluwerAcademic Publishers Group. p. 153–179.
3. Richard E. Schantz and D.C. Schmidt, *Middleware for Distributed Systems: Evolving the Common Structure for Network-centric Applications*, in *Encyclopedia of Software Engineering*. 2001, Wiley & Sons.
4. Banavar, G., et al., *A Case for Message Oriented Middleware*, in *Distributed Computing*, P. Jayanti, Editor. 1999, Springer Verlag Kg. p. 1–18.
5. Mark Hapner, et al., *Java Message Service*. Version 1.0.2b, http://java.sun.com/products/jms/docs.html, 2001
6. *IBM website*. http://www.ibm.com,
7. *Tibco website*. http://www.tibco.com,
8. *Arjuna website*. http://www.arjuna.com
9. Lee, S.J., M. Gerla, and C.C. Chiang. *On-Demand Multicast Routing Protocol*. in *Wireless communications and networking conference*. 1999. New Orleans; LA: IEEE Operations Center.
10. Wu, C. and Y. Tay, *AMRIS: A Multicast Protocol for Ad Hoc Wireless Networks*. Milcom, 1999. **1**: p. 25–29.
11. Garcia-Luna-Aceves, J.J. and E.L. Madruga, *The Core-Assisted Mesh Protocol*. Ieee Journal on Selected Areas in Communications Sac, 1999. **17**(8): p. 1380–1394.
12. Elizabeth M. Royer and C.E. Perkins, *Multicast Ad hoc On-Demand Distance Vector (MAODV) Routing*. Work in progress, 2000
13. Kunz, T. and E. Cheng. *On-Demand Multicasting in Ad-Hoc Networks: Comparing AODV and ODMRP*. in *Distributed computing systems*. 2002. Vienna: IEEE Computer Society.
14. Lee, S.J., et al. *A Performance Comparison Study of Ad Hoc Wireless Multicast Protocols*. in *Computer communications; IEEE INFOCOM 2000*. 2000. Tel Aviv, Israel: Ieee.
15. Liao, T., *WebCanal: a multicast Web application*. Computer Networks and Isdn Systems, 1997. **29**(8/13): p. 1091–1102.
16. Sonja Buchegger and J.-Y.L. Boudec. *Performance Analysis of the CONFIDANT Protocol*. in *MobiHoc*. 2002. Lausanne, Switzerland.
17. S-J Lee, William Su, and M. Gerla, *On Demand Multicast Routing Protocol (ODMRP) for Ad Hoc Networks*. http://www.ietf.org/proceedings/00jul/ID/ manet-odmrp-02.txt, Work in progress, 2000
18. *The Java Network Simulator (JNS)*. http://jns.sourceforge.net,
19. Arjuna-MS for MANETs webpage http://www.cs.ncl.ac.uk/people/einar.vollset/home.formal/arjunamanet.html, 2002
20. Chandra, R., V. Ramasubramanian, and K. Birman, *Anonymous Gossip: Improving Multicast Reliability in Mobile Ad-Hoc Networks*. International Conference on Distributed Computing Systems, 2001. **21**: p. 275–283.
21. Park, V.D. and M.S. Corson, *A Highly Adaptive Distributed Routing Algorithm for Mobile Wireless Networks*. Ieee Infocom, 1997. **3**: p. 1405–1413.

22. Ko, Y.B. and N. Vaidya. *GeoTORA: A Protocol for Geocasting in Mobile Ad Hoc Networks*. in *Network protocols*. 2000. Osaka, Japan: IEEE Computer Society.
23. M.J. Fischer, N.A. Lynch, and M.S. Paterson, "Impossibility of Distributed Consensus with one faulty Process," Journal of the ACM, Vol. 32, No. 2, pp. 374–382, April 1985
24. Einar Vollset, "Extending an enterprise messaging system to support mobile devices", MSc thesis, University of Newcastle upon Tyne, September 2002.

Dynamic Device Access for Mobile Users

Dirk Kutscher and Jörg Ott

Technologiezentrum Informatik (TZI), Universität Bremen,
Postfach 330440, 28334 Bremen, Germany,
{dku,jo}@tzi.uni-bremen.de

Abstract. The Dynamic Device Association (DDA) framework presented in this paper provides service discovery and secure device access for mobile users. We discuss scenarios and requirements for enabling mobile users to find and access co-located (mobile) multimedia services on demand. As example, we use the dynamic discovery of IP telephony services in a network that an application on a mobile device can locate, personalize and control on behalf of a user. We examine existing protocols such as SLP and UPnP and derive a different, more general solution that addresses the issues of scalability for service announcements and security for device access. We outline the DDA approach with its five phases: service discovery, selection, device association, application protocol operation, and device dissociation. We show how the DDA approach can be used to bootstrap arbitrary application protocol sessions and present a concrete protocol specification and its implementation.[1]

1 Introduction

Mobile users increasingly rely on the ubiquitous availability of applications ranging from paging and voice communications to Internet access and web browsing to full-fledged multimedia collaboration and entertainment. However, the typical mobile device is not able to implement all the applications of potential interest to a user, at least not equally well. This mostly stems from its form factors and the respective incompatible requirements for different types of applications (e.g. text processing with a large screen and keyboard vs. voice communications with minimal weight and size). Hence, typical users carry numerous devices, that overlap in some parts of their functionality and are complementary in others: notebook computers, tablet PCs, PDAs, and cellular phones comprise just a subset of the variety of devices available today.

To overcome the inherent limitations of individual mobile devices, they may be combined with one another to create a more powerful working environment. And they may be complemented by functionally richer, more powerful, or just specialized ones: "personal" appliances such as desktop computers and telephones in the office space or at home as well as "public" devices such as PCs in Internet cafés, speaker phones in conference rooms, or video presentation and communication equipment in dedicated meeting rooms. One notable example is the use of IP telephones in conjunction with PDAs, where the latter support configuring and controlling the former: by means of

[1] A long version of this paper is available from `http://www.dmn.tzi.org/dda/`.

M. Conti et al. (Eds.): PWC 2003, LNCS 2775, pp. 53–64, 2003.

some infrared interface [ABL02] or through third party call control.[2] Users may log in to any such IP phone device within their domain and take control of it; i.e. they become able to place calls with their privileges under their identity and may receive calls placed to their phone number or URI and have all their personal settings loaded into the phone upon login.

In this IP telephony scenario, an employee is able to move around in the office and take control of an arbitrary phone by means of a portable device (provided she possesses the necessary permissions). This concept can be generalized in several ways: 1) IP telephony can be extended to arbitrary services, i.e. a framework for device access should not be restricted to a certain class of applications and associated protocols. 2) Devices providing certain services do not need to be found "manually" by the user but can rather be discovered automatically. 3) While numerous service discovery protocols are available today, those mostly address rather closed and static deployment environments (see also section 3). The closed service location environment is opened up to support "external" users and even enable public services.

In this paper, we combine ideas from auto-configuration, service discovery, and component integration to design and implement a protocol framework for Dynamic Device Association (DDA). DDA includes discovering, potentially locating, and associating with devices for authorized users; allows for bootstrapping application protocols; and provides means for dissociation after use. All DDA components are based upon standardized protocols as far as possible. In section 2, we present a general outline of the problem and identify the various steps of a generic solution. In section 3, we review related work for each of the necessary steps based upon which we devise a solution and present our system design in section 4. We conclude this paper with a summary and a discussion of open issues as well as future work in section 5.

2 Conceptual Overview

In a sample mobile user scenario, Alice and her visitor Bob meet in a conference room of Alice's company. Alice brings her laptop, Bob his PDA, both use the available WLAN. The conference room is equipped with a SIP-based conference phone and a regular SIP phone. Alice's laptop and Bob's PDA find two devices offering telephony services and obtain their locations and labels. Alice attaches to both devices and uses the conference phone to place a call to Carol with whom they are supposed to have a tele-chat. Alice and Bob use the speakerphone for the voice conversation and they use Alice's laptop to add a slide presentation to the call. During the teleconference the speakerphone is not accessible for others. Bob is only allowed to access the regular SIP phone to which he also attaches. Both may receive incoming calls through the regular SIP phone, but only Alice may place outgoing ones. After the conference call completes, Alice and Bob dissociate from the two phones and leave.

From this scenario, we can roughly identify the following five phases of dynamic device association:

[2] E.g.,by using CTI solutions such as TAPI [MS99] or SIP third party call control [RPSC02].

1. Service and Device Discovery
   Initially, the mobile devices need to find services offered by other devices. The discovery obviously involves an identification of the services, their availability (free vs. in use), their location information (to determine physical proximity), and a rendezvous address.
2. Device selection
   Once a set of suitable devices has been found, the user may select a particular device. This selection process may also be fully automated (by user preference, physical proximity, or some other algorithm).
3. Service and device association
   As soon as the user has picked a particular device, an association process is invoked by the mobile device. The selected service is contacted and an authentication procedure is carried out, after which the application protocol is bootstrapped: all necessary configuration parameters are exchanged and the application protocol is initialized.
4. Application protocol operation
   The application protocol is run in the context of the dynamically established association. This may involve all kinds of interactions between the mobile and the associated device.
5. Service and device dissociation
   When the associated device is no longer needed, the user's device dissociates from the device, freeing all allocated resources, and potentially making the device fully available again to the public.

Different protocols may be employed to implement the necessary mechanisms for each of the above phases. The following section discusses related work in this area, particularly regarding phases 1 and 3.

3 Related Work

Service discovery and the establishment of communication session between peers is addressed by multiple protocols and architectures, including but not limited to the *Service Location Protocol* (SLP, [GPVD99]), *Universal Plug and Play* (UPnP, [MS00]), *Salutation* [SAL99] and *Jini* [SUN01]. We will compare SLP and UPnP in the following.

3.1 Service Location Protocol

The *Service Location Protocol* (SLP, [GPVD99]) is a framework for service discovery and service selection in IP networks. It is intended for automating the configuration of applications that want to use services such as network printers or remote file systems in an administrative domain.

SLP is a very lightweight protocol: Essentially, an SLP user agent (a client application that is looking for a service) sends a request for a service and receives responses from service agents (entities that advertise services) or directory agents (entities that aggregate service advertisements from multiple service agents and can answer requests on their behalf). The request can either be directed to a service agent or directory agent or can be sent via multicast to a standard multicast group using a well-known port. Service agents never advertise their service using multicast or broadcast announcements. In order to increase scalability for multicast requests, a user agent can add the addresses of service agents that have answered before to a *previous responders list*. A service agent that sees its address in this list of a multicast request does not answer the request. Directory agents are intended for larger networks in order to enhance scalability.

User agents can specify three types of information in service requests: scope identifiers, a service type[3] and a query predicate. A filter predicate can optionally be specified to query a service with respect to its attributes. After receiving responses to service requests a user agent can query the service's attribute explicitly by sending an attribute request. SLP also provides optional authentication but does not provide confidentiality.

SLP provides a solution to locating services, not to associating with the services. Service association (i.e., allocating a service resource, exchanging confidential access credentials) has to be done in a second step, with a different protocol.

3.2 Universal Plug and Play

Universal Plug and Play (UPnP, [MS00]) is an architecture for device control in local networks and relies on different protocols for functions such as service discovery and description, transmission of control commands and event notifications. It provides peer-to-peer communication between different types of devices without the need for manual configuration.[4] To allow these devices to interwork, they do not only need to locate each other but must also dynamically learn their capabilities and be enabled to exchange information without knowing in advance which devices are present. The following functions can be distinguished:

Device discovery: UPnP's *Simple Service Discovery Protocol* (SSDP, [GCL+99]) relies on periodic unsolicited service advertisements from devices *and* on explicit service requests from clients (*control points*). Devices periodically send service advertisements containing information about the device type, an identifier, a URI for more information about the device and a duration, for which the advertisement is valid. There is no mechanism in SSDP to scale the announcement interval with the number of UPnP devices.

Device description: A UPnP service description is an XML document that describes the service's interface (in terms of methods and variables that can be inspected and modified. After discovering the device, a control point retrieves the corresponding service description by sending an HTTP GET request.

[3] The service type is a fragment of an SLP *service URI*, a URI that can be used to specify the type and the address of a service, e.g., service:printer:lpr://printer.example.com.

[4] E.g., in home networks where devices from different vendors are connected to an IP network in an ad-hoc fashion.

Device control: UPnP relies on the *Simple Object Access Protocol* (SOAP, [BEK^{+}02]) for remote procedure calls and variable manipulation.[5] Neither SOAP-1.1 or UPnP in general define any security mechanisms to authenticate control points and to provide integrity and confidentiality of the communication.

Event notification: UPnP allows control points to subscribe to certain state variables of a service relying on the *General Event Notification Architecture Base* (GENA, [CaYG00]) that specifies the methods `SUBSCRIBE`, `UNSUBSCRIBE` and `NOTIFY`.

While SLP carefully addresses the issue of scalability, e.g., by minimizing the use of multicast messages, UPnP has obvious deficits: The service advertisement rules do not allow UPnP to scale to larger environments with many services. In addition, the control features of UPnP are completely insecure, lacking authentication of control points, message integrity and confidentiality. This makes UPnP difficult to use except for controlled environments.

4 System Design

When we compare SLP and UPnP we can state that SLP provides mechanisms for service discovery and selection in static enterprise networks without addressing the service association problem. It focuses on simplicity and scalability with respect to the number of user agents. We believe that SLP's exclusive usage of service requests by user agents can lead to sub-optimal behavior in the case of ad-hoc communication and mobility. UPnP is intended as a complete solution for service discovery and device control through SOAP. It is targeted at service discovery in "dynamic" environments where devices can be connected dynamically, hence relying on periodic service announcements, however without properly addressing scalability.

These observations have led us to develop a new design for a dynamic device association framework for dynamic ad-hoc environments that addresses the necessary scalability and security issues and is not tied to any particular application protocol. Figure 1 depicts the DDA process schematically. The concepts we present in the following refer to the five phases of device association that we have discussed in section 2.

4.1 Service and Device Discovery

When discovering services a user agent is interested in the following information: the existence of devices (in range) and their availability, the characteristics of the services they offer and their rendezvous parameters.

In principle, the discovery of existing and available services, is addressed by SLP and SSDP. However, to enable mobile user agents to discover (possibly dynamically changing) services offered by (potentially mobile) devices, the SLP mechanisms are not directly applicable:

5 SOAP is a specification of how to represent remote procedure calls, including their parameters and responses, in XML documents, which are (usually) transmitted using HTTP.

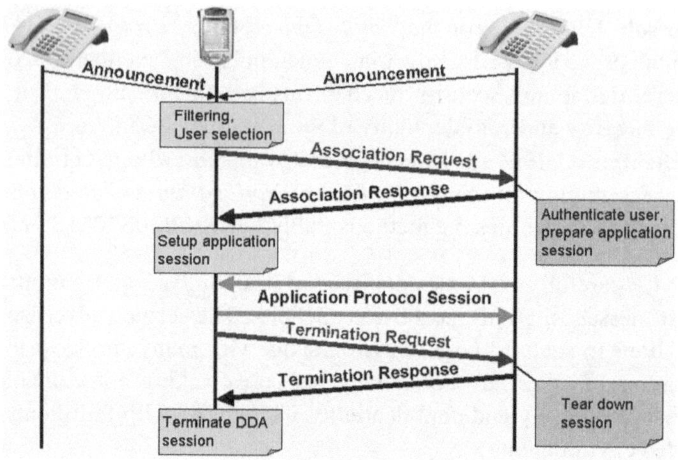

Fig. 1. The DDA process

For mobile devices that roam between networks, we cannot assume a static enterprise network with constant connectivity. A mobile device can enter and leave the scope of services quite frequently requiring the user agent to initiate new queries every time it enters a new network. Relying on the query-response model would also prevent the user agent from timely detecting that a service is no longer reachable. In cases of intermittent connectivity a user agent would thus have to send out queries periodically to validate its view of the available services in a network.

A query-response model does not allow a mobile device to build up and maintain an ephemeral "directory" of available services: this would require either iterating through all conceivable service names (which appears infeasible) or using wildcard queries at the risk of response implosions. Finally, relying on directory agents announcing their availability is not a suitable fallback: some wireless ad-hoc networks may be affected by the *hidden terminal problem*: this could lead to situations where a mobile device can communicate with the directory agent but not with the corresponding service agents.

As a solution, we propose an announcement-based scheme, where services actively announce their availability and user agents receive and filter announcements depending on the service description. This approach allows user agents to discover services more easily, especially in environments with mobile user agents and ad-hoc-networking characteristics. SSDP also relies on an active announcement model but does not provide a rate-adaptation scheme and thus does not scale to large numbers of services. In our approach service agents announce themselves by multicasting advertisements, and service agents participate in a *rate adaptation process*: Each service agent observes the announcements of other service agents per time interval and adapts its own transmission rate so that the total announcement data rate roughly remains constant, regardless of the number of service agents. Service agents also announce discontinuation of their service. User agents receive service announcements but can optionally also send explicit service requests, and service agents include their current estimate of the group size in their announcements in order to inform user agents and other agents about their view of the group.

Using active announcements with a well-defined rate-adaptation scheme has several advantages: User agents do not have to poll to detect the availability of services, instead they receive announcements automatically when connecting to a network. User agents can also detect when services are no longer available by monitoring the group size and calculating the retransmission interval themselves, automatically expiring services that cease sending announcements after some time.

Instead of static service descriptions, we propose the use of a *soft-state* approach, where the latest announcement is conveying the current service description, which is more practical for also accommodating dynamic parts in the service description such as service availability.

In order to provide user agents with enough meaningful information as a basis for the service selection, the service descriptions should include at least the following information in addition to the service type:

- the session protocol that is used to access the service;
- the current availability of the service and its geographic location;
- service attributes that describe the capabilities of the services; and
- a service URI (the "rendezvous point") to associate with a service.

The DDA framework provides all this information in the periodic service announcements as well as in responses to explicit service queries, thus differing from SLP's two-stage request-response procedure. For implementing the announcement-based, scalable DDA service discovery procedure, we have selected the *Session Announcement Protocol* (SAP, [HPW00]).[6] SAP announcements carry a description of a communication session, including transport parameters that are required to join the session. Today, SAP is almost always used with the Session Description Protocol (SDP, [HJ98]): a simple text-based format that can be described as a key-value scheme, where a key is a single character.

Figure 2 is a sample DDA service announcement.[7] The *session level* section (up to the first 'm=' line) contains general information about the device; each *media level* section describes a service and provides details how to access it. The media field indicates the media type `dda-control`, and the protocol field specifies the used application session protocol, e.g., `HTTP`, `HTTPS`, `SIP`, or `MBUS`. Figure 2 illustrates the use of SDP for HTTPS and Mbus. The service URI is specified in the attribute `dda-connect`. The following session and media attributes are used (the attributes are always prefixed with `dda`): `device-type` to specify the announcement type and the session protocol, `device-id` to distinguish multiple devices of the same type, `device-location` to allow for device selection based on geographic locations, `stats` to indicate the number of visible service agents in a given scope, and `device-status` to specify the availability status of a device.

[6] SAP is a multicast-based protocol for the advertisement of multicast multimedia sessions. The protocol provides scalability with respect to the number of sessions that are announced in total and the amount of bandwidth that is used, which is achieved by adapting the announcement interval to the total number of SAP announcements.

[7] Some lines that are not related to DDA have been omitted.

```
a=app:dda mbus
a=dda-device-id:32778be73ef9823097d22957b3e5809a
a=dda-device-location:MZH%205160
a=ip-phone:SIP dku's%20phone Example.com%20Phone-X
a=dda-stats Local 42
m=dda-control:443 HTTPS -
a=dda-connect:https://10.1.2.3/connect
a=dda-device-type:ip-phone
a=dda-device-status:AVAILABLE
```

Fig. 2. Sample DDA service announcement (abbreviated)

The example in figure 2 also provides an application specific attribute "ip-phone" that is used to specify parameters for this special device type.

4.2 Device Selection

The actual device selection can be realized in different ways, depending on the application, the network characteristics, and user settings: The user agent can be configured with a filter expression to consider only services of a specified type and with certain attributes. A preference-based ranking may be included as well, and physical proximity may be used as a criteria for automated device selection.

If the device selection process cannot be automated, the list of currently available services is presented to the user. Mobile devices may continuously monitor their environment and present a (structured) service directory to the user from which she can select services to access.

4.3 Service and Device Association

After a user agent has selected an appropriate service, the actual device association takes place. The user contacts the service URI, authenticates itself to the service and requests the service session parameters. The following requirements can be identified for these steps:

Authentication: Some services such as telephony and printing services may only be made available to authorized users. Therefore, DDA must support authentication of user identities. The specific mechanisms depend on the available infrastructure and the application scenario.

For corporate environments with a set of well-known users and an appropriate security infrastructure, public-key based mechanisms as well as shared secrets (i.e. passwords) may be used. Guests may be authenticated using one-time (or one-day) credentials which may take either shape. With an inter-domain public-key infrastructure, guests may also be authenticated as individuals.

```
a=dda-connect:https://10.1.2.3/connect/4367-abc4-9786
a=dda-disconnect:https://10.1.2.3/disconnect/4367-abc4-9786
a=dda-lease:600
m=control 47000 MBUS -
c=IN IP4 224.224.224.224
a=mbus:HASHKEY=(HMAC-MD5-96,T21/U6/ORLxKF/0a)
a=mbus:ENCRYPTIONKEY=(NOENCR)
a=mbus:SCOPE=LINKLOCAL
```

Fig. 3. Sample DDA session description (abbreviated)

Our DDA scheme supports certificate-based authentication as well as password-based authentication. If no personalized authentication is possible, access passwords can be provided, e.g., at a help desk.

Confidentiality and integrity: After the user has been authenticated the parameters for the actual session protocol have to be negotiated. Since this may include sensitive information such as transport parameters that should not be disclosed and keying material for securing the service session itself, this data exchange may need to be secured with respect to confidentiality and message integrity. Confidentiality is achieved by encrypting the communication; integrity is accomplished by relying on hashed message authentication codes (HMACs).

Session protocol parameter description: One or more services can be described in a session description. The DDA framework is not tied to any specific session protocol, and hence it must be possible to describe session parameters for different protocols such as SIP [RSC$^+$02], Mbus [OPK02], HTTP [FGM$^+$97], and SOAP [BEK$^+$02]. Independent of the specific protocol in use, the following parameters are described: an explicit lease duration, a service URI for re-associating and a service URI for dissociating.

We have mapped the service association step to an HTTP GET request: The user agent requests the session configuration, possibly providing authentication credentials for the user. After the user has been authenticated, the service agent provides the configuration data in the HTTP response. We have considered two mechanisms for user authentication: The user agent can either connect to the server using HTTP or HTTP/TLS. In case of HTTP/TLS, the user agent and the service agent can authenticate themselves using certificates. Alternatively, e.g., if a public key infrastructure is not available and certificates cannot be validated, the user can be authenticated through HTTP digest authentication.

Figure 3 depicts a sample session description.[8] The parameters for the session are described in a media level section of the SDP description. The `m=` line must specify the type `control`, followed by a port number (if applicable) and a protocol identifier. The session level section includes the lease time (`a=dda-lease`), a URI for lease-renewal (`a=dda-connect`), and may also provide a URI (`a=dda-disconnect`) to explicitly

[8] Some lines that are not related to DDA have been omitted.

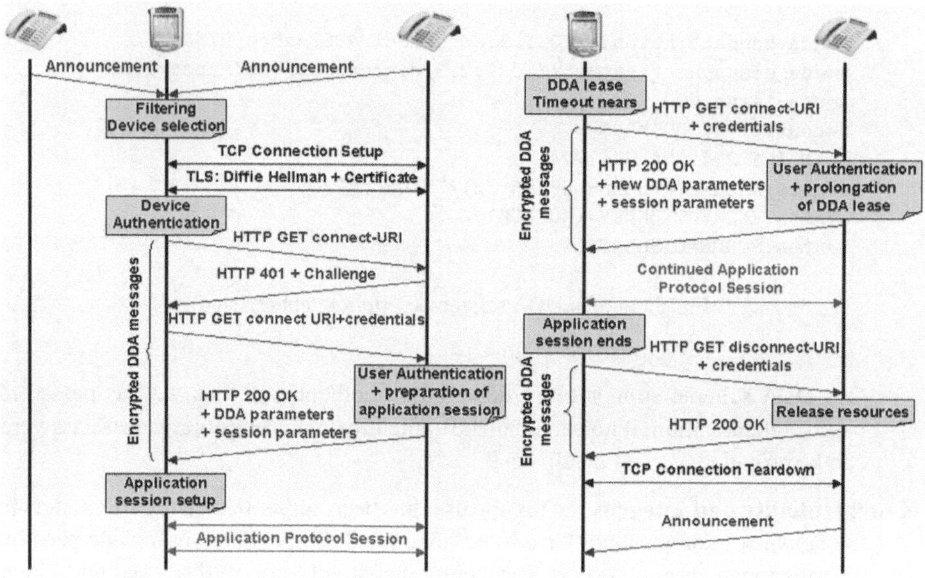

Fig. 4. Full sequence of implemented DDA interactions

terminate the application session. All other parameters are given as session protocol-specific attributes. The session description uses the attribute `mbus` that is used multiple times to specify different Mbus communication parameters.[9]

Figure 4 shows the complete operation of a DDA process: the user selects an IP phone, connects to it using TLS, thereby authenticating the device and establishing a secure communication link. HTTP digest authentication is used to verify the mobile user's identity and the application session parameters are conveyed in the HTTP body. The application protocol operates and, when the lease expiration time nears, the DDA process is re-invoked to refresh the lease. Eventually, the application session and the device association are terminated.

4.4 Application Protocol Operation

After the mobile device has obtained the session description it can start using the corresponding service. Depending on the application protocol in use, the two associated devices may continue operation in more or less tight coupling: Some protocols (such as Mbus) support liveness monitoring of peers: this allows one device to notice that the other has disappeared and terminate the application session as well as the device association. Other protocols, e.g., HTTP based protocols, do not support liveness detection, which means that services could be blocked indefinitely if no other mechanisms exist to terminate application sessions and device associations.

[9] We have defined DDA fields for the required Mbus parameters (see [OPK02]).

We therefore propose the usage of *leases*, i.e., explicitly limited usage durations: a service agent specifies a maximum lease time and, if a user agent wishes to use the service beyond this time, it has to re-associate with the service agent using the corresponding URI and obtain a new lease (with repeated authentication and authorization and potentially new parameters). If the user agent does not re-new the lease the service agent can assume the session has terminated when the lease expires.[10]

4.5 Service and Device Dissociation

For service dissociation, we have to consider similar issues as discussed in the previous section: If an application protocol supports the notion of application sessions and provides mechanisms for their tear-down, those mechanisms may be used as a hint to also terminate the device association.[11] For other protocols, DDA-based mechanisms for dissociation and for dealing with failure situations are required. The DDA support for terminating device associations is hence twofold: first, the service agent provides a dissociation URI to be used by the client for explicit termination. This is complemented by the lease concept which ensures that service agents eventually may terminate sessions: implicitly in case the user agent has disappeared or explicitly by denying a re-association attempt of a user agent.

5 Conclusion

The Dynamic Device Association concept generalizes the ideas of service location and bootstrapping of communication sessions. The service location and selection functionality is usable in different network environments: statically configured enterprise networks, ad-hoc-networks, and networks with both mobile user agents and mobile service agents. The device association is not limited to specific application protocols and provides secure authentication to dynamically located services. We have presented some sample scenarios for which DDA can be useful, e.g., the dynamic association and control of "public" IP telephony systems.

DDA builds on some concepts that have been implemented in other protocols before – we have especially considered SLP and UPnP and have observed some shortcomings. These observations have led to a different approach with a scalable service announcement mechanism and a generic, secure service association procedure.

We have implemented these concepts in a protocol specification and have applied the DDA protocol to the dynamic association of IP phones using PDAs and laptops with "DDA browsers". The PDA applications can locate available IP phones, personalize and control these devices. The combination of SAP and HTTP with SDP turned out to be a good compromise yielding a scalable, secure, and extensible solution that meets the requirements of our target application scenarios.

---

[10] The concept of leases is used by other protocols as well such as DHCP [Dro97] and the Jini architecture [SUN01].

[11] Examples are the `mbus.bye()` command and the `RTCP BYE` packet.

For future enhancements, we are considering to develop suitable mechanisms for scoping and automatically selecting devices of interest (e.g. using physical proximity), and we would like to investigate the practical issue of segmented networks and firewalls within enterprises and their influence on dynamic device association. Finally, support for additional application protocols will be added.

References

[ABL02] V. Azondekon, M. Barbeau, and R. Liscano. Service selection in networks based on proximity confirmation using infrared. International Conference on Telecommunications (ICT), Beijing, China, 2002.

[BEK+ 02] Don Box, David Ehnebuske, Gopal Kakivaya, Andrew Layman, Noah Mendelsohn, Henrik Frystyk Nielsen, Satish Thatte, and Dave Winer. *Simple Object Access Protocol (SOAP) 1.1*, May 2002.

[CaYG00] Josh Cohen and Sonu Aggarwal and Yaron Y. Goland. General event notification architecture base: Client to arbiter. available online at http://www.upnp.org/, September 2000. Internet Draft, Work in Progress.

[Dro97] Ralph Droms. Dynamic host configuration protocol. RFC 2131, March 1997.

[FGM+ 97] Roy T. Fielding, Jim Gettys, Jeffrey C. Mogul, Henry Frystyk Nielsen, and Tim Berners-Lee. Hypertext transfer protocol – http/1.1. RFC 2068, January 1997.

[GCL+ 99] Yaron Y. Goland, Ting Cai, Paul Leach, Ye Gu, and Shivaun Albright. Simple service discovery protocol/1.0, operating without an arbiter. available online at http://www.upnp.org/, October 1999. Internet Draft, Work in Progress.

[GPVD99] Erik Guttmann, Charles Perkins, John Veizades, and Michael Day. Service location protocol, version 2. RFC 2608, June 1999.

[HJ98] Mark Handley and Van Jacobsen. Session description protocol. RFC 2327, April 1998.

[HPW00] Mark Handley, Colin Perkins, and Edmund Whelan. Session announcement protocol. RFC 2974, October 2000.

[MS99] Microsoft Corporation. *IP Telephony with TAPI 3.0*, 1999.

[MS00] Microsoft Corporation. *Universal Plug and Play Device Architecture*, June 2000.

[OPK02] Jörg Ott, Colin Perkins, and Dirk Kutscher. A message bus for local coordination. RFC 3259, April 2002.

[RPSC02] Jonathan Rosenberg, Jon Peterson, Henning Schulzrinne, and Gonzalo Camarillo. Best current practices for third party call control in the session initiation protocol. Internet Draft draft-ietf-sipping-3pcc-02.txt, Work in progress, December 2002.

[RSC+ 02] Jonathan Rosenberg, Henning Schulzrinne, Gonzalo Camarillo, Alan Johnston, Jon Peterson, Robert Sparks, Mark Handley, and Eve Schooler. Sip: Session initiation protocol. RFC 3261, June 2002.

[SAL99] The Salutation Consortium. *Salutation Architecture Specification V2.1*, 1999.

[SUN01] Sun Microsystems. *Jini Architecture Specification Version 1.2*, December 2001.

A Checkpointing Algorithm for Mobile Computing Environment

Pushpendra Singh and Gilbert Cabillic

Aces Projet, INRIA-Rennes,
Campus Universitaire de Beaulieu
35042 Rennes, France
{Pushpendra.Singh,Gilbert.Cabillic}@irisa.fr

Abstract. Mobile computing is going to change the way, computers are used today. However mobile computing environment has features like high mobility, frequent disconnections, and lack of resources, such as memory and battery power. Such features make applications, running on mobile devices, more susceptible to faults. Checkpointing is a major technique to confine faults and restart applications faster. In this paper, we present a coordinated checkpointing algorithm for deterministic applications. We are using anti-messages along-with selective logging to achieve faster recovery and reduced energy consumption. Our algorithm is non-blocking in nature and avoids unnecessary computation. We ask only minimum number of processes to take the checkpoint and also take in to account the limited storage available at mobile devices.

1 Introduction

Mobile environment has features like high mobility, frequent disconnections, and lack of resources such as memory and battery power; such features make applications, running on mobile devices, more susceptible to faults. Thus providing good fault-tolerant mechanisms is necessary to ensure reliable working of computing devices. Coordinated Checkpointing is a major technique to provide fault-tolerance.

In this paper, we present a coordinated checkpointing algorithm for deterministic applications. Coordinated checkpointing has two phases. In the first phase, checkpoint initiator asks other processes to take a checkpoint. If every process takes a checkpoint, then the initiator enters into the second phase and asks processes to make their recent checkpoint permanent. However, in the first phase, if a process denies to take a checkpoint then whole checkpointing process is aborted. In coordinated checkpointing, last checkpoints of every process form a consistent state; thus it is free from domino effect [1] and problem of storing number of checkpoints.

Recently proposed checkpointing algorithms [2,3] do not take into account the nature of application and insist on avoiding orphan messages [3] to have a correct recovery. We have found out that for deterministic applications, their

M. Conti et al. (Eds.): PWC 2003, LNCS 2775, pp. 65–74, 2003.

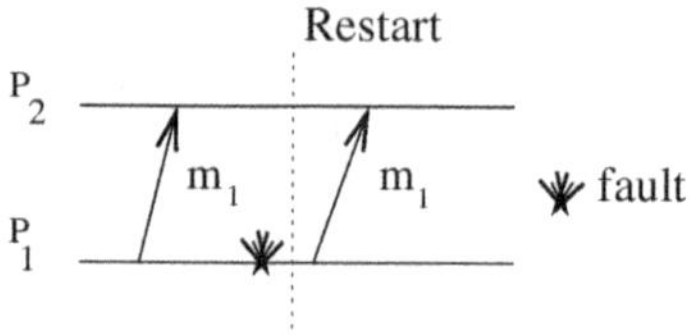

Fig. 1. Assumption of same processing

approach results in repeated computation and slow recovery. We have developed our algorithm only for deterministic applications and orphan messages are present in the checkpointing states of processes in our algorithm. However, we show that we have a faster and more energy-aware recovery than [2,3]. We have used anti-messages [4] to make our algorithm non-blocking; but, we store anti-messages only at the receiver's end, and only for messages that can become orphan [3], which we call, *selective logging.* In our algorithm, only the processes dependent on each other take checkpoints and not every process; a process i is dependent on process j, if it has received at least one message from process j since its last checkpoint [5].

For deterministic applications, we assume the model of Johnson-Zwaenepoel [6] i.e. if two processes start in the same state, and both receive the identical sequence of inputs, they will produce the identical sequence outputs and will finish in the same state. The state of a process is thus completely determined by its starting state and by the sequence of messages it has received. For example in Fig. 1, suppose P_1 has a permanent checkpoint before sending m_1 to P_2 and after sending of m_1 some fault occurs at P_1. So when P_1 restarts from its last checkpoint, it will send m_1 again to P_2.

The rest of the paper is organized as follows: section 2 provides necessary background. In section 3, we present our checkpointing algorithm. An analysis of algorithm and comparison with other works is done in section 4. We give our future direction and conclude in section 5.

2 Preliminaries

1. *System Model:* Mobile Computing environment consists of fixed base stations, known as mobile service stations (MSS), and mobile hosts (MH). A MSS is responsible for MHs present in its cell Communication between MH and its corresponding MSS is through reliable *FIFO* wireless channel. A MH can directly talk only to the MSS responsible for it. Communication between a MH and another MH or MSS is only through the MSS responsible for that MH. MSSs are connected together through a fixed wired network, communicating through a reliable *FIFO* wired channel.

   Our distributed computation model consists n processes $\{P_1, P_2, \ldots, P_n\}$. The processes run on same (or different) MH(s) or MSS(s) with fail-stop computation model. Processes do not share a global memory or global clock. However every process has access to the stable memory of its MH or MSS.

Data stored on stable storage is shielded from failures. The processes interact with each other by message passing and no other means. Delays in communication channels are arbitrary but finite.

2. *Anti-message:* When a process receives a message, it can store an anti-message for it. An anti-message [4] is same as the original message. The function of anti-message is to annihilate duplicate original message, if sent again. For example in Fig. 1, P_2 can save an anti-message for m_1, when m_1 was sent first time. Now, if P_1 restarts, after a fault, and sends m_1 again, because of deterministic nature, then P_2 can delete m_1 with its anti-message.
3. *Selective Logging :* Our logging is selective in nature because we do logging only during the checkpointing's phase one and phase two and only for possible orphan messages, as explained later in section 4. This logging is very much limited in nature in comparison to [6], where every message is logged, since last checkpoint.

3 The Checkpointing Algorithm for Deterministic Applications

To explain our approach clearly, we assume only one checkpoint initiator at a single time.

3.1 The Basic Idea

The basic idea of algorithm is as follows: in the first phase, a process starts checkpointing as checkpoint initiator by sending requests to processes over which it depends. When a process receives a checkpoint request, it takes the checkpoint, propagates the checkpoint request and continues its processing. During the checkpointing, when P_i sends a computation message to P_j, it also sends id of current checkpoint initiator, if P_i has taken a checkpoint, otherwise a NULL value. When P_j receives the message, it checks the value of checkpoint initiator. If it is not NULL then P_j knows that checkpointing is going on and it might receive a checkpoint request later; thus it saves an anti-message for the received message, before processing it. However if value of checkpoint initiator is NULL, it simply processes the message. After a process has taken the checkpoint, whenever it receives a computation message from another process, it stores it (before processing) in a serial order to previously received such messages. An anti-message is also stored for same message. This storing is done only during the checkpointing to insure same processing in the recovery phase. In the second phase, checkpoint initiator asks processes to make their checkpoint permanent. The checkpoints stored in first phase are made permanent and checkpointing process ends.

The saved anti-messages, messages and checkpoints form the complete package for recovery. If a fault occurs, processes start from their last checkpoint and executes the messages, they had stored during checkpointing, in the same serial order as they were received. If during the course of recovery a process receives

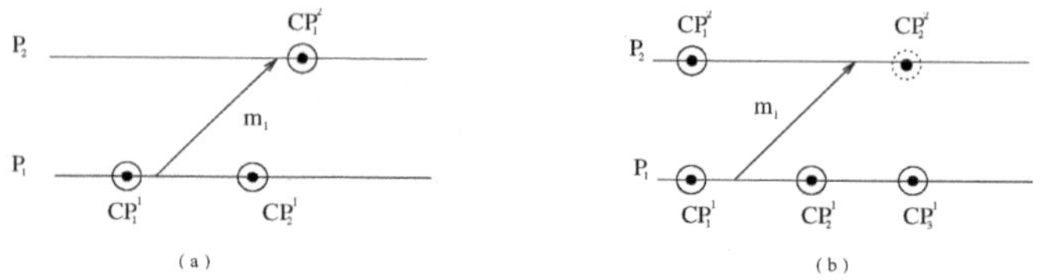

Fig. 2. Existing Problems

a message for which it has an anti-message, it annihilates the message without processing it.

3.2 An Improved Version

In our basic scheme, processes might keep storing unnecessary messages and anti-messages. For example in Fig. 2(a), if P_1 later takes another checkpoint CP_2^1, storing of anti-message of m_1 at P_2 becomes unnecessary because now P_1 will not rollback to CP_1^1.

We solve this problem by passing the information of last checkpoint to other processes through a variable *cp_number*. Every process has a variable *cp_number*, denoted by cp_number_i for P_i. The cp_number_i contains the number of last permanent checkpoint of P_i. When a process sends another process a computation message, it will also send value of its *cp_number*. When P_j receives the message from P_i, it checks the value of cp_number_i. If P_j has stored anti-messages for sender P_i, with a value of cp_number_i less than newly received cp_number_i, it will delete all such anti-messages.

Besides it, in our algorithm, a process asks other processes to take the checkpoint based on dependency information [5]. But this dependency information may be outdated, thus a process can be forced to take unnecessary checkpoints [2]. For example in Fig. 2(b), if P_2 decides to take CP_2^2, it will ask P_1 to take a checkpoint based on the dependency caused by message m_1. However this dependency information is outdated because P_1 has taken other checkpoints CP_2^1 and CP_3^1, which removes the old dependency caused by m_1.

To solve this problem, every process has a boolean array $message\_sent$, denoted by $message\_sent_i$ for P_i; which is initialized to zero. When P_i sends a message to P_j since its last checkpoint state, it will set value of $message\_sent_i[j]$ to 1. $message\_sent_i$ will be reset to zero after a successful checkpoint of P_i. So when P_j receives a checkpoint request from P_i, it will check the value of $message\_sent_j[i]$. If the value is 1, it will take the checkpoint and will propagate the checkpointing request, otherwise, it will ignore the request and will send P_i a *cp_ignore* message. On receiving a *cp_ignore* message, P_i will not include P_j in second phase of checkpointing.

Cao-Singhal [2] have tried to solve the same problem using an array of *checkpointing sequence number (csn)*, which is same as our *cp_number*. However to maintain the information of *csn*, they need an integer array, which requires more storage than our boolean array; besides, $message\_sent$ is also used in reducing the number of system messages as explained later.

Data Structures

1. checkpoint Number: A process gives a unique serial number to the checkpoint taken by it i.e. i^{th} checkpoint taken by P_k is denoted as CP_i^k. After a successful checkpoint, checkpoint number is stored permanently at a process.
2. cp_finish: It will be used to find out the end of first phase of checkpointing. We will use following technique to know the completion: every process will have a variable *cp_finish*. A process will increase *cp_finish* by 1 with every checkpoint request sent and will decrease by 1 with every checkpoint request received. When the process has taken its checkpoint it will send value of its *cp_finish* to the initiator. Initiator will add all the *cp_finish*, received by it, with its own *cp_finish*. When the sum will become 0 , it will understand that every process has taken the checkpoint and it will start the second phase.
3. dependency array: Every process has a n-bit array, which is initialized to zero [5]. When P_i receives a message from P_j, since its last checkpoint, it sets j^{th} bit of its dependency array to 1. *Dependency array* is again reinitialized to zero when a process completes a successful checkpoint. This array is used to maintain the information of dependency between processes, so that only processes dependent on each other take checkpoints.
4. message_sent array: Every process has a n-bit array, which is initialized to zero. *message_sent* array will be used to remove unnecessary checkpoint requests as explained earlier and also in reducing number of system messages as follows: a logical AND operation on *dependency array* and *message_sent* array will give the number of checkpoint requests a process will receive during a checkpointing process. So when a process receives first checkpoint request, it can calculate the final value of *cp_finish* and send it to the checkpoint initiator. Later when it receives other checkpoint requests, it can simply ignore them without notifying checkpoint initiator. Without *message_sent* array a process would have had to notify the initiator in response of every checkpoint request, as it is done in [5], otherwise *cp_finish* would not add up to 0.
5. cp_initiator: Every process has a field of *cp_initiator*, denoted by $cp\_initiator_k$ for P_k. This field contains *id* of the checkpoint initiator, while checkpointing is going on, otherwise this field is set to NULL.

Algorithm: To explain our approach clearly, we assume only one checkpoint initiator at a single time, however our approach can be extended to multiple initiators by extending the data structures appropriately.

1. Our distributed computation model consists n processes $\{P_1, P_2, \ldots, P_n\}$.
2. Initiator will take a checkpoint and then it will send checkpoint requests to the processes, over which it depends using its dependency array. It also sends its dependency array, its *id* as initiator, and its *cp_number*.
3. When a process receives a checkpoint request, it will check the validity of request, using its *message_sent*. If the request is valid, it will take a checkpoint, will propagate the checkpointing request, and will continue its normal processing. It will store the value of checkpoint initiator in its *cp_initiator*

field so that later if it receives a duplicate request, it will discard the request. If the request is not valid then it will discard the request and will notify the initiator.

4. When a process propagates a checkpointing request, it sends the request only to the processes over which it depends and they have not already got the request from initiator. This information is obtained by doing a logical AND operation on the dependency array, received with checkpoint request, and its own dependency array. Besides, when process sends new checkpoint request it will send total number of processes, for whom request has been sent by doing a logical OR operation on dependency array received in checkpoint request and its own dependency array [2]. It will also send the value of *cp_initiator*.
5. If a process receives a computation message, it will do one of the two things:
   (a) If the process had previously received a checkpoint request and have taken a checkpoint then it will store the message with *cp_number* and *id* in a serial order to previous similar messages, before processing. An anti-message will also be stored for this message.
   (b) If the process is not aware of any checkpointing process going on then it will check the checkpoint initiator field of the message.
      i. If the value of the checkpoint initiator field is NULL, it will simply process the message.
      ii. Otherwise, it will store an anti-message with *cp_number* and *id* for the message before processing it
6. After taking a checkpoint in the first phase, every process will send the initiator a checkpoint completion message alongwith its cp_finish.
7. Initiator will add all cp_finish, received by it, to its own and when its cp_finish will become 0, it will start the second phase.
8. In the second phase, initiator will ask participating processes to make their checkpoints permanent and will make its own checkpoint permanent.
9. Databases like dependency array, message_sent array, etc., will be reinitialized to 0. New permanent checkpoint numbers will be assigned.
10. This will end the checkpointing process.
11. When a process receives a computation message, containing higher *cp_number* of a sender, for which the process has stored anti-messages or message with a less *cp_number*; it will delete all such anti-messages.

Recovery: For recovering from failure, every process will begin from its last stored checkpoint state. A process will do its processing as it was doing before failure. If the process had stored some messages, it would process them in the same serial order as they had arrived. When a process receives a message for which it has an anti-message, it will delete the message with its anti-message without processing.

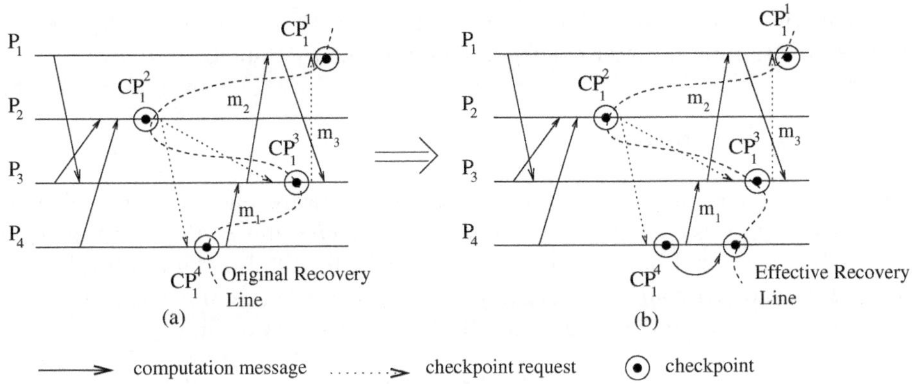

Fig. 3. Example

3.3 Example

In Fig. 3(a), P_2 is the initiator and it depends upon P_3 and P_4 because it has received messages from them in past. P_2 starts checkpointing and sends checkpoint requests to P_3 and P_4 and increases *cp_finish*$_2$ by 2. P_4 receives the request, takes a checkpoint, decreases its *cp_finish* by 1 (because it has received one request but sent none), sends it to P_2 and continues its processing. P_4 sends message m_1 to P_3 with checkpoint initiator *id* equal to 2. When P_3 receives m_1, it finds a not NULL value for checkpoint initiator. It stores the anti-message for m_1 before processing and continues its processing. P_3 sends m_2 with checkpoint initiator field NULL. P_1 processes m_2. P_3 receives checkpoint request, takes a checkpoint, sends a checkpoint request to P_1 over which it depends. Its *cp_finish* remains 0 (because it has received one request and sent one), which it sends to P_2. P_1 sends m_3 to P_3 with checkpoint initiator field NULL. Since P_3 have taken a checkpoint, it stores m_3, giving it a serial number; it also stores an anti-message for m_3 before processing it. Meanwhile checkpoint requests reach P_1, it takes the checkpoint, decreases its *cp_finish* by 1 and sends it to P_2. P_2 adds all the *cp_finish*'s received by it to its own and as it becomes 0, P_2 enters into the second phase and asks every process to commit their checkpoints and completes the checkpointing. Checkpoints CP_1^1, CP_1^2, CP_1^3, CP_1^4, message m_3, and anti-messages of m_1 and m_3 form complete recovery package.

If some fault occurs, system starts from last saved checkpoints i.e. CP_1^1, CP_1^2, CP_1^3, and CP_1^4. P_4 sends message m_1 to P_3 as it has done before failure. When P_3 receives m_1, it annihilates m_1. Later P_3 process m_3 from its log and system progresses.

Here we see that P_3 annihilates m_1 (which is an *orphan message*), without processing, thus hiding its presence. In other words we have basically shifted the checkpoint state of P_4 to after the sending of m_1, as shown in Fig. 3(b); which is a complete consistent state. Thus we recovered as if there was no *orphan message*.

Later, if P_4 takes another checkpoint CP_2^4 and sends a message m_4 to P_3, P_3 will delete the anti-message of m_1. Similarly stored message m_3 and its anti-message will also be deleted when P_3 will take CP_2^3 or when it will receive a message containing a larger checkpoint number of P_1 than 1.

3.4 Advanced Recovery

We can enhance our recovery to a more energy-aware one by using following scheme: every message will have two parts viz. *header* and *content*. In the beginning only *header* will be sent to the receiver, which will also contain total size of *content* with usual routing information. If a process receives a message header for which it has an anti-message, it will do one of the two things:

1. Continue receiving it and later delete it with its anti-message.
2. Depending on the size of content, if the receiver thinks that it is a waste to receive the message just to be deleted later, for example for big messages. It will send "not to send" message to the sender for this message.

When a sender receives a "not to send" message from the receiver, it will not send the *content*. This enhancement involves use of an extra message viz "not to send". Generally extra messages are advised to be avoided given the argument that they involve extra energy consumption. However receiving and sending of a message, which is to be discarded later, also involves consumption of energy. For big messages say 1 Mb, this energy consumption can be more than energy consumed in sending one extra message. Besides it "not to send" message is very lightweight compared to big size computation messages, thus a flexible approach can be adopted by receiver, if it thinks that it will be wise to send an extra message, it will do so, otherwise it will simply receive the message and discard it.

4 Analysis of Algorithm and Comparison with Other Works

We have made our algorithm for deterministic application to take advantage of the nature of application in saving resources such as energy and memory of a mobile device. In this section we will analyze our algorithm for similar reasons.

1. Storage : As a part of our algorithms, we store checkpoints and do some message logging. But this message logging is done only for a brief period i.e. between phase 1 and phase 2 of the checkpoint algorithm and only when the sending process has not taken a checkpoint but receiving process has taken a checkpoint. We store anti-messages only for the messages that can become orphan. Anti-messages are not stored for the messages, where receiver has taken the checkpoint or sender has not taken the checkpoint like in Fig. 3, anti-message is stored only for m_1 and not for m_2. As mentioned in our enhanced recovery scheme, for anti-messages we can store only the message header, which will not consume much of storage.

Unnecessary anti-messages are also deleted as part of algorithm. Since our algorithm is based on coordinated checkpointing scheme, so only last checkpoint is needed to be stored.

Previous work has been done on the same approach by [7,6]. However, they log every message since last checkpoint and also block the processing till logging is not done for a message. The overload of such logging is very high for mobile devices because of limited storage available.

2. Energy : The algorithm saves energy by avoiding the same calculation that it has already done. For example in Fig. 3, m_1 is an orphan message. Other checkpointing algorithms [8,9,2] avoid orphan message by taking checkpoint prior to processing the message. Which basically means that at the time of recovery such messages are again processed by receiver, because they had not been included in the checkpoint. This repeated execution of same messages consumes energy.

   In our approach receiver will discard such messages, thus we save energy by avoiding the same computation unlike [8,9,2]. Besides it, with our enhanced recovery scheme, whole sending and receiving of a big message can be avoided by using an extra lightweight message. It will result in energy saving by avoiding the communication as well.
3. Execution time : The arguments of energy saving also holds true for execution time for deterministic recovery. Since a receiving process does not compute the same messages, thus has less average time for other messages; this results in faster recovery. Again taking example of Fig. 3, suppose a fault occurs after the reception of m_3. During the recovery, our algorithm would not compute m_1 and other such messages, unlike [8,9,2].

Recently developed algorithms [8,9,2] do not take advantage of deterministic nature of application and applies same method for every kind of application. We think that such approach is not good for small wireless devices, such as PDA's, working in mobile environment. Because for such devices, even small savings of energy and storage are substantial; thus we advocate use of different techniques suitable to the natures of applications to achieve different gains [10].

5 Conclusion and Future Work

In this paper, we presented a checkpointing algorithm suitable to deterministic application. We used anti-message technique [4] with selective logging to achieve our aim. With the help of anti-messages, our approach avoids same computation during the recovery phase; so it provides a faster recovery and saves energy. We have limited our logging to a small period thus without causing extra overhead on mobile device for storage. We find that our algorithm results in longer utilization of device by conserving energy then previously proposed algorithms [2, 11]. However our algorithm is limited to the deterministic applications, which we find a limitation in evolving mobile computing field.

Presently we are working on developing an algorithm for non-deterministic applications. In our approach, we are using *successive checkpoint* approach to achieve our aim [12]. *Successive checkpoint* is an extra checkpoint taken to avoid orphan message. However our approach is totally different than taken by [2], because our successive checkpoint will never be discarded unlike their *mutable checkpoint*. Thus taking a successive checkpoint will not result in a energy wastage.

Our Future work will also explore use of the surrounding resources, present in mobile computing environment, to reduce memory and energy-consumption for our present checkpointing algorithm.

References

1. Elnozahy, E., Johnson, D., Wang, Y.: A survey of rollback-recovery protocols in message-passing systems. Technical Report CMU-CS-96-181 (1996)
2. Cao, G., Singhal, M.: Mutable checkpoints: A new checkpointing approach for mobile computing systems. IEEE Transactions on Parallel and Distributed Systems **12** (2001)
3. Manivannan, D., Singhal, M.: Failure recovery based on quasi-synchronous checkpointing in mobile computing systems. Technical Report OSU-CISRC-796-TR36, Dept. of Computer and Information Science, The Ohio State University (1996)
4. Myers, D.R.J.A.: Virtual time. ACM Transactions on Programming Language and Systems **7** (1985) 404–425
5. Prakash, R., Singhal, M.: Low-cost checkpointing and failure recovery in mobile computing systems:. IEEE Transactions on Parallel and Distributed Systems **7** (1996) 1035–1048
6. Johnson, D.B., Zawenepoel, W.: Sender-based message logging. In Proceedings of 17th International Symposium on Fault-Tolerant Computing (1987) 14–19
7. Johnson, D.B., Zwaenepoel, W.: Recovery in distributed systems using optimistic message logging and checkpointing. (1988) 171–181
8. Manabe, Y.: A distributed consistent global checkpoint algorithm for distributed mobile systems. 8th International Conference on Parallel and Distributed Systems (2001) 125–132
9. Cao, G., Singhal, M.: Low-cost checkpointing with mutable checkpoints in mobile computing systems. Proceedings of 18th International Conference on Distributed Computing Systems (1998) 462–471
10. Singh, P., Cabillic, G.: Fault tolerance and availability in mobile computing environment. selected to appear in PDPTA (2003)
11. Manabe, Y.: A distributed first and last consistent global checkpoint algorithm. 12th International Conference on Information Networking (1998)
12. Singh, P., Cabillic, G.: Successive checkpointing approach for mobile computing environment. selected to appear in ICWN (2003)

Testing Mobile Wireless Applications

Ichiro Satoh

National Institute of Informatics
2-1-2 Hitotsubashi, Chiyoda-ku, Tokyo 101-8430, Japan
Tel: +81-3-4212-2546, Fax: +81-3-3556-1916
ichiro@nii.ac.jp

Abstract. A framework is described that can be used to build and test application-level software for wireless mobile computing. It emulates the physical mobility of wireless devices by using the logical mobility of software-based emulators of the devices and target software. Since each emulator is implemented as a mobile agent, it can dynamically carry the target software to each of the sub-networks to which its device is connected on behalf of the device, permitting the software to interact with other servers in the current sub-network. That is, it can test software designed to run on a wireless device in the same way as if the software were disconnected from the network, moved with the device, and reconnected to and operated on another network. Also described are the lessons learned from exploiting the framework in developing typical software for wireless devices.

1 Introduction

The development of software for portable computing devices is very difficult due to the limited computational resources these devices have, even if they are standalone. Recent advances in networking technology have enabled portable computing devices to link up with servers through wireless networks, such as IEEE802.11b and Bluetooth, to access information from them, and to delegate heavy tasks to them. A typical IEEE802.11b wireless LAN consists of more than one base station (i.e. access point), whose typical radio area is within at most a few hundred meters, connected through a local area network. When a user moves from location to location, his/her mobile computing device may be disconnected from the current network and reconnected to another network. Several researchers has explored mechanisms to transparently mask variations in mobility at the network or system level, such as Mobile-IP [14] and host mobility [19]. These approaches have been designed to enable packet delivery while a mobile device is away from its home base.

However, the notion of mobility transparency is not always suitable for wireless applications, such as user navigation systems in a city or museum and printing services. Such applications need to access local servers on a local area network in the current location. That is, a change in the network and location implies movement away from the servers currently in use, toward new ones. Therefore, software development for mobile computing devices that use short range wireless networks is often tedious and extremely susceptible to change. To construct a correct application, it must be tested in all the networks to which the device could be moved and hooked up to. Unfortunately, the task

M. Conti et al. (Eds.): PWC 2003, LNCS 2775, pp. 75–89, 2003.

of testing software for mobile computing devices has attracted little attention so far. This is a serious impediment to its growth beyond mere laboratory prototypes.

To overcome this problem, a software testing approach suitable to wireless devices, including PDAs and wireless appliances, is needed. We introduced a framework, called Flying Emulator, for developing software running on portable computing devices in our earlier paper [17]. The key idea of the framework is to offer a mobile agent-based emulator of a mobile computing device. The emulator performs application-transparent emulation of its target device for application software written in the Java language. Furthermore, since the emulator is implemented as a mobile agent, it can carry its software to remote networks according to patterns of physical mobility and test the software inside the environments of those networks. However, the framework presented in the previous paper had no mechanism for simulating wireless networks or the small form factor displays and controls of mobile wireless devices such as PDAs. Therefore, it is not always suitable for testing applications for small wireless devices, in particular PDAs and wireless appliances. The goal of this paper is to enhance the existing framework to solve these problems. The enhanced framework can have a mechanism to simulate characteristics of wireless networks by using Java bytecode rewriting technique. That is, it can automatically replace Java classes for network processing in applications by customized classes, which enable the developer to control features of wireless networks such as network disconnection. It also provides a graphical front end for target devices so that it is easy enough for end-users of wireless devices to use in testing applications and evaluating contents.

The remainder of this paper is organized as follows. Section 2 discusses requirements in testing wireless applications and then outlines a framework for testing wireless applications. Section 3 presents the design and implementation of the framework and Section 4 demonstrates the usability of the framework through two real-world examples. Section 5 surveys related work and Section 6 provides a summary and discusses some future issues.

2 Approach

The framework presented in this paper aims testing network-dependent application-level Java-based software, including application-level protocols designed to run on mobile devices, such as PDAs and notebook PCs, and which may often access servers on local networks in the device's current location either through short-range wireless networks such as the IEEE802.11b or Bluetooth networks. This framework does not address mobile phones because wireless networks for mobile phones provide a global view in the sense that they offer continuous access to services and resources through a land-based network, even when the device's location changes. Emulation of the performance of wireless networks, such as bandwidth and connectivity latency, is also beyond the scope of the framework.

2.1 Requirements

To test applications designed to run on wireless devices, the framework should satisfy the following requirements.

Network-dependency and interoperability: Cooperation among mobile computing devices and servers within a domestic or office network is indispensable because it complements various features missing in the device. As a result, the appropriateness of the software running on the device not only depends on its internal execution environment but also on the external environments, including servers, provided by the network that it connects to. Moreover, testing the interoperability of various devices tends to be tedious, since there are countless varieties of devices with which the target device can interface.

Mobility and disconnection: Wireless access points, so-called *hotspots*, are being installed in various places, such as airports, hotels, and cafes. While a wireless device roams among the radio cells of the base stations within a hotspot, it can continue to access servers provided within the hotspot as well as global networks such as the Internet. On the other hand, when a device moves outside of the area of the current hotspot, it is disconnected and cannot always enter the area of another hotspot, because the areas of hotspots are still sparse. Such devices must often sleep to save battery life and avoid the risk of accidental damage while moving.

Ease of use: The framework should be simple enough for end-users of wireless devices to use in testing. It must be able to run on servers without any custom hardware and enable easy operation of applications using a graphical user interface displayed on a stationary computer in front of the developer. Many applications have their own GUIs. The framework should enable the developer to test his/her target applications including their GUIs.

Spontaneous and plug-and-play management: When a wireless mobile device is reconnected, it may have to detect servers on the network to perform its task. To achieve this, several middleware systems, such as Jini [2] and Universal Plug and Play (UPnP) [12], have been used to manage devices. These mechanisms use multicast communications whose packets can be transmitted to only the hosts within specified sub-networks. Therefore, the software target to run on a wireless device must be tested inside the sub-networks to which the device can be connected.

2.2 Overview of the Framework

It is difficult to build and debug software designed wireless mobile devices within the devices themselves, because it has a less powerful processor with less memory and a limited user interface with a clamped keyboard and small screen. A popular solution is to use a software-based emulator for the target device. However, existing emulators are not always usable for the development of software dependent on resources provided in networks, because an emulator running on a standalone computer cannot simulate all the resources provided in the networks to which its target device may connect. The simplest way to solve this problem is for the developer to actually carry a workstation running an emulator of the target device (or the device itself) and to attach it to the local networks in the current location. This is of course troublesome for the developer and consequently should only be resorted to in the final phase of software development.

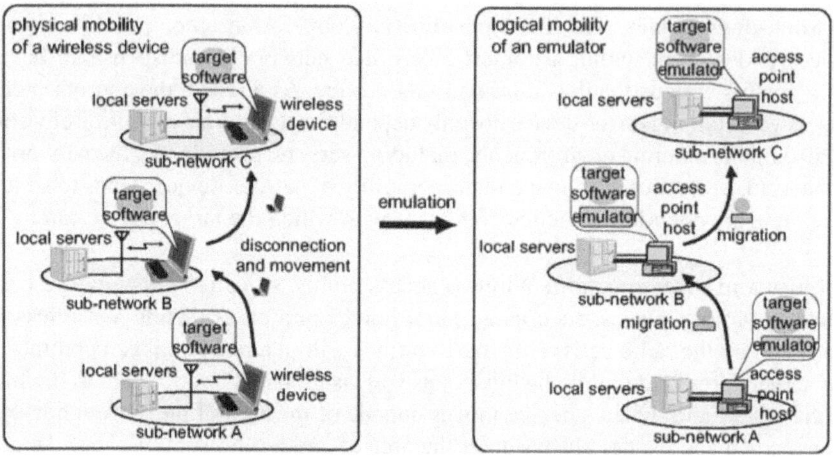

Fig. 1. Correlation between physical and logical mobilities

Our framework aims to solve these problems through the use of a software-based emulator that can simulate the internal execution environment of its target device, like the approaches taken in previous works. The key idea of the framework is to emulate the physical mobility of a wireless device between networks by using the software's logical mobility, which has been designed to run on the device between the networks, as shown in Figure 1. The framework constructs a software-based emulator as a mobile agent, which can travel from host to host under its own control. When a mobile agent-based emulator and its target software moves among networks, it transfers the code and state of the software to the destination network. The carried applications can access servers provided in the destination and continue their processes as if they had been physically moved with the target device.

Each mobile agent is simply a logical entity that must thus be executed on a computer. Therefore, we assume that each network to which the device may be moved and attached has more than one special stationary host, called an access-point host that offers a runtime system for mobile agents. Each access-point host has a runtime environment that enables applications running in a visiting emulator to connect to local servers in its network.

3 Design and Implementation

The current implementation of this framework is based on a Java-based mobile agent system called MobileSpaces[15].[1] As Figure 2 shows, the framework has the following three components:

– A *mobile agent-based emulator* that can carry the target software to specified access-point hosts on remote networks on behalf of a target wireless device.

[1] The framework itself is independent of the MobileSpaces mobile agent system and can thus work with other Java-based mobile agent systems.

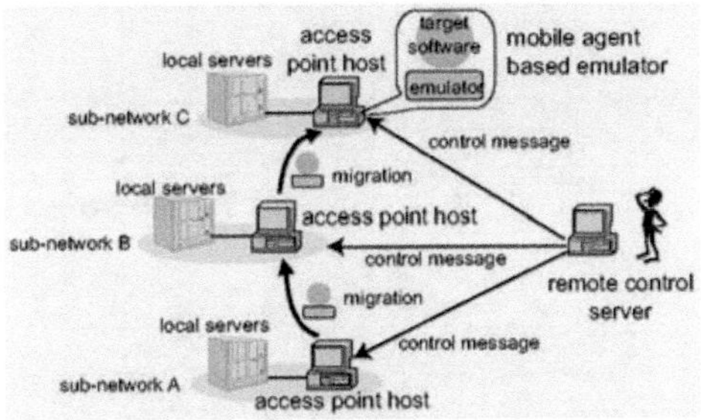

Fig. 2. Architecture of the framework.

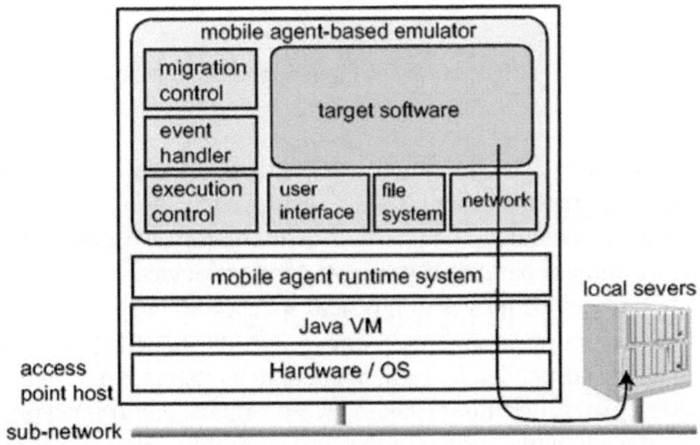

Fig. 3. Mobile agent-based emulator running on an access-point host.

- *Access-point hosts* that are allocated to each network and enable the software carried by an emulator to connect with various servers running on the network.
- A *remote-control server* that is a front-end to the whole system, enabling the moving emulator and its target software to be monitored and operated by remotely displaying their GUIs on its screen.

In addition, we provided a runtime system that runs on a wearable computer and supports the execution of the tested software. As the framework is constructed independently of the underlying system, it can run on any computer with a JDK 1.1- or 1.2-compatible Java virtual machine, including Personal Java, and the MobileSpaces system.

3.1 Mobile Agent-Based Emulator

Our mobile agent-based emulator can carry and test software designed to run on its target wireless device. Figure 3 shows the structure of a mobile agent-based emulator running on an access-point host.

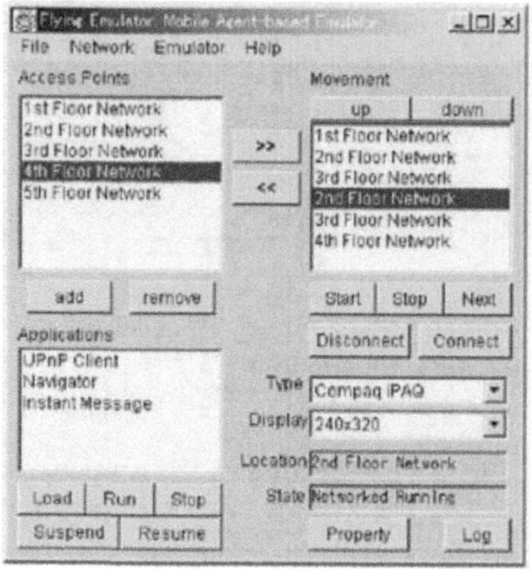

Fig. 4. User interface for controlling mobile agent-based emulators.

Emulation of Mobility: The developer can interactively control the movement of the emulator through the graphical user interface displayed the remote-control server as Figure 4 shows. Also, each emulator can have its own itinerary, a list of hosts corresponding to the physical movement pattern of its target wireless device.

When a wireless device moves in physical space, it may still be running. However, our emulator cannot migrate over networks when its inner applications are running because they must be suspended and marshaled into a bitstream before being transferred to the destination. To solve this problem, we designed our framework to divide the life-cycle state of each application into the following three phases: networked running, isolated running, and suspended. In the networked running state, the software is running in its emulator on an access-point host and is allowed to link to with servers on the network. In the isolated running state, the software is still running but is prohibited from communicating with any servers or devices on the network. This means that the device is disconnected from the network. In the suspended state, the emulator stops its target software and maintains the execution states, such as program variables, for the software by marshaling itself into a bit stream along with the states and codes of its target software.

When an emulator is suspended or migrated over networks, it can marshal itself into a bit stream along with the heap blocks and codes of its target software since it is implemented as a mobile agent. The emulator also dispatches certain events to its target software to explicitly restart (or stop) its activities and acquire (or release) the computational resources of the current host when the life-cycle state of the software is changed. In addition, our framework can provide each mobile wireless device with lightweight middleware to monitor the environment of the device, such as network connectivity and location, and dispatch certain events to its target as a mobile agent-based emulator corresponding to the device.

Emulation of Wireless Networking: When anchored at an access-point host, each emulator can directly inherit most network resources from the host, such as `java.net` and `java.rmi` packages. In the current implementation, a moving emulator cannot have

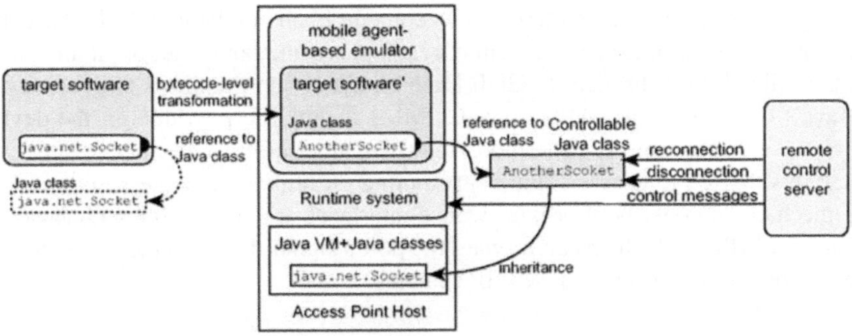

Fig. 5. Bytecode transformation for customizing a class for network operation.

its own network identifier, such as an IP address and port number, but this is not a serious problem because our target software is a client-side program as mentioned previously. Applications running on an emulator can interact with other applications running on different emulators and servers on the current sub-network and the Internet if the sub-network is connected to the Internet.

The current implementation simply maps the wireless device TCP/IP stack onto the desktop TCP/IP stack to simulate IP connectivity. Some issues in wireless device use can be simulated as well as to the extent that the medium of a desktop computing environment allows; but there may be other issues, like network disconnection, latency, and bandwidth. This framework has a mechanism to simulate characteristics of wireless networks. The mechanism overrides Java's classes for network operation, such as `java.net.Socket` and `java.net.ServerSocket`, with customized classes that emulate that characteristics of wireless networks by using a bytecode rewriting technique. Our bytecode rewriting tool is based on Byte Code Engineering Library (BCEL) [5], which enables bytecode manipulations of Java classes and is also entirely written in Java and does not have to extend the Java virtual machine. Our mechanism detects certain classes in target applications and then transforms them into the corresponding customized classes when the original classes are loaded at runtime. The current implementation of this framework provides customized TCP socket classes that are subclasses of `java.net.Socket` and `java.net.ServerSocket` and can be explicitly disconnected and reconnected by the remote control server as shown in Figure 5. Moreover, the developer can easily define customized classes to specify other characteristics of wireless networks.

Emulation of Computing Environment: The framework assumes that its target software will be Java application programs. Accordingly, the Java virtual machine can actually shield the target software from many features of the hardware and the operating system of mobile wireless devices. Each emulator permits its target software to have access to the standard classes commonly supported by the Java virtual machine as long as the target device offers them. In addition, the current implementation of our emulator supports several typical resources of mobile wireless devices. Each emulator maintains a database to store files. Each file can be stored in the database as a pair consisting of its file/directory path name pattern and its content. Each emulator provides basic primitives for file operation, such as creation, reading, writing, and deletion; it allows a user to insert files into it through its GUI. Each emulator can permit its target software to be Java's communication APIs (Java COMM) if they are provided on the device on which the emulator runs. Furthermore, the framework offers a mechanism that enables its target software to access equipment running on remote computers via serial ports. The mechanism consists of proxies whose interfaces are compatible with Java's communication APIs and which can forward the port's signals between the emulator and the remote-control server through TCP/IP channels.

Emulation of User Interface: The user interfaces of most handheld computers are limited by their screen size, color, and resolution, and they may not be equipped with traditional input devices such as a keyboard or mouse. Each emulator can explicitly constrain the size of the user interface being used from its inner applications by using a set of classes for the visible content of the MobileSpaces system, called MobiDoc [16]. Also, it can have images of the physical user interface of the target device as it would appear to the end-user. Typical handheld devices will include a screen that may allow content to be displayed. Therefore, the screen is seamlessly embedded into the pictures of the device, and the basic controls of the device can be simulated through mouse-clickable buttons.

Our framework enables the whole user interface of a device, including the graphical user interface of target applications, to be displayed on the screen of the remote control server and operated from the standard input devices of the server, such as a keyboard and mouse. This mechanism is constructed on the Remote Abstract Window Toolkit (RAWT) developed by IBM [8]. This toolkit enables Java programs that run on a remote host to display GUI data on a local host and receive GUI data from it. The toolkit can be incorporated into each access-point host, thus enabling all the windows of applications in a visiting emulator to be displayed on the screen of the remote control server and operated using the keyboard and mouse of the server. Therefore, the developer can always test his/her target applications, including their GUIs, within a desktop computing environment and the access-point hosts do not have to offer any graphics services and user-input devices. The current implementation of the framework supports emulators for three kinds of computing devices: standard notebook PCs, pen-based tablet PCs, and palm-sized PDAs. Figure 6 (A) shows a screenshot of the remote control server and (B) is a picture of a tablet PC running a user navigator application. The left window in Figure 6 (A) is the window of a mobile agent-based emulator of the tablet-PC, where the emulator tests the application.

3.2 Access-Point Host

We assume that more than one access-point host is allocated in each network, to which the wireless device may be attached. As previously mentioned, the framework is built on the MobileSpaces mobile agent system. Each access-point host is a server or workstation offering a MobileSpaces runtime system for executing the mobile agent-based emulator

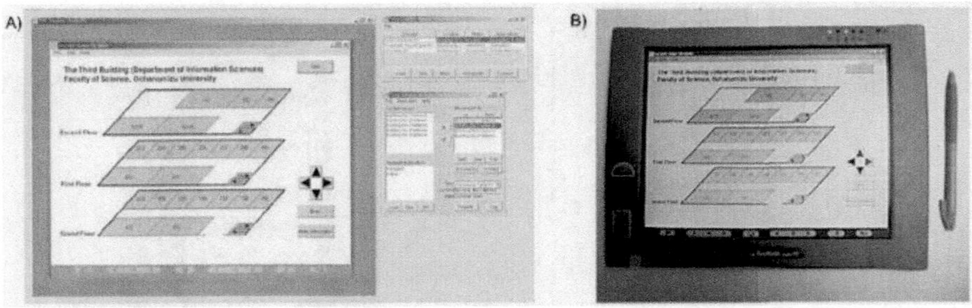

Fig. 6. (A) Screenshot of remote control server and (B) Picture of a tablet PC

and migrating it to another access-point host. The host does not need any custom hardware. When an agent is transferred over a network, the runtime system stores the state and codes of the agent, including its software, in a bitstream defined by Java's JAR file format, which can support digital signatures for authentication. The MobileSpaces runtime system supports a built-in mechanism for transmitting the bitstream over networks by using an extension of the HTTP protocol. In almost all intranets, there is a firewall that prevents users from opening a direct socket connection to a node across administrative boundaries. Since this mechanism is based on a technique called HTTP tunneling, emulators can be sent outside a firewall as HTTP POST requests, and responses can be retrieved as HTTP responses.

3.3 Remote-Control Server

The remote-control server is a control entity responsible for managing the whole system. It can run on a standard workstation that supports Java and does need any custom hardware. It can always track the locations of all the emulators, because each access-point host sends certain messages to the control server whenever a moving emulator arrives or leaves. Moreover, the server acts as a graphical front end for the system, enabling the developer to freely instruct moving emulators to migrate to other locations and terminate, through its own graphical user interface. It also can monitor and record the status of all access-point hosts by periodically multicasting query messages to them.

4 Application

To demonstrate the utility of our framework, we used it to test two typical mobile wireless applications.

4.1 Testing of Network-Dependent Software

Portable computing devices have been used in several other projects to provide user navigation in a city [1,4]. Similarly, we developed a prototype navigation system for a building of the National Institute of Informatics using PDAs with IEEE 802.11b wireless

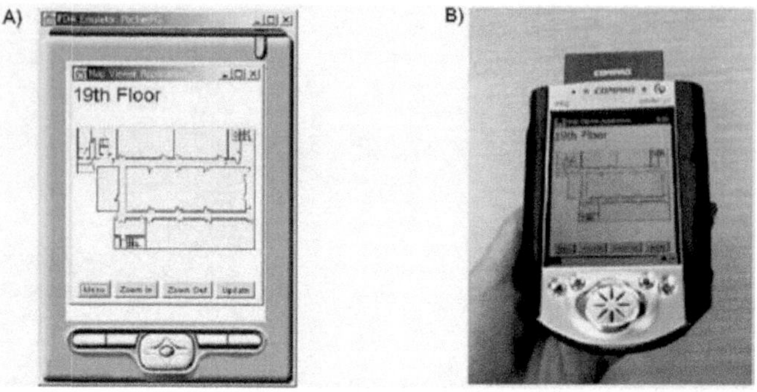

Fig. 7. (A) the window of a mobile agent-based emulator with a map viewer application running on an access point host displayed on the screen of remote control server and (B) the same application running on a PDA.

LAN connectivity. Each floor in the building has its own local area networks and one or more wireless LAN base stations. The system provides each visitor with a PDA that obtains location-dependent information from servers allocated on the sub-network of the current location via IEEE802.11b wireless networks. As a visitor moves from floor to floor, the PDA automatically displays a map of the current floor. To test the system, we constructed a mobile agent-based emulator for the PDA. The emulator can migrate a map viewer application designed to run on the PDA to the sub-network of another floor and enable the application to access the local database of the floor and display suitable maps.

Figure 7 (A) shows the window of the map viewer application tested in the emulator forwarded from an access point host to the remote control server by using the RAWT toolkit and (B) is a picture of the target PDA (Compaq iPAQ) running the map viewer application. As illustrated in Figure 7 (A) and (B), both the application running on the emulator and the application running on the target device can present the same navigation information. That is, the tested application can run in the target device in the same way as it was executed in the emulator. Furthermore, this example shows that the framework can provide a powerful method for testing not only application software for portable computers but also for creating location-dependent contents, such as map and annotations about locations. Moreover, by using the RAWT toolkit, this framework enables a content creator to view the location-dependent information that should be displayed on the PDA on the screen of his/her stationary computer. Also, since the emulator can define its own itinerary among multiple access points, it can easily and exactly trace the mobility of each visitor and test the contents displayed on the screen of his/her PDA.

4.2 Testing of Multicast-Based Protocol

Since wireless devices move from network and network, they need to be managed in an ad-hoc manner. Universal Plug and Play (UPnP) [12] is a powerful infrastructure

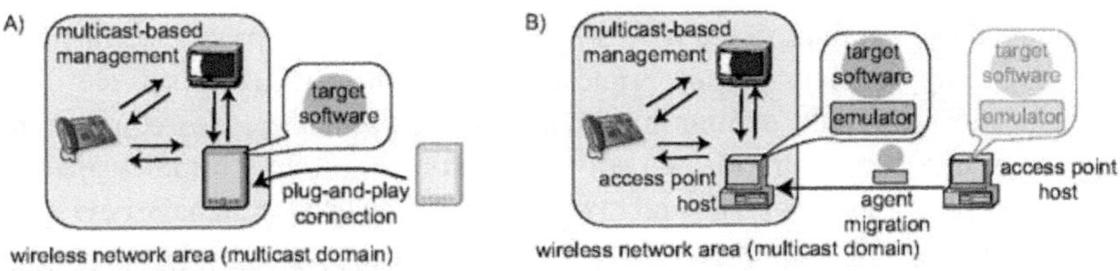

Fig. 8. Emulation of (A) the plug-and-play operation of a mobile device by (B) the migration of the emulator for the device between access-point hosts.

for enabling a device to dynamically join a network, obtain an IP address, convey its capabilities upon request, and learn about the presence and capabilities of other devices. Using it, we easily tested the interoperability of UPnP-aware Java application software designed to run wireless devices and a subset implementation of the UPnP protocol in our previous project [13]. UPnP uses a multicast-based management protocol, called simple service discovery protocol (SSDP), with which a device can announce its presence to others as well as to discover other devices or services. For example, a joining device multicasts messages to advertise its services to the UPnP's control points. Since such multicast messages are available within the domain of specified sub-networks, UPnP-aware software designed to run on a device must operate within the domain to receive the messages. Therefore, we constructed a mobile agent-based emulator as a carrier for the software. When the emulator arrives at an access-point host within the domain, the software it carries can multicast advertisement-messages to hosts in the domain and can receive search-messages multicasted from other devices in the domain as if the emulator's target were joined to the domain, as shown Figure 8. We demonstrated that the software in an emulator running on a host can interact with software in another emulator running on a different host as well as on the same host through UPnP protocols. In addition, software tested successfully in the emulator could still be run in the same way on the device without modifying or recompiling it. Our framework can thus provide a powerful methodology for testing the interoperability of protocols, within limited specified sub-networks, for reasons of security and reduced network traffic.

5 Related Work

As mentioned above, software-based emulators of portable computing devices facilitate development and testing of standalone software running on the devices, but most existing emulators do not support wireless devices in the sense that they cannot simulate the whole network in which the target device will interact. An extreme solution is to actually carry around portable devices and attach them to local networks at the destinations, but this would be extremely troublesome for developers and content creators. Another solution is to let the target software run on a local workstation and link to remote devices and servers through networks, e.g., the InfoPad project at Berkeley [11] and Lancaster University's network emulator [6]. However, accomplishing this in a responsive and reliable manner is difficult, and the emulators cannot remotely access all the services and resources that are available only within the local networks because of security protection mechanisms. Moreover, the approach is inappropriate in testing software using service discovery protocols. Since a mobile computing environment is dynamic, we require zero user configuration and administration. To solve this problem, several middleware systems, such as Jini [2] and Universal Plug and Play [12], are often used to manage devices. These systems use multicast communications to find their management servers and devices, but the multicast messages can only be transmitted to hosts within specified sub-networks. Therefore, the software that is to run on ubiquitous computing devices must be tested within the sub-networks to which the devices may be connected.

There have been various attempts to apply mobile agent technology [3,7,10], including the mobile code approach, to wireless mobile computing, but their goals are to handle network disconnection using the mobility of agents, instead of any testing of software for mobile computing. Our previous framework presented in [17] lacks a mechanism for simulating the user interface of wireless devices and the characteristics of wireless networks. It also does not support multicast networking, so that it cannot be used to test wireless appliances managed in a plug-and-play manner.

6 Conclusion

We have described a framework for building and testing application-level Java-based software designed to run on wireless devices. The framework provides software-based emulators for target software by incorporating a Java virtual machine. Since these emulators are constructed as mobile agents, they can carry software on behalf of the target device to networks that the device may be moved and connected to. That is, testing software is provided with the services and resources provided through its current network as if the software were being executed on the target device when attached to the network. Software successfully tested in the emulator can be run in the same way on the target device without being modified or recompiled. Moreover, the framework allows emulators to easily simulate other characteristics of wireless networks by using a runtime bytecode rewriting technique, Our early experience indicated that we can greatly reduce the time required to develop software for wireless devices using the framework.

There are, however, further issues that need to be resolved. Security is one of the most serious concerns in mobile agent technology. However, since our framework is used in the development phase instead of the operation phases, this issue is not as serious as it is in other mobile agent-based applications. Nevertheless, we plan to devise schemes to guarantee security and to control access, since the current implementation relies on the JDK 1.1 security manager. Since our framework can complement existing software-development methods for wireless computing discussed in Section 5. Therefore, we are interested in developing a tool to integrate our approach with other methods. The location-aware mobile agent infrastructure we developed incorporates RF-based and infrared-based tag sensors [18] and the framework we propose should be able to support these sensors.

References

[1] G.D. Abowd, C. G. Atkeson, J. Hong, S. Long, R. Kooper, and M. Pinkerton, "Cyberguide: A Mobile Context-Aware Tour Guide". ACM Wireless Networks 3, pp.421–433. 1997.

[2] K. Arnold, A. Wollrath, R. Scheifler, and J.Waldo, "The Jini Specification". Addison-Wesley, 1999.

[3] G. Cabri, L. Leonardi, F Zambonelli, "Engineering Mobile Agent Applications via Context-Dependent Coordination", pp.1039–1055, IEEE Transaction of Software Engineering, Vol. 28, No. 11, November 2002.

[4] K. Cheverst, N. Davis, K. Mitchell, and A. Friday, "Experiences of Developing and Deploying a Context-Aware Tourist Guide: The GUIDE Project", Proceedings of ACM/IEEE Conference on Mobile Computing and Networking (MOBICOM'2000), pp.20–31, 2000.

[5] M. Dahm, "Byte Code Engineering Library", http://jakarta.apache.org/bcel/index.html

[6] N. Davies, G. S. Blair, K. Cheverst, and A. Friday, "A Network Emulator to Support the Development of Adaptive Applications", Proceedings of USENIX Symposium on Mobile and Location Independent Computing, USENIX, 1995.

[7] A. Fuggetta, G. P. Picco, and G. Vigna, "Understanding Code Mobility", IEEE Transactions on Software Engineering, 24(5), 1998.

[8] International Business Machines Corporation, "Remote Abstract Window Toolkit for Java", http://www.alphaworks.ibm.com/, 1998.

[9] J. Jing, "Client-Server Computing in Mobile Environments", ACM Computing Survey.

[10] B. D. Lange and M. Oshima, "Programming and Deploying Java Mobile Agents with Aglets", Addison-Wesley, 1998.

[11] M. Le, F. Burghardt, and J. Rabaey, "Software Architecture of the Infopad System", Workshop on Mobile and Wireless Information Systems. 1994.

[12] Microsoft Corporation, "Universal Plug and Play Device Architecture Version 1.0" June, 2000. http://www.upnp.org/UpnPDevice_Architecutre_1.0.htm

[13] T. Nakajima, I. Satoh, and H. Aizu, "A Virtual Overlay Network for Integrating Home Appliances", Proceedings of International Symposium on Applications and the Internet (SAINT'2002), pp.246–253, IEEE Computer Society, January, 2002.

[14] C. Perkins, "IP Mobility Support", Internet Request For Comments RFC 2002, 1996.

[15] I. Satoh, "MobileSpaces: A Framework for Building Adaptive Distributed Applications Using a Hierarchical Mobile Agent System", Proceedings of International Conference on Distributed Computing Systems (ICDCS'2000), pp.161–168, IEEE Computer Society, April, 2000.

[16] I. Satoh, "MobiDoc: A Framework for Building Mobile Compound Documents from Hierarchical Mobile Agents", Proceedings of Symposium on Agent Systems and Applications / Symposium on Mobile Agents (ASA/MA'2000), Lecture Notes in Computer Science, Vol.1882, pp.113–125, Springer, 2000.

[17] I. Satoh, "Flying Emulator: Rapid Building and Testing of Networked Applications for Mobile Computers", Proceedings of Conference on Mobile Agents (MA'2001), LNCS, Vol.2240, pp.103–118, Springer, December, 2001.

[18] I. Satoh, "SpatialAgents: Integrating of User Mobility and Program Mobility in Ubiquitous Computing Environments", to appear in Wireless Communications and Mobile Computing (Accepted), Vol. 3, John Wiley, 2003.

[19] A. C. Snoeren and H. Balakrishnan, "An End-to-End Approach to Host Mobility", Proceeding of Conference on Mobile Computing and Networking (MobiCom'02), pp.155–166, ACM Press, 2002.

Appendix: Application Program

As mentioned previously, each application, which can be tested in our mobile agent-based emulators, is composed of more than one mobile agent-based component. However, typical Java software units, including Java applets and Java beans, can be easily modified to such components by implementing the following listener interface.

```
interface ApplicationListener
  created() // invoked after creation
  terminating() // invoked before termination
  networked() // invoked after network enabled
  isolated() // invoked after network disconnected
  suspending() // invoked before suspension
  resumed() // invoked after resumption
}
```

The above interface specifies callback methods invoked by the emulator and the runtime system on the target device when the life-cycle states of an application, such as networked running state, isolated running state, and suspended state as shown in Figure 9. Each application must define proper processes in each of these methods to hook and handle such changes. For example, suppose that a mobile agent-based emulator is just about to migrate from its current host to another host. As shown in Figure 10, an application contained in the emulator is notified through the following process:

1. The `isolated()` method of the application is invoked to handle the disconnection from the network, and then the application must release resources, such as sockets

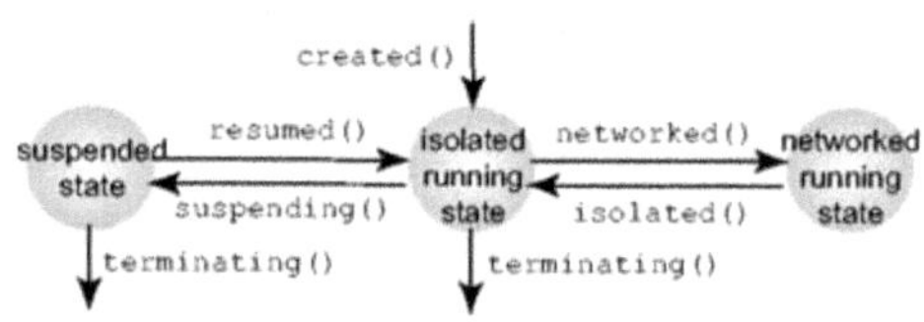

Fig. 9. Callback method invocations in the life-cycle state-transition.

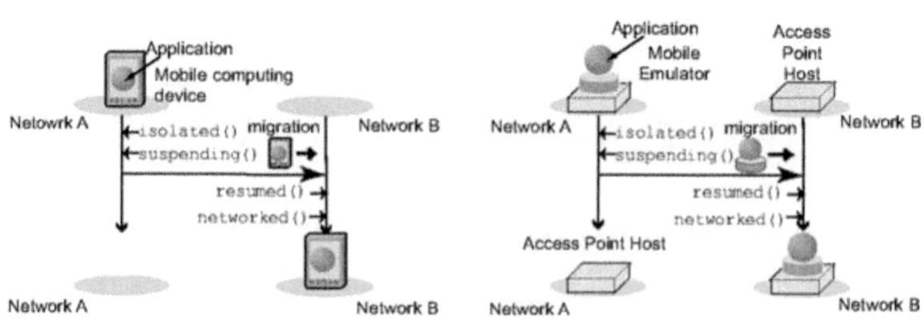

Fig. 10. The movement of a computing device and the migration of the corresponding mobile agent-based emulator.

and rmi's remote references, which are captured by the application and be prohibited from connecting to any servers.

2. Next, the `suspending()` method of the application is invoked to instruct it to do something, for example, closing its graphical user interface, and then the application is marshaled into a bit-stream.
3. The emulator migrates to the destination as a whole with all its inner applications.
4. After the application is unmarshaled from the bit-stream, its `resumed()` method is invoked to do something, for example, redrawing its graphical user interface.
5. After the `networked()` method is invoked, the application is permitted to connect servers on the current networks.

TAPI: Transactions for Accessing Public Infrastructure

Matt Blaze[1], John Ioannidis[1], Sotiris Ioannidis[2], Angelos D. Keromytis[3], Pekka Nikander[4], and Vassilis Prevelakis[5]

[1] AT&T Labs – Research
{mab,ji}@research.att.com
[2] CIS Department, University of Pennsylvania
sotiris@dsl.cis.upenn.edu
[3] CS Department, Columbia University
angelos@cs.columbia.edu
[4] Nomadic Lab
pekka.nikander@nomadiclab.com
[5] CS Department, Drexel University
vp@drexel.edu

Abstract. This paper describes TAPI, an offline scheme intended for general Internet-based micropayments. TAPI, which extends and combines concepts from the *KeyNote Microchecks* and OTPCoins architectures, encodes risk management rules in bank-issued users' credentials which are in turn used to acquire small-valued payment tokens. The scheme has very low transaction overhead and can be tuned to use different risk strategies for different environments and clients.

Keywords: Micropayments, trust management, wireless networks, access control

1 Introduction

Traditional electronic payment systems impose a low bound on the value of each transaction due to the associated processing and clearing overhead. For small-value transactions, this overhead dominates the value of the transaction itself, making the use of such a system uneconomical. Various schemes have been proposed, aiming to reduce overheads so as to handle payments of fractions of a cent. These systems must cope with problems of scale, risk, and trust. It is important to have mechanisms that can scale to millions of transactions while maintaining acceptable levels of risk.

However, cryptographic and other computational operations have non-negligible cost. Thus, we need to minimize the crypto operations by aggregating them in larger transactions. Our observation is that we can take advantage of any locality of reference exhibited by micropayments, *i.e.,* a user paying for a service from a web site is more likely to purchase additional services from the same site. For applications where this holds, we can amortize the cost of many micropayments over a larger payment.

We present a mechanism that allows multiple partial charges against a payment authorization. By splitting a micropayment transaction into a number of partial transactions (minipayments) for smaller amounts, up to the amount of the original micropayment, we can accommodate multiple purchases within the original (single) transaction. Thus, we

M. Conti et al. (Eds.): PWC 2003, LNCS 2775, pp. 90–100, 2003.

spread the cost of the transaction over a number of distinct purchases. A similar approach is being used by some vendors: multiple small credit card transactions are aggregated and presented as a single transaction to the credit card company. Typically, special agreements that cover liability and specify dispute handling policies need to be in place before this can be used. We built a system where dispute handling can easily be managed, *i.e.,* the merchant or the user can prove (or disprove) that a particular minipayment occurred, and thus limit the exposure to fraud.

We discuss a case study involving per-packet charging in a wireless network. In Section 2, we describe the background for our case study and arrive at the requirements for the charging scheme. We then describe the key features of our partial micropayment architecture by presenting a detailed example, namely a microcheck payment framework that is based on the KeyNote [1] trust-management system and a mechanism for making partial payments from a single microcheck. We briefly discuss our implementation in Section 3, and give an overview of related work in Section 4.

1.1 Motivation

The massive reduction in cost of wireless LAN (WiFi) base-stations has resulted in the gradual deployment of wireless LANs in public places by commercial operators, who want to charge for access. In most existing installations, the user must establish credit with the site operator (usually through a Web portal) before being allowed access to the network. As the density of WiFi coverage increases, the requirement for separate authentication with each provider becomes more onerous. Ideally, the user should be able to move between WiFi networks and access the Internet with authentication and payment done automatically. The rigidity of current payment methods (including the inability of the payment infrastructure to handle small payments) forces the network operators to charge for access in large time slots (*e.g.,* on a daily or weekly basis). The use of micropayments would allow the operator to be much more flexible. For example, the operator may wish to charge for each packet sent or received by the user.

Even if operators do not charge real money for the services offered, it is sometimes desirable to have some type of accountability, to detect infrastructure abusers (to avoid the "tragedy of the commons"). Such schemes still depend on some type of user registration and accounting; the missing part is the translation to the real-world concepts (*i.e.,* money). A micropayment scheme with "play" money can be used to provide this accountability. In order to be able to perform this type of charging, we need a system that satisfies the following requirements: First, it must be able to handle very small payments. Second, it should not require a user-initiated login procedure; instead, be able to receive payment dynamically. Third, it should not require on-line authentication with the user's credit institution in order to minimize connection overhead and accommodate situations where the user's credit institution is temporarily inaccessible.

To satisfy these requirements, we employ two different techniques: (a) The KeyNote trust-management System that establishes trust between the user, the service provider and the user's credit institution, using the architecture we originally described in [2], and (b) we introduce the concept of *OTP Coins* that allow single microchecks to be broken into smaller payment increments. We present these two techniques in detail.

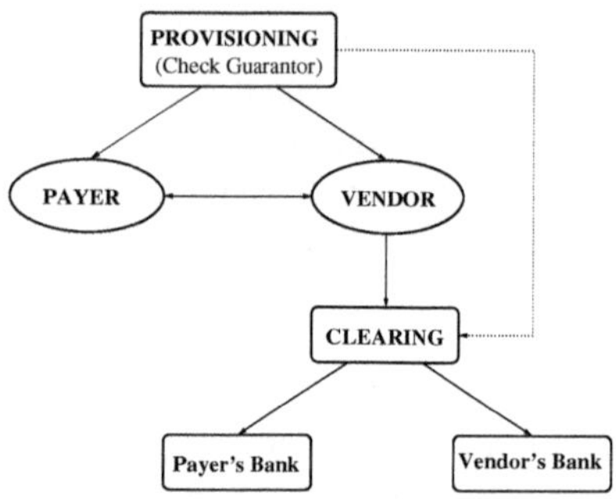

Fig. 1. Microbilling architecture diagram. We give the generic terms for each component, and in parentheses the corresponding players. Arrows represent communication between two parties: Provisioning issues credentials to Payers and Merchants; these communicate to complete transactions; Merchants send transaction information to Clearing, which verifies the transaction and posts the necessary credits/charges or arranges money transfers. Provisioning and Clearing exchange status information about Payer and Merchant accounts.

1.2 KeyNote Microchecks

The micropayments system introduced in [2] forms the basis of our approach. The general architecture of this microbilling system is shown in Figure 1. We consider an environment where Merchants and Payers sign up for service with a Provisioning Agent (PA). Merchants interact with Payers through the Merchant Payment Processor (MPP). The Clearing and Settlement Center (CSC) for reconciling transactions may be a separate entity, or may be part of the PA.

The PA issues KeyNote [1] credentials to *Payers* and *Merchants,* that describe the conditions under which a Payer is allowed to perform a transaction, and the fact that a Merchant is authorized to participate in a transaction. When a Payer wants to buy something from a Merchant, the Merchant encodes the details of the proposed transaction into an *offer,* which is sent to the Payer. To proceed, the Payer issues to the Merchant a microcheck for this offer. The microchecks are encoded as KeyNote credentials that authorize payment for a specific transaction. This credential is effectively a check signed by the Payer and payable to the Merchant. The conditions under which this check is valid match the Merchant's offer. Part of the offer is a nonce, which maps payments to specific transactions and prevents double-depositing of microchecks by the Merchant.

To determine whether he will be paid, the Merchant passes the offer description and the Payer's key along with the Merchant's policy (that identifies the PA key), the Payer credential (signed by the PA) and the microchecks credential (signed by the Payer) to his local KeyNote compliance checker. If the compliance checker authorizes the transaction, the Merchant is guaranteed that Provisioning will allow payment.

If the transaction is approved, the Merchant stores a copy of the microcheck along with the payer credential and associated offer details for later settlement. Otherwise, depending on their network connectivity, either the Payer or the Merchant can request a transaction-specific credential that can be used to authorize the transaction. This approach, if implemented transparently and automatically, provides a continuum between online and offline transactions tuned to the specific risk and operational conditions.

Periodically, the Merchant will 'deposit' the microchecks and associated transaction details to the CSC, which may or may not be run by the same entity as the PA, but must have the proper authorization to transmit billing and payment records to the PA for the customers. The CSC receives payment records from the various Merchants; these records consist of the Offer, the KeyNote microcheck, and the credential from the payer sent in response. In order to verify a microcheck, the CSC goes through the same procedure as the Merchant did when accepting the microcheck. If the KeyNote compliance checker approves, the check is accepted and the account balances adjusted.

The main advantage of this architecture is the ability to encode risk management rules for micropayments in user credentials. Other electronic systems have focused on preventing fraud and failure, rather than on managing it. As prevention mechanisms are often too expensive for micropayments, risk management seems particularly attractive.

1.3 OTP Coins

Electronic coins based on One Time Passwords (OTP) are another fundamental aspect of our approach. While the microchecks manage the risks of single transactions, OTP coins allow the cost of a microcheck to be distributed even more thinly, effectively making it possible to divide a microcheck transaction into hundreds of smaller, partial transactions. This approach is especially suitable for paying for access time, e-content, or other kinds of "continuous" goods, *i.e.,* goods that can be sold by some measure.

The basic approach, without microchecks, was outlined in [3]: an OPIE [4] OTP account was sent to the Client, who used the passwords to pay for wireless Internet access. The system was based on the IEEE 802.1x protocol, running OPIE over TLS.

When combined with microchecks, the Merchant spells out the OTP terms in the offer, *e.g.,* it might state he provides wireless Internet access time at \$0.001 per 5 seconds when bought in lots of 100 5 second units. That is, he offers 100 pieces of 5 second access time units for the price of \$0.1. If the Client accepts the offer, she generates a random number H_{100}, calculates a hash function over it 100 times, forming a reverse hash chain $H_{100}, H_{99}, ..., H_1, H_0$, where $H_i = hash(H_{i+1})$, and embeds the result H_0 into the microcheck she sends to the Merchant. The Merchant stores the hash value H_0 (called H_{check}) along with the number of remaining valid tokens, 100. At this point, the Merchant has sold to the Client 100 OTP coins, only valid with that Merchant. However, the construction allows the Client to be charged only for the actual amount spent.

When the Client wants to use the coins, she sends the next hash value to the Merchant. That is, she first sends H_1, then H_2, *etc.* The Merchant checks that the received hash value gives the previously stored value, *i.e.,* that $H_{stored} = hash(H_{received})$. If so, she decrements the number of remaining valid tokens, and stores the new received value. Thus, we have established a convention where a single OTP password represents the value for a certain commodity, *e.g.,* for 5 seconds of access time. Once the commodity has been used up, the Merchant asks for the next token, to continue service.

Once the Client has used all coins or stops using more coins, the Merchant possesses a hash value H_N where N is the number of coins used. When he deposits the microcheck to the CSC, he also sends these numbers. The CSC computes $H_{check} = hash^N(H_N)$

and compares this to the number stored in the microcheck. If they match, it can be certain that the Client has indeed bought N units of the good.

2 Architecture

We describe the TAPI architecture through an example use in pay-per-use 802.11 access. We then give a brief security analysis of our architecture.

2.1 Example Usage Scenario

As an example, we show how the system can be adapted to a public wireless Internet access using Wireless LANs. We begin with a client that has signed up with an acceptable Provisioning agent. Here, access points subsume the role of the Merchant and users play the role of the Payer. As a result of this registration process (which happens offline), the user is issued with a *clearing check,* signed with the PA's public key:

```
Authorizer: PA_KEY
Licensees: PAYER_KEY
Conditions: app_domain == "Internet Access" &&
      currency == "USD" &&
      &amount < 2.51 && date < "20031231" -> "true";
Signature: ...
```

Wireless LAN Authentication. The IEEE 802.1x standard [5] defines a means to authenticate clients in an Ethernet-like network, *e.g.,* it allows authenticating devices starting to use WLAN or a corporate LAN for Internet access. In practice, the standard defines how to run the IETF standard Extensible Authentication Protocol (EAP) [6] over raw Ethernet frames. The encapsulation is called EAP over LAN (EAPoL) [5].

Since we use the standard EAP protocol, it is possible to use any or all of its subprotocols. However, since neither EAP or EAPoL provide any cryptographic protection themselves, the security of the system depends on the security of the underlying network and on the properties of the EAP subprotocol. Thus, the risks and the protections must be matched to provide the desired level of security.

When 802.1x is used, there are two kinds of client hosts: authenticated and unauthenticated. In a wired LAN, the clients are usually distinguished based on the port: a physical port is either authenticated or not. In a shared medium, *e.g.,* Wireless LAN (WLAN), the distinction is usually based on the Layer 2 addresses. It may be possible to falsify or "steal" a MAC address, depending on the actual implementation. In the case of public WLAN, where no encryption is used, the only protection is the relative difficulty of using a MAC address at the same time another client is using it.

Buying OTP coins. Whenever a new client host wants to join a LAN that uses IEEE 802.1x, the access-point attempts to run EAPoL. The status of the client is kept unauthenticated as long as the client fails to authenticate through EAPoL. In our case, we provide unauthenticated clients limited access so that they can buy OTP coins, used for the actual EAPoL level authentication (see below). That is, any unauthenticated client is served (via DHCP) a private IP address. This address can be used only locally.

The client uses the MPP protocol to purchase a pile of OTP coins. In the simplest case (and lacking any special-purpose protocol for purchasing coins), a simple web interface can be used with the user. When the user contacts the captive portal, he sees a web page that encodes the details of the Merchant offer, *e.g.*:

```
merchant = "ADK'S WIRELESS"
currency = "USD"
product = "Internet Access"
date = "20020916"
packets_per_coin = "100"
coins_per_dollar = "10000"
amount = "2"
nonce = "eb2c3dfc860dde9a"
```

The user examines the details of the request and, if acceptable, authorizes a payment to the merchant by issuing the appropriate KeyNote microcheck:

```
Authorizer: PAYER_KEY
Licensees: "ADK's Internet"
Conditions: app_domain == "Internet Access" &&
        currency == "USD" && amount == "2" &&
        packets_per_coin == "100" &&
        coins_per_dollar == "10000" &&
        first_coin == "c637bf92f9f371dfa09\
                  59bc467d04b91c2ea1b29" &&
        nonce == "eb2c3dfc860dde9a" &&
        date == "20001227" -> "true";
Signature: ...
```

The microcheck also contains the value for the first OTP Coin. This coin is not actually used, but serves as the beginning of the OTP chain. The next time the client needs to authorize a payment, she will use the next coin in the chain, *i.e.*, “310b86e0b62b82856-2fc91c7be5380a992b2786a”. The user sends this microcheck and its guaranteeing check, issued by the PA, to the access point. The latter verifies the integrity of the credentials and determines (by invoking KeyNote) whether the CSC will honor the payment terms.

Using OTP coins. Once the Client has acquired a set of OTP coins, she runs the standard 802.1x EAPoL protocol with the local access point. The access point requests a user identifier from the client, who answers with a string identifying the microcheck used for buying the OTP coins, and the merchant the coins where bought from. The access point then contacts the back-end authenticator (the Merchant). The microcheck fingerprint indicates the correct unused OTP coin pile.

Once the back-end authenticator receives the identity response, it checks the OTP coin pile and sends an OPIE request, requesting for the next unused OPIE password, *i.e.,* OTP coin. The Client responds with the next unused coin, H_{i+1}. The back-end authenticator checks the coin, records it as used, and replies with an EAP SUCCESS message. As the access point receives the EAP SUCCESS message from the back-end authenticator, it changes the status of the client into authenticated, and passes the message to the client. When the client receives the SUCCESS message, she releases her current IP address and requests a new one with DHCP. Since she is now authenticated, she gets a new IP address that she can use to communicate with the outside world. Alternatively, the client could have received a valid IP address which was appropriately filtered by the access point; on success, the relevant filters are simply removed.

Before the OTP coin is used up, the back-end authenticator sends a new OPIE request to the client. If the client wants to continue, she replies with the next OTP coin. On the other hand, if the client does not want to continue access for any reason, she simply does not respond to the request. Thus, if the client goes off-line, the access point changes the status of the client's MAC address into unauthenticated.

Clearing. Periodically, the access point provides all these microchecks along with the related transaction records to the CSC, which uses this information to verify the transaction and charge/credit the relevant accounts. The user's device (laptop, PDA, *etc.*) may also keep a record of all transactions, which can be used in case of a charge dispute. CSCs communicate with PAs to indicate the status of Payers' and Merchants' accounts. Part of the transaction records include the last OTP coin received from a user, and its serial number. The CSC can verify its validity, by repeatedly hashing it the appropriate number of times and comparing the result with the initial OTP coin included in the microcheck. Thus, the CSC can respectively debit the Merchant's account and credit the Payer's account for the appropriate amount. In case of dispute, the exact usage can be determined by verifying the credentials and the OTP coin chain. Assuming the underlying cryptography is not broken, the results are non-repudiable.

2.2 Security Analysis

The security of the system can be broken into two parts: one that relates to the security at the network level, and one that refers to the security of the payment mechanisms.

WLAN security. Wireless LANs are known to be notoriously insecure. However, their insecurity depends heavily in the way they are used. In our example case, where WLAN is used for providing public Internet access, the operator is mainly interested in collecting the access fees, while the clients are interested in getting the service they pay for. Other security concerns that the users may have (e.g. privacy) can be taken care of at an upper layer, and fall beyond the scope of this paper. Consequently, the main threats we are interested in are: (a) someone gaining access without paying, and (b) someone paying but not gaining access. Naturally, these threats may occur at the same time, through an attacker "stealing" access that another user has paid for.

Thus, it is certainly *possible* for an attacker to cause an authenticated client to disconnect from the network, and start using its MAC address. However, the access point is likely to detect the event and may require immediate re-authentication as the MAC address re-connects to the network. If the value of the OTP coins is low enough, *e.g.,* just a few seconds of access time, the gain for the potential attacker is small. Finally, it should be straightforward to detect an attacker that repeatedly steals MAC addresses.

A more powerful attacker can set up a phony access point. If he lures other clients to send OTP coins to it, he can then use these coins to pay towards the real access point, effectively riding free. The victim clients are unlikely to notice anything, since they still get the service they expect. The real access point may not notice anything either, depending on its sophistication. While this attack can be made harder, *e.g.,* by including the Client's MAC address in the microchecks, the simple nature of the OTP coins makes it impossible to block the attack altogether. However, given the current status of WLAN deployment, the cost of the attack compared to the benefits gained seems to be high enough to render the attack academic. For proper security, either the OTP coins must be replaced with something more sophisticated (and costly), or the underlying network must be secured. The additional cost should be evaluated against the expected risk and cost of fraud, and implemented only if economically viable.

Payment Framework Security. When dealing with electronic payments we must ensure that fraudulent transactions cannot take place, *e.g.,* the merchant should not be able to forge an OTP coin, nor should the user be able to deny that she has spent one.

The scheme requires that we select a non-reversible hash function. Thus, the merchant can verify that H_{n-1} was derived from H_n, but is unable to produce H_{n-1} given H_n. Similarly, if the client sends a number other than H_{n-1}, the merchant will detect that and revoke the service. Similarly, if the merchant produces H_{n-1}, the client cannot claim that she has not sent it. An extensive discussion of the security of the KeyNote microcheck architecture may be found in [7]. A key observation is that the low value of the checks and the need for light-weight verification mechanisms favor the use of credential expiration (with short lifetimes) over the use of a more heavyweight revocation mechanism such as credential revocation lists.

3 Implementation

We have implemented the IEEE 802.1x protocol and the OTP coins in the FreeBSD operating system [3]. Our initial performance measurements indicate that the effect on payload performance is negligible: a typical EAPoL transaction is performed in less than two seconds, making it possible to support re-authentication every 5-10 seconds.

Our 802.1x implementation consists of kernel code that implements the basic framing functions for the authentication protocol, plus a number of user level programs. The user-level programs implement the individual EAP subprotocols, and in particular the EAP OTP authenticator and supporting modules. These rely on a small new library, *libeap.* They also utilize the *libskey* library present in FreeBSD. To make it easy to buy OTP coins, we use a captive portal to allow users to download the 802.1x and EAP OTP implementations. Our MAC filter module forward to a web server packets arriving from unauthenticated users. Thus, it is possible to create a situation where the only services provided to an unauthenticated client are DHCP and the captive web server.

We are currently working on implementing the full-fledged MPP protocol on top of EAP, without the need for a captive portal. Users can specify their payment policies using

KeyNote (or some other front-end mechanism, which is then translated to KeyNote). On receipt of an offer from an access point, KeyNote is called to determine whether the terms are acceptable. If so, a microcheck is automatically issued, and the necessary OTP coins are generated and used without user interaction. If the offer is not acceptable, the user is notified and presented with the offer.

4 Related Work

IEEE 802.1x Security. IEEE 802.1x [5] is a forthcoming standard for authenticating and authorizing users in Ethernet like local area network (LAN) environments. It is primarily meant to secure switched Ethernet wireline networks and IEEE 802.11 based WLANs. In the typical usage scenarios, the network requires user authentication before any other traffic is allowed, *i.e.,* even before the client is assigned an IP address. This allows corporations to strictly control access to their networks. It is important to note that 802.1x implements only authentication and MAC address based access control. Since MAC spoofing is fairly easy, the resulting system may not be secure enough.

[8] argues that 802.1x security is flawed since it does not provide per-packet integrity and authenticity. Depending on the settings, this may allow session hijacking, enabling an attacker to take over a MAC address that belongs to a legitimate, authenticated user. In a shared medium such as 802.11, authentication should be tightly integrated with a link-level integrity system using different session keys for different clients.

Electronic Cash and Micropayments. NetBill [9] is a transactional payment protocol with many advanced features (atomicity, group membership, pseudonyms, *etc.*) that requires communication with the NetBill server for each transaction, thus exhibiting the same drawback with respect to micropayments as the simpler online protocols already mentioned. Other general payment protocols [10,11,12] suffer the same problem.

Digital cash-based systems (*e.g.,* [13]) do not directly address the issue of double-spending (fraud). Some e-cash systems use online checking, thus negating the off-line operation capability. Others rely on detection after the fact. This drawback is manifest in several micropayment protocols [14,15,16,17]. While double-spending is a problem in all off-line systems, none of these protocols address the issue of risk management.

NetCents [18] and Millicent [19] are scrip-based off-line-friendly micropayment protocols. As the monetary unit used in these protocols is vendor-specific, double-spending is made difficult. A hidden assumption is that merchants have "total information". If there are many points of sale, continuous communication and synchronization is required between the different points, negating the benefits of off-line operation.

MiniPay [20] was developed primarily for use with a web browser, with a lot of effort gone into the user interface aspect. Risk management is implemented as a decision to perform an online check with the billing server based on the total spending by the customer that day, and some parameter set by the merchant. The billing provider cannot customize the risk-management parameters per-customer or per-merchant. Fileteller [7] uses a scheme similar to ours for buying and selling network-based storage.

5 Summary and Concluding Remarks

We presented TAPI, a simple offline electronic payment scheme intended for general Internet-based micropayments. TAPIallows multiple partial charges on a single payment authorization by splitting a micropayment transaction into a number of minipayments for smaller amounts (totaling the amount of the original micropayment), thus accommodating multiple purchases within the original transaction.

To demonstrate our design in practice, we implemented the IEEE 802.1x protocol along with the OTP coins in the FreeBSD operating system. Our case study involved per-packet charging in a wireless network. Initial performance measurements indicated that the effect on payload performance was negligible and that the typical EAPoL transaction takes less than 2 seconds on average. We are in the process of implementing the full-fledged MPP protocol on top of EAP, without the need for a captive portal.

References

1. Blaze, M., Feigenbaum, J., Ioannidis, J., Keromytis, A.D.: The KeyNote Trust Management System Version 2. Internet RFC 2704 (1999)
2. Blaze, M., Ioannidis, J., Keromytis, A.D.: Offline Micropayments without Trusted Hardware. In: Proceedings of the Fifth International Conference on Financial Cryptography. (2001)
3. Nikander, P.: Authorization and charging in public wlans using freebsd and 802.1x. In: Proceedings of the Annual USENIX Technical Conference, Freenix Track. (2002)
4. N. Haller, C. Metz, P.N.M.S.: A One-Time Password System. RFC 2289, IETF (1998)
5. : IEEE Draft P802.1X/D11: Standard for Port based Network Access Control (2001)
6. L. Blunk, J.V.: PPP Extensible Authentication Protocol (EAP). RFC 2284, IETF (1998)
7. Ioannidis, J., Ioannidis, S., Keromytis, A., Prevelakis, V.: Fileteller: Paying and Getting Paid for File Storage. In: Proceedings of the Sixth International Conference on Financial Cryptography. (2002)
8. Mishra, A., Arbaugh, W.A.: An Initial Security Analysis of the IEEE 802.1x Standard. Technical Report UMIACS-TR-2002-10, Computer Science Department, University of Maryland (2002)
9. Cox, B., Tygar, D., Sirbu, M.: NetBill security and transaction protocol. In: Proceedings of the First USENIX Workshop on Electronic commerce, USENIX (1995)
10. Neuman, C., Medvinsky, G.: Requirements for network payment: The Netcheque prospective. In: Proceedings of IEEE COMCON. (1995)
11. Bellare, M., Garay, J., Herzberg, A., Krawczyk, H., Steiner, M., Tsudik, G., Waidner, M.: iKP – A Family of Secure Electronic Payment Protocols. In: Proceedings of the First USENIX Workshop on Electronic Commerce, USENIX (1995)
12. Foo, E., Boyd, C.: A Payment Scheme Using Vouchers. In: Proceedings of the Second International Conference on Financial Cryptography. Number 1465 in Lecture Notes in Computer Science, Springer-Verlag (1998) 103–121
13. Chaum, D.: Achieving Electronic Privacy. Scientific American (1992) 96–101
14. Rivest, R., Shamir, A.: PayWord and MicroMint. CryptoBytes (**2**) 7–11
15. Jutla, C., Yung, M.: Paytree: amortized signature for flexible micropayments. In: Proceedings of the Second USENIX Workshop on Electronic Commerce, USENIX (1996)
16. Hauser, R., Steiner, M., Waidner, M.: Micro-payments based on ikp. In: Proceedings of the 14th Worldwide Congress on Computer and Communication Security Protection. (1996)

17. Tang, L.: A Set of Protocols for MicroPayments in Distributed Systems. In: Proceedings of the First USENIX Workshop on Electronic Commerce, USENIX (1995)
18. Poutanen, T., Hinton, H., Stumm, M.: NetCents: A Lightweight Protocol for Secure Micropayments. In: Proceedings of the Third USENIX Workshop on Electronic Commerce, USENIX (1998)
19. Manasse, M.S.: The Millicent protocols for electronic commerce. In: Proceedings of the First USENIX Workshop on Electronic Commerce, USENIX (1995)
20. Herzberg, A.: Safeguarding Digital Library Contents. D-Lib Magazine (1998)

On the Coexistence of UWB with Fixed Wireless Access Systems*

Romeo Giuliano, Gianluca Guidoni, Franco Mazzenga, and Francesco Vatalaro

University of Rome - Tor Vergata and *RADIOLABS*
Viale del Politecnico 1, 00133 Rome - ITALY
{giuliano,guidoni,mazzenga,vatalaro}@radiolabs.it

Abstract. Ultra Wide Bandwidth (UWB) spread-spectrum techniques are getting an increased importance due to their many desirable features such as high bit rates support and low power consumption. UWB is considered one enabling technology for advanced wireless local area networks providing integrated multimedia services to nomadic users over hot-spot areas. The assessment of the interference caused by UWB devices on already existing systems is fundamental to guarantee not conflicting coexistence and, therefore, the acceptance of this new technology worldwide. In this paper we use simulation to analyze the coexistence issues between and indoor UWB system and an outdoor fixed wireless access (FWA) system in the $3.5 \div 5.0$ GHz frequency range. We consider a realistic UWB system architecture and we show that in many cases UWB system can coexist with FWA without causing dangerous interference.

1 Introduction

Ultra Wide Bandwidth (UWB) spread-spectrum techniques are gaining increased importance due to their many desirable features such as supported high bit rates and low power consumption [1]. In order to protect existing services against UWB interference the Federal Communications Commission (FCC) has restricted the UWB operating frequencies and allowed UWB transmissions under the operating conditions stated by frequency-power masks issued for each UWB application/device. As a result, UWB communications are confined to the frequency range of $3.1 \div 10.6$ GHz.

The UWB technology is based on the transmission of very short duration pulses and the information signal is spread over a very wide bandwidth (of the order of several GHz) [2]. UWB is one viable candidate for short-range indoor radio communications supporting very high bit rates services and applications. It enables the implementation of innovative wireless local area networks providing multimedia services to nomadic users over hot-spot areas. Thus the assessment of the possible interference caused by UWB devices is of fundamental importance to guarantee not conflicting coexistence between UWB devices and any other

* Work done for *RADIOLABS*, Consorzio Università Industria - Laboratori di Radiocomunicazioni, Italy within the IST ULTRAWAVES European Project

M. Conti et al. (Eds.): PWC 2003, LNCS 2775, pp. 101–113, 2003.

system and to obtain acceptance of this new technology worldwide. To meet the UWB interference requirements on the existing narrow-band systems UWB low transmission-power and short-range operations have to be ensured. The ITU document [3] contains a preliminary analysis on the possible interference of indoor UWB terminals onto an outdoor Fixed Wireless Access (FWA) system. However results in [3] are based on assumptions leading to pessimistic conclusions about the impact of UWB on FWA systems. Main misleading assumptions deal with the selection of the free space channel model for both indoor and outdoor propagation and to the procedure used to evaluate UWB interference on the FWA receiver. In particular in [3] it is assumed that each UWB terminal continuously transmits at its maximum power without accounting for (possible) power control. In this paper we analyze the coexistence problems between indoor UWB system and an outdoor FWA system through computer simulation. In order to compare our results with those already presented in the current literature we keep many of the original assumptions in [3]. However we account for additional UWB system features such as: the UWB system architecture, more realistic propagation models, UWB terminal activity factor. We demonstrate that in many practical cases UWB system can coexist with FWA without dangerous interference. The paper is organized as follows. In Section 2 we summarize the main characteristics of the considered FWA systems and we assess the FWA interference requirements. In Section 3 we describe the selected scenarios and we define the UWB system features introducing the main parameters describing its usage modes. In Section 4 we define the assessment methodology introduced in this work. Simulation results presented and discussed in Section 5. Finally conclusions are given in Section 6.

2 Interference Limits Requirements

To assess the limits of UWB interference on the existing FWA systems we start from the ITU-R requirements in [4] concerning the interference from unwanted emissions from sources other than fixed service (FS) or services sharing the same band on primary bases. As indicated in [4] the total interference to a fixed service link is partitioned as:

- 89 % for the intra service interference;
- 10% for the co-primary services interference;
- 1% for the aggregation of the following interference sources:
  - Emissions from radio services which share frequency allocations on a non-primary basis;
  - Unwanted emissions (i.e. out-of-band and spurious emissions such as energy spread from radio systems, etc.) in non-shared bands;
  - Unwanted radiations (e.g. ISM applications);

As indicated in [5] and [6] the previous percentages apply to the performance objectives. In addition, this degradation allowance refers to the aggregation of the whole secondary services transmitters and unwanted signals. UWB systems are commonly classified as secondary services therefore this criterion is applied. The ITU-R has defined the following long and short term interference criteria for not conflicting coexistence.

Long term criteria (20% of the time)

- For co-primary sharing a margin degradation of 0.5 dB, equivalent to a $I/N = -10$ dB, leads to a performance degradation of 10 %.
- For secondary service interference and unwanted emissions, the performance degradation shall not exceed 1%. For these services it was concluded that a $I/N = -20$ dB is the adequate criteria since it represents a margin degradation of 0.04dB.

The previous two statements indicate generic objectives and implicitly assume that interference has spectral characteristic similar to white noise. However in the UWB case, due to the pulsed characteristic of UWB signals separate considerations would be needed for both average (RMS) and peak interference objectives within the FS receiver bandwidth. It should be observed that the allocation of 99% of the interference margin to intra-service and/or co-primary services (89% + 10%) might also be too pessimistic with today's technology and in more realistic conditions higher margins than 1%, in particular I/N much closer to 0 dB, could be tolerated by fixed links for secondary services. In the rest of the paper we mention the limits and results both considering the long-term interference with the conservative $I/N = -20$ dB and with $I/N = -10$ dB.

Short term criteria (0.0001% of time)

Short-term criteria in [7] gives allowance for a positive (in dB) I/N ratio to happen for very short percentage of time. Positive I/N ratio can be related to peak interference. Similarly to [3] this criterion will not be considered in this paper.

The FWA system considered in this paper includes two communicating devices: a FWA Central station (CS) positioned on the roof of the building, and a FWA Terminal Station (TS) close to the building. The most relevant characteristic to our interference study is the FWA receiver noise parameter. This can be expressed as:

$$N_A = -144 + 10\log_{10}(RX_{BW}) + NF \quad (1)$$

where RX_{BW} is the 3dB receiver bandwidth (in MHz) and NF is the receiver noise figure (dB).

Considering a FWA receiver operating at 3.5GHz, we assume $NF = 5$dB, and using (1) we obtain $N_{A50} = -122dBW$ for $RX_{BW} = 50$MHz and $N_{A14} = -127.5$dBW for $RX_{BW} = 14$MHz. Considering FWA systems operating in the 3.6÷4.2 GHz and 4.4÷5.0 GHz we assume $NF = 6$dB and for $RX_{BW} = 50$MHz we obtain $N_{A50} = -121$dBW.

Starting from the calculated FWA receiver noise characteristics, we obtained the limits on the allowed UWB inter-system interference indicated in Table 1.

Table 1. Summary of UWB aggregate interference objectives for coexistence with FS-ITU-R objectives

FWA features	Interference limits
FWA CS ($B_W = 14$ MHz)	$\leq -$ 159dBW/MHz @ I/N= -20dB
	$\leq -$ 149dBW/MHz @ I/N= -10dB
FWA CS ($B_W = 50$ MHz)	$\leq -$ 159dBW/MHz @ I/N= -20dB
	$\leq -$ 149dBW/MHz @ I/N= -10dB

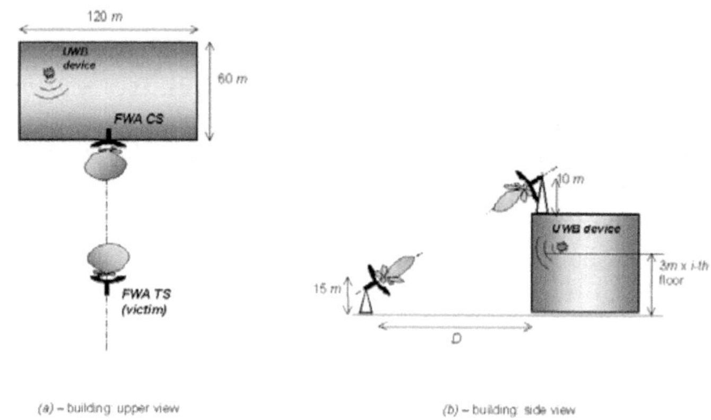

Fig. 1. Hot spot reference scenario for FWA - UWB system coexistence.

Also the corresponding values for the less conservative, but still very safe, protection levels $I/N = -10$dB were included. To study the effects of UWB on the FWA we first evaluate the power of the UWB interference at the FWA receiver $I_{UWB \to FWA}$. The power spectral density of the UWB interference is obtained by dividing $I_{UWB \to FWA}$ for the bandwidth of the UWB signal. The result is then compared with the values in Table 1.

The interference $I_{UWB \to FWA}$ is evaluated as:

$$I_{UWB \to FWA} = \sum_n P_{T_n} L(d_n), \qquad (2)$$

where P_{T_n} is the power transmitted by the n-th UWB terminal and $L(d_n)$ is the path loss between the n-th UWB terminal and the FWA receiver located at relative distance d_n. The path loss $L(d)$ accounts for both indoor and outdoor propagation and antenna gains.

3 Selected Scenario

To evaluate (2) we consider an UWB system located in a commercial/industrial building (hot-spot scenario). We assume that UWB system interferes with the

Table 2. Main parameters of the considered UWB system

Simulator feature	Alternatives
Outdoor Propagation Model	Free Space, ITU-R P.1411-1 Cost231-Hata model/3GPP
Indoor Propagation Model	Free space, ITU-R P.1238-2, Cassioli, Molish Win, (CMW)[8]
Power Control	Used for every UWB device
Activity Factor	1%, 2%, 5%, 10%
Bit Rates	2, 25, 110 Mb/s

FWA system in the proximity of the considered building. The FWA-TS antenna is steered in order to optimally point to the FWA-CS. We assume the building has 10 isolated floors and we neglect the UWB intra-floor interference. The reference scenario is plotted in Figure 1. The UWB system contains a mixed population of devices characterized by different bit rates. UWB devices are located inside building and communicate among them according to a master-slave architecture. To evaluate UWB interference on FWA we introduced more accurate models for indoor and outdoor propagation, power control, terminals activity factors less than unity. The antenna radiation patterns and other geometrical parameters such as the FWA distance from the building, floor dimensions were assumed as in [3]. The parameters and models used for analysis are summarized in Table 2. The number of possible scenarios obtained from the combinations of the parameters in Table 2 and that should be explored is very large. However many scenarios could result as unrealistic. As an example it could be noted that using 100% activity factors contrasts with the typical deployment scenarios envisaged for UWB devices where they will be active only for a fraction of time. In particular it is estimated that in any case the aggregate percentage of time that any single device transmits is to be considered not exceeding 10% of the time. For brevity, in the following we restrict our analysis to some important and realistic cases.

4 Interference Calculation

To evaluate $I_{UWB \to FWA}$ in (2) we used the software simulator developed within the IST ULTRAWAVES European project [9]. We consider an UWB system based on a Code Division Multiple Access (CDMA) technique with multiple parallel radio paths for upstream (slave to master) and downstream (master to slave) communications. The simulated area is assumed to be rectangular and UWB masters are located on a regular grid. Slaves can be arbitrarily and uniformly located within each floor according to a uniform distribution. The simulator is snapshot based and its operating principle is illustrated in the flowchart of Fig.2. In each *outer loop* iteration a new random scenario is generated. Assuming the positions of the masters scenario generation consists in randomly placing the slaves in the service area. The *inner loop* is used to simulate the power control procedure. For each inner iteration the power of each transmitter is increased or

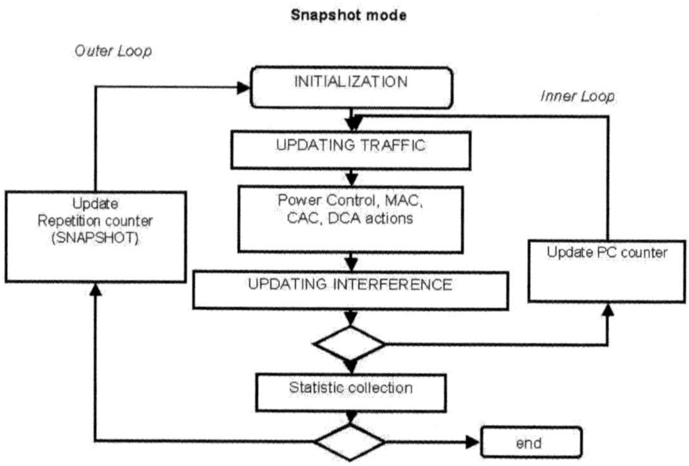

Fig. 2. Simulator flowchart

decreased in accordance to the comparison of the carrier-to-interference ratio, C/I, calculated for each slave by the master device to which the slave is connected to, and the reference carrier-to-interference ratio, $(C/I)_{Target}$. The interference I measured for each master receiver is calculated in each inner iteration accounting for the active UWB transmitters in the area and the propagation losses. The inner procedure is halted when the power transmitted by each slave remains constant or has reached the maximum. After reaching this steady condition all the relevant parameters such as the power level transmitted by each user, the users in outage etc., are collected. More than 10000 snapshots were generated and the statistical behavior of the collected parameters was analyzed. The mobility of the UWB devices was not considered in this paper.

The simulator parameters are now discussed in detail.

- *Source Traffic Model*
  We consider traffic sources characterized by different bit rates and activity factors. The simulation results presented in the next Section will consider different combinations of source traffic models.
- *Transmitted Power and Interference*
  The maximum transmitting power for each UWB terminal is −6.2dBm and can be easily calculated starting from the maximum allowable power spectral density (i.e. −41dBm/MHz), assuming a UWB signal bandwidth (W_{UWB}) of 3GHz and a flat UWB spectrum. The total interference I in each master receiver is:

$$I = I_{Intra} + I_{FWA \to UWB} + \eta \tag{3}$$

  where I_{Intra} is the intra-system interference caused by the UWB devices; η is the thermal noise power and $I_{FWA \to UWB}$ is the inter-system interference due to FWA on the UWB system. The $I_{FWA \rightarrow UWB}$ is accounted for by increasing the noise figure in the UWB receiver by 5dB. The $(C/I)_{Target}$ used in the power control procedure for each UWB source was evaluated as

$$\left(\frac{C}{I}\right)_{\text{Target}} = \left(\frac{E_b}{N_0}\right)_{\text{Target}} - PG \tag{4}$$

where $PG = W_{UWB}/R_b$ and R_b is the source bit rate. The UWB reference $(E_b/N_0)_{Target}$ was always set to 4dB in each scenario and for each bit rate.

– *Pathloss models: indoor propagation*
To analyze the dependence of the UWB interference from the selected channel model, we repeated the simulations assuming two different path-loss models: the simple free space model and the Cassioli-Molish-Win (CMW) model [8] repeated below:
  - *Free space propagation model*

$$A_{FS}(d) = c_0 + c_1 \log_{10}(d) \quad [dB] \tag{5}$$

  where c_0= pathloss at 1m = $20 \cdot \log_{10}(\frac{4\pi}{\lambda})$= 43.32dB at 3.5GHz;$c_1 = 20$ dB/m
  - *CMW model*

$$A_{CMW}(d) = \begin{cases} 0 & d \in [0,1] \\ c_0 \cdot \log_{10}(d) & d \in [1, d_{break}] \\ c_1 + c_2 \cdot \log_{10}\left(\frac{d}{d_{break}}\right) & d > d_{break} \end{cases} \tag{6}$$

  it is a dual slope model where $c_0 = 20.4$dB/m, $c_1 = 20.4 \log_{10}(d_{break})$, $c_2 = 74$dB/m and $d_{break} = 11m$.

The selected path loss models represent two extremes among those indicated in Table 1. In fact, neglecting possible tunnel effects, the path loss exponents of the two selected models vary between 2 and 7.4 thus accounting for many intermediate values of the path loss exponents of other models.

– *Antenna models*
We assumed that any UWB device has an omnidirectional antenna on the horizontal plane with gain 0dBi. For the FWA-TS, we use the antenna model indicated in [10]. It is a sectorial antenna with sectors of 90° and a main lobe gain of 16dBi. Its radiation pattern envelopes are reported in [10].

Call admission control strategies (CAC) strategies enhance performance and allow to reduce the interference both in the UWB system and other narrowband systems. For example an UWB slave could be stopped from transmitting when its power level requirements are judged too high by the master. The beneficial effects of CAC strategies were not considered in this paper.

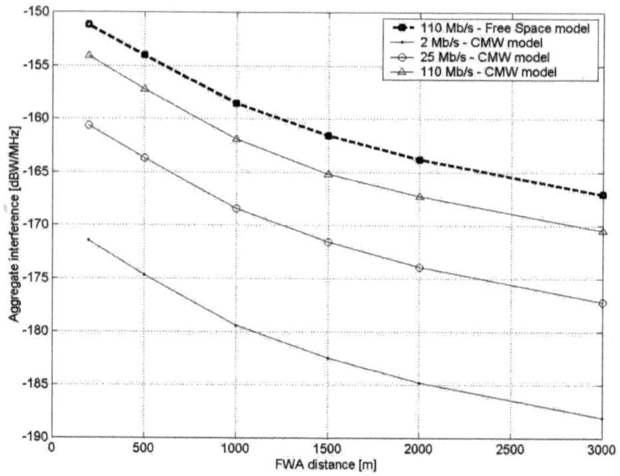

Fig. 3. UWB aggregate interference vs FWA distance - Single Service Case - Free Space and CMW channel models - 32 masters per floor, activity factor 1

5 Simulations Results

Simulations results were obtained considering different combinations of the system parameters indicated in Table 2. In this paper only upstream (slave-to-master) transmissions were considered. To validate our simulator we first re-implemented the scenario in [3] and we considered an UWB terminal density of 450 users per floor. The floor dimensions where 120×60 m^2) and the vertical distance between floors is 3m (see Fig.1). Every UWB device is unrealistically set to transmit 100% of the time at its maximum power level (i.e. -6.2 dBm for $W_{UWB} = 3$GHz). The UWB devices were positioned over a grid at a relative distance of 4m. We used the free space path loss model for both indoor and outdoor links. The results obtained are in very good agreement with those presented in [3] and for brevity are not reported. In the following subsections we illustrate the simulation results in more realistic scenarios considering both single and multiple bit rate services.

5.1 Single Service Cases

We assume that UWB devices transmit at the same bit rate: 2, 25 or 110 Mb/s. The density of the UWB terminals is varied and both free space and CMW models have been considered. The building penetration loss is neglected [1].

[1] For a correct evaluation of interference at least 10 dB should be considered bringing the target aggregate interference to -139 dBW/MHz @ $I/N = -10$ dB (or -149 dBW/MHz @ $I/N = -20$ dB)

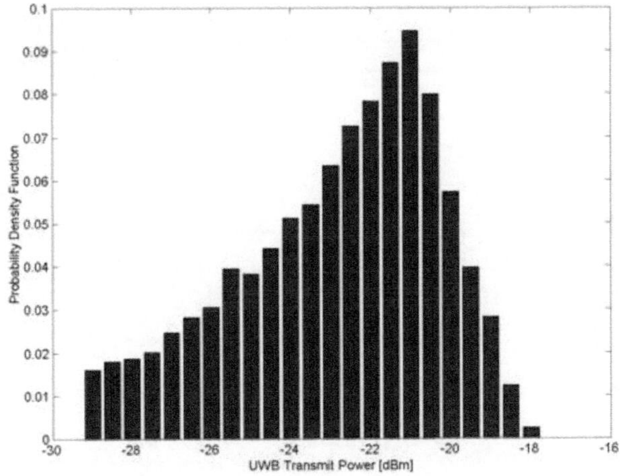

Fig. 4. Power distribution for (power controlled) devices transmitting at 110 Mb/s

In Fig. 3 we plot the UWB aggregate interference density vs the distance of the FWA-TS from the building. We considered the free space and the CMW channel models for indoor propagation while we always adopted the free space channel model for outdoor. A density of 500 UWB devices per floor was considered. From the results in Fig.3 it can be noticed that due to power control action the aggregate interference (dBW/MHz) is lower than the ITU requirements in Table 1 in any conditions and for practical FWA distances from the building. Including 10 dB due to building penetration loss, the target aggregate interference is fully respected, even at these extreme high densities.

In Fig. 4 we plot the histogram of the power transmitted by slaves operating at 110Mb/s. It can be observed that no UWB users are in outage and that due to power control action the transmitted power is always well below than the FCC limit of −6.2dBm.

Having proved the compatibility between the UWB and the FWA systems with different service classes and for densities higher than those indicated in [3], we now vary the activity factor to the maximum levels foreseeable for an UWB terminal from 1% up to 10%.

In Fig.5 and Fig.6 we plot the aggregate interference density for different FWA receiver distance with transmitters at 110 Mb/s and for different activity factors. As expected, results in Fig.5 and Fig.6 show the raise in the aggregate interference levels with the increase of activity from 1% to 5% and then to the maximum expected of 10%. Always from the results in Fig.5 and Fig.6 even in the unrealistic case of 500 UWB devices per floor and even increasing the activity factor to 10%, in most practical cases the effects on the FWA system are not harmful.

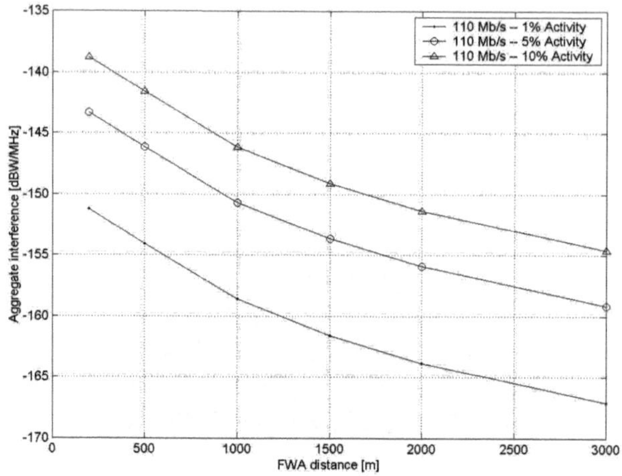

Fig. 5. Aggregate interference density for different FWA receiver distance with transmitters at 110 Mb/s, free space indoor model, Activity factors of 1%, 5% and 10%

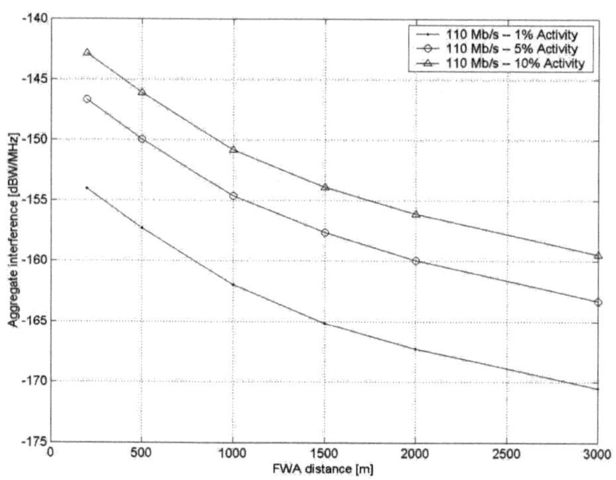

Fig. 6. Aggregate interference density for different FWA receiver distance with transmitters at 110 Mb/s, CMW indoor model, Activity factors of 1%, 5% and 10%

5.2 Multi-service Cases

In this Section we consider a scenario where UWB terminals simultaneously operate at different bit rates. The first set of simulations considered slaves with two different bit rates: 2 Mb/s (50% of slaves) and 110 Mb/s (50% of slaves - possibly unrealistic). The system parameters are unchanged but the devices

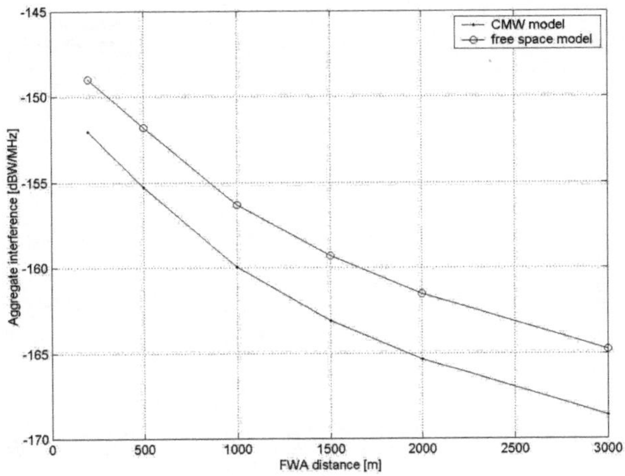

Fig. 7. Aggregate interference density vs FWA receiver distance - UWB services 110 Mb/s (50%)+2 Mbit/s (50%)

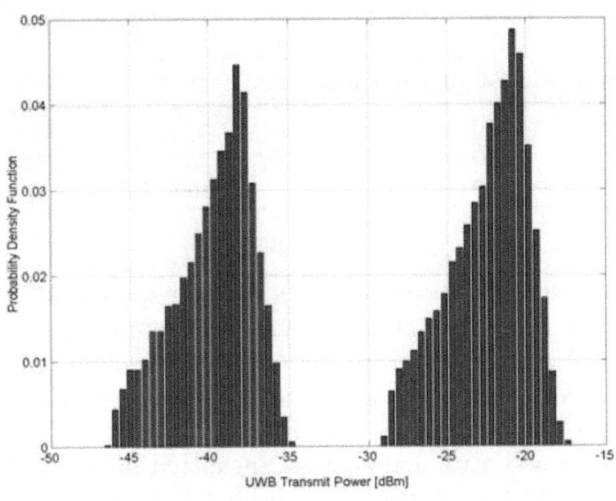

Fig. 8. Power distribution for power controlled devices transmitting at 110 Mb/s and 2 Mbit/s (50%)- indoor free space channel model

density is reduced to 300 UWB terminals per floor to obtain a more realistic scenario. Slaves are power controlled and have an activity factor of 5%.

In Fig.7 we plot the aggregate interference density vs the FWA receiver distance from the building. Different indoor propagation models were considered. From the results in Fig.7 it can be observed that full compatibility between

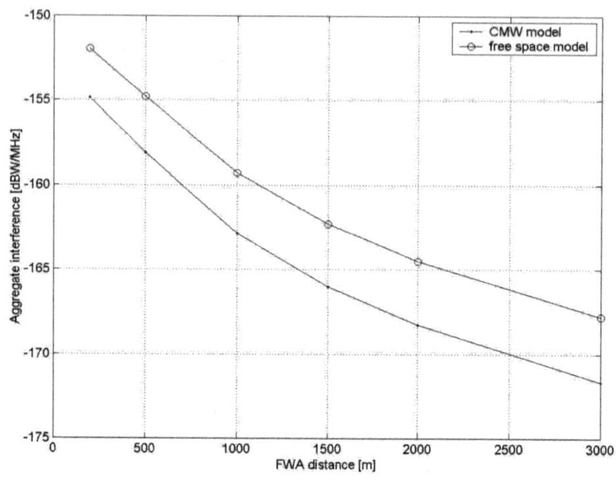

Fig. 9. Aggregate interference density vs FWA receiver distance - UWB services 110+Mb/s(20%)+25 Mb/s(30%)+2 Mbit/s(50%)-Power control ON - free space and CMW indoor propagation model

Table 3. Traffic Parameters for mixed bit rate simulation: 2+25+110 Mb/s.

Population [% of the total]	Activity Factor	Data Rate
20	2%	110 Mb/s
30	5%	25 Mb/s
50	10%	2 Mb/s

UWB and FWA services is ensured both for free space and CMW models. In fact, for practical FWA distances the aggregate interference is well below the targets in Table 1 that however are met even at very close FWA distances.

In Fig.8 we plot the histogram on the UWB transmitted power in the case of terminals operating at 2 Mb/s and 110Mb/s.

Thanks to power control action the transmitted power is always well below than the FCC limit of −6.2dBm. As expected the average power transmitted by the 110 Mb/s terminals is larger than the 2 Mb/s. This is due to the lower coding gain available for the higher bit rate services.

Finally in Fig.9 we plot the aggregate interference by considering a UWB mixed bit rate scenario as a function of the FWA receiver distance. We considered 300 UWB users per floor and 32 UWB master devices. Traffic parameters used to obtain the results in Fig.9 are indicated in Table 3. From the results in Fig.9 it can be observed that full compatibility between UWB and FWA services is ensured both for free space and CMW models.

6 Conclusions

The assessment of the potential interference caused by UWB devices is a fundamental research topic to ensure not conflicting coexistence between UWB devices and any other existing system. Any study needs to go beyond simplistic models and should consider realistic scenarios. As many other wireless local area network systems, UWB terminals will be in sleep mode for the large majority of times, will not run continuously and will not emit constantly at the maximum allowed power. More accurate path loss models were considered to evaluate the UWB aggregate interference on FWA. The results presented in this paper show that considering basic system features such as the UWB system architecture, the activity factor, the power control and including realistic propagation models, even in the extreme densities proposed there is no risk for the FWA operations. Further margins not even considered in this paper still exist such as admission control techniques, antenna polarization mismatches, deep NLOS and multiple trough-wall indoor losses. The analysis procedure introduced in this paper can be easily extended to other significative scenarios.

References

1. R. A. Scholtz, R. Weaver, E. Homier, J. Lee, P. Hilmes, A. Taha, and R. Wilson, "UWB radio deployment challenges", in Proc. IEEE Int. Symp. on Personal, Indoor and Mobile Radio Communications, vol. 1, pp. 620–625, Sept. 2000. London, UK.
2. M. Ho, L. Taylor, and G. Aiello, "UWB technology for wireless video networking," in Int. Conference on Consumer Electronics, ICCE, pp. 18–19, 2001.
3. ITU-R M18-20A5-1R0-SE24-UWB - annex5-1- FS- Updated UWB and FS coexistence report (Annex 5-1), updated version of: "*Preliminary Study on Coexistence between UWB and the Fixed Service in Bands from 1 to 6 GHz*"
4. ITU-R F.1094-1 (10/95) "Maximum allowable error performance and availability degradations to digital radio-relay systems arising from interference from emissions and radiations from other sources"
5. ITU-R F.1397: "Error performance objectives for real digital radio links used in the international portion of a 27500 km hypothetical reference path at or above the primary rate"
6. ITU-R F.1491: "Error performance objectives for real digital radio links used in the national portion of a 27500 km hypothetical reference path at or above the primary rate"
7. ITU-R F.758-2 (05/00) "Considerations in the development of criteria for sharing between the terrestrial fixed service and other services"
8. D. Cassioli, Moe Z. Win and Andreas F. Molisch, "The ultra-wide bandwidth indoor channel: from statistical model to simulations", IEEE Journal on Selected Areas on Commun. – Special issue "Channel and propagation models for wireless system design," Aug. 2002
9. "ULTRA Wideband Audio Video Entertainment System (ULTRAWAVES)" European Project – IST-2001-35189
10. ITU-R F.1336: "Reference radiation patterns of omnidirectional, sectoral and other antennas in point-to-multipoint systems for use in sharing studies in the frequency range from 1 to about 70 GHz"

Improving Wireless Access Control Schemes via Adaptive Power Regulation

Francesca Cuomo and Cristina Martello

Dpt. Info-Com, University of Rome "La Sapienza" ,
Via Eudossiana, 18 - 00184 Rome (Italy)
{cuomo,martello}@infocom.uniroma1.it

Abstract. Two challenging issues for future wireless communication systems are the support of *Quality of Service* (QoS) and the definition of flexible access schemes to be used in heterogeneous systems (public cellular systems, wireless local area and ad-hoc networks). This paper deals with the class of *Radio Resource Control* (RRC) based on power regulation, where typical PHY layers parameters (powers, interference, etc.) are combined with upper layers ones to i) increase radio channel utilization; ii) perform admission control of heterogeneous wireless links; iii) maintain negotiated QoS levels expressed as a function of the *Signal-to-Interference-Ratio* (SIR). In the recent literature power control is mainly employed to dynamically adapt the transmission of concurrent links to the varying system conditions (due to channel quality, mobility) still guaranteeing the perceived QoS. In this framework, we investigate how the design of a suitable power control can be used in the support of traffic requiring some QoS guarantees. The paper presents a paradigm where all the RRC operations are performed in a distributed way, i.e., independently for each link involved in a communication. The distributed approach assures that the strategies are quite flexible to be used both in cellular-like systems and in ad-hoc networks.

1 Introduction

One challenging issue in the design of future wireless communication systems is the possibility to provide *Radio Resource Control* (RRC) schemes combining, in accordance to a cross-layer approach, PHY, MAC and RLC information in order to optimize the use of the radio resource and satisfy user requirements. This paper deals with the class of RRC based on *PoWer Control* (PWC) where typical PHY layers parameters (power, interference, etc.) are combined with upper layers ones to increase radio channel utilization and to support *Quality of Service* (QoS). PWC is the paradigm to accommodate in a wireless system users heterogeneous both in QoS requirements (data rate, packet error rate, etc.) and in perceived quality of the respective wireless channel [1][2][3].

In the recent literature PWC is envisioned as the means to adapt dynamically the transmission of concurrent links to the varying system conditions (due to channel quality and mobility) still guaranteeing the desired QoS. In this framework we investigate how the design of a suitable PWC scheme can be used for

M. Conti et al. (Eds.): PWC 2003, LNCS 2775, pp. 114–127, 2003.

the support of traffic requiring QoS guarantees that can be expressed, in a very general way, as a soft state of the *Signal-to-Interference-Ratio* (SIR); i.e., for each link it is negotiated that the SIR will not drop below a given threshold. We remark that the SIR is a simple metric that implicitly includes QoS parameters like the *Bit Error Rate* (BER), data rate (e.g., processing gain in CDMA systems) and consequently packet transmission time. RRC based on PWC assures that the system resource allocation is adapted to the current environment state.

The paper aims at presenting the frame of PWC exploited for the QoS support in a system where all the RRC operations are performed in a distributed way, i.e., independently for each link involved in a communication. The distributed approach assures that the strategies are quite flexible to be used in both an infrastructured and an ad-hoc network.

The remainder of this paper is organized as follows. Section 2 briefly recalls some literature on distributed power controlled access schemes while in Section 3 we present our proposal (named *Distributed Power Regulated Admission Control* scheme, D-PRAC); in particular we describe the admission control rules and we discuss issues concerning the adaptivity to system's dynamics as well as implementation aspects. A comparison of D-PRAC with other approaches found in literature is presented by a performance analysis (Section 4). Finally, Section 5 concludes the paper.

2 Admission Control Schemes Based on Power Control

Quality of service based on adapting transmission power allows to:

1. establish and maintain links by adapting power levels on the basis of current interference and link extension;
2. make links achieving differentiated QoS levels in terms of rate and target SIR (see for example [4]);
3. mitigate interference and thus channel reuse.

In classical cellular systems (e.g., UMTS) PWC is mainly employed as a mechanism to uniform the power levels of different links concurrent to the same Base Station [5]. Here we extend the PWC adoption to perform the *Admission Control* (AC) of links requiring some QoS in a wireless network and to maintain the desired QoS during the link lifetime. Furthermore, it can be shown that a suitable PWC can improve the overall system performance and can guarantee an efficient channel reuse that is a key aspect when referring to networks lacking of spectrum (cell) planning schemes.

We focus on distributed AC, where each link will autonomously decide its own admission by collecting information regarding the system it is going to enter. In particular the receiver typically has the task of measuring the perceived interference and reporting the result to the transmitter [6]. Besides measurements, some proposals (e.g., [7]) provide also for signaling broadcasted by each device (typically the transmitter) to all others as further information used to coordinate the power selection process of a number of links. Depending on the multiple access technique, PWC schemes can be applied in single channel systems (e.g., 802.11) and in multi-channel ones (e.g., CDMA). However, it can be shown that in both cases the problem of power controlled access can be formalized in very similar terms [8]. The main difference between the two is that, in case of multi-channel, interference is mitigated by a processing gain and thus resource reuse is higher. The need of making a number of links coexist, leads to the conclusion that each one should regulate its own power in order not to disturb too much the other neighbor communications while keeping to preserve its desired SIR. In [9] it has been investigated how the introduction of PWC in the access procedure can enhance the system performance in terms of radio channel utilization efficiency.

Two main classes of distributed power controlled access schemes can be identified:

1. based exclusively on measurements of the surrounding environment;
2. based on both measurements and signaling among different links.

As for methods based just on measurements, proposals in this class are [6][8][10]. The common idea of these works is that a link can extract information on the system just by performing measurements. In particular in [10] a local (hence distributed) probing scheme is developed aiming at discovering admissibility of the new link just by "comparing" interference measurement results before and after activating the transmission of the link: a link evaluates the impact of its entrance on the others (which will adapt to the new interference) and decides whether it is admissible. In [8] it is proposed a distributed PWC algorithm which protects active links with respect to the interference brought by new access attempts. Each link just applies a rule of power updating on the basis of current SIR measure: the algorithm has the desirable feature to guarantee the maintenance of the target SIR during updating. In spite of their advantageous lack of inter-link signaling, a strong limit of such works consists in not considering the power-constrained problem (i.e., radio devices have a bound on the maximum power they can emit). This problem can not be overcome when links operate only on the basis of measurements. A link evaluates the possibility to enter the system on the basis of its own impact on the others and of their consequent reaction, but in a power constrained system this reaction in terms of additional interference can be less due to power saturation of some device. As a consequence the admission decision could wrongly success causing the drop of power saturated links.

Other approaches face and overcome the problem of power constraint by means of explicit inter-link signaling [7][11][12]. The adopted paradigm consists in protecting links by maintaining a *margin* with respect to the target SIR. This margin is managed in order to allow the admission of new links avoiding to reconfigure (in particular increase) current transmission powers. These strategies result more robust since maintenance of a margin implies that each link can tolerate positive extra interference due to new accesses and also absorbs link quality variations due to "unpredictable" phenomena (e.g., radio channel fluctuations and devices' mobility). Thus a *soft QoS* level is maintained since the changed link conditions imply SIR variations above the target SIR thanks to the margin. Obviously a tradeoff exists between setting large margins accommodating a high number of soft QoS flows or fixing the margins at very low values thus reducing system flexibility and the number of contemporary flows. In the case of margin adoption, an explicit inter-link signaling is required since a link willing to be established must regulate its transmission power in accordance to the margins of its neighboring links. Margins exchange could be done periodically (like in [7]) or on demand (like in [11]).

A main difference between schemes based only on measurements and schemes based on both measurements and signaling concerns the way of managing powers each time a new link tries to be established: the power management can be either global or incremental [11]. Typically schemes based only on measurements fall into the class of global power management, since all powers are continuously adapted to current interference; anyway not all the proposals can guarantee consistence with the SIR requirements during the updating phase (like in [10]). Instead schemes based on both measurements and inter-link signaling can be either global or incremental although proposals found in literature do not provide for power reconfiguration and are in the class of the incremental approaches.

In this paper we propose a strategy for distributed admission control based on both measurements and inter-link signaling. In order to privilege the robustness of the AC and the complexity of an eventual implementation still aiming at a flexible and efficient system, we pursue an incremental-like approach with the additional feature of margins reconfiguration during the link lifetime.

3 A Distributed Power Regulated RRC Based on Margins

3.1 Access Model and Problem Statement

The considered reference model is constituted by a number N of single-hop interfering links. In a general case, either single or multi channel interface, the following expression can be provided for the SIR (denoted as γ) of the i-th link:

$$\gamma_i = \frac{P_i \cdot g_{ii}}{\sum_{j=1, j \neq i}^{N} P_j \cdot g_{ij} + \eta_i} \tag{1}$$

where g_{ij} represents a kind of "cross-correlation effect" between the i-th and the j-th link. In general g_{ij} is the total path gain form the j-th link's transmitter to the i-th link's receiver and accounts for the distance attenuation as well as for shadowing and multi-path fading effects; in case of multi-channel (like CDMA) it also includes the processing gain. P_i is the i-th link's transmitter emitted power and η_i is the noise power at the i-th link's receiver.

In the framework of a power regulated access, the AC problem can be formulated in the following way. Without loss of generality, we assume that the link requirements are expressed in terms of SIR target γ^T and - in case of CDMA systems - of desired rate, this latter in turn implies that some code with a certain processing gain is required for transmission. We consider that N-1 links are already active satisfying their QoS requirements. The problem of evaluating the access of an N-th link consists in finding, if it exists, a proper set of transmission powers $\{P_i\}_{i=1,...,N}$ such that $\gamma_i \geq \gamma_i^T$ for each link $i = 1, ..., N$. The problem can be expressed in the following matrix form allowing to define the well-known condition for existence of a feasible solution:

$$\begin{cases} (\mathbf{I} - \mathbf{F}) \cdot \mathbf{P} \geq \mathbf{h} \\ \mathbf{P} \qquad\qquad\; \geq \mathbf{0} \end{cases} \tag{2}$$

where:

- $\mathbf{I}$ is the $N \times N$ identity matrix;
- $\mathbf{P}$ is the column vector $(N \times 1)$ of transmission powers;
- $\mathbf{F}$ is a $N \times N$ matrix which depends on the current topology; in particular $F_{ii} = 0$ and $F_{ij} = \gamma_i^T \cdot \frac{g_{ij}}{g_{ii}}$;
- $\mathbf{h}$ is a column vector $(N \times 1)$ and is essentially related to noise powers $(h_i = \gamma_i^T \cdot \eta_i / g_{ii})$.

The existence of a feasible solution of problem (2) is expressed as a condition for the maximum modulus eigenvalue of $\mathbf{F}$, ρ_F, and is $\rho_F < 1$.

In other terms, the N links can coexist satisfying their QoS requirements depending on the topology they are structured in. If a solution of problem (2) exists, then the minimum power solution (called *Pareto-optimal* solution) is:

$$\mathbf{P}^* = (\mathbf{I} - \mathbf{F})^{-1} \cdot \mathbf{h} \tag{3}$$

which has the property that, with every other solution, all links' power levels are not lower that their corresponding Pareto-optimal levels. In particular any other solution can be expressed as $\mathbf{P}^* = (\mathbf{I} - \mathbf{F})^{-1} \cdot (\mathbf{h} + \epsilon)$ where ϵ is a column vector of N real positive values.

3.2 Approach Based on Margins (D-PRAC)

As stated before, the method we adopt is based on making active links keep a margin with respect to their desired SIR γ^T. In order to maintain this margin we introduce the *Maximum Extra Interference* (MEI) an active link can sustain still maintaining its SIR not below γ^T. If MEI is greater than 0 the link perceives a SIR above γ^T, while a MEI equal to 0 means that the link operates at the target SIR. The scheme we propose, D-PRAC, proceeds in a step-by-step way by evaluating the access of a new link given a number of already active ones. Once verified the admissibility of the new link, D-PRAC will maintain the links at power levels $\mathbf{P}^* < \mathbf{P} \leq \mathbf{P}^{\mathbf{dv}}_{\mathbf{max}}$, $\mathbf{P}^{\mathbf{dv}}_{\mathbf{max}}$ being the column vector of the devices' maximum powers. Thus, in general, the scheme will not be able to configure $\mathbf{P}$ at the Pareto-optimal levels.

The D-PRAC main steps are:

Starting point: N-1 links are active at transmission powers $\mathbf{P^{(N-1)}}$ (where N-1 represents the length of the vector) such that they can tolerate some positive MEI level.

Access attempt: the access of the N-th link can take place at power P_N within a range ($P_{min,N}$, $P_{max,N}$) where $P_{min,N}$ is the minimum transmission power allowing to reach the desired QoS and $P_{max,N}$ is the maximum transmission power compliant with (i.e., not exceeding) active links' MEIs and with the device maximum power $P^{dv}_{max,N}$ (the computation of these parameters is explained in Section 3.3). The admission rule compares $P_{min,N}$ and $P_{max,N}$ in order to check whether they limit a feasible range of power values.

Power/MEI selection: if the access can take place (i.e. $P_{min,N} \leq P_{max,N}$), a criterion must be adopted to select the transmission power. Depending on the chosen transmission power, the new link will acquire a certain MEI level as well as it will consume a certain amount of the other links' MEIs. Thus a major component of the admission rule is definitely the power/MEI selection criterion which affects the radio resource utilization efficiency. We recall that the acquired MEI accounts for the device power constraint since represents the maximum additional interference the link could sustain by keeping its power not above P^{dv}_{max}.

Power/MEI Selection Criterion. We introduce a power/MEI selection criterion which aims at: i) adapting power configuration to environment variations; ii) guaranteeing location fairness with respect to access possibility, iii) avoiding block states that could be overcome with a suitable MEI reconfiguration action. Let us consider a sequence of links acquiring an initial MEI at the access. Due to changing conditions (such as link accesses and releases, radio channel variations and mobility) the active links' MEIs will vary dependently also on the reciprocal distances. At a certain time the AC may fail due to some too low - or even null - MEI, nevertheless the considered link might be admissible after having performed a suitable power reconfiguration. While in [11] every new link enters the network acquiring a constant MEI, on the basis of the previous observation, here we introduce a more flexible definition of the margin approach by providing for an access with balanced and partially reconfigurable MEIs.

Here we deal with the problem of selecting the transmission power of a new N-th link within the existing range ($P_{min,N}$, $P_{max,N}$). The main point is the trade-off existing between the new link's and the already active ones' MEIs. The larger the new link's transmission power the larger the MEI it acquires and the larger the other links' MEI consumption. This trade-off can be graphically represented by the plot of MEIs as functions of power P_N, in particular in Figure 1 we show the MEIs of two active links (i-th and j-th) and the N-th attempting the access. We observe that $P_{min,N}$ corresponds to the minimum power above which a non-negative MEI_N is acquired while the value $P_{max,N}$ is the one annulling the first MEI in the system. The optimal working point consists in the power $P_{opt,N}$ allowing to maximize the minimum MEI which indeed constitutes a bottleneck for further accesses. In fact, the greater the maximum allowed power the higher the probability of a successful access for a further link. Thus an unbalanced MEI configuration would imply that at least one large MEI could be reduced in favor of links with less MEIs.

Another main observation regards the fact that not only MEIs' balancing but also their common value impacts the possibility of accessing the network. Let us assume that all MEIs are equal and that a new link tries to enter the network. Then, no matter the reciprocal link distances, there is enough room for the new link only if enough MEI has been acquired to absorb the new entrance. Since during lifetime of the network MEIs will generally evolve to unbalanced values and will be also consumed by new accesses, a MEI reconfiguration process is introduced in order to partially re-balance the system. According to the incremental philosophy such reconfigurations are not global but partial, i.e., they are locally performed by a subset of links (possibly only one at time) thus resulting in sub-optimal power configurations although they are very simple to be implemented. The general structure of the reconfiguration process is defined by the following steps:

Starting point: N links are active at transmission powers $\mathbf{P}^{(\mathbf{N})}$.

Link selection: according to a suitable criterion, a link - say the k-th link ($1 \leq k \leq N$) - is selected for MEI reconfiguration.

Power/MEI reconfiguration: since the k-th link is already active, there actually exists a range of suitable powers ($P_{min,k}$, $P_{max,k}$) and it is currently adopting a transmission power $P_{curr,k}$ within this range. On the basis of current other links' MEIs it computes and selects for transmission the optimal power $P_{opt,k}$ which maximizes the minimum MEI. This situation is described in Figure 2 where it is assumed that the k-th link corresponds to the one currently perceiving the maximum MEI. This power reconfiguration process is simple since it is performed within MEIs' constraints. As for the criterion of choice of the link to be reconfigured, it actually represents an open point and affects effectiveness of the admission control scheme. In general we can observe that, like in the previous example of Figure 2, reduction of higher MEIs results in improvement of access possibility since it implies a power decreasing which in turn results in an increasing of lower MEIs (e.g., MEI_j in the figure).

3.3 D-PRAC Protocol and Implementation Issues

To clarify the D-PRAC mechanism, we specify some implementation issues. First we show how the parameters P_{min}, P_{max}, P_{opt} are computed and what information they are derived from. Then we provide the signaling exchange needed for the support of D-PRAC and address the main problems concerning its implementation. P_{min} is the minimum power that allows the link to reach the desired QoS level and can be computed by imposing that exactly the SIR target is achieved:

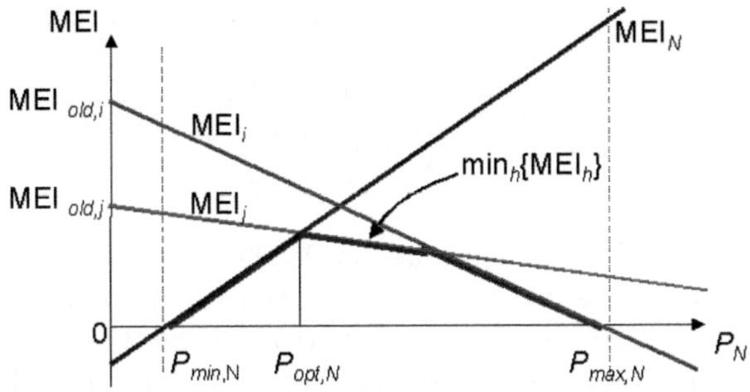

Fig. 1. MEIs as functions of the new link's transmission power.

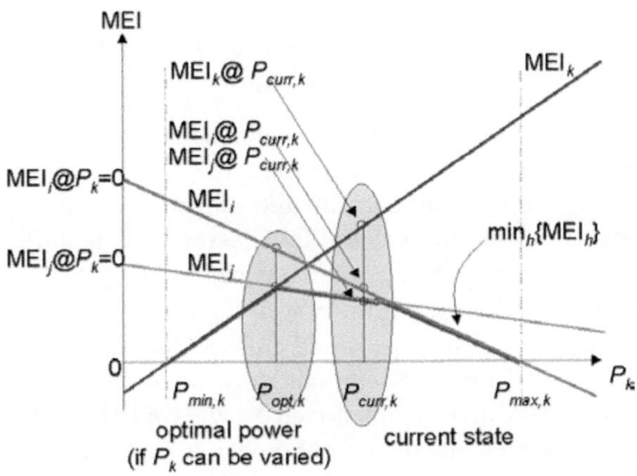

Fig. 2. MEIs as functions of an active link's transmission power under reconfiguration.

$$\frac{P_{min,N} \cdot g_{NN}}{\sum_{j=1}^{N-1} P_j \cdot g_{Nj} + \eta_N} = \gamma_N^T \Rightarrow P_{min,N} = \gamma_N^T \cdot \frac{\sum_{j=1}^{N-1} P_j \cdot g_{Nj} + \eta_N}{g_{NN}} = \gamma_N^T \cdot \frac{I_{NN}}{g_{NN}} \tag{4}$$

Thus, besides γ_N^T, $P_{min,N}$ essentially depends on the interference level I_N at the receiver and the link extension. The link's receiver shall measure the first quantity and estimate the second one by cooperation with the relevant transmitter, as we will specify in the following.

As for $P_{max,N}$, it represents the maximum power that at most annuls one of current MEIs and is calculated according to the following expression (where also the device power constraint is considered):

$$P_{max,N} = \min\{P_{max,N}^{dv}, \min_{j \neq N}\{\frac{MEI_j}{g_{jN}}\}\} \tag{5}$$

According to (5), $P_{max,N}$ can be derived if neighbor links' MEIs are known as well as their receivers' distance from the link's transmitter. As for the optimal power $P_{opt,N}$, the new N-th link can calculate it on the basis of MEIs and the interference measured I_N:

$$P_{opt,N} = \min_{i \neq N} \{ \frac{MEI_i + I_N}{g_{iN} + g_{NN}/\gamma_N^T} \} \tag{6}$$

A major component of the mechanism is the signaling of MEIs which must be updated by links' receivers as their variations come about and broadcasted to neighbor transmitters. To this end, two basic methods can be adopted: a periodic broadcasting and an on-demand signaling. Besides, the MEI level can be either explicitly signaled by a data packet or conveyed by a tone whose power level is inversely proportional to MEI (see [7]). In the following we assume a periodic broadcasting. The protocol main steps performed by two radio entities A and B willing to communicate respectively as transmitter and receiver, are the following (qualitatively described in Figure 3):

step 1 A contacts B by sending a specific message; the reception of this message triggers in B interference measurements and allows B to estimate the path gain by comparing transmitted and received power relevant to this message; moreover, in the contact message, A includes the desired SIR so that B can compute P_{min} in accordance to (4).

step 2 B notifies to A the minimum power P_{min}.

step 3 In the meanwhile A listens to and collects broadcasted MEIs from neighboring devices. In case of explicit data packets are adopted, they must also include the used transmitted powers so that A can estimate the path gain between itself and the various receivers. In case of tones, A will directly extract the quantity $\frac{MEI_i}{g_{Ai}}$ from the i-th tone's received power. Thus A calculates the maximum power P_{max} as in (5).

step 4 A compares P_{min} and P_{max}: if $P_{min} \leq P_{max}$ the access can take place and the transmission power P_{opt} is selected according to the above balancing criterion on the basis of the knowledge of MEIs.

step 5 A acknowledges to B the access result and in case of success it notifies the selected transmission power. B calculates its own MEI and signals to neighbors the relevant level.

As transmission starts, the other links' receiver will update their MEIs. A crucial point concerns MEI signaling. In particular the QoS assurance could be affected by not complete and/or inaccurate acquisition of MEIs values; in this case estimation of P_{max} would be incorrect and the new link could select a transmission power exceeding some MEI and thus causing the relevant drop. Another critical point arising in defining D-PRAC implementation is represented by the problem of concurrent access/reconfiguration attempts. As for access, if two links evaluating the admission rule on the basis of the acquired MEIs succeeded the access procedure at the same time, they might cause an amount of interference level overdoing current MEIs. In a soft QoS context these two events can be overcome by making links tolerate additional quality variations included in a negotiated range. Finally, the selection of MEI when no other links

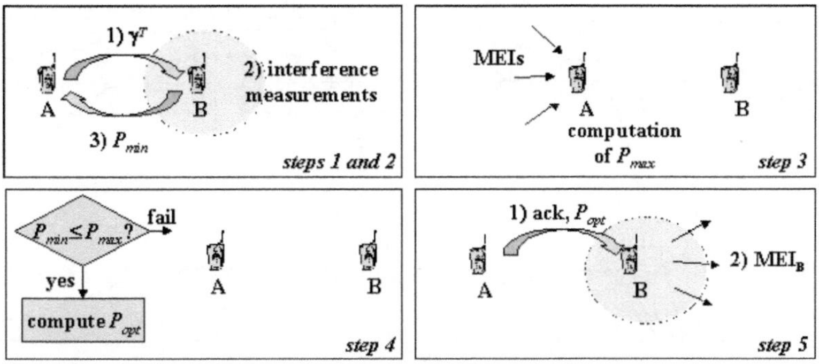

Fig. 3. Main steps of the D-PRAC scheme.

are active (say MEI0) must be specified. In general the selection of MEI0 affects the amount of available resource for new subsequent accesses. Thus a criterion of choice of MEI0 could be to set it at the value allowing to accommodate a given amount of communications before a block occurs referring to an expected situation for further access requests. More in detail the first link should select a power P_1 depending on the MEI0 by:

$$\frac{P_1 \cdot g_{11}}{MEI0 + \eta_1} = \gamma_1^T \tag{7}$$

MEI0 should be calculated in order to sustain up to a given number N0 with a given γ^T. The estimation can be done only by referring to the expected value of the path gain (g_{ex}) and accounts for the maximum power allowed to each new link which is equal to $\frac{MEI0}{N0 g_{ex}}$.

$$\frac{g_{ex}\frac{MEI0}{N0 g_{ex}}}{\eta_0 + g_{ex} \cdot P_1} = \gamma^T \tag{8}$$

Thus P_1 and MEI0 are computed as solutions of the equation system constituted by (7) and (8); due to the power constrain the actual adopted P_1 will be the minimum between P_{max}^{dv} and the computed solution and the resulting MEI0 can be found by equation (7).

4 Performance Results

4.1 General Behavior of Access Schemes Based on Margins

We aim at showing the impact of a given MEI configuration for a set of active links on the probability of new access success.

We considered a topology of N links for different values of N and we set their transmission parameters so that they acquired given MEI levels. In particular we considered a square area of $100m \times 100m$, a central frequency of 1 GHz, a

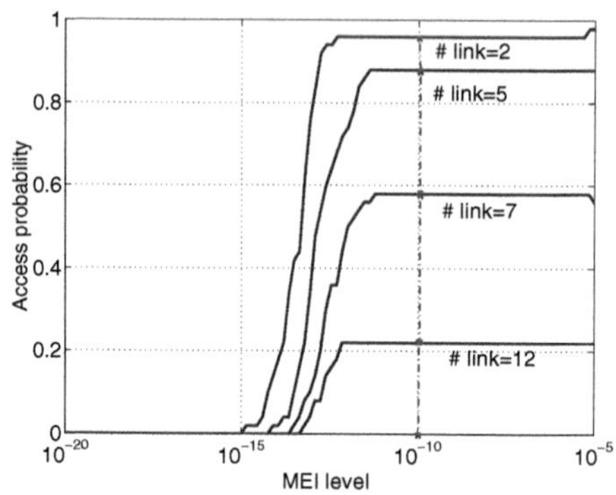

Fig. 4. Access probability as a function of a common MEI level.

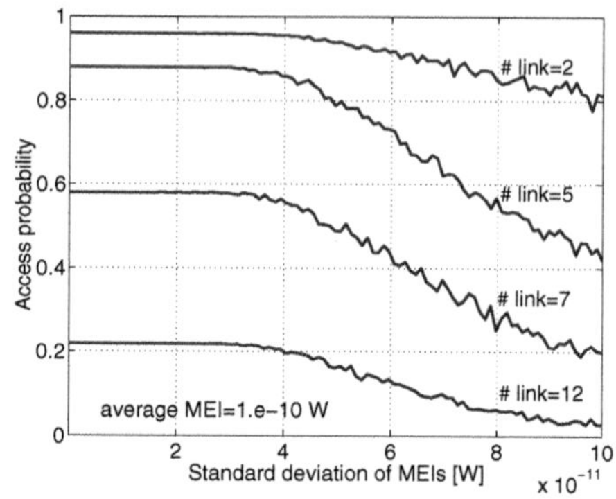

Fig. 5. Access probability as a function of the MEI standard deviation.

geometric factor of attenuation equal to 2, a processing gain equal to 200, γ^T=5, P_{max}^{dv}=1 W and η=2 e-12 W.

Figure 4 shows the results relevant to the access probability for a new link obtained when a common MEI level (reported in the abscissa) is set in all the active links. Different numbers of active links are considered. The main observation concerns the behavior of the access probability as a step function of the MEI level. Active links have to maintain enough MEI in order to permit the access to a new link. As we try to force a MEI level above the step point, the access probability saturates due to the existence of maximum transmission power reflecting in an upper bound on the acquirable MEI. The access probability saturation level gets lower as the number of active links increases due to the greater link density in the considered area.

Figure 5 reports the access probability obtained when active links' MEI levels are set according to a fixed average value, equal to 1.0 e-10 W highlighted also in Figure 4, and with a variable standard deviation for different values of the number of links. This plot shows the interesting behavior that a MEI configuration with sparse MEI levels degrades the access probability for new links and thus confirms that a margin based strategy aiming at balancing MEIs will outperform schemes which do not care of keeping MEIs balanced.

4.2 Comparison Among Different Access Strategies

We first present an example showing the advantage and difference of the proposed AC with respect to the scheme of constant initial MEIs proposed in [11] (here named *Constant MEIs* approach - C-MEI). As for D-PRAC, in our simulations we did not introduced the power reconfiguration process. Let us consider a sequence of five links willing to activate. The MEIs in correspondence of the receiver of each active link are shown (the values are normalized with respect to MEI0). Table 1 and 2 refer respectively to D-PRAC and C-MEI. Note that a MEI equal to infinite characterizes a non-active link. In particular the considered

Table 1. Example of 5 links activated by applying D-PRAC.

step	MEI_1	MEI_2	MEI_3	MEI_4	MEI_5
a	1	∞	∞	∞	∞
b	0.78	0.78	∞	∞	∞
c	0.68	0.77	0.72	0.68	∞
d	0.37	0.76	0.71	0.67	0.37

Table 2. Example of 5 links activated by applying C-MEI.

step	MEI_1	MEI_2	MEI_3	MEI_4	MEI_5
a	1	∞	∞	∞	∞
b	0.74	1	∞	∞	∞
c	0.58	0.99	0.99	1	∞
d	**0**	0.98	0.98	0.98	0.71

topology is represented at four steps corresponding to the access of the first link (a), of the second (b), of the fourth (c), and of the fifth (d). In case D-PRAC is adopted (Table 1), the entering link acquires a normalized MEI equal to the minimum one (see for example link 4 with respect to link 1 in (c)). Instead, in case of C-MEI (see Table 2), each new link enters acquiring a constant maximum MEI, i.e., MEI0. Thus at the step (b), with D-PRAC, the two links achieve equal MEIs while with C-MEI the second achieves a normalized MEI equal to 1 causing the maximum reduction on MEI of link 1. In Table 2 at step (d) the previous unbalanced MEI condition leads to a block of the network for future accesses since link 1's MEI reduces to zero due to the access of link 5 at the maximum power. The disadvantage of unbalanced MEIs is evident in the fact that the other links 2, 3, 4 and 5 continue maintaining useless high MEIs while their reduction would increase MEI of link 1 and unblock the network.

We generated a number of topologies with N links, for different values of N, randomly located in a square area of $100m \times 100m$, in order to estimate the probability of successfully activating all the links. The adopted parameters are the same as in the previous section. Furthermore, we do not introduce devices' mobility.

We have investigated the impact of the choice of MEI0 on performance for both algorithms D-PRAC and C-MEI. In Figure 6 the probability of activating N links is reported as a function of MEI0 for different values of N. Two opposite effects arise: on one hand higher MEI0 values increase the successful probability (first part of the curve), on the other hand (second part of the curve) links acquire higher MEIs thus generating higher interference that future links will have to overcome with higher transmission powers, and this avoids further improvement of the access probability. The saturation of the two curves is due to the maximum transmission power which bounds the maximum initial MEI that can be actually acquired. The behavior in case of either D-PRAC or C-MEI is slightly different since in case of D-PRAC the curve settles in faster than C-MEI. This effect is due to the fact that as concerns D-PRAC MEI0 is just the initial MEI acquired by the first link while then the MEIs evolve according to the balanced mechanism shown in Figure 1, while, as far as C-MEI is concerned, MEI0 represents the initial maximum extra interference tolerated by *each* entering link. Figure 6 also suggests a possible criterion of reconfiguring MEIs in D-PRAC. When the network evolves towards values of MEIs located in the first part of the curve, the reconfiguration process should make it move to higher MEIs allowing to achieve a

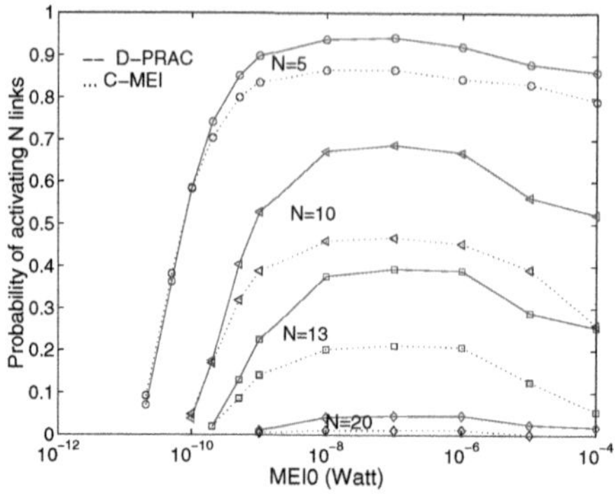

Fig. 6. Probability of activating N links as a function of MEI0.

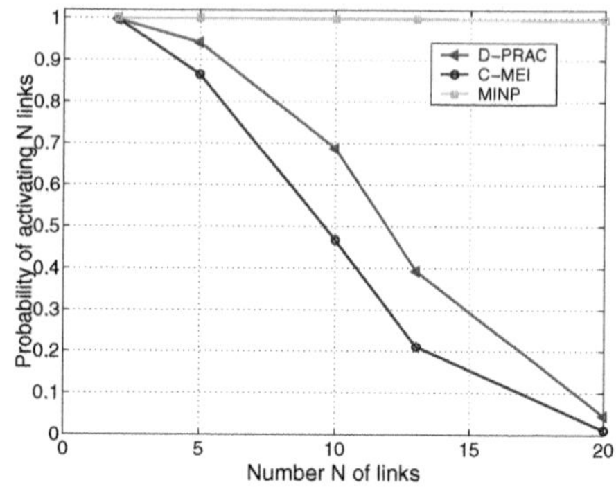

Fig. 7. Probability of activating N links as a function of N.

near-maximum access probability for future links. Obviously the problem stands in identifying the working point corresponding to the "elbow" of the curve.

We compared the D-PRAC approach to an ideal scheme reconfiguring powers to the Pareto-optimal solution, i.e. the minimum power (in the following MINP) that can be assumed as reference for the performance analysis. We assume that this scheme behaves ideally rejecting links that would force some link's drop due to the maximum power saturation. The obtained results are reported in Figure 7 for the three strategies D-PRAC, C-MEI and MINP. The MEI0 value for D-PRAC and C-MEI is 1.0 e-8 W. The MINP curve sets the maximum probability that can be achieved by any strategy. We can observe that D-PRAC outperforms C-MEI; besides, since we have not introduced any reconfigurable mechanism yet, we can expect further improvement making D-PRAC curve closer to MINP when one or more links undertake a reconfiguration. When comparing D-PRAC and C-MEI, let us remind that the two strategies require the same kind and amount of extra information and thus of signaling overhead. Instead, as for the application of MINP in a distributed scenario, some proposals exist based on probing the radio channel before and after the access (see [10]); however in this case the algorithm neglects the bound on the transmission power of a device implying drop of active links.

5 Conclusions

In this paper, we deal with a Radio Resource Control based on distributed power regulation. The system supports QoS flows requiring a given SIR level. Our approach is based on the maximum extra interference that can be tolerated by a QoS flow and can be applied in a distributed system where the optimal "minimum power approach" seems not applicable due to the high complexity required to reconfigure the links. D-PRAC is quite general and may be applied in "CDMA" oriented systems based on multichannel-multirate paradigms. Unlike previous proposals, our margin-based approach allows balancing links' parameters to reduce system block probability, typical of incremental admission control approaches. We discuss the advantages of the balanced margin approach by means of qualitative examples and selected performance results. Finally, we give some guidelines for the design of an RRC protocol, which needs only the exchange of few parameters over broadcast channels.

References

1. T.S. Rappaport, A. Annamalai, R.M. Buehrer and W.H. Tranter, "Wireless communications: past events and a future perspective", IEEE Communications Magazine, vol. 40, no.5, pp. 148–161, May 2002.
2. A. Chandra, V. Gummalla and J.O. Limb, "Wireless Medium Access Control Protocols", IEEE Communications Surveys, Second quarter 2000, pp. 2–15, 2000.
3. N. Bambos and S. Kandukury, "Power-controlled multiple access schemes for next-generation wireless packet networks", IEEE Wireless Communications, pp. 58–64, June 2002.
4. D. Kim, "Rate-Regulated Power Control for Supporting Flexible Transmission in Future CDMA Mobile Networks", IEEE J. Select. Areas Commun., vol. 17, no. 5, pp. 968–977, May 1999.
5. D.M. Novakovic and M.L. Dukic, "Evolution of the Power Control Techniques fo DS-CDMA toward 3G Wireless Communication Systems", IEEE Communications Surveys, Fourth Quarter 2000, pp. 2–15, 2000.
6. M. Xiao, N.B. Shroff and E.K.P. Chong, "Distributed Admission Control for Power-Controlled Cellular Wireless Systems", IEEE/ACM Transactions on Networking, vol. 9, no. 6, pp. 790–800, December 2001.
7. J.P. Monks, V. Bharghavan and W.W. Hwu, "A Power Controlled Multiple Access Protocol for Wireless Packet Networks", Proc. INFOCOM 2001, pp. 219–228, 2001.
8. N. Bambos and S.C. Chen, "Channel Access Algorithms with Active Link Protection for Wireless Communication Networks with Power Control", IEEE/ACM Trans. On Networking, vol. 8, no. 5, pp. 583–597, October 2000.
9. C. Martello and D. Bocchetta, "Power Controlled MAC Protocols for Wireless Ad-Hoc Networks", Proc. of European Wireless 2002, pp. 319–326, February 2002.
10. C. Zhu and M.S. Corson, "A Distributed Channel Probing Scheme for Wireless Networks", IEEE INFOCOM 2001, pp. 403–411, 2001.
11. S. Lal and E.S. Sousa, "Distributed Resource Allocation for DS-CDMA-Based Multimedia ad hoc Wireless LAN's", IEEE J. Select. Areas Commun., vol. 17, no. 5, pp. 947–967, May 1999.
12. S. L. Wu, Y.C. Tseng, J.P. Sheu, "Intelligent Medium Access for Mobile Ad-hoc Networks with Busy Tones and Power Control", IEEE J. Select. Areas Commun., vol. 18, no. 9, pp. 1647–1657, September 2000.

Achieving Service Differentiation and High Utilization in IEEE 802.11*

Vasilios A. Siris** and Matina Kavouridou

Institute of Computer Science (ICS)
Foundation for Research and Technology - Hellas (FORTH)
P.O. Box 1385, GR 711 10 Heraklion, Crete, Greece
{vsiris,mkavouri}@ics.forth.gr

Abstract. Service differentiation in wireless LANs is a growing demand, due to their increasing use for applications with different requirements and the scarcity of wireless channel resources. The contribution of this paper is twofold: First, we investigate how various parameters of the IEEE 802.11 MAC protocol affect service differentiation, in terms of both throughput and delay. Second, we propose a new approach for providing service differentiation, while achieving high wireless network utilization. Simulation investigations demonstrate that our approach can effectively adapt to varying network conditions. The approach can be implemented at the access point of a wireless LAN, while the wireless stations need only to support the emerging IEEE 802.11e standard.

Keywords: weighted fairness, throughput monitoring, wireless LANs

1 Introduction

The wireless LAN (WLAN) area is a field of wide development over the last few years. The two main WLAN standards are the ETSI High Performance European Radio (HIPERLAN) and the IEEE 802.11 WLAN, with the latter dominating the market. In IEEE 802.11, the primary media access mechanism is the CSMA/CA (Carrier Sense Multiple Access with Collision Avoidance) algorithm, which currently lacks support for service differentiation.

The capacity of wireless networks is much smaller than that of fixed networks. Moreover, unlike fixed networks, there is a limited ability for increasing the wireless channel capacity. At the same time, there is an increasing use of wireless networks for multimedia and delay sensitive applications, and an increasing deployment of wireless LAN hotspots. For all these reasons, it is becoming increasingly important to support service differentiation in wireless LANs.

The contribution of this paper is twofold. First, we investigate mechanisms for supporting service differentiation in 802.11 networks in terms of both throughput

* Part of this work was supported by British Telecommunications (BT), UK.
** Corresponding author: tel.: +30 2810 391726, fax: +30 2810 391601. Also with the Department of Computer Science, University of Crete.

M. Conti et al. (Eds.): PWC 2003, LNCS 2775, pp. 128–137, 2003.

and delay. Moreover, we investigate the weighted fairness of the various schemes, by presenting the results as the ratio of throughput achieved for the corresponding ratio of values of the control parameter. Second, we propose and investigate a new approach for supporting service differentiation, while at the same time achieving high wireless network utilization. The approach involves dynamically adjusting parameters of the 802.11 MAC protocol, to track varying conditions of the network load. The approach can be implemented solely at the access point of an 802.11 network, without requiring changes to the wireless stations; the latter need only to support the emerging IEEE 802.11e standard.

This rest of the paper is organized as follows. In Section 1.1 we briefly describe the CSMA/CA protocol, including the extensions being developed for IEEE 802.11e. In Section 1.2 we review related work on supporting service differentiation in wireless LANs. In Section 2 we investigate how various parameters of the 802.11 MAC affect service differentiation, in terms of both throughput and delay. In Section 3 we present and investigate a new approach for achieving service differentiation while attaining high network utilization. Finally, in Section 4 we present some concluding remarks and identify areas for further investigation.

1.1 IEEE 802.11

The IEEE 802.11 MAC supports two access modes: *Distributed Coordination Function* (DCF) and *Point Coordination Function* (PCF). In the first, which is the primary mode and best suited for traffic without strict delay requirements, wireless stations contend for the wireless channel at every frame transmission. The basic scheme for DCF is *Carrier Sense Multiple Access with Collision Avoidance* (CSMA/CA). A collision can be caused by two or more stations trying to transmit at the same time. After each frame transmission, the sender waits for an acknowledgment (ACK) from the receiver. If no ACK is received, a collision must have occurred and the frame is retransmitted.

Frames can have different priorities by varying the time interval, called *interframe spacing* (IFS), the channel must be sensed idle prior to their transmission. In DCF, two IFSs are defined: Short IFS (*SIFS*) and DCF IFS (*DIFS*), where *SIFS* is smaller than *DIFS*. Hence, an ACK which uses *SIFS* has a higher transmission success probability than a new data frame, which uses *DIFS*.

The objective of the *collision avoidance* part of CSMA/CA is to avoid simultaneous frame transmissions right after the channel is sensed idle. This can be achieved if, prior to transmitting a frame, a station waits for the channel to be idle for some random backoff time. This backoff time is selected randomly from the interval $[0, CW-1] \cdot \text{slot\_time}$, where CW is the contention window, whose initial value is $CWmin$. After each unsuccessful transmission, the contention window is doubled up to some maximum value $CWmax$.

An important activity of IEEE is its work towards 802.11e, which aims to improve the media access mechanism and to add support for service differentiation [1]. The extensions include the Enhanced DCF (EDCF) algorithm. Although all the details have not yet been finalized, EDCF supports different classes, with different values for $CWmin$, $CWmax$, *DIFS* (now called Arbitration IFS - *AIFS*), and the persistence factor *PF*, which determines the increase of the contention window after a collision. Smaller values of *CWmin*, *CWmax*, *AIFS*, or *PF* correspond to a higher priority.

1.2 Related Work

Next we review some representative related work, without attempting to perform an exhaustive survey of the area. The work in [2] investigates different backoff increase rates and *DIFS* intervals for providing service differentiation. Simulation experiments show that such schemes work well for UDP traffic, but not so well for TCP traffic. The work of [3] investigates how the above mechanisms, in addition to *CWmin* differentiation, can be adjusted to support per flow differentiation, rather than per station differentiation. In both works, differentiation is in terms of throughput, and the wireless channel utilization is not considered.

Another direction involves the development of new distributed access algorithms for supporting service differentiation [4,5,6,7,8]. The algorithm proposed in [5] tries to achieve fair bandwidth distribution by setting the frame size and the initial value of the backoff interval based on weights. In the work of [6] each wireless station adjusts its contention window based on a fairness index, which measures the degree of fairness of the station's throughput, relative to some reference value. In earlier work, the MACAW protocol [4] adjusts the backoff timer using a multiplicative increase/additive decrease algorithm.

The work in [7] deals with fairness and wireless channel utilization, and proposes a scheme for setting the backoff interval for different traffic classes, based on their weights and the estimated number of contending stations. The work in [8] introduces virtual MAC and virtual source algorithms that passively monitor the radio channel to determine if the channel can support new service requests in terms of delay and loss. The schemes described in the last two paragraphs all require changes to the MAC layer running at the wireless stations.

2 Service Differentiation Mechanisms

In this section we investigate how parameters of IEEE 802.11, and 802.11e in particular, affect service differentiation in terms of both average throughput and delay. The parameters we consider are 1) the maximum frame size, 2) *DIFS* (interval the channel must be sensed idle prior to frame transmission), and 3) *CWmin* (minimum value of the contention window). Other parameters that can affect service differentiation include *CWmax* (the maximum value for the contention window) and the persistence factor *PF*. These parameters will affect service differentiation in the presence of collisions, which should be avoided since they decrease the channel utilization. Hence, we do not discuss them further.

Our experiments were performed using the ns-2 simulator [9]. For the experiments investigating *DIFS* and *CWmin* differentiation, we used the EDCF modules developed by Atheros Communications. The procedure for dynamically adjusting the *CWmin* values was implemented on top of these modules. For all

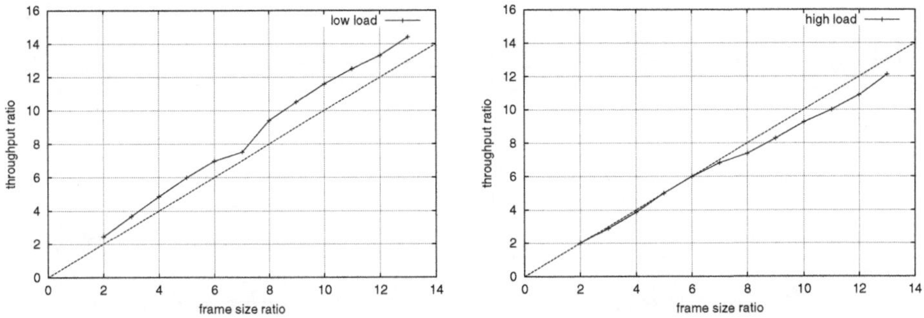

(a) Low load (= 1.5 Mbps)

(b) High load (= 2 Mbps)

Fig. 1. Service differentiation with varying maximum frame size.

experiments the channel capacity was 11 Mbps, expect for the first concerning the maximum frame size where it was 2 Mbps.

2.1 Differentiation Based on Maximum Frame Size

We first consider varying the maximum frame size that a station is allowed to transmit. The traffic used was constant bit rate (CBR) sources over UDP connections. The wireless LAN considered had capacity 2 Mbps, and carried 4 sources with an aggregate traffic of 1.5 Mbps. Fig. 1(a) shows that the ratio of throughput is approximately proportional to the ratio of frame sizes; each point in the graph corresponds to an experiment where the frame payload size for 2 of the sources was 300 bytes, and for the other 2, a multiple of this value. Indeed, it appears that the source with a larger frame size receives proportionally higher throughput compared to a source with a smaller frame size. This is not the case when the load is increased, as shown in Fig. 1(b), which was for 10 sources with an aggregate rate of 2 Mbps; this figure shows that sources with large frame sizes receive somewhat less capacity than that suggested by the ratio of frame sizes.

Note that varying the frame size cannot achieve delay differentiation. To achieve such differentiation, we need to use one of the other two parameters, *DIFS* or *CWmin*, which we investigate next.

2.2 Differentiation Based on *DIFS*

Next we investigate the service differentiation that is achieved with different values of *DIFS*, or *AIFS* as the inter-frame spacing interval is called in 802.11e. Our experiments involved both UDP and TCP traffic. The default value of *DIFS* is 34 μs; this value corresponds to the class with highest priority. The lower priority classes have a *DIFS* value larger than the default by some number of time slots, with each slot time equal to 9 μs.

Fig. 2(a) shows the differentiation in terms of throughput, for 5 and 10 CBR sources over UDP connections with aggregate rate 20 Mbps and 40 Mbps, respec-

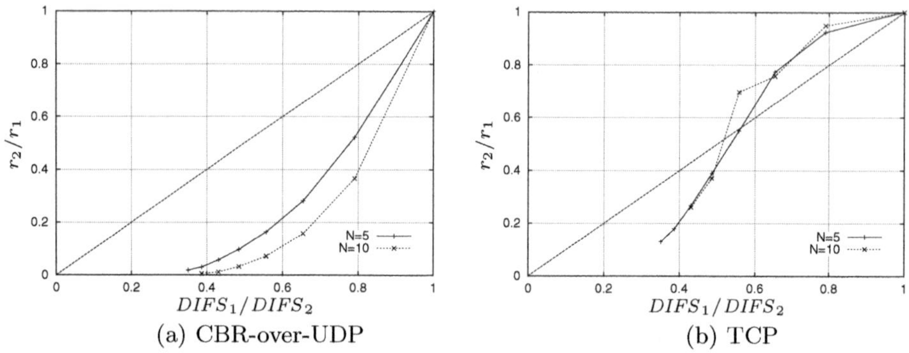

Fig. 2. Throughput ratio r_2/r_1 as a function *DIFS* ratio $DIFS_1/DIFS_2$, for UDP and TCP traffic, and $N = 5$ and $N = 10$ sources.

tively, and payload size 500 bytes. Observe that the dependence of the throughput is far from being inversely proportional to the value of *DIFS*. Fig. 2(b) shows that throughput differentiation in the case of TCP traffic is quite different than that for UDP traffic; indeed, for small values of *DIFS*, the effect of increasing *DIFS* is small. Also observe in Fig. 2(a) that for UDP traffic the number of sources affects the dependence of the throughput ratio on the *DIFS* ratio.

Fig. 3(a) shows how the average delay for voice traffic depends on the number of slots added to the default value of *DIFS*. The traffic consisted of 2 UDP connections carrying voice traffic, and 4 CBR sources with an aggregate bandwidth of 30 Mbps. The modelled voice traffic was assumed to be G.729 encoded: a 60 byte packet is generated every 20 ms during the talk spurt period, and the length of talk spurt periods is exponentially distributed with mean 352 ms, whereas the length of silence (inactive) periods is exponentially distributed with mean 650 ms. Fig. 3(a) shows that, as expected, the average delay increases with the number of slots; indeed, the rate of increase is larger for a larger number of slots added to *DIFS*. Also shown in the figure is the 90% confidence interval, estimated from 10 independent runs of the experiment.

Fig. 3(b) shows the average delay of TCP traffic, in the case of 2 TCP connections, and 3 CBR sources with total bandwidth 30 Mbps. Comparison with Fig. 3(a) shows that the addition of up to 6 slots results in an average delay that is the same as in voice-over-UDP traffic. The addition of more than 6 slot times results in a higher delay for voice-over-UDP traffic. Hence, the average delay appears to be less affected by *DIFS* for TCP traffic, compared to UDP traffic.

2.3 Differentiation Based on *CWmin*

Next we investigate the differentiation achieved by varying the minimum contention window *CWmin*. Fig. 4(a) shows the throughput differentiation, for 5 and 10 CBR sources with aggregate rate 20 Mbps and 40 Mbps, respectively, and payload size 500 bytes. Observe that the ratio of throughput is approximately

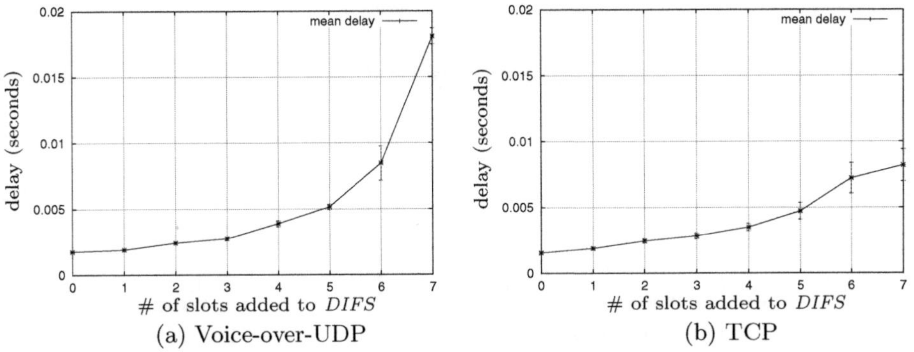

Fig. 3. Delay differentiation for different *DIFS* values.

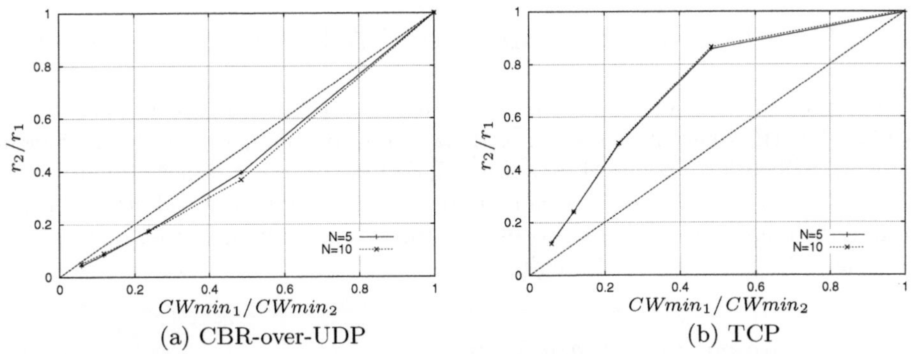

Fig. 4. Throughput ratio r_2/r_1 as a function *CWmin* ratio $CWmin_1/CWmin_2$, for UDP and TCP traffic, and $N = 5$ and $N = 10$ sources.

inversely proportional to the *CWmin* ratio. Fig. 4(b) shows that the differentiation in the case of TCP traffic is less effective; see also [3]. Indeed, increasing the value of *CWmin* from 16 to 32 results in a throughput decrease of only 15%. Also observe in Figs. 4(a) and 4(b) that the effect of the number of connections on the throughput ratio as a function of the *CWmin* ratio is minimal, unlike *DIFS* differentiation with UDP traffic, Fig. 2(a).

Fig. 5(a) shows the average delay of voice traffic as a function of *CWmin*, for 16, 32, 64, 128, and 256 time slots, in the case of 2 UDP connections carrying voice traffic, and 3 CBR sources with an aggregate bandwidth of 30 Mbps. As expected, the delay increases with increasing *CWmin*. Fig. 5(b) shows how the average delay of TCP traffic depends on *CWmin*, in the case of 2 TCP connections, and 3 CBR sources with an aggregate bandwidth of 30 Mbps. First observe that the delay in the TCP case is an order of magnitude smaller than in the UDP case. Furthermore, observe that for values of *CWmin* up to approximately 64 time slots, there is no increase of the average delay; indeed in this range the delay decreases with increasing *CWmin*. For values of *CWmin* above 64 time slots, the average delay increases with increasing *CWmin*.

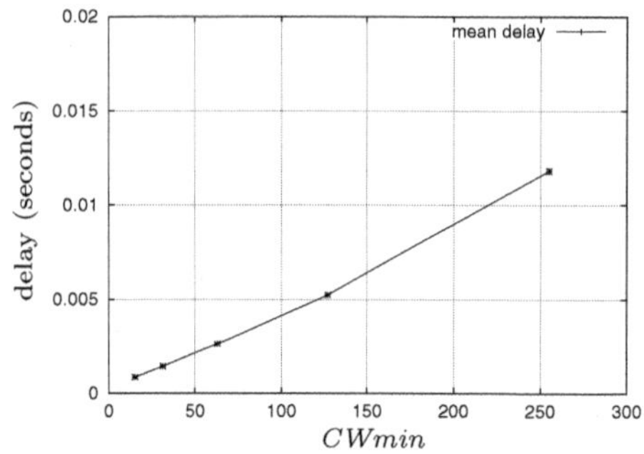

(a) Voice-over-UDP

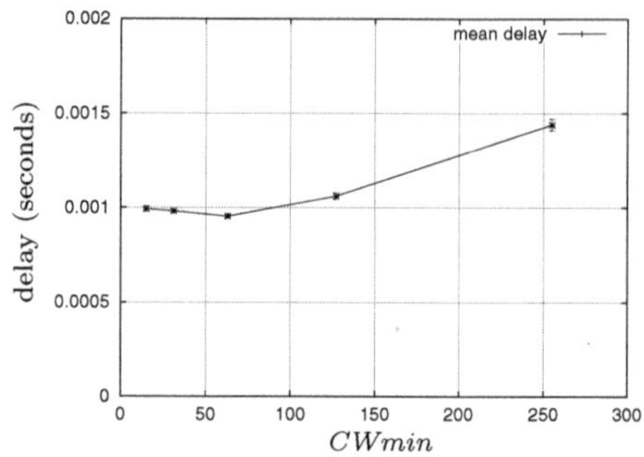

(b) TCP

Fig. 5. Delay differentiation for different *CWmin* values.

3 Algorithm for Assignment and Adjustment of *CWmin*

In the previous section we investigated the service differentiation that can be achieved with different values of *CWmin*. The optimal, in terms of efficient network utilization, values of *CWmin* will depend on the number of contending stations. In this section we describe an approach for assigning *CWmin* values to sources belonging to different classes, and for dynamically adjusting these values to achieve high aggregate throughput.

3.1 Assignment of *CWmin* Values

Assume that different classes are associated with different weights, where a larger weight is assigned to a higher priority class. Based on the results of Section 2.3, the values of *CWmin* can be assigned inversely proportion to the weight of each class. Hence, if ϕ_i is the weight for class i, the value of the minimum contention window $CWmin_i$ for class i is calculated using[1]

$$CWmin_i = \left\lfloor SF \cdot \frac{Li}{\phi_i} \right\rfloor , \tag{1}$$

where SF is some scaling factor and L_i is the frame size; the frame size is added so that differentiation depends solely on the weight. In the case of collisions, the usual exponential backoff algorithm of 802.11 is performed.

The above assignment of *CWmin* values for different classes, based on the class' weight and the frame size, is similar to the calculation of the backoff interval in the scheme proposed in [5]; our approach differs in that we assign different values of *CWmin* rather than the backoff interval, and the behaviour of the exponential backoff algorithm in 802.11 remains the same. This presents

[1] More precisely, the left-hand side should be $CWmin_i - 1$, since the backoff is selected from $[0, CW - 1] \cdot \text{slot\_time}$. However, because we consider $CWmin \geq 16$, the difference in practice is insignificant.

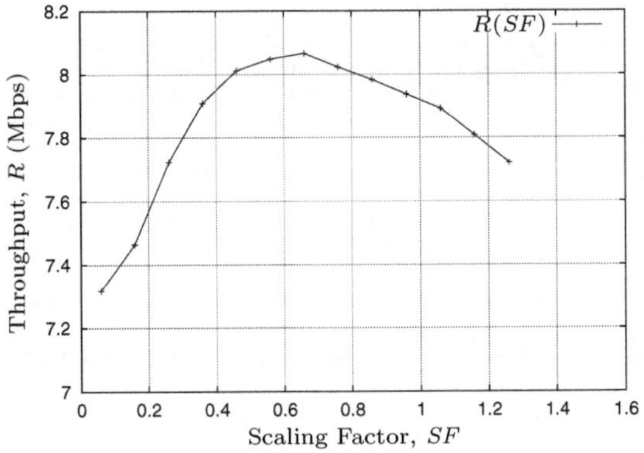

Fig. 6. Aggregate throughput as a function of the scaling factor. The measurement interval is $T_m = 10$ seconds and the scaling factor step is $\Delta_{SF} = 0.1$.

an important advantage, since our approach can be implemented solely at the access point of a WLAN, without requiring any changes at the wireless stations, assuming these support the IEEE 802.11e standard.

In the assignment of *CWmin* values, the value of the scaling factor is important, since it affects the channel utilization. Next we describe and evaluate an approach for adjusting the scaling factor, and subsequently the values of *CWmin*, based on monitoring the actual throughput of the WLAN.

3.2 Adaptive Recalculation of *CWmin*

The dependence of the aggregate throughput on the scaling factor is shown in Fig. 6; an identical result is presented in [5]. Observe that the throughput initially increases with the scaling factor up to a maximum value, after which it starts to decrease. Such a behaviour can be explained as follows: When the scaling factor is smaller than the optimum, the values of *CWmin* are small, resulting in many collisions, hence the utilization and throughput is low. On the other hand, when the scaling factor gets too large, the values of *CWmin* are large, resulting in a large percentage of idle times, hence the wireless channel is underutilized.

Another important observation is that the optimum scaling factor is different for different network loads; this motivates the need to recalculate the scaling factor, hence the values of *CWmin*, adaptively when the network load changes. From the shape of Fig. 6, this can be achieved by measuring the average throughput, and moving the scaling factor in the direction that increases the aggregate throughput. The details of the proposed procedure is shown in Fig. 7. The procedure includes three parameters: the interval T_m over which the throughput is measured, the scaling factor step size Δ_{SF}, and the percentage a used for deciding when to search for a new optimal scaling factor. A detailed discussion of how these factors affect the performance of the algorithm is contained in [10].

The above procedure for adjusting the scaling factor can be implemented at the access point. Moreover, each time the scaling factor changes, the access

Step 1. The scaling factor is set to some initial value SF_1, and the aggregate throughput R_1 is measured at the access point (AP) over an interval T_m
Step 2. $SF_2 := SF_1 + \Delta_{SF}$, and the aggregate throughput R_2 is measured at the AP
Step 3. If $R_2 > R_1$ then SF *increases* with step Δ_{SF} while the aggregate throughput increases else if $R_2 < R_1$ then SF *decreases* with step Δ_{SF} while the aggregate throughput increases
Step 4. Let SF^* be the optimal scaling factor, and R^* the throughput when Step 3 ends
Let $SF_1 := SF^*$
For the same scaling factor, the AP continuously measures the average throughput R_1
If $R_1 < aR^*$, where $a \in (0, 1)$ then goto Step 2

Fig. 7. Procedure for adjusting the scaling factor SF.

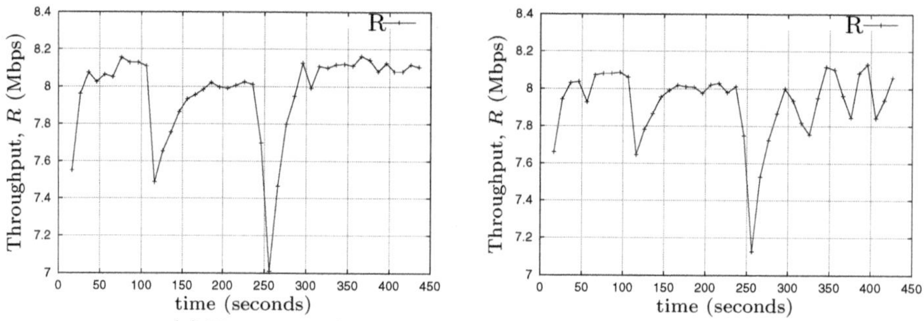

Fig. 8. Results for exponential and pareto traffic sources. Exponential: average "on" 800 ms, "off" 200 ms. Pareto: same average on/off, and shape parameter 1.5.

point would recalculate, using Eq. 1, the value of $CWmin_i$ for each class i, and distribute it to the wireless stations; this communication can utilize the particular procedure that will be used in IEEE 802.11e.

3.3 Simulation Results

In this section we demonstrate the operation of the procedure described in the previous section for dynamically adjusting $CWmin$. The values for the three parameters of the procedure we consider are the following, see [10] for a justification of this selection: throughput measurement interval $T_m = 10$ seconds, scaling factor step size $\Delta_{SF} = 0.1$, and throughput decrease threshold $a = 0.95$.

Initially, the WLAN has 13 stations, each producing exponential (Fig. 8(a)) or pareto (Fig.8(b)) traffic with rate 1 Mbps. At time 130 seconds, 20 more identical stations enter the WLAN; at time 240 seconds, 23 of the stations depart, leaving 10 stations in the WLAN. Observe that the procedure for adjusting $CWmin$ kicks in whenever the number of stations changes, and tries to maintain a high throughput. The maximum throughput is approximately 8 Mbps, with the exact value depending on the number of stations. Indeed, a smaller number of stations can achieve a larger aggregate throughput; this is due to the smaller probability of collisions when there are fewer stations. The results from these figures show that, for the particular parameter values, the procedure's convergence time is of the order of 10s of seconds, which is sufficient in environments where the number of wireless users change over timescales larger than a few minutes.

4 Conclusion

In this paper we first presented simulation experiments on service differentiation, in terms of both throughput and delay, using various 802.11 MAC layer parameters, namely the maximum frame size, the minimum contention window, and the DCF inter-frame spacing interval. Second, we proposed a simple yet effective procedure for adjusting the minimum contention window based on actual throughput measurements. An important advantage of our approach is that it can be implemented solely at the access point; the wireless stations are only required to support the emerging IEEE 802.11e standard. Our initial experiments show that the approach is quite robust, and can effectively adjust the contention window to achieve high aggregate throughput. Further investigations focus on quantifying the tradeoffs of the procedure's parameters; such information will be useful for tuning these parameters in a real environment. Furthermore, the procedure's reactivity should depend on the timescales over which the network conditions change significantly; identifying these timescales can assist in tuning the procedure's degree of reactivity.

References

1. IEEE: Draft Supplement - Part 11: MAC Enhancements for QoS, IEEE 802.11e/D2.0 (2001)
2. Aad, I., Castelluccia, C.: Differentiation mechanisms for IEEE 802.11. In: Proc. of IEEE INFOCOM'01 (2001)
3. Aad, I., Castelluccia, C.: Remarks on per-flow differentiation in IEEE 802.11. In: Proc. of European Wireless 2002
4. Bharghavan, V., Demers, A., Shenker, S., Zhang, L.: MACAW: a media access protocol for wireless LANs. In: Proc. of ACM SIGCOMM'94 (1994)
5. Vaidya, N.H., Bahl, P., Gupta, S.: Distributed fair scheduling in a wireless LAN. In: Proc. of ACM MOBICOM'00 (2000)
6. Bensaou, B., Wang, Y., Ko, C.C.: Fair medium access in 802.11 based wireless ad-hoc networks. In: Proc. of ACM MOBIHOC'02 (2002)
7. Qiao, D., Shin, K.G.: Achieving efficient channel utilization and weighted fairness for data communications in IEEE 802.11 WLAN under DCF. In: Proc. of IEEE/IFIP IWQoS'02 (2002)
8. Veres, A., Campbell, A.T., Barry, M., Sun, L.H.: Supporting service differentiation in wireless packet networks using distributed control. IEEE J. Select. Areas in Commun. **19** (2001) 2094–2104
9. UCB/LBNL/VINT: Network Simulator - ns (version 2). (`http://www-mash.cs.berkeley.edu/ns.html`)
10. Siris, V.A., Kavouridou, M.: Achieving service differentiation and high utilization in IEEE 802.11. Technical Report No. 322, ICS-FORTH (2003)

Optimized Power Allocation and Signal Shaping for Interference-Limited Multi-antenna “Ad Hoc” Networks

Enzo Baccarelli and Mauro Biagi

INFO-COM Dept., University of Rome,
via Eudossiana 18, 00184, Rome (Italy)
FAX no. +39 (0)64873300
enzobac@infocom.uniroma1.it
Phone.no. +39 (0)644585466
m.biagi@ieee.org
Phone.no. +39 (0)644585471

Abstract. This paper deals with optimized Multiple-Input Multiple Output (MIMO) channel estimation and ensuing information throughput conveyed by pilot-based multi-antenna systems affected by both spatially colored Multiple Access Interference (MUI) and *errors* in the available channel estimates. The architecture of the Minimum Mean Square Error (MMSE) MIMO channel estimator is derived and the related analytical conditions for the optimal design of space-time training sequences are provided. Afterwards, closed form expressions for the maximum information throughput sustained by the considered systems for Gaussian distributed input signals are given and, then, a novel powers' allocation algorithm for the asymptotical achievement of the system capacity is developed. Considerations about optimized space-division MAC strategies are also provided.

1 Introduction and Goals

Due to fast increasing demand lastly experienced for pervasive high-throughput Wireless “ad-hoc” LANs, Multi-Antenna systems seem to be adequate as innovative technological approach to guarantee reliable performance and also power saving. In this context, we propose an innovative algorithm for power allocation in Rayleigh flat fading environment for Multi-Antenna transceiver so to exploit spatial diversity in order to maximize mutual information also in the presence of Multi Access interference. The transmission scheme we based on is continuous and does not required orthogonal access (e.g. orthogonal TDMA, FDMA or CDMA). Therefore, in this work we focus on the ultimate information throughput conveyed by pilot-based wireless MIMO systems equipped with *imperfect* channel estimates at both transmit and receiver and impaired by *spatially colored MUI.* Specifically, main contributions of this work may be summarized as follows. First, we develop the optimal MMSE channel estimator for pilot-based MIMO systems impaired by colored MUI. Second, we provide the analytical

M. Conti et al. (Eds.): PWC 2003, LNCS 2775, pp. 138–152, 2003.

properties characterizing optimized space-time training sequences and then we show as these last are related to the statistics of spatial MUI. Third, we develop closed-form analytical expressions for computing the information throughput sustained by considered MIMO system for the case of Gaussian distributed input signals and then we point out several asymptotical operating conditions guaranteeing achievement of the corresponding system capacity. Fourth, we propose an iterative algorithm for the optimized power allocation when *imperfect* channel estimates are available at both transmit/receiver. Before proceeding few words about the adopted notation. Capital letters indicate matrices, lower-case underlined symbols denote vectors while characters overlined by arrow $^{\rightarrow}$ denote block-matrices and block-vectors. Apexes *, ^{T}, † mean conjugation, transposition and conjugate-transposition respectively while lower-case letters will be used for scalar quantities. Again, det $[A]$ and $Tra[A]$ mean determinant and trace of matrix $A \doteq [\underline{a}_1 \ldots \underline{a}_m]$, while $vect(A)$ indicates the (block) vector obtained by the ordered stacking of the column of matrix A. Finally, I_m is the (mxm) unit matrix, $||A||_E$ is the Euclidean norm of matrix A, $A \otimes B$ is the Kronecker product of matrix A by matrix B, $\underline{0}_m$ is the m-dimensional zero-vector, lg denotes natural logarithm and $\delta(m,n)$ is the Kroenecker delta.

2 System Modeling

The considered application scenario is that of emerging local wireless "ad-hoc" networks [18,20] where a (large) number of uncoordinated transmit-receiving nodes simultaneously communicate over a limited hot-spot cell and then give arise to mutual multiple access interference [18]. Simply stated, it is composed by a transmitter unit equipped with t≥1 antennas communicating to a receiving unit equipped with r≥1 antennas via a radio channel impaired by both slow-variant flat Rayleigh fading and additive multiple access interference induced by adjacent nodes active over the same hot-spot cell. Path gain h_{ji} from transmitter antenna i to receiver on j may be modelled as a complex zero-mean unit-variance proper complex variable (r.v.) [5,6,7,8] and, for sufficiently spaced apart antennas, these path gains $\{h_{ji} \in \mathbb{C}^1, \ 1 \leq j \leq r, \ 1 \leq i \leq t\}$ may be considered[1] uncorrelated. Furthermore, for low-mobility applications as those serving nomadic users over hot-spot cells, path gains $\{h_{ji}\}$ may be also assumed time-invariant over $\mathrm{T} \geq 1$ signalling periods, after which they change to new *statistically independent* values held for another T signalling periods and so on. We assume that the coded and modulated streams radiated by transmitter antennas are split into packets composed by $\mathrm{T} \geq 1$ slots, where first $\mathrm{T}_L \geq 0$ slots are used by Rx for learning the MUI statistics (See Sect.II.1), second $\mathrm{T}_{tr} \geq 0$ slots are employed for estimating the path gains $\{h_{ji}\}$ of the forward MIMO channel (see Sect.II.2) and, finally, last $\mathrm{T}_{pay} \triangleq \mathrm{T} - \mathrm{T}_{tr} - \mathrm{T}_L$ slots convey payload data (see Sect.II.3). Thus, after indicating as R_C (nats/slot) the information rate of

[1] For hot-spot local area applications, antenna spacing of the order of $\lambda/2$ suffices for meeting the above assumption [15].

the employed space-time encoder, the spectral efficiency η (nats/sec/Hz) of the described system equates

$$\eta = \frac{\mathrm{T}_{pay}}{\mathrm{T}} \cdot \frac{\mathrm{R}_C}{\Delta_s \mathrm{B}_w}, \tag{1}$$

where Δ_s (sec.) and B_w are slot duration and RF bandwidth of radiated signal, respectively.

2.1 Learning Phase

During the learning phase, no signals are radiated by transmitter Tx so to allow corresponding receiver Rx to learn statistics of the impairing MUI. More in particular, the r-dimensional (complex column) vector $\underline{\dot{y}}(n) \triangleq [\underline{\dot{y}}_1(n)...\underline{\dot{y}}_r(n)]^T$ collecting the outputs of the r receiving antennas over the n-th slot of the learning phase may be modeled as

$$\underline{\dot{y}}(n) \triangleq \underline{\dot{d}}(n) \equiv \underline{\dot{v}}(n) + \underline{\dot{w}}(n). \tag{2}$$

The first component accounts for the receiver thermal noise and then $\{\underline{\dot{w}}(n) \in \mathbb{C}^r, 1 \leq n \leq \mathrm{T}_L\}$ may be modeled as a zero-mean, proper complex, spatially correlated and temporally white, Gaussian sequence with covariance matrix equal to

$$\mathrm{E}\{\underline{\dot{w}}(n)(\underline{\dot{w}}(m))^\dagger\} = N_0 \mathrm{I}_r \delta(m,n) \tag{3}$$

where N_0 (watt/Hz) is the level of the receiving thermal noise. Since the second component $\{\underline{\dot{v}}(n)\}$ in (2) accounts for the MUI due to multiple co-located transmitting nodes active over the same hot-spot cell, it is likelihood to model $\{\underline{\dot{v}}(n) \in \mathbb{C}^r\}$ as a zero-mean temporally white, spatially colored proper Gaussian sequence, whose covariance matrix

$$\mathrm{K}_v \triangleq \mathrm{E}\{\underline{\dot{v}}(n)(\underline{\dot{v}}(m))^\dagger\} \equiv \begin{bmatrix} c_{11} & \dots & c_{1r} \\ c_{12}^* & \dots & c_{2r} \\ \vdots & \vdots & \vdots \\ c_{1r}^* & \dots & c_{rr} \end{bmatrix} \tag{4}$$

remains unchanged over time intervals at least equal to the duration of an overall packet. However, this last may change from a packet to another so that it is reasonable to assume that both Tx and Rx nodes are not aware of the covariance matrix of the overall disturbance:

$$\mathrm{K}_d \triangleq \mathrm{E}\{\underline{\dot{d}}(n)(\underline{\dot{d}}(m))^\dagger\} \equiv \mathrm{K}_v + N_0 \mathrm{I}_r, \tag{5}$$

at the beginning of each transmitted packet. However, since the received signals $\{\underline{\dot{y}}(n)\}$ equate the MAI ones $\{\underline{\dot{f}}(n)\}$ during the learning phase, laws of large numbers guarantees that an unbiased and consistent (e.g, asymptotically exact) estimate $\overset{\vee}{K}_d$ of the (a priori) unknown covariance matrix K_d is given by the following relationship:

$$\overset{\vee}{K}_d = \frac{1}{\mathrm{T}_L} \sum_{n=1}^{\mathrm{T}_L} \underline{\dot{y}}(n)(\underline{\dot{y}}(n))^{\dagger}. \tag{6}$$

2.2 Training Phase

On the basis of the allowable MUI covariance matrix K_d, during the training phase the Tx transmitter node is able to perform the optimized shaping of the deterministic pilot streams $\{\widetilde{x}_i(n) \in \mathbb{C}^1, T_L + 1 \leq n \leq T_L + T_{tr}\}, 1 \leq i \leq t$, to be used for estimating the (a priori unknown) path gains $\{\mathrm{h}_{ji}\}$ of the MIMO (forward) channel. In particular, the (Δ_s-sampled) signals $\{\widetilde{y}_i(n) \in \mathbb{C}^1, T_L + 1 \leq n \leq T_L + T_{tr}\}, 1 \leq j \leq r$, measured at the output of j-th receiving antenna during the training phase may be modelled as

$$\widetilde{y}_j(n) = \frac{1}{\sqrt{t}} \sum_{i=1}^{t} \mathrm{h}_{ji}\widetilde{x}_i(n) + \widetilde{d}_j(n),\ T_L + 1 \leq n \leq T_L + T_{tr}, 1 \leq j \leq r, \tag{7}$$

where the corresponding overall disturbance

$$\widetilde{d}_j(n) = \widetilde{v}_j(n) + \widetilde{w}_j(n),\ T_L + 1 \leq n \leq T_L + T_{tr}, 1 \leq j \leq r, \tag{7.1}$$

is independent from path gains $\{h_{ji}\}$ and exhibits the *same* statistics previously detailed in (4), (5) for the learning phase. Thus, after assuming the (usual) constraint

$$\frac{1}{t} \sum_{i=1}^{t} ||\widetilde{x}_i(n)||^2 = \widetilde{P},\ T_L + 1 \leq n \leq T_L + T_{tr}, \tag{8}$$

about the average power $\widetilde{P}$ radiated by transmitting antennas over each slot of the training phase, the corresponding signal to interference-plus-noise ratio (SINR) $\widetilde{\gamma}_j$ measured at the output of *j-th* receiving antennas equates (see eqs.(7), (8))

$$\widetilde{\gamma}_j = \widetilde{P}/(N_0 + c_{jj}),\ 1 \leq j \leq r \tag{8.1}$$

where $N_0 + c_{jj}$ is the j-th diagonal entry of the MUI matrix K_d in (5). Therefore, the T_{tr}xr (complex) samples gathered at the outputs of the r receive antennas during overall training phase may be organized into the (T_{tr}xr) observed matrix $\widetilde{Y} \triangleq [\underline{\widetilde{y}}_1 ... \underline{\widetilde{y}}_r]$ given by [6,7]

$$\widetilde{Y} = \frac{1}{\sqrt{t}} \widetilde{X} H + \widetilde{D}, \tag{9}$$

where $\widetilde{X} \triangleq [\underline{\widetilde{x}_1} ... \underline{\widetilde{x}_t}]$ is the (T_{tr}xt) matrix of (deterministic) pilot symbols, $H \triangleq [\underline{h}_1 ... \underline{h}_r]$ is the (txr) complex matrix composed by path gains $\{h_{ji}\}$ and the

(T_{tr}xr) matrix $\widetilde{D} \triangleq [\underline{\widetilde{d}}_1 ... \underline{\widetilde{d}}_r]$ collects the disturbance samples $\{\widetilde{d}_j(n)\}$ in (7) experienced during the training phase. Obviously, from (8) it follows that the pilot matrix $\widetilde{X}$ in (9) must satisfy the following second order constraint:

$$Tra[\widetilde{X}\widetilde{X}^{\dagger}] = tT_{tr}\widetilde{P}. \tag{9.1}$$

As detailed in Sect.III, training observations in (9) are employed by the receiver node Rx for computing the MMSE matrix estimate $\hat{H} \triangleq E\{H|\widetilde{Y}\}$ of the MIMO channel H. In turn, at step $n = T_L + T_{tr}$ (e.g.,at the end of training phase) this estimates $\hat{H}$ are communicated back to the transmitter via an (ideal) feedback link.

2.3 Payload Phase

Thus, on the basis of available K_d and $\hat{H}$ matrices and actual message $\mathfrak{M}$ to be communicated, the transmitting Tx node suitably shapes the (random) signal streams $\{\phi_i(n) \in \mathbb{C}^1, T_L + T_{tr} + 1 \leq n \leq T\}$, $1 \leq i \leq t$, to be radiated by t transmitting antennas during the payload phase. The corresponding (sampled) signals $\{y_j(n) \in \mathbb{C}^1, T_L + T_{tr} + 1 \leq n \leq T\}$, $1 \leq j \leq r$ measured at the outputs of r receiving antennas may be modelled as

$$y_j(n) = \frac{1}{\sqrt{t}} \sum_{i=1}^{t} h_{ji}\phi_i(n) + d_j(n), \; T_L + T_{tr} + 1 \leq n \leq T, \; 1 \leq j \leq r, \tag{10}$$

where the sequences $d_j(n) \triangleq v_j(n) + w_j(n)$, $1 \leq j \leq r$, account for the overall disturbance experienced during the payload phase. They still exhibit the *same* statistics previously detailed in (4),(5) are assumed independent from both path gains $\{h_{ji}\}$ and payload streams $\{\phi_j\}$. Therefore, after assuming that this last meets the (usual) power constraint

$$\frac{1}{t} \sum_{i=1}^{t} E\{||\phi_i(n)||^2\} = P, \; T_L + T_{tr} + 1 \leq n \leq T \tag{10.1}$$

the resulting SINR γ_j measured at the output of the j-th receiving antenna during the payload phase equates[2] (see eqs.(5), (10))

$$\gamma_j = P/(N_0 + c_{jj}), \; 1 \leq j \leq r. \tag{10.2}$$

Furthermore, from (10) we also deduce that column vector $\underline{y}(n) \triangleq [y_1(n)... y_r(n)]^T$ with r elements collecting the outputs of the r receiving antennas over n-th payload slot is linked to the (tx1) column vector $\underline{\phi}(n) \triangleq [\phi_1(n)...\phi_t(n)]^T$ of the corresponding signals radiated by Tx node as in

$$\underline{y}(n) = \frac{1}{\sqrt{t}} H^T \underline{\phi}(n) + \underline{d}(n), \; T_L + T_{tr} + 1 \leq n \leq T, \tag{11}$$

[2] We point out that our model explicitly for the different power levels $\widetilde{P}$ and P possibly radiated by t transmitting antennas during training and payload phases respectively.

where $\{\underline{d}(n) \triangleq [d_1(n)...d_r(n)]^T,\ T_L + T_{tr} + 1 \le n \le T\}$ is the temporally white Gaussian sequence of the disturbance with spatial covariance matrix still given by K_d in (5). Furthermore, directly from (10.1) it follows that the txt spatial covariance matrix $R_\phi \triangleq E\{\underline{\phi}^\dagger \underline{\phi}\}$ of the t-dimensional signal radiated during each slot must meet the power constraint

$$Tra[R_\phi] \equiv E\{\underline{\phi}^\dagger \underline{\phi}\} = tP,\ T_L + T_{tr} + 1 \le n \le T. \tag{11.1}$$

Finally, after stacking the T_{pay} observed vectors in (11) into the corresponding (T_{pay}rx1) block vector $\underline{y} \triangleq \left[\underline{y}^T (T_L + T_{tr} + 1) ... \underline{y}^T (T)\right]^T$, we may compact the T_{pay} relationship (11) in the following one:

$$\underline{\overrightarrow{y}} = \frac{1}{\sqrt{t}} \left[I_{T_{pay}} \otimes H\right]^T \underline{\overrightarrow{\varphi}} + \underline{\overrightarrow{d}} \tag{12}$$

where the (block) covariance matrix of the corresponding disturbance block vector $\underline{d} \triangleq \left[\underline{\overrightarrow{d}}^T (T_L + T_{tr} + 1) ... \underline{d}^T (T)\right]^T$, equates

$$E\{\underline{\overrightarrow{d}}(\underline{\overrightarrow{d}})^\dagger\} = I_{T_{pay}} \otimes K_d \tag{12.1}$$

while the block vector $\underline{\phi} \triangleq \left[\underline{\phi}^T (T_L + T_{tr} + 1) ... \underline{\phi}^T (T)\right]^T$ of random signals transmitted during overall payload phase is constrained as in (see 11.1):

$$E\{\underline{\overrightarrow{\phi}}^\dagger \underline{\overrightarrow{\phi}}\} = T_{pay} tP. \tag{12.2}$$

3 MMSE MIMO Channel Estimation in the Presence of Spatially Colored MUI

Since in [9] it is proved that the MMSE estimates $\hat{H} \equiv [\hat{\underline{h}}_1 ... \hat{\underline{h}}_r] \triangleq E\{H|\widetilde{Y}\}$ of the MIMO channel H in (9) is a *sufficient statistic* for the ML detection $\hat{\mathfrak{M}}$ of the transmitted message $\mathfrak{M}$, no information loss is paid by the here considered receiving architecture that it is composed by a MIMO MMSE channel-estimator cascaded to an ML detector of the transmitted message $\mathfrak{M}$. A suitable application of the Principle of Orthogonality leads to the following expression for the MMSE estimates $\hat{\underline{H}}_j \triangleq E\{\underline{h}_j|\widetilde{Y}\}$ of the j-th column of matrix H when both $\widetilde{Y}$ and the transmitted pilot matrix $\widetilde{X}$ are optimally employed at the receiver side:

$$\hat{\underline{h}}_j = \frac{1}{\sqrt{t}} \left[\underline{e}_j^T K_d^{-1/2} \otimes \tilde{X}^\dagger\right] \left[\frac{1}{t}\left(K_d^{-1} \otimes \tilde{X}\tilde{X}^\dagger\right) + I_{rT_{tr}}\right]^{-1} \left(K_d^{-1/2} \otimes I_{T_{tr}}\right) vect(\tilde{Y}),$$
$$1 \leqslant j \leqslant r \tag{13}$$

where r-dimensional (columnn) vector $\underline{e}_j$ is the (usual) j-th unit vector of $\mathbb{R}^r$ [13], $\mathrm{K}_d^{-1/2}$ indicates the positive square root of matrix K_d^{-1} [13] while $vect(\widetilde{Y})$is the $\mathrm{r}T_{tr}$-dimensional column vector given by the ordered stacking of the columns observed matrix $\widetilde{Y}$ in (9). Furthermore, the cross-correlation matrices of the columns of the resulting MMSE error matrix $\mathfrak{E} \equiv [\underline{\varepsilon}_1 ... \underline{\varepsilon}_1] \triangleq H - \hat{H}$ are given by

$$E\left\{\underline{\varepsilon}_j \left(\underline{\varepsilon}_i\right)^{\dagger}\right\} = \delta(j,i)I_t - \frac{1}{t}\left(\underline{e}_j \otimes I_t\right)^{\dagger}\left(K_d^{-1/2} \otimes \tilde{X}\right)^{\dagger}\left[\frac{1}{t}\left(K_d^{-1} \otimes \tilde{X}\tilde{X}^{\dagger} + I_{rT_{tr}}\right)\right]^{-1} \cdot$$

$$\cdot \left(K_d^{-1/2} \otimes \tilde{X}\right)\left(\left(\underline{e}_i \otimes I_t\right)\right), 1 \leqslant j, i \leqslant r \qquad (14)$$

so that the resulting total mean square error $\sum_{tot}(\tilde{x}) \triangleq ||\mathfrak{E}||_E^2$ equates (see (14) for j=i)

$$\sum\nolimits_{tot} = \sum_{j=1}^{r} Tra\left[\underline{\varepsilon}_j \underline{\varepsilon}_j^{+}\right] = rt +$$

$$-\frac{1}{t}\sum_{j=1}^{r} Tra\left[\left(\underline{e}_j \otimes I_t\right)^{\dagger}\left(K_d^{-1/2} \otimes \tilde{X}\right)^{\dagger}\left[\frac{1}{t}\left(K_d^{-1} \otimes \tilde{X}\tilde{X}^{\dagger}\right) + I_{rT_{tr}}\right]^{-1} \cdot\right.$$

$$\left.\cdot \left(K_d^{-1/2} \otimes \tilde{X}\right)\left(\underline{e}_j \otimes I_t\right)\right] \qquad (15)$$

3.1 Condition for the Optimal Training

Since the total mean square error in (15) depends on the matrix $\tilde{X}$ employed for the training, the key-problem becomes how to choose $\tilde{X}$ do to minimize (15) under power constraint (9.1). By fact, an application of Cauchy inequality [13] leads to the following condition characterizing optimal $\tilde{X}$.

Proposition 1. A training matrix $\tilde{X}$ minimizes the total squared error (15) under power constraint (9.1) if and only if it meets the following relationship

$$K_d^{-1} \otimes \tilde{X}^{\dagger}\tilde{X} = aI_{rt} \qquad (16)$$

where the positive scalar a equates

$$a \triangleq \frac{T_{tr}\tilde{P}}{r} Tra[K_d^{-1}] \qquad \diamond\,(16.1)$$

4 Conveyed Information Throughput

The block-fading model introduced in Sect.II for the forward MIMO channel guarantees that this last is information stable [16] so that the corresponding Shannon's capacity C fixes the maximum information throughout conveyable in a reliable way during the payload phase. Following quite standard approaches [14], the capacity C of the MIMO channel (12) can be expressed as

$$C = \{C(\hat{H})\} \equiv \int C(\hat{H})p(\hat{H})d\hat{H}, \quad (nats/payload\ slot), \qquad (17)$$

where the random variable

$$C(\hat{H}) \triangleq \sup_{\overrightarrow{\underline{\phi}}:E\{\overrightarrow{\underline{\phi}}^{\dagger}\overrightarrow{\underline{\phi}}\}\leq tT_{pay}P} \frac{1}{T_{pay}} I\left(\overrightarrow{\underline{y}};\overrightarrow{\underline{\phi}}|\hat{H}\right), \tag{18}$$

is the capacity of the MIMO channel (12), (12.2) *conditioned* on the realization $\hat{H}$ of the channel estimates actually available at both transmitter and receiver. Finally, $I(\cdot;\cdot|\cdot)$ in (18) is the mutual information conveyed by the MIMO channel (12) during the payload phase. Unfortunately, barring two limit cases of PCSI [1,2,4,5] and NCSI [6,7], pdf of the input signals $\overrightarrow{\underline{\phi}}$ achieving sup in (18) is currently *unknown* even for the simpler case of spatially white MUI. However, it is known that Gaussian distributed input signals achieve sup in (18) not only when condition of PCSI is approached [1,2,4,5] but also for $0 < \sigma_{\varepsilon}^2 < 1$ when length T_{pay} of the payload/phase (largely) exceeds number of t of transmit antennas (see [7] about this asymptotic important result). Therefore, motived by above considerations, in the following we focus on the evaluation of (18) for Gaussian distributed input signals. In this case the T_{pay} components $\{\underline{\phi}(n) \in \mathbb{C}^t, T_L + T_{tr} + 1 \leq n \leq T\}$ in (11) of the overall signal vector $\overrightarrow{\underline{\phi}}$ in (12) are modeled as uncorrelated zero-mean proper complex Gaussian vectors with correlation matrix R_{ϕ} meeting (11.1). Obviously, the corresponding information throughput

$$C_G(\hat{H}) \triangleq \frac{1}{T_{pay}} \sup_{Tra[R_{\phi}]\leq Pt} I\left(\overrightarrow{\underline{y}};\overrightarrow{\underline{\phi}}|\hat{H}\right) \tag{19}$$

conveyed by the MIMO channel (12) generally falls below $C(\hat{H})$ in (18), so that we have $C_G(\hat{H}) \leq C(\hat{H})$. However, above inequality is satisfied as equality when at least one of the above cited operating conditions is met. Therefore, passing now to deal with evaluation of $C_G(\hat{H})$ in (23), we have the following first property.

Proposition 2. Let us assume the training matrix $\tilde{X}$ meeting the relationship (16) and also assigned the spatial correlation matrix R_{ϕ} in (11.1) of the payload streams radiated by t transmit antennas. Thus, the resulting conditional mutual information $I\left(\overrightarrow{\underline{y}};\overrightarrow{\underline{\phi}}|\hat{H}\right)$ in (19) supported by the MIMO channel (12) admits the following closed-form expression:

$$I\left(\overrightarrow{\underline{y}};\overrightarrow{\underline{\phi}}|\hat{H}\right) = T_{pay} \lg\det\left[\left(I_r + \frac{1}{t}K_d^{-1/2}\hat{H}^T R_{\phi}\hat{H}^* K_d^{-1/2} + \sigma_{\varepsilon}^2 P K_d^{-1}\right)\cdot \left(I_{rt} + \frac{\sigma_{\varepsilon}^2 T_{pay}}{t}(K_d^{-1}) \otimes R_{\phi}\right)^{-1/T_{pay}}\right] \tag{20}$$

when at least one of the conditions below listed is met:

a) both T_{pay} and t are large; (20.1)

b) σ_{ε}^2 vanishes; (20.2)

c) all SINRs γ_j, $1 \leq j \leq r$, in (10.2) vanish. $\diamond$(20.3)

4.1 Optimized Power Allocation in the Presence of Colored MUI

Therefore, according to (19), we must proceed to the power-constrained maximization of conditional throughput (20). For this purpose, first let us indicate as

$$K_d = U_D \Lambda_D U_D^{\dagger}, \tag{21}$$

the Singular Value Decomposition (SVD) of MUI spatial covariance matrix K_d, with

$$\Lambda_D \triangleq diag\{\mu_1, ..., \mu_r\} \tag{21.1}$$

indicating the corresponding (rxr) diagonal matrix of the *magnitude-ordered* singular values. Second, after introducing the (txr) matrix

$$A \triangleq \hat{H}^* K_d^{-1/2} U_D, \tag{22}$$

accounting for the *combined* effects of imperfect channel estimates $\hat{H}$ and MUI spatial coloration K_d, let us denote as

$$A = U_A D_A V_A^{\dagger} \tag{22.1}$$

the related SVD, where U_A and V_A are unitary matrices, while

$$D_A \triangleq diag\{K_1, ..., K_s, \underline{0}_{t-s}\}, \tag{22.2}$$

is the corresponding (txr) diagonal matrix collecting the $s \triangleq \min\{r, t\}$ *magnitude-ordered* singular-values $k_1 \geq k_2 \geq ... \geq k_s > 0$ of matrix A. Finally, for future convenience, let us also introduce the following dummy positions:

$$\alpha_m \triangleq \frac{\mu_m k_m^2}{t(\mu_m + P\sigma_\varepsilon^2)}, \;\; 1 \leq m \leq s; \;\; \beta_l \triangleq \frac{\sigma_\varepsilon^2 T_{pay}}{t\mu_l}, \;\; 1 \leq l \leq r. \tag{23}$$

Thus, the application of the Kuhn-Tucker conditions [14, eqs.(4.4.10), (4.4.11)] allows us to evaluate the optimized transmit powers $\{P^\star(m),\ 1 \leq m \leq t\}$ achieving the constrained sup in (19) as detailed in the following proposition.

Proposition 3

Let us assume that at least one of the above operating conditions (20.1), (20.2), (20.3) is fulfilled. Thus, for $m = s+1, ..., t$, the powers achieving sup in (19) *vanish* while for $m = 1, ..., s$ they are computed according to the following two relationship

$$P^\star(m) = 0, \text{ when } k_m^2 \leq \left(1 + \frac{\sigma_\varepsilon^2 P}{\mu_m}\right)\left(\frac{t}{\rho} + \sigma_\varepsilon^2 Tra[K_d^{-1}]\right), \tag{24}$$

$$P^\star(m) = \frac{1}{2\beta_{min}}\left\{\beta_{min}\left[\left(1 - \frac{r}{T_{pay}}\right)\rho - \frac{1}{\alpha_m}\right] - 1 + \right.$$

$$+\sqrt{\{\beta_{min}\left[\left(1-\frac{r}{T_{pay}}\right)\rho-\frac{1}{\alpha_m}\right]-1\}^2+4\beta_{min}\left(\rho-\frac{1}{\alpha_m}-\frac{r\rho\beta_{min}}{\alpha_m T_{pay}}\right)}\Big\},$$

when

$$k_m^2 > \left(1+\frac{\sigma_\varepsilon^2 P}{\mu_m}\right)\left(\frac{t}{\rho}+\sigma_\varepsilon^2 Tra[K_d^{-1}]\right). \tag{25}$$

Furthermore, the nonnegative scalar parameter ρ in (24), (25) is set so to satisfy the power constraint (see eq.(11.1))

$$\sum_{m\in\Im(\rho)} P^\star(m) = Pt, \tag{26}$$

where

$$\Im(\rho) \triangleq \{m=1,...,s : k_m^2 > \left(1+\frac{\sigma_\varepsilon^2 P}{\mu_m}\right)\left(\frac{t}{\rho}+\sigma_\varepsilon^2 Tra[K_d^{-1}]\right)\}, \tag{26.1}$$

is the (ρ-depending) indexes set fulfilling inequality (25). Finally, the corresponding optimized spatial correlation matrix for the radiated signals is aligned along right-eigenvectors of matrix A in (22.1) according to

$$R_\phi(opt) = U_A diag\{P^\star(1),...P^\star(s),\underline{0}_{t-s}\}U_A^\dagger, \tag{27}$$

so that the resulting maximized throughput in (19) admits the following (simple) closed-form expression:

$$C_G(\hat{H}) = \sum_{m=1}^{r}\left(1+\frac{\sigma_\varepsilon^2}{P}\mu_m\right)+\sum_{m=1}^{s}\left[\lg(1+\alpha_m P^\star(m))-\frac{1}{T_{pay}}\sum_{l=1}^{r}\lg\left(1+\beta_l P^\star(m)\right)\right]\diamond \tag{28}$$

About above reported power allocation, some remarks are in order. First, an exploitation of the (truncated) expansion: $\sqrt{1+x}\approxeq 1+0.5x$ allows us to rewrite (24), (25) in the following form for vanishing σ_ε^2:

$$lim_{\sigma_\varepsilon^2\to 0}P^\star(m) = \max\left\{0,\rho-\frac{t}{k_m^2}\right\},\ m=1,..,s, \tag{29}$$

and this last agrees with the water-filling like power allocation previously reported in [1,2] for the case of *perfect* channel estimation. Second, when σ_ε^2 approaches unit then channel estimates $\hat{H}$ vanish so that no information is available at both transmit and receiver about actual values assumed by MIMO channel path gains $\{h_{ji}\}$. In this limit case, we arrive at the following limit expression for the sustained throughput $C_G(\hat{H})$ in (19):

$$\lim_{\sigma_\varepsilon^2\to 1} C_G(\hat{H}) \triangleq C_G(0) = \sum_{m=1}^{r}\lg\left[\frac{\left(1+\frac{P}{\mu_m}\right)}{\left(1+\frac{T_{pay}}{\mu_m}\right)^{1/T_{pay}}}\right],\ (nats/payload\ slot) \tag{30}$$

Since this last holds for large t and T_{pay} regardless of radiated power P, relationship (30) directly *supports* the conjecture originally reported in [7] about the capacity-achieving property retained for *large* T_{pay} by the input Gaussian pdf *even* in application scenarios with non-coherent detection.

4.2 A Numerical Algorithm for Implementing the Optimized Power Allocation

Passing now to shortly consider numerical implementation of presented Power Allocation formulas, the first step for computing (24), (25) relies on evaluating ρ value satisfying relationship (26). Although this computation resists closed-form analytical evaluation in the simpler case of PCSI [1,2], nevertheless we observe that cardinality $|\Im(\rho)|$ of the indexes' set in (26.1) vanishes at $\rho = 0$ and then increases for growing ρ. This means that the solution of (26) may be found by implementing a (very) simple iterative procedure that starts with $\rho = 0$ and then progressively increases the current value of ρ by a pre-assigned step-size Δ until summation in (26) equates the power constraint Pt. The resulting algorithm for the numerical evaluation of the optimized powers (24), (25) is summarized in Table I and several performance plots obtained via its computer-implementation will be presented in Sect.VI.

Table 1. Pseudo-code for a numerical implementation of the optimized power allocation algorithm.

1. Compute and order eigenvalues of MUI covariance matrix K_d;
2. Compute SBD of matrix A in (22.1) and order its singular values;
3. Set $P^{\star}(m) = 0,\ 1 \le m \le t$;
4. Set $\rho = 0$ and $\Im(\rho) = \emptyset$;
5. Set step size Δ;
6. While $\left(\sum_{m \in \Im(\rho)} P^{\star}(m) < Pt\right)$ do
7. Update $\rho = \rho + \Delta$;
8. Update set $\Im(\rho)$ via eq. (26.1);
9. Compute powers set $\{P^{\star}(m),\ m \in \Im(\rho)\}$;
10. end;
11. Compute optimized powers $\{P^{\star}(m),\ 1 \le m \le s\}$ via eqs. (24), (25);
12. Compute optimized shaping matrix $R_{\phi}(opt.)$ via eq.(27);
13. Compute conveyed throughput $C_G(\hat{H})$ via eq.(28);

5 Numerical Results and Performance Comparisons

Although joined pdf of the (rxt) elements of the channel estimates $\hat{H}$ is known [9], nevertheless the corresponding expectation

$$C_G \triangleq E\{C_G(\hat{H})\}, \tag{31}$$

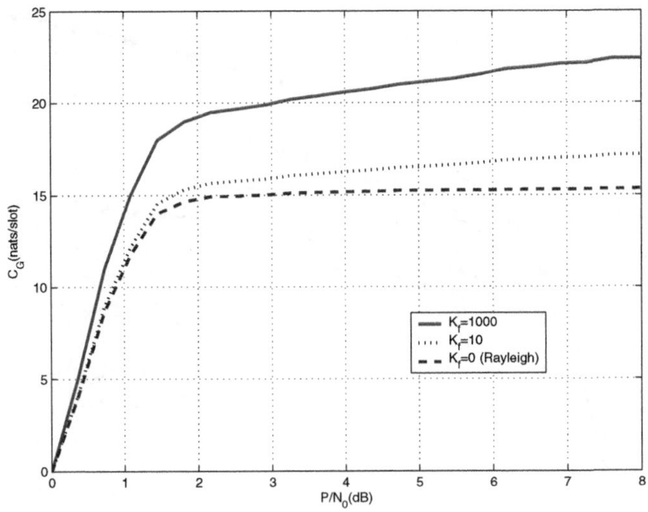

Fig. 1. Sensitivity of the Throughput C_G conveyed by the reference link Tx0 $\rightarrow$ Rx0 on the Ricean Factor k in (38) for $T_{pay} = 80$, r=t=10, $\sigma_\varepsilon^2 = 0.1$).

of the conditional throughput in (28) resists closed-form analytical evaluation even in the simplest case of spatially white MUI with vanishing σ_ε^2 [4,5,17 and reference therein]. Thus, as in [1,2,4] in this Section we resort to a Monte-Carlo approach for computing expectation based on sample-average of 10,000 independent realizations of conditional throughput $C_G(\hat{H})$. Furthermore, all the reported numerical plots refer to a hexagonal network with N_0 *set to unit* and various values of the power level P radiated by transmit nodes. Plots of Fig.1 allow us to appreciate the effect of the Ricean Factor k_f on the average throughput conveyed by the reference link for $T_{pay} = 80$, r=t=10 and $\sigma_\varepsilon^2 = 0.1$. From these last plots we conclude that when interference presents Rayleigh features the system performances fall short. Although in these years MAI-mitigation capability offered by multi-antenna systems based on smart-like technology has been often claimed [8,15,18], nevertheless a still open question concerns comparison of information throughputs C_G conveyed by here considered MAI-impaired systems with those guaranteed by orthogonal MAI-free TDMA (or FDMA) based access techniques.

By fact, till now no firm evidence of superiority of an access technique over the other one is available in the literature, specially for the application scenarios here considered where typical SINRs values are of the order of few dBs so that multiuser detection strategies based on iterative subtractive cancellations of MAI tend to fall short [22]. To gain some (preliminary) insight about this important question, we have numerically evaluated the average information throughout $C_{TDMA} = E\{C_{TDMA}((\hat{H}))\}$ (nats/ payload slot) conveyed by the reference link when a MAI-free TDMA-based access is implemented.

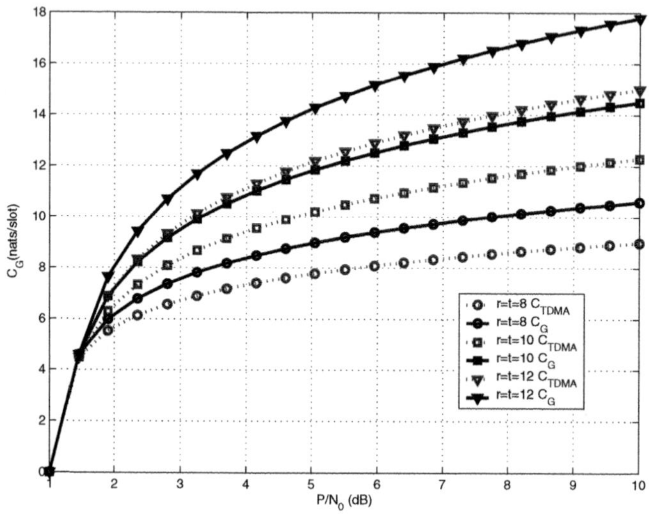

Fig. 2. C_{TDMA} throughput comparisons for the reference link Tx0 → Rx0 for $T_{pay} =$ 80, k=1000, $\sigma_\varepsilon^2 = 0.1$).

The obtained numerical plots of Fig.2 refer to the network with $T_{pay} = 80$, $\sigma_\varepsilon^2 = 0.1$, $k = 1000$ and values of number of transmit/receiving antennas ranging from 4 to 12. Although C_G has been evaluated in the worst-MAI condition, nevertheless plots of Fig.2 show that C_G outperforms corresponding C_{TDMA}, specially at low radiated power level P and for transceivers equipped with large number r=t of transmit/receiving antennas. Thus, we may conclude that when r and t increase the proposed spatial shaper allows us to achieve Channel Capacity that is higher than the one achieved by orthogonal access methods.

6 MAC Implications and Conclusions

The underlying conclusion seems to be that allowing multiple users to collide in the time-frequency plan leads to a more effective utilization of the spatial dimension of the receivers that, by fact, over-compensates throughput loss induced by collisions experienced in the time-frequency domain. Based on this conclusion, a more effective MAC strategy may be to allow for some forms of hybrid multi-access scheme where (a moderate level of) collisions are tolerable in the time-frequency domain so to fully exploit the space-division access capability of the multi-antenna system. An example of non-orthogonal access technique may be OFDMA with *tone-sharing* among multiple users, where the degree of tone-sharing increases with the number of transmit/receive antennas. This conclusion demands for new MAC paradigm and related MAC design criteria where the spatial MAC capability of the Multi-Antenna system is carefully exploited.

References

1. F.R. Farrokhi, G.J. Foschini, A. Lozano, R.A. Valenzuela,"Link-Optimal BLAST processing with Multiple-Access Interference", VTC2002, pp.87–91.
2. F.R. Farrokhi, G.J. Foschini, A. Lozano, R.A. Valenzuela,"Link-Optimal Space-Time Processing with Multipole Transmit and Recieve Antennas", IEEE Comm. Letters, vol.5, no.3, pp.85–87, March 2001.
3. S.L. Marple Jr., Digital Spectral Analysis with Applications, Prentice Hall, 1987.
4. C.-N. Chuan, N.D.S. Tse, J.M .Kahn, R.A. Valenzuela,"Capacity-scaling in MIMO wireless Systems under correlated Fading", IEEE Trans. on Inform. Theory, vol.48, no.3,pp.637–650, March2002.
5. G.J. Foschini, M.J. Gans,"On limit of wireless communicqations in fading environment when using multiple antennas", Wireless Pers. Comm., vol.6, no.3, pp.311–325, June 1998.
6. B. Hassibi, T.L. Marzetta, "Multiple-Antennas and Isotropically Random Unitary Inputs: the received Signal Density in closed form", IEEE Trans. on Infrom. Theory, vol.48, no.6, pp.1473–1485, June 2002.
7. T.L. Marzetta, B.M. Hochwald, "Capacity of a mobile Multiple-Antenna Communication link in Rayleigh flat fading", IEEE Trans. on Inform. Theory, vol.45, no.1, pp.139–157, January 1999.
8. R.D. Murch, K.B. Letaief,"Antenna System for broadband Wireless Access", IEEE Comm. Mag., pp.637–650, March 2002.
9. J.-C. Guey, M.P. Fitz, M.R. Bell, W.-Y. Kuo,"Signal Design for Transmitter Diversity Wireless Communication System Over Rayleigh Glat fading channels", IEEE Trans. on Comm., vol.47, no.4, pp.527–537, April 1999.
10. J. Baltarsee, G. Fock, H. Meyr,"Achievable Rate of MIMOChannels with data-aided channel-estimation and perfect interleaving", IEEE Journ. on Selected Areas in Comm., vol.19, no.12, pp.2358–2368, Dec.2001.
11. E. Baccarelli, M. Biagi, A. Fasano, "Optimized Design and Performance of Multiple-Antenna 4th Generation WLANs for Partially-Coherent Decoding", European Wireless 2002 Conference Proceedingd, Florence 25–28 February 2002, vol2, pp.858–864
12. T.M. Cover, J.A. Thomas, Elements of Information Theory, Wiley, 1991.
13. P. Lancaster, M. Tismetesky, The theory of Matrices, 2nd Ed., Academic press, 1985.
14. R.G. Gallagher, Information heory and Reliable Communication, Wiley, 1968.
15. G.T. Okamoto, Smart Antennas Systems and Wireless LANs, kluwert 2001.
16. S. Verdú, T.S. Han,"A general Formula for channel Capacity", IEEE Trans. on Inform. Theory, vol.40, no.6, pp.1147–1157, July 1994.
17. D.W. Bliss, K.W. Forsythe, A.O. Hero, A.F. Yegulalp,"Environmental Issues for MIMO Capacity", IEEE Trans. on Signal Proc., vol.50, no.9, pp.2128–2142, Sept. 2002.
18. A. Santamaria, F.J. L.-Hernandez, Wireless LAN Standards and Applications, Artech House, 2001.
19. H. Sampath, S. Talwar, J. Tellado, V. Erceg, A. Paulraj, "A 4th Generation MIMO-OFDM Broadband Wireless System: Design, Performance and Field Trials results", IEEE Comm.Mag., pp.143–149, Sept.2002.
20. C.E. Perkins, Ad Hoc Networking, Addison Wesley, 2000.
21. E. Baccarelli, M. Biagi, "Error Resistant Space-Time Coding for Emerging 4G-WLANs", presented at Wireless Communication and Networking Conference (WCNC) 2003, 16–20 March 2003, New Orleans, Lousiana, U.S.A.

22. S. Verdú, Multiuser Detection, Cambridge Uni. Press, 1998.
23. A. Lozano, A.M. Tulino,"Capacity of Multiple-Trasmit Multiple-Receive Antenna Architecture", IEEE Trans. on Inform. Theory, vol.48, no.12, pp.3117–3128, Dec.2002.
24. E. Baccarelli, M. Biagi,"A New Family of Space-Time Codes for Multi-Antenna Systems with Imperfect Channel Estimation", (INVITED PAPER), IEEE International Symposium on Advances in Wireless Communications (ISWC'02) conference proceedings, Victoria, BC, Canada September 23–24, pp.83–84.

On the Performance of a Medium Access Scheme Based on EY-NPMA in a High Bitrate Wireless LAN

G. Dimitriadis and F.-N. Pavlidou

Department of Electrical and Computer Engineering
Aristotle University of Thessaloniki
54124 Thessaloniki, Greece
{gedimitr,niovi}@eng.auth.gr

Abstract. Wireless LANs provide an efficient and inexpensive way for the creation of networks that do not constrain the users' mobility. Furthermore, advanced physical layer techniques have allowed wireless LANs to support bitrates that until recently were attainable only in wired networks. The growing penetration of WLANs in the market, as well as the wider spectrum of applications that these kinds of networks are called to support, demands the introduction of new methods for efficient medium access. In this paper, we propose and evaluate the performance of a medium access scheme that is based on the EY-NPMA protocol for medium access, showing through simulation the good characteristics of the proposed scheme.

1 Introduction

Wireless LANs provide an efficient and inexpensive way for the creation of networks that do not constrain the users' mobility. Spontaneous, hassle free communication has become possible through ad hoc networks, while infrastructure based wireless LANs have permitted true flexibility and freedom without demanding installation of new cables or retrofitting. Furthermore, advanced physical layer techniques have allowed wireless LANs to support bitrates that until recently were attainable only in their wired counterparts. The growing penetration of WLANs in the market, as well as the wider spectrum of applications that these kinds of networks are called to support, demands the introduction of new methods for efficient medium access, since it is the MAC who defines how efficiently the available raw bandwidth is used. In this paper we present and evaluate the performance of of a modified version of the EY-NPMA medium access scheme, a protocol that is part of the HIPERLAN [1] wireless networking standard. The rest of the paper is structured as follows. In section 2 we provide a brief presentation of the background on medium access for wireless LANs. In section 3, we present the proposed medium access scheme, while sketching the mechanics of the base EY-NPMA protocol. In section 4, the simulation results are presented and finally, section 5 concludes the paper.

M. Conti et al. (Eds.): PWC 2003, LNCS 2775, pp. 153–162, 2003.

2 Background Work

A medium access scheme must possess three characteristics in order to be an efficient one. It must ensure that the common medium is fairly shared between the nodes that consist the network, provide explicit quality of service (QoS) guarantees for different classes of traffic and finally maximize the medium utilization by minimizing the wasted capacity that is caused by packet collisions and/or overhead. Furthermore, all of the above should be implemented in a design that is both robust and simple. In the field of wireless LANs, there are two dominating families of protocols for medium access. The first one includes those that are based on contention, while the other includes the protocols that are based on the paradigm of dynamic assignment.

Contention based medium access schemes represent a paradigm that has been traditionally connected with Local Area Networks (mostly because of the widespread deployment of IEEE 802.3 - Ethernet). According to protocols belonging to this family, packet transmissions take place in a completely stochastic way, with minimal or totally absent coordination between the nodes participating in the network. Consequently, there exists a probability that multiple transmissions take place simultaneously, resulting in erroneous reception. To reduce such occurences, the concept of carrier sensing is widely employed, since it prevents network nodes from transmitting, when they sense that a transmission is already in progress. For technical reasons, the variant of carrier sensing that uses collision detection (CSMA/CD) cannot be applied in wireless networks, so in order to minimize the impact of collisions, the technique of collision avoidance (CSMA/CA) is widely used, initially proposed by Karn in [2]. According to this scheme, packet collisions are restricted only between small Request-To-Send, Clear-To-Send packets (RTS/CTS), while the actual data packets are transmitted collision-free. Furthermore, this handshake between transmitter and receiver effectively mitigates the hidden terminal problem [3].

Contention based medium access schemes are particularly well fit for bursty traffic (which is the case for most data communications), but generally they are not well suited for providing quality of service guarantees for different traffic classes. Regarding throughput, as a rule contention based protocols perform best under light to medium traffic loads, since the rate of collisions increases sharply as the traffic load exceeds a threshold. However, in the last few years, the importance of service differentiation has triggered intense research activity on embedding quality of service (QoS) capabilities in contention based medium access schemes. Protocols that are QoS aware include Blackburst proposed by Sobrinho and Krishnakumar in [6], DFS (Distributed Fair Scheduling) [7], the under standardization by the 802.11e task group medium access protocol EDCF (Enhanced Distributed Coordinator Function) [5], as well as the aforementioned EY-NPMA protocol. On the other hand, DCF (Distributed Coordinator Fuction) of the original IEEE 802.11 [4] standard for wireless LANs is insensitive to different traffic classes.

Medium access schemes following the dynamic assignment approach gain access to the common medium either via an arbitrator (e.g. a terminal that

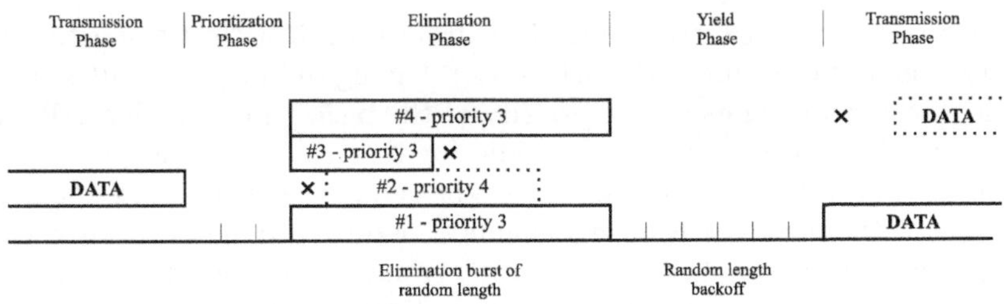

Fig. 1. EY-NPMA's synchronized channel access cycle

polls each station in a round-robin fashion) or through a well-defined process that ensures that only one terminal at a time has permission to gain access to the channel (e.g. reservation or token passing schemes). An important difference from contention based medium access protocols is that with dynamic assignment schemes there is usually a need for a management entity, undertaking the role of polling, creating tokens when they get lost and validating reservations. In infrastructure based wireless LANs, the access point (AP) is the obvious candidate for such a role, but in ad hoc networks a node is dynamically chosen to become channel coordinator, usually through clustering techniques. Compared to contention based medium access schemes, the performance of dynamic assignment protocols does not deteriorate when the traffic load increases, while the deterministic access to the channel inherently provides quality of service support. On the other hand, the fact that each station must explicitly acquire access permission (i.e. get polled or capture the token), introduces a latency that is mostly notable in cases of light traffic, while traffic load asymmetries tend to deteriorate the performance of these protocols. Also, the hidden terminal problem proves to be more difficult to alleviate in dynamic assignment protocols, rather than in contention based ones. Protocols that follow the dynamic assignment paradigm are the medium access layers of the HIPERLAN/2 [8] and Bluetooth [9] standard, as well as the PCF (Point Coordinator Function) access mode of the IEEE 802.11 standard.

3 Protocol Model

EY-NPMA stands for Elimination-Yield Non-pre-emptive Priority Multiple Access. It is a contention based protocol that has been standardised under ETSI's HIPERLAN standard for wireless LANs. EY-NPMA provides excellent support for different classes of traffic regarding quality of service and demonstrates very low collision rates. In the following subsections, we will present the inner workings of the base EY-NPMA scheme and after that the proposed modified scheme.

3.1 EY-NPMA

According to EY-NPMA, the channel access cycle consists of four distinct phases: Prioritization, Elimination, Yield and Data Transmission. In prioritization, EY-

NPMA recognizes five distinct priorities from 0 to 4, with 0 being the highest priority. The cycle begins with each station having data to transmit sensing the channel for as many slots as the priority of the packet in its buffer. All stations that succesully sense the channel as idle for the whole interval proceed to the next phase, the elimination phase. Those that do not, exit the contention process and wait for the next synchronized channel condition to make another attempt. During the elimination phase, each station transmits an energy burst of random length. These bursts ensure that only the stations having the highest priority data at a time proceed to the elimination phase. As soon as a station finishes bursting, it immediately senses the channel. If the channel is sensed as idle, the station proceeds to the next phase. Otherwise, it leaves the cycle. During the yield phase, the stations that survived the two previous ones, back off for a random number of slots. The station that backs off for the shortest interval eventually gets access of the channel for data transmission. All other stations sense the beginning of the transmission and refrain from transmitting. Each phase reduces the number of stations that remain into the contention process, so that (hopefully) the station that will commence transmitting data in a given time will be unique. In figure 1, we present a typical synchronized channel access cycle. Solid line boxes represent actual transmissions, while dashed line boxes represent projected transmissions that did not take place because the station left the contention process. The X marks show when and why a station left the cycle.

Through this four-phases cycle, EY-NPMA manages to provide a low and quasi-constant rate of collisions. The parametres chosen in the HIPERLAN standard (maximum number of slots for bursting and backing off, etc) aimed at a target collision rate of 3.5% for a population of 256 simultaneously contending nodes, a figure that guaranteed that for most practical situations communications would be collision-free. However, this important merit of EY-NPMA turns out to also be its most severe disadvantage. In order to achieve such a low rate of collisions, a large number of slots is allocated to the elimination and yield phase and thus are being experienced as overhead. As advances in the physical layer lead to higher bitrates, this effect becomes even more intense. Typical values for the duration of each slot are around 10 μs, a value which at high bitrates becomes a significant fraction of the time needed to transmit the actual data payload. Because of the wireless environment, but also for technical reasons, there is a lower limit to the slot duration. Propagation delay and delay spread demand that the slot duration for both elimination and yielding exceeds a certain threshold, while especially for bursting, rise and fall times of each burst place this threshold even higher. In the following subsection, we describe a medium access scheme based on the original EY-NPMA protocol, that manages to achieve a more efficient utilization of the shared medium.

3.2 Proposed Scheme

The proposed medium access scheme features a different structure for the prioritization phase and a mechanism for dynamically promoting the packet priority

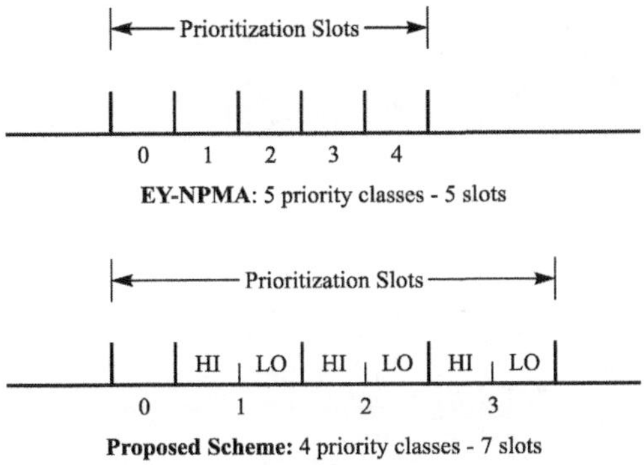

Fig. 2. Structure of the prioritization phase for both schemes

for packets that have survived the elimination process, but did not survive the yield phase. Instead of 5 priority classes, we propose a scheme with 4 priorities. The three lower ones consist of two subclasses, namely a low and a high. The structure of the prioritization slots for the two schemes is depicted in figure 2. A data packet of priority x that is placed in the transmission buffer of the MAC contoller by an upper layer is automatically labeled as being x-low priority. Let x-low be the highest priority when a channel access condition occurs and a number of N stations enter the contention process with all N stations commencing bursting at the same slot. At the end of the elimination phase, a fraction of the initial population, N_s, will have survived elimination and will choose a random number of slots to backoff. At this point all N_s stations switch their packets priorities from x-low to x-high. At the next channel access cycle, through the prioritization phase only the N_s stations will enter the elimination phase, in contrast to the at least $N-1$ stations that would enter contention according to the base EY-NPMA scheme. Access cycles at x-low priority will be postponed, until there are no more x-high packets, which will happen at the end of at least N_s cycles. When there are no more packets of x-high priority, an access cycle of x-low will follow, and the whole process will be repeated.

Furthermore, since at x-high priority level there are much fewer contending nodes, a reasonable rate of collisions can be achieved with fewer slots dedicated to elimination and yielding, than in the case of x-low priority. Hence, by demanding the all x-high cycles employ fewer cycles for the two contention resolution phases, we are led to better medium utilization figures, since for the same data payload the access cycles become shorter. The cost of this modification is the reduction of the traffic classes by one and the addition of two extra slots for the prioritization phase. Consequently, when a big population of stations wants to send data in the same base priority class x, according to the base EY-NPMA scheme in each access cycle this big population will content for channel access. On the contrary, with our modification the whole population contents for one cycle, while for a number of subsequent cycles only a subset of these stations will participate in the contention process. Furthermore, the protocol's behaviour to the base priority

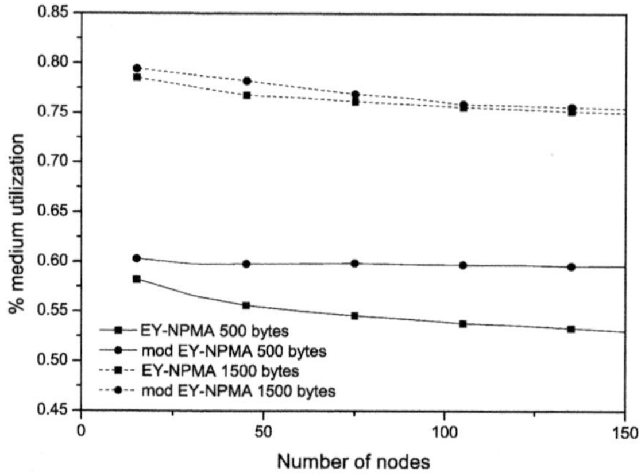

Fig. 3. Medium utilization vs number of contending nodes

classes does not change by our modification. That is a priority 3 packet will always be of lower priority than a priority 2 packet, no matter what subclass – low or high – the packets happen to be.

4 Simulation

4.1 Simulation Scenario

The simulation trials for both the base and the modified EY-NPMA scheme were conducted using a custom simulator developed by the authors in C++. For the needs of this work, two scenarios were examined and evaluated. The first one considered the performance of a high bitrate wireless LAN as the number of contending nodes increased, while the second dealt with the transport capabilities of the same wireless LAN for two different traffic classes as the offered load increased. For both scenarios we assumed a fully connected topology, that is a network where between each pair of nodes there exists a bidirectional link. Consequently, with the above assumption we ruled out the existence of any hidden nodes and also avoided routing of data packets as would be the case in a multihop wireless network. The channel capacity was set to 20 Mbps, while in the case of the base EY-NPMA protocol the maximum number of elimination slots (bursting) was set to 9 and the maximum number of slots for the yielding phase (backoff) was set to 6. For the proposed scheme, 5 slots were allocated to elimination and 6 slots to yielding for x-low priority packets, while for x-high priority cycles, 3 slots were allocated for elimination and 4 slots for yielding. The duration of a slot for both phases was set to 10 μs.

For the first set of experiments, an increasing number of nodes contend for channel access. A high rate packet generator of priority 1 is attached to each node, so that their transmission buffer is never empty. For this scenario, we

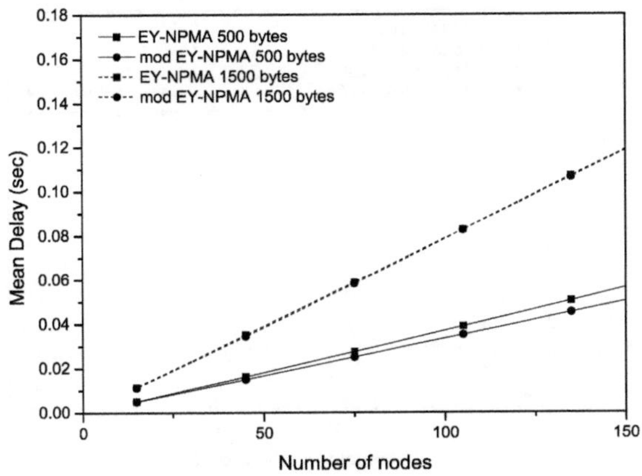

Fig. 4. Mean packet delay vs number of contending nodes

experimented with two packet sizes, the first being 500 bytes and the second 1500 bytes. The network performance was evaluated using two metrics, namely medium utilization and mean packet delay. Medium utilization is defined as the percentage of time that is used for the successful transmission of actual data frames, while the metric of mean delay expresses the time needed between placement of a data packet in the transmission buffer and positive acknowledgement of its reception. In the second scenario, we examine the performance of a network consisting of 100 nodes, under a variable offered load. The network population is divided into two groups: a group of 40 nodes generating high priority data packets (priority 1), and a group of 60 nodes generating low priority data packets (priority 2). For both priority classes, the packet size is equal to 500 bytes, while the packet generation rate for high priority traffic is 5 time the correspondent rate for low priority traffic. Consequently, for any given value of aggregate offered load, 23% is due to low priority sources and 77% is due to high priority sources. For this scenario, the network performance was measured also in terms of medium utilization and mean packet delay.

4.2 Simulation Results

In figures 3 and 4, we present the results of the first scenario. Figure 3 depicts the medium utilization for a variable number of simltaneously contending nodes. The base EY-NPMA scheme shows a poorer performance than the modified medium access scheme for both 500 and 1500 bytes of packet size. Furthermore, the difference in performance increases as data packets become shorter. In the case of 1500 bytes long data packets, the modified EY-NPMA scheme shows an increase in medium utilization of 2.4%, while for 500 bytes long packets, the correspondent increase is somewhat less than 10%. This phenomenon is accounted to the importance of the slot duration relative to the time needed for a data frame transmission. The proposed scheme employs fewer slots for the

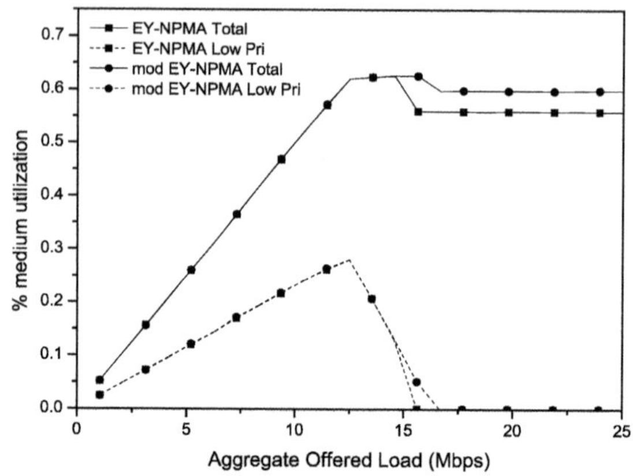

Fig. 5. Medium utilization vs aggregate offered load

contention resolution process than the base EY-NPMA protocol. The impact of these savings is more evident in the case of short data packets, where the slot duration is a more significant fraction of the time needed for the transmission of the actual data packet, than in the case of long data packets.

In this figure, three out of four curves show a decreasing trend. As was expected, in the case of the base EY-NPMA scheme, our simulations showed that the collisions rate remained quasi-constant while the node population increased. The decrease in medium utilization is owed to the fact that when the number of contending nodes increases, the average length of the elimination phase increases too, since each node randomly chooses a number of slots for bursting according to a geometric distribution. This increase of the elimination phase is translated to increases in overhead and hence to worse medium utilization. In the case of the modified EY-NPMA scheme, the number of nodes that survive the elimination phase is larger, since we allocated fewer slots for bursting. This had two conflicting results: In x-low there was a higher collisions probability, but more data frames were transmitted in x-high, achieving data transmissions with favorable collisions probability in shorter cycles. In the case of short data packets, these two phenomena reach an equilibrium and hence the medium utilization is almost independent of the number of contending nodes. In the case of long data frames, however, the effect of the increased collisions rate is slightly more intense and a decrease in medium utilization is experienced.

In figure 4, we present the mean delay experienced by a packet between acceptance by the MAC layer and positive reception acknowledgement. For both packet sizes, the modified EY-NPMA scheme showed an increase in performance compared to the base medium access protocol. For the same reasons that were reported in the previous paragraphs, there are more significant differences in performance in the case of short packets, rather in the case of packets that are 1500 bytes long. The increase in mean packet delay as the population size

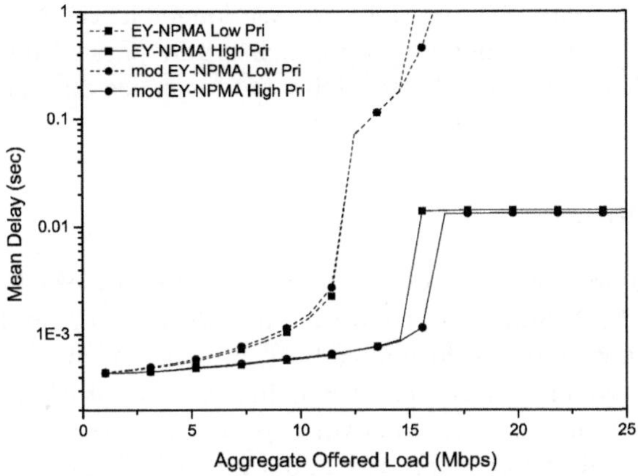

Fig. 6. Mean packet delay vs aggregate offered load

increases shows a linear behaviour for both schemes. This is primarily attributed to the fact that the same capacity is shared by a larger number of nodes and not to an increase in collisions, something that is also confirmed by the results of figure 3.

Figures 5 and 6 depict the results gained from the second scenario simulations. In figure 5 the medium utilization versus variable aggregate load is presented. For each scheme the curves of total medium utilization and medium utilization due to low priority traffic are drawn on the graph. The medium utilization by the high priority traffic can easily be deducted by substracting the two curves. In this figure the medium utilization for both schemes initially increases linearly as the offered load increases. However, as hinted by figure 3, in the case of the base EY-NPMA scheme the network reaches its saturation point earlier than in the case of the modified scheme, while in steady state EY-NPMA also shows a worse performance. Beyond the saturation point, low priority traffic is gradually phased out, until eventually it is completely starved. However, in the case of the modified EY-NPMA scheme, starvation of low priority traffic occurs later, than in the case of the base protocol.

In figure 6, we present the mean packet delay for the same scenario. We see that for both schemes the curves of low and high priority data follow a similar trend. When the network operates beneath the saturation point, packets experience a mean delay that is less than 5 ms. We also notice that beneath the saturation point, the mean delay of the proposed scheme is slightly higher than in the case of the base EY-NPMA scheme, something that is especially evident in the case of low priority traffic. This is owed to the fact that even though both classes have the same priority (1 and 2), in the case of the modified EY-NPMA protocol nodes commence bursting 1 or 2 slots later, for high and low priority traffic respectively, than in the case of the base EY-NPMA scheme. Since there is no intense contention at this point of operation, almost all cycles are at x-low and hence the access cycles are longer in the case of the modified scheme. However, as the network approaches and passes the saturation point the modified scheme provides better delay figures, since an increasing number of shorter cycles are done in x-high priority.

5 Conclusions

In this paper we presented and evaluated the perfomance of EY-NPMA in a high bitrate wireless LAN and also proposed a modification of the base scheme that manages to achieve good performance both in terms of medium utilization and mean packet delay. The guaranteed low collisions rate of EY-NPMA comes at a cost that becomes higher, as the network speed increases. The proposed scheme reaches a balance between overhead and rate of collisions resulting in better utilization of the available capacity. Simulation trials for two different scenarios confirm the positive impact of these modifications. However, these first results also show that the proposed scheme has still room for improvement. Our future work will primarily concentrate on optimizing the proposed scheme and also compare it to other QoS aware medium access protocols.

References

[1] European Telecommunications Standards Institute (ETSI), ETSI EN 300 652 V1.2.1: Broadband Radio Access Networks (BRAN): HIgh PERformance Local Area Network (HIPERLAN) Type I: Functional Specification, 1998

[2] P. Karn, "MACA – A New Channel Access Protocol for Packet Radio," ARRL/CRRL Amateur Radio Ninth Conputer Networking Conf., pp. 134–140, 1990

[3] F. Tobagi, L. Kleinrock, "Packet Switching in Radio Channels: Part II - the Hidden Terminal Problem in Carrier Sense Multiple-Access and the Busy-Tone Solution," IEEE Transactions in Communications, vol. 23, no. 12, pp. 1417–1433, 1975

[4] The Institute for Electrical and Electronics Engineers (IEEE), IEEE Std 802.11 – Wireless LAN Medium Access Control (MAC) and Physical Layer (PHY) specifications, 1999

[5] M. Benveniste et al., EDCF Proposed Draft Text. IEEE working document 802.11-01/131r1, 2001

[6] J. Sobrinho, A. Krishnakumar, "Quality-of-Service in Ad Hoc Carrier Sense Multiple Access Networks," IEEE Journal on Selected Areas in Communications, vol. 17, no. 8, pp. 1353–1368, 1999

[7] N. Vaidya et al., "Distributed Fair Scheduling in a Wireless LAN," Sixth Annual International Conference on Mobile Computing and Networking, USA, 2000

[8] European Telecommunications Standards Institute (ETSI), DTS/BRAN-0020004-1 V0.m: Broadband Radio Access Networks (BRAN): HIgh PERformance Local Area Network (HIPERLAN) Type II: Data Link Control (DLC) Layer: Part 1: Basic Transport Functions, 1999

[9] Bluetooth Special Interest Group, Specification of the Bluetooth System - Core, V1.0B, 1999

WILMA: An Open Lab for 802.11 HotSpots*

Extended abstract

R. Battiti[1], M. Brunato[1], R. Lo Cigno[1], A. Villani[1], R. Flor[2], and G. Lazzari[2]

[1] Dipartimento di Informatica e Telecomunicazioni – Università di Trento
Via Sommarive 14, 38050 Povo, Trento, Italy
{battiti,brunato,locigno,villani}@dit.unitn.it
[2] ITC-IRST – Via Sommarive 18, 38050 Povo, Trento, Italy
{flor,lazzari}@itc.it

1 WILMA at-a-Glance

WILMA (Wireless Internet and Location Management Architecture) is an on-going research project based in Trento, Italy, whose aim is the study of the management of 802.11–based HotSpot Networks and the value added services that can be provided through such an infrastructure.

The founding partners of WILMA [9] are ITC-Irst [10], the DIT [11] at the University of Trento, and Alpikom [12] a local telecom provider. One of the main achievements of the project is the creation of an Open Laboratory (Open Lab) where other institutions can join the experiments. For instance, an agreement with the Municipality of Trento was reached for the realization of several Hot Spots in the City buildings and historical center.

What makes WILMA different from other experimental 802.11 based services (see for instance projects like MegaBeam [13], Telia HomeRun [14] or Boingo [15]), is the stress on research, innovation and a new architectural concept. The WILMA network is based on a layered structure as shown in Fig. 1. The owners of the access points are not necessarily the content or Internet connectivity providers, indeed, they seldom are. Roaming is guaranteed throughout all the access points, independently from the infrastructure provider. The only similar ongoing project is, to the best of our knowledge, the Stockholm Open Network [16], where the service and network architecture resemble the one envisioned by WILMA, though many differences can be identified, such as, for instance, the different approach used for IP addresses management and authentication procedures.

2 Network Architecture

One of the main novelties of the project is the independence of the physical wireless infrastructure providers, the HSPs in Fig. 1, from the service and content providers, the ISPs. This is indeed one of the key issues from the point of

* The WILMA project (www.wilmaproject.org) is supported by the Province of Trento under Grant N. 437, issued on March 3, 2002.

M. Conti et al. (Eds.): PWC 2003, LNCS 2775, pp. 163–168, 2003.

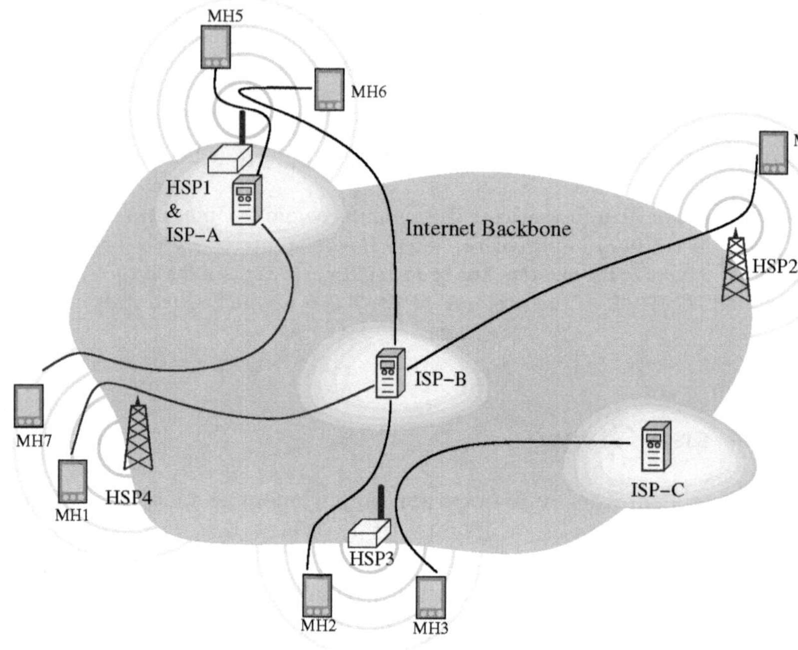

Fig. 1. Network and Service Architecture envisioned in the WILMA project

view of the network and service architecture, because it poses several new issues concerning security, privacy, AAA functions, pricing, etc.

The choice of this multi-actor infrastructure is not casual, but it is based on the deep belief that it is the only possible architecture that will ensure success to 802.11–based commercial networks. The Internet experience has shown that a clear separation between connectivity, which is technology related and can benefit from the quick pace of technology evolution, and services, which are software based and evolve with a much slower time scale, is a necessary condition for success in a competitive environment. The same Internet, on the other hand, has also shown that business without a pricing and investment strategy is not possible. The hundred and more years of telephony, fixed and mobile, teach the other part of the story: Pricing and planning strategies pays off, but service–specific networks cannot handle the burden of evolution. Besides, mobile networks have shown that only a standard, open solution as GSM can be really successful in the telecommunication market. One of the key points in GSM success is roaming, that guarrantees ubiquitous service and fosters competition, since it enables the ingress in the market of new operators without requiring the installation of a completely new infrastructure.

The challenges with wireless access on non-licensed bands go one step beyond both Internet and 2-3G cellular networks, since it is not possible for operators to furnish wide-scale access infrastructure with a constant quality. Hence, the idea of a new architecture, bases on local HotSpot infrastructure Providers (HSP) and global Internet Service Providers (ISP), that can be the traditional telecom and Internet providers. HSPs are a large number, possibly privates covering specific sites (airports, trains and train stations, gas stations and rest areas on motorways, etc.), or municipalities offering coverage as a basic infrastructure on the territory, just like potable water or sewage. ISPs on the other hand are few and provide global coverage services. Mobile Hosts (MH) access the global infrastructure through an HSP, but the entity that ensure that the user is enabled to access the network are the global ISPs. This scenario is scalable, business oriented and conceptually appealing, but it raises a large number of still unresolved problems that must be studied and experimented.

3 Research Areas

The main research topics that are addressed in the WILMA project concern planning, security, protocols and innovative services. They can be summarized as follows (not necessarily in importance order).

Network Management. The management of complex networks includes many aspects, but when wireless networks are involved, mutual trust, authentication, authorization and accounting (AAA) are the top-list priorities. In a network scenario as envisioned by WILMA, where HSPs and ISPs are different entities, the management becomes of paramount importance.
Similarly to the Virtual Operators proposed in [1], ISPs in WILMA manage users accessing the network through the HSPs. In its present phase the project provides for a single ISP, played by the researchers at DIT, while IRST and Alpikom act as HSPs only. The connectivity between HotSpots is provided by Alpikom directly or is obtained through IP tunneling. The second phase of the project is open to other institutions that would like to experiment this role.
As already mentioned pricing plays a key role in the success of provisioned networks. In order to be competitive, both ISPs and HSPs need simple and effective mechanisms to control the congestion and guarantee the QoS, while keeping the business profitable. Price-based policy for the access control can be a means to enforce QoS and profit, for instance in [2] we explore a policy that controls the hot spot traffic by dynamically determining the access cost as a function of the current load in the hot spot.

Location Aware Services. In wireless networks, the physical location of a user is an important parameter, whose utility ranges from routing and content-delivery optimization, to location-dependent service provisioning. The estimation of the position of a mobile host by measuring the strength of beacon radio signals received by the network interface is a novel method open to research, that may make location independent, in some cases, from satellite based systems like GPS [3,4].
The WILMA city-wide open lab offers the chance of studying the benefits of new value-added services on a large scale, unavailable to purely indoors

implementations. For instance, cooperation with public institutions enables the experimentation of systems that support human decisions both in leisure environments (tourism, recreation) and in critical contexts (emergency management), while other subjects will collaborate in testing context-aware applications that adapt their behavior to the user's location (e.g., home or work).

In the prospected framework, access points serve a double purpose: they provide connectivity, and at the same time they are used to determine the location of the mobile systems, offering to the user a context-dependent interface even without being actually connected to the network.

An example application is `PILGRIM` (Personal Item Locator and General Recommendation Index Manager) [5,6], a system that provides a list of interesting web links to users, where the measure of interest to a link for a particular user depends on his position. The system is based on the *collaborative filtering* paradigm: the interest metric depends on previous user actions, and no parameters need to be set by the system manager.

Security. Security means both encryption of data and dependable resource use, hence interacting deeply with the network management. The main difference between security related topics and AAA and network management-related problems is that security is concerned with the semantics of data, while AAA operations are normally not concerned with it.

One of the key points within this project regards the study of service differentiation based on security needs. Most of the multimedia traffic (voice and video) generally require a mild security enforcement (GSM-like encryption is generally sufficient). On the other hand, security requirements of data traffic may range from zero (e.g., downloading of the information regarding the monument in front of you) to applications whose requirements go well beyond secrecy and privacy, like for instance all credit card based transactions, where the user must also be granted about the generalities of the counterpart.

Network Planning. W-LANs are being deployed widely, but rarely with careful planning procedures. Access points (APs) are placed where needed, with the only constraint that power and network plugs must be close by. While 'optimal' for fast bootstrap, this strategy will hardly lead to well organized Hot Spots. Within the WILMA project several fundamental issues are studied. Among them:

- Optimal positioning of APs as a function of the environment and traffic requirements;
- Dimensioning of the wireless and wired infrastructure;
- Interference between neighboring and/or partially overlapped systems;
- Planning for localization, i.e., with the additional constraint that APs positions should provide a good structure of the signal strength to support the positioning algorithms.

Protocols and QoS Provisioning. A carefully engineered network requires some form of admission control in order to provide some degree of QoS, and suitable protocols to provide, implement and negotiate QoS and SLAs. A

broadband wireless network will also support multimedia communications, with their different QoS requirements. To guarantee different QoS requirements in a wireless network one needs suitable scheduling and admission control algorithms [7]. Besides, differentiation is needed for pricing and accounting purposes, as a function of the traffic types [8], thus also complementing the researches on pricing carried out in the framework of network management.
WILMA is studying and addressing these topics with an incremental and experimental approach, tackling problems as they arise in the experimental phase, though also fundamental research on admission techniques and scheduling is carried out within the project.

4 Ongoing and Future Work

WILMA Open Lab presently consists of a first HotSpot within the University of Trento, and a second HotSpot at ITC-IRST that will be interconnected to the network management soon. In the next few months HotSpots scattered in the Trento area will be connected through the Alpikom infrastructure, supporting a seamless Open Lab where services are experimented, and data supporting research and incremental service deployment are collected and elaborated. Meanwhile, smaller scale experiments will start in selected areas, such as Trento Library, providing free public access to the Internet and allowing the collection of important measures and informations about user needs and habits.

One of the initial goals of the project is the experimentation of Open Software solutions to implement the functionalities identified above, both using and customizing existing tools and developing new ones.

Service differentiation experiments are medium term goals, allowing the study of viable business models that ensure both free (or very cheap) access to basic services and income-providing value added ones.

The long term goal is handling the infrastructure used for research and experiments, together with the expertise accumulated during the project, to private enterprises, thus paving the road for the widespread introduction of nomadic services based on standard and open architectures.

References

1. J. Zhang, J. Li, S. Weinstein, N. Tu, "Virtual Operator based AAA in Wireless LAN Hot Spots with Ad-hoc Networking Support," *Mobile Computing and Communication Review*, Vol. 6, N. 3, pp. 10–21, Jul. 2002
2. R. Battiti, M. Conti, E. Gregori, M. Sabel, "Price-based Congestion-Control in Wi-Fi Hot Spots," *Proc. WiOpt'03*, INRIA Sophia-Antipolis, France, Mar. 3–5, 2003,
3. M. Brunato, Cs. Kiss Kalló, "Transparent Location Fingerprinting for Wireless Services," *Proc. of Med-Hoc-Net 2002*, Cagliari, Italy, 2002
4. R. Battiti, Th. Le Nhat, A. Villani, "Location-Aware Computing: a Neural Network Model for Determining Location in Wireless LANs," *Technical Report DIT-02-0083*, Dipartimento di Informatica e Telecomunicazioni, Università di Trento, Oct. 2002

5. M. Brunato, R. Battiti, "PILGRIM: A Location Broker and Mobility-Aware Recommendation System," *Proc. of IEEE PerCom 2003, First IEEE Annual Conference on Pervasive Computing and Communications*, Fort Worth, TX, USA, Mar. 2003
6. M. Brunato, R. Battiti, "A Location-Dependent Recommender System for the Web," *Proc. of MobEA 2003 Workshop*, Budapest, Hungary, May 20 2003
7. S.K. Das, R. Jayaram, N.K. Kakani, S.K. Sen, "A Call Admission and Control Scheme for Quality-of-Service (QoS) Provisioning in Next Generation Wireless Networks," *ACM/Baltzer Journal on Wireless Networks (WINET)*, Vol. 6, pp. 17–30, 2000
8. G.V. Záruba, I. Chlamtac, S.K. Das, "A Prioritized Real-Time Wireless Call Degradation Framework for Optimal Call Mix Selection," *ACM/Baltzer Journal on Mobile Networks and Applications (MONET)*, Vol. 7, No. 2, pp. 143–151, 2002

URLs and Web Sites

9. The Wilma Project – `http://www.wilmaproject.org`
10. Istituto Trentino di Cultura; Istituto per la Ricerca Scientifica e Tecnologica `http://www.itc.it/IRST`
11. Dipartimento di Informatica e Telecomunicazioni (DIT), Università di Trento `http://www.dit.unitn.it`
12. Alpikom S.p.A, Conectiong & COM, Trento, Italy `http://www.alpikom.it/index.html`
13. Megabeam Italia S.p.A., Milano, Italia – `http://www.megabeam.it`
14. Telia HomeRun, Telia Mobile, Telia, SE – `http://www.homerun.telia.com`
15. Boingo Wireless, Inc., USA – `http://www.boingo.com/`
16. Stockholm Open Net, Stockholm, SE – `http://www.stockholmopen.net`

MobileMAN: Mobile Metropolitan Ad Hoc Networks*

Marco Conti[1], Silvia Giordano[2], Gaia Maselli[1], and Giovanni Turi[1]

[1] IIT Institute, CNR, Via G. Moruzzi 1, 56124 Pisa, Italy
{marco.conti,gaia.maselli,giovanni.turi}@iit.cnr.it
[2] SUPSI, Switzerland (CH)
silvia.giordano@die.supsi.ch

Abstract. MobileMAN is a project funded by the Future and Emerging Technologies arm of the IST Programme of the the European Commission. This project investigates the potentialities of the Mobile Ad hoc NETworks (MANET's) paradigm. Specifically, the project aims at defining and developing a metropolitan area, self-organizing, and totally wireless network, called Mobile Metropolitan Ad hoc Network (MobileMAN). The main technical output of this proposal can be summarized as follow. i) Development, validation, implementation and testing of the architecture, and related protocols, for configuring and managing a MobileMAN. ii) Physical implementation of this architecture for lowers layers (i.e., wireless technologies). iii) Integration of applications on top of our self-organized network. iv) Validation of the self-organizing paradigm from the social and economic standpoint.

1 Introduction

The future information society is expected to rely heavily on wireless technology. Advances in wireless communication will enable a radical new communication paradigm based on self-organized information.

This new networking environment is composed by a group of wireless devices, called nodes, that must cooperatively realize the functionalities usually provided by the network infrastructure (e.g. routing, forwarding, network management). Such systems are known as mobile ad hoc networks [1], or infrastructureless networks. The Internet Engineering Task Force community refers to them as MANETs [2]. The MobileMAN project [3] will advance the MANET paradigm focusing on communications in a (possibly large) metropolitan area, called Mobile Metropolitan Ad Hoc Network (MobileMAN). A MobileMAN is an autonomous, self-organized, wireless network, solely composed by wireless nodes like, for example, next generation mobile phones, emerging Personal Digital Assistants (PDAs), and laptops equipped with wireless interfaces.

* This work was partially funded by the Information Society Technologies programme of the European Commission, Future and Emerging Technologies under the IST-2001-38113 MOBILEMAN project.

M. Conti et al. (Eds.): PWC 2003, LNCS 2775, pp. 169–174, 2003.

Table 1. Project partners

Partner	Country	Research areas
IIT-CNR	Italy	Wi-Fi, Networking, Transport, Middleware
EURECOM	France	Cooperation, Security
UCAM-CLAB	United Kingdom	Middleware, Economic Issues
HUT	Finland	Networking
NETikos	Italy	Applications
SUPSI	Switzerland	Wi-Fi, Cooperation, Social Aspects

In a MobileMAN the lack of centralized points leads to the necessity of distributing all the network functions to the available nodes, that must cooperate and provide services to each other. While adjacent nodes communicate directly, far away nodes rely on intermediate nodes acting as relays.

A MobileMAN is not intended as a replacement for current infrastructure based (wired and wireless) networks. It is intended to complement them, and to enable new application scenarios in which a centralized infrastructure is impossible, undesirable, or unnecessary. In addition, we also envisage intercommunication between a MobileMAN and the Internet. This can be provided by nodes which have simultaneous access to the MobileMAN and to the Internet, perhaps via special access points, or via user terminals with multiple network interfaces (e.g., a PDA equipped with both Bluetooth and GPRS interfaces). Applications supported by a MobileMAN might range from text messaging systems (evolving from the very popular SMS), up to more demanding multimedia (voice and video) services. Eventually, they might even range from simple interactive games to ones that truly merge real, and virtual worlds.

The project objectives are twofold. On the *technical* aspect, the project aims at inventing and demonstrating solutions that make the self-organisation paradigm effective. On the *business and social* aspect, it aims at exploiting the self-organisation paradigm for supporting innovative applications, which will improve people's living standards.

The main expected technical output of MobileMAN regards the development, validation, implementation, and testing of the MobileMAN architecture. Another expected result is the integration and validation of popular services (such as SMS and chatting) on top of our self-organised network, as well as, the extensions of these services into new realms, such as multimedia messaging, spontaneous electronic collaboration, and wireless interactive games. Finally, a socio-economic evaluation of the effectiveness of the mobile ad hoc paradigm, and its market access will be provided.

Table 1 lists all the project participants, with the associated main research areas within the MobileMAN project framework.

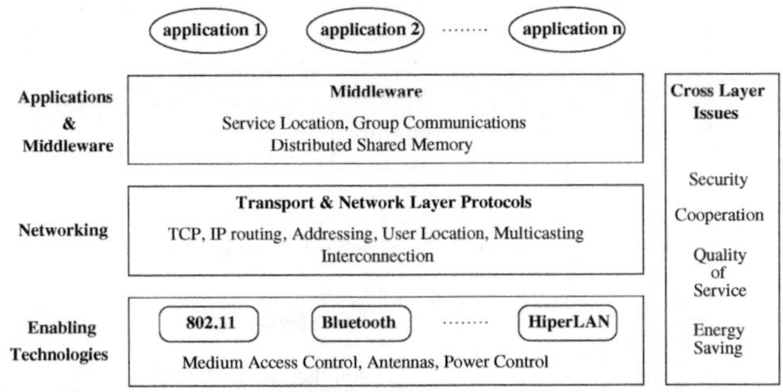

Fig. 1. MANET reference architecture.

2 MANET Architecture

Classical networking architectures layer protocols and network responsibilities. Even the MANET architecture groups the ad hoc networking activities into three main areas: enabling technologies, networking, middleware and applications (see Figure 1). This layered approach leads the research efforts to target mainly isolated components of the overall network design (e.g., routing, MAC, power control). In this model, each layer in the protocol stack is designed and operated separately, with interfaces between layers that are static and independent of the individual network constraints and applications. However, as shown in Figure 1, in a MANET some functionalities cannot be assigned to a single layer. For example, energy management, security and cooperation, quality of service, cannot be completely implemented in a single layer, but they are implemented by combining and exploiting mechanisms of all layers.

The layered approach has proved to be effective in wired networks where the conditions are stable. In this case, it guarantees for example that the TCP/IP protocol stack can operate on a large set of heterogeneous technologies. This principle does not apply anymore when wireless technologies are included in the Internet world (see the problems related to TCP over wireless links [4]) and becomes less efficient when ad hoc networks are taken into consideration. Without any doubt, efficient solutions for ad hoc networks require a more strict cooperation among protocols belonging to different layers.

Specifically, we envisage a complete loop in the design/behavior of protocols belonging to different layers. Low layer protocols will be designed taking into account the requirements/characteristics of high-layer protocols. In addition, high layer protocols will react to signalling generated by low layer protocols, as well as low layer protocols will adapt their behavior according to the information collected above them. We think that, in ad hoc networks, efficient solutions can be designed by further strengthening cross-layer interdependencies. For example, adopting multi-path forwarding, TCP acknowledgements can be used to estimate the reliability of different paths. In the following we present the cross-layered MobileMAN architecture.

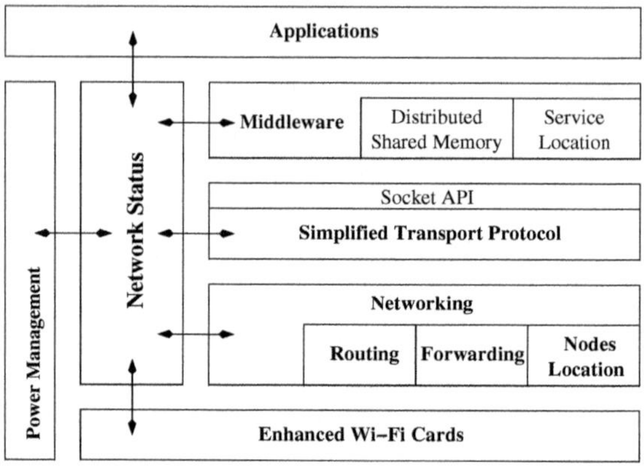

Fig. 2. MobileMAN architecture.

3 Cross-Layered MobileMAN Architecture

Information sharing avoids duplicate efforts for collecting data about the network status. For example, in [5] a strict cross-layer design principle is applied to the routing and the middleware layers, to facilitate multimedia data accessibility for various applications at the end-systems. Active communication between the layers is achieved through signalling, regarding the condition of data transmission.

A cross-layer protocol design supports adaptivity across multiple layers of the protocol stack. The different timescales of the network variations suggest that each layer should attempt to compensate for variation at that level. If adapting locally is unsuccessful, information should be exchanged with other layers for a more general response. For example, let us assume that a weak link occurs on a multi-hop path. Firstly, the physical link protocol can try to fix the problem by increasing the transmission power. However, if the link weakness is caused by some physical obstacle, it is better for a higher layer to respond by, for example, delaying packet transmissions (e.g., until the mobile node leaves the tunnel) or by identifying an alternative route. Similarly, if nodes in the network are highly mobile, link characteristics and network topology will change rapidly. Informing the network layer of highly mobile nodes might change the routing strategy from unicast to broadcast in the direction of the intended user. This integrated approach to adaptive networking is the biggest challenge in adaptive protocol design. The above examples support a cross layering design, as it is emerging in the framework of the MobileMAN project [6]. The kernel of a MobileMAN network node (see Figure 2) is a shared memory, called *Network Status* that is a repository of information collected by network protocols at all layers. Each protocol can access this shared memory to read and write collected information. Hence, layers efforts in duplicating network-status information are minimized, facilitating inter-layer cooperation.

Hereafter we discuss the protocol stack for the ad hoc node, with special attention to exploiting cross-layer interactions.

Wi-Fi. A physical implementation of an enhanced IEEE 802.11 wireless technology will be done by improving the existing standard, dealing with the multi-hop organization of the MobileMAN network [7]. Specifically, the cross-leyering functionality will be exploited importing a 2-3 hops horizon knowledge of the neighborhood, built up at the routing layer. This will allow a fair channel scheduling that could take into account the hidden and exposed node issues.

Routing, forwarding, and nodes location. Our claim is to consider the routing problem according to the cross-layering principle, so that routing data is useful also at other layers. For example, the knowledge of node locations can be useful for identifying the closer node that is implementing a service, without requiring the middleware to discover the same location once more. From these observations, our reference choice for routing is based on a link-state protocol. According to the Fisheye policy [8], a node propagates link state information to other nodes in the network, basing on how far away (in terms of hops number) they are. Closer nodes will exchange link state information more frequently than far away nodes. The primary consequence of this choice is that a node has a precise view of the 2-3 hops network around it, and each node has the knowledge of all the possible paths to a destination. These features help in implementing mechanisms of multi-path reliable forwarding [9], and of security/cooperation enforcing [10,11,12,13]. Considering a possibly large MobileMAN network, the 2-3 hops network knowledge mentioned above is not sufficient to route packets between far away nodes. In this case, a node location service [14] with a geographic forwarding procedure become necessary.

Simplified Transport Protocol. The main goals in the design of a Simplified Transport Protocol, are to provide a reliable and connection-oriented service that minimizes useless data re-transmissions by analysing and reacting appropriately to the different phenomena happening at the below layers (e.g., route failures, route changes, congestions) [15]. For this purpose, the Simplified Transport Protocol could exploit information reported by the routing and Wi-Fi layers in the Network Status.

Middleware. The middleware layer generally provides context abstractions, able to hide complex details to application programmers. In mobile ad hoc environments, this trend has to be reversed to context-awareness. The MobileMAN cross-layered architecture aims at supporting this aspect. Finally, recent proposals [17,18] emphasise data sharing among nodes and support for disconnected operations. To achieve this, concepts like tuple spaces, mobile agents, and event based communications have been adopted. While these technologies provide advanced abstractions, efficient implementations are still open issues. Specifically, solutions must be devised to support a distributed shared memory, service discovery, and group communication.

References

1. I. Chlamtac, M. Conti, and J. Liu. Mobile Ad Hoc Networking: Imperatives and Challenges. In *Ad Hoc Networks*, No. 1(1) 2003.
2. http://www.ietf.org/html.charters/manet-charter.html
3. http://cnd.iit.cnr.it/mobileMAN
4. H. Elaarag. Improving TCP Performance over Mobile Networks. *ACM Computing Surveys*, Vol. 34, No. 3, pp. 357–374, Sept. 2002.
5. K. Chen, S.H. Shah, and K. Nahrstedt. Cross-Layer Design for Data Accessibility in Mobile Ad Hoc Networks. In *Wireless Personal Communications* No. 21, pp. 49–76, 2002.
6. M. Conti. Cross Layering in MANETs' Design: the MOBILEMAN Approach. *The first NSF – COST Workshop on Networking*, Chania, Crete, June 2003.
7. R. Bernasconi, I. Defilippis, S. Giordano, and A. Puiatti. An enhanced MAC architecture for multi-hop wireless networks. To appear in *Proceedings of the Personal Wireless Communications Conference 2003 (PWC 2003)* , Sept. 2003.
8. G. Pei, M. Gerla, and T.W. Chen. Fisheye State Routing in Mobile Ad Hoc Networks. In *Proceedings of the 2000 ICDCS Workshops*, Taipei, Taiwan, Apr. 2000.
9. M. Conti, E. Gregori, and G. Maselli. Towards Reliable Forwarding for Ad Hoc Networks. To appear in *Proceedings of the Personal Wireless Communications Conference 2003 (PWC 2003)*, Sept. 2003.
10. S. Giordano, and A. Urpi. Self-Organized and Cooperative Ad Hoc Networking. In *Mobile Ad Hoc Networking*, S. Basagni, M. Conti, S. Giordano, I. Stojmenovic (Editors), IEEE Press and John Wiley and Sons, Inc., New York, 2003.
11. P. Michiardi and R. Molva. A game theoretical approach to evaluate cooperation enforcement mechanisms in mobile ad hoc networks. In *Proceedings of WiOpt '03*, Sophia-Antipolis, March 2003.
12. A. Urpi, M.A. Bonuccelli and S. Giordano. Modelling Cooperation in Mobile Ad Hoc Networks: a Formal Description of Selfishness. In *Proceedings of WiOpt '03*, Sophia-Antipolis, March 2003.
13. J. Crowcroft, R. Gibbens, F. Kelly and S. Östring. Modelling Incentives for Collaboration in Mobile Ad Hoc Networks. In *Proceedings of WiOpt '03*, Sophia-Antipolis, March 2003.
14. G. Turi. A Context-Aware Location Service for Metropolitan Ad Hoc Networks. To appear in *Proceedings of the Personal Wireless Communications Conference 2003 (PWC 2003)*, Sept. 2003.
15. G. Anastasi, and A. Passarella. Towards a Novel Transport Protocol for Ad Hoc Networks. To appear in *Proceedings of the Personal Wireless Communications Conference 2003 (PWC 2003)*, Sept. 2003.
16. C. Mascolo, L. Capra , and W. Emmerich. Middleware for Mobile Computing (A Survey). In *Networking 2002 Tutorial Papers*, editors E. Gregori, G. Anastasi, S. Basagni. Springer. LNCS 2497.
17. Amy L. Murphy, Gian Pietro Picco, and Gruia-Catalin Roman. Lime: A Middleware for Physical and Logical Mobility. In *Proceedings of the 21st International Conference on Distributed Computing Systems (ICDCS-21)*, May 2001.
18. C. Mascolo, L. Capra, S. Zachariadis, and W. Emmerich. XMIDDLE: A Data-Sharing Middleware for Mobile Computing. *Int. Journal on Personal and Wireless Communications.*

Wireless Networking for Virtual Immersive COMmunications: The VICOM Project

F. Davide[1], A. Detti[2], E. Gregori[3], and F. Vatalaro[4]

[1] Telecom Italia Learning Services
Rome, Italy
Fabrizio.Davide@telecomitalia.it
[2] CNIT – University of Rome "Tor Vergata",
Italy
andrea.detti@uniroma2.it
[3] IIT Institute of the Italian National Council,
Pisa, Italy
enrico.gregori@iit.cnr.it
[4] University of "Rome Tor Vergata",
Electronic Dept., Italy
vatalaro@ing.uniroma2.it

Abstract. The paper describes the Virtual Immersive Communication (VICOM) project, and reports the state of art with respect to the wireless networking issues. The project vertically exploits the enabling technologies for Immersive communications enriched with virtual reality components. The research topics are common driven toward the final project goal, consisting in proof of concept demonstrators: Mobility in Immersive Environments (MIE) and Virtual Immersive Learning (VIL).

1 Introduction

The increasing relevant role of multimedia mobile communications, joined with the emerging technologies on audio, video and sensing interfaces, makes challenge their integration, in order to realize novel form of human interactions at distance. Thus it is possible to create a sense of being in the remote place and to make natural, simple and continuous the interactions with the ambient. The first characteristic is often referred as *telepresence*, while the other is referred *immersive service*.

With this regard, some building blocks necessary to realize these services, consist in: i) multi-modal service interfaces aware of the user context; ii) wireless and sensor networks; iii) middleware platforms for mobile applications.

The multi modal interfaces allow the user to interact with the system in natural ways (e.g., voice, gesture, etc.); moreover, they output such realistic multimedia contents that create the sense of telepresence. The awareness of the user context saves in the retrieval of information from the user, therefore simplifies the service accessibility [2]; the previous ones joined with the continuous opportunity in dialing with the system, give to the user the sensation of the continuous presence of the service around him; i.e. "immersivity".

M. Conti et al. (Eds.): PWC 2003, LNCS 2775, pp. 175–181, 2003.

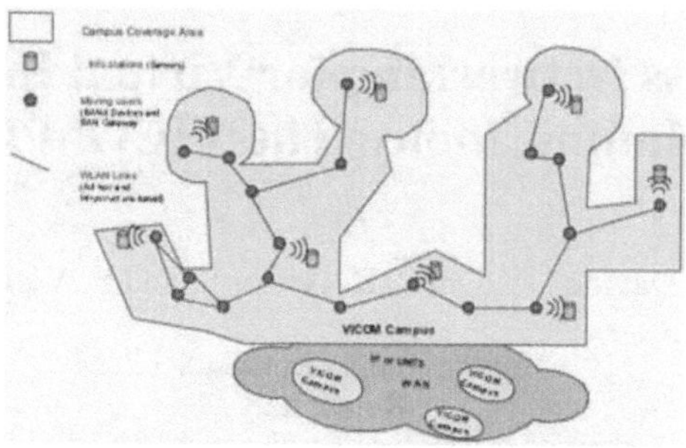

Fig. 1. VICOM network reference scenario

In these research areas, some italian partners, (universities, governemnet research centers and private companies), started the Virtual Immersive COMmunication Project (VICOM)[1] [1]. The project goal is to study and development solutions for *ambient intelligence* systems and to prepare two demonstrative scenarios. These scenarios are named Mobility in Immersive Environment (MIE) and Virtual Immersive Learning (VIL), and are aimed at stressing complementary services aspects; respectively, "*immersivity*" and "*telepresence*".

To provide the immersivity functionalities, MIE scenario mainly exploits advanced mobility issues. It is assumed that the user wears a set of devices forming a body area network (BAN). One of these performs the gateway functions towards the surrounding Personal Area Network (PAN) and with Wireless Local Area Network (WLAN). In order to provide a seamless connectivity in different PAN/WLAN environment, the BAN gateway is able to reconfigure its low layers (up to the data link), via software radio mechanisms. These services will provided within an area of limited coverage, e.g., a campus. Communications between BANs is obtained both with access base stations (infrastructure-based WLAN) and with ad hoc solutions, see Fig. 1.

The resulting service coverage defines the VICOM Campus. Within the campus, a set of sensors provides the user contextual. Finally, the campuses are together interconnected by means of IP Wide Area Network (WAN).

VIL scenario foresees that the users join a tele-training session, enriched by virtual reality contents, and exploit the telepresence concept.

In the next sessions, we report the network topics dealt within the VICOM project in order to support the proposed demonstrators.

2 VICOM Networking: State of Art

In both scenarios, we have to integrate the several wireless networking technologies (for BAN, WLAN end Sensors) to provide connectivity among user within the room

---

[1] The VICOM Project is funded by the Italian Ministry of Instruction University and Research (MIUR). It is three years long, and it is started in the end of 2002. The partnership is formed by: the CNIT universities consortium; the Pisa and Bologna units of the National Research Center (CNR); the "Politecnico di Milano"; the ISCTI national institute and Telecom Italia Learning Services.

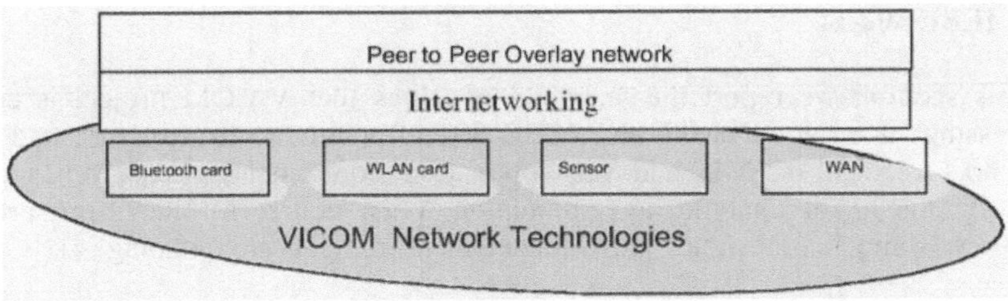

Fig. 2. Sketch of VICOM protocol architecture of the fixed network devices (WAN), of the gateway sensor devices, of the BAN gateway (Bluetooth and WLAN IEEE 802.11)

or the campus and connectivity with remote user via Internet; anyway, the different scenario requirements split the research directions. Due to the wideness of service coverage, in the MIE scenario the focus is placed on solution for mobility as Quality of Service is not critical. On the other hand in the VIL scenario we have the opposite situation: Quality of Service management is a strong requirement (we have high bandwidth requirement from the application) while mobility is not critical due to the limited coverage of the services (e.g., an equipped room).

The target of the VICOM project is to set up two proof-of-concept demonstrators; hence, the research topics are strongly tied with the off-the-shelf network technologies. The networking research is focused on the Bluetooth technology for the BAN/PAN, on 802.11 for the PAN/WLAN and on a not yet defined technology for sensor networks. As consequence, from a technology point of view, VICOM faces internetworking problems with four network types (sensor, Bluetooth, IEEE 802.11 and WAN). The networking and QoS strategies within the specific area will be customized according to the technology features and employment (Fig. 2). A coarse customization, instead, will be performed at the transport layer, where the issues coming out for the TCP unsuitability with the wireless link will be probably solved in a technology independent fashion, by developing a "wireless TCP" suitable both for Bluetooth and IEEE 802.11 and whenever needed adopting indirect-TCP solutions. Nevertheless, for those devices that do not require to widely be accessible (such as the membership of a BAN with the exclusion of the BAN gateway), it is not sure that those IP compliant transport and routing techniques will be necessary, and so other possible technology-specific protocols "non-ip compliant" have to be taken into account.

At least for the wireless gateway, we envisage a possible protocol differentiation up to the transport level that requires some bridging layer functions to limit the negative impact of wireless technologies on the need-to-end performances. For this reason we will introduce some proxy nodes in our networks. In addition we need some middleware services. The most important is the many to many multimedia communications, which is a basic requirement of cooperative work applications, that is obtained constructing a peer-to-peer overlay network.

In the following sections, we report the VICOM research lines more tied with the wireless networking issues and describe the topics of the current research.

2.1 IEEE 802.11

In this section, we report the technical novelties that VICOM project is currently addressing in order to enable the IEEE 802.11 wireless infrastructure and ad hoc networks to fulfill the MIE and VIL application requirements. These topics are dealt both by means of analytic and simulation tools; hence, for the more promising solutions, some "on-the-field" performance measurements are on going [4].

2.1.1 Ad hoc Routing Protocols

For the applications envisioned in the VICOM scenarios, and more specifically for MIE, it is important to have mechanisms, which make it possible to establish data communications even beyond the coverage of the fixed infrastructure. It is therefore appealing to use the ad hoc networking technology, where nodes flexibly collaborate in delivering the information to the final destination. Due to its natural broadcast characteristics, the IEEE 802.11 is a promising and available technology to realize an ad hoc network. With this regard, a VICOM research line is currently focused on routing in ad hoc networks, with special effort on energy efficiency of the considered schemes [5] and in contrast with the intuitive goal of choosing the shortest paths, actually accomplished by the well know AODV and DRS ad hoc routing protocols.

2.1.2 Quality of Services Mechanisms

As previously stated, most of the VICOM services require the delivery of multimedia data, like streaming video and audio. This feature needs the support of time-bounded services by the infrastructure or by ad hoc network. Nevertheless, the legacy IEEE 802.11 standard was designed to support only best-effort services. Therefore, a careful analysis is coming on to verify the applicability of two mechanisms, both discussed in the latest version of the 802.11e draft standard: the Enhanced Distributed Coordination Function (EDCF) and the Hybrid Coordination Function (HCF).

2.2 Bluetooth

Although IEEE 802.11 technology furnish higher bandwidth respect to the Bluetooth (BT) one and so, it appears as more suitable for multimedia applications, we have to take in mind that IEEE 802.11 drains a significant amount of energy and processing resources often not available on small devices as PDA [6]. On the contrary, Bluetooth owns an high electronic integrability, a low power drain and a low device processing requirement being a lot of the computation directly implemented on the radio chip; that is more or less witnessed by the market penetration of Bluetooth in small and low performance devices such as PDA, mobile phones and earphones. For these reason the Bluetooth technology is the more promising candidate for the BAN, but it has to be considered also in the realization of "simple" device PAN, with or without infrastructure.

In the following, we report the main Bluetooth research topics currently dealt in VICOM project.

2.2.1 Scatternet Formation Algorithms

While the ad hoc communication are quite natural in IEEE 802.11, this doesn't occurs in Bluetooth; hence, suitable scattered formation protocols have to be developed [7]. Both scatternet formation protocols that require the radio vicinity of all nodes, for BAN and infrastructured PAN, and protocols working in the more general multi-hop scenario, for ad hoc PAN, are taken into account. Moreover, major focus is placed on distributed techniques, in the sense that the protocols are executed at each node with limited knowledge of the surrounding topology.

2.2.2 Handover and Fast Connection Restablishment

Most of Bluetooh communications are based on L2CAP connections, unlike with the connectionless nature of IEEE 802.11 MAC layer. This difference, joined with the slowness of the inquiry procedure [8], leads the need for the Bluetooth research line to investigate on fast and reliable handover and in fast connection reestablishment protocols. The former may be accomplished for those devices that sojourn within the radio coverage during the movement and are able to simultaneously belong to two or more piconet (i.e. are able to perform the scatternet functionalities); hence, they can hand over from one piconet to the other without breaking the connectivity. In the other cases, e.g. in presence of shadowing or of device not quite able to manage advanced scatternet functionality, fast connection restablishment mechanisms may be candidate to avoid the high layer protocol timeout (e.g. TCP, HTTP and so on).

2.2.3 Ad hoc Routing Protocols

As done in the IEEE 802.11 research line, Ad hoc Routing Protocols that take into account of Bluetooth technology characteristics are dealt in VICOM project. Currently, the major focus is based on the optimization of the following protocols: the Broadcast Incremental Power (BIP), the Broadcast least unicastcost (BLU) and the Broadcast Link-based MST (BLiMST) algorithm.

2.3 Sensor Networks

During the description of VICOM scenario, we show the opportunity in using sensor networks to retrieves the relevant contextual information. Sensor networks are inherently resource poor, i.e., they are fundamentally constrained in terms of CPU power, energy, memory. This calls for new design in routing techniques since, e.g., complicated routing protocols cannot be loaded onto sensor nodes due to the lack of memory to store the code and possibly due to the excessive amount of computations they may involve. Also, storing locally large routing tables is impractical and may not be feasible due to memory constraints [9]. As consequence, the current VICOM research on sensor networks are focused on energy efficient routing protocols that differs form the Bluetooth and IEEE 802.11 ones in terms of lower complexity to the detriment of routing performance, which, however may be acceptable due to the small network load.

2.4 Upper Layer Protocol

Above the routing layer, the transport and the overlay protocols form what we call upper layer protocols. These protocols may support the applications layer, both directly or through the middleware level.

2.4.1 Wireless TCP

Wireless links are characterized by a bit error rate, which is much higher than in wired links, and segment losses may occur due to link errors rather than congestion [10]. The transmission rate decreases caused by segment losses due to link errors are unnecessary and only result in a decrease of the throughput performance. This is the major cause of the scarce TCP efficiency in all wireless communication environments. VICOM research are facing this issues with two approaches: i) adaptation of TCP congestion control algorithm to the wireless environment; ii) decrease in the wireless link bit error rate through the use of appropriate error correcting schemes and power management.

2.4.2 Peer to Peer Overlay Networking

Several VICOM applications are inherently many to many and require different performance constrains. Due to the limited processing capacities of the ad hoc network devices, it appears not much practical to implement a lot of routing enhancement, which optimally solve the specific applications requirements. Starting from this consideration, it is foreseeable that the network layer of devices will offer basic unicast routing functionalities (and in the best case, some coarse QoS mechanisms), while on-demand implementing the routing enhancements only for the application that it is currently running. Those enhancements are tightly connected with the application need, therefore their natural place on the protocol architecture is between the transport and the application layer. We refer this intermediate layer as overlay.

With this regard, VICOM research is focused on mechanisms for setup suitable peer to peer overlay networks facing with the many to many communication involved in collaborative work applications. Those mechanisms will take into account of the contextual information regarding the network and the application itself. Currently, overlay network solutions for MPEG 4 streaming over ad hoc network are under investigation.

References

[1] www.vicom-project.it

[2] A. Detti, F. Davide, G. Cortese, "eCASA: an Easy Context Aware System Architecture", to appear in Proc. *IEEE Vehicular Technology Conference 2003*

[3] R. Bruno, "802.11", VICOM Project, *Technical report of first semester of WP4 activity*, section 1.1, www.vicom-project.it

[4] G. Anastasi, E. Borgia, M. Conti, E. Gregori, "IEEE 802.11 Ad Hoc Network: Perfornamce Measurements", *IIT Internal Report*, January 2003

[5] L. Badia, M. Zorzi, "On the Construction of Broadcast and Multicast Trees in Wireless Networks – Global vs. Local Energy Efficiency," in Proc. *IEEE GLOBECOM2002*, Taiwan, November 2002

[6] F. Cuomo, "Bluetooth", VICOM Project, *Technical report of first semester of WP4 activity*, section 2.1, www.vicom-project.it

[7] S. Basagni, R. Bruno, and C. Petrioli, "A performance comparison of scatternet formation protocols for networks of Bluetooth devices," in Proc. *IEEE International Conference on Pervasive Computing and Communications*, PerCom 2003, Forth Worth, TX, March 23–26 2003, pp. 341–350.

[8] A. Detti, P. Loreti, F. Mazenga, F. Vatalaro, "Performance Analysis of Bluetooth Inquiry and Page Procedures", in Proc. *European Conference on Wireless Technology 2002 (ECWT2002)*, 26–27 September – Milan

[9] M. Zorzi, "Sensor Networks" VICOM Project, *Technical report of first semester of WP4 activity*, section 1.2, www.vicom-project.it

[10] G. Morabito, "End-to-End Protocols" VICOM Project, *Technical report of first semester of WP4 activity*, section 1.3, www.vicom-project.it

DARWIN: Demonstrator for an Adaptive and Reconfigurable Wireless IP Network*

Guillem Femenias, Josep Lluis Ferrer, and Loren Carrasco

Departament de Matemàtiques i Informàtica, Universitat de les Illes Balears
Cra. de Valldemossa km. 7,5, Palma de Mallorca 07122, Spain
guillem.femenias@uib.es, dijjfg@tomir.uib.es, loren@ipc4.uib.es

Abstract. Project DARWIN is working towards the specification and demonstration of a flexible broadband WLAN radio access system based on an IP network platform. In a sense, DARWIN can be considered as an evolution of IEEE 802.11 and HIPERLAN/2. This paper presents the main innovative ideas behind this project, that can be summarized as: a flexible network architecture providing extended range ad-hoc and infrastructure modes of operation, enhanced air interface (low-density parity check (LDPC) codes, adaptive modulation, smart antennas combined with space-time block codes), enhanced wireless link and data link control functions, and an enhanced security architecture.

1 Introduction

Within the Spanish ICT (Information and Communications Technology) initiative, project DARWIN is working towards the definition and demonstration of a flexible broadband WLAN radio access system based on an IP network platform. DARWIN systems are intended to provide short range, wireless access to multi-media services over IP, operating in the 5 GHz band and providing bit rates up to 100 Mbps. In a sense, DARWIN can be considered as an evolution of IEEE 802.11 and HIPERLAN/2 and, as such, the main innovative ideas behind it can be summarized as:

- a flexible network architecture providing extended range ad-hoc and infrastructure modes of operation;
- enhanced air interface, through the use of low-density parity check (LDPC) codes, adaptive modulation algorithms, smart antennas at both transmitter and receiver combined with space-time block codes, etc;
- enhanced wireless and data link control functions that fulfil the requirements on Quality of Service (QoS) and IP support dictated by upper layers; and
- an enhanced security architecture.

In the first phase of the project specifications of the different system layers have been generated. In a second phase the functionality of the system is being

* This work has been supported in part by the Ministerio de Educación y Ciencia, Spain, and FEDER (Fondo Europeo de Desarrollo Regional), under grant TIC2001-0287.

M. Conti et al. (Eds.): PWC 2003, LNCS 2775, pp. 182–187, 2003.

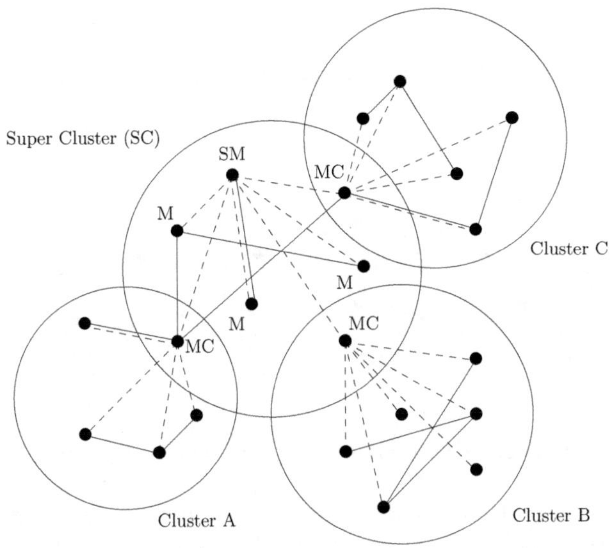

Fig. 1. DARWIN subnet

tested through the use of simulation tools. The third phase of the project is intended to provide as a result a prototype implementation of the system.

2 Flexible Network Architecture

One of the main features of DARWIN is its flexibility in network configuration. A DARWIN network can be defined as a group of DARWIN devices which have in common the same NET-ID. Within a DARWIN network we can define a subnet as a set of synchronized DARWIN devices using the same carrier frequency. A subnet is created when a Darwin Super Master (DSM) starts to generate valid Broadcast Control CHannel (BCCH), and allows other DARWIN devices to associate with the network. All devices of a subnet shall be synchronized to the frequency chosen by the DSM, and access the channel using the MAC frame structure dictated by the DSM. The selection of the DSM is dynamic, and seamless switching of the DSM responsibility from one DSM-capable wireless terminal to another is possible. Apart from the DSM, all the other devices in a subnet belong to one of the following hierarchical categories: Darwin Master (DM), Darwin Master of a Cluster (DMC) or Darwin Slave (DS). A DM is a device that is directly associated to the DSM and does not have any other device associated with it. A DMC is a DMC-capable device that is associated to the DSM and has other devices associated with it. A DS is a device that is associated to the DSM through the intermediation of a DMC. A DMC together with its associated DSs constitute a cluster and the DSM, the DMs and the DMCs of a subnet constitute a super-cluster.

The control plane of DARWIN is centralized for all general features. That is, only the DSM, either directly or through the mediation of a DMC, can instruct

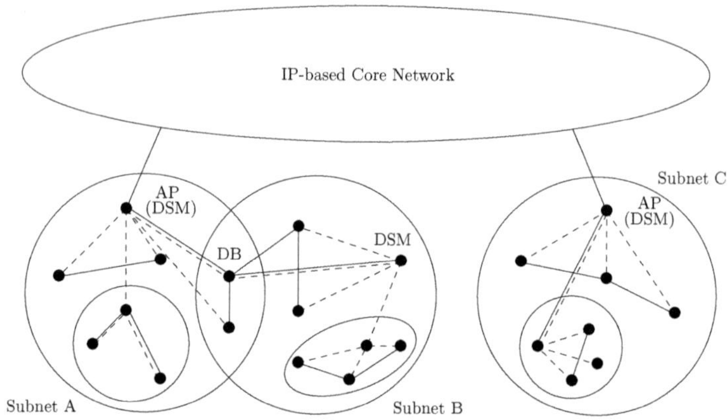

Fig. 2. General architecture of a DARWIN network

a DARWIN terminal to do something. However, distributed control is also made possible for some ad-hoc features by introducing logical control channels, which can be used for direct exchange of control messages between terminals. In the user plane, DARWIN makes extensive use of Direct Link (DiL) user connections. Direct Mode is used, provided that two wireless terminals can reach each other directly. A link quality calibration process helps to track the connectivity between any two devices by measuring the associated RF link quality. The DSM (or DMC) is used as a user data relay for a pair of terminals, provided that they cannot reach each other directly. Even this user data relaying is performed during the DiL phase between the wireless terminals and the DSM (or DMC).

Apart from providing connectivity between wireless terminals (ad-hoc network), DARWIN can also provide connectivity of wireless terminals to access points (AP) that bridge traffic to wired IP networks (infrastructure based network). In this case, the role of DSM will be played by the AP and DARWIN will work as a seamless extension of other networks, so wired IP network nodes will see DARWIN nodes as other network nodes.

In order to increase the capacity or the coverage area of a single subnet, multiple subnets *operating on different frequencies* can be deployed. Each subnet is under the control of its own DSM and works independently of the other subnets. Dynamic Frequency Selection (DFS) is used to enable dynamic selection of the RF channel. Different subnets can be interconnected either by a fixed network through the APs (infrastructure based network), or by DARWIN Bridges (DB) (ad-hoc network). Potentially any DB-capable DM device in a subnet can be configured as a bridge to another subnet.

3 Physical Layer Enhancements

DARWIN physical layer [1] is very similar to that adopted by IEEE 802.11 and HIPERLAN/2. Parameters like channel raster, OFDM subcarrier spacing, cyclic prefix duration, burst structure, ... coincide with those used in IEEE 802.11 and

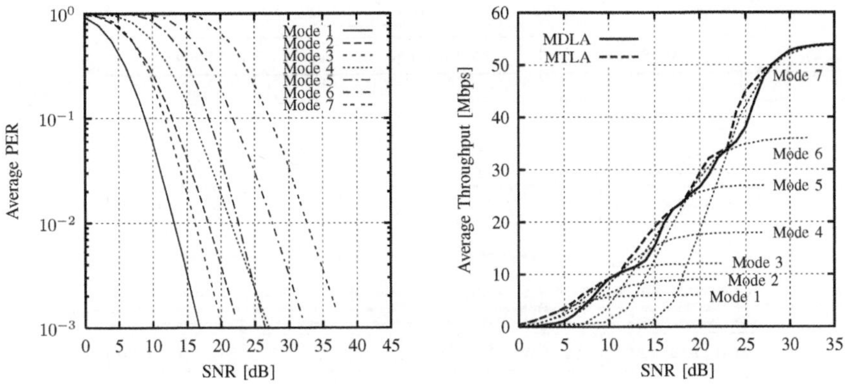

(a)

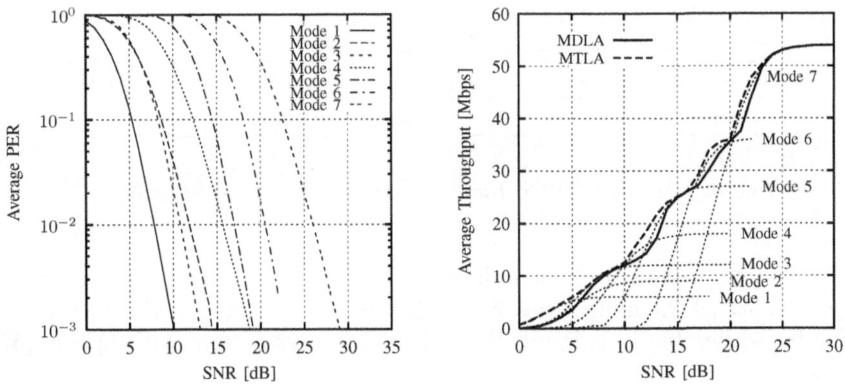

(b)

Fig. 3. Average PER and average throughput of (a) conventional (punctured ($R_c = 1/2, K = 7$) convolutional code) and (b) enhanced (punctured (864,432) irregular LDPC code combined with Alamouti's STBC) systems. It has been assumed the transmission of 54-bytes packets over ETSI BRAN Channel A. Two link adaption algorithms have been implemented: minimum delay (MDLA) and maximum throughput (MTLA).

HIPERLAN/2. However, in order to fulfil the requirements of QoS dictated by upper layers, the physical layer of DARWIN incorporates a series of enhancements with respect to IEEE 802.11 and HIPERLAN/2 systems. In particular, some of the techniques that are being investigated within DARWIN project are:

- use of adaptive coding strategies based on low-density parity check codes;
- use of adaptive modulation schemes;
- use of smart antennas at both transmitter and receiver combined with space-time block codes.

Preliminary results (see Fig. 3) show that these techniques can provide performance improvements in terms of packet error rate (PER) and throughput of the system. By providing a more reliable link, these enhancements result in a

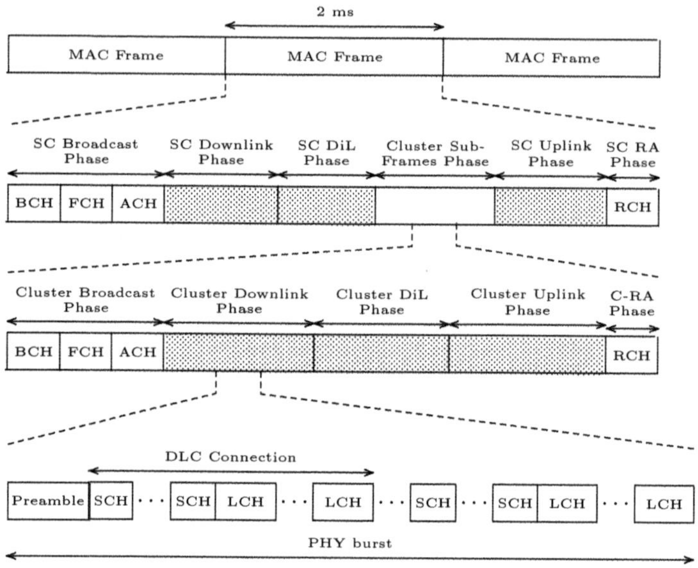

Fig. 4. Basic DARWIN frame structure (non-sectorized antennas).

significant improvement of the link control functions and, at the same time, contribute to one of the main objectives of DARWIN project, i. e. the consecution of an IP-based flexible broadband WLAN radio access system with QoS support.

4 Data Link Control Layer Enhancements

The DLC of DARWIN consists of three functional entities:

- Radio Link Control (RLC), responsible for the exchange of control functions' information between DSMs and DMs (or DMCs) and between DMCs and DSs. Supported DARWIN RLC control functions are: association control, radio resource control, DLC control, security management, QoS control, address management and configuration management.
- Error Control (EC), when applied to a user connection, is responsible for transmission error detection and recovery, and is based on ARQ.
- Medium Access Control (MAC), responsible for the implementation of DARWIN MAC protocol.

The DARWIN MAC protocol [2] is based on a dynamic TDMA/TDD scheme and its flexibility is, also, one of the main features of DARWIN project. Transmissions in downlink, uplink and direct link, are structured on the air interface as a sequence of MAC frames, each one of 2ms duration. Six major variable-length phases can be identified in a typical DARWIN MAC frame as follows:

- *Super Cluster Broadcast phase*, in which control information is broadcast from the DSM to all the terminals in the super cluster.

- *Super Cluster Downlink phase*, reserved for user control information and data transmission from the DSM to the terminals in the super cluster.
- *Super Cluster Direct Link phase*, reserved for user control and data direct transmissions between two terminals associated with the DSM and operating in direct mode.
- *Cluster Sub-frames phase*, in which DARWIN MCs are allocated a variable-duration time slot that is also divided in five phases: cluster broadcast phase, cluster downlink phase, cluster direct link phase, cluster uplink phase and cluster random access phase.
- *Super Cluster Uplink phase*, reserved for user control information and data transmission from the terminals in the super cluster to the DSM.
- *Super Cluster Random Access phase*, consisting of one or more random channels, in which random access is allowed for transmission of control information from DMs or DMCs to the DSM.

The DARWIN MAC structure is able to extend the coverage of a DSM providing a 'relay' function for terminals that are outside the range of a DSM. The gain in communication range is paid with a degradation in system capacity. Planning strategies that result in an acceptable trade-off between these parameters are now being investigated, as well as powerful inter-layer scheduling structures that achieve a coupling between IP and DLC, providing QoS support.

5 Security Services

DARWIN incorporates all the basic security services, with a high degree of flexibility, allowing to negotiate services and algorithms to be used [3]. As a clearly differential element regarding HIPERLAN/2 and IEEE 802.11, DARWIN allows the use of public key infrastructure for authentication services. In relation to key exchange, besides Diffie-Hellman and pre-shared key, DARWIN establishes an scheme based on RSA. For confidentiality service DARWIN uses strong algorithms and secure key generation processes, based on the SSL model. DARWIN provides stream ciphers and block ciphers, and stablishes separated keys for both directions of communication. Finally, the integrity service is not implemented in HIPERLAN/2, and in IEEE 802.11 is linked to the confidentiality service. In DARWIN approach, integrity and confidentiality are independent services.

References

[1] DARWIN TR WG1.1 V1.0. Physical (PHY) layer, 2002.
[2] DARWIN TR WG2.1 V1.0. DARWIN MAC layer specification, 2003.
[3] DARWIN TR WG3.1 V1.0. DARWIN security services, 2003.

μPROS – Micro Protocol Based Development of Mobile Ad Hoc Networks

Alexander Geraldy and Reinhard Gotzhein

Computer Networks Group
Computer Science Department, University of Kaiserslautern
Postfach 3049, 67653 Kaiserslautern, Germany
{geraldy,gotzhein}@informatik.uni-kl.de

1 Project Context

Communication systems are an integral part of large, distributed systems, they form the basis for operating systems and applications. The recent evolution of wireless communication and user mobility has led to a new main research focus, aimed at offering farther advanced and more flexible communication services while improving the effective use of communication resources. This requires, among other things, customized and flexible mechanisms for unicast and multicast routing as well as for quality of service provision. Customization of communication systems is closely related to an application area, and calls for new protocol engineering techniques that are based on reuse in order to manage the development effort. In the project, customization of protocols for mobile ad hoc networks in the Ambient Intelligence domain (AmI, [1]) will be addressed. Ambient Intelligence is a vision that inherits the dynamic topology and automatic configuration from mobile ad hoc networks, and exceeds ad hoc networks by new applications based on the integration of "intelligent" network devices into everyday's objects. By this large-scale information integration, the AmI vision aims at easing the peoples' life.

2 Project Framework

The project is carried out by the Computer Networks Group at the Technical University of Kaiserslautern in Germany, in cooperation with several partners, and is publicly funded. The project staff currently consists of 2 full time researchers, 2 PhD students, and 6 master students. Financial support is received from several sources and covers full time research positions as well as PhD grants, which are funded by the German Science Foundation, the government of the province Rheinland-Pfalz, and the University of Kaiserslautern. Infrastructural support is obtained via membership in the research network "Ambient Intelligence", a Center of Excellence funded by the provincial government. There is close cooperation with Concordia University in Montreal/Canada

M. Conti et al. (Eds.): PWC 2003, LNCS 2775, pp. 188–191, 2003.

and the Fraunhofer Institute for Experimental Software Engineering in Kaiserslautern/Germany. The project is planned for a duration of 5 years, and has started in October 2002. Being in a starting phase at the moment, the project is expected to double in size in 2004.

3 Project Objectives

The main objective of the project is the provision and application of engineering approaches and techniques for the development of adaptive, integrated mobile ad hoc networks. In the core of the project is the construction and the use of a library of specialized design components called "micro protocols" for (unicast and multicast) routing and quality of service provision in mobile ad hoc networks, as well as a micro protocol framework for their composition. A micro protocol [2] is a self-contained, ready-to-use building block covering a single protocol functionality. Micro protocols are then composed to form complex communication systems. The selection of micro protocols is based on an assessment of existing routing and quality of service mechanisms, augmented with measures for their adaptation and integration in the Ambient Intelligence domain. This domain is characterized by a high degree of mobility as well as diverse communication requirements. To formally define micro protocols, the design language SDL (Specification and Description Technique), which is widely used in industry, will be applied.

4 Challenges

There are a number of challenges that will be addressed in the course of the project, including:

- **Increase of productivity and quality in the communication systems development**
  Reuse in the protocol engineering domain has not been very successful in the past. Yet, reuse is a key to increasing both quality and productivity. In a previous project, we have shown how to foster reuse in protocol design through specific design patterns called "SDL patterns". The challenge in this project will be to apply SDL patterns in order to define communication components called "micro protocols", and to provide a framework to compose them. Furthermore, it needs to be investigated how validation and implementation techniques can benefit from these reuse techniques.

- **Advanced services for heterogeneous applications**
  Most research in ad hoc networks is aimed at IP networks, which provide general-purpose communication services for common applications. In a heterogeneous environment, applications require more specialized and diverse services, which can only be provided by special-purpose, customized communication systems. It is an open issue how to design and implement heterogeneous communication functionalities in an adaptive, integrated manner.

- **Adaptive routing and quality of service provision**
  Routing as well as quality of service provision are major challenges in mobile ad hoc networks. This is due to the constantly changing topology and the sensitivity to noise. For instance, there is no single routing mechanism that yields best results in all situations. Therefore, techniques that take the current topology, channel quality, and traffic situation into account for short-term as well as for long-term adaptation are needed.

5 Project Tasks

The following is an excerpt of the list of tasks that have been defined for the project:

- **Provision of micro protocol design libraries for selected protocol functionalities**
  In the context of mobile ad hoc networks, both multi-hop routing and quality of service provision are of specific interest. There exists a variety of mechanisms for these functionalities, which are to be assessed and compared using discrete simulation techniques. Based on the outcome, suitable mechanisms will be selected and encased into micro protocols. This will lead to design libraries of routing and quality of service micro protocols, which can then be composed into customized communication systems in the AmI domain.
- **Definition of a general micro protocol design framework**
  In general, a component framework is understood as a rigid system skeleton to which components are attached in precisely defined places. In this sense, layered architectures are communication system skeletons. For the composition of micro protocols, this view needs to be generalized: a micro protocol framework is defined by a set of composition operators and mechanisms to build flexible, hierarchical communication systems from micro protocols. Thus, synchronization and causal relationships between micro protocols are established on a high level of abstraction.
- **Design of adaptation schemes for routing and quality of service provision**
  Based on the micro protocol library, new adaptive protocols will be devised. We will focus on the design of the adaptation decision process (reliable in-system replacement of functionalities), each node's knowledge about the network, metrics, and cost functions. Based on these parameters, a run-time adaptation process that must be efficient (network and CPU load) and reliable (no loops or oscillation) will be designed. The adaptation approach will be analyzed by implementing adaptation schemes, and by studying them in the AmI domain. In particular, we will investigate how to substitute protocols properly, and how to design the decision process for substitutions. We focus on adaptive protocols for routing and quality of service provision.

- **AmI case studies**
  In cooperation with other research groups of the AmI Center of Excellence, several applications and their customized communication systems will be developed. At the moment, the choice of applications is subject to further study. Possible candidates can, e.g., be found in the consumer domain ("Intelligent House"). Further tasks that are to be addressed in a later project stage are related to the effective implementation of micro protocols and their composition. First results on how to design micro protocols with SDL [2], to automatically implement micro protocol designs [3], and how to integrate quality of service mechanisms [4] are readily available.

References

1. K. Ducatel, M. Bogdanovicz, F. Scapolo, J. Leijten, J-C. Burgelman: *That's what friends are for. Ambient Intelligence (AmI) and the IS in 2010.* Innovations for an e-Society, Challenges for Technology Assessment, Institute for Technology Assessment and Systems Analysis (ITAS), 2001
2. R. Gotzhein, F. Khendek, P. Schaible: *Micro Protocol Design: The SNMP Case Study,* in: Telecommunications and Beyond: The Broader Applicability of SDL and MSC, E. Sherrat (Ed.), LNCS 2599, Springer, 2003, pp. 61–73
3. P. Schaible, R. Gotzhein: *Development of Distributed Systems with SDL by Means of Formalized APIs,* 11th International SDL Forum "System Design", July 1–4, 2003, Germany
4. J. Brandt, R. Gotzhein, R. Grammes, B. Schürmann: *Chatroom over Powerline: Systematical Development of a QoS-Integrated Distributed System,* SFB 501 Report 04/2003, Computer Science Department, University of Kaiserslautern, 2003

Simulation Results and a Proof-of-Concept Implementation of the FleetNet Position-Based Router
Extended Abstract

Hannes Hartenstein[1], Holger Füßler[2], Martin Mauve[2], and Walter Franz[3]

[1] NEC Europe Ltd. Network Laboratories
D-69115 Heidelberg, Germany
hannes.hartenstein@ccrle.nec.de
[2] University of Mannheim
Department of Mathematics and Computer Science
D-68131 Mannheim, Germany
{fuessler,mauve}@informatik.uni-mannheim.de
[3] DaimlerChrysler Research and Technology
D-89081 Ulm, Germany
walter.franz@daimlerchrysler.com

1 Introduction

The FleetNet project[1] [1,2,3] develops a platform for inter-vehicle communications based on ad hoc networking principles. With an ad hoc network established by moving vehicles, a realm of applications – ranging from exchange of emergency warnings or sensor data to unicast communication between passengers of different vehicles and to integration of the vehicular ad hoc network with the Internet through static or mobile FleetNet gateways – could be enabled in a low-latency, robust and low-cost fashion.

Vehicular ad hoc networks show a high level of mobility and, correspondingly, a high degree of dynamics with respect to topological changes. Therefore, multi-hop communication represents a serious challenge. Since in a vehicular ad hoc network communication end-points might be addressed by specifying a geographic region (geocast) in addition to standard IP addressing, positional information has to be used in any case for forwarding packets. GPS (and in the future Galileo) receivers will soon become a standard piece of equipment in vehicles and can be used to support routing tasks. In previous publications we have shown on the basis of simulations that routing methods that make use of geographic information can outperform other routing approaches in the case of vehicular ad hoc networks. Example results will be given in Section 2.

As a 'proof-of-concept' as well as to compare simulation results with reality, we currently have set up a real-world test network consisting of 6 `Smart`^{TM} cars equipped with GPS receivers, IEEE 802.11b wireless LAN NICs, and our FleetNet router. In this extended abstract we outline the testbed architecture of this 'FleetNet Demonstrator'

[1] The FleetNet Project is partly funded by the German Ministry of Education and Research (BMBF) under contract no. 01AK025. Partners are DaimlerChrysler AG, NEC Europe Ltd., Siemens AG, Robert Bosch GmbH, Temic GmbH, and the Universities of Mannheim, Braunschweig, Hannover and Hamburg-Harburg.

M. Conti et al. (Eds.): PWC 2003, LNCS 2775, pp. 192–197, 2003.

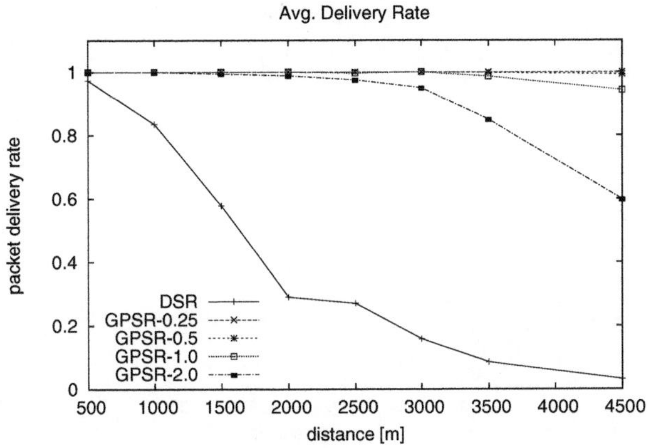

Fig. 1. A comparative study via NS-2 simulations of position-based routing (greedy part of greedy perimeter stateless routing, GPSR) and DSR. Shown is the achieved packet delivery ratio versus the multi-hop communication distance for the case of a highway scenario. The transmission range was set to 250m. The beaconing interval was set to 0.25, 0.5, 1.0, and 2.0 seconds, respectively.

(FND) on both, hardware and software side and some first insights and experiences with the demonstrator in Section 3 and 4. The goal of this extended abstract is to present a current snapshot of the project with respect to position-based routing. A final evaluation of the FleetNet Demonstrator is underway. Related work has studied a real-world Dynamic Source Routing-based vehicular ad hoc test network [4] and single-hop characteristics with two vehicles [5].

2 Simulation Results

In position-based unicast routing, a packet header does not only indicate the destination's ID, e.g., IP address, but also the destination's approximate current geographic position. To acquire a destination node's position, a location service (see [6]) is used. As location service we use a the *reactive location service* (RLS) [7] that essentially has translated the DSR route discovery procedure to the domain of position-based ad hoc routing. At each hop packets are then forwarded in a greedy manner to the neighbor that is closest to the indicated destination's position. Each node is aware of its neighbors within transmission range through periodic 'beacon' messages. Several metrics for greedy routing as well as recovery strategies for failures of greedy routing are presented in [6]. The main advantage of position-based routing over classical topological approaches is given by the fact that there is no need for route setup and maintenance. Instead, packets are forwarded 'on the fly'. Thus, position-based routing deals very well with mobility. To 'prove' this claim, we have conducted simulation experiments based on realistic vehicle movements on a highway and a city scenario. In [8] we studied highway scenarios and showed that position-based routing outperforms Dynamic Source Routing (DSR) due to better handling of mobility and smaller header sizes (see Figure 1). In [9] we studied city

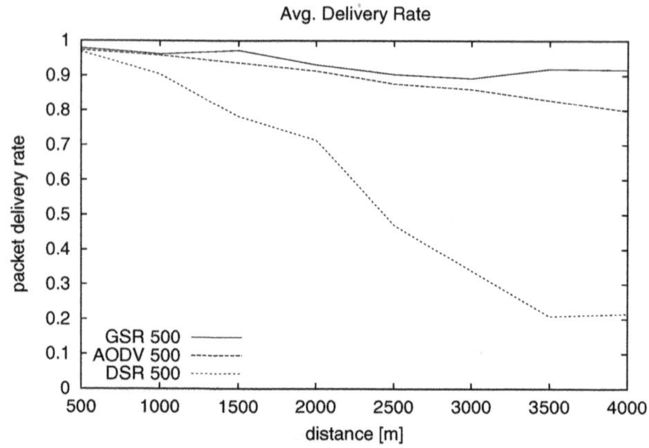

Fig. 2. A comparative study via NS-2 simulations of position-based routing making use of digital maps, DSR and AODV. Shown is the achieved packet delivery ratio versus the multi-hop communication distance for the case of a city scenario. The transmission range was set to 500m.

scenarios and compared position-based routing with DSR and with Ad Hoc On-Demand Distance Vector routing (AODV) showing that position-based routing outperforms the other approaches when a digital map of the city is available (see Figure 2). However, without a digital map position-based routing in a city environment suffers severely from the fact that frequently no greedy route is available to the destination and recovery strategies might fail due to radio obstacles.

3 Testbed and Implementation Architecture

In the 'real-world' test network, each of the six `Smart`^{TM} cars is equipped with a Linux-based FleetNet router that forwards packets coming from either the 802.11b network interface card (NIC) or from the FleetNet car area network (FCAN), see Figure 3a. We use external planar antennae with a gain of 4 dBi. Connected to the FCAN is a GPS-receiver with a navigational system. To support global monitoring of the 802.11-based ad hoc network, each car is in addition equipped with a GPRS NIC (Figure 3b).

The router is implemented as a user-space daemon that uses packet sockets in promiscuous mode on the wireless side. The FleetNet routing layer represents a '2.5 layer' approach by putting the FleetNet routing header between 802.11 MAC header and IP header (Figure 4). In order not to grep all packets coming from the FCAN, `netfilter` is used to select packets destined for other car subnets requiring wireless transport. For IP packets sent by the FCAN and addressed to another vehicle, a location service is used to determine the current geographic position of the vehicle corresponding to the destination IP address in case the position is not known. As location service we use again the *reactive location service* (RLS) [7]. When a required position is not known, a location request is flooded through the network. Corresponding packets for this destination are queued until the location reply is received. The forwarding decisions of the router

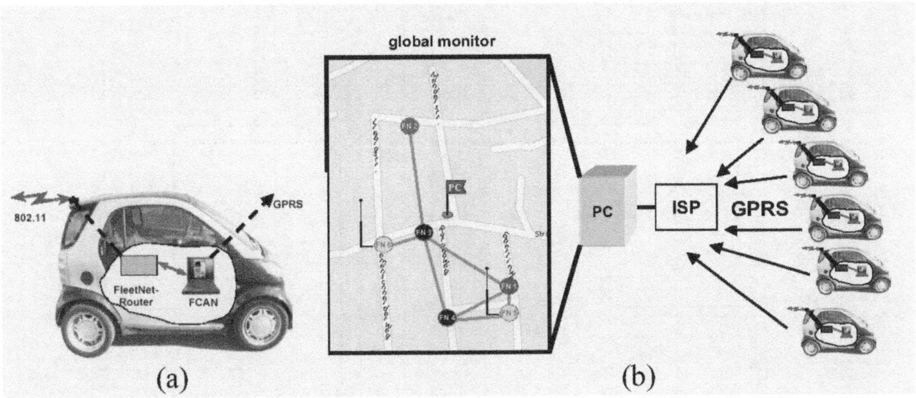

Fig. 3. (a) Each Smart car is equipped with a Linux-based FleetNet Router connected to the FleetNet car area network; (b) A global view of the ad hoc network is generated by having each car sending its local state via GPRS to a global monitor server.

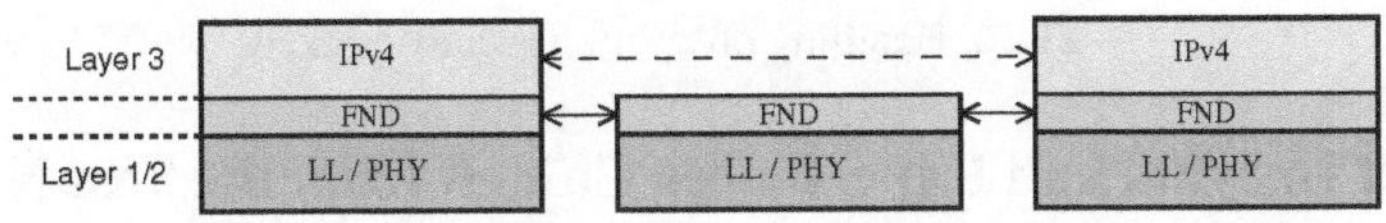

Fig. 4. The 2.5 layer approach taken in the 'proof-of-concept' implementation.

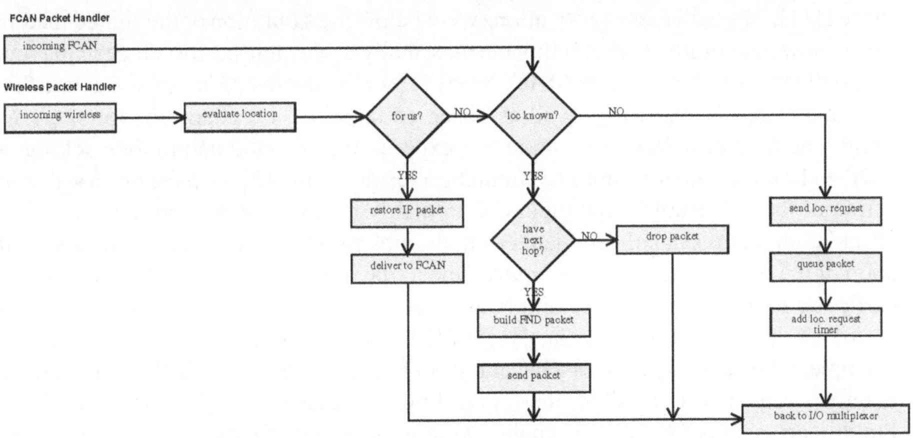

Fig. 5. Unicast position-based routing logic.

are based on greedy position-based routing as outlined in [10]: a packet is given to the neighbor that is closest to the final destination's position. Information on the positions of the neighbors is gained through periodic beacons each node sends out as one-hop broadcasts. The routing logic including location requests/replies is given in Figures 5 and 6.

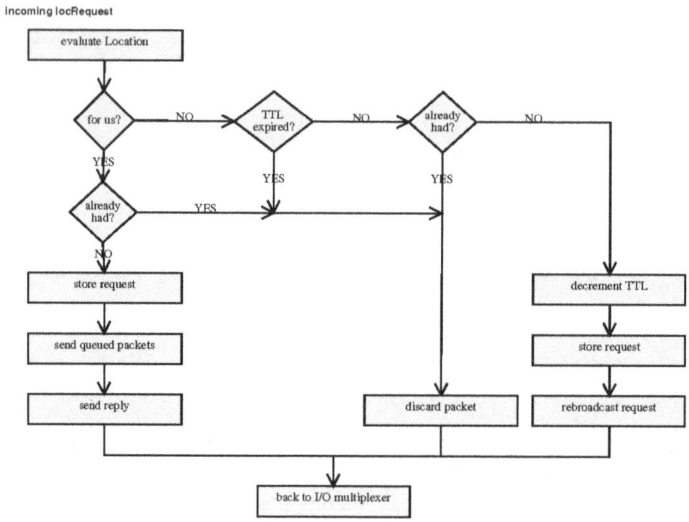

Fig. 6. Handling of location request packets.

4 First Insights and Experiences Gained from the Test Network

Various applications like broadcasting of emergency warnings as well as unicast multi-hop communication between two vehicles have been successfully tested and demonstrated [11]. Since all these applications were below the saturation of the shared medium, we are now quantitatively evaluating the maximally achievable performance of the router in combination with the 802.11 MAC. As in [12] we are interested in assessing the achievable multi-hop throughput as well as observed latency for a chain-like topology with chain lengths 2 to 6. We have conducted experiments in static and mobile settings for UDP and TCP traffic. Contrary to simulation results as in [12] we have observed degradation of the achievable throughput due to highly changing link conditions and due to obstacles that aggravate the hidden terminal problem. The statistical analysis and evaluation of the measurements is underway. Due to the highly changing radio conditions, in particular in mobile scenarios, simple averages over several runs do not lead to meaningful results without a proper analysis of the 'third variables' of the experiments. For example, when a node receives a beacon from a far away node to which communication is only 'occasionally' feasible, the forwarding of packets might be seriously affected. Preliminary results for the static chain topology show, e.g., for the case of a chain length of 4 nodes, IEEE 802.11 with 2 Mbps, and bidirectional communication between the two outer nodes, that a throughput of 200 kbps and round-trip-times varying between several hundred milliseconds and 3 seconds can be achieved.

5 Outlook

The movements, connectivity graphs and radio conditions are all recorded during the test runs in addition to the actual measurements on throughput and latency. These data allow for ‘replay’ within a simulator, and will help us to compare simulation results with reality and to enhance simulations. While the amount of work in setting up tests with a vehicular ad hoc networks of six cars is already considerable, we plan to increase the number of cars to several tens of vehicles in order to get insight into multi-hop communication behaviour over larger areas. In addition, we plan to migrate to 802.11a NICs as the current ASTM standardisation process has selected 802.11a as a basis for its standard E2213-02 [13]. On the routing side, we will further investigate geocast routing strategies and will perform simulations of a new position-based forwarding strategy, called *contention-based forwarding* [14], on top of realistic vehicle movement patterns.

References

1. Franz, W., Eberhardt, R., Luckenbach, T.: Fleetnet – internet on the road. In: Proceedings of the 8th World Congress on Intelligent Transportation Systems. (2001) Sidney, Australia.
2. Hartenstein, H., Bochow, B., Ebner, A., Lott, M., Radimirsch, M., Vollmer, D.: Position-aware ad hoc wireless networks for inter-vehicle communications: The FleetNet project. In: Proc. ACM MobiHoc ’01, Long Beach, California (2001)
3. The FleetNet project. http://www.fleetnet.de
4. Maltz, D.A., Broch, J., Johnson, D.B.: Experiences designing and building a multi-hop wireless ad hoc network testbed. Technical Report CMU-CS-99-116, School of Computer Science, Carnegie Mellon University (1999)
5. Singh, J.P., Bambos, N., Srinivasan, B., Clawin, D.: Wireless LAN Performance under Varied Stress Conditions in Vehicular Traffic Scenarios. In: Proceedings of the IEEE 56th Vehicular Technology Conference (VTC’02 Fall). Volume 2., Vancouver, Canada (2002) 743–747
6. Mauve, M., Widmer, J., Hartenstein, H.: A Survey on Position-Based Routing in Mobile Ad-Hoc Networks. IEEE Network **15** (2001) 30–39
7. Käsemann, M., Hartenstein, H., Füßler, H., Mauve, M.: A Simulation Study of a Location Service for Position-Based Routing in Mobile Ad Hoc Networks. Technical Report TR-07-002, Department of Computer Science, University of Mannheim (2002)
8. Füßler, H., Mauve, M., Hartenstein, H., Käsemann, M., Vollmer, D.: A Comparison of Routing Strategies for Vehicular Ad Hoc Networks. Technical Report TR-02-003, Department of Computer Science, University of Mannheim (2002)
9. Lochert, C., Hartenstein, H., Tian, J., Füßler, H., Herrmann, D., Mauve, M.: A Routing Strategy for Vehicular Ad Hoc Networks in City Environments. In: Proc. IEEE Intelligent Vehicles Symposium, Columbus, OH (2003)
10. Karp, B.N., Kung, H.T.: GPSR: Greedy Perimeter Stateless Routing for Wireless Networks. In: Proc. ACM/IEEE MobiCom ’00, Boston, Massachusetts (2000) 243–254
11. Enkelmann, W.: Fleetnet – applications for inter-vehicle communications. In: Proc. IEEE Intelligent Vehicles Symposium. (2003)
12. Li, J., Blake, C., DeCouto, D.S.J., Lee, H.I., Morris, R.: Capacity of Ad Hoc Wireless Networks. In: Proc. ACM/IEEE MobiCom ’01, Rome, Italy (2001) 61–69
13. Standard specification for telecommunications and information exchange between roadside and vehicle systems. ASTM Designation: E 22123 – 02e1.
14. Füßler, H., Widmer, J., Käsemann, M., Mauve, M., Hartenstein, H.: Contention-based Forwarding for Mobile Ad Hoc Networks. Accepted for publication in the Elsevier Ad Hoc Networks Journal (2003)

EYES – Energy Efficient Sensor Networks

Paul Havinga[1], Sandro Etalle[1], Holger Karl[2], Chiara Petrioli[3], Michele Zorzi[4], Harry Kip[5], and Thomas Lentsch[6]

[1] University of Twente, The Netherlands
p.j.m.havinga@utwente.nl
[2] Technical University of Berlin, Germany
[3] Rome University "La Sapienza", Italy
[4] CNIT, Italy
[5] Nedap N.V., The Netherlands
[6] Infineon Technologies, Austria

Abstract. The EYES project (IST-2001-34734) is a three years European research project on self-organizing and collaborative energy-efficient sensor networks. It will address the convergence of distributed information processing, wireless communications, and mobile computing.
The goal of the project is to develop the architecture and the technology which enables the creation of a new generation of sensors that can effectively network together so as to provide a flexible platform for the support of a large variety of mobile sensor network applications.
This document gives an overview of the EYES project.

1 Introduction

The vision of ubiquitous computing requires the development of devices and technologies, which can be pervasive without being intrusive. The basic components of such a smart environment will be small nodes with sensing and wireless communications capabilities, able to organize flexibly into a network for data collection and delivery. Realising such a network presents very significant challenges, especially at the architectural and protocol/software level. Major steps forward are required in the field of communications protocols, data processing, and applications support. The EYES project (IST-2001-34734) is a three years European research project on self-organizing and collaborative energy-efficient sensor networks. It will address the convergence of distributed information processing, wireless communications, and mobile computing.

The goal of the project is to develop the architecture and the technology which enables the creation of a new generation of sensors that can effectively network together so as to provide a flexible platform for the support of a large variety of mobile sensor network applications.

1.1 Ubiquitous Computing

"Ubiquitous Computing", a phrase that the late Mark Weiser (1952 – 1999) described in 1988 as "the calm technology, that recedes into the background of our

M. Conti et al. (Eds.): PWC 2003, LNCS 2775, pp. 198–201, 2003.

lives", matures from the vision of the Nineties to the reality of the young millennium, enabling increasing mobility and interaction of services and applications in a large variety of areas in daily life. Recently, we have seen rapid progress in developing a new off-the-desktop computing paradigm that moves towards the notion of a pervasive, wearable, unobtrusive, disappearing, or invisible computer. Sensors are tiny devices capable of capturing physical information, such as heat, light or motion, about an environment. Rapid advances in micro-electro-mechanical systems (MEMS), digital circuitry, and wireless communication have enabled a new generation of tiny, inexpensive, networked sensors. Embedding millions of sensors into an environment creates a digital skin or wireless network of sensors. These massively distributed sensor networks, communicate with one another and summarize the immense amounts of low-level information to produce data representative of the overall environment. From collaboration between (large) groups of sensor nodes, intelligent behaviour can emerge that surpasses the limited capabilities of individual sensor nodes.

Collaborative, smart sensor networks present information in a qualitative, human-interpretable form, which allows people (or computers) to respond intelligently. Sensor networks will change the way we work and live. Sensor network systems will enhance usability of appliances, and provide condition-based maintenance in the home. These devices will enable fundamental changes in applications spanning the home, office, clinic, factory, vehicle, metropolitan area, and the global environment. Sensor node technology enables data collection and processing in a variety of situations, for applications, which include environmental monitoring, context-aware personal assistants (tracking of location, activity, and environment of the user) home security, machine failure diagnosis, medical monitoring, and surveillance and monitoring for security.

Due to the high volume of energy constrained sensors in such networks and their use in possibly harsh environments, replacement of batteries in sensor nodes is not practical in these systems, their lifetime is however required to extend over years. Even though preliminary works have identified some of the requirements for sensor network implementation stressing the role of energy-efficient design, and have proposed preliminary solutions for some of the critical issues such as energy-efficient routing, some important areas such as self-configurable and re-configurable energy-efficient architectures, maintainability, security, as well as application-dependent issues like reliability, have been at least partially overlooked by the current research. Also an integrated approach to energy-efficient design is lacking and would be extremely beneficial for sensor network design.

2 Objectives

We believe that the solution of these problems at their core depends on highly distributed, reliable, and secure information systems that can evolve and adapt to radical changes in their environment, delivering information services that adapt to the people and services that need them. These systems must easily and naturally integrate devices, ranging from tiny sensors and actuators to handheld information appliances. Such devices must be connected by short-range wireless networks as well as by high-bandwidth backbones. Data and services must be secure, reliable, and high-performance, even if part of the system is down, disconnected, under repair, or under attack. The system must configure, install, diagnose, maintain, and improve itself - this applies especially to the vast numbers of sensors that will be cheap, widely dispersed, and even disposable. Finally, it must allow vast quantities of data to be easily and reliably accessed, manipulated, disseminated, and used in a customized fashion by users.

The goal of the EYES project is to develop the architecture and the technology, which enables the creation of a new generation of self-organizing and collaborative sensors that can effectively network together so as to provide a flexible platform for the support of a large variety of mobile sensor network applications.

We will also show the feasibility of the concepts and technologies developed by means of a prototype in which a sizable network of sensors will be built and some example applications demonstrated.

3 Description of the Work

The technical work will focus mostly on architectural, protocol and software issues. This area is in fact believed to be the true bottleneck in current sensor networks. The challenges to face in developing new technologies for sensor networks are the need for the nodes to be smart, self-configurable, capable of networking together, and the inherent poverty of resources of the nodes themselves. The main thrust of the work will therefore be directed towards the development of new architectural schemes and communications protocols and algorithms at multiple layers, taking into account those specific features. In particular, schemes, which are able to work efficiently in the presence of limited energy, processing power and memory, will be developed.

4 Architecture

In our approach we define two distinct key system layers of abstraction: the sensor and networking layer, and the distributed services layer. Each layer provides services that may be spontaneously specified and reconfigured.

- The sensor and networking layer contains the sensor nodes (the physical sensor and wireless transmission modules) and the network protocols. Ad-hoc routing protocols allow messages to be forwarded through multiple sensor nodes taking into account the mobility of nodes, and the dynamic change of topology. Communication protocols must be energy-efficient since sensor nodes have very limited energy supply. To provide more efficient dissemination of data, some sensors may process data streams, and provide replication and caching.

- The distributed services layer contains distributed services for supporting mobile sensor applications. Distributed services co-ordinate with each other to perform decentralized services. These distributed servers may be replicated for higher availability, efficiency and robustness. We have identified two major services. The lookup service supports mobility, instantiation, and reconfiguration. The information service deals with aspects of collecting data. This service allows vast quantities of data to be easily and reliably accessed, manipulated, disseminated, and used in a customized fashion by applications.

On top of this architecture applications can be built using the sensor network and distributed services.

5 Milestones and Expected Result

Both theoretical research results and the implementation of a prototype will be produced. Specifically, we will:

- Create new technologies for communications, networking and service provision which take explicitly into account the constraints of sensor networks
- Propose and test a new architecture for flexible support of mobile sensor network applications
- Test ideas by building a prototype sensor network in which to implement the proposed schemes thereby verifying their effectiveness.

6 Project Details

- Project Reference: *IST-2001-34734*
- Start Date: *2002-03-01*
- End Date: *2005-02-28*
- Project budget: *4.730 M-Euro*
- Project URL: *http://eyes.eu.org/*

7 Participants

- Universiteit Twente, CTIT, The Netherlands
- Infineon Technologies, Austria
- Consorzio Nazionale Interuniversitario per le Telecomunicazioni, Italy
- N.V. Nederlandsche Apparatenfabriek "Nedap", The Netherlands
- Universita degli Studi di Roma "La Sapienza", Italy
- Technische Universitat Berlin, Germany

NetMoAd: Scalability, Dependability, and Flexibility in Ad hoc Networks

Gísli Hjálmtýsson

Network Systems and Services Laboratory
Reykjavík University, Reykjavík, Iceland
gisli@ru.is

Abstract. Wireless ad-hoc networking is increasingly seen as critical part of ubiquitous self-configuring networking environments, including home and sensory networks and networks required to support ideas of pervasive computing. While basic principles of networking remain intact the large number of networked elements many of which exhibit mobility while being limited in power and other aspects introduces challenges not seen in traditional networks. The complexity of these future networks is further aggravated as increasing flora of applications exploit increasing diversity in of network and user equipment. Ad-hoc networking requires these devices to play a dual role of an end-system acting in its own interests and also the role of a router forwarding package on behalf of other devices. We foresee a wide range of operating environments from highly volatile – with rapidly moving mobiles and fluttering availability, to semi stable sensory environments – where node failures or external changes are the primary sources of network volatility.

1 Framework

We have recently started a project on ad-hoc networking focused on scalability and dependability and flexibility by targeting three main areas: reachability and routing, aggregation of routing information and on nodal architecture and operating system support for the duality required for the networked elements. While based on our own prior work and that of others, the project is in its early stages with limited amount of results so far. This abstract briefly describes the project, our approaches and our preliminary results.

The research is conducted at the Network Systems and Services Laboratory at Reykjavík University, lead by Dr. Gísli Hjálmtýsson, professor of Computer Science. The effort constitutes a new direction for our research on network and router architectures, and novel network systems and services. In particular, we aim to build on our prior and ongoing work on active and programmable networking [1,2] and self-configuring services [3,4], rapid restoration [5,6,7], and service dependability [8]. In addition to Professor Hjálmtýsson, Prof. Heimir Þór Sverrisson and two Ph.D. students are involved at this early stage of the project.

The project is currently supported by a grant from the Icelandic Centre for Research, with the aim of applying for full funding to the European Commission project. Currently this project is done by Reykjavík University, but part of our current

M. Conti et al. (Eds.): PWC 2003, LNCS 2775, pp. 202–206, 2003.

funding is attributed to building a collaboration group and to connect our efforts with related work performed elsewhere, particularly for the purposes of joining forces to seek EU funding.

2 Objective

Immediate objective of the project is to evolve to the level of maturity and group size (collaboration) to become viable project to be funded by the European Commission. Technical project objectives are to achieve scalability in routing, to achieve dependability in highly mobile ad-hoc environments, and to understand the role of nodal architecture in these two.

3 Challenges

We foresee as a primary challenge to achieve aggregation in network information, particularly routing/forwarding information. Also, whereas reliability is well defined in traditional transport networks, finding ways to make connectivity and availability of network services dependable over a volatile ad-hoc network appears to be a significant challenge.

4 Task Performed to Tackle Challenges

Initially, we will perform a significant amount of analysis and simulation to explore the range of potential options and deepen our understanding of the underlying problem principles. At the same time, we will implement and experiment with relevant node OS support to rationalize the practicality of these potential options. Our goal is that the result of this initial result will be concrete protocol and mechanism proposals, validated with analytical models, simulation and implementation on network elements; moreover, that the implementation and experimentation will result a set of requirements for system support.

5 Project Organization

Wireless ad-hoc networking is increasingly seen as critical part of ubiquitous self-configuring networking environments, including home and sensory networks and networks required to support ideas of pervasive computing. While basic principles of networking remain intact the large number of networked elements many of which exhibit mobility while being limited in power and other aspects introduces challenges not seen in traditional networks. The complexity of these future networks is further aggravated as increasing flora of applications exploit increasing diversity in of network and user equipment. Ad-hoc networking requires these devices to play a dual role of an end-system acting in its own interests and also the role of a router forwarding package on behalf of other devices. We foresee a wide range of operating environments from highly volatile – with rapidly moving mobiles and fluttering availability, to semi stable sensory environments – where node failures or external changes are the primary sources of network volatility.

5.1 Scalability of Reachability and Routing

The first focal point of our effort is reachability and routing. While in a (semi-) stable environments automatic allocation of addresses may be optimized to aggregate routing, in more volatile settings address allocation is reduced to ensuring uniqueness thus effectively eliminating route aggregation. In small static environments it is viable for each node to keep a routing entry for every host. This is in fact common in modern layer-2 devices that self-configure a MAC layer forwarding table by observing the transiting traffic. However, this approach is infeasible either in highly volatile environments or in networks having a large number of elements. We expect that the networks needed for pervasive computing and ubiquitous networking to be both, encompassing millions of elements in small areas many of whom exhibit fluttering availability and/or mobility. This is the inverse of what destination based routing achieves on the Internet. Volatility causes information to become stale rapidly, with staleness increasing with distance from origination. Volatility and scale conspire to increase the intractability of maintaining the routing information. Maintaining millions of routing entries in a low powered low cost device is not viable.

We are experimenting with novel and existing proposals to manage routing and forwarding information in large-scale volatile network environments. Some existing proposals employ source based routing where the sender specifies every node on the path to the intended destination. Intermediate nodes process the source route header to extract information about hosts on the path, creating a routing table of destinations. In addition to scalability problems of maintaining routing paths to each destination, source based routing fails to exploit the improvement in information closer to the destination. Indeed, this is one of the attractive attributes of destination based routing on the Internet, where routing is based on coarse grained (aggregate) information far from destinations, but becomes gradually refined at routers closer to the destination.

To achieve aggregation and avoid the need for detailed information to be maintained throughout the network we are experimenting with a combination of featherweight flows and virtual topologies (overlays). A particular problem as the number of network elements grows is that without route aggregation, the number of routes maintained at each node is likely to far exceed the number of flows active at the node at any given time. To combat this problem we break large networks into smaller regions, maintaining per-host information for nearby hosts but keeping track of regions rather than systems and paths in more distant parts of the network. While in many ways similar to self-configured IGP areas, the regions are constantly changing and may be overlapping. Featherweight flows, allow a fast creation of a tunnels/connections between regions and systems. A lightweight connection setup may be source routed or may carry a "report-route" type option to propagate information about hosts and connectivity. This way a tradeoff based on scale and volatility allows individual nodes to dynamically select a point in the spectrum from fully host based routing information to more aggregate virtual topology style routes.

5.2 Dependability of Connectivity and Network Services

The second element of our effort is dependability of network connectivity. In wired networks where physical failures in connectivity are relatively rare, dependability translates into configuration management, dynamic routing updates, fast restorations and hot-spot diffusion. The wireless ad-hoc networks that we are studying require self-configuration of both node-local facilities and of inter-nodal state maintenance, in particular routing protocols. While automatic configuration avoids the misconfigurations that threaten the reliability of current production networks, self-configuration of routing protocols at the scale of interest is inherently tricky.

In more volatile environments maintaining basic connectivity is an ongoing effort of significant complexity. Fast restoration is replaced with continuing local search for the currently best available channel. However, whereas local scanning and rapid soft hand-offs can maintain local connectivity, rapid local restoration in wired networks assumes very low likelihood of multiple failures. This assumption is not valid in ad-hoc wireless networks. To achieve dependability we have devised a scheme similar to the one used to achieve scalability in routing, where network elements are grouped into regions. Within a region, the network nodes monitors the network connectivity to neighboring nodes, constructing a profile of dependability for these nodes. Using these profiles nodes select the most dependable nodes to form a set of virtual access points for that region. When routes fail, fast restoration and hot-spot diffusion exploit these virtual access points as intermediaries until new information restores better routes to replace those lost.

5.3 Nodal Architecture and Operating Systems Support

The third focus of our effort is on nodal architecture and operating system support for ubiquitous and ad-hoc networking. Over the last few years we have researched and implemented architecture for an active and programmable network node, and experimented with it as a network server, high-performance router, and optical-networking controller, with applications such as media streaming, multicast, and adaptive congestion control. The architecture supports a range of dynamically programmable activity, ranging from pure control plane functionality, to limited access the datapath, to intercepting and processing every datagram. The flexibility of this platform makes it attractive as a basis for ad-hoc networked elements and ubiquitous networking. These attributes make our nodal architecture an ideal platform to realize and experiment with the spectrum of choices in solving the issues detailed below, while gaining understanding of the requirements that ubiquitous networking imposes on the operating systems of network elements.

Our initial assessment is that our platform and more generally active and programmable networking approaches seem effective and beneficial in ubiquitous networking and pervasive computing environments. Favorable bandwidth to processing ratio – relatively low bandwidth compared to processing power – affords additional per packet processing when required. Programmability provides adaptability to the elements, allowing in-field adaptation and software upgrades. Our architecture has proven effective either as a networked end-system or as a router; instantiated as the nodal OS for ubiquitous network element and sensory equipment it now provides both functions.

References

[1] Gísli Hjálmtýsson, "The Pronto Platform - A Flexible Toolkit for Programming Networks using a Commodity Operating System," in the proceedings of OpenArch 2000, Tel Aviv, Israel, March 2000.

[2] Gísli Hjálmtýsson, Heimir Sverrisson, Björn Brynjúlfsson and Ólafur R. Helgason, "Dynamic packet processors - A new abstraction for router extensibility," in the proceedings of OpenArch 2003, San Francisco, April 2003.

[3] Gísli Hjálmtýsson, Björn Brynjúlfsson and Ólafur Ragnar Helgason, "SLIM – Self-configuring Lightweight Internet Multicast," RU Technical Report. In preparation to be submitted for publication.

[4] Gísli Hjálmtýsson, Björn Brynjúlfsson and Ólafur Ragnar Helgason, "Overcoming last-hop/first-hop problems in IP multicast," RU Technical Report. Submitted for publication.

[5] Gísli Hjálmtýsson, Jennifer Yates, Sid Chaudhuri and Albert Greenberg, "Smart Routers . Simple Optics: An Architecture for the Optical Internet," IEEE/OSA Journal of Lightwave Technology, December 2000.

[6] Gísli Hjálmtýsson, Jennifer Yates and Sid Chaudhuri, "Restoration Services for the Optical Internet," in the proceedings SPIE.s Terabit Optical Networking: Architecture, Control and Management Issues conference, Boston, MA, November 2000.

[7] Gísli Hjálmtýsson, Panagiotis Sebos, Graham Smith, and Jennifer Yates, "Simple IP Restoration for IP/GbE/10GbE optical networks," Postdeadline paper PD-36, OFC 2000, Baltimore, MD, March 2000.

[8] Nikos Anerousis and Gísli Hjálmtýsson, "Service Level Routing on the Internet," in the proceedings of Globecom'99, Rio de Janeiro, Brazil, December 1999.

The SONG (SOlutions for Next Generation Mobile Systems) Project*

Sándor Imre, László Pap, Ferenc Balázs, János Horváth Cz., Róbert Schulcz, and Sándor Szabó

Budapest University of Technology and Economics
Department of Telecommunications
Mobile Communications and Computing Laboratory
Magyar Tudósok krt.2, H-1117. Budapest, Hungary
ipmobility@mcl.hu

Abstract. The SONG project was formed in 2001, when research team of Mobile Communications and Computing Laboratory (MCL) at Budapest University of Technology and Economics (BUTE) joined the Inter-University Centre for Telecommunications and Informatics (ETIK). ETIK consists of academic research teams form BUTE and Eötvös Lóránd University of Sciences (ELTE) and industrial partners among the leaders of information technology in Hungary. Industrial members are: Ericsson Communications Systems Hungary Ltd., KFKI Computer Systems Sc., Hungarian Telecommunications Sc., Sun Microsystems Hungary Ltd., Westel Mobile Telecommunications Sc., Compaq Computer Hungary Ltd., Antenna Hungária Sc., Computer Science and Automation Research Institute of the Hungarian Academy of Sciences.

1 Introduction

The main mission of ETIK is to perform internationally recognised research and development activities in the area of Internet-based telecommunications and mobile telecommunication systems, which advance the operation of the Internet and high-speed data-communication networks; and to integrate different communication technologies and services, like traditional telecommunications, Internet and wireless systems. In order to achieve these goals ETIK launches projects with 3 years of duration. Research activities of ETIK are in close connection with the Ph.D. Schools of BME and of ELTE, therefore Ph.D. and undergraduate students are involved into the R&D projects. The achievements of the projects are evaluated twice a year during the ETIK workshops where the projects hold scientific presentations and demonstrations of practical results (e.g. software tools, test beds, etc.). In this paper we present a short survey of the ETIK project entitled SONG.

* The SONG project is powered by ETIK.

M. Conti et al. (Eds.): PWC 2003, LNCS 2775, pp. 207–212, 2003.

2 Goals and Research Areas

The main goal of the SONG project is to explore solutions next generation mobile systems, analyse their performance by means of simulations and analytical calculations. Besides already known methods, our research team also develops new ideas and implements test beds in the following areas (see below).

2.1 Hierarchical Mobility Management Optimisation: Optimal Network Topology and Agent Selection

We reviewed the modifications introduced by HMIPv6 and RegReg6, along with their detailed protocol operations and special features. Then, a thorough comparative analysis of the two extensions was given, including the aspects of regional signalling differences, the process of registrations, the issues of distributing regional mobility agent functionality, modifications to MIPv6, etc. [1]

HMIPv6 and RegReg6 were found obviously advantageous for mobile nodes that are roaming far away from their communication partners. Therefore, they are essential components for Mobile IPv6 networks, enabling more seamless mobility and a reduced signalling load outside the visited domains. However, the simulation results confirmed the suppositions that not only the approaches but also the performance of both protocols is comparatively equal. This similarity has probably historical roots in a common idea. The reason why this original idea parted in two separate but parallel directions might be that HMIPv6 and RegReg6 are related to different companies (Ericsson and Nokia). However, there is definitely no need for two separate protocols if the differences are minor – this might also be justified by the fact that the IP Routing for Wireless/Mobile Hosts Workgroup has recently removed RegReg6 from the list of managed protocols.

Nevertheless, both HMIPv6 and RegReg6 possess some unique and useful features. [2] Therefore, the subsistent of the two extensions should not be either HMIPv6 or RegReg6, but a protocol that combines the conceptions of these. A feasible solution could take HMIPv6 as a basis and equip it with the Previous Access Router sub-option and a Binding Cache for each visited domain router to enable crossover router selection if large domain hierarchies are considered. Even more important, a mechanism for mobility agent selection should be worked out for domains with distributed agent functionality, according to the aspects listed previously [3].

2.2 Application of Different Mobility Models

In case of several problems (like CAC, location update and paging strategies) the optimal solution can only be found, if the future number and bandwidth demand of mobile terminals can be predicted precisely for each radio cell. In the literature several mobility models can be found, but the suitability (e.g. what parameters should be measured in a real network, in order to determine the model's parameters) and accuracy of these models are not examined. We have developed a new Markov mobility model especially appropriate for accurate prediction of the future number of mobile users in an urban environment [4]. The other advantage of our model is its simplicity, making it useful for numerical analysis.

2.3 Accuracy of Call Admission Control Algorithms in Mobile Environment

The effectiveness of resource reservation based CAC algorithms depends highly on future user location and bandwidth predictions. We are trying to describe analytically the relationship between the optimal prediction distance and the user's velocity [4].

2.4 Optimisation of IPv6 Address Autoconfiguration Methods

We are focusing on probably the most time-consuming part of the handover process: if the mobile terminal has to obtain a new IP address, which will identify the given mobile node uniquely on the new subnetwork. IPv6 provides an easy way to obtain an IP address automatically: the address autoconfiguration feature in IPv6 is responsible for the IP address distribution. We are studying the performance of the two kinds of address autoconfiguration methods (stateful and stateless). We suggest carrying out Duplicate Address Detection (DAD) parallel with AAA and CAC procedures, so the handover process can be completed more rapidly.

2.5 Mobility Support Based on IP Anycast Addresses

The use of anycast addresses is not prevalent; however the use of such addresses inside a micromobility area can overcome the problem of frequent address changes. Currently we are developing a test bed to implement the details of a mobility supporting method based on anycast address utilisation.

2.6 The Application of IP QoS Mechanisms in Mobile Wireless Environment

The current trends and future visions of next generation infocom networks are all common in employing IP in the wireless access and in the backbone too, so the QoS provisioning techniques of wired IP networks have to be extended to mobile scenarios also. We compare different IP QoS techniques (eg. DiffServ and IntServ) from the viewpoint of mobile wireless environment [5].

2.7 Binding Cache Size and Lifetime Optimisation

Every mobile device in IPv6 can always be addressed with its home address. When the mobile device is not attached to its home network, it obtains a temporary IP address – a care-of address – from the foreign network it is currently attached to. In order to be able to receive packages in this case the mobile always informs its home agent – a router in its home subnetwork – about its current care-of address. Correspondent nodes can send packages directly to the care-of address if they know it, otherwise they send them to the home address and the home agent forwards them to the mobile. The association between the home address and the care-of address is called binding. In IPv6 networks, every node contains a so-called Binding Cache to store binding information about mobile devices.

With the limited capability of mobiles and network overhead caused by triangle routing the optimisation of the binding cache's size and the binding entries' lifetimes is very important. Our simulation demonstrates this issue in different network scenarios. We investigate different statistics like end-to-end delay time, rate of packets sent via triangle routing, rate of packet loss, handover frequency, etc.

We used our simulator to examine the Binding management methods of IPv6, which aim at optimising the network for less triangle routing (packets sent across Home Agent). First we examined the effect of the Binding Cache's size on the rate of triangled packets, then the effect of the binding entries' lifetimes on the rate of triangled packets. We traced only the data request and spontaneously sent packets, because these are the only packets in our network to cause triangling [6].

2.8 DSP Implementation of Independent Component Analysis Algorithm of Noisy Underdetermined Systems

In wireless networks due to the parallel communication the interference load is heavy. Therefore the received signal streams could become not understandable for the receiver, which makes the interference filtering or cancelling very important. This task is much more difficult than for example the noise filtering because the parameters of interfering signals are equal or nearly equal. Because of the problem of data collision a question may arise: is it possible to offer a method, which is able to retrieve useful information form the data packages suffering from interference, and able to choose the required data stream from the received signal. In this situation it is conceivable that in case of two or more workstations transmit signals to the same receiver at the same time, the receiver could interpret all of the datastreams or a fraction of these at least. Should the new method be blind, none or minimal a-priori information about the source signals is necessary, it will be further advance. A method that only needs to fulfill the former requirements are independent component analysis (ICA) [10] and one of its important applications the blind source separation (BSS).

2.9 Dynamic Subchannel Allocation in OFDM Systems

The Orthogonal Frequency DivisionMultiplexing (OFDM) is a multicarrier modulation technique (MCM) which is the principle of transmitting data by dividing the stream into several parallel bit streams, each of which has a much lower bit rate, and by using these substreams to modulate several carriers. One subcarrier is used only by one user at pure OFDM systems so the subchannels are shared between the subscribers. The dynamic allocation provides facilities that the base station reserves the best subchannels to the users at downlink transmission. This method is based on observation of the channel characteristics, the number of bits to be transmitted and the transmission power in each subcarrier. In a TDD system the base station (BS) can estimate the instantaneous channel characteristics of all downlinks based on the beforehand uplink transmission.

3 Project Outputs

New, internationally published efficient algorithms and methods for the above listed problems are the scientific output of the project (see reference list below and www.mcl.hu).

The most important practical output of SONG is the complete IP mobility simulator program suite. Our goal was to create a flexible simulator for optimising and developing new IP mobility solutions. The simulator is capable of studying various kinds of IP mobility problems and proposals and can be used to study realistic IP implementation plans, to study the performance of versatile IP mobility management proposals and for academic purposes. Most of our already published results are achieved with the help of this simulator.

The simulations were created using the OMNeT++ simulation environment [7]. OMNeT++ (Objective Modular Network Testbed in C++) is an object-oriented, modular, discrete event simulator. It can be used for various simulation purposes such as modelling the behaviour of communication protocols, computer networks, distributed systems, etc. It is an open-source tool and can be paralleled with other open-source tools like Parsec, NS, Ptolemy, as well as with commercial products like OPNET or COMNET III.

An OMNeT++ model consists of hierarchically nested modules, with an arbitrary depth of nesting level. This allows simulating entities in various resolutions, for instance, simulating a network that consists of nodes, where each node contains layers, each layer has processes, etc. Entities of the simulation are referred to as modules. A module can be either compound or simple. A compound module is a larger building block since it can contain several other modules, which can be compound and simple as well. Simple modules implement the actual functionality of an entity. The user programs the actual algorithm in C++. All features of the object-oriented concept can be utilized, like polymorphism, inheritance, etc. Component libraries containing reusable modules are also available.

Modules are interconnected with parametric connections. They communicate with each other by sending messages, either through a path defined by their connections or directly. A module can also send a message directly to itself - this functionality is used to implement timers. Messages can carry arbitrary complex data structures.

The topology of the simulated network is described in a simple textual file with the help of the NED (Network Description) language. NED files can be created in various ways: edited by the user, generated with drag-and-drop method using the graphical GNED tool of OMNeT++, or generated with a text-processing tool like perl or awk; or with a combination of the above. The created topology can be changed even during the simulation by way of dynamic module and connection creation.

We have also developed a user friendly Graphical User Interface (GUI) connected to the IP mobility Simulator. In the beginning, the handling of the simulator was difficult, like setting parameters was not easy, because the values could only be modified in the source NED and C++ code, and, of course, it assumed the knowledge of both languages.

The main requirements for the GUI were as follows. To enable users to make test scenarios easily, components (base station, mobile station) and walls should be placed in a planning field and should restrict movements of the mobile stations. Furthermore, users should be able to assign arbitrarily routers to base stations and make any kind of router network topology. We developed the GUI in JAVA 2 because this language supports well the development of graphical user interfaces (with java.awt pack), it is quite robust and platform free.

The simulator is also used for education purposes at BUTE.

References

[1] Róbert Schulcz, Sándor Imre, László Pap, "Qualitative Analysis and Comparison of HMIPv6 and RegRegv6 Protocols", SoftCOM 2002, International Conference on Software, Telecommunications and Computer Networks, October, 2002, Croatia

[2] Róbert Schulcz, Sándor Imre, László Pap, "Quantitive Analysis and Comparison of HMIPv6 and RegRegv6 Protocols", Eunice 2002, September, 2002, Norway

[3] Vilmos Simon, Árpád Huszák, Sándor Szabó, "Hierarchical Mobile IPv6 and Regional Registration Optimization", ELMAR 2003, 2003, Croatia

[4] Alexandrosz Burulitisz, Balázs Rózsás, Sándor Szabó, dr. Sándor Imre, "On the Accuracy of Mobility Modelling in Wireless Networks", SoftCOM 2002, International Conference on Software, Telecommunications and Computer Networks, October, 2002, Croatia

[5] Gábor Zsolt Bilek, István Dudás, Róbert Milesz, Sándor Szabó, dr. Sándor Imre, "Providing QoS in IP Micromobility Networks" Workshop of the WWW2003 conference, 2003, Budapest, Hungary

[6] S. Imre, R. Schulcz, Cs. Csegedi, Sz. Vajda: "IPv6 Macromobility Simulation Using OMNeT++ Environment" International Teletraffic Seminar, Telecommunication Network and Teletraffic Theory, January 29 – February 1, 2002., St. Petersburg, Russia, ISBN 5-89160-028-5, pp. 180–193.

[7] OMNeT++ simulation environment webpage, http://whale.hit.bme.hu/omnetpp/

[8] Satyabrata Chakrabarti, Amitabh Mishra, "QoS Issues in Ad Hoc Wireless Networks.", IEEE Communications Magazine. vol. 39, no. 2, February 2001, pp. 142–148.

[9] Tien-Chung Tien, Shambhu J. Upadbyaya, "A Local/Global Strategy Based on Signal Strength for Message Routing in Wireless Mobile Ad-hoc Networks", IEEE, 2000

[10] A. Hyvrinen, "Survey on independent component analysis," Neural Computing Surveys, 1999.

[11] Z. Németh, F. Balázs, S. Imre, "Comparison of overcomplete independent component analysis algorithms in ad-hoc environment," International Conference on Signal Processing, Pattern Recognition and Applications [SPPRA 2002],June 25–28, 2002, Crete, Greece, pp. 13–18, 2002.

[12] T. Balog, F. Balázs, S. Imre, and L. Pap, "Analysis of extended ofdm-cdma system," 6th Int. OFDM-Workshop, Hamburg, pp. 24–1:24–4, September. 2002.

Mobile Ad Hoc Communication Issues in Ubiquitous Computing – The Smart-Its Experimentation Platforms

Albrecht Schmidt[1], Frank Siegemund[2], Michael Beigl[4], Stavros Antifakos[3], Florian Michahelles[3], and Hans-W. Gellersen[5]

[1] Media Informatics Group, University of Munich, Germany
albrecht.schmidt@informatik.uni-muenchen.de
[2] Distributed Systems Group, ETH Zurich, Switzerland
[3] Perceptual Computing and Computer Vision Group, ETH Zurich, Switzerland;
{siegemund,antifakos,michahelles}@inf.ethz.ch
[4] TecO, University of Karlsruhe, Germany
michael@teco.edu
[5] Computing Department, Lancaster University, UK

1 European Project "Smart-Its"

As we become familiar with the idea of connecting PDAs, notebook PCs and wall-sized screens, as put forward in [1], ubiquitous computing in everyday environments creates yet new challenges for mobile ad-hoc networking. When looking at less powerful mobile and stationary devices that act as sensors or actors, new requirements for communication evolve.

1.1 Project Vision

The European Project "Smart-Its" [2] is interested in a far-reaching vision of computation embedded in the world. In this vision, mundane everyday artefacts become augmented as soft media, able to enter into dynamic digital relationships. In our project, we approach this vision with the development of "Smart-Its" - small-scale embedded devices that can be attached to everyday objects to augment them with sensing, perception, computation, and communication. We think of these "Smart-Its" as enabling technology for building and testing ubiquitous computing scenarios, and we use them to study emerging functionality and collective context-awareness of information artefacts.

1.2 Project Partners

Smart-Its is a collaboration of Lancaster University in the UK, ETH Zurich in Switzerland, University of Karlsruhe in Germany, Interactive Institute in Sweden, and VTT in Finland. The project is part of the European initiative "The Disappearing Computer", and funded in part by the Commission of the European Union, and by the Swiss Federal Office for Education and Science.

M. Conti et al. (Eds.): PWC 2003, LNCS 2775, pp. 213–218, 2003.

1.3 Wireless Communication

A key issue in the Smart-Its project is to explore different options for wireless communication in ubiquitous computing scenarios. In this project presentation we concentrate on this aspect. In particular, we report requirements found in such environments and show ways to implement ad-hoc networks of augmented everyday objects that possess only very limited resources and computing capabilities.

2 Ad Hoc Communication Issues in Ubiquitous Computing Environments

From the extensive studies of settings and scenarios in wearable computing, mobile communities, and intelligent environments the different requirements for communication in ubiquitous computing have been extracted. In this section we elaborate on the most important ones.

- **Local Broadcast**
  For many applications a mechanism that offers a broadcast to all devices in a local vicinity is important. Often, the identity of the communication partner is of minor interest; the mere fact that another object is close by is sufficient to regard it as a potential communication partner. In many cases the basic radio link offers this feature, which however is often hidden by higher protocol layers.
- **Radio-signal strength indication and control**
  Support for adjusting transmission levels and accessing information about the received signal strength at runtime is of prime interest in many Ubicomp scenarios. These features can be used to dynamically adapt an application to the current radio environment or to limit the transmission range and thereby to control the number of potential communication partners.
- **Stateless vs. stateful communication**
  Especially in networks where communication partners are likely to be unreliable, it is desirable that communication is not trapped in a certain state when a device fails to respond according to a certain protocol. The use of time-out mechanisms may introduce unacceptable latencies. For example, it is questionable whether connection-oriented as opposed to pure broadcast-based communication is necessary in simple Ubicomp environments.
- **Power issues**
  For mobile computing devices, such as laptops, PDAs or mobile phones, power is of major importance. For artifacts deeply embedded in the environment, power becomes one of the most critical issues [3]. One obvious measure of power efficiency is the power expenditure per transmitted bit (unit: Joule/bit). However, this measure must not ignore the time and power spent during device discovery and connection establishment.
- **Instant communication**
  Many applications rely on mechanisms for instant communication between devices that are in common vicinity. Therefore, it is crucial for the communication subsystem to establish relationships between devices almost instantaneously and to provide short start-up times.

- **Unidirectional communication**
  For particular classes of devices unidirectional communication is sufficient. In particular, simple sensing devices often only send information to other devices but never need to process any incoming messages. Whereas for simple display devices just receiving information could be desirable.
- **Bandwidth**
  Many of the applications found in the analysis require very low bandwidth communication. In the design of a communication system bandwidth requirements of such systems should be regarded, too. In many applications it is acceptable to trade bandwidth to save energy, however, a lower bandwidth generally results in longer communication times (longer on-times) that could counter the saving effect. Therefore, in the design the optimum of bandwidth vs. energy consumption should be found.
- **Common Time base**
  Providing a distributed common time for all participants in a network is important for distributed perception and data fusion. In general it is of particular interest to know when exactly a sensor acquired certain data.

3 The Smart-Its Experimentation Platforms

In the course of the Smart-Its project [2] we develop small-scale embedded devices that can be attached to everyday objects to augment them with sensing, perception, computation, and communication capabilities. The general device architecture is depicted in figure 1. Each module has an autonomous power supply, dynamically attachable and deployable sensors, a microprocessor unit, and RF-communication modules.

As communication is a vital part in the development of these devices, the design decision regarding the communication technology has far reaching implications on scenarios and applications that can be supported and built based on these devices.

To assess communication issues and to evaluate different technologies, we have developed two complementary devices; one is based on standard Bluetooth-radio technology whereas the other is based on a custom RF-communication stack.

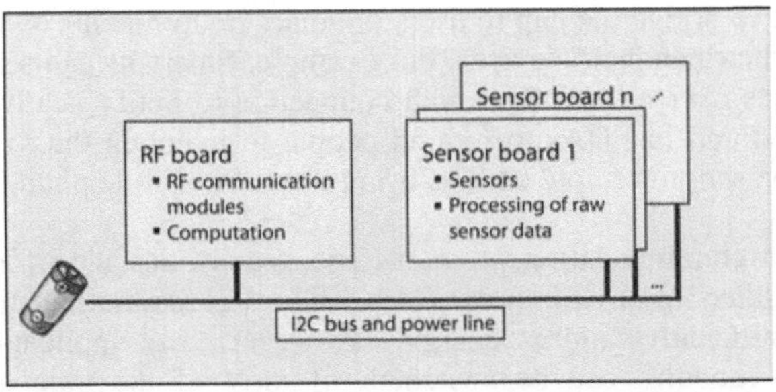

Fig. 1. Basic Smart-Its component-based device architecture.

Fig. 2. Bluetooth-based Smart-It module with acceleration/ temperature sensor board attached.

3.1 Bluetooth Smart-It

The Bluetooth Smart-It, depicted in figure 2, is designed around an 8-bit RISC microcontroller unit with embedded memory and an Ericsson Bluetooth module [4, 5]. The modules offer a generic Host Controller Interface (HCI) to the lower layers of the Bluetooth protocol stack, while the higher layers of the protocol as well as applications must be implemented on the host system. We developed protocol stacks for some of the higher layers of the Bluetooth standard. The software supports data exchange on the L2CAP (Logical Link Control and Adaptation Protocol) and RFCOMM (the serial port emulation of Bluetooth) layers. Furthermore, objects such as calendar entries or business cards as well as commands for accessing the capabilities of mobile phones can be transmitted over standardized Bluetooth profiles that are supported by our implementation.

The main reason for choosing Bluetooth as a candidate communication technology for Smart-Its, was that it provides a degree of interoperability necessary to easily integrate augmented objects into existing computing environments. As the number of consumer devices such as PDAs, Laptops, cellular phones, and digital cameras that are equipped with Bluetooth modules increase rapidly, Bluetooth-based Smart-Its are also well suited to bridge the gap to users of smart environments by building upon the capabilities of their handheld devices. For example, Smart-Its can send business cards to mobile phones over a local Bluetooth connection to notify nearby users. Calendar entries can be stored in PDAs to remind people to maintain their objects, and SMS messages can be sent to remote entities using a nearby mobile phone as mobile access point [6].

Compared to communication protocols specifically designed for wireless sensor networks and Ubicomp environments (cf. section 3.2), communication via Bluetooth consumes significantly more energy. However, in applications where the communication modules can be switched off most of the time and the need for communication be recognized by considering local sensor readings only, the Bluetooth-based Smart-Its are well suited for realizing typical Ubicomp application scenarios.

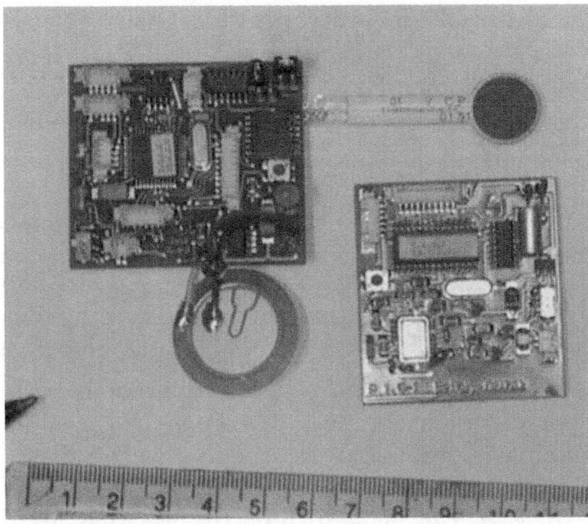

Fig. 3. On the left a fully equipped Smart-Its sensor board is shown. On the right a Smart-Its custom RF-communication board is depicted.

3.2 Simple-RF Smart-It

When implementing the Simple-RF Smart-It we focused on an approach that is complementary to Bluetooth (See [7] for technical details). This implementation is especially suited for applications where lowest power consumption, lower bandwidth and very fast connection times are needed. The hardware implementation is based on the TR1001 transceiver from RFM running on the European 868.35 MHz ISM band (figure 3) providing 125 kBit/s bandwidth for shared bi-directional data transfer providing a connection-to-network time of less than 12 milliseconds in average.

The main goal of this solution was to address the major requirements for an ad-hoc network intended to support communication of (small) context messages. The protocol is not designed for short range wireless multimedia communication, but specially adopted to run on very small devices with very limited amounts of energy. The RF Smart-Its protocol implements a one-hop peer-to-peer stateless datagram protocol with collision avoidance. Smart-Its broadcast information to other devices in a range of 1 to 10 meters. Physical range of the broadcast signal and therefore the resulting set of communication partners are selectable on the module for each communication packet. For addressing of communication partners different variants are available, such as addressing by the type of object (e.g. this message is for all chairs in the vicinity), by the context in which objects are in (e.g. all object that are moved within the vicinity), and also by ID based addressing.

As power consumption was one major concern in the design of the modules (as it also is for Bluetooth) several design decisions have been taken accordingly. Minimizing the time the transceiver is on and consumes energy is a key step beyond minimizing the actual power consumption of the RF-module. Based on a common time-base and transmission start time bound to a slot (transmission could start roughly every 13ms in a slot of about 1ms) means that in order to listen to communication a RF-Modules has to be on only very little time (in our implementation only one millisecond within 13ms which results in a reduction of the power consumption by a factor of about ten). This results in a run-time of up to one year with one battery set in some settings.

4 Conclusions

In the Smart-Its project we investigate different forms of wireless communication to suit the needs of ad-hoc communication in Ubiquitous computing environments. Analysing the requirements in such environments we explored different communication platforms – one based on Bluetooth and another based on a simple slotted protocol implemented on a microcontroller using a low power RF-Transceiver.

We realized there is no “one-fits-all” solution in this domain. In applications where interoperability is paramount and access to commercial user devices is required a Bluetooth based solution is preferable. In scenarios where energy is crucial and the amount of data communicated is small a simple RF solution has significant advantages. Also in scenarios where little information is exchanged in brief encounters (e.g. two people on bikes passing each other) it is crucial to have a system that allows rapid communication set-up, which was included in the custom RF solution.

Acknowledgement. We like to acknowledge the contributions of Oliver Kasten to the “Smart-Its”-Bluetooth platform. The work presented here was funded in part by the Commission of the European Union under contract IST-2000-25428, and by the Swiss Federal Office for Education and Science (BBW 00.0281). The project ‘Smart-Its’ is part of the Disappearing Computer initiative, http://www.disappearing-computer.org.

References

1. Weiser, M.: The Computer of the 21st Century. Scientific American 265, 3, September 1991, pp. 66–75.
2. Smart-Its, EU Project. Disappearing Computer Initiative, http://www.smart-its.org. 2002.
3. Gellersen, H-W., Schmidt, A. and Beigl, M.: Multi-Sensor Context-Awareness in Mobile Devices and Smart Artefacts. ACM journal Mobile Networks and Applications (MONET), Vol. 7, No. 5. October 2002.
4. Beutel, J., Kasten, O.: A Minimal Bluetooth-Based Computing and Communication Platform. Technical Note, May 2001, http://www.inf.ethz.ch/vs/publ/
5. Kasten, Oliver, Langheinrich, Marc: First Experiences with Bluetooth in the Smart-Its Distributed Sensor Network. Workshop on Ubiquitous Computing and Communications,In: Proc. PACT 2001, Oct. 2001
6. Siegemund, F., Rohs, M.: Rendezvous Layer Protocols for Bluetooth-enabled Smart Devices, ARCS 2002, International Conference on Architecture of Computing Systems, Karlsruhe, Germany, 2002.
7. TecO, University of Karlsruhe. Smart-its Project page. http://smart-its.teco.edu/artefacts/prototype/prototype.html. 2002.

IST Project: BroadWay – The Way to Broadband Access at 60 GHz

Athanasios Vaios and Ioannis Stavrakakis

University of Athens,
Department of Informatics & Telecommunications
Panepistimiopolis, Ilissia, 157 84, Athens, Greece
{avaios,istavrak}@di.uoa.gr

BroadWay (IST-2001-32686) is an Information Society Technologies project. It is a three-year project, which started in 2002, and it is partly funded by the Commission of the European Community. The partners involved in the project are the Motorola Labs (France), TNO-FEL (Netherlands), Intracom and the National & Kapodistrian University of Athens (Greece), IMST, IRK and Dresden University of Technology (Germany) and FARRAN (Ireland).

BroadWay aims to develop a tightly integrated platform for hot spots supporting private (e-home, entertainment, business) and public (fast outdoor downloading) applications. This platform will be able to cope with very dense user environments satisfying the user expectations in terms of the available throughput. BroadWay objectives include the definition, development and demonstration of the components of a hybrid dual frequency system based on HIPERLAN/2 OFDM high spectrum efficiency technology at 5 GHz and an innovative fully ad-hoc extension at 60 GHz, named HIPERSPOT. The HIPERSPOT system architecture will be based on hardware extensions of HIPERLAN/2 in order to ensure easy reconfigurability and allow for backward compatibility with the 5 GHz technology. It will support two compatible classes of mobile terminals: one of lower cost that will target the same transmission speed as HIPERLAN/2 (25 Mbps or even less data rate) and one providing a significantly higher data rate (more than 100 Mbps) than the maximum possible in HIPERLAN/2 by exploiting the larger bandwidth available at 60 GHz. HIPERSPOT will be equipped with a novel modified multi-carrier transmission scheme enhancing the robustness of transmissions at 60 GHz. Main goals of BroadWay are to offload the 5 GHz radio band in dense deployment areas, to exactly focus radio beams and to allow for self-organizing autonomous operation at the new frequency.

BroadWay is obviously part of the 4G scenario, as it complements the wide area infrastructure by providing a new hybrid air interface technology working at 5 GHz and at 60 GHz. This air interface is expected to be particularly innovative as it addresses the new concept of convergence between wireless local area network and wireless personal area network systems. The concept of BroadWay extends and complements existing 5 GHz broadband WLAN systems into the 60 GHz range allowing for terminal mobility while providing much higher capacity (achieving data rates exceeding 100 Mbps) and increased

M. Conti et al. (Eds.): PWC 2003, LNCS 2775, pp. 219–221, 2003.

privacy. For the first time a spectral efficient broadband modulation scheme at 60 GHz will be presented. The tight integration between both types of system (HIPERLAN/2 and HIPERSPOT) into the BroadWay frame will result in a wider acceptance and lower cost for both of them through massive silicon reuse. BroadWay provides a very high data rate mode for wireless broadband services, flexibility in order to adapt to changing environments and traffic requirements, spectral efficiency, efficient support of IP-based backbone networks and services, compatibility with existing communications systems and support of ad-hoc networking. One of the innovative aspects of the system is that it provides a solution to the problem of the inter-cell interference of the 5 GHz wireless networks that are expected to proliferate in the near future due to the expected increase of service demand and user density. This is achieved by proposing a system vision that already integrates current technologies by offering a bridge between the 5 GHz and 60 GHz spectrum. This way, offloading the 5 GHz radio spectrum from data traffic is possible through seamless switching to 60 GHz using the new HIPERSPOT extension of HIPERLAN/2 when in range of coverage.

BroadWay includes innovative work in the physical and baseband layers as well as in the link and convergence layer architecture. Concerning the Radio Frequency module, BroadWay targets the development of a low cost integrated front-end architecture with dual 5/60 GHz frequency operation. For its high integration capabilities, the front-end is modelled using QMMIC (Quantum Monolithic Microwave Integrated Circuit) multi-functional blocks. QMMIC is based on the combination of conventional MMIC (Monolithic Microwave Integrated Circuit) used on hybrid HEMPT (High Electron Mobility Pseudomorphic Transistor) technology and Resonant Tunnelling Diodes. It addresses the problem of 60 GHz smart antenna as well as the coexistence of two such systems in one small antenna. Planar antenna technology is required to develop miniaturized systems for use in both the HIPERLAN/2 system and the HIPERSPOT extension at 60 GHz. Due to the frequencies addressed, the size of the antenna will be small, which allows for the implementation of smart antennas not only in the base stations but also directly in the handheld equipment. According to the project objectives to offload an existing WLAN system (HIPERLAN/2) and extend this system with additional features, an enhancement of HIPERLAN/2 baseband algorithms is required to cope with the extended system requirements for the additional BroadWay modes. Backward compatibility is ensured to reuse HIPERLAN/2 hardware and software building blocks. The algorithm complexity will be kept as low as possible to provide for a potential mass-market solution. Regarding the link and convergence layer architecture, enhancements in HIPERLAN/2 protocol stack are proposed and evaluated to adapt to the BroadWay environment; these enhancements are validated through simulations to allow for a system that concordantly works in both 5 GHz and 60 GHz.

Focusing on the networking part of the conducted work, major challenges imposed by the dual mode of operation are to be addressed. BroadWay introduces two modes of operation: the cellular one at 5 GHz and the ad-hoc one at 60 GHz. At any time instant, one 5 GHz and several 60 GHz channels (based on the channel bandwidth and spacing) can be used inside the HIPERLAN/2 cell. The Access Point operates at 5 GHz and at 60 GHz simultaneously, whereas the mobile terminals can only operate at one frequency band at a time - meaning that they must switch from one mode of operation to the other. First of all, the support of ad-hoc functionality at 60 GHz in BroadWay by incorporating existing ad-hoc algorithms is not effective. Preliminary tests in the 60 GHz band have shown that communication in this frequency is strongly limited due to the Line Of Sight constraint. Although higher speeds can be achieved, the communication range cannot exceed 15-20 meters at most. In addition, indoor or outdoor environments as well as the mobility of users affect the stability of 60 GHz links. Simulations have shown that as the number of hops to reach a destination using the 60 GHz band increases, the lifetime of the path drastically decreases. Due to the short-range nature of the links at 60 GHz, it is inefficient to consider the construction of paths of more than 2-3 hops. Moreover, there are several frequency channels, which a mobile terminal can be tuned at. Thus, the design of a new protocol is needed to cater to the system's complexity and enable the establishment of 5 GHz and 60 GHz links. To support routing in the BroadWay system, it is necessary to determine which nodes operate at which frequency band as well as to establish efficient routes at 60 GHz. Since distributed routing protocols used for pure ad-hoc networks are not applicable in the BroadWay environment due to short-living 60 GHz links, QoS demands and system's peculiarities, a centralized routing scheme is considered where the Access Point is responsible for making routing decisions at both frequencies and allocating the needed resources, based on information collected during the neighborhood discovery process. This process provides the AP with 60 GHz topology information (every mobile terminal's one-hop away neighborhood and the quality of the corresponding links when operating at 60 GHz). Since the 60 GHz links may typically last for only a few seconds, the information collected by the neighborhood discovery process needs to be refreshed frequently. This would require an additional overhead that should not deteriorate the performance of the overall system, but contribute to offloading the 5 GHz band by enabling the establishment of credible 60 GHz paths. Among the critical issues that affect the performance of BroadWay is the switching time required not only for changing band (5/60 GHz) but also for switching between the various 60 GHz channels, the scheduling of transmissions inside the HIPERLAN/2 cell, the synchronization between the frames at 5 GHz and 60 GHz, and the signaling burden to incorporate the ad-hoc functionality at 60 GHz. The format of the HIPERLAN/2 frame needs to be enhanced to cater to the system's functionalities and a new frame at 60 GHz needs to be defined to enable the communication at the new frequency.

An Adaptive Radio Link Protocol to Improve TCP Performance over Correlated Fading Wireless Channels

Haitao Lin[1], Sajal K. Das[1], and Hee Yong Youn[2]

[1] Center for Research in Wireless Mobility and Networking (CReWMaN)
Department of Computer Science and Engineering
University of Texas at Arlington
Arlington, TX 76019, USA
Telephone:+1 817-272-7409
Fax:+1 817-272-3784
{hlin,das}@cse.uta.edu
http://crewman.uta.edu

[2] School of Information and Communication Engineering
Sungkyunkwan University
Jangangu Chunchundong 300
Suwon, Korea, 440-746
Telephone:+82 31-290-7147
Fax:+82 31-290-7231
youn@ece.skku.ac.kr
http://csn.skku.ac.kr

Abstract. Improving Internet service performance is essential for the future wireless networks. The correlated fading channels are found very unfriendly to TCP performance in such networks. Moreover, the variation in fading rate brings further challenge to wireless networks when supporting Internet applications, most of which runs TCP as the transport layer protocol. In this paper, we first analytically study the performance of different RLP (Radio Link Protocol) retransmission schemes in supporting TCP. An analytic metric called *RLP retransmission efficiency* is introduced to evaluate performance of different RLP retransmission schemes under correlated fading channels. With analytic and simulation results, we show that a single retransmission scheme does not yield the best performance over different rates of fading. Based on this observation, we propose an Adaptive Radio Link Protocol (ARLP) that has a fading rate detector to estimate the channel fading rate. Based on the estimated channel fading rate, ARLP adaptively adjusts its retransmission schemes to achieve the optimal data transmission performance. Simulation results show that ARLP can improve both the performance of a particular mobile host and the total throughput of wireless networks.

M. Conti et al. (Eds.): PWC 2003, LNCS 2775, pp. 222–236, 2003.

1 Introduction

TCP performance over wireless channels has been studied extensively in recent years. The major effort is focused on avoiding TCP's congestion control mechanism from being initiated when packet losses happen over unreliable wireless links. Proposals addressing this problem can be divided into three classes: i) Splitting the TCP connection into a wireline segment and a wireless segment [1][2]; ii) designing new TCPs such as W-TCP [11] or modifying normal TCP as to differentiate between losses over wireless links and real congestions; iii) Using local retransmissions in link layer or MAC layer so that most packet losses are recovered before TCP times out. In this way, TCP is presented with a much more reliable link. The unnecessary initiations of TCP's congestion control mechanism are avoided and throughput is improved. Due to the inherent problems in the first and second classes of mechanisms to address this problem, only local retransmission solutions are commercially implemented. In cdma2000 (IS2000) systems, a link layer protocol called Radio Link Protocol (RLP) is introduced to recover data losses over wireless links [14]. Similar protocol called RLC (Radio Link Control) is utilized in UMTS system [12]. The commercial implementations of local retransmission schemes imply their important roles in solving TCP over wireless problem, and any improvement to their performance is very important to the future wireless networks.

On the other hand, the local retransmission is a particular implementation of Automatic Repeat reQuest (ARQ). For example, the RLP used in cdma2000 is a negative acknowledgement (NACK) based Selective-Repeat ARQ, with finite number of retransmissions. It has been proposed by many researchers that the link layer protocol over wireless channels should be adaptive to the varying channel characteristics as wireless channels experience time-varying fading most of the time. Many adaptive ARQ schemes have been proposed [3][4][7][13]. In [3], three adaptive hybrid error control schemes are proposed and compared. These schemes differ in two aspects: how many Reed-Solomon (RS) code segments are used for each packet and how a packet with uncorrectable errors are retransmitted. Hlaing [4] proposed to use XOR-ing to do bit error rate (BER) estimation, and to use Frequency Hoping to counteract long deep fading. Paul and Mani [7] proposed to use adaptive MAC frame length to improve the link layer protocol performance over time-varying fading channels. In general, among all these works, adaptiveness is incorporated into link layer protocol in several ways. First, the coding rate can be adaptive: the general idea is that for high bit-error-rate (BER) channels, use more robust error correction coding, which results in lower effective data rates; for low BER channels, use less robust coding, with higher effective data rate. Second, the use of ARQ can be adaptive: for low BER channels, use Forward Error Correction (FEC) coding to correct the errors, which yield higher throughput in this environment; for high BER channels, use ARQ to recover the erroneous data since ARQ is more efficient in this situation. Third, the size of the atomic transmission unit can be adaptive: it is desirable to use larger frame size over low BER channels; while using smaller frame size over high BER channels yields better performance. In this paper, we explore another possibility of adaptiveness: the way to do retransmissions in

ARQ. By introducing this adaptiveness, we solve the problem of how to optimize the ARQ retransmission efficiency in a time-varying correlated fading channel, especially when the rate of fading varies. This problem is extremely important as it is closely related to the problem of how to further improve the link layer performance in supporting TCP. As mentioned, the local retransmission schemes like RLP are nothing but particular implementations or variants of the standard ARQs.

In the RLP specified in cdma2000 [14], when an RLP frame is found in error and can not be recovered by FEC, the frame is retransmitted for a finite number of times, and each time the number of copies of the retransmitted frame can be different. For example, for the first time of retransmission, two copies of the lost frame can be sent in two consecutive time slots. If both of these copies do not get through and time out happens, another 3 copies of this frame are transmitted. If all of these retransmissions fail, RLP gives up and let upper layers like TCP recover the lost data. This particular way of doing retransmission is called the RLP scheme (1,2,3), which involves first a one-copy transmission, then a two-copy retransmission, followed by a three-copy retransmission. There are several other retransmission schemes, such as (1,1,1,1,1,1) and (1,1,2,3), etc. For detailed information on RLP, refer to [14].

In this paper, we first show that over correlated fading channels of different characteristics, RLP retransmission schemes perform differently and have different retransmission efficiency. More specifically, scheme (1,2,3) performs better over fast fading channels, where the channel states of two consecutive frames have low correlation; while scheme (1,1,1,1,1,1) performs better over slow fading channels, where the channel states of two consecutive frames have high correlation. We present both analytical metric and experimental results to support this finding because of its significance: it serves as the primary motivation and the basic supporting fact of our proposed Adaptive Radio Link Protocol (ARLP).

ARLP enhances the link layer protocol performance over correlated fading channels where fading rate varies. The protocol consists of a simple but efficient sequential test algorithm to determine the fading rate of the channel, and a modified RLP receiver to respond to the detected fading rate changes with corresponding retransmission scheme that performs best with the detected fading rate. As a result, the ARQ retransmission efficiency is optimized.

The rest of this paper is organized as follows: Section 2 discusses the performance of two typical RLP retransmission schemes, their difference and implications. Section 3 presents the proposed adaptive RLP, including sequential test algorithm and the adaptive retransmission algorithm. Section 4 evaluates the performance of the proposed protocol ARLP and compares it with the standard RLP. Section 5 concludes the paper.

2 RLP Performance Study

We start by briefly introducing the wireless correlated fading channel model. The two-state first order Markov chain has been widely adopted as a mathematical tool to model the wireless fading channels because of its accuracy and its ease

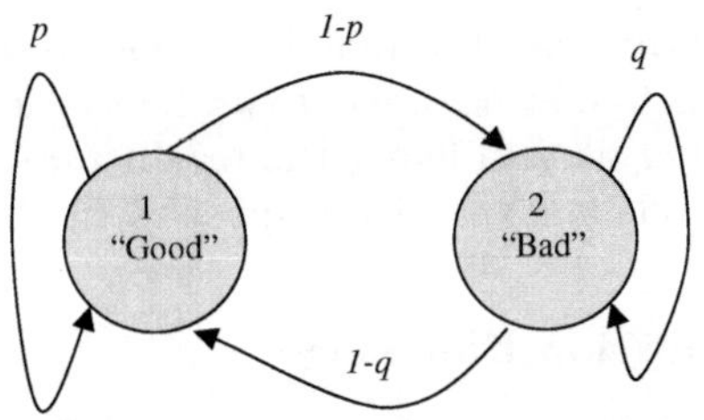

Fig. 1. Two-state Markov Chain

to mathematically adopt into analysis. When the wireless channel is subject to correlated multipath fading, the probabilities of getting an error among the bits being transmitted are not independent and identically distributed (i.i.d.) [8]. Instead, there are high correlations between consecutive bits being transmitted. If we assume the channel state does not change during the transmission of a bit, which is true for high speed data transmission or slow moving speed, the wireless channel can be modeled as a two-state Markov chain [9][10]. If we further assume that the channel state during transmission of a link layer frame does not change, the two-state Markov chain can be applied to the frame level channel error modeling and has been shown to be an accurate model [10]. In this section, we investigate the system performance based on this channel model.

2.1 Correlated Fading Channel Model

The frame error model can be approximated by a simple two-state Markov chain [10] as shown in Fig. 1. Assuming that the channel state does not change during the transmission of a link layer frame, we define the channel state during transmission of frame t as S_t, where $S_t \in \{1, 2\}$. Value 1 and 2 correspond to the channel state "good" and "bad", respectively, where "good" channel state means the frames transmitted in such state are correctly received by the receiver, and "bad" channel means the frames transmitted in such state are not correctly received. Defining the transition probability $M_{ij} = \text{Prob.}\{S_t = j | S_{t-1} = i\}$, the transition probability matrix of the Markov chain is given by:

$$M = \begin{pmatrix} p & 1-p \\ 1-q & q \end{pmatrix} \tag{1}$$

For a Rayleigh fading channel, given the fading margin F, the Doppler frequency f_d and the symbol (A symbol is an RLP frame here) length T, the transition matrix M of the Markov chain, as well as the average frame error rate (FER) P_E can be calculated. Because of limited space, we omitted the details of how to obtain M and P_E from F and f_dT, which can be found in [8][10] . Note that in the above model, the Frame Error Rate (FER) is solely decided by F, the fading margin, while the fading rate of the wireless channel depends on the product of f_d and T. Large f_dT means fast fading rate and short error burst, while small value of f_dT implies slow fading rate and long error burst. When $f_dT = 1$, the model reduces to i.i.d channel model. Note that $f_d = f(\frac{u}{c})$, where f is the carrier frequency, v is the relative speed of mobile station to base station, and c is the speed of light. For a particular channel, f can be taken as a constant; when the link layer frame length T is also fixed, the product $f_d T$ is solely dependent on v, the speed of the mobile station with respect to the base station.

2.2 RLP Retransmission Efficiency

A typical link layer ARQ used in wireless networks is Radio Link Protocol (RLP). The RLP is introduced into CDMA (IS95 and IS2000) systems in order to solve the TCP performance problem in wireless environment. Instead of letting TCP take care of the lost packets, the RLP tries to recover the lost link layer frames to avoid the unnecessary initiation of congestion control mechanism in upper TCP. The RLP is a negative acknowledgement (NACK) based Selective Repeat (SR) ARQ protocol. When a frame is detected to be in error or lost, copies of the lost frame are retransmitted. There are several optional retransmission schemes [8], including (1,1,1,1,1,1) and (1,2,3). Consider scheme (1,2,3) for example: when RLP at the receiver side finds out that a frame is in error (or missing), it sends back two NACKs requesting for retransmission of this frame. A timer is set for this missing frame. The sender retransmits two copies of this frame. When the timer expires (times out) for the first time and this frame is still not received, three NACKs are sent back, the timer is reset, and each NACK received at the sender side triggers a retransmission of the missing frame so that three copies of the missing frame are transmitted. If the timer expires again, the RLP stops retransmission of this frame and sends whatever it has to the upper layer. For more details on RLP, refer to [14][16].

In this section, we study the retransmission efficiency of different RLP retransmission schemes both analytically and with simulation results. Note that an RLP frame can experience more than one transmission attempt before it is successfully received. For example, for scheme (1, 2, 3), a frame can be successfully received in any of the six RLP transmission attempts. We say frames arriving at different RLP transmission attempts in different "phases". Formally, an RLP frame is in *phase* i if it is received correctly for the first time at the ith RLP transmission.

The probability of a frame in phase i is given by:

$$P_h(i) = \begin{cases} 1 - P_E & \text{if} \quad i = 1 \\ S_i(M^{d(i)})_{21} & \text{if} \quad 2 \le i \le |scheme| \\ S_i & \text{if} \quad i = |scheme| + 1 \end{cases} \tag{2}$$

where $M^{d(i)}$ is matrix M to the $d(i)$th power, and M_{ij} is the element at ith row, jth column of matrix M. $|scheme|$ is the total number of RLP transmissions a frame can experience, which is dependent on particular RLP retransmission scheme. For example, for scheme (1,2,3), $|scheme| = 1 + 2 + 3 = 6$. When $i = |scheme| + 1$, $P_h(i)$ means the frame is finally aborted by RLP since it has been transmitted for the maximum number of times that is allowed. S_i is the probability that all previous transmission attempts fail, which is recursively given by:

Table 1. $d(i)$ for RLP scheme (1,2,3) and (1,1,1,1,1,1), assuming RTT of 4 frames and RLP timeout value of 13 frames

i	$d(i)$ for (1,2,3)	$d(i)$ for (1,1,1,1,1,1)
1	0	0
2	9	9
3	10	23
4	25	37
5	26	51
6	27	65
7	35	73

$$S_i = \begin{cases} 0 & \text{if} \quad i = 1 \\ P_E & \text{if} \quad i = 2 \\ S_{i-1}(M^{d(i-1)})_{22} & \text{if} \quad 3 \leq i \leq |scheme| + 1 \end{cases} \tag{3}$$

For a frame in phase i, for $1 \leq i \leq |scheme|+1$, $d(i)$ is the time that has passed (in unit of RLP frames) from the first reception of a frame to the time the frame is first correctly received or finally aborted by RLP. Note that different RLP retransmission schemes have different values of $d(i)$. Table 1 gives the values of $d(i)$ for scheme (1,2,3) and (1,1,1,1,1,1). Assuming that d_t is the propagation delay from RLP sender to RLP receiver, $X(i) = d(i) + d_t$ is the delay seen by a frame that is in phase i. Thus the average delay of an RLP frame is:

$$E(X) = \sum_{i=1}^{|scheme+1|} Ph(i)X(i) \tag{4}$$

In order to study how different RLP schemes perform over correlated fading channels, and also to study the performance of different RLP retransmission schemes in terms of their efficiency in supporting TCP transmission, some analytic metric is needed. Because of the complexity in developing pure analytical model for TCP throughput with RLP retransmissions and correlated fading channel conditions, we use an indirect analytical metric, which is simple but provide enough insight into RLP's ability to support TCP in correlated fading channel.

Here we discuss the efficiency of two typical RLP retransmission schemes: (1,2,3) and (1,1,1,1,1,1). To develop a metric to evaluate the RLP performance of supporting TCP, both the probability of successful recovery and the time it takes for RLP to recover have to be taken into account. We first define P_r to be the probability that, given the original transmission is in error, a frame is recovered during RLP retransmissions successfully. The complement probability $\overline{P_r}$ is given by:

$$\overline{P_r} = \prod_{i=2}^{|scheme|-1} (M^{d(i)})_{22} \tag{5}$$

Note that the average delay $E(X)$ decides the speed of RLP recovery, while recovery rate P_r decides the RLP's recovery capability.

Both of $E(X)$ and P_r are important to the performance of RLP in supporting TCP connections. Hence, we take into consideration both of these two factors for defining the RLP retransmission efficiency EF_r. On the other hand, these two factors are not of the same magnitude, namely $0 < P_r < 1$, while$E(X) \geq 4$ assuming $d_t \geq 8$. It is not appropriate to use their direct ratio as the efficiency definition, because it magnifies the effect of $E(X)$ and ignoring the significance of P_r. we define EF_r as

$$EF_r = \frac{P_r}{\eta + E(X)'} \tag{6}$$

where η is a tunable factor so that the relative significance of P_r and $E(X)$ can be tuned. According to the simulations, we found that $\eta = 10$ is appropriate for our purpose of evaluating the RLP scheme in supporting TCP. $E(X)'$ is the normalized delay based on the maximum delay under consideration.

2.3 Analytical and Simulation Results

Fig. 2 shows the analytical results of EF_r of two schemes under different channel conditions. First, it is observed that in fast fading channels ($f_dT = 1$, or i.i.d. channel), the RLP retransmission efficiency is better than in slow fading channels ($f_dT = 0.01$). This implies that slow fading channels, with high correlations of channel states between frames, are very unfriendly to SR-ARQs like RLP. Second, we observed a very interesting characteristic of RLP: when the degree of correlation is high (slow fading), scheme (1,1,1,1,1,1) has better retransmission efficiency. While in low correlated channels (fast fading), scheme (1,2,3) has better retransmission efficiency. To understand the reason behind this phenomenon, we look into the channel characteristics implied by slow and fast fading. Slow fading, i.e., large f_dT, means longer error burst length. In other words, if a frame is in error, there is a high possibility that the frame following the error frame is in error too (note that this probability is M_{22} in the channel model). As a result, in scheme (1,2,3), retransmitting 2 or 3 copies of the same frame in consecutive frames does not help much to improve the recovery probability. It has almost the same efficiency as the scheme (1,1,1) in this case. Scheme (1,1,1,1,1,1) is better because it just retransmits one copy of the lost frame at a time, thus significantly reducing the correlation between two retransmitted copies and having better recovery probability, in turn higher retransmission efficiency. When the degree of correlation is low, scheme (1,2,3) has better efficiency. Because of the low correlation, recovery probability of scheme (1,2,3) and (1,1,1,1,1,1) are almost the same. But scheme (1,2,3) has shorter average delay, which means it causes less delay in delivering the frames to upper layer, which makes the scheme (1,2,3) perform better in this case.

We also study the RLP retransmission efficiency through simulations, and look at the RLP performance from another angle, TCP throughput, which is directly dependent on the RLP retransmission efficiency. The simulation is based

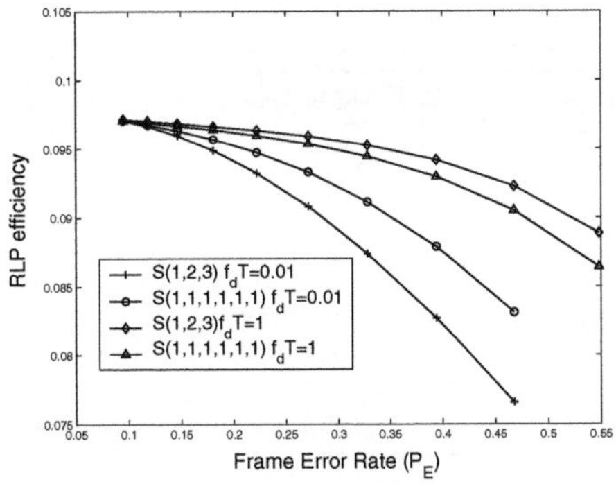

Fig. 2. RLP Retransmission Efficiency (EF_r) vs. P_E

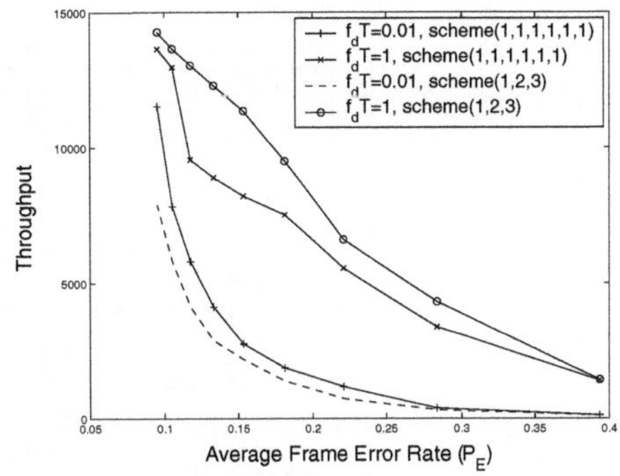

Fig. 3. TCP throughput in correlated channels vs. P_E

on Network Simulator 2 (NS2) [20]. Since NS2 does not provide RLP implementations, we developed RLP based on [14]. Other simulation parameters are: the RLP frame length is 5 ms, i.e., an RLP frame is sent every 5 ms. The wireless link bandwidth is assumed 2 Mbps, which is the common rate that 3rd generation wireless networks (such as IS2000 and UMTS) can provide. FTP is used as the application, which basically means the application always has enough data to transfer. TCP is used as the transport layer protocol with SACK option.

Fig. 3 shows the TCP throughput from simulation experiments. The TCP throughput shows the same characteristics as discussed above: RLP's overall performance is better in fast fading channels than in slow fading channels. In particular, in slow fading channels, scheme (1,1,1,1,1,1) performs better, while in fast fading channels, scheme (1,2,3) performance better. More importantly, the simulation results indicate that EF_r, the RLP retransmission efficiency, is a reasonable analytic metric for comparing the performances of different RLP schemes under correlated fading channels, especially when the supported transport layer protocol is TCP.

3 An Adaptive RLP

As we have pointed out, a single transmission scheme in RLP does not always yield the best performance. Sometimes, a mobile station may operate in highly correlated fading channels, while other times, it may operate in low correlated fading channels. This gives rise to the question of how we can achieve the best possible performance both in slow fading and fast fading channels with the RLP retransmission schemes that is most appropriate for the channel. The answer is to adaptively change the retransmission scheme based on the measured fading rate of the channel. Here we propose an adaptive RLP that embodies such strategy, which is called *Adaptive Radio Link Protocol* (ARLP). It is based on the RLP specifications of IS2000 systems [14].

Fig. 4 shows the basic functional units of the proposed ARLP. Because of the asymmetry of traffic load in uplink and downlink, most of bulk data transfer

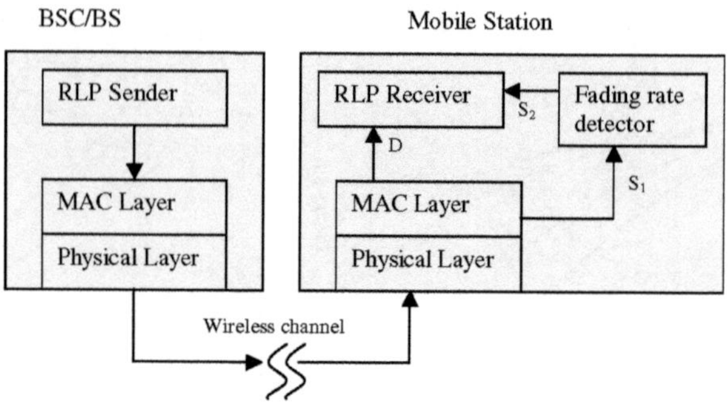

D: frame in RLP frame format, sent by MAC layer if it is correctly decoded and checked by MAC.
S_1: Signaling sent by MAC layer, indicating the current frame is "good" or "bad"
S_2: Signaling sent by Fading Rate Detector, indicating "fast" or "slow" fading characteristic of current channel

Fig. 4. Functional Units of ARLP

happens in the downlink direction, thus making the performance of down link more important in wireless networks for data. We will explain the application of the proposed ARLP mainly in downlink direction. Uplink can be treated similarly. If system complexity is a concern, the standard RLP can be used in uplink because the uplink traffic load is expected to be much lighter than downlink. The the performance improvement of utilizing ARLP in uplink may not be significant.

To detect the fading rate and its changes, some additional signaling is needed between the MAC layer and link layer. The MAC layer continuously decodes the frames received over the air interface from the base station. When a frame is decoded, no matter whether the frame is intended for this particular mobile station or not, the MAC sends a simple binary signal to the fading rate detector in link layer notifying the result of decoding: good frame or bad frame. This is different from most of the existing infrastructures where the MAC simply discards the frame if the decoding result indicates an error frame or a frame not for it. In such an infrastructure, the link layer is not aware of the discarded frames. As a result, most link layer adaptive protocols use ACK or NACK coming from receiver as the only source for channel estimate. However, for RLP, NACK is only sent when the RLP receiver finds gap(s) in the receiver buffer. This causes inefficient channel estimate and large delay in detecting change in channel characteristics, as we can easily see that a NACK received by sender indicates the channel condition almost a round trip time ago (or even longer because of bursty errors). Another reason justifying this inter-layer signaling is that we propose to do the channel estimate on the receiver side. The signaling between MAC layer and link layer in receiver significantly facilitates channel estimate. As proved in many earlier works, though the additional inter-layer interaction violates the principle of limiting the coupling of different layers of protocols as much as possible, sometimes it is necessary for wireless networks where efficiently utilizing the limited resources (e.g. bandwidth) is so important that it is worth to make a compromise.

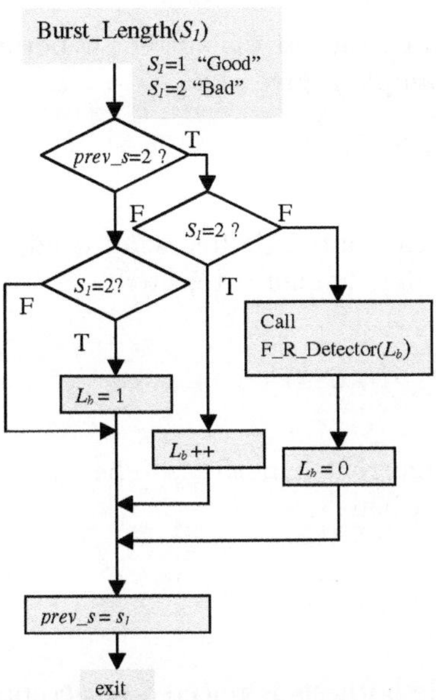

Fig. 5. Burst Length Sampling

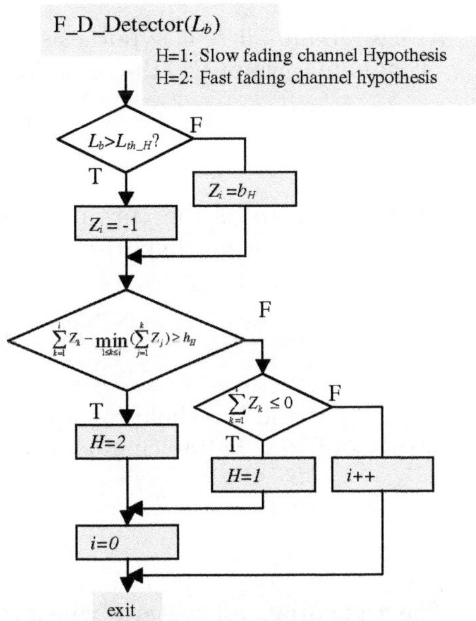

Fig. 6. Fading Rate Sequential Test

The purpose of fading rate detection is to estimate the channel fading rate, which is decided by the product of f_d and T in the correlated fading model. However, f_dT is not a directly measurable parameter for a wireless receiver. Therefore, we have to resort to a measurable parameter that has close relationship with f_dT. The average length of frame error bursts $L_b = \frac{1}{1-q}$, is a monotonically decreasing function of f_dT, when the average frame error rate P_E is fixed. Remember q is the conditional probability Prob.$\{S_t =$ "*bad*"$|S_{t-1} =$ "*bad*"$\}$. Since L_b is a measurable parameter in receiver, we use it as the main source of channel fading rate estimate.

There are several test techniques to estimate characteristics of a random variable generated by a stochastic process. The problem of detecting the average burst length can be directly transformed into such a problem: given a sequence of frame error burst lengths, how can we tell if this is a fast fading channel or a slow fading channel? We choose the sequential test technique to do this detection because of its advantages over other adaptive detection techniques [5]. Unlike other test techniques, the sequential test does not have a fixed sample size. Also, it has short response time. Our algorithm consists of two parts: burst length calculation and fading rate detection. Fig. 5 and Fig. 6 respectively show the pseudocodes for these two functions: Burst_Length(S_1) and F_D_Detector(L_b). When the $(i-1)$th frame is good, and the current ith frame is bad, a new burst of error frames begins. The burst length counter L_b keeps counting the number of continuous error frames until a good frame is detected. The length of this burst is then fed into the fading rate detector as a sample.

The fading rate detector is basically a sequential testing with two hypothesis: "fast fading channel" and "slow fading channel", which corresponds to short burst length and long burst length. Each burst length sample is scored based on its length and current hypothesis of the fading rate. If the current hypothesis is "slow fading channel", the score for each sample is given by:

$$Z_i = \begin{cases} -1 \text{ if } L_b > L_{th1} \\ +b \text{ if } L_b \leq L_{th1} \end{cases} \tag{7}$$

After the score of the current sample is calculated, if following condition is satisfied, the current hypothesis of slow fading channel is rejected:

$$\sum_{k=1}^{i} Z_k - \min_{1 \leq k \leq i} (\sum_{j=1}^{k} Z_j) \geq h_1 \tag{8}$$

where h_1 is the threshold for rejecting the current hypothesis. The current hypothesis of slow fading channel is accepted when

$$\sum_{k=1}^{i} Z_k \leq 0 \tag{9}$$

The sequential test restarts whenever the hypothesis is rejected or accepted. If the current hypothesis is rejected, the hypothesis is changed to the opposite one, which in our case is "fast fading channel", otherwise, current hypothesis is kept. The sequential test with "fast fading channel" hypothesis is done similarly to the test for "slow fading channel" hypothesis we have describe above. The current hypothesis, an indicator of current fading rate, is made a binary variable accessible to RLP receiver as the output of the fading rate detector. According to the value of this variable, the RLP receiver decides which retransmission scheme should be used for currently not correctly received frames. If the hypothesis is slow fading channel, scheme (1,1,1,1,1,1) is used; otherwise, scheme (1,2,3) is used. Note that the difference between these retransmission schemes is when to send NACKs and how many NACKs to send, which are actually decided by the RLP receiver. This is the reason why the channel fading rate estimate is done on the receiver side.

We set the default hypothesis as "fast fading channels" when the link is established. One may note that if the overall frame error rate (P_E) is low, even in slow fading channels, the error frame burst length may not be long. Thus our fading rate estimate, solely based on average burst length could be wrong. However, this false estimate is not going to hurt the performance of the adaptive proposal, ARLP. Low P_E indicates a fairly good channel condition. As we can see from the analysis and simulation results, in this case, the TCP throughput is indifferent to the retransmission scheme, because retransmission is rarely used.

4 Performance

4.1 Response Time

The time interval between a change in channel fading rate to the time the ARLP detects that change is defined as the *response time*. Fig. 7 shows the simulation results of average respond time to a typical channel fading rate change (from $f_dT = 1$ to $f_dT = 0.01$). We have observed that response time varies between 5 to 30 seconds. Also, the higher the P_E is, the shorter the response time is. This response time is reasonable if we take into account that the sample here is the burst length, which is available only when a burst finishes. The interval between bursts can range from less than one second to several seconds long and certain number of samples are need to make an accurate estimate. This is the reason why a channel estimate mechanism can detect the change of BER in less than a second, while detecting fading rate change needs several seconds. On the other hand, this response time is fast enough for the purpose of fading rate detection. Though the signal strength may vary fast, the fading rate is unlikely to vary very fast. The variation of SIR (Signal to Interference Ratio) is caused by the mobility, while the variation of fading rate is caused by the variation of the speed of mobile stations. Note that fading rate is decided by f_dT, which is proportional to the speed of the mobile station. It is reasonable to assume that most of the mobile terminal users are not likely to vary their speed in a large range in a short period. Instead, most users tend to maintain their speed for a relatively long time (for at least several seconds). If they change the speed (e.g. from walking to riding a car), they are also likely to maintain that changed speed for a relatively long time. The response time of several seconds is fast enough to catch the change and respond with the appropriate retransmission schemes.

4.2 TCP Performance Improvement

We study the performance improvement of system with ARLP over systems with standard RLP through simulation experiments. We use the same simulation environment as described in Section 2.2.

We first consider the scenario of one mobile user: a mobile user first walks for 80 seconds with a speed of 5 km/hour (about 3.10 miles/hour), which corresponds to $f_dT = 0.02$, then the user gets on his car and moves for another 80 seconds with a speed of 96 km/hour (about 60miles/hour), which corresponds to $f_dT = 0.4$. This scenario represent a typical fading rate change, thus is a good test for the proposed ARLP. The bulk data transmission happens during the whole process. Three systems are considered: a system of the proposed ARLP, a system of the standard RLP with scheme (1,2,3) and a system of standard RLP with scheme (1,1,1,1,1,1). Fig. 8 compares the TCP throughput of these three systems. We define the performance improvement as

$$\delta = \frac{T_{ARLP} - \max(T_{123}, T_{111111})}{\max(T_{123}, T_{111111})} \tag{10}$$

where T_{ARLP}, T_{123}, T_{111111} are throughput of ARLP, standard RLP with scheme (1,2,3) and standard RLP with scheme (1,1,1,1,1,1) respectively. Fig. 9 (a) shows that the performance improvement can be as much as 10% for the simulated scenario, which is a significant improvement considering that it is achieved at a fairly high frame error rate of 0.12.

In the second scenario, we consider the total TCP throughput improvement of the system. We assume a system with 100 users in full rate transmission. 20%

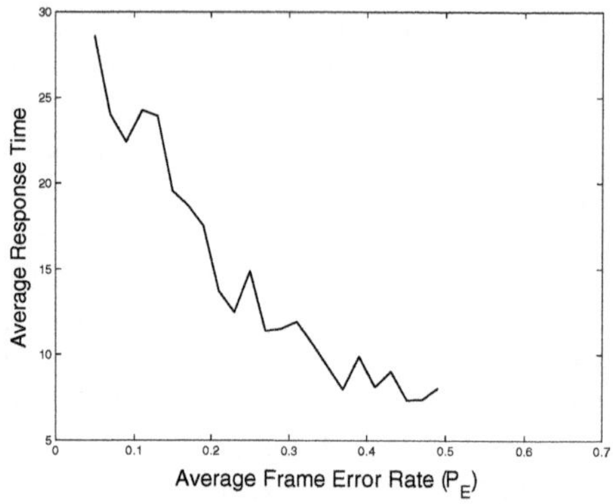

Fig. 7. Response Time vs. P_E

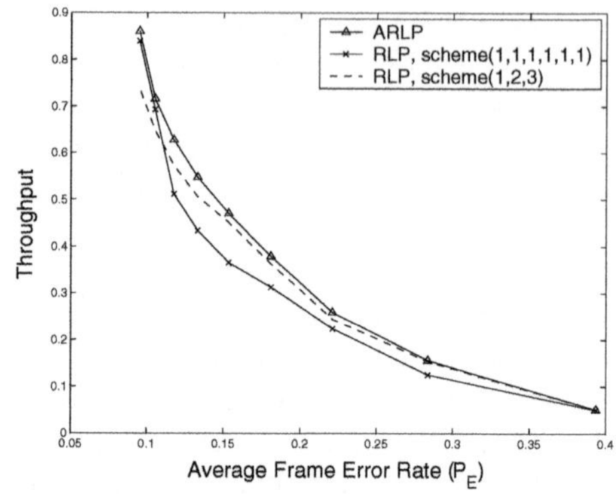

Fig. 8. Performance Comparison

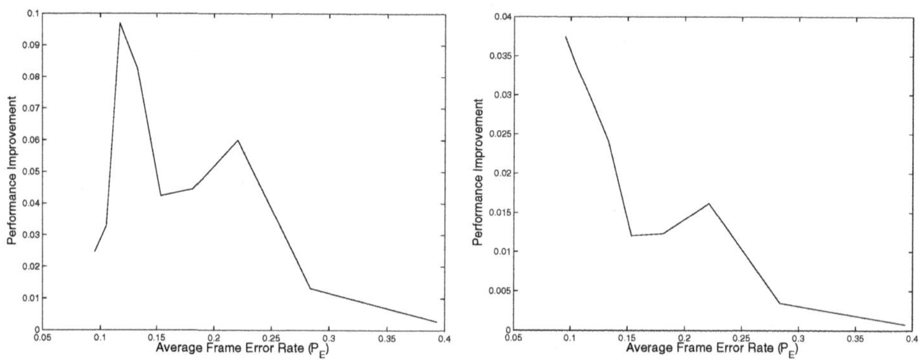

(a) Single User

(b) System

Fig. 9. ARLP Performance Improvement

of the users are pedestrians with a speed of 5km/hour, while 80% of the users are in vehicles with a speed of 96km/hour. The same three systems as in single user scenario are considered. Throughput improvement has the same form of definition as (10), with T_{ARLP}, T_{123}, T_{111111} representing the throughput of all users of each system instead of the throughput of a single user. Fig. 9 (b) shows the performance improvement δ, with respect to average frame error rate P_E. It shows that the throughput of the system can also be improved up to 3.8% with ARLP.

It is observed that the most significant improvement always happens in moderate average frame error rate. This can be explained by noting that on one hand, the proposed ARLP works by optimizing retransmission schemes, which can only be effective when there are considerable errors, one the other hand, when the error rate is too high ($P_E > 0.5$), link layer recovery becomes inefficient and not much improvement can be expected.

5 Conclusion

In this paper, we study the performance of different RLP retransmission schemes, especially scheme (1,2,3) and scheme (1,1,1,1,1,1), over correlated fading channels. We proposed a mathematically tangible metric: retransmission efficiency to evaluate the performance difference. We show that a single retransmission scheme does not result in best performance when the fading rate of correlated fading channels changes. To solve this problem, we propose a new adaptive RLP protocol ARLP, which uses sequential test technique to estimate the channel correlation, and adjusts its retransmission scheme according to the detected fading rate. Simulation results show that our ARLP can improve the TCP throughput performance when the user changes his/her moving speed, which results in changes in fading rate.

References

1. A.V Bakre and B.R. Badrinath, "I-TCP: indirect TCP for mobile hosts," *International Conference on Distributed Computing Systems*, 1995 pp. 136–143.
2. A.V Bakre and B.R Badrinath, "Implementation and performance evaluation of Indirect TCP," *IEEE Trans. on Computers*, Vol. 46 Issue 3, 1997 pp. 260–278.
3. S. Choi and K.G. Shin, "A class of adaptive hybrid ARQ schemes for wireless links," *IEEE Trans. on Vehicular Technology*, Vol. 50 Issue: 3, 2001 pp. 777–790.
4. H. Minn, M. Zeng and V.K. Bhargava, "On ARQ scheme with adaptive error control," *IEEE Trans. on Vehicular Technology*, Vol. 50 Issue: 6, 2001 pp. 1426–1436.
5. M. Rice and S.B. Wicker, "A sequential scheme for adaptive error control over slowly varying channels," *IEEE Trans. on Communications*, Volume: 42 Issue: 2, 1994, pp. 1533–1543.
6. R. van Nobelen, M. Seshadri, J. Whitehead and S. Timiri, "An adaptive radio link protocol with enhanced data rates for GSM evolution," *IEEE Personal Communications*, Volume: 6 Issue 1, 1999 pp. 54–64.
7. P. Lettieri and M. B. Srivastava, "Adaptive frame length control for improving wireless link throughput, range, and energy efficiency," *INFOCOM* 98, Volume: 2, 1998 pp. 564–571.
8. A. Chockalingam and G. Bao, "Performance of TCP/RLP protocol stack on correlated fading DS-CDMA wireless links," *IEEE Transactions on Vehicular Technology*, Volume: 49, 2000 pp. 28–33.
9. H.S.Hang, "On verifying the first-order Markovian assumption for a Rayleigh fading channel model," *Proc IEEE ICUPC*, 1994, pp. 160–164.
10. M. Zorzi, R.R. Rao and L.B. Milstein, "On the accuracy of a first-order Markov model for data block transmission on fading channels," *IEEE ICUPC'95*, pp. 211–215.
11. P. Sinha, N. Venkitaraman, R. Sivakumar and V. Bharghavan, "WTCP: A Reliable Transport Protocol for Wireless Wide-Area Networks," *ACM Mobicom*, Seattle. WA. August 1999.
12. 3G Partnership Project, Release 99
13. M. Rice and S.B. Wicker, "Adaptive error control for slowly varying channels," *IEEE Transactions on Communications*, February-April 1994 pp. 917–926.

14. TIA/EIA/IS-707-A-2.10, "Data Service Options for Spread Spectrum Systems: Radio Link Protocol Type 3," January 2000.
15. S. Souissi and S.B. Wicher, "A diversity combining DS/CDMA system with convolutional encoding and Viterbi Decoding," *IEEE Trans. on Vehicular Technology*, Vol44, NO.2, May 1995 pp. 304–312.
16. J. M. Harris and M. Airy, "Analytical model for radio link protocol for IS-95 CDMA systems," *Vehicular Technology Conference*, 2000. Vol. 3 pp. 2434–2438.
17. M. Abramowitz and I. A. Stegun, "Bessel Functions J and Y." §9.1 *Handbook of Mathematical Functions with Formulas, Graphs, and Mathematical Tables*, 9th printing. New York: Dover, pp. 358–364, 1972.
18. M. Schwartz, W. R. Bennett and S. Stein, *Communication Systems and Techniques*, New York: McGraw-Hill, 1966
19. H. Lin and S. K. Das, "TCP performance analysis of CDMA systems with RLP and MAC layer retransmissions," *10th IEEE International Symposium on Modeling, Analysis and Simulation of Computer and Telecommunications Systems* (MASCOTS 2002), pp. 313–320.
20. The Network Simulator – ns-2, http://www.isi.edu/nsnam/ns/

Novel Delayed ACK Techniques for Improving TCP Performance in Multihop Wireless Networks

Eitan Altman[1] and Tania Jiménez[2]

[1] INRIA, BP93
2004 Route des Lucioles,
06902 Sophia Antipolis Cedex, France
altman@sophia.inria.fr
http://www-sop.inria.fr/mistral/personnel/Eitan.Altman/me.html
[2] C.E.S.I.M.O., Facultad de Ingeniería,
Universidad de Los Andes,
Mérida, Venezuela
tjimenez@ula.ve

Abstract. We study in this paper TCP performance over a static multi-hop network that uses IEEE 802.11 protocol for access. For such networks it has been shown in [6] that TCP performance is mainly determined by the hidden terminal effects (and not by drop probabilities at buffers) which limits the number of packets that can be transmitted simultaneously in the network. We propose new approaches for improving the performance based on thinning the ACK streams that competes over the same radio resources as the TCP packets. In particular, we propose a new delayed ACK scheme in which the delay coefficient varies with the sequence number of the TCP packet. Through simulations we show that the ACK thinning allows to increase TCP throughput substantially more than previous improvement methods.

1 Introduction

Various mobility-induced aspects of TCP performance over ad hoc networks have been studied in the past, such as link breakage and routing failures and have motivated proposals for improvements of TCP:
(1) ELFN (Explicit Link Failure Notification) [7,11]: This allows TCP to interact with routing protocols and to freeze its timers when a route failure is detected
(2) Fixed RTO: An important source of degradation of TCP performance is the exponential backoff of TCP during timeouts due to disconnections, which may result in very long silence periods when connection is available again. In [5] it is proposed to disable the exponential backoff.
(3) ATCP proposed in [10] does not change TCP but implements another layer between the network and transport layers.
(4) TCP DOOR: TCP protocol is changed so as to be able to better react to out-of-order packets that may occur frequently in Ad-hoc networks [17].

M. Conti et al. (Eds.): PWC 2003, LNCS 2775, pp. 237–250, 2003.

(5) The ATRA framework [4]: consists of a set of MAC and routing layers mechanisms that reduce route failures, predict route failures before they occur and minimize the latency for route error propagation.

Most of the above work is summarized in [12]. Other work has been devoted to comparison of the performance of TCP under various Ad-Hoc protocols [3,5].

In this paper we focus on a static multihop network that uses the IEEE 802.11 protocol for access, see [6,15]. For such scenarios it has been shown in [6] that TCP performance is mainly determined by the hidden terminal effects (and not by drop probabilities at buffers) which limits the number of packets that can be transmitted simultaneously in the network (this is called the "spatial reuse"). In particular, for the chain topology that we study in this paper, the spatial reuse factor of the network has been shown in [6] to have some limit which was around 1/4, which means that given h nodes, only around $h/4$ can transmit packets simultaneously. This allowed the authors to conclude that there is an optimal size for the maximum window of TCP which is close to the value of $h/4$. Any increase in the maximum window size results in decreasing TCP throughput, although the loss in performance is not large (around 20%).

Two techniques have been proposed in [6] to improve the performance, one based on RED type technique and the other on adaptive spacing at the link layer. The reported gains in TCP throughput were between 5% and 30%.

In practice the number of nodes cannot be assumed to be known and can also vary in time. Thus the value of the maximum window for TCP for optimal operation may not be known, and we may have to use a large value for reasonable performance in cases there is a possibility that the number of nodes is large. Our first goal in this paper is to obtain a performance of TCP which is good for a large range of number of nodes; this is not possible with standard TCP where we know that good performance can be obtained only for a given number of nodes n by fixing the maximum window size to around $n/4$. Our second goal is to further obtain improvement in TCP performance with minimal changes in TCP.

To that end, we observe that the bottleneck of the system is the spatial reuse: the number of packets that can be transmitted simultaneously. Our aim is thus to decrease the flow of ACKs so as to give more bandwidth to TCP data packets. We note that although TCP ACKs are much smaller (40bytes) than TCP data packets (typically between 500 bytes and 1kbyte), their transmission requires the same signaling overhead of the 802.11 MAC, i.e. the RTS and CTS packets that precede each transmission of a packet and an ACK packet (of the MAC layer) for acknowledging successful transmission.

To reduce the ACK flow, we use the delayed ACK option of TCP, in which an ACK is generated for every d TCP packets, with one exception: if the first packet (of an expected group of d packets) arrives at the destination, then after some time interval (typically 100ms) if d packets have not yet arrived, then an acknowledgement is generated without further waiting. The standard delayed ACK option has the value $d = 2$ (see RFC 1122). We observe significant improvement of standard delayed ACK with respect to the nondelayed option. This then

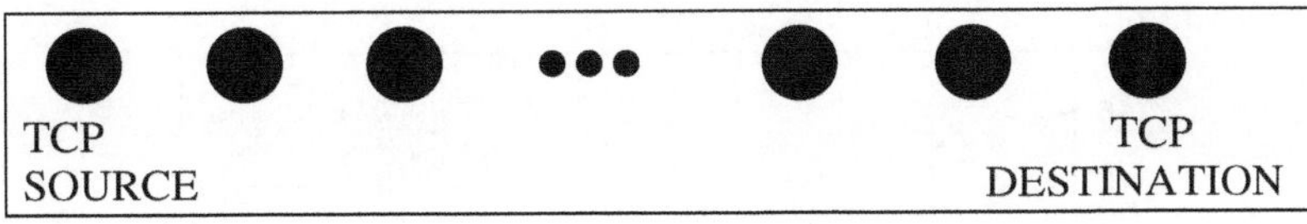

Fig. 1. The chain topology

motivates us to modify the standard delayed ACK option and to take $d > 2$. Using simulations in ns [1], we test this approach for both permanent as well as short TCP connections. We show that this allows to increase TCP throughput by around 50%, which is substantially more than the previous improvement methods. We then propose a modification in the delayed ACK approach which we call the "dynamic delayed ACK", where the parameter d gradually increases with the sequence number of the acknowledged packet. This allows us to have more ACKs when the window size is small, thus enabling to increase rapidly the throughput, and less ACKs when the window is large (and during which TCP is more vulnerable to losses).

The structure of the paper is as follows. We present in the next section the simulation scenario. Then we present in Section 3 our findings on performance of persistent TCP connections. In Section 4 we introduce our dynamic delayed ACK modification; through simulations of short TCP connections, we compare this proposal to standard TCP (without delayed ACK) and to TCP with delayed ACK with different fixed parameters d. We end with a concluding section.

2 The Simulation Scenario

We use ns2 [1] to simulate the chain scenario of [6] that consists of n nodes over a line separated by a distance of 200m, as seen in Figure 1. For each wireless node, the transmission range is 250m, the carrier sensing range is 550m and the interference range is about 550m.

We use the standard two-ray ground propagation model, the IEEE802.11 MAC, an omni-directional antenna model of ns [1] and an interface queue length of 50 at each node. We tested the NewReno version of TCP, which is the most deployed one. We tested four scenarios: 3,9,20 and 30 nodes. The cases of 3 and 9 nodes required 150 sec per simulation (to obtain stationary behavior). The other cases required 1500 sec per simulation. A TCP data packet is taken to be of size 1040 bytes. We used the AODV (Ad hoc On-Demand Distance Vector) routing algorithm (see [14]) in our simulations.

3 Simulation Results for Persistent TCP Connections

3.1 Effect of Delayed Ack on TCP Throughput

Our simulation results for $n = 9, 20$ and 30 nodes are summarized in Figures 2-4, respectively. In all those cases the hidden terminal effect (rather than buffer

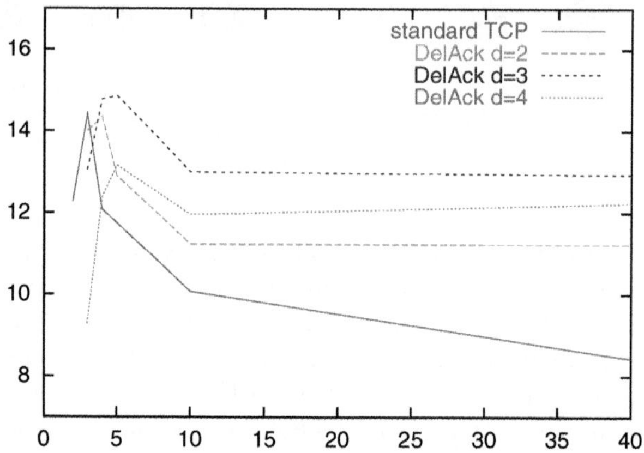

Fig. 2. Throughput in pkt/sec for $n = 9$ as a function of the maximum window size

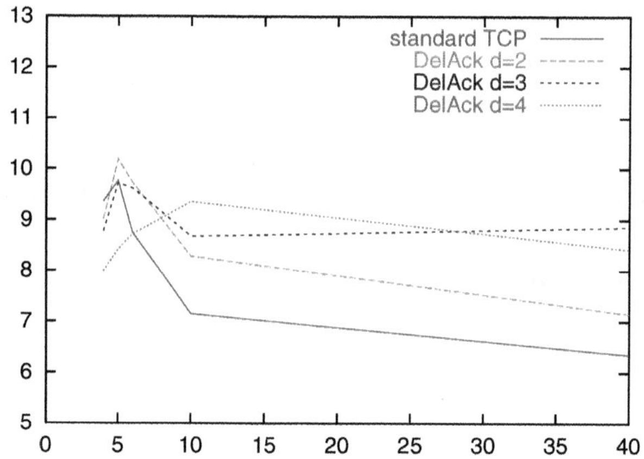

Fig. 3. Throughput in pkt/sec for $n = 20$ as a function of the maximum window size

overflow) causes losses (the hidden terminal problem is well explained in [6]). We shall later examine the case of $n = 3$ in which we do not have TCP losses.

We see that the standard Delayed Ack option ($d = 2$) slightly outperforms the standard TCP (yet with another value of maximum window size) for $n = 9$, and largely outperforms (more than 10%) the standard TCP for $n = 30$. A further improvement is obtained by the Delayed Ack with $d = 3$ (for both $n = 9$ as well as $n = 20$). But the most important improvement that we see is that all delayed ACK versions are better than the standard TCP for maximum window sizes of more than 10, with the options of $d = 3$ or $d = 4$ outperforming the standard delayed ACK option. For $n = 9$, the Delayed ACK version with $d = 3$ is seen to yield between 30% to 40% of improvement over standard TCP for

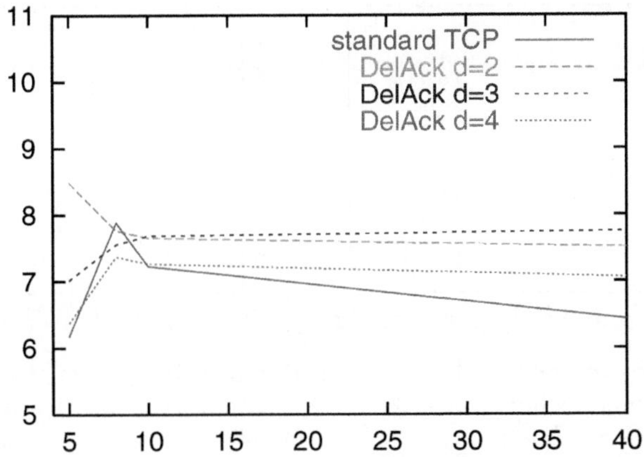

Fig. 4. Throughput in pkt/sec for $n = 30$ as a function of the maximum window size

any maximum window sizes larger than 10; in that range it also outperforms standard TCP by 20%-30% for $n = 20$ and by $6\% - 20\%$ for $n = 30$. The version $d = 4$ performs even better for $n = 20$ for maximum windows between 10 to 25. An even better performance of delayed ACK can be obtained by optimizing over the timer duration of the Delayed Ack options, as we shall see later.

Yet the most important conclusion from the curves is the robustness of the Delayed Ack options. In practice, when we do not know the number of nodes, there is no reason to limit the maximum window size to a small value, since this could deteriorate the throughput considerably. When choosing large maximum window, the delayed ACK versions considerably outperform standard TCP. They achieve almost the optimal value that the standard TCP could achieve if it knew the number of nodes and could choose accordingly the maximum window.

For a fixed small size of maximum window size, the Delayed Ack option does not outperform the standard TCP version since most of the time, the window size limits the number of transmitted TCP packets to less than d, which means that the delayed ACK option has to wait until the timer (of 100ms by default) expires before generating an ACK; during that time the source cannot transmit packets.

Next, we plot the window size evolution for $n = 9$ for standard TCP and for TCP with delayed ACK option with $d = 3$. The window size is sampled every 0.1 sec.

We see that although the maximum window size is 2000, the actual congestion window does not exceed the value of 13. We see that from the figures that in standard TCP, losses are more frequent and more severe (resulting in timeouts) whereas the $d = 3$ version of delayed ACK does not give rise to timeouts.

In Figure 7 we present the evolution of the congestion window size for standard TCP with maximum window size of 3 for the case of 9 nodes. We know from [6] that a maximum size of between 2 and 3 should indeed give optimal

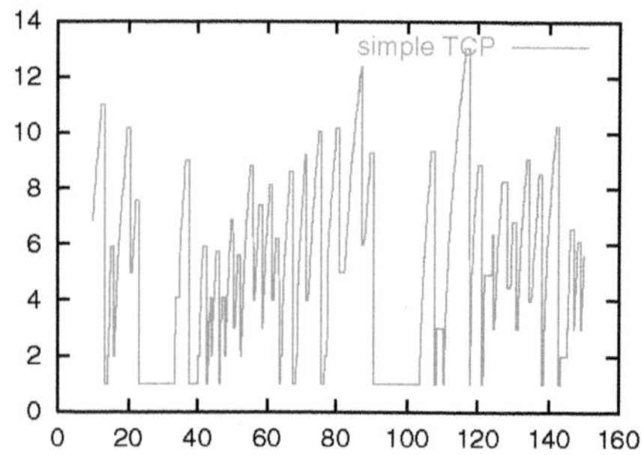

Fig. 5. Window size evolution for standard TCP with maximum window of 2000

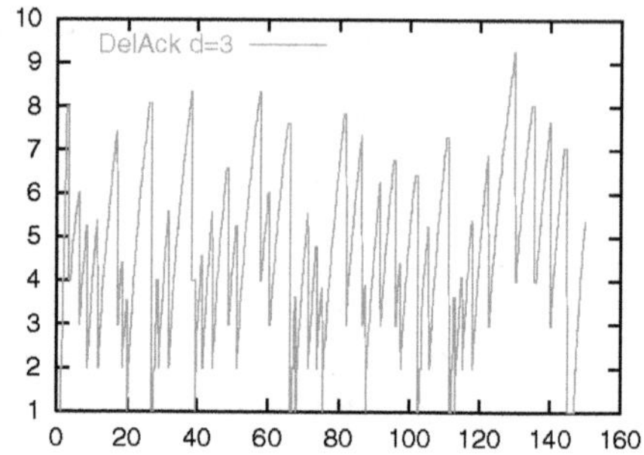

Fig. 6. Window size evolution for DelAck TCP with $d = 3$, with maximum window of 2000

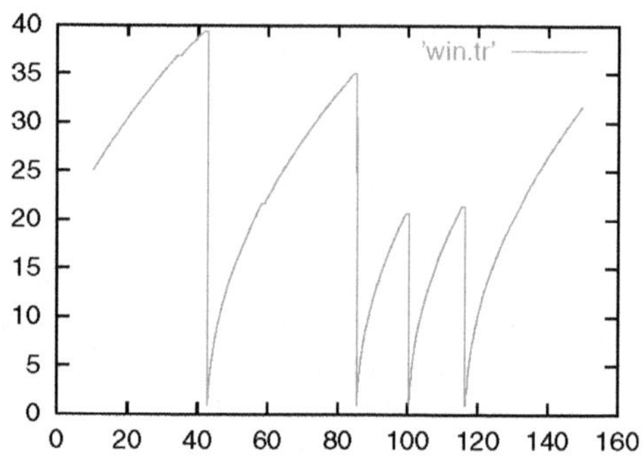

Fig. 7. Window size evolution for standard TCP (delayed ACK disabled) with 9 nodes and maximum window size of 3

performance (and this is confirmed in Figure 2). We see in Figure 7 that there are almost no losses (the window is seen to decrease only 4 times, in contrast to a much larger number in previous figures). Note that the actual window size is the minimum between the congestion window (depicted in the Figure) and the maximum window size (whose value here is 3).

3.2 Varying the Delayed Ack Time Interval

In the previous Figures, all versions using delayed Acks had the default interval of 100msec (as explained in the Introduction). Next, we vary the interval length and check its impact on throughput, see Fig. 8. We consider the delayed ACK version with $d = 3$.

We see that the default value performs quite well, although for small maximum windows, shorter intervals perform slightly better, whereas with large max-

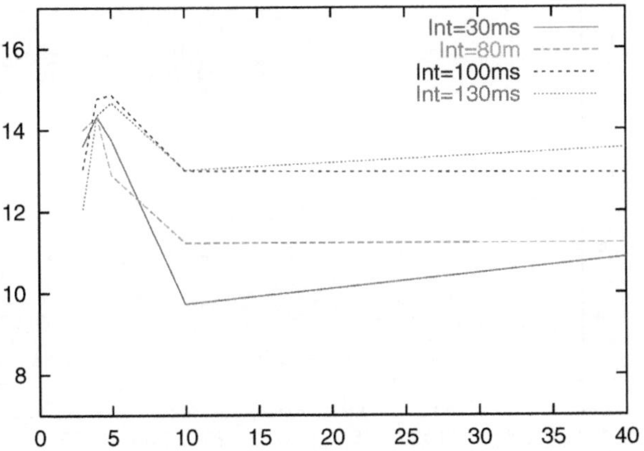

Fig. 8. The influence of Delayed Ack interval on TCP throughput, as a function of the maximum window size. $d = 3$

Table 1. Number of transmitter packets during 149sec for $n = 3$ as a function of the maximum window size

	Standard TCP	*Delayed Ack Versions*		
WinMax	Standard	$d = 2$	$d = 3$	$d = 4$
3	6068	6602	6763	2699
2000	6094	6565	6779	6888

imum window, a larger interval (130ms) is slightly better. We tried to further increase the time interval beyond 130ms but then the throughput decreased.

3.3 $n = 3$: No Hidden Terminals

Finally, we consider the case of $n = 3$ nodes. In that case the hidden terminal phenomenon does not occur anymore (due to the RTS/CTS handshake mechanism, see [6]), and we have not observed TCP losses during the simulation for any value of window size. Even then, delayed ACKs can be used to improve considerably the performance. This is illustrated in Table 1 that gives the number of TCP packets successfully received within 149 sec for $n = 3$. Since there are no losses, then as long as d is greater than the max window, we expect to improve the performance as d gets larger, since TCP packets compete with less ACKs. This is indeed confirmed in Table 1. The improvement that increases from 10% to 15% as d grows from 2 to 4, does not depend on the maximum window (as long as it is greater than d). However for $d = 4$ we see, as can be expected, that we get a bad performance for a maximum window of 3, since the destination always needs to wait till the 100ms interval of the Delayed Ack option expires in order to send an ACK (since the windows allows for sending only 3 data packets).

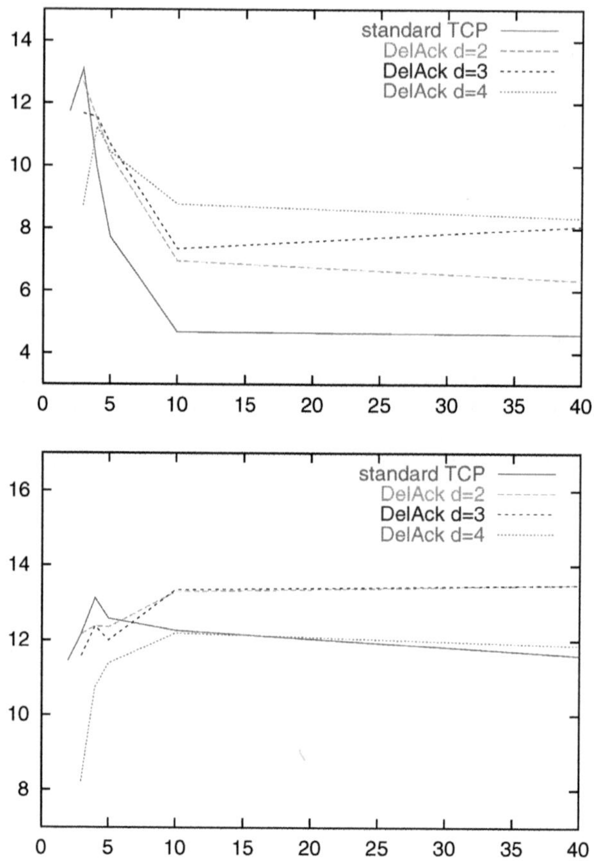

Fig. 9. Throughput in pkt/sec for $n = 9$ as a function of the maximum window size, DSR (up) and DSDV (down) routing protocol

3.4 Other Routing Algorithms

As we already mentioned, TCP performance is affected by the fact they have to share the channel with the ACK packets. But ACK are not the only packets that compete for the radio channel. Also packets involved in route discovery and in route maintenance take a share of the radio resources, so it is not surprising that different ad-hoc routing algorithms may result in different performance of TCP even in our simple static chain topology. We shall thus study TCP performance under various routing protocols.

We repeat the simulations for $n = 9$ with the DSR (Dynamic Source Routing) protocol [8] as well as for the DSDV (Destination-Sequenced Distance-Vector) [13], and depict the throughput as a function of the maximum window sizes in Figure 9.

We make the following observations:

- In general, the throughputs obtained under AODV are of the same order as under DSDV, and both are larger than under DSR.
- In DSR, a large degradation in throughput for standard TCP (without delayed ACK) is observed as the maximum window size increases from its best value of 3. The degradation here is from around 13pkt/sec to 4.6pkt/sec (a decrease by 65%) whereas for the AODV routing we obtained (as in [6]) a degradation of only 40%.
- In DSDV, there is a very small degradation of the throughput for standard TCP as the maximum window size increases from its best value of 4 (note also that the best window size is 4, here which is slightly larger than in the value of 3 obtained for the other routing protocols).
- For DSR, the improvement in TCP throughput with the delayed ACK versions of $d = 3$ and $d = 4$ with large maximum windows (which may be necessary when we do not know ahead what the number of hops is) is even larger than in AODV. It is almost twice that of standard TCP, where as in the AODV it was around 1.5 times larger. In DSDV we also obtain an improvement but it is lower than for AODV (around 17% of improvement).

4 Short TCP Transfers and Dynamic Delayed Ack

In a recent paper [9], the authors show that conclusions drawn from simulating permanent TCP connections can be qualitatively quite different than those obtained from simulations of transfers that have large time scale variability. The latter is obtained by replacing the infinite source model by ones in which transferred files have heavy-tailed distributions.

4.1 Simulated Scenario

We use TCP connections whose size has a Pareto distribution. Recall that for Pareto distribution, $E[size] = \beta k/(\beta - 1)$ where $\beta = 1.5$ is our shape parameter and where the average file size that we took is $E[size] = 30kbytes$; thus the parameter k equals in our case to 10000. The complementary distribution of a file size is $Pr(size > s) = (k/s)^{\beta}$. The time between the end of a connection till the beginning of the next connection has an exponential distribution with average of half a second. This characterization of TCP sessions is compatible with measured Internet traffic, see e.g. [16].

In addition to the throughput, we now consider the average session delay (the time between the opening of a session till the last ack is received at the source). This performance measure of course does not have a meaning in the previous setting of a persistent TCP connection.

All simulations last 1500sec and use the AODV routing protocol. TCP data packet sizes are again 1000bytes plus 40bytes of headers (for IP and TCP). We used the same generators and seeds for simulations with different parameters so as to have always the same input traces (sizes of file transfers and times between transfers).

4.2 The Dynamic Delayed Ack

We already saw in the previous section that the delayed ACK performs better when the window size is large, since sufficiently packets arrive so that we do not have to wait till the Delayed Ack time interval (of 100ms) expires before generating an ACK.

We therefore introduce a modification of the Sink behavior in TCP that is based on an adaptive version of the delayed ACK, in which the parameter d changes dynamically according to the sequence number of the acknowledged packet; it increases gradually from 1 to 4 as the sequence number increases. More precisely, we define three thresholds: $l1, l2, l3$ such that $d = 1$ for packets with sequence N numbers smaller than $l1$, $d = 2$ for packets with $l1 \leq N < l2$, $d = 3$ for $l2 \leq N < l3$ and $d = 4$ for $l3 \leq N$.

This will then be tested in our framework of short TCP connections. Note that we could expect TCP to perform even better if the decision on the size of d could depend on the window size (and/or whether we are in slow start or in congestion avoidance phase, as was proposed in [2]) rather than on the sequence number. This would then allow to use dynamic delayed Acks also for persistent TCP connections. However, this information is not available at the sink and would require further complications that would either involve extra signaling either at the transport layer or perhaps at a higher layer.

4.3 Simulation Results

We examine in this section the influence of different delay ACK schemes on three performance measures of the TCP connections: the throughput, the losses and the average session delay. The comparisons are done for various values of maximum window sizes. We consider the case of 9 and 20 nodes respectively, and depict the simulation results in figure 13.

The threshold that we used for the dynamic delayed Ack approach are $\{2, 5, 9\}$ which gave the best results for both throughput and delays and for both $n = 9$ and $n = 20$ nodes. The dynamic Delayed Ack was only tested for maximum window sizes of 10 or more for $n = 9$ and for 15 or more for $n = 20$ (as its goal is to be able to perform with large value of maximum window so as to be useful when we have no knowledge that allows us to choose the optimal value of the maximum window size).

In terms of throughput, we see in Figure 10 again that delayed ACK approaches have better performance than the standard TCP: any version of delayed ACK gives better throughput than the standard TCP for maximum window size larger than 7 (for $n = 9$) and larger than 10 (for $n = 20$). For large window sizes, the new dynamic DelAck has the best performance. The standard delayed ACK ($d = 2$) outperforms the standard TCP for any value of maximum window size in the case of $n = 20$, and almost all values of maximum window size in the case of $n = 9$.

Next, we consider the losses. We see from Figure 11 that almost for all values of maximum window, the number of losses is largest for the case of standard

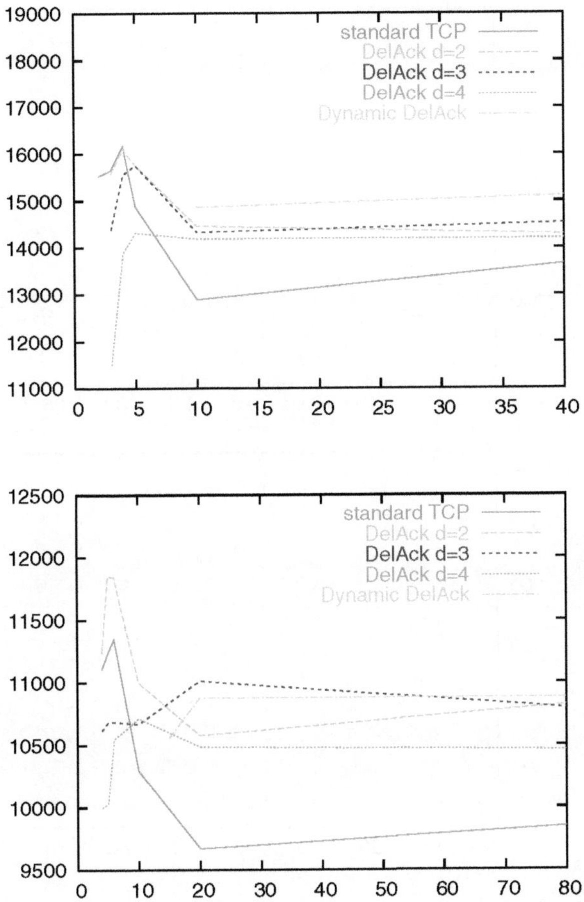

Fig. 10. No. of transferred packets (not including retransmission) during 1500sec for $n = 9$ (up) and $n = 20$ (down) as a function of the maximum window size

TCP, and it *decreases* in the delayed ACK versions as d increases. The smallest number of losses are obtained in our dynamic DelAck scheme.

Next we observe the delay in Figures 12-13. We see that the standard delayed ACK ($d = 2$) outperforms the standard TCP (without delayed ACK) for any value of maximum window for both $n = 9$ and $n = 20$. For large window size, the improvement is of around 18% for $n = 9$ and 22% for $n = 20$. Again, the best performance is obtained by the dynamic DelAck scheme for maximum window size larger than 10 (with $n = 9$) and larger than 20 (for $n = 20$).

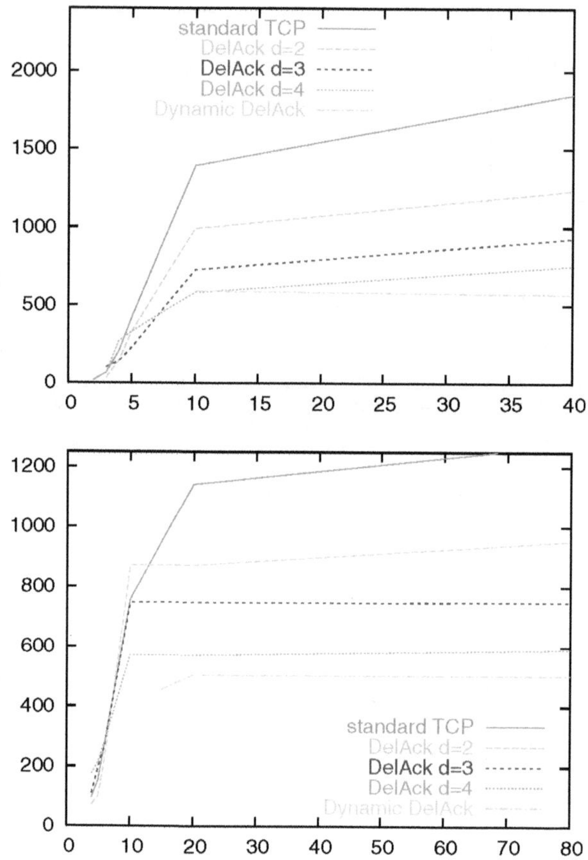

Fig. 11. No. of losses of packets during 1500sec for $n = 9$ (up) and $n = 20$ (down) as a function of the maximum window size

5 Concluding Remarks

We have shown in this paper that Delaying Ack (as defined in RFC 1122) improves the performance of TCP over static multi-hop networks that uses IEEE802.11 as MAC. Increasing the number packets that are acknowledged by an ACK to $d > 2$ (which is not recommended in RFC 1122) further improves the throughput. The improvements are due to the fact that ACKs and TCP packets contend over the same channel, so decreasing the throughput of ACK can improve the throughput of TCP. When we do not know the number of nodes in the system, we cannot limit the maximum window size to a small value, since this could result in a dramatic decrease of throughput if the number of nodes is large. Large values for maximum windows have to be used, for which the Delayed Ack options with $d > 2$ have been shown to yield an improvement of between 20% to 40%.

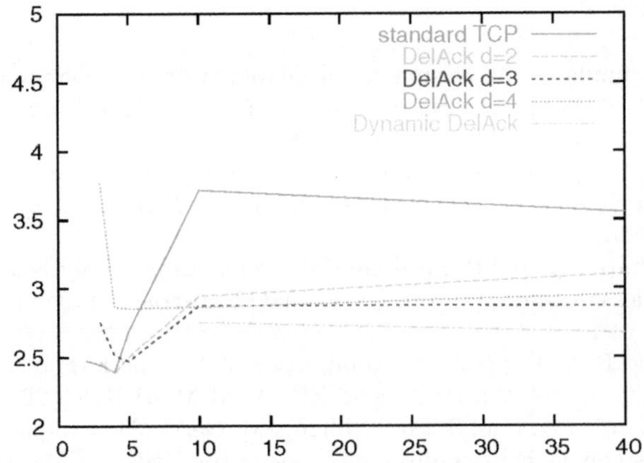

Fig. 12. Average session delay for $n = 9$ as a function of the maximum window size

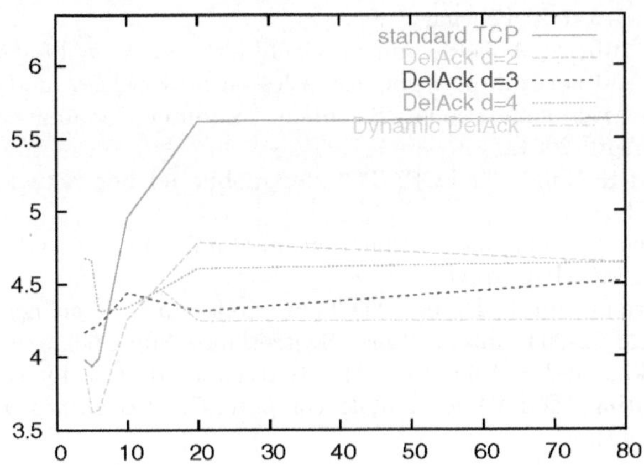

Fig. 13. Average session delay for $n = 20$ as a function of the maximum window size

Yet the advantage of more ACKs is that they allow to increase rapidly the window size when it is small. Therefore further improvement can be expected if the value of d increases with the window size. Since the window size is not available at the sink, we proposed instead a dynamic DelAck scheme in which the value d increases with the sequence number of the acknowledged packet. We showed that its performances in terms of delay, loss probabilities and throughput outperform those of delayed ACK schemes with fixed d for large values of maximum window.

References

1. Network Simulator, ns version 2.1, available at http://www.isi.edu/nsnam/ns/
2. M. Allman, "On the generation and use of TCP acknowledgements", ACM Computer Communication Review, Oct. 1998.
3. A. Ahuja, S. Agrawal, J. P. Singh and R. Shorey, *Performance of TCP over different routing protocols in mobile ad-hoc networks*, *Proceedings of IEEE VTC*, Tokyo, Japan, 2000.
4. V. Anantharaman and R. Sivakumar, "A microscopic analysis of TCP performance over wireless Ad-hoc networks", (extended abstract) *Proc. of ACM Sigmetrics*, 2002.
5. T. D. Dyer, R. V. Boppana, "A comparison of TCP performance over three routing protocols for mobile Ad Hoc networks", ACM MOBIHOC, 2001.
6. Z. Fu, P. Zerfos, H. Luo, S. Lu, L. Zhang, M. Gerla, "The impact of multihop wireless channel on TCP throughput and loss", Proc. *IEEE INFOCOM* 2003. Available on www.cs.ucla.edu/wing/publication/publication.html
7. G. Holland and N. Vaidya, "Analysis of TCP performance over mobile Ad Hoc networks", *ACM Mobicom*, Seattle, Washington, 1999.
8. D. B. Johnson, D. A. Maltz, Y.-C. Hu, J. G. Jetcheva, "The Dynamic Source Routing Protocol for Mobile Ad Hoc Networks (DSR)", IETF MANET Working Group INTERNET-DRAFT, available at http://www.ietf.org/internet-drafts/draft-ietf-manet-dsr-07.txt 21 February, 2002.
9. Y. Joo, V. Ribeiro, A. Feldmann, A. C. Gilbert and W. Willinger, "TCP/IP traffic dynamics and network performance: a lesson in workload modeling, flow control, and trace-driven simulations", Sigcomm Computer Communications Review, vol 31 No. 2, April 2001.
10. J. Liu and S. Singh, "ATCP: TCP for mobile ad hoc networks", *IEEE JSAC*, **19**(7), 2001.
11. J. P. Monks, P. Sinha and V. Bharghavan, "Limitations of TCP-ELFN for Ad Hoc networks", MOMUC 2000.
12. R. de Oliveira and T. Braun, "TCP in wireless mobile ad hoc networks", Tech. Report IAM-02-003, univ. of Bern, Switzerland, July, 2002.
13. C. E. Perkins and P. Bhagwat, "Highly dynamic destination-sequenced distance-vector routing (DSDV) for mobile computers", *Proceedings of SIGCOMM*, pp. 234–244, 1994.
14. C. E. Perkins and S. R. Das, "Ad hoc On-Demand Distance Vector (AODV) Routing", IETF MANET Working Group INTERNET-DRAFT, available on http://www.ietf.org/internet-drafts/draft-ietf-manet-aodv-11.txt, 19 June, 2002.
15. V. Ramarathinam and M. A. Labrador, "Performance analysis of TCP over static ad hoc wireless networks", *Proc of ISCA 15th International Conf on Parallel and Dist. Computer Systems (PDCS)*, 410–415, Sept 2002.
16. B. Sikdar, S. Kalyanaraman and K. S. Vastola, "An Integrated Model for the Latency and Steady-State Throughput of TCP Connections", *Performance Evaluation*, v.46, no.2–3, pp.139–154, September 2001.
17. F. Wang and Y. Zhang, "Improving TCP performance over mobile Ad-Hoc networks with out-of-order detection and response", *ACM MOBIHOC*, June 2002.

Performance Evaluation of Transport Protocols with Local Mobility Management

Nikos Georganopoulos* and A. Hamid Aghvami

Centre for Telecommunications Research,
Kings College London, Strand,
London WC2R 2LS, UK
n.g@ieee.org
hamid.aghvami@kcl.ac.uk

Abstract. There is an increasing interest to design a full IP Access Network that can provide IP connectivity to mobile users. UDP and TCP transport layer protocols are designed to deliver data across wired networks. Mobility of users in such a wireless system results in transport layer protocols facing a very different packet loss and delay pattern. The use of an efficient mobility management protocol is essential to the performance of the protocols and the relevant applications running on top. In this paper[1] the performance of UDP and TCP transport layers is evaluated on a wireless IP access network running the BCMP localised mobility management protocol, through simulations using the ns-2 simulation platform. Data and multimedia applications are employed to test their performance. UDP and various TCP variants are tested and the objective is to evaluate their performance in such a system and to examine the effect of the mobility of users.

1 Introduction

There are currently numerous new data applications being developed and provided in the Internet. At the same time important applications like voice and video are to be offered also over IP with the use of developed QoS mechanisms. Furthermore, the philosophy behind the IP protocol stack is regarded to be better than that of traditional networks such as ISDN or ATM that are connection oriented employing cell or frame switching. Finally the growth of the Internet is rapid and is extending in terms of data traffic volumes, users and geographical locations. All these reasons prompt the use of IP for a wireless access network that will provide a vast number of services to its users, on the move. As a result there is currently ongoing research effort to design such a system that can provide IP services and applications to mobile users through a mobile-fixed converged backbone network. These systems could be considered as next generation cellular system offering full IP connectivity to its users.

* Currently with: Toshiba Research Europe Ltd, 32 Queen Square, Bristol BS1 4ND, UK.

1 Part of this work has been performed in the context of the IST consortium BRAIN (IST-1999-100 50) and MIND (IST-2000- 28584), which are partly funded by the European Commission. The authors would like to thank their colleagues for their contribution in this work.

M. Conti et al. (Eds.): PWC 2003, LNCS 2775, pp. 251–260, 2003.

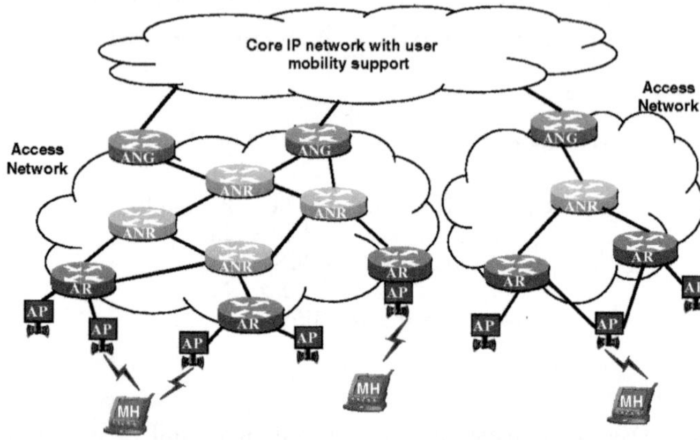

Fig. 1. Access Network Reference Architecture, including one or more Access Network Routers (*ANR*) and is connected to the Core Network (*CN*), a 'standard' IP network, through an Access Network Gateway (*ANG*). Access Router (*AR*) are special ANRs residing on the edge of an AN and connected to one or more Access Points (*AP*), a layer-2 device that is connected to one or more ARs and offers the wireless link connection to the Mobile Host (*MH*). An AR offers IP connectivity to MHs, acting as a default router to the MHs it is currently serving

1.1 Access Network-Local Mobility Management Domain

There is thus a strong motivation behind the design of an IP wireless access network system. A high level description of such a network topology can be found in [1] where definitions of the different system entities are given. Such a system will manage mobility of the users locally inside its administrative domain through the use of a Localised Mobility Management (LLM) [2] solution. Figure 1 shows a general diagram of such wireless access network architecture. The main network components in the architecture are the following [1]. The Access Network (AN) (Fig. 1) is a full IP network.

An LMM solution will be running inside the AN on specific ANRs that will be responsible for tracking the movement of the MHs at the edge of the network and ensuring that packets are timely delivered to every MHs' current point of attachment to the network.

1.2 Transport Layer Protocols

User Datagram Protocol (**UDP**) is a connectionless transport protocol that provides the fastest way for applications to send messages and is used by time sensitive multimedia applications. Transmission Control Protocol (**TCP**) provides a reliable data-transfer and is used for bulk data traffic and interactive data applications. TCP is the mostly used transfer protocol and supports the transfer of over 90% of all traffic in the public Internet today. Different variants of TCP have been proposed and are considered in this study. These include Tahoe, Reno, Vegas, NewReno, Sack and Fack. Support of wired Internet will be inherit in such a system, thus rendering necessary the evaluation of the performance of transport layer protocols and the effect of the mobility of users

2 Mobility in the Wireless IP Access Network

In theory, transport protocols should be independent of the technology of the underlying layers and particular the transmission media at the physical layer. However the wireless IP-based AN will be designed to provide IP services and applications to MHs, over the air where user mobility is of primary concern. There are thus two main characteristics of the AN that will have an effect on the performance of the transport protocols:

- **Host (terminal) mobility:** Packets can get lost as hosts move and handover between different ARs residing at the edge of the network. The path that packets follow in the AN towards the MHs changes and is updated to track their movement. Packets can then be lost or reordered during this update and the applications will suffer. The impact of host mobility on the transport protocol performance is mainly studied and presented in this paper, together with the interaction with the chosen mobility management solution that is running in the AN.
- **Wireless channel:** Packets will be lost because of bad performance of the wireless channel (fast and slow fading). This will have an impact on applications running on top of the transport protocols. The architecture of the network considered in this study is independent of the chosen air-interface, which is though dependent on the actual system implementation. This aspect thus is not considered in this study. There are various other proposals to improve the performance of transport protocols over wireless links.

2.1 A Localised Mobility Management Solution

The mechanism used to track the movement of the MHs at the edges of the AN is called LMM protocol. To manage handovers between domains a global mobility management protocol is needed like Mobile IP (MIP) [3]. An LMM protocol is needed to provide either fast (minimal packet delays) or smooth (minimal loss) handovers. Furthermore, local mobility should be managed as transparently as possible to the nodes outside of the access network. These solutions include protocols like Hierarchical MIP, Cellular IP, HAWAII.

In this study a new LMM proposal, called BRAIN Candidate Mobility Management Protocol (BCMP) [4] is employed. Only limited number of nodes need be modified to support BCMP. The network consists of legacy IP routers with added mobility aware functionality in just two types of nodes. Anchor Points (ANPs) own and allocate IP addresses, authenticate users, maintain records, and tunnel packets towards MHs. ARs terminate tunnels from ANPs and forward packets to/from mobile hosts. By supporting the ability for the ANG to change, the BCMP provides a resilient and scalable solution. The main protocol procedures dealing with mobility are:

- **Handover and Path Updates:** This function includes an optional handover preparation phase to ensure fast and smooth change of ARs by performing a planned handover. The handover execution procedure is the same whether or not there has been a preparation phase (unplanned handover). The ANP is also notified about the new location of the MH and the path is updated.

• **Inter-Anchor Handover:** With this function the protocol allows (but does not mandate) the network operator to request that a MH changes ANP if it has moved far away from its original ANP. This improves routing efficiency in the AN in exchange for exposing mobility toward the Internet.

Furthermore login, logout and paging procedures are defined but are not relevant to the study presented here.

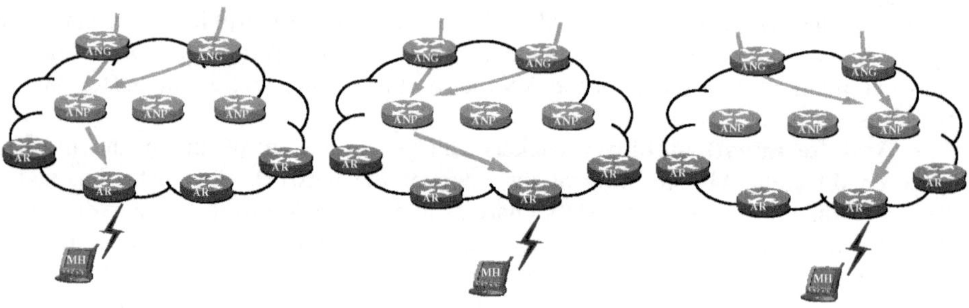

Fig. 2. BCMP Operation

3 Experimental Set-Up

The evaluation of the performance of transport layer protocols has been performed using the ns-2 simulation platform [5]. As mentioned above different network topologies are employed. These include a tree topology and a mesh topology. Regarding BCMP [4] has shown the best configurations of the protocol entities to get the best performance in terms of handover delay and application performance. This configuration requires to put the ANPs in the middle of the AN, that is nodes 10, 11, 12 in the tree topology and nodes 14, 16,18 in the mesh topology.

Nodes 0 to 5 represent various CHs and node 6 in this study represents the MHs' HA that forwards the packets to the current address of the MH. One of the properties of the BCMP is that it is independent of the global mobility protocol but for this study the use of MIP is assumed. Although there is no mobility between different ANs, the MH can change its assigned address inside the AN when changing ANPs something that will trigger a global mobility event. This also depends on the actual size and physical topology of the AN. Furthermore the protocol supports multiple ANGs, which in the mesh topology are in nodes 7 and 9. In the tree topology there is only one ANG in node 7. A static routing protocol is employed in the AN. Regarding the physical characteristics of the network, the following table summarises the different values used. These values are the same for both network topologies.

There are 24 MHs attached to the AN, only 2 of which will be moving in the AN, performing handovers. Moving MHs will be moving at the edge of the AN performing a total of 100 handovers every 1 second. It is assumed that cell coverage of the areas of the ARs is overlapping, thus ensuring a graceful physical handover from one AR to the other. The wireless channel is assumed to be perfect, with no packet loss occurring.

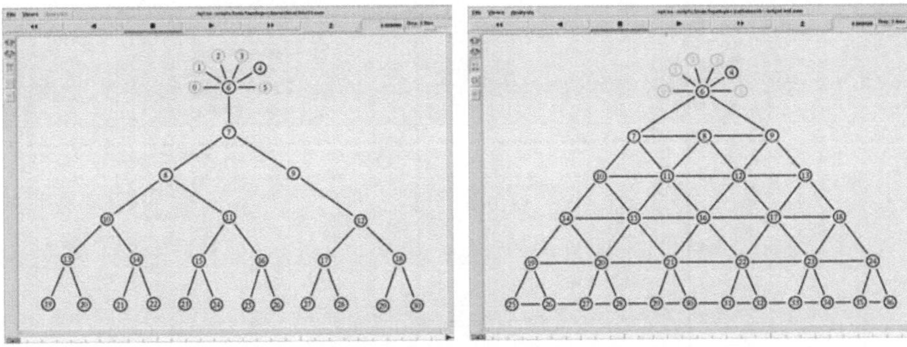

Fig. 3. Tree and Mesh Topologies

Table 1. Network Physical Characteristics

Link	Bandwidth	Delay
CH(s) – HA	10 Mbytes	5 ms
HA – ANG(s)	100 Mbytes	5 ms
AN wired links	20 Mbytes	1 ms
AN wireless links	10 Mbytes	0.5 ms

Various applications will be supported by the MHs, which will produce a mixture of realistic traffic throughout the duration of the simulations. These applications will be running over UDP and TCP. This will enable to study the effect of mobility on the transport protocols.

• **UDP Applications:** VoIP is one application implemented, which is a two-way communication between the MH and the CH. Streaming video is another application that is used over (RTP over) UDP, to transmit a movie from a Video Server to the MH. The employed video model is a trace-driven one, using a real movie trace file (Jurassic Park I). Finally a videoconference application (over RTP over UDP) is also simulated and used to generate traffic load. For such an application real video traces are used again.

• **TCP Applications:** Web browsing is one application that relies on TCP to transfer information from a web server to the MH. For this application, a web-browsing model, included in the libraries of ns, is used. File transfer with FTP is the main TCP application that is used during the simulation. This application is run throughout the duration of the simulation, to achieve a maximum file size transfer at the end of the simulations. The achieved throughput is controlled by TCP, as it will be illustrated later in the document. The different variables of the TCP connections can also be set for the simulation. The packet size of the TCP packets is 1480 bytes, and the maximum window is varied to plot graphs with varied advertised window size.

4 Performance Evaluation

This section will describe the results of the performed simulations. Results for UDP and TCP will be given in two different sections starting with UDP.

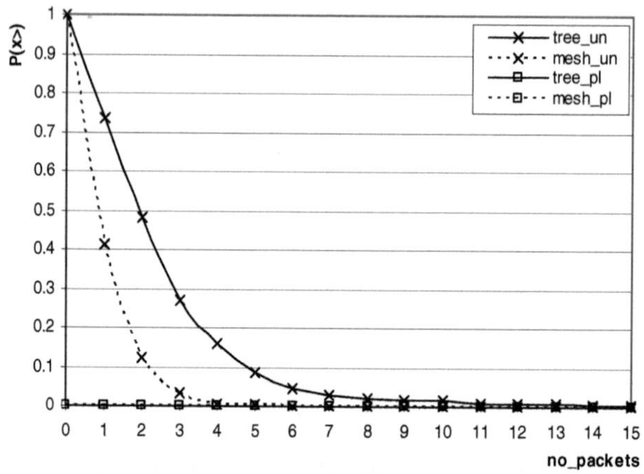

Fig. 4. Packet Loss Probability per Handover

Analysis of Experimental Results for UDP: The graph of Fig. 4 plots the probability of packet loss per handover because user mobility. It can be seen that when performing planned handovers user mobility can be completely hidden, as there is zero packet loss in both topologies. In the case of unplanned handovers packets can be lost and in the case of tree topology there is a high probability of even up to 5 packets been lost during a handover going up to even 14 lost packets. In this case the topology has a big impact on the number of lost packets since in the tree topology, for specific handover the handover messages have to go through the ANG before reaching the old or new AR (handover between nodes 26 and 27 in Fig. 3). In the case of the mess topology only 1 or 2 packets have a high loss probability with 5 the higher value that can occur.

One metric is the number of packets with higher than a specific end-to-end delay variance value, where delay variance is the difference between the time the packet arrives at the destination and the time the packet was generated at the sender plus the time it takes to forward the packet through the network links end-to-end. This is similar to IPPM's One-Way Delay Metric defined in [7] minus the end-to-end forwarding delay, including transmission and propagation delay, thus leaving the queuing and mobility induced delays. This metric thus shows the number of packets that will be lost depending on the playout delay offset of the specific application and its decoding process. In the mesh topology case results show that the end-to-end delay variance is increased by around 0.5 ms for up to 5% of the packets with smaller values for planned that unplanned handovers. In the tree topology, up to 10% of the packets face a higher end-to-end delay variance even up to 4 ms, with higher values for planned than unplanned handovers. Actual graphs are not included due to limited space. Note that the forwarding delay through the network is 15-16 ms, resulting in an end-to-end delay variance ranging from 10% to 40% of that delay.

Next graphs in Fig. 5 show the probability of the delay jitter, which is the relative difference between the arrival of two consecutive packets in sequence. This is similar to IPPM's IP Packet Delay Variation Metric described in [8] and assuming that the packets are in sequence.

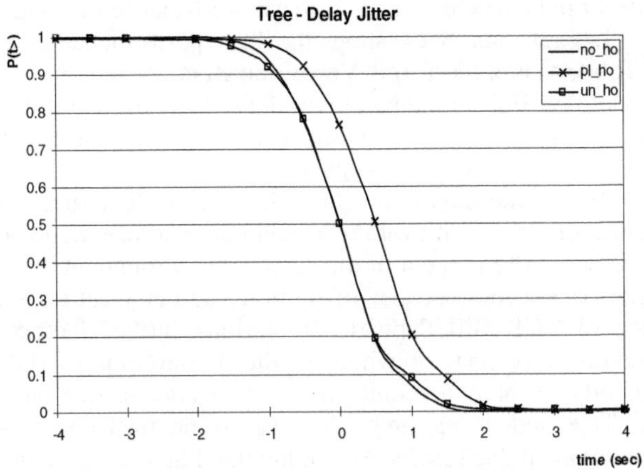

Fig. 5. Packet Delay Jitter Probability

This metric is important for some multimedia applications as they use this metric in the decoding process and ignore packets if they are late compared to the previous packet in the sequence. This metric can take negative values as the difference between the inter-arrival time between two packets in sequence can be smaller than the difference between their relevant generation time. However, what is important is the positive side, since packets with a value of delay jitter higher that the delay offset of the application will be ignored. The experiment showed that in the mesh topology mobility of the users does not have an effect on the delay jitter as the three plots are almost the same. However in the tree topology when planned handovers are performed, packets have a higher probability of larger delay jitter. This is because of the fact that packet during planned handovers have to be forwarded from the current AR to the new AR, so that they don't get lost and this redirection can include various more links thus increasing the jitter delay.

Analysis of Experimental Results for TCP: When evaluating the performance of TCP and since it guarantees delivery of all packets the major metric is the achieved throughput of the connections. Although the performance of the TCP variants has been compared before and this comparison has been the base for designing more variants, this has happened considering only the problem of congestion in the network. However considering mobility modifies the problem since the network is not congested and the packet loss pattern is very different. Thus, the performance evaluation performed here will show the effect of mobility on TCP and its variants.

The experiment showed that without mobility, with this network configuration, up to a window size value of 30 the TCP variants perform identical since congestion doesn't occur anywhere in the network. For higher max window values there is a small differentiation because of congestion with all the TCP variants achieving 98% to 100% of the 10Mbps throughput, which is the bandwidth of the CH to HA link. When mobility is included and planned handovers are performed the results are similar, showing that again planning a change of AR can hide the mobility of the users and eliminate packet loss.

However in the mesh topology (Fig 6) and in the case of unplanned handover the performance of the various TCP variants is differentiated since packets are lost during handover (up to 30 max window) and furthermore because of congestion (more than 30 max window). Fack and Sack show the best performance followed closely by NewReno and Tahoe that outperforms Vegas and Reno.

In the tree topology (Fig. 7) the handover delay is larger and thus more packets are potentially lost, which results in a higher differentiation in the performance of the TCP flavours. Like in the mesh topology Fack and Sack show the best performance achieving 98%, 99% of the maximum throughput. NewReno and Tahoe show a very good performance of 95% achievable throughput whereas Reno and Vegas suffer achieving only 65% to 70% percent of the achievable throughput.

Assuming that packet loss because of mobility and congestion is the main reason, the performance of TCP still depends on various further factors like how many consecutive packets were lost, or where in the whole window did this packet loss occur. Consequently, is hard to come to absolute conclusions when evaluating the performance of TCP and its variants. However some firm conclusions can be well supported and more available results are not included here because of limited space.

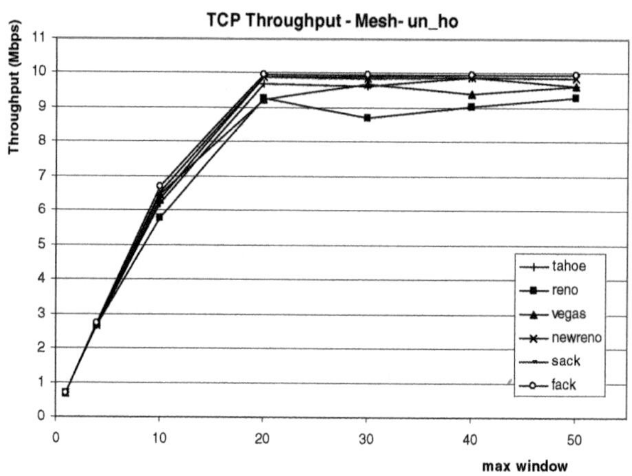

Fig. 6. TCP Throughput – Mesh Topology

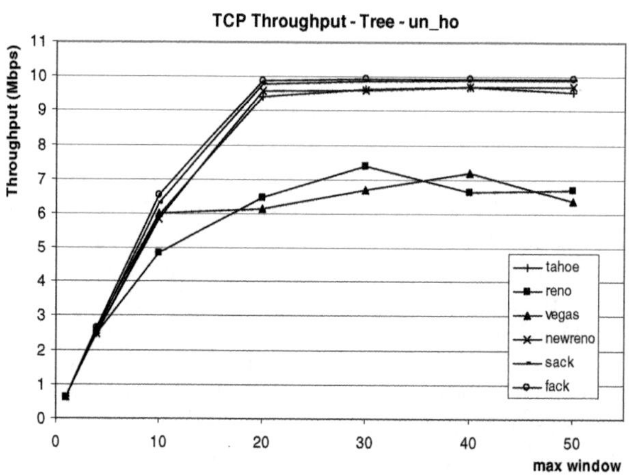

Fig. 7. TCP Throughput – Tree Topology

5 Conclusions and Future Work

This article has presented the results of the study on the performance of standard transport layer protocols over a wireless IP access domain with host mobility locally managed with the use of BCMP. A software experimental platform has been built based on ns-2, using standard data and real-time multimedia applications running on MHs moving inside the access network and performing handovers between ARs. The topology of the access network has also been varied including mesh and tree topology. IP transport layer including UDP and TCP with its various flavours have been employed.

The focus of the study was on the effect of the mobility of the users inside the access network. It has been shown that mobility may have an effect on the performance of the protocols. Detailed graphs were not included due to limited space. When trying to evaluate these results the following conclusions can be drawn:

• Successful planned handovers can hide mobility entirely from UDP and TCP. Packet loss is eliminated, not affecting UDP applications and aiding TCP to achieve maximum throughput, irrespective of the variant used. This is irrespective of the topology of the network, as mesh and tree topologies show the same performance. End to end delay variance is minimised and slightly increased packet delay jitter can only occur in the tree topology.

• In terms of network topology, mesh topology shows an excellent performance from all aspects of the system for both UDP and TCP. Of course mesh topology involves more nodes and links making it more expensive. Furthermore more efficient routing protocols (QoS routing) can be used to further improve the performance, thus increasing the complexity of the system. Tree topology, which is smaller and simpler, also shows a very good performance especially when planned handovers can be successfully executed by the MHs.

• When comparing the various TCP variants, interesting results can be drawn when studying the case of packet loss occurring because of unplanned handovers only or together with congestion. **Sack** and **Fack** with the use of selective ACKs yield the best performance, which is expected since these techniques use a different type of ACK that can acknowledge reception of multiple packets in a window. Bearing in mind that in this mobility scenario multiple packet loss per window is very probable, very good performance is achieved, yet the complexity of the algorithm and the size of the ACK packets are increased. **Reno**, which is currently widely used in the Internet, yields the worst performance, around 65% of the maximum possible throughput, together with **Vegas** that also yields a very poor performance. This is justified since Reno is often forced to wait for a retransmission timeout to recover from two or more packets dropped from the same window of data [9], which is what happening in this scenario. Regarding Vegas, although its flow control mechanism is very well designed and no congestion ever occurs, it suffers when trying to recover from packets that are lost because of handovers. **NewReno** also exhibits very good performance from the family of TCPs that uses a standard receiver, i.e. one that sends a single ACK per received packet. So NewReno with small modifications to the basic Reno implementation does give a big improvement on its performance. Finally **Tahoe**, which is the basic implementation without Fast Recovery, yields a very good overall performance, outperforming Reno and Vegas. This could be justified as when facing multiple packet loss regularly (every handover) Tahoe TCP by halving its widow size and re-transmitting all the packets starting from the first lost, manages to transmit more packets in the long run of the experiment. Tahoe has shown similar good performance in a similar study involving a different localised mobility protocol in [10].

More detailed analysis and graphs could justify the results presented here. Further, work in this area should include the use of a realistic wireless channel and the study of further TCP versions designed for wireless systems like M-TCP, Snoop protocol, I-TCP and others. Finally, layer-4-aware layer-2 protocols can be developed to improve the performance TCP and UDP over wireless links.

References

1. J. Manner, M. Kojo, C. Perkins, T. Suihko, P. Eardley, D. Wisely, R. Hancock, N. Georganopoulos, "Mobility Related Terminology" IETF draft, draft-ietf-seamoby-mobility-terminology-04.txt, (work in progress), April 2003.
2. C. Williams, "Localized Mobility Management Requirements" IETF draft, draft-williams-mobileip-lmm-requirements.txt, (work in progress), July 2001.
3. C. Perkins, "IP Mobility Support for IPv4", IETF RFC 3220, January 2002.
4. K. Ceszei, N. Georganopoulos, Z. Turanyi, A. Valko, "Evaluation of the BRAIN Candidate Mobility Management Protocol", Proceeding IST Summit '01, September 2001.
5. G. Almes, et al, "A One-way Packet Loss Metric for IPPM", IETF RFC 2680, Sep. 1999.
6. C. Demichelis, P. Chimento, "IP Packet Delay Variation Metric for IPPM", IETF draft, <draft-ietf-ippm-ipdv-09.txt> (work in progress), April 2002.
7. G. Almes, et al, "A One-way Delay Metric for IPPM", IETF RFC 2679, September 1999.
8. ns-2 website, www.isi.edu/nsnam/
9. K. Fall and S. Floyd, "Simulation-based Comparisons of Tahoe, Reno and SACK TCP", Computer Communications Review, July 1996.
10. A. Delgado, A. Mihailovic, N. Georganopoulos, A. H. Aghvami, "Adaptation of Transport Protocols for an IP-Micromobility Scheme", Proc. ICC '01, June 2001.

Analysis and Measurement of TCP/IP Performance over GPRS Networks

L. Benetazzo, M. Bertocco, C. Narduzzi, and R. Tittoto

Università di Padova, Department of Information Engineering,
Via Gradenigo 6/a – 35131 Padova, Italy
Phone: +39 049 827 7500; Fax: +39 049 827 7699
{luigino.benetazzo,matteo.bertocco,claudio.narduzzi,
ronny.tittoto}@unipd.it

Abstract. The paper presents an experimental analysis of some link layer characteristics of GPRS (General Packet Radio Service) mobile networks and discusses their relationship to TCP/IP performances. Mobile networks implement sophisticated mechanisms in order to provide mobility and capacity in a wide area. Unfortunately, large delays along the radio path may be introduced, thus reducing throughput in data transfers. The paper focuses its attention on a set of phenomena (excess queuing, delay spikes, redundant retransmissions and bandwidth oscillations) that were observed in real mobile GPRS systems. Analyses are carried out on a set of TCP traffic traces captured by passive monitoring at the main protocol interfaces. Several aspects related to TCP performances and throughput are then discussed and results obtained are compared with simulation studies and characterisations already available in the literature. Optimal network configurations and tuning of various parameters are also discussed.

1 Introduction

The need for heterogeneous interoperability of IP (Internet Protocol) across broadly different technologies has attracted considerable research interest, in particular with regards to wireless technology. While the 3G mobile standard [1] represents an example of a full-scale IP implementation designed to provide various customer services, 2.5 generation mobile wireless networks, such as GPRS [2] and EDGE [1], provide data services and Internet accessibility through the IP protocol, in addition to more traditional voice services on circuit-switched networks.

GPRS networks are deployed world-wide and their performance evaluation and optimisation is an active research area. In particular, the use of the TCP protocol [3], [4] over such networks has stimulated a number of studies and analyses aimed at the improvement of performances in heterogeneous networks. TCP performances are strictly related to network characteristics and are known to be often affected by implementation details and network configuration. For these reasons, it is important to assess how well different services are or can be implemented, by measuring significant network parameters. Furthermore, continuous network monitoring is required to detect possible dimensioning problems or failures during normal network operation.

M. Conti et al. (Eds.): PWC 2003, LNCS 2775, pp. 261–275, 2003.

Several aspects of network behaviour have been analysed in [5], and in the references therein, by means of multi-layer tracing at various interfaces. The analysis allows to understand network performance and to discover possible improper configurations. Network characterisation and further TCP performance analysis have also been presented in [6], where specific aspects, such as link buffering, are investigated. The approach followed by the authors in [5] and [6] was to implement the full GPRS connection, keeping both endpoints within reach in their laboratory. In this way, artificial traffic could be generated in a controlled manner.

From the viewpoint of a mobile network operator, traffic can be traced by probing only at some intermediate interfaces along the TCP connection path, since both endpoints of the connections are unreachable. Hence, operators usually have no access to the whole end-to-end connection and have to evaluate overall TCP performances by processing data gathered at their own monitoring points. This is known as passive monitoring and, in spite of some difficulties, it is a powerful tool for network troubleshooting.

This paper presents an analysis of traces captured by passive monitoring of real GPRS traffic at the interfaces typically available to a network operator. The possibility of detecting and quantifying relevant behaviours is discussed and demonstrated, by showing how various TCP anomalous behaviours due to mobile link characteristics can be evidenced. Results are compared with previously published work, where relevant. The proposed analysis could provide useful tools to improve quality of service from the end user point of view.

The next section introduces GPRS link layer characteristics, whereas Sect. 3 describes several details regarding the measurement set-up and observed phenomena on real networks. Final remarks presented in Sect. 4 conclude the paper.

2 GPRS Link Layer Characteristics

Limited data rates, high latency, delay variation, bursty losses and bandwidth oscillations are important GPRS link layer characteristics, as described in [7]. Bandwidth-Delay Product (BDP) values between 4 and 10 kBytes are typical, therefore these networks can be classified as LTNs (Long Thin Networks) [8].

To allow a better understanding of the issues that will be discussed in this paper, some network elements and a number of protocols are briefly introduced. Fig. 1 illustrates the most relevant elements of a GPRS network from a performance monitoring point of view. During a data transfer between a mobile terminal (MT) and an Internet server, packets are processed by the Gateway GPRS Support Node (GGSN), the Serving GPRS Support Node (SGSN), the Base Station Controller (BSC) and the Base Transceiver Station (BTS). GGSN handles connectivity to external networks, whereas SGSN deals with mobility over a single routing area and controls the data flow toward BSC over the *Gb* interface. Finally BSC handles the radio resource scheduling, medium access and data transmission toward MTs over the *Abis* interface. A detailed description of the GPRS system can be found in [2].

Fig. 2 shows the user plane GPRS protocol stack. Protocols that will be considered in this paper have been evidenced with a different colour. The Radio Link Control protocol (RLC) provides acknowledged or unacknowledged data transfer between MTs and BSCs. The Logical Link Control protocol (LLC) provides packet transfer between MTs and SGSNs; it can operate in both acknowledged or unacknowledged mode, though most GPRS networks operate in the unacknowledged mode. Maximum size of the LLC frames can be configured up to 1556 bytes. The Base Station Subsystem GPRS Protocol (BSSGP) provides data control flow between BSC and SGSN; in fact, it is used by BSCs to communicate buffer status and leak rate for the downlink flow. Finally, MT and Internet hosts communicate by means of the TCP/IP protocol suite.

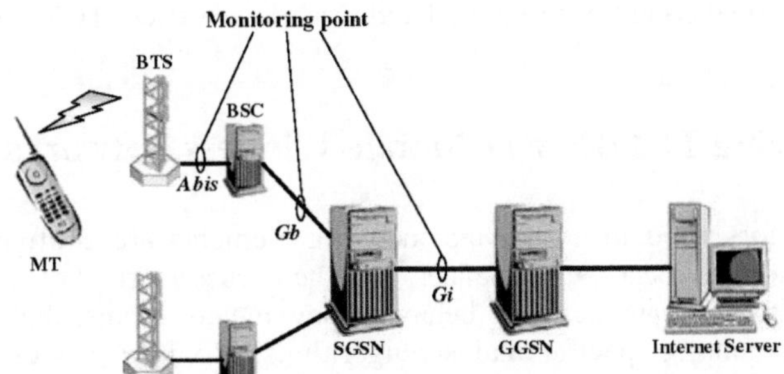

Fig. 1. GPRS network elements involved in a data connection and measurement probe points location

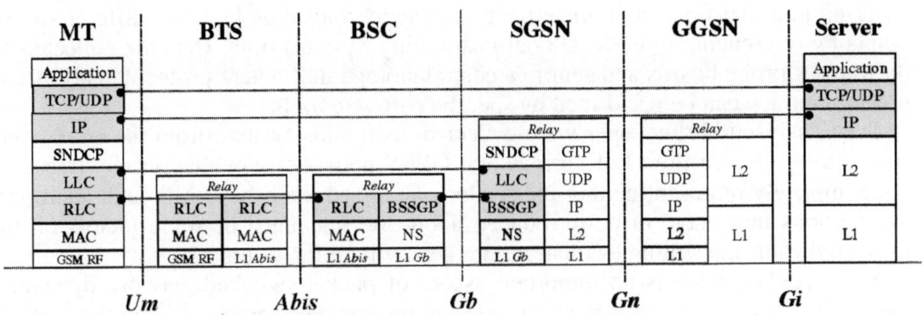

Fig. 2. The GPRS protocol stack

Mobile wireless networks exhibit several differences from traditional fixed networks which, in some cases, affect TCP performances. Data rate is the most important parameter to take into account. Typical data rates for 2.5G systems (GPRS and EDGE) are between 10kbps and 40kbps. In addition, actual data rates change dynamically: users connected to a cell can ask for resources or release them, thus increasing or reducing the available bandwidth, which is also influenced by the distance from the base station. Furthermore, users can move from a cell to another, experiencing abrupt changes in bandwidth. These phenomena, often referred to as bandwidth oscillations, may cause TCP malfunctions and degrade throughput.

Latency is another important characteristic. For GPRS radio links this is typically high, due to the mechanisms implemented to provide mobility, capacity in a wide area and reliability along the radio path. For the latter purpose, the RLC protocol [9] uses the acknowledged mode, implementing specific ARQ (Automatic Repeat reQuest) and FEC (Forward Error Correction) techniques, which ensure reliability at the cost of increased latency.

Delay along a wireless path also exhibits large variations, due to changing radio conditions and distances between MTs and BTSs. Delay variations compromise data transfer quality and appear like jitter at higher level protocol layers. Latency can increase abruptly and these variations could generate TCP malfunctions such as redundant retransmissions, as will be seen in the next sections.

GPRS networks do not exhibit high packet loss rates and values typically found in normal operational conditions (e.g. no handover) do not affect TCP performances.

3 Monitoring TCP/IP over Mobile Wireless Networks

Mobile operators need to ensure that network elements are configured properly in order to guarantee best performances for their customers. For this reason it is fundamental to evaluate network behaviour by means of distributed measurement devices implementing specific analysis algorithms. Monitoring is usually carried out by passive measurements, i.e., by observing existing network traffic, rather than by generating dedicated test flows. Tektronix NET-GPRS is a monitoring system that enables operators to troubleshoot, optimise and manage GPRS networks by means of protocol analysis, multi-interface multi-protocol procedure trace and statistical measurement. It can monitor different network interfaces by means of several probes, ensuring minimal timestamp uncertainty and synchronisation between different probe points by referencing to GPS (Global Positioning System) time. Data are collected by the various probe boards and sent to a central storage unit where protocol analysis and procedure trace can be performed by specific software tools.

Data analysed in this paper were collected, using this system, from the *Gi*, *Gb* and *Abis* interfaces (see Fig.1) of various real GPRS networks. For this reason, traces do not completely represent neither the sender side, nor the receiver. Although additional assumptions may have to be introduced about the relevance of some factors on the measured values, resulting evaluations are generally quite accurate.

Since packet delay is an important aspect of packet-switched network dynamics, the analysis presented in this paper focuses on transport protocol behaviour and is based on the measurement of RTT (Round Trip Time). Experimental results are discussed with the aim of relating possible causes for RTT variations to buffering and flow control procedures.

A set of over 800 traces have been analysed, accounting for over 500000 LLC frames. Useful information, including TCP sequence numbers, IP addresses, port numbers, flags, etc., are extracted from header packets by the NET-GPRS software tools. The capture time is indicated by a timestamp associated to each packet. Traces are seen as transport level flows, defined by different (source IP address, source port number, destination IP address, destination port number) 4-ples.

Main data services accessed through GPRS networks are web browsing (HTTP and HTTPS), email (POP3 and SMTP) and file transfers (FTP etc.). Preliminary packet analysis showed that TCP accounts for 83% of the total IP traffic. UDP is also frequently used (17%), mainly for Wireless Application Protocol (WAP) services. As reported in Table 1, the best part of the captured traces is related to data downloads. This is confirmed by packet size analysis that showed a large number of small IP packets, of size between 40 and 60 bytes, in the uplink direction and IP packets larger than 500 bytes in downlink.

Table 1. Percentage of TCP connections for the different type of services.

	HTTP HTTPS	POP3	SMTP	FTP	Others
TCP traces	78,95%	6,62%	2,60%	9,45%	2,38%

Since data transfers in GPRS networks suffer from large delays in the downlink direction, we focused our analysis on large data downloads from Internet servers to MTs. For these traces it is possible to evaluate the partial RTT experienced along the radio tract by evaluating the time elapsed between the capture of a packet and the corresponding acknowledgement (same sequence number) observed in the opposite direction. This is what a sender usually does, for instance, to determine its own retransmission time-out (RTO). In this case, however, the measurement is performed using packets collected at the *Gb* interface; the partial RTT evaluated in this way can give a good estimate of the time elapsed in the BSC, the BTS, the MT and along the air path, allowing to keep trace of the evolving characteristics of the radio path alone.

In order to correctly interpret the behaviour of transmitters and receivers in the analysed connections, a preliminary examination was carried out in the attempt to detect the various TCP implementations used by the hosts involved. For the majority of traces behaviour was very similar way to that expected for SACK and New Reno implementations [10]. Predictably, some differences were observed, and can be related to proprietary implementations of TCP in specific operating systems.

3.1 Partial Round Trip Time Measurements

Several partial RTT values were obtained for each trace considered in the analysis. Each captured acknowledgement in the uplink direction was considered together with its corresponding packet and a partial RTT was calculated, as the difference between the respective timestamps, in every non-ambiguous [11] case.

Fig. 3 presents a histogram of the measured RTTs obtained analysing a set of over 200 TCP connections. These traces are all HTTP downloads of 43 kbytes and have been captured in a test plant with perfect radio conditions. As can be noticed, values are typically greater than 1 s, with a mean RTT of 1.5 s. Values are considerably high if compared with normal Internet wired connections, anyway they agree with measurements previously reported in the literature [6].

By comparison, Fig. 4 shows measured partial RTTs in the case of bulk download traces gathered from real traffic. In this case different services are considered, such as HTTP, FTP and POP3. The majority of the MTs analyzed uses 4 or 3 timeslots in downlink and 1 timeslot in uplink. Also, code scheme 2 (CS-2) encoding is widely used since it provides good performances if compared with CS-3 and 4. For these mobiles the bandwidth-delay product (BDP) is approximately equal to 10 kbytes.

The figure refers to 4795 measures divided in two main groups: the first group, represented by white bars, refers to all RTT estimates obtained from TCP traces with a receiver advertised window up to the BDP (10 kbytes in this case), which is suggested as the reference value [7]. The second group, represented by black bars, refers to the remaining traces, where the receiver advertised window was greater than the BDP. It should be emphasised that the histogram has been obtained by analysing only large bulk transfers of at least 20 kbytes; in this way, it is ensured that the size of the transmitter congestion window for the second group of traces is as close as practically possible to the maximum allowed value.

As can be seen from the figure, large window sizes imply larger RTT values. For long TCP sessions the transmitter fills the buffer located at the BSC excessively, leading to partial RTTs that can be 4 to 6 times longer. This confirms the excessive queuing phenomenon described in [6], where a number of consequences are briefly described (RTT inflation, inflated retransmission timer value, problems of stale data and high recovery time).

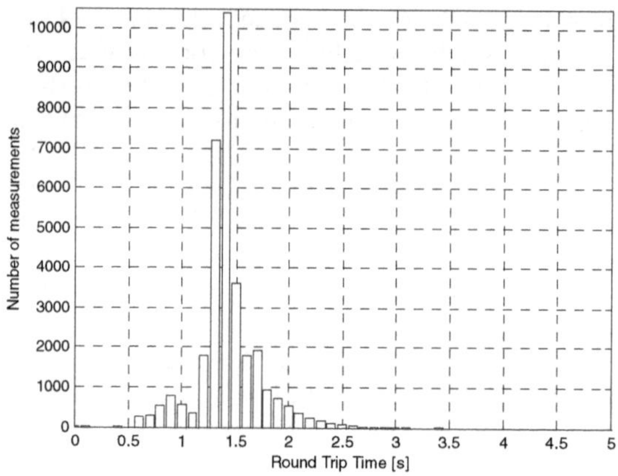

Fig. 3. Partial RTT evaluated analyzing a set of over 200 HTTP downloads of 43 kbytes

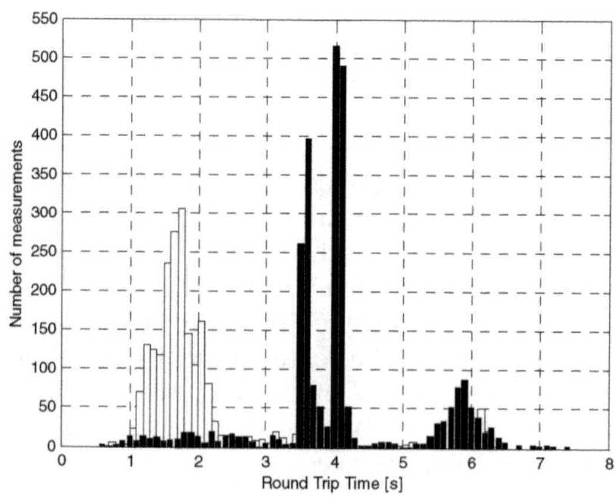

Fig. 4. Partial RTT evaluated analyzing a set of over 120 TCP downloads of at least 20 kbytes. Black bars refers to measurements collected from traces with advertised window greater than 10 kbytes, white bars refers to traces with advertised window up to 10 kbytes.

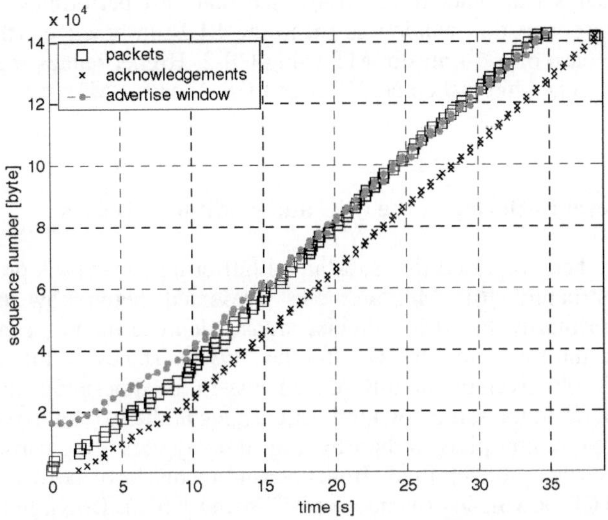

Fig. 5. A TCP trace example (advertised window equal to 16 kbytes); delay gets larger as the congestion window increases

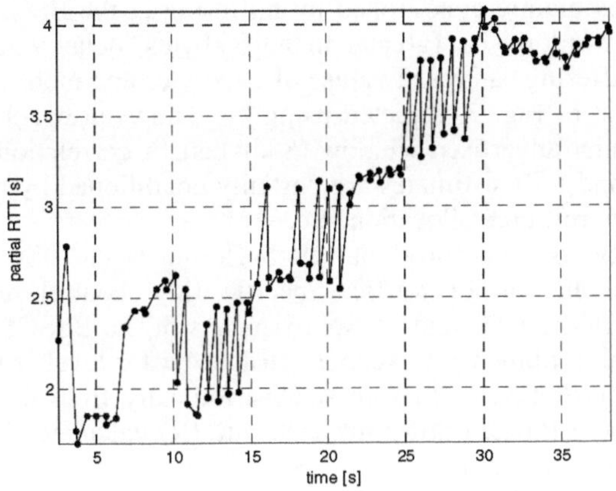

Fig. 6. Partial RTT estimates for the trace of Fig. 5

Fig. 5 shows a TCP trace, depicted using squares to represent packets and "X"s to represent acknowledgements. The x-axis shows the packet timestamp (capture time) expressed in seconds. The y-axis shows the packet sequence number expressed in bytes. Sequence numbers and timestamps are scaled, so that the starting value appears to be always equal to 0 for each connection. The dot-dash line represents the receiver advertised window.

Fig. 6 presents the corresponding partial RTT measurements as functions of the capture time. It can be noticed that delay gets larger as the transmitter increases its congestion window. The maximum allowed size is set to 16 kbytes, therefore the congestion window can reach values greater than the optimal BDP.

Further analysis of other traces confirmed that best performances are achieved if the maximum congestion window is set to the BDP; in practice, 10 kbytes is a good value for the case of a 4-timeslot MT using CS-2. Higher values must be used in the case of higher encoding schemes, CS-3 and CS-4, or for MTs that can support more timeslots.

3.2 Link Level Buffering at the BSC and Additional Delays

It has already been reported that substantial buffering capacity is provided within the GPRS infrastructure [6]. Measurements presented here show that the element involved is primarily the BSC. Precise transit delay estimates can be obtained by monitoring at the same time the *Gb* and the *Abis* interfaces. In this case delay within the BSC is simply given by the difference between the timestamps that the monitoring device associates to the same downlink LLC frame at the two interfaces.

In Fig. 7 two scatter plots of the BSC transit delay versus the partial RTT evaluated at the *Gb* interface are reported. Both measurements have been obtained analysing 500-kbytes TCP downloads (4-timeslot MT using CS-2). Dots refer to the case of a large window size (advertised window equal to 64 kbytes). As can be noticed, partial RTTs and correspondent delay queues are extremely high and highly correlated. In fact, a correlation coefficient equal to 0,98 is obtained as a remarkable proof that round trip delay is mainly conditioned by buffering at the BSC. For real traces with large congestion window size (greater than 10 kbytes) delay values of about 3 or 4 s are typical and buffering can reach values of 24 kbytes and more.

The second plot of Fig. 7, depicted using '+', shows how correlation decreases in the case of a smaller advertised window (8 kbytes). A correlation coefficient equal to 0,74 is obtained and RTT estimates are partially conditioned by other aspects such as radio transmission, resource allocation, etc.

The GPRS network constantly controls buffering at the BSC for each single MT and for the total amount of traffic. Specific flow control messages (FCM) are exchanged between the BSC and SGSN by means of the BSSGP protocol. Basically, the flow control algorithms try to keep a constant buffer level at the BSC by delaying LLC frames at the SGSN. FCMs are sent periodically from the BSC to the SGSN, informing it of the estimated downlink leak rate (R) calculated by the RLC protocol and of the maximum buffer size (B_{max}).

Detailed analysis of the current flow control implementations showed that better performances are achieved using small size LLCs. In fact, with LLCs of approximately 500 bytes the system ensures less variable delays at the BSC buffer, reducing jitter seen by higher layer protocols. Since GPRS networks provide transparent link layer fragmentation, performed both by SNDCP and RLC protocols, these size limitations do not need to reflect on the IP MTU (Maximum Transfer Unit). Further considerations regarding IP datagram maximum size will be discussed later in this paper.

Buffering at the BSC is not the only cause of high latency in GPRS networks. For example, considering Fig. 7 it can be seen that packets spend approximately 1 s along the radio tract and at the MT. This has to be taken into account when computing the partial RTT observed at the *Gb* interface. A rough assessment of the way this delay builds up can be given as follows.

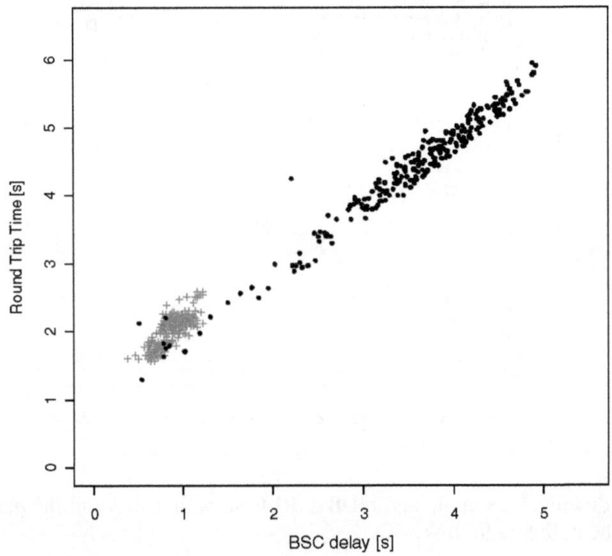

Fig. 7. Scatter plot of the delay queue versus the partial RTT evaluated at the *Gb* interface; dots, '.' refers to a trace with large TCP transmitter window values; '+' refers to a trace with TCP transmitter window limited to 8 kbytes

Data of Fig. 7 refer to TCP connections using IP datagrams of 1500 bytes. Since CS-2 and 4 timeslots are used for both cases, data packet radio transmission in the downlink direction takes approximately 350 ms [12], [13]. A corresponding 52-byte TCP acknowledgement is transmitted uplink, using a single timeslot, in approximately 50 ms. The remaining 600 ms are due to the use of delayed acknowledgement (200 ms), packet buffering and processing time at the various network elements; a delay contribution may also come from radio resource requests that periodically occur for the uplink transmission, as discussed later on.

Another important issue regarding TCP over GPRS is related to delayed acknowledgements. In fact, considering current performances of 4-timeslot MTs, practical considerations reported in [14] suggest that a small IP MTU (e.g. 576 bytes) should be used. However, analysed traces have shown that, even with these MTU sizes, the IP datagram can in some cases occupy the radio link for longer than the delayed ACK timeout, causing a considerable increase of the total RTT. A simple solution could be to acknowledge every segment [8]. It is the authors' opinion that ack-every-segment can be applied for the total duration of the TCP connection in a GPRS network. In fact, this rather drastic solution could also partially solve the problem related to periodic radio resource request and the consequent acknowledgement compression described in the next subsection.

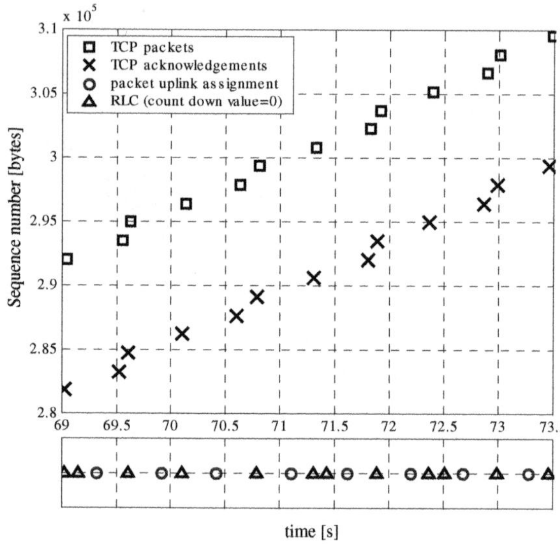

Fig. 8. A TCP download example and relative RLC signaling showing the periodic release and assignment of the uplink radio link.

3.3 RTT Variability

Partial RTT estimates obtained from the majority of the traces analysed in this paper exhibit large variability. This can be observed, for instance, in Fig. 6, where RTTs oscillate about the local mean value with an amplitude equal to 1 s. This behaviour is very common in real GPRS traces and can be defined as transport layer jitter. The causes of this phenomenon are various and are not only related to flow control mechanisms discussed above (Sub. 3.2).

A not uncommon phenomenon, also observed in [5] and [6], is acknowledgement compression, a situation where acknowledgements seem to be released in bursts. The main cause of this behaviour is related to the frequent release of the uplink radio resources by the MT. Resources on the uplink air interface are released by applying the countdown procedure [9]. The mobile station continuously keeps track of the number of RLC data blocks in the transmission queue and, when there is only a certain number left, initiates the countdown procedure. This is done by decreasing the so-called countdown value (CV) for each RLC block sent uplink, the very last block being sent on reaching zero. If data have been correctly received, the BSC then releases the radio link.

It is important to note that the countdown procedure is irreversible. When a MT has only small IP packets to transmit in uplink, such as TCP acknowledgements, the count down procedure is frequently started, so that radio link releases and requests alternate during the whole length of the connection. An example is shown in Fig. 8, where basic RLC signalling is plotted along the TCP trace. Triangles represent uplink RLC data blocks with CV equal to 0: they indicate a subsequent radio link release performed by the BSC. Packet Uplink Assignments (PUAS), depicted using circles, are always present after these RLC blocks, proving that the monitored MT requested back the radio resources. This mechanism is the main cause of acknowledgement compression and large RTT variability.

Other causes for RTT variations are periodic radio resource uplink losses and reassignments and burst reorganisations due to buffering and fragmentation performed by the RLC protocol.

An example of burst reorganisation is given in Fig. 9 where the burst structure of the transmitted TCP packets, as observed at the monitoring point, does not correspond to that of TCP acknowledgements returning from the receiver (MT). In view of the behaviour of TCP receivers, acknowledgements can be assumed to reflect the actual burst structure of packets as they are received. The figure shows that the last packets of a TCP burst can be buffered and finally depart with the first packets of the following burst. This phenomenon is mainly related to the RLC window size [9] used in the downlink direction and can be observed for older GPRS MTs. Analysis on real traffic has shown that this anomalous behaviour is extremely rare in recent networks where RLC window size is typically equal or greater than 192 radio blocks.

Both phenomena described here, together with high delay variation due to buffering at the BSC, are significant causes of bad TCP retransmission timeouts (RTO) and redundant retransmissions (RR) described in detail in the following section.

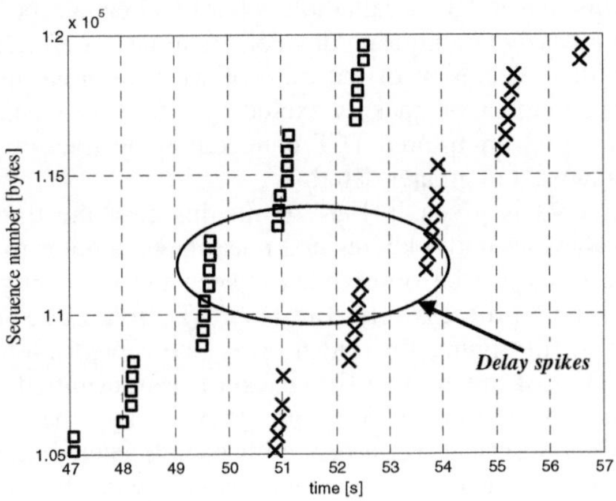

Fig. 9. Burst reorganisations due to buffering and fragmentation

3.4 Redundant Retransmissions

Several simulation results are available in the literature [15] regarding TCP/IP behaviour over GPRS networks. The most common phenomenon observed during these simulations regards redundant retransmissions (RRs) caused by high round trip delays and bad retransmission time-outs. Simulations show that the principal cause of RTO expiration are delay spikes, which are usually associated to dynamic allocation of resources and mobility related mechanisms. Analysis of real traces presented in this paper showed that a third cause is related to the phenomena described in Sect. 3.2 and 3.3.

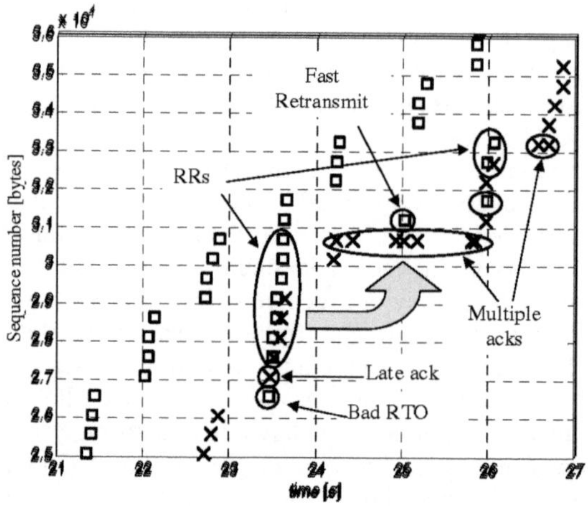

Fig. 10. Typical GPRS bad retransmission time-out and successive related RRs

In order to determine the exact percentage of RRs, it would be necessary to track both sender and receiver traces [16]. Unfortunately, this is not possible for a mobile operator and for this reason RRs evaluation is hard and cannot be sufficiently precise if implemented in a passive monitoring device. A simple yet effective solution for this problem could be the calculation of the ratio between the total number of sent TCP packets and the net number of packets excluding retransmissions. This ratio can be computed easily for each monitored TCP connection and traces exhibiting high ratio values should be analysed in greater detail.

An example of RRs is given in Fig. 10. In this case the transmitter congestion window reaches values around 5 kbytes and phenomena similar to the ones described in the previous sections can be observed. Between the beginning of the connection and 22 s the TCP sender progressively adjusts its RTO to a value approximately equal to 2 s. As shown in the figure, this situation is interrupted after a delay spike that occurs at around 23 s. At this point a first packet is retransmitted after a RTO; this is actually a redundant retransmission, as proved by the fact that a "late acknowledgement" is observed just a few milliseconds later. In this case, because of burst reorganisation, the acknowledgement arrived with the first packet of the successive burst. Typically, after a RTO the sender enters the slow start phase so that, for each successively received acknowledgement, it retransmits successive packets. A total number of 7 retransmitted packets have been reached in Fig. 9. Since the packets had already been received at the MT, multiple acknowledgements are generated, as evidenced by the "x" between 24 s and 26 s. These make the sender enter the fast retransmit phase, whereby other packets are unnecessarily retransmitted.

While the results reported agree with the behaviour described in the literature, they show that delay spikes and RRs can also be related to burst reorganisation, rather than to resource allocation and mobility. Other traces showing similar phenomena have been analysed. It must be pointed out that, after the RRs, the transmitter reverted to a stable and regular transmission. This is due to the fact that the sender enters the fast retransmit phase only when three multiple acknowledgements are received. In other words, even a large number of RRs compromises TCP performances only in a relatively limited period, since TCP mechanisms ensure a further regulation in subsequent traffic.

RRs inevitably compromise TCP performances [16]. In order to develop non-invasive tools capable of correct performance evaluation, RRs need to be considered since they represent important aspects related to TCP/IP over mobile networks.

3.5 Bandwidth Changes

Users can get variable resources as a function of the radio path characteristics, distance and number of MTs simultaneously accessing the same cell. These variations can be observed also by monitoring TCP/IP traffic. Captured traces related to real GPRS traffic exhibit these behaviours. Examples are given in Fig. 11. In this case the traces shown refer to large file transfers and present abrupt bandwidth changes and high delay variations.

Other traces with similar characteristics have been observed, especially regarding long file transfers lasting more than 30 seconds. This behaviour is strongly related to the mobile mechanisms implemented to share the limited resources available along the radio path. Handoff operations can also lead to abrupt delay variations with total RTT that can reach even 10 s and longer.

When link characteristics change abruptly, long delay spikes and packet loss are frequently observed. However, in the majority of cases analysed during this work, these phenomena were not accompanied by redundant retransmissions. The GPRS traces presented also demonstrate how bandwidth oscillations can occur frequently during a single download. It can be noticed that, because of the fast variations, it would be difficult to evaluate bottleneck bandwidth, especially when monitoring TCP/IP traffic by a non-invasive technique.

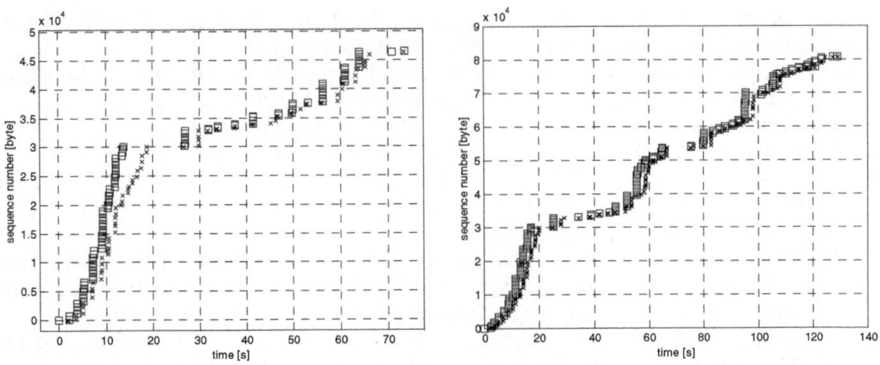

Fig. 11. Bandwidth oscillations due to resource sharing and handoff procedure, real GPRS traffic.

4 Conclusions

This article discusses a set of phenomena that can be observed by analysing real TCP traces over new generation mobile networks. The work points to some problems related to TCP/IP and GPRS wireless links. The purpose is to provide experimental evidence of behaviours that have already been noticed in the literature by means of simulations, but also to describe some peculiar behaviours in detail and propose optimal network configurations that allow to avoid them.

Further work has to be done in order to develop measurement techniques that could help network operators discover possible anomalies and improve performances. In particular, the relationship between delay variability and flow control is currently under closer investigation. The intention is to achieve a more precise evaluation of this type of phenomena.

Traces reported here refer to real GPRS traffic obtained from various mobile operators. Results are in agreement with the analyses in [5], [6], proving that passive monitoring can effectively provide operators with the tools for an accurate assessment of network performances and optimisation. It is hoped to extend the analysis to UMTS traces as soon as possible, with the intention of discovering differences and improvements related to this new technology.

Acknowledgements. The authors would like to acknowledge the support granted by Tektronix Padova S.p.a. for the studies reported in this work. The useful suggestions of the anonymous reviewers are gratefully acknowledged.

References

1. H.Kaaranen, A.Ahtianinen, L.Laitinen, S.Naghian, V.Niemi: UMTS Networks: Architecture, Mobility and Services. John Wiley & Sons, 2001.
2. G.Brasche, B.Walke: Concepts, Services, and Protocols of the New GSM Phase 2+ General Packet Radio Service. IEEE Communications Magazine, August 1997.
3. J.B.Postel: Transmission Control Protocol. RFC 793, September 1981.
4. W.R.Stevens: TCP/IP Illustrated, Volume 1: The Protocols. Addison-Wesley, 1994.
5. A.Gurtov, M.Passoja, O.Aalto, M.Raitola: Multi-Layer Protocol Tracing in a GPRS Network. Proceedings of the IEEE Vehicular Technology Conference (Fall VTC2002), Vancouver, Canada, Sepember 2002
6. R. Chakravorty and I. Pratt: Performance Issues with General Packet Radio Service. Journal of Communications and Networks (JCN) – Special Issue on "Evolving from 3G deployment to 4G definition", pages 266–281, Vol. 4, No. 2, December 2002
7. H.Inamura, G.Montenegro, R.Ludwig, A.Gurtov, F.Khafizov: TCP over Second (2.5G) and Third (3G) Generation Wireless Networks. IETF RFC 3481, February 2003.
8. G.Montenegro, S.Dawkins, M.Kojo, V.Magret, N.Vaidya: Long Thin Networks. IETF RFC 2757, January 2000.
9. 3GPP TS 04.60: Radio Link Control/ Medium Access Control (RLC/MAC) protocol. Release 1999, January 2002.
10. K.Fall, S.Floyd: Simulation-based Comparisons of Tahoe, Reno, and SACK TCP. Computer Communication Review, Vol. 26 No. 3, pp. 5–21, July 1996.
11. V.Paxson, M.Allman: Computing TCP's Retransmission Timer. RFC 2988, Proposed Standard, November 2000.

12. M.Meyer: TCP Performance over GPRS. IEEE Wireless Communications and Networking Conference, New Orleans, LA, September 1999.
13. U.Vornefeld: Analytical Performance Evaluation of Mobile Internet Access via GPRS Networks. In Proc. of the European Wireless '02, pp. 712–718, Florence, Italy, February 2002.
14. S.Dawkins, G.Montenegro, M.Kojo, V.Magret: End-to-end Performance Implications of Slow Links. BCP 48, IETF RFC 3150, July 2001.
15. A.Gurtov, R.Ludwig: Evaluating the Eifel Algorithm for TCP in a GPRS Network. In Proceedings of European Wireless, Florence, Italy, February 2002.
16. V.Paxson: End-to-End Internet Packet Dynamics. IEEE/ACM Transactions on Networking, Vol.7, No.3, pp. 277–292, June 1999.

Exact Probabilistic Analysis of the Limited Scheduling Algorithm for Symmetrical Bluetooth Piconets*

Gil Zussman[1], Uri Yechiali[2], and Adrian Segall[1]

[1] Department of Electrical Engineering
Technion – Israel Institute of Technology,
Haifa 32000, Israel
{gilz@tx,segall@ee}.technion.ac.il
http://www.comnet.technion.ac.il/{~gilz,segall}
[2] Department of Statistics and Operations Research
School of Mathematical Sciences
Tel Aviv University,
Tel Aviv 69978, Israel
uriy@post.tau.ac.il
http://www.math.tau.ac.il/~uriy

Abstract. Efficient communication in Bluetooth scatternets requires design of intra and inter-piconet scheduling algorithms, and therefore numerous algorithms have been proposed. However, due to inherent complexities of the Bluetooth Medium Access Control (MAC), the performance of such algorithms has been analyzed mostly via simulation. We show that a piconet operated according to the limited (pure round robin) scheduling algorithm is equivalent to a *1-limited polling system* and utilize methods developed for the analysis of such systems to analyze this algorithm. We present *exact* analytic results regarding symmetrical piconets with bi-directional traffic. Our results differ from the corresponding analytical results presented as exact in [12],[13],[14],[16],[19], and [21]. We show that [14] actually presents *approximate* results, as it ignores important dependencies incorporated in the piconet operation model. Consequently, [14] *underestimates* the intra-piconet delay, in some cases by more than 50%. We also indicate that for similar reasons the analyses presented in [12],[13],[15],[16],[17],[18],[19],[20], and [21] seem to provide only approximate results.

Keywords: Bluetooth, Scheduling, Polling, Queueing, Limited, Time Division Duplex, Personal Area Network (PAN)

1 Introduction

Bluetooth is a Personal Area Network (PAN) technology, which enables portable devices to connect and communicate wirelessly via short-range ad-hoc networks [2],[3]. The basic network topology (referred to as a piconet) is a collection of slave devices operating together with one master. A multihop ad-hoc network of piconets in which some of the devices are present in more than one piconet is referred to as a

* This research was supported by a grant from the Ministry of Science, Israel.

M. Conti et al. (Eds.): PWC 2003, LNCS 2775, pp. 276–290, 2003.

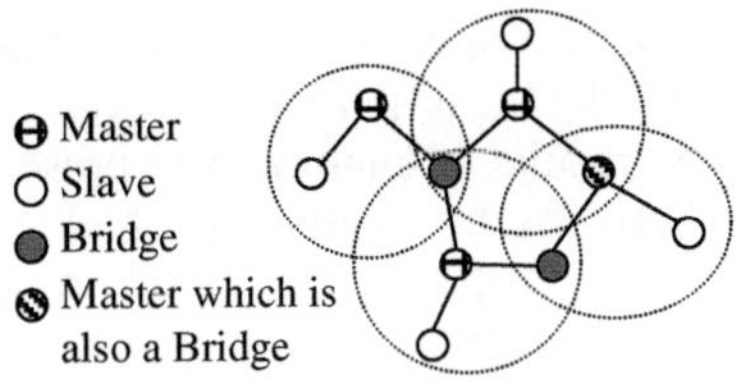

Fig. 1. An example of a Bluetooth scatternet

scatternet (see for example Fig. 1). A device that is a member of more than one piconet (referred to as a bridge) must schedule its presence in all the piconets in which it is a member (it cannot be present in more than one piconet simultaneously).

In the Bluetooth specifications [2], the capacity allocation by the master to each link in its piconet is left open. The master schedules the traffic within a *piconet* by means of polling and determines how the bandwidth capacity is to be distributed among the slaves. Efficient *scatternet* operation requires determining the link capacities that should be allocated in each piconet, such that the network performance will be optimized [27],[28]. The required link capacities should be allocated by *inter-piconet* scheduling algorithms. These algorithms schedule the presence of the bridges in different piconets. Numerous heuristic *intra* and *inter-piconet* scheduling algorithms have been proposed (e.g. [4],[5],[6],[7],[22] and references therein).

Analytical performance evaluation of intra and inter-piconet scheduling algorithms has great importance, since it may provide insight on their design and optimization. However, as mentioned in [5], due to the special characteristics of the Bluetooth Medium Access Control (MAC), the operation model of most scheduling regimes differs from those of classical polling models. Accordingly, in the past most of the proposed scheduling algorithms have been evaluated via simulation.

In this paper we focus on the *limited (pure round robin) scheduling algorithm.* We show that when all packets are 1-slot long[1], the piconet can be analyzed as a *TDMA* (Time Division Multiple Accesses) system. Then, we show that when all packet sizes are used the piconet is equivalent to a *1-limited polling* system[2]. The problem of computing exact mean delays in general 1-limited polling systems has not been resolved yet [8], but we derive exact analytic results for a *symmetrical* piconet with *bi-directional* traffic. We note that due to the equivalence to a polling system, approximate results can be obtained for more complex systems by utilizing the vast amount of research dedicated to 1-limited polling (see [8] for a review).

Recent papers [12],[13],[14],[16],[17],[18],[19],[20],[21] have *claimed to provide exact analytic results* regarding the performance of various intra and inter-piconet scheduling regimes. The analyses there are based on the theory of M/G/1 queue with vacations (introduced in [9], see also [25]). Since our exact results differ from these results, we show that the closed form solutions exhibited in [14] for the *limited* (pure round robin) scheduling algorithm are actually approximate solutions, as they are based on unsatisfied assumptions leading to inaccurate probability generating functions. In particular, we argue that important dependencies incorporated in the piconet operation model are ignored when applying the results of the M/G/1 queue with

---

1 In Bluetooth piconets, the channel is slotted and the packets lengths are 1, 3, and 5 slots [2].

2 A polling system consists of several queues served by a single server according to a set of rules (polling scheme) [1, p. 195],[8],[24],[26].

vacations to the analyzed system. We also indicate that the analyses in [12],[13],[15], [16],[17],[18],[19],[20], and [21] are based on similar models, and therefore seem to provide only approximate results. Finally, we present numerical results that demonstrate the difference between the results in [14] and our exact results.

We note that Miorandi et al. [10] have presented an *approximate* analysis of the limited scheduling algorithm for a piconet with *asymmetrical* traffic. Their findings support our observation that the analysis of the limited algorithm in [14] can serve only as an approximation.

Due to space constraints, we do not elaborate on piconets with *unidirectional traffic* and on the *exhaustive scheduling algorithm*. However, we note that in [29] we have shown that a piconet with *unidirectional* traffic operated according to the *exhaustive* scheduling algorithm can be modeled as an *exhaustive polling system* and derived exact analytic and numerical results regarding intra-piconet delays. It should be noted that those results also apply to piconets with asymmetrical traffic. We have also shown that a piconet with *unidirectional* traffic operated according to the *limited* scheduling algorithm can be modeled as a *1-limited polling system*. Then, we have shown that in a piconet with only uplink traffic in which all arrival rates are statistically equal, the mean delays for the *limited* and *exhaustive* regimes are *equal*. This observation has been extended for any arbitrary Time-Division-Duplex system, operated in a similar manner to a Bluetooth piconet, in which the packets are not necessarily 1, 3, and 5 slots long (as required by the Bluetooth specifications [2]). Finally, we have outlined the complexities in analyzing a piconet with *bi-directional* traffic operated according to the *exhaustive scheduling algorithm*.

To the best of our knowledge, the results presented in this paper and in [29] are the only available *correct exact* analytic results regarding the performance of Bluetooth scheduling algorithms.

The rest of the paper is organized as follows. Section 2 gives a brief introduction to the Bluetooth technology, while Section 3 presents the model. In Section 4 we analyze the limited scheduling algorithm and discuss its analysis in [12],[13],[14],[15], [16],[17],[18],[19],[20], and [21]. In Section 5 we present numerical results and in Section 6 we summarize the main results and discuss future research directions.

2 Bluetooth Technology

In a *piconet* one unit acts as a *master* and the others act as *slaves* (a master can have up to 7 slaves). Bluetooth channels use a Frequency-Hop/Time-Division-Duplex (FH/TDD) scheme in which the time is divided into 625-μsec intervals called *slots*. The master-to-slave transmission starts in even-numbered slots, while the slave-to-master transmission starts in odd-numbered slots. Masters and slaves are allowed to send 1, 3 or 5-slot *packets*, which are transmitted in consecutive slots. Packets can carry synchronous information (voice link) or asynchronous information (data link).[3] Information can only be exchanged between a master and a slave, i.e. there is no direct communication between slaves.

[3] We note that we concentrate on networks in which only data links are used.

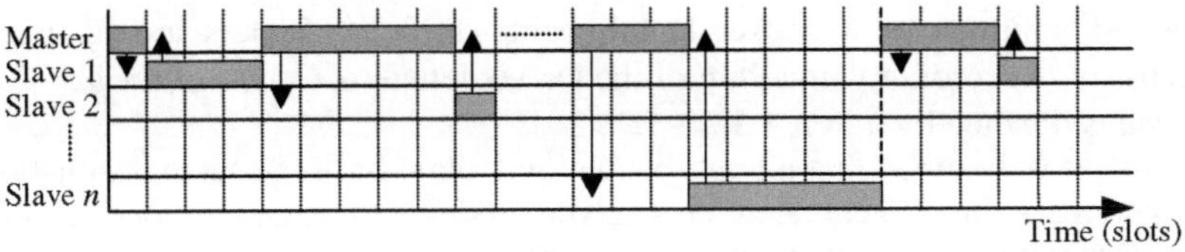

Fig. 2. An example of the Time Division Duplex (TDD) scheme in a Bluetooth piconet

A slave is allowed to start transmission in a given slot if the master has addressed it in the preceding slot. The master addresses a slave by sending a data packet or a 1-slot *POLL packet* (if it has no data to transmit). The slave must respond by sending a data packet or a 1-slot *NULL packet* (if it has nothing to send). We refer to the master-to-slave communication as *downlink* and to the slave-to-master communication as *uplink*. An example of the TDD scheme in a piconet with n slaves is given in Fig. 2.

The master schedules the traffic within a *piconet* according to an *intra-piconet* scheduling algorithm. Various intra-piconet scheduling algorithms have been recently proposed. For example:

- Limited Round Robin (Pure Round Robin [5]) – The master communicates with the slaves according to a fixed cyclic order. At most a single packet is sent in each direction (downlink and uplink) every time a master-slave queue pair is served.
- Exhaustive Round Robin – The master communicates with the slaves according to a fixed cyclic order. The master does not switch to the next master-slave queue pair until both the downlink (master-to-slave) and the uplink (slave-to-master) queues are empty.

In this paper, we focus on the limited algorithm.

In a *scatternet*, a unit (referred to as a *bridge*) can participate in two or more piconets, on a time-sharing basis, and even change its role when moving from one piconet to another. Namely, a bridge can be a slave of a few masters or a master in one piconet and a slave in another piconet. Fig. 1 above illustrates an example of a scatternet including bridges from these two types. The presence of a bridge in different piconets has to be controlled by an *inter-piconet* scheduling algorithm.

3 The Model

To facilitate our claims we describe the piconet model presented in [14] and use similar notation.

The number of nodes is denoted by m (accordingly, the number of slaves is $m-1$). We assume that each node has an infinite buffer. It is assumed that the traffic into each node is a compound Poisson process generating bursts (batches) of packets according to a Poisson arrival process with rate λ (bursts/slot). The probability generating function (PGF) of the burst (batch) size (number of packets in a burst) is denoted by $G_b(x)$. Its mean and second factorial moment are denoted by $\overline{B}$ and $\overline{B}^{(2)}$. We will show that the results presented in [14] are inaccurate even for the simplest case in which the traffic is non-bursty i.e. the burst size is always 1. To that end, in the rest of the paper we assume that $G_b(x) = x$.

The probabilities of a packet length being 1, 3, or 5 slots are p_1, p_3, and p_5, respectively. Accordingly, the PGF of the packet length is $G_p(x) = p_1x + p_3x^3 + p_5x^5$ and the mean is denoted by $\overline{L} = p_1 + 3p_3 + 5p_5$.

In [14], it is assumed that all packets within a burst have the same destination node. Furthermore, a burst generated at a given node is intended to one of the other $m-1$ nodes with probability $1/(m-1)$. As a node, the master generates traffic intended for the slaves and in addition routes packets between the slaves. Under these assumptions, the burst arrival rate to each uplink (slave-to-master) queue is $\lambda_u = \lambda$ and the burst arrival rate to each downlink (master-to-slave) queue is $\lambda_d = \lambda$ (i.e. the model is symmetrical). Notice that the arrival process to the uplink queues is Poisson whereas the arrival process to the downlink queues, being dependent on the scheduling regime, is, in general, not Poisson.

Simplifying the above, we assume that the master is the destination of all packets generated at the slaves (i.e. the master *does not* route packets between slaves). On the other hand, we assume that packets are generated at every downlink queue according to a Poisson arrival process with rate λ. Due to the assumption regarding the Poisson arrival process, the analysis of this scenario is simpler than the analysis of the scenario described above in which the master *does* route packets. Therefore, the results regarding the access delay obtained in [14] should also hold for this scenario. However, we show that this is not the case.

Finally, three related performance indicators are defined:

- Access delay – The time a packet has to wait in the uplink queue before it is served (denoted by Wa).
- Queueing delay at the master – The time a packet has to wait at a downlink queue before it is served (denoted by Wm).
- End to end delay – The total time a packet spends in the master's and the slave's queues not including the service times (denoted by We = Wa + Wm).

The mean values of the delay functions described above are denoted by $\overline{W}_a$, $\overline{W}_m$ and $\overline{W}_e$.

4 Analysis of the Limited (Pure Round Robin) Regime

In this section we show that a piconet operating according to the limited regime and in which all packets are 1 slot long is equivalent to a TDMA system. Then, we consider a "standard" piconet with packet sizes of 1, 3, and 5 slots, and formulate it as a 1-limited *polling* system. We obtain exact analytical results which are *different* from those derived in [14] and indicate that [14] applies the results of the M/G/1 queue with vacations *without* considering the dependencies between the queues.

4.1 Analysis as a TDMA System

Consider a piconet operated in the limited regime in which *all packets are 1 slot long* (i.e. $p_1 = 1$). In such a piconet, a single slot is allocated to each downlink and uplink in every cycle. Therefore, the piconet can be analyzed as a *TDMA system* [1, p. 194] with a cycle length of $2(m-1)$ slots. Every slot in the cycle is allocated to one of the

$2(m-1)$ downlinks and uplinks. The computation of the delay in a TDMA system is based on the analogy with the M/D/1 queue with deterministic service time and vacation length both equal to $2(m-1)$ [1, p. 194]. Accordingly, applying [1] eq. (3.58), where the number of queues is $2(m-1)$ and the total arrival rate is $2(m-1)\lambda$, we obtain the mean access delay and queueing delay at the master (in slots):

$$\overline{W}_a = \overline{W}_m = \frac{m-1}{1-2(m-1)\lambda}. \tag{1}$$

For this simple scheduling regime, the result regarding the access delay obtained in eq. (5) in [14] coincides with our result (1).

The model presented in Section 3 is symmetrical with respect to the slaves (the arrival rates to all master and slaves queues are equal). However, in a TDMA system, the queuing behavior of one user is independent of the queuing behavior of other users. Thus, analytic results can be obtained for an asymmetrical piconet. We denote the arrival rate to slave i by λ_u^i and the arrival rate to the master of packets intended for slave i by λ_d^i. We denote the access delay of packets in the uplink queue of slave i by W_a^i and the queueing delay at the master of packets intended to slave i by W_m^i. In this (deterministic) case every link can be independently analyzed as an M/D/1 queue. Thus, the mean access delay (in slots) is again derived from eq. (3.58) in [1]:

$$\overline{W}_a^i = \left(\frac{2(m-1)\lambda_u^i}{2\left(1-2(m-1)\lambda_u^i\right)}\right)2(m-1) + \frac{1}{2}\cdot 2(m-1) = \frac{m-1}{1-2(m-1)\lambda_u^i}. \tag{2}$$

A similar equation describes the mean queueing delay at the master of packets intended to slave i ($\overline{W}_m^i$). In this case $\overline{W}_m^i$ replaces $\overline{W}_a^i$ and λ_d^i replaces λ_u^i.

4.2 Analysis as a 1-Limited Polling System

Since in most Bluetooth applications the length of the packets varies, we now consider a "standard" piconet operated in the limited regime with *1,3, and 5-slot packets*. We show that such a piconet can be modeled as a *1-limited polling system*[4] with $2(m-1)$ queues and present a closed form expression for the mean access delay in a piconet with symmetrical traffic.

In a piconet operated according to the limited scheduling regime, even if the master has nothing to send to a specific slave, one slot is used during the downlink communication (by the POLL packet). Similarly, even if the slave has nothing to send, one slot must be used during the uplink communication (by the NULL packet). In order to model the piconet as a 1-limited *polling* system we utilize the fact that data packets are at least one slot long. Thus, when data packets are sent at least one slot must be used.

We define an equivalent *1-limited gated* polling[5] system with the following characteristics: (a) There are $2(m-1)$ customers and a single server, (b) the server

---

[4] In a 1-limited polling system, at each visit of the server to a queue only the *first* packet in the queue is served. The server incurs a *switchover* time when it shifts from one queue to another [1, p. 201],[8],[24].

[5] The system is referred to as the *limited gated* polling system, since only a message that is found in the *beginning* of the switchover time is served.

serves the customers in a fixed cyclic order, (c) the server incurs a *switchover* time of 1 slot when it shifts from one customer to another, (d) when the server serves a customer, at most a single packet is served. If at the *beginning* of the switchover the queue is empty, the server completes the switchover and immediately switches to the next customer, and (e) the packet service times are 0, 2, and 4 slots.

This polling system is equivalent to a Bluetooth piconet operated according to the limited scheduling algorithm. Namely, for the same arrival process and packet length distribution (i.e. $p_0 = p_1$, $p_2 = p_3$, and $p_4 = p_5$), the delay (time until the *end* of service) in both systems is equal. This property is demonstrated in Fig. 3, which illustrates an example of the operation of a piconet composed of a master and two slaves and of the equivalent polling system. When the master starts transmitting to the first slave it has a 3-slot data packet in the downlink queue. In the equivalent system, this packet is represented by a 1 slot of switchover and 2 slots of data. The first slave has nothing to send, and therefore it responds with a NULL packet. A 1 slot of switchover represents this packet in the equivalent system. Then, the master sends a 1-slot data packet to the second slave. It is represented in the equivalent system by a 1 slot of switchover and *0 slots of data.* The rest of the transmissions (including a POLL packet) can be seen in the figure.

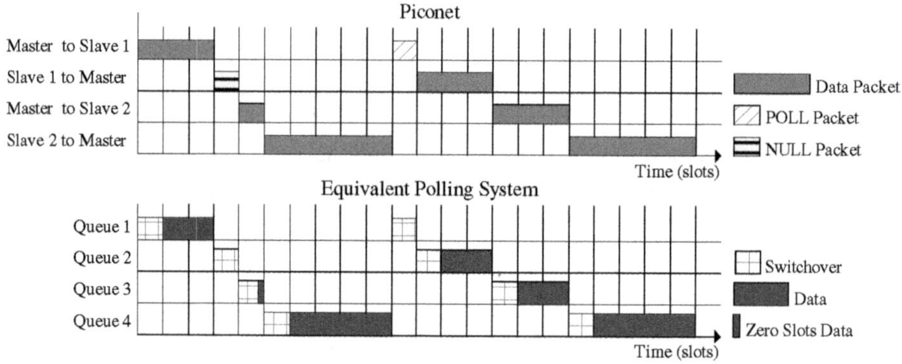

Fig. 3. An example of the operation of a piconet and of the equivalent polling system

In order to obtain the access delay (W_a) in a piconet, one has to deduct the Bluetooth packet length (L) from the delay (time until the *end* of service) in the equivalent polling system. Alternatively, if one obtains the waiting time (the time until the service *starts*) in the equivalent polling system, a single slot has to be deducted in order to obtain the access delay in a Bluetooth piconet. This results from the fact that when Bluetooth data packets are sent, some of the data is actually sent during the "switchover" time, as it is defined in the equivalent polling system.

We now focus on symmetrical systems in which the arrival rates to all queues are equal. By applying the model for a symmetrical *limited gated* polling system described in [1, p. 201] we can obtain the mean waiting time of a packet in a queue. The service time of a k-slot (k = 1,3,5) data packet is defined as $k - 1$ slots. Moreover, the waiting time in [1] is defined as the time a packet waits until its service starts. Thus, in order to obtain the mean access delay, one has to deduct 1 slot from the expression for the waiting time in [1], eq. (3.77). Accordingly, we apply [1] eq.

(3.77), where the number of queues is $2(m-1)$, the total arrival rate is $2(m-1)\lambda$, the switchover time is one slot with zero variance, the traffic intensity is $\rho = 2(m-1)\lambda(\overline{L}-1)$, and the second moment of the service time (denoted in [1] as $\overline{X^2}$) is $4p_3 + 16p_5$. Deducting 1 time unit (i.e. 1 slot), we obtain the mean access delay and the mean queueing delay at the master (in slots):

$$\overline{W}_a = \overline{W}_m = \frac{1+(m-1)\{1+2\lambda(p_3+6p_5-1)\}}{1-2(m-1)\lambda\overline{L}} - 1. \quad (3)$$

Notice that in this system it must hold that $2(m-1)\lambda\overline{L} < 1$. We shall refer to $2(m-1)\lambda\overline{L}$ as the load in the system.

As a special case, consider a piconet in which only 1-slot packets are used (i.e. $p_1 = 1$, $p_3 = 0$, and $p_5 = 0$). For such a piconet, eq. (3) reduces to eq. (1), which represents the delay in a piconet with 1-slot packets. Moreover, the result given by (3) was verified by two independent simulation models based on OPNET (for more details regarding the design of the simulation models, see [6][6] and [10]). For example, Fig. 4 compares the exact mean access delay ($\overline{W}_a$) (computed according to (3)) to the average access delay computed by simulation[7] in a piconet with 4 slaves in which the probabilities of 1, 3, and 5-slot packets are equal. For each load value, the results have been computed after 230,000 slots using the model presented in [6] or after 48,000 to 2,400,000 slots (depending on the load) using the model presented in [10].

Finally, we note that the equivalence between a piconet operated according to limited algorithm and a 1-limited polling system can be exploited in order to obtain exact and approximate results for more complex systems. Namely, using the equivalence property and methods for analyzing 1-limited polling systems (see for example [8]), one may analyze piconets with bursty arrival process ($G_b(x) \neq x$), asymmetrical traffic, and intra-piconet routing. For example, Miorandi and Zanella [11] have recently used this property in order to analyze piconets with asymmetrical traffic and to obtain approximate results which outperform the results in [10] and [14].

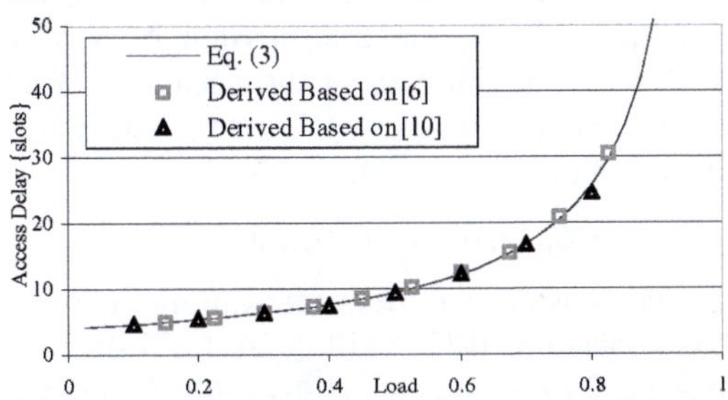

Fig. 4. The mean access delay (computed according to (3)) and the average access delay values computed by simulation in a piconet with 4 slaves in which $p_1 = p_3 = p_5 = 1/3$

[6] In [6] the delay is defined as the time until the whole packet is received by the destination.

[7] The simulation results computed by the model presented in [10] have been obtained by Carlo Caimi from the University of Padova.

4.3 Examination of the Analysis as an M/G/1 Queue with Vacations [14]

The theory of M/G/1 queue with vacations[8] [9],[25] is used in [14] in order to analyze a piconet operated in the limited regime in which the packets are 1, 3, and 5 slots long. We now briefly describe that analysis and point out that the direct use of the model of M/G/1 queue with vacations to analyze a 1-limited polling system, *without taking into consideration the dependencies between the queues*, leads to inaccurate results and may serve only as an approximation. We assume, for simplicity, that the traffic is non-bursty ($G_b(x) = x$) and that packets are generated at every master-to-slave (downlink) and slave-to-master (uplink) queue according to a *Poisson* arrival process with arrival rate λ.

The *piconet service cycle time* X_c is defined in [14] as the time (number of slots) required for the master to serve all the slaves once. The PGF, the mean, and the second moment of the cycle time are denoted by $G_{X_c}(x)$, $\overline{X}_c$, and $\overline{X}_c^2$. It is claimed that since the model is symmetric, it is sufficient to consider a *single master-slave channel* (accordingly, we refer to the considered slave as slave i). It is mentioned that the probability that a downlink queue is not empty is $P_d = \lambda \overline{X}_c$ and that the probability that an uplink queue is not empty is $P_u = \lambda \overline{X}_c$. In [14], eq. (1) and (2), the PGFs of the durations of the downlink and uplink communications are presented as:

$$G_d(x) = (P_d p_1 + (1 - P_d))x + P_d p_3 x^3 + P_d p_5 x^5 , \tag{4}$$

$$G_u(x) = (P_u p_1 + (1 - P_u))x + P_u p_3 x^3 + P_u p_5 x^5 , \tag{5}$$

while the PGF of the cycle time is presented as:

$$G_{X_c}(x) = (G_d(x) G_u(x))^{m-1} . \tag{6}$$

The model of M/G/1 queue with vacations is used to analyze the system. Accordingly, the service time of a single uplink queue (i.e. of the queue at slave i) in the vacation model is equal to the cycle time (X_c). A vacation starts when the master polls slave i and finds its uplink queue to be empty. As we understand, the vacation is composed of the NULL packet returned by slave i, the service of the other $m - 2$ uplink and downlink queues, and the transmission of a packet to slave i in the downlink queue. V_l denotes the duration of the vacation period and its PGF, mean, and second moment are denoted by $G_{V_l}(x)$, $\overline{V}_l$, and $\overline{V}_l^2$. The PGF of the duration of the vacation period is provided in [14], eq. (3):

$$G_{V_l}(x) = x G_d(x) (G_d(x) G_u(x))^{m-2} . \tag{7}$$

Finally, the access delay is derived from the waiting time in a batch arrival system $M^{[x]}/G/1$ with multiple vacations ([25, p. 143], eq. (3.21a)). Since we assume that the arrival process is non-bursty (i.e. $G_b(x) = x$, $\overline{B} = 1$, $\overline{B}^{(2)} = 0$), eq. (5) in [14] reduces to [25, p. 123], eq. (2.14a):

---

[8] According to the model of M/G/1 queue with multiple vacations, the server begins a vacation every time the system becomes empty. If the server returns from a vacation to find the system not empty, it starts working immediately and continues until the system becomes empty again. If the server returns from a vacation to find an empty system, it begins another vacation immediately [25].

$$\overline{W}_a = \frac{\lambda \overline{X_c^2}}{2\left(1-\lambda \overline{X}_c\right)} + \frac{\overline{V_l^2}}{2\overline{V}_l}. \tag{8}$$

We now describe a few problems in the model presented above. For the clarity of presentation, the description of each problem ignores the existence of other problems.

1. *The cycle length described in eq. (6) is inappropriate for use as a service time in a vacation model.* Consider the instant when the server returns to slave i from a vacation or when it completes "service" (i.e. the master completes a cycle) there, and at least one packet is found in the uplink queue. In such a case, the server will not take a vacation and a service period will start. According to [14], the PGF of the service time is defined as the PGF of the cycle time (Xc) presented in (6). It is composed of the duration of the communication in the uplink queue of slave i and the durations of the communication in the rest of the uplink and downlink queues.

   When the service starts, there is obviously a data packet in the queue of slave i. Hence, the PGF of the duration of the uplink communication of slave i is Gp(x) (defined in Section 3) and not Gu(x) (presented in eq. (5)). Therefore, one of the terms in eq. (6) should be replaced. Consequently, eq. (6) cannot be used as a service time in a vacation model[9].
2. *The uplink and downlink communication periods composing a cycle are not independent.* The PGF of the cycle length presented in (6) is a multiple of the PGFs of the uplink (Gu(x)) and the downlink (Gd(x)) communication periods. Thus, an underlying assumption in the derivation of $G_{X_c}(x)$ is that all the downlink and uplink communication periods are independent. However, the length of a communication period depends on the existence of a packet in the corresponding queue (if the queue is empty, the length is 1 slot). The existence of a packet depends on the lengths of the preceding uplink and downlink communication periods, since long periods increase the probability of a packet arrival. Hence, the lengths of the periods composing a cycle do not seem to be independent, and therefore the derivation of the cycle length in (6), where independence is assumed, is inaccurate[10].
3. *The cycle time depends on the length of the vacation or the cycle that precedes it.* The vacation model described by Takagi [25] is used in order to compute the mean access delay, presented in eq. (8). One of the important assumptions made in [25, p. 111] is that: "Messages arrive in the system according to a Poisson process of fixed rate and have service times with independent and identical distribution. These service times are independent of the arrival process, and each service time is independent of the sequence of vacation periods that precede that service time."

   However, the service time in the model described in [14] is taken as the cycle time (Xc) and is composed of uplink and downlink communication in m – 1 queues. The length of each of the communication periods depends on the existence of packets in the corresponding queue. If the cycle follows a long vacation or cycle, the probabilities that the queues are not empty will increase, and thereby the probability of a long cycle will increase. On the other hand, if the cycle follows a short vacation or cycle, the probability of a short cycle will increase. Thus, the

[9] This difference is critical for small piconets where only a few components compose $G_{X_c}(x)$.

[10] It seems that the mean cycle length derived from combining (4), (5), and (6) is correct. However, for the vacation model the second moment of the cycle length is also required.

requirement for independent service times, and for independence between vacation lengths and service times, does not seem to hold. Therefore, eq. (8) can provide only an approximation for the delay.

In Section 4.1, we have analyzed a piconet operated in the limited regime in which *all packets are 1 slot long*. In such a piconet the duration of a communication period is deterministic (1 slot), and therefore, there are no dependencies between the cycle and vacation lengths as well as within a cycle. Thus, as we have mentioned, for this simple regime, the result regarding the access delay obtained in eq. (5) in [14] coincides with our result (1). However, due to problems 1, 2, and 3, described above, in a "standard" piconet (in which the packets are 1, 3, and 5 slot long) the results presented in [14] differ considerably from our results.

Recall, that we assume that the master is the destination of all packets generated at the slaves (i.e. the master *does not* route packets). On the other hand, in [14] it is assumed that the master routes some of the traffic between the slaves. Accordingly, the mean value of the *end-to-end delay* is defined as the sum of the access delay and the queueing delay at the master. A method for computing the mean queueing delay at the master is described at the end of Section 2 in [14]. For non-bursty traffic, the delay computed according to that method is equal to the access delay described in (8) (i.e. $\overline{W}_a = \overline{W}_m$). Thus, it seems that an underlying assumption is that the arrival process of packets to the master from the slaves can be treated as Poisson.

This assumption probably follows the analysis of *polling systems with probabilistic routing* (e.g. [23]). The analysis of such systems is based on an important assumption that the *service times of a packet in different queues are independent*. However, this is not the case in a piconet (for example, a 1-slot packet sent from the slave to the master cannot become a 3-slot packet when it is forwarded to another slave). Thus, the computation of the end-to-end delay is inaccurate not only because of the inaccuracies in the computation of the access delay but also due to the assumptions made regarding the arrival process of packets which require routing.

We note that Miorandi et al. [10] present an *approximate* analysis of the limited scheduling algorithm by using the tool of probabilistic routing and assuming that the various resulting flows are independent. They state that "*the assumption of independent flows, although providing good results at low traffic load, leads to substantial mismatch with the simulation results as the system gets close to stability limit*".

Finally, we note that assumptions, similar to the ones indicated in this section, also appear in [12],[13],[15],[16],[17],[18],[19],[20], and [21]. For example,

- In [12] the limited scheduling algorithm is analyzed in a similar methodology to the analysis described in [14]. For instance, eq. (5) in [12] presents the mean waiting time. However, the calculation of this waiting time ignores the dependency between the service time and the vacation length preceding it (see item 3 above).
  Moreover, according to eq. (7) in [12] the arrival rate must be lower than $\lambda_0 = (2(m-1)(\overline{L}-1))^{-1}$. If the arrival rate approaches this value, the cycle length defined in [12] approaches infinity. However, since the longest possible cycle is composed of only 5-slot packets, the maximal cycle length is 2(m – 1)·5 slots.
  Although both [12] and [14] deal with the limited scheduling regime, the vacation length described in eq. (4) in [12] differs from the vacation length described in eq.

(3) in [14]. In both cases the use of the vacation model leads to approximate results disregarding the exact vacation length.

- The analysis of the limited scheduling algorithm in [13] and [16] is very similar to the analysis in [14]. The only difference is that in [13] and [16] it is assumed that the master does not generate traffic and it only routes packets between the slaves.
- In [15],[19], and [21] the performance of scatternets composed of two piconets connected through a bridge is analyzed. The performance of the scatternets is analyzed for exhaustive and limited intra-piconet scheduling algorithms. The limited algorithm is analyzed in a similar manner to the analysis in [14].
- The analysis of the exhaustive regime in [14] is also based on the theory of M/G/1 queue with vacations. In [29] we show that the PGF of the time to exhaust the queues derived according to [14] differs from the correct PGF. Moreover, we argue that due to the reasons discussed above, the use of the model of M/G/1 queue with vacations in order to analyze the exhaustive regime leads to approximate results. The same remarks apply to works [17],[18],[19],[20], and [21].

5 Numerical Results

In this section we present exact numerical results computed according to the analysis in Section 4.2. Then, we demonstrate the difference between our results and those presented in [14]. It turns out that in some cases the results obtained according to [14] *underestimate* the mean access delay by more than 50%. We guess that the noticeable difference between analytical and simulation results in high arrival rates, indicated in [12], is due to the inaccuracy of the analytical results there.

Fig. 5 illustrates the exact mean access delay (computed according to (3)) in piconets with various numbers of slaves in which the probabilities of 1, 3, and 5-slot packets are equal ($p_1 = p_3 = p_5 = 1/3$). The figure presents the delay (in slots) as a function of the load in the system (defined in Section 4.2 as $2(m-1)\lambda\overline{L}$).

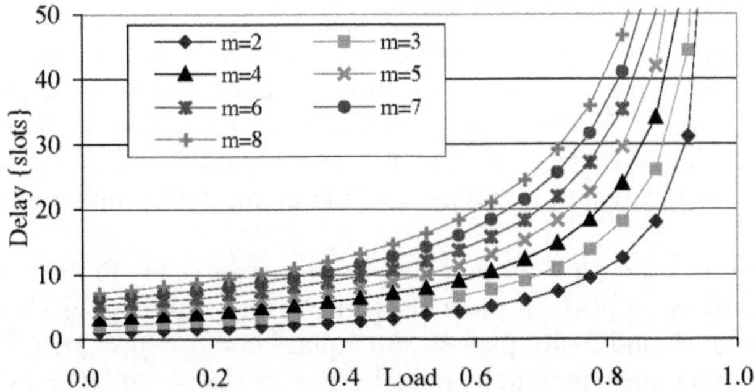

Fig. 5. The exact mean access delay (computed by (3)) in piconets in which $p_1 = p_3 = p_5 = 1/3$

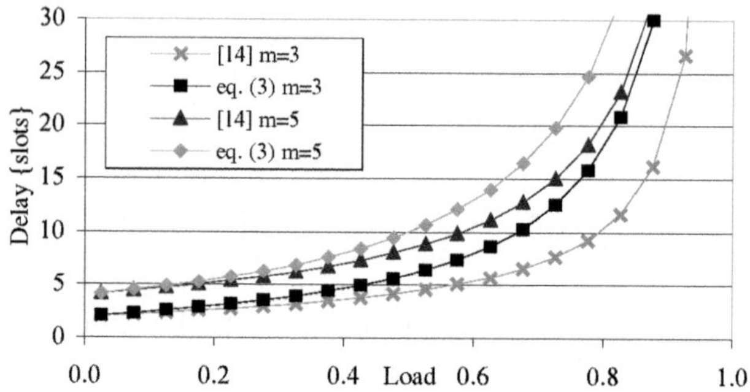

Fig. 6. The mean access delay derived according to [14] and the exact mean access delay (computed by (3)) in piconets with 2 and 4 slaves in which all packets are 5 slots long ($p_5 = 1$)

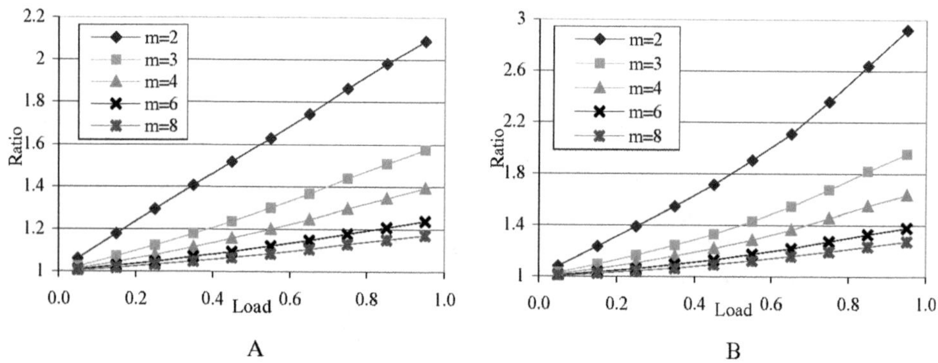

Fig. 7. The ratio of the exact mean access delay (obtained by (3)) to the mean access delay derived according to [14] in piconets in which (A) $p_1 = p_3 = p_5 = 1/3$ and (B) $p_5 = 1$

Fig. 6 compares the mean access delay computed according to [14] to the mean access delay computed according to our model (i.e. according to (3)) when all packets are 5 slots long ($p_5 = 1$). The delay is depicted as a function of the system load in a piconet with 2 slaves ($m = 3$) and in a piconet with 4 slaves ($m = 5$). The figure demonstrates that the results obtained in [14] significantly underestimate the access delay.

Fig. 7-A presents the ratio of the exact mean access delay to the mean access delay computed according to [14], in piconets with various numbers of slaves in which the probabilities of 1, 3, and 5-slot packets are equal ($p_1 = p_3 = p_5 = 1/3$). Fig. 7-B presents the same ratio in piconets in which all packets are 5 slots long ($p_5 = 1$).

6 Conclusions

This work presents an analytical study of the limited (pure round robin) scheduling algorithm for Bluetooth piconets, and examines the analytical study of this algorithm in [14]. We have modeled a piconet in which all packets are 1 slot long operated according to the limited scheduling algorithm as a TDMA system. Then, we showed that a piconet operated according to the limited scheduling algorithm is equivalent to a 1-limited polling system, and derived exact analytic results for symmetrical systems. These results differ from those obtained in [14] which can actually be viewed as approximate results. We have argued that [12],[13],[15],[16],[17],[18],[19],[20], and [21] seem also to present approximate results. Finally, we have provided numerical examples that illustrate the difference between the exact results and those presented in [14].

Future study will focus on utilizing the equivalence between a piconet and a 1-limited polling system along with the vast amount of research on 1-limited polling in order to obtain good approximate results for asymmetrical piconets with complex traffic patterns. One of the first attempts in that direction has been recently made in [11] where approximate results that outperform those in [10] and [14] have been presented. Moreover, due to the inherent complexities in obtaining the PGF of the time to exhaust the queues at the master and a given slave in the gated and exhaustive regimes (presented in [29]), it seems that there is no closed form expression for the delay under such regimes. Thus, a major future research goal is to obtain a good (at least approximate) analysis of such regimes.

Acknowledgments. We thank Nir Naaman and Daniele Miorandi for helpful discussions, and Ronen Kofman and Carlo Caimi for obtaining simulation results. We also thank the anonymous reviewers for their helpful comments.

References

1. Bertsekas, D.P., Gallager, R.: Data Networks. Prentice-Hall Inc., New Jersey (1992)
2. Bluetooth SIG: Specification of the Bluetooth System – Version 1.1. (Feb. 2001)
3. Bray, J., Sturman, C.: Bluetooth 1.1 Connect without Cables. Prentice-Hall Inc., New Jersey (2001)
4. Bruno, R., Conti, M., and Gregori, E.: Bluetooth: Architecture, Protocols and Scheduling Algorithms. Cluster Computing, Vol. 5 (Apr. 2002) 117–131
5. Capone, A., Gerla, M., Kapoor, R.: Efficient Polling Schemes for Bluetooth Picocells. Proc. IEEE ICC'01 (June 2001)
6. Har-Shai, L., Kofman, R., Zussman, G., Segall, A.: Inter-Piconet Scheduling in Bluetooth Scatternets. Proc. OPNETWORK 2002 (Aug. 2002)
7. Johansson, N., Korner, U., Johansson, P.: Performance Evaluation of Scheduling Algorithms for Bluetooth. Proc. IFIP TC6 Int. Conf. on Broadband Comm. (Nov. 1999)
8. Levy, H., Sidi, M.: Polling Systems: Application, Modeling, and Optimization. IEEE Trans. on Comm., Vol. 38 (Oct. 1990) 1750–1760
9. Levy, Y., Yechiali, U.: Utilization of the Idle Time in an M/G/1 Queueing System. Management Science, Vol. 22 (Oct. 1975) 202–211
10. Miorandi, D., Caimi, C., Zanella, A.: Performance Characterization of a Bluetooth Piconet with Multi-Slot Packets. Proc. WiOpt'03 (Mar. 2003)

11. Miorandi, D., Zanella, A.: An Approximate Delay Analysis for Limited-1 Polling in a Bluetooth Piconet. Working Paper (May 2003)
12. Misic, J., Misic, V.B.: Queueing Analysis of Bluetooth Piconet Traffic. Proc. SPECTS'02 (July 2002)
13. Misic, J., Misic, V.B.: On Delay Times in a Bluetooth Piconet: The Impact of Different Scheduling Policies. Proc. EurAsia ICT 2002, Lecture Notes in Computer Science, Vol. 2510 (eds: Shafazand M.H., Tjoa A M.). Springer-Verlag (Oct. 2002) 891–898
14. Misic, J., Misic, V.B.: Modeling Bluetooth Piconet Performance. IEEE Comm. Letters, Vol. 7 (Jan. 2003) 18–20
15. Misic, J., Misic, V.B.: Bridges of Bluetooth County: Topologies, Scheduling, and Performance. IEEE J. on Selected Areas in Comm., Vol. 21 (Feb. 2003) 240–258
16. Misic, V.B., Misic, J.: On Bluetooth Piconet Traffic Performance", Proc. IEEE PIMRC'02 (Sep. 2002)
17. Misic, V.B., Misic, J.: Minimizing End-to-End Delays in Bluetooth Scatternet with a Slave/Slave Bridge. Proc. ICCCN'02 (Oct. 2002)
18. Misic, V.B., Misic, J.: Queueing Analysis of Delays in Bluetooth Slave/Slave Bridge. Proc. IEEE/ACM MASCOTS'02 (Oct. 2002)
19. Misic, V.B., Misic, J.: Bluetooth Scatternet with a Master/Slave Bridge: a Queueing Theoretic Analysis. Proc. IEEE GLOBECOM'02 (Nov. 2002)
20. Misic, V.B., Misic, J.: Performance of Bluetooth Bridges in Scatternets with Exhaustive Service Scheduling. Proc. HICSS-36 (Jan. 2003)
21. Misic, V.B., Misic, J.: Performance of Bluetooth Slave/Slave Bridge. Telecommunication Systems, Vol. 22 (Jan. – Apr. 2003) 221–239
22. Racz, A., Miklos, G., Kubinszky, F., Valko, A.: A Pseudo Random Coordinated Scheduling Algorithm for Bluetooth Scatternets. Proc. ACM MOBIHOC'01 (Oct. 2001)
23. Sidi, M., Levy, H., Fuhrmann, S.: A Queueing Network with a Single Cyclically Roving Server. Queueing Systems, Vol. 11 (1992) 121–144
24. Takagi H.: Analysis of Polling Systems. MIT Press (1986)
25. Takagi, H.: Queueing Analysis: a Foundation of Performance Evaluation, Vol. 1 – Vacation and Priority Systems. North-Holland (1991)
26. Yechiali, U.: Analysis and Control of Polling Systems. In: Performance Evaluation of Computer and Communication Systems (eds: Donatiello L., Nelson R.). Springer-Verlag (1993) 630–650
27. Zussman, G., Segall, A.: Capacity Assignment in Bluetooth Scatternets – Analysis and Algorithms. Proc. IFIP-TC6 Networking 2002, Lecture Notes in Computer Science, Vol. 2345 (eds: Gregori E. et al.). Springer-Verlag (May 2002) 411–422
28. Zussman, G., Segall, A.: Capacity Assignment in Bluetooth Scatternets – Optimal and Heuristic Algorithms. To appear in: ACM/Kluwer Mobile Networks and Applications (MONET), Vol. 9, No. 1 (Feb. 2004)
29. Zussman, G., Yechiali, U., Segall, A.: On the Exact Analysis of Bluetooth Scheduling Algorithms. Submitted for Publication (available from http://www.comnet.technion.ac.il/~gilz/pub_files/BT_ex_li.pdf) (May 2003)

A Probabilistic Topology Unaware TDMA Medium Access Control Policy for Ad Hoc Environments*

Konstantinos Oikonomou[1] and Ioannis Stavrakakis[2]

[1] INTRACOM S.A., Development Programmes Department,
19.5 Markopoulou Avenue, Paiania 190 02
Athens, Greece
Tel: +30 210 6677023, Fax: +30 210 6671312
okon@intracom.gr

[2] University of Athens,
Department of Informatics & Telecommunications
Panepistimiopolis, Ilissia 15 784
Athens, Greece
Tel: +30 210 7275343, Fax: +30 210 7275333
istavrak@di.uoa.gr

Abstract. The design of an efficient Medium Access Control (MAC) is challenging in ad-hoc networks where users can enter, leave or move inside the network without any need for prior configuration. Chlamtac and Farago have proposed a topology unaware TDMA-based scheme, suitable for ad-hoc networks, while Ju and Li have proposed an enhanced version that maximizes the minimum guaranteed throughput. Both approaches consider a deterministic policy for the utilization of the assigned scheduling time slots. In this work it is shown that this deterministic policy fails to utilize non-assigned slots that would result in collision-free transmissions even under heavy traffic conditions. A simple probabilistic policy is proposed, capable of utilizing the non-assigned slots according to an access probability, fixed for all users in the network. An analytical study establishes the conditions under which the probability of success for a specific transmission under the probabilistic policy, is higher than that under the deterministic. The dependence of both policies on the topology density is shown and a simple topology density metric is introduced as well. Simulation results show that there exists a suitable range of values for the access probability for which the probabilistic policy outperforms the deterministic and show how this range is affected by the variations of the topology density.

1 Introduction

Nowadays, the user demand for modern applications anytime and anywhere cannot be fulfilled by the *traditional* networks (wired, optical, infrastructure-

* This work has been supported in parts by the IST program under contract IST-2001-32686 (BroadWay).

M. Conti et al. (Eds.): PWC 2003, LNCS 2775, pp. 291–305, 2003.

based mobile, etc.). These networks are infrastructure-based and require prior configuration in order for a user to be allowed to use the network resources. On the other hand, *ad-hoc* networks are *infrastructureless* and a node can enter, leave or move inside the network without any need for prior configuration. The infrastructureless nature of the ad-hoc networks provides flexibility to the user but at the same time introduces new challenges. The idiosyncrasies of the ad-hoc networks make the design of an efficient Medium Access Control (MAC) a challenging problem.

Several MAC protocols have been proposed for ad-hoc networks. CSMA/CA has been employed as the main mechanism, [1], [2], [3], whereas many others have employed the Ready-To-Send/Clear-To-Send (RTS/CTS) mechanism, [4], [5], in addition to CSMA/CA, to avoid the *hidden/exposed terminal* problem.

TDMA-based MAC protocols have also been proposed for ad-hoc networks. S-TDMA - proposed by Kleinrock and Nelson, [6], is capable of providing collision-free scheduling based on the exploitation of noninterfering transmissions in the network. Chlamtac and Farago, [7], have proposed an original TDMA-based scheme for topology transparent scheduling. Their scheme exploits the mathematical properties of polynomials with coefficients from finite Galois fields, to randomly assign scheduling time slot sets to each node of the network. Collisions are allowed, but it is guaranteed that for each node at least one time slot in a frame would be collision-free, [7]. Ju and Li, [8], have proposed another scheme that maximizes the minimum guaranteed throughput. However, certain factors, such as the topology density and the particular random assignment of the polynomials, have not been taken into account.

In this paper the general approach proposed in [7] and [8] is considered and the idea of allowing the nodes to utilize (according to a common access probability) scheduling slots not originally assigned (according to the rules in [7], [8]) to them, is introduced. As it is shown in this paper, the proposed policy achieves a higher throughput under certain conditions (that are studied here), when the benefit of utilizing otherwise idle slots outweighs the loss due to collisions induced by the introduced controlled interference.

In Section 2 a general ad-hoc network is described and some key definitions are introduced. The proposed policy (to be referred to as the *Probabilistic Policy*) is motivated and introduced in Section 3; the one introduced in [7], [8] is also described and is referred to as the *Deterministic Policy*. In Section 4 the case of a specific transmission between two given neighbor nodes, is considered. A study is presented establishing the conditions (topology density and polynomial assignment) under which the Probabilistic Policy achieves higher success probability, for the specific transmission; it is also shown that the set of two-hop neighbor nodes influence the probability of success, for a specific transmission. In Section 5 the system throughput is discussed and the motivation for the introduction of a topology density metric and a categorization of topologies is presented in Section 6. The simulation results, presented in Section 7, show that there exists a range of values of the access probability used to utilize probabilistically the otherwise idle slots, that result in a higher throughput. Furthermore, the simulation results show that this range of values decreases almost exponentially, as the density of the topology increases, as it is expected from the analysis in Section 4. Section 8 presents the conclusions.

2 System Definition

An ad-hoc network may be viewed as a time varying multihop network and may be described in terms of a graph $G(V, E)$, where V denotes the set of nodes and E the set of links between the nodes at a given time instance. Let $|X|$ denote the number of elements in set X and let $N = |V|$ denote the number of nodes in the network. Let S_u denote the set of neighbors of node u, $u \in V$. These are the nodes v to which a direct transmission from node u (transmission $u \rightarrow v$) is possible. Let D denote the maximum number of neighbors for a node; clearly $|S_u| \leq D$, $\forall u \in V$.

Suppose that a node u wants to transmit to a particular neighbor node v in a particular time slot i. In order for the transmission $u \rightarrow v$ to be successful, two conditions should be satisfied. First, node v should not transmit in the particular time slot i, or equivalently, no transmission $v \rightarrow \psi$, $\forall \psi \in S_v$ should take place in time slot i. Second, no neighbor of v - except u - should transmit in time slot i, or equivalently, no transmission $\zeta \rightarrow \chi$, $\forall \zeta \in S_v - \{u\}$ and $\chi \in S_\zeta$, should take place in time slot i. Consequently, transmission $u \rightarrow v$ is corrupted it time slot i if at least one transmission $\chi \rightarrow \psi$, $\chi \in S_v \cup \{v\} - \{u\}$ and $\psi \in S_\chi$, takes place in time slot i.

The transmission(s) that corrupts transmission $u \rightarrow v$ may or may not be successful itself. Specifically, in the presence of transmission $u \rightarrow v$, transmission $\chi \rightarrow \psi$, $\chi \in S_v \cup \{v\} - \{u\}$ and $\psi \in S_\chi \cap (S_u \cup \{u\})$, is corrupted. If $\psi \in S_\chi - (S_\chi \cap (S_u \cup \{u\}))$, then transmission $\chi \rightarrow \psi$ is not affected by transmission $u \rightarrow v$.

3 Scheduling Policies

Under the Deterministic Policy, [7], [8], each node $u \in V$ is randomly assigned a unique polynomial f_u of degree k with coefficients from a finite Galois field of order q ($GF(q)$). Polynomial f_u is represented as $f_u(x) = \sum_{i=0}^{k} a_i x^i (\text{mod } q)$ [8], where $a_i \in \{0, 1, 2, ..., q-1\}$; parameters q and k are calculated based on N and D, according to the algorithm presented either in [7] or [8]. For both algorithms it is satisfied that $k \geq 1$ and $q > kD$ or $q \geq kD + 1$ (k and D are integers). The access scheme considered is a TDMA scheme with a frame consisted of q^2 time slots. If the frame is divided into q subframes s of size q, then the time slot assigned to node u in subframe s, $(s = 0, 1, ..., q-1)$ is given by $f_u(s) \text{mod } q$ [8]. Let the set of time slots assigned to node u be denoted as Ω_u. Consequently, $|\Omega_u| = q$. The deterministic transmission policy, [7], [8], is the following.

The Deterministic Policy: Each node u transmits in a slot i only if $i \in \Omega_u$, provided that it has data to transmit.

Depending on the particular random assignment of the polynomials, it is possible that two nodes be assigned overlapping time slots (i.e., $\Omega_u \cap \Omega_v \neq \emptyset$). Let $C_{u \to v}$ be the set of overlapping time slots between those assigned to node u and those assigned to any node $\chi \in S_v \cup \{v\} - \{u\}$. $C_{u \to v}$ is given by (1).

$$C_{u \to v} = \Omega_u \cap \left(\bigcup_{\chi \in S_v \cup \{v\} - \{u\}} \Omega_\chi \right). \tag{1}$$

Let $R_{u \to v}$ denote the set of time slots i, $i \notin \Omega_u$, over which transmission $u \to v$ would be successful. Equivalently, $R_{u \to v}$ contains those slots not included in set $\bigcup_{\chi \in S_v \cup \{v\}} \Omega_\chi$. Consequently,

$$|R_{u \to v}| = q^2 - \left| \bigcup_{\chi \in S_v \cup \{v\}} \Omega_\chi \right|. \tag{2}$$

$R_{u \to v}$ is the set of *non-assigned eligible* time slots for transmission $u \to v$, that if used by transmission $u \to v$, the probability of success for the particular transmission could be increased. The increased probability of success for transmission $u \to v$ does not necessarily increase the average probability of success of all transmissions in the network (throughput); the presence of transmission $u \to v$ in a slot i, $i \notin \Omega_u$, may corrupt another, otherwise successful, transmission $\chi \to \psi$. Then, transmission $\chi \to \psi$ will not be a successful one, even though $u \to v$ will be.

Theorem 1. *$|R_{u \to v}|$ is greater than or equal to $q(k-1)D$.*

Proof. Notice that $\left| \bigcup_{\chi \in S_v \cup \{v\}} \Omega_\chi \right| \leq (|S_v| + 1)q$, since $|\Omega_\chi| = q$, $\forall \chi \in V$. From Equation (2) it is concluded that $|R_{u \to v}| \geq q^2 - (|S_v| + 1)q$, or $|R_{u \to v}| \geq q(q - |S_v| - 1)$. Given that $D \geq |S_v|$, $|R_{u \to v}| \geq q(q - D - 1)$. Since $q \geq kD + 1$ (see [7], [8]), $q - D - 1 \geq (k-1)D$. Consequently, $|R_{u \to v}| \geq q(k-1)D$. □

From Theorem 1 it is obvious that for $k > 1$, $|R_{u \to v}| > qD$. Consequently, the number of non-assigned eligible slots may be quite significant for the cases where $k > 1$ (this case corresponds to large networks, [8]). Even for the case where $k = 1$, $|R_{u \to v}| \geq 0$, that is, $|R_{u \to v}|$ can still be greater than zero. For those nodes for which the set of overlapping slots is not the largest possible $\left(\text{i.e., } \left| \bigcup_{\chi \in S_v \cup \{v\}} \Omega_\chi \right| < (|S_v| + 1)q\right)$, $|R_{u \to v}|$ is greater than zero, even for $k = 1$. Furthermore, if the neighborhood of node v is not dense, or $|S_v|$ is small compared to D, then $|R_{u \to v}|$ is even higher.

In general, the use of slots i, $i \in R_{u \to v}$, may increase the average number of successful transmissions, as long as $R_{u \to v}$ is determined and time slots $i \in R_{u \to v}$ are used efficiently. The determination of $R_{u \to v}$ requires the existence of a mechanism for the extraction of sets Ω_χ, $\forall \chi \in S_v$. In addition, the efficient use of slots in $R_{u\to v}$ by node u, requires further coordination and control exchange with neighbor nodes χ, whose transmissions $\chi \to \psi$, with $R_{\chi\to\psi} \cap R_{u\to v} \neq \emptyset$, may utilize the same slots in $R_{\chi\to\psi} \cap R_{u\to v}$ and corrupt either transmission $u \to v$ or $\chi \to \psi$, or both.

Moreover, under non-heavy traffic conditions, there exist a number of idle slots, in addition to those in $R_{u\to v}$, not used by the node they are assigned to. In order to use all non-assigned time slots without the need for further coordination among the nodes, the following probabilistic transmission policy is proposed.

The Probabilistic Policy: Each node u always transmits in slot i if $i \in \Omega_u$ and transmits with probability p in slot i if $i \notin \Omega_u$, provided it has data to transmit.

The Probabilistic Policy does not require specific topology information (e.g., knowledge of $R_{u\to v}$, etc.) and, thus, induces no additional control overhead. The access probability p is a simple parameter common for all nodes. Under the Probabilistic Policy, all slots $i \notin \Omega_u$ are potentially utilized by node u: both, those in $R_{u\to v}$, for a given transmission $u \to v$, as well as those not in $\Omega_u \cup R_{u\to v}$ that may be left by neighboring nodes under non-heavy traffic conditions. On the other hand, the probabilistic transmission attempts induce interference to otherwise collision-free transmissions. The following section establishes the conditions under which the loss due to the induced interference is more than compensated for by the utilization of the non-assigned time slots.

4 Specific Transmission Analysis

In this section both policies are analyzed for a specific transmission (transmission $u \to v$). The analysis assumes heavy traffic conditions; that is, there is always data available for transmission at each node, for every time slot. Let $P_{i,u\to v}$ denote the probability that transmission $u \to v$ in slot i is successful. Let $P_{u\to v}$ be the average probability over a frame for transmission $u \to v$ to be successful during a time slot. That is,

$$P_{u\to v} = \frac{1}{q^2} \sum_{i=1}^{q^2} P_{i,u\to v}, \tag{3}$$

where q^2 is the frame size, in time slots.

Under the Deterministic Policy, $P_{i,u\to v} = 0$, $\forall i \notin \Omega_u$. For $i \in \Omega_u$ there are two distinct cases: for $i \in C_{u\to v}$, $P_{i,u\to v} = 0$, while for $i \notin C_{u\to v}$, $P_{i,u\to v} = 1$ (note the if $i \in C_{u\to v}$ then $i \in \Omega_u$ as well). Since $|\Omega_u| = q$, it is evident that under the Deterministic Policy the average over a frame probability of success for transmission $u \to v$ (denoted by $P_{D,u\to v}$), is given by

$$P_{D,u\to v} = \frac{q - |C_{u\to v}|}{q^2}. \tag{4}$$

Under the Probabilistic Policy, it is evident that $P_{i,u\to v} = 0$, for $i \in C_{u\to v}$, as well as for $i \notin \Omega_u$ and $i \notin R_{u\to v}$. On the other hand, $P_{i,u\to v} = (1-p)^{|S_v|}$, for $i \in \Omega_u$ and $i \notin C_{u \to v}$, whereas $P_{i,u \to v} = p(1-p)^{|S_v|}$, for $i \in R_{u \to v}$ (note that if $i \in R_{u \to v}$ then $i \notin \Omega_u$). Consequently, $P_{i,u \to v} = (1-p)^{|S_v|}$ for $q - |C_{u \to v}|$ time slots, while $P_{i,u \to v} = p(1-p)^{|S_v|}$ for $|R_{u \to v}|$ time slots. As a result, under the Probabilistic Policy the average over a frame probability of success for transmission $u \to v$ (denoted by $P_{P,u \to v}$), is given by

$$P_{P,u \to v} = \frac{q - |C_{u \to v}| + p|R_{u \to v}|}{q^2}(1-p)^{|S_v|}. \tag{5}$$

The term $\frac{q-|C_{u \to v}|}{q^2}(1-p)^{|S_v|}$ is equal to $P_{D,u \to v}$, decreased by the factor $(1-p)^{|S_v|}$ that is due to the interference introduced by the probabilistic transmission attempts. The term $\frac{p|R_{u \to v}|}{q^2}(1-p)^{|S_v|}$ is the gain due to the use of the non-assigned eligible slots $R_{u \to v}$. The aforementioned reduction of $P_{D,u \to v}$ due to the interference, is possible to be more than compensated for by the gain due to the use of the non-assigned eligible slots, resulting in $P_{P,u \to v} > P_{D,u \to v}$. This is investigated in the sequel.

It is evident that $P_{P,u \to v} = P_{D,u \to v}$, when $p = 0$. For $p > 0$, $P_{P,u \to v}$ may or may not be greater than $P_{D,u \to v}$, depending on the values of p, q, $|S_v|$, $|R_{u \to v}|$ and $|C_{u \to v}|$. The scope of the following analysis is to determine the conditions under which $P_{P,u \to v} > P_{D,u \to v}$.

Theorem 2. *$P_{P,u \to v} = P_{D,u \to v}$ for $p = 0$ and $P_{P,u \to v} < P_{D,u \to v}$ for $1 \geq p > 0$, provided that $|R_{u \to v}| \leq (q - |C_{u \to v}|)|S_v|$.* □

The proof of Theorem 2 can be seen in Appendix 1.

Theorem 3. *Provided that $|R_{u \to v}| > (q - |C_{u \to v}|)|S_v|$ is satisfied, $P_{P,u \to v} > P_{D,u \to v}$ for $p \in (0, p_{max,u \to v})$, for some $0 < p_{max,u \to v} < 1$.* □

The proof of Theorem 3 can be seen in Appendix 3.

According to Appendix 3 there exists a maximum value for $P_{P,u \to v}$ for a particular value of $p \in (0, p_{max,u \to v})$ $(\equiv p_{0,u \to v})$. In Appendix 2 it is shown that there exists a value $p_{s,u \to v}$, such that $p_{0,u \to v} < p_{s,u \to v} < 1$, for which the second derivative of $P_{P,u \to v}$ with respect to p is zero.

It should be noted that $p_{max,u \to v}$ is not easy to be calculated analytically from Equation (5). On the other hand, it can be calculated using numerical methods, such as the well-known Newton-Raphson method.

5 On the System Throughput

The analysis presented so far has established the conditions for which the probability of success under the Probabilistic Policy is higher than that under the Deterministic Policy, for a specific transmission. Since the same value of p is assumed to be adopted for all nodes under the Probabilistic Policy it is possible that this common value results in different comparative performance under the Deterministic and the Probabilistic Policies, for different transmissions. First, it may be that for some transmissions Theorem 2 holds and thus the Probabilistic Policy can never outperform the Deterministic one. Second, it may be that for some transmissions $\chi \to \psi$ Theorem 3 holds but the common p is outside the range $(0, p_{max,\chi\to\psi})$ and thus the Probabilistic Policy induces a lower probability of success. Finally, for some transmissions $\chi \to \psi$ Theorem 3 may hold and p is within the range $(0, p_{max,\chi\to\psi})$ and thus the Probabilistic Policy outperforms the Deterministic one. From the aforementioned three cases it is clear that the system throughput (averaged over all transmissions) under the Probabilistic Policy (denoted by $P_{P,succ}$) may or may not outperform that under the Deterministic Policy (denoted by $P_{D,succ}$) for a given value of p.

The following theorem shows that for large networks ($k > 1$, [8]) $|R_{u\to v}| \geq (q - |C_{u\to v}|)|S_v|$ holds for every transmission $u \to v$. For $k = 1$ the condition also holds for any transmission $u \to v$ for which $|S_v| \leq D/2$.

Theorem 4. *For $k > 1$, $|R_{u\to v}| \geq (q - |C_{u\to v}|)|S_v|$, for every transmission $u \to v$. For $k = 1$ $|R_{u\to v}| \geq (q - |C_{u\to v}|)|S_v|$, provided that $|S_v| \leq D/2$.*

Proof. From Theorem 1 and for any transmission $u \to v$, it is concluded that $|R_{u\to v}| \geq qD$, for $k > 1$. Since $q \geq q - |C_{u\to v}|$ and $|S_v| \leq D$, it is concluded that $(q - |C_{u\to v}|)|S_v| \leq qD$, and, consequently, $|R_{u\to v}| \geq (q - |C_{u\to v}|)|S_v|$ holds for $k > 1$.

In [7], [8] it is established that $q \geq D + 1$ for $k = 1$. From Theorem 1 and for any transmission $u \to v$, $|R_{u\to v}| \geq q(q - |S_v| - 1)$ or $|R_{u\to v}| \geq (q - |C_{u\to v}|)(q - |S_v| - 1)$, since $q \geq q - |C_{u\to v}|$. To show that $|R_{u\to v}| \geq (q - |C_{u\to v}|)|S_v|$ it suffices to show that $q - |S_v| - 1 \geq |S_v|$ or $q \geq 2|S_v| + 1$. Since $q \geq D + 1$, it suffices to show that $D + 1 \geq 2|S_v| + 1$ or $|S_v| \leq D/2$. □

From theorems 2, 3 and 4, it is shown that for $k > 1$ there exists a range of values of p such that $P_{P,\chi\to\psi} \geq P_{D,\chi\to\psi}$, for any transmissions $\chi \to \psi$. Consequently, for $k > 1$, there exists a range of values for p such that $P_{P,succ} \geq P_{D,succ}$.

The analysis presented in the previous section (as well as in Theorem 4) has shown an obvious connection between the number of the neighbor nodes of node v and the probability of success for transmission $u \to v$. From equations (1) and (2) it is concluded that $|C_{u\to v}|$ increases and $|R_{u\to v}|$ decreases as $|S_v|$ increases. Under the Deterministic Policy (see Equation (4)) the probability of success for transmission $u \to v$ decreases linearly as $|C_{u\to v}|$ increases (as $|S_v|$ increases). Under the Probabilistic Policy (see Equation (5)) as $|S_v|$ increases the probability of success: (a) decreases linearly as $|C_{u\to v}|$ increases; (b) increases linearly as $|R_{u\to v}|$ decreases; and (c) decreases exponentially $\left(\text{term } (1-p)^{|S_v|}\right)$ as $|S_v|$ increases. Thus, it seems that an increase in $|S_v|$ has a more negative impact on the probability of success under the Probabilistic Policy than under the Deterministic. Consequently, it is expected that: (a) the Probabilistic Policy (with access probability p) outperforms the Deterministic Policy for a certain topology $G(V, E)$ with parameters N and D and low topology density, while the reverse can be true for the same parameters N and D and access probability p but a higher topology density; (b) the value of p that results in a higher probability of success for the Probabilistic Policy decreases, to reduce the impact of the interference caused by the increased number of probabilistic attempts by the increased number of neighbors. The maximum system throughput achieved under the Probabilistic Policy under some p is expected to decrease as the topology density increases.

From the above discussion, it is evident that the topology density affects strongly the probability of success under the Probabilistic Policy. In the next section, a topology density metric is introduced to be used for the categorization of each topology $G(V, E)$ for a given pair N, D. In Section 7, the relation between the range of suitable values of p (range of values for which $P_{P,succ} \geq P_{D,succ}$) and the topology density metric is investigated as well as the relation between the maximum value of $P_{P,succ}$ and the particular metric.

6 Topology Density Analysis

It has been shown that a specific transmission $u \to v$ is influenced (is corrupted or successful) by the transmission set $\Phi_{u \to v} \cup \Theta_{u \to v}$. Consequently, node v as well as any node $\chi \in S_v - \{u\}$ influence transmission $u \to v$. Let the set of nodes S_u be referred to as the set of *one-hop* neighbor nodes of node u with respect to transmission $u \to v$, the set of nodes $S_v - \{u\}$, $\forall v \in S_u$, be referred to as the set of *two-hop* neighbor nodes of node u and the union of both previous sets to be referred to as the *aggregate two-hop* set of neighbor nodes of node u with respect to transmission $u \to v$. Consequently, the aggregate two-hop set of a node influences the transmissions of that particular node. Based on this observation, the topology density is defined as it is presented next.

For a given pair N and D numerous topologies $G(V, E)$ exist. It is possible to categorize these topologies depending on the number of the aggregate two-hop nodes. Let *Calculated*$(G(V, E))$ denote the total number of aggregate two-hop nodes, for all nodes in the network, calculated for topology $G(V, E)$.

$$Calculated(G(V,E)) = \sum_{u \in V} \sum_{v \in S_u} |S_v - \{u\}| = \sum_{u \in V} \sum_{v \in S_u} (|S_v| - 1). \tag{6}$$

Let $Upper(N, D)$ and $Lower(N, D)$ be defined as follows.

$$Upper(N, D) = ND^2, \tag{7}$$

$$Lower(N, D) = \begin{cases} D^2 + D & N - D = 1 \\ D^2 + D + 4 & N - D = 2 \\ D^2 - 3D + 4N - 4 & N - D \geq 3. \end{cases} \tag{8}$$

Theorem 5. *For a topology $G(V, E)$ with parameters N and D, the following holds: $Lower(N, D) \leq Calculated(G(V, E)) \leq Upper(N, D)$.*

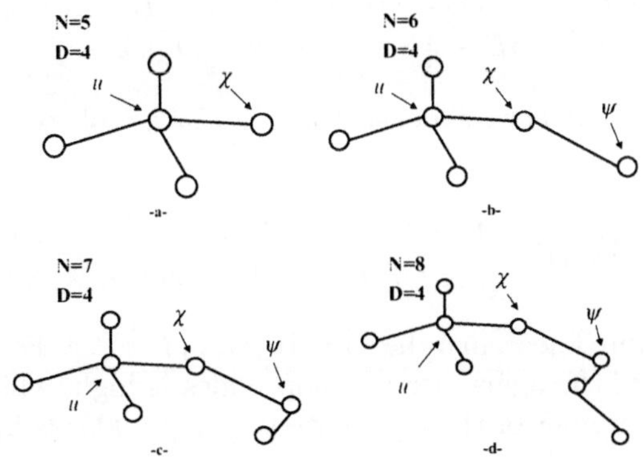

Fig. 1. Least dense topology for $D = 4$ and $N = 5, 6, 7, 8$ respectively.

Proof. The upper bound for $Calculated(G(V, E))$ is derived by considering the most dense topology $G(V, E)$ for which $N = |V|$. The most dense topology corresponds to the case for which $|S_u| = D, \forall u \in V$. Consequently, for each of the N nodes, there are D one-hop neighbors, and $D(D-1)$ two-hop neighbors. Thus, the total aggregate two-hop nodes in the network are $N(D(D-1)+D) = ND^2$. As a result, $Upper(N, D) = ND^2$. Note that in the case in which the most dense topology for a pair N and D is not possible to have D neighbor nodes for every node, $Upper(N, D) < ND^2$ still holds.

The lower bound for $Calculated(G(V, E))$ corresponds to the total number of aggregate two-hop nodes of the least dense topology possible for a given pair N, D. The latter is a topology in which there is only one node u with $|S_u| = D$ one-hop neighbor nodes, $D - 1$ nodes are one-hop neighbors of u but have no other one-hop neighbor except node u, and one node χ is one-hop neighbor of node u and the initiator of a line of nodes. No loops are present in the topology (see Figure 1).

In case $N - D = 1$ (see Figure 1.a for the corresponding least topology) each node other than u has only one one-hop neighbor node; node u. Consequently, there are D aggregate two-hop nodes for node u, as well as D for the rest D nodes. As a result, $Lower(D, N) = D^2 + D$, for $N - D = 1$.

In case $N - D = 2$ (see Figure 1.b for the corresponding least dense topology) node χ has another one-hop neighbor node, node ψ, expect u. Consequently, there are $D + 1$ aggregate two-hop nodes for node χ, and 2 for node ψ. For node u, there are $D + 1$ aggregate two-hop nodes, and for each of the $D - 1$ remaining nodes (for which their only one-hop neighbor node is u), D aggregate two-hop nodes. As a result $Lower(D, N) = (D - 1)D + (D + 1) + (D + 1) + 2$ or $Lower(D, N) = D^2 + D + 4$, for $N - D = 2$.

In case $N - D \geq 3$ (see Figures 1.c and 1.d for the corresponding least dense topology) node χ is the initiator of a line of nodes. The number of nodes in this line is equal to $N - D - 1$. For the two nodes at the end of the line there is $3 + 2$ aggregate two-hop nodes, while for each one of the $N - D - 3$ nodes there are 4.

As a result $Lower(D,N) = (D-1)D + (D+1) + (D+2) + 4(N-D-3) + 3 + 2$ or $Lower(D,N) = D^2 - 3D + 4N - 4$ for $N - D \geq 3$. $\square$

Let $W(G(V,E))$ denote a density metric for topology $G(V,E)$ defined as follows;

$$W(G(V,E)) = \frac{Calculated(G(V,E)) - Lower(N,D)}{Upper(N,D) - Lower(N,D)}. \tag{9}$$

It is evident from Theorem 5 that $0 \leq W(G(V,E)) \leq 1$. For $W(G(V,E))$ close to 1 the number of the aggregate two-hop nodes is high, while for $W(G(V,E))$ close to zero the number of the aggregate two-hop nodes is low.

7 Simulation Results

For the simulation purposes four different topology categories are considered. The number of nodes in each topology category is $N = 100$, while D is set to 5, 10, 15 and 20. These four topology categories are denoted as D5N100, D10N100, D15N100 and D20N100 respectively. Parameters q and k are determined according to the algorithm presented in [8]. Time slot sets are assigned randomly to each node, for each particular topology. The particular assignment is kept the same for each topology category throughout the simulations. Different topologies that correspond to different topology density values $W(G(V,E))$ are considered for each topology category.

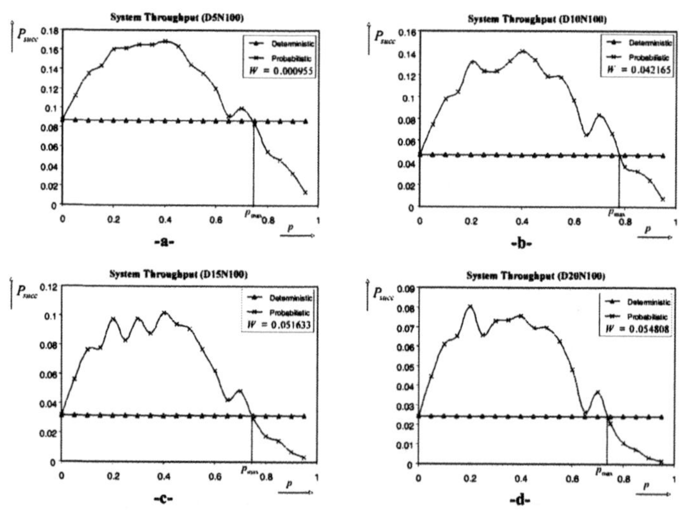

Fig. 2. P_{succ} under the Deterministic as well as the Probabilistic Policy, for small topology density values.

Note from Theorem 1 that for $k > 1$ the number of non-assigned eligible time slots is expected to be higher for a higher value of k and Theorem 3 always holds as it has been proved in Theorem 4. Consequently, it is expected that for $k > 1$ all transmissions will achieve a higher throughput under the Probabilistic Policy for any value of p in the range $(0, p_{max})$, where p_{max} depends on the specific transmission's environment (see Theorem 3). When $k = 1$, the Probabilistic Policy will outperform the Deterministic Policy for those transmission for which $|S_v| \leq D/2$ and may not for others. That is, the lower the topology density the more transmissions will satisfy the condition $|S_v| \leq D/2$ and the wider the range of values of p for which the Probabilistic Policy will outperform the Deterministic Policy.

The simulation results presented demonstrate the performance for $k = 1$ (the resulting value for k is equal to 1 for the four topology categories, [8]), that is the case that the number of non-assigned eligible time slots is expected to be rather small and, thus, the effectiveness of the Probabilistic Policy rather low.

The simulation results show that: (a) there exists a range of values for $0 \leq p \leq p_{max}$ $(0 < p_{max} < 1)$ such that $P_{P,succ} \geq P_{D,succ}$; (b) this range decreases as the topology density, measured by $W(G(V,E))$, increases as expected, suggesting that the introduced topology density metric $W(G(V,E))$ is an effective one; (c) $P_{P,succ}$ decreases as $W(G(V,E))$ increases.

Figures 2 and 3 depict simulation results for the system throughput (P_{succ}), under both the Deterministic and the Probabilistic Policies, as a function of the access probability p. In Figure 2, $W(G(V,E))$ is close to zero, while in Figure 3, $W(G(V,E))$ is as high as possible (close to 0.8). In all three sets of simulations it can be observed that the system throughput achieved under the Deterministic Policy is constant with respect to p. Under the Probabilistic Policy there exists a range of values for p $(0 \leq p \leq p_{max} < 1)$ such that $P_{P,succ} \geq P_{D,succ}$.

For values of p close to 1 the throughput is almost zero. This is expected since the interference induced due to the high access probability is rather high; it may also be concluded from Equation (5).

A careful consideration of the results in figures 2 and 3 shows that p_{max} decreases as $W(G(V,E))$ increases as expected. This trend is shown more clearly in Figure 4, where the values for p_{max} derived in the simulations are plotted as a function of $W(G(V,E))$. Notice that p_{max} decreases almost exponentially as $W(G(V,E))$ increases, as expected from the discussion in Section 5. The results in Figure 4 demonstrate the effectiveness of the introduced topology density metric $W(G(V,E))$ in capturing the topology density and determining the range of probabilities p under the Probabilistic Policy that lead to a higher system throughput than that under the Deterministic Policy. Notice that for a particular value of $W(G(V,E))$, any value of p below the curve results in $P_{P,succ} \geq P_{D,succ}$. Based on the results in Figure 4, one could measure or estimate the value of $W(G(V,E))$ and depending on whether it is low or high select a larger or smaller probability p.

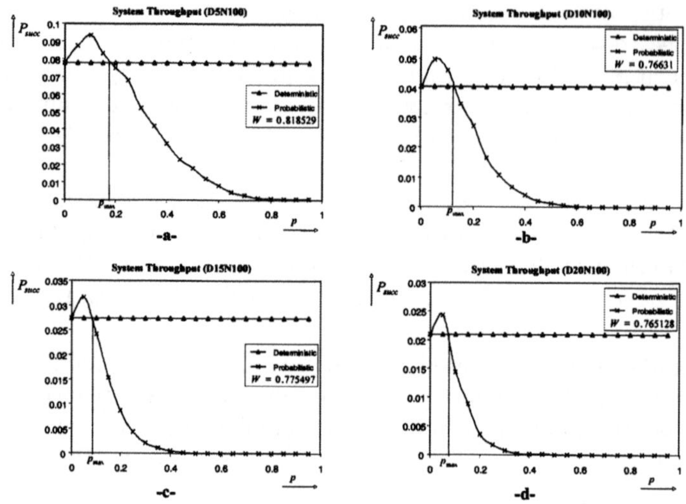

Fig. 3. P_{succ} under the Deterministic as well as the Probabilistic Policy, for high topology density values.

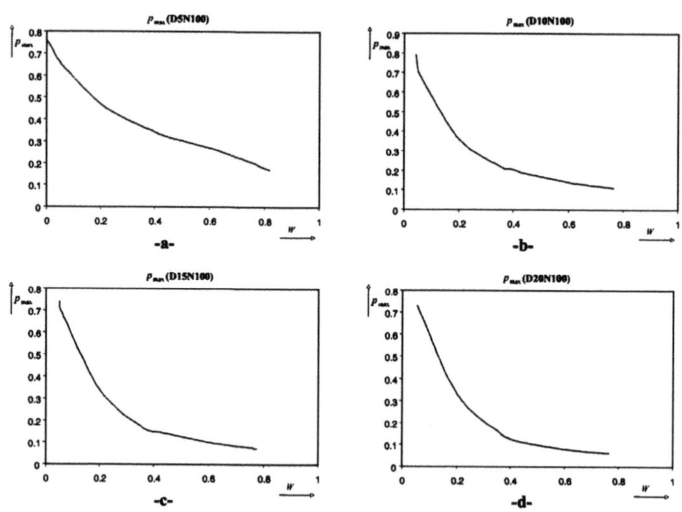

Fig. 4. p_{max} parameterized by $W(G(V,E))$.

8 Summary and Conclusions

In this paper the inherent inefficiencies of the Deterministic (slot assignment) Policy in an ad-hoc network, proposed in [7] and [8], are investigated and the Probabilistic (slot assignment) Policy is introduced in an effort to improve the achieved network throughput. The basic idea behind the proposed policy is to use (with some probability p) slots not assigned to a node under the assignment scheme in [7], [8]. The study in this paper has been carried out under heavy traffic conditions, which are expected to minimize the benefits of the Probabilistic Policy that eventually tries to utilize slots non-assigned to anybody or not used by others.

A common probability p is assumed for all transmissions in the network - as it would practically be the case - and the system throughput is considered, as shaped by all transmissions (which may or may not achieve a higher throughput under the Probabilistic Policy). First, it is shown that the networks for which the assigned polynomials have degree $k > 1$ (see [7], [8]) - which is the case for large networks - a nonzero range for the common probability p exists under which the Probabilistic Policy outperforms the Deterministic Policy. Second, for $k = 1$ (typically for rather small networks) sufficient conditions are established regarding the two-hop neighborhood under which the conditions for the existence of a range $(0, p_{max})$ of the access probability are satisfied and, thus, such transmissions will achieve a higher throughput under the Probabilistic Policy; thus, such transmissions will contribute a higher value toward the system throughput achieved under the Probabilistic Policy. The latter conditions show that the size of the two-hop neighborhood affects the performance of of a specific transmission (the smaller the size the higher the achieved throughput) and suggest that the system throughput achieved under the Probabilistic Policy is affected by the overall one- and two-hop (aggregate two-hop) neighborhood size of the network. These observations have led to the idea of introducing the topology density metric $W(G(V,E))$ and use it to infer how large the aggregate two-hop neighborhood is for a certain topology. As $W(G(V,E))$ decreases it is expected under the Probabilistic Policy that both the range of effective access probabilities $(0, p_{max})$ and the achieved system throughput will increase.

Simulation results have been derived for four network topology categories $\Big(\text{for four pairs } (N,D)\Big)$ and the for two values of the corresponding value of $W(G(V,E))$ for each one of them; $k = 1$ and heavy traffic conditions have been assumed, both of which are expected to induce only a small advantage of the Probabilistic Policy over the Deterministic Policy, compared to the cases of networks with $k > 1$ and non-heavy traffic conditions. The derived results have supported the claims and expectations regarding the comparative advantage of the Probabilistic Policy over the Deterministic Policy, as well as the dependence of the former on the introduced topology density metric.

References

1. IEEE 802.11, "Wireless LAN Medium Access Control (MAC) and Physical Layer (PHY) specifications", Nov. 1997. Draft Supplement to Standard IEEE 802.11, IEEE, New York, January 1999.

2. V. Bharghavan, A. Demers, S. Shenker, and L. Zhang, "MACAW: A Media Access Protocol for Wireless LAN's", Proceedings of ACM SIGCOMM'94, pp. 212–225, 1994.
3. C.L. Fullmer, J.J. Garcia-Luna-Aceves, "Floor Acquisition Multiple Access (FAMA) for Packet-Radio Networks", Proceedings of ACM SIGCOMM'95, pp. 262–273, 1995.
4. P. Karn, "MACA – A new channel access method for packet radio", in ARRL/CRRL Amateur Radio 9th Computer Networking Conference, pp. 134–140, 1990.
5. J. Deng and Z. J. Haas, "Busy Tone Multiple Access (DBTMA): A New Medium Access Control for Packet Radio Networks," in IEEE ICUPC'98, Florence, Italy, October 5–9, 1998.
6. R. Nelson, L. Kleinrock, "Spatial TDMA, A collision-free Multihop Channel Access Protocol", IEEE Transactions on Communications, Vol. COM–33, No. 9, September 1985.
7. I. Chlamtac and A. Farago,"Making Transmission Schedules Immune to Topology Changes in Multi-Hop Packet Radio Networks", IEEE/ACM Trans. on Networking, 2:23–29, 1994.
8. J.-H. Ju and V. O. K. Li, "An Optimal Topology-Transparent Scheduling Method in Multihop Packet Radio Networks", IEEE/ACM Trans. on Networking, 6:298–306, 1998.

Appendices

Appendix 1 *Proof of Theorem 2*

From Equation (5) it is concluded that $P_{P,u\to v} = P_{D,u\to v}$ for $p = 0$.

The first derivative of $P_{P,u\to v}$ with respect to p is calculated in Appendix 4. For $|R_{u\to v}| \leq (q - |C_{u\to v}|)|S_v|$, the first derivative is zero for two values of p: $p = 0$ and $p = 1$. For any other value $p \in (0, 1)$ the first derivative is always negative (see Appendix 4) and therefore, the global maximum corresponds to $p = 0$ ($P_{P,u\to v} = P_{D,u\to v}$) whereas, the global minimum corresponds to $p = 1$ ($P_{P,u\to v} = 0$). Consequently, for any value of $p \in (0, 1]$, $P_{P,u\to v} < P_{D,u\to v}$.

Appendix 2 *On the Existence of* $p_{s,u\to v}$

$\frac{d^2 g(p)}{d^2 p} = 0$ for $p = 1$ and $p = \frac{2|R_{u\to v}|-(q-|C_{u\to v}|)(|S_v|-1)}{|R_{u\to v}|(|S_v|+1)} \equiv p_{s,u\to v}$. Since $0 < p \leq 1$ (0 is not included since it is the case for which $P_{P,u\to v} = P_{D,u\to v}$), in order for $p_{s,u\to v}$ to be a valid root it is required that $0 < p_{s,u\to v} \leq 1$. $p_{s,u\to v} > 0$ when $2|R_{u\to v}| > (q - |C_{u\to v}|)(|S_v| - 1)$. Note that $p_{s,u\to v} \leq 1$, when $p_{s,u\to v} > 0$. Notice that $2|R_{u\to v}| > (q - |C_{u\to v}|)(|S_v| - 1)$ is always met if $|R_{u\to v}| > (q - |C_{u\to v}|)|S_v|$ and, in this case, $p_{s,u\to v} > p_{0,u\to v}$. To show the latter it suffices to show that $\frac{2|R_{u\to v}|-(q-|C_{u\to v}|)(|S_v|-1)}{|R_{u\to v}|(|S_v|+1)} > \frac{|R_{u\to v}|-(q-|C_{u\to v}|)|S_v|}{|R_{u\to v}|(|S_v|+1)}$ or $2|R_{u\to v}| - (q - |C_{u\to v}|)(|S_v| - 1) > |R_{u\to v}| - (q - |C_{u\to v}|)|S_v|$ or $2|R_{u\to v}| - (q - |C_{u\to v}|)|S_v| + q - |C_{u\to v}| > |R_{u\to v}| - (q - |C_{u\to v}|)|S_v|$ or $2|R_{u\to v}| + q - |C_{u\to v}| > |R_{u\to v}|$ or $|R_{u\to v}| + q - |C_{u\to v}| > 0$. The latter always holds.

Appendix 3 *Proof of Theorem 3*

From Equation (5) it is concluded that $P_{P,u\to v} = P_{D,u\to v}$ for $p = 0$ and $P_{P,u\to v} = 0$ for $p = 1$. Consequently, the range of values for which $P_{P,u\to v} > P_{D,u\to v}$ includes neither 0 nor 1.

For $0 < p < 1$ and $|R_{u\to v}| > (q - |C_{u\to v}|)|S_v|$ the first derivative of $P_{P,u\to v}$, with respect to p, is zero (see Appendix 4), when $p = \frac{|R_{u\to v}|-(q-|C_{u\to v}|)|S_v|}{|R_{u\to v}|(|S_v|+1)}$ $(\equiv p_{0,u\to v})$. For $p = p_{0,u\to v}$ (the first derivative is zero) the second derivative is negative (see Appendix 4). Consequently, $p_{0,u\to v}$ corresponds to a maximum value for $P_{P,u\to v}$. For every value p, $0 < p \le p_{0,u\to v}$, the first derivative is always positive and consequently, $P_{P,u\to v} > P_{D,u\to v}$.

On the other hand, for every value $p_{0,u\to v} < p < 1$ the first derivative is always negative. For $p \to 1$, $P_{P,u\to v} \to 0$ and given that $P_{P,u\to v}$ is a continuous function of p, there exists a value $p_{0,u\to v} < p < 1$, such that $P_{P,u\to v} = P_{D,u\to v}$. Let $p_{max,u\to v}$ denote that value of p.

Finally, it is evident that for any value $p \in (0, p_{max,u\to v})$, $P_{P,u\to v} > P_{D,u\to v}$, provided that $|R_{u\to v}| > (q - |C_{u\to v}|)|S_v|$.

Appendix 4 *Complement of Proofs of Theorems 2 and 3*

For convenience, the following function $g(p)$ is considered

$$g(p) = (q - |C_{u\to v}| + |R_{u\to v}|p)(1-p)^{|S_v|} = P_{p,u\to v}q^2.$$

The first and second derivative of $g(p)$ with respect to p are given by:

$$\frac{dg(p)}{dp} = \Big(|R_{u\to v}| - (q - |C_{u\to v}|)|S_v| - |R_{u\to v}|(|S_v|+1)p\Big)(1-p)^{|S_v|-1}.$$

$$\frac{d^2 g(p)}{d^2 p} = -|S_v|\Big(2|R_{u\to v}| - (q - |C_{u\to v}|)(|S_v|-1) - |R_{u\to v}|(|S_v|+1)p\Big)(1-p)^{|S_v|-2}.$$

$\frac{dg(p)}{dp} = 0$ for $p = 1$ and $p = \frac{|R_{u\to v}|-(q-|C_{u\to v}|)|S_v|}{|R_{u\to v}|(|S_v|+1)} \equiv p_{0,u\to v}$. Since $0 < p \le 1$, in order for $p_{0,u\to v}$ to be a valid root it is required that $0 < p_{0,u\to v} \le 1$. $p_{0,u\to v} > 0$ when $|R_{u\to v}| > (q - |C_{u\to v}|)|S_v|$. Note that $p_{0,u\to v} \le 1$, when $p_{0,u\to v} > 0$.

If $|R_{u\to v}| < (q - |C_{u\to v}|)|S_v|$ then $\frac{dg(p)}{dp} < 0$ for any value of p. Therefore the maximum value for $g(p)$ is assumed for $p = 0$ and it is $g(0) = q - |C_{u\to v}|$.

If $|R_{u\to v}| > (q - |C_{u\to v}|)|S_v|$ then for $p = p_{0,u\to v}$ the second derivative is equal to

$$\left.\frac{d^2 g(p)}{d^2 p}\right|_{p=p_{0,u\to v}} = -|S_v|(q - |C_{u\to v}| + |R_{u\to v}|)(1-p)^{|S_v|-2}.$$

Since $(q - |C_{u\to v}|) + |R_{u\to v}| > 0$ and $(1-p)^{|S_v|-2} > 0$ for $p = p_{0,u\to v}$, it is concluded that $\left.\frac{d^2 g(p)}{d^2 p}\right|_{p=p_{0,u\to v}} < 0$. As a result $g(p)$ assumes a maximum value at $p_{0,u\to v}$.

Power-Saving in Wi-Fi Hotspots: An Analytical Study*

G. Anastasi[1], M. Conti[2], E. Gregori[2], and A. Passarella[1]

[1] University of Pisa,
Dept. of Information Engineering
Via Diotisalvi, 2 - 56122 Pisa, Italy
{g.anastasi,a.passarella}@iet.unipi.it
[2] CNR - IIT Institute
Via G. Moruzzi, 1 - 56124 Pisa, Italy
{marco.conti,enrico.gregori}@iit.cnr.it

Abstract. Wi-Fi hotspots are one of the most promising scenarios for mobile computing. In this scenario, a very limiting factor is the scarcity of mobile-device energetic resources. Both hardware and software architectures of current devices are very inefficient from this standpoint, mainly the networking subsystem. This work analyzes a power-saving network architecture for the mobile-Internet access through Wi-Fi hotspots. Specifically, this solution supports any kind of best-effort network applications, since it is application-independent. In this paper we derive a complete analytical model of the power-saving system when applied to mobile Web access. Furthermore, we use this model to compare our solution with a well-known approach, i.e., the Indirect-TCP. The comparison is performed by considering two performance figures: the energy saved in downloading a Web page and the related transfer-time. The results show that, in the average, our solution saves up to 78% of the energy. Furthermore, the power-saving system introduces an additional average transfer-time of 0.4 sec, and hence it does not significantly affect the QoS perceived by the users. Finally, we assess the sensitiveness of the power-saving system with respect to Internet key parameters, such as the available throughput and the RTT.

Keywords: Wi-Fi, Power Saving, Web, Mobile Internet, Analytical Models.

1 Introduction

In this work we analyze a power-saving network architecture for 802.11 "Wi-Fi" hotspot environments. This is today one of the most promising scenarios for mobile computing, and is rapidly becoming a key business area. In the typical

* This work was carried out under the financial support of the Italian Ministry for Education and Scientific Research (MIUR) in the framework of the Projects: FIRB-PERF and FIRB-VICOM.

M. Conti et al. (Eds.): PWC 2003, LNCS 2775, pp. 306–320, 2003.

deployment of such scenario, Internet Service Providers guarantee wireless Internet access in a limited-size environment, such as a campus or a mall (i.e., a "hotspot"). Wireless coverage is achieved by means of Access Points which build a 802.11 WLAN. Moreover, Access Points are connected to the Internet through a standard high-speed LAN. Mobile users subscribe a contract with an ISP, and are allowed to access the Internet on-the-move inside the hotspot.

Integrating mobile devices in such an environment is still an open research issue. Since mobile devices are battery-fed, their energetic resources are a very critical factor. Furthermore, the development of energy-aware solutions for mobile networking is a very pressing requirement [2,3,4,11,12,17,18]. Specifically, the wireless interface may drain up to 50% of the total energy spent by a mobile device [17]. Moreover, legacy Internet protocols (such as TCP/IP) are very inefficient from this standpoint [1,2]. Due to the specific consumption pattern of 802.11 wireless interfaces [14], the optimal power-saving strategy consists in transferring data at the maximum throughput allowed on the wireless link, and switching the wireless interface off (or to put it in a "doze" mode) whenever it is idle [2,3,4,16,17,18]. As it is clear, the key point of such an approach is the identification of idle periods within the network traffic pattern.

In this paper we provide an analytical study of the power-saving networking solution for Wi-Fi hotspots that we developed in [4]. This solution follows an Indirect-TCP approach [7,8], and operates at the transport and middleware layers. Furthermore, it is application independent in the sense that it does not rely on any a-priori knowledge of the application behavior. Specifically, our system traks dynamically the behavior of the network application(s) running on the mobile host. Based on the application behavior, it predicts time intervals during which the wireless interface will be idle, and switches it off accordingly.

In [4] we tested our system by considering the Web application. Therefore, in this paper we provide an analytical model of the system behavior when it is used to support mobile Web access. However, both the power-saving system in [4], and the models provided here are quite general and can be used with any kind of best-effort applications.

The first step of our study consists in building a model of the traffic generated by a typical Web user. Then, we provide closed formulas that describe the average behavior of the power-saving system when used to support such a traffic. Moreover, we compare our solution with a pure Indirect-TCP approach. The comparison is carried out by focusing on two main performance figures, i.e., i) the energy saved in downloading a Web page, and ii) the related additional transfer-time.

After assessing the model accuracy against results presented in [4], we exploit the model to deeply analyze the power-saving system. The results show that, under the Internet conditions experienced in [4], our solution saves up to 78% of the energy spent when using the pure Indirect-TCP approach. Moreover, the average additional transfer-time is less than $0.4sec$, and hence we can conclude that the power-saving system does not significantly affect the QoS perceived by the users.

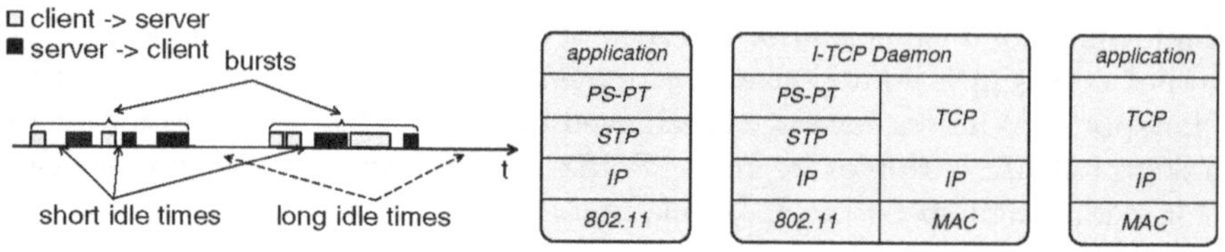

Fig. 1. Snapshot of a typical best-effort data exchange (left) and the power-saving network architecture (right).

Finally, by using our model, we perform a sensitiveness analysis to investigate the dependence of the power-saving system on two Internet key parameters, i.e., the throughput and the Round Trip Time (RTT) between the Web client and server. The results show that power-saving is mainly affected by throughput variations. Specifically, the energy saving varies from 48% to 83% when the throughput increases from 0 to ∞, i.e., from the lower to the upper theoretical bound. On the other hand, the additional transfer-time is a slightly increasing function of RTT. However, the average additional transfer-time is always less than $0.5sec$.

The paper is organized as follows. Sections 2 and 3 present the power-saving system and its analytical model. Section 4 is devoted to the validation of this model. Finally, Sect. 5 presents the sensitiveness analysis.

2 Reference Environment

2.1 Power-Saving Network Architecture for Best-Effort Traffic

As highlighted in the previous section, our power-saving architecture supports any kind of best-effort applications. In such applications data-transfer phases are characterized by *bursts* separated by *idle phases* during which data are locally processed. Figure 1-left shows a snapshot of a typical data exchange.

Specifically, bursts can be seen as made up of data packets. Packets are separated by *short idle times*, while bursts are separated by *long idle times.* Short idle times are due to automatic interactions between computers, while long idle times are related to human reaction times. Therefore, short idle times are typically shorter than long idle times. A typical cut-off value used in the literature is $1sec$.

Our power-saving system relies upon the dynamic estimation of (short and long) idle times based on the traffic behavior observed in the past. Specifically, short idle times are estimated by exploiting the Variable-Share Update algorithm [15], while a binary exponential backoff, starting from $1sec$, is used for long idle times.

In detail, the estimation algorithm works as follows. When a short idle time begins, the Variable-Share Update algorithm provides an estimate, $t_i^{'}$, of the actual idle time, t_i. If $t_i^{'}$ occurs to be less than t_i, the estimate is updated with the 90th percentile of the short idle times, throughout referred to as k. Finally, if t_i too occurs to be greater than k, the algorithm assumes that t_i is a *long* idle time. Therefore, it provides the next updates by using a binary exponential backoff starting from $1sec$.

The network interface of the mobile host is managed according to these estimates. It is worth noting that the network interface has a transient in getting on (throughout referred to as t_{so}) during which it drains power from the battery but is not available for exchanging data. Therefore, for idle times less than t_{so} it is energetically convenient to leave the network interface on. A complete description of the algorithm is provided in [4], and is here omitted for the sake of space.

To integrate the estimation algorithm described above in a real Wi-Fi system, we defined the network architecture shown in Fig. 1-right. This power-saving architecture exploits the Indirect-TCP model, since it splits the transport connection between the (mobile) client and server at the Access Point. In addition, the transport protocol between the mobile and the Access Point is a *Simplified Transport Protocol* (STP in the figure) that is tailored to the characteristics of the wireless link. As shown in [8], this solution outperforms the standard Indirect-TCP architecture [7].

In our architecture the idle-time estimation algorithm is implemented in the *Power-Saving Packet Transfer* (PS-PT) protocol. The PS-PT protocol follows a simple master-slave scheme. When there are no more data to be exchanged (i.e., when an idle time occurs), the PS-PT module at the Access Point generates a short idle time estimate and decide whether it is convenient to switch off the network interface at the mobile host. If so, it sends a "shutdown" command to the mobile host including an indication of the time interval during which it should remain disconnected. Possible data destined to the mobile host that become available while it is disconnected are buffered at the Access Point. When the mobile host reconnects, it polls the Access Point to obtain either new data or an update of the idle-time estimate. Updates are derived according to the algorithm described above. Upon deriving the update, the Access Point decides whether it is convenient to the mobile host to switch the wireless interface off. Finally, if the mobile host generates new data while it is disconnected, the wireless interface is immediately switched on and data are sent. Details about the network architecture shown in Fig. 1-right can be found in [4].

2.2 Performance Figures

To evaluate the performance of our system, we define two indexes. We measure the energy saving achieved by the power-saving architecture – with respect to a pure I-TCP architecture – by means of the I_{ps} index, defined as

$$I_{ps} = \frac{C_{ps}}{C_{I-TCP}} \quad . \tag{1}$$

Specifically, C_{ps} is the energy spent to download a set of Web pages when using the power-saving system, C_{I-TCP} is the energy spent to download the

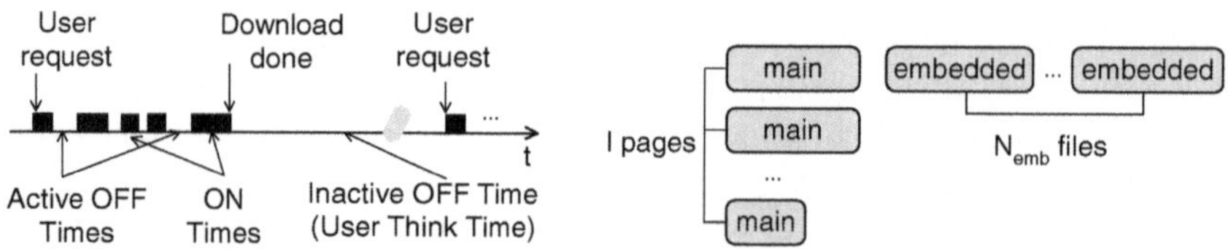

Fig. 2. Web-page download (left) and scheme of the basic block (right).

same set of Web pages when using a pure Indirect-TCP approach[1]. Due to the 802.11 wireless interface consumption patterns, the energy spent is proportional to the time interval during which the wireless interface remains switched on [14]. Therefore, we express both C_{ps} and C_{I-TCP} in seconds.

Finally, it must be noted that the power-saving system can introduce an additional delay to the Web-page transfer-time. Specifically, additional delays may be introduced when idle-time estimates are greater than the actual idle times. Web users might perceive the additional transfer-time as a degradation of the QoS, and hence it is important to quantify it. To this end, we use the I_{pd} index, defined as

$$I_{pd} = URT_{ps} - URT_{I-TCP} \ , \tag{2}$$

where URT is the User Response Time, i.e., the time elapsed to download a Web-page, between the user request and the page rendering at the mobile host.

3 System Model

The description of the power-saving system (Sect. 2) shows that several parameters affect the system behavior. In particular, the throughputs on the wireless and wired networks, the accuracy of the idle-time estimates, the application-level traffic profile play an important role. To clearly understand their influence on the system performance, we describe the average behavior of the power-saving system by means of an analytical model. Since the system performance is related to the application-level traffic, it is necessary a preliminary characterization of the traffic generated, in the average case, by a generic Web-user.

3.1 Web-User Model

To characterize a client access to a Web server, it is sufficient to focus on a single Web-page download (see Fig. 2-left).

As is well known a Web page consists of a main file and zero or more embedded files (e.g., figures). All files composing a Web page are transferred during the ON Time interval while in the Inactive OFF Time (or User Think Time)

[1] It should be noted that, throughout our work, we use an Indirect-TCP architecture with a simplified transport protocol (i.e., the STP) between the mobile host and the Access Point, as discussed in [8].

Table 1. Parameters that define the Web-user average traffic profile.

Definition	Symbol	Value
Probability that a Web page contains embedded files	p_{emb}	0.44
Average number of embedded files in a Web page	N_{emb}	1.50
Average size of the embedded files (bytes)	$\overline{D}_{emb}$	6348
Average size of the main files (bytes)	$\overline{D}_{mf}$	17496
Average User Think Time (seconds)	$\overline{UTT}$	3.25

the user reads the content of the downloaded Web page. Finally, during Active OFF Times the browser parses a piece of the main file and sends the request for the next embedded file(s)[2]. Active OFF Times are typically less than User Think Times, since they are due to automatic interactions between computers.

In [4] we used the SURGE simulator to simulate a typical Web user. The SURGE is based on the statistical model of the Web traffic presented in [9,10, 13]. Therefore, we exploit this model to derive the parameters that define the *average* traffic profile generated by the generic Web-user (see Tab. 1).

Specifically, the Web session of the simulated user can be thought of as the continuous download of the same *basic block*. The basic block consists of a set of Web pages defined in such a way that the traffic generated for downloading them meets the statistics of Tab. 1. In detail, the basic block is made up of $l \triangleq \lceil 1/p_{emb} \rceil$ Web pages. The first page contains the main file and N_{emb} embedded files, while the others pages are composed by the main file only. The dimension of each main file is $\overline{D}_{mf}$, except for the main file of the last page, whose dimension is $\overline{D}_{mf} \cdot (l - 1/p_{emb})$ (for instance, if $1/p_{emb}$ is 3.4, the basic block is made up of 4 pages, the dimension of the last main file being $0.4 \cdot \overline{D}_{mf}$). The dimension of each embedded file is $\overline{D}_{emb}$. Finally, the user waits $\overline{UTT}$ seconds before downloading the next page. Figure 2-right shows a scheme of the basic block.

Idle Times Characterization. With respect to the reference network architecture shown in Fig. 1-right, we assume that the application layer generates the average traffic profile shown in Fig. 2-right. The power-saving system exploits idle times in this traffic to save energy. The characterization of these idle times is thus a necessary step to model the energetic behavior of the power-saving system.

It is worth recalling that Web utilizes TCP as the transport protocol. In particular, HTTP/1.1 uses persistent connections. As discussed in [6], we can reasonably assume that the download of the basic block occurs over a unique, steady-state, TCP connection. Moreover, in average, data transfer on a steady-state TCP connection can be seen as the transfer of fixed-size groups of TCP segments, separated by *RTTs* [19]. Based on these remarks, we can conclude that *short* idle times in the user traffic profile are produced by the TCP behavior, and can thus be considered as samples of the *RTT* between the client and the

[2] Several files can be requested at the same time, by exploiting the pipelining mechanism of the HTTP.

Table 2. Symbols used throughout the paper.

Definition	Symbol
Total dimension of the basic block	B
Number of pages in the basic block	l
Average available Internet throughput	$\overline{\gamma}$
Average available wireless-link throughput	$\overline{\gamma}_{wl}$
Switching-on transient interval of the wireless interface	t_{so}
Average number of switching-on events during the basic block download	S
Average number of switching-on events during a short idle time	S_1
Average number of short idle times during the basic block download	r
Average number of switching-on events during a UTT	S_2
Average number of switching-on events to send the embedded-file(s) request	S_3
Short idle-time sample	t_i
Short idle-time estimate	$t_i^{'}$
Upper bound of the idle-time distribution	M
90th percentile of short idle-times	k
Error of the idle-time estimator	e

Web server. On the other hand, *long* idle times (i.e., idle times between bursts) can be considered as samples of the User Think Time.

Hereafter we focus on a single basic-block download to derive the analytical model of the power-saving system. Specifically, we provide a model for evaluating both the energy consumption and the additional URT introduced. For the reader's convenience, we summarize in Tab. 2 the parameters used throughout the paper.

3.2 Energy Consumption Model

We are now in the position to evaluate the energy spent to download a single basic block, by using either the power-saving system or the pure Indirect-TCP approach. In the following we claim a set propositions that allow us to provide closed formulas for I_{ps} and I_{pd}, respectively. Proofs of these propositions are omitted for the sake of space. They can be found in [6].

The energy spent when using an Indirect-TCP approach (i.e., C_{I-TCP}) is the total time required to download the basic block, since no power-saving strategy is used. Therefore, the following theorem holds.

Theorem 1. *The energy spent to download a single basic block by using a pure Indirect-TCP approach is*

$$C_{I-TCP} = \frac{B}{\overline{\gamma}} + l \cdot \overline{UTT} = \frac{\overline{D}_{mf} \cdot 1/p_{emb} + \overline{N}_{emb} \cdot \overline{D}_{emb}}{\overline{\gamma}} + l \cdot \overline{UTT} \ , \qquad (3)$$

where $\overline{\gamma}$ is the average available Internet throughput between the client and the Web server and B is the total size (in bytes) of the basic block.

On the other hand, the energy consumed when using the power-saving system is made-up of two components. The first one corresponds to the time required to transfer the basic block over the wireless link. The second one is due to the wireless network interface at the mobile host that consumes t_{so} seconds every time it is switched on. Therefore, the following lemma hold.

Lemma 1. *The energy spent by using the power-saving system is*

$$C_{ps} = \frac{B}{\overline{\gamma}_{wl}} + t_{so} \cdot S \triangleq \frac{B}{\overline{\gamma}_{wl}} + r \cdot S_1 + l \cdot S_2 + S_3, \tag{4}$$

where $\overline{\gamma}_{wl}$ is the average available throughput allowed by the wireless link, and S is the average number of switching-on events during the download of a basic block.

It is worth noting that, thanks to the network architecture design, $t_{so} \cdot S$ is the only energy contribution that is introduced by the power-saving mechanisms. The Lemmas presented belows allow us to derive a closed formula for this contribution.

Firstly, S includes three components, i.e., i) the number of swithing-on events within *short* idle times (i.e., $r \cdot S_1$ in (4)); ii) the number of switching-on events within *long* idle times (i.e., $l \cdot S_2$ in (4)); and iii) the number of switching-on events within the download of the first page of the basic block, between the end of the main-file download and the beginning of the embedded-files download (i.e., S_3 in (4)). Hence, S can be evaluated as shown in (4).

According to the characterization of short idle-times (see Sect. 2), the r term in (4) can be seen as the number of $RTTs$ within the download of a single basic block, i.e. r is equal $B / (\overline{\gamma} \cdot \overline{RTT})$. S_1 can be derived by considering the algorithm for estimating short idle-times, and the distributions of short idle-times and their estimates (i.e., t_i and $t_i^{'}$, respectively). Details are reported in [6].

Furthermore, the S_2 term in (4), can be derived as follows. In the initial part of a User Think Time (1 sec), the network interface is managed according to the estimator of *short* idle times (see Sect. 2.1). Then, it is managed according to a binary exponenential backoff. Hence, the following lemma holds.

Lemma 2. *The average number of switching-on events within a User Think Time is*

$$S_2 = H + \left\lceil \log_2 \overline{UTT} \right\rceil ,$$

where H is the average number of switching-on events within the first second of the User Think Time.

A formula for H can obtained by following the same approach used to derive S_1, and is thus omitted (see [6] for details). Finally, since the PS-PT protocol detects a short idle-time after the last packet of the main file, the S_3 term of (4) can be evaluated as follows

Lemma 3. *The average number of switching-on events between the end of the main-file download and the beginning of the embedded-files download is*

$$S_3 = p\left(t_i' > t_{so}\right) .$$

The above lemmas allow us to prove the following theorem.

Theorem 2. *The energy spent to download a single basic block by using the power-saving system is*

$$C_{ps} = \frac{B}{\overline{\gamma}_{wl}} + t_{so} \cdot \left\{ \frac{B}{\overline{\gamma} \cdot \overline{RTT}} \cdot S_1 + l \cdot \left(H + \lceil \log_2 \overline{UTT} \rceil\right) + p\left(t_i' > t_{so}\right) \right\} . \quad (5)$$

3.3 Additional URT Model

In the HTTP/1.1 version, the download of a Web page can be seen as the sequence of two non-overlapping transactions between the client and the server, each of which starts with a client request. During the first transaction the client downloads the main file, while in the second transaction it fetches all embedded files, if any. Due to the TCP behavior, each transaction consists of several groups of back-to-back TCP segments separated by idle times, as discussed in Sect. 2. When an idle-time estimate occurs to be too long, the Access Point stores TCP segments received from the Web server in a local buffer, waiting for the mobile host to reconnect (see Sect. 2.1). In this way, the power-saving system introduces an additional time to the delay experienced by that group of TCP segments. It is worth noting that the additional delay introduced in a whole transaction is only the delay related to the *last* group of TCP segment in that transaction.

It must be pointed out that the Access Point is usually connected to the Internet through an high-speed LAN (e.g., a 100-Mbps fast-Ethernet). Hence, the time interval required by the Access Point to receive a group of TCP segments is typically negligible. As a consequence, the additional delay introduced for a group is well approximated by the additional delay introduced for the first TCP segment within the group, i.e., the error of the idle-time estimator (throughout referred to as e). Finally, if the transaction starts when the wireless interface of the mobile host is off, it is delayed of further t_{so} seconds.

Based on these remarks, we can evaluate the average additional delay introduced in a Web-page download ($\overline{I_{pd}}$) as follows:

Theorem 3. *The average additional delay introduced by the power-saving system when downloading of a Web page is*

$$\overline{I_{pd}} = t_{so} + \overline{e} + \left\{ t_{so} \cdot p\left(t_i' > t_{so}\right) + \overline{e} \right\} \cdot p_{emb} , \quad (6)$$

where $\overline{e}$ is the average value of e and p_{emb} is the probability that a Web-page contains embedded files.

In detail, $t_{so} + \overline{e}$ is the additional delay introduced in the transfer time of the main-file. If the Web-page contains embedded files, a further delay is introduced.

This delay is $\overline{e}$ if the wireless interface is initially switched on, $t_{so} + \overline{e}$ otherwise. Thus, to evaluate $\overline{I_{pd}}$ we need to derive the idle-time estimator error, e whose formula is provided by the following theorem. As above, the related proof can be found in [6], and is here omitted.

Theorem 4. *The average error of the short idle-time estimator is*

$$\begin{array}{l}\overline{e} = \frac{1}{2}\left(\frac{M^2-t_{so}^2}{4M} \cdot u\left(M,\, t_{so}\right) + 0.9 \cdot \frac{k^2-t_{so}^2}{4M} \cdot u\left(k,\, t_{so}\right) + \right. \\ \left. + \; 0.1 \cdot \frac{2sec-M-k}{2} \cdot \chi\left(k,\, t_{so}\right)\right)\end{array} , \qquad (7)$$

$$where\ u\left(x,\, y\right) = \begin{cases} 1\ if x \geq y \\ 0\ otherwise \end{cases},\ and\ \chi\left(k,\, t_{so}\right) = \begin{cases} 1\ if\, 1sec - k > t_{so} \\ 0\ otherwise \end{cases}.$$

4 Model Validation

In [4] we implemented a prototype of the power-saving network architecture. We used a real-Internet path to connect the Access Point and the Web server on the fixed host and simulated an 11-Mbps wireless link between the Access Point and the mobile host. We performed an extensive set of experiments, and each experiment provided us with a sample of I_{ps} and $\overline{I_{pd}}$. We replicated the experiments by considering different wireless-link throughputs, starting from 11 Mbps down to 2 Mbps. We observed very small variations in energy saving since the bottleneck between the client and the server is always the wired part of the path (see [5] for details on the methodology).

The model validation is carried out by comparing the analytical results for I_{ps} and $\overline{I_{pd}}$ provided by the model with the average hourly results obtained from the prototype[3]. Specifically, for each hour of the day, i) we instantiate the model parameters with the average values experienced by the prototype; ii) we derive the model predictions for I_{ps} and $\overline{I_{pd}}$; and finally iii) we compare these predictions with the prototype results. It is worth noting that in each experiment the Web user generates a traffic that meets the statistics shown in Tab. 1. Furthermore, according to the remarks discussed in Sect. 2, we use $2 \cdot \overline{RTT}$ as the maximum value for both t_i and $t_i^{'}$. The $\overline{RTT}$ experienced in the experiments is almost independent on the specific hour. Hence, the M parameter in (7) does not vary over different hours. Finally, S_1, S_2 and S_3 values are almost identical over the whole day, and $\overline{\gamma}_{wl}$ is assumed equal to $11Mbps$. Therefore, in our testbed, the only parameter varying over different hours is the throughput experienced in the (wired) Internet, i.e., $\overline{\gamma}$. Table 3 reports the values of the other parameters used to validate our analytical model.

Figure 3 shows the hourly average values for I_{ps} and $\overline{I_{pd}}$ measured by using the prototype. We also plot the I_{ps} and $\overline{I_{pd}}$ figures derived from the analytical model.

As far as the I_{ps} index (left-side plot), the model and the prototype provide very close results: the difference is always less than 9% of the prototype results.

[3] As discussed in [5], I_{ps} and $\overline{I_{pd}}$ samples collected within the same hour (also in different days) are i.i.d.

Table 3. Parameters used to validate the analytical model.

Definition	Symbol	Value
Average number of switching-on events in a short idle time	S_1	1.55
Average number of switching-on events in a *UTT*	S_2	5
Number of switching-on events to send the embedded files request	S_3	1
Network RTT between the client and the Web server	$\overline{RTT}$	0.3 sec
Switching-on transient interval of the wireless interface	t_{so}	0.1 sec
Indicator function of k and t_{so} relative values	$\chi(k, t_{so})$	1

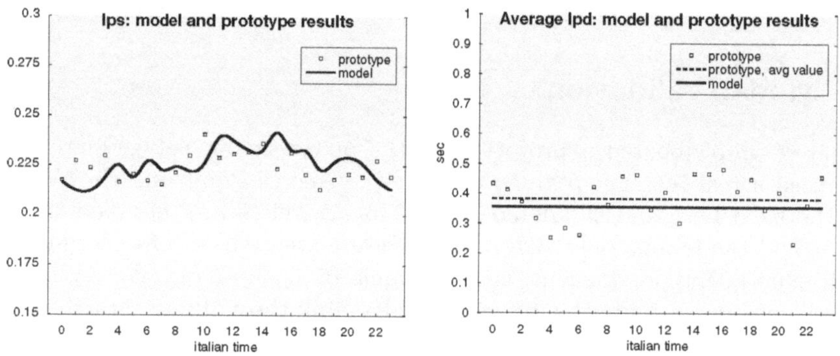

Fig. 3. Hourly average I_{ps} and $\overline{I_{pd}}$ obtained from the model and by the prototype, respectively.

Furthermore, we have compared the daily average of I_{ps} values with the corresponding analytical predictions provided by the model. The results obtained (not reported here) show that the difference between the model and the prototype is less than 1% of the prototype average daily value.

As far as the $\overline{I_{pd}}$ index (right-side plot), the results show that experimental values vary during the day, i.e., the prototype is sensitive to variations of $\overline{\gamma}$. Specifically, $\overline{\gamma}$ depends on two factors, i.e., i) congestion in the Internet, that reduce the TCP window size; and ii) variations of the *RTT* between the client and the server. As discussed in Sect. 3.3, the additional URT is affected by *RTT* variations. However, we have no sufficient information to include the precise *RTT* pattern in the analytical model. As a consequence, the $\overline{I_{pd}}$ model allows us to measure the additional URT related to the average *RTT*. Specifically, if we compare the model prediction with the daily average value of $\overline{I_{pd}}$, the difference is about 7% of the prototype results (the average *RTT* over the whole day is 0.3*sec*).

The above results show the accuracy of our analytical model. In the next session we use this model to investigate the sensitiveness of the power-saving system to two Internet key parameters, i.e. the available throughput and the *RTT*.

5 System Sensitiveness

5.1 Throughput Analysis

As noted in the previous section, the Internet throughput (i.e., $\overline{\gamma}$) depends on both the network RTT and the TCP window size. Since the I_{ps} index depends on $\overline{\gamma}$ (see (3) and (5)), it follows that it is influenced by both the RTT and the TCP window size. On the other hand, the $\overline{I_{pd}}$ index is only affected by variations in the RTT, as shown by (6) and (7). Therefore, below we analyze I_{ps} as a function of $\overline{\gamma}$, and $\overline{I_{pd}}$ as a function of RTT.

Firstly, we derive the I_{ps} index as a function of $\overline{\gamma}$. Sarting from (3) and (5), after some algebraic manipulations, $I_{ps}(\overline{\gamma})$ becomes:

$$I_{ps}(\overline{\gamma}) = \frac{a\overline{\gamma} + b}{c\overline{\gamma} + d} , \tag{8}$$

where a, b, c and d are constant terms. Figure 4-left shows I_{ps} as a function of $\overline{\gamma}$[4].

It clearly appears that when $\overline{\gamma}$ increases I_{ps} decreases and, hence, the power-saving system saves more energy. This result is somehow counter-intuitive, since one would expect that the best power saving is achieved when $\overline{\gamma}$ is at its lowest value. In this case the overall idle time during the download of the basic block is at its maximum value. However, the I_{ps} behavior in the above plot can be explained as follows. As shown in (3) and (5), variations of $\overline{\gamma}$ affect both C_{I-TCP} and C_{ps}. In the Indirect-TCP architecture, when $\overline{\gamma}$ increases, the time needed to fetch the basic block from the Web server (i.e., $B/\overline{\gamma}$) decreases, and C_{I-TCP} decreases accordingly. On the other hand, the dependence of C_{ps} on $\overline{\gamma}$ is as follows. As highlighted in Sect. 4, $\overline{\gamma}$ is strictly related to the the TCP window size. Since RTT is almost stable, large TCP windows mean high $\overline{\gamma}$ values, while narrow TCP windows correspond to low $\overline{\gamma}$ values. Furthermore, if the TCP window size increases, the number of $RTTs$ needed to fetch the basic block (i.e., r in (4)) drops, since more bytes are downloaded in a single RTT. Equations (4) and (5) show that if r decreases, the number of switching-on events (i.e., S) decreases. Therefore, we can conclude that the more $\overline{\gamma}$ increases, the more C_{ps} decreases. Since both C_{I-TCP} and C_{ps} benefit from increases of $\overline{\gamma}$, the I_{ps} pattern is defined by the parameters a, b, c and d. Specifically, in the Internet configuration we experienced, when $\overline{\gamma}$ increases, C_{ps} decreases more than C_{I-TCP} does, and hence I_{ps} drops.

As a final remark, it is worth noting that Fig. 4-left highlights the theoretical lower and upper bounds of I_{ps}. Specifically, in the Internet configuration we experienced, I_{ps} ranges between 0.517 (when $\overline{\gamma} = 0$) and 0.168 (when $\overline{\gamma}$ approaches ∞). Therefore, with respect to the I-TCP architecture, our power-saving system guarantees energy savings that are always above 48%, and raise up to 83%. However, if we focus on realistic throughput values (i.e., between $50Kbps$ and $1Mbps$), energy savings are almost stable, since they vary between 68% and 82%.

[4] Clearly $\overline{\gamma} > 0$ is the only $\overline{\gamma}$ range that makes sense in (8).

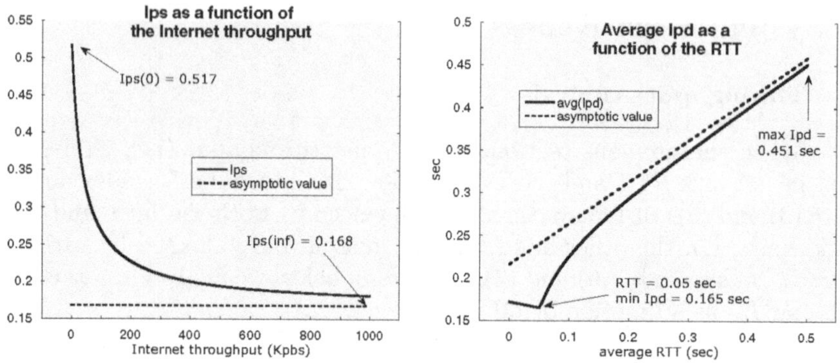

Fig. 4. I_{ps} as a function of the Internet throughput $\overline{\gamma}$ (left-side plot) and $\overline{I_{pd}}$ as a function of the $\overline{RTT}$ value (right-side plot).

5.2 *RTT* Analysis

In this section we analyze the dependence of $\overline{I_{pd}}$ on the average Round Trip Time, i.e., $\overline{RTT}$. As a preliminary step, it is necessary to define the range of valid $\overline{RTT}$ values. Specifically, our power-saving system defines $1sec$ as the upper bound of short idle-times. Therefore, $1sec$ is also the upper bound of both the short idle-time and estimate distributions (i.e., $M \leq 1sec$). Since in our model M is equal to $2 \cdot \overline{RTT}$, $\overline{RTT}$ must be less than $0.5sec$. On the other hand, we can use 0 as the lower bound – or, more precisely, as the theoretical lower limit – of $\overline{RTT}$.

Also, it must be pointed out that t_{so} is the lower bound of $\overline{I_{pd}}$. Specifically, even if the estimator error is always equal to 0, the mobile host switches the wireless interface on at least once every Web-page download (i.e., when the user sends a new Web page request). Therefore t_{so} represents an additional URT that can never be eliminated when using our power-saving system.

Finally, for the sake of simplicity, hereafter we assume that $t_{so} = 0.1sec$ holds. Therefore, since t_{so} is equal to $0.1sec$ and $\overline{RTT}$ is less than $0.5sec$, $\chi(k, t_{so})$ is always equal to 1. However, the analysis can be easily extended to any t_{so} value.

From (6) and (7) we derive the plot shown in Fig. 4-right. In this plot we can observe three regions corresponding to i) M less than t_{so}; ii) M between t_{so} and $t_{so}/0.9$; and, iii) M greater than $t_{so}/0.9$. The second region is very small and, actually, can be hardly distinguished in Fig. 4-right.

In the first region, the $\overline{RTT}$ is so short that $t_i^{'}$ is always less than t_{so}. Hence, the wireless interface remains idle for k seconds and then (if necessary) it is switched off for the next $1 - k$ seconds. This latter value also defines the estimation error. Therefore, in this region, both $\overline{e}$ and $\overline{I_{pd}}$ are decreasing functions of $\overline{RTT}$, and $\overline{I_{pd}}$ reaches its minimum value ($\min \overline{I_{pd}} = 0.165sec$, when $\overline{RTT} = 0.05sec$). On the other hand, in the second and third regions $t_i^{'}$ may be greater than t_{so}. Therefore, $\overline{e}$ and $\overline{I_{pd}}$ become increasing functions of $\overline{RTT}$, and hence $\overline{I_{pd}}$ reaches its maximum value ($\max \overline{I_{pd}} = 0.451sec$, achieved when

$\overline{RTT} = 0.5sec$). The complete analysis of Fig. 4-right is provided in [6], and is here omitted for the sake of space.

As a final remark, it is worth noting that the $\overline{I_{pd}}$ figure in the third region can be well approximated by a linear increasing function, that grows as $0.487 \cdot \overline{RTT}$. Therefore, we can conclude that increases of $\overline{RTT}$ have a moderate impact on the additional URT.

6 Conclusions

In this work we have derived an analytical model of the power-saving architecture developed in [4]. This solution is tailored to Wi-Fi hotspot scenarios, and is aimed at reducing the energy consumed by a mobile host running non-real-time network applications. The analytical model has been used to analyze the performance of our system when used to support mobile Web access. The results have shown that, under the Internet conditions experienced by the prototype, the power-saving system saves up to 78% of the energy consumed by using a pure Indirect-TCP approach. Furthermore, the additional transfer-time of a Web page is about $0.4sec$, and, hence, Web users do not perceive the presence of the power-saving system as a significant degradation of the QoS.

Furthermore, we have performed a sensitiveness analysis with respect to two Internet key parameters, i.e., the throughput on the wired network and the RTT. This analysis has shown that, when focusing on typical throughput ranges (i.e., between $50Kbps$ and $1Mbps$), the energy saved is always greater than 68%. Finally, we have found that the average additional transfer-time is a slightly increasing function of the average RTT. However, we can conclude that the power-saving never affects the QoS perceived by Web users, since the average additional transfer-time is always less than $0.5sec$.

References

1. S.Agrawal, S.Singh, "An Experimental Study of TCP's Energy Consumption over a Wireless Link", 4th European Personal Mobile Communications Conference, February 20–22, 2001, Vienna, Austria.
2. G. Anastasi, M. Conti, W. Lapenna, "A Power Saving Network Architecture for Accessing the Internet from Mobile Computers: Design, Implementation and Measurements", The Computer Journal, Vol. 46, No.1, 2003, pp. 3–15.
3. G.Anastasi, M.Conti, E.Gregori and A.Passarella, "A power saving architecture for web access from mobile computers", Proc. 2nd IFIP TC-6 Networking Conf. (Networking 2002), 2002, Pisa, Italy, LNCS# 2345, pp. 240–251.
4. G. Anastasi, M. Conti, E. Gregori and A. Passarella, "Balancing Energy Saving and QoS in the Mobile Internet: An Application-Independent Approach", Proceedings of the 36th Hawaii Int. Conf. on System Sciences (HICSS–36), January 03.
5. G.Anastasi, M.Conti, E.Gregori and A.Passarella, "Performance Comparison of Power Saving Strategies for Mobile Web Access", Performance Evaluation, to appear.

6. G.Anastasi, M.Conti, E.Gregori and A.Passarella, "Power-Saving in Wi-Fi Hotspots: an Analytical Study (Extended Version)", Tech. Rep. DII03–5, Dept. of Information Engineering, University of Pisa, May 2003 (available at http://www.ing.unipi.it/~o783499/research/docs/PWC_WiFi_ext.pdf).
7. A.Bakre, B.R.Badrinath, "Implementation and Performance Evaluation of Indirect TCP", IEEE Trans. on Computers, Vol.46, No.3, March 1997.
8. H. Balakrishnan, V. N. Padmanabhan, S. Seshan, R. H. Katz, "A Comparison of Mechanisms for Improving TCP Performance over Wireless Links", IEEE/ACM Trans. on Networking, Vol. 5, N. 6, December 1997.
9. P.Barford e M.Crovella, "Generating Representative Web Workloads for Network and Server Performance Evaluation", Proc. of ACM SIGMETRICS '98, pp. 151–160, June 1998.
10. P.Barford, A.Bestavros, A.Bradley e M.Crovella, "Changes in Web Client Access Patterns", World Wide Web (Special Issue on Characterization and Performance Evaluation), 1999.
11. L.Bononi, M.Conti and M.Donatiello, "A distributed mechanism for power saving in IEEE 802.11 wireless LANs", Mobile Networks Applic. (MONET), 2001, Vol. 6, pp. 211–222.
12. R.Bruno, M.Conti and E.Gregori, "Optimization of efficiency and energy consumption in p-persistent CSMA-based wireless LANs", IEEE Trans. Mobile Comput., 2002, Vol. 1, 10–31.
13. M.Crovella e A.Bestavros, "Self-Similarity in World Wide Web Traffic: Evidence and Possible Causes", IEEE/ACM Trans. on Networking, Vol.5, No.6, pp.835–846, Dec. 1997.
14. L.M. Feeney and M. Nilsson, "Investigating the energy consumption of a wireless network interface in an ad hoc networking environment", Proceedings of the 20th IEEE Computer and Communications Societies Conf. (INFOCOM 2001), 2001.
15. M. Herbster and M.K. Warmuth, "Tracking the best expert", Proceedings of the 12th Int. Conf. on Machine Learning, 1995, pp. 286–294.
16. IEEE standard for Wireless LAN – Medium Access Control and Physical Layer Specification, P802.11, November 1997.
17. R.Kravets e P.Krishnan, "Power Management Techniques for Mobile Communication", Proc. of the 4th ACM/IEEE Int. Conf. on Mobile Computing and Networking (Mobicom'98).
18. R.Krashinsky and H.Balakrishnan, "Minimizing Energy for Wireless Web Access with Bounded Slowdown", Proc. 8th Int. Conf. on Mobile Computing and Networking (Mobicom 2002), 2002.
19. M. Mathis, J. Semke, J. Mahdavi and T.Ott, "The macroscopic behavior of the TCP Congestion Avoidance Algorithm", Computer Communication Review, Vol. 27, N. 3, July 1997.

A Queuing Analysis of Packet Dropping over a Wireless Link with Retransmissions

Rachid El Azouzi and Eitan Altman

INRIA B.P 93, 2004 Route des Lucioles
06902 Sophia Antipolis Cedex France
{relazouz,altman}@sophia.inria.fr

Abstract. One of the most important performance measure for real time applications is the packet loss probability. In a wireless environment losses are first due to noisy channels (fading, shadowing etc). In order to improve the radio link, one often retransmits packets that have not been well received (using the Automatic Retransmission reQuest - ARQ). This however may lead to queuing phenomena due to retransmissions and to losses of packets due to buffer overflow. We present a queuing analysis in order to compute the combined effect of the noisy channel, the retransmissions and the buffer overflow. We use advanced spectral methods of quasi-birth and death process in order to analyze the system for general multistate Markov channels and provide an approximation approach based on the theory of singular perturbation.

1 Introduction

The study of the performance of wireless channels and their effect on the quality of service is a major research issue as it provides tools for design of link layer protocols and dimension link layer buffers, see e.g. [12,13] and references therein. We study in this paper wireless channels that use retransmission of lost packets: the Automatic Retransmission reQuest (ARQ) protocol. We note however that retransmissions cause extra delay which results in larger queues and therefore in possible losses due to buffer overflow. In order to study both the delay statistics as well as the global loss process, we provide a queuing analysis of the radio link.

Our starting point is a link model provided in [12]. The radio channel considered in [12] is modeled using a two state Markov chain: a perfectly "good" state in which transmission succeeds with probability one, and a perfectly "bad" state in which transmission fails with probability one. Our first goal is to allow for a much more rich class of models for the channel state. The need for richer models is motivated by the following considerations:

- It has been shown in [9,11] that more than two states are needed for describing some radio channels; in fact it was shown that one needs typically at least four states.
- Even in the "best" channel state it might be too optimistic too expect losses to occur with probability zero. Conversely, it may be too pesimistic to assume that in a bad state all packets are systematically lost.

M. Conti et al. (Eds.): PWC 2003, LNCS 2775, pp. 321–333, 2003.

– Assume that for a given packet loss probability p we assign a single state of the channel. This means (if a discrete time model as in [12] is used) that the time during which a packet suffers a loss probability of p has a geometric distribution. If we wish to be able to model other distributions then more than one state should be assigned to a given loss probability!

Our analysis approach is based on modeling of the system as a two dimensional Markov chain, where the first corresponds to the channel state and the second to the number of packets queued at the buffer. In order to obtain the steady state distribution of this system we make use of recent advanced spectral methods of analysis of quasi birth and death (QBD) processes [5,6]. These methods allow us to represent the steady state solution as a simple function of the eigenvectors and eigenvalues of some matrices. For the case that the channel is described by a two dimensional chain, we obtain an *explicit solution* which has not been available in [12] even for the simpler special case of "perfectly" good and "perfectly" bad channel states.

The second objective of the paper is to study a singular perturbation solution approach for approximating the steady state solution. This approach allows to obtain simple approximations for the case that the channel state evolves much slower than the queue size. In fact, if the speed ratio is given by a parameter ϵ, then the singular perturbation approach allows to represent the steady state probabilities as a Taylor series of ϵ. The first term in the series already gives a very good approximation when ϵ is sufficiently small, and this term is much simpler to compute than by using the previous exact approach.

The structure of the paper is as follows. We present the model in section 2. Section 3 provides a detailed spectral analysis and give the steady state channel-queue length for the finite queue model. We concluded by sample numerical results. In section 4 We study a singular perturbation solution approach for approximating the steady state solution and we compare the two methods presented in section 3 and section 4 by numerical results.

2 Model

We consider the following queuing system. In each slot, a packet is generated at the transmitter according to a Bernoulli process with arrival probability $p > 0$. The channel state can be described by the variable $l \in E := \{1, 2, .., L\}$, where $l = i$ denote that the packet can be transmitted with probability q_i, with $q_i \in [0, 1]$. If a transmission fails we assume that it will be attempted again in the next time unit. As in the first model studied in [12], we assume that the number of retransmissions of a packet by the ARQ protocol is not bounded: retransmissions will be attempted until a success.

$$\Lambda = \begin{pmatrix} \lambda_{11} & \lambda_{12} & \cdots & \lambda_{1L} \\ \lambda_{21} & \lambda_{21} & \cdots & \lambda_{2L} \\ \vdots & \vdots & \vdots & \vdots \\ \lambda_{L1} & \lambda_{22} & \cdots & \lambda_{LL} \end{pmatrix} \tag{1}$$

be the transition matrix for the error process, where $\sum_{j=1}^{L} \lambda_{ij} = 1$. We shall denote $\nu = [\nu_1, \nu_2.., \nu_L]$ the probability distribution of the transition matrix Λ. We assume throughout that Λ is irreducible and aperiodic.

The state of the system at time n is described by $X(n) = (l(n), i(n))$, where $l(n)$ is the state of channel in slot n and $i(n)$ is the number of packets in the queue at the beginning of slot t. The arrival process in slot m is independent of $X(n)$ for all $m \geq n$. Arrivals are assumed to occur at the end of the slot so that, in particular, they cannot depart in the same slot.

The system has a finite queue of size B. A loss occurs whenever an arriving packet finds the queue full. It can be shown that the Markov channel evolution and the assumption of independent arrivals imply that $X(n)$ is a Markov chain with value in a finite state space $\mathcal{B} = \{(i,j),\ i = 1,..L,\ j = 0,..,B\}$

Since we consider discrete time, we have to make some convention on the order of events within each time unit. We assume that at each time unit, if the queue is nonempty then first a transmission is attempted, and only if it succeeds then there is a departure; then if there have been an arrival during the slot then it joins the queue. We call this a service-arrival order. This order means in particular that if the queue is full at the beginning of the slot then an arriving packet during the slot need not be lost: it is only lost if the transmission has failed.[1]

The transition probability $L(B+1) \times L(B+1)$ matrix of the Markov chain $X(n)$ are given by

$$P = \begin{pmatrix} B_0 & B_1 & 0 & \cdots & & & \\ A_2 & A_1 & A_0 & 0 & \cdots & & \\ 0 & A_2 & A_1 & A_0 & 0 & \cdots & \\ \vdots & \ddots & \ddots & \ddots & \ddots & & \\ \vdots & & \ddots & \ddots & \ddots & A_1 & A_0 \\ 0 & \cdots & & & 0 & A_2 & F_1 \end{pmatrix} \begin{matrix} 0 \\ 1 \\ \vdots \\ \vdots \\ \vdots \\ B \end{matrix}$$

P consists of $B+1$ block rows and block columns. The matrices B_0, B_1, A_0, A_1, A_2, and F_1 are all $L \times L$ non-negative matrices, where $B_0 = \bar{p}\Lambda$, $B_1 = p\Lambda$,$A_0 = diag(p\bar{q}_i, i = 1,,..,L)\Lambda$, $A_1 = diag(pq_i + \bar{p}\bar{q}_i, i = 1,..,L)\Lambda$, $A_2 = diag(\bar{p}q_i, i = 1,..,L)\Lambda$ and $F_1 = diag(pq_i + \bar{q}_i, i = 1,..,L)\Lambda$. Since Λ is irreducible and aperiodic, and since $p > 0$, the Markov chain $X(n)$ is aperiodic and it contains a single ergodic class. Hence it has a unique stationary (or steady state) probability distribution defined as

$$\pi(i,j) = \lim_{n \to \infty} Pr(l(n) = i, i(n) = j);\quad i = 1, 2, .., L;\ j = 0, 1, .., B$$

[1] If in contrast, a real system operates in the arrival-service order i.e. first an arrival occurs (and is lost if the queue is full) and then a transmission is attempted, then the expressions for the steady state probability obtained for the service-arrival order are still valid but they apply to the state of the queue just *after* the arrival occurred in the arrival-service order system. From this it is then easy to derive the steady state distribution at the beginning of a time slot.

The objective of this analysis is to determine the blocking probabilities, the expected delays, and in fact the whole probability distribution $\pi(i,j)$ in terms of the parameters of the system. $\pi(i,j)$ is the unique solution to the equation

$$\pi = \pi P, \quad \pi \bar{e}_L = 1 \tag{2}$$

where the notation $\bar{e}_k$, $k \geq 0$ is used to denote a column vector of ones of length $k(B+1)$. For convenience, we partition the stationary vector π as $[\pi_0, \pi_1, ..., \pi_B]$, where π_k of size $1 \times L$, will called the solution vector of level k, $0 \leq k \leq B$.

3 Analysis of the System

3.1 Spectral Analysis

In the analysis a special role is played by the number of states of the server for which the transmission probability is zero or one. We denote these numbers by

$$m_0 := |\{i : q_i = 1\}|, \quad m_2 := |\{i : q_i = 0\}|$$

The structure of the transition probabilities would be similar to the one studied [6] if we did not have the first row and column. We handle this problem by reducing the state space, by removing the level zero and in the second phase, we construct the stationary probability of the reduced chain by using the spectral analysis. This will allow us to obtain the steady state probabilities of the original system.

Let $\mathcal{B}_1$ be subset of $\mathcal{B}$, where $\mathcal{B}_1 = \{(i,j)/i = 1,..,L; j = 1,..,B\}$. We define the restricted process $\{X_n^0, n \geq 0\}$ as follows: $X_n^0 = X(s_n)$ for $n \geq 0$, so that X_n^0 is the actual state visited when the Markov chain is not for the nth time in level 0.

Lemma 1. *The restricted precess X_n^0 is an irreducible positive Markov chain on the state $\mathcal{B}_1$. Its transition is given by*

$$P^0 = \begin{pmatrix} C1 & A_0 & 0 & \cdots & & & \\ A_2 & A_1 & A_0 & 0 & \cdots & & \\ 0 & A_2 & A_1 & A_0 & 0 & \cdots & \\ \vdots & \ddots & \ddots & \ddots & \ddots & & \\ \vdots & & \ddots & \ddots & \ddots & A_1 & A_0 \\ 0 & \cdots & & & 0 & A_2 & F_1 \end{pmatrix} \begin{matrix} 1 \\ 2 \\ \vdots \\ \vdots \\ \vdots \\ B \end{matrix}$$

where $C_1 := A_1 + A_2(I - B_0)^{-1}B_1$. Its stationary probability vector is proportional to $\pi^0 := [\pi_1, \pi_2, .., \pi_B]$, i.e., we have that $\pi^0 P^0 = \pi^0$.

Proof See Lemma 10.1.1 in [8]. ∎

The matrix C_1 records the probability, starting from the level 1 of returning to the level 1, before reaching the level 2. The lemma implies that the stationary probability vector may be determined by the following equations:

$$\pi_0 = \pi_1^0 A_2 (I - B_0)^{-1},\ \pi_1^0 = \pi_1^0 C_1 + \pi_2^0 A_2, \tag{3}$$

$$\pi_i^0 = \pi_{i-1}^0 A_0 + \pi_i^0 A_1 + \pi_{i+1}^0 A_2,\ \pi_B^0 = \pi_{B-1}^0 A_0 + \pi_B^0 F_1,\ \pi \bar{e}_L = 1 \tag{4}$$

If $\sum_{i=0}^{L} \nu_i q_i \neq p$, then the steady-state probability vector can be written as a sum of two matrix-geometric terms,

$$\pi_j^0 = x_1 R^{j-1} + x_L S^{B-j},\ \ j = 1, 2, .., B \tag{5}$$

where the rate matrices R and S are respectively the minimal nonnegative solutions to the quadratic matrix equations (for details see Naomov [5]).

$$A_0 + R(A_1 - I) + R^2 A_2 = 0, \quad S^2 A_0 + S(A_1 - I) + A_2 = 0 \tag{6}$$

Next we provide a simpler representation of (5). To that aim we introduce some definitions and two Lemmas. Define the quadratic matrix polynomials $T(z)$ and $\bar{T}(z)$:

$$T(z) := A_0 + z(A_1 - I) + z^2 A_2,\ \ \bar{T}(z) := A_2 + z(A_1 - I) + z^2 A_0 \tag{7}$$

The solution (3)-(4) is closely related to the roots and the left nullvector of $T(z)$. Let (α, u) be a root-nullvector pair of $T(\alpha)$, thus satisfying the equation

$$uT(\alpha) = 0,\ \ det[T(\alpha)] = 0 \tag{8}$$

We observe that the non-zero roots of $det[T(z)]$ are the reciprocals of non zero roots of $det[\bar{T}(z)]$ with the same corresponding left nullvector. We shall analysis the nature of the roots and the nullvectors of $T(z)$

Lemma 2. *1. If α is an eigenvalue of R with algebraic multiplicity m, then α is a root of the polynomial* $\det[T(z)]$ *and its multiplicity is at least m. And if u is an eigenvector of R corresponding to the eigenvalue α, then u is a left nullvector of $T(\alpha)$.*
2. If $\alpha \neq 0$ and $1/\alpha$ is an eigenvalue of S with algebraic multiplicity m, then α is a root of the polynomial $\det[T(z)]$ *and its multiplicity is at least m. And if u is an eigenvector of S corresponding to the eigenvalue $1/\alpha$, then u is a left nullvector of $T(\alpha)$.*
3. The left nullspace of R and A_0 coincides, and so do the left nullspace of S and A_2

Proof See the full of the paper [4]. ∎

Remark 1. Recall m_0 and m_2 are the number of state $i \in \{1, 2, .., L\}$ for which $q_i = 1$ and $q_i = 0$. Then after some permutation of the rows in matrix A_i,

$i = 0, 2$, the matrix A_i $i = 0, 2$ can be writing as the following block triangular matrix[2]

$$A_i = \begin{pmatrix} A_i^1 & A_i^2 \\ \mathbf{0} & \mathbf{0} \end{pmatrix} \tag{9}$$

The eigenvalues of A_i, $i = 0, 2$ are those of A_i^1 with those of matrix null $\mathbf{0} \in M_{m_i}$, counting multiplicities. Then if A_i^1, $i = 0, 2$ has eigenvalue *zero* with multiplicity n_i, 0 is eigenvalue of A_i with algebraic multiplicity r_i where $r_i = m_i + n_i$, $i = 0, 2$.

In the sequel, we assume that the sum of all principal minus of $T(z)$ of order $L - n_0$ is nonzero. We deduce from these results the following

Lemma 3. *(i)- The polynomial* $det[T(z)]$ *has a root with geometric multiplicity* r_0 *located at* $z = 0$*, and (ii)-* $2L - r_0 - r_2$ *roots. For their location we distinguish two cases:*
1. If $p < \sum_{i=1}^{L} \nu_i q_i$, $L - r_0$ *of them lie in disc* $|z| < 1$*, one at* $z = 1$*, and* $L - 1 - r_2$ *in disc* $|z| \geq 1$ *with* $z \neq 1$.
2. If $p > \sum_{i=1}^{L} \nu_i q_i$, $L - 1 - r_0$ *of them lie in disc* $|z| < 1$*, one at* $z = 1$*, and* $L - r_2$ *in disc* $|z| \geq 1$ *and* $z \neq 1$

Proof see the full version of the paper [4] ∎

Now, we return to the steady-state for the restricted process X_n^0. It was already showed that the matrices R and S have a full set of eigenvectors. Consequently the expression (5) can thus rewritten in spectral-expansion form :

$$\pi_j^0 = \sum_{k=1}^{L} \beta_k (\phi_k)^{j-1} w_k + \sum_{k=L+1}^{2L} \beta_k (1/\phi_k)^{B-j} w_k, \ 1 \leq j \leq B \tag{10}$$

with $\phi_1, .., \phi_L$ the eigenvalues of the matrix R and $w_1, w_2, .., w_L$ the corresponding left eigenvectors, and $1/\phi_{L+1}, .., 1/\phi_{2L}$ are the eigenvalues of S with corresponding left eigenvectors $w_{L+1}, .., w_{2L}$. If $m_2 + n_2 > 0$, then we set the corresponding ϕ_k equal to ∞ and we write $1/\phi_k = 0$. Moreover by convention we set $0^0 = 1$.

Here the vectors π_j, $j = 1, 2, .., L$ are know if β_k's are known. We derive an alternative set of equations which uniquely determines the coefficients β_k by combining the relation (10) with the equations (3), (4) and the normalization condition.

3.2 Examples and Numerical Results

In the full version of the paper [4], we illustrate how to apply this method to a channel that can be described by two states.

[2] Suppose that there exists $i \in \{1, 2, .., L\}$ such that $q_i = 0$ (resp $q_i = 1$). Then all entries in row i of the matrix S (resp R) equal 0, so that 1_i with the ith entry equal to 1 and all other entries equal to 0 is a left nullvector of S (resp R), Indeed, when the state of server is i in which $q_i = 0$, no packet can be transmitted. Hence starting from (j, i), $\forall j$, no state state of level $j - 1$ can ever be visited before another state in the level j is visited.

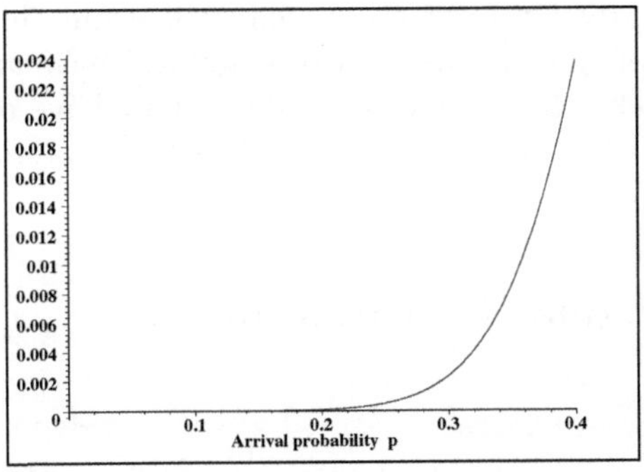

Fig. 1. The dropping probability of a packet as function of the arrauval probability p for $q_1 = 1$, $q_2 = 0$, $\lambda_{11} = 0.3$, $\lambda_{22} = 0.5$ and buffer size $B = 6$

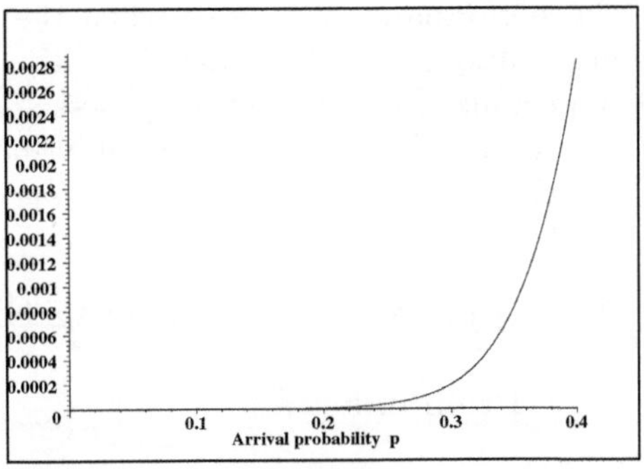

Fig. 2. The dropping probability of a packet as function of the arrival probability p for $q_1 = 1$, $q_2 = 0.2$, $\lambda_{11} = 0.3$, $\lambda_{22} = 0.5$ and buffer size $B = 6$

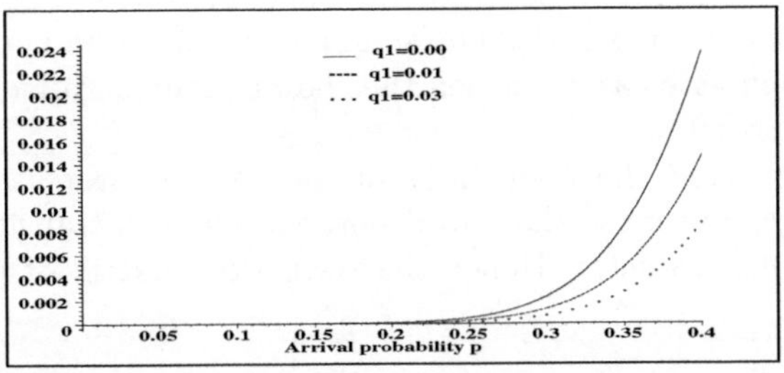

Fig. 3. The dropping probability of a packet as function of the arrival probability p for $q_1 = 1$, $\lambda_{11} = 0.3$, $\lambda_{22} = 0.5$ and buffer size $B = 10$ and $q2 = 0, 0.01, 0.03$

Numerical investigation. In this paragraph, we present some numerical results obtained based on the analysis described in previous section 3. We investigate the relationship between the dropping probability and arrival rate p. In particular, we compare the values of the performance measures for two different schemes: The first schemes is perfectly good/absolutly bad state in which the transmission succeeds with probability 1 or 0, and second scheme is a perfectly good/fairly bad in which the transmission succeeds with probability 1 or 0.2.

In figures 1 and 2, the dropping probability is plotted versus the arrival rate for first scheme (resp. scheme 2).

As shown in figure 1-2, the performance measures (dropping probability) for two schemes are different. It is true that the modification on the transmission in bad state will influence the performance , but more surprisingly, the difference between the performance measures is of the order 10^{-1}. This actually confirms that using the simple model of scheme 1 (e.g. [12,14,13]) does not adequately capture the process of packet errors. In order to verify this, we calculate the performance measures with a small change in the transmission probability q_2 in the region near 0 for scheme 1 (see figure 3), we observe a large effects on the performance measures when the arrival rate is more than 0.3. From this observation, it is clear that the values of the transmission probabilities have an important role to model the process packet errors. We conclude that we have to be careful when using the simplify model to represent the wireless the analytic system.

4 Approximations Using Singular Perturbation

4.1 Preliminaries

Singular perturbation is a powerful approach do obtain simple approximation of steady state probabilities of Markov chains in which we can identify two time scales: there are several group of states such that there are rare transitions between groups and frequent transitions within each group (see e.g. [2,3,10]) We shall apply this method to the case when the transitions of the queue size occur much more frequently than those of the channel states. This will typically be the case in high speed channels. The singular perturbation approach can also be useful for the opposite case in which the channel states evolve much faster than the queue size process.

We thus represent the transitions of the channel state as $\Lambda(\epsilon) = I + \epsilon\Lambda_1$ where Λ_1 is a generator (so that $\Lambda(\epsilon)$ represents indeed transition probabilities) and ϵ is a small parameter. Hence the transition matrix of the Markov chain $X(n)$ becomes

$$P(\epsilon) = Q_0 + \epsilon Q_1$$

where Q_0 is the probability transition matrix of the unperturbed Markov chain corresponding to strong interaction, and Q_1 is the generator corresponding to weak interaction, i.e.,

$$Q_0 = \begin{pmatrix} \bar{p}I & pI & 0 & \cdots & & & \\ \bar{A}_2 & \bar{A}_1 & \bar{A}_0 & 0 & \cdots & & \\ 0 & \bar{A}_2 & \bar{A}_1 & \bar{A}_0 & 0 & \cdots & \\ \vdots & \ddots & \ddots & \ddots & \ddots & & \\ \vdots & \ddots & \ddots & \ddots & \ddots & \bar{A}_0 & \\ 0 & \ldots & & 0 & \bar{A}_2 & \bar{F}_1 & \end{pmatrix} \text{ and } Q_1 = \begin{pmatrix} \tilde{B}_0 & \tilde{B}_1 & 0 & \cdots & & & \\ \tilde{A}_2 & \tilde{A}_1 & \tilde{A}_0 & 0 & \cdots & & \\ 0 & \tilde{A}_2 & \tilde{A}_1 & \tilde{A}_0 & 0 & \cdots & \\ \vdots & \ddots & \ddots & \ddots & \ddots & & \\ \vdots & \ddots & \ddots & \ddots & \ddots & \tilde{A}_0 & \\ 0 & \ldots & & 0 & \tilde{A}_2 & \tilde{F}_1 & \end{pmatrix}$$

where $\bar{A}_2 = diag(\bar{p}q_i, i = 1.., L)$, $\bar{A}_1 = diag(pq_i + \bar{p}\bar{q}_i, i = 1.., L)$, $\bar{A}_0 = diag(p\bar{q}_i, i = 1.., L)$, $\bar{F}_1 = diag(pq_i + \bar{q}_i, i = 1.., L)$, $\tilde{B}_0 = \bar{p}(\Lambda_1)$, $B_1 = p(\Lambda_1)$, $\tilde{A}_2 = diag(\bar{p}q_i, i = 0, 1.., L)\Lambda_1$, $\tilde{A}_1 = diag(pq_i\bar{p}\bar{q}_i, i = 1.., L)\Lambda_1$, $\tilde{A}_0 = p\,diag(p\bar{q}_i, i = 1.., L)\Lambda_1$, and $\tilde{F}_1 = diag(pq_i + \bar{q}_i, i = 1.., L)\Lambda_1$.

Let $\pi^\epsilon(k, i)$, $i = 1..L$, $j = 0, 1, .., B$, be the probability distribution of the state of the channel and the number of packets in the system in steady-state.

This is exactly the perturbation problem under the assumption that there exists a ergodic class (i.e., has exactly one closed communicating set of states), and Q_0 contains L subchains.

The partition of state set of Markov chain associated with the probability matrix Q_0 depends of the number of service for which the transmission probability is zero or one. Indeed, if $m_0 = m_2 = 0$, the L recurrent classes ξ_i, $i = 1,, L$ are given by

$$\xi_i = \{(i, j),\ j = 0, 1, ..B\}$$

and the unperturbed chain admits no transient state (block diagonal structure). Now if $m_0 > 0$ or $m_2 > 0$, let $E_B = \{i/\ q_i = 0\}$ and $E_G = \{i/\ q_i = 1\}$. In this case, the L recurrent classes are given by : $\xi_i = \{(i, 0), (i, 1)\}$ if $i \in \{j/\ q_j = 1\}$, $\xi_i = \{(i, B)\}$ if $i \in \{j/\ q_j = 0\}$ and $\xi_i = \{(i, j)\ j = 0, 1, ..B\}$ otherwise. The set of transient states T is given by $T = \{(i, j)/j \neq 0, 1$ and $q_i = 1\} \cup \{(i, j)/j \neq B$ and $q_i = 0\}$. Without loss of generality, the channel state in which the transmission probability is zero or one, are described by the variable $l \in \{1, .., r_0\}$ where $r_0 = m_0 + m_2$.

4.2 Taylor Series Expansion for Steady State

The above assumptions still implies that the stationary probability π^ϵ has the following property

Lemma 4. *The stationary probability π^ϵ of the perturbed chain has a Taylor series expansion : $\pi^\epsilon = \sum_{n=0}^{\infty} \pi^{(n)} \epsilon^n$. Moreover $\pi^{(n)}$, $n \geq 0$ are recursively calculated as follows:*

$$\pi^{(0)}[I - Q_0] = 0, \quad \sum_{(i,j)} \pi^{(0)}(i, j) = 1 \tag{11}$$

$$\pi^{(n)}[I - Q_0] = \pi^{(n-1)} Q_1, \quad \sum_{(i,j)} \pi^{(n)}(i, j) = 0 \tag{12}$$

Proof see Schweitzer [10], ∎

We consider the unperturbed Markov chain associated with the transition probability matrix Q_0. There exists a partition of states set of Markov chain $X(n)$ into of family of L recurrent classes ξ_i, $i = 1, .., L$. To each class $\bar{\xi}$ is associated the invariant measure (row vector) $m_{\bar{\xi}}$ of the recurrent subchain defined on the class $\bar{\xi} \in \bar{S} = \{\xi_1, .., \xi_L\}$.

Let us consider the natural numbering the states of $\mathcal{B}$ after the grouping defined by the partition $\bar{S}$. With this numbering, the transition probability matrix Q_0 has the following block structure:

$$Q_0 = \begin{pmatrix} P_1 & \cdots & 0 & 0 & \cdots & 0 & 0 & \cdots & 0 \\ \vdots & \ddots & \vdots & & \ddots & & & \ddots & \\ & & P_r & 0 & \cdots & 0 & 0 & \cdots & 0 \\ 0 & \cdots & & P_{r+1} & \cdots & 0 & 0 & \cdots & 0 \\ & \ddots & \vdots & \vdots & \ddots & \vdots & & \ddots & \\ 0 & \cdots & 0 & 0 & 0 & P_L & 0 & \cdots & 0 \\ F_1 & \cdots & 0 & 0 & \cdots & 0 & Y_1 & \cdots & 0 \\ \vdots & \ddots & \vdots & \vdots & \ddots & \vdots & \vdots & \ddots & \vdots \\ 0 & \cdots & F_r & 0 & \cdots & 0 & 0 & \cdots & Y_r \end{pmatrix}$$

where we distinguish three cases : For $i \notin \{1,..,r\}$, P_i is a $(B+1)\times(B+1)$ matrix :

$$P_i = \begin{pmatrix} \bar{p} & +p & \cdots & & \\ \bar{p}q_i & pq_i + \bar{p}\bar{q}_i & p\bar{q}_i & & 0 \\ 0 & \ddots & \ddots & \ddots & \\ \vdots & & \ddots & pq_i + \bar{p}\bar{q}_i & p\bar{q}_i \\ 0 & & \cdots & \bar{p}q_i & pq_i + \bar{q}_i \end{pmatrix}$$

For $i \in E_G$, P_i is 2×2 matrix, F_i is $(B-1)\times 2$ and Y_i is a $(B-1)\times(B-1)$ matrix :

$$P_i = \begin{pmatrix} \bar{p} & p \\ \bar{p} & p \end{pmatrix}, \quad F_i = \begin{pmatrix} 0 & \bar{p} \\ \vdots & 0 \\ & \vdots \\ 0 & 0 \end{pmatrix}, \quad Y_i = \begin{pmatrix} p & 0 & \cdots & 0 \\ \bar{p} & p & \cdots & 0 \\ \vdots & \ddots & \ddots & \\ 0 & \cdots & \bar{p} & p \end{pmatrix}$$

For $i \in E_B$, the matrix $P_i = 1$, F_i is a B vector given by $F^T = (0,..,0,p)$ and Y_i is a $B\times B$ matrix, where

$$Y_i = \begin{pmatrix} \bar{p} & p & \cdots & 0 \\ 0 & \ddots & \ddots & 0 \\ \vdots & \cdots & \bar{p} & p \\ 0 & \cdots & 0 & \bar{p} \end{pmatrix}.$$

Let us define the matrix $\bar{Q}_1 = (\bar{Q}_1)_{\xi\xi'}$ by: $(\bar{Q}_1)_{\xi_i\xi'_j} := \sum_{k=1}^{B} m_{\xi_i}(k) \sum_{k'=1}^{B}(Q_1)_{(i,k),(j,k')}$. We note that $\bar{Q}_1$ is a generator of an aggregated Markov chain on a state space $\bar{S}$. Now let $\bar{\nu} = [\bar{\nu}_1,..,\bar{\nu}_L]$ the probability distribution of the aggregated Markov chain. Note in view of definition of $\bar{Q}_1$, we observe that the generator of aggregated Markov chain is Λ_1. Hence $\bar{\nu}$ is the unique stationary distribution of ergodic Markov chain $l(n)$, i.e.,

$$\bar{\nu}\Lambda_1 = 0, \quad \nu\mathbf{1} = 1. \tag{13}$$

Following Delebecque[3], the unperturbed stationary probability is given by

$$\pi^{(0)}(i,j) = \bar{\nu}_i m_{\xi_i}(j) \tag{14}$$

where m_{ξ_i} is the probability distribution of the recurrent class which given by the following lemma.

Lemma 5. *For each $i = 1, 2, .., L$, the stationary distribution m_{ξ_i} of recurrent class ξ_i is given by:*
1- If $i \notin E_G \cup E_B$, $m_{\xi_i}(j) = \frac{p^j(1-q_i)^{j-1}}{q_i^j(1-p)^j} m_{\xi_i}(0)$, $j = 1, .., B$, where $m_{\xi_i}(0) = \frac{q_i - p}{q_i - p\alpha^B}$ with $\alpha = \frac{p(1-q_i)}{q_i(1-p)}$.
2- If $i \in E_G$, $m_{\xi_i}(0) = 1 - p$ and $m_{\xi_i}(1) = p$.
3- If $i \in E_B$, $m_{\xi_i}(B) = 1$.

Proof see the full version of the paper [4]. ∎

We shall denote q_{ξ_i} the right eigenvector corresponding to the one eigenvalue of the matrix P_i. Recall that the right eigenvector provide the probability to end in the class ξ_i starting from i. In our model, the vector q_{ξ_i} is given as follows: For each $j = 0, 1, .., B$, $q_{\xi_i}(k,j) = 1$, if $k = i$ and $q_{\xi_i}(k,j) = 0$ otherwise.

Following the analysis in [1, Chap. 3], we can obtain the stationary distribution of the perturbed system by another equivalent system with the matrix coefficients of smaller dimension.

Theorem 1. *A solution of the equations (11)-(12) with the normalization conditions is given by the following recursive formulae:*
1. $\pi^{(0)} = \bar{\nu}M$, where $\bar{\nu}$ is the stationary distribution of aggregated Markov chain $\Lambda_1 \in \mathbb{R}^{L+1 \times L+1}$ and $M \in \mathbb{R}^{(L+1)\times(L+1)(B+1)}$ is a matrix whose rows are stationary distribution of the Markov chain Q_0 and $N \in \mathbb{R}^{(L+1)(B+1)\times(L+1)}$ is a matrix of eigenvectors corresponding to the zero eigenvalue of the unperturbed generator $Q_0 - I$.
2. $\pi^{(n)} = \pi^{(n-1)}Q_1 H_0 + \pi_1^{(n)}M$, $\pi_1^{(0)} = \nu$, $\pi_1^{(n)} = \pi_1^{(n-1)}MQ_1H_0Q_1NH_1$, where H_0 is a deviation matrix of the unperturbed Markov chain $H_0 = [I - Q_0 + NM]^{-1} - NM$ and H_1 is a deviation of the aggregated Markov chain $H_1 = [-\Lambda_1 + \Lambda^*]^{-1} - \Lambda^*$, where Λ^* is the ergodic projection of transition matrix $\Lambda - I$, i.e., $\Lambda^* = \mu_L \nu$, with $\mu_L = [1, .., 1]^T \in \mathbb{R}^{L+1}$.

Observe that the stationary distribution is obtained here more efficiently than by using the spectral analysis given in section 3. In this reduced system, we need only to compute the stationary distribution of aggregated Markov chain Λ since the stationary distributions of each recurrent class of the unperturbed matrix transition Q_0 are given by Theorem 1.

4.3 Examples and Numerical Results

Consider the same example as the one presented in subsection 3.2. Also without loss of generality, we assume that $q_1 \neq q_2$. Since the channel state describes by two states, the Markov chain associated with probability matrix Q_0 has 2

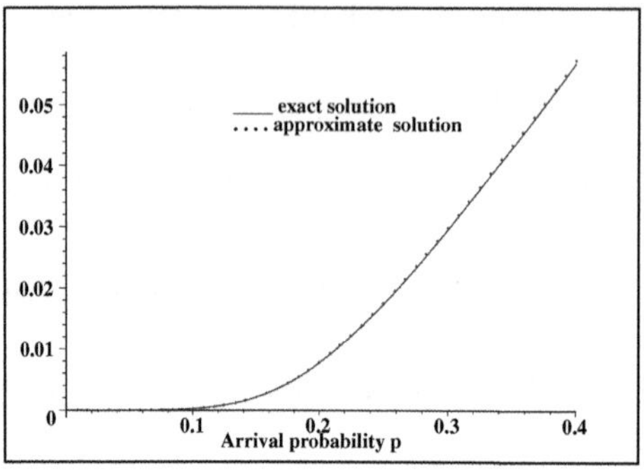

Fig. 4. Dropping probability versus arrival probability for $\epsilon = 10^{-3}$, $a = 1/5$, $b = 1/2$, $q_1 = 1$, $q_2 = 0$ and $B = 5$. Comparison of two methods: spectral analysis (exact solution) and approximate solution.

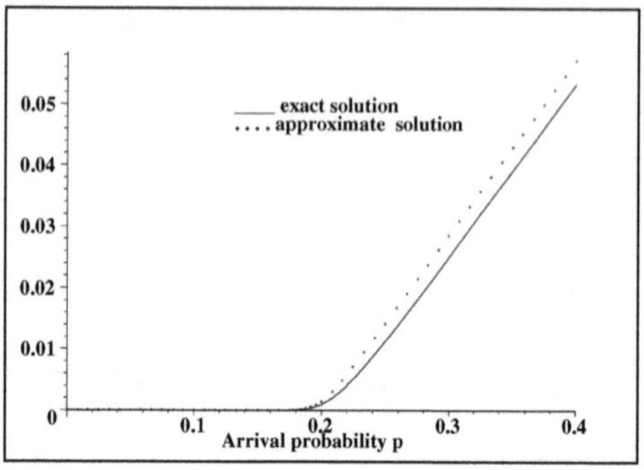

Fig. 5. Dropping probability versus arrival probability for $\epsilon = 10^{-3}$, $a = 1/5$, $b = 1/2$, $q_1 = 1$, $q_2 = 0.2$ and $B = 30$. Comparison of two methods: spectral analysis (exact solution) and approximate solution.

recurrent classes and transient states if $m_0 > 0$ or $m_2 > 0$. The stationary probability distribution of the aggregated Markov chain Λ_1 is given by $\bar{\nu}_1 = \frac{b}{a+b}$, $\bar{\nu}_2 = \frac{a}{a+b}$, where $\Lambda_1 = \begin{pmatrix} -a & +a \\ b & -b \end{pmatrix}$

Now, we are going to illustrate how to apply this method and obtain a simple approximation for steady state solution. This approximation can be presented by first term in the Taylor series of ϵ in which the first term is given by $\pi^{(0)} = [\bar{\nu}_1, \bar{\nu}_2] \begin{bmatrix} m_{\xi_1} \\ m_{\xi_2} \end{bmatrix}$

In figures 4-5 we compare the dropping probability for various values of the buffer size and for $a = 1/5$, $b = 1/2$, $q_1 = 1$ and $q_2 = 0.2$, as obtained via the two methods presented in section 3 and section 4.

From plots in figures 4-5, we observe that the first term in the Taylor series gives a good approximation when ϵ is small ($\epsilon = 10^{-3}$). Moreover, we observe that when the buffer size increases the error between the exact solution and the approximation solution increases. An intuitive explanation is as follows: when we use the first term of the Taylor series, we get the limiting behavior as ϵ tends to zero, i.e. as the transitions of the queue sizes occur much faster than the transitions of the channel state. However, as the buffer size increases, the probability of reaching its boundaries become smaller (it converges to zero when the buffer size converges to infinity for the arrival probability that we chose) which means that the time between transitions to the state of full queue do not occur much faster than the transitions of the channel state.

References

1. K. E. Avrachenkov, "Analytic Perturbation Theory and its Applications", Thesis in University of South Australia, 1999. Available at http://www-sop.inria.fr/mistral/personnel/K.Avrachenkov/moi.html
2. P.J. Courtois, "Decomposability: queueing and computer system applications", Academic Press, New York, 1977.
3. F. Delebecque, "A reduction process for perturbed Markov chains," SIAM J. App. Math., vol. 43, pp. 325–350, 1983.
4. R. El Azouzi, and E. Altman, "Queueing Analysis of Link-Layer Losses in Wireless Networks", available at http://www-sop.inria.fr/mistral/personnel/Rachid.Elazouzi/moi.html
5. V. A. Naoumov. Matrix-multiplicative approach to Quasi-Birth-and-Death processes analysis. In: Matrix-analytic Methods in Stochastic Models- Lecture Notes in Pure And Applied Mathematics. Eds. A.S. Alfa and S.R. Chakravarthy. Marcel Dekker, New York (1996), 87–106.
6. R. N. Queija, "Processor-Sharing Models for Intergrated-Service Networks", PhD Thesis, Eindhoven University of Technology, January 2000, ISBN 90-646-4667-8.
7. P. Lancaster, Theory of Matrices, Academic Press, New York, 1969.
8. G. Latouche and R. Ramaswami. "Introduction to Matrix Analytic Method in Stochastic Modeling". Society for Industrial and Applied Mathematics, 1999.
9. M. Rahman, M. Bulmer and M. Wilkinson, "Error models for land mobile satellite channels", *Australian Telecommunication Research*, Vol. 25 No 2, pp. 61–68, 1991.
10. P.J. Schweitzer, "Perturbation theory and finite Markov chains", *J. Appl. Prob.*, v.5, pp.401–413, 1968.
11. B. Vucetic and J. Du, "Channel modeling and simulation in satellite mobile communication systems", *IEEE J. on Selected Areas in Communnications*, Vol. 10, No. 8, pp. 1209–1218, 1992.
12. M. Zorzi, "Data-Link dacket dropping models for wireless local communication" IEEE Tran on Vehicular tachnology, vol. 51, No. 4, July 2002.
13. M. Zorzi and R. R. Rao, "Lateness probability of a transmission scheme for error control on a two-state Markov channel", IEEE Tran Commun vol 47, pp. 1537–1548, Oct, 1999.
14. M. Zorzi, R. R. Rao and L. B. Milstein, Onthe accuracy of a first-order Markov model for data transmission on fading channels: Proc. IEEE ICUPC, pp. 211–215, Nov, 1995.

Models for Non-intrusive Estimation of Wireless Link Bandwidth

Jian Zhang[1], Liang Cheng[2], and Ivan Marsic[1]

[1] Rutgers University, Department of Electrical and Computer Engineering
Piscataway NJ 08854, USA
{jianz,marsic}@caip.rutgers.edu
http://www.caip.rutgers.edu/disciple/

[2] Lehigh University, Department of Computer Science and Engineering
Bethlehem, PA 18015, USA
cheng@cse.lehigh.edu

Abstract. Dynamics of link bandwidth of a wireless link, which changes frequently and abruptly due to the dynamic channel sharing, fading, and mobility, is of interest to adaptive network applications and communication protocols. This paper presents a novel approach to estimate wireless link bandwidth based on radio signal-to-noise ratio (SNR). Unlike traditional methods that send probe packets, our method is non-intrusive to the wireless network since in IEEE 802.11 wireless local area networks, SNR information is provided by the physical layer for the MAC- and upper layers' functionality. Theoretical analysis and experimental observation indicate a nonlinear relationship between SNR and the wireless bandwidth. Based on this, nonlinear models using neural network and Bayesian inference methods are proposed and evaluated on data collected in 802.11b wireless networks. The effectiveness of our method under various environments and scenarios has been studied.

1 Introduction

Adaptation is widely recognized as key to overcoming the resource constraints in mobile computing systems [1,9]. Only through effective adaptation can mobile applications overcome mobile resource constraints. A key mobile resource is wireless link bandwidth, and its measurement is important in order to devise effective adaptation techniques. Due to dynamic channel sharing, fading, and user mobility, the bandwidth of a wireless link changes frequently and abruptly. Since wireless link generally exhibits characteristics such as large transmission latency, high packet loss rate, and dynamic link bandwidth, it is challenging to estimate wireless link bandwidth accurately and efficiently.

There exist a number of methods for bandwidth estimation. Most of them can be categorized into two groups [12]. One group is a variant of pathchar [5,8] based on the round-trip times for single packets. The disadvantage of this group is heavy overhead of the probing traffic. The other is a variant of packet-pair [2, 3,13,16,17] based on the packet-pair model. This group imposes lower overhead

M. Conti et al. (Eds.): PWC 2003, LNCS 2775, pp. 334–348, 2003.

relative to the first one. In [11], a multi-packet model is presented to unify the one-packet and packet-pair model and a packet-tailgating technique is proposed.

However, these methods are intrusive to the network since they introduce overhead traffic by sending probe packets during the estimation process. The overhead is not desirable, especially in wireless communication environments where network bandwidth and battery energy are scarce and precious. In addition, due to the dynamics of the wireless link, there is a need for more frequent bandwidth measurement, thus consuming even more resources than in wired networks. Therefore, a non-intrusive bandwidth estimation method is required for wireless networks.

In the existing work on non-intrusive or passive methods, e.g., [11], instead of sending own probe packets, the traffic of other applications running on the same host is observed. However, these methods are inaccurate because they cannot control the size and transmission time of the packets. E.g., [11], estimates the bandwidth based the time interval between packet pair arrivals, which may be inaccurate for small packets due to the coarse granularity of the system clock, thus degrading the estimation performance.

Our objective is to monitor the wireless link as it appears on top of the MAC layer. *Wireless link bandwidth* is defined as the effective transmission bandwidth of a wireless link or *saturated throughput* that can be achieved at the wireless link. The techniques reviewed above measure the nominal or "ideal" channel bandwidth, since they do not account for high error rate and lengthy retransmission time in the MAC layer.

In this paper we propose a non-intrusive method that uses RF signal-to-noise-ratio (SNR) information to dynamically estimate the wireless link bandwidth. The SNR information is provided by most IEEE 802.11 [7] wireless card drivers for the MAC layer and upper layers' functionality. It is updated by each received frame, which is not necessarily a data frame, but could also be a management or control frame. Another salient characteristic of our method is that the estimation accuracy is not influenced by packet sizes, compared to other traditional methods such as packet-pair techniques.

The radio propagation channel exhibits many different forms of channel impairments [18]. Common ones include distortion due to multipath delay spread, Doppler spread, RF (intracell and intercell) interference, shadow fading, ambient noise, etc. It is to be expected that the wireless bandwidth will follow the variations in the received signal power. Experimental results for an IEEE 802.11b network shown in Fig. 1(a) corroborate this intuition. The SNR is read from the network interface card and the actual bandwidth is measured by sending saturated probe traffic, as explained in Section 3 below. Fig. 1(b) indicates that the relationship is nonlinear. The high bandwidth variability visible in Fig. 1(a) is due to the rapid variations in the channel characteristics, which cause packet loss and retransmission. The high spikes represent the nominal bandwidth, i.e., without errors and MAC-layer retransmissions; the low spikes represent the worst case of errors and retransmissions.

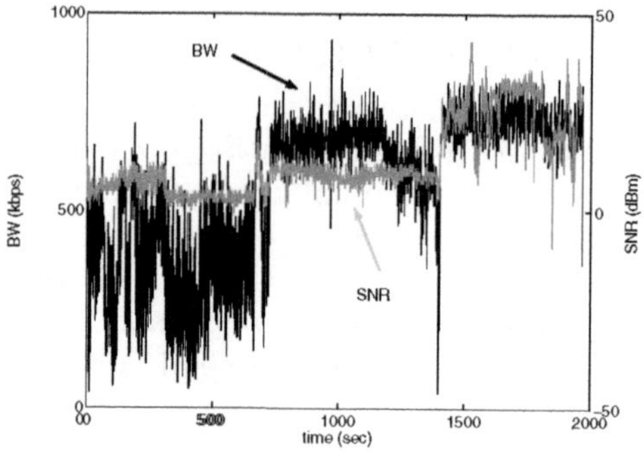

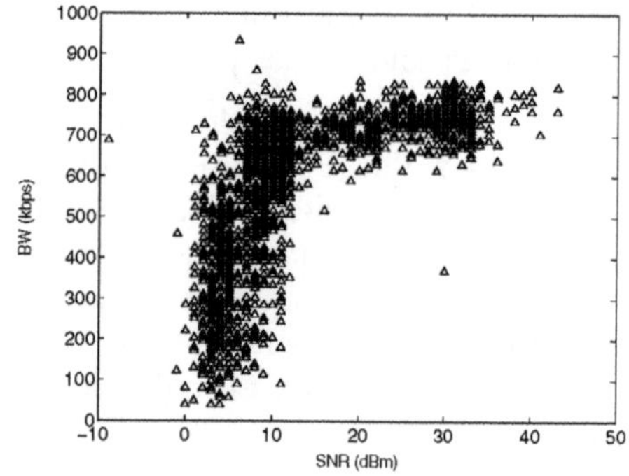

Fig. 1. (a) Example of measured wireless link bandwidth and SNR data. (b) SNR-bandwidth relationship for a typical dataset clearly shows nonlinearity.

Note that our method provides a single-hop rather than end-to-end bandwidth, because SNR only reveals the quality of the immediate wireless link. Nonetheless, this is sufficient for the following reasons. In infrastructure networks, the wireless links are almost always located in the first or last hops. In such networks, the wireless-hop link is usually the weakest link, and its bandwidth information is critical for providing QoS and improving end-to-end performance in such networks [15]. A piecewise mechanism, which considers separately wireless and wired networks, can also be employed [4]. In mobile ad-hoc networks, knowing the bandwidth of individual links is sufficient to determine the best routing path. Link bandwidth information can be exchanged in link-state packets along with topology information.

The rest of the paper is organized as follows. Section 2 presents the theoretical model that shows a relationship of the wireless link bandwidth and SNR. Section 3 describes experiments to collect the data for the empirical bandwidth estimation models and Section 4 presents two empirical models. The effectiveness of the models under different scenarios and environments is discussed in Section 5. Lastly, Section 6 concludes the paper.

2 Theoretical Relationship between SNR and Wireless Link Bandwidth

Here we first demonstrate analytically the relationship between the SNR and wireless link bandwidth. The analysis is based on baseband SNR, which is different from what the 802.11 wireless device drivers provide, i.e., SNR of RF signals. The difference does not invalidate the derived relationship, though.

Signal-to-noise ratio (SNR) in analog and digital communications is a measure of signal strength relative to the background noise. If the signal strength in microvolts is V_s, and the noise level, also in microvolts, is V_n, then the SNR, in decibels, is given as $\text{SNR} = 20\log_{10}(V_s/V_n)$. A corresponding measure in a wireless communication environment is the received bit-energy-to-noise ratio, denoted as $\text{SNR}(t) = \alpha(t)^2(E/N)$, where $\alpha(t)$ is a Rayleigh-distributed random process in the flat fading channel model [18], E is the average bit energy of the received signal, and N is the noise power density.

The bit error rate (BER) is the percentage of bits in error relative to the total number of bits received in a transmission. In the wireless communications, the relationship between BER and SNR depends on the channel modulation scheme [10], such that $\text{BER}(t) = Q[\text{SNR}(t)]$, where $Q[\cdot]$ is a nonlinear operator. E.g., in the $(\pi/4)$QPSK modulation scheme, $Q[\cdot]$ can be expressed as

$$Q(p) = 1\sqrt{2\pi}\int_{\sqrt{2p}}^{\infty} e^{-\frac{x^2}{2}}\,dx$$

The relationship between packet error rate (PER) and BER depends on the channel coding scheme. Assume that there is no error-correction coding applied and the number of bits in a packet is n, then

$$\text{PER}(t) = 1 - \prod_{i=1}^{n}[1 - \text{BER}(t_i)]$$

where t_i is the time instant the receiver received the ith bit. If the wireless channel fading variation is very slow compared to the packet transmission time (true for walking and driving speeds), then PER can be approximated as $\text{PER}(t) = 1 - [1 - \text{BER}(t)]^n$. Thus $\text{PER}(t)$ gives the probability of packet loss at time t on the wireless channel and can be viewed as $\text{PER}(t) = P[\text{BER}(t)]$, where $P[\cdot]$ is a nonlinear operator.

The relationship between the wireless link bandwidth and PER depends on MAC layer protocols, such as automatic retransmission (ARQ) scheme. Generally a packet is kept in the transmitter queue until it is successfully received and acknowledged by the receiver. According to the ARQ scheme, a new packet will not be sent to the receiver until either (*i*) a timer times out, or (*ii*) the previous packet has been successfully received and acknowledged by the receiver and the sender received the acknowledgement. Therefore the channel efficiency η, which is the percentage of time the channel is used to effectively communicate data between the transmitter and the receiver, can be derived as follows.

Assume that t_p and t_a are the times to transmit a packet and to transmit an ACK, respectively. Furthermore, t_{proc} and t_{prop} are the packet processing time at the end-hosts and the packet propagation time across the channel, respectively. Let t_{to} and t_s denote timeout value of the timer and the minimum time between successive packet transmissions at the sender in case of no packet error, respectively. Both t_{to} and t_s are set differently for different ARQ schemes. Then define t_s^* as the maximum value of t_s across all ARQ schemes, which is $t_s^* = t_p + 2t_{proc} + 2t_{prop} + t_a$. Define x as the total time taken for a successful transmission of a packet and its ACK, which is a random variable depending on the dynamics of the PER. Thus,

$$E|x| = (1 - PER)t_s + PER(t_{to} + E|x|) = t_s + (PER \times t_{to})/(1 - PER)$$

Table 1. Parameters for channel efficiency.

ARQ scheme	t_s	t_{to}	η
S&W	t_s^*	t_s^*	$(1 - PER) \times t_p/t_s^*$
GBn	t_p	$n \times t_p$	$(1 - PER)/(1 - PER + n \times PER)$
SRP-infinity	t_p	t_p	$1 - PER$

The channel efficiency can be expressed as follows:

$$\eta(\text{ARQ, PER}) = \frac{t_p}{E[x]} = \frac{(1 - \text{PER})t_p}{(1 - \text{PER})t_s + \text{PER} \times t_{to}} \tag{1}$$

For stop-and-wait (S&W), go-back-n (GBn), and selective repeat with infinite buffer (SRP-infinite) schemes, the values of the parameters are set according to the Table 1.

The above analysis assumes that the transmitter sends a frame to the receiver once the communication channel is detected as idle. However, in some protocols, such as IEEE 802.11, after the transmitter detects the idle wireless channel, it still has to wait for a random period of time before sending out the frame. Thus (1) can be generalized as:

$$\eta(\text{ARQ, PER}) = \frac{t_p}{E[x]} = \frac{(1 - \text{PER})t_p}{t_0 + (1 - \text{PER})t_s + \text{PER} \times t_{to}}$$

where t_0 is the random time before the transmitter sends out the frame during the idle channel period. Then the *wireless link bandwidth* can be expressed as:

$$B(t) = \eta[\text{ARQ, PER}(t)] \times C$$

where C is the wireless link capacity. The relationship between the wireless bandwidth and the signal strength can be denoted as (2) and it can be simplified as (3):

$$B(t) \overset{\eta[\cdot]}{\longleftrightarrow} \text{PER}(t) \overset{P[\cdot]}{\longleftrightarrow} \text{BER}(t) \overset{Q[\cdot]}{\longleftrightarrow} \text{SNR}(t) \tag{2}$$

$$B(t) = f[\text{SNR}(t)] \tag{3}$$

where $f[\cdot] = \eta[P[Q[\text{SNR}(t)]]]$ or $f = \eta \circ P \circ Q$. Therefore, (3) shows that there is a well-defined relationship between the SNR and the wireless link bandwidth.

Since the equation from RF signal SNR to BER depends on how the wireless receiver implements components of synchronization, interference handling, equalization, and demodulation, it is not possible in practice to build the exact theoretical model for bandwidth estimation. For this reason, below we consider empirical methods for bandwidth estimation. Nonetheless, the above analysis is important to back our intuition about the existence of the relationship.

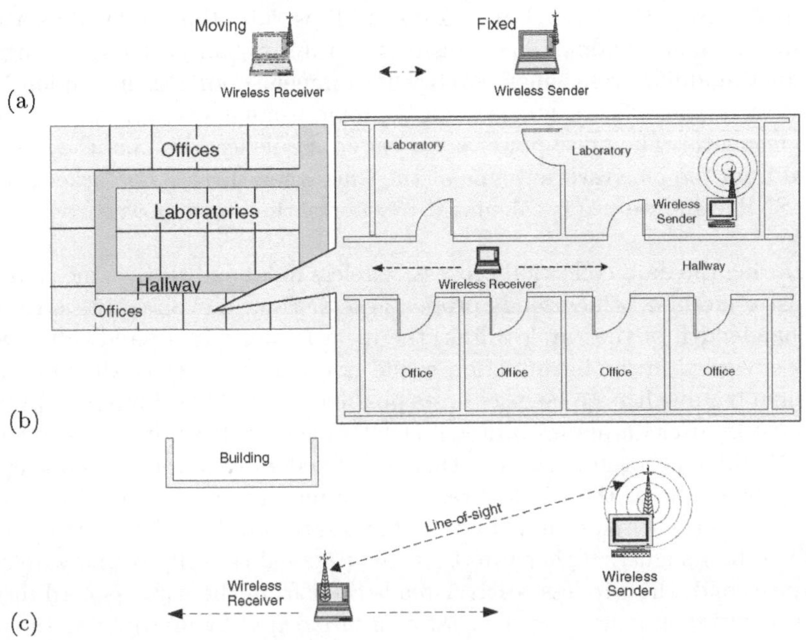

Fig. 2. Layouts of the wireless network for experiments. (a) Topology of the wireless/mobile LAN. (b) Physical layout of the wireless/mobile LAN (Indoors case). (c) Physical layout of the wireless/mobile LAN (Outdoors case).

3 Experiments for Training-Data Collection

Before presenting the empirical bandwidth estimation models, we first describe the experiments for collecting the data that will be used to train the models and verify the estimates. Fig. 2 shows the layouts of our experimental IEEE 802.11b wireless/mobile LANs. In the indoors case (Fig. 2(b)), the mobile laptop travels around the fixed wireless sender along the hallway and enters the offices and the labs. During each session a dataset of about 500 samples is recorded. The maximum distance between the receiver and the sender reaches 100 feet. Generally, there is no line-of-sight (LOS) communication between the sender and the mobile laptop. In the outdoors case (Fig. 2(c)), the mobile receiver travels with varying speeds around the sender with a line-of-sight wireless connection.

The experiments were performed in the peer-to-peer (ad-hoc) mode for 802.11 wireless LAN. About 20 datasets of SNR and corresponding wireless link bandwidth recorded at the mobile receivers have been collected. Various IEEE 802.11b wireless cards are used in the experiments, such as Sony VAIO, Orinoco Gold, and Compaq WL 110 cards. The times of day for the data collection are randomly chosen over the period of experiments. To avoid other traffic influencing the recordings, our testbed is isolated and has only one sender and one receiver operating on a fixed channel. The recordings here are based on one single transmission rate that has been set manually to 1Mbps. In reality, the raw bit rate is automatically set to a level of 1 Mbps, 2 Mbps, 5.5 Mbps or 11Mbps according to current signal quality. The transmission rate is changed by switching among different modulation schemes. We briefly comment upon this in Section 5 below.

Most of the 802.11b wireless cards provide signal strength and noise power independently. The noise power is measured as silence level, i.e. the power measured from the receiver's antenna at the time when there is no packet coming in. The SNR value we use is calculated by subtracting noise power from the signal power.

During the data collection process, wireless link bandwidth is measured by an intrusive method, which sends probe packets. Since our objective is to monitor the bandwidth of the wireless link, the methods for wired bandwidth measurement reviewed in the introduction would not work here, since they measure the nominal bandwidth. To be as close as possible to the MAC layer, the probe traffic must be user datagram protocol (UDP) packets. We use a constant bit rate (CBR) traffic generator [14] that runs on a fixed wireless sender and keeps sending probe traffic to the wireless receiver. To measure the maximum throughput that reflects the wireless link bandwidth, we saturate the link by setting the CBR traffic rate a slightly higher rate than the physical capacity of the wireless link. Every second, the wireless receiver reads SNR from the wireless card driver. At the same time, it calculates the *saturated throughput* by multiplying the number of packets it has received during the 1 sec period with the sum of the sizes of IP and UDP headers and the UDP payload. This is recorded as the actual link bandwidth for a given SNR. An example measurement is shown in Fig. 1(a). A 1-second sampling period is chosen only for computational convenience. While doing estimation, the sampling rate can be higher than once per second depending on the requirements of the application and it can increase up to the rate of updating the SNR information, i.e., the frame arrival rate. The recorded SNR and bandwidth data are used to perform model identification.

4 Empirical Nonlinear Models

System identification is about modeling the relationship between inputs and outputs. When building a model for recorded datasets, we try to build a model such that the relationship of its output (estimated bandwidth) to its input (SNR) matches what is exhibited by most of the points in the datasets. When the model is used in estimation, real-time measured SNRs are fed into the model and the generated outputs are estimates of bandwidth. In order to evaluate the accuracy of the model, the estimated values are compared to the actual bandwidth, which is again measured by sending probe packets as in Section 3 above.

We use absolute mean error and average relative error as indices for performance evaluation. The average relative error is defined as $\left(\sum |y_i - \hat{y}_i|\right) / \sum y_i$, where y_i and $\hat{y}_i$ are the desired and estimated outputs at i^{th} point, respectively. Since both the empirical data (Fig. 1(b)) and the analytical model (Eq. (3)) indicate that the relationship is nonlinear, we employ nonlinear estimation methods. Two methods, neural networks and statistical inference, are described below.

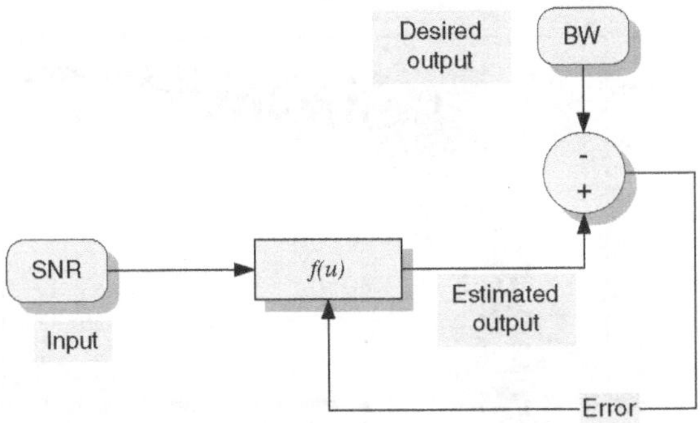

Fig. 3. Error-correction learning model for neural networks.

4.1 Back-Propagation Neural Network (BPNN)

Back-propagation neural network (BPNN) has been applied successfully to solve some difficult problems of input-output nonlinear mapping, curve fitting, etc. [6]. For BPNN, the nonlinear function $f(u)$ consists of a series of units combining linear operations (weights and bias) and nonlinear operations (sigmoidal functions). These units are called perceptrons and organized layer by layer. The network is trained in a supervised manner with an algorithm known as the error back-propagation algorithm based on error-correction learning rule. The training process is illustrated in Fig. 3, where $f(u)$ is given as an initial model. A set of input values and corresponding desired output values are used as a training dataset. The difference between desired output values and the output of $f(u)$ is considered an error and propagated back to the model. Based on the errors, the learning algorithm adjusts the parameters in function $f(u)$ and finally minimizes the sum of square errors of all training points, i.e., $\sum (y_i - \hat{y}_i)^2$.

The reason we choose BPNN for nonlinear modeling is that the relationship of measured bandwidth vs. SNR, shown in Fig. 4, resembles a sigmoidal function. Therefore, with a simple structure, i.e., small number of perceptrons and layers, and a short training time, the BPNN can learn to represent the relationship. Moreover, the computation complexity of the estimation is also low. We may notice in Fig. 4 that the SNR-to-bandwidth mapping is not one-to-one, i.e., there could be different bandwidth values corresponding to a single SNR value. Given a set of training points, a BPNN learns a one-to-one sigmoidal mapping that minimizes the sum of square errors.

In this paper, a 1-4-1 BPNN is used for modeling, i.e., one input, 4 hidden nodes, and 1 output. The model is first trained on a certain dataset and then it is used to generate estimates by feeding other datasets of recorded SNRs. The estimates on each dataset are compared with the corresponding actual measured link bandwidth, and errors are presented in the form of relative and absolute mean value. This is shown in Fig. 5. The relative error between the estimated bandwidth and the actual measured bandwidth in Fig. 5(a) is 14.81%. Fig. 5(b)

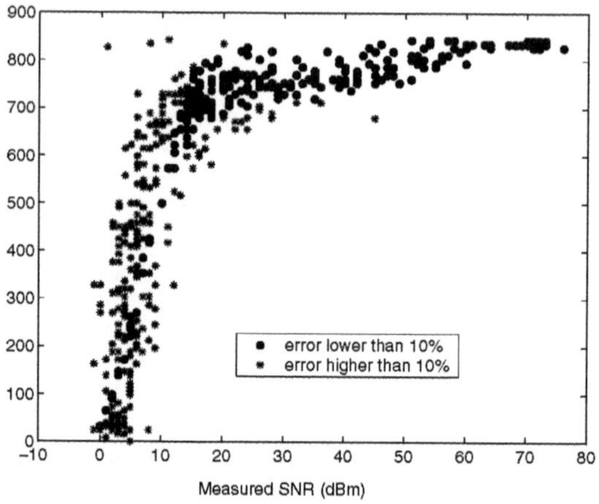

Fig. 4. Least sum of square errors fitting obtained by BPNN.

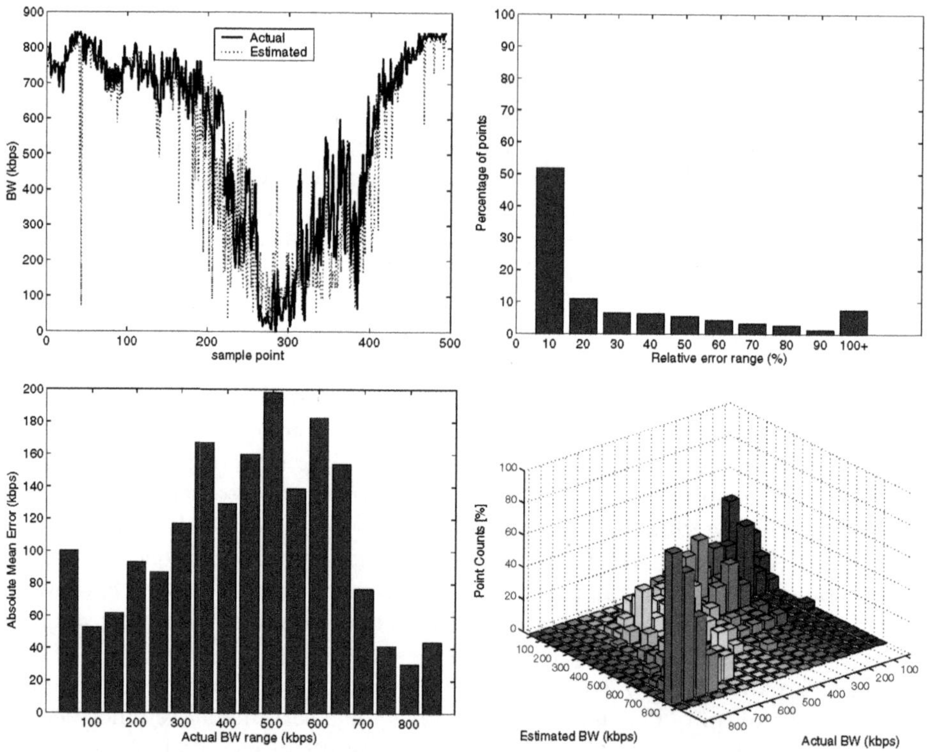

Fig. 5. Performance of BPNN model: (a) Comparison of measured and estimated bandwidths; (b) The distribution of relative errors; (c) Mean absolute error for estimations; (d) 2-D histogram.

shows the distribution of estimates in different relative error ranges. We may notice that for about 50% of estimates relative error is less than 10%. For very low bandwidths, even a small absolute error means a large relative error. Therefore, Fig. 5(c) shows the mean absolute errors for different values of actual bandwidth. The largest mean absolute error is around 200 Kbps and it is located in the bandwidth range from 320 Kbps to 600 Kbps, which means bandwidths in that region are more difficult to track by SNRs. Lastly, Fig. 5(d), shows the model performance as a 2-D histogram. Each bar counts the number of different estimates for a given actual bandwidth. The bars close the diagonal (estimated = actual) are the tallest, indicating that the model works well for this dataset.

We repeated testing the BPNN model by training it on 5 datasets and applied it to do estimation on 5 different datasets. The average relative error obtained is 24.94% and the standard deviation of error is 13.57%.

4.2 Statistical Model: Bayesian Inference

Considering the collected SNR data as observations and the corresponding measured link bandwidth as output states, we can build a Bayesian inference model. Firstly, we divide the ranges of input data and output data into small regions, and represent them by a finite state space $s_1, s_2, ..., s_n$ and an observation space $o_1, o_2, ..., o_m$. Then, based on the historical data, a Bayesian inference model is built by calculating the priority distribution: $Pr(o_i|s_j)_{i=1,2,...,m;j=1,2,...,n}$. This gives the probability of a certain SNR value at region o_i observed in the condition that real or measured bandwidth is at state s_j, based on collected data.

To perform the estimation, we use the following Bayes rule to update the probabilities, which represent the new level of belief in all possible outcome states after an observation:

$$Pr(s_j|o^*)_{j=1,2,...n} = \frac{Pr(s_j)Pr(o^*|s_j)}{\sum_{i=1}^{n} Pr(s_i)Pr(o^*|s_i)}$$

where o^* is the current observation; $Pr(s_j)$ is the prior probability and it is initialized by assuming the each state has the same probability in the beginning.

The output of the Bayesian inference model provides information of the probability distribution of all possible states of estimated bandwidth. We calculate the expected value of the probability distribution and take it as the estimated bandwidth. Similar to the BPNN model, one dataset with both measured SNR and measured bandwidth is used to train the model, i.e., to build the priority distribution. Then, the estimations are made on other datasets and result compared with the measured actual bandwidth. The training and estimation datasets used to obtain Fig. 6 are the same as the ones used with the neural network model. Here, the relative error is 15.70%.

We repeated testing the Bayesian model by training it with 5 datasets and applying it to do estimation on 5 other datasets. The average relative error obtained is 26.07% and standard deviation of error is 15.43%. This is close to what we obtained with the BPNN model. Fig. 7 explains the similarity, since the

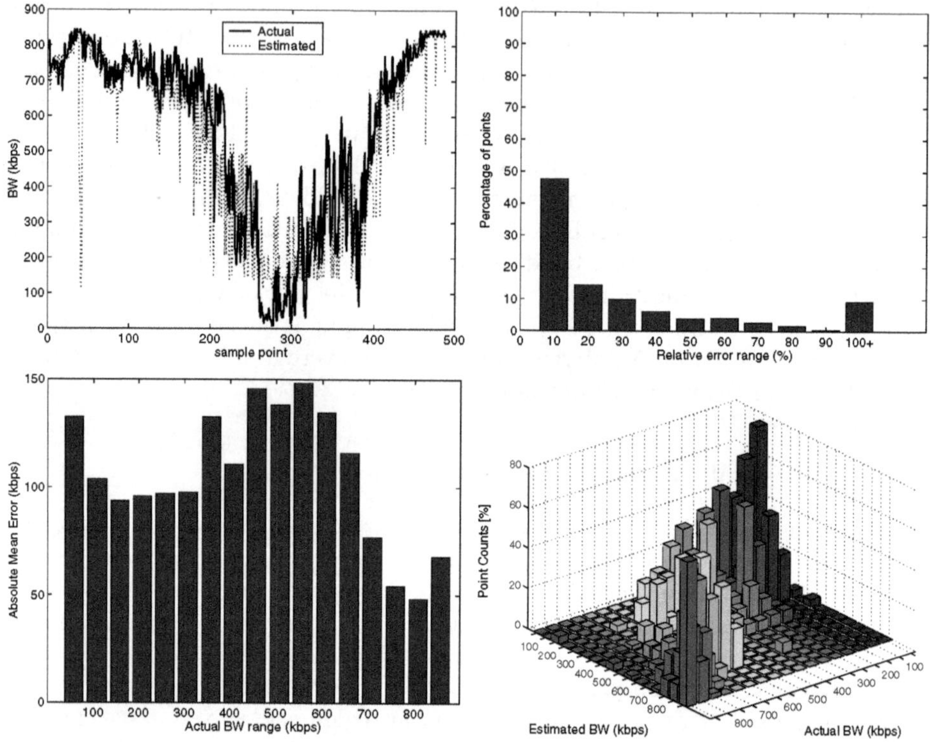

Fig. 6. Performance of the Bayesian inference model: (a) Comparison of measured and estimated bandwidths; (b) The distribution of relative errors; (c) Mean absolute error for estimations; (d) 2-D histogram.

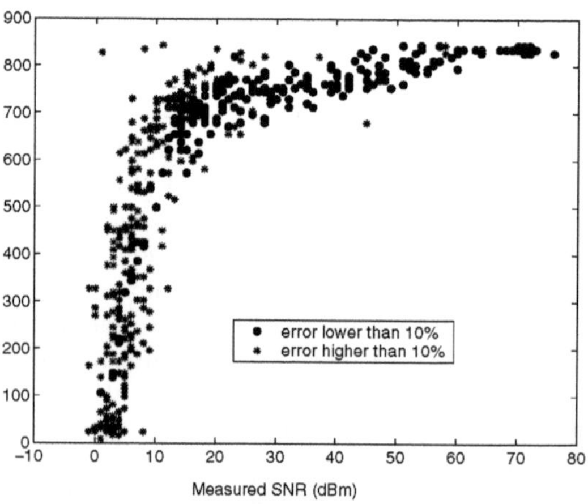

Fig. 7. The Bayesian model mapping resembles the least sum of square errors fitting.

Table 2. Performance comparison of neural network and Bayesian inference models. Shown are the error rates for each. The performance on some of the datasets is significantly worse than on others, e.g., Test data III. This is since the datasets are not balanced, so some datasets contain more points in the "bad" region (SNR < 15 dBm).

BPNN **Bayes**	**Test data I**	**Test data II**	**Test data III**	**Test data IV**	**Test data V**
Training data I	15.32% 16.08%	15.98% 17.80%	44.69% 56.22%	22.74% 24.06%	22.46% 21.03%
Training data II	15.91% 15.96%	15.48% 16.18%	45.31% 49.97%	24.09% 21.83%	23.66% 20.62%
Training data III	15.48% 15.70%	15.82% 16.18%	46.85% 49.97%	23.62% 21.83%	23.77% 20.62%
Training data IV	14.69% 15.87%	16.00% 19.77%	52.28% 60.05%	19.39% 18.78%	20.54% 18.98%
Training data V	15.15% 15.78%	16.10% 20.91%	61.71% 62.02%	17.20% 17.84%	19.20% 17.82%

mapping curve obtained by the Bayesian model resembles the least-square-error curve obtained by the BPNN in Fig. 4.

The performances of these two models are compared in Table 2. In the columns list the training datasets and the rows list the estimation datasets. Relative errors caused by BPNN model and Bayesian model corresponding to different datasets are shown respectively in the upper line and lower line of each cell. The table shows their performances parallel each other.

The performance of both models is particularly weak at the steep part of the SNR-bandwidth curve, for SNRs lower than 15 dBm, Figs. 4 and 7, where the mapping is obviously not one-to-one. We tried to improve this performance by designing the models with multiple input variables. One possibility is to use the signal and noise strengths as two independent inputs, rather than a single SNR input. Another possibility is to use past values of SNR or bandwidth some of their statistical properties as additional input variables for the model. So far these efforts only resulted in minor performance improvements. However, we still believe that the model performance at this part of the curve can be improved only through additional input data and this is a topic of our current research.

It may be argued that these models are suitable only to the extent that they estimate a "break point" between a good and bad link. However, although accurate bandwidth estimation for low SNRs is hard to achieve, having at least average bandwidth estimate is useful for many applications. In addition, for the case of automatic rate selection (11 Mbps, 5.5 Mbps, 2 Mbps, and 1 Mbps), the "break point" is not as easy to identify since the performance deteriorates rather gradually. Even if it could be located, it would not provide sufficient information for link characterization for bandwith-sensitive applications.

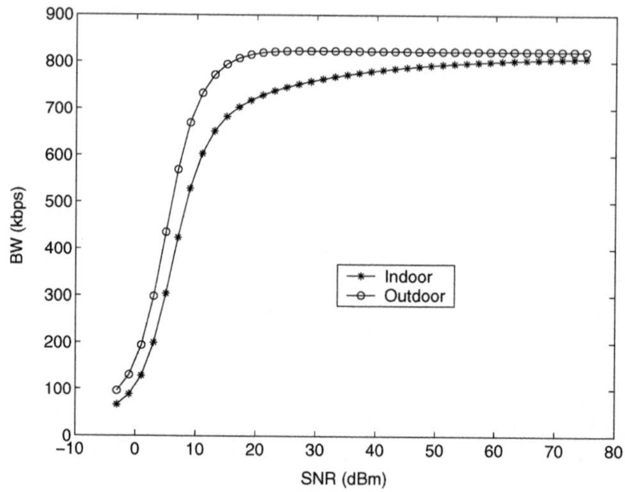

Fig. 8. The least sum of square errors fittings for indoors and outdoors datasets.

5 Model Effectiveness under Different Scenarios

Since both BPNN and Bayesian models are built or trained offline on previously collected data and there is no actual bandwidth information available at runtime when doing estimation, it is important to know whether the environmental changes could cause a performance deterioration. The first experiment was in training the models on data measured in one building and using them for estimation in a different building. The average relative error is 17.6%.

Fig. 8 shows the input-output curves for the models trained on all indoor collected data and all outdoor collected data. Obviously, there is a slight difference and this difference contributes to the performance degradation when using a model trained on one environment for estimation in another one:

Estimation → **Training ↓**	**Indoor**	**Outdoor**
Indoor	24.94%	33.04%
Outdoor	32.98%	31.97%

A possible solution is to employ a dual-model mechanism. When mobile nodes are located outdoors, the system employs the model trained on outdoors data and vice versa for indoors. In this case, it is important to automatically recognize the environment change and switch to a proper model and this is a possible future research topic.

Experiments similar to those reported above for peer-to-peer (ad-hoc) mode were also performed for the infrastructure (access point) mode. The results remain virtually the same if one of the endpoints is connected to the access point over a wire. If both endpoints communicate wirelessly over the access point, the performance changes due to channel sharing and different spatial relationships to the access point. This case needs to be investigated further.

In addition the original 802.11-DS standard, 802.11HR offers additional 5.5 Mbps and 11 Mbps for high rate transmission [7], where a different modulation (CCK) is adopted. To extend our models to a multi-rate situation, we repeated the experiments with non-fixed transmission rates. Experiment results show that the method performs well in multi-rate case and the average relative error in our tested cases is 18.58%. Future research includes extending the method to other wireless LAN standards, such as 802.11g.

Since our method estimates the link bandwidth at the MAC layer, it can be used with different network or transport protocols, including TCP. However, in such a case, the estimated bandwidth does not necessarily predict the throughput that will be achieved by the transport protocol. We evaluated the estimation models developed here with TCP traffic. The measured TCP throughput sharply falls off for low estimated bandwidth because of the congestion control mechanisms reacting to highly increased packet loss under a low link bandwidth. This relationship needs to be further investigated.

6 Conclusions

The dynamics of the wireless link bandwidth is of interest to adaptive applications and protocols in wireless communication environments. This paper presents a novel approach to estimate wireless link bandwidth based on radio signal-to-noise ratio. Its salient feature is that it provides a passive, non-intrusive method for wireless link bandwidth estimation, which avoids the use of probe packets as in traditional methods. The method is evaluated in IEEE 802.11b wireless local area networks under different environmental conditions. Two nonlinear models, BP neural network and a statistical model, are used and their performance is studied and compared. Both BPNN and Bayesian inference models achieve similar accuracy in bandwidth estimation. Future work includes improving the accuracy of the estimation method under different scenarios and employing it in real-world applications to provide quality-of-service. Another interesting parameter related to signal strength is the link latency, which we also plan to estimate. Also, the scenarios with multiple senders and receivers will be studied.

Acknowledgements. The research is supported by NSF Contract No. ANI-01-23910, US Army CECOM Contract No. DAAB07-02-C-P301, a research grant from Cisco, Inc., and by the Rutgers Center for Advanced Information Processing (CAIP).

References

1. Badrinath, B., Fox, A., Kleinrock, L., Popek, G., Reiher, P., Satyanarayanan, M.: A conceptual framework for network and client adaptation. *Mobile Networks and Applications (MONET)* **5** (2000) 221–231
2. Bolot, J. C.: End-to-end packet delay and loss behavior in the internet. *Proc. ACM SIGCOMM'93 Conf. Applications, Technologies, Architectures, and Protocols for Computer Communications* (San Francisco CA, September 1993) 289–298
3. Carter, R. L., Crovella, M. E.: Measuring bottleneck link speed in packet-switched networks. *Performance Evaluation* **27/28** (1996) 297–318
Also appears as Technical Report TR-96-006, Department of Computer Science, Boston University, Boston, MA, March 1996
4. Cheng, L., Marsic, I.: Piecewise network awareness service for wireless/mobile pervasive computing. *Mobile Networks and Applications (MONET)* **7** (2002) 269–278
5. Downey, A. B.: Using pathchar to estimate internet link characteristics. *Proceedings of the ACM SIGCOMM'99 Conference on Applications, Technologies, Architectures, and Protocols for Computer Communications* (Cambridge MA, August/September 1999) 241–250
6. Hertz, J., Krogh, A., Palmer, R. G.: *Introduction to the Theory of Neural Computation.* Addison-Wesley Publ. Co., Reading, MA, 1991
7. IEEE Computer Society, LAN MAN Standards Committee: IEEE Standard 802.11–1999, 802.11 Wireless LAN Medium Access Control (MAC) and Physical Layer (PHY) Specifications (1999)
8. Jacobson, V.: Pathchar – A tool to infer characteristics of internet paths. Mathematical Sciences Research Institute (MSRI) Math Awareness Week (Mathematics and the Internet), April 21, 1997
Online presentation and software at: ftp://ftp.ee.lbl.gov/pathchar
9. Katz, R. H.: Adaptation and mobility in wireless information systems. *IEEE Personal Communications* **1** (1994) 6–17
10. Kim, Y. Y., Li, S.: Modeling multipath fading channel dynamics for packet data performance analysis. *Wireless Networks* **6** (2001) 481–492
11. Lai, K., Baker, M.: Nettimer: A tool for measuring bottleneck link bandwidth. *Proc. 3rd USENIX Symposium on Internet Technologies and Systems* (San Francisco CA, March 2001) 122–133
12. Lai, K., Baker, M.: Measuring link bandwidths using a deterministic model of packet delay. *Proc. ACM SIGCOMM 2000 Conf. Applications, Technologies, Architectures, and Protocols for Computer Communications* (Stockholm Sweden, August 2000) 283–294
13. Lai, K., Baker, M.: Measuring bandwidth. *Proc. Conf. on Computer Communications (IEEE INFOCOM'99)* (New York NY, March 1999) 235–245
14. Naval Research Laboratory (NRL): The Multi-Generator Toolset MGEN-3.2 User's Guide, Washington, DC
Online at: http://manimac.itd.nrl.navy.mil/MGEN/MgenUserGuide.html
15. Qiu, L., Bahl, P., Adya, A.: The effect of first-hop wireless bandwidth allocation on end-to-end network performance. Proc. NOSSDAV'02 (Miami FL, May 2002)
16. Paxson, V.: End-to-end internet packet dynamics. *IEEE/ACM Trans. Networking* **7** (1999) 277–292
17. Paxson, V.: Measurements and Analysis of End-to-End Internet Dynamics. Ph.D. Dissertation, Computer Science, University of California, Berkeley, CA, April 1997
18. Rappaport, T. S.: *Wireless Communications Principles and Practice.* 2nd edition. Prentice-Hall, Englewood Cliffs, NJ, 2002

Supporting Proactive Location-Aware Services in Cellular Networks

Axel Küpper[1], Florian Fuchs[2], Michael Schiffers[1], and Thomas Buchholz[1]

[1] Munich Network Management Team, University of Munich, Munich, Germany
{kuepper,schiffers,buchholz}@informatik.uni-muenchen.de
[2] Department of Informatics, Munich University of Technology, Munich, Germany
fuchs@cs.tum.edu

Abstract. Location-aware services (LAS) adapt content and service execution to the user's current physical location. In today's cellular networks, only reactive LASs are offered. The user has to explicitly request the service before the network determines her current location and starts service execution. However, in the future the user would like to see proactive LASs, which are automatically triggered as soon as she enters a predefined point of interest. In contrast to reactive LASs, this requires that the network performs a continuous user tracking, which entails expensive point-to-point signaling at the air interface.
However, proactive LASs can be supported more efficiently if the mobile terminal locates itself. The terminal can derive its own location by processing location information which is broadcasted by the network or by a dedicated localization infrastructure like GPS. The network does not need to know where the user is, unless she has entered a point of interest. This paper follows this idea and presents the Location-Aware Mobility Architecture (LAMA), a platform for deriving and processing location information at the mobile terminal.

1 Introduction

"Pst, you have 'Buy new album of Madonna' on your To-Do list. This album is on offer at the record store on your right. Don't you want to drop in?" This could be a notification generated by a so-called *location-aware service* (LAS), which is offered by a service provider and delivered by the operator of a cellular network like GSM or UMTS through a *mobile terminal* (MT). In contrast to conventional mobile services, LASs adapt service execution or content delivery to the user's current physical location and thus allow a new dimension in service personalization.

Today's LASs are offered as *Reactive LASs* (RLAS) only. Location-aware information is only passed on demand to the user. She has to know by herself or at least assume that there may be a *point of interest* (e.g., a record store) nearby before going online to establish a service session and to request the LAS in order to retrieve the desired information. However, from the point of view of the user,

M. Conti et al. (Eds.): PWC 2003, LNCS 2775, pp. 349–363, 2003.

Proactive LASs (PLAS) are desired, which are automatically initialized as soon as she enters a pre-defined point of interest (as described in the example above). This requires that the user fixes her points of interest prior to service usage in a *location profile* when subscribing for a PLAS. Such a location profile may be derived from the user's individual preferences, for example "interest in vinyl records". From these preferences, a PLAS service provider derives points of interest and assembles them in the user's location profile. In this example, this could be the locations of record stores offering rare vinyl records. When the user approaches one of her points of interest, she is automatically alarmed by the service provider and can seize the opportunity of passing by. While RLASs are widely extended in cellular networks in the meantime, PLASs have been barely realized so far, which is due to the technical limitations of positioning methods used in these network. Nevertheless, PLASs are increasingly gaining momentum in numerous research and standardization activities, see [1], [2], and [3] for example.

From this description, a more detailed comparison between RLASs and PLASs can be derived. RLASs are invoked by the user from a front-end application of her mobile device, for example a WAP browser. Upon service invocation, the network derives the user's location and service execution is triggered. However, PLASs usually must be triggered at service components located at the fixed-network site, e.g., at a WAP gateway, as soon as the user enters a point of interest. In contrast to RLASs, the network does not need to derive the user's location just once, but must locate the user continuously, which is called *user tracking* in the following. In addition, each location information derived in this way must be checked against the user's location profile, which is referred to as *location monitoring* in the following.

Obviously, user tracking is the crucial point when realizing PLASs. To keep the network informed about a user's location requires dedicated *point-to-point signaling* between the network and the MT. Using RLAS, the overhead associated with point-to-point signaling can be neglected, as the user needs to be located only once or twice. However, if executed on a regular basis and for a substantial amount of subscribers, the signaling traffic dramatically increases and significantly burdens the air interface. In [1], this problem is referred to as "the dilemma of cell-based proactive location-aware services". Furthermore, whenever the network obtains new location information, location monitoring must be activated, which causes further signaling and additionally occupies computational resources.

To efficiently support PLASs, we propose the *Location Aware Mobility Architecture* (LAMA). The basic idea behind LAMA is that the network does not need to know where the user is, unless she has reached one of her points of interest. This implies that the MT is always aware of its own location and that it knows the user's points of interest. If and only if the user enters a certain point of interest, LAMA then signals this event to associated service components at the fixed-network site and triggers service execution in this way. Thus, LAMA combines user tracking with location monitoring at the MT. For an efficient user tracking, LAMA utilizes localization by *broadcast signaling*, where location information is broadcasted in each cell of the network. From this information, LAMA can derive the current location and thereby avoids dedicated point-to-point signaling between the MT and the network.

The remainder of this paper is organized as follows. The subsequent section gives an overview of available technologies for localization, which are called *positioning methods*, and discusses their pros and cons with regard to PLASs. Section 3 introduces the system design of LAMA, i.e., its components and the interactions between them. Section 4 describes our LAMA prototype realization and demonstrates the deployment of LAMA by means of an application scenario. LAMA has been evaluated in a field test, which is described in Sect. 5. Finally, Sect. 6 briefly covers related work, and Sect. 7 gives an outlook on further work.

2 Positioning Methods

A positioning method comprises the infrastructure, physical fundamentals (i.e., the propagation delay of signals), and the technical process for deriving a user's location. Each tracking process applies a certain positioning method and fixes its parameters, such as the frequency of positioning and the user it is applied for. Cellular systems offer various alternatives for positioning, see [4] and [5] for example, but not all of them are applicable for PLASs.

Table 1 gives an overview of positioning methods applied in GSM and UMTS networks and lists some of their features. It is distinguished between *network* and *terminal-based positioning*, which denotes whether the location is derived at the network or at the MT. A further classification distinguishes in which mode of operation positioning is done. If it is executed in the context of an existing or forthcoming connection between the MT and the network, i.e., if the MT is busy, it is referred to as *online positioning.* If the MT is merely registered with the network, but no connection exists, i.e., if the MT is idle, it is called *offline positioning.* Note that only offline positioning is applicable for PLASs, because a connection is not established before the user has entered one of her pre-defined locations.

Furthermore, Tab. 1 shows for each positioning method the kind of signaling (broadcast vs. point-to-point), the accuracy, and the format location information is delivered in. A location format describes how location is represented and is usually based on a certain reference system. If this reference system is a coordinate system, location is available in a *physical format.* Typical examples are the *World Geodetic System 1984* (WGS84), *Universal Transversal Mercator* (UTM) and *Gauß-Krüger* (GK) coordinate systems. A *symbolic format* describes an object's location by abstract names or identifiers. In cellular networks, usual symbolic formats are *cell* and *location-area identifiers* (CIs and LAIs).

As can be derived from this table, some methods are based on broadcast signaling, for example LAI, CI, and GK-broadcast positioning. LAI-broadcast refers to the partitioning of an operator's coverage area into *location areas*, each represented by an unambiguous LAI. A location area comprises a sufficient number of cells and each cell, in turn, is assigned another identifier, the CI. Each

Table 1. Overview of positioning methods and their features

Positioning method	Net-work	Ter-minal	On-line	Off-line	Signaling	Accuracy	Format	Appl. for PLAS
LAI-broadcast		×	×	×	broadc.	5–15km	symbolic	yes
LA-update	×		×	×	p2p	5–15km	symbolic	yes
Paging	×			×	p2p	500m–10km	symbolic	no
CI-broadcast		×	×	×	broadc.	500m–10km	symbolic	yes
Cell-update	×		×		p2p	500m–10km	phy./symb.	no
GK-broadcast		×	×	×	broadc.	500m–10km	physical	yes
TA	×	×	×		p2p	10m–10km	phys./symb.	no
TOA	×		×		p2p	20m	physical	no
E-OTD		×	×	×	broadc.	20m	physical	yes
GPS		×	×	×	broadc.	20m	physical	yes

base station permanently broadcasts the LAI and CI of its location area and cell on a signaling channel. A MT is permanently listening to these channels, be it in idle or in busy mode, and is therefore always aware of these identifiers. Unfortunately, LAI and CI are symbolic formats, which is not very convenient for LASs. To overcome this lack, base stations can additionally broadcast location information in a physical format. For example, this is done in the German o2 network, where each base station transmits its own GK-coordinates using the SMS cell broadcast service [6].

To enable the network to locate the user, dedicated point-to-point signaling is required. This operation is called *location update* and is initialized each time the MT recognizes that either the LAI or the CI has changed. In the former case, it is called *location-area update* (LA update), in the latter *cell update.* Cellular networks only apply LA updates, which cause a moderate signaling overhead in contrast to cell updates. Because cell updates are not applied, the network must page the user's MT to determine her current cell. *Paging* is another positioning method that is based on point-to-point signaling. Like LA-updates, it is executed in the context of location management if a mobile-terminated call arrives for a user. For a deeper introduction into location management of cellular networks, see [7] and [8].

The positioning methods presented so far show only a moderate accuracy, which is not adequate for many LASs. Higher accuracy can be achieved if CI-broadcast or cell-update positioning are combined with the *timing advance* (TA) parameter, which refers to the distance between an MT and the serving base station. It is actually needed for synchronizing the TDMA medium access between MTs located at different distances to the serving base station, but can also be utilized for refining the accuracy of positioning. The distance is measured by means of the signal propagation delay and transmitted from the serving base station to the MT. It requires a dedicated point-to-point signaling and is only applicable for online-positioning.

A further improvement of accuracy can be achieved with triangulation. It is applied in the *Global Positioning System* (GPS), where the current location is derived by signal propagation delays between at least three satellites and a GPS receiver. The principle of triangulation can also be realized by the cellular network, where the signal propagation delays to or from at least three neighboring base stations are processed. The triangulation method is called *Time of Arrival* (TOA) if the measurements and calculations for determining the location are done in the uplink, i.e., by the network. If it is done in the downlink by the MT, it is called *Enhanced Observed Time Difference* (E-OTD). TOA is only applicable if the MT is busy and requires point-to-point signaling. This is not necessary if GPS and E-OTD are used, because all parameters required to calculate the location are broadcasted in the downlink.

To draw a conclusion, user tracking should be supported by positioning methods based on broadcast signaling. This kind of signaling easily enables a MT to locate itself and, in contrast to point-to-point signaling, is independent of the number of subscribers for which user tracking is performed. Furthermore, some of the broadcast positioning methods are needed for location management anyway and therefore cause no additional overhead. As a consequence, possible candidates for user tracking in LAMA are LAI, CI, and GK-broadcast positioning as well as E-OTD and GPS. However, it must stressed that the power consumption of MTs increases with positioning accuracy. For example, E-OTD and GPS are based on complex calculations for triangulation, which significantly exhaust the batteries of MTs, especially if they are continuously performed for user tracking. Therfore, E-OTD and GPS should only be deployed if high accuracy is requested by the associated PLAS.

3 The LAMA System Design

LAMA is an integrated platform which combines user tracking and location monitoring at the MT. Because of this integration and because LAMA preferably deploys positioning methods which are based on broadcast signaling, LAMA significantly reduces signaling traffic in comparison to approaches where user tracking and location monitoring are performed at the fixed-network site. If location monitoring has detected that the user has entered a point of interest, LAMA notifies an associated PLAS service component about this event and triggers service execution in this way. This service component can either be located at the MT or at the fixed-network site.

LAMA has been designed with the following design goals in mind:

1. *Configurability:* This design goal primarily focuses on the set of positioning methods installed at a LAMA-enabled MT. The availability of a certain positioning method depends on the capabilities of the MT, the underlying wireless system, and the operator the user is registered with. As a consequence, it is desired that LAMA, i.e., the set of positioning methods used by LAMA, can be tailored on an individual basis according to this environment.
2. *Transparency:* The tracking process should be transparent to both the service developer and the user. The service developer should easily and rapidly create new services, without taking care for positioning methods supported by the

respective operator, system, or MT. On the other hand, the user should not be concerned with aspects of tracking once she had passed through the necessary configurations of the system, e.g., determination of her preferred positioning methods. This makes the system easy to use.
3. *Generality:* The interfaces for accessing LAMA should be very general so that it is not restricted to specific PLASs, but can be used universally.
4. *Portability:* LAMA should be portable in order to be used in conjunction with mobiles of different types, e.g., mobile phones, PDAs, or a combination of both.
5. *Extensibility:* Although LAMA has initially been designed for operation in second and third generation cellular networks, it should also cope with proprietary and standardized positioning methods used in present and future wireless systems. This is very important, because it is expected that cellular technologies like UMTS will converge with local wireless systems like IEEE 802.11. Therefore, LAMA must be easily extensible in order to support converging networks.

Following these design goals, LAMA has been decomposed into three subsystems: the *location monitor*, the *location tracker*, and the *location converter*. The location tracker performs the user tracking and hides the associated technical details of the tracking process and the used positioning methods from the other subsystems. It determines the MT's location by using one or several positioning methods and delivers it to the location monitor.

The location tracker consists of one or more *trackers* and a central *tracker controller* which coordinates the different trackers (see Fig. 1a). Other subsystems access the location tracker only via the tracker controller. A tracker implements a certain positioning method and delivers the derived location to the tracker controller. In addition, it associates location information with quality parameters, such as accuracy (average error) and up-to-dateness (time stamp). LAMA can adopt any terminal-based positioning method and provides the corresponding trackers. In correspondence to Tab. 1, Fig. 1a shows trackers for LAI-broadcast, CI-broadcast, GK-broadcast, TA, E-OTD, and GPS positioning. If other positioning methods are supported by the underlying mobile environment, LAMA can be easily extended by appropriate trackers.

The life cycles of trackers are coordinated by the tracker controller. This component initializes and terminates trackers and processes the location information they deliver. To access different types of trackers in a unified way, each tracker has to implement the general `Tracker` interface. This interface provides operations to start and stop the tracking process, which are invoked by the tracker controller and which are implemented in the trackers according to the special hardware and software of the positioning method they are based on. Every time one of the trackers measures a new location, it notifies the tracker controller by calling its `locationTracked()` method.

From the locations received in this way, the tracker controller selects the location with the highest accuracy and up-to-dateness and delivers it to the location monitor. The location monitor observes a user's location received by the location

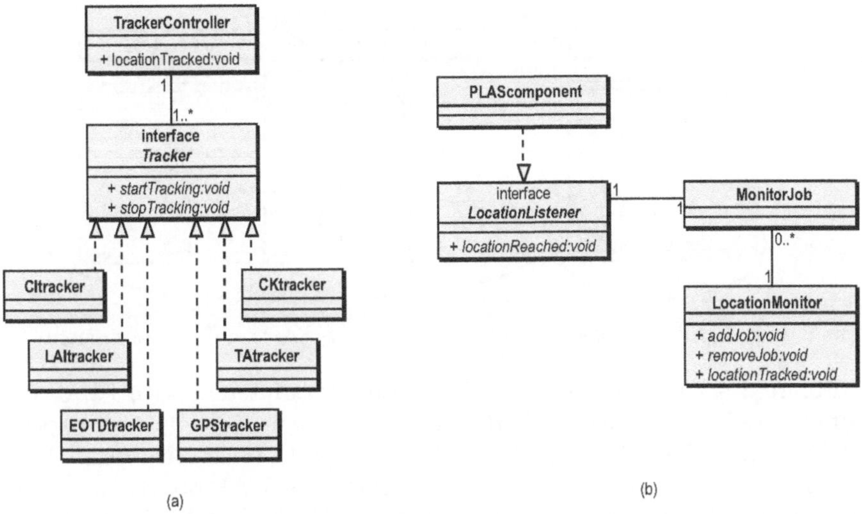

Fig. 1. Object design of the LocationTracker subsystem

tracker on behalf of a PLAS and notifies the associated PLAS service component as soon as a certain location has been entered. The object design of the location monitor is depicted in Fig. 1b. The location monitor manages several monitor jobs, each created by a PLAS service component and each containing one or several points of interest of a user who has subscribed for that PLAS. In order to be notified when the current location matches with one of the points of interest, the calling PLAS component must implement the `LocationListener` interface. Location listeners add new monitor jobs to the location monitor by calling the `addJob()` method and they remove jobs by calling the `removeJob()` method. Every time the location monitor receives updated location information from the location tracker, all monitor jobs are successively activated and executed in order to check the received location against the predefined points of interest. In case of a matching, the location listener of the associated PLAS component is notified that a certain point of interest has been entered. This is done by invoking the `locationReached()` method of the `LocationListener` interface.

The location converter allows to convert location information between different formats, for example from the GK to the WGS84 coordinate system and vice versa. It is used by the location tracker if a tracker does not support a certain physical or symbolic format as required by the invoking LAS. The location converter can also be used by the location monitor if it is necessary to convert a location profile provided by a PLAS. If a transformation between physical and symbolic formats is needed, the location converter at the MT must be connected to a remote directory. For example, if a PLAS requires the location in GK-coordinates and the CI-tracker, delivering locations as CIs, is the only available tracker, then the associated GK-coordinates must be obtained from the remote directory, which performs the mapping between these formats. In order to limit the number of such remote invocations, intelligent caching mechanisms are needed which buffer multiple mappings at the MT.

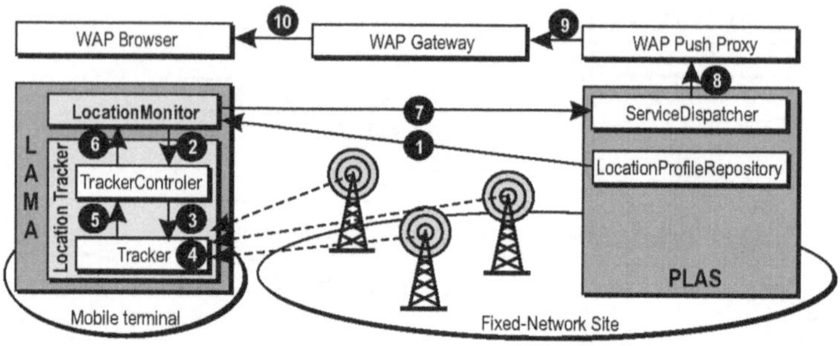

Fig. 2. Interactions between LAMA and a WAP service environment

Figure 2 shows how these LAMA components interact with PLAS components located at the fixed-network site (the location converter is disregarded in this example). The *Wireless Application Protocol* (WAP) has been chosen as the underlying service platform, which consists of a WAP browser at the MT as well as a WAP gateway and a push proxy at the fixed-network site. Generally, LAMA can be deployed independently of any service platform. For example, the WAP components can be replaced by an SMS infrastructure or by toolkits intended for UMTS. Furthermore, for the interactions between the MT and the fixed-network site the *General Packet Radio Service* (GPRS) is used.

In this scenario it is assumed that a location profile has already been generated for a particular user and stored in a location profile repository at the fixed-network site. This repository invokes the `addJob()` method at the location monitor with the location profile and the `LocationListener` interface of the service dispatcher as parameters (1). The location monitor activates the tracker controller (2), which, in turn, starts one or several trackers (3). Each time a tracker derives a new location (4), it returns it to the tracker controller (5), which forwards it to the location monitor (6). The location monitor then checks the location against the user's location profile. In case of a matching, it invokes the `LocationListener` interface of the service dispatcher at the fixed-network site and transfers the point of interest the user has reached (7). The next steps are not LAMA specific, but depend on the used service platform. The dispatcher might dynamically generate location-dependent content and combine it with a specific URL. This URL is then forwarded over a push proxy (WAP uses the term *push* instead of *proactive*) (8) and a WAP gateway (9) to the user's WAP browser (10), which can then request this URL.

4 Prototype Realization

Generally, today's MTs are characterized by limited computational resources, marginal storage, and proprietary interfaces. To evaluate LAMA according to usability, configurability, and portability, a prototype has been realized. The hardware configuration, i.e., the MT, consists of a Palm Vx handheld and a Siemens S35i mobile phone, which are connected via the serial interface. It is of course desired to use a single device with integrated radio facilities, why

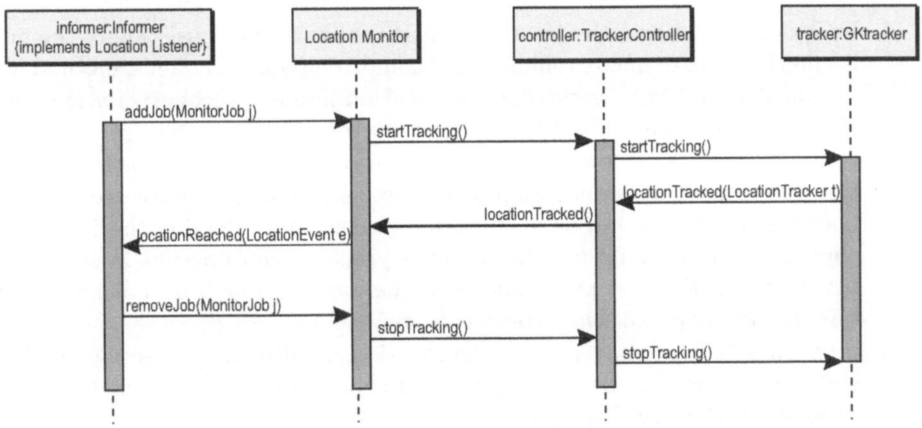

Fig. 3. Sequence diagram of the interaction between the LAMA components and Informer

another version of LAMA for the IPAQ handheld is in preparation as well. We have focused on GK-broadcast and GPS positioning and implemented the corresponding trackers. This configuration has been deployed in the German o2 network, where GK-broadcast is available. The mobile phone is used to receive GK-broadcast messages and to interact with LAMA and PLAS components located at the fixed-network site. GK-broadcast messages are requested from the handheld over the serial interface using AT commands according to [9]. Communication with remote LAMA and PLAS components can be done over SMS or GPRS bearer services.

To deploy LAMA at different types of MTs, we have focused on a Java-based solution for the software configuration. Basically, the *Java 2 Micro Edition* (J2ME) seems to be an appropriate platform, but suffers from the missing of some important device-dependent functions. Therefore, we have chosen the WebSphere Studio Developer (WSDD) platform from IBM [10]. WSDD is nearly J2ME-compliant, offers the desired functions for different hardware configurations, and is available for most of the popular handheld operating systems.

To easily test LAMA, an application called *Informer* has been implemented as well. Informer notifies the user about landmarks in his close proximity and, if desired, provides additional information about these landmarks. A landmark consists of the location in GK-coordinates, the name of the point of interest, and a short description. For example: 3690021, 5341076, "Munich Olympic Tower", "Landmark of the Olympic Summer Games 1972 in Munich". A landmark can be set at the current location. Alternatively, a landmark location profile, which could be, for example, offered by a service provider for selected cities, can be downloaded from a remote server. Informer is realized as a J2ME application and imports the LAMA classes. Informer is an application for *Personal Information Management* (PIM), like calendars and to-do lists, and is primarily executed at the MT. It interacts with the LAMA system in the following way (see figure 3):

1. Informer implements the interface `LocationListener` consisting of the public method `void locationReached(LocationEvent e)`. This method will be called by LAMA every time the location monitor recognizes a matching of the current location with a point of interest.
2. Informer creates a new job for the location monitor by creating a monitor job with a reference to the corresponding object implementing the `LocationListener` interface. In a subsequent step, it adds the job to the current location monitor. The tracking process is automatically started by LAMA when the first job is added to the location monitor. Tracking is automatically stopped, when there is no more job.
3. A job has to be removed from the location monitor by the service component which has initialized this job. This is done by calling the method `void removeJob(MonitorJob j)`.

5 Field Test Evaluation

To evaluate the LAMA concept in general and the accuracy of the GK-tracker in particular, a field test including 81 radio cells of the o2 network in Munich has been carried out (see Fig. 4). For this purpose, the sample application Informer has been arranged with a location profile containing popular landmarks of Munich.

During the field test, LAMA worked with a high reliability and displayed neighboring landmarks with an acceptable accuracy (when used in conjunction with the GK-tracker). In addition to the GK tracker, a GPS tracker has been implemented and both of them have been deployed in parallel. Due to simplicity, we have decided to choose GK-coordinates as our reference format, making it necessary to translate the WGS84 coordinates (delivered by the GPS tracker) using the location converter. From an economical point of view, a further advantage is that third party providers can offer PLASs without the need to enter into cooperation with network operators. However, this is in contrast to the efforts of operators, who probably want to gain revenues by selling location information.

LAMA has only been tested with the landmarks of Munich, i.e., with a manageable number of points of interest. However, MTs are equipped with limited storage and processing capabilities, which is why scalability problems have to be expected if the number of points of interest to be checked by the location monitor increases a certain order of magnitude. For example, confronting the location monitor with all landmarks of Germany or even the entire world would overload the system and would be inefficient anyway, because the users' mobility patterns are dominated by local movements. It is therefore obvious to charge the location monitor only with points of interest of local relevance and to automatically refresh the monitor jobs in dependence on the user's current location (see also Sect. 7).

To evaluate the accuracy of GKtracker the following data has been logged during the field test : GK-coordinates sent out by the base stations (*base-station coordinates*) and the actual coordinates of the MT when new GK coordinates

Fig. 4. Geographic overview of the field test

arrive (*arrival coordinates*). The arrival of new GK coordinates is referred to as *location update* in the following. The accuracy of the GK-tracker has been determined by calculating the *positioning error*. This error refers to the distance in meters between the arrival coordinates of a location update and its associated base-station coordinates. Each cell has been entered for several times from different directions. Correspondingly, *maximum*(MaxPE), *minimum* (MinPE), and *average positioning errors* (APE) could be derived for each of the 81 cells. In addition, the *location-update distance* (LUD) has been evaluated, i.e., the distance in meters between the arrival of two consecutive location updates.

Figure 5 shows the area around Munich Central Station, a clipping of the entire field test, and the measured base station and arrival coordinates. Table 2 contains the associated values for MaxPE, MinPE, APE, and LUD. The average MaxPE of all cells covered by the field test amounts to 595 meters, whereas the average APE is 429 meters. The smallest MinPE that has been derived is

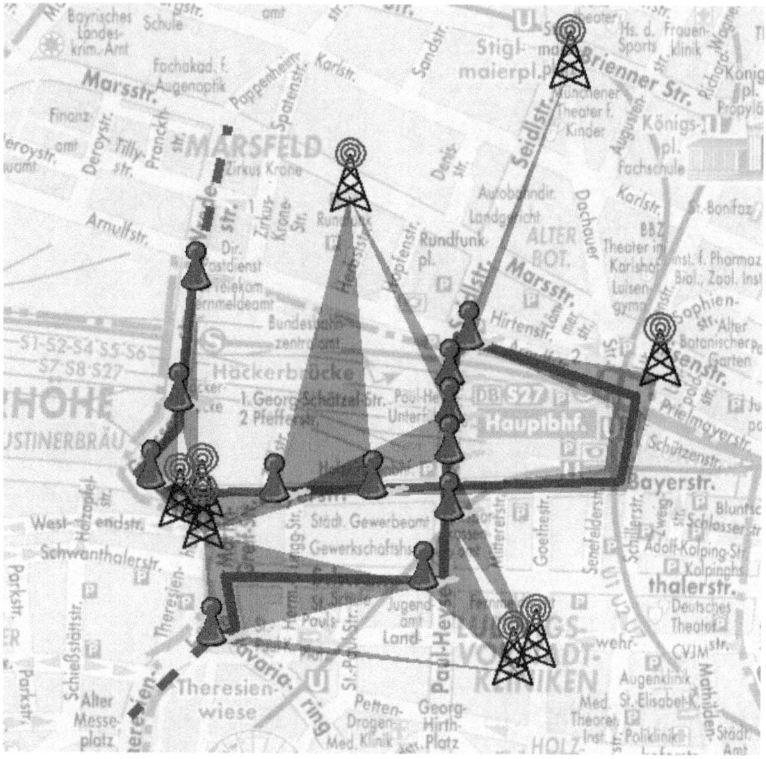

Fig. 5. Graphical evaluation showing the area around Munich Central Station

98 meters. The average LUD amounts to 345 meters. Obviously, the positioning error depends on the density of base stations in a certain area. Usually, this density is much higher in urban areas than in suburban or rural areas. However, the results achieved by the field test can be considered to be representative for cellular network in urban areas.

To draw a conclusion, the field test has demonstrated that a terminal-based approach of user tracking in conjunction with location monitoring is a promising way to realize PLASs. Currently, this approach is further refined in that trackers installed at an MT are coordinated in an adaptive way in dependence on the user's current distance from a point of interest.

6 Related Work

One of the pioneers in location awareness is the *Active Badge* system which has been developed in 1992 [11]. It enables the localization of individuals in buildings by determining the location of their Active Badge. An Active Badge sends out an infrared beacon every 10 seconds which is detected by stationary sensors installed within a building. Thus, one can say that Active Badge realizes network-based positioning. Principally, this approach can be adopted to realize

Table 2. Measured distances corresponding to the area shown in figure 5

Cell	MinPE	MaxPE	APE	LUD
50	641	641	641,0	0
51	223	469	346,0	541
52	262	355	308,5	183
53	345	451	398,0	118
54	541	578	559,5	86
55	423	500	461,5	95
56	596	670	633,0	76
57	400	641	520,5	1033
58	618	646	632,0	204
59	149	165	157,0	250
60	100	231	165,5	200
61	206	434	320,0	255

PLASs. However, its implementation in cellular networks would cause a significant signaling overhead if MTs of a substantial amount of subscribers would frequently send out these beacons.

The *Guide* project has focused on another application and developed a tour guide for the city of Lancaster [12]. Unlike Active Badge, beacons containing location information are sent out by the network and received by the terminal. The system has been built upon an IEEE 802.11 infrastructure, why the relevant places to be covered by the tour guide have to be equipped with IEEE 802.11 access points. Apart from the limitation that Guide focuses on a certain positioning method, its designers have restricted to reactive services only.

As interactive data-oriented services for cellular networks have gained more and more attraction in the recent years, an increasing number of projects have started making services in these networks location-aware. The *Deep Map* project has developed a system for GSM to exchange location information between members of a tourist group [13]. The system is based on traces which contain the coordinates of representative tours through the area to be covered by the service. These traces have to be fixed in advance and stored on a central server. The current location of a group member can then be predicted by the server. In order to obtain the initial location or to correct the prediction, the terminal needs to perform location updates, i.e., the current location is measured by GPS and transferred to the server. However, the number of these location updates can be dramatically reduced according to a system without location prediction. Although this approach can also be adopted for PLASs, there is too much overhead for the acquisition of representative tours, especially for operators serving a national-wide coverage area.

One of the first platforms that combines different positioning methods has been developed by the *Nexus* project [2]. Nexus proposes an integrated platform which incorporates different positioning methods for indoor and outdoor positioning as well as a spatial data model for representing the physical world. Thus, Nexus is very close to the approach followed by LAMA. The *Lol@* project has developed another platform for location-aware services [3]. Unlike the aforementioned platforms, Lol@ explicitly addresses PLASs. It defines a network-centric *Gateway Mobile Location Center* (GMLC) which performs the user tracking and triggers content delivery as soon as the user enters a point of interest. The GMLC is assisted by positioning methods located either inside the MT or at the network. However, location monitoring is executed by the GMLC at the fixed-network site and thus, dedicated point-to-point signaling between MTs and the GMLC is needed to support PLASs.

7 Outlook

Generally, the costs of user tracking increase with the accuracy of the underlying positioning method. It is therefore intended to add more intelligence to LAMA. Positioning methods should be automatically activated and deactivated in dependence on their costs, accuracy requirements of the used PLAS, and the user's current distance from a point of interest. For example, it would be desired to switch from a positioning method with moderate accuracy to one with higher accuracy if the user approaches a point of interest and if this degree of accuracy is really required for the associated PLAS. LAMA is currently extended to provide these functions.

Furthermore, ongoing work concentrates on an extension of the location monitor in order avoid scalability problems at the MT, which occur if too many points of interest have to be checked against the current location. It is intended that the monitor jobs only contain a subset of points of interest that are located within a certain distance to the user's current position or within her city or region. Thus, the location monitor itself becomes location-aware, and monitor jobs must be automatically refreshed in dependence on the user's location.

Acknowledgment. The authors wish to thank the members of the Munich Network Management (MNM) Team for helpful discussions and valuable comments during the final phase of the paper. The MNM Team directed by Prof. Dr. H.-G. Hegering and Prof. Dr. C. Linnhoff-Popien is a group of researchers of the University of Munich, the Munich University of Technology, and the Leibniz Supercomputing Center of the Bavarian Academy of Science. For more information see `http://wwwmnmteam.informatik.uni-muenchen.de`.

References

1. Fischmeister, S., Menkhaus, G.: The Dilemma of Cell-based Proactive Location-Aware Services. Technical Report TR-C042, Software Research Lab, University of Constance (2002)
2. Klinec, D., Volz, S.: NEXUS – Positioning and Communication Environment for Spatially Aware Applications. In: International Archives of Photogrammetry and Remote Sensing. Volume XXXIII of Inter-Technical Commission, IC 10., Amsterdame, Netherlands (2000)

3. Popischil, G., Stadler, J., Miladinovic, I.: A Location-based Push Architecture using SIP. In: Proceedings of the 4th International Symposium on Wireless Personal Multimedia Communications (WPMC '01, Aalborg, Denmark (2001)
4. 3GPP TS 02.71: Location Services; Service Description, Stage 1 (2001)
5. 3GPP TS 03.71: Location Services; Functional Description, Stage 2 (2002)
6. 3GPP TS Group Terminals 03.41: Technical Realization of Cell Broadcast Service (CBS) (1998)
7. Tabbane, S.: Location Management Methods for Third-Generation Mobile Systems. IEEE Communications Magazine (1997) 72–84
8. Zhang, J.: Location Management in Cellular Networks. In: Handbook of Wireless Networks and Mobile Computing. John Wiley & Sons (2002) 27–49
9. 3GPP TS Group Terminals 07.07: AT Command Set for GSM Mobile Equipment (ME) (1998)
10. : (http://www.ibm.com/embedded/)
11. Want, R., Hopper, A., Falcao, V., Gibbons, J.: The Active Badge Location System. ACM Transactions on Information Systems (1992) 91–102
12. Davies, N., Cheverst, K., Mitchell, K., Efrat, A.: Using and Determining Location in a Context-Sensititve Tour Guide. IEEE Computer (2001) 35–41
13. Walther, U., Fischer, S.: Metropolitan Area Mobile Services to Support Virtual Groups. IEEE Transactions on Mobile Computing **1** (2003) 96–110

Implementation of a Scalable Context-Aware Computing System*

Tamer Nadeem, Adel Youssef, Suman Banerjee, Moustafa Youssef, Sulabh Agarwal, Kevin Kamel, Andrzej Kochut, Christopher Kommareddy, Pankaj Thakkar, Bao Trinh, A. Udaya Shankar, and Ashok Agrawala

MIND Lab, UMIACS and Department of Computer Science
University of Maryland, College Park, MD 20742, USA
{nadeem,adel,suman,moustafa,sulabh,kamelkev,kochut,
kcr,thakkar,bao,shankar,agrawala}@cs.umd.edu

Abstract. Context-aware computing involves the automatic tailoring of information and services based on the current location of the user. In this paper, we describe our experience in implementing Rover, a system that enables location-based services, as well as the traditional time-aware, user-aware and device-aware services. To achieve system scalability to very large client sets, Rover servers are implemented in an "action-based" concurrent software architecture that enables fine-grained application-specific scheduling of tasks. We have demonstrated its feasibility through implementations for both outdoor and indoor environments on multiple platforms.

Keywords: ubiquitous, context-aware, location-based, operational-law.

1 Introduction

Context-aware computing requires the automatic tailoring of information and services based on the current context of the user. The *context* of the user typically consists of a set of user-specific parameters including his location, the characteristics of the access device and interface, and the interests of the user, usually represented in an user profile. The different technology components needed to realize context-aware computing are present today, powered by the increasing capabilities of mobile personal computing devices and the increasing deployment of wireless connectivity.

What has hindered its ubiquitous deployment is the lack of system-wide integration of these components in a manner that scales with large user populations. In our prior work [2], we had described an architecture of such a context-aware computing system, called *Rover*, and had discussed its applications, devices, and

* This work was supported in part by the Maryland Information and Network Dynamics (MIND) Laboratory, its Founding Partner Fujitsu Laboratories of America, and by the Department of Defense through a University of Maryland Institute for Advanced Computer Studies (UMIACS) contract.

M. Conti et al. (Eds.): PWC 2003, LNCS 2775, pp. 364–374, 2003.

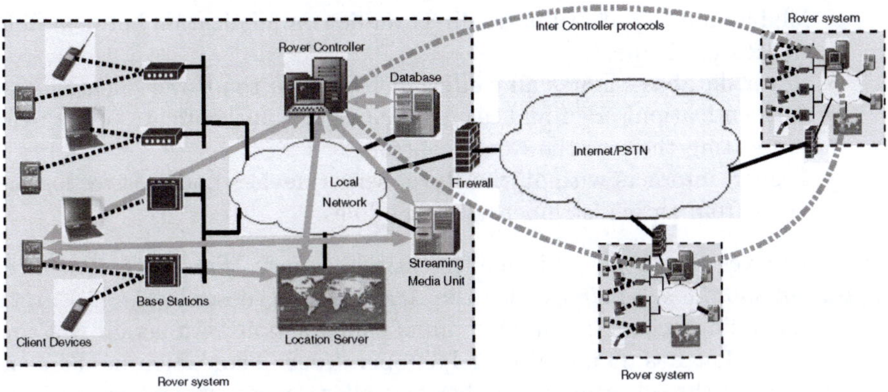

Fig. 1. Physical architecture of the Rover System

users. In this paper, we focus on our implementation experience of this system for both indoor and outdoor environments. Rover enables services with characteristics: a) Location-aware, in addition to the more traditional notions of time-aware, user-aware, and device-aware. b) Available via a variety of wireless access technologies and devices. c) Scales to a very large client population.

In the next section we present an overview of the Rover system. In Section 3 we describe the our implementation of the system. We explain the functionality of the system in Section 4. In Section 5, we briefly comment on our experiments and analysis to test system scalability. We describe some related projects in Section 6 and finally conclude in Section 7.

2 Overview of Rover

The Rover system, depicted in Figure 1, consists of the following entities:

- *End-users* of the system. Rover maintains a *user profile* for each end-user, that defines specific interests of the user and is used to customize the content served.
- *Rover-clients* are the client devices through which users interact with Rover. Rover maintains a *device profile* for each device, identifying its capabilities and thus, the functionality available at that device.
- *Wireless access infrastructure* provides wireless connectivity to the Rover clients.
- *Servers* implement and manage the various services provided to the end-users. The server system consists of the following set of devices:
  - Rover controller: is the "brain" of the Rover system. It provides and manages the different services requested by the Rover clients. It schedules and filters the content sent to the clients based on user and device profiles and their current locations.
  - Location server: is a dedicated unit responsible for managing the client device location services within the Rover system.

- Media manager: coordinates direct audio communication between different Rover clients.
- Rover database: stores all content delivered to the Rover clients.
- Authentication server and security manager[1]: authenticates users before connecting them to the Rover controller.
- Logger: interacts with all the Rover server devices and receives log messages from their instrumentation modules.

Rover achieves scalable performance using a new *fine-grained real-time application-specific scheduling* called the *Action Model* (described in [2]). In this model, scheduling occurs in "atomic" units called actions. An action is a small piece of code that has no intervening I/O operations. The Action model avoids the overhead of thread context switches and allows more efficient scheduling of execution tasks.

The Rover system represents a single domain of administrative control that is managed and moderated by its Rover controller. A large domain can be partitioned into multiple administrative domains each with its own Rover system, much like the existing Domain Name System . For this multi-Rover system, we define protocols that allow interaction between the domains to ease user roaming. In the next section we describe our prototype of the Rover system that has been implemented in the University of Maryland campus.

3 System Implementation

In our prototype implementation all system components were developed on Linux-based platforms in C++ and they can easily be ported to any other operating system as well.

Rover Controller
Scalability and reliability are the most important factors required in such system. It is implemented using the action model. Each independent transaction of the Rover controller with the clients is called a *server operation.* A server operation consists of a sequence (or more precisely, a partial order) of actions interleaved by asynchronous I/O events.

Rover Database
The database is implemented using a standard SQL language (MySQL[2]). It consists of two different components — the relational data and the spatial data. All attribute-based data, e.g. user and client states, are stored in the relational component. All spatial information, e.g. floor plan of a building, are stores in the spatial component in a hierarchical structures. Vector-based image is used to represent spatial data.

[1] The Authentication server is not shown in Figure 1.
[2] Any other relational database could replace MySQL.

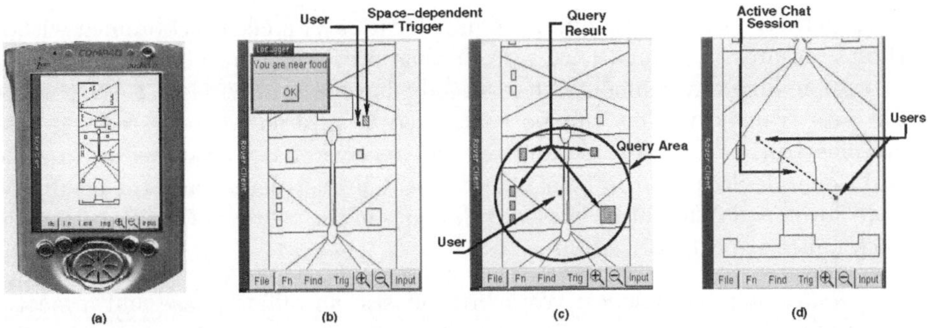

Fig. 2. Rover client screen shots taken from a demonstration at the McKeldin mall of the University of Maryland campus. (a) Rover client running the client software showing the mall map. (b) A notification to the client about a nearby food stall. The user associated with the client had previously set a trigger notification request when he is close to a food stall. (c) The user had issued a query operation about the sites of interest in his vicinity. On receiving the response from the Rover system, the client has highlighted the relevant sites. (d) An active chat session between this user and another user is marked as a dotted line connecting both users.

Rover Clients
The Rover clients are implemented on the Compaq IPAQ devices running the familiar distribution of Linux for handheld devices. The user interface is implemented using GTk++ (Figure 2).

Media Manager
Media manager enables Rover clients to establish direct audio communication channels, independent of the Rover server system. We have defined a protocol called *Click & Connect* that allows users to easily initiate voice communication with each other. The protocol is so named because an user can directly click on the icon representing another user to initiate voice communication. The Media Manager arbitrates the handshaking process between two clients to provide a degree of organization.

Wireless Infrastructure
University of Maryland has a IEEE 802.11 wireless network widely deployed across the campus, which we used for all wireless communication to and from the Rover clients. We also deployed some additional base stations in the campus to increase the reachability of Rover in areas not covered by the university wireless network.

Location Server
The location server is responsible for storing and managing user locations in the Rover system. The system is designed to work in both indoors and outdoors environments. We have experimented with RF-based systems that infer the location of a device based on the signal strength of received RF signals of IEEE 802.11 wireless LAN frames.

In our RF-based techniques, the user location of a client is obtained without the use of any additional hardware. It thus provides more ubiquitous coverage in campus-like environments that already have a rich wireless LAN coverage for data transport. This can be contrasted to alternative Infra-red tag-based systems [10,1] or ultra-sonic emitter and receiver based systems [9] in which additional devices need to be attached to the infrastructure and the clients. We have developed different RF-based techniques in the context of the Rover system. Techniques are categorized into:

- *Radio-map Techniques:* Work in 2 phases: an offline phase and a location determination phase. During the offline phase, the signal strengths received from the access points, at selected locations in the area of interest, are gathered as vectors and tabulated over the area. During the location determination phase, the vector of samples received from each access point is compared to the radio-map and the "best" match is returned as the estimated user location. We used two methods to calculate the best match:
  - K-Nearest Neighbors (KNN): A voting mechanism based on a distance function is used to estimate the best user location.
  - Probabilistic Clustering-based: The Baye's theorem is used to select the most probable location of the user. Refer to [13] for more details.
- *Model-based Techniques:* The relation between the signal strength received from an access point and the distance to this access point is captured by some function (model). By using three or more access points, the user location is estimated. Two methods are used:
  - Minimum Triangulation: Given each access point i, the distance (d_i) between the receiver, located at x, y, z and the access point, located at x_i, y_i, z_i, is modeled as:

    $$d_i^2 = (x - x_i)^2 + (y - y_i)^2 + (z - z_i)^2 = \frac{k}{v_i}$$

    where v_i is the strength of the received signal, and k is a constant.
  - Curve Fitting: The received signal power is modeled as:

    $$PL(d)[dB] = A + B\log(d)$$

    $$A = PL(d_0)[dB] - 10n\log(d_0),\ and\, B = 10n$$

    where PL is the received power, d is the three dimensional path length to the transmitter, d_0 is a reference distance, and n represents the path loss exponent. Using curve fitting techniques, A and B are estimated for each access points.

For indoor environments, we found that radio-map based techniques achieve better accuracy than model-based techniques. This is because the relation between the signal strength and distance in indoor environments is complicated by to the multi-path effect and other phenomena which are difficult to capture by simple models. On the other hand, model-based techniques have the advantage of not depending on the calibration process required to build the radio-map. This advantage favors model-based techniques in outdoor environments, where the relation between signal strength and distance can be captured by simple functions and the coverage area is large making building the radio-map a time consuming process.

Security Manager
The Rover system inherits the vulnerability of the IEEE 802.11 wireless communication system such as *unauthorized access*, *non confidential*, and *non integrity*. Therefore, we have implemented different security mechanisms to protect client interactions.

- *Client-Server Security:* A session based security protocol, Secure Sockets Layer (SSL) is used to secure the channels between the clients and the server.
- *Chat (peer-peer) Security:* The direct audio chat sessions we employ the Data Encryption Standard (DES), which is an inexpensive symmetric key cryptographic system. The session key is transferred using the SSL-protected secure channel.

4 System Functionality

Rover provides different capabilities to the users, which can be categorized as follows:

- *System Admin Operations* are available only to the authorized system administrator. These set of operations are used to monitor, manage, and query users, devices, and Rover system.
- *User Access Operations* are the basic set operations that every user avails to access the Rover system.
- *Trigger Operations* allow users to set context-specific alerts. The triggers are activated based on user interests and depend on current time and/or location of the user. When the trigger condition is satisfied the Rover server system sends appropriate notification to the particular user (Figure 2(b)).
- *Query Operations* allow users to acquire information about different aspects of the system and the environment. Figure 2(c) shows a client screen shot in response to a client query on sites of interest in its vicinity.
- *Location Update Operation* inform the server system about the client's location.
- *Audio Chat Operations* enable direct audio communication between clients. Audio chat between clients is initiated with the coordination of the Media Manager. Once an audio chat is initiated, the clients interact directly with each other without intervention of the Rover server system (Figure 2(d)).

5 System Performance

To asses the performance and scalability of the Rover System we take two approaches: a) *Active Monitoring* where we instrumented the controller to collect

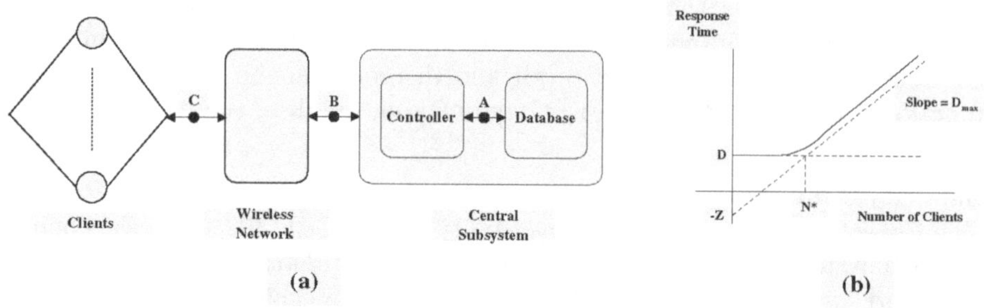

Fig. 3. Passive monitoring analysis: (a) Performance model for passive monitoring. (b) Typical asymptotic bounds.

different performance statistics (e.g. queue lengths for each component, the response time for each operation, etc.). b) *Passive Monitoring* which is described in this paper.

5.1 Analysis Using Passive Monitoring

In this approach, no instrumentation code is introduced in the server system. Instead, we use a client load generator to stress test the server and observe two different metrics — the response time obtained by individual clients and the number of clients that can simultaneously served by the system without significantly impacting the performance.

We model the Rover system as a single-server multi-client system as shown in Figure 3(a). The Rover System is modeled as a central subsystem consisting of two devices, the Rover controller and the Rover database, and N terminal subsystems. Each of the N terminals is a client of the Rover System and perform the cycle of issuing a request, waiting for the response and processing the response (think time). Wireless network models the communication channel between the server and the clients. Since we are interested in assessing the performance of the Rover system we do not explore the affects of communication channel in this paper.

Using a technique called *operational law* [5] to analyze such systems, it can be shown that the response time observed by clients increase marginally with increasing the number of clients up to a critical client population. Let D denotes the time required by the system to process a single client operation, and $D_{\max}$ denotes the time required at the bottleneck server of a multi-server system. For the single server model, $D_{\max} = D$. If N^* indicates the critical number of clients that the system can support without impacting the response time for the clients and Z the *think time* used by the clients between operations, then operational analysis suggests that:

$$N^* = \frac{D + Z}{D_{\max}} \tag{1}$$

The graphical representation of N^* is shown in Figure 3(b).

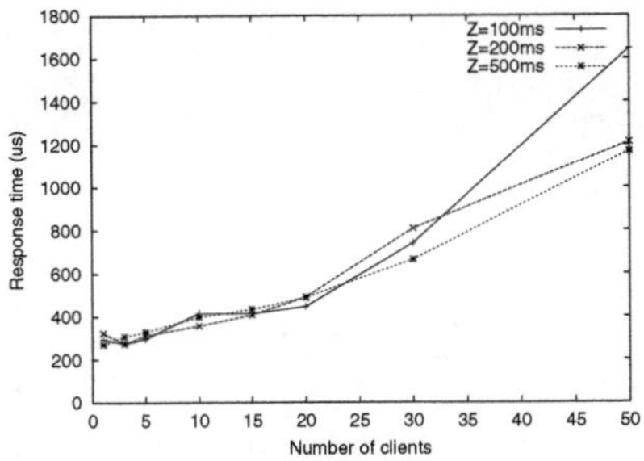

(a)

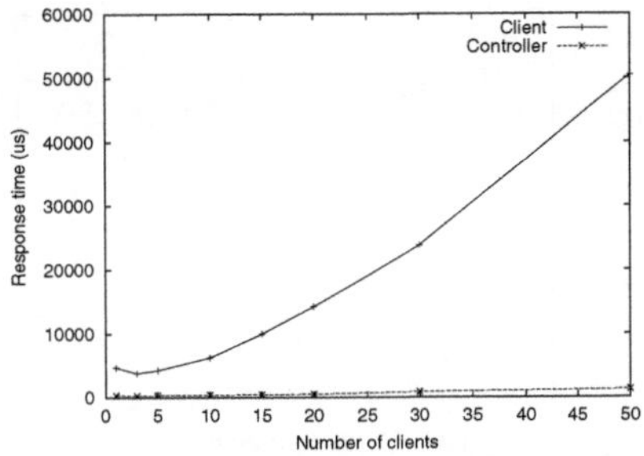

(b)

Fig. 4. *GetAllLoginUsers* operation: (a) Controller response time, and (b) Response time when Z=200ms.

5.2 Experiment Configuration

The central subsystem runs on Pentium IV 1.5 GHz desktop machine with 256 MB of RAM running the Linux OS with kernel version 2.4.7. A second machine is used to behave as a set of clients (client loader). The client loader runs on a Pentium III 800 MHz laptop with 128 MB of RAM and running a Linux OS with kernel version 2.4.2. The client machine uses 802.11b wireless network to connect to the network.

The response time for each operation were collected as observed at the database, the controller and the client (points A, B, C respectively in Figure 3(b)). Instead of collecting response time for each of the system operations, we experimented with three different operations representing three different categories:

1. *GetAllLoginUsers:* Gets the position of all users who are logged into the system. This operation is controller intensive and does not involve the database.
2. *VectorMap:* Gets the vector map of an area. This operation is computationally intensive at the database side.
3. *Locate:* Locates the object containing the given point. This operation involves the database though it is not very computationally intensive at either the database or the controller sides.

5.3 Results and Discussion

GetAllLoginUsers

Figure 4(a) shows the response time, at the controller, plotted against the number of clients in the system for different think times ($Z = 100ms, 200ms, 300ms$). The total service demand time, D of the controller is observed to be around 300 microseconds. For only one device in the system $D_{max} = D$. Using Equation 1 where $D = 300$ microseconds and a think time $Z = 200$ milliseconds, we get N^* to be approximately 667 requests. Hence the server can support 667 requests

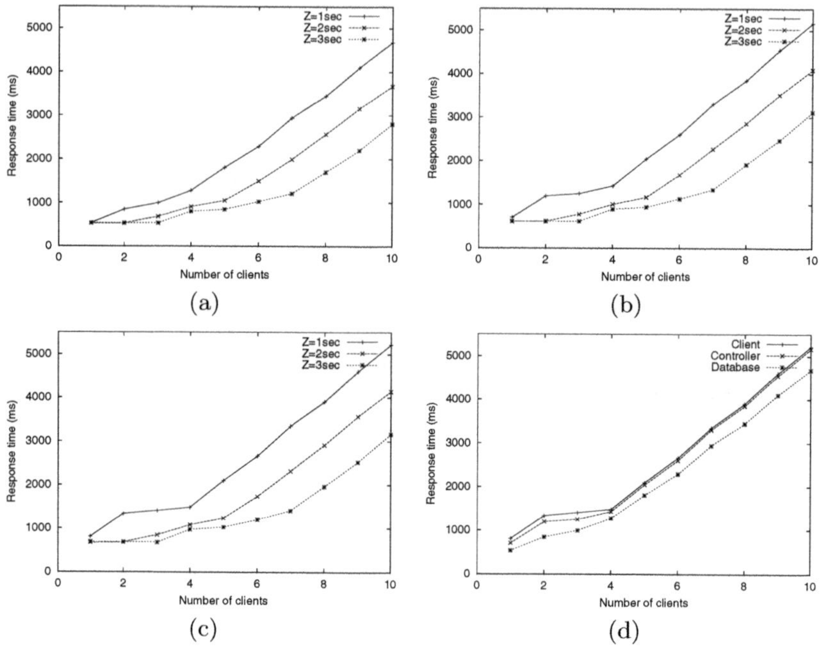

Fig. 5. *VectorMap* operation: (a) Database response time, (b) Controller response time, (c) Client response time, and (d) Response time when Z=1s.

without any significant delays. In an actual deployment, the think time would be of the order of 10's of seconds and that would give an even higher value of N^* (of the order of thousands of clients).

Figure 4(b) shows the response time behavior of both the controller and the client when the think time is Z = 200 milliseconds. As the graph shows the controller graph stays almost horizontal as the number of clients are increased which shows the controller can handle a large number of clients. On the other hand, the client graph grows with the number of clients. This can either be due to the effect of the wireless hop involved or the processing involved at the OS level. The performance of the 802.11b wireless network has not been taken into account and is left for future work.

VectorMap

Figure 5 shows the response time at the different Rover components. For the database the total service demand time is observed to be around 0.5 seconds. Using equation 1, we can predict the knee-point to be at $N^* = 3$ with a think time $Z = 1$ second. We should note that the *VectorMap* operation is an infrequent operation and has been used only to assess the performance of the system in the extreme case. In an actual deployment, the duration between subsequent *VectorMap* operation requests (Z) would be in the order of minutes. The difference in the database response and the controller response (Figure 5(d)) could be

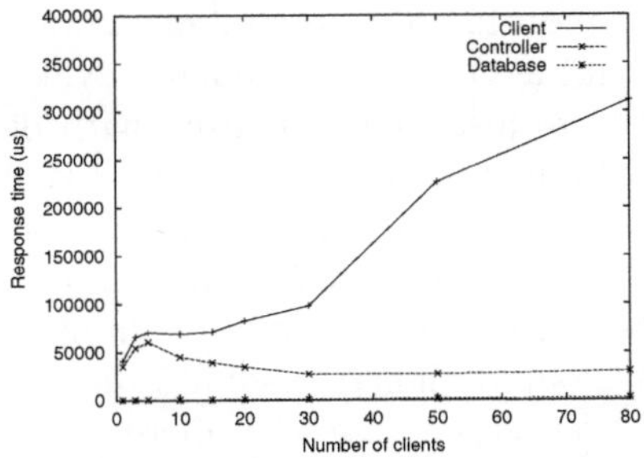

Fig. 6. Response time of *Locate* (Z=200ms)

explained by the fact that at the controller all the data is touched and a copy is created for debugging purpose.

Locate

Similar to the analysis of the previous operations, we show the response time at the database, the server and the client for a think time of 200 milliseconds in Figure 6.

6 Related Work

There are several ongoing research efforts in the area of context-aware applications. The Active Badge system [10] uses infrared based special badges to locate current people in each room. Georgia Tech's conference assistant [4] was designed to assist conference attendees in choosing specific presentations to attend based on their profile. AT&T sentient computing system [7] location based system is based on ultrasonic measurements. The Cyberguide [1] project is a context-aware tourist guide prototype based on infrared technology. Other personal assistant projects include: office assistant [11], GUIDE system [3], CMU's smart sight [12], ComMotion [6] project, and HP's Cooltown [8]. All the context-aware applications described in this section were developed as a framework to assist individual users. Rover, *in contrast*, defines a system framework in which such applications can be built. It allows direct interaction between the users and as well as between the users and the environment in a scalable manner. Rover system architecture enables easy instantiation of new applications into the system by appropriate definition of server operations and client interactions.

7 Conclusions and Future Work

Rover is currently available as a deployable system using specific technologies, both indoors and outdoors. Our final goal is to provide a completely integrated system that operates under different technologies, and allows a seamless experience of location-aware computing to clients as they move through the system.

With this in mind, we are continuing our work in a number of different directions. We are experimenting with a wide range of client devices, specially the ones with limited capabilities. We are also experimenting with other alternative wireless access technologies including a Bluetooth-based LAN. We are also working on the design and implementation of a multi-Rover system.

We believe that Rover Technology will greatly enhance the user experience for many different context-aware applications in different environments. Our initial experience indicate that our designed system scales to large user populations and the benefits of the system increase with increase in the number of users.

References

1. G.D. Abowd, C.G. Atkeson, J. Hong, S. Long, R. Kooper, and M. Pinkerton. Cyberguide: A mobile context-aware tour guide. *Wireless Networks*, 3(5), October 1997.
2. S. Banerjee, T. Nadeem, A. Shankar, and A. Agrawala et al. Rover technology: Enabling scalable location-aware computing. *IEEE Computer*, 2002.
3. K. et al. Cheverst. Experiences of developing and deploying a context-aware tourist guide: The lancaster guide project. In *Proc. 6th Ann. Int'l Conf. MobileComputing and Networking (Mobicom 00)*, New York, 2000.
4. A.K. Dey, M. Futakawa, D. Salber, and G.D. Abowd. The conference assistant: Combining context-awareness with wearable computing. In *Proceedings of the 3rd International Symposium on Wearable Computers (ISWC '99)*, San Francisco, California, October 1999.
5. R. Jain. *The Art of Computer Systems Performance Analysis*, chapter 4. John Wiley and Sons, New York, 1991.
6. N. Marmasse and C. Schmandt. Location-aware information delivery with commotion. In *Proceedings of Second International Symposium on Handheld and Ubiquitous Computing, HUC 2000*, Bristol, UK, September 2000.
7. Steve Hodges Joe Newman Pete Steggles Andy Ward Andy Hopper Mike Addlesee, Rupert Curwen. Implementing a sentient computing system. *IEEE Computer Magazine*, 34(8), 2001.
8. S. Pradhan, C. Brignone, J.H. Cui, A. McReynolds, and M.T. Smith. Websigns: Hyperlinking physical locations to the web. *IEEE Computer*, 34(8), 2001.
9. N.B. Priyantha, A. Chakraborty, and H. Balakrishnan. The cricket location-support system. In *Proceedings of ACM Mobicom*, August 2000.
10. R. Want, A. Hopper, V. Falco, and J. Gibbons. The active badge location system. In *ACM Transactions on Information Systems*, January 1992.
11. H. Yan and T. Selker. Context-aware office assistant. In *Proceedings of the 2000 International Conference on Intelligent User Interfaces*, New Orleans, LA, January 2000.
12. J. Yang, W. Yang, M. Denecke, and A. Waibel. Smart sight: a tourist assistant system. In *3rd International Symposium on Wearable Computers*, San Francisco, California, October 1999.
13. M. Youssef, A. Agrawala, and A. Shankar. Wlan location determination via clustering and probability distributions. *IEEE International Conference on Pervasive Computing and Communications (PerCom) 2003*, March 2003.

SPREADing the Web

Paul Couderc and Michel Banâtre

IRISA/INRIA
Campus Universitaire de Beaulieu
35042 Rennes Cedex, France
{Paul.Couderc,Michel.Banatre}@irisa.fr
http://www.irisa.fr/aces

Abstract. This paper demonstrates the use of a physical/spatial computing model (SPREAD) to build contextual web (or wap) information systems. The approach presented consists in spreading information items in the environment (through short distance wireless technologies); then, sensing these data allows to integrate a context description, which can be used to help users co-navigate seamlessly the web as they move. The paper contributes an unified definition of the concept of context, applicable to both real (physical) and virtual (information system) environments. The underlying SPREAD system is also presented, as well as its potential for dynamic web content driven by physical entities.

1 Introduction

Today, ubiquitous access to information system is a reality. Systems like WAP effectively allow access to Web or Web-like information systems almost anywhere on a mobile phone or a PDA. However, we think these systems lack a crucial feature to qualify as truly pervasive information systems : context-awareness and spontaneous operation.

Context-awareness is not new, and has been extensively discussed in the academic domain. The basic motivation supporting this concept is that the most precious resource in mobile computing is user-attention: the user is focusing on his "real-life" activity, and interactions with the device are a disturbance which should be limited to the minimum. The idea behind context-awareness is to sense information from the environment of the user to guess its situation. The concept of context-awareness was born in the pioneer project Parctab at the beginning of the 1990's [13,11], and many similar works followed [2]. However, context-awareness does not seem to emerge beyond the research domain.

It has been already argued [10,2] that it is the lack of models, architectures and tools which prevents context-awareness to be enabled to its full potential. The design of context-aware applications remains too difficult. The notion of context usually considered is often quite fuzzy and difficult to deal with from an information system point of view.

In this paper, we propose a different approach based on the concept of spatial information systems [5], offering several advantages. We also propose a uni-

M. Conti et al. (Eds.): PWC 2003, LNCS 2775, pp. 375–384, 2003.

fied definition of context which is convenient to both physical and virtual environments. This notion of context allows to interpret mobility from a new and exciting perspective. Moreover, the concept of spatial information systems is supported by a well-understood programming model sharing similarities with LINDA [7]. This model is implemented in SPREAD (Spatial PRogramming Environment for Ambient computing Design), which allows extremely easy application developments for the spatial aspects of the context.

The paper is organized as follows : section 1 proposes to reconsider the notion of context, from the perspective of both real and virtual environments. Section 2 describes how the notion of context in the web can be enhanced with new aspects, like a spatial dimension. The SPREAD programming model is described in section 3. Section 4 concludes the paper by highlighting the innovation of our approach, and by giving some research perspectives.

2 Context-Awareness Revisited

The importance of context-awareness has been identified since the beginning of ubiquitous computing. The Parctab project exhibited context-sensitive applications as early as 1994 [11]. However, this notion of context is still not clearly defined and understood. Some important questions, which can be easily answered in the particular case of a given application, lacks general answers and require a deeper understanding:

What does the context represent, and for who (end user, application, system ...)? What kind of information constitutes "context", how does the context relate to the notion of mobility, how does the context should be represented and used from the application perspective?

2.1 What is "Context" ?

Context in the Real World. The first works focused on demonstrating the interest of exploiting context-awareness at the application level, enabling what is now called "pervasive computing". The considered notion of context was related to the situation of a human user in the physical environment. Basically, the idea is to guess the situation of the user in order to allow the application to operate more seamlessly. Many works followed Parctab on this principle, like the shopping companion "Shopnavi" [9] which helps a user to track shops, prices, and products using positioning and visual identification techniques, or the tourism applications CyberGuide [1] and GUIDE [3] which enhance a city visit with contextual information and navigational help.

For these applications, "context" can be defined as any information characterizing the situation of the user in the physical environment. This may include time, location, nearby physical entities (people, objects, machines, animals. . .)

All these information contribute to help understanding a situation from an application perspective : for example, a navigation application deals with location information, and thus the current location of the user is part of the context to which this application should be aware of. In this way, the context is defined "from the bottom" by a set of information (context attributes) which constitutes it. A problem in this way of defining the context is its incompleteness and lack of generality: the definition refers to the notion of "situation", which is subjective and dependent on the considered environment and applications. The definition does not help to answer some of the fundamental questions we presented earlier. In particular, it does not address the fundamental relationships between context and a mobile entity, and how an information system should integrate the notion of context.

We insist on the importance of this problem because ubiquitous computing involves a growing integration between information systems and physical environment : it may be seen as a dissolution of the information system in the real world. This trend requires to carefully examine the notion of context not only in the physical environment ("situation" of the user) but also for computing processes associated with real-world entities which evolve in the information system.

Context in Information Systems. Notions of context also exist in information systems. For example, the context of a computing process is a point in its execution graph. Another example is a link to a web page, which represents a citation context. A set of words can also be interpreted as a context, which can be found in relevant documents from a collection. It is this notion of context which is used by search engines. A last example is the context of an e-mail, which is a rich set of information: the date when it was received, the sender identity, the thread (or subject) it belongs to, etc.

As we can see, the notion of context in information system refers to the concept the location of an information item or document, relatively to a set of other information items. Thus, the context can be seen as *a set of references to an information item*, defining a location in an information space.

An important aspect to examine about the notion of context is its relationship in respect to the mobility. In the real world moving physically effectively changes the current context. A similar notion of mobility exists in information systems, when a process is addressing data. For example, a web browser is an agent which allows a human users to "move" inside the web space by jumping from document to document along web links, being each time exposed to a new context. This important notion of mobility inside information system was discussed by Dix and al. in [6].

Because ambient computing tends to blend the frontier between the real (physical) and virtual worlds, we think that a unified notion of context which could be applied to both kind of environments is desirable.

2.2 Unified Context

From the previous descriptions, we think that the notion of context refers to two main concepts: the first one is the idea of a "location" of an entity in a given space. The considered spaces may be of any kind: the physical space, the time, the space of idea, the space of web pages, the space of words from a language etc.

The second aspect which appears essential to us is the notion of "relativity": the notion of context involves relationships with other elements to define the location. The context cannot be considered as an "absolute location" (like a location defined by coordinates), but a relative location charaterized by the *proximity* of other elements.

Thus, we can define the context as the location of a (mobile) entity inside the spaces it belongs to, relatively to others objects inside these spaces.

We will not detail *Spaces*, which depend on the considered environment. We just need to think of them as mathematical spaces, populated with objects where relationships between these objects depend on the specific kind of spaces considered. For example, a computer memory can be seen as a one-dimensional address-space, populated with machine-word objects.

The notion of relative position means that objects of the space are comparable in some way (according to some dimension). Consider for example the physical space: one object can be compared to surrounding objects according to the euclidian distance.

More precisely, the context of an entity e is defined by a set of information items
$C = \{x_1, x_2, ..., x_n\}$ such as $\forall i, d(e, x_i) \leq T_i$

Where d is a distance function, giving a metric of the relative proximity of e against x_i, and T_i are thresholds defining the limits of the context.

The parameter T_i could be used to dynamically reduce or expand the limit of the context, and thus the set of objects considered as relevant. In the case of the physical space, this parameters is supposed to be set by the user as the notion of distance is meaningful for him.

The interesting aspect of this definition is that an entity may belong to several spaces (or one space with multiple dimensions), which is especially important in the ubiquitous computing model, where information systems define information spaces related to the physical space. For example, a PDA or a cell-phone includes both attributes related to the physical environment (like the current date/time and location), and attributes related to an information system (for example, the URI of a Web page associated with the current location).

Distance functions may be defined to represent either the distance according to one dimension, or over several dimensions. For example, one may define a distance function to sort relevant web pages accordingly to both physical distance and semantic relevance : $d(e, x_i) = d_{sem}(e, x_i)/d_{phy}(e, x_i)$.

2.3 Context in the Web

To build context-aware Web systems, and according to the previous definition of context, we have to precise the structure of the space considered. In the case of the Web, we can identify two common dimensions: the topical dimension, and the "temporal" dimension.

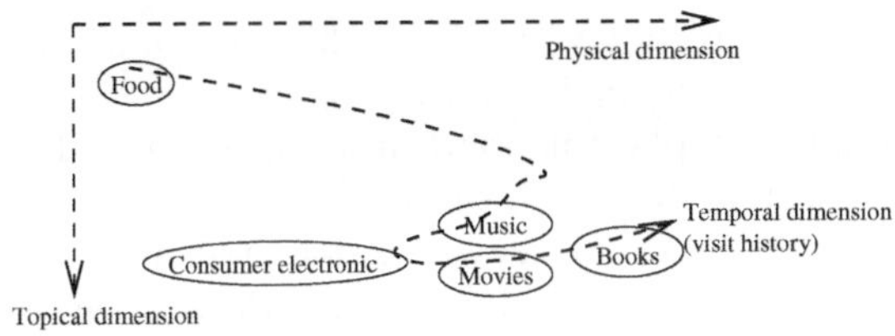

Fig. 1. Web contextual dimensions of a user visiting a shopping center

The first one refers to the topical proximity between Web pages. It may be measured by the number of hyperlink hops separating two pages, or by other means like the textual proximity between the pages (note that we could also consider two distinct dimensions, corresponding respectively to the hyperlink proximity and the textual proximity).

The temporal dimension corresponds to the information space defined by the history of the visited pages, which we all know and navigate using the "back" and "next" buttons of the Web browser.

To link the Web with the real world, we need to add a spatial dimension to the context, and associate Web pages or links with physical volumes. In such a Web system, implicit links between information are defined by the physical proximity. When navigating on such a Web system, the entity (human user, or agent) can change of navigation mode at any time, by simply switching to another dimension. Consider for example someone visiting a shopping mall enabled with such a web system; the user walks as usual to look for his favorite products, the browser navigates in spatial mode reflecting the links to nearby products. The user discovers an interesting item, but needs to have more information about the product: he just clicks the link corresponding to the product, making him "dive" into the virtual web by switching the dimension of navigation to topical. From this point, the user navigates on the Web (for example until he reaches the page of a review of the product).

The temporal dimension may be used to backtrack along both topical and physical dimensions. The figure 1 illustrates a web information space of a shopping center, with the pages positioned according to the different contextual dimensions (including the physical dimension). The picture shows for example that food is far away from cultural products in both physical and topical dimensions, but close in the visit history of the user: music is visited one hop after food).

3 Building and Navigating the Web Physically

The perspective of such a multi-dimensional Web browsing is quite exciting. Physical mobility becomes an integral part of Web browsing (the user is effectively walking on new kind of web links), while the already existing dimensions of the web enhances perception of the real world. From the perspective of our definition, the context of the mobile entity (the Web browser, which is itself an agent of the user) is its position inside a three dimensional (topical, physical and temporal) space. The context is thus a kind of cursor in the physical and information spaces where the mobile entity evolves.

While conceptually simple, this kind of applications will only emerge when location dependent web systems will become massively available. We propose a "physical" approach to the problem, which does not involve a logical representation of the physical space (like a Geographic information system).

3.1 Principle

In the physical approach, the data representing the context is implicitly specified in the physical space, by the physical arrangement of objects [5,4]. In such systems, control flow and information processing are directly done through physical objects placement and mobility, like in a mechanical machine, except that instead of processing physical forces it processes information.

The approach has two main advantages :

First, it is not needed to maintain the coherency between a logical representation of the physical world and the reality, which is crucial when the number of physical objects associated with information system grows, or when the objects are highly mobile.

Second, building and managing the information system is done implicitly by the management of physical objects. No work on an intermediary logical representation of the physical space is required, because the logic of the information system is represented in the physical space.

3.2 The SPREAD Programming Model

In our model, a spatial information system is constituted of a set of information, each information filling a sphere in the physical space. Computing in this so-called physical information system is supported by a the SPREAD middleware, which adopted a LINDA-like tuple-space approach [7].

In SPREAD, the contextual aspects of applications are programmed with a minimalist API, yet powerful. Four primitives are supported: `rd` and `capture` to read a single (respectively a set of) tuple(s), `out` to put a tuple in a specified physical area relatively to the publisher entity (actually a specified range, due to technological limitations), and finally `drop` to remove a tuple from the physical space. For more information about SPREAD, please refer to [4].

Note that when tuple range is infinite, our the primitives `rd` and `out` have the same semantic as LINDA ones because the tuple space is global and independant of physical objects location.

3.3 Spreading the Web in the Physical Space

We used the physical information system approach to realize our contextual web system. The idea is to "spread" data defining a web context directly in the physical space. The data are called contextual clues. These contextual clues are

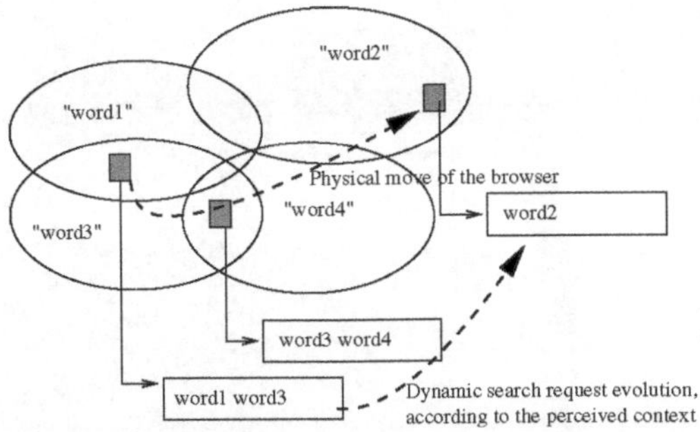

Fig. 2. Contextual analogy

published by physical objects, to cover a specified area. At a given point in the space, an application perceives a set of clues which contributes to define the context.

We used two types of clues: direct pointers, and textual clues.

Direct Pointers. In this mode, a physical object represents one or more web links (URIs). The object rays in a sphere around him (of a specified range) the URIs, which may be perceived by nearby entities (depending on the range). The resulting structure is a set of URIs filling different area in the physical space. These area may correspond to places, objects, machines, peoples and so on. As an entity moves, it navigates along the spatial dimension of the Web, implicitly existing through the surrounding objects.

Textual Clues. Direct pointers are not always adequate to describe a Web context. In many cases, a more fuzzy specification is required. For example, someone may only want to publish a specification of his interests, instead of a direct pointer to his personal Web page.

We propose the concept of contextual analogy to allow similar specification through the physical space : the proximity along the spatial dimension is considered analog to the proximity along another dimension (in this case, topical or textual dimension). Given this principle, we can fill the physical space with different clues defined by words or piece of text. A given point of the physical space will be covered by zero, one or more clues, thus defining spatially a web search request like shown on figure 2.

As an entity moves, the context integrated by the application from the visible clues changes according to the surrounding objects, dynamically building the current search request. Note that is also possible to combine clues from both the physical space and a user specified request or profile to further refine spatial browsing. We experimented the concept with textual clues, but we may imagine other forms.

Fig. 3. Internet Explorer driven by Webwalker in spatial mode

3.4 Programming the Web Application with SPREAD

We will now illustrate the use of SPREAD to build the contextual web application described in the previous section.

Filling the Physical Space with Links or Clues. Spreading the existing Web into the physical space is as easy as disposing in the space physical devices enabled with our system. These devices integrate a simple application which just issues `out(<t>,range)` operations for each URI or clues associated with the area surrounding (which depends on the specified range) the device. The tuples are formatted as follows:

- For URI, a tuple has the form :
  `<"URI", "http://...">`
- For textual clues, a tuple has the form :
  `<"TEXT_CLUE", "a piece of text">`

Prototype Overview. The idea exposed previously has been implemented and experimented : the contextual web application has been developed with the SPREAD system, and experimented inside our laboratory (figure 3) and the large technological museum of *La Cité des Sciences*, Paris. We give here an overview of the prototype architecture.

As we said earlier, the physical objects which may participate in the spatial information system should integrate autonomous computing and communicating ability to make the physical information system possible. We use iPAQ 3600 serie PocketPC PDAs running WinCE 3.0 and IEEE 802.11B wireless LAN PCMCIA cards.

The contextual Web browser (webwalker) is implemented as a local web proxy interfaced with Microsoft Pocket Explorer. This proxy is a SPREAD applications which integrates contextual clues from the surrounding objects. Depending on the mode of browsing, the clues (URI) are either forwarded directly to the standard web browser, or used to constitute a dynamic search request.

4 Conclusion

We conclude this paper by a highlighting some strong points of our approaches, and giving some research perspectives.

The idea of location sensitive Web browser was first introduced by Mobisaic [12]. The concept was refined further by the Cooltown project [8] which proposes an architecture to bridge the web to the physical world through wireless URI broadcast and sensing. The contextual Web we presented innovates in two ways :

The multi-dimensional context definition we proposed *integrates* the different aspects of the context, *both in the existing web (virtual) and the physical world.* This definition also allows to interpret mobility in this multi-dimensional space.

The second innovation resides in the physical approach we used: the information system is built from the physical space, by the contribution of many clues which are integrated to define contexts. Contextual interactions are all expressed in terms of a spatial programming model. The power of the model allows much more "spatial web applications" than browsing, like for example contextual notification in a museum or a shopping center when the user is near a bookmarked context.

From a broader perspective, the model proposed by SPREAD makes possible a radically new way of programming context-aware applications. It also allows to envisage a new concept of computing machines where information flow and actions synchronization is implicitly controlled by the placement and movements of smart mobile objects. The simplicity, power of expression and completeness of the LINDA-like programming model are strong advantages for designing and analyzing context-aware systems and spatial information systems.

An important topic of further research would be to re-consider the analysis of web accesses traces from the perspective spatial dimension. If spatial web browsing emerges, many results depending on analysis of web access locality in existing dimensions (temporal or topical) are likely to be re-considered. The spatial dimension will add complexity to these analysis as it may involve dimensions "switches", or multi-dimensional mobility : for example, a user may be browsing along the topical dimension while moving spatially.

References

1. G. D. Abowd, C. G. Atkeson, J. Hong, S. Long, R. Kooper, and M. Pinkerton. Cyberguide: A mobile context-aware tour guide. *ACM Wireless Networks*, 3, 1997.
2. Guanling Chen and David Kotz. A survey of context-aware mobile computing research. Technical Report TR2000-381, Dept. of Computer Science, Dartmouth College, November 2000.

3. K. Cheverst, N. Davies, K. Mitchell, and A. Friday. Experiences of developing and deploying a context-aware tourist guide: The guide project. In *MOBICOM'2000*, 2000.
4. P. Couderc and M. Banâtre. Ambient computing applications: an experience with the spread approach. In *Hawaii International Conference on System Sciences*, January 2003.
5. Paul Couderc. *Mobilité contextuelle dans les systemes d'information.* PhD thesis, Université de Rennes-1, 2001.
6. Alan Dix, Tom Rodden, Nigel Davies, Jonathan Trevor, Adrian Friday, and Kevin Palfreyman. Exploiting space and location as a design framework for interactive mobile systems. *ACM Transactions on Computer-Human Interaction*, 7(3):285–321, 2000.
7. David Gelernter. Generative communication in Linda. *ACM Transactions on Programming Languages and Systems*, 7(1):80–112, 1985.
8. T. Kindberg, J. Barton, J. Morgan, G. Becker, D. Caswell, P. Debaty, G. Gopal, M. Frid, V. Krishnan, H. Morris, J. Schettino, B. Serra, and M. Spasojevic. People, places, things: Web presence for the real world, 2000.
9. K. Nagao and J. Rekimoto. Agent augmented reality : A software agent meets the real world. In *Second International Conference on Multiagent Sytems*, December 1996.
10. J. Pascoe. Adding generic contextual capabilities to wearable computers. In *The Second International Symposium on Wearable Computers*, pages 92–99, October 1998.
11. B. N. Schilit, N. Adams, and R. Want. Context-aware computing applications. In *Proceedings of the Workshop on Mobile Computing Systems and Applications*, Santa Cruz, CA, December 1994.
12. G. M. Voelker and B. N. Bershad. Mobisaic – an information system for a mobile wireless computing environment. Technical Report TR-95-04-01, 1995.
13. M. Weiser. Some Computer Science Issues in Ubiquitous Computing. *Communication of the ACM*, 36(7):75–83, July 1993.

A Service Discovery Model for Wireless and Mobile Terminals in IPv6

Bilhanan Silverajan, Jaakko Kalliosalo, and Jarmo Harju

Dept. of Information Technology,
Tampere University of Technology,
P.O. Box 553, FIN-33101 Tampere, Finland
{bilhanan,kalliosa,harju}@cs.tut.fi

Abstract. As the mobility and the number of affordable, powerful, and highly portable devices becoming networked increases, so will the amount of networked services offered, managed and discovered. In this paper, we discuss the need and use of service discovery mechanisms in future fixed and mobile networks. In particular, we choose to focus on future enterprise networks that we anticipate would need to support both fixed and wireless terminals with IPv6 and Mobile IPv6 access. We describe service discovery mechanisms using the Service Location Protocol (SLP) in IPv6, and propose extensions to the protocol for utilisation by visiting mobile devices using Mobile IPv6, that will allow service discovery to be performed without breaking compatibility to standard SLP. The extensions introduce 2 new Agents for mobility detection and overcoming limitations of multicast usage by Mobile Nodes in foreign networks. Finally we describe our SLP implementation for IPv6.

1 Introduction

The much anticipated demand in supporting mobility in modern networks for pervasive and nomadic computing is rapidly being realized, spurred by the availability today of affordable, powerful, and highly portable computers. Next-generation networks are widely expected to support user, terminal and session mobility together with mobile hosts possessing location and context awareness. Closely tied to these expectations are issues regarding service location. As the mobility and the number of these devices becoming networked increases, so will the amount of networked services offered, managed and discovered. Often, manual configuration and service discovery administration for these services may not be possible without increasing the complexity of administration, especially in mobile and ad-hoc environments.

In general, the mechanism of locating a specific resource, object or service can be performed in two ways [1]:

The first is a passive, "Lookup" method which is initiated by a seeker. It requires the existence of some directory or other agent to answer the request. Successful lookup may be achieved by exact name or address, or by some matching criteria. Lookup may be done in a statically configured environment.

M. Conti et al. (Eds.): PWC 2003, LNCS 2775, pp. 385–396, 2003.

The second is an active, more dynamic "Discovery" method, performed without the assistance of an intermediate entity. Entities use discovery mechanisms to spontaneously locate each other without previous configuration or knowledge of other entities. In many cases in the absence of configuration, discovery is often the first step to locate lookup services for subsequent interaction.

Today, "service location" implies substantially more than simply locating network printers or web servers. There is a widespread need for service location methods ranging from basic bootstrapping operations at low-level system operations such as network-based boot-up for diskless computers, to high-level demands for locating objects and architecture-specific services needed by application-level programs in distributed systems. As an example, [2] demonstrates how the CORBA Naming Service [3], a traditionally statically configured lookup service for CORBA [4] applications was enhanced with dynamic service discovery by using the Service Location Protocol (SLP) [5]. The Naming Service itself was implemented with an LDAP [6] directory service that also supported LDAP-based lookups for accessing entries.

When considering mobile terminals in conjunction with wireless and mobile networks, a naturally inherent need arises for service location, because terminal and user mobility would be high. Location awareness for computing needs, as well as context preservation while roaming, would become important factors. All kinds of local network and application services would need to be dynamically discovered, used, or even offered, and the likelihood is small that these mobile terminals possess prior knowledge of foreign visited networks.

In this paper, we choose to focus on future enterprise networks that we anticipate would need to support both fixed and wireless terminals with IPv6 and Mobile IPv6 access. We describe service discovery mechanisms using the Service Location Protocol (SLP) in IPv6, and propose extensions to the protocol for utilisation by visiting mobile devices using Mobile IPv6, that will allow service discovery to be performed without breaking compatibility to standard SLP.

Section 2 quickly introduces SLP in its present form for IPv4 networks. Section 3 then discusses the use of SLP in IPv6 and how we propose to extend its functionality for use over Mobile IPv6. Section 4 describes our IPv6 SLP implementation and Section 5 presents our conclusions.

2 Overview of SLP

Although SLP has been standardized several years ago, its usefulness is only just becoming apparent. In IPv4, SLP has formed the basis of several modern discovery mechanisms today in AppleTalk, Novell Netware and Sun Microsystem's Jini technology. An open-source implementation of SLP, called OpenSLP [7] is available in addition to Sun Microsystem's own version, shipped with Solaris 8 and 9. Test versions for research purposes also exist [8].

SLP standardization efforts in the Internet Engineering Task Force (IETF) have produced several Requests for Comments (RFCs). The initial protocol version and its components for IPv4 were described in 1997 [9]. Then, a second version of the protocol was introduced in 1999 [5]. Among other things, SLPv2 rectified race conditions that existed in SLPv1, introduced clearer usage for service scopes, and added LDAP compatibility for string encoding of attributes and search filters. Though SLP v2 introduced slight incompatibilities with SLPv1, it provided a migration path away from the original by allowing a limited level of backward compatibility; the general components and overall capabilities, however, have been largely preserved. Today, SLPv2 remains the more prevalent of the two versions in use for service discovery in IPv4 networks. It is this version that all subsequent references to SLP in this paper are based on.

Briefly, using SLP involves introducing 3 types of agents in a network: The Service Agent (SA), User Agent (UA), and Directory Agent (DA). The UA is used by client applications on their behalf to discover the locations of services, by querying the network with SLP request messages containing a "service:" string, which may be formulated as a URI, and returning responses received from the network to its applications.

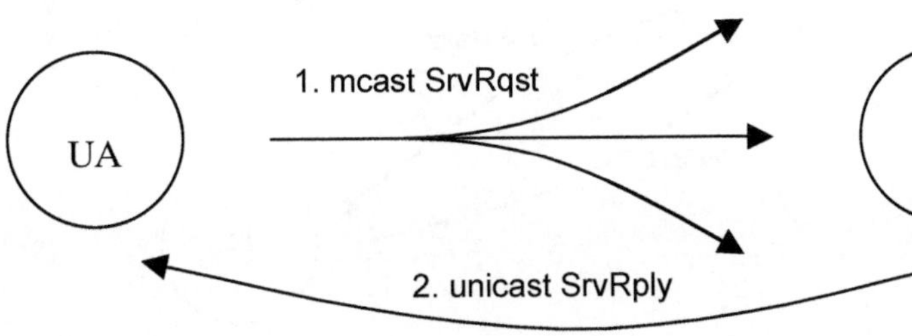

Fig. 1. Service discovery with UA and SA

Generally, SAs tends to reside on the same hosts as server applications, but they could reside elsewhere in the network too. Server applications register and unregister their service with SAs with a Service Type string URL. For example, a print server might register itself with an SA with the URL

"service:printer:http://printerlocation.mycompany.com".

The SA then listens for UAs multicasting or unicasting requests for a printing service, such as "service:printer" or "service:printer:http", and provides unicast replies containing the location of the printer. SAs also unicast SA Advertisement messages on the network if solicited by a UA. Figure 1 depicts a User Agent multicasting a service request, and receiving a unicast response.

The DA is an optional element in SLP. If the DA is present in the network, it periodically announces its presence with DA Advertisement multicast messages. The main role of the DA is to provide a centralized service for all service announcements in the very large network, so that a point of single contact exists for a UA trying to discover various services. Thus, SAs are required to register their services with DAs they discover from DA advertisements. UAs also interact directly with the DA instead of SAs, if one is present in the network. Figure 2 demonstrates how DA Advertisements, SA Service Registrations and UA Service Requests proceed in a network.

In SLP, UAs, SAs, and DAs can be flexibly configured statically (manually) to find each other prior to start-up, or dynamically to use multicast and broadcast discovery mechanisms for Request messages. All multicast messages use the administratively scoped multicast address 239.255.255.253. Responses however, travel directly point-to-point.

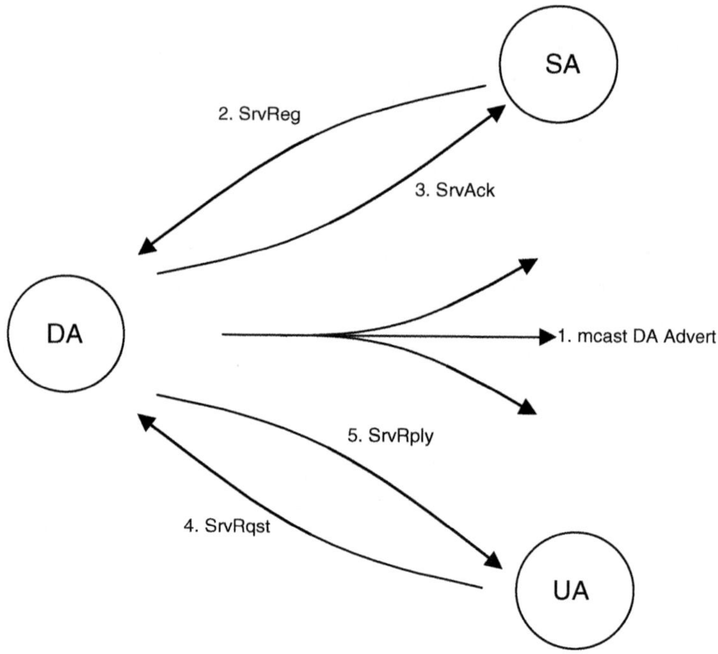

Fig. 2. Service discovery with UA, DA and SA

Apart from simple service request and reply messages, the protocol also defines messages that allow defining various other attributes, service types, and security enhancements through the use of digital signatures for message content verification. Optional extensions such as attribute lists are also specified [10].

3 Using SLP in IPv6 and Mobile IPv6

In 2001, the use of SLP in IPv6 networks was standardized [11], although at the time of this writing, despite its popularity in IPv4 networks, commercial and open-source SLP implementations for IPv6 remain few or non-existent.

Using IPv6 as a network protocol for SLP introduced some major differences from IPv4 in the way the protocol works. Service requests using broadcast methods are replaced with link-local multicast. Also, instead of a fixed multicast address for SLP messages, IPv6 multicast addresses are dynamically calculated using a hash algorithm for the various "service:" types. The addresses range from FF0X:0:0:0:0:0:1:1000 to FF0X:0:0:0:0:0:1:13FF, and DA advertisements use FF0X:0:0:0:0:0:0:123. The value of X can be 1, 2 or 5 and thus defines the scope of the service to be either node-local, link-local or site-local. Lastly, the notion of an SLP zone is introduced. A zone is a connected region of topology of a given scope. For example, the set of links connected by routers within a particular site, and the interfaces attached to those links, comprise a single zone of site-local scope. Agents are not allowed to propagate advertisements or respond to requests across different zones.

In contrast to fixed IPv6, Mobile IPv6 allows an effective way for computers to remain connected to the Internet with one IP address, while roaming across different subnets or organizational networks [12]. Each mobile node (MN) is always identified by its home address, regardless of its current point of attachment. While situated away from its home, a mobile node is also associated with a Care-Of Address (COA).

3.1 MN Usage of SLP in Home Network

A mobile node that is connected to its home link functions in exactly the same way as a fixed IPv6 node. No modification is thus needed in its interaction with other nodes in its home network. Thus, it is able to use all the features of SLP, different classes of scoped multicast, and use and offer services with the UA, SA and DA functionalities as specified in [11].

3.2 MN Usage of SLP in a Foreign Network

When MNs roam across networks using Mobile IPv6 wirelessly or otherwise, we cannot assume that they are always able to rely on static configuration for rapidly locating services in their immediate vicinity. Dynamic discovery is needed, especially in foreign networks the MN has no previous knowledge of. Using SLP would serve this purpose well for locating local site-based services, whenever the MN moves into a new network.

Mobility detection is thus useful in this context for SLP to perform discovery automatically and immediately when the MN moves across into a new network. However, Mobile IPv6 currently shields this from the upper layers, aiming to keep applications unaware of the terminal mobility, which presents a technical difficulty to SLP.

The current Mobile IPv6 specification prescribes two completely different, but equally acceptable choices for all Mobile IPv6 implementations to support routing multicast packets to and from MNs in foreign networks [12]: In the first method, the MN becomes a direct member to the local multicast router in the foreign network. In the second method, the MN remains a member of its home network and all multicast packets are bi-directionally tunnelled to and from its home agent (HA).

From the perspective of direct site-local multicast by the MN, these two methods are clearly incompatible with each other. The second method would also breach SLP zone enforcement rules for discovering services in a foreign network. There is no way for an MN to clearly determine beforehand which multicast behaviour is in force, since it is implementation dependent. Also, currently none of the present Mobile IPv6 implementations support either of the above 2 methods. Thus proper and consistent multicast usage by the MN when it is not in its home network might very possibly become limited to only the link-local scope, where packets are guaranteed to reach all nodes in the same link, be it in the home network or foreign network.

To overcome the above-mentioned difficulties, we propose extending SLP with 2 new agents. When using Mobile IPv6 and roaming, an SLP Visiting Agent (VA) resides in the MN performing link-local discovery of an SLP Access Agent (AA), sitting on the local link in the foreign network that helps the MN and VA find site-local services.

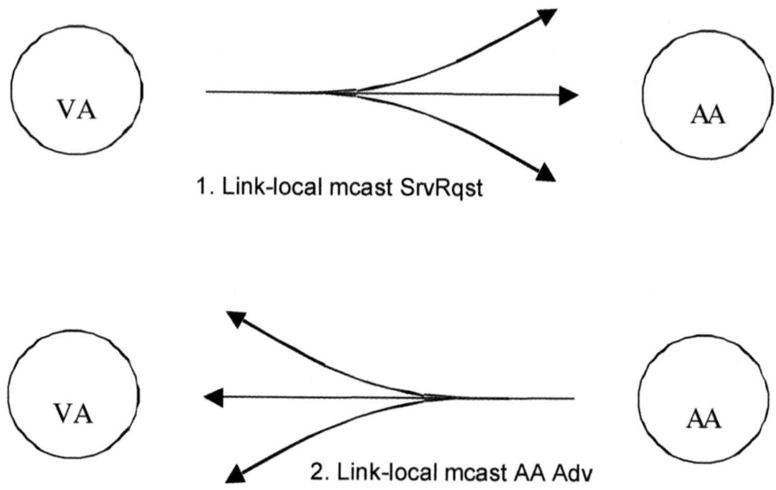

Fig. 3. Active AA Discovery

The VA can either actively solicit, or passively listen for, periodic AA advertisements in the foreign network. This is similar to how UAs and DAs interact in a fixed network, with the exception that multicast always uses link-local scope, and replies also use multicast instead of unicast. This allows other VAs present in other MNs to detect the AA quicker without resorting to active discovery or creating a potential multicast avalanche in the link. Fig 3 shows how active AA discovery is performed when a VA enters into a new network.

AA advertisements contain a "service:access-agent:<address>" URL using the multicast address FF02:0:0:0:0:0:1:1259. This address is calculated using the hash algorithm described in section 4.1 of RFC 3111 [11]. In addition to the source address, AA Advertisements can also piggy-back the address of a local site-wide DA, if one exists in the foreign network.

MN mobility detection is performed by the VA by observing changes in the AA Advertisements. When the source address in the advertisements change, the VA can determine whether the MN has moved or not by checking its Care-Of Address (COA). If the COA has not changed, it implies the MN has not moved, but a new AA has instead become active. In this case, all interaction now proceeds with the new AA. On the other hand, if AA Advertisements become completely absent, the VA confines itself to link-local services only after determining that unicast reachability to the AA and DA (if one is present) has failed.

Because the AA sits on a fixed node, it can perform proper multicast routing for site-local discovery on behalf of the VA. It can therefore relay whatever site-local Service Requests the VA might have, and forward responses it receives from the network back to the VA. The AA also serves to announce all services that the VA may wish to advertise to the foreign network. Viewed from this perspective, the VA and AA can together be seen as a decoupled hybrid User/Service Agent-pair working on behalf of applications in the MN to both utilize and provide services locally in the foreign network.

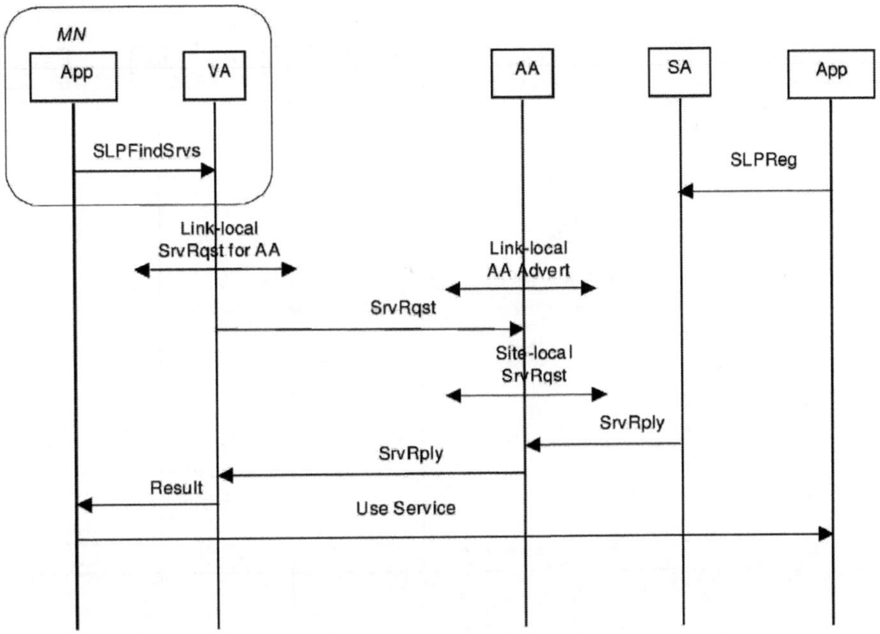

Fig. 4. MN using services in foreign network

Fig 4 illustrates the sequence of events taking place when an MN enters a new network and wants to use site-local services. Here, the VA performs an active AA discovery. Upon receiving an AA Advertisement, the VA can check if there is a DA present in the foreign network. If a DA is present, the VA may optionally choose to communicate directly with it to find site-local services in the event of a communication failure with the AA. In this example, the foreign network has no DA present, so all site-local service discovery proceeds via the AA. The AA behaves as a normal UA, and performs site-local multicast discovery on behalf of the VA to discover services directly from local SAs. The received replies are then forwarded back to the VA.

Fig 5 describes what happens when the MN enters into a new network and wants to also provide site-local services. In this foreign network, an organization DA is present which announces its presence with periodic site-local DA Advertisement multicast messages. This is noted by the AA, which piggybacks the advertisements onto its own periodic AA Advertisements. The VA first performs mobility detection tests to discover whether the MN has moved. Once it ascertains that the new AA Advertisements do indeed originate in a new network, it then proceeds to unregister all previously registered services from the old AA in the previous network (depicted in dotted lines) using unicast messages.

The VA then begins to register services with the current AA. The AA now emulates a normal SA mechanism. If no DA was present, it would perform SA Advertisements and respond to UA multicast service requests. Here, it registers the VA's services with a site-local DA with the service type URL containing the original location as specified by the VA. Service Acknowledgement packets are returned by the AA to the VA once service registration in the foreign network is successful.

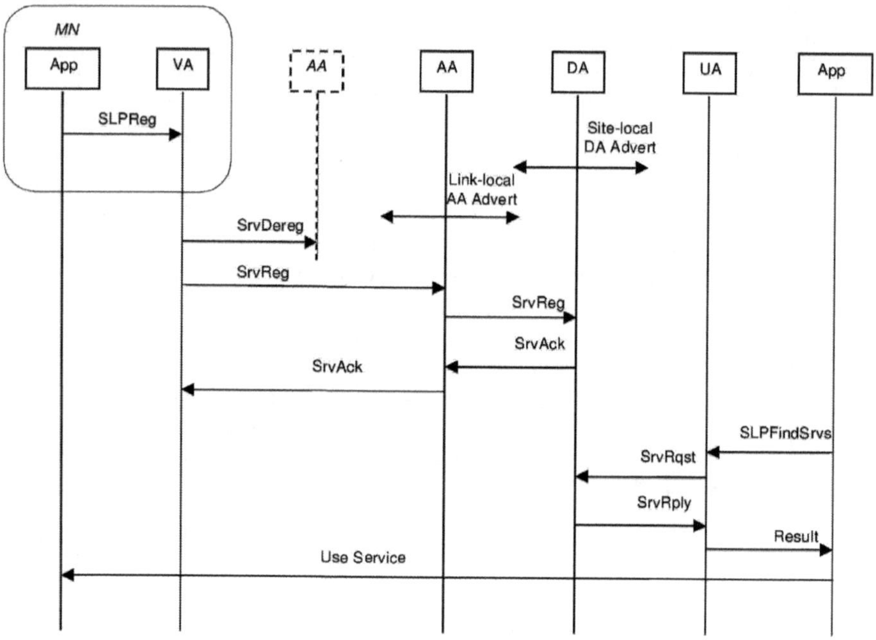

Fig. 5. MN providing services in foreign network

In general, the VA always listens on the link-local scope for service requests, and the AA on the site-local scope. The VA interacts directly with a DA only if an AA is unreachable with both unicast and multicast methods. If there is no local AA, the VA can do a link-local multicast DA discovery. If successful, it registers its services with the DA and can listen for service requests on the link-local scope. It also can use site-local services that it finds from the DA. If no AA or DA is present on the link, the VA only listens for service requests and uses services found with link-local multicast. This is summarized in Table 1.

Table 1. Agent Presence and Service Scope

		AA present	DA present	Absent
VA Service Scope	Service Usage	Site-Local	Site-Local	Link-Local
	Service Provision	Site-Local	Site-Local	Link-Local

4 SLP Implementation for IPv6

This section illustrates our SLP UA, SA and DA implementations for IPv6. These agents were implemented with the use of DOORS [13], a publicly available event-based C++ framework. DOORS facilitates the making of portable and advanced communication software ranging from socket and protocol implementations to object oriented CORBA applications. It abstracts applications into a set of stateful event-driven tasks communicating via asynchronous messages transmitted and received via their ports.

Figure 6 illustrates a simplified UML diagram depicting the classes making up the overall architecture developed for the UA, SA and DA. The SLPPeer class is a common class that is used by all 3 agents, implementing the basic PDU encoding and decoding methods for all the SLP messages.

DAMain and DAMainStateMachine implement the DA functionality. All service registrations are currently stored in memory within a C++ STL vector container, but with a little effort, its functionality can be extended to support the integration of an LDAP-based back end to store service registrations using the schema template defined in RFC 2609 [14].

Applications needing SLP UAs use the UAMain objects. For each service location request received from the application, the UAMain object creates a new UAConn object. The newly created UAConn subsequently sends the request either to the DA residing in the network directly, or to the proper multicast group. The UAConn object is also responsible for resending the request if necessary.

The architecture of the SA is similar to that of the UA. Applications send service registration messages to the SAMain object, which creates new SAReg objects. Each SAReg object is responsible for 1 service registration, and it joins to the correct multicast group to respond to any SLP requests it receives.

Although the UAMain and SAMain tasks are responsible for routing an incoming message to the proper child task for handling, the UAConn and SAReg objects directly use their parents' corresponding socket connections and ports for sending messages to the upper layers, applications or other SLP Agents. As the child tasks are able to communicate with their applications and the network without encumbering their parent tasks, this simplifies implementation and significantly reduces messaging overhead.

The UDP6Task object is supplied by the DOORS framework, and provides a uniform and simple message-based interface to send and receive IPv6-based UDP datagrams to and from the network. It is capable of understanding unicast and multicast, and is able to join and listen to multiple multicast groups. The PTask class, also part of the DOORS framework, is a base class containing commonly functionality for protocol development, such as State Machine handling functions and specialised methods which understand communication through User and Provider Service Access Point and Peer Ports. All the five main task classes implementing the three agents of the protocol, are derived from the DOORS PTask class.

5 Conclusions

SLP is a simple and elegant protocol that can be flexibly configured to custom-fit the needs of service location in various types of enterprise networks. Active work will continue with our research and implementation efforts towards extending SLP to support dynamic discovery for mobile terminals within core IPv6 networks.

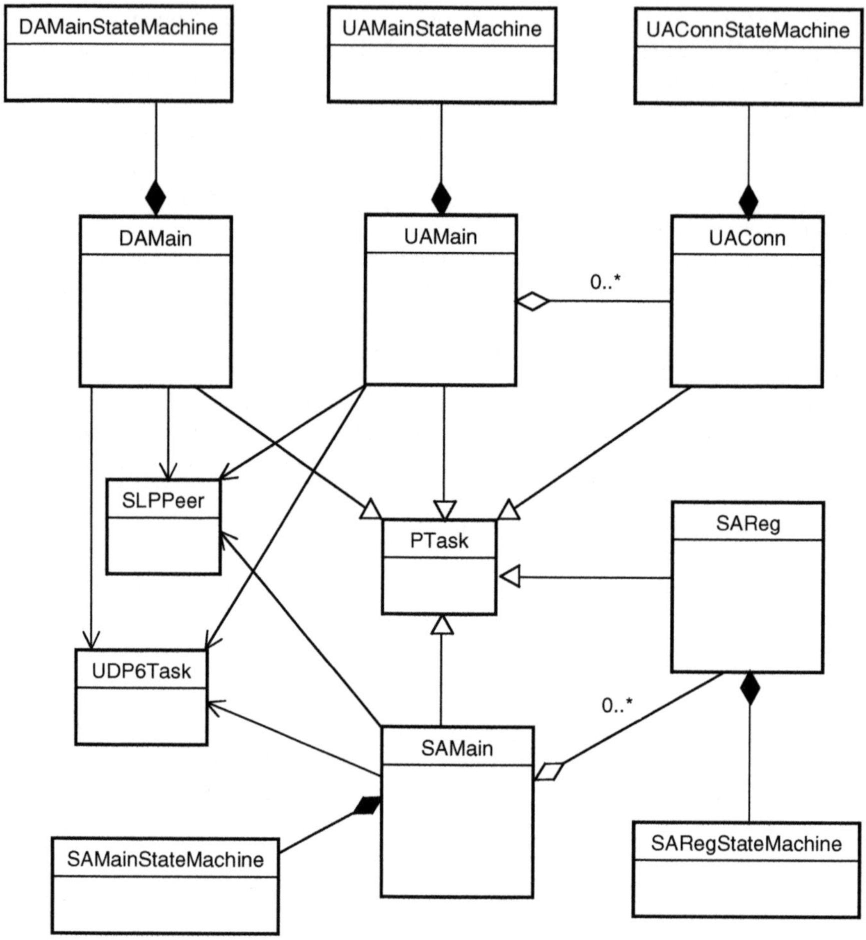

Fig. 6. DOORS SLP design for IPv6

Effort is also being spent on understanding the influence of mobile computing paradigms in IPv6 networks on the evolution of distributed object oriented systems and architectures, particularly with regards to the role of lightweight and flexible service location and discovery methods for both clients and servers. This is especially interesting because, owing to the smaller granularity and size of object-oriented services and servers, there is a high likelihood that they would reside in portable devices, making these server applications just as mobile as client applications.

In IPv4, SLP is a mature and popular protocol that has been put to good use, and interaction with many types of applications and services is well understood. IPv6 itself boasts an undeniably rich and flexible feature set for fixed and mobile networks. However, experience has yet to be gained from using SLP in IPv6. The protocol modifications that RFC 3111 [11] introduces for IPv6 changes the fundamental methods of how dynamic discovery and scoping are done, and consequently render it incompatible with its IPv4 counterpart.

In effect, SLP in IPv6 can be regarded as a new protocol whose behaviour and supporting network requirements need to be thoroughly investigated. With the mobility extensions we propose to SLP, MNs will in future easily be able to tap into an SLP deployment in IPv6 networks, by neither breaking any compatibility with fixed IPv6 services nor introducing the need for a completely new service discovery architecture. RFC specification guidelines of not allowing SLP Advertisement and Request/Reply packets to cross SLP zones, are preserved too.

Static configuration of the VA within the MN for service discovery is unnecessary, and this does not also enforce any requirement as to how the agents such as the AA discover other agents in the fixed network, or should be configured.

Measurements have yet to be made as to the optimal duration between AA Advertisements. The need to rely on link-local multicast is required only within the context of the initial AA discovery, since all subsequent communication between the AA and VA use unicast messages. However, in order not to populate every single link in the foreign network with an AA, sane network engineering practices must be applied to deploy AAs only in subnets which are expected to serve MNs and provide services. Also, the widespread use of IEEE 802.1Q Virtual LAN (VLAN) technology in many networks allow much flexibility in segmenting networks logically using VLAN identification numbers (VLAN IDs) without having to be physically on the same link to receive link-local messages; many switches already in common use today support multiple simultaneous VLAN IDs. For example, this could be used to let one AA belong to several VLANS, allowing link-local multicast to reach all of them.

Finally, SLP uses digital signatures for content verification of messages. There is nothing in the proposed model which weakens or strengthens this technique. Agents may verify digital signatures provided in advertisements, but the responsibility of authenticating Mobile Nodes in foreign networks lies with Mobile IPv6.

References

1. Robert E. McGrath: Discovery and Its Discontents: Discovery Protocols for Ubiquitous Computing, Presented at Center for Excellence in Space Data and Information Science, NASA Goddard Space Flight Center, April 5, 2000
2. Bilhanan Silverajan, Joona Hartman, Jani Laaksonen: Investigating Service Discovery, Management and Network Support for Next Generation Object Oriented Services. Proceedings of Smartnet 2002, Saariselkä Finland April 8–10, 2002
3. OMG: Naming Service Specification, February 2001
4. OMG: CORBA v3.0: The Common Object Request Broker: Architecture and Specification, July 2002
5. IETF RFC 2608, "Service Location Protocol, Version 2", June 1999
6. IETF RFC 2251, "Lightweight Directory Access Protocol (v3)", Dec 1997
7. The OpenSLP Project, http://www.openslp.org
8. Christian Bettstetter and Christoph Renner, "A Comparison of Service Discovery Protocols and Implementation of The Service Location Protocol", Proceedings of 6th Open European Summer School EUNICE 2000, Enschede, The Netherlands, September 13–15, 2000
9. IETF RFC 2165, "Service Location Protocol", June 1997
10. IETF RFC 3059, "Attribute List Extension for the Service Location Protocol", February 2001

11. IETF RFC 3111, "Service Location Protocol Modifications for IPv6", May 2001
12. IETF Mobile IP Working Group: Mobility Support in IPv6, draft-ietf-mobileip-ipv6-22.txt, May 26, 2003
13. Bilhanan Silverajan, Ilkka Karvinen, Joona Hartman, Jani Laaksonen: Enterprise-level Integration and Interoperability in Future Networks with DOORS Middleware. IFIP WG6.7 Workshop and EUNICE 2002 Summer School on Adaptable Networks and Teleservices, Trondheim, Norway September 2–4, 2002
14. IETF RFC 2609, "Service Templates and Service: Schemes", June 1999

Comparing SOAP Performance for Various Encodings, Protocols, and Connections

Jaakko Kangasharju, Sasu Tarkoma, and Kimmo Raatikainen

Helsinki Institute for Information Technology
PO Box 9800, 02015 HUT, Finland
Tel: +358 50 384 1518, Fax: +358 9 694 9768
{jaakko.kangasharju,sasu.tarkoma,kimmo.raatikainen}@hiit.fi

Abstract. SOAP is rapidly gaining popularity as the Web service protocol. At the same time, small mobile devices with wireless access, in particular to the Internet, are becoming more prevalent. At first look, it would seem that SOAP as a protocol consumes quite a lot of network bandwidth and processor time. Therefore its suitability for small devices and wireless links needs to be evaluated. This paper presents two optimizations that can be applied to typical uses of SOAP, message compression and persistent connections, and measures their performance in some common situations. Asynchronous messaging with SOAP is also treated briefly. The measurements indicate that a suitable compression scheme can save bandwidth substantially, and that the protocols underlying the typical use of SOAP can be improved considerably in the presence of unreliable high-latency networks.

Keywords: Measurement Of Wireless And Mobile Systems, Mobile And Wireless Applications, Web Services Over Mobile And Wireless Networks

1 Introduction

Web services are a rising phenomenon in the network service world, and SOAP is the protocol of Web services. One of SOAP's benefits is that, being XML, it is more human-readable than binary protocols and therefore easier to debug. Another often-mentioned benefit is the use of the ubiquitous HTTP as a tunneling protocol, which e.g. allows SOAP messages to penetrate firewalls.

A visible trend in future communications is a large increase in the use of mobile terminals. These include small devices with limited processing power and a wireless connection to larger networks. Any future ubiquitous communication protocol should preferably be usable even on these devices. The main difficulties in wireless communication compared to fixed links are significantly lower bandwidth, which limits the amount of data that can be sent, and higher latency, which limits the number of round trips a protocol can effectively make within a communication.

The most common version of SOAP currently in use is SOAP 1.1 [9] while SOAP 1.2 [10] is being prepared by the World Wide Web Consortium. The most

M. Conti et al. (Eds.): PWC 2003, LNCS 2775, pp. 397–406, 2003.

common underlying protocol used for SOAP messages is HTTP 1.0 [4]. SOAP 1.1 requires messages to be encoded as XML documents, which are not designed to be space-efficient. While SOAP messages are fundamentally one-way, a common use of SOAP is for synchronous RPC, which requires the application to wait for a network round trip for each message sent. Additionally, HTTP 1.0 supports only a single exchange per TCP connection, so each invocation requires an additional round trip to first establish a connection. Thus typical current use of SOAP would seem to accentuate the main problems in wireless communication.

The size of XML documents can possibly be overcome by compressing them in some manner. Unfortunately, SOAP messages are often quite small, so typical compression algorithms, such as the popular Gzip [5] supported by many HTTP clients and servers, will not manage very well. There are indications that a binary representation of XML, such as WAP Binary XML [11], could do better than traditional compression in this case [7].

The latency problems caused by the need to reopen connections when using HTTP 1.0 could be solved simply by moving to HTTP 1.1 [6], which supports persistent connections. The SOAP programming framework could also offer asynchronous operations, which could eliminate the effects of high latency for client applications not needing synchronicity. Asynchronous operations could, for example, be implemented on top of HTTP's synchronous request-response, but it would also be possible to move to full asynchronicity even at the transfer protocol level.

The problems of TCP over wireless links are well known [1] and several improvements to TCP have been proposed [3]. Our work here is more concerned on the impact of different TCP usage patterns on application performance, rather than gaining improvements by modifying TCP itself. Apart from the compression approaches mentioned above, a good XML-specific compressor is XMill [8], which is, however, better suited to large documents. The MPEG-7 standard [2] also includes a specification for a binary XML format.

In this paper we examine performance implications of alternatives to the HTTP protocol binding for SOAP. The alternatives include use of plain TCP, Gzip compression of XML documents, persistent TCP connections, and cache-based tokenization of SOAP message items. The rest of the paper is organized as follows: In Sect. 2 we describe what tests were run and in which environments, in Sect. 3 we present our results, and Sect. 4 gives the conclusions based on the results, as well as plans for extending this work in the future.

2 Test Description

A test system with easily exchangeable protocol implementations was set up to measure the effects of the two above-mentioned optimizations, compression of messages and persistent connections. Tests were run on several different connections to see how link quality degradation affects protocol performance. These tests were intended to identify the benefits and drawbacks of different optimizations.

Table 1. The message exchange scenarios used in the tests

Scenario	Description
Deploy	The client sends a 391-byte XML document to the server in a SOAP message (total size 655 bytes) and receives a 31-byte XML document back in a SOAP message (total size 296 bytes)
Stress	The client sends, as quickly as possible, 17472[a] times a 258-byte XML document in a SOAP message (total size 652 bytes), each message receiving a 195-byte XML document back in a SOAP message (total size 464 bytes)

[a] 17472 equals $2 \times 24 \times 364$; the test system was based on a calendaring application, so the Stress scenario measures the cost of filling every half-hour of a year (minus a day)

The SOAP framework used in testing was Apache Axis 1.0 running on Debian GNU/Linux 3.0 with Linux version 2.4.18. The machines used for testing were one desktop PC and two laptop PCs with the desktop PC having a fixed network connection and the laptop PCs having a choice of fixed and wireless connections. The desktop PC had a 1333 MHz AMD Athlon processor and 512 MB of main memory. The laptop PCs's model was HP Omnibook 500 with a 500 MHz Intel Pentium III processor and 512 MB of main memory.

During testing the machines were in normal multi-user operation, but with no other significant computation proceeding at the same time. Apache Axis is implemented in the Java programming language; the Java versions used were Java 2 SDK 1.4.1 on the desktop and Java 2 SDK 1.3.1 on the laptops, both from Sun Microsystems. All network traffic generated by the tests was captured with the Ethereal network traffic analyzer and saved for closer analysis.

Two basic scenarios, shown in Table 1, were designed. The intent of the Deploy scenario was to measure the performance of a single typical invocation. Accordingly, the invocation consisted of an Apache Axis deployment descriptor, which is used to initialize a service with Axis. The Stress scenario was designed to measure the performance of the system under a heavy load.

The protocols shown in Table 2 were intended mainly for testing the effects of compression and persistent connections. The single asynchronous protocol was only included as a rough start; it is not suitable for actual SOAP use, since it is unable to correlate different messages together without major changes, and is therefore incapable of proper two-way communication. None of these protocols (apart from HTTP) is really designed for serious use. However, their performance characteristics should be indicative of those of actual protocols with similar designs.

As can be seen from the descriptions, each new protocol is a minor modification of some other protocol. This is to maximize the chances that differences in the protocol performance numbers are actually the result of the changes made and not just random chance. For example, it was noticed that leaving the application-level message buffer too small for the actual message caused many-fold worsening in processing times; this was a clear difference between the needs of the compressing and the non-compressing protocols.

Table 2. The SOAP transfer protocols used in the tests

Protocol	Description
HTTP	The standard SOAP-over-HTTP binding shipped with Axis; synchronous request-response is implemented with HTTP's corresponding messages; HTTP 1.0 is used so a single connection is used only for a single interaction
PTCP	Messages consist of a content length in decimal followed by the actual SOAP content as XML with messages sent directly over a TCP connection; each connection is closed after a single request-response interaction; shipped with Axis for demonstration purposes
TCPZ	The PTCP protocol modified to compress each message's content with Gzip prior to sending
Pers	The PTCP protocol modified to keep connections open to permit several interactions to be carried out over a single TCP connection
Perz	The Pers protocol modified to compress each message's content with Gzip prior to sending
Bper	The Pers protocol modified to compress each message's content with an XML tokenizer prior to sending
Pera	The Pers protocol modified not to send response messages to permit the client to continue processing immediately without needing to wait for the server's response

The tokenizer used for Bper is a simple binary format inspired by WAP Binary XML [11]. It represents each XML element and attribute name as binary values, caching new names for later use. Standard element and attribute names used in SOAP messages are pre-cached. Integers, booleans, and dates are represented in a compact binary form, where the length of an integer's or a date's representation depends on its size.

The implementation of the tokenizing in Bper is quite suboptimal. Due to unfamiliarity with the Axis implementation at the time, it had to generate the binary representation from the message's XML representation. With the tokenizer implementation this necessitated parsing the XML at the sending end in addition to the ordinary parsing done at the receiving end. Therefore the main interest in the Bper protocol is the amount of data sent, and not the execution time.

The main intent of the experimentation was to measure SOAP's suitability for wireless environments. Therefore the Stress scenario, where differences should show more markedly, was run over four different connections, as described in Table 3, to measure the effect the network's degradation would have. The precise configuration of each connection is shown in Fig. 1. The Wireless LAN card used was Nokia D211 operating in ad hoc mode. Round trip times for the connections were measured with the ping program, i.e. ICMP Echo packets, and are given in minimum/average/maximum format, or as a single number where there was no variation.

Table 3. The connections used in the Stress scenario

Connection	Description
Localhost (LH)	Client and server on the same machine, communicating over the localhost network interface; round trip time 0.0 ms
LAN	Client and server on different machines, both connected to the same LAN; round trip time 0.1 ms
WLAN	Client and server on different machines, both connected to the same Wireless LAN; round trip time 2.3/2.5/4.4 ms
Routed WLAN (RW)	Client and server on different machines with a five-hop network route between them, the first hop of which is a Wireless LAN link; round trip time 9.2/13.8/25.8 ms

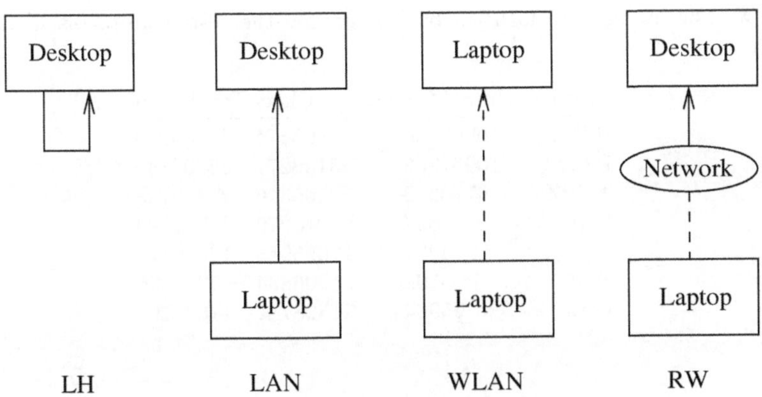

Fig. 1. Test configuration for the different connections; solid lines indicate localhost and wired connections, dashed lines indicate wireless connections

3 Test Results

In both scenarios, measurements were made of the amount of data each tested protocol sent over the network. The measured amount was further split into pure TCP packets (typically containing SYN, FIN, or ACK) and packets containing actual data. The count for data packets includes the IP and TCP headers. Total execution time was measured in the Stress scenario at the application level on the server side over all four tested connections. There was no noticeable fluctuation in the amount of data sent with different connections.

The amounts of data sent in the Deploy scenario are shown in Table 4. There is more variation in the results than would be expected, especially concerning the amount of pure TCP packets. This is explained by noting that the total number of packets differs among protocols, mostly depending on how much application-level buffering each protocol does. The amount of data sent by the Pera protocol is low only because it does not send a response message; if it did, its amount of data should be approximately equal to that of PTCP and Pers.

Table 4. The amount of data (in bytes) sent by the tested protocols in the single-request-response Deploy scenario

Protocol	Pure TCP	Data packets	Total data
HTTP	628	1399	2027
PTCP	628	1160	1788
TCPZ	560	678	1238
Pers	684	1165	1849
Perz	628	681	1309
Bper	560	583	1143
Pera	628	796	1424

Table 5. The amount of data (in bytes) sent by the tested protocols in the Stress scenario

Protocol	Pure TCP	Data packets	Total data
HTTP	10969736	27365474	38335210
PTCP	10967868	23138277	34106145
TCPZ	9773372	12566356	22339728
Pers	3824	21018279	21022103
Perz	4232	12655943	12660175
Bper	5592	5206834	5212426
Pera	95964	11528744	11624708

The amount of actual data sent is what would be expected. HTTP sends more data than the other protocols, since it has more extensive message headers. A protocol usable in practice would definitely require more headers than the simplistic message length included in PTCP (and by extension, all the others).

The effect of compression is relatively minor: Gzip compresses by approximately 40% and the binary representation by 50%. When taking also in the account the proportion pure TCP data takes of the communication, the compression does not have much advantage in this case.

The amounts of data sent in the Stress scenario are shown in Table 5. As mentioned earlier, these amounts remained effectively static independent of the connection used, so these numbers apply to all connections.

In this case the advantages of persistent connections are clearly visible, with the persistent protocols piggybacking almost all of their TCP ACKs in data packets. This permits the uncompressing Pers protocol to achieve a lower amount of total sent data than even the compressing TCPZ.

Again, as before, the amount of data sent by the Pera protocol is somewhat misleading. By not sending response messages, it manages to cut down the amount of sent data by half without any compression. If there were response messages, its amount in data packets should approach that of PTCP and Pers.

Table 6. Average execution times in milliseconds with mean deviations measured in the Stress scenario over the Localhost and LAN connections

Protocol	LH Total	LH Single	LAN Total	LAN Single
HTTP	130336±0.09%	7.460	145049±1.06%	8.302
PTCP	106698±0.13%	6.107	155599±0.66%	8.906
TCPZ	158209±0.11%	9.055	204916±0.67%	11.728
Pers	92819±0.12%	5.312	128290±0.12%	7.343
Perz	142039±0.21%	8.130	198882±0.10%	11.383
Bper	190944±1.35%	10.929	265647±0.26%	15.204
Pera	40556±0.87%	2.321	36178±2.82%	2.071

The messages sent in this scenario are only slightly larger than the ones in the Deploy scenario, so the performance of Gzip is also only slightly better, with compressed messages being approximately 50% of the size of the originals.

The binary representation's utilization of XML item caching permits it to compress messages to under 25% of their original size in this kind of repetitive scenario. While in this case the messages are highly similar, there would also be similarities in attributes and SOAP headers in more general cases, which should still permit this representation to have an advantage over Gzipped XML.

The XML item caching is not without its drawbacks, though. It requires that no messages are lost or reordered during transit, and that both ends can keep their caches synchronized, so for example application restarts are not acceptable. In this test scenario the reliability in messaging is achieved through TCP and a non-mobile terminal, and the simplicity and single purpose of the test applications preclude the possibility of caches falling out of synchrony. When the client is mobile, however, messages are typically lost during an address-changing handover. In addition, it is possible that the application protocol on top of the transport layer does not feed the messages upward sequentially. Therefore the protocol would need to ensure neither of these happens in actual situations.

The measured execution times, both total time taken and average time per invocation, for the Localhost and LAN connections are shown in Table 6. There is not much difference between these two connection types, and from the deviations it can be seen that there is little variation in measured times from one execution to another. The only noteworthy issue is that the execution time for the Pera protocol does not increase, indicating that the network is not its bottleneck at these bandwidths.

The times for the other two connections, WLAN and Routed WLAN, are shown in Table 7, for both total time and average per-invocation time as before. The measurements show considerable fluctuation depending on when measurements were made. From the deviations it can be seen that the times for the persistent protocols fluctuate wildly. This is an artifact of the Wireless LAN; its performance characteristics varied quite a lot depending on the measurement time.

The degradation in performance of the non-persistent protocols is particularly striking, which is even better visible in Fig. 2 summarizing the average

Table 7. Average execution times in milliseconds with mean deviations measured in the Stress scenario over the WLAN and Routed WLAN connections

Protocol	WLAN Total	WLAN Single	RW Total	RW Single
HTTP	509549±24.07%	29.164	2126201± 1.30%	121.692
PTCP	457714±10.95%	26.197	2253455± 4.01%	128.975
TCPZ	475750±13.96%	27.229	2195746± 2.56%	125.672
Pers	255564± 6.02%	14.627	393324±42.54%	22.511
Perz	417271±26.04%	23.882	391899±33.86%	22.430
Bper	415225± 6.62%	23.765	416045±27.32%	23.812
Pera	81396± 1.46%	4.659	113035±20.68%	6.469

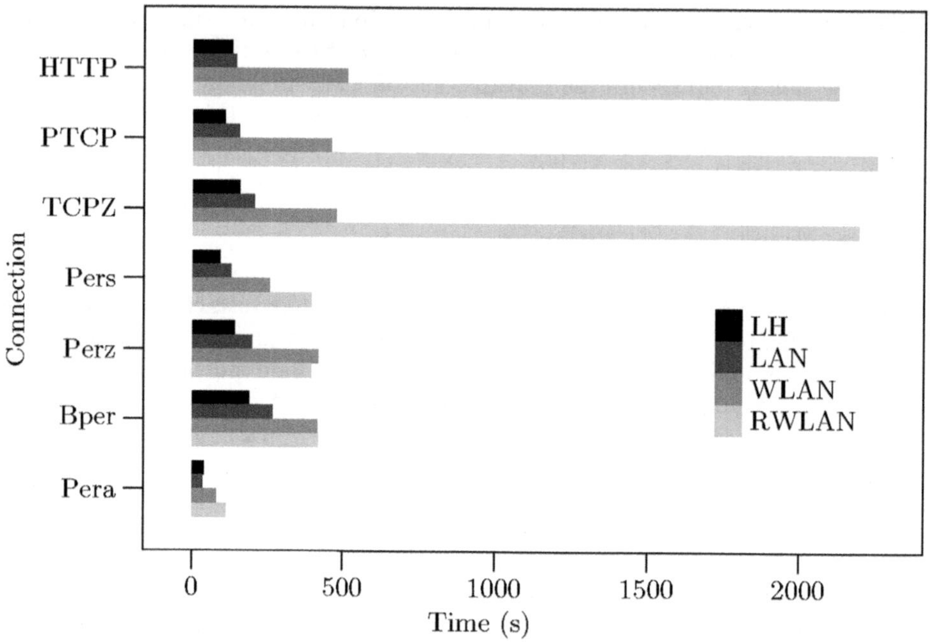

Fig. 2. Average times for all protocols and connections

times for all protocols and connections. The main cause of this degradation is the fickle nature of the Wireless LAN, which proved to be quite willing to drop ACKs in response to connection-opening SYNs, causing several-second delays as the SYN sender's timeout triggered.

The performance of the compressing persistent protocols can even be seen to improve when moving from the WLAN connection to the Routed WLAN connection. This is most probably an artifact of the fickleness of the Wireless LAN. However, since the performance of the compressing protocols in Routed

WLAN is similar to that of Pers, it would seem to imply that the time taken for compression is not a significant factor in the runtime anymore.

4 Conclusions

From the measurements we can see that the default way of using SOAP is significantly suboptimal, especially under heavy communication load in slower networks. In the test scenarios the main problem was the need to open new connections for new invocations, mainly because Wireless LAN bandwidth was not a bottleneck when invocations were synchronous. However, the effect of bandwidth reduction is clearly visible with the asynchronous Pera protocol.

The generic Gzip compression algorithm manages to decrease message size only by about 50%. Observations made during testing reveal that compressing and decompressing also consume significant amounts of processor time, which is directly visible in measured execution times on Localhost and LAN. In the Stress scenario the binary format manages to decrease message size by approximately 75%.

The processor time used is especially significant when considering the power of devices typically having wireless links, such as phones or PDAs. Preliminary experimentation seems to indicate that the extra processing time incurred by binary compression can be mostly eliminated by directly generating the compressed format, as would be expected. Furthermore, significant speedups have been achieved by parsing a binary XML format directly without going through the textual form [7].

Keeping connections open is clearly a benefit in the Stress scenario due to the heavy load the scenario places on the network. The effects become more pronounced with increased connection latency and unreliability, as would be expected. Persistence of connections would also benefit interactive applications communicating with a small number of servers, since it eliminates the round trip needed in TCP connection establishment. This could, and in the future probably will, also be achieved by using HTTP 1.1 with its built-in persistent connections.

The performance of the Pera protocol shows that further reducing the effects of latency is possible for applications able to send several messages before needing possible responses to any of them. This could include e.g. event-based systems, where no application-level response is returned to the event generator. However, this is more of an issue for the programming framework and not the protocol, since an asynchronous interface can be built on top of a synchronous protocol if needed.

The latency becomes even more of a problem when considering wireless technology used in mobile phones, such as GPRS. Measurements show that the average latency for a GPRS-based connection equivalent to the Routed WLAN is between 800 and 900 ms with large variations. This kind of connection would make it completely unsuitable to use a synchronous interface if it could at all be avoided. A GPRS connection also has a much lower bandwidth than Wireless LAN. Preliminary experiments would seem to indicate that the Stress scenario with the Pera protocol would take over an hour to run with a GPRS-based connection.

Future plans are to extend these tests to also cover a GPRS-based connection. For these tests it would also be useful to have a new protocol designed to implement the optimizations shown most beneficial by the above results. The future tests are also expected to cover a new scenario based on some real-world situation with a heavy network load.

References

1. Elan Amir, Hari Balakrishnan, Srinivasan Seshan, Randy Katz. Efficient TCP over networks with wireless links. In *Proceedings of the Fifth IEEE Workshop of Hot Topics in Operating Systems*, May 1995.
2. Olivier Avaro, Philippe Salembier. MPEG-7 Systems: Overview, *IEEE Transactions on Circuits and Systems for Video Technology*, 11(6):760–764, June 2001.
3. Hari Balakrishnan, Venkata Padmanabhan, Srinivasan Seshan, Randy Katz. A comparison of mechanisms for improving TCP performance over wireless links, *IEEE/ACM Transactions on Networking*, 5(6):756–769, December 1997.
4. Tim Berners-Lee, Roy Fielding, Henrik Frystyk Nielsen. *RFC 1945: Hypertext Transfer Protocol – HTTP/1.0*, May 1996. `http://www.ietf.org/rfc/rfc1945.txt`
5. Antaeus Feldspar. *An Explanation of the DEFLATE Algorithm*, August 1997. `http://www.gzip.org/deflate.html`
6. Roy Fielding, James Gettys, Jeffrey Mogul, Henrik Frystyk Nielsen, Larry Masinter, Paul Leach, Tim Berners-Lee. *RFC 2616: Hypertext Transfer Protocol – HTTP/1.1*, June 1999. `http://www.ietf.org/rfc/rfc2616.txt`
7. Marc Girardot, Neel Sundaresan. Millau: an encoding format for efficient representation and exchange of XML over the Web. In *Ninth International World Wide Web Conference*, May 2000. `http://www9.org/w9cdrom/154/154.html`
8. Hartmut Liefke, Dan Suciu. XMill: an efficient compressor for XML data. In *Proceedings of the 2000 ACM SIGMOD International Conference on Management of Data*, May 2000.
9. World Wide Web Consortium (W3C). *W3C Note: Simple Object Access Protocol (SOAP) 1.1*, May 2000. `http://www.w3.org/TR/SOAP/`
10. World Wide Web Consortium (W3C). *W3C Proposed Recommendation: SOAP Version 1.2 Part 1: Messaging Framework and SOAP Version 1.2 Part 2: Adjuncts*, May 2003. `http://www.w3.org/TR/2003/PR-soap12-part1-20030507/` and `http://www.w3.org/TR/2003/PR-soap12-part2-20030507/`
11. World Wide Web Consortium (W3C). *W3C Note: WAP Binary XML Content Format*, June 1999. `http://www.w3.org/TR/wbxml/`

Channel Capacity of MIMO Wideband CDMA System under the Imperfect Channel Estimation and Near/Far Effect

Jae-Sung Roh[1], Hyung-Rae Park[2], and Sung-Joon Cho[2]

1 Dept. of Information & Communication Eng.,
SEOIL College, Seoul, Korea
jsroh@seoil.ac.kr
2 School of Electronics, Telecommunication and Computer Eng.,
Hankuk Aviation Univ.,
Kyonggi-do, Korea
{hrpark,sjcho}@mail.hangkong.ac.kr

Abstract. The channel capacity of multiple-input multiple-output (MIMO) wideband CDMA system with coherent RAKE receiver is considered. General multi-path intensity Nakagami fading and multiple-access near/far interference channel are assumed. The analysis for system performance shows that the error of channel estimation significantly degrades BER performance and can be effectively suppressed by RAKE receiver and MIMO schemes. Also, an attempt for comparing the channel capacity on different performance improvement schemes has been made. In particular, the channel capacity of MIMO system is compared with single-input single-output (SISO) system. The MIMO complex spreading CDMA system with path correlation yields better performance with respect to channel capacity than a SISO system with i.i.d. input and output. And a discussion on the multiple-access near/far interference is also included, which illustrates that it can be effectively limited by power control and channel estimation schemes.

1 Introduction

To provide higher data rates for end users, as well as to accommodate more users over wireless channels in the next generation communication systems, wideband CDMA has become the focus of current research interests. Two of the important features of wideband CDMA systems are the use of complex spreading and user-dedicated pilot channel [1]-[5]. Coherent reception requires the knowledge of channel characteristics, which are time varying in fading environments [6]. A conventional and effective method to accomplish this task is to use a separate pilot channel or insert pilot symbols in data symbols.

Generally, multi-path is viewed as an undesirable feature of wireless communications. Recently, to increase the spectrum efficiency and the link reliability, multiple-input multiple-output (MIMO) scheme is devised to exploit multi-path in a scattering wireless channel. Previous results about MIMO channel [7], [8] confirm that the enormous capacity gain is obtained by using the multi-element antennas. In

M. Conti et al. (Eds.): PWC 2003, LNCS 2775, pp. 407–416, 2003.

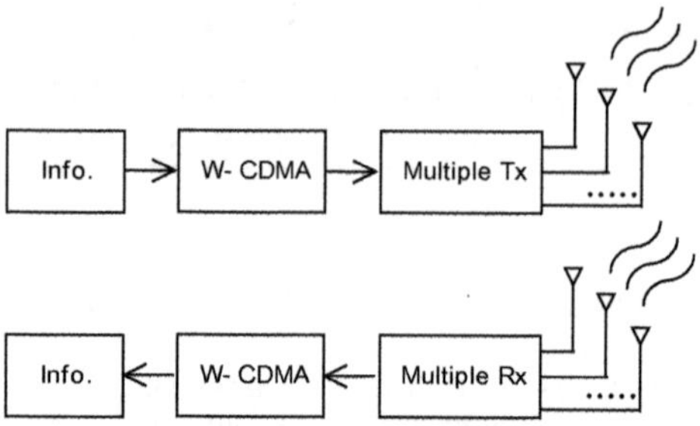

Fig. 1. System structure of MIMO W-CDMA.

order to improve the channel capacity, the signals at various elements must be uncorrelated. But multi-path components arrive from a wide range of azimuth angles in a heavy scattering channel. If the paths are correlated due to inappropriate spacing or mutual coupling effects, the channel capacity becomes substantially smaller.

In this paper, the channel capacity of MIMO wideband CDMA system with imperfect channel estimation and multiple-access near/far interference is evaluated. The channel capacity improvement due to the RAKE receiving and multiple antenna schemes is also investigated in order to achieve reliable performance of wideband CDMA system.

2 System Model

The system structure of MIMO wideband CDMA system is shown in Fig. 1, where M transmitting antennas send the wideband CDMA signal over scattering wireless channel to M receiving antennas at each symbol time.

2.1 Transmitter Model

In the reverse link of wideband CDMA systems, where an I/Q code-multiplexed pilot is utilized, spreading consists of two operations. The first is channelization operation, in which pilot and data symbols on I- and Q-branches are independently multiplied with an orthogonal variable spreading factor code and transformed into a number of chips. The second operation is scrambling, where the resultant signal are further multiplied by a complex valued scrambling code. This spreading scheme is called complex spreading, as illustrated in figure 2(a).

The transmitted signal of the k-th user can be written as [1]

$$\begin{aligned} S_j(t) = \sqrt{E_{D,j}} \sum_{n=-\infty}^{\infty} \{ \\ &\cdot \left[\sqrt{g_j} C_{P,j}(n) S_I(n) - D_j(m) C_{D,j}(n) S_Q(n) \right] \\ &\cdot h(t - nT_c) \cos(\omega_o t + \phi_j) \\ &\cdot - \left[\sqrt{g_j} C_{P,j}(n) S_Q(n) - D_j(m) C_{D,j}(n) S_I(n) \right] \\ &\cdot h(t - nT_c) \sin(\omega_o t + \phi_j) \ \} \end{aligned} \tag{1}$$

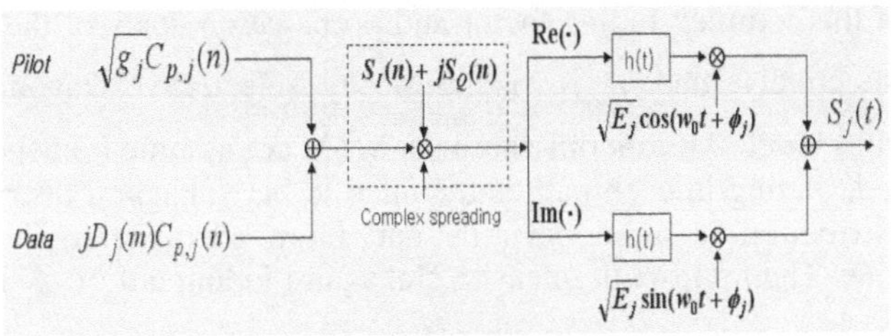

(a) Transmitter model

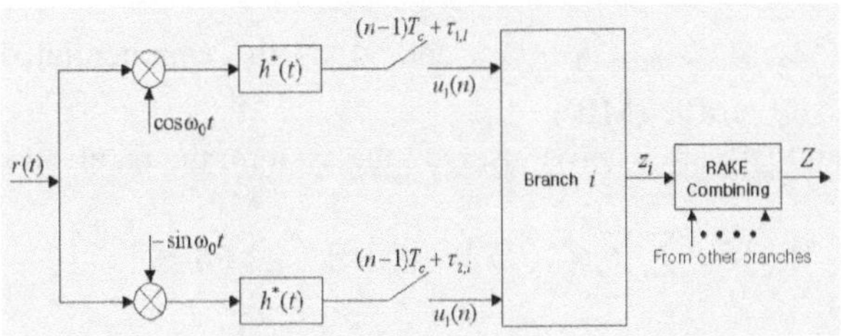

(b) Receiver model

Fig. 2. Transmitter and RAKE receiver model of the k-user.

where $E_{D,j}$ is chip energy of data channel, g_j is power ratio of pilot channel to the data channel, $D_j(m)$ is data symbols for the *j*-th user, ϕ_j is carrier phase, $C_{P,j}(n)$ and $C_{D,j}(n)$ are orthogonal channel codes for pilot and data symbols, respectively, $S_I(n)$ and $S_Q(n)$ are real and imaginary parts, respectively, of the cell-specific scrambling sequence, $h(t)$ is impulse response of the pulse shaping filter truncated by the length of AT_c for practical systems, where $A > 1$, and T_c is the chip interval. And, the power of the transmitted signal is expressed as

$$P_j = (1 + g_j) E_{D,j} / T_c = E_j / T_c \tag{2}$$

where E_j is the chip energy for both data channel and dedicate pilot channel of the *j*-th user.

2.2 Channel and Receiver Model

The complex low-pass equivalent impulse response of a multi-path fading channel can be written as

$$h_j(t) = \sum_{l=0}^{L-1} \alpha_{j,l}(t) \delta[t - \tau_{j,l}(t)] e^{j\theta_{j,l}(t)} \tag{3}$$

where L ($L \geq 1$) is the number of resolvable propagation paths. For the sake of simple notation, it is assumed that all users have the same number of multi-paths. $\alpha_{j,l}(t) e^{j\theta_{j,l}(t)}$

and $\tau_{j,l}(t)$ are the complex fading factor and propagation delay of the *l*-th path of the *j*-th user, respectively, and $\alpha_{j,l}(t)$ is frequency selective Nakagami-distributed in MIMO channel model. All random variables in (3) are assumed independent for *j* and *l*. The amplitude fading in each path is assumed to be Nakagami distributed since the Nakagami distribution is more versatile and more adequate to describe different fading situations. The multi-path intensity Nakagami fading pdf of ζ is given by

$$f(\zeta)=\left(\frac{m_T}{\Omega_T}\right)^{m_T}\frac{\zeta^{m_T-1}}{\Gamma(m_T)}\exp\left(\frac{-m_T\zeta}{\Omega_T}\right) \tag{4}$$

where $\Omega_T=\sum_{i=0}^{N_R-1}\Omega_o e^{-i\delta}$, $m_T=\sum_{i=0}^{N_R-1}m_i$, and δ is the exponential decay ratio of the multi-path intensity profile (MIP).

Assuming that there are *J* active users in the system, the received signal is given by

$$r(t)=\sum_{j=1}^{J}\sum_{l=0}^{L}\alpha_{j,l}(t)S_j[t-\tau_{j,l}(t)]e^{j\theta_{j,l}(t)}+\eta(t) \tag{5}$$

where $\eta(t)$ is the AWGN with double-side power spectrum density $\eta_0/2$.

3 Channel Capacity Evaluation

3.1 Effect of Channel Estimation

From the analysis results of W-CDMA system, it has been seen that the channel estimation using pilot symbols suffers from multiple-access interference and multi-path fading. In most publications on coherent demodulation, perfect channel estimation is assumed for the purpose of simplification. Therefore, we considered the channel estimation error in the following performance analysis. In order to mitigate multi-path fading effect, a RAKE receiver with pilot symbol aided coherent demodulation and maximum ratio combining (MRC) is employed. The RAKE receiver structure is shown in figure 2(b), where the number of branches is less or equal to the number of resolvable paths. Assuming that the *i*-th path delay $\tau_{1,i}$ can be accurately estimated for the reference user (*j*=1), each path that corresponds to a RAKE branch gives an output component. The outputs of all branches are added together to form the decision statistic.

Assuming that the fading of each path is independent of each other and the output in each branch of the RAKE receiver is independent of each other, the summed output $Z(m)=\sum_{i=0}^{N_R-1}z_i(m)$ is a Gaussian variable with mean and variance given by

$$\begin{aligned}&E\left\{Z(m)\big|_{\alpha 1,i(mP_N),\,i=0,1,\dots,N_R-1}\right\}\\&=E\left\{Z(m)\big|_{\zeta}\right\}\\&=\sum_{i=0}^{R_f-1}E\{z_i(m)\}\end{aligned} \tag{6}$$

and

$$\begin{aligned}&\operatorname{var}\left\{Z(m)\big|_{\alpha 1,i(mP_N),\, i=0,1,\ldots,N_R-1}\right\}\\&=\operatorname{var}\left\{Z(m)\big|_{\zeta}\right\}\\&=\sum_{i=0}^{R_f-1}\operatorname{var}\left\{z_i(m)\right\}\end{aligned} \tag{7}$$

where R_f is the number of RAKE finger.

Therefore, conditioned on the instantaneous multi-path fading amplitude ζ of the reference user, the equivalent signal-to-noise plus interference ratio (SNIR) is expressed as

$$\gamma_{eq1}=\left\{\begin{aligned}&\frac{(1+g)}{\zeta}\left(1+\frac{1}{gN_P}\right)\left(\frac{1}{E_b/N_o}+\frac{\Delta\Omega_T J}{2P_N}\right)\\&+\left(\frac{2N_R(1+g)^2}{\zeta^2 gN_P}\right)\left(\frac{1}{E_b/N_o}+\frac{\Delta\Omega_T J}{2P_N}\right)^2\end{aligned}\right\}^{-0.5} \tag{8}$$

where $\Delta = 2/3$ for rectangular pulse shaping filter, Ω_T is channel parameter given by $\Omega_T=\sum_{l=0}^{L-1}E\{\alpha_{j,l}^2(mN)\}$, and E_b/N_o is the signal-to-noise ratio.

3.2 Effect of Multiple-Access Near/Far Interference

As well known, DS/CDMA system is susceptible to multiple-access near/far interference, which occurs when the base station input include one or more other CDMA signals that are stronger by adjusting the transmitted power of mobile users so that the base station gets the same power from the received signal of each transmission. In this paper, we consider the effects of multiple-access near/far interference on performance of the wideband CDMA system with complex spreading scheme that has the channel estimation error. Modifying the previous equation (8), the equivalent SNIR include multiple-access near/far interference is expressed as

$$\gamma_{eq2}=\left\{\begin{aligned}&\frac{(1+g)}{\zeta}\left(1+\frac{1}{gN_P}\right)\left(\frac{1}{E_b/N_o}+\frac{\Delta\Omega_T}{2P_N}\sum_{j=1}^{J}\frac{P_j}{P_o}\right)\\&+\left(\frac{2M(1+g)^2}{\zeta^2 gN_P}\right)\left(\frac{1}{E_b/N_o}+\frac{\Delta\Omega_T}{2P_N}\sum_{j=1}^{J}\frac{P_j}{P_o}\right)^2\end{aligned}\right\}^{-0.5} \tag{9}$$

where P_o is the power controlled reference power.

3.3 Channel Capacity of MIMO CDMA System

Consider a CDMA system with M transmitting antennas and M receiving antennas operating in scattering and path correlated Nakagami fading channel with different power profile. Assuming that the transmitted signal is comprised of M statistically independent, equal power components, the expression for channel capacity in bps/Hz is given as [7], [8]

$$C_{M\times M}(\gamma)=\log_2 det\left(\overline{I_M}+\frac{\gamma}{M}\overline{HH^*}\right) \tag{10}$$

where det denotes determinant of a matrix, $\overline{I_M}$ is the identity matrix of dimension M where M is the number of transmitting and receiving elements. The superscript * denotes conjugate transpose. Due to the random nature of the $M\times M$ matrix channel transfer function $\overline{H}$, the channel capacity $C_{M\times M}$ is also a random quantity. In the case of $M\times M$ parallel transmission of each path, and the correlation coefficient between any two received paths is ρ, the channel capacity of MIMO wideband CDMA system is obtained as

$$C_{M\times M}(\gamma_{eq},\rho)=M\log_2\left(1+\frac{\gamma(1-\rho)}{M}\right)+\log_2\left(1+\frac{\rho M\gamma}{M+\gamma(1-\rho)}\right) \tag{11}$$

The received signal is corrupted by AWGN and multiple-access interference signal with statistically independent components of power σ_n^2 and σ_i^2, respectively. The quantity γ_{eq} is independent of M and is a random variable due to Nakagami fading channel. The average channel capacity of MIMO wideband CDMA system, $\overline{C_{M\times M}(\rho,\gamma)}$ is simply the expected value of $C_{M\times M}(\rho,\gamma)$ and is given by

$$\begin{aligned}\overline{C_{M\times M}(\rho,\gamma)}&=\int_0^\infty\!\!\ldots\!\int_0^\infty C_{M\times M}(\rho,\gamma_i)\times\frac{m_f^{m_f}\gamma_i^{m_f}}{\Gamma(m_f)\gamma^{m_f}}\exp\left(\frac{-m\gamma_i}{\gamma}\right)d\gamma_i\\&=\frac{M\,m_f^{m_f}}{\Gamma(m_f)\,\gamma^{m_f}}\left[\begin{aligned}&\int_0^\infty\!\!\ldots\!\int_0^\infty\log_2\left(1+\frac{\gamma_i(1-\rho)}{M}\right)\times\gamma_i^{m_f}\exp\left(\frac{-m\gamma_i}{\gamma}\right)d\gamma_i\\&+\frac{1}{M}\int_0^\infty\!\!\ldots\!\int_0^\infty\log_2\left(1+\frac{\rho M\gamma_i}{M+\gamma_i(1-\rho)}\right)\times\gamma_i^{m_f}\exp\left(\frac{-m\gamma_i}{\gamma}\right)d\gamma_i\end{aligned}\right]\end{aligned} \tag{12}$$

where γ_i is the instantaneous SNR and m_f is the Nakagami fading parameter. In order to simplify the numerical evaluation, it is assumed that each user has independent but the same fading characteristics. In general, the integral of equation (12) can be computed numerically using computer software. When the parameters set as $K=1$, $M=1$, $M_C=1$, $m_f=1$, and $\rho=0$, equation (12) reduces to the well-known Shannon capacity formula in Rayleigh fading channel [9].

4 Numerical Results and Discussion

In this section, the effects of different parameters on the channel capacity of wideband CDMA system with complex spreading, imperfect channel estimation by dedicated pilot symbols, and multiple-access near/far interference are investigated by numerical calculations. The multi-path fading is assumed to be Nakagami distributed, and the exponential decay ratio of MIP model is considered. Unless noted otherwise, the decay ratio of MIP is 0.2, the number of multi-path is 8, the LPF length is 32, and the power ratio of in-phase and quadrature branches is 0.3.

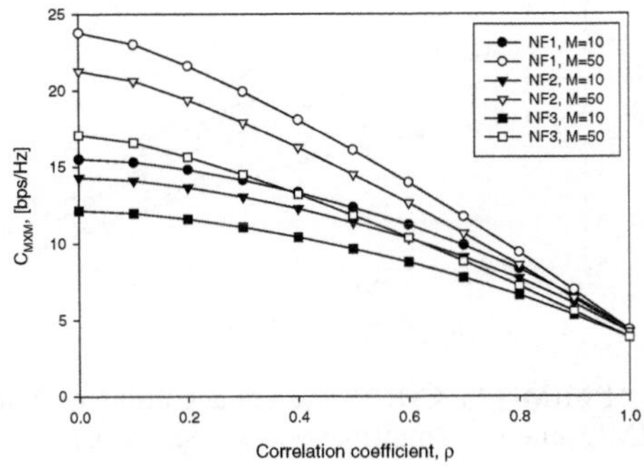

Fig. 3. Channel capacity of MIMO W-CDMA system according to multiple-access interference model and number of MIMO antennas (multiple-access user is 10).

Figure 3 shows the channel capacity of W-CDMA system with imperfect channel estimation as a function of correlation coefficient ρ for different number of antennas, i.e., $M = 10, 50$. Bit-energy-to-noise power density, the number of RAKE fingers, and the number of multiple-access users are set to 15 dB, 8, and 10, respectively. It can be seen that the MIMO scheme takes advantage of the multi-path fading and multiple-access user interference channel and gives better performance as the number of MIMO antennas increases. Also, Fig. 3 illustrates the channel capacity according to the multiple-access near/far interference model and path correlation. In this figure, we select the three kinds of near/far interference model, and NF 1 model show the perfect power control case. According to Fig. 3, NF2 model gives better performance than NF 3 model. This is because NF 2 model has a small variation of power control error than NF 3 model. In order to simplify the performance analysis, we assumed that multiple-access near/far interference model has discrete power and users distribution.

Figure 4 shows the effects of multiple-access near/far interference model and MIMO antennas according to the correlation coefficient ρ when the number of multiple-access users is set to 30. As expected, it is shown that as the number of multiple-access users increase, the channel capacity of MIMO W-CDMA system decrease, and the difference of channel capacity are small according to the multiple-access near/far interference model.

Figure 5 shows the channel capacity of MIMO W-CDMA system with perfect power control according to the parameters N_R and E_b/N_o in multi-path Nakagami fading and multiple-access near/far interference model. The number of Tx/Rx antennas, number of multiple-access user, and path correlation coefficient are set to 10, 10, and 0.4, respectively. In the range of $E_b/N_o \rangle 20\,dB$, the capacity increase for four curves is little. Specially, the difference of channel capacity is same according to the RAKE finger at any E_b/N_o.

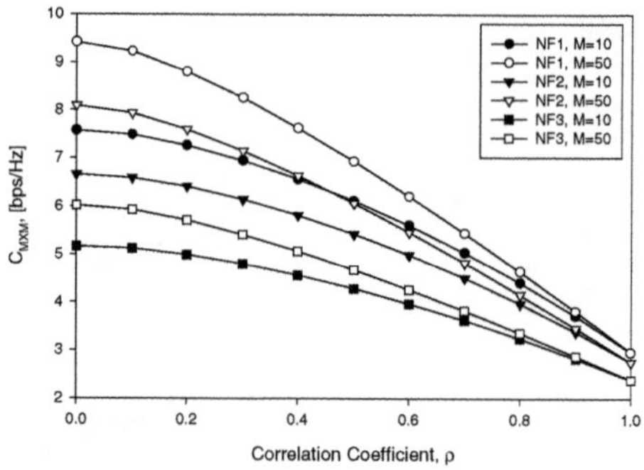

Fig. 4. Channel capacity of MIMO W-CDMA system according to multiple-access interference model and number of MIMO antennas (multiple-access user is 30).

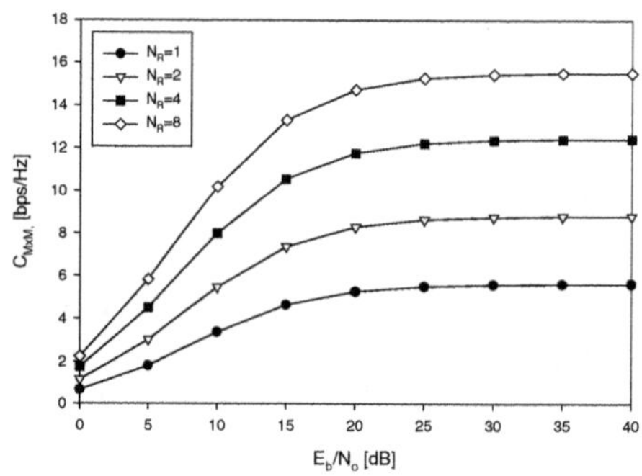

Fig. 5. Channel capacity of MIMO W-CDMA system according to the number of RAKE finger and E_b/N_o at perfect power control case.

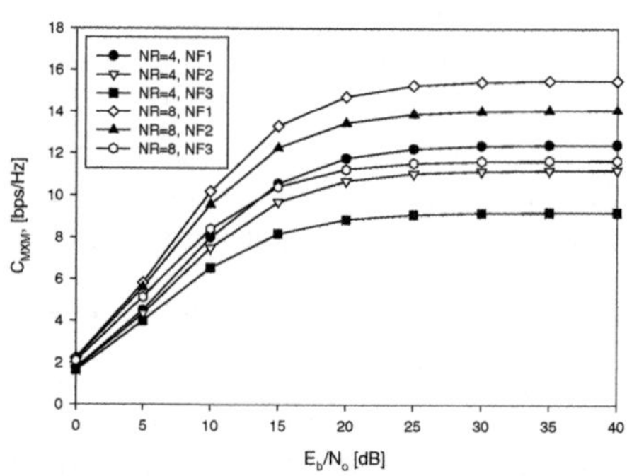

Fig. 6. Channel capacity of MIMO W-CDMA system according to multiple-access interference model, the number of RAKE finger, and E_b/N_o.

Figure 6 shows the channel capacity of MIMO W-CDMA system according to the number of RAKE finger and multiple-access near/far interference model. The number of Tx/Rx antennas, number of multiple-access user, and path correlation coefficient are set to 10, 10, and 0.4, respectively. For small E_b/N_o, the channel capacity increases linearly with E_b/N_o in the range of $E_b/N_o \langle 10dB$. But, in the range of $E_b/N_o \rangle 20dB$, the channel capacity increases according to the multiple-access near/far interference model.

5 Conclusion

In this paper, the channel capacity of MIMO wideband CDMA system with imperfect channel estimation and multiple-access near/far interference in path-correlated Nakagami fading channel is evaluated. The following conclusions have been drawn. The effects of multi-path fading, imperfect channel estimation error, and multiple-access interference can be suppressed by RAKE receiver, I/Q complex spreading and multiple Tx/Rx antennas scheme. The system capacity degrades significantly due to the multiple-access near/far interference. However, this interference can be effectively limited by power control and channel estimation scheme. Employing the multi-element arrays at both transmitter and receiver can lead to increase the spectral efficiency for a given total transmit power in a scattering channel. From the results, the MIMO wideband CDMA with RAKE receiver scheme improves the equivalent SNIR from the multiple-access users in MIMO wireless channel. The channel capacity of the proposed MIMO wideband CDMA system may be as high as 4~5 times that with the conventional SISO system for the case of $M = 10$ and $0 \langle \rho \langle 0.4$.

Acknowledgements. This work was supported by the Korea Science & Engineering Foundation (KOSEF) and the Kyonggi Province through the Internet Information Retrieval Research Center (IRC) of Hankuk Aviation University.

References

1. J. Wang and J. Chen, "Performance of wideband CDMA systems with complex spreading and imperfect channel estimation," IEEE J. Select. Areas Commun., vol. 19, no. 1, pp. 152–163, Jan. 2001.
2. F. Adachi, M. Sawahashi, and H. Suda, "Wideband DS-CDMA for next generation mobile communications systems," IEEE Commun. Mag., vol. 36, pp. 56–69, Sept. 1998.
3. 3G TS 25.213 version 3.3.0, "Spreading and modulation (FDD)," 3GPP TSG-RAN, 2000-06.
4. L. Staphorst, M. Jamil, and L. P. Linde, "Performance evaluation of a QPSK system employing complex spreading sequences in a fading environment," in Proc. IEEE VTS 50th Vehicular Technology Conf., pp. 2964–2968, Sept. 1999.
5. T. G. Macdonald and M. B. Pursley, "Complex processing in quaternary direct-sequence spread-spectrum receivers," in Proc. IEEE Military Communications Conf., pp. 494–498, Oct. 1998.

6. T. Eng and L. B. Milstein, "Coherent processing in quaternary direct-sequence spread-spectrum receivers," IEEE Trans. Commun., vol. 43, pp. 1134–1143, Feb./Mar./Apr. 1995.
7. G. J. Foschini, "Layered space-time architecture for wireless communication in a fading environment when using multiple antennas," Bell Labs Tech. J., pp. 41–59, Autumn, 1996.
8. S. L. Loyka and J. Mosig, "Channel capacity of n-antenna BLAST architecture," Electr. Lett., vol. 36(7), pp. 660–661, March 2000.
9. W. C. Y. Lee, "Estimate of channel capacity in Rayleigh fading environment," Trans. on Veh. Technol., vol. 39, no. 3, Aug. 1990.

Performance Evaluation of Partially Coherent MC/DS-CDMA System with MOC Sequence

Jae-Sung Roh[1] and Sung-Joon Cho[2]

[1] Dept. of Information & Communication Eng.,
SEOIL College, Seoul, Korea
jsroh@seoil.ac.kr

[2] School of Electronics, Telecommunication and Computer Eng.,
Hankuk Aviation Univ.,
Kyonggi-do, Korea
sjcho@mail.hangkong.ac.kr

Abstract. This paper deal with the mutually orthogonal complementary (MOC) sequences to assigning a spreading sequences to each user and propose the partially coherent equal gain combined multicarrier direct-sequence code-division multiple access (MC/DS-CDMA) system. And, we analyze the reverse link capacity and BER performance in Rayleigh fading plus multiple-access interference (MAI) channel, and evaluate the effect of phase error on receiver and transmission activity in a cell. Each user is assumed to have a distinct set of spreading sequences, with a different spreading sequence for each carrier in each user's set. By selecting MOC sequences, MAI of asynchronous MC/DS-CDMA system can be eliminated when compared to systems employing a single spreading sequence to each carrier for a particular user in the phase noise channel, and either data rate or channel capacity can be increased.

1 Introduction

Recently, there has been great interest in applying multi-carrier techniques to obtain diversity effect in communications systems. One example is MC/DS-CDMA [1]-[3], in which each of the M_C carriers is multiplied by a spreading sequence unique to each user. This MC/DS-CDMA system has a number of desirable features, including narrow-band interference suppression and a lower required chip rate than that of a single-carrier system occupying the same total bandwidth. The lower required chip rate is a result of the fact that the entire bandwidth is divided equally among M_C frequency bands. In addition, it is easier to implement the parallel receiver architecture of a number of carriers than a larger order RAKE [4].

The MC/DS-CDMA system using MOC sequences does have some advantages when compared to the single DS-CDMA system. Main advantage is that the reduction in MAI reduces the effect of the near-far problem and the autocorrelation sidelobes. Therefore, the MC/DS-CDMA system can support more users and more information symbols for a fixed error probability constraint. And hence increase the data rate achievable by a single user. And the disadvantage of MC/DS-CDMA system using

M. Conti et al. (Eds.): PWC 2003, LNCS 2775, pp. 417–425, 2003.

MOC is that the system is not as resistant to frequency-selective fading. However, even with this disadvantage, the system appears well suited to the fiber optical channels or Rician channels with a strong line-of-sight path.

In this paper, we introduce a MC/DS-CDMA that employs a set of spreading sequences for each user. Each user applies a different sequence to each sub-carriers. By selecting these sequences to be MOC sequences, MAI can be eliminated in the ideal phase-coherent channel, and either data rate or capacity can be increased. In the receiver, despreading is accomplished on a carrier-by-carrier basis. Hence, excluding MAI and noise, the output of the matched filter corresponding to a particular carrier channel is the autocorrelation function of the corresponding spreading sequence. The MAI part of the output of the matched filter is the summation of cross correlation functions between the intended user's and unintended users' spreading sequences. After adding the output of all matched filters, the autocorrelation sidelobes and MAI are zero by the defining property of MOC sequences. In this paper, the capacity and BER performance of MC/DS-CDMA system combined MOC sequences and EGC scheme for the reverse link (mobile-to-base) over Rayleigh fading channel is analyzed. The MOC sequences can provide high data rate and capacity for multi-users. The impacts of phase error of receiver and transmission activation are also considered.

2 Partially Coherent MC/DS-CDMA System with MOC Sequences

The relation of spectrum of the transmitted signal for single-carrier and multi-carrier system is given by

$$W_{MC} = M_C \cdot W_{SC} \tag{1}$$

where, M_C is the number of carriers, and we assume a strictly band-limited chip waveform with bandwidth W_{SC}. The symbol duration is $T = M_C \cdot T_C$, where the chip period T_C is M_C times larger than that of single-carrier systems. Similarly, $E = M_C \cdot E_C$ is the bit energy, where E_C is the chip energy. Without loss of generality, we assume user $k = 1$ is the intended user. The phase and delay of user $k = 1$ are assumed to be zero without loss of generality. Users $k = 2, \cdots, K$ are interfering users. They are assumed to be independent.

2.1 Transmitter

Let $d_h^{(k)}$ be the data stream for the k th user, and let $\left\{C_i^{(k)}, i = 1, 2, \cdots, M_C\right\}$ be one of the mutually orthogonal complementary sets of sequences of length L_S for the user k. The transmitted signal of user k in the i th branch is given by

$$S_i^{(k)} = \sum_{n=-\infty}^{\infty} d_h^{(k)} c_{n,i}^{(k)} \phi(t - nT_C - \tau^{(k)}) \tag{2}$$

$$k = 1, \cdots, K, \qquad i = 1, \cdots, M_C$$

where $h = \lfloor n/L_S \rfloor$, $\tau^{(k)}$ is the delay for user k, and the chip waveform $\phi(t)$ has Fourier transform $\Psi(f)$. We assume $\Psi(f)$ is strictly band-limited.

Here, we describe mutually orthogonal complementary sets of sequences. Each of the M_C sequences assigned to an individual user (one sequence per carrier) is distinct. So each user has a unique set of spreading sequences, and each of the spreading sequences in a user's set is different. Binary and multiphase complementary sets of sequences have been used in radar applications for reducing both range and doppler sidelobes, allowing the detection and resolution of objects that would otherwise be hidden in the range sidelobes of large nearby scatterers. One complementary set of sequences is a mate of the other, and they are MOC sets, because the sum of cross correlation functions is zero. However, to illustrate the basic ideas of the MC/DS-CDMA system and to simplify the analysis, we only consider binary MOC sets in the proposed system.

2.2 Receiver

The chip-matched filters i.e., ideal bandpass filters are used to separate the M_C multi-carrier frequency bands. Matched fillers matched to $c_i^{(k)}$, $i = 1, 2, \cdots, M_C$ are then used for despreading in the receiver. The M_C matched filler outputs are then summed and sampled. Note that it is natural to use a fast Fourier transform to perform this processing.

We assume the bit rate is the same for our system and the system described in [1]. We assume perfect symbol and chip synchronization for user $k = 1$, and we evaluate the performance of the first user. It is standard to assume $d_h^{(k)} c_{n,i}^{(k)}$, $k = 2, \cdots, K$ are independent sequences. We first look at the ith branch of the receiver for user $k = 1$.

From the figure 1, the chip-matched filter output is given by

$$Y_i(t) = S_i(t) + I_i(t) + N_i(t) \tag{3}$$

where $S_i(t)$ is the signal component, $I_i(t)$ is the MAI term, and $N_i(t)$ is the noise passing through bandpass and lowpass filters. After matched filtering and sampling, we have decision statistic $Z_i = S_{Z_i} + I_{Z_i} + N_{Z_i}$.

We use equal gain combine techniques at the receiver, i.e.,

$$Z = \sum_{i=0}^{M_C} Z_i \tag{4}$$

Denote $\overline{\alpha} = (\alpha_{1,i}, \cdots, \alpha_{1,M_c})$. Then

$$E[Z \mid \overline{\alpha}] = L_S \sqrt{E_C} \sum_{i=1}^{M_C} \alpha_{1,i}, \tag{5}$$

$$\text{var}[N_Z] = \sum_{i=1}^{M_C} \text{var}\left[N_{z_i}\right] = M_C L_S \eta_0 / 2, \tag{6}$$

and

$$\text{var}[I_Z] = \sum_{i=1}^{M_C} \text{var}\left[I_{Z_i}\right] = \sum_{i=1}^{M_C} \left[L_S R_{I,i}(0) + 2 \sum_{i=1}^{M_C - 1} R_{I,i}(lT_C) \sum_{j=1}^{M_C - 1} c_{j,i}^{(1)}, c_{j-l,i}^{(1)} \right] \tag{7}$$

where $R_{I_i}(\tau)$ is the autocorrelation function of interference.

By the correlation properties of complementary sequences at any user k, $\sum_{i=1}^{M_C} c_{n,i}^{(k)} c_{n+l,i}^{(k)} = M_C \delta(l)$, and when a sinc function with unit energy $\left(E_C = 1\right)$ is used as the chip waveform, the conditional SIR_{MOC-MC} can then be written as

$$SIR_{MOC-MC} = \frac{E^2[Z \mid \overline{\alpha}]}{\text{var}[I_Z] + \text{var}[N_Z]} = \frac{2M_C^2 E_C \mu \cos^2\theta}{M_C L_S (K-1) E_C + M_C L_S \eta_0} \tag{8}$$

where $\mu = \sum_{i=1}^{M_C} \alpha_{1,i}^2$ is a chi-square random variable with $2M_C$ degrees of freedom.

As we stated it before, $C^{(k)}(\lambda)$ is an all-zero sequence. Therefore, the total $MAI = \sum_{k=2}^{K} MAI_k$ is zero no matter what data bits the other users are transmitting. In addition, the MAI is independent of the chip waveform. The receiver output signal for user 1 has a very narrow peak at $t = T$. At all other times, the output is zero. The error probability is

$$P_g(e) = \frac{1}{\sqrt{2\pi}} \int_{\sqrt{SIR_{MOC-MC}}}^{\infty} \exp\left(\frac{-x^2}{2} \right) dx \tag{9}$$

The bit-error probability remains the same as the number of interfering users increases, assuming perfect carrier synchronization. Of course, the number of interfering users cannot increase arbitrarily, as there must be MOC sequences to accomplish them.

3 Evaluation of Partially Coherent MC/DS-CDMA with MOC Sequences

3.1 Capacity of MC/DS-CDMA with MOC Sequences in EG Combined Rayleigh Fading Channel

We consider two channels in this paper. First, we assume the system operates over an imperfect carrier-phase noise channel. Second, we analyze the system's performance in the Rayleigh slow fading channel. For user k, the impulse response of the i th frequency band is $\alpha_{k,i}e^{j\beta_{k,i}}$, where the $\alpha_{k,i}$ are independently, identically distributed (i.i.d.) Rayleigh random variables with unit second moment, so average received signal power is equal to transmitted signal power, and the $\beta_{k,i}$ are i.i.d. uniform random variables over $[0,2\pi)$.

The channel capacity of Gaussian noise environment was an upper bound of the maximum transmission rate, and it can be expressed as

$$C = B\log_2(1+S/N) \tag{10}$$

where B is the channel bandwidth in Hertz and S/N is the signal to noise power ratio. This formulation is known as the Shannon-Hartley law. The S/N ratio of (10) for the fading channel is a random variable which should be replaced by SIR_{MOC-MC} of (8) for the reverse link. Therefore, the channel capacity is as follow

$$\overline{C}_{fading} = \int_0^\infty B\log_2\left(1+SIR_{MOC-MC}(\mu)\right)\cdot f(\mu)d\mu \tag{11}$$

where SIR_{MOC-MC} is expressed in (8) and $f(\mu)$ is the pdf of Rayleigh distribution and is given by [5], [6]

$$f(\mu)=\frac{\mu^{M_C}e^{-\mu/2}U(\mu)}{2^{M_C}(M_C-1)!} \tag{12}$$

where $U(\mu)$ is a unit function.

3.2 BER of Partially Coherent MC/DS-CDMA System with MOC Sequences in MAI Channel

The base station receiver with partially coherent correlation type contains a random phase error. The random phase error assumed to be generated from PLL. The fading bandwidth is assumed to be much smaller than the loop bandwidth of PLL. At the output of the PLL, we have

$$\hat{\theta} = \theta + \Delta\theta \tag{13}$$

where the phase error $\Delta\theta$ have Tikhonov density function [7], [8]

$$f(\Delta\theta) = \frac{\exp\left(R\cos(\Delta\theta)\right)}{2\pi I_o(R)}, \qquad -\pi \le \Delta\theta \le +\pi \tag{14}$$

where $I_0(\cdot)$ is the zeroth-order modified Bessel function of the first kind and R is the loop SNR in PLL. The loop SNR R is proportional to system signal-to-noise ratio of E_b/N_0, i.e., $R = \rho * E_b/N_0$. When the loop SNR R is exceed 10 dB, $\cos(\hat{\theta})$ can be approximated by its expected value with respect to $\Delta\theta$. And this case not incur significant error. Therefore

$$\cos(\hat{\theta}) \approx E_{\Delta\theta}\left[\cos(\theta + \Delta\theta)\right] \tag{15}$$

where $E_{\Delta\theta}$ denotes the expectation with respect to the phase errors $\Delta\theta$.

The average error probability of partially coherent MC/DS-CDMA with MOC sequences in equal gain combined multipath Rayleigh fading channel is obtained as

$$P(e\,|\,k) = \int_0^\infty \int_{-\pi}^{\pi} P_g(e\,|\,k, \Delta\theta, \mu) f(\Delta\theta) f(\mu) d\theta d\mu \tag{16}$$

where $f\left(\Delta\theta\right)$ has the same form as Eq.(14).

Because of not every user in the cell is always transmitting simultaneously, the effect of transmission activity is include in the performance analysis. The probability that k out of K interferers are active can be described by a binomial distribution.

$$P(k) = \binom{K}{k} a^k \left(1-a\right)^{K-k} \tag{17}$$

where a is the transmission activity factor. Therefore, the average error probability is

$$P(e) = \sum_{k=1}^{K} P(e\,|\,k) P(k) \tag{18}$$

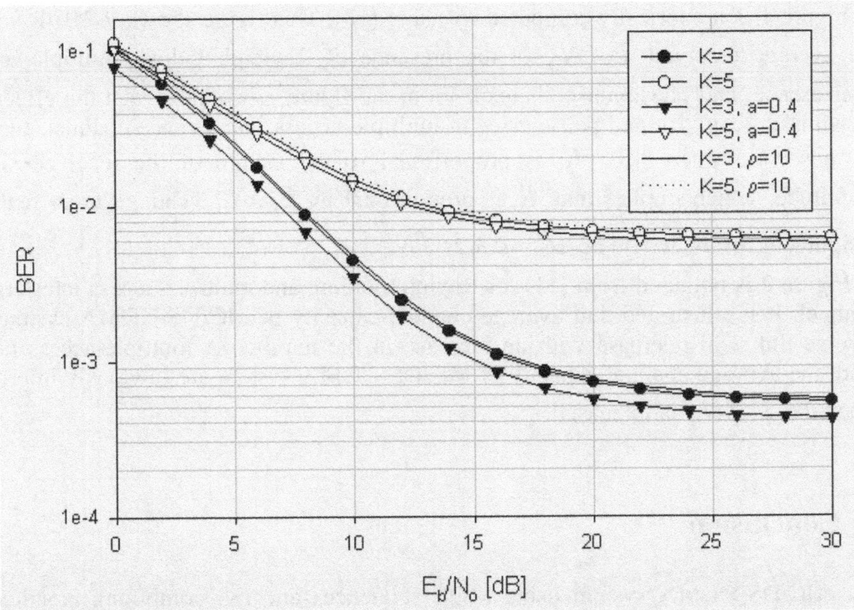

Fig. 1. Average error probability of MC/DS-CDMA using MOC sequences over the Rayleigh fading channel ($M_C = 8,\ L_S = 8$).

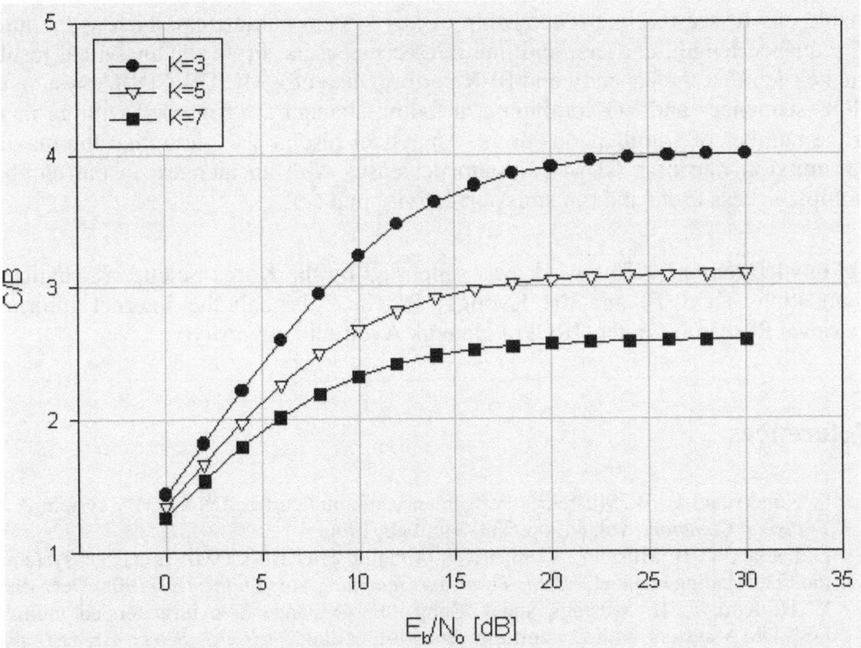

Fig. 2. Channel capacity of MC/DS-CDMA using MOC sequences according to the multiple-access user ($M_C = 8,\ L_S = 8$).

Figure 1 is numerically computed by using (18). This figure shows the BER versus the average received E_b/N_0 in the presence of Rayleigh fading, multiple-access interference, and phase noise channel. From the figure 1, we know that the effects of transmission activity and phase error in multiple-access interference channel. In fact, for a second-order PLL, R is proportional to the square of the received signal amplitude, which implies that R is proportional to E_b/N_0. Thus, a more realistic assignment for R would be $R = \rho * E_b/N_0$.

Figure 2 is obtained from (11) for Rayleigh fading and multiple-access interference channel. It is illustrated that average channel capacity of MC/DS-CDMA system for reverse link will decrease with an increase in the number of multiple-access users. And, the average channel capacity of MC/DS-CDMA system are obviously improved when the E_b/N_0 is increased.

4 Conclusion

The MC/DS-CDMA system using MOC sequences and EG combining is suited to phase noise and fading channels. In such channel, the effects of phase error and transmission activity on the system performance of a CDMA system is examined and quantified for reverse links. The closed form of SIR_{MOC-MC} derived in this paper can enable one to see the interrelationship of key system parameters, such as the number of sequence length, carriers, and multiple-access users. From the analytical results, it can be seen that the capacity and BER performance of a MC/DS-CDMA system using MOC sequences and EG combining in fading channel are degraded with an increase in the number of simultaneous users. Analytical results also show that the maximum transmission rate of a CDMA system decreases with an increase in the number of multiple-access users and transmission activity in a cell.

Acknowledgements. This work was supported by the Korea Science & Engineering Foundation (KOSEF) and the Kyonggi Province through the Internet Information Retrieval Research Center (IRC) of Hankuk Aviation University.

References

1. S. Kondo and L. B. Milstein, "Performance of multicarrier DS CDMA systems," *IEEE Trans. on Commun.*, vol. 44, pp. 238–246, Feb. 1996.
2. D. Lee and L. B. Milstein, "Comparison of multicarrier DS-CDMA broadcast systems in a multipath fading channel," *IEEE Trans. on Commun.*, vol. 47, pp. 18971904, Dec. 1999.
3. Y. H. Kim, L. B. Milstein, and I. Song, "Performance of a turbo coded multicarrier DS/CDMA system with nonuniform repetition coding," *IEEE J. Select. Areas Commun.*, vol. SAC–19, pp. 1764–1774, Sept. 2001.

4. S. M. Tseng and M. R. Bell, "Asynchronous multicarrier DS-CDMA using mutually orthogonal complementary sets of sequences," *IEEE Trans. on Commun*, vol. 48, no. 1, Jan. 2000.
5. S. Sampei, *Applications of digital wireless technologies to global wireless communications*, Prentice Hall, 1997.
6. J. Proakis, *Digital Communications*, New York: McGraw-Hill, 1989.
7. T. Eng and L. B. Milstein, "Partially coherent DS-SS performance in frequency selective multipath fading," *IEEE Trans. Commun*, vol. 45, pp. 110–118, Jan. 1997.
8. G. L. Stuber, *Principle of Mobile Communication*, KAP, 2001.

Throughput Competitiveness of WCDMA Channel Assignment

Jiun-Horng Cherng, Fen-Fen Shueh, and Wen-Shyen Eric Chen

Institute of Computer Science, National Chung-Hsing University
250, Kuo-Kuang Road, Taichung, 402, Taiwan, Republic of China
Phone: +886-4-2284-0497 ext 917, Fax: +886-4-2359-7915
{cherng,fenfen,echen}@cs.nchu.edu.tw

Abstract. Efficient and effective resource utilization is important in the bandwidth allocation of the communication network, especially when the resources are scarce such as the radio channels in the wireless links. Channel assignment in Wideband CDMA transmission schemes, i.e., the bandwidth allocation in the 3G wireless links, can be transformed into the Orthogonal Variable Spreading Factor code assignment to the connection requests. The code assignment problem presents a typical on-line feature that the decision to accept/reject the request has to be made at the arrival time of the request without the knowledge of future requests. An algorithm for an on-line problem is considered *competitive* if its performance is within some constant fraction of the performance of any other algorithm for the same input sequences. The study in deriving a competitive algorithm is valuable as it guarantees a lower bound of the system performance in all circumstances. This paper presents an on-line competitive algorithm called Left_Right_Aggressive (LRA) for the OVSF code assignment problem in the wireless links regarding system throughput as the performance indicator. With preemption allowed, the proposed algorithm achieves the constant competitive ratio of 2/(1-2β), where β is the fraction of the maximum requested bandwidth to the system capacity. The simulation results demonstrate that it is promising to achieve better throughput performance and lower preemption ratio in request patterns of Poisson distribution.

1 Introduction

High-rate multimedia data transmissions and multiple services to users are significant characteristics in the third-generation (3G) mobile communication systems. International Mobile Telecommunications-2000 (IMT-2000) sets up the air interface objectives [1] to support wideband data services as high as 2Mb/s for limited, and 144Kb/s to 384Kb/s for full coverage and mobility, while keeping the same transmission quality as in fixed telecommunication network [2]. Orthogonal Variable Spreading Factor (OVSF) [3] codes were proposed and adopted by IMT-2000 as the channelization codes on the forward radio link for the Wideband CDMA [4,5] radio transmission scheme.

M. Conti et al. (Eds.): PWC 2003, LNCS 2775, pp. 426–436, 2003.

The efficient and effective utilization of the scarce radio resources is the main theme for data transmissions over the air, and system throughput, which is the aggregated occupied bandwidth over time, is usually regarded as the performance indicator. Since the connection requests arrive one by one over time, it manifests the typical on-line feature [6,7], i.e., the system has to make the acceptance/rejection decision at the arrival time of the request without the advanced knowledge of the future requests. To achieve higher bandwidth utilization and/or lower blocking rate of the system, the preemptive feature [8,9,10] is deployed as to provide more "valuable" connections at the expense of less profitable ones already in services, when the total bandwidth requested exceeds the system capacity so that not all requests can be served at the same time. When and how to reject/preempt connections, and simultaneously guarantee the throughput performance, are more challenging in the on-line problem domain as there is no contribution to the throughput performance if the request is rejected/preempted. Authors in [11] demonstrated that an average 20% gain of throughput efficiency can be achieved if reassignment scheme is deployed.

We present an on-line preemptive algorithm Left_Right_Aggressive (LRA) for OVSF code assignment in this paper that can achieve the constant competitiveness in the system throughput aspect. Algorithm LRA is an improvement of algorithm LR [10] in reducing the unnecessary preemption and increasing the system throughput. For any $\beta < 1/2$ where β is the ratio of the maximum bandwidth request to the system capacity, the algorithm is $2/(1 - 2\beta)$-competitive. Our simulation demonstrates that the LRA algorithm is capable of achieving both higher throughput and at the same time keeping the low preemption ratio.

The remainder of the paper is organized as follows. Section 2 delineates the background and the related work of OVSF code assignment, on-line problem and competitiveness, and preemption. In Section 3, the proposed algorithm LRA, with both preemptive and code reassigning features, is presented and proved to achieve the constant competitive ratio. The comparison of the throughput among LRA algorithm and others are demonstrated in Section 4. We present the discussion and the future work in Section 5.

2 Preliminaries

We describe the background and the related work of OVSF code assignment in the first subsection, followed by the introduction and definition of on-line problem and competitiveness. Finally, the preemptive operation and related work is delineated.

2.1 OVSF Code Assignment

Three Direct Sequence Code Division Multiple Access (DS-CDMA) transmission schemes [12,13,14], namely multi-code (MC-CDMA) method, OVSF code method, and hybrid method, had been proposed to achieve variable and high

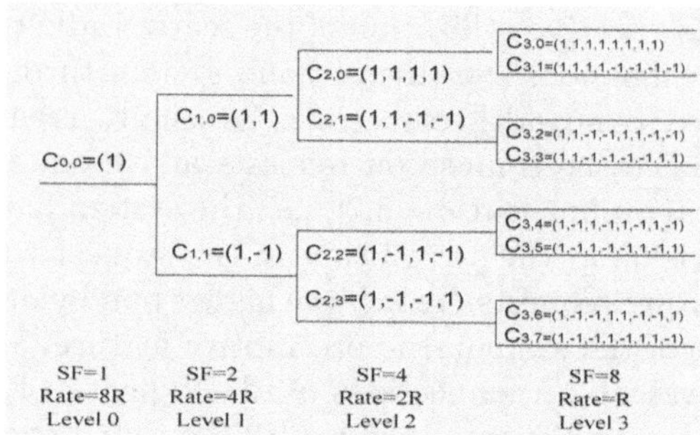

Fig. 1. An OVSF code tree

data-rates as required by 3G air interface objectives. The OVSF codes, a modified Hadamard transformation, were adopted by IMT-2000 as the spreading code generation scheme on the forward radio link as the channelization codes.

Through the recursive method in orthogonal code tree generation [13], the OVSF code tree with 4 levels (level 0 $\sim$ level 3) is shown in Figure 1. $C_{L,S}$ uniquely identifies an OVSF code in the code tree, where Lindicates the level of the code and S is the sequence number in that level. For a $(B+1)$-level OVSF code tree, the spreading factor (SF) of codes in level L is 2^L. Let the codes in the last level (level B), be capable of carrying R data-rate information, the codes in level L can carry data with rate $2^{B-L}R$.

The spreading codes are generated recursively as follows. Let $C_L = \{C_{L,S} | S = 0 \text{ to } 2^L - 1\}$ denotes a matrix representation of the set of 2^L binary spreading codes with spreading factor 2^L. The code length of the Lth level is 2^L chips and the code is used for transmitting symbols at the rate of 2^L times lower than the chip rate. Assuming that the $\bar{C}$ is the binary complement of C, the C_L codes can be easily constructed from C_{L-1}as $C_L = \begin{bmatrix} C_{L-1} & C_{L-1} \\ C_{L-1} & \bar{C}_{L-1} \end{bmatrix}$.

The OVSF code tree is basically a full binary tree. In the OVSF code tree, all codes are orthogonal to each other, except the pairs that one code is the ascendant (descendant) of the other. It is thus evident that one code in level L can be assigned if and only if no code on the path from root to this code or any code in the subtree of this specific code is assigned. The restriction is called the *orthogonality constraint.* The request is blocked and rejected even when the system has the adequate bandwidth capacity, and the scenario is called *code blocking* in differentiating from the *system blocking* which results from the inadequacy of system capacity to serve the request.

A simple example in Figure 2 illustrates the impact of code blocking. In Figure 2(a), a request with data rate 2R cannot be accepted because any code in level 2 is orthogonal to the existing codes of $C_{1,0}$, $C_{3,4}$, and $C_{3,6}$. It is accepted only after the reassignment, say code $C_{3,6}$ is reassigned to $C_{3,5}$, and Code $C_{2,3}$ is available to be used as shown in Figure 2(b).

The authors in [11] devised a dynamic code assignment method on OVSF codes. Under the code constraint property, they proposed a code assignment

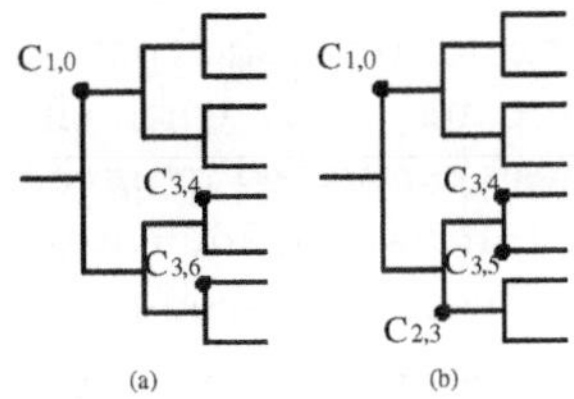

Fig. 2. OVSF code assignment (a) code blocking, (b) code reassignment

scheme intending to minimize the number of necessary reassignments of the occupied codes. The detailed description of the algorithm was not given, especially in the recursive searching steps. The comparison between more heuristic garbage collection methods with the recursive one was studied in [15], and the result implied that garbage collection is well suited for code reassignment when OVSF code blocking occurs.

Some research efforts focused on code assignments and reassignments for QoS purpose or by MC-CDMA and hybrid schemes [16,17,18,19]. Another approach [20] was the protocol design in supporting bursty data traffic by statistical multiplexing in the forward radio link of W-CDMA instead of assigning a fixed-rate code to a user.

2.2 Online Problem and Competitiveness

The main characteristic of the on-line problem is to make the acceptance/rejection decision at the time of the request without the knowledge of the future requests. A competitive algorithm for an on-line problem was discussed in [7]. It states that an algorithm for an on-line problem is called *competitive* if its performance is within some constant fraction of the performance of any other algorithm on the same input sequence, even an off-line algorithm. Let $P_A(S)$ and $P_B(S)$ represent the performance function of algorithms A and B respectively on a input sequence of requests S. For $c \geq 1$, algorithm A is called *c-competitive* if the following conditions are satisfied: (a) A is an on-line algorithm; (b) B is any other algorithm, even the off-line algorithm; and (c) A constant d exists for any sequence S such that $P_A(S) \geq 1/c * P_B(S) + d$.

Authors in [8] also demonstrated that, for the connection admission problems, if the holding time of the request is unknown when the request arrives, it is impossible to find a c-competitive algorithm if throughput is considered as the performance function. In this paper, we assume that the holding time of the request (the duration of each request) becomes known when the request arrives at the system.

2.3 Preemption

For real-time tasks, preemption techniques are often deployed. Some "weighting functions" or "values" are set for comparison in the preempting processes. The bandwidth request preemption in [8,9] considered several forms of values such as the uniform profit for all requests, routing path counts and duration of the request. The values are gained only when the requests are completed, and no value counts if the request is rejected or preempted as the penalty of rejection/preemption. The study in [8] also indicates that if no penalty is paid for preemption, it is trivial to achieve the 1-competitive algorithm.

When throughput is considered as the performance function, the "Left_Right" (LR) strategy [10] which deploys preempting operations in the bandwidth allocation problems achieves the constant competitive ratio of $2/(1-2\beta)$, where β is the fraction of the maximum requested bandwidth to the bandwidth of the channel, i.e., system capacity. This is in contrast to the poor logarithmic competitive ratio that the non-preemptive algorithm [21] can obtain. The authors in [10] showed that the traditional throughput (bandwidth times the duration of the request) considerations as the parameter of preemption was not desirable, no matter in "greedy" or "double the gain" situations.

An alternative method considering both the periods of time of the requests that have been served and will be served by the system was suggested, yet this strategy is too conservative in that it may unnecessarily preempt connection requests. Further discussion and improvement will be entailed in the next section.

3 LRA Scheme for OVSF Code Assignment

3.1 LR Strategy

For an on-line communication system, there are times that the proper preemption of requests is allowed for obtaining more valuable "profit". Let the profit be the system throughput, and let the request k is of 3-tuple form (a_k, d_k, r_k), where a_k and d_k specifies the arrival time and duration of request k, and r_k is the data-rate requirement. The goal is to derive an algorithm that, with preemption allowed, is capable of achieving the guaranteed system throughput. The authors in [10] suggested that, with system capacity W and $\beta = \max\ \{r_k/W\}$ for all k, the procedure LR as depicted in Figure 3, can achieve $2/(1-2\beta)$-competitive. That is, its throughput performance is at least $(1-2\beta)/2$ times of the best algorithm that may not exist.

The procedure starts, upon arriving of request k, by sorting all requests currently in the system in the ascending order of the starting times, and then selecting the top ones of requests into a set L such that the total bandwidth of those requests do not exceed half of system capacity. The same is done by sorting the ending times of the requests in descending order and selecting the top ones into a set R with the same $W/2$-bandwidth constraint. Those requests that are neither in L nor R can be preempted or rejected.

The process is demonstrated using an example in Figure 4(a). Set system bandwidth W=10, and $\beta = 0.4$. Let requests 3, 4, 7 be in services while request 8 arrives at time t=13. Each request is with the input parameters (8,12,4), (9, 9, 2), (12, 5, 2), and (13, 6, 4), respectively, with total occupied bandwidth 8 at t=13. According to the procedure LR, L={3}, R={3}, and P = {4,7,8},

```
Procedure LR
Let Z' be the set of requests currently in service;
Let P be the set of requests that can be preempted or rejected.
W is the system bandwidth;
Upon the request k with bandwidth request r_k, DO:
BEGIN
   Z = Z' U {k};
   Sort Z by increasing order of starting time;
   L = top requests of Z such that the total bandwidth is at most W/2;
   Sort Z by decreasing order of ending time;
   R = top requests of Z such that the total bandwidth is at most W/2;
   P = Z - (L U R);
   Preempt/reject the requests in P;
END
```

Fig. 3. Procedure LR for determining the requests to be rejected and/or preempted

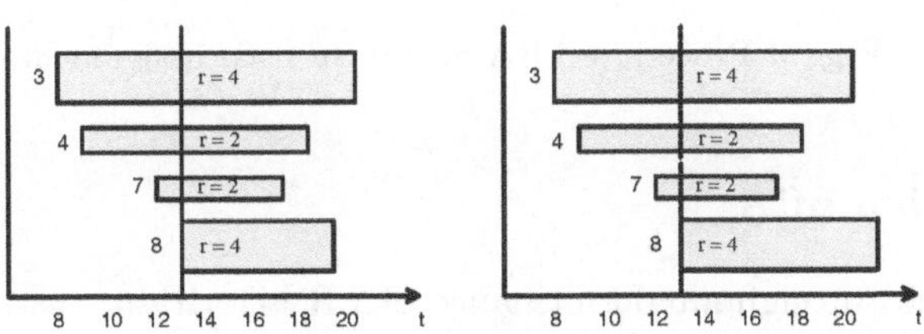

Fig. 4. Request 8 arrives at time t=13 while requests 3, 4, and 7 are in service

the request 4, 7 are preempted and request 8 is rejected. The procedure LR guarantees that for all the time periods of request 4, 7, and 8, the occupied bandwidth ratio is always greater than (1/2 - β). The requests that are rejected and/or preempted have no contribution to the throughput.

Figure 4(b) illustrates another conservative behavior of procedure LR. Let the input sequence be the same as in Figure 4(a) with the exception that the input parameter of request 8 is (13, 8, 4). In this example, L={3}, R={8}, and requests 4 and 7 are preempted. As can be easily verified that request 4 is preempted unnecessarily since input sequence {3, 4, 8} can coexist in the system for the time period t $\in$ [8, 21] without exceeding the system capacity.

The LR technique, trying to keep the requests that have been served and will be serviced for longer period, is similar to the concern of the requests' holding times. If the duration of the request is large enough, it will certainly be kept in either L or R, even though there is no guaranteed competitive ratio. In Section 4, the simulation is also conducted for the DUR algorithm which, upon request k's arrival, sorts the duration of the requests in descending order and selects those long-duration requests with total bandwidth less than or equal to W, and preempts/rejects those unselected.

```
Procedure LRA for OVSF code assignment
Let Z' be the set of requests currently in service;
Let P be the set of requests that can be preempted or rejected.
W is the system bandwidth;
U is the system bandwidth currently occupied.
Upon the request k with bandwidth request r_k, DO:
BEGIN
 IF ( r_k > W - U ) THEN                    //insufficient system resource (i)
  Z = Z' U {k};
  Sort Z by decreasing order of ending time;
  R = top requests of Z such that the total bandwidth is at most W/2;
  IF request k is not included in R THEN goto end;              // (ii)
  V = totoal bandwidth aggregated in R;
  Sort Z by increasing order of starting time;
  L = top requests of Z such that the total bandwidth is at most W-V; // (iii)
  P = Z - (L U R);
  Preempt/reject the requests in P;
  Release the OVSF codes previously assigned to requests in P;
 ENDIF
  Assign request k by OVSF code assignment procedure;
END
```

Fig. 5. Procedure LRA for OVSF code assignment

3.2 Algorithm LRA

To improve the throughput performance of LR algorithm, the more aggressive one called LRA (Left_Right_Aggressive) is designed and depicted in Figure 5. The bandwidth utilization is considered only in two conditions: when request k is rejected, or when request k is accepted while others are preempted.

The procedure LRA, upon the arrival of request k with bandwidth request r_k, checks the occupied bandwidth U and decides if the system has enough capacity. If the resource is available, the request is assigned an OVSF code by the OVSF code assignment procedure that will be presented later. If there is insufficient resource, the process begins sorting the ending times of all the requests in service. If request k is not included in set R, as shown in the example of Figure 4(a), the request k alone is rejected without affecting other requests in service. Since request k has the earliest ending time compared to all other requests currently in the system and definitely the latest starting time among them, the total bandwidth ratio of all others requests in service exceeds (1- β) for the period t $\in [a_k, a_k+d_k]$ for request k, the bandwidth utilization is thus guaranteed to be greater than (1- β) in this period.

If request k is included in R, as example shown in Figure 4(b), and let the total bandwidth in R be V, the set L could accommodate the aggregated bandwidth up to $(W - V)$ instead of (1/2 - β)W.

As emphasized in Section 2.1, the OVSF code assignment of the $(B+1)$-level OVSF code tree may encounter the code blocking that prohibits the allocation of bandwidth to the request. In this paper, the First-Fit algorithm is carried out to find an available code $C_{L,S}$ from S $=$ 0 in level L when request bandwidth r $= 2^{B-L}R$, where R is the basic data rate. If no available code found, the garbage collection operation is then executed from the bottom level B, up to level $(L-1)$ so that it is guaranteed to get an available code in level L for request k. These operations result in code reassignment in compensation for the gain in throughput performance.

The LRA algorithm thus guarantees to achieve the competitive ratio of $1/(1/2 - \beta) = 2/(1 - 2\beta)$, yet this is only the worst-case scenario, the real implementation achieves much higher value as demonstrated in the next section.

4 Simulation and Results

Throughput performance, as well as the preemption ratio, is compared among the three algorithms: LR algorithm, LRA algorithm, and DUR (preemption algorithm based on the duration of the requests.) Event driven simulation is performed. Each run produces 30,000 requests with different patterns of data rate requirement. System bandwidth W is assumed 256R for an OVSF code tree with B=8, which complies with the IMT-2000 specification. The data rate r $\in$ [R, $2R$, $4R$, $8R$, $16R$, $32R$, $64R$]. Five data patterns are generated: UR (Uniform Rate) with equal probability in each data rate, and the other four rate patterns P_8, P_16, P_32, and P_64, with Poisson distribution of rate request with mean rate $8R$, $16R$, $32R$ and $64R$ respectively.

4.1 Parameters

Input parameters are λ, μ, and r, and important output values are throughput, the preemption ratio, and the reassignment ratio. They are defined as follows:

λ : new request rate. The inter-arrival time of each request is assumed exponentially distributed with mean value of $1/\lambda$.

μ : the mean service rate of each connection. $1/\mu$ is the mean service time (duration) of each connection, and is assumed of exponential distribution.

λ/μ : the ratio of request rate to the service rate, and is called the *traffic load* of the system.

r : bandwidth request. r is assumed R, $2R$, $4R$, $8R$, $16R$, $32R$, and $64R$. The frequency on different rate depends on the rate patterns.

Throughput: the aggregated product of requested bandwidth and the duration of the requests that are completed without being rejected/preempted.

Preemption ratio: the fraction of preempted requests to the accepted requests. It indicates the ratio of requests that were previous accepted to the system but later preempted with no contribution to the system throughput.

Reassignment ratio: the number of reassignment per accepted request. The code reassignment is executed when the code blocking occurs during the OVSF code assignment procedure. This ratio indicates how often one request will be reassigned a new OVSF code.

4.2 Results

The system throughput is the major goal of the preemption algorithms. The LR and LRA algorithms are both competitive while DUR algorithm is not. The

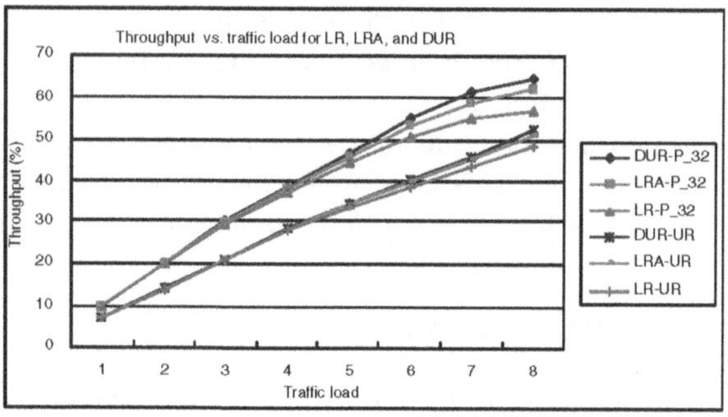

Fig. 6. Throughput comparison vs. traffic load for LR, LRA and DUR algorithms

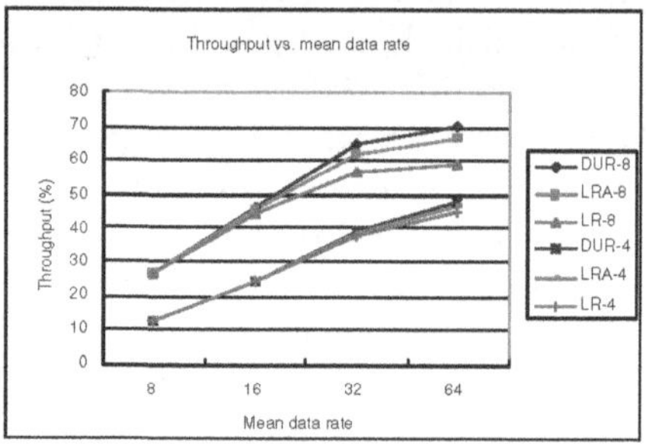

Fig. 7. Throughput comparison vs. mean data rate for LR, LRA and DUR algorithms.

throughput comparison of the three algorithms using the uniform data pattern (UR) and Poisson data rate with mean 32 (P_32) is illustrated in Figure 6. The throughput performance of LRA is better than LR, about 5% throughput improvement when traffic load approaches 8 in P_32 data pattern is gained, and 2% throughput improvement gained for uniform rate. The improvement increases as traffic load and mean data rate increase. The DUR algorithm performs quite well under the Poisson traffic, even though there is no guarantee for other traffic patterns. The throughput of DUR algorithm is higher than LRA for about 1~2% as shown in Figure 6.

Figure 7 depicts the throughputs of the three algorithms when traffic loads (λ/μ) are 4 and 8 respectively, as the mean data rates grow from $8R$ to $64R$ (P_8, P_16, P_32, and P_64). The LRA algorithm performs better than LR, and the difference grows as the mean data rate increases. Again the DUR algorithm performs slight better than the LRA algorithm.

High preemption ratio indicates the inefficiency of the algorithm since it accepts the requests but easily abandons them later. The preemption ratios of the three algorithms under different traffic load are demonstrated in Figure 8. While the DUR algorithm performs well in throughput aspect, it is quite poor in this category. The preemption ratio is close to 20% when traffic load approaches between 7 and 8 in the P_32 data pattern. The LRA algorithm, on the other hand, keeps the ratio below 13% even for when traffic load is 8. The LRA algorithm

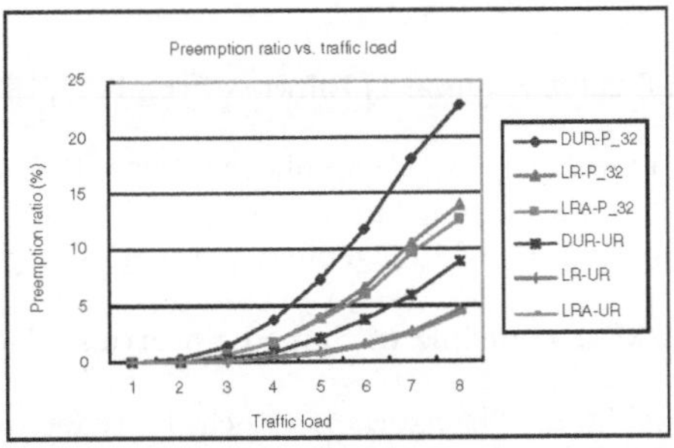

Fig. 8. Preemption ratio comparison vs. traffic load for LR, LRA and DUR algorithms

Table 1. Reassignment ratios under uniform data rate pattern

Traffic load	LR Algorithm	LRA Algorithm	DUR Algorithm
3	0.0028	0.0029	0.0030
4	0.0092	0.0097	0.0121
5	0.0175	0.0178	0.0200
6	0.0312	0.0325	0.0387
7	0.0475	0.0510	0.0572
8	0.0686	0.0733	0.0851

obtains lower preemption ratio than LR algorithm, and at the same time, higher throughput performance.

Reassignment ratios (the fraction of reassigned requests to the accepted requests) for different algorithms under the uniform data rate pattern are shown in Table 1. As the value are all very low, the LR algorithm outperforms the other two as the system may find available OVSF code more easily since the bandwidth utilization is comparatively lower.

5 Conclusions and Future Work

The competitiveness for an on-line program is valuable in that it guarantees the performance in all input sequences. To achieve better throughput performance, it is sometimes suitable to reassign and/or preempt the previous accepted requests. This paper presents a modified version of competitive algorithm called LRA. The LRA algorithm for OVSF code assignment problem is proved and demonstrated to improve the original LR algorithm by reducing the unnecessary preemptions to the requests, and also achieving higher throughput performance than the theoretical value.

Traffic patterns other than Poisson distribution may cause different behavior of the three algorithms in discussion. This is the interesting topic to work into. As MC-CDMA and hybrid method other than OVSF code assignment schemes are paid much attention, the further studies of the throughput competitive analysis in both non-preemption and preemption scenarios are also worthwhile.

References

1. Special Issue, IMT-2000: Standards Efforts of the ITU, IEEE Pers. Commun., vol. 4, Aug. 1997
2. R. Prasad and T. Ojanpera, "An Overview of CDMA Evolution Towards Wideband CDMA," IEEE Communications Surveys, vol. 1, no. 1, 4-th Quarter, pp. 2–29, 1998
3. TR25.922, "Radio Resource Management Strategies," 3GPP, V3.1.0, pp. 25–27, March 2000
4. A. J. Viterbi, CDMA: Principle of Spread Spectrum Communications, Addison-Wesley, 1995
5. A. Baier, U.-C. Fiebig, W. Granzow, W. Koch, P. Teder, and J. Thielecke, "Design Study for a CDMA-based Third Generation Mobile Radio System," IEEE J. Sel. Areas Commun., vol. SAC–12, pp. 733–43, May 1994
6. D. D. Sleator and R. E. Tarjan, "Amortized Efficiency of List Update and Paging Rules," Communications of ACM, vol. 28, pp. 202–08, 1985
7. M. Manasse, L. McGeoch and D. D. Sleator, "Competitive Algorithms for On-line Problems," Proc. ACM Symposium on Theory of Computing, pp. 322–33, May 1988
8. J. A. Garay and I. S. Gopal, "Call Preemption in Communication Networks," Proc. INFOCOM '92, vol. 3, pp. 1043–50, 1992
9. J. A. Garay, I. S. Gopal, S. Kutten, Y. Mansour, and M. Yung, "Efficient On-line Call Control Algorithms," Proc. 2nd Israel Symp. on the Theory and Computing Systems, pp. 285–93, 1993
10. A. Bar-Noy, R. Canetti, S. Kutten, Y. Mansour, and B. Schieber, "Bandwidth Allocation with Preemption," Proc. ACM Symp. on Theory of Computing, pp. 616–25, May 1995
11. T. Minn and K.-Y. Siu, "Dynamic Assignment of Orthogonal Variable Spreading Factor Codes in W-CDMA". IEEE J. Sel. Areas Commun., vol. 18, no. 8, pp. 1429–40, Aug. 2000
12. C.-L. I and R. D. Gitlin, "Multi-code CDMA Wireless Personal Communications Networks," Proceedings of ICC'95, vol. 2, pp. 1060–64, 1995
13. F. Adachi, M. Sawahashi, and K. Okawa, "Tree-structured Generation of Orthogonal Spreading Codes with Different Length for Forward Link of DS-CDMA Mobile Radio," Electronics Letters, vol. 33, no. 1, pp. 27–28, Jan. 1997
14. S. H. Kim., S. H. Shin, and K. S. Kwak, "A Hybrid Multi-rate Scheme for WCDMA," Proceedings of TENCON' 99, vol. 2, pp. 1224–27, Dec. 1999
15. F. Shueh, and W.-S. E. Chen, Minimizing Reallocation on OVSF Channel Assignment," in preparation
16. W.-T. Chen, Y.-P. Wu, and H.-C. Hsiao, "A Novel Code Assignment Scheme for W-CDMA Systems," Proceedings of VTC'2001 Fall, vol. 2, pp. 1182–86, 2001
17. R.-G. Cheng and P. Lin, "OVSF Code Channel Assignment for IMT-2000," Proceedings of VTC'2000 Spring, vol. 3, pp. 2188–92, 2000
18. F. Shueh, Z.-E. P. Liu, and W.-S. E. Chen, A Fair, Efficient, and Exchangeable Channelization Code Assignment Scheme for IMT-2000, "IEEE ICPWC'2000, pp. 429–33, Dec. 2000
19. F. Shueh and W.-S. E. Chen, "Code Assignment for IMT-2000 on Forward Radio Link," Proceedings of VTC'2001 Spring, vol. 2, pp. 906–10, 2001
20. A. C. Kam, T. Minn, and K.-Y. Siu, "Supporting Rate Guarantee and Fair Access for Bursty Data Traffic in W-CDMA," IEEE J. Sel. Areas Commun., vol. 19, no. 11, pp. 2121–30, Nov. 2001
21. B. Awerbuch, Y. Azar, and S. Plotkin, "Throughput-Competitive On-line Routing," IEEE Proc. Foundations of Computer Science, pp. 32–40, 1993

Radio Planning and Optimization of W-CDMA Systems

Edoardo Amaldi, Antonio Capone, Federico Malucelli, and Francesco Signori

Politecnico di Milano,
Piazza Leonardo da Vinci 32 ,
20133 Milano, Italy
{amaldi,capone,malucell,signori}@elet.polimi.it

Abstract. Radio planning and optimization of 3G wireless systems are key issues for service providers which need to control the huge costs involved. In this paper we describe algorithms and a mathematical programming model for locating and configuring base stations aiming at maximizing coverage and minimizing cost. The present model considers both uplink and downlink direction, studied separately in [1,3,4]. The Tabu Search algorithm we propose exploits solutions provided by the uplink model and a simple power control mechanism to drastically reduce the computational time required to find an approximate solution of the mixed uplink and downlink model. Experimental results obtained for realistic instances with voice as well as data traffic are reported and discussed.

1 Introduction

Most of third generation (3G) wireless access systems are based on W-CDMA (Wideband Code Division Multiple Access), a more flexible but also more complex radio access scheme [10]. In recent years these systems have been attracting considerable attention from a technological as well as a scientific point of view. The challenge is to enhance today's mobile telecommunication systems (e.g., GSM) by providing increased capacity, data transmission capabilities and a wide range of new multimedia services.

Due to the peculiarities of W-CDMA, the radio planning problem cannot be subdivided into a coverage problem and a frequency allocation problem like it is the case for planning second generation cellular systems with a TDMA-based access scheme [11,12]. Indeed, in W-CDMA the bandwidth is shared by all active connections and no actual frequency assignment is strictly required. Moreover, the area actually covered by a base station (BS) also depends on the signal quality constraints, usually expressed in terms of Signal-to-Interference Ratio (SIR), and on the traffic distribution [6]. Since SIR values depend on emission powers, the specific power control mechanism and the power limitations must be taken into account.

To address the network planning problem for systems with W-CDMA air interface, the following information are usually supposed to be known: i) a set of

M. Conti et al. (Eds.): PWC 2003, LNCS 2775, pp. 437–447, 2003.

candidate sites where BSs can be installed, ii) a set of possible configurations of each base station (rotation, tilt, height), iii) the traffic distribution estimated by using empirical prediction models and iv) the propagation description based on approximate radio channel models or ray tracing techniques. To take into account SIR constraints, two power control (PC) models are considered: a power-based PC model which assumes that emission powers are adjusted to guarantee a target received power and a SIR-based PC model that assumes that emission powers are adjusted to guarantee a SIR target value to all active links.

In previous work [1,3] we proposed discrete optimization models and algorithms to support decisions on where to install new BSs and which antenna configuration to select for the uplink direction. Uplink turns out to be more critical than the downlink direction in the presence of symmetrical traffic such as voice calls (see e.g. [10]). Since 3G systems are also intended to provide data services which are expected to have a substantial impact on the downlink direction (e.g., web-browsing) and to yield asymmetrical traffic, the downlink direction has been addressed in [4].

In this paper we describe a complete mathematical programming problem which includes uplink and downlink constraints as well as other relevant issues for BS configuration such as antenna height, tilt, and sectors orientation. Moreover we also consider the constraints due to pilot signals which are emitted by base station at a constant power level to broadcast system information and to allow mobile terminal to select best serving station. The problem formulation considers the SIR-based PC but it can be easily modified to model a power-based PC just assigning a proper value to power variables.

Besides providing a deeper insight into the different aspects of the UMTS base station location and configuration problem, considering and studying separately the uplink and downlink direction models as well as power-based and SIR-based PC mechanisms yields a considerable improvement from the computational point of view.

After describing the model in Section 2, a Tabu Search algorithm which provides good solution in a reasonable amount of time is briefly presented in Section 3. A procedure that allows to reduce the computations needed to determine the emitted powers with a SIR-based PC mechanism is also described. In Section 4 results obtained for medium-size realistic instances generated using classical propagation models are reported and discussed. Section 5 contains some concluding remarks.

2 Model with SIR-Based PC for Uplink and Downlink Directions

Assume that a set of *candidate sites* $S = \{1, \ldots, m\}$ where a BS can be installed, is given and that an installation cost c_j is associated with each candidate site (CS) j, $j \in S$. A set of *test points* (TPs) $I = \{1, \ldots, n\}$ is also given. Each TP $i \in I$ can be considered as a centroid where a given amount of traffic d_i (in Erlang) is requested and where a certain level of service (measured in terms of SIR) must be guaranteed. The required number of simultaneously active connections for TP i, denoted by u_i, turns out to be a function of the traffic demand, i.e., $u_i = \phi(d_i)$. The actual definition of the function ϕ is a degree of freedom of the planning process. It can simply correspond to the average number of active connections or to the number of simultaneous connections not exceeded with a given probability p. The connection activity factor can also be considered.

The propagation information is supposed to be known, either computed by using prediction tools (e.g. Hata's models or ray tracing) or obtained by actual measurements. In the case of omnidirectional BSs or of directive BSs with a uniform horizontal diagram let g_{ij}, $0 < g_{ij} \leq 1$, be the propagation factor of the radio link between TP i, $1 \leq i \leq n$, and a CS j, $1 \leq j \leq m$. In the sequel matrix $G = [g_{ij}]$, $1 \leq i \leq n, 1 \leq j \leq m$, denotes the gain between TP i and BS j.

Assuming a SIR-based PC mechanism, for each connection the emission power is adjusted so as to guarantee a signal quality level that is equal to a target value SIR_{tar} [9].

In the W-CDMA UMTS base station location and configuration problem one wishes to select a subset of candidate sites within the set S where to install directive BSs as well as their configuration, and to assign the TPs to the available BSs so as to maximize the traffic covered and/or minimize the installation costs while taking into account the signal quality requirements in terms of SIR and the power limits on the BSs.

To construct a mixed integer programming overall model, let us define the two classes of variables:

$$y_j = \begin{cases} 1 \text{ if a BS is installed in } j \\ 0 \text{ otherwise} \end{cases}$$

for $j \in S$ and

$$x_{ij} = \begin{cases} 1 \text{ if TP } i \text{ is assigned to BS } j \\ 0 \text{ otherwise} \end{cases}$$

for $i \in I$ and $j \in S$. Suppose we consider directive BSs with three identical 120 degree sectors and with an omnidirectional antenna diagram along the horizontal axis. Let the index set $I_j^\sigma \subseteq I$ denote the set of all TPs i that fall within the sector σ of the BS installed in CS j. Obviously, for each j, $I_j^1 \cup I_j^2 \cup I_j^3 = I$ and the index sets I_j^σ with $\sigma = 1, 2, 3$ are disjoint.

Since we wish to maximize the total traffic covered and minimize the total installation cost and those two aspects are clearly contrasting, the following objective function is maximized:

$$\sum_{i=1}^{n} \sum_{j=1}^{m} u_i \, x_{ij} - \lambda \sum_{j=1}^{m} c_j \, y_j \tag{1}$$

where $\lambda \geq 0$ is a trade-off parameter between maximizing coverage and minimizing costs.

Each TP $i \in I$ can be assigned to at most one BS: $\sum_{j \in S} x_{ij} \leq 1$.

Note that by restricting the assignment variables x_{ij} to take binary values, it is required that in every feasible solution all active connections must be assigned to a single BS. Moreover, a TP i can be assigned to a CS $j \in S$ only if a BS has been installed in j: $x_{ij} \le y_j$.

We also need to introduce an explicit continuous variable p_i to denote either the power emitted by TP i or the power received at each TP i from the BS it is assigned to, depending on which direction (uplink or downlink) is considered. Since the received powers must be nonnegative and there exists an upper bound on the maximum power P_{max} that a BS or a TP can assign to each connection, we have:

$$0 \le p_i \le P_{max}^{up} \tag{2}$$

for uplink and

$$0 \le p_i \le \sum_{j \in S} P_{max}^{down}\ g_{ij}\ x_{ij} \tag{3}$$

for downlink.

For the downlink direction we also consider the powers of the pilot signals $\hat{p}_j$ associated to each BS installed in j, $j \in S$. They are supposed to be constant and equal to: $\hat{p}_j = \delta\ P_{tot}$, where δ is equal to 0.15 and P_{tot} is the total power assigned to each BS j.

Moreover, the limit on the total power that each BS j can emit is accounted for by the following inequality:

$$\sum_{i \in I} \frac{p_i}{g_{ij}}\ x_{ij} + \hat{p}_j \le P_{tot}\ y_j \tag{4}$$

Given the SIR-based PC mechanism, for each pair of TP i in I and CS j in S we consider the following SIR constraints:

$$\frac{p_i\, g_{ij}\, x_{ij}}{\sum\limits_{h \in I_j^{\sigma_{ij}}} u_h\, g_{hj} \sum\limits_{t=1}^{m} p_h\, x_{ht} - p_i\, g_{ij} + \eta} = SIR_{tar}\ x_{ij} \tag{5}$$

$$\frac{p_i\ x_{ij}}{\alpha\left(\sum\limits_{\substack{k \in I_j^{\sigma_{ij}} \\ x_{kj}=1}} u_k g_{ij} \frac{p_k}{g_{kj}}\ x_{kj} + \hat{p}_j - p_i\right) + \sum\limits_{\substack{l \in S \\ l \ne j}} \hat{p}_l \sum\limits_{\substack{k \in I_l^{\sigma_{ij}} \\ x_{kl}=1}} u_k g_{il} \frac{p_k}{g_{kl}}\ x_{kl} + \eta_0} = SIR_{tar} x_{ij} \tag{6}$$

where for any CS l, the index set $I_l^{\sigma_{il}}$ denotes the set of all TPs in I that fall within the sector σ_{il} of the BS installed in l, which contains TP i. For any single connection between a BS located in CS j and a TP i falling in one of its sectors (denoted by σ_{ij}), the numerators of the left-hand-side terms correspond to the power of the relevant signal emitted by or received at TP i while the denominators amount to the total interference due to all other active connections

in the system. Thus, constraints (5) and (6) make sure that the SIR of any active connection between BS j and TP i (i.e., $x_{ij} = 1$) is equal to SIR_{tar}. Note that, since we consider directive BSs with uniform antennas diagram along the horizontal axis, the propagation gains g_{ij} do only depend on the distance between TP i and CS j and not on the sector σ_{ij} in which TP i falls.
Moreover, we have to consider an additional SIR constraint regarding the pilot signal from BS j to TP i:

$$\frac{\hat{p}_j \, x_{ij}}{\alpha \left(\sum_{\substack{k \in I_j^{\sigma_{ij}} \\ x_{kj}=1}} u_k g_{ij} \frac{p_k}{g_{kj}} \, x_{kj} \right) + \sum_{\substack{l \in S \\ l \neq j}} \hat{p}_l \sum_{\substack{k \in I_l^{\sigma_{ij}} \\ x_{kl}=1}} u_k g_{il} \frac{p_k}{g_{kl}} \, x_{kl} + \eta_0} = SIR_{tar} \, x_{ij} \quad (7)$$

If a power-based PC mechanism is considered, the power emitted from any any TP i or BS j (depending on which direction is considered) is no longer a continuous variable p_i, but is adjusted so as to guarantee a received power respectively at BS j or at TP i equal to a target value P_{tar}. Therefore, the quality constraints (5, 6) amount to imposing an upper bound on the number of connections that can be assigned to that BS. As new users are added to the system, the SIR values of all the other users decrease until one falls below the lowest acceptable quality level SIR_{min}.

The SIR-based PC mechanism is more complex than the power-based one since the power emitted by each BS or by each TP depends on that emitted by all the others, but it is also more efficient since it allows for the use of lower powers. Therefore, from a planning prospective, assuming a power-based PC mechanism instead of a SIR-based one leads to a conservative dimensioning which may allocate more radio resources than necessary; moreover, while in uplink direction both mechanisms give good quality results (as shown in section 4), in downlink direction the power-based PC mechanism turns out to be totally inadequate (see [4]).

See also [7] for a discussion of various modelling aspects.

3 Grasp and Tabu Search Algorithms

Since even the UMTS BS location problem restricted to uplink direction is NP-hard [2], we have developed Greedy Randomized Adaptive Search Procedures (GRASP) and Tabu Search algorithms to find good approximate solutions in a reasonable amount of time.

GRASP is a simple heuristic in which a randomized greedy procedure is applied a predefined number of times and the best solution found is returned. In each run of GRASP, one starts from an empty set of active BSs and at each iteration one randomly selects an available CS (in which to install an additional BS) from a set of available CSs which yield the best improvements in the objective function.

Tabu Search [8] is a meta-strategy for guiding local search procedures through the solution space of an optimization problem towards good approximate solutions. In our Tabu Search algorithms the following 'moves' are considered to explore the solution space: removing a BS, installing a new BS, removing an existing BS and installing a new one (swap). At each step, a new current solution is obtained by carrying out the best available move even though it may worsen the objective function value. To prevent cycles and to try to escape from local optima, some moves are forbidden for a certain number of iterations (they are added to a Tabu list). The best solution found during the iterations is stored and returned after a predefined maximum number of steps. As initial solution of the Tabu Search algorithm we consider the solution (i.e., the set of active BSs and the assignment of TPs to active BSs) obtained with GRASP.

Looking for approximate solutions of the SIR-based model is very intensive computationally even for instances with a moderate number of TPs. Indeed, just computing the transmitted powers corresponding to a given assignment of TPs to BSs involves inverting a $n \times n$ matrix. This operation has to be done each time a single transmitted power has to be changed: clearly, this produces a great computational effort. Moreover, there is no straightforward way of handling power constraints (2) and (3): indeed, if some of the calculated powers p_i do not satisfy them, the solution cannot be considered as feasible. To cope with the above-mentioned problems, we adapted a recently proposed iterative method to find the power levels which allows a substantial reduction of computational time [5].

We assume that new users arrive one at a time; before the new arrival, all users achieve their SIR_{tar}. The new user i is given a starting power level high enough to overcome the current interference, i.e. (as an example for uplink direction)

$$p_i = \frac{SIR_{tar}}{g_{ij}} \left(\sum_{h \in I_j^{\sigma_{ij}}} u_h \, g_{hj} \sum_{t=1}^{m} p_h \, x_{ht} - p_i \, g_{ij} + \eta \right) \tag{8}$$

and the iteration starts only if p_i is less than P^{up}_{max}. Otherwise, it's no worth trying since the power levels are monotonically increasing during the iterations. After each iteration step, we bound the transmitted powers between 0 and P^{up}_{max}. If all the links achieve their SIR_{tar} within $maxiter$ steps (e.g. 5-10), then the new connection is supported. The full algorithm looks as follows:

```
Set p_i according to (12)
feasible = false
maxiter = 5, tol = 0.95
if p_i < P_max
 p_l(0) = p_l, forall l
 for n = 1:maxiter
  Calculate SIR_l using equation (8), forall l
  Update the power levels p_l(n)= min{P_max, SIR_tar*p_l(n-1)/SIR_l(n-1)}
  if any (p_l == P_max) then
```

```
     break
  else if SIR_l > SIR_tar*tol, forall l then
     feasible = true,
     break
  end if
 end for
end if
if (feasible == true) then
 Admit new user
else
 Block the user
end if
```

Of course a similar algorithm is implemented for downlink direction.

4 Some Computational Results

In previous work [2,4] we discussed the importance of power control mechanisms in UMTS network planning. Indeed, the SIR depends on the received powers which in turn depend on the transmitted ones: therefore, these powers have to be adjusted by the PC mechanism so as to minimize interference and guarantee quality.

As shown in [2,4], in uplink direction the power-based PC mechanism gives good results, even if it is more conservative than the SIR-based one. Yet, the latter needs extremely high computational time, which is in uplink direction almost 10 times higher than the power-based one. Typical results obtained with a single iteration of GRASP and Tabu Search for small-size realistic instances with 390 TPs and 20 CSs are reported in Table 1.

As explained in [4], in downlink direction the difference between the results obtained with SIR-based PC mechanism and power-based one becomes more relevant than in uplink direction: with the addition of the pilot signals this difference is still more increased, as shown in Table 2.

This substantial difference is mainly due to the wide range in which the emitted power levels can vary in downlink direction, thus generating high intra-cell and inter-cell interference. In the presence of a power-based PC mechanism the emission powers, which are selected to guarantee a P_{tar} level at the receiving end, are often much higher than needed to yield SIR values equal to SIR_{min}, thus producing high interference levels. On the contrary, with a SIR-based PC the emission powers are selected in order to have a signal quality exactly equal to SIR_{tar}, thus keeping the interference levels under limited values; this in turn leads to solutions having excellent coverage with a limited number of BSs.

Due to the above considerations, we focused our attention on the uplink and downlink model with SIR-based PC.

In the first two columns of Table 3 we compare the results obtained with a single iteration of GRASP and with a single iteration of GRASP followed by Tabu Search on realistic small instances including pilot signals and with variable

Table 1. Results obtained for uplink with the models assuming power-based and *SIR*-based PC: n=390, m=20

SIR-based uplink		**Power-based uplink**	
Served TPs	**# of BSs**	**Served TPs**	**# of BSs**
368	7	368	8
368	7	368	9
390	9	385	9
376	7	376	9
390	7	389	9
390	8	390	10
390	7	390	9
390	7	387	8
390	8	390	11
386	7	380	7

Table 2. Results obtained for downlink with the models assuming power-based and *SIR*-based PC: n=390, m=20

SIR-based downlink		**Power-based downlink**	
Served TPs	**# of BSs**	**Served TPs**	**# of BSs**
385	5	175	7
383	6	194	9
390	5	190	10
390	4	160	8
390	5	203	11
390	4	180	9
390	5	201	11
390	5	210	10
389	4	196	10
390	5	203	9

sector orientations. We consider $m = 20$ CSs and $n = 780$ TPs, equally subdivided in two sets: one producing voice call traffic for uplink direction, the other receiving traffic on downlink direction (see www.elet.polimi.it/res/3Gplanning for some example instances). As expected, the Tabu Search algorithm produces solutions of higher quality than the ones obtained with GRASP: a greater number of users is served with less BSs.

In the third column, solutions produced by an uplink power-based model and subsequently subjected to the uplink and downlink model constraints are shown. These solutions are very similar to the ones directly produced with the complete model (i.e. uplink and downlink together): this is a significant result for at least two reasons. First of all it shows that in a UMTS network the most relevant direction from a planning point of view is without doubt the uplink one (see [4]). This is primarily due to the low intra-cell interference that TPs experiment in the downlink direction because each BS uses orthogonal spreading codes. Moreover,

Table 3. Results obtained with the model including uplink and downlink SIR-based constraints and with starting solutions provided by the simplified uplink power-based model; section orientation is also considered

Uplink and downlink GRASP			Uplink and downlink Tabu search			Power-based uplink		
TPs dw	TPs up	# of BS	TPs dw	TPs up	# of BS	TPs dw	TPs up	# of BS
383	363	7	390	370	8	384	372	8
380	369	8	390	390	9	390	389	9
390	389	11	390	390	8	390	390	9
390	375	8	390	388	8	390	387	10
390	390	9	390	390	7	390	390	8
390	390	8	390	390	7	390	390	9
390	389	10	390	390	8	390	390	8
390	390	11	390	390	8	390	390	7
390	390	8	390	390	7	390	390	8
390	385	9	390	388	8	390	388	8

Table 4. Results obtained for downlink assuming SIR-based PC with pilot signal and without pilot signal: n=390, m=20

SIR-based downlink with pilot signal		SIR-based downlink without pilot signal	
Served TPs	# of BSs	Served TPs	# of BSs
385	5	390	4
383	6	390	4
390	5	390	3
390	4	390	3
390	5	390	3
390	4	390	3
390	5	390	4
390	5	390	4
389	4	390	3
390	5	390	3

from a computational point of view, the second way of finding good solutions for an uplink and downlink scenario is obviously much less intensive than producing them with the entire model: indeed, the algorithm used for the uplink model with power-based PC is about 50 times faster than the one used for the overall model.

Finally, Table 4 shows the impact of the introduction of pilot signals in 3G planning: these signals considerably reduce the available power at BSs, thus worsening the quality of the results obtained without them. Moreover, pilot signals themselves have to satisfy some SIR-constraints (see 7): again, this produces a degrading of the quality of the results.

5 Concluding Remarks

We have presented a mathematical programming model to optimize the location and configuration of BSs in UMTS networks taking into account both uplink and downlink directions, with pilot signals included, and assuming a SIR-based PC mechanism. GRASP and Tabu Search algorithms have been proposed to find good approximate solutions within a reasonable amount of computational time. To make them computationally effective, we adapted a recently proposed iterative method to find the power levels of the signals emitted by TPs or BSs.

Unlike for the uplink model with power-based PC, in downlink the more accurate model with SIR-based PC is required. Moreover, the results obtained with the combined model including uplink SIR-based as well as downlink SIR-based constraints indicate that by starting from the solutions provided by the simplified uplink model with power-based PC, computing times are reduced by a factor of fifty without heavily affecting the solutions' quality. Besides, the introduction of pilot signals obviously worsen the solutions' quality in downlink direction but makes the results more realistic.

Finally, we can assert that the fact of considering and studying separately the uplink and downlink direction models as well as a power-based PC mechanism or a SIR-based one has yielded a striking improvement from the computational point of view and has given us a better insight into the overall UMTS BS location and configuration problem.

References

1. E. Amaldi, A. Capone, and F. Malucelli. Optimizing UMTS radio coverage via base station configuration. In *The 13th IEEE International Symposium on Personal, Indoor and Mobile Radio Communications (PIRMC'02)*, volume 1, pages 315–319, 2002.
2. E. Amaldi, A. Capone, and F. Malucelli. Planning UMTS base station location: Optimization models with power control and algorithms. *IEEE Transactions on Wireless Communications*, 2002.
3. E. Amaldi, A. Capone, F. Malucelli, and F. Signori. UMTS radio planning: Optimizing base station configuration. In *Proceedings of IEEE VTC Fall 2002*, volume 2, pages 768–772, 2002.
4. E. Amaldi, A. Capone, F. Malucelli, and F. Signori. Optimization models and algorithms for downlink UMTS radio planning. In *Proceedings of IEEE Wireless Communications and Networking Conference (WCNC'03)*, 2003. To appear.
5. M. Berg. Radio resource management in bunched personal communication systems. Technical report, Kungl Tekniska Hogskolan (KTH), Stockholm, Sweden, 2002.
6. E. Berruto, M. Gudmundson, R. Menolascino, W. Mohr, and M. Pizarroso. Research activities on UMTS radio interface, network architectures, and planning. *IEEE Communications Magazine*, pages 82–95, February 1998.
7. A. Eisenblatter, A. Fugenschuh, T. Koch, A. Koster, A. Martin, T. Pfender, O. Wegel, and R. Wessaly. Modelling feasible network configurations for UMTS. Technical Report 02-16, Konrad-Zuse-Zentrum fur Informationstechnik (ZIB), Berlin, Germany, 2002.

8. F. Glover and M. Laguna. *Tabu search.* Kluwer Publishers, 1997.
9. S.A. Grandhi, J. Zander, and R. Yates. Constrained power control. *Journal of Wireless Personal Communications*, 1(4), 1995.
10. H. Holma and A. Toskala. *WCDMA for UMTS.* John Wiley and sons, 2000.
11. C.Y. Lee and H.G. Kang. Cell planning with capacity expansion in mobile communications: A tabu search approach. *IEEE Trans. on Vehicular Technology*, 49(5):1678–1690, 2000.
12. R. Mathar and Michael Schmeink. Optimal base station positioning and channel assignment for 3g mobile networks by integer programming. *Annals of Operations Research*, 107:225–236, 2001.

Soft Output Bit Error Rate Estimation for WCDMA

Lodewijk T. Smit[1], Gerard J.M. Smit[1], Johann L. Hurink[1], and André B.J. Kokkeler[2]

[1] Department of Electrical Engineering, Mathematics & Computer Science
University of Twente, Enschede, The Netherlands
L.T.Smit@utwente.nl
[2] Ericsson, Eurolab, Enschede, The Netherlands

Abstract. This paper introduces a method that computes an estimation of the bit error rate (BER) based on the RAKE receiver soft output only. For this method no knowledge is needed about the channel characteristics nor the precise external conditions. Simulations show that the mean error of the estimation is below 2%, with only a small variance. Implementation issues for a practical use of the method are discussed.

Keywords: BER estimation, WCDMA

1 Introduction

In this paper we introduce a method to compute an estimation of the bit error rate (BER) of a wireless channel. The presented method is used to estimate the current quality of the wireless channel using the data received by a RAKE receiver [8]. This information can be used to adapt the receiver to obtain the desired Quality of Service (QoS) for a given application or protocol with minimal computational effort. This reduction in computational effort can be translated to a reduction in energy consumption for a mobile terminal or to a reduction in the amount of resources for a base station.

Applications or protocols demand a certain QoS that translates to a certain quality (BER) of the wireless link. In general, two principles are used to determine the quality of the output of the RAKE receiver. First, known sequences of (pilot) symbols are transmitted in parallel with the data, so the BER can be determined. Second, models are used which require the current status of the environment as input to compute the BER. In this paper, we use an alternative (third) method. We apply statistical methods on the soft output of the RAKE receiver, to compute the BER without additional knowledge of the current environment or transmission of extra pilot symbols.

To illustrate our approach we give an example of the application of our BER estimation algorithm in a wide-band code division multiple access (WCDMA) system [6]. The output of the RAKE receiver is used as input to an (adaptable) forward error correction (FEC) turbo decoder as depicted in Figure 1. The used turbo FEC operates on a sequence of bits, grouped in a block. Given the number of errors per block, we can predict whether the used forward error decoder is able to correct the received block [9] or not. In our adaptive system [10], the spreading factor used by the WCDMA transmission is decreased until the limit of the error decoding capacity of the FEC decoder is reached. Decreasing the spreading factor leads to a higher bitrate, which has two main

M. Conti et al. (Eds.): PWC 2003, LNCS 2775, pp. 448–457, 2003.

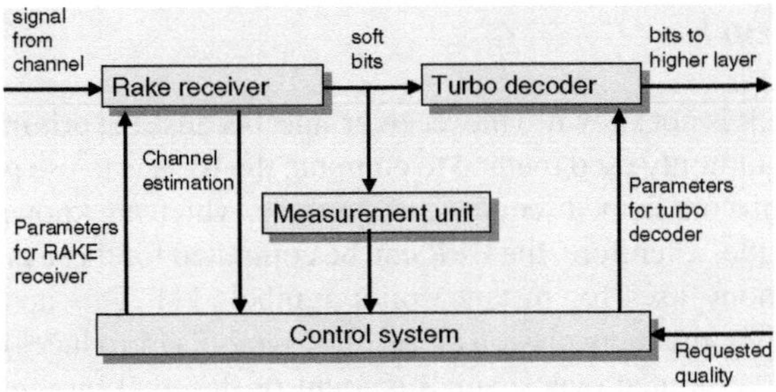

Fig. 1. The Control System of The Terminal

advantages. First, a certain amount of data is transmitted as fast as possible, providing a good QoS for the user. Second, the transmitter and receiver can be switched off earlier, saving power, which is especially useful for a mobile handheld terminal.

The frame error rate (FER) after the FEC decoder also gives an indication of the quality of the received signal. However, the BER estimation after the RAKE receiver gives much more information about the quality. There are two reasons for that. First, we not only know whether the turbo decoder is able to correct the received frame or not, but we also know whether the quality of the received frame is near the turbo decoder error correcting capacity or there is room for improvement. In this way we can predict how much the quality of the output of the RAKE receiver should be improved or decreased (e.g. by changing the spreading factor) so that the turbo decoder is just able to correctly decode the received frames. Second, we can predict what will happen when we change parameters. For example, if we plan to change the spreading factor of the RAKE receiver, we are able to predict the resulting BER and FER. Thus, we can predict whether the quality of the output of the RAKE receiver after reduction of the spreading factor is still good enough such that the turbo decoder can still correct most of the frames and what the consequences are for the QoS (e.g. latency, throughput, BER).

Our BER estimation algorithm gives detailed information about the quality of the signal and as we know the characteristics of the forward error decoder, we can made a careful trade-off between the different parameter settings that are possible at physical layer (RAKE receiver), link layer (FEC decoder) and transport layer (e.g. retransmissions) of the network protocol stack. This cross-layer approach ensures a global optimization, with potential higher savings compared to optimizations performed per individual layer of the network protocol stack.

In this way, we can minimize the energy consumption and/or the use of resources at run-time, while satisfying an adequate QoS, which is requested by the end user of the system.

Section two describes related work. Section three derives and explains the method for BER estimation. Section four presents simulation results giving the difference between the real BER and the estimated BER. Section five discusses how to implement the presented method in hardware, followed by the conclusions in the last section.

2 Related Work

In general, the BER is not known at the receiver side, because the original transmitted data is unknown. A commonly used method to compute the BER is to use pilot symbols. Pilot symbols represent a predefined sequence of symbols, which are known at the transmitter and the receiver side. Therefore, the BER can be computed for these pilot symbols. Third generation telephony uses for instance pilot symbols [1]. This approach has several disadvantages. First, the transmission of the pilot symbols introduces overhead. Second, the BER is only computed over a small amount of the total bits that are transmitted. Third, the BER of the pilot symbols may differ from the BER of the data.

Another approach is to model the channel with all the known effects, e.g. [7]. A state of the art article on this area is [3]. Using this method it is possible to achieve accurate BER estimations for the modeled channel. However, the actual properties of the channel and the modeled effects can differ significantly from the constructed model. Also, effects that are not modeled can happen in real situations. In practice, it is not possible to model all the different effects that cause the disturbance of the wireless channel. Estimation of the exact quality of the signal of the wireless channel is therefore impossible.

Our approach differs significantly from the two mentioned approaches. We only use the soft output from the rake receiver, and require no additional information about the channel. In our opinion, it doesn't matter which physical effect is responsible for the degradation of the signal to determine the BER. The advantage is that an accurate estimation can be made independent of the unpredictable dynamic changing external environment.

3 BER Estimation

In an ideal situation, without disturbance of the channel, the output of the soft value of the rake receiver is equal to the used sf (spreading factor) for a transmitted bit with value one. Similarly, for a bit with value zero (represented by minus one), the soft output of the rake receiver is $-sf$ This perfect situation is shown in Figure 2.

In case of disturbance of the channel, the sampled chip values are no longer exactly equal to one or minus one, but can be higher of lower. A lot of external causes may be responsible for this disturbance. Most effects that change the signal can be modeled with a normal distribution. For example, AWGN behavior and the fact that the spreading codes of other channels are not perfectly orthogonal, can be modeled with a normal distribution. A few effects, e.g. fading, do not behave like a normal distribution. However, the central limit theorem [5] states that regardless of the type of distribution, the distribution will approximate a normal distribution, if the number of samples is large (>30). Therefore, we can approximate the values of the soft values of the output of the RAKE receiver with a normal distribution. One soft output value is composed of different chip values. If the number of chips per bit is higher, a better approximation of the normal distribution is made.

Figure 3 shows the expected normal distribution behavior for the soft output values of the RAKE receiver for a pretty good channel. When the channel becomes worse, the mean will not change (significantly), but the standard deviation will increase, as shown in

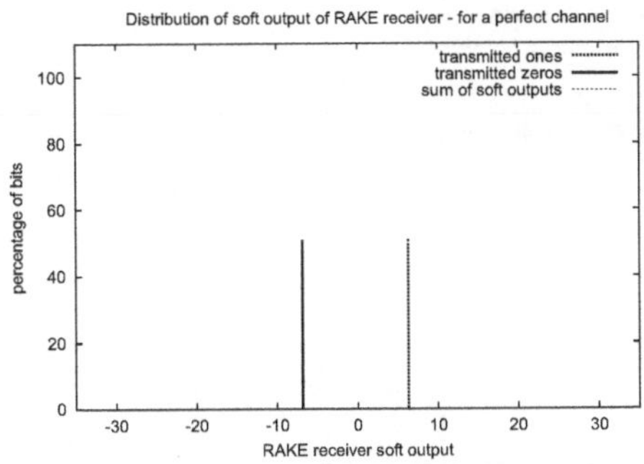

Fig. 2. Perfect channel

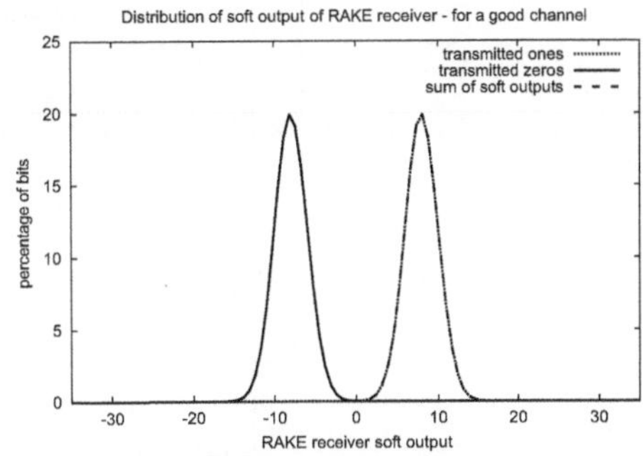

Fig. 3. Good channel

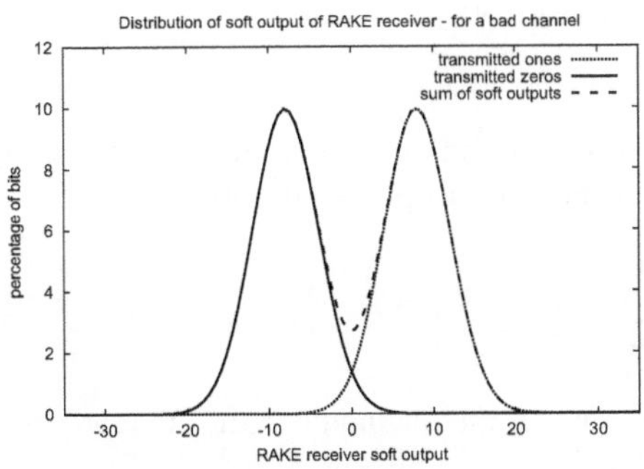

Fig. 4. Bad channel

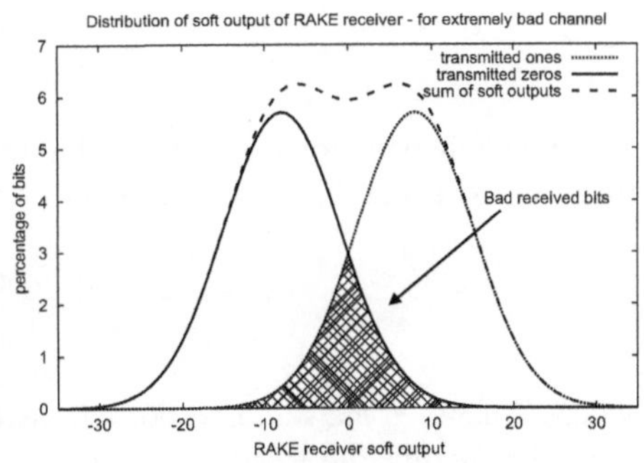

Fig. 5. Very bad channel

Figure 4. Some bits are received incorrect in this figure. All soft values > 0 are considered to be transmitted ones and all soft values < 0 are considered to be transmitted zeros. Figure 5 shows the effect for an extremely bad channel. As can be seen from the figure, the two distributions are heavily mixed up. Every bit with value one that is received with a negative soft output is received incorrectly and also the positive soft output for a transmitted bit with value zero is received incorrectly. The marked area in Figure 5 is the probability that a bit is received incorrectly.

In reality, the distribution is not as ideal as the distribution shown in Figures 2 to 5. Figure 6 shows the soft output values of one transmitted block (3012) bits. To plot the distribution, all the soft values are rounded to the nearest integral number to make classes. Figure 6 shows the distributions for the transmitted ones and zeros.

Unfortunately, the receiver can not determine whether a soft value belongs to the 1-distribution or to the 0-distribution. The soft output of the RAKE receiver is the addition of the 1-distribution and the 0-distribution, which is also plotted in Figures 2 to 6 as a dotted line.

Our goal is to predict the bit error rate (BER), i.e. the size of the marked area in Figure 5. Let X(Y) denote the distribution of the soft output values of the transmitted zeros (ones). Using these distributions, the BER can be expressed by:

$$BER = pP(X \geq 0) + (1-p)P(Y \leq 0). \tag{1}$$

where p denotes the probability that a zero is transmitted.

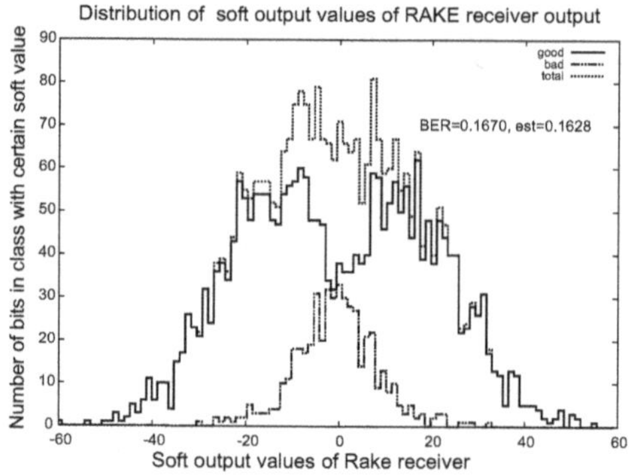

Fig. 6. Output of RAKE receiver for bad channel

Since both distributions are mirrored to the zero axis and due to the mentioned assumption, X and Y can be expressed in terms of a standard normal distribution:

$$X = \sigma Z - \mu. \tag{2}$$

$$Y = \sigma Z + \mu. \tag{3}$$

where Z denotes the standard normal distribution, μ the mean and σ the standard deviation. Using this, the BER reduces to:

$$BER = P(X \geq 0) = P(Z < \frac{\mu}{\sigma}) = \Phi(-\frac{\mu}{\sigma}). \tag{4}$$

where $\Phi(z)$ is the function that gives the area of the standard normal distribution to the right of z, i.e. the probability that a value is smaller than z. The function $\Phi(z)$ is widely available in tabular form.

We want to get a prediction of μ and σ based on the soft output values of the RAKE receiver. Using the soft output values, we derive estimates $\widehat{\mu}$ and $\widehat{\sigma}$ for μ and σ respectively.

Note that if there are only effects with a normal distribution (like disturbance of other users, AWGN, etc), $\widehat{\mu}$ will be equal to the spreading factor. However, for other effects (e.g. fading effects like Doppler), the $\widehat{\mu}$ can differ significantly from the spreading factor.

As mentioned before, the received soft output values of the RAKE receiver do not correspond to the distribution X and Y, but to a distribution W, which results from the combination of the distributions X and Y (with probability p we get distribution X and with probability $(1-p)$ distribution Y). For W we have:

$$P(W \leq w) = pP(X \leq w) + (1-p)P(Y \leq w). \tag{5}$$

Based on measured results for W and using moments of distributions, it is possible to estimate the characteristic values μ and σ of the distributions X and Y, which together form distribution W (see [11]). If r is a positive integer, and if X is a random variable, the rth moment of X is defined to be $m_r(X) \equiv E(X^r)$, provided the expectation exists, see [4]. For a standard normal distribution, the first, second, third and fourth moments

are respectively zero, one, zero and three. The first and third moment of Z are zero and can not be used to compute the two unknown variables $\widehat{\mu}$ and $\widehat{\sigma}$. Therefore the second and fourth moment of W are used.
The second moment of W is:

$$m_2(W) = p(E(X^2)) + (1-p)(E(Y^2)). \tag{6}$$

Scrambling (used in almost every wireless communication system) ensures that approximately an equal number of ones and zeros are transmitted. This means that $p \approx \frac{1}{2}$. Setting $p = \frac{1}{2}$, and using equations (2), (3) and the moments of the standard normal distribution, equation (6) becomes:

$$m_2(W) = \mu^2 + \sigma^2. \tag{7}$$

therefore,

$$\sigma^2 = m_2(W) - \mu^2. \tag{8}$$

The fourth moment of W is:

$$m_4(W) = p(E(X^4)) + (1-p)(E(Y^4)). \tag{9}$$

With p = $\frac{1}{2}$, this equation becomes:

$$m_4(W) = \mu^4 + \binom{4}{2}\mu^2\sigma^2 E(Z^2) + \sigma^4 E(Z^4). \tag{10}$$

Substituting the moments of Z gives:

$$m_4(W) = \mu^4 + 6\mu^2\sigma^2 + 3\sigma^4. \tag{11}$$

Replacing σ^2 with (7) and simplifying yields:

$$\mu^4 = \frac{3}{2}(m_2(W))^2 - \frac{1}{2}m_4(W). \tag{12}$$

So,

$$\mu = \sqrt[4]{\left|\frac{3}{2}(m_2(W))^2 - \frac{1}{2}m_4(W)\right|}. \tag{13}$$

Using Equations (8) and (13) in combination with the estimation of the second and the fourth moments of W based on the individual samples $V_1..V_n$ of the output of the rake receiver, the Formulas (14) and (15) can be derived for the estimators $\widehat{\mu}$ for the mean and $\widehat{\sigma}$ for the standard deviation:

$$\widehat{\mu} = \sqrt[4]{\left|\frac{3}{2}\left(\frac{\sum_{i=1}^{n} V_i^2}{n}\right)^2 - \frac{1}{2}\frac{\sum_{i=1}^{n} V_i^4}{n}\right|} \tag{14}$$

$$\widehat{\sigma} = \sqrt{\left|\frac{\sum_{i=1}^{n} V_i^2}{n} - \widehat{\mu}^2\right|} \tag{15}$$

Finally, the BER estimation can be computed with:

$$\widehat{BER} = \Phi\left(-\frac{\widehat{\mu}}{\widehat{\sigma}}\right) \tag{16}$$

4 Results

In our simulation environment we performed several simulations with a realistic time-variant channel. In successively simulations, the number of simultaneously transmitting users, the number of paths and the amount of added white Gaussian noise (AWGN) is changed. All simulations uses blocks with 1000 randomly generated turbo encoded bits, making a block size of 3012 bits. For each received block, the real BER is determined and compared with the estimated BER. The reported estimation error is the absolute difference between the estimated BER and the real BER (expressed in %); i.e.,

$$est\ error = |BER_{EST} - BER_{REAL}| * 100\%. \tag{17}$$

The estimation error is reported as the absolute difference, because the relative difference can be very high with a low BER. For example, if 2 errors (BER=0.002) are estimated for a block with 1000 bits and the block contains 1 error (BER=0.001) the relative difference is large, while the absolute difference is only 0.1%. For our application, we are interested in the absolute difference.

In Figure 7 the mean estimation error is depicted, as function of classes with a width of 0.01 of the real BER of the received block (e.g., the estimation error of all blocks with a real BER in the range [0.15, 0.16) are summed up and divided through the number of blocks in the class to get a mean estimation error). For a specific case, the presented results show that the estimation is better for a lower spreading factor. Having a specific BER, a lower spreading factor means a better channel than the same BER for a higher spreading factor. Therefore, the estimation works better for a better channel. In addition to the average estimation error, information about the variance in the estimation is relevant, because the estimation will be worthless if the variance is too high. In Figure 8, the estimation error for $sf = 8$ is depicted. Beside the mean of the estimated BER also the variance is given. For each BER class, the interval $[\mu - \sigma, \mu + \sigma]$ is given. Given this figure, we can conclude that, for a real BER below 0.2, a good prediction is possible with an error of at most 2%. We are not interested in BER > 0.2, because blocks with a BER > 0.2 can not be corrected by a FEC decoder (e.g. a turbo decoder). The same kind of simulations have been performed for different scenarios, e.g. Rayleigh fading channels, different amount of users, different amount of paths, etc. The achieved results were similar to the ones given in the Figures 7 and 8.

4.1 External Validation

To validate the results of our simulation and to verify the proper working of our algorithm, Ericsson Eurolab in Enschede did an additional set of simulations. They got only Formulas (14) – (16) without additional information. Ericsson used their UMTS simulator and tried to estimate the BER with the Formulas (14) – (16) and the soft output

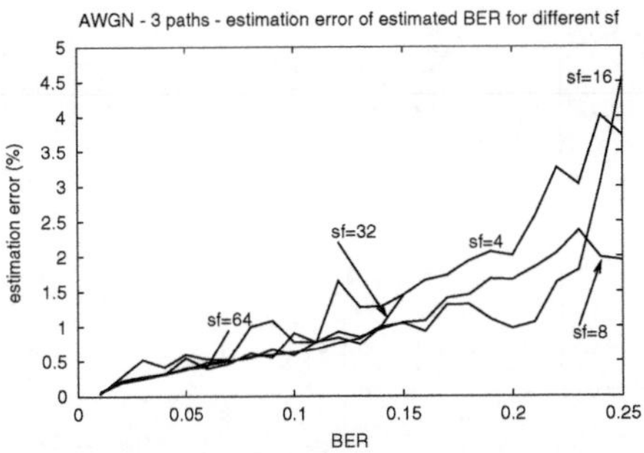

Fig. 7. Estimation Error, for different Spreading Factors

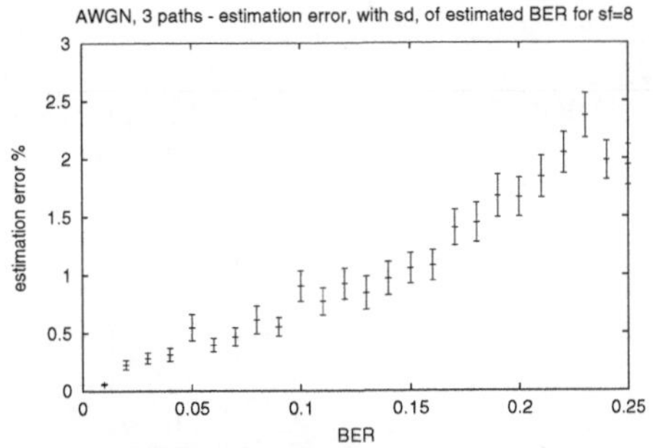

Fig. 8. Estimation Error and its Variance displayed for $sf = 8$ ($[\mu - \sigma, \mu + \sigma]$ for each class)

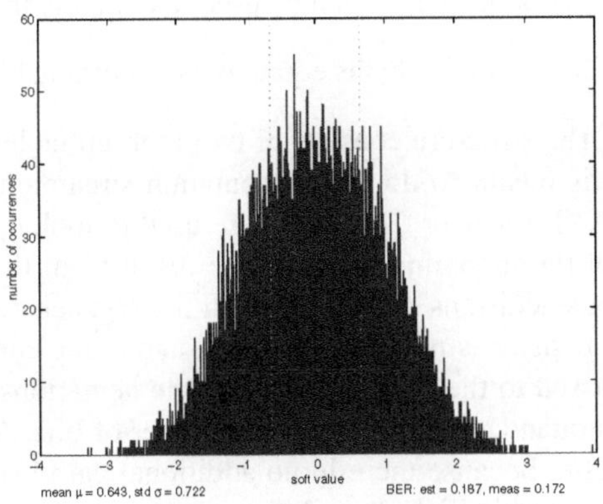

Fig. 9. DCH, real. channel, 120k samples

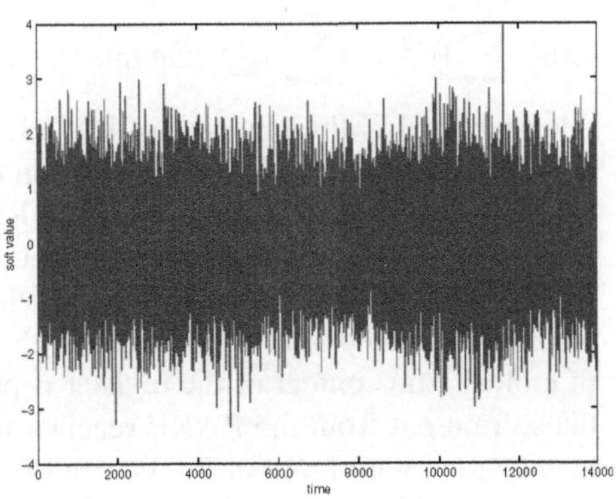

Fig. 10. DCH, real. channel, data

of their simulator. Two different channels have been simulated: AWGN and an Ericsson proprietary channel that is very realistic with multiple users, multiple paths, power variations, etc. Figure 9 shows the soft output values of the RAKE receiver. The estimated BER is 0.187 and the real BER is 0.172. Figure 10 shows the accompanying diagram of the received power. The estimation error for the Ericsson proprietary channel was about 1.7% and the estimation error for the AWGN channel was even lower. As expected, the BER estimation algorithm gives less accurate results when power control is disabled. However, the whole performance of WCDMA depends on a good power control.

5 Implementation

The proposed method is simple and the involved Formulas (14) and (15) can be implemented easily on an ALU (e.g. an ARM). In this section we give some considerations how a real implementation can be made on an ASIC or reconfigurable architecture.

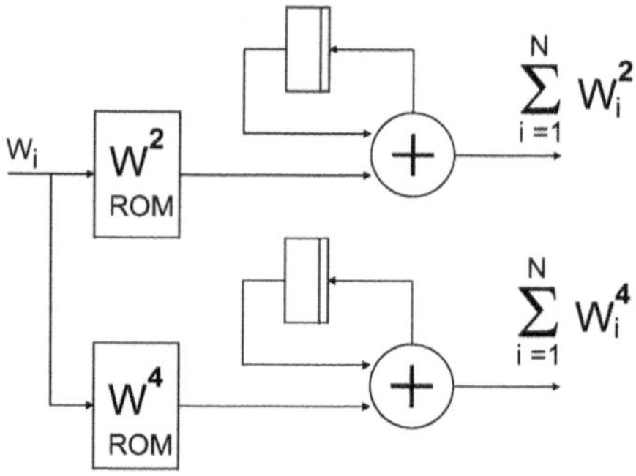

Fig. 11. Hardware support for BER estimation

Figure 11 shows a very simple hardware support, which can be used to compute the terms $\sum_{i=1}^{n} W_i^2$ and $\sum_{i=1}^{n} W_i^4$, that must be done at a speed that is equal to the incoming bit rate (maximal 2Mbit/s in case of UMTS). The structure consists of two look-up tables (LUT), two adders and two registers and is meant to do the computation streaming, while the samples are coming from the RAKE receiver. The LUTs are used to look up the power of two and the power of four of the incoming sample. The result from the LUT is added to the subtotal of the previous additions that is stored in the register. At begin of the reception of a new block, the register is initialized to zero, and at the end of a block, the content of the register is passed to the output. In real implementations, the soft output from the RAKE receiver is quantized with a limited number of bits. A quantization with more of 8 bits is not useful, because there in no additional gain [2]. Even with 6 bits quantization, there is no observable SNR degradation. Supposing 6 bits that represent a signed soft value, the LUTs can be limited to $2^5 = 32$ entries because the sign bit can be ignored. This proposed structure can be implemented in dedicated hardware or an FGPA.

The remainder of the computation of the formula can be done after finishing the computation of the summation. Note that the speed of this computation can be much lower, because this has to be done only once per received block. Therefore, this computation can be done by a general purpose processor, like an ARM. If everything has to be done in dedicated hardware or FPGA, division by n and multiplication by $\frac{1}{2}$ and $\frac{3}{2}$ can be done by shifting, if only the first 2^k samples of all samples of a block are used, where k is as large as possible. The square roots can be stored in a LUT.

6 Conclusions

A method is introduced that makes an estimation of the bit error rate based on statistical analysis of the output of the RAKE receiver soft output only, without prior knowledge about the channel model and all external influences. Simulations show that the mean error of the estimation is below 2%, having only a small variation. Implementation issues for use of the method in practice are discussed.

Acknowledgements. This research is conducted within the Chameleon project (TES.5004) supported by the PROGram for Research on Embedded Systems & Software (PROGRESS) of the Dutch organization for Scientific Research NWO, the Dutch Ministry of Economic Affairs and the technology foundation STW. We would like to thank dr. W.C.M. Kallenberg for his support.

References

1. http://www.3gpp.org.
2. Becker, J., Pionteck, T., Glesner, M.: Simulation, prototyping and reconfigurable hardware realization of cdma rake-receiver algorithms for flexible mobile transceivers. In: Proc. of ERSA'01. (2001) 119–126
3. Cheng, J., Beaulieu, N.C.: Accurate ds-cdma bit-error probability calculation in rayleigh fading. IEEE transactions on wireless communications **1** (2002) 3–15
4. Dudewicz, E.J., Mishra, S.N.: Modern Mathematical Statistics. John Wilsey & Sons, Inc. (1988) ISSN: 0271-6232.
5. Mann, P.S.: Introductory Statistics. 2 edn. John Wiley & Sons (1995) ISBN: 0-471-31009-3.
6. Milstein, L.B.: Wideband code division multiple access. IEEE Journal on Selected Areas in Communications **18** (2000) 1344–1353
7. Morrow, R.K.: Accurate cdma ber calculations with low computational complexity. IEEE Transactions on Communications (1998) 1413–1417
8. Price, R., Green, P.: A communication technique for multipath channels. In: Proceedings of the IRE. Volume 46. (1958) 555–570
9. Smit, L.T., Smit, G.J., Havinga, P.J., Hurink, J.L., Broersma, H.J.: Influences of rake receiver/turbo decoder parameters on energy consumption and quality. In: Proc. of 2002 International Conference On Third Generation Wireless and Beyond. (2002) 227–235
10. Smit, L.T., Smit, G.J., Havinga, P.J., Hurink, J.L., Broersma, H.J.: Run-time control for software defined radio. In: proceedings PROGRESS 2002 workshop. (2002)
11. Tan, W., Chang, W.: Some comparisions of the method of moments and the maximum likelihood in estimating parameters of a mixture of two normal densities. Journal of the American Statistical Association **67** (1972) 702–708

The Convergence of Control, Communication, and Computation*

Scott Graham and P.R. Kumar

Department of Electrical and Computer Engineering, and
Coordinated Science Laboratory
University of Illinois at Urbana-Champaign
1308 West Main Street
Urbana, IL 61801
USA
{srgraham,prkumar}@uiuc.edu
http://black1.csl.uiuc.edu/~prkumar/

Abstract. The convergence of communication and computation over the past two decades has given us the Internet. We believe that the next phase of the information technology revolution will be the convergence of control, communication, and computation. This will provide the ability for large numbers of sensors, actuators, and computational units, all interconnected wirelessly or over wires, to interact with the physical environment. We argue that in the proliferation of this "convergence," a critical role will be played by the architecture. We describe an experimental Convergence Testbed at the University of Illinois, outline the architectural challenges, and our efforts in this direction.

1 Introduction

Over the past two decades we have seen the convergence of communication and computation, which has given us the Internet. Worldwide there are over 150 million internet hosts [1], and over 600 million users [2]. Indeed networked computers nowadays are critical not only for their computation capabilities but also for their communication capabilities. This phase of the information technology revolution has provided us the ability to exchange information in the form of email or to browse each other's webpages.

We anticipate that the next phase of the information technology revolution will provide us the ability to actively interact with the environment and alter it.

* This material is based upon work partially supported by USARO under Contract Nos. DAAD19-00-1-0466 and DAAD19-01010-465, DARPA under Contract Nos. N00014-01-1-0576 and F33615-01-C-1905, AFOSR under Contract No. F49620-02-1-0217, DARPA/AFOSR under Contract No. F49620-02-1-0325, and NSF under Contract No. NSF ANI 02-21357. Any opinions, findings, and conclusions or recommendations expressed in this publication are those of the authors and do not necessarily reflect the views of the above agencies. Captain Graham is studying under U.S. Air Force sponsorship through the AFIT/CI program.

M. Conti et al. (Eds.): PWC 2003, LNCS 2775, pp. 458–475, 2003.

Such interaction will require sensing the environment and acting on it, and will be achieved by interconnecting sensors and actuators with computation elements, and providing all with communication capability.

Two technological trends making this feasible are the growth in embedded computers and wireless networking. About 98% of all microprocessors sold are embedded, and their percentage is growing [3]. They are present in cell-phones, watches, stereos, microwaves, washing machines, wireless thermometers, cordless phones and answering machines. They exist in pocket video games, VCRs, DVD players, printers, and components of computer systems. Automobiles typically have multiple such processors.

Currently these embedded devices function in an isolated way and are not significantly interconnected. The cost of wires alone is comparable to the cost of many of these devices, not to mention installation costs. Thus, toasters are not connected to alarm clocks. However we may be on the cusp of a wireless revolution. Wi-Fi (IEEE 802.11x) has experienced double-digit growth since 2000 [4], and is now installed as a default on several makes of computers. Lower cost wireless connectivity is possible with Bluetooth available at a $6 per chipset cost to manufacturers [5]. Extrapolating these trends in wireless communication, we can envision a time, not far off, in which wireless connectivity is a commodity. With each embedded device functioning as a sensor or an actuator, and each wirelessly connected with others, the future could well see orchestras of sensors and actuators playing over the ether in vast interconnected control systems. Indeed, the Berkeley Motes [6] already provide a combination of sensing, wireless communication, and computation, all in a package with a small spatial footprint and low energy usage.

In short we anticipate the convergence of control with communication and computation. But how will these systems interoperate? How will they be interconnected, physically, in applications, and in theoretical frameworks? These are the issues addressed in this paper. We argue that the architecture of these systems will play a critical role in their emergence and proliferation. We describe a testbed for convergence at the University of Illinois. We elaborate on the issues that arise, and outline an architectural solution being pursued by us to realize the twin goals of reliability and minimal design and deployment time.

2 Convergence Towards a More Holistic Theory

The aforementioned technological developments are leading to accompanying changes in research directions which are aimed at a more integrated view of systems theory. Though it may not be completely accurate to put too clear a historical marker, it can be said that the last half of the twentieth century was the age of developing the individual areas of control, communication, and computation. von Neumann's idea of a stored program (1944) and the ENIAC (1946) are about a half century old, and roughly mark the beginning of the age of computers. Wiener's second World War work, embodied in his "Yellow Peril" book (so known for the color of its cover and its perceived incomprehensibility) dates to 1949. Shannon's foundational information theory was published in 1948. Kalman's work on providing a foundation for state-space control theory dates to around 1960. In signal processing the seminal work of Cooley and Tukey is slightly more recent, around 1965.

However, we anticipate that the next few decades will witness the development of a more integrated system theory combining all these areas. For example, signal/image processing methods with information theoretic performance assessment and connections are already emerging [7,8]. Networking is seeing the confluence of computer science with more traditional communications research conducted in Electrical Engineering departments.(INFOCOM, for example, is jointly organized by the IEEE Computer and IEEE Communications Societies). Communication and control have a long history of involvement, dating back to the work of Wiener and Nyquist.

In the future, at the theoretical level (but not necessarily at the architectural level), issues such as addressing messages, and combining sensory inputs, while computing based on locally available data, will all be seen simply as tradeoffs in the context of design of a larger system.

As we enter this age of convergence, several challenges arise. In this paper we highlight one of them; how to design such "systems" (to use the EE term), or the "application" (to use the CS term), in an environment of constant change where new features are always being added, and hardware is constantly being changed, all without necessitating major recoding, and with a view to minimizing the designer's time. A closely related issue is what sort of software infrastructure (or "middleware" to use the CS term) is needed to facilitate the rapid development and deployment of systems.

Consider the IP stack in networking. It is present in all computers and has provided essential communication services by making interconnections transparent to the user. But what will the "IP stack" equivalent be for distributed and interconnected embedded systems? A suitable architectural construct will need to provide the appropriate services for sensors, actuators, computation, and communication to work together. Such infrastructure code should self-organize, taking care of details such as which computation is running on which host, and relieve the designer from mundane details such as IP addresses and the problem of start-ups, etc. The software should provide the right abstractions and interfaces to application programmers, and a rich set at that, so that they can concentrate on developing applications.

We can today build one-of-a-kind systems to fit just about any single use case, although such systems could be, and often are, enormously expensive. But that is not the vision we are projecting. Our vision is broader. The convergence of communication with computation is today ubiquitous because of several forces. We contend that similar forces are at play which will lead to the confluence of control with communication and computation also becoming ubiquitous. Moreover, the resulting systems will need to adapt to changing uses.

Our goal, thus, is to move from an era of carefully hand-crafted systems to mass production of interconnectable devices, with easy to configure interfaces,

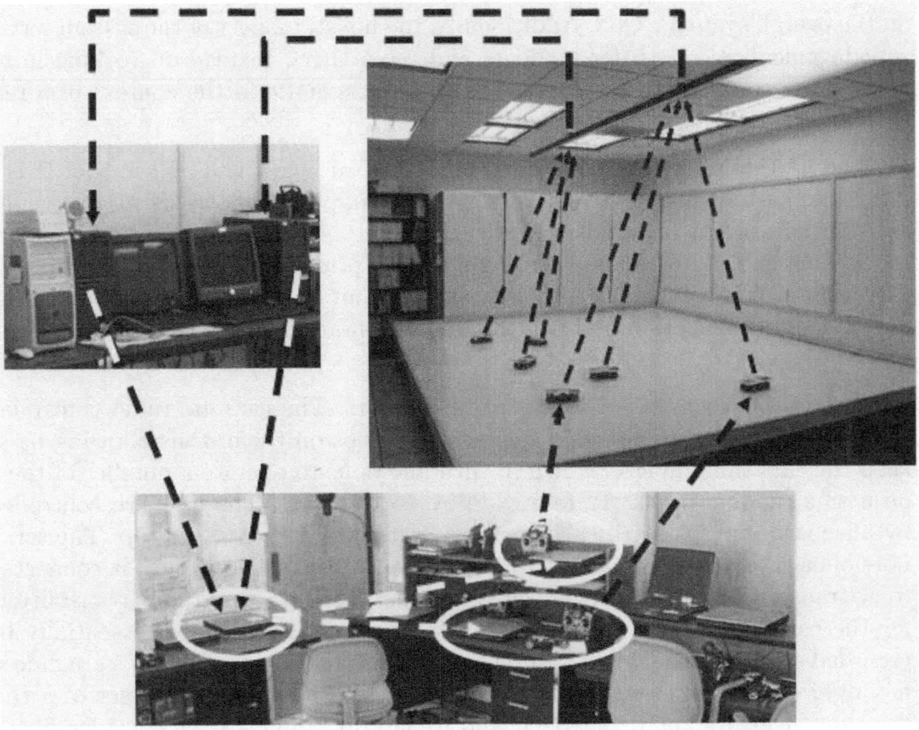

Fig. 1. Convergence Laboratory Testbed at the University of Illinois

such that systems which feature the convergence of control with communication and computation are routinely deployed with short design and development time, while incorporating flexibility to meet changing needs.

3 A Testbed: The IT Convergence Lab

To investigate these issues, we have set up an "IT Convergence Lab" at the University of Illinois, which features a testbed, as shown in Figure 1.

There are several reasons for the use of an experimental testbed. First, it represents a complete system, as we see below. Thus, while researchers often, and with good reason, focus in depth on a particular aspect of the overall system (say, the control law used, or the routing protocol, or the image processing algorithm), working with the system in its entirety educates us about all aspects of the system. Thus we are able to identify what is the critical bottleneck in the system at any given stage in the design. This may vary at different stages of evolution, as has been witnessed by us, between the search strategy used in predictive control, to the cycle time of computation in the image processing algorithm which needs to locate cars under non-uniform lighting, etc. Thus we are able to assess the impact of the choice of a particular strategy for a subsystem on the overall system's QoS. Additionally, the holistic aspect of the system serves a pedagogical role both for students and researchers. Instead of working in an abstract setting, the testbed forces us to be pragmatic in the context of a real system.

The particular choice of the testbed described below is useful in that it is a malleable system, which is at the same time simple. It allows us to investigate substantial aspects of middleware development as well as application development. Our goal is to develop and explore the principles important to the proliferation of control, communication, and computation, and to provide at least one concrete application and working implementation which incorporates these principles.

Our testbed consists of a fleet of fifteen cars. The cars are radio controlled (RC), which allows us to avoid mounting laptops on them, thus allowing us to keep the cars small in size, which in turn allows us to operate a number of them on a small indoor track, 12 feet wide by 16 feet long. The cars are controlled by off-board laptops, with each car having its own dedicated laptop. The serial port of each laptop is connected to a microcontroller, which in turn is connected to a transmitter which drives the particular RC car. The radio transmitters for the cars use separate non-interfering frequencies, and should essentially be regarded as dedicated "wires," emulating a scenario where the laptops are indeed mounted on the cars and connected to them by wires. The entire set of a car, its transceiver, its microcontroller, and its laptop, could just be considered as a single "car unit".

For communication between computers, the lab uses an ad-hoc wireless network, comprised of IEEE 802.11 PCMCIA cards, which carries the control and data packets used in the application. For monitoring and as a diagnostic backup, there is also a dedicated wired network consisting of an Ethernet switch and a small hub. The hub connects two VisionServers to a DataServer and the network switch.

The function of the vision system is to provide feedback to individual cars in the form of position and orientation information. Two overhead cameras continually monitor the platform on which the cars are mounted, and serve as the only sensors in the system. The video feed from each camera is sent over a cable to its dedicated desktop PC, called a VisionServer, where image processing is done to determine the orientations and locations of all the cars. To accomplish this, we have placed six color patches on each car's "roof," in distinct patterns. This color coding allows the vision system to distinguish cars. The vision system segments each of the pixels in a frame into predetermined colors, then searches for groups of colors, and identifies cars through their color patterns. When patterns are correct, the vision system then extracts the location of the color patches and determines the position and orientation of the car. The position information from the two desktops is sent over an Ethernet cable to a DataServer laptop, which in turn is connected to all the other laptops (controlling the cars) by the ad hoc wireless network. Figure 1 shows the complete "loop" on the testbed.

For simplicity and malleability, we employ dedicated laptops (all running Linux) for each radio controlled car. For now, the controller for each car runs on its dedicated laptop. However, as described in the paper [9] describing the software infrastructure and middleware aspects of this project, the next phase of our software infrastructure development will involve automatic migration of code so that the controller for a given car could be running on any of several laptops. For example, computations can move to locations which have maximum computational resource availability or require minimal communication resources. Or, to enhance reliability without sacrificing efficiency, multiple controllers may run on one laptop with other laptops standing by in case of controller failures.

Thus the entire system features multiple sensors (currently two vision systems, with more that can be added as necessary), multiple actuators (the fifteen cars), and multiple computational resources, with loops closed over an ad hoc wireless network.

We refer to systems of this type as Federated Control Systems [9]. In principle, we should be able to replace the cars with airplanes, vision systems with GPS or other sensors, and have an air traffic control system. The architecture should be the same. Or, should we replace the VisionServers with thermometers, door sensors, motion detectors, and smoke detectors, and the cars with heating controllers, sprinkling systems, lighting controllers, etc, the resulting system should be an easily re-configurable total home control system.

But before such systems are common, they must be inexpensive, easy to use, and useful.

4 The Importance of Proliferation

Proliferation is important for proliferation. There is positive feedback. Individual one-of-a-kind systems may serve to fill a particular need. But when these systems are mass-produced, they become inexpensive, and the demand for their use increases. This in turn leads to improvements, which further increases demand, driving down cost, and so on. Many of the eventual uses may be of limited value, and hence would never support large scale costs of development on their own, but when the costs come down and are amortized over a huge number of applications, these lower value needs will begin to drive the market.

Our goal is to address system design challenges arising when we move towards this level of demand in the vision of convergence. We contend that the critical ingredient involved in realizing this goal is architecture.

5 Importance of Architecture

Ultimately, the usefulness of an overarching design depends on how well it adapts to the particular needs of an individual context, while at the same time capturing the essence of that which is common across all the usages. The particular useful variations of a product may not be known to the designers beforehand.

We believe that successful proliferation will depend on providing the right abstractions and architecture for use by designers. We begin by providing some examples of technologies which have successfully proliferated.

We contend, and this may be controversial, that the success of the Internet is primarily architectural, and secondarily algorithmic, though protocols such as TCP have played a most important role. Consider the OSI model of networks, even though in practice it is not followed precisely. This model separates the various functions of communication into layers of abstractions, giving a specific purpose to each layer and hopefully enabling it to perform at that layer. (There are many cross-layer design issues as well that deserve attention; see [10]). Services at a layer can be oblivious to lower layers, and hence can focus on that portion of the design which has been delegated to them. In order to interoperate, we must, of course, provide interfaces between layers which are well-defined and understood by both sides of the interface. With well-defined interfaces above and below each layer in the protocol stack, it becomes possible to make changes to an intermediate layer without affecting the other layers in the system. This allows for incremental evolution. It gives the design longevity.

Consider the alternative of merging all the layers so the implementation runs faster. Such a system would not have longevity since any small change would necessitate the redesign of the entire system. It would simply not be conducive to proliferation.

In addition to longevity, the overall scheme must provide a rich enough set of abstractions to support individual uses. It must provide an architecture that allows a designer to visualize where different aspects of a problem are solved and where they are located.

Indeed, architecture is important for proliferation of technologies, in general. Valiant [11] claims that the success of serial computation is due to the von Neumann bridge, and contends that it is the lack of a von Neumann bridge that is one of the reasons for the failure of certain efforts in parallel computation.

In communication, we contend that the separation of source coding from channel coding has played a major role in the proliferation of digital communication. Indeed we argue that this structural result established by Shannon [12] has been more important than the precise characterization of channel capacity.

Similarly, more important than the precise values of the feedback gains is the closed-loop architecture of a control system. The very separation of the overall system into a portion that cannot be modified, the "plant," and a portion that can be, the "controller," is important, and is obvious only in retrospect. Indeed, simulation software often does not make this distinction, with the result that implementing certain policies may require digging into code where the plant and controller interactions are intermingled.

These examples illustrate the fundamental and far reaching influence of architecture on systems and motivate the desire to address the architecture appropriate for the convergence of control, communication, and computation.

6 Application Architecture for Ever Evolving Systems

We now turn to the issue of concern in the paper – how to design a large system featuring a multiplicity of sensors and actuators.

Our approach is motivated by several fundamental considerations. The overarching issues are:

(i) The ever changing nature of a system,
(ii) The complexity of the design.

7 The Need for the Incremental Evolution of the Design of Complex Systems

The first consideration is that the design of a large system is always in flux. It is never at an end. As a system is built, new features are always added.

In the early mass production of WWII aircraft, US automobile manufacturers assumed that automobile assembly-line methods would translate to aircraft manufacturing, without a strong understanding of the additional complexity of aircraft and the manufacturing precision required. Frequently, design changes were required even before the first aircraft would come off the line. Rather than change the assembly line, the fixes were often done in separate modification centers. Even then, further changes were often made at front-line bases [13].

Similarly, it is erroneous to design today's large and complex systems under the assumption that software is easy to change, and therefore adaptable. The ability of a system to adapt to changing requirements depends heavily on the overall architecture of the system and the nature of the changes. Only if the system is well-designed, with flexible architecture, can one hope that the resulting system will be adaptable.

An important driver of change is "feature bloat," though we do not use the phrase in a pejorative sense. But it must be carefully managed. One starts with a modest goal, and an eye toward future changes, and completes it reliably. Then one inserts additional functionality to make the system more useful. Indeed, this is an ever present feature of many software projects. (Successive versions of Microsoft Word are just one prominent example).

Similarly, viewed from the usage end, customers do not always know what they want or need at the beginning of a design cycle. Upon experiencing a new capability, they may envision slight variations that would make the capability more useful. Apparently small changes can, however, have large unintended negative effects as they ripple through the design of a complex system. Systems should thus be well-designed a priori, to the extent possible, so as to be able to incorporate this inevitable feature bloat, and insulate the risk of feature failure from other parts of the system which must be reliable.

Incremental development may also be necessary from an economic point of view in the proliferation and mass adoption of a technology. A system under development for an extended period of time will not produce any financial support for the developer during the development phase. Thus, for large development efforts, it is useful to build the system in smaller increments, each of which provides an increase in functionality. This produces continuous revenue, making the proliferation phase financially viable.

Incremental development also provides useful feedback in the design and application of the system. As increments are tested, identified problems can be resolved before future increments suffer from the need for redesign. We can see this principle in the early development of our testbed. In the beginning, we simply worked to get a single car running in open loop, according to a preplanned sequence of speed and steering commands. In this phase, the cars were found to be too slow, and the motors unreliable. We did not need to have an entire system working to discover this. Moreover, this discovery led to changes in the motors and gearboxes which would have changed all of the calibration data for each car. We had not yet invested time calibrating every car, thus early feedback helped to avoid this time consuming task for the remaining cars. We were also initially concerned about slack in the steering mechanism and hence the repeatability of the cars performance. Several open-loop demonstrations proved that the cars were sufficiently repeatable to meet our needs, thereby avoiding a redesign of the steering which we had thought necessary.

Of course, incremental upgrades must be relatively simple to incorporate at each stage. Moreover, it is useful to be able to "roll back" if an upgrade fails in some fashion. This ability to "undo" is a challenge to system design, but provides much needed flexibility to designers and users alike.

8 Design Goals for Application Architecture

Thus, we contend that complex system design must be regarded as a continuing process. Our purpose is to address the issue of design of such a system so as to meet two goals:

(i) Reliability.
(ii) Minimizing the time to design and deploy a new feature.

The need for reliability is now well accepted. Indeed, rather than "high performance," the focus of much of current software research is on "reliability." It may be referred to as robustness, or fault-tolerance, or perhaps security as well, highlighting particular aspects of system reliability. In any case, reliability is now entrenched as a primary performance criterion.

The second focus is on cost—in terms of human time. It is important to reduce both the time to design a system, as well as the time to deploy it. In fact, in our approach these go hand in hand. By abstracting certain aspects of systems in appropriate ways, we aim to realize our ultimate vision that these systems be "mass produced" rather than "hand crafted." Our process aims to reduce the process to one of designing an individual block and then easing its implementation through a process of selecting among interoperable components and setting appropriate constraints. Thus we aim to make one-of-a-kind systems affordable and hence useful.

Our design process views the overall system as a composition of "decision-making modules," segregating levels of decision making to broadly conform to commonly used and well understood methods for control system design, and adopting an evolutionary approach that we call "Collation." This provides support for reliability and evolution at many architectural layers.

9 The Levels and Modules of Decision Making

At the highest architectural level of any application is the system goal, which may change over time. A clean architectural design separates goals at all levels of decision making from the means to accomplish them.

High level goals are further instantiated or translated as they percolate through the layers of the design to elements of the system that are more aware of information that pertains to the optimization of the goals. One common refinement of goals is by time scale decomposition [14], though there are also other possibilities such as, for example, spatial decomposition.

As one moves across layers, one finds perhaps several steps of control refinement and data abstraction. As a concrete illustration, an air traffic controller need not know the precise settings of the throttle on a particular aircraft, which may however be of utmost interest to the pilot. Similarly, the pilot may be unaware of aircraft movement on the ground at the destination airport, even though such movement will ultimately affect the pilot's operating conditions. The air traffic controller need only know the plane's position, airspeed, and flight plan in order to accomplish the higher level goal of ensuring flight safety, while the pilot need only know if the flight plan is still acceptable to the controller, and need not know about the flight plans of any other plane. Of course, a fail safe mechanism is in place locally. Upon detecting an oncoming plane, a pilot will change course for safety without clearing it with the air traffic controller who probably does not have data at the level of refinement the pilot needs, or the ability to make a decision fast enough to help.

From decision theory [15], we can regard any decision making unit as subject to three "inputs." First is the "goal," which could, for example, be specified as a cost function to be optimized. Second is the "model." This provides the basis for deciding what will be the result of actions, and thus allows us to choose between different actions when seeking to optimize the cost function. Third is the "information" available to the decision making unit. This could be noisy measurements of say the locations of the cars, etc. The output of the decision making module can be regarded as the optimal (or near optimal, or satisfactory) choice of an action which does well by the goal, based on the model of the environment, and the available measurements.

To illustrate these issues more concretely, in the Convergence Lab Testbed we have a centralized top level planner called the ScheduleServer which is responsible for generating collision-free timed trajectories for each car along routes

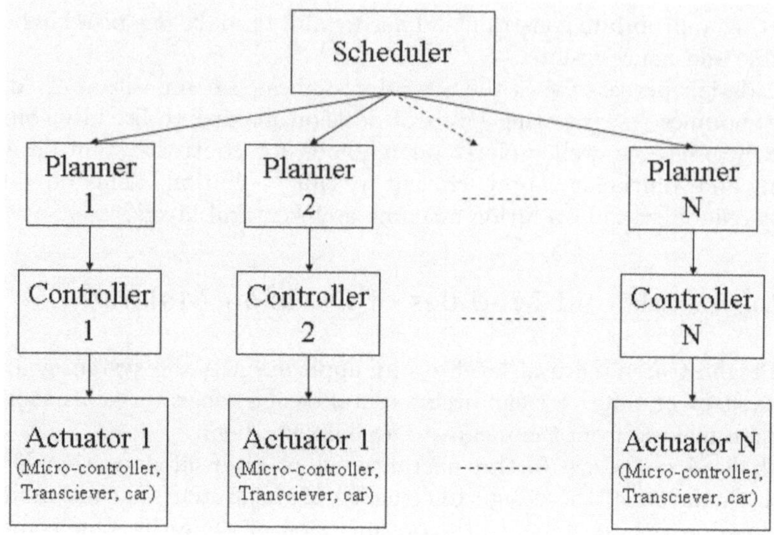

Fig. 2. Goal decomposition in the testbed

which represent the high level goals of the system [16]. The goals may be specified merely as a triple comprising an origination location, an intermediate way-point, and a final destination on the track. The server must then be able to determine feasible paths which satisfy the constraints of the track, and then schedule collision-free trajectories for each car.

The output of this plan is a set of timed way-points for each car. This is given to the middle level in the task architecture, which is distributed. Specifically, each car has it's own mid-level controller called the Planner.

The ScheduleServer monitors the locations of the cars and determines when something has gone wrong and re-plans, sending new trajectories which replace the old ones. Figure 2 shows the goal decomposition/refinement on the testbed.

The Planner in normal mode merely passes the trajectory on to the low level real-time controller which dutifully attempts to track the trajectory. However, the Planner has also been given access to position and orientation information for each of the other cars, or obstacles, on the track. By monitoring their movement, the Planner can perform additional functions such as following another car in a formation or in a pursuit-evasion scenario. It may also provide collision avoidance by predicting potential collisions, which may occur even with collision-free trajectories due to misbehaving cars, malicious cars, or simply dead batteries. Upon detection of imminent collision, the Planner may simply stop the car to prevent the collision, or instead plan an alternative path based upon some criterion. This recovery planning then represents a mid-level goal. It is not the high level goal, but represents refinement of a higher level goal as a result of additional information. Moreover, in the face of unpredicted behavior, it is able to prevent system failure.

The low level real-time controller is myopic. In our case, we have traded cheap computational power for algorithmic complexity. The real-time controller uses a linearized model of the car and examines thousands of potential control sequences, comparing the predicted results of each of them with the desired result, choosing the one with the lowest cost, i.e., the one which most closely follows the desired trajectory. Note that at this level, the goal is merely to conform to the given trajectory. There is no understanding of higher goals such as collision avoidance. Separation of concerns in this hierarchical structure greatly simplifies the design, while providing flexibility at the same time.

It should be mentioned that there is really one more lower layer. The microcontroller which sends signals to the car stores a sequence of controls, which for fail-safe reasons is always terminated with a stop command. Failure to receive an update from the low level real-time controller is thus handled appropriately in the hardware/firmware level of the system. Such bottom end failsafe mechanisms are necessary in safety critical applications and could even be implemented through run-time invariance checking in micro-processors.

Using a proper task architecture, we have separated responsibility among the various parts of the system. Incorporating new tasks is greatly simplified. For example, if one of the cars represents an ambulance, and we want it to have priority, then the only part of the system which is affected is the very top level ScheduleServer which must be altered to give priority to it.

10 Reliability and Dependence

As noted earlier, the first attribute of "performance" may well be reliability, which in turn has many dimensions.

In any system, certain portions of the system may depend on other parts for its operation. A cellular phone clearly "depends" critically on its antenna to complete a call. However, some dependencies are not inherently necessary, but creep in as design or implementation dependencies[1]. For example, a laptop may only "use" wall power to operate but does not "depend" on it since it has a battery. Complex systems exhibit complex dependencies which are difficult to trace, and pose challenging troubleshooting problems. Many design or implementation dependencies could perhaps be converted into "use" relationships. Consider the power steering system in a car. If the power steering fails, the power assist to the driver is gone, but the steering wheel remains physically connected to the movement of the front wheels, allowing a driver to continue to drive safely, although requiring increased effort.

To understand dependencies, it is useful to visualize the space of errors and categorize them. We will enumerate a few for purposes of this discussion.

Execution. This includes all forms of system crashes, all segmentation faults, all power failures, deadlocks, livelocks, infinite loops, and any other faults which

[1] We are grateful to Professor Lui Sha for educating us about what the requirements for reliability are in practice, and how to address them.

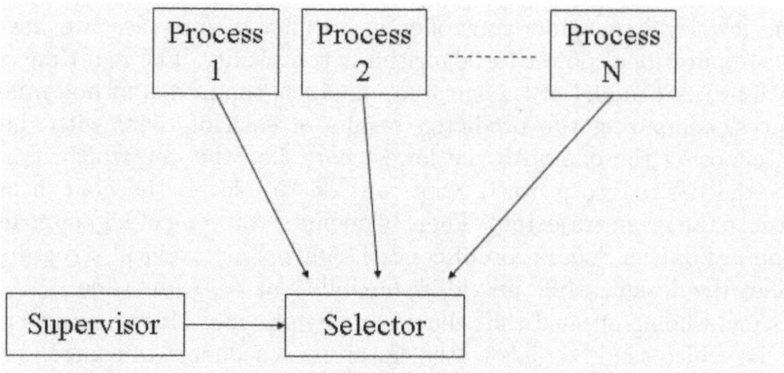

Fig. 3. Collation architecture

prevent a process from executing. There is a great deal of research on the prevention of execution errors. We are interested in how to continue to operate in the presence of such errors.

Timing. This includes any operation which does not return the result within the deadline required. It is important in systems which interact with their environment.

Semantic. This means that while the function proceeded and returned a result by the deadline, its value was not correct in some sense. This could simply be a design flaw, and is frequently ascribed to the application or domain expert.

11 Collation for Evolution

In keeping with traditional principles of functional programming, our design is as modular as we are able to achieve. Thus the focus is on code reuse rather than rewrite. When adding new features, our goal is to "insert" functionality rather than revamp the existing architecture.

The ability to "undo" an action while word processing a document provides tremendous flexibility. Similarly, instead of being critically dependent on correct operation of a more complex implementation, we can merely use it when it is satisfactory, and revert to a simpler version when it is not. It is therefore reasonable to deduce that large complex systems must incorporate the ability to switch between components when they fail or when increased functionality is desirable, while maintaining system integrity in the face of faults, failures, and changes in operational environments. Such capability must be built in, and even itself upgradeable perhaps. Figure 3 presents the architectural construct, or design pattern, of "Collation."

The Convergence Testbed has experienced many system failures and system upgrades as well as changes in the operational environment. As these changes have occurred, we have observed the utility of a process of "Collation" as a design pattern and have incorporated it wherever feasible. An example concerning the

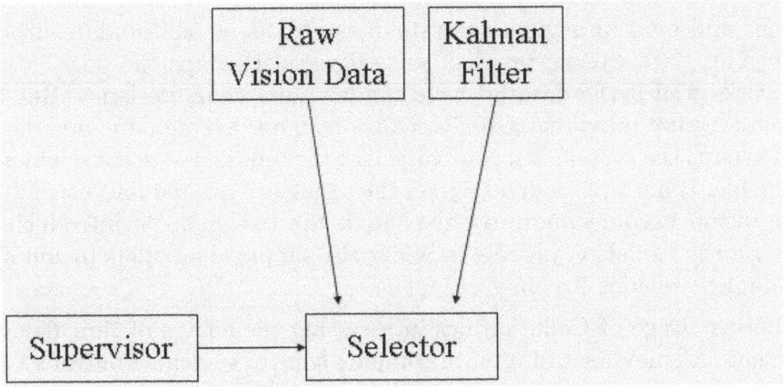

Fig. 4. Kalman filter inserted with Collation

incorporation of a vision data filter illustrates this method. In an early version of the testbed, with just one car running, and just one camera, the vision system was not responsible for identifying the car, but just reporting its position and orientation. A reliable system for this functionality was in place, and the real-time controller was able to use the raw vision data reliably for its operation. Moreover, there was no need for a centralized store of vision data; therefore, the DataServer was not yet implemented. The system performed its task of following predetermined trajectories quite well in this early version. However, at a later stage, to improve the smoothness of trajectory following, it was decided to add a Kalman filter. This was done by adding it as a parallel block to an existing direct breakthrough. During the debugging phase, the existing position and orientation information which was "reliable" but not "very accurate" was used to monitor the Kalman filter's output. Figure 4 illustrates the Collation process applied to the Kalman filter.

This idea of Collation can be cast into the Simplex architecture of [17]. Sha, *et al.* [18] have considered the use of a simple reliable controller as a backup to a complex, unreliable controller. This method is based on using simplicity to control complexity. The key notion is that the simple controller, previously established to be reliable in some way, can always maintain stability of the system and meet certain safety parameters provided that the system state is within a well defined operating region, as for example, the basin of attraction of its Lyapunov function. Then a smaller region is defined within which the complex controller is given authority over the system. A supervisory process, which must also be reliable, observes the system state in order to determine if and when the complex controller will cause the system state to move outside the basin of stability of the simple controller. When this occurs, the supervisor switches control to the simple controller, thereby maintaining stability.

Collation encompasses more than redundant safety systems. It extends to what we call "data fusion." Consider multiple data sensors in a system. An aircraft avionics system may receive position information from GPS, land based beacons, and inertial navigation system, as well as manual updates from a navigator. This data can be fused in several ways. Perhaps the data is averaged. This works well if the data all have similar noise characteristics. But if one of the sources were providing grossly erroneous data, averaging is not the best fusion. Perhaps the system can compare all of the inputs and vote on which sources are reliable. If multiple sources agree, then they are trusted and used. Of course, as the fusion becomes more sophisticated, the likelihood of introducing errors grows. So the Collation process provides the simple algorithms to run alongside the complex versions just in case.

Another usage of Collation lies in assessing the effects of time-delays in the incremental deployment of a more complex control system. Control systems are generally sensitive to timing. Delays introduced into a stable control loop can even render it unstable. A system designed with the ability to switch between a stable version of a process and an experimental version, can accommodate online development and testing safely. Consider, as a simple example, a filter located somewhere along a control feedback loop. When a more sophisticated filter is being entertained, the additional processing required for it may introduce additional delay, which could render it worse than the original simpler design. By applying the Collation design pattern, we are able to first program another version of the simple filter that includes the additional delay, without any algorithmic changes, and use the Collation process to switch between the original and the delayed versions, monitoring the system for undesirable effects. Once we have tested this sufficiently, we may then install the full functionality of the complex filter and run it in place of the delayed version of the original filter. Because of the supervisor, we can make these changes at run-time (in real-time) without bringing down the system. Moreover, the original filter is still in place, ready to be used in the event of undesirable behavior of the complex filter. So Collation facilitates incremental operational testing by allowing low risk online upgrade.

Yet another place where Collation is useful is in "planning." Multiple plans can be generated and evaluated, and the plan with the best performance can be implemented. One example of this in the Testbed is in the mid-level Planner which continuously monitors the vision data, predicting where cars will be in the next several steps and comparing the current trajectory with those positions in order to predict future collisions. Upon detection of a potential collision, the Planner may create several alternative plans, perhaps a path to the left of the collision, and perhaps one to the right. These alternate paths are then checked for collisions, and if one is deemed successful, it is used. If not, the desired behavior is to come to a stop, and the Planner accordingly stops the low level controller, thereby avoiding a potential collision. Figure 5 shows Planning inserted via Collation.

These examples illustrate the fundamental ability to connect to, and select among, multiple sources of data or control. Properly implemented, this functionality provides for evolution, rollback or undo, and reliability. It provides a separation of decision criteria, or rules, from the execution of the criteria. We

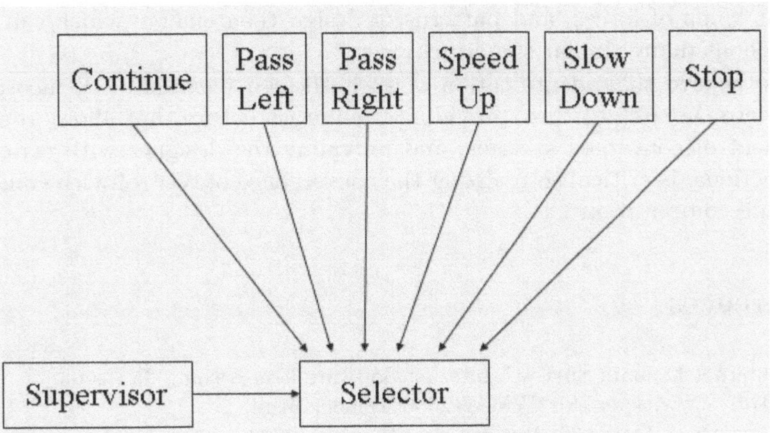

Fig. 5. Planning as Collation

can implement implementing the Selector as a separate process from the other components, we can create the Simplex Architecture for reliable on-line system upgrade. By choosing among many pieces of source code for compilation, the process can be called software configuration management. When multiple external components can connect to a single component (which can in turn provide its output to multiple external components) and each component has local "intelligence" contained in the Supervisor to govern its action, we can realize an interoperable Federated Control System. If such "intelligence" includes rules for fail-safe operation, then the system has fail-safe at that level of abstraction.

Collation bears a resemblance to object oriented design. In aggregation or hierarchical composition, each element of the Collation architecture can be a trivial one, or a very complex system of its own, or something in between. Thus, Collation exhibits a self-similar nature useful for hierarchical construction and decomposition.

The design pattern that we call Collation combines widely used principles into a useful architectural construct. Whether or not a system realizes the benefits of connection to and selection among multiple sources depends upon the software design of the underlying infrastructure.

A movie showing several applications of the functioning system is available on the Testbed website [19].

12 Concluding Remarks

The architect Christopher Alexander argues in [20] that in any system underlying patterns of use exist. The principles, or forces, which drive the use of that system must be understood, in order to design it in such a way that it comes alive, which is to say that it fulfills the purpose of its creation. Many systems may experience similar kinds of forces, and patterns of design then emerge which can be seen throughout many similar structures.

We believe that identification of such "design patterns" [21], incorporating them into the architecture, providing the infrastructure that allows routine deployment of converged systems, and providing the designer with a rich set of abstractions, is critical to realizing the convergence of control with communication and computation.

References

1. "Internet Domain Survey," Internet Software Consortium, Jan 2003. http://www.isc.org/ds/WWW-200301/index.html.
2. "How Many Online?", Nua Internet Surveys, 2003. http://www.nua.ie/surveys/how_many_online/.
3. J. Stankovic, "VEST: A Toolset For Constructing and Analyzing Component Based Operating Systems for Embedded and Real-Time Systems," University of Virginia TRCS-2000-19, July 2000.
4. K. Carter, A. Lahjouji, N. McNeil, "Unlicensed and Unshackled: A Joint OSP-OET White Paper on Unlicensed Devices and Their Regulatory Issues," May 2003, http://hraunfoss.fcc.gov/edocs_public/attachmatch/DOC-234741A1.doc.
5. V. Lipset, "In-Car Bluetooth To Grow Beyond Telephony, Study Says" May 23, 2003 http://www.thinkmobile.com/Everything/News/00/67/32/.
6. "Wireless Sensor Networks," CrossBow Technology Inc, http://www.xbow.com/Products/Wireless_Sensor_Networks.htm.
7. S. S. Pradhan and K. Ramchandran, "Distributed source coding using syndromes (DISCUS): Design and construction," *Proceedings of the IEEE Data Compression Conference (DCC)*, Snowbird, Utah March 1999.
8. R. Shukla, P. L. Dragotti, M. N. Do, M. Vetterli, "Rate-distortion optimized tree structured compression algorithms for piecewise smooth images (448 kB)," *IEEE Transactions on Image Processing*, Jan. 2003, submitted.
9. G. Baliga and P. R. Kumar, "Middleware Architecture for Federated Control Systems," IEEE Distributed Systems Online, June 2003, http://dsonline.computer.org/0306/f/bal.htm.
10. V. Kawadia, P.R. Kumar, "A Cautionary Perspective on Cross Layer Design," Technical Report, CSL, University of Illinois, Jun 28, 2003. http://black1.csl.uiuc.edu/~prkumar/ps_files/Cross_Layer.ps.
11. L. G. Valiant, "A Bridging Model for Parallel Computation." *Communications of the ACM*, vol. 33, no. 8, August 1990.
12. C. E. Shannon, "A Mathematical Theory of Communications," Bell System Technical Journal, vol. 27, pp. 379–423 and 623–656, July and October, 1948 (2 parts).
13. J. Rumerman, "The American Aerospace Industry During World War II," *US Centennial of Flight Essay*, 2003, http://www.centennialofflight.gov/essay/Aerospace/WWII_Industry/Aero7.htm.
14. S. B. Gershwin, *Manufacturing Systems Engineering*, Prentice-Hall, Englewood Cliffs, NJ 1994.
15. D. Blackwell and M. Girschick, *Theory of games and statistical decisions*, Wiley, New York, NY, 1954.

16. A. Giridhar and P. R. Kumar, "Scheduling Traffic on a Network of Roads," Technical Report, CSL, University of Illinois, Apr 2003. http://black1.csl.uiuc.edu/~prkumar/ps_files/trafficpaper.ps.
17. L. Sha, R. Rajkumar, and M. Gagliardi, "The Simplex Architecture: An Approach To Building Evolving Industrial Computing Systems," *Proceedings of the International Conference on Reliability and Quality in Design*, pp. 122–126, Seattle, Washington, Anaheim, CA, ISSAT Press, March 16–18, 1994.
18. L. Sha, R. Rajkumar, and M. Gagliardi, "Evolving Dependable Real Time Systems," *Proceedings of IEEE Aerospace Conference*, Vol. 1, pp. 335–346, Aspen, Colorado, IEEE Computer Society Press, February 3–10, 1996.
19. "The Convergence Laboratory testbed," University of Illinois, http://black1.csl.uiuc.edu/~prkumar/testbed/.
20. C. Alexander, *The Timeless Way of Building*, Oxford University Press, Oxford, UK, 1979.
21. E. Gamma, R. Helm, R. Johnson, and J. Vlissides, *Design Patterns: Elements of Reusable Object-Oriented Software*, Addison-Wesley, Boston, MA, 1995.

Self-Organising Node Address Management in Ad Hoc Networks*

Stephen Toner and Donal O'Mahony

Networks & Telecommunications Research Group (NTRG) Trinity College, Dublin 2, Ireland
Tel:+353 6082336, Fax: +353-1-6772204
{stephen.toner,donal.omahony}@cs.tcd.ie

Abstract. Whilst much effort has been put into the creation of routing algorithms to handle all sorts of mobility scenarios in ad-hoc networks, other fundamental issues, such as the addresses used by nodes, haven't been dealt with adequately. This addressing problem has recently attracted increased attention and a few proposals have been made, though often these schemes only work in limited scenarios. In this paper we present an autoconfiguration protocol designed to work in a mobile ad hoc network (MANET). The scheme allows joining nodes to dynamically obtain addresses, and has been designed to efficiently manage addressing, and to handle such scenarios as the merging and partitioning of networks. We discuss an implementation used within an emulated environment and a real self-organising ad-hoc network.

1 Introduction

An ad-hoc network is one where nodes collaborate to allow communication without the required presence of network infrastructure. Their dynamically changing membership and topology means that specialised routing protocols are required. The lack of manual management means that auto-configuration is a highly desirable goal. Whilst research in ad-hoc routing protocols has been strong in recent years, the same intensity has not been applied to other important related areas, such as node addressing. However these routing protocols typically rely on nodes having a unique address, and ignore this vital issue. Often nodes are assumed to have addresses configured a priori, but this is impractical and not easily accomplished. Instead we propose a dynamic scheme for assigning and managing addresses within ad-hoc networks. Nodes require a unique address for packets to be delivered to the "correct" destination, and due to the routing side effects that may arise from nodes using duplicate addresses.

In fixed IP-based networks hosts use IPv4 or IPv6 addresses, which have a hierarchical element to their structure. An IP address typically has 2 purposes: It identifies the node, and encodes routing information. This second point is vitally

* This material is based upon work supported, in part, by the European Office of Aerospace Research and Development, Air Force Office of Scientific Research, Air Force Research Laboratory, under Contract No. F61775-01-WE052.

M. Conti et al. (Eds.): PWC 2003, LNCS 2775, pp. 476–483, 2003.

important when nodes are mobile. Traditionally this mobility has been handled in two ways. DHCP uses servers to auto-configure nodes with a topologically correct address as they move, so normal IP routing can be used. Also Mobile IP allows nodes to maintain a static identity based on the "home address", so a node is always contactable via a static address (albeit inefficiently).

Unfortunately the use of the same combination of DHCP and Mobile IP within an ad-hoc network may prove impractical, and may be even impossible. Because there may be no infrastructure available, DHCP, which relies on an address server, is unable to provide a solution to the addressing problem. In fact it is actually unnecessary to configure a topologically correct address as, unlike in the fixed network, the address used in an ad-hoc network has no routing purposes (for most routing protocols). Addresses are simply used as a means of identification within the ad hoc network, and so must simply be unique within the network. Instead the ad-hoc routing protocol handles the routing, typically using a flat address space.

An obvious solution to this problem is to provide each node with a permanent unique identifier that could be used to identify the node within the MANET. *A hardware-based solution,* where every piece of hardware has a permanent unique identifier introduces a number of problems:

- Whilst 48-bit IEEE MAC addresses are the closest thing we have to this, and are designed to be unique, existing MAC addresses aren't guaranteed to be unique [7, 3]. There are known instances where network interface cards (NICs) have been issued with duplicate MAC addresses. Also not all hardware has an IEEE MAC address, as it may be using different wireless technologies.
- It would require a change in existing software, as often 32-bit addresses are used.
- Interworking between fixed and ad-hoc networking would be complicated
- The fact that the identity of the node can be determined from its address raises privacy concerns.
- The larger the address used, the greater the per-packet overhead. This becomes more of an issue in source routing protocols where a list of addresses is placed in each packet.

Mobile-IP based solutions (e.g. MIPMANET [6]) typically require nodes to have a permanent home address, which is used as the unique address within the ad-hoc network. All existing hardware then requires a permanent address, and the increased address size that would be required (probably requiring the migration to IPv6 due to the shortage in IPv4 addresses) would require software changes, and create extra overheads. It would also create new issues if the ad-hoc network was to become connected to the global network, with the selection of a care-of address either through the selection of an appropriate foreign agent (as would be the case if IPv4 were used), or the configuration of a local care-of address. For example with MIPv6 the mechanism for auto-configuring a care-of address [7] still requires Duplicate Address Detection (DAD), even though the address has the MAC address embedded in it. Performing this DAD in an ad-hoc network is not a trivial matter, and a new scheme is required to manage this autoconfiguration of addresses.

If we discount the use of a static address then a scheme is required to allow nodes to autoconfigure an address within an ad-hoc network. Current solutions typically have limitations, working only in certain scenarios. The scheme we present provides an ad-hoc autoconfiguration mechanism, that isn't reliant on the MAC address and works in an ad-hoc environment. It is based on the use of IPv4 addresses, but this could easily be extended to use IPv6 addresses. It aims to provide a solution that minimises both traffic and time requirements for configuration and management. We also briefly examine how these networks could be connected to the fixed network.

2 Related Work

Perkins et al [1] have proposed a simple Duplicate Address Detection (DAD) based scheme whereby nodes choose a tentative address and send a request to this address. The absence of a reply is assumed to indicate that this address is free for use. However there are a few limitations to this approach.

- In the event of a partitioned network, or the coalescing of two networks, the address would not be guaranteed to be unique and the procedure would have to be repeated again after merging.
- No mechanism is suggested to detect the merging of networks, and even if such a mechanism were in place, every single node in both networks would have to redo this DAD test, potentially causing a disruptive broadcast storm. Whilst this problem is recognised, no solution is suggested.
- Timeout-based approaches suggest that the interval required for a packet to successfully traverse the entire network can be calculated, whereas the nature of ad-hoc networks makes this difficult.
- Joining the network takes a long period, as a number of retries are necessary to ensure that all nodes received the broadcast, and time must be allowed for any node to respond.

Vaidya's proposal [2] aims to ensure packets are delivered to the "correct" node even if two nodes are using the same address. [2] argues that "Strong" duplicate address detection (i.e. showing "correct" behaviour where the existence of duplicates can be detected within a bounded interval) is impossible. The basis of this argument is that an ad-hoc network may become partitioned and so the detection of duplicates within these two partitions is infeasible. Whilst this may be true, we argue that an ad-hoc network that has two partitions can be treated as two distinct ad-hoc networks. A node's address is only required to be unique within the same partition as only these nodes are reachable. This proposal requires the modification of existing routing protocols to incorporate their scheme, something that we would rather avoid.

In MANETconf [3] joining nodes select an existing member (the "initiator") of the network to obtain an unused address. This floods the network with a request for a specific address and all nodes reply with positive or negative responses, if this address is in use. This scheme displays a few limitations:

- If 2 nodes were to try to concurrently join the network, in the same region, their requests would be indistinguishable. The "initiator" node has no means to distinguish between the nodes when communicating, and so both nodes could select the new address.
- The requirement for the nodes to reply positively (suggesting the address is free) creates extra traffic and is superfluous as regards the particular address allocation at hand (a single negative response will override all these responses), and is just used as a means to detect nodes having left.

- To detect merging and resolve duplicate addresses it is proposed that two previously distant nodes should exchange information about their partition when they come together (suggesting some sort of proactive beaconing). In the event of conflict the node with the least TCP connections is expected to acquire a new address, though the scheme provides no details on how this would be determined.

Patchipulusu [4] uses a leader to identify the group, and nodes joining the network receive sequential addresses, with the newest member taking over the role as leader. Each node periodically sends an update beacon message to the nodes with the next and previous addresses so that node losses can be detected. Any node that becomes separated for a period must acquire a new address. Similarly when two networks merge, all the nodes in the network with the least number of nodes must acquire a new address. This scheme potentially requires a lot of unnecessary address changes, causing loss in routing information, loss in connectivity of connections, and disrupts communication.

3 Our Approach

Our proposal is based on the election of a "leader" within each network partition. This leader:

- provides identification for the network so that the merging of networks can be detected.
- acts in a manner similar to a DHCP server, whereby it hands out addresses to new nodes.
- allows for an efficient merging of networks.

This leader maintains a list of all addresses in use, and new nodes must apply to it for an appropriate address. One of the fundamental considerations of an approach like this is that the mechanism should cope easily with the loss of the leader node.

Each node has a 32-bit address (for compatibility with the existing IPv4 network), an extra 32-bit unique identifier (UID), and a network identifier (NID). In the event of the ad-hoc network becoming joined to the fixed network, the IP network will not use this auto-configured address to identify the node (Later in this section we outline our global connectivity scheme). Only this 32-bit address is used elsewhere, for routing etc., with the other identifiers (UID and NID) only used within the addressing protocol to uniquely identify nodes, and network merges.

Joining

A node wishing to join chooses a temporary random address from a reserved range. Whilst using this address the node does not participate in the routing protocol. Due to the limited number of addresses available in this range the possibility of duplicates being chosen is reasonably high, but the address is only used in forming a link-local address with an existing member of the network, and must only be unique within the physical range of this existing member. To supplement this, a node also chooses the UID. Any identifier could have been chosen, for example the MAC address would have been ideal, but a random 32-bit value was selected for simplicity. The address management scheme uses this tuple {address, UID} to determine if a node is actually a different node using the same address, or just the same node rejoining the network (possibly due to the temporary segmentation of the network).

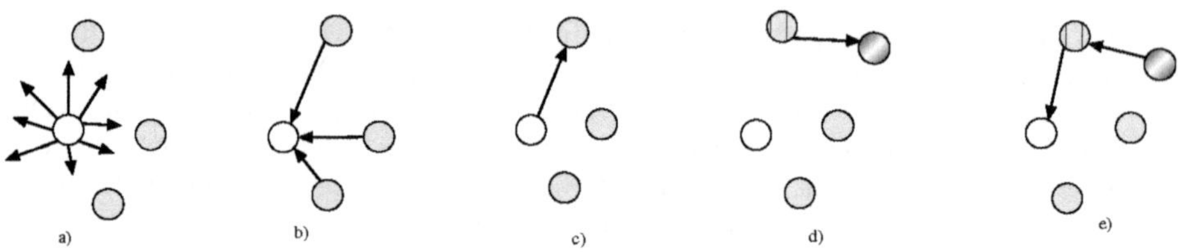

Fig. 1. Joining the network

The node broadcasts the *join* message, containing the temporary address and UID. All nodes in range reply, and the joining node selects one to act as an "agent" (this is a node already in the network and so can participate in the routing protocol) to obtain an address (preference is given to leader nodes if they are directly within range as this reduces the necessary traffic). This agent node stores the address details and sends a request to the leader node, containing a unique reference number for this request. The leader selects an address that isn't currently in use (this can be done immediately as the leader has a list of all addresses in use and so doesn't need to flood the network to determine if the address is free), and returns this to the agent, along with a copy of the current table of addresses in use. The agent then returns this to the joining node, after adding the appropriate destination, by translating from the reference number in the response to the temporary address the corresponding node was using.

Leader Election

A fundamental principle of ad-hoc networking is that any node is liable to fail, so we distribute a copy of the table to each joining node, along with a corresponding version number. When a node joins it is added to the table, and the version number is incremented. The election of a new leader is then reduced to the task of simply locating the holder of the latest version. Whilst this may not contain all the entries that the original master table held it is guaranteed to hold all the addresses of the currently connected nodes. The last node to join will hold the latest version of the table.

The leader proactively floods a beacon periodically throughout the network. This addition of a proactive element is realistically unavoidable as nodes must become aware that adjacent nodes do not belong to the same network. This could be accomplished through one of two mechanisms: either every node periodically broadcasts a message that isn't repeated, or a single node (the leader) broadcasts a message that is to be flooded throughout the entire network. However, if the leader floods the network, then the loss of this node can also be detected readily. To cope with network losses a threshold number of flooded messages may be missed before an election is initiated.

The election should cause a minimal amount of traffic. If a node realises that the leader is absent, it waits a random period before flooding an *election* message through the rest of the network stating its intention to take over as the new leader. This message contains the version number of the table held by the node, the node's address, and the last beacon message received. Nodes hearing this message will react accordingly: It compares the version of the table it holds, and the table indicated by the prospective leader. If the node's own version is greater it becomes the new provisional leader and starts a new election. If the initiating node had a more recent version, then it is marked as the provisional leader and this election message is flooded onwards. If both versions are the same, the conflict is resolved by choosing the node with the highest address as the provisional leader. In this way every node should independently and correctly determine the new leader.

When a node initiates an election it initiates a timer. If no other messages are received from more "eligible" nodes then the node assumes that it is has been elected as the new leader and floods the network with a *leader_elected* beacon containing the new information. All other nodes will be waiting for this message, as its election message should have been flooded throughout the network. If any node doesn't receive this message within a threshold period then it will reinitiate the election.

Network merging

The merging of the two networks can be detected through the presence of more than one leader. A node receiving a beacon from a foreign leader (containing a different NID and leader address) message must inform his current leader. Only the two leaders of the respective networks need to communicate, comparing their lists of currently used addresses to discover if any duplicate addresses are in use. These duplicate nodes will then have to change their address by re-applying for a new address.

Communication between the leaders is complicated by the fact that the path through which they communicate cannot be guaranteed to be free of duplicate addresses, as it passes through two address spaces. For example Figure 2 shows a path where two nodes along the path are using the same address. Therefore the leaders communicate through the nodes that received the foreign *leader_beacon.* For example, in Figure 2 node 23 is used so that two distinct partial paths are generated (M1 and M2) with the addresses within each guaranteed to be unique.

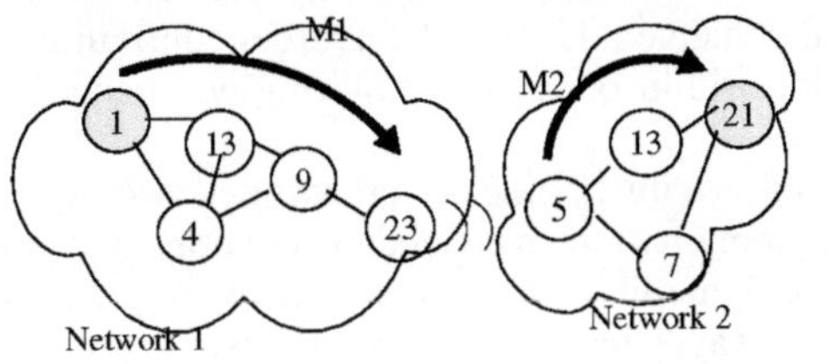

- Node 23 heard a leader_beacon from the 2nd network, and delivered this to its leader (Node 1)
- This encapsulates its packets, and sends to this node (i.e. Node 23, along path M1), which in turn injects the inner packet into the other network
- This is then delivered (along M2) to the leader of the second network.

Fig. 2. Merging Networks

Address Reuse: To prevent "address leakage", whereby nodes leave the network without freeing the address for reuse, a management mechanism, whereby infrequent periodic requests test for the presence of nodes, was introduced. Although this address leakage wouldn't be a problem where the address space is large enough relative to the number of nodes, it does create a problem whereby the address table could become prohibitively large.

Routing Traffic Isolation: When two networks merge, there will be a short period of vulnerability during which adjacent nodes belong to different networks, and so should not participate in routing with the other network until addresses have been checked for duplicates. By adding the NID to each packet this can be easily detected, and prevented. It also means that network merges can be detected more readily than having to wait on the periodic network beacon.

Table Size: One potential problem with the scheme is the fact that in large networks the size of the table containing the used addresses could become prohibitively large. Mechanisms for limiting the size are being examined whereby addresses may not be completely random, but may follow some sequence.

Global Connectivity: So far we have considered the ad-network as an isolated entity. However it would be desirable for this network to be able to interoperate with the fixed network should a possible connection be available. We are currently using a NAT-based solution to accomplish this. Related work, based on the TRIAD [8] scheme, aims to identify nodes globally using a name rather than an address. The operation of this scheme is outside of the scope of this paper.

4 Implementation

In the Networking and Telecommunications Research Group (NTRG) at Trinity College Dublin we have developed a flexible testbed, where components are assembled using a layered architecture. Layers have been developed performing a variety of networking functions, with working implementations provided for a number of ad-hoc routing protocols. Testing was performed with DSR (Dynamic Source Routing). The JEmu emulator [5] was used extensively to emulate realistic movement scenarios. This aided the development, as well as throwing up some interesting observations and design considerations. For example, when nodes were programmed to follow a random mobility pattern, it was found that the network often split for short periods and then might become reconnected. It was important therefore that merging and splitting of networks could be performed quickly and efficiently, and that nodes weren't unnecessarily made change address. The address configuration layer was implemented as a separate layer within our stack, requiring no changes to existing routing protocols.

Emulation scenarios were run with up to thirty nodes following a variety of mobility patterns for testing. This implementation provided validation that the protocol does handle mobility in a realistic dynamic environment, where merging and partitioning could occur, and where message loss was evident. By artificially constricting the available address range to between 51 and 100 for testing (with 1-50 used for the initial temporary address), the chance of duplicate addresses occurring was hugely increased, providing an adequate test for the protocol's ability to handle duplicate addresses. Initial testing using virtual nodes (using the JEmu emulator) indicated that the protocol could operate in realistic scenarios. The frequency of the beaconing by the Leader node had a large impact on performance, and had to be chosen to provide a balance between the need for rapid resolving of network partitioning and merging, and to avoid excess traffic.

Next the implementation was tested with real nodes (on iPaqs running Windows CE, and on laptops, using 802.11b as the wireless medium). The Dublin Ad-hoc Wireless Network (DAWN) network is a project underway in Trinity College aiming to create a useful operational ad-hoc network, and all mobile nodes in the network use this scheme. A number of applications have been developed to run across this network, including point-to-point telephony and instant-messaging applications. It contains both mobile nodes, and fixed machines that provide a permanent population for testing and connectivity. To avoid a centralised system for allocating addresses, this protocol has been used so nodes can enter the network and obtain an address without the requirement for any pre-configuration.

5 Conclusions

The proposed address autoconfiguration mechanism provides a dynamic protocol for mobile ad-hoc nodes, and allows the creation of a self-organising network. The proposed solution has a number of advantages over other proposals.

- It works with existing hardware and software.
- It allows nodes to quickly join the network, without having to wait for timers to expire to indicate that no other node is using the same address.
- Merging is extremely efficient, as only two nodes need communicate, rather than requiring every node to individually validate its address by flooding the entire network.
- Unlike some related protocols, nodes maintain the first address received wherever possible. This stability helps maintain cached routing information, and minimises network disruption.
- It recovers quickly from the failure of any node.
- It is independent of both routing protocol and the underlying wireless medium (e.g. there is no requirement for a unique MAC address).

This scheme provides a locally unique address within a MANET. Further work will look at evaluating this protocol in more detail. We are currently examining the issue of providing connectivity to the Internet through the use of a NAT-based mechanism for IPv4. We will also examine possible security considerations.

References

[1] Perkins et al. *"IP Address Autoconfiguration for Ad Hoc Networks"* – Internet Draft (Nov. 2001).

[2] Nitin Vaidya *"Weak Duplicate Address Detection in Mobile Ad Hoc Networks"*, ACM International Symposium on Mobile Ad Hoc Networking and Computing (MobiHoc), June 2002.

[3] S. Nesargi and R. Prakash , *"MANETconf: Configuration of hosts in a mobile ad hoc network"*, in INFOCOM, 2002.

[4] P. Patchipulusu, *"Dynamic Address Allocation Protocols for Mobile Ad Hoc Networks"*, M.Sc. thesis.

[5] Flynn, J. et al., *"A Real-Time Emulation System for Ad Hoc Networks"*. Communication Networks and Distributed Systems Modelling and Simulation Conference (CNDS 2002), Texas, 2002.

[6] U. Joönsson, F. Alriksson, T. Larsson, P. Johansson, and G. Q. M. Jr. *"MIPMANET - mobile IP for mobile ad hoc networks"*, MobiHoc, August 2000.

[7] S. Thomson and T. Narten, "IPv6 Stateless Address Autoconfiguration, RFC 2462," Internet Engineering Task Force, Zeroconf Working Group, December 1998.

[8] D. Cheriton and M. Gritter. "TRIAD: A new next generation Internet architecture", March 2000.

DiSUS: Mobile Ad Hoc Network Unstructured Services

Paul Fergus, Anirach Mingkhwan, Madjid Merabti, and Martin Hanneghan

Networked Appliances Laboratory
School of Computing and Mathematical Sciences
Liverpool John Moores University,
Byrom Street, Liverpool L3 3AF, UK.
{cmppferg,cmsaming,M.Merabti,M.B.Hanneghan}@livjm.ac.uk

Abstract. Mobile devices are situated in an environment that consists of infrastructure and ad hoc networks. Unlike infrastructure networks, ad hoc networks are highly dynamic, wireless and do not have any centralised administration. These two environments accommodate two types of services: one type is registered and discovered using toolsets found within structured networks, e.g., JINI and UDDI, and the other type of service is situated within a network that has no structure, as in P2P. Our work highlights a key solution for providing unstructured services in a mobile ad hoc environment. This paper shows how mobile ad hoc systems can construct and semantically discover unstructured services in their environment. We describe the Distributed Semantic Unstructured Services (DiSUS) framework; the protocols used and describe the functionality of the prototype system we have developed.

1 Introduction

Visualise a high street shopping area, which is a simple outdoor environment. The street is full of shops, restaurants, street vendors and other people. We pop in and out from one store to another (structured services), buy a quick snack from a street vendor – here today gone tomorrow (unstructured services) – and greet people we know (neighbour nodes). All of these activities happen within our focal view. Mobile devices within real-world environments have to work the same way as our shopping area analogy. This provides users of mobile devices, with the ability to interact and use services in the same way we interact with shops and people within our real environment.

Information Space [8] is the concept of integrating information and services from the environment the user has access to. By considering the Mobile User as the centre of the surrounding information and services we find that, in reality, the environment the user moves into provides both structured and unstructured services. The ability to select and use both of these types of services to offer the maximum flexibility for the user is of paramount importance.

The need for an integrated Information Space will require the unification of wired and wireless networks and their services. In particular, the challenge is to bring together services within mobile ad hoc networks and infrastructure networks like the Internet [9]. Service nodes provide services throughout the Information Space by using Gateways that interconnect infrastructure networks and mobile ad hoc networks together.

M. Conti et al. (Eds.): PWC 2003, LNCS 2775, pp. 484–491, 2003.

We describe services within the Information Space as structured and unstructured and define these as follows:

- *Structured Services* use third party services to register and advertise services the peer provides, e.g., Directory Services, Proxy Services and Naming Services. These kinds of services typically have complex structures, such as network connectivity, database access and multimedia functions.
- *Unstructured Services* provide services independent of any third party intervention. This concept is based on a simple service definition, such as a kiosk that provides quick information, a TV remote control that simply changes the channel or a file-sharing application that exchanges digital content.

There are an increasing number of structured services available to users over the Internet and mobile ad hoc networks, yet unstructured services remain far behind. Internet-based structured services like JINI [12] and UDDI are already well defined; however they are incapable of providing services within dynamically changing network environments. This limitation can be overcome by situating services within the Information Space, using decentralised P2P concepts [14].

In addition to using a distributed ad hoc network, our Distributed Semantic Unstructured Services (DiSUS) framework contributes additional knowledge to this area by enabling peers to discover services dynamically based on semantic mappings between queries and distributed services.

The DiSUS framework enables every peer within the network to be a stranger. An individual with a notebook does not have to register with any centralised registry – we envisage that the individual will connect the device to the P2P network irrespective of where they are in an ad hoc fashion.

In the following section, we discuss the requirements for unstructured services in mobile ad hoc networks; in section 3 we provide an overview of the DiSUS framework and its components followed by an explanation of how the prototype works; section 4 documents our conclusions and our proposed future work.

2 System Requirements

Before we can distribute and discover unstructured services, it is paramount that our framework addresses a number of key requirements. The very nature of P2P networks means that peers will come and go over time, therefore we cannot guarantee the availability of unstructured services at any given moment. The challenge is to enable unstructured services to exist in such a non-deterministic environment and allow mobile ad hoc peers within the P2P network to effectively expose and discover unstructured services. We describe these requirements as follows:

- Interoperability: The framework must support all types of services, and not be restricted to specific services like Gnutella [4].
- Decentralisation: Services have to be completely decentralised; every peer that joins the P2P network must be capable of reaching any other peer without the help of a centralised third party.
- Unstructured services in Information Space: Services must exist in infrastructure and mobile ad hoc networks.
- Dynamic environments: Services have to be able to work in dynamically changing environments [16].

- Platform Independence: To support a variety of services and devices, a service has to be lightweight and capable of communicating across different platforms for example by using W3C standards [15].
- Ubiquity: The deployment of services must include a variety of mobile devices ranging from workstations to mobile phones.
- Intelligence: Service location must support discovery using semantic languages and toolsets that have inferential capabilities [6, 7].
- On-demand Services: Services will not run continuously and will be invoked as and when they are required [13].

3 Distributed Semantic Unstructured Services (DiSUS) Framework

In this section we describe our Distributed Semantic Unstructured Services (DiSUS) Framework, which implements several subsystems to distribute and semantically discover services within a P2P network. It provides a description of subsystem interaction and overall framework capabilities.

3.1 Framework Design

The requirements described in Section 2 are addressed within the DiSUS framework using three core components. These three components implement a P2P network, a knowledge base with inferential capabilities that allow us to semantically represent and query distributed peer services and a Web container capable of dynamically hosting Web Services as illustrated in Fig. 1. The integration of these components comprises the DiSUS framework and provides the foundation for our research.

The framework consists of two types of peer service that can choose to implement any, all or none of these components dynamically. These peers in turn are described as a Simple-Peer and a Specialised-Peer.

The DiSUS framework describes three peer services, which communicate with each other using standardised XML messaging wrapped in FIPA-ACL [11] compliant objects, as illustrated in Fig. 1. The two peers labelled Specialised-Peer-A and Specialised-Peer-B are peer services that have the ability to host Web Services, semantically match queries with local services on the peer machine and propagate queries within the P2P network. The peer labelled Simple-Peer-C is a simple peer and has limited capabilities. This type of peer service does not have the capability to host Web Services or respond to queries propagated from neighbouring peers. This peer service propagates its own queries and invokes discovered Web Services.

In this framework Web Services are hosted locally on peers and are described within the peer's knowledge base best describing, conceptually, what the Web Service means. This process enables the peer to capture the capabilities the service offers. Information about the Web Service itself is not entered into the knowledge base automatically, but rather a knowledge engineer relies on his/her understanding of the services capabilities and manually adds this information to the knowledge base. This is a temporary measure, which enables us to test our solution and we are currently

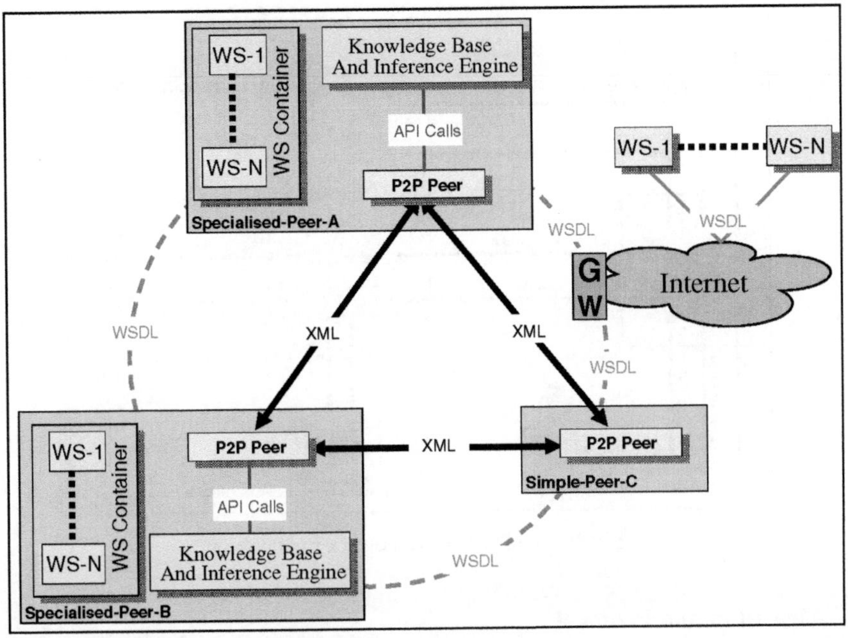

Fig. 1 Distributed Semantic Unstructured Services (DiSUS) Framework

working on algorithms to automate this process [2]. The challenge is to automatically create these representations and enable mapping processes between information contained in the knowledge base and the service itself.

The Web Services themselves are developed and hosted within distributed Web Service Containers. Invocation of Web Services is achieved in conformance with the WSDL 1.0 specification [15] – this ensures that any compliant Web Service can be invoked by a peer irrespective of where the service is located, providing it has been conceptually described within a peer's knowledge base.

Knowledge base servers reside on each Specialised-Peer, which ensures that the knowledge within the peer network is not centralised. This requirement allows us to move away from centralised registries such as JINI and UDDI and enables Web Service descriptions to be distributed within knowledge bases throughout Specialised-Peers connected to the P2P network. This ensures that there is no central point of failure and that service descriptions are distributed throughout the network – if a peer becomes unavailable you only lose the services on that peer.

We believe this framework to be very flexible and it provides distinct advantages over other areas of research such as the Service Location Protocol (SLP) [5] and Avancha, S. et al. [1]. Our framework has the ability to extend its reach outside an organisational LAN and access services contained within a peer network and semantically match queries with distributed services. Furthermore the DiSUS framework is protocol agnostic and we envisage that we will be capable of enabling interoperability between different protocols such as Bluetooth, WiFi and HomeRF. Our understanding is that [5] and [1] are incapable of achieving this level of functionality.

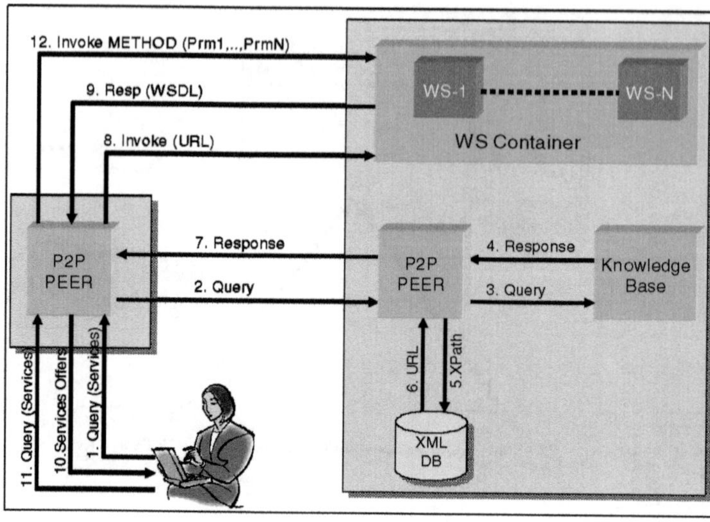

Fig. 2. The DiSUS Framework in operation

3.2 Algorithm and Protocol

This section describes the systematic interaction between subsystems in the DiSUS framework. It shows how services are semantically discovered, selected and invoked.

Fig. 2 describes the interaction between the user and two peer services. This process begins when the user selects a service category from the user interface. After selection a query message is retrieved from an XML configuration file and wrapped in a XML message. This message is propagated to all peers on the network. When a Specialised-Peer receives a message it extracts the queries from the XML message and passes it to its knowledge base to determine semantically if it has a service corresponding to the received query.

It does not matter what name we give to services because services are not retrieved by keyword but rather on their capabilities. Services that have the same capabilities, but are named differently should be conceptually equivalent, therefore they will be returned irrespective of what they are called.

If the peer has a service, the abstract name of the service is retrieved from the knowledge base and used to extract the Web Service's relative URL from an XML file as illustrated in Fig. 2. We programmatically retrieve the IP address dynamically from the peer machine at runtime and construct the URL by combing the IP address and Port number with the relative URL extracted from the XML file.

Before a Web Service is invoked, the peer service retrieves the WSDL file and programmatically determines what functions the service offers. Typically, the functions might be displayed to the user who then selects the required function (method) they want to use, or the program might intelligently invoke the most suitable service. This is outside the scope of this research. Before the service method is invoked the user is prompted to enter the required parameters. The Web Service processes this method invocation and a query response is created and wrapped within an XML message. This message is returned to the peer that initiated the query and presented to the user.

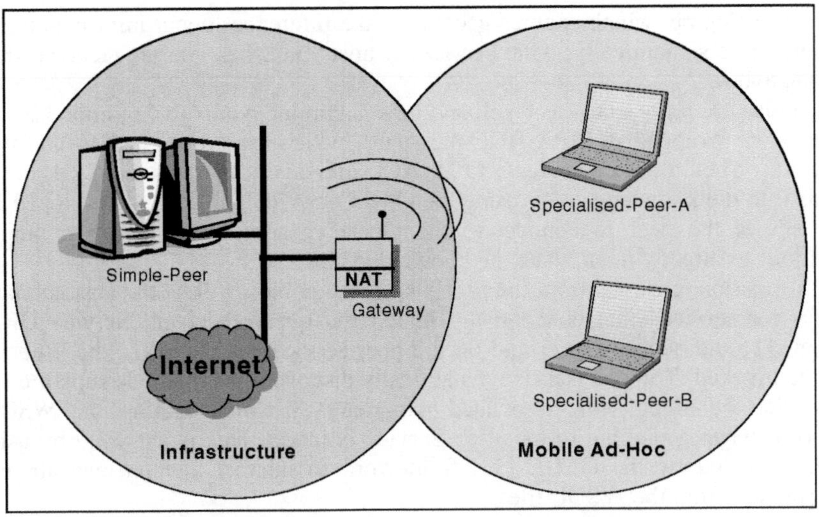

Fig. 3. DiSUS Prototype system configuration

3.3 Prototype

The prototype we have created hosts several Web Services, which are distributed and discovered within three Windows XP peers connected in a JXTA [16] P2P network on a Wireless LAN (802.11b). Two Notebooks where used and one Desktop (see Fig. 3).

Although the machines used within our prototype are fairly high powered we envisage deployment of our prototype to much smaller devices such as PDAs and Mobile Phones, however our current research has not addressed this issue.

All software was constructed using Java based on the 1.4 JRE. We chose three toolsets to construct our framework – JXTA was used to implement the P2P network, OpenCyc [10] was used to implement the knowledge base and Inference Engine, and GLUE was used to implement the Web Service container. In conjunction with the three toolsets described, the FIPA-ACL [11] standard was used to standardise message passing between peer services.

The test environment consists of two Specialized-Peers that host three Web Services each and provide OpenCyc services to semantically process queries received from the peer network. The third peer is a Simple-Peer and provides no services except the minimal functionality required, as described in section 3.

In the test environment, the Specialised-Peer-A hosts the following Web Services; PF Book service; the Restaurant Finder service and the Postcode Checker service. The Specialised-Peer-B hosts the AM Book service; the ABC Restaurant Finder service and the Thesaurus service.

The Web Service descriptions in our prototype are represented in the OpenCyc Knowledge Base using the built-in language provided by OpenCyc, which is CycL [10]. This is a language based on Description Logics and allows information to be represented taxonomically based on classifications obtained from specific domains. As well as a means of representing information, OpenCyc provides a built-in

Inference Engine, which enables peers to make inferences about information in order to determine semantically, what services it hosts, matches queries received from the peer network.

Queries are represented as CycL and have a similar syntax to Common Lisp. These queries are wrapped in FIPA-ACL messages, which are propagated around the peer network. When a peer receives a FIPA-ACL message it extracts the CycL query and sends it to the knowledge base using the OpenCyc API. This API is developed in Java and allows the peer to connect to the OpenCyc server and invoke a number of functions to directly manipulate the knowledge base.

Information retrieved from the peer's knowledge base will be the abstract names of URLs for services that conceptually match the query received. Service URLs are returned to the querying peer and once a peer service has the URL, the Web Service can be invoked. The peer service dynamically discovers the methods supported by the Web Service along with associated parameters by interrogating the WSDL file received when a binding to the Web Service is made. This is achieved by using the APIs provided by the GLUE [13] framework to interact and retrieve all required information from the WSDL file.

4 Conclusion and Future Work

In this paper we have described the DiSUS framework and explained how it can support unstructured services in Mobile Ad hoc Networks. We defined the requirements such a framework needs to address and described the prototype we developed. Within our research into the DiSUS framework our primary goal was to implement a prototype that could ground our ideas and provide a framework on which to build. We believe we have achieved this and have successfully implemented a solution that enables us to semantically discover distributed Web Services in an ad hoc network.

The construction phase required a great deal of manual work to conceptually create service descriptions and insert them into the knowledge base as well as formulating the queries used to conceptually discover and invoke services. Although we are currently working on algorithms to automate this process [2], it initially enabled us to develop a level of confidence that illustrated that our ideas are viable and uncovered key areas we need a better understanding of. This is very much a work in progress framework, which in this paper forms the base framework for future work.

Within this paper we successfully addressed a number of key issues - firstly we addressed the issue of semantically representing descriptions of Web Services and illustrated how these descriptions can be used to semantically determine if peers contain services that match the query being processed. Secondly we illustrated how unstructured services could be discovered within a dynamic environment devoid of any centralisation or any third party registries. We successfully embedded services within a distributed P2P network and proved that we could retrieve services without have any pre-knowledge of their existence and employ on-demand service invocation.

Within this paper we acknowledge that we need to address a number of issues. Firstly we need to better understand how we can conceptually match natural language queries with knowledge contained in a peer's knowledge base. We are currently addressing this issue and we are working on an algorithm that can determine the semantics of unstructured information using distributed dynamic ontologies. We believe that our approach will provide an efficient means of addressing semantic interoperability between disparate knowledge structures. Secondly we need to automatically evolve the conceptual capabilities contained within the peer's knowledge base by collaboratively evolving knowledge structures based on a general consensus received from the peer network. We have in part, developed an algorithm to achieve this by using evolutionary programming techniques, however this is still a work in progress [3].

References

[1] S. Avancha, Joshi, A., Finin, T., Enhancing the Bluetooth Service Discovery Protocol, University of Maryland Baltimore County: Baltimore, MD 21250, Technical Report, TR-CS-01-08, 2001.

[2] P. Fergus, Mingkhwan, A., Merabti, M., Hannegham, M. Capturing Tacit Knowledge in P2P Networks, to appear in 4th EPSRC Annual Postgraduate Symposium on the Convergence of Telecommunications, Networking and Broadcasting (PGNet2003). 2003. Liverpool, UK

[3] P. Fergus, Mingkhwan, A., Merabti, M., Hanneghan, M., Distributed Emergent Semantics in P2P, John Moores University: Liverpool, Technical Report, TR-CMS-2003-04-002, 2003.

[4] The Gnutella Protocol Specification v0.4. 2001, Accessed: 23-05-2003, http://www9.livewire.com/developer/gnutella_protocol_0.4.pdf.

[5] E. Guttman, Service Location Protocol Automatic Discovery of IP Network Services. IEEE Internet Computing, 1999. Vol. 3(No. 4): p. 71–80.

[6] A. Maedche, Staab, S. Services on the Move - Towards P2P-Enabled Semantic Web Services, Proceedings of the 10th International Conference on Information Technology and Travel & Tourism. 2003. Helsinki: Springer, p. 124–133.

[7] S. A. McIlraith, Son, T. C., Zeng, H., Semantic Web Services. IEEE Intelligent Systems, 2001. 16(2): p. 46–53.

[8] A. Mingkhwan, Merabti, M., Askwith, B. Interoperability of Structured and Unstructured Services in Personal Mobility Information Space, European Wireless 2002. 2002. Florence, Italy

[9] A. Mingkhwan, Merabti, M., Askwith, B., Hanneghan, M. Global Wireless Framework, European Personal Mobile Communications Conference (EPMCC'03). 2003. Glasgow, Scotland

[10] OpenCyc Project. 2002, Cycorp, Inc., Austin, TX, USA, Accessed: 5-02-03, http://www.opencyc.org.

[11] M. Schalk, Liebig, T., Illmann, T., Kargl, F. Combining FIPA ACL With DAML+OIL - A Case Study, Proceedings of the Second International Workshop on Ontologies in Agent Systems. 2002. Bologna, Italy

[12] JINI Technology. 2002, Accessed: 23-05-2003, http://www.jini.org/.

[13] GLUE. 2003, The Mind Electric, Accessed: 5-02-03, http://www.themindelectric.com.

[14] Framework for Peer-to-Peer Distributed Computing in a Heterogeneous, Decentralized Environment. 2002, Sun Microsystems, Inc., Accessed: 28th November 2002, http://wwws.sun.com/software/jxta/mdejxta-paper.pdf.

[15] World Wide Web Consortium. 2003, W3C, Accessed: 5-02-03, http://www.w3c.org.

[16] B. J. Wilson, (2002), "JXTA", New Riders Publishing, Indianapolis.

Performance Evaluation of Base-Station-Assisted Link State Routing Method for Mobile Ad Hoc Networks

Hiroaki Morino[1], Tadao Saito[1,2], and Mitsuo Nohara[2]

[1] Research and Development Initiative Chuo University
1-13-27 Kasuga Bunkyo-ku Tokyo 112-8551, Japan
Tel: +81-3-3817-1626 Fax: +81-3-3817-1606
morino@m.ieice.org
[2] TOYOTA InfoTechnology Center Co., Ltd.

Abstract. For communication systems of automobiles, hot spot networks using wireless LAN are attractive in that broad bandwidth is available at low cost. Further, inter-vehicle ad hoc networks organized by cars are expected to cooperate with the hot spot network and effectively extend the coverage of the hot spot network. The authors have proposed base-station-assisted link state routing method(BALSR)[1] that is effective for large scale inter-vehicle ad hoc networks. This paper presents performance evaluation of BALSR, and it shows that BALSR significantly reduces the number of conrol packets in the ad hoc network compared to conventional link state routing method.

Keywords: Inter-vehicle mobile ad hoc network, link state routing method, wireless LAN base station

1 Introduction

Recently, broadband wireless communication systems including 3G cellular system , hot spot networks by wireless LAN and so on have emerged, and capacity of communication systems of automobiles is increasing from tens of kilo bit/sec to hundreds of kilo bit/sec. Especially, hot spot networks are attractive for the availability of broad bandwidth at the lower cost. Although coverage of hot spot networks is limited, it will be effective when some terminals in the coverage receive data from the hot spot network and they share received data with other terminals out of the coverage using the inter-vehicle ad hoc network. This is called the information delivery system in this paper. In this system, performance of throughput will be largely affected by the routing method of inter-vehicle ad hoc network. For the routing method, mainly the following three conditions will be required; (1) Many communication pairs are handled efficiently (2) The method works with low control overhead in the case that the ad hoc network is large scale whose size is several kilometers (3) The route between the information sender terminal to the information receiver terminal can be constructed in a few seconds in order that the system efficiently works.

M. Conti et al. (Eds.): PWC 2003, LNCS 2775, pp. 492–497, 2003.

There have been many proposals of routing methods for mobile ad hoc networks. For example, MANET(Mobile Ad-hoc NETworks) Working Group of IETF (Internet Engineering Task Force) discusses mainly two types of routing methods. One is the proactive method, where every terminal maintains routing table by periodic exchange of routing information, and OLSR[2] and TBRPF[3] are examples of this method. The other is the reactive method, where only terminals that have data to send construct routes on demand. AODV[4] and DSR[5] are examples of this method. In these methods, however, each terminal has to relay packets of routing information more frequently as the number of hops of the network gets large. In the wireless LAN based ad hoc network, it is possible that excessive amount of control packets causes congestion that leads to throughput decrease of the network. From this viewpoint, we have proposed base-station-assisted link state routing method (BALSR).

This paper presents performance evaluation of BALSR, and it shows that BALSR reduces the number of control packets in ad hoc mode wireless LAN in inter-vehicle communications.

2 Overview of BALSR

2.1 Overview of the Information Delivery Systems Using Hot Spot Network and Inter-vehicle Communication

Figure 1 illustrates overview of the information delivery system using the hot spot network and the inter-vehicle ad hoc network. In this figure, the car (a), which is out of coverage of the hot spot network, wants to receive some data from the network. It is called information receiver terminal. Car (b) is one of terminals newly coming into the coverage of the hot spot network, and it is called information sender terminal. The procedures of the system for information delivery between these two terminals are as follows.

(1) The information sender terminal begins to join the inter-vehicle ad hoc network and constructs routes to other all of terminals that already joined the network.
(2) When the route between the information receiver terminal and the information sender terminal is established, the information receiver terminal sends request to the information sender terminal for receiving data from the hot spot network.
(3) The information sender terminal receiver the requested data from the hot spot network and sends the data to the information receiver terminal.

2.2 Operations of BALSR

Figure 2 shows an overview of the inter-vehicle ad hoc network realized by BALSR. In this system, wireless LAN base stations are provided along the road for routing control of the network. Each terminal has two wireless LAN interfaces; one is ad hoc mode wireless LAN for inter-vehicle communication and

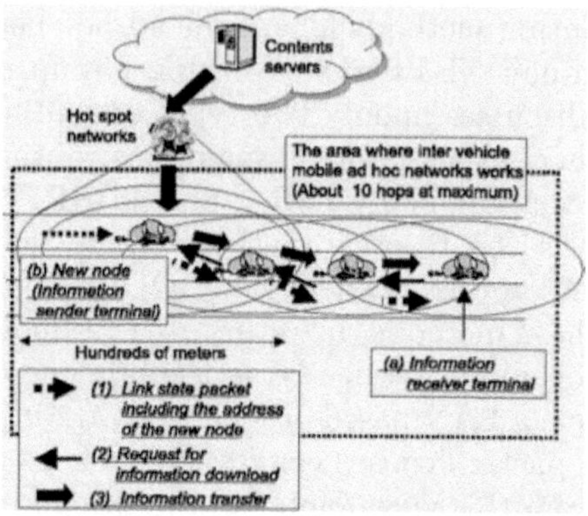

Fig. 1. Information delivery system using the hot spot network and inter-vehicle mobile ad hoc networks

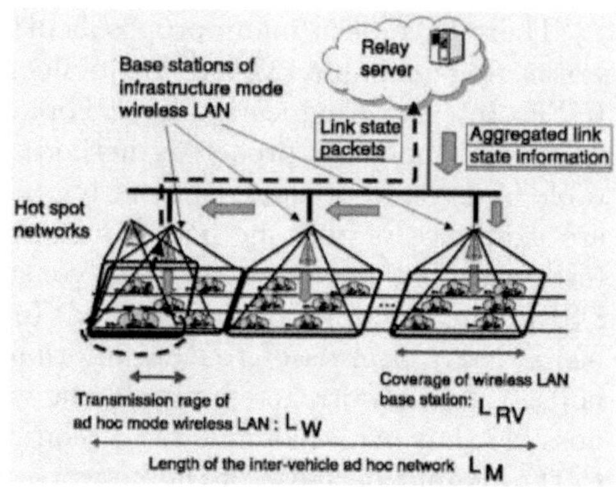

Fig. 2. Overview of the base-station-assisted link state routing method

another is infrastructure mode wireless LAN to communicate with base stations. Infrastructure mode wireless LAN is configured so that it has larger transmission range and lower bit rate compared to ad hoc mode wireless LAN. Functions of wireless LAN base stations described here can be also realized by base stations of cellular system in the case that the cellular system supports point-to-multipoint communications. Operations of the system are as follows. Each terminal sends the hello packet in the certain interval T(hello) by ad hoc mode wireless LAN and recognizes neighbor terminals. Then the terminal generates the link state packet in the certain interval T(ls) that includes link status information between itself and neighbor terminal, and it sends it to the relay server via wireless LAN base stations. The relay server receives link state packets sent by terminals, aggregates link state information in these packets, and sends the information to all terminals via wireless LAN base stations to all terminals at the certain period. The period at which the relay server sends link state information to all terminals is called packet relay period, and it is denoted by T(relay).

In this method, the number of control packets handled in ad hoc mode wireless LAN is significantly reduced, and throughput will be improved.

3 Performance Evaluation

3.1 Evaluation Model

The Model of Car Movement and Data Packets Transmission between Cars. The model of car movement is shown in Fig.3, where each car runs at a constant speed and keeps a constant distance to a car ahead. The model of data packets transmission between cars is as follows. Among cars that newly enter the area of the ad hoc network, predetermined numbers of cars are selected as information sender terminals. For each of these terminals, a car most distant from it is selected as an information receiver terminal, and the information sender terminal sends data packets to the information receiver terminal by unicast for 20 second, at a sending rate of 10 packet/sec. When 20 seconds have passed, the information sender terminal selects another information receiver terminal and repeats the same procedure until it reaches at the end of the area of the network.

Parameters of Wireless LAN and Routing Method. In the model of BALSR shown in Fig.2, parameters of wireless LAN are as follows.

- Ad hoc mode wireless LAN
  - Bit rate : 2Mbps
  - Transmission range L_W : 150-300m
- Infrastructure mode wireless LAN
  - Bit rate : 128kbps
  - Coverage of base station L_{RV} : 1km

Parameters of ad hoc mode wireless LAN are determined to model IEEE 802.11b 2Mbps mode. On the other hand, parameters of infrastructure mode wireless LAN are determined by assuming that the system has larger spreading factor than IEEE 802.111b. In the ad hoc mode wireless LAN, packets sent from a terminal arrive at the next terminal when the distance between two terminals is equal or less than L_W. It is assumed that packets are lost when distance is larger than L_W, and lost packets are not retransmitted. It is also assumed that any packet loss caused by collision does not occur Routing methods evaluated are BALSR and OLSR. In both methods, interval of hello packets T(hello) is 2 seconds. Interval of link state packets T(ls) is 2 seconds at minimum and 6 seconds at maximum respectively. Packet relay period T(relay) in BALSR is set as 100msec. In BALSR and OLSR, parameters are configured so that routes between terminals within two hops are constructed in mesh topology, and routes between terminals distant by more than two hops are constructed in tree topology.

At last, the way of utilization of wireless LAN base stations in BALSR and OLSR is as follows. Though wireless LAN base stations are provided mainly for control of the ad hoc network, it is assumed that they are connected to content servers shown in Fig.1 and each terminal in both method receive information from content servers via them.

3.2 Results

Relationship between Data Packet Delivery Ratio and Transmission Range. Figure 4 shows data packets delivery ratio in BALSR. Data packets delivery ratio is defined as the ratio of data packets that arrived at the information receiver terminal among data packets sent from information sender terminal. Since BALSR and OLSR has almost the same data characteristics, the figure only shows results of BALSR. The figure shows that data packet delivery ratio in the network with the same value of maximum number of hops(calculated by

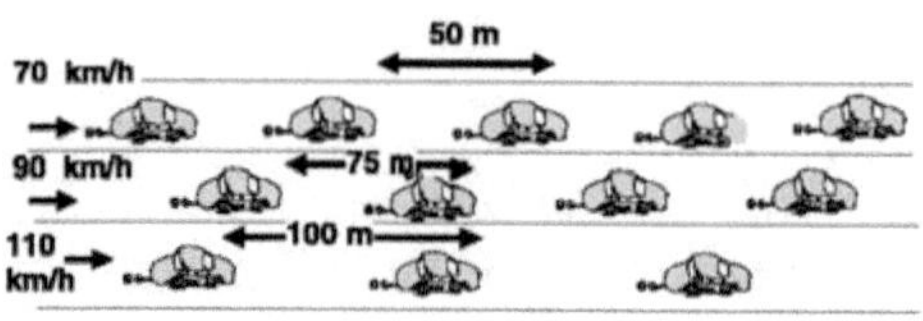

Fig. 3. Car movement model in the sytem shown in Fig.2

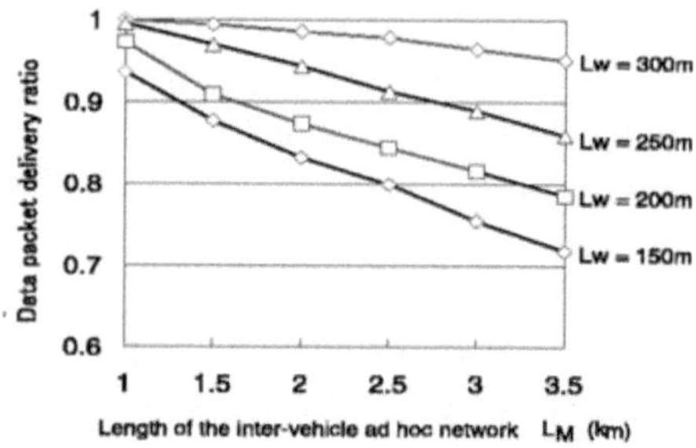

Fig. 4. Data packet delivery ratio in BALSR

L_M/L_W)is improved as L_W increases. This is by the effect that routes between terminals within two hops are configured to be constructed in mesh topology, as described in Section 3.1. If a route fails between some two terminals within two hops by topology change, an alternative route can be immediately established among redundant routes already established in mesh topology. From this results, it can be said that L_W should be set as large value as possible to improve data packet delivery ratio in large scale network.

Control Overhead. Figure 5 shows the number of conrol packets that ad hoc mode wireless LAN in each terminal and each wireless LAN base station sends and receives in BALSR and OLSR. In OLSR, it increases as L_M increases, and it reaches 76 packets/sec when L_M=3km. On the other hand, BALSR reduces the number of control packets in each terminal to 22 packets/sec, which is about 30 % of that of OLSR. Fig.6 shows amount of information of control packets each terminal and each wireless LAN base station sends and receives in BALSR and OLSR. BALSR reduces amount of information of control packets handled in each terminal by about 50kbps compared to OLSR. On the other hand, BALSR consumes about bandwidth of each wireless LAN base station by about 25kbps. Therefore, it can be said that total bandwidth that control packets consume in ad hoc mode wireless LAN and wireless LAN base stations in BALSR is almost the same as that of OLSR.

From these results, it can be concluded that BALSR is effective in that it reduces the number of control packets in ad hoc mode wireless LAN compared to OLSR. It is also shown that total bandwidth consumed by control packets in ad hoc mode wireless LAN and wireless LAN base station in BALSR is almost the same as OLSR.

4 Conclusion

This paper presents performance evaluation of BALSR in large scale inter-vehicle ad hoc networks under the conditions that the inter-vehicle ad hoc network works in order to effectively extend the coverage of the hot spot network.

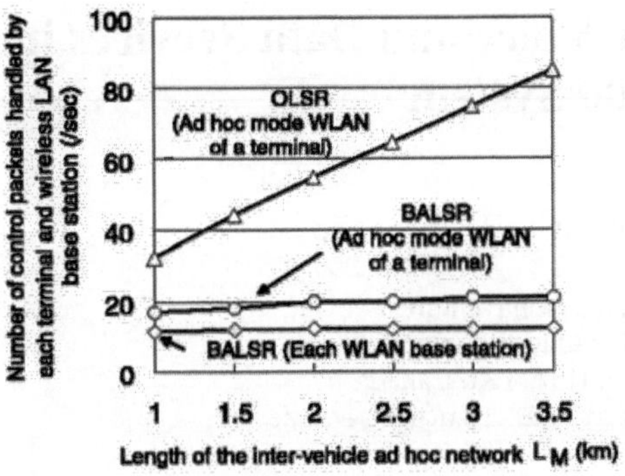

Fig. 5. The number of control packets handled by each terminal and each wireless LAN base station

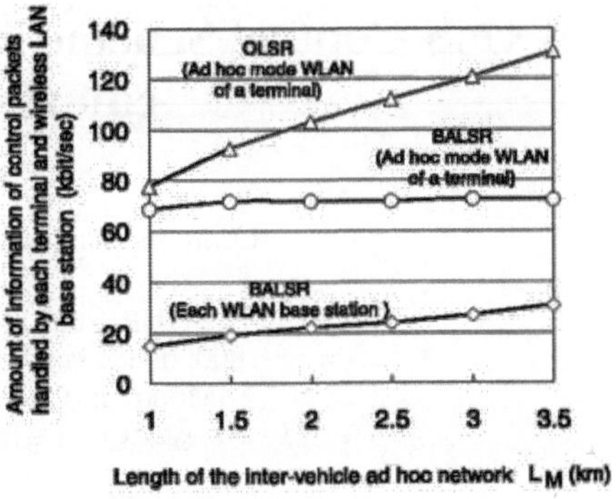

Fig. 6. Amount of information of control packets handled by each terminal and each wireless LAN base station

Firstly, data packet delivery ratio is evaluated under the condition of various transmission ranges of ad hoc mode wireless LAN. It is found that BALSR achieves data packet delivery ratio above 90 % in the case that the length of the ad hoc network is 3 km, when the range is equal to or greater than 300 m. Secondly, the number of control packets that ad hoc mode wireless LAN in each terminal sends and receives is evaluated. It is shown that BALSR reduces the number of control packets to 30% of that of OLSR in the case that the length of ad hoc network is 3 km. This reduces load of ad hoc mode wireless LAN in each terminal, and it will lead to improve throughput of the ad hoc network. For future work, throughput performance of BALSR will be further evaluated in detail by considering wireless LAN media access control.

Acknowledgements. The authors wish to thank Prof. Shigehiko Naoe (Chuo University), Mr. Tadao Mitsuda (TOYOTA InfoTechnology Center Co., Ltd) and Prof. Takeshi Hattori (Sophia University) for their valuable comments.

References

1. H. Morino et al., “An efficient proactive routing method for mobile ad hoc networks using peer-to-peer and cellular communication system,” PWC 2002 Oct 2002.
2. Phillipe Jacquet et al., “Optimized Link State Routing Protocol,” Proc. of IEEE INMIC 2001.
3. Richard Ogier et al., “Topology Dissemination Based on Reverse-Path Forwarding,” IETF Internet-Draft <draft-ietf-manet-tbrpf-07.txt> Mar 2003.
4. C. Perkins et al., “Ad hoc On-demand Distance Vector (AODV) Routing,” Internet-Draft <draft-ietf-manet-aodv-10.txt> Jan 2002.
5. David B. Johnson et al., “The Dynamic Source Routing Protocol for Mobile Ad Hoc Networks (DSR)”, Internet-Draft <draft-ietf-manet-dsr-07.txt> Feb 2002.

A QoS Control Scheme for Voice and Data Services in cdma2000 System

Omneya Issa and Jean-Charles Grégoire

INRS-EMT, Place Bonaventure,
800, de la Gauchetière Ouest, bureau 6900
Montréal (Québec), H5A 1K6 Canada
{issa,gregoire}@inrs-EMT.uquebec.ca

Abstract. The 3GPP2 standards gave many new features to cdma2000 to be able to support voice and data services, and established a layered structure for this network, but they left out QoS control as an open question. The aim of this paper is to present a QoS control scheme that exploits some of these new features. The proposed scheme consists of resolving contention between admitted services, scheduling the service packets and efficiently multiplexing data blocks on physical channels according to their QoS requirements. Its performance is evaluated using voice and data applications. It is compared with the performance of existing cdma2000 scheme in both indoor and outdoor environments.

1 Introduction

The new design of cdma2000 gives rise to a challenge concerning the use of these new capabilities and the control of QoS to get the maximum service satisfaction. Furthermore, the standards left out the QoS control as an open question. A QoS control scheme that exploits all these new features is needed. It must be able to resolve contention between service entities, schedule the service packets according to their QoS requirements and efficiently multiplex data blocks on physical channels. In fact, a form of priority scheduling is mentioned in the RLP standard [1]. That is, control messages have the highest priority followed by retransmitted data blocks, and, at the bottom, the data blocks that are being sent for the first time.

Some previous research efforts on cdma2000 [2] considered data burst scheduling between mobile users waiting to be admitted. However, the QoS control is left unspecified. [3] also proposed a framework for flow control in cdma2000. The framework deals with the bandwidth as a whole; it does not partition it into distinct traffic channels that differ in characteristics such as BER. Other researchers have proposed scheduling schemes for CDMA-based systems. But their work conforms rather to a hybrid TD/CDMA system, comparable to the forward packet data control function of 1XEV-DV extension of cdma2000 [1].

This paper proposes a QoS control scheme for the service connections that are admitted by the BS and hence the matching dedicated physical channels can be established. Section 2 describes the design components. Section 3 presents the simulation model and the design performance is investigated section 4.

M. Conti et al. (Eds.): PWC 2003, LNCS 2775, pp. 498–504, 2003.

2 Proposed QoS Control Scheme

The cdma2000 MAC layer contains a QoS control entity that is left unspecified. This section presents the design of a QoS control scheme that we propose for this undefined entity. This scheme controls the transmitted packets, both signalling and data packets, across their path through the layers based on requested QoS levels agreed to during service negotiation phase and the corresponding established physical channels. The QoS parameters of the service connections consist of the maximum delay (MaxD), the minimum throughput (MinTh) and the maximum loss [1].

Since the standards do not specify any particular classification, we choose to divide the traffic in four classes: *control messages* (CM) including messages such as IP routing, service signalling and handoff messages; b*ounded delay* (BD) for traffic with a delay bound such as voice services; m*inimum throughput* (MT) for traffic that requires minimum throughput such as *ftp* and *http* downloads; and b*est effort* (BT) for the traffic that does not have any QoS requirements. Each class has its queue; when a connection is admitted, its packets are forwarded to their matching class queue.

It is mentioned in cdma2000 standards that priority levels can be used to provide over-the-air QoS, yet no specific priority assignment is defined. So we decide to assign a priority to each class shown in figure 1 and use a priority-based scheduler.

Our design deploys shaping of service traffic before being scheduled for helping the design to better commit to the agreed QoS and simplifying the design of the scheduler. The shaper of a data service with fixed-size packet simulates a CBR source by setting the bucket size B to one packet and token rate r to the agreed service rate. A buffer is used to store packets that cannot pass because there is no waiting token. However, a voice service generates bursty traffic. A source model that matches the voice behavior is used by [4]. It assumes a two-state, on-off, Markov model; when on, voice user sends at the peak rate P. The transition rate from on to off is α, and the transition rate from off to on is λ. Based on this model, [4] deduced a relation (1) between r and B for N users in function of P, α, λ and loss rate L. So we can determine B of a voice connection as a function of r.

$$[(\ln L)/N]\, r^2 - [P \ln L + (\alpha + \lambda)\, B]\, r + \lambda N P B = 0 \tag{1}$$

BE connections do not need shaping and the control messages pass without shaping, as well. Note that packets, that are released, pass through the RLP entity. The RLP entity fragments the packets into 266-bit PDUs (Protocol Data Unit), called data blocks (DBs), which are then forwarded to the class queue of their service connection.

Our priority-based scheduler is not sufficient to handle the different QoS needs of each class. So each class must have its own scheduling policy. Note that the overall scheduling scheme follows the standard RLP's. The CM queue has FIFO policy except that the handoff messages have higher priorities. An earliest deadline first policy is adopted in the BD queue; the packet deadline is the packet arrival time plus MaxD. When this queue is selected for service, all blocks whose deadlines have expired are dropped. A FIFO policy is adequate for MT queue since the throughput guarantee is handled by the admission phase. In this queue, if a block has the same arrival time as a RLP retransmitted block, the retransmitted one will be scheduled first. The DBs of BE queue are sorted according to their connection priority (if specified).

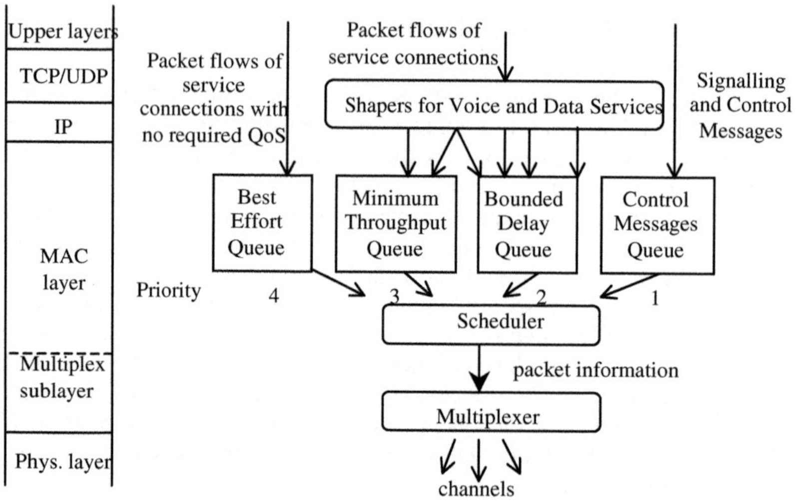

Fig. 1. QoS Control Scheme.

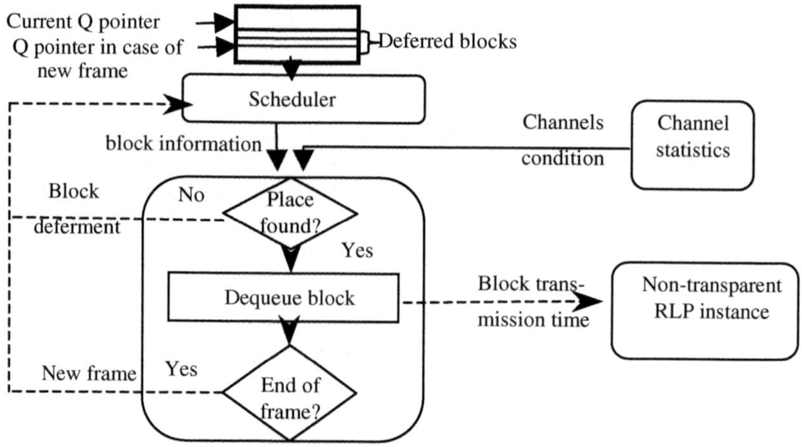

Fig. 2. The multiplexer.

The multiplexer is responsible for multiplexing DBs onto the adequate channels. The standards only specify a multiplex option that defines the maximum number of DBs and the allowed block size for a channel. They do not specify the way a channel is selected for DB transmission. In our design, figure 2, the multiplexer first checks the presence of a place on a channel supporting the BER required by the DB connection or a lower BER. A place on a channel is not found when the channel is reported in error by the statistics gathering done at this sublayer. The DB transmission time is reported to the RLP to avoid a useless generation of a duplicate DB in case of the reception of an early duplicate NAK for this DB. When the multiplexer fails to transmit a DB, the latter is not discarded; it is left at the head of the queue. When the multiplexer signals a new frame, the scheduler begins with the DBs that were deferred within the last frame and are still at the head of the served queue. This compensates for the DBs that should have been sent in the previous frame.

3 Simulation Model

The proposed scheme is implemented using the ns-2 over both the MS and the BS nodes. We simulate a MS connected to a BS in a cmda2000 network and communicating with a remote host on a fixed network. The propagation delay is 3 μs on the wireless network, and 100ms on the fixed network. The voice is modelled as in [4], α is set to 1/1.2s, λ to 1/1.8s, L to 10^{-3}, P to 26.6kbps, r to 13.3kbps, B to 180 packets and packet size is 50 bytes. The FTP application from ns-2 is used, packet size is 512 bytes. The web browsing traffic is modelled as in [5]. Simulation models for both the cdma2000 forward and reverse link are created according to the standards. Slow fading is modelled by lognormal shadow fading, table 1. Fast fading is modelled by the superposition of 2 Rayleigh-distributed faded paths. Additive white Gaussian noise is added to simulate the effect of overall interference. The radio configuration of reverse and forward channels is 4 and 5 respectively. The BER of the fundamental (FCH) and supplemental channel (SCH) is 10^{-3} and 10^{-5} respectively. The frame length is 20 ms.

Table 1. Parameters of the two simulated environments.

	Indoor Environment	**Outdoor Environment**
Model	In building	Vehicular
Place	Office (soft partition)	Shadowed urban area
Velocity	3 km/h	100 km/h
Path loss exponent	1.6	3
Shadowing deviation	9.6	4

4 Simulation Results

The simulation incorporates a FTP upload generated by the MS, 3 web sessions and a voice conversation. The MaxD of the voice is 100ms, the MinTh of FTP varies from 26.6kbps to 113.7kbps. Note that the effective channel rate is lower than the nominal data rate of the standards because of the presence of headers in the transmitted frame. FTP and web sessions use non-transparent RLP while voice has a transparent one. The satisfaction of each application is measured according to its most pertinent QoS parameter. For FTP, the throughput is measured (Tm). For the voice, the percentage of discarded DBs (Pdp) due to expiration is presented. For web sessions, the average response time (RT) experienced by web pages is measured, that is, the time between the launching of the web page and the arrival of the entire page objects.

To investigate the performance of the proposed scheduler, without the effect of multiplexer, only one reverse channel is established with a rate that varies from 28.8kbps and 115.2kbps, done by varying the transmission power and the target SIR. The DBs of the 3 applications are transmitted on this channel. A forward 57.6kbps

SCH is opened for TCP ACKs and web sessions replies. Figure 3(a,b) shows that the proposed scheduler raises the Tm of FTP in indoor environments with a difference of 1-8kbps comparing to the scheduling of RLP standard. In outdoor environments, Tm is lower (2-6kbps) because the channels experienced more successive frame errors.

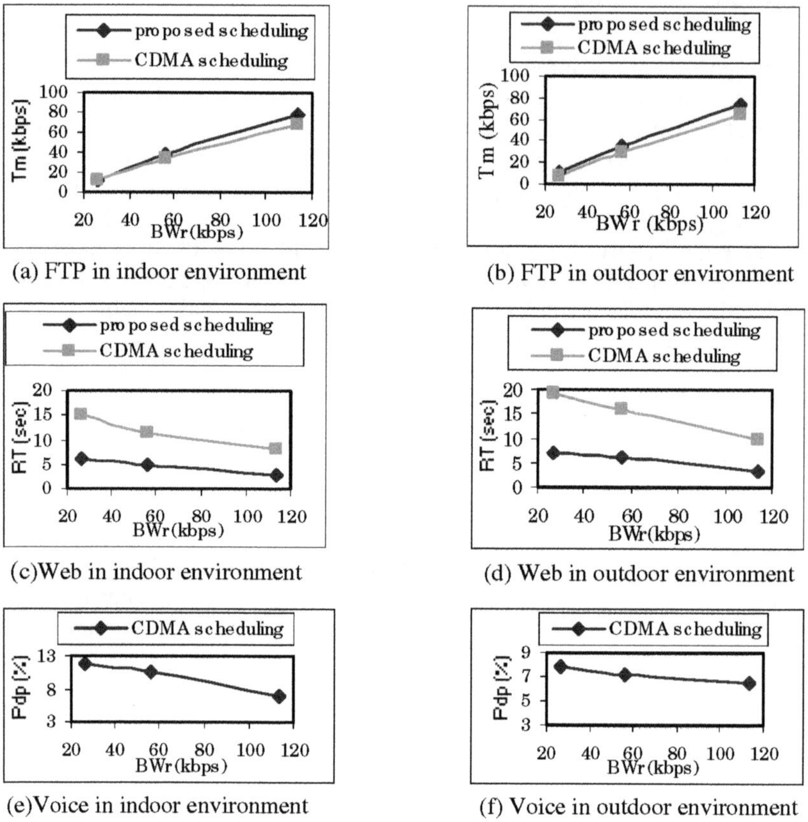

Fig. 3. Scheduler performance.

The difference in web RT using the proposed scheduler vs. RLP scheduling is 9s that goes down to 5s when doubling the reverse bandwidth (BWr) in indoor environment. In outdoor environment, this difference goes up to 12s and decreases to 7s when doubling BWr, figure 3(c,d). When deploying our proposed scheduler, no voice packet is discarded because those packets have the priority of transmission. Whereas, in cdma2000 scheduling case, the new packets are served on a FIFO-basis, so voice blocks may wait till expiration. In indoor environments, Pdp is 11.5-7%. In outdoor environments, Pdp goes from 8% to 6.5% because FTP packet generation rate is lower, due to successive errors, figure 3(e,f). Note that the discarded voice blocks remain in queue for longer time which results in web request packets and FTP packets waiting longer before being transmitted, which increases the TCP estimation of round-trip time (RTT) and thus, reduces FTP Tm and increases web RT.

Due to channel errors and its effect on TCP connections, Tm does not reach the required one, although the channel bandwidth is adequate to hold the required rate. That is, some frames are transmitted empty. When transmitting a voice conversation on a 14.4kbps channel, adequate for the average voice rate 13.3kbps, 20% of the frames are empty; those frames should carry DBs discarded due to expiration. Those idle frames can be saved by using the cdma2000 variable channel rate capability.

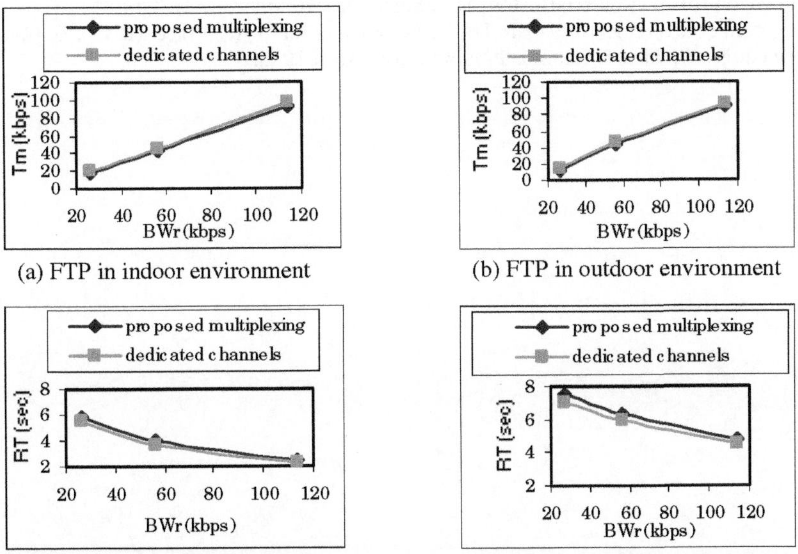

(a) FTP in indoor environment (b) FTP in outdoor environment

(c)Web in indoor environment (d) Web in outdoor environment

Fig. 4. Multiplexer performance.

Another solution would be to give other services the possibility to transmit their DBs on the same channel, and thus, avoid opening new dedicated channels for those services. This solution favours the proposed multiplexing scheme. Figure 4 shows the results of simulating a FTP service and 3 web sessions that require a BER of 10^{-5} whereas the voice service needs a 10^{-3} BER. 2 reverse channels are opened, one 14.4kbps FCH channel and one SCH channel whose rate varies from 28.8kbps to 115.2kbps. A forward 57.6kbps SCH is opened as well. The BER of the FCH only fits the voice service, while the BER of SCH fits the 3 applications. To measure the effect of multiplexing on the applications' performance, we compare it with their performance when opening a dedicated channel adequate for each. When compared with the throughput of dedicated channel, the Tm of FTP using the proposed multiplexing is only affected by a difference of 0.5kbps in low rates that goes up to 2kbps in 113.7kbps rate. The web RT rose by 0.4s in low BWr and by 0.2s when reaching a BWr of 115.2kbps. The voice Pdp is about 20% in the case of dedicated channel while no packet is discarded when using the proposed scheme. Those differences are comparatively the same in both indoor and outdoor environments. This shows that we can avoid the opening new dedicated channels for applications like web sessions, email and telnet and the expiration of voice packets by multiplexing the DBs of several services on the same channel when it fulfills their QoS requirements.

References

1. Cdma2000 standards at http://www.3gpp2.org/Public_html/specs/index.cfm.
2. Nanda et al. Adaptation techniques in wireless packet data services. IEEE Comm.Mag. 2000.
3. Comaniciu, N. et al. QoS Guarantees for 3G CDMA Systems via Admission and Flow Control. VTC'2000, Boston, MA, September 2000.
4. J. Rosenberg. Engineering VBR Traffic Parameters for WAN Voice. (ICC '97), Canada.
5. The cdma2000 ITU-R Candidate Submission. April 1998.

Push Driven Service Composition in Personal Communication Environments

Justinian Oprescu, Franck Rousseau, Laurentiu-Sorin Paun, and Andrzej Duda

LSR-IMAG Laboratory
BP. 72, 38402 Saint Martin d'Héres, France
{Justinian.Oprescu,Franck.Rousseau,Laurentiu-Sorin.Paun,
Andrzej.Duda}@imag.fr
http://www-lsr.imag.fr

Abstract. Our current mode of communication is mainly based on the *pull* model in which the user requests information or initiates a communication stream. We believe that a personal communication environment should behave according to the *push* model in which a source of data or the network infrastructure takes care of preparing communications and proposing them to the user. In this proactive way of operation, the only user intervention is to choose the right communication flow. In this paper we present the service discovery and composition in Omnisphere, a personal communication environment for wireless appliances. All elements in Omnisphere are considered as services that can be dynamically discovered and composed to form complex communication applications. Based on user preferences, device capabilities, and context Omnisphere makes use of existing discovery protocols such as SLP, Jini, UPnP, or DNS-SD to discover relevant services. Service descriptors provide further information on the data types generated on output or accepted on input. All this information allows Omnisphere to configure services and user applications so that the user can benefit from complex communication applications composed on demand.

1 Introduction

We present the service discovery and composition in Omnisphere [1], a personal communication environment for wireless appliances. The goal of Omnisphere is to mediate communications on behalf of the user in a proactive way. Many of our current applications are governed by the *pull* model in which the user requests information or initiates a communication stream (Web, e-mail etc). In this interactive mode of operation, the user is an active part and must initiate the process of information acquisition. In the *push* model, it is the source or the network infrastructure that proposes the information to the user in a similar way to the SIP protocol which invites the user to a communication session. Such a mode of operation is at the base of Omnisphere: it takes care of preparing communications and proposing them to the user. In this proactive way of operation, the only user intervention is to choose the right communication flow.

M. Conti et al. (Eds.): PWC 2003, LNCS 2775, pp. 505–510, 2003.

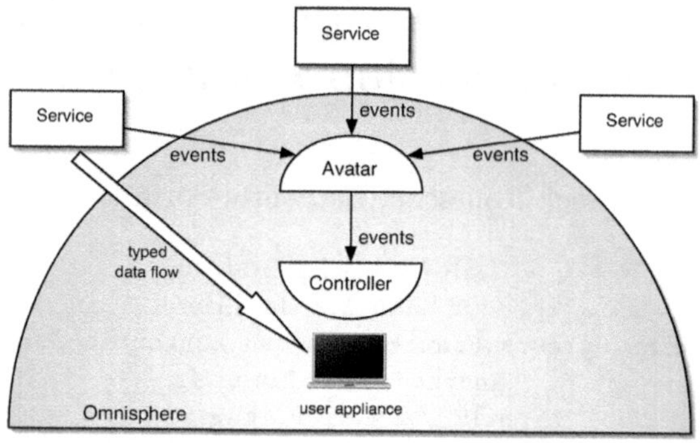

Fig. 1. Omnisphere architecture

For common applications we transform information sources that require user intervention into sources that notify the user about some events and then the user may decide to consume the proposed information or not (e.g. a NotifyMail service checks the user mailbox for new messages and notifies the user about important messages, the user may then read a message using a standard mailer and the POP protocol). The role of Omnisphere is to mediate notification events according to the user preferences, discover required services, configure them, and compose them into complex communication applications.

In the rest of the paper, we focus on the push driven service composition in Omnisphere. We consider that communication chains in a personal communication environment can be dynamically managed when all components follow the same paradigm – "everything-is-a-service": all basic services, sources, transcoders or filters, as well as user applications are viewed as services from the discovery point of view. Their descriptors can be found and used along with other information such as the user preferences to compose complex communication applications. Due to the space limitation we skip over the related work, obviously there are several projects related to our work.

The paper is organized as follows. Section 2 presents the architecture of Omnisphere. Section 3 describes our approach to service discovery and composition. Finally, we present conclusion and future work (Section 4).

2 Architecture of Omnisphere

Omnisphere is a communication and information universe surrounding wireless appliances (more detailed description can be found elsewhere [1]). It mediates various communication flows to relieve the user from any cumbersome details of finding the right service components, configuring, and composing them based on user preferences, device capabilities, and the current context. *Omnisphere* comprises the following elements:

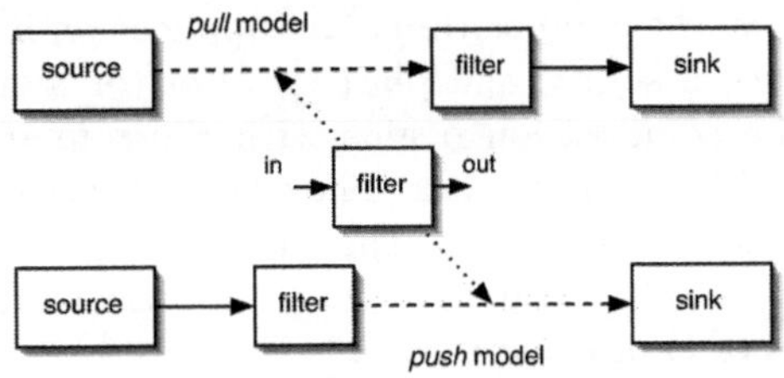

Fig. 2. Composition of basic services

- *Ambient services.* An ambient service is an abstract view of a communication component. It represents a high-level concept that allows us to construct complex services out of primitive ones by connecting them with *typed data flows.* An ambient service encapsulates some processing, provides a signaling interface for generating events and accepting a configuration commands, and publishes a service description.
- *Typed data flows.* A typed data flow is an abstract view of communication between ambient services. It allows to associate the type of data with a given flow.
- *User avatar.* An avatar represents the user in Omnisphere. Its role is to mediate event notifications concerning different data flows. It receives events from ambient services and acts according to the information on user preferences, the current context, and device capabilities to either notify the user or set up communications. It monitors the presence of user devices, manages services available at the given Omnisphere, and discovers services required for the user. An instance of the user avatar exists at each Omnisphere in which the user is present.
- *Omnisphere controller.* It runs on a user appliance and takes care of events sent to the user. When needed, it will configure and run applications used by the user to consume communication flows. When requested by the user, it will initiate the discovery process of available services in the present Omnisphere. Upon the arrival in a new place, it will register the device within a given Omnisphere and authenticate the user.

3 Service Discovery and Composition in Omnisphere

We can compose services in two ways. The first one is *pull driven*—a sink service looks for intermediate filter services and a source service to provide an end to end communication flow. An example of such a composition is RealPlayer that looks for a streaming source. The second way is *push driven*—it makes the network infrastructure the active part of the communication: a source service looks for a suitable sink to deliver a flow in a proactive manner, e.g. an e-mail notifier looks for a mailer on a PDA to inform the user about e-mails upon entering a personal communication environment. Figure 2 illustrates these approaches with two chains of services: the composition of an application in the pull model starts with a sink that finds a source and places a filter (or transcoder) if needed. In the push model, the composition begins with the source that finds a suitable sink with an optional filter in between. The graph of service composition may be more complex than the simple chains shown in the figure. Although we provide both composition methods, we believe that the push model corresponds better to the requirements of Omnisphere.

The user avatar is the central element in the operation of Omnisphere. It takes care of all the operations that should be performed in the network infrastructure on behalf of the user. In this way most of the operations are delegated into the network infrastructure and are automated as much as possible. When an appliance enters Omnisphere, it authenticates itself so that based on the User ID and Appliance ID, Omnisphere can retrieve the `User Preferences` and `Device Capabilities`. `User Preferences` define what are the most common needs of the user in terms of services and different types of data and `Device Capabilities` provide information about the characteristics of appliances (screen resolution, processor speed, network interface). It also uses `Context` that adds some location specific information such as the geographic location, available devices, type of environment—public or private, and all other information related to the close neighborhood of an appliance. It then creates an avatar that discovers relevant services using a service discovery protocol. The discovered services are matched with the required characteristics provided by `User Preferences`, `Device Capabilities`, and `Context`. They are then configured and activated. If needed the avatar contacts the Omnisphere controller on the user appliance to run the user applications that receive data flows. To take into account state changes, for example when an appliance becomes inactive, or a new appliance enters Omnisphere, the service bindings are re-evaluated.

Such a discovery process relieves appliances, which may have limited resources, from the operation that may consume scarce resources and may require the availability of different discovery protocols on the appliance. It may also save critical radio channel resources, because only selected information is provided to the appliance.

3.1 Example

We present below an example of e-mail notification. Alice wants to use a mailer such as Netscape or Eudora at the office. Instead of running them all the time to periodically check her mailbox via POP, she wants the mailer to be run upon the arrival of a new mail and configured to use ssh tunnels with the company POP and SMTP servers. When she leaves the office, she wants to receive e-mail notifications as SMS messages on her cellular phone for some urgent and important messages. When attending a meeting, she wants to be notified on her PDA to decide whether it is worthwhile to read a mail. Upon arrival at home, she wants to read the mail on the home PC by using the SMTP server of her ISP.

Let us start with `User Preferences` for Alice.

Communication	Attributs	Preferred User Applications
`Mail`	`domain=imag.fr` `transport=secure`	`laptop-netscape, pda-popup, gsm-sms`

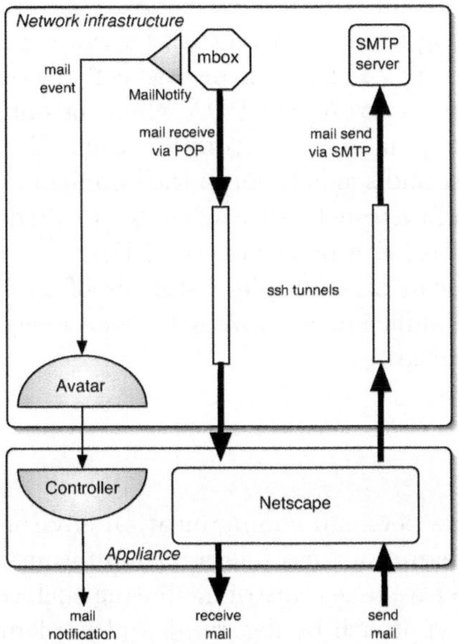

Fig. 3. Mail example

This information means that Alice wants to use her mailbox in the `imag.fr` domain and her preferred applications are the usual Natscape mailer on her laptop, a popup dialog window for deciding whether to read a mail, or a notifier that sends a SMS message to her cell phone. Each of these applications can be configured with some additional information such as the port number or the cell phone number.

The composition of ambient services is presented in Figure 3. A user mailbox exists on a server in `domain=imag.fr` and there is a service MailNotify able to check for a new mail of a given user. If Alice is in her office, Omnisphere looks for services that generate type `/mixed/e-mail:incoming` (we define a simple hierarchy of types similar to the DNS names). The query returns the addresses of incoming mail servers (POP, IMAP, HTTPS). Then, Omnisphere looks for a service that generates type `/mixed/e-mail:event`. In our example, the query returns the MailNotify service. A query for `/mixed/e-mail:outcoming` yields an outgoing local SMTP server.

User applications such as Netscape register with the service discovery protocol so that Omnisphere can find them. If several applications are discovered, in the case for example when Alice is in the office with her laptop and a PDA, Netscape is chosen as the preferred user application. At this point all the elements are discovered and Omnisphere can configure them and compose to form the complex Mail application.

When Alice goes to a meeting room, the avatar detects that she is not longer present in the office and stops the mailer. According to the preferences, it discovers a SMS notifier, configures it with the Alice's cell phone, and forwards notification events coming from the MailNotify service. In this way, Alice receives automatically an excerpt of the mail on her cell phone. Then, Omnishpere in the meeting room detects the Alice's PDA when she enters the room. It requests the migration of the avatar to the current location. The avatar stops forwarding mails to the cell phone and sends them to the Omnisphere controller on the PDA which pops up a window to ask Alice what to do with the incoming mail. She may decide to read it using a mailer on the PDA.

In general, there can be multiple instances of discovered services. In this case, we need to add sufficient attributes to `User Preferences` in order to help resolving such ambiguities.

4 Conclusion

We believe that future personal communication environments will be based on the *push driven* paradigm: sources of data push information towards the user and the network infrastructure takes care of mediating such communication flows. We experiment with this approach by designing and implementing Omnisphere. We have already prototyped the main elements of Omnisphere: the user avatar and the Omnisphere controller. The avatar uses the DNS-SD for service discovery—we have modified the Apple Rendezvous implementation of DNS-SD to take into account our definition of data types and wildcard service queries.

Our current work concerns seamless handover of the user between different places, the detection of the user movement, and implementing more complex scenarios such as *follow me video*—the user can continuously watch a video stream presented on different appliances when moving between different places.

References

1. F. Rousseau, J. Oprescu, L-S. Paun, and A. Duda. Omnisphere: a Personal Communication Environment. In *Proc. Thirty-Sixth Annual Hawaii International Conference on System Sciences (HICSS-36)*, Big Island, Hawaii, 2003.

Performance Analysis and Modelling of an OSA Gateway

Jens Andersson, Christian Nyberg and Maria Kihl

Department of Communication Systems
Lund Institute Of Technology, Sweden
Box 118 221 00 Lund
Telephone +46 46 222 91 58
Telefax +46 46 14 58 23
{jens.andersson,christian.nyberg,maria.kihl}@telecom.lth.se

Abstract. It is foreseen that you in the future should be able to use the same services independent of where you are positioned or which terminal that is used. The open service architectures provide these opportunities. Open Service Access (OSA) is an example of such an architecture and it is part of the specification delivered by 3GPP. This paper explains the OSA architecture and presents a model of an OSA gateway. Further, it discusses and proposes some feasible overload control mechanisms for the gateway. The behaviour of the mechanisms is investigated through simulation.

1 Introduction

During the last years there has been a change in service architectures towards so called open service architectures. One of the first open service architectures that was successfully developed is PARLAY specified by the Parlay group. In Parlay a set of standard application interfaces (APIs) is defined. These will enable applications residing outside the network to access and control network resources.

Open Service Access (OSA) is the service architecture that is proposed for the 3G networks. OSA is based on the concept of PARLAY, and is developed by the 3GPP [4]. It is foreseen that there will be a great demand of services and in order to respond to this demand the pace of the development has to speed up.

One common problem for all service architectures is what actions to take if the control nodes become overloaded. Overloaded nodes leads to long waiting times for service. If the waiting times get too long, customers will abandon the request for service and perhaps make a retry. These abandoned requests consume valuable processing time. In the worst case, an overloaded node will only be processing abandoned requests for service. Thereby the need of an overload control mechanism is obvious.

Overload control has been around for some decades. In Wildling [7] the protection of telephone exchanges is discussed. One paper on overload control in IN is Kihl [8].

Very few papers have been published on load issues for open service architectures. However, the performance of a Parlay gateway is analysed in Melen [9].

In this paper, we investigate overload control mechanisms for the OSA service architecture. We propose a queuing model for the most critical nodes in the architecture and investigate different ways of measuring load and rejecting customers.

M. Conti et al. (Eds.): PWC 2003, LNCS 2775, pp. 511–517, 2003.

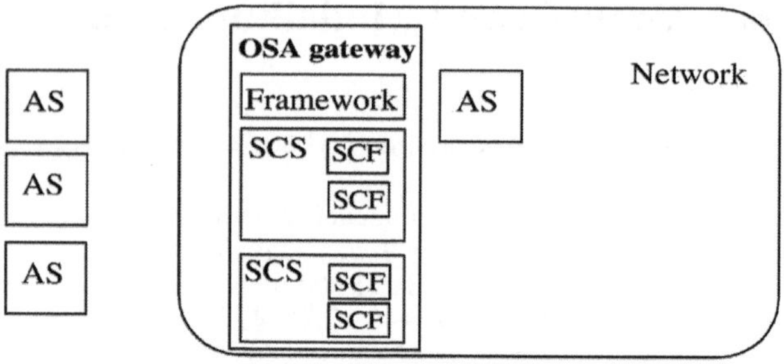

Fig. 1. The architectural picture of an OSA architecture

The paper is organized as follows: in section 2 a description of OSA is given. In 3 the simulation model is presented. The proposals for overload control mechanisms can be found i section 4 followed by the results and discussion in section 5. Finally we draw some conclusions in section 6.

2 Open Service Access (OSA)

From the beginning OSA was an acronym for Open Service Architecture, but it has been re-termed to Open Service Access. OSA offers an increased security and integrity enabling the operators to open up their networks to independent software developers and service providers. Thereby the number of feasible service providers has increased.

2.1 Architecture

OSA consists of three parts, the Application Servers (AS:s), the Service Capability Servers (SCS:s), and the Framework. Fig. 1. shows one possible configuration of an OSA architecture. The part referred to as the OSA gateway can be built on several physical entities. In Fig. 1. the Framework and both the SCS:s constitute the OSA gateway.

The AS:s host the applications. An application is usually triggered by the dialling of a special number or by some kind of HTTP request. The AS:s can be physically positioned inside or outside the network they are communicating with. An example of a typical OSA application in a 3G network is an “application initiated call” proposed in [3]. The sequence diagram of this service is shown in Fig. 2.

In an OSA architecture there can be one or several SCS:s, see Stretch [5]. The SCS provides network functionality to the applications via one or several SCF:s. An SCF consists of several narrow functions, which together makes it possible to utilize the network capability. Examples of SCF:s are Call Control, Mobility and Charging SCF. For example the Call Control SCF provides functionality to establish different kinds of calls to a mobile user. The Framework can be seen as a separate SCS providing the applications with basic mechanisms, like authentication before accessing the network functionalities or the possibility to find out which SCF:s that are provided by the SCS:s. It is important to notice that there is always exactly one framework in an OSA gateway.

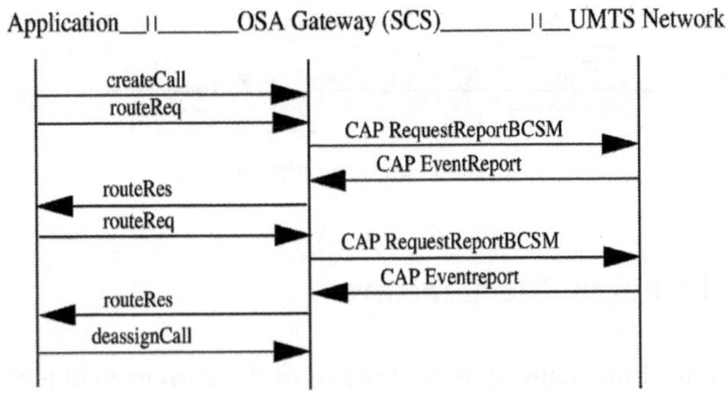

Fig. 2. Message sequence diagram for an application initiated call

2.2 Overload Control in OSA

In an OSA architecture the AS:s and the SCS:s are especially sensitive to overload. It is possible for both the AS:s and for the SCS:s to have overload control.

The overload related functionality is managed by the Framework as described in the specifications [2]. Information about the load condition in the SCS:s and the AS:s can be exchanged between the AS:s and the Framework. This gives the opportunity to control the load either from the AS or from the Framework.

There are three load levels, 0, 1 and 2 corresponding to normal load, overload and severe overload respectively. Nothing is said about how the load levels should be set or what actions they should cause. The actions should be defined in the load management policy, which is created via contract writing.

3 Simulation Model

We have developed a model consisting of one AS and a gateway containing one SCS and a Framework, Fig. 3. Each new application call is authenticated by the Framework. The execution time for this is assumed to be negligible, thus the gateway can be modelled as only the SCS. In the AS the application described in Fig. 2. is implemented. Of course in a real system there will be many applications with different behaviour. However, one is enough to create an overload situation and to evaluate the behaviour of an OSA gateway. The arrivals of the application calls are modelled as a Poisson process with the rate calls each second. The SCS is modelled as a one server queue with capacity of serving 100 application calls per second. The capacity of the AS is dimensioned so the overload will appear in the SCS and thereby the AS can be seen as a delay.

In Fig. 2. it is shown that each service has to execute in the network twice. The first time is modelled as a delay of 10 ms and the second is modelled as an exponentially distributed delay with mean 2 s. The other service times are set as follows: If a message in the SCS results in a new message the execution time is 2 ms, else 1 ms. Each delay in the AS in Fig. 3. is 1 ms.

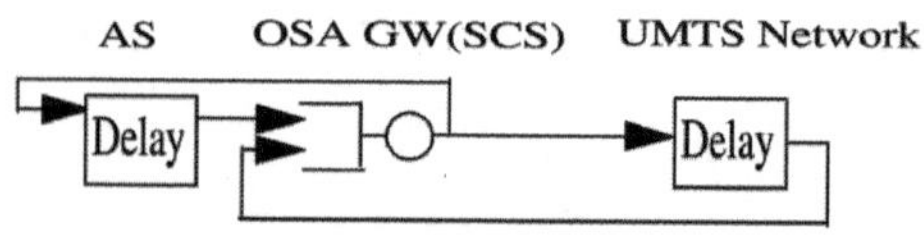

Fig. 3. The simulation model

4 Overload Control Mechanisms

An overload control mechanism should measure the current load and reject new calls if necessary. In our model the Framework rejects new application calls.

When the gateway is overloaded the waiting times get too long. In [6] a maximal delay of 100 ms is proposed and that value will be used here. If a completed application call has had a mean delay in the OSA gateway longer than 100 ms, it is said to be an expired call.

The main objectives for the overload control mechanism in this paper are to maximize the throughput and minimize the number of expired calls. To do this the number of calls in the gateway should fluctuate as little as possible so that the server is kept busy as much of the time as possible at the same time as the queue length should be kept short.

4.1 Measurement Methods

Two ways of measuring the load, A and B, are proposed below. In both cases, the measured load level at an arrival is compared to a threshold. If the measured load level is above the threshold at five consecutive arrivals or departures the load level is increased (if smaller than 2). If it is below the threshold at five consecutive arrivals or departures it is decreased (if it is larger than 0).

Method A measures the total number of application calls in the SCS, network and AS. Method B measures the number of calls in just the SCS. Calls in the network and AS will sooner or later come back to the SCS and demand processing. Method A takes this into account, B does not.

To estimate the threshold values when A is used we set T_{tot}=E(total time in system for an application call), T_{scs}=E(total time in the SCS) and x_{scs}=E(service time in the SCS for each application call). If is the threshold it must satisfy

$$\left(\hat{A} \cdot \frac{T_{scs}}{T_{tot}}\right) = \frac{100}{x_{scs}} \Rightarrow \hat{A} = \frac{100 \cdot T_{tot}}{x_{scs} \cdot T_{scs}} \tag{1}$$

to satisfy the requirement so that calls not are expired.

When method B is used the threshold can be calculated as

$$\left(\hat{B} = \frac{100}{x_{scs}}\right) \tag{2}$$

Table 1. Threshold parameters used in the simulations

Measurement method	Static method load level 1	Static method load level 2	Dynamic method load level 1	Dynamic method load level 2
Method A	190	210	190	200
Method B	40	45	30	35

4.2 Rejecting Methods

This paper proposes two methods for rejecting calls, the static method and the dynamic method. Both methods use Percent blocking [1] where R_f=P(a call is rejected).

The static method works like this: when load level is 0, R_f is 0, when load level is 1, R_f is set to 0.5 and when load level is 2, R_f is set to 1.

The dynamic method tries to stabilize the measured load just below the threshold. When load level 1 is reached R_f is increased by 0,1. If load level 1 remains after X seconds, R_f is increased one more time etc. If load level 2 is reached, R_f is increased by 0.4 in the same way. If the load level is 0, R_f is decreased by 0,1 every X:th second. Of course R_f must always be in the interval [0, 1].

In our simulations X is 25*E(total execution time in the SCS for one application call).

5 Results and Discussion

The rejecting and measurement methods will be compared in this section. The comparisons will be done with both constant and varying average arrival rates, λ. The threshold values are chosen such that the fraction of expired services never exceeds 0.5%. The used threshold values are shown in Table 1.

5.1 Comparisons of Measurement Methods

In the steady state case when we let λ keep the same value during a long interval method B gives a better throughput. We also conclude that method A is more sensitive to changes of the threshold values. If the thresholds are lowered to decrease the rate of expired calls there is a sharp decrease of the throughput when method A is used.

However, steady state arrival rates are not very realistic. It seems more probable that the value of λ is rather bursty and shifts at random times. Fig. 4. shows the simulation results when λ is randomly varying between the discrete values 0, 50, 100 and 150 calls per second and the times between changes of λ are exponentially distributed with mean 2.0 seconds. The upper plot shows how λ is varying over 200 seconds with the mean 81.6 calls per second. In the plots it can be discerned that method B is better than method A, because of the smaller variations in the number of application calls in the SCS. The mean throughput values corresponding to each of the graphs starting from the upper are 60.3, 57.6, 63.5 and 60.4 respectively.

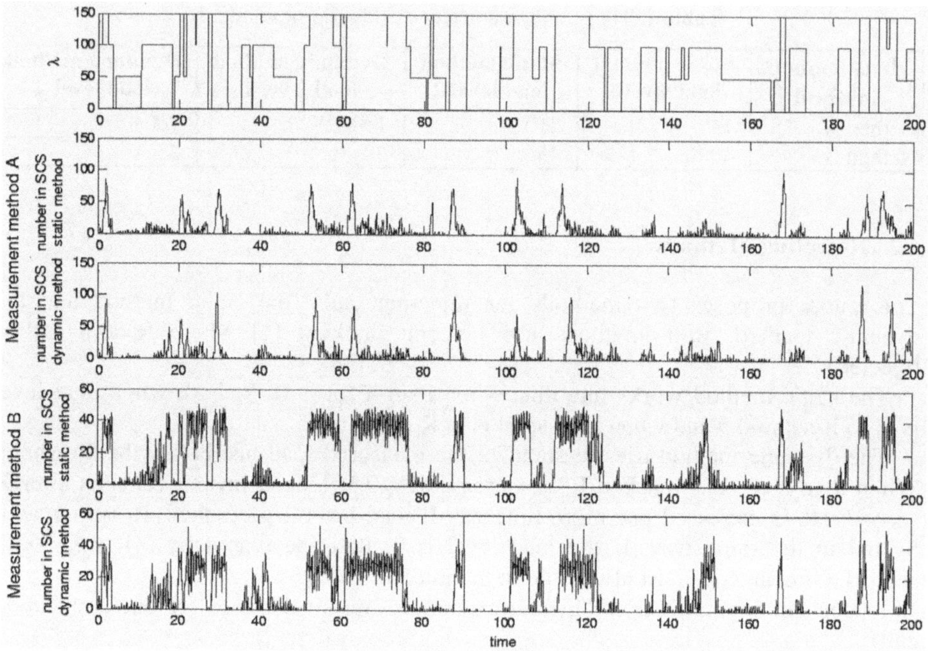

Fig. 4. The number of services in the SCS is plotted as a function of time when is varying as the top plot. In each plot different measurement method or rejecting method is used

5.2 Comparisons of Rejecting Methods

In the steady state case the throughputs are about the same irrespective of which rejecting method that is used. However, when λ shifts it can be discerned that the static method has a better behaviour from transients point of view in Fig. 4. When there is a change from $\lambda = 0$ to $\lambda = 150$ it can be discerned how the dynamic method has a slow reaction. This means that the static method fulfils the requirements just as well as the dynamic method and it seems to have a better behaviour concerning transients.

6 Conclusions

In the OSA architecture it has not been defined how to measure the load or how to react on an overload situation. In this paper it is shown that the throughput is larger when the number of calls in the SCS is used as a measure of the load than when the total number of active calls are used as a measure. We have also proposed two rejection methods of which the static method seems to have the best behaviour.

References

1. Berger, A.: "Comparsion of Call gapping and Percent blocking for overload control in distributed switching systems and telecommunications networks", IEEE Transactions on Communications, vol. 39, 1991, pp 407–414
2. ETSI standard 201 915-3 v1.3.1 "OSA API; Part 3: Framework", july 2002
3. ETSI standard 201 915-4 v1.3.1 "OSA API; Part 4: Call Control SCF", july 2002
4. The 3GPP home page, "www.3gpp.org"
5. Stretch, R.M.: B T Technol J. The OSA API and other related issues, Vol. 19 No 1, Jan 2001, pp 80–87
6. Eurescom Technical Information, "Non-functional aspects and requirements related to Parlay/OSA products", june 2002
7. Wildling, K., T. Karlstedt: "Call Handling and Control of Processor Load in SPC-systems", ITC 9, Torremolinos 1979
8. Kihl, M, Nyberg, C: "Investigation of overload control algorithms for SCPs in the intelligent network", Communications IEE Proceedings, vol. 144, pp 419–423, 1997
9. Melen, R., Moiso, C., Tognon, S.: Performance evaluation of an Parlay gateway, "http:// exp. telecomitalialab.com/pdf/06-MOISO4.pdf", 2001

Experimental Results of 802.11a Wireless LAN System over Optical Fiber

I. Haroun, F. Gouin, L. Boucher, and L. Bouchard

Communications Research Centre Canada (CRC),
Ottawa, Ontario, Canada K2H 8S2
{ibrahim.haroun,francois.gouin,luc.boucher,luc.bouchard}@crc.ca

Abstract. In this paper, the rationale of using wireless over optical fiber to support broadband wireless deployments is presented and initial measured results derived from a proof-of-concept experiment are described. The primary objective of this project is to investigate the technical feasibility of using an optical/wireless infrastructure capable of delivering broadband multi-media traffic to and from subscribers. In such a scheme, the optical fibers are used to route the broadband optically modulated signals to distribution centers, where the RF signals get detected and wirelessly transmitted to each remote user. An experimental test setup using an 802.11a access point over optical fiber was developed in order to evaluate the potential of this technology merge.

1 Introduction

The initial objective of this project was to explore strategies to provide high capacity mobile broadband wireless access for the future, enabling the support of a wide range of applications under different propagation and mobility environments. Some of the limitations involved in providing broadband wireless access in a mobile environment are described in the background section of this paper. The rationale of using wireless over optical fiber as part of the solution to implement mobile broadband communications is then presented. In section 3, a possible architecture for wireless over optical fiber transmission is described. In order to evaluate the potential of the solution presented (the RF and optical fiber technologies merge), an experimental proof-of-concept setup using an 802.11a access point over optical fiber was put in place and measurements were conducted. This is reported in section 4. Conclusions and future work are presented in section 5.

2 Background

2.1 Limitations in Providing Broadband Wireless Access

One of the key issues in broadband wireless access is the ability to transmit high peak data rates to individual terminals. Apart from the obvious requirement of having enough spectrum space, there are a few fundamental limitations on how high the data

M. Conti et al. (Eds.): PWC 2003, LNCS 2775, pp. 518–525, 2003.

rate can go over wireless links. Factors influencing the data rate include RF power, signal distortion, mobility, etc. [1-2]:

- RF Power: Reliable data transmission requires a certain minimum amount of energy for each binary bit of information. Then, transmitting data faster requires higher power. Some of the methods that can be employed to increase the average received power include: transmitting more power, reducing the distance between the transmitter and receiver and/or using higher antenna gain.
- Signal Distortion: In a multipath radio propagation environment, the receiver intercepts several copies of each transmitted symbol. As the symbol transmission rate increases, the symbol duration gets shorter and consequently the different copies of a symbol overlap with other symbol causing the well-known phenomenon of Inter-Symbol-Interference (ISI). The ISI problem cannot usually be corrected by increasing the power. It can be corrected to some extent by employing equalization techniques or techniques such as spread spectrum (e.g. CDMA -Code Division Multiple Access) or frequency division multiplexing (e.g. OFDM - Orthogonal Frequency Division Multiplexing), each technique having its limitations.
- Mobility: Another physical problem is the ability of the system to support terminal mobility. If terminals are allowed to move from location to location, then the system must be able to provide coverage over the service area.

The key requirements to providing broadband wireless access in urban areas include the needs to support: very high density of terminals, high data rate per unit area, various modes (e.g. fixed, nomadic and mobile) and the requirement to be economically deployable on a large scale [2,3].

2.2 Why RF over Optical Fiber?

A broadband wireless solution in urban areas with wide range of radio services must support fixed, nomadic and mobile terminals. In order to maintain high data rates everywhere and be spectrum efficient, the radio cells must be small [3]. The small cells translate into a very large number of base stations. Hence, for a successful deployment, it is necessary to have economical and effective transport and distribution schemes. With such a large number of base stations, the fiber distribution loop becomes an attractive solution, with the following supporting justifications [2,3]:

- One of the main justifications is to combat the physical limitations of large cells. Having small cells allows one to: (1) shorten the distance between bases and mobiles; (2) reduce the signal distortion by reducing the range of delay spread and scattering and (3) provide an effective means of deploying site-specific radio coverage and also (4) facilitate effective growth strategies.
- With a large number of cells (or base station antennas), the wireless over optical fiber concept solves two problems: (1) it reduces the cost of base stations since many functions can be removed from the base station and placed at some central processing locations. The huge bandwidth capacity of the fiber is used to communicate these functions between the processing units and the base stations

and (2) the use of fiber greatly simplifies the communications among base stations.

- Since the wireless over optical fiber concept allows for removing most of the intelligence from the base stations and placing them at a few central points, better management of the radio resources is possible. Algorithms can be developed to match the radio resources to cell requirements and to adaptively change the radio resource allocations among adjacent cells.
- One of the key advantages of the "wireless over optical fiber" concept is the fact that fiber optic local distribution is gaining considerable momentum. Cable and other wired access operators are cooperating to extend the fiber optic broadband capacity as close as possible to the end user. The concept is termed: "fiber-to-the-curb". If and when this happens, the wireless operators could use advantageously this metropolitan fiber infrastructure to support large-scale deployments for mobile broadband access.

The use of wireless over fiber is not a new concept. Since the first investigations of optical-RF interactions several decades ago, the applications now cover areas such as antenna remoting, optically-controlled phased array antenna systems, and microwave signal processing. However, the optical and optoelectronic components used for optically distributing RF signals have in the past been very expensive and, as a result, the technology has only been used in very limited deployments. Advances in fiber optic technology now provide more affordable modulating laser devices and optical detectors.

These developments lead to the possibility of utilizing wireless over optical fiber to carry wireless LAN (WLAN) signals, therefore enabling the delivery of broadband multi-media traffic to and from subscribers on a large scale [3,4]. An experimental test setup using 802.11a access points over optical fiber was developed in order to demonstrate the potential of this technology merge. The test set-up and the measured results are described in Section 4.

3 Wireless over Optical Fiber Architecture

Figure 1 illustrates the infrastructure of a single cell of a WLAN over optical fiber system. In this scheme, the base station (also called access point) transceiver of each cell is equipped with an Electric-to-Optic converter (E/O) in the receive path, and an Optic-to-Electric converter (O/E) in the transmit path. A possible architecture for multiple base stations using wireless over optical fiber transmissions is illustrated in Figure 2. All the digital signal processing functions such as modulation and demodulation, and frequency assignment are performed at the central office.

In the downstream path (from the access point to the customer), an optical detector is required to convert the optical signal to an RF signal. The RF signal will need to be amplified in order to support the needed EIRP (Effective Isotropic Radiated Power). For example, the 802.11a standard [5,6] for North America uses three power levels for its 300 MHz bandwidth, the first 100 MHz (5.15–5.25 GHz) is restricted to a maximum output power of 50 mW, the second 100 MHz (5.25–5.35 GHz) is restricted to maximum power of 250 mW, and the third 100 MHz (5.725–5.825 GHz) is restricted to a maximum output power of 1 Watt for outdoor applications. The

802.11a standard for North America also specifies channel spacing of 20 MHz and guard band spacing of 30 MHz at the band edges in order to meet FCC (US) and IC (Canada) spectral mask requirements.

One of the main advantages of 802.11a systems is its robustness against multi-path fading and increased spectrum efficiency because of the use of OFDM modulation [7,8]. WLAN 802.11a systems currently provide high-speed Internet access (up to 54 Mbps) for indoor environments such as public buildings, shopping malls, airports, and similar applications. These systems can be modified to support larger coverage outdoor applications.

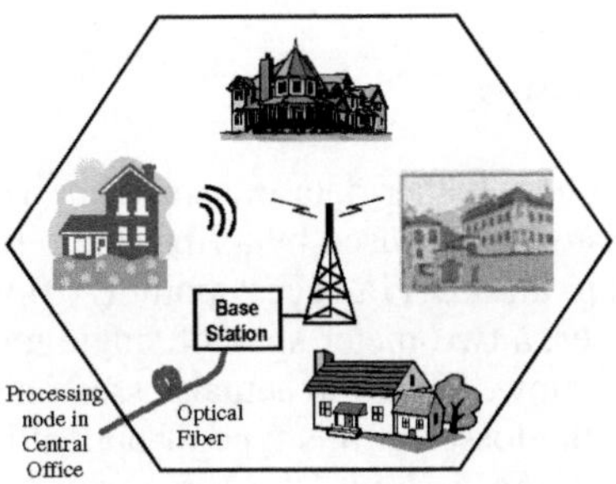

Fig. 1. Single Cell Architecture

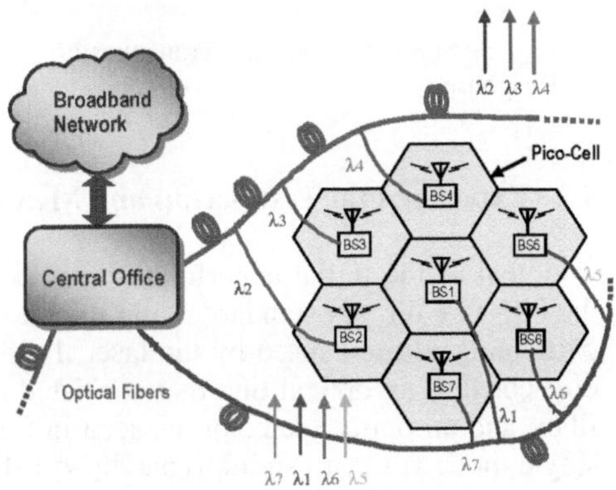

Fig. 2. Possible Architecture for Multiple Cells

4 Proof-of-Concept

4.1 Optoelectronic Components Selection

The 802.11a WLAN standard operates in the frequency range between 5.150 and 5.825 GHz. A few sources and photo-receivers that were relatively easily available and capable of operating over that frequency range were evaluated. These fiber optic links were first evaluated in terms of transmission as a function of frequency. A network analyzer was used for these measurements, as shown in Figure 3. The optical link transmission response of one of the source/photo-receiver selected for this experimentation (the HP834 series) is shown in figure 4. The transmission of the optical link was measured over the range from 400 MHz to 10 GHz. As seen, the RF transmission is attenuated by approximately 8.5 dB at 5 GHz and decreases smoothly to -10 dB at 6 GHz, the range of interest for 802.11a. Of course, the variation in the RF transmission is even smaller over any given 802.11a RF channel, which is only 20 MHz wide. The losses are mainly due to the conversion from RF to optical and back to RF.

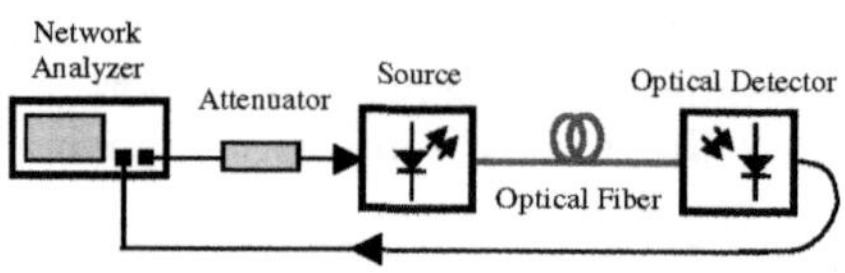

Fig. 3. Setup to Measure Transmission Response

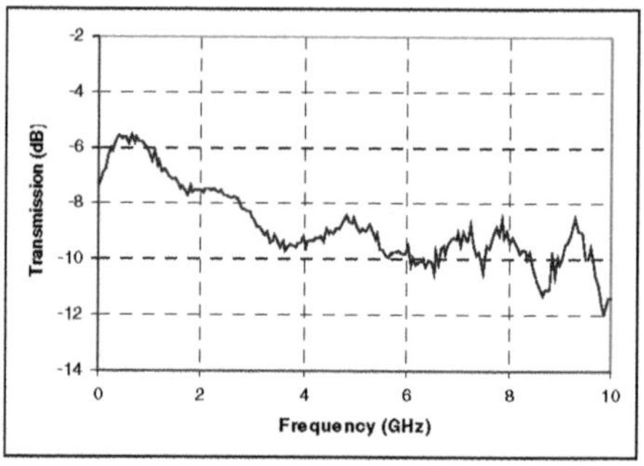

Fig. 4. Optical Link Transmission Response

4.2 Experimental Test Set-up and Measured Results

The goal of the initial experiments was to determine whether data transmission using the 802.11a WLAN standard could tolerate the distortions caused by a fiber optic link, particularly those caused by the laser. In this setup, an 802.11a access point (AP) was connected to an optical link that consisted of a laser, a two-meter span of single mode fiber, and an optical detector, as seen in figure 5 below. Since the actual losses on the single mode fiber are small, typically < 1 dB/km, the losses of this type of optical link are primarily due to conversion at its end points. As a result, the two-meter span should give a good indication of some of the degradations that can be expected from the optical conversion.

The laser used was an "HP 83403C Lightwave source" emitting at 1.5 μm and the detector was an "HP 83411D Lightwave receiver". Both are specified as having a bandwidth of 300 MHz to 6 GHz. The 802.11a access point was a D-Link DWL-5000AP.

In order to establish a baseline for this experiment, the RF spectrum of the signal transmitted from the access point towards the optical link was first measured. The access point (AP) was configured for 54 Mbps transmission rate and 5.28 GHz channel frequency. As shown in figure 5, the output of the optical detector was connected to a spectrum analyzer to monitor the 802.11a signal at the output of the access point (1) and after transmission over the fiber (2). Figure 6 compares these 2 points, i.e. the measured spectrum at the input and output of the 2-meter optical link.

From the measured results, one observes a drop in the signal level by about 8.5 dB in the central part of the 20 MHz channel. This drop is in agreement with the response of the HP 834 series optical link shown previously in Figure 4. One can also observe that there is no significant degradation in the shape of the channel response.

Figure 7 shows the RF spectrum at ½ meter distance away from the transmitting antenna (test set-up shown in figure 5 (3)). The measured results indicate a drop in the signal level due to the propagation path-loss, but the shape of the channel spectrum did not degrade. With increasing distance between the transmitting and receiving antennas, the shape of the RF spectrum starts to degrade. In order to compensate for the optical link losses and to maintain the system required BER, an RF amplifier will be required to boost the signal level at the output of the optical link.

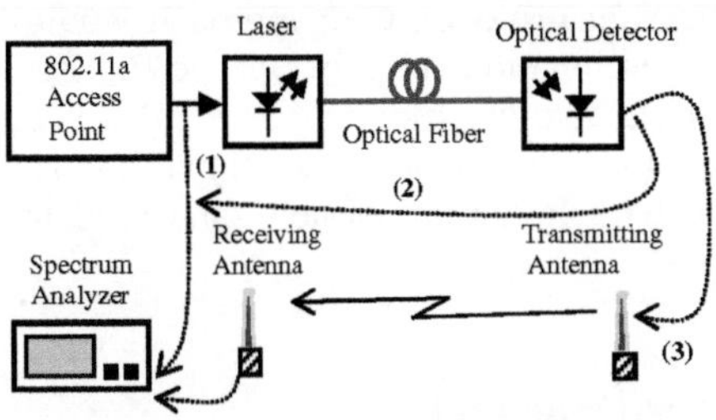

Fig. 5. 802.11a over Optical Fiber Test Setup

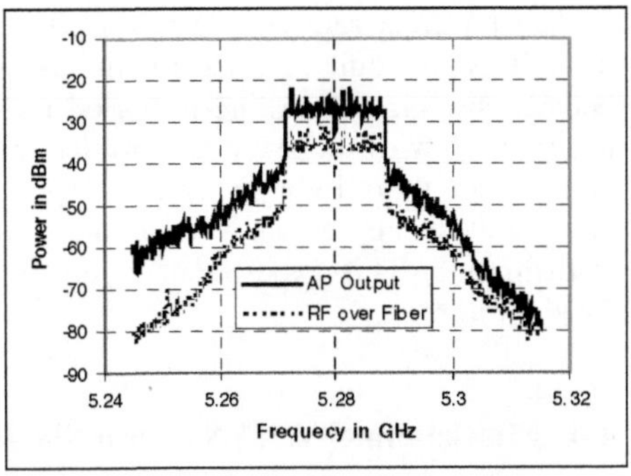

Fig. 6. RF Spectrum before and after Transmission over the Optical Link

Finally, the spectrum analyzer shown in the test set-up of Figure 5 (3) was replaced with a laptop computer equipped with an 802.11a card, in order to establish a link with another laptop computer at a remote distance. Data files transfer between the access point and the remote laptop computer were established and successful.

4.3 Bi-directional Optical Fiber Link

Once the successful transmission of 802.11a over optical fiber was demonstrated, another optical-link was added in the reverse direction. A Miteq MDD optical link was used for the up-stream link because the source includes a pre-amplifier which amplifies the received RF signal before modulating the laser thereby improving the signal-to-noise ratio. The down-stream link used the HP 834 series as described before. Using this test set up, successful transfer of data files to and from the access point, via optical links was demonstrated. This experiment demonstrated that the optical links were able to successfully transmit and receive WLAN 802.11a signals.

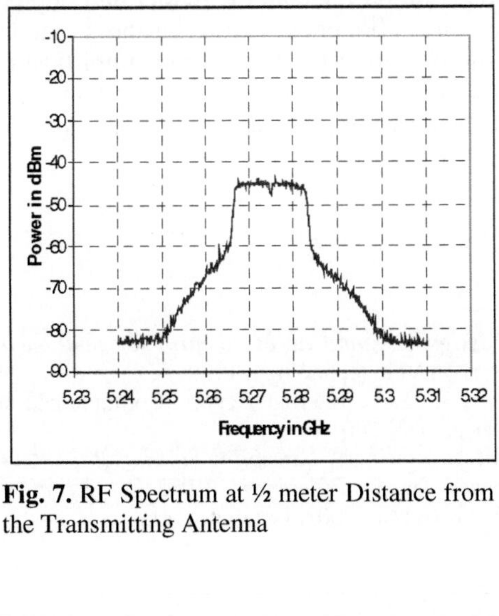

Fig. 7. RF Spectrum at ½ meter Distance from the Transmitting Antenna

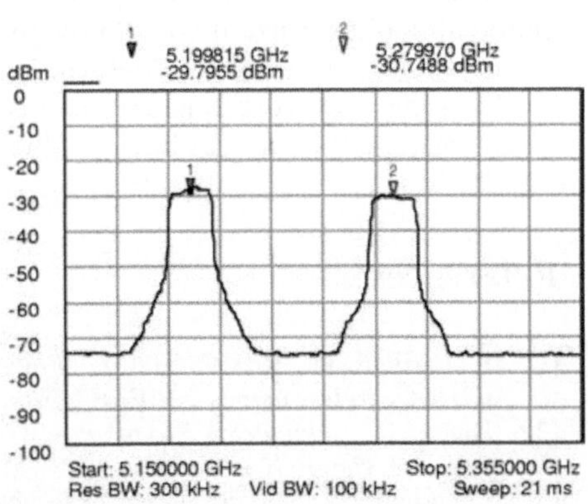

Fig. 8. Two-OFDM Channels – RF Spectrum at the Output of the Optical Link

The bit-error-rate (BER) is typically used to characterize the quality of a digital link. It is difficult to measure the BER directly when using an access point (AP) because the error handling is buried in the transfer protocol. One approach would be to do away with the AP and cardbus adapter combination (to generate the RF signal) and replace them by a vector signal generator capable of producing an OFDM signal. A vector spectrum analyzer could then be used to display the signal after the transmission. This experiment would further characterize this system, and is planned for the future.

4.4 Multichannel WLAN over a Single Fiber-Optic Link

In order to determine that the link is free of intermodulation products, the impact of transmitting more than one RF channel over a single optical wavelength was investigated. In this experiment, one channel was generated by an 802.11a access point (D-Link DWL-5000AP) and the other channel was obtained from an Agilent E4438C Vector Signal Generator (using the Agilent 802.11a signal studio software). The two channels were combined and applied to an optical link (Miteq MDD). The output of the optical link was fed to a spectrum analyzer to investigate any inter-modulation that might occur due to carrying multiple channels. The measured spectrum at the output of the optical link is shown in Figure 8. As seen the resulting spectrum showed no apparent inter-modulation distortion when two OFDM signals were carried over the optical link.

5 Conclusions and Future Work

Initial measured results using 802.11a WLAN over optical fiber demonstrates that the 802.11a signals can be transmitted over optical fiber without causing undue degradation to the signal characteristics. Merging of the optical fiber and the WLAN (802.11a) technologies could provide a good solution for both the indoor and outdoor broadband access, while improving spectrum efficiency and robustness against multipath fading. Also, it could provide a significant cost reduction and simplification to the wireless base-station deployments for future generation wireless systems such as 4G. Additional work is needed and is planned to complement these results, in order to demonstrate the full feasibility of this approach.

References

[1] P. Larijani and R.H.M. Hafez, "*Power assignment and access control for multimedia CDMA wireless networks*" IEEE transactions on Veh. technology, 1999.

[2] Hafez, R., "*Universal Broadband Wireless Access*", Internal Report, Communications Research Centre (CRC), Ottawa, Canada, December 2001.

[3] I. Haroun, G. Chan, R. Hafez, L. Bouchard, L. Boucher, "*Feasibility Study of Radio Over Optical Transmission Systems for Increased System Efficiency*", IEEE Int. Symp. on Electromagnetic Compatibility EMC 2003, 11-16 May 2003, Istanbul, Turkey.

[4] Al-Raweshidy, H., Komaki, S., "*Radio over Fiber Technologies for Mobile Communications Netw*orks", Norwood, MA, Artech House, 2002.
[5] http://standards.ieee.org/getieee802/
[6] Bob O'Hara, Al Petrick "*The IEEE 802.11 Handbook: a Designer's Companion*", IEEE Press, 1999.
[7] Heiskala, J., Terry, J., "*OFDM Wireless LANs: A Theoretical and Practical Guide* ", SAMS Publishing, 2002.
[8] Van Nee, R., Prasad, J., "*OFDM For Wireless Multimedia Communications*", Artech House Publishing, 2000.

Last Interaction Based Paging in Mobile Cellular Networks

I.Z. Koukoutsidis and M.E. Theologou

National Technical University of Athens,
School of Electrical and Computer Engineering,
9 Heroon Polytechneiou st., Zographou 157 73, Athens, Greece
koukou@telecom.ntua.gr

Abstract. In a mobile network, the paging operation is used to indicate the subscriber's position in order to establish a connection with another user, calling from a fixed or mobile environment. The currently employed scheme, blanket paging, incurs a significant cost in radio bandwidth utilization and is cost-effective only in small networks. In this paper, we introduce a new strategy that combines an optimal partitioning of the location area with a model to predict user movements, based on the additional information of the cell where the last interaction between the network and the terminal took place. We study the performance of the strategy under delay bounds and different mobility and call arrival characteristics. It is found that the new scheme further minimizes signaling costs and enhances previous sequential paging algorithms.

1 Introduction

Tracking the location of a mobile terminal (MT) is one of the fundamental operations in a wireless system. Current mobile networks implement this operation by dividing the system coverage area into geographical location areas (LAs). Each LA consists of a number of cells whose base stations are connected to the same service switching center (Mobile Switching Center, MSC). The MSC is itself connected to the telephone network (PSTN). The mobile network is always aware of the LA the user currently resides in. Upon an LA crossing, it is the responsibility of an MT to initiate a location update (LU) or registration procedure to inform the network of the change and update the appropriate databases [1]. When an incoming call or messaging service arrives for a specific subscriber, the network must determine the exact cell where he currently resides in order to deliver the service. For this, a search is conducted in all the cells of the registered location area, during which the network sends polling messages and awaits for a response from the right cell. The process is called *paging* (PG) and is shown graphically in Fig. 1.

This scheme has been eloquently named *blanket paging* because it covers every possible cell where the terminal might be located. The location update and paging procedures both define a *location management* (LM) scheme. The LM scheme described above is simple and easy to implement, however it is not cost-effective, especially under heavy traffic conditions and high user densities. A significant amount of bandwidth and processing power (both at the air-interface and the fixed

M. Conti et al. (Eds.): PWC 2003, LNCS 2775, pp. 526–540, 2003.

network) is consumed for the location updates, and unnecessary signaling load is generated for paging all cells in a location area.

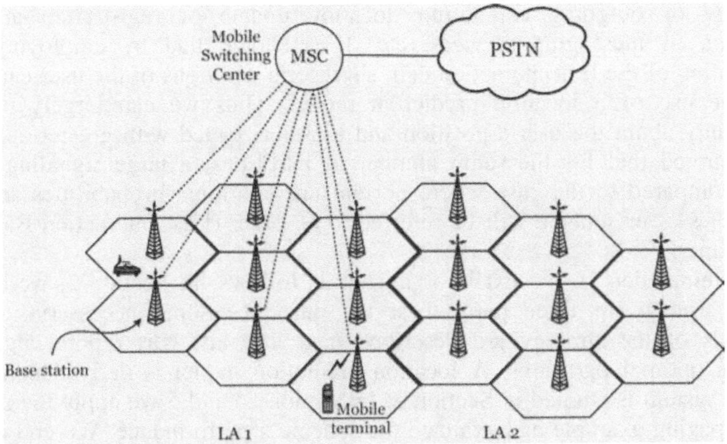

Fig. 1. Paging procedure in a mobile cellular network (the figure shown here is simplified, as other components involved with the paging procedure are omitted, such as the SS7 network and HLR/VLR databases. Additionally, the BSC/RNC is omitted, since its primary function is to maintain the radio resources of its base stations and is not directly involved in paging)

In general, the objective to minimize the total signaling cost of the location update and paging procedures constitutes a complex optimization problem: "Given the system area size and shape, the mobility behavior of a specific terminal and the incoming call characteristics, determine an optimal LA design and corresponding LU and PG mechanisms such that the total signaling cost incurred by location management procedures is minimized, and the delay for finding a mobile user is kept under a reasonable constraint". In [2], it was shown that the problem in general is *NP*-complete, as it is reduced to a *min-cut clustering* problem. In this way, an exact solution is amenable only for small networks. A heuristic algorithm applied to the problem can be found in [3], where a sub-optimal LA design is acquired by initially merging cells to groups by means of a greedy method and then exchanging cells between the groups until no further reduction in the signaling cost can be attained.

Assuming that a favorable partition of the coverage area into location areas is acquired, we can further reduce total costs by employing a different paging strategy. In a foundation work in [4], Rose and Yates formally showed the intuitive idea that polling groups of cells sequentially in order of decreasing location probabilities minimizes paging cost. The penalty paid for reducing the number of paging messages is the increase in the average delay for finding a mobile user, which is proportional to the number of groups the location area is divided into. The strategy described above is also referred to as *selective paging* in [5]. A similar algorithm presented in [6] is called *multi-step paging*, where each group of cells is called a paging area (PA). Optimal sequential paging is extensively studied in [7], where dynamic programming algorithms for partitioning network areas are derived.

Our work is motivated by the ideas in [4,7]. The aim is also to derive optimal partitions to improve the paging mechanism, but this time using conditional location probabilities from the time of the last interaction with the network to the next incoming call arrival. A *last interaction* (LI) is defined as any transaction that can provide incidental information regarding the cell position of a mobile user, such as incoming or outgoing call setup, location update or registration at switch-on, execution of messaging services, etc. It is shown that by employing the extra information of the last interaction cell, short-term positions of the user can be derived with the use of a location prediction model. Thus we can largely diminish the uncertainty about the user's position and a user is paged with greater accuracy. It is demonstrated that for the same number of partitions, a large signaling saving can occur compared to the case where normal unconditional probabilities are used. For conciseness, our scheme will be referred to as LIBP (Last Interaction Based Paging) in the paper.

The remainder of the text is organized as follows. In Section 2, we describe the paging strategy in three parts: first we introduce some necessary notation; the essentials of the strategy are described next; the last part reports algorithms for deriving optimal partitions. A location prediction model is derived mathematically and the system is studied in Section 3. In Sections 4 and 5 we apply the analysis to a simple paging example and evaluate the scheme's performance. We end in Section 6 with a brief review of the main points of our strategy and the major conclusion.

2 Paging Strategy

2.1 Preliminary Definitions and Notation

Let us denote by $S=\{1,2,\ldots,n\}$ the set of n cells in a location area, such that a mobile is guaranteed to be in one of those cells at the time of a call arrival. We suppose that a model for conditional user location can be acquired either mathematically or by gathering statistics. The model describes a time-varying conditional distribution $p(i\,|\,u,t)$, which gives the probability of a user being located in cell i at time t, provided he was in cell u at $t=0$. The probability takes distinct values for every $i\in S$, such that $\sum_{i\in S} p(i\,|\,u,t)=1$. By ordering these conditional probabilities in decreasing values, we can apply an algorithm to determine the optimum number of paging areas N and their associated sizes, n_i $(i=1\ldots N)$, to minimize the number of messages sent subject to a delay constraint. For $N>1$, the size of PA_i is $0<n_i<n$, and it holds that

$$\forall\, i\neq j \;\rightarrow\; PA_i \cap PA_j = \varnothing$$

$$\text{and } \sum_{i=1}^{N} n_i = n$$

,i.e. the partitions are mutually exclusive and collectively exhaustive. The conditional location probability of a user in paging area PA_i is then defined as the sum of individual location probabilities, as

$$q_i(u,t)=\sum_{j\in PA_i} p(j\,|\,u,t)\,. \tag{1}$$

2.2 Algorithm Description

After a last interaction event where the position of an MT in the location area is updated in the network database, a timer is set on the network side. When an incoming call request arrives for a specific user, the conditional location probabilities $p(i \mid u)$ are generated by the system for the time instance of the arrival, t_a. Based on these probabilities, cells are grouped into paging areas according to an optimal partitioning scheme.

Assuming that $PA_1, PA_2, \ldots, PA_N$ is an optimal N-partition of the set S, then the paging algorithm works by first sending polling messages to all cells in PA_1. If there is a response from one of the polled cells, the first step is successful and the algorithm stops. Otherwise, a second step is conducted that queries cells in the second partition PA_2. In case of another failure, the third partition is queried, then the third, and so on. Each step in the algorithm corresponds to a polling cycle, from the time the paging message is sent until a response is received. If the mobile is not in one of the cells in the paging area, then the polling cycle corresponds to the duration of a timeout interval. Assuming the cost for polling one cell is 1 (e.g. measured in bits/message), the average paging cost based on this algorithm is

$$C = \sum_{i=1}^{N} \sum_{j=1}^{i} n_j \cdot q_i(u, t_a), \tag{2}$$

i.e. it is the average number of paged cells. The average delay, in number of polling cycles is

$$d = \sum_{i=1}^{N} i \cdot q_i(u, t_a). \tag{3}$$

The main effort in implementing this scheme is required at the network to calculate the conditional location probabilities at the time of an incoming call arrival –based on the last interaction cell– and derive the optimal partitions. This is not inhibitive considering that modern networks have increased computational capabilities. In addition, a small extra cost occurs from setting the timer anew each time a transaction between the user and the network occurs, and storing and retrieving the identity of the last interaction cell when necessary. This information is actually the base station id, and is usually transmitted but not exploited in message transactions between the MSC and cell base stations.

2.3 Optimal Partitions

The objective of a partitioning algorithm is to choose the optimum N and n_i to minimize the number of paging messages subject to a delay constraint. In [7], the authors have shown that an optimal partition (OP) of a set S must be *non-increasingly ordered*. This means that for $i, j \in S$ with $i \in PA_k$, $j \in PA_l$ such that $k \leq l$, it must hold that $p(i \mid u, t_a) \geq p(j \mid u, t_a)$. From now on, we consider all cells in the location area to be numbered in order of decreasing conditional location probabilities. Hence, an optimal N-partition will consist of contiguously numbered cells, that is

$$PA_1 = \{1,2,\ldots,n_1\}\ , PA_2 = \{n_1+1,\ldots,n_1+n_2\},\ldots,\ PA_N = \{n_{N-1}+1,\ldots,n_{N-1}+n_N\}\ .$$

In this work we only consider OPs subject to a *worst case delay* constraint. For worst case delay, the total number of partitions equals the delay upper bound D, i.e. $N = D$. A dynamic programming algorithm can be constructed according to [7] which provides the partition into N zones that minimizes the paging cost. Let $g[n', N']$ be the minimum average paging cost achievable by partitioning the first n' cells into N' paging areas. Then we have the following recursion:

$$g[n',N'] = \min_{j=N'-1}^{n'}\left(g[j,N'-1] + n'\sum_{i=j+1}^{n'} p(i \mid u,t_a) \right). \tag{4}$$

The recursive formula exhibits the fact that an optimal partition of size N' that consists of n' numbered cells can be found by the minimum of an expression that allocates $j \le n'$ cells to an optimal partition of size $N'-1$ and the overhead $n'-j$ cells to the last partition N'. The initial conditions are

$$g[j,1] = j\sum_{i=1}^{j} p(i \mid u,t_a)\ . \tag{5}$$

With the help of these conditions we can apply the recursive equation to gradually build the two-dimensional matrix $g_{n\times N}$. Then by looking up the entries in the table, partition sizes can be found for increasing number of partitions. The time complexity of this algorithm is $O(Nn^2)$, which is the time to calculate all entries in the table.

Alternatively to the dynamic programming approach, we can employ a heuristic algorithm introduced in [8] to find the optimal allocation of cells to paging areas. If p_j^s and p_j^l denote the smallest and largest probability cell, respectively, in a partition j for conditional probabilities calculated at time instant t_a, then the algorithm starts with an initial partition and proceeds to move partition boundaries left or right, depending on the following conditions:

- *Backward Boundary Condition*: if the total conditional location probability of a partition j, $q_j(u,t_a) > p_j^s(u,t_a)\cdot(n_{j+1}+1)$, then the cell with probability p_j^s will be moved to the next PA, i.e. the partition boundary is moved left.
- *Forward Boundary Condition*: if the total conditional location probability of a partition j, $q_j(u,t_a) < p_{j+1}^l(u,t_a)\cdot(n_{j+1}-1)$, then the cell with probability p_{j+1}^l will be moved to the previous PA, i.e. the partition boundary is moved right.

The algorithm checks all partitions in the area until forward and backward boundary conditions have been satisfied. This is the strategy we employ to derive optimal partitions in the paper. Practically this algorithm is very effective; for small scale problems and depending on the form of the location probabilities, one can cleverly choose the initial partition so that relatively few boundary movements are necessary to achieve the optimal configuration.

Partitioning under average delay constraints is not considered here, as this is very difficult for $N > 2$. However, a similar polynomial time dynamic programming algorithm can be used to derive optimal paging areas in the average delay case [7].

For the case of $N = 2$ it is relatively easy, and a corresponding algorithm can be found in [9].

3 Mathematical Model

In order to model the proposed scheme mathematically, we assume that the incoming call interarrival time is an exponentially distributed random variable t_a with rate λ_a. Then the probability density function (pdf) is given by

$$f_{t_a}(t) = \lambda_a \cdot e^{-\lambda_a t}. \tag{6}$$

The residence time of a mobile user in a cell i is also assumed to be exponentially distributed with mean $1/\lambda_{m_i}$ and density $f_{m_i} = \lambda_{m_i} \cdot e^{-\lambda_{m_i} t}$. We can safely postulate that in most cases, a user movement direction depends only on the cell he is currently located. That, together with the cell dwell times being exponentially distributed, permits us to model the system as a continuous time *Markov chain*, as shown in Fig. 2.

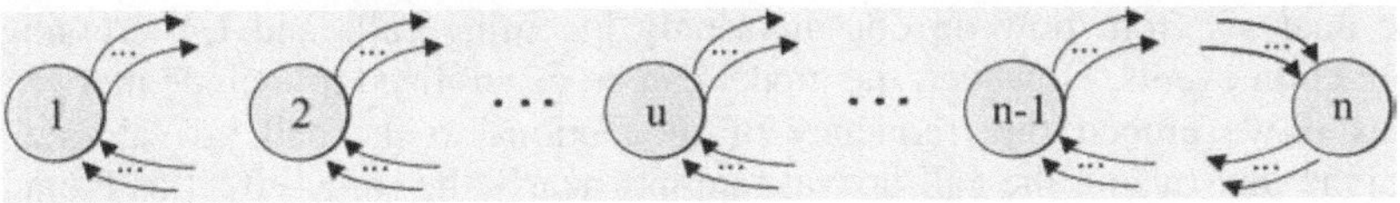

Fig. 2. Markov chain diagram for modeling MT movement in a location area

Each state of the diagram corresponds to a cell in the LA. The transition rates depend on the movement rates λ_{m_i} and the transition probabilities p_{ij}, $\forall\, i, j \in S$. The latter provide the routing probability that an MT currently in cell i will move towards an adjacent cell j. Accordingly, rates q_{ij} are calculated as

$$q_{ij} = \lambda_{m_i} p_{ij},$$
$$q_{ii} = -\sum_{j \neq i} q_{ij}, \tag{7}$$

for $i, j = 1, 2, \ldots, n$. The p_{ij} probabilities are influenced by the topological structure of the LA and the mobility pattern of the terminal. Here we follow a *symmetric random walk* model, so that a mobile moves with equal probabilities to each of its neighboring cells. Moreover, for convenience the mean residence times in every cell will be taken equal, $\lambda_{m_i} = \lambda_m \;\forall\, i$, although different values can be applied provided the required data is in hand. It is worth noting that the estimation of the mean residence time is rather difficult since its value depends on the individual behavior and mobility pattern of a mobile user, as well as the size and shape of each cell. For implementing the LIBP scheme, estimates may be acquired either by a user-supplied profile, or by gathering statistics regarding the crossings of the MT from certain points in the network. These would be logged periodically either via the existing structure or an autonomous sub-network, and communicated to the MSC/VLR. Secondarily, we note that a model usually classifies objects into classes. Since it would be too arduous to estimate mobility parameters for each and every subscriber of a network provider, we may say that the mean residence time and call rate are representative of a given class of subscribers.

Using expressions in (7), we create the transition rate matrix (generator matrix), $\mathbf{Q}=[q_{ij}]$, $i, j \in S$. Then by solving the system $\pi \cdot \mathbf{Q} = 0$ with $\sum_{i\in S}\pi_i = 1$, we obtain the steady-state probabilities for the continuous time case. A probability π_i denotes the proportion of time the mobile spends in state i, *provided that* it resides in the specified LA. Then equivalently, using the PASTA theorem [10], π_i is the probability that an incoming call will find a user in cell i. Therefore, steady-state probabilities can be used to derive optimal partitions for a general unconditioned case, where no knowledge of the last interaction cell is employed.

If the additional information of the last interaction cell is available, we can calculate the likelihood of reaching state (cell) i after a period t, based on the transition matrix $\mathbf{Q}$. If X_t is the state of the Markov chain at time t, by applying the Chapman-Kolmogorov equations it is possible to derive the following property [11]:

$$p(i \mid u,t) = \Pr\{X_t = i \mid X_0 = u\} = e^{t\mathbf{Q}}(u,i) \quad (8)$$

The analysis from now on considers only incoming calls and LA crossings as last interaction events. However, the model can be generalized to include more transaction types if we embed the frequency of interactions to the call arrival rate. We can imagine that raising the call arrival rate has nearly the same effect as the addition of more interactions in the model.

The timing diagram describing a possible sequence of events as the mobile moves in the network area is shown in Fig. 3.

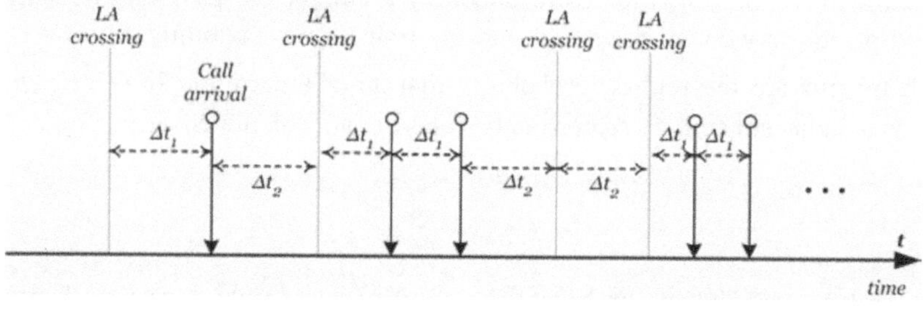

Fig. 3. Timing diagram of modeled user-network interactions during MT movement

Each one of the events in the figure is a random renewal point and corresponds to an update of the user position in the network database. The periods $\Delta t_1, \Delta t_2$ define the roaming intervals from the most recent interaction to the next event —a call arrival or normal registration event respectively. Any incoming call request for a specific subscriber arrives at the end of a *paging interval.* This is defined here as the period between the last interaction and the next call arrival. Since we always assume that a call will arrive *while* the MT is in the LA, this is equal to Δt_1 *if and only if* $\Delta t_1 < \Delta t_2$. Due to the memoryless property of the exponential, the distribution of Δt_1 is the same as the interarrival distribution, even if the last interaction was an LA crossing. Consequently, the distribution of a paging interval is

$$f_p(t) = f_{\Delta t_1}(t \mid 0 \le t < \Delta t_2) = \frac{\lambda_a \cdot e^{-\lambda_a t}}{\int_{t_2=0}^{\infty}\int_{t_1=0}^{t_2} \lambda_a \cdot e^{-\lambda_a t_1} \cdot f_{\Delta t_2}(t_2) dt_1 dt_2}. \qquad (9)$$

From eqs. (8), (9) using the continuous version of the total probability theorem [12], the probability of the user residing in cell i when the next call arrives is

$$p(i \mid u) = \int_0^{\infty} p(i \mid u,t) f_p(t) dt = \int_0^{\infty} e^{t\mathbf{Q}}(u,i) \frac{\lambda_a \cdot e^{-\lambda_a t}}{\int_{t_2=0}^{\infty}\int_{t_1=0}^{t_2} \lambda_a \cdot e^{-\lambda_a t_1} \cdot f_{\Delta t_2}(t_2) dt_1 dt_2} dt. \qquad (10)$$

The distribution of Δt_2, $f_{\Delta t_2}$ is the distribution of the residence time in a location area, when the user initiates his movement from the cell of the most recent interaction. Suppose the MT resides in the n-cell LA for a period T, during which it visits m cells, and the time spent at each cell is t_i, $i \in [1, m]$. Then the total sojourn time $T = t_1 + t_2 + \ldots + t_m$ has density function

$$f_{LA}^{(m)}(T) = \int_{t_1=0}^{T} \int_{t_2=0}^{T-t_1} \cdots \int_{t_m=0}^{T-t_1-\ldots-t_{m-1}} f_m(t_1) f_m(t_2) \ldots f_m(t_m) dt_m \ldots dt_2 dt_1, \qquad (11)$$

i.e. is the convolution of $f_m(t_1), \ldots, f_m(t_m)$.

In order to calculate the number of cells visited until exiting the LA, we can use an embedded version of the Markov chain in Fig. 2, with stepwise transition probabilities p_{ij} and an absorbing state E to mark transitions to cells in the external boundary of the area. If $\mathbf{P} = (p_{ij})$ is the $(n+1) \times (n+1)$ transition matrix, then its power is defined recursively as

$$\mathbf{P}^{[k]} = \begin{cases} \mathbf{P}, & if\ k = 1 \\ \mathbf{P} \times \mathbf{P}^{[k-1]}, & if\ k > 1 \end{cases}, \qquad (12)$$

where an element $p_{i,j}^{[k]}$ of $\mathbf{P}^{[k]}$ is the probability that an MT initially in cell i moves into cell j after k steps. Likewise, $p_{u,E}^{m}$ is the probability that a mobile exits the LA after m steps. It follows that the probability that an MT exits the LA in *exactly* m steps after the terminal was last seen in cell u is given by

$$p_{m,u,E} = \begin{cases} p_{u,E} & , if\ m = 1 \\ p_{u,E}^{(m)} - p_{u,E}^{(m-1)}, & if\ m > 1 \end{cases} \qquad (13)$$

With the help of (11), the theorem of total probability gives us the density function of the residence time in the LA, conditioned on the last interaction:

$$f_{\Delta t_2}(t) = \sum_{m=0}^{\infty} p_{m,u,E} \cdot f_{LA}^{(m)}(t). \qquad (14)$$

To avoid the difficult algebraic computation of $f_{\Delta t_2}(t)$, we may approximate the distribution of a paging interval with an unconditional pdf, so that (10) becomes

$$p(i \mid u) = \int_0^{\infty} e^{t\mathbf{Q}}(u,i) \lambda_a e^{-\lambda_a t} dt. \qquad (15)$$

This approximation is valid when the last interaction cell is an inner cell in the LA and/or the *call-to-mobility ratio* (CMR) λ_a / λ_m takes relatively high values. In this case, the probability of an incoming call arriving before the user exits the LA approximates unity. On the other hand, when the last interaction cell is near the LA border or the CMR is low, the approximation is not very accurate. Then $\Pr\{0 \le t < \Delta t_2\} < 1$ and higher values near the LI cell are expected. As it is shown in the next section, because of this fact paging costs calculated using the simplified formula (15) provide a lower bound on the signaling reduction achieved with the proposed scheme.

It is worth noting that since we restrict the model to a location area, the total number of states in the Markov chain is not too large and the algorithm maintains a relatively low complexity, both in space and time. Fast numerical methods may be used to facilitate the computation of the matrix exponential in (8) for application to a system with limited processing capabilities. A reference is cited in [13] for further details. We also remark that a similar analysis like the one presented above can be made when the cell residence times have a general distribution; for interested readers, details are included in [14].

Besides the framework presented here, any other model that derives a predictive probability of user location can be used to examine the algorithm. For instance, a natural extension of the presented work could be an isotropic diffusion model [15], where the last interaction considers the exact position in the *xy*-plane and drift velocities are $\upsilon_x = \upsilon_y = 0$. Further, a Gauss-Markov model can be used as in [16]. Such a model describes a mobile's movement pattern more accurately, as it better captures the essence of the correlated behavior in time.

4 Application

We apply the analysis above to a simple paging example. The location area consists of 31 cells in a hexagonal configuration. Cells are numbered from 0 to 30, as shown in the simplified representation in Fig. 4. Our aim is to derive optimal partitions of the LA based on the call arrival and movement characteristics. The symmetric random walk model discussed in the previous section is followed and equation (15) is used to derive the conditional location probabilities. Results are derived for several values of the call-to-mobility ratio. Tables 1-3 depict the allocations of cells to optimal partitions, for N=2, 3 and 4, in the general OP case and when LI cell=15 in the LIBP scheme, for different CMR values. A higher number of partitions is not considered; as it was demonstrated in [5] for an LA of similar size, the advantages gained from paging cost reduction for higher values of N become less significant compared to the increase in the average delay. The delay in locating an MT and thus establishing a call connection is a quality of service parameter for a subscriber, therefore we want to sustain it beneath acceptable levels. Moreover, it will be demonstrated that for common CMR values the behavior of the LIBP scheme does not change by adding more partitions.

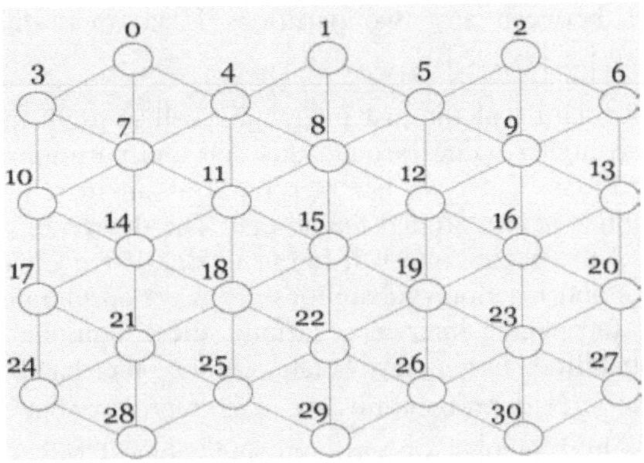

Fig. 4. Simplified view of the 31-cell hexagonal location area in our example

Table 1. Optimal partitions of the location area in the general OP case

Partitions	Cell Selection
N=2	{(21,18,14,22,7,11,15,8,12,19,9,16,25,4,23); (5,17,10,26,29,13,20,28,24,3,0,1,2,6,30,27)}
N=3	{(21,18,14,22,7,11,15,8,12); (19,9,16,25,4,23,5,17,10,26); (29,13,20,28,24,3,0,1,2,6,30,27)}
N=4	{(21,18,14,22,7,11,15); (8,12,19,9,16,25,4); (23,5,17,10,26,29,13,20); (28,24,3,0,1,2,6,30,27)}

Table 2. Optimal partitions of the location area in the LIBP scheme (LI cell=15, CMR=10^{-1})

Partitions	Cell Selection
N=2	{(15,22,8,18,11,12,19,14,25,16,21,7); (4,5,9,23,26,29,1,17,10,13,20,28,0,2,24,3,6,30,27)}
N=3	{(15,22,8,18,11,12,19); (14,25,16,21,7,4,5,9,23,26,29); (1,17,10,13,20,28,0,2,24,3,6,30,27)}
N=4	{(15,22,8,18,11,12); (19,14,25,16,21,7,4); (5,9,23,26,29,1,17,10); (13,20,28,0,2,24,3,6,30,27)}

Table 3. Optimal partitions of the location area in the LIBP scheme (LI cell=15, CMR=1)

Partitions	Cell Selection
N=2	{(15,22,8,19,18,12,11); (25,5,4,16,14,26,29,21,9,7,23,1,20,13,17,10,28,2,0,30,24,6,3,27)}
N=3	{(15); (22,8,19,18,12,11); (25,5,4,16,14,26,29,21,9,7,23,1,20,13,17,10,28,2,0,30,24,6,3,27)}
N=4	{(15); (22,8,19,18,12,11); (25,5,4,16,14,26,29,21,9,7); (23,1,20,13,17,10,28,2,0,30,24,6,3,27)}

By observing the tables, we remark that for the unconditioned case paging areas have almost equal sizes. This happens because the steady-state probabilities do not differ very much from one cell to another. As it has been proven in [7], "if each cell has equal probability of user location then the *N*-partition of an *n*-cell LA which minimizes the average cost of paging is balanced such that the difference in the number of cells between any two partitions is no more than one", specifically $\forall\, i \in \{1,2,\ldots,N\}$, $\lfloor \frac{n}{N} \rfloor \le n_i \le \lceil \frac{n}{N} \rceil$.

When the information of the last interaction cell is used, the derived conditional probabilities have higher values around this cell and partition sizes become smaller for the first partitions. This can be interpreted as an increased "concentration" of location probabilities at and around the LI cell. The degree of concentration depends on the duration of the paging interval. In our model, if the CMR is small, the paging interval increases and location probabilities are less concentrated around the LI cell. Theoretically, if the paging interval is infinite, the conditional probabilities become equal to the probabilities in the steady state. On the other hand, when the CMR takes higher values, the $p(i \mid u)$ probabilities are much more concentrated in a small portion of the LA. For a high number of partitions and a high CMR, the first partition may consist of only one cell (usually, the LI cell) as is shown in Table 3. The concentration of user location probabilities with increasing values of CMR compared to the steady-state probabilities is shown in the stack diagram of Fig. 5.

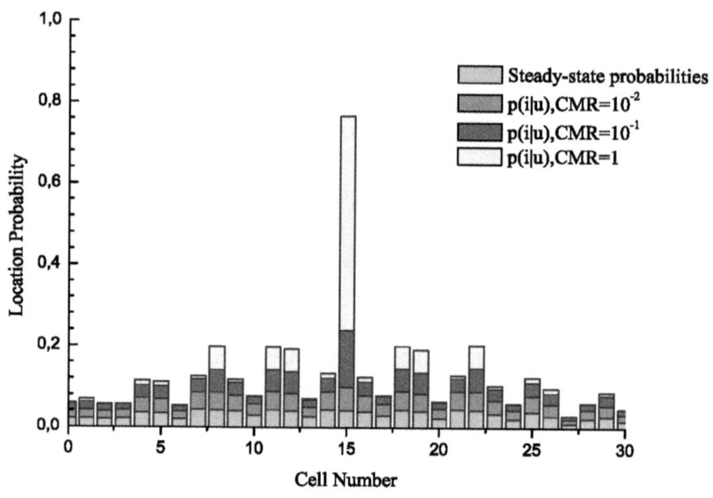

Fig. 5. Concentration of user location probabilities with increasing values of CMR in the LIBP scheme, when LI cell=15, compared to the steady-state probabilities

The more concentrated the location probabilities are, the higher is the signaling load reduction by the application of an optimal partitioning scheme. This is an essential conclusion first drawn explicitly in [7], and constitutes the rational basis of our work. It also explains why the simplified formula (15) in our LA model provides a lower bound on the signaling savings achieved with the proposed method. If the condition that a call arrives while an MT is in the LA were taken into consideration, the individual probabilities would be higher around a smaller number of cells. Therefore the savings using the approximate formula are always less than the exact savings.

5 Performance Evaluation

To evaluate the performance of the proposed method, we calculate the cost in number of paging messages to reach a user and the resulting average paging delay, for the LA of Fig. 4. Graphic results are shown in Figs. 6,7. We have considered two cases of last interaction cells: the first when LI cell=15, the center cell in the area, and second for LI cell=1, a cell near the border of the LA.

The results in terms of paging cost verify the following:

- In all cases, as the call-to-mobility ratio increases, the relative mobility of the terminal decreases or, equivalently, the paging interval diminishes. Thus the location uncertainty is reduced and the signaling savings increase. In Fig. 6(a), for *N*=2 partitions the percentage saving is 9.68% for CMR=0.1, 38.16% for CMR=0.5, 51.56% for CMR=1 and 63.72% for CMR=3. For higher values of the call-to-mobility ratio results are not shown here, however it is evident that more gains are to be expected, until eventually the curve reaches the saturation point of *C*=1.
- As the number of partitions increases, the saving in messaging cost is higher. Initially the most significant reduction occurs when we move from the conventional blanket polling strategy to a 2-partition scheme. Then the slope of the curve is smaller for higher number of paging areas; hence our gains are also smaller, until there is no improvement by increasing the number of partitions. Beyond *N*=3 the reduction is very slight for the LIBP scheme with high CMR values.
- The LIBP scheme is always better than the general optimal partitioning strategy with unconditional location probabilities. Even for small CMR values where the location database is not updated very often or the mobile moves rapidly, the saving in signaling cost from the general OP case can be worthwhile. For CMR=0.1 in Fig. 6(a), the percentage saving is 9.68% for *N*=2, 15.07% for *N*=3 and 15.22% for *N*=4.
- The LIBP scheme produces better results quicker than the general OP scheme, so that more gains can be achieved with a smaller price to pay for increasing delays. In the most commonly studied case where CMR is around 1, the optimum value of *N* that can achieve significant traffic reduction with acceptable delay is $N < 4$, smaller or at least equal than the one for the general unconditioned OP case.
- When the LI cell is near the border of the LA, the paging cost is lower for the same number of partitions and the same CMR. By comparing Figs. 6(a)-(b) we note that, for example, when *N*=2 and CMR=1, the average decrease is 51.56% when LI cell=15 and 61.79% when LI cell=1. This happens because a near border cell has always fewer adjacent cells or fewer 'escape routes' to move, therefore chances are that it will be closer to its origination point at the end of a paging interval. So, in correspondence with our central idea, the conditional probabilities are more concentrated around the LI cell and savings come increased.

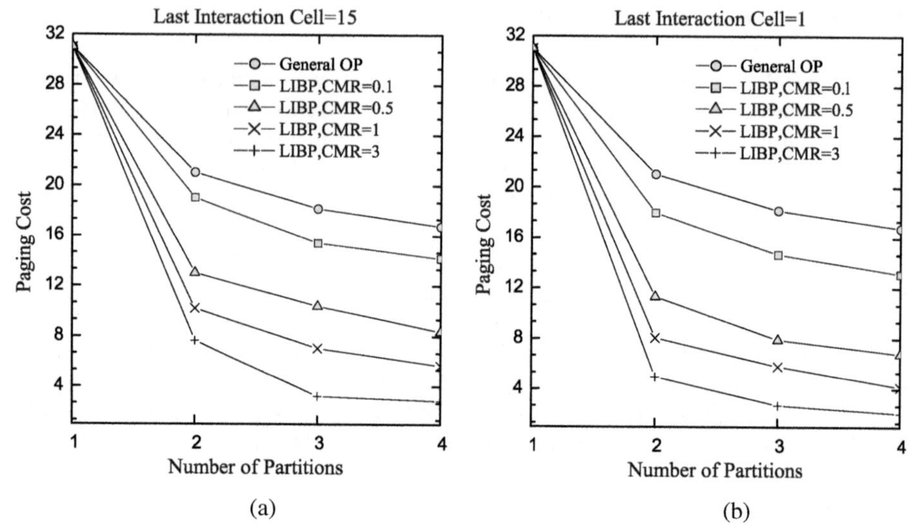

Fig. 6. Paging cost vs. number of partitions for the general OP scheme and the LIBP algorithm with different values of the call-to-mobility ratio, when (a) LI cell=15 and (b) LI cell=1

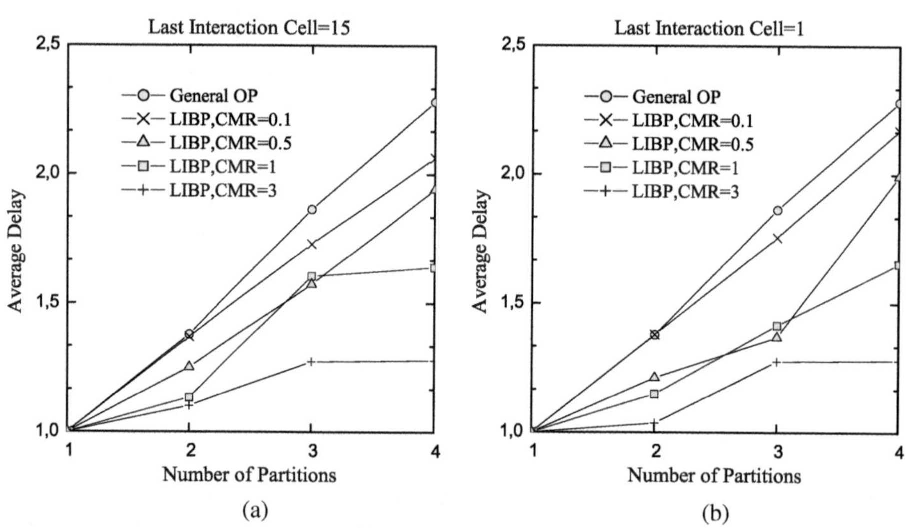

Fig. 7. Average delay for the general OP scheme and the LIBP algorithm with different values of the call-to-mobility ratio, when (a) LI cell=15 and (b) LI cell=1

Regarding the metric of average delay to locate a mobile user, we note that the LIBP scheme is again superior to a general OP scheme. For instance, in Fig. 7(a), when N=3 the delay for the LIBP scheme is 1.60 polling cycles when CMR=1, while for the general OP case it amounts to 1.86 cycles. This reduction can be quite significant when translated to real time. As an example, if the duration of a polling cycle is 500 ms, the extra delay when 3 partitions are used is 300 ms for the LIBP scheme and 430 ms for the unconditioned strategy. Thus the overhead delay deteriorates performance much less when the LIBP scheme is employed. Moreover, as for the average cost, the LIBP scheme performs better for higher CMR values. This happens because the location database is updated more often, and thus the probability of finding the user in one of the first steps increases.

We also notice that delay increases approximately linearly for a scheme using the unconditional location probabilities. This is in accordance with previous results in [5,7]. However, in the LIBP scheme this is not true for high CMR; a proportionally greater or lower increase may be observed by adding one more partition. This depends on the exact allocation of cells –and hence location probabilities– to paging areas, which is not symmetrical for increasing number of partitions. In addition, we note that choosing a different starting cell does not affect the results clearly as in the case of paging cost. Delays may increase when the last interaction cell is near the border of the LA.

Finally, when the CMR is very high, a mobile terminal is almost certainly in the last interaction cell, with a small probability to be found elsewhere. In our test cases, when CMR>10, the LI cell probability was >0.9. For such values, partitioning the LA into more than two paging areas is almost useless and does not improve the performance of the algorithm or affect delays. Then an optimal partition of the LA would be first to poll the last interaction cell, and secondly the remaining cells in the area.

6 Summary and Conclusion

This paper has presented a novel paging strategy that helps in reducing the signaling load generated by the search operations in a wireless network. The new strategy enhances current partitioning algorithms by exploiting the location information acquired through recent interactions with the system. It employs a mathematical or statistical model to predict conditional location probabilities based on the cell where the last interaction between the network and the terminal took place. The derived location probabilities are more concentrated in a small portion of the location area, which results in greater savings by the application of an optimal partitioning algorithm. Numerical results in a simple LA paradigm have shown that significant savings with smaller delays can be accomplished compared to a conventional partitioning scheme, especially for high usage or low mobility subscribers. Our scheme is particularly flexible in that all or some transaction types may be used to update the system or terminal database about the last interaction position, depending on implementation, protocol and processing cost issues. However, the more frequent transactions are, the higher are the gains of the LIBP scheme.

A symmetric random walk model was used to describe MT movements in a location area and the obtained results correspond to lower bounds on the specific system performance. More realistic mathematical or statistical models can be employed in compliance with the general framework described in the paper, which will help in further attesting the effectiveness of the proposed scheme.

References

1. Akyildiz, I.F., Ho, J.S.M.: On Location Management for Personal Communications Networks. IEEE Communications Magazine **9** (1996) 138–145
2. Ali, S.Z.: Location Management in Cellular Mobile Radio Networks. In Proc. 13th IEEE PIMRC, Lisbon, Portugal, Sept. 2002, 745–749
3. Plehn, J.: The design of Location Areas in a GSM-network. In Proc. 45th IEEE Veh. Tech. Conf., Chicago, USA, July 1995, 871–875
4. Rose, C., Yates, R.: Minimizing the Average Cost of Paging Under Delay Constraints. Wireless Networks **1** (1995) 211–219
5. Abutaleb, A., Li, V.O.K.: Paging Strategy Optimization in Personal Communication Systems. Wireless Networks **3** (1997) 195–204
6. Markoulidakis, J.G., Sykas, E.D.: Performance Bounds of a Multiple-Step Paging Strategy in Future Universal Mobile Telecommunication Systems. Int. J. Wireless Inf. Networks **3** (1995) 133–147
7. Krishnamachari, B., Gau, R.-H., Wicker, S.B., Haas, Z.J.: Optimal Sequential Paging in Cellular Networks. To appear in Wireless Networks
8. Wang, W., Akyildiz, I.F., Stüber, G.L.: An Optimal Paging Scheme for Minimizing Signaling Costs Under Delay Bounds. IEEE Comm. Letters **2** (2001) 43–45
9. Koukoutsidis, I.Z., Demestichas, P.P., Papaioannou, P.I., Theologou, M.E.: Optimal Decision Strategies for Paging in Wireless Networks. Submitted to Wireless Personal Communications (2003)
10. Wolff, R.W.: Stochastic Modeling and the Theory of Queues. Prentice-Hall, Englewood Cliffs, NJ (1989)
11. Feldman, R.M., Valdez-Flores, C.: Applied Probability and Stochastic Processes. PWS Publishing Co., Boston (1996)
12. Papoulis, A., Pillai, U.S.: Probability, Random Variables and Stochastic Processes. 4th Int. edn. McGraw-Hill, New York (2002)
13. Stickel, E.U.: An Algorithm for Fast High Precision Computation of Matrix Exponential and Logarithm. Analysis **10** (1990) 85–95
14. Koukoutsidis, I.Z., Theologou, M.E.: A Combination of Optimal Partitioning and Location Prediction to Assist Paging in Mobile Cellular Networks. Submitted to Int. J. Wireless Information Networks (2003)
15. Lei, Z., Rose, C.: Probability Criterion Based Location Tracking Approach for Mobility Management of Personal Communications Systems. In Proc. IEEE Globecom, Phoenix, USA, Nov. 1997, 977–981
16. Liang, B., Haas, Z.J.: Predictive Distance-Based Mobility Management for PCS Networks. In Proc. 18th IEEE Infocom, New York, USA, March 1999, 1377–1384

Multi-period Channel Assignment

Hakim Mabed[1], Alexandre Caminada[1] and Jin-Kao Hao[2]

[1] France Télécom R&D, 6 Avenue des Usines, BP 382, 90007 Belfort, France
{hakim.mabed,alexandre.caminada}@francetelecm.com
Tel: (+33)03.84544309, Fax (+33)0384544396.
[2] University of Angers, 2 Bd Lavoisier, 49045 Angers Cedex, France
jin-kao.hao@univ-angers.fr

Abstract. The well-known fixed channel assignment scheme for cellular networks is not flexible enough to follow the evolution of traffic. This paper introduces a multi-period channel assignment model. In addition to the usual objective of minimizing the interference, the model integrates another requirement to minimize the transition cost from a frequency plan to another one. Several heuristic solution approaches are also proposed. Experimental results on real data are presented to compare the multi-period model and the fixed model, and to assess the effectiveness of the proposed solution algorithms.

Keywords: Multi-period channel assignment, optimization, genetic search, tabu search.

1 Introduction

In a GSM network [9], the geographical area is partitioned into cells, each one served by a single base station. To ensure communications occurring on their cells, stations require a certain number of frequencies depending on the expected traffic load. In other words, lightly loaded cells are assigned fewer channels than heavily loaded ones. The mobile network operators dispose of a very limited number of frequencies to cover all the network area. For this reason, frequencies reusing is indispensable to increase the network capacity. Channel assignment consists in assigning the available frequency spectrum to the stations of the network in order to satisfy their demands and to minimize the interference. Interference is caused by the presence of overlapping areas between cells where several signals of good quality are received.

The quality of communications in cellular networks depends closely on how the available frequency spectrum is managed. Because of its implementation simplicity, fixed channel assignment (*FCA*) is largely used in today's GSM mobile networks. In this case, a subset of nominal frequencies is definitively allocated to each base station. However, the main inconvenient of *FCA* is that it is not adaptive to traffic variation. In fact, usually frequency plan dimensioning is based on an over-sizing of traffic data [1][6][17] and unused channels in lightly loaded cell are not reassigned to heavily loaded ones. To overcome this handicap, many alternative strategies have been adopted as dynamic channel assignment [1][3], hybrid channel assignment [15] and channel borrowing [16]. Usually those techniques perform badly in heavy traffic or require additional signaling loads to ensure channel readjustment [8].

M. Conti et al. (Eds.): PWC 2003, LNCS 2775, pp. 541–554, 2003.

This paper presents a channel assignment model noted *MCA* for Multi-period Channel Assignment [12], which associates simplicity and adaptability. In this case, the frequency-planning problem consists in finding a *sequence of frequency plans* following the traffic evolution for a number of time periods. Each frequency plan is conceived to fit the traffic situation at the period in which it is operational. Two reasons make the problem more complicated. First, in addition to the classical criteria dealing with interference, the transition cost caused by frequency plan change must be minimized. Second, the multi-period character of the problem increases its combinatorial complexity. To cope with this complexity, we propose several optimization techniques based on a genetic tabu search algorithm and we compare their performances against the *FCA* scheme in terms of lost traffic.

The paper is organized as follows. In next section we formally describe the *MCA* model and we give a set of definitions used in the remainder of the paper. Section 3 describes in details the basic genetic tabu search algorithm used to solve the *FCA* problem. Section 4 presents how genetic tabu search algorithm is readapted to *MCA* model. Section 5 is dedicated to the experimental tests carried out in order to assess the *MCA* model.

2 Multi-period Channel Assignment Problem

In fixed channel assignment, a single frequency plan is built in order to be permanently operational even if the traffic evolves in time. The key word is then the robustness of the frequency plan over time. To that end, modelers use an aggregation of traffic data, for example traffic at second busy hour to evaluate the quality of frequency plans [10].

In the case of multi-period channel assignment, we assume that traffic evolution follows a cyclical scheme. According to the desired scale level, one cycle is divided into periods of equal duration (hours, days...). We assume also that the traffic load is known on every cell for each period. The objective is then to find a sequence of frequency plans. Each frequency plan is built with the objective to minimize the interference recorded at the associated period. In addition, the frequency plan must meet another requirement to minimize transition costs between frequency plans. Transition cost measures the required effort or damage caused by the frequency plan changes. Several aspects can be taken into account to measure the transition cost between two frequency plans: (a) Minimizing the number of changed frequencies between two frequency plans; (b) minimizing the number of stations affected by the changes or (c) minimizing the traffic load affected by the changes. In this work, the number of changed frequency is taken as the transition criterion.

2.1 Basic Notations

We introduce here the basic notations and definitions, which will be used in the continuation of the paper:

- N: The number of stations.
- $\{S_1, \ldots, S_N\}$: The set of stations composing the network.
- m_i / $i \in [1..N]$: The number of frequencies required by the station S_i.

- F: The number of available frequencies.
- np: The number of studied time periods.
- $2B_i$: The period on which the traffic load on the station S_i reaches its second greatest value.

Interference damage between stations depends on several factors such as inter-channel separation between used frequencies, signal powers... It is also largely depending on traffic intensity on these stations. The impact of traffic on interference is twofold. As interferer station, traffic load describes the utilization rate of frequencies and hence impacts on the quantity of generated interference. As interfered station, traffic intensity reflects the importance of the area covered by the station and consequently the interest of interference reduction on this area.

Let us note by:

- $I_{i,j,d}$: The interference damage between S_i and S_j caused by a pair of frequencies distanced by d channels.
- $I^{2B}_{i,j,d}$: The interference damage between S_i and S_j measured according to traffic load on station S_i at the period $2B_i$ and on station S_j at the period $2B_j$.
- $I^{p}_{i,j,d}$: The interference damage between S_i and S_j measured according to traffic situation at the period p.
- Assignment $f_{i,k} \in [1..F]$ corresponds to the kth "permanent" frequency assigned to the station S_i.
- Assignment $f^{p}_{i,k} \in [1..F]$ corresponds to the kth frequency assigned to the station S_i at the period p.
- A frequency plan is either a vector of "permanent" assignments $< f_{1,1}, \ldots, f_{1,m_1}, \ldots, f_{1,m_N}, \ldots, f_{N,m_N} >$ (for FCA) or a vector of "temporary" assignments $< f^{p}_{1,1}, \ldots, f^{p}_{1,m_1}, \ldots, f^{p}_{N,1}, \ldots, f^{p}_{N,m_N} >$ (for MCA).
- A sequence is a vector of temporary frequency plans $\langle \langle f^{1}_{1,1}, \ldots, f^{1}_{1,m_1}, \ldots, f^{1}_{N,m_N} \rangle, \ldots, \langle f^{np}_{1,1}, \ldots, f^{np}_{1,m_1}, \ldots, f^{np}_{N,m_N} \rangle \rangle$.

2.2 Problem Formulation

In the fixed channel scheme, a single frequency plan is constructed on the basis of $I^{2B}_{i,j,d}$ values. The objective of the optimization is to find the vector $< f_{1,1}$, $\ldots, f_{1,m_1}, \ldots, f_{1,m_N}, \ldots, f_{N,m_N} >$ which minimizes the total interference depicted by the function F_{2B}.

$$F_{2B} = \sum_{i=1,j=1}^{N,N} \sum_{k=1,l=1}^{m_i,m_j} I^{2B}_{i,j,|f_{i,k}-f_{j,l}|} \tag{1}$$

Where the double sum in the formula 1 measures the total of interference over the network, caused by used frequencies. The frequency plan thus worked out will be permanently operational.

By opposition, in multi-period channel assignment, the objective is to find a sequence of frequency plans corresponding to $f^{p}_{i,k}$ values, which minimizes the two functions:

$$F_{\Sigma} = \sum_{p=1}^{np} \sum_{i=1,j=1}^{N,N} \sum_{k=1,l=1}^{m_i,m_j} I^{p}_{i,j,|f^{p}_{i,k} - f^{p}_{j,l}|} \tag{2}$$

$$C_{\Sigma} = \sum_{p=1}^{np-1} \sum_{i=1}^{N} \sum_{k}^{m_i} IND(f^{p}_{i,k} \neq f^{p+1}_{i,k}) \tag{3}$$

$$IND(condition) = \begin{cases} 1 \; if \;\; condition \; is \; true \\ 0 \; otherwise \end{cases}$$

The function F_{Σ} represents the sum of interference recorded over all the time periods whereas the function C_{Σ} depicts the transition cost between frequency plans composing the sequence.

3 Genetic Tabu Search for Fixed Channel Assignment

The multi-period frequency assignment problem can be seen as an extension of the fixed channel assignment, requiring the generation of a sequence of frequency plans instead of a single frequency plan. For this reason, we describe first the algorithm serving to generate a single frequency plan. This algorithm is also used in Section 5 to compare the fixed channel assignment model and the multi-period frequency assignment model.

Many algorithms based on metaheuristics have been proposed for the fixed channel assignment problem [4][7][11][5]. We present here a hybrid genetic tabu search algorithm that is described in details in [10]. This article doesn't aim to study the performance of such algorithm but tries to show the relevance of MCA scheme and how the FCA algorithms can be readapted to this model.

The FCA algorithm starts from a population of individuals corresponding to frequency plans. The algorithm makes evolve the frequency plans iteratively. At each generation, the algorithm selects two frequency plans from the population and applies a crossover operator to them. The two new generated frequency plans are then improved using a Tabu Search based mutation.

3.1 Crossover Operator

As crossover operator, we adopt the geographical crossover described in [10][13]. The principle is this one: we randomly choose a reference station S_R and we build the set of its neighbors $V(S_R)$ composed of interfering stations S_i (i.e. $\exists d / I_{R,i,d} > 0$). The parts of the frequency plans corresponding to $V(S_R) \cup \{S_R\}$ are then exchanged between the two parents.

Geographic crossover allows the conservation of the building blocs present in the parent chromosomes. This is made by swapping information related to the local resolution of interference between stations. This operator is generalized later to multi-period assignment (see §4.1.3).

3.2 Tabu Search Based Mutation

After crossover, the two new frequency plans are improved by a tabu search based mutation. The idea is to apply a cycle of local search to the new frequency plans. More concretely, we associate to each assignment $f_{i,k}$ of the individual, a value called *violation score* measuring the contribution of that assignment to the recorded interference. Equation 4 gives the function serving to calculate the violation score of the assignment $f_{i,k}$. At each cycle of the local search operator, one assignment is chosen on the basis of the violation scores and its value is changed. The new frequency value corresponds to the best one which is not tabu. After the change, the new and the old value are considered tabu for this assignment.

$$SCORE_{i,k} = \sum_{j=1}^{N}\sum_{l=1}^{m_j} I^{2B}_{i,j,|f_{i,k}-f_{j,l}|} \tag{4}$$

Notice that such a tabu management contributes to two different roles. The element (i, k, f_{old}) avoids the recurrence of visited solutions, whereas the element (i, k, f_{new}) prevents the remainder individuals from re-exploring the same search area since the tabu list is shared by all population individuals.

After mutation the new frequency plan are inserted in the population in replacement of another one. The replaced frequency plan is chosen on the basis of its fitness. More precisely, individuals of bad fitness have more chance to be replaced.

The algorithms below describe the main procedure of the genetic tabu search algorithm as well as the tabu based mutation procedure.

```
TabuSearchOperator(Frequency plan fp)
Begin
    Best_fp:=fp;
    CalculateScores(fp);
    for iter:=1 to TSML {Tabu Search based mutation length}
            (i,k):=SelectAssignment(fp);{on the basis of violation scores}
            f_old:= fp[i,k];
            f_new := SelectBestFrequency(fp, i, k); {which is not tabu}
            AddToTabuList(i,k,f_old); AddToTabuList(i,k,f_new);
            fp[i,k]:= f_new;
            UpdateScores(fp);
            If BetterThan(fp,Best_fp) then Best_fp=fp; End if
    End for
End.
```

```
Genetic Tabu Search
Begin
    P:=RandomInitPopulation(Pop_size);
    For g:=1 to NbGenerations
            (p1,p2):=SelectParents(P)
            with a Pc probability do (f1,f2):=Crossover(p1,p2)
            otherwise f1:=p1; f2:=p2;
            f1:=TabuSearchOperator(f1); f2:=TabuSearchOperator(f2);
            (v1,v2):=SelectVictims(P);
            ReplaceBy(v1,f1); ReplaceBy(v2,f2);
    End for
End.
```

4 Genetic Tabu Search for Multi-period Channel Assignment

For the purpose of finding multi-period channel assignment, we have designed and experimented different optimization techniques. Each technique presents a particular manner to readapt FCA algorithms (in our case the Genetic Tabu Search) for the resolution of the MCA problem. These techniques can be roughly classified into two classes: direct optimization and decomposed optimization.

4.1 Direct Optimization

The multi-period character of the problem increases its combinatorial complexity. In direct optimization, the problem is considered in its totality without restriction on search space. In other words, search space will correspond to all the sequences of the form:

$$\left\langle \left\langle f_{1,1}^{1},\ldots,f_{1,m_1}^{1},\ldots,f_{N,m_N}^{1}\right\rangle,\ldots,\left\langle f_{1,1}^{np},\ldots,f_{1,m_1}^{np},\ldots,f_{N,m_N}^{np}\right\rangle\right\rangle$$

The optimization algorithm generates the different frequency plans composing the optimal sequence in a competing way. It is then necessary to readapt search operators of the basic algorithm.

4.1.1 Objective Function

To assess the fitness of a sequence, two criteria are considered: the total of interference recorded over time periods F_Σ and the total of transition cost C_Σ. The quality of each frequency plan in the sequence is calculated regarding to the other plans. Therefore, choices made on a part of the sequence may lead to other changes in the entire sequence.

The interference and transition criteria (§2.2) are aggregated into a single objective function. A threshold value S_Σ is defined as the maximal tolerated number of changes in the sequence. Exceeding this threshold the sequence quality is penalized with a very high value M. The objective function takes then the following form:

$$F = F_\Sigma + M \times IND(C_\Sigma > S_\Sigma) \text{ where } M \text{ is a very high value} \quad (5)$$

4.1.2 Initial Population

Generation of the initial population passes through a pre-optimization phase. For each sequence of the initial population, we choose iteratively one period *p*. An optimization phase is launched to generate a frequency plan, *fp*, well adapted to that period with the objective function given in equation 6. Then the frequency plan *fp* is fixed during all periods forming a sequence <*fp*,..,*fp*> which is inserted in the initial population. This process is reiterated for the other individuals of the initial population.

$$F_p = \sum_{i=1,j=1}^{N,N} \sum_{k=1,l=1}^{m_i,m_j} I^p_{i,j,|f_{i,k}-f_{j,l}|} \;/\; p \in [1..np] \quad (6)$$

4.1.3 Crossover Operator

Considering the effectiveness of the geographical crossover, a multi-period version of this operator should be interesting. The objective is to allow both spatial and temporal configuration exchange between sequences. In other words, the frequency plan evolution in a part of the network is grafted into another sequence. To that end, a reference station is randomly selected and the set of its neighbors is built. Then the corresponding parts in the two parent sequences are exchanged. The crossover working is schematized in the following figure.

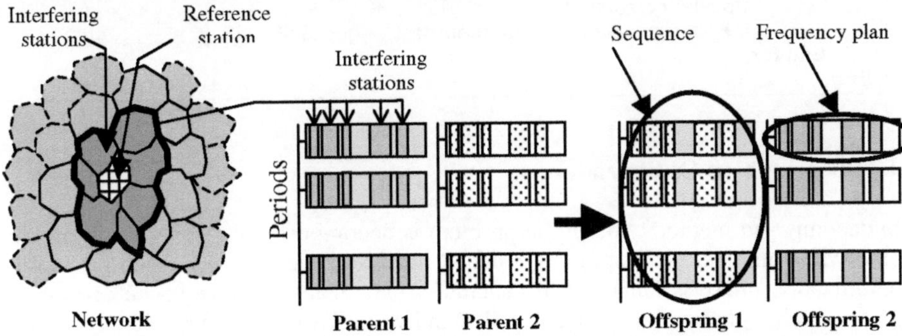

Fig. 1. Crossover operator for multi-period channel assignment

4.1.4 Mutation Operator

Two variants of the previous tabu search based mutation operator are implemented. The first variant (M1) changes the value of a single assignment $f^p_{i,k}$. First, a period *p* is randomly chosen and the violation score of each assignment of the considered period is calculated using the formula 7. Then an assignment $f^p_{i,k}$ is selected with a probability proportional to its violation score and the best not tabu value is attributed to it.

$$SCORE_{i,k} = \sum_{p=1}^{np} \sum_{j=1}^{N} \sum_{l=1}^{m_j} I^{p}_{i,j,|f^{p}_{i,k} - f^{p}_{j,l}|} \tag{7}$$

The second variant (M2) resets to the same value all the assignments $f^{p}_{i,k}$ ($p \in [1..np]$). The working scheme is the same as in mutation (M1) except that the new value is attributed to all assignments of the same position as $f^{p}_{i,k}$. These two variants are used in a competing way with probabilities Pm_1, $1\text{-}Pm_1$. The algorithm below depicts the Tabu Search based mutation of direct optimization. The main procedure is the same as in fixed assignment except that manipulated individuals are sequences.

```
TabuSearchOperator(Sequence seq)
Begin
    Best_seq:=seq;
    p=Random(np);
    CalculateScores(Seq[period]); {seq(period) corresponds to the frequency
                                          plan of the period p}
    for iter:=1 to TSML {Tabu Search based mutation length}
            (i,k):=SelectAssignment(fp);{on the basis of violation scores}
            f_old:= seq[p,i,k];
            f_new := SelectBestFrequency(seq,p, i, k); {which is not tabu}
            with a Pm1 probability, do
                    AddToTabuList(p,i,k,f_old); AddToTabuList(p,i,k,f_new);
                    seq[p,i,k]:= f_new;
            else do
                    AddToTabuList(p,i,k,f_old); AddToTabuList(TOUT,i,k,f_new);
                    for each per=1..np do seq[per,i,k]:= f_new;
            end with
            UpdateScores(seq);
            If BetterThan(seq,Best_fp) then Best_seq=seq; End if
    End for
End.
```

4.2 Decomposed Optimization

In decomposed methods, the initial problem is decomposed into several sub-problems of lower complexity, leading to reduced search space. On each sub-problem, an optimization phase is launched to generate a part of the final sequence of frequency plans. Each optimization phase handles individuals of frequency plan type. Three decomposed algorithms are implemented. Details of their implementation are given here below.

4.2.1 Step by Step Optimization

The optimal sequence of frequency plans is built in an iterative manner. At each iteration, one period is considered according to its chronological order. A frequency plan is then generated (by optimization) to fit the traffic situation at this period and to minimize transition cost from previous frequency plan. The final solution corresponds then to the set of those frequency plans. Note that the part of the optimal sequence already built can't be readjusted in further iterations. We give hereafter the different steps followed by the method. The value S designates the maximal tolerated change threshold between two consecutive frequency plans in the sequence. This threshold serves to aggregate the two partial functions F_p (equation 6) and C_p (described in the algorithm).

> Find the values $f_{i,k}^{1}$, which minimize: F_1
> For each period $p \in [2..np]$
> Find the values $f_{i,k}^{p}$, which minimize:
> $$F_p + M \times IND(C_p > S), \text{ where } C_p = \sum_{i=1}^{N} \sum_{k}^{m_i} IND(f_{i,k}^{p} \neq f_{i,k}^{p-1})$$

The final solution will correspond to the sequence

$$\left\langle \left\langle f_{1,1}^{1}, \ldots, f_{1,m_1}^{1}, \ldots, f_{N,m_N}^{1} \right\rangle, \ldots, \left\langle f_{1,1}^{np}, \ldots, f_{1,m_1}^{np}, \ldots, f_{N,m_N}^{np} \right\rangle \right\rangle.$$

4.2.2 Sequential Optimization

The idea is to use the robust frequency plan generated by fixed channel assignment method as a starting point for search. More precisely, an initial optimization phase using the function F_{2B} is performed producing a robust frequency plan. The different frequency plans composing the sequence are constructed iteratively in chronological order of periods exactly as in step-by-step optimization. The first frequency plan corresponding to the initial period is generated starting from the robust frequency plan (with respect to the transition cost criterion). We give hereafter the details of sequential optimization algorithm.

> Find the value $f_{i,k}^{0}$, which minimize F_{2B}
> For each period $p \in [1..np]$
> Find the values $f_{i,k}^{p}$, which minimize: $F_p + M \times IND(C_p > S)$

4.2.3 Parallel (or Simultaneous) Optimization

The iterative aspect of sequential optimization makes it slow. To overcome this inconvenient, a parallel variant of this technique is proposed. In this case, the frequency plans associated with the different periods are constructed starting from the robust plan in parallel. To explain this difference we give the working scheme of this parallel optimization, the parallel algorithm being implemented under PVM (Parallel Virtual Machine) system.

> Find the value $f_{i,k}^{0}$, which minimize F_{2B}
> For each period $p \in [1..np]$ **do simultaneously**
> Find the values $f_{i,k}^{p}$, which minimize:
> $$F_p + M \times IND(C'_p > S), \text{ where } C'_p = \sum_{i=1}^{N} \sum_{k}^{m_i} IND(f_{i,k}^{p} \neq f_{i,k}^{0})$$

5 Experimental Tests

The objective of this section is twofold. On the one hand, we compare the performance of the implemented multi-period optimization techniques. On the other hand, we compare the quality of solutions generated by the multi-period model with those produced by the FCA model. Results of multi-period and fixed channel assignment are compared from two points of view. The first is based on objective functions (Formulas 1 to 3). The second adopts operator's point of view and compares the solutions according to the lost traffic.

5.1 Benchmark Problems

Tests are carried out on both fictitious[1] and real data. The first problem, B-63, represents a fictitious problem instance with 63 stations, 30 available frequencies and 6 periods. The second instance, D-639, corresponds to a real world problem. The network is composed of 639 stations with 62 frequencies and traffic data during 13 hours (periods). The third instance, BM-120, is another real world problem with 120 stations and 62 available frequencies. *BM-120* is dedicated to study the performance of *MCA* for large-scale traffic data. Traffic evolution is thus studied over one week, day by day.

5.2 Comparison between Multi-period Channel Assignment Techniques

Four multi-period optimization algorithms, described before, are compared. Those algorithms correspond to direct optimization, step by step optimization, sequential optimization and parallel optimization. Table 1 gives the results obtained by each technique for the two problems B-63 and D-639. We run each algorithm 5 times on every problem. Only the best solution is reported for each algorithm.

Two implementations of the direct optimization technique are presented. The first uses only the mutation operator *M1*. The second uses in a simultaneous way the two mutation operators *M1* and *M2*. For each technique we give the name, the objective and eventually the mutation operator used. Obtained solutions are compared according to interference (F_p) and transition (C_p) cost at each period as well as their sum over time.

From columns (2) and (3) we remark the effectiveness of using the two mutation operators in cooperative way. By using only the *M1* operator, the transition cost reaches quickly the threshold S_Σ and hence slows down the algorithm evolution.

Step-by-step technique gives bad results. This can be explained by the absence of a global vision. In fact, at each phase, step-by-step algorithm optimizes the frequency plan according to the traffic situation at the associated period without taking into account the future evolution of traffic.

However, the main observation is that decomposed approaches, represented in table 1 by columns (4) and (5), give the best results. We notice also that results of sequential and parallel-decomposed optimization are very close.

[1] By fictitious data, we mean a real network whose traffic data are artificially modified.

Table 1. Comparison between the different multi-period channel assignment techniques

		(1)		(2)		(3)		(4)		(5)	
Name		Step by step decomposed		Direct		Direct		Sequential decomposed		Parallel decomposed	
Objective		F_p+*IND*(C_p>S), S=30, 50		F_Σ+*IND*(C_Σ>S_Σ), S=130, 600		F_Σ+*IND*(C_Σ>S_Σ), S=130, 600		F_p+*IND*(C_p>S), S=30, 50		F_p+*IND*(C'_p>S), S=30, 50	
Mutation				M1		M1+M2					
Cost		F_p	C_p	F_p	C_p	F_p	C_p	F_p	C_p	F_p	C_p
B-63	P0	62340	0	68828	0	57110	0	51569	0	51322	0
	P1	59740	29	66587	27	57431	38	49123	25	49012	27
	P2	53525	30	58725	23	47318	37	41837	20	41750	23
	P3	52350	30	55135	34	43402	15	40922	27	41033	19
	P4	62282	30	61645	15	52139	21	48144	25	48281	28
	P5	63484	30	65181	21	56880	15	51650	6	51584	10
	Total	**353721**	**149**	**376104**	**120**	**314280**	**126**	**283245**	**103**	**282982**	**107**
D-639	7:00	17289	0	18014	0	16026	0	16663	0	16458	0
	8:00	47376	50	48363	50	45832	50	43268	50	42875	50
	9:00	80311	50	83892	50	75913	50	73962	50	73784	50
	10:00	99960	50	103470	50	89749	50	85750	50	85912	48
	11:00	106898	50	109018	50	98944	50	94713	50	94799	50
	12:00	103367	50	105913	50	95013	49	92103	38	92355	46
	13:00	87468	50	89354	50	82832	49	77725	50	77623	50
	14:00	88088	50	91325	50	83549	48	77904	50	77789	50
	15:00	92669	50	95102	50	90114	50	82921	44	82805	46
	16:00	104715	49	106822	50	99322	50	92023	49	91987	47
	17:00	124727	48	128341	50	119444	48	110814	45	110612	50
	18:00	138748	50	143755	50	136617	44	128595	33	128904	30
	19:00	124642	50	126015	50	125029	41	119212	45	119447	42
	Total	**1216258**	**597**	**1249384**	**600**	**1158387**	**579**	**1095653**	**554**	**1095350**	**559**

5.3 Comparison between Fixed and Multi-period Solutions for D-639 Problem

To compare fixed and multi-period channel assignment, we have run the fixed channel assignment algorithm (Section 3) five times on the D-639 problem. In the tables 2 and 3, we compare the best solution found by the *FCA* with the multi-period solution found by the parallel decomposed optimization (column 5 in table 1). This comparison is made on the basis of objective function (table 2) and lost traffic (table 3). In table 3, we give the lost traffic (in Erlang) at each period as well as the total of lost traffic for the two compared solutions. We use for that, the quality evaluator of PARCELL©[2]. Results show a reduction of lost traffic reaching sometime 8% by using the *MCA* model. Notice that in table 2, transition cost for fixed solution is usually zero since there is a single frequency plan.

2 Engineering tool for design of mobile radio network, ORANGE society all rights reserved.

Table 2. Comparison between fixed and multi-period channel assignment in terms of objective function

Name		Fixed		Parallel decomposed	
Objective		F_{2B}		F_p+*IND*(C'_p>50)	
Cost		F_p	C_p	F_p	C_p
D-639	7:00	17944	0	16458	0
	8:00	47065	0	42875	50
	9:00	78944	0	73784	50
	10:00	92338	0	85912	48
	11:00	102398	0	94799	50
	12:00	96356	0	92355	46
	13:00	84723	0	77623	50
	14:00	86766	0	77789	50
	15:00	94820	0	82805	46
	16:00	104131	0	91987	47
	17:00	121026	0	110612	50
	18:00	138218	0	128904	30
	19:00	125507	0	119447	42
	Total	**1189467**	**0**	**1095350**	**559**

Table 3. Comparison between fixed and multi-period channel assignment in terms of lost traffic for the D-639 problem

Periods	**Traffic**	**Fixed**	**Multi-period**	**Gain**
7h-8h	504	1.60	1.47	8.12%
8h-9h	1161	4.90	4.77	2.6%
9h-10h	1746	9.58	9.57	0.1%
10h-11h	2015	10.65	11.27	
11h-12h	2168	12.30	12.19	0.9%
12h-13h	2092	11.42	11.45	
13h-14h	1861	9.84	9.39	4.5%
14h-15h	1944	10.30	9.75	5.3%
15h-16h	1972	11.15	10.30	7.6%
16h-17h	2160	12.77	12.25	4%
17h-18h	2486	15.75	14.40	8.5%
18h-19h	2745	19.01	17.83	6.2%
19h-20h	2696	16.71	15.82	5.3%
Total	**25550**	**145.98**	**140.46**	

Table 4. Comparison between fixed and multi-period channel assignment for BM-120 instance

Days	**Fixed solution** F_{2B}		**Multi-period solution** F_p+*IND*(C'_p>50)	
	F_p	C_p	F_p	C_p
June, Monday 24	39219	0	38708	31
June, Tuesday 25	40171	0	39647	21
June, Wednesday 26	40140	0	39866	15
June, Thursday 27	44537	0	44226	19
June, Friday 28	42958	0	42504	24
June, Saturday 29	30543	0	28890	50
June, Sunday 30	23221	0	21306	50

5.4 Comparison between Fixed and Multi-period Solutions for Large-Scale Traffic Data (BM-120 Problem)

In tables 4 and 5, we compare two solutions generated for the BM-120 instance. The first solution is generated using the *FCA* model, and the second using the parallel decomposed algorithm for the multi-period model. As for *D-639* problem, we compare these two solutions in terms of objective function (table 4) and lost traffic quantity (table 5). The first observation is that, during the weekend, frequency plan adaptation requires more changes. For both Saturday and Sunday, the change threshold is reached. This can be explained by the great difference between the traffic situation during the weekend and the remainder days. This observation results in table 4, where we note a great quality improvement during the weekend (the gain is of 8% and 11.4%) in the *MCA* model.

Table 5. Lost traffic recorded for fixed and multi-period solution for the BM-120 instance

Days	Traffic	Fixed	Multi-period	Gain
June, Monday 24	183	3.62	3.60	0.55%
June, Tuesday 25	246	3.65	3.35	8%
June, Wednesday 26	340	3.76	3.71	1%
June, Thursday 27	364	4.28	3.97	7%
June, Friday 28	340	4.07	3.98	2%
June, Saturday 29	338	2.66	2.44	8%
June, Sunday 30	322	1.76	1.59	11.4%
Total	**21133**	**145.98**	**140.46**	

6 Conclusion

In this paper we have proposed a multi-period channel assignment (*MCA*) model for GSM mobile networks. In addition to the classical minimization interference criterion, we introduced another optimization criterion based on the transition cost from the frequency plan of a period to the plan of another one. Compared with the fixed channel assignment model, the proposed model has the advantage of being flexible and adaptive to traffic evolution.

Based on the MCA model, we have developed several optimization techniques to find a sequence of frequency plans for a given time periods. These solution techniques are adapted from a hybrid Genetic Tabu Search algorithm for fixed channel assignment. We proposed two ways of generating a solution for the MCA model: direct optimization in which the best sequence of frequency plans is sought directly; and decomposed optimization in which the whole solution is built by finding frequency plans for each individual periods.

Several experiments on three realistic data sets have been carried out. These data sets include both fine grained (hour by hour) and large scale (day by day) time steps. Experimental results have led to the following observations. First, comparing the different optimization techniques for the *MCA* model on these data sets shows that the sequential and parallel implementation of the decomposed optimization give frequency plans of better quality in terms of the two optimization criteria (global interference and transition cost between frequency plans). Second, when comparing solutions obtained using the *MCA* model and the *FCA* model, one observes that the multi-period model leads to frequency plans of lower interference. Third and most importantly, thanks to the multi-period model, the lost traffic is always reduced, reaching sometimes a gain of communications up to 11.4%. This last point is especially beneficial from an operator's operational point of view.

Finally, let us mention two possible improvements for multi-period frequency assignment. As to the model itself, other optimization objectives may be taken into consideration (as mentioned in Section 2). As to solution techniques, an interesting alternative to the penalty-based aggregation approach used in this study is a true multi-criteria optimization approach that would be certainly worth of investigation.

References

1. Baier, K. Bandelow "Traffic engineering and realistic network capacity in cellular radio networks with inhomogeneous traffic distribution" Proc. of IEEE VTC, 1997, P 780–784.
2. T. H Chan, M. Palaniswani, D. Everitt "Neural network-based dynamic channel assignment for cellular mobile communication systems" IEEE Transaction on Vehicular Technology, Vol 43, N° 2, 1994, P 279–288.
3. L. Chen, S. Yoshida, H. Murata "A dynamic channel assignment algorithm for voice and data integrated TDMA mobile radio" Proc. of IEEE VTC, 1996, P 213–217.
4. R. Dorne, J-K. Hao "An evolutionary approach for frequency assignment in cellular radio networks" IEEE International Conference on Evolutionary Computation, 1995, P 539–544.
5. M. Duque-Anton, D. Kunz, B. Rüber "Channel assignment for cellular radio using simulated annealing" IEEE Transaction on Vehicular Technology, Vol 42, N° 1, 1993, P 14–21.
6. D. Grillo, R. A. Skoog, S. Chia, K. K. Leung. "Teletraffic Engenieering for Mobile Personal Communications in ITU-T Work: The Need to Match Practice and Theory" IEEE Personal Communications, Vol 12, 1998, P 38–58.
7. J-K. Hao, R. Dorne, P. Galinier "Tabu search for frequency assignment in mobile radio networks" Journal of Heuristics 4, 1998, P 47–62.
8. Katzela ,M. Naghshineh "Channel assignment schemes for cellular mobile telecommunication systems: A Comprehensive Survey" IEEE Personal Communications, June 1996, P 10–31.
9. W. Lee, "Mobile communications design fundamentals" Wiley Series in Telecommunications: 1992.
10. H. Mabed, A. Caminada, J-K. Hao "A dynamic traffic model for frequency assignment" Parallel Problem Solving from Nature PPSN VII, Lecture Notes in Computer Science 2439, 2002, P 779–788.
11. S. Matsui, I. Watanabe, H. Tokoro "A parameter-free genetic algorithm for fixed channel assignment problem with limeted bandwidth" Parallel Problem Solving from Nature PPSN VII, L Lecture Notes in Computer Science 2439, 2002, P 789–799.
12. K. Murray, D. Pesch "Adaptive radio resource management for GSM/GPRS networks" First joint IEI/IEE Symposium on Telecommunications Systems Research, 27 Nov 2001.
13. D. Renaud, A. Caminada "Evolutionary Methods and Operators for Frequency Assignment Problem" SpeedUp Journal 11(2), 1997, P 27–32.
14. P. Reininger, S. Iksal, A. Caminada, J. Korczak "Multi-stage optimization for mobile radio network planning" Proc. IEEE Vehicular Technology Conference, 1999, P 2034–2038.
15. J. Tajima, K. Imamura "A strategy for flexible channel assignment in mobile communication systems" IEEE Transaction on Vehicular Technology, Vol 37, N° 2, 1988, P 92–103.
16. V. Wille, H. Multimaki, S. Irons "A practical approach to channel borrowing for microcells in GSM systems" Proc. IEEE VTC, 1998, P 144–148.
17. J. Zander. "Radio Resource Management in Future Wireless Networks: Requirements and Limitations" IEEE Communications Magazine, 1997, P 30–36.

Supporting UMTS QoS in Wireless LANs

Nikos Passas[1], Dimitris Skyrianoglou[1], and Apostolis K. Salkintzis[2]

[1] University of Athens
Department of Informatics & Telecommunications
Communication Networks Laboratory
Panepistimiopolis, 15764, Athens, Greece
{passas, dimiski}@di.uoa.gr
[2] Motorola
Global Telecom Solutions Section
32 Kifissias Av., 15125, Athens, Greece
salki@motorola.com

Abstract. This paper focuses on how the UMTS Quality of Service (QoS) architecture can be supported in Wireless LANs (WLANs), deployed as alternative access networks to the UMTS core. More specifically, the paper addresses the provision of UMTS QoS over WLANs and proposes a solution based on the introduction of the Wireless Adaptation Layer (WAL). The WAL provides means for extending the UMTS QoS in WLAN environments, independently of the underlying WLAN technology. Moreover, the WAL configuration to support the provisioning of UMTS QoS is discussed. In this context, a mapping between UMTS QoS attributes and WAL modules is proposed and justified.

1 Introduction

The Universal Mobile Telecommunications System (UMTS) is expected to allow the operators and service providers to extend the set of available applications offered to the user; nevertheless it will not be capable of supporting all the advanced bandwidth-demanding applications. Even in indoor environments, where a 2 Mbps channel will be offered, applications such as high-quality video will not always be feasible. Users will have to wait for the fourth generation systems (4G), which promise to increase the available bandwidth by offering a variety of broadband access systems. One way to extend the set of applications offered by UMTS to hot-spots and improve the QoS, is the interworking of UMTS with wireless local area networks (WLANs). Technology in the area of WLANs has been evolved during the last few years to offer efficient communications at rates in the range of tens of Mbps. Two of the major standards for WLANs today are the IEEE 802.11 [1] and the ETSI HIPERLAN/2 [2].

WLAN technologies can complement rather than compete with UMTS systems, offering alternative high performance access systems. UMTS/WLAN interworking can combine the wide-area benefits of second- and third- generation mobile systems, including unlimited roaming and mobility, with the additional throughput and capacity of WLANs in strategic local areas. An essential requirement for UMTS/WLAN tight interworking [3] concerns the compatibility of the QoS offered

M. Conti et al. (Eds.): PWC 2003, LNCS 2775, pp. 555–567, 2003.

by the two systems. Different WLANs technologies offer different capabilities for QoS, making it difficult to guarantee the QoS agreed between the user and the UMTS network. In this paper, we address this requirement by introducing an intermediate layer that accepts the UMTS user profiles, as expressed by specific QoS attributes, and performs certain actions in order to guarantee the committed QoS when a WLAN is used as a UMTS access network.

The paper is organized as follows. In section 2, a brief outline of the QoS defined for UMTS is presented. Section 3 discusses issues and alternatives for UMTS/WLAN interworking. In section 4, a solution for QoS interworking is described, focusing on traffic classes and attributes mapping. Finally, section 5 presents our conclusions.

2 QoS in UMTS

The QoS architecture in UMTS is based on the concept of *Bearer Service*, following a layer-based approach. As stated in [4], in order to realize a certain level of QoS, a Bearer Service with clearly defined characteristics and functionality has to be set up from the source to the destination of the service. The QoS architecture of UMTS follows a layer-based approach. Each layer consists of one or more bearer services that are concatenated in order to provide the desired QoS provision on an end-to-end basis (see Figure 1).

The UMTS Bearer Service provides QoS support within the UMTS network, i.e., from the Mobile Terminal (MT) to the Core Network (CN) Gateway. UMTS Bearer Service is further divided into the *Radio Access Bearer Service*, which provides confidential transport between MT and CN Iu Edge Node, and the *Core Network Bearer Service*, which connects the CN Iu Edge Node with the CN Gateway and through this with the external network.

2.1 UMTS QoS Classes and Attributes

Four different QoS classes are defined for the UMTS bearer service, taking into account the nature of the traffic produced by the various applications that are used over UMTS, as well as the restrictions and characteristics of the wireless interface.

The four classes are defined as follows [4]:

- **Conversational Class**. It includes real-time services such as speech and videoconferencing. Services belonging to the conversational class have strict transfer delay constraints.
- **Streaming Class**. This class applies to services where a stream of real-time data is destined to a user (e.g., video or audio streaming). The flow of data is unidirectional while the destination is usually a human. In contrast to the conversational class, delay constraints are not very strict, but the need for preserving the time relation (delay variation) remains important.
- **Interactive Class**. Applications such as web browsing or database retrieval fall into this class. There is an interactive scheme involved with the end-user requesting data from a remote host or equipment. In this case, a key aspect is the round-trip delay, which should be kept within reasonable limits.

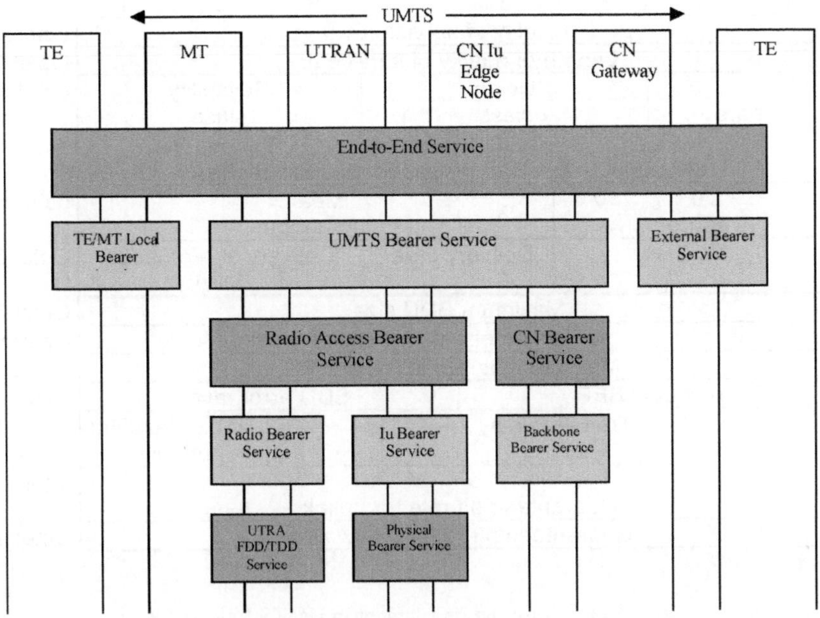

Fig. 1. UMTS QoS Architecture

- **Background Class**. This class includes services such as e-mail, or file downloading. All these services are characterized by the fact that the end-user is expecting the data within a large period of time, therefore time constraints do not normally apply. However the content of the data packets should be preserved so the bit error rate should be kept low.

Figure 2 presents the QoS Information Element of UMTS Release 99, used to specify the QoS attributes for UMTS Bearer Service [4]. More specifically, the QoS Information Element includes the following set of QoS attributes:

- **Traffic Class:** It defines the type of application (class) for which the bearer service is optimized. UMTS defines four traffic classes, as described earlier.
- **Maximum Bit-rate:** It is the maximum number of bits delivered by UMTS and to UMTS at a particular Service Access Point (SAP) within a certain period of time divided by the duration of the period. The traffic is conformant with the maximum bit-rate as long as it follows a token bucket algorithm where token rate equals maximum bit-rate and bucket size equals maximum SDU size.
- **Guaranteed Bit-rate:** It is the guaranteed number of bits delivered by UMTS at a particular Service Access Point (SAP) within a certain period of time divided by the duration of the period (provided that there are data to deliver). The traffic is conformant with the guaranteed bit-rate as long as it follows a token bucket algorithm where token rate equals guaranteed bit-rate and bucket size equals Maximum SDU size.
- **Delivery Order:** Indicates whether the bearer service should provide in-sequence delivery of application data packets, referred to as Service Data Units (SDUs).
- **Maximum SDU Size:** Indicates the maximum SDU size.

<table>
<tr><th>8</th><th>7</th><th>6</th><th>5</th><th>4</th><th>3</th><th>2</th><th>1</th><th></th></tr>
<tr><td colspan="8">Quality of service IEI</td><td>octet 1</td></tr>
<tr><td colspan="8">Length of quality of service IE</td><td>Octet 2</td></tr>
<tr><td colspan="2">0 0
spare</td><td colspan="3">Delay
class</td><td colspan="3">Reliability
class</td><td>octet 3</td></tr>
<tr><td colspan="4">Peak
throughput</td><td>0
spare</td><td colspan="3">Precedence
class</td><td>octet 4</td></tr>
<tr><td colspan="3">0 0 0
spare</td><td colspan="5">Mean
throughput</td><td>octet 5</td></tr>
<tr><td colspan="3">Traffic Class</td><td colspan="2">Delivery order</td><td colspan="3">Delivery of erroneous
SDU</td><td>Octet 6</td></tr>
<tr><td colspan="8">Maximum SDU size</td><td>Octet 7</td></tr>
<tr><td colspan="8">Maximum bit rate for uplink</td><td>Octet 8</td></tr>
<tr><td colspan="8">Maximum bit rate for downlink</td><td>Octet 9</td></tr>
<tr><td colspan="4">Residual BER</td><td colspan="4">SDU error ratio</td><td>Octet 10</td></tr>
<tr><td colspan="6">Transfer delay</td><td colspan="2">Traffic Handling
priority</td><td>Octet 11</td></tr>
<tr><td colspan="8">Guaranteed bit rate for uplink</td><td>Octet 12</td></tr>
<tr><td colspan="8">Guaranteed bit rate for downlink</td><td>Octet 13</td></tr>
</table>

Fig. 2. QoS Information Element in UMTS Release 99

- **SDU Format Information:** Defines a list of possible exact SDU sizes. This information is usually used by the UMTS radio access network in order to achieve spectral efficiency and lower transfer delay.
- **SDU Error Ratio:** It indicates the acceptable fraction of SDUs that can be lost or detected as erroneous.
- **Residual Bit Error Ratio:** Indicates the undetected bit error ratio in the delivered SDUs. If no error detection is requested, residual bit error ration indicates the bit error ratio in the delivered SDUs.
- **Delivery of Erroneous SDUs:** Indicates whether SDUs detected as erroneous shall be delivered or discarded.
- **Transfer Delay:** Defines maximum delay for the 95th percentile of the distribution of delay for all delivered SDUs during the lifetime of a bearer service.
- **Traffic Handling Priority:** Specifies the relative priority of an interactive-class bearer with respect to other interactive-class bearers.
- **Allocation/Retention Priority:** Specifies the relative importance of allocating and retaining a UMTS bearer compared to other UMTS bearers. It is used for admission control during congestion periods.

For each traffic class within UMTS a different set of QoS attributes is applicable. Table 1 shows which parameters are applicable to each traffic class.

Table 1. QoS attributes defined for each Traffic Class

Attributes	**Conversational**	**Streaming**	**Interactive**	**Background**
Maximum Bitrate	X	X	X	X
Guaranteed Bitrate	X	X		
Delivery Order	X	X	X	X
Maximum SDU Size	X	X	X	X
SDU Format Information	X	X		
SDU Error Ratio	X	X	X	X
Residual Bit Error Ratio	X	X	X	X
Delivery of erroneous SDUs	X	X	X	X
Transfer Delay	X	X		
Traffic Handling Priority			X	
Allocation/Retention Priority	X	X	X	X

3 UMTS Interworking with WLANs

Three types of interworking have been identified by ETSI [5] and the choice between them is mainly a trade-off between minimizing additional functionality and increasing the seamlessness of interworking [6]:

- **No Coupling:** The UMTS and WLAN are used as completely independent networks and the user has to sign up for different contracts in each network. Only Internet services are available to the user in the WLAN, through a standard IP router. This type of interworking enables a rapid introduction of WLAN in UMTS and is suitable for all WLAN technologies. On the other hand, it provides poor transition performance and complete separation between UMTS and Internet services.
- **Loose Coupling:** The main difference with the no coupling case is that the WLAN and UMTS networks share the same subscribers database, stored in the core UMTS network. Authentication, Authorization, and Accounting (AAA) functions in the WLAN access directly the UMTS control functions, in order to perform the

corresponding operations. On the other hand, there is complete separation between the services the terminal can access in the UMTS and the WLAN.

- **Tight Coupling:** The WLAN is considered by the core UMTS network as another access network (such as UTRAN, GERAN, etc.). This enables the QoS and security functions of UMTS to be reused for WLAN, through proper adjustments. Handovers are possible for UMTS connections, while the user only needs to subscribe to one network operator. Access to the Internet is also supported, but through the UMTS core in this case. This type of interworking offers improved transition performance and complete service interworking, but it requires full control and user data translation.

A possible configuration for a UMTS/WLAN interworking architecture is depicted in Figure 3. The WLAN consists of a number of Access Points (APs) that are interconnected through a Distribution System (e.g., an 802.3 Ethernet cable). The WLAN is connected:

- to the UMTS core network through the Inter-Working Unit (IWU), which is responsible for translating and adapting signalling (loose and tight coupling only) and data (tight coupling only) from one system to the other, and,
- to the Internet, through a standard IP router (no and loose coupling only).

A possible scenario could involve a dual mode mobile terminal that leaves the coverage area of UMTS and enters the coverage area of a WLAN (e.g., in a trade center or an airport). It would be desirable for the MT to maintain a QoS level within the coverage area of WLAN and enjoy an extended set of services, with the same or possibly improved level of QoS support, including both UMTS and Internet services.

In the next section, we focus on the tight coupling scenario, trying to provide solutions for mapping of the UMTS traffic classes and QoS attributes to specific WLAN behavior.

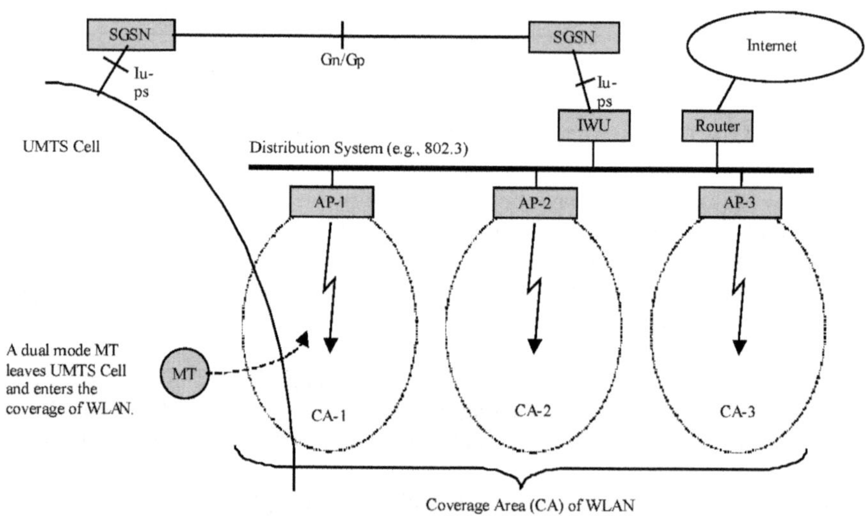

Fig. 3. UMTS-WLAN Tight-Coupling Interworking Configuration

4 The Use of the WAL in UMTS/WLAN Interworking

4.1 The Wireless Adaptation Layer (WAL)

In the communication of WLAN with both the UMTS and the Internet, QoS support is required in order to guarantee a specific end-to-end QoS for all services. Different WLAN technologies offer different or incompatible QoS capabilities, making it essential to design a unique QoS supporting scheme, applicable to all. It is clearly desirable to use the same scheme for both UMTS and Internet communications, in order to minimize complexity.

A unique QoS supporting scheme has already been proposed for WLAN communications with the Internet, in the context of the European Commission funded project WINE, referred to as the Wireless Adaptation Layer (WAL) [8]. The WAL is an intermediate layer, between the IP and the WLAN Data Link Control (DLC) layers, providing a QoS mechanism adaptable to both the capabilities of the underlying WLAN technology and QoS requirements. It provides a uniform interface to IP, while being independent of the underlying wireless network technology. The WAL incorporates a set of functional modules that can be dynamically combined and adapted to the special characteristics of the wireless link and the transport protocol. Here we extend the WAL's usage to a UMTS environment, in order to support the tight coupling interworking.

The concept of the WAL is depicted in Figure 4. The central "intelligence" of the WAL is the **WAL Coordinator**. In the downstream, it accepts data packets from the upper layers and decides on the appropriate WAL chain. In the upstream, the WAL Coordinator receives data packets from the lower layers and passes them through the same WAL chain in the reverse order.

The **QoS** module provides flow isolation and fairness guarantees through traffic shaping and scheduling. Other modules that are considered are:

- **ARQ module (ARQ)**: Can be used to improve packet error rate of non-real-time traffic (for example interactive or background) through retransmissions of erroneous packets.
- **FEC module (FEC)**: Its use is recommended for real-time-traffic (usually conversational and streaming) because it can reduce packet error rate without increasing delays like ARQ.
- **Fragmentation module (FRM)**: Can decide on the recommended fragment length based on channel conditions and traffic parameters.
- **Header Compression module (HC)**: This module can be used to reduce the overall header overhead, leading to increase of the available bandwidth.

The WAL architecture is not restrictive concerning the number or kind of modules. One of the main considerations is openness in order to introduce new modules where necessary. A detailed description of the WAL can be found in [8].

In order to be compatible with the UMTS QoS scheme, UMTS Service Bearers correspond in the WAL layer to the so-called *"WAL Bearers"*. In this way, the Radio Bearer Service of Figure 1 can be divided in two layers, the WAL layer and the standard WLAN DLC layer (e.g., 802.11 DLC). Every WAL Bearer can be treated in a different way in the WAL layer, based on the QoS for the corresponding UMTS Service Bearer and the capabilities of the standard Radio Bearer Service (i.e., WLAN

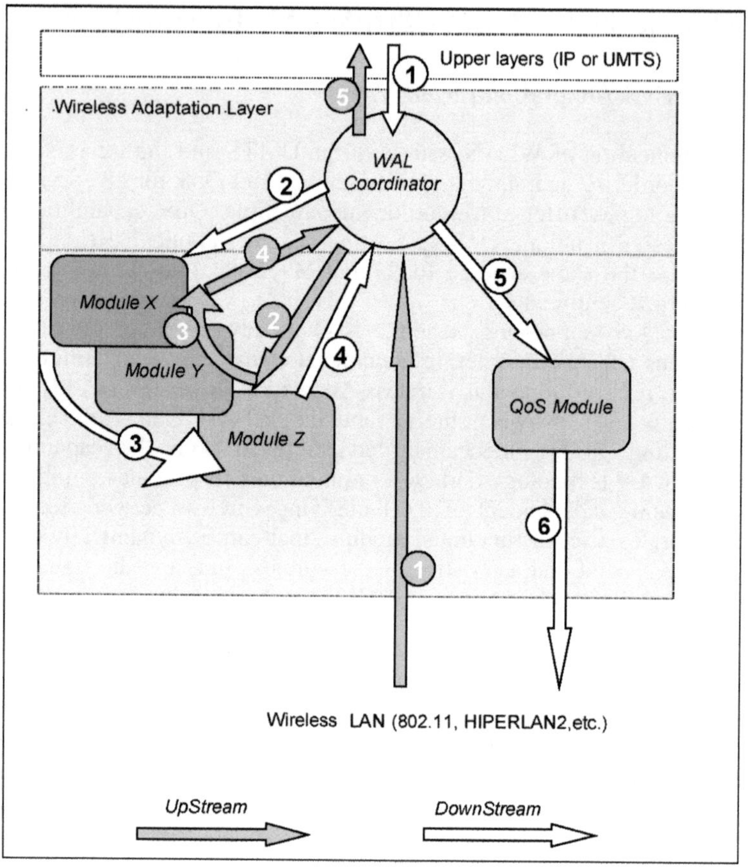

Fig. 4. The WAL design

technology). Treatment is characterised by the sequence of WAL modules X/Y/Z (referred to as a *WAL Chain*) that the packets of a WAL Bearer will pass through, as well as the specific parameters of these modules. These parameters can be adjusted dynamically by the WAL Coordinator, based on time-varying channel or traffic conditions.

4.2 Mapping of UMTS QoS Attributes to WAL Parameters

The WAL can be appropriately configured in order to be compatible with the communication of WLAN with both the UMTS (tight coupling) and Internet (no- or loose-coupling). Here we study the former case of UMTS/WLAN QoS interworking; the latter case has been investigated in [9].

Seen from the UMTS QoS architecture perspective, the WAL (together with the underlying WLAN) should implement a Radio Bearer Service. In this respect, it is necessary to map the QoS attributes of the Radio Bearer Service onto the appropriate semantics and attributes of the WAL modules. Since the WAL has a large degree of parameterization in terms of modules and module parameters, it leaves enough space for mapping of the UMTS QoS. Below we discuss on how the UMTS QoS attributes described in section 2, can be mapped onto WAL modules. The discussion focuses on the user plane. Future work will include the procedures for configuration of the WAL bearers through the control plane, for example when a PDP context is created. Although we give a high-level approach of the mapping, it is shown that all the UMTS QoS attributes can be utilized by one or more WAL modules, and adjust their performance per bearer. The specific usage is part of each module's operation.

Traffic Class

Each UMTS traffic class is mapped to a specific WAL chain (i.e., sequence of WAL modules) that can efficiently service this class. A possible mapping can be as shown in Figure 5. The choice of the modules in each chain is based on the nature and requirements of each UMTS QoS Class, such as delay and loss requirements. Delay sensitive classes, such as Conversational or Streaming, cannot use ARQ, since this would add extra delay; they instead use FEC for error correction. ARQ is used in the not so demanding Interactive Class. Background Class is not using error control, since the service offered to it is considered as best-effort. Possible errors can be recovered end-to-end through retransmissions. Header Compression and Fragmentation are used in all classes, since they can improve the performance in the wireless medium, provided proper configuration. Additionally, the order of the modules should be chosen so that the preceding modules do not cause conflicts or degrade the operation of the succeeding ones. For example, the QoS module is always placed last in every WAL chain, so as to guarantee that the scheduled packets will not experience any further delays due to processing or queuing.

Maximum Bit-Rate

It can be used within the transmitting WAL QoS module, so as to impose constraints on the maximum transmission rate for a particular WAL Bearer. Knowing the maximum bit-rate of the bearer, the QoS module can ensure that this is never violated by bursts of packets, causing QoS degradations to other bearers. The incorporation of Maximum Bit-rate within the QoS module can be achieved with the use of a token bucket, according to the definition of this attribute in [4].

Guaranteed Bit-Rate

This attribute can be used both at the transmitting and the receiving side of a WAL Bearer. At the transmitting side, it can be mapped on a specific parameter of the QoS module that determines the mean transmission rate of the WAL Bearer that should be policed. Appropriate traffic shaping can also be performed, according to this attribute. At the receiving side, it can be used by the WAL Coordinator, in order to determine the mean bandwidth offered to a specific bearer. In case the WAL fails to provide the Guaranteed Bit-rate for a bearer service (due to excess traffic or low channel quality), the WAL Coordinator at the receiving side may request an adjustment of the parameters of the QoS module at the transmitting side, so as to provide extra bandwidth to the bearer service or change the parameters of other modules (e.g., FEC, ARQ, Fragmentation) in order to improve the performance of the wireless link.

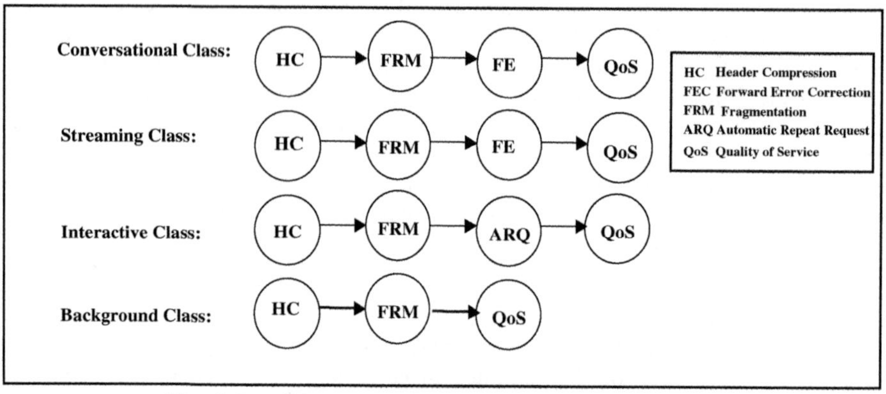

Fig. 5. Possible Module Chains for each UMTS QoS Class

Delivery Order

This attribute can be applied both to the ARQ module and the WAL Coordinator, which is responsible for delivering the packets to the upper layers. In case the Delivery Order attribute is set to "yes", either the WAL Coordinator or the ARQ module (for Interactive Class only) need to maintain a buffer where the out-of-order packets are stored until the preceding packets have been received from the lower layers and delivered to the upper layers. In case the delivery order is set to "no", the packets can be immediately forwarded to the upper layers.

Maximum SDU Size

It can be directly applied to the Fragmentation Module to set the maximum size of fragments that can be applied. The fragmentation module should adapt the fragments' size to the channel conditions without violating this parameter. When the channel is in a good state, the Fragmentation module prefers long SDU sizes, to increase the throughput and decrease the transmission overhead, but should never exceed the maximum SDU size. When the channel condition deteriorates, the Fragmentation module sets shorter SDU sizes, which are less vulnerable to channel errors. Maximum SDU size can also be used by the WAL Coordinator for admission control and traffic policing [4], and, the Header Compression module for determining the optimal compression ratio.

SDU Format Information

Similar to the Maximum SDU Size, this attribute can be utilized by the Fragmentation module and the Header Compression module, in order to optimize their operation by determining the allowed sizes of the SDUs.

SDU Error Ratio

This attribute can be used as a parameter to the FEC or the ARQ module. In order to preserve the agreed SDU Error Ratio, FEC may need to utilize a more powerful CRC scheme, by using longer error correction codes. The ARQ module may have to decide to adapt the parameters of the ARQ scheme (e.g., increase the number of allowed retransmissions, or enlarge the window size for a specific bearer), in order to maintain error ratio below this value. This attribute can also be utilized by the WAL Coordinator as a threshold for triggering an adjustment of the WAL module parameters.

Residual Bit Error Ratio
Since bit error detection/correction is performed within the FEC module, this attribute can be used to select the appropriate error code. It determines the "undetectable" errors and therefore corresponds to the efficiency of error detecting schemes. As in the SDU Error Ratio case, the Residual Bit Error Ratio can also be used by the WAL Coordinator, in order to trigger adjustment of WAL module parameters.

Delivery of Erroneous SDUs
This attribute can be used by the FEC module or the WAL Coordinator, to decide on how to treat erroneous SDUs. FEC can mark the erroneous SDUs while the WAL Coordinator may decide whether they are forwarded to the upper layers or discarded, based on this attribute.

Transfer Delay
It can be directly used by the QoS module to schedule and prioritize the transmission of SDUs, so as not to violate this tolerance. It can also be utilized by the ARQ module in order to determine the number of retransmissions that are allowed for a specific SDU before it exceeds its delay deadline or set the window size and the duration of the window reset period. SDUs that do not succeed to maintain this value should be discarded.

Traffic Handling Priority
This attribute can be used by the QoS module during scheduling for determining the priority of packets belonging to different WAL bearers. According to [4], it is used as an alternative to absolute guarantees (e.g., Transfer Delay) and thus the two attribute types (relative/absolute) cannot be used together for a single bearer. Based on relative priorities, this attribute may lead to simpler and faster scheduler implementations, reducing the overall cost.

Allocation/Retention Priority
It can be used by the WAL Coordinator upon admission control and instantiation of a new WAL bearer, to determine the importance of this bearer. In situations where resources are reduced (e.g., due to channel quality deterioration), the WAL Coordinator may use this attribute to prioritize bearers with high allocation priority over bearers with lower priorities. If bearer discard is needed, lower priority bearers should be chosen first.
Table 2 shows a possible mapping of all UMTS QoS attributes to WAL modules, based on the above discussion. The specific use of the attributes is part of the operation of the modules.

Table 2. Mapping of UMTS attributes to WAL modules

WAL Modules	WAL Coordinator	ARQ	FEC	Header Compression	Fragment ation	QoS
Maximum Bitrate	X					X
Guaranteed Bitrate	X					X
Delivery Order	X	X				
Maximum SDU Size	X			X	X	
SDU Format Information				X	X	
SDU Error Ratio	X	X	X			
Residual Bit Error Ratio	X		X			
Delivery of erroneous SDUs	X		X			
Transfer Delay		X				X
Traffic Handling Priority						X
Allocation/Retention Priority	X					

5 Conclusions

In this paper, we discussed a framework for QoS interworking between UMTS and WLANs. The approach presented here is based on the WAL, an intermediate shim layer between the upper layers and the WLAN, aiming at improving its performance.

The WAL features a modular architecture that can be easily extended and adapted to meet both the different requirements from the upper layers and the different QoS capabilities of the lower layers. Moreover, the WAL operates transparently to both the upper and lower layers, thus it does not require any modifications or enhancements of the existing technologies. It has also the ability to adapt to the changing channel conditions by the dynamic adaptation of the WAL module parameters. The paper presented a mapping of all the UMTS QoS attributes, as included in the QoS Information Element of UMTS, to specific WAL modules, in order to parameterize their operation per bearer. It was shown that every QoS attribute can be utilized by one or more WAL modules.

Conclusively, the WAL appears as a generic and flexible solution for the UMTS/WLAN QoS interworking. Future work will describe in more detail the operation of each WAL module, based on UMTS QoS attributes, and evaluate the performance of the WAL against no-WAL architectures.

References

1. "IEEE Standard for Wireless LAN Medium Access Control (MAC) and Physical Layer (PHY) Specifications", P802.11, Nov. 1997.
2. ETSI TS 101 761-1 V1.2.1 (2000-11) "Broadband Radio Access Networks (BRAN); HIPERLAN Type2; Data Link Control (DLC) Layer; Part 1: Basic Data Transport Functions", available at http://www.etsi.org.

3. Salkintzis et al., "WLAN-GPRS Integration for Next Generation Mobile Data Networks," IEEE Wireless Communications, vol. 9, no. 5, pp. 112–124, Oct. 2002.
4. 3GPP TS 23.107 V5.5.0, "Universal Mobile Telecommunication System (UMTS; QoS Concept and Architecture", 2002, available at http://www.3gpp.org.
5. ETSI TR 101 957 V1.1.1, "Broadband Radio Access Networks (BRAN); HIPERLAN Type2; Requirements and Architectures for Interworking between HIPERLAN/2 and 3rd Gebration Celluler Systems", 2001, available at http://www.etsi.org.
6. J. Berkwist, J. Engren, O. Eriksson "WLAN as a Complement to UMTS", Master Thesis, available at http://www.s3.kth.se/~matthias/current/Ericsson.pdf.
7. 3GPP, "Universal Mobile Telecommunication System (UMTS; Core Network Protocols – Stage 3", TS 24.008 v.3.11.0, 1999, available at http://www.3gpp.org.
8. P. Mähönen, T. Saarinen, N. Passas, et. al, "Platform-Independent IP Transmission over Wireless Networks: The WINE Approach", IEEE Personal Communications Mag. Vol.8 No.6, Dec. 2001.
9. P. Mähönen, T. Saarinen, N. Passas, et. al, "Platform-Independent IP Transmission over Wireless Networks: The WINE Approach", IEEE Personal Communications Mag. Vol.8 No.6, Dec. 2001.
10. D. Skyrianoglou, N. Passas, and S. Kampouridou, "A DiffServ-based Classification Scheme for Internet Traffic Over Wireless Links", in Proc. ICWLHN 2001, Singapore, Dec. 2001.

"On the Integration of MPEG-4 Streams Pulled Out of High Performance Mobile Devices and Data Traffic over a Wireless Network"

Spyros Psychis, Polychronis Koutsakis, and Michael Paterakis

Electronics & Computer Engrg. Dept. &
Telecommunication Systems Institute
Technical University of Crete
73100, Chania, Greece
{psycho,polk,pateraki}@tsinet.gr

Abstract. In this work we propose and evaluate mechanisms for the multiplexing and the integrated delivery of Video and Data Traffic over a Wireless Cellular High Speed Packet Switched Network. We envision a system where the Mobile Terminals are considered to be high performance devices with extended storage capabilities which can act like cache memories streaming multimedia material. We focus on the uplink channel and we thoroughly investigate the system's performance under a variety of possible loads which consist of actual MPEG-4 streams and Data Traffic.

Keywords: Wireless Packet Switched Networks, TDMA, MAC, Media Access Control, MPEG-4 streams

1 Introduction

The coexistence of wired and wireless communication networks is a hint that none of the two network types is able to service, on its own, all of the users needs. The basic advantage of wireless communication is that it allows the users to move, even during the time they use the network services. The goal of wireless communication networks is to allow user access to the capabilities of the wired networks at any time without regard to location or mobility. Thus, problems and issues that until now appeared only on wired networks now must be solved on wireless and vice versa. And of course the main concern in wireless communication is the same like in wired: scaling, in other words how to extend the broadband frontier to the end user. Therefore, the system capacity needs to grow as much as the demands of the various services grow.

System capacity can be increased by:

a. using a cellular structure with a cell size as small as possible (microcells) in order to reduce transmitting power and to increase frequency reuse. Microcell diameters are usually of the order of a few hundred meters; therefore the round-trip propagation delay within a microcell is negligible (of the order of 1 μs or less),

b. using efficient medium access control (MAC) protocols to exploit the variations in access and service required by the disparate user sources.

M. Conti et al. (Eds.): PWC 2003, LNCS 2775, pp. 568–577, 2003.

Within the microcell, spatially dispersed mobile terminals (MTs) share a radio channel that connects them to a fixed base station or a wireless hotspot. The base station allocates the channel resources (transmission scheduling), delivers feedback information and serves as an interface to the mobile switching center (MSC), which provides access to the fixed network infrastructure and the Internet.

Another important problem is the optimization of the usage of the system capacity.

During the last five years a lot of work has been done on various caching schemes that permit the fair and effective sharing of the network resources [1,2,3]. Most of them have been deployed especially for the retrieval and delivery of high quality stored multimedia content. And since the notion of broadband access is transferred to the wireless field, most of these caching architectures must be extended. In this work, Mobile Terminals are considered to be high performance devices with extended storage capabilities which can act like cache memories, streaming multimedia material to other MT's (in the same or in adjacent micro cells) in order to reduce client start up latencies and improve the network performance. Even the whole micro cell can be faced as a cache farm from the rest of the network. Therefore, we envision high quality stored video streaming on the uplink channel (MT's to the BS) in order for these streams to be delivered for playback to their destinations.

The delivery of high quality stored video material is a service with very strict QoS requirements. Consequently the design of MAC strategies that integrates this class of service with others, which often require contradictory QoS guaranties, is necessary.

As mentioned earlier, in this work we focus on the uplink (mobiles to base station) channel, where a MAC scheme is required in order to resolve the source terminals contention for channel access.

As for the downlink channel, since the base station is the sole transmitter on that channel, it is in complete control of the downstream traffic, using TDMA to relay information to the users.

2 MPEG-4 and Media Streaming

MPEG-4 is an ISO standard that provides a standardized and worldwide-accepted technological framework that enables the integration of the production, storage distribution and content access paradigms. MPEG-4 leverages existing digital video content by supporting the MPEG-1 and MPEG-2 coding standards. Furthermore, it enables richer development of new digital video applications and services. An MPEG-4 scene is composed by a number of audiovisual objects, which can be either static or time-varying in nature. [4]

MPEG-4 Visual provides a natural video coding algorithm that is capable of operation from 5 kbps with a spatial resolution of QCIF (144x176 pixels). It is ITU-T H.263 compatible, in the sense that an MPEG-4 video decoder also correctly decodes an H.263 bit stream. The algorithm scales up to bit rates of some Mbit/s and optimization has been carried for ITU-R 601 resolution pictures (288x720@50Hz and 240x720@59.94 Hz). The recently developed Studio Profile can operate at over 1 Gbit/s. In addition MPEG-4 has a so-called Fine Granularity Scalability mode that allows transmission of the same video content at different bit rates from one coded version of the movie [10].

Table 1.

Movie Name	Mean Bit rate (Kbps)	Peak Bit rate (Kbps)
News	720	3400
Parking Lot Cam	790	2800
Silence of the Lambs	580	4400
The Simpsons	1300	8800
Soccer	1100	3600
Star Wars	280	1900

Movie Name	Mean Bit rate (Kbps)	Peak Bit rate (Kbps)
News	720	3400
Parking Lot Cam	790	2800
Silence of the Lambs	580	4400
The Simpsons	1300	8800
Soccer	1100	3600
Star Wars	280	1900

In our study, we use the trace statistics of actual MPEG-4 streams from [11]. The video streams that we used, were carefully selected in order to cover a broad range of movie characteristics. Table 1 shows the movies we used together with their mean and peak bit rates.

Figures 4 through 9 (in page 9) show the histograms of the bit rate of each movie. Horizontal axis shows for each video frame the corresponding bit rate quantized at 1 Kbps sectors. Vertical axis shows the number of video frames that correspond to a specific bit rate. Histograms delineate the bit rate behavior for each movie; a spread distribution corresponds to a bursty stream. Notice that the axes are not necessarily under the same scale.

All the streams are encoded at 25 video frames per second, which means that a new video frame is generated every 40 msecs. Since the video streams are pulled out from Mobile Terminals that act like caches and not dedicated cache engines with optimization or smoothing functionalities, no packet shaping mechanisms were used. Video content is delivered to the network in the initial form that was encoded. In our study we have made the assumption that the maximum transmission delay for video packets is 40 msecs, with packets being dropped when this deadline is reached. That is, all video packets of a VF must be delivered before the next VF arrives. The allowed video packet dropping probability is set to 0.0001 from [5].

3 Data Traffic

We adopt the data traffic model based on statistics collected on email usage from the Finish University and Research Network (FUNET) [6]. The model simulates email data traffic. The probability distribution function f(x) for the length of the data messages was found to be well approximated by the Cauchy (0.8,1) distribution. The message inter-arrival time distribution for the FUNET model is exponential. We assume an upper limit on the average data message delay equal to 2 seconds.

4 Frame Structure, BS Scheduling, and Terminal Actions

The uplink channel time is divided into time frames of equal length. Each frame consists of two types of intervals. These are the request intervals and the information intervals. Each of these intervals is divided into a number of time slots.

Within an information interval, each slot accommodates exactly one, fixed length, packet that contains video or data information and a header. Each request slot is subdivided into two mini-slots and each mini-slot accommodates exactly one, fixed length, request packet. By using more than one mini slot per request slot, a more efficient usage of the available request bandwidth is possible, [7]. The base station broadcasts a short binary feedback packet at the end of each mini-slot indicating only the presence or absence of a collision within the mini-slot (collision (C) versus non-collision (NC)). Upon successfully transmitting a request packet the terminal waits until the end of the corresponding request interval to learn of its reservation slot (or slots). If unsuccessful within the request intervals of the current frame, the terminal attempts again in the request intervals of the next frame. A terminal with a reservation transmits freely within its reserved slot [5].

We adopt the idea that video and data terminals can share the request slots (first by the video terminals, and after the video contention period, by the data terminals). The reason we require video terminals to contend in order to reserve information slots is that we are dealing with high quality stored video content which, in contrast to videoconference streams, generates highly bursty information. Video requirements change frequently over time and it is impossible to ensure that their demands will reach the BS successfully if we use techniques similar to the ones described in [7],[8]. The two-cell stack reservation random access algorithm [9] is used to resolve the collisions, among video request packets.

After the end of the video contention period, data terminals without a reservation and with message ready for transmission, begin their contention in order to reserve information slots. In the case that two or more data request packets collide, the two-cell stack random access algorithm [14] is used to resolve the collision, due to its operational simplicity, stability and high throughput.

The BS allocates channel resources at the end of the corresponding request interval, and follows a different allocation policy for video terminals than that for data terminals.

Video terminals have higher priority in acquiring the slots they demand. If a full allocation is possible, the BS then proceeds to allocate any still available information slots to the requesting data terminals. Since we assume here that data slot reservations can not be preempted in favor of video terminals waiting for transmission, we choose not to allocate more than one slot per frame to a requesting data terminal. Otherwise, if a full allocation is not possible, the BS grants to the video terminals as many of the slots they requested as possible (i.e., the BS makes a partial allocation). The BS allocates the earliest available information slots to the video terminals, which, if needed, keep these slots in the following channel frames, until the next video frame (VF) arrives.

There are two reasons for adopting the partial allocation policy, described above, to the video users. The first is that in our system there are 10 available reservation slots (20 minislots) in each channel frame, which practically means that almost all of the possible collisions among video terminal request packets can be resolved within the contention period of one channel frame. The second reason is that the video terminals are transmitting high quality bursty video. The size of a video frame is usually of the order of a few kilobytes, which means that after segmentation it corresponds to a few hundred channel cells (of ATM size). Especially in MPEG-4 coding, the correlation factor between two adjacent video frames, is extremely high. Therefore, even in the case that a few tens of cells do not reach the destination; the decoder can successfully reconstruct the video frame. Additionally, MPEG-4 incorporates very capable error correction techniques and can therefore suffer relatively higher packet loss rates compared to other media standards without serious degradation of the playback quality. Consequently, we try to conform to the QoS requirement on packet dropping while spreading the packet losses in time. This spreading of losses can lead to an imperceptible blurring during video playback, which is expected to be acceptable when watching a movie, instead of concentrating the lost cells in time, and consequently receiving video "hiccups" and severe losses in the decoder.

When a video terminal wants to decrease its bit rate, it releases all the slots that were previously allocated to it and are no longer necessary. The BS realizes the change in the bit rate when the first currently unnecessary reserved slot is released by the video terminal (and is observed empty by the BS in the current frame), and consequently allocates the newly released slots to any other requesting terminals, after the end of the request interval of the next frame

When the bit rate of a video terminal increases then the terminal must enter the contention process in order to acquire the additional information slots it needs. In such case, the terminal releases all of its currently reserved information slots before entering the contention period. When it passes the contention period successfully, then the required information slots are reserved and the video terminal can freely transmit within them. The rationale behind this policy is fairness. We prefer to let each video terminal compete with other video terminals at an equal basis (since video has absolute priority), in order to "spread" the video packet losses uniformly.

5 System Parameters

The channel rate is 20Mbps (from [12]). As mentioned before, in this study we investigate the case where the system traffic includes video and data. Nevertheless, high capacity wireless channels will often be used for integrating voice, video and data traffic, and the frame duration is usually chosen to be equal to the time a voice terminal needs to generate a new voice packet [13].

Table 2.

Movie		λ	Throughput %
A	News	5.1	86.47266
B	Parking Lot Cam	4.85	83.91547
C	Silence of the Lambs	4.78	79.1234
D	The Simpsons	3.1	68.67518
E	Soccer	5.05	92.85171
F	Star Wars	5.85	88.76075
Average		4.788333	83.29986

Assuming a speech codec rate of 32 Kbps and that the information packet length is equal to the size of an ATM cell, yields the channel frame duration of 12 ms. The 12 ms of frame duration accommodate 566 slots (556 information slots plus 10 reservation slots). Consequently, the above channel's information payload transmission speed is slightly above 18Mbps (because of the ATM headers encapsulation and the existence of reservation slots).

Finally, given that the packet length is equal to the size of an ATM cell, it can be seen that the average data message length is equal to 80 packets.

6 Simulation Results and Discussion

An extensive simulation study was carried out. Each run simulated one hour of actual network activity (300005 channel frames). We simulated the system under all possible movie loads from 1 to 6 movies. For each load, all different combinations were examined (63 in total), from 5 times each (Monte Carlo method). The lack of similar work in the literature (i.e., proposed mechanisms for handling Video Streams of this quality, together with performance evaluation of such mechanisms) prevents us from making result comparisons.

Table 2 and Figure 1 show the results for each movie separately. "The Simpsons" turned out to be the most demanding stream in contrast with "Star Wars" which was the less demanding, although the highest channel throughput appears with the Soccer stream.

Figure 2 and Table 3 show the average results of the simulation runs under all possible movie loads. As the number of video terminals increases, the aggregate bit rate gets smoother, since the superposition of the movies is known from the literature to always be less bursty than a single movie, and the average bandwidth of the superposition is close to the sum of the mean bit rates of the movies. This is proven once more in our study, as the case of the system accommodating only one movie is shown to provide the smallest throughput than any other case of movie load.

The λ parameter corresponds to the maximum data message arrival rate (expressed in messages per frame) sustained by the system (i.e. such that the video packet dropping probability <0.0001 and the average data message delay < 2 seconds).

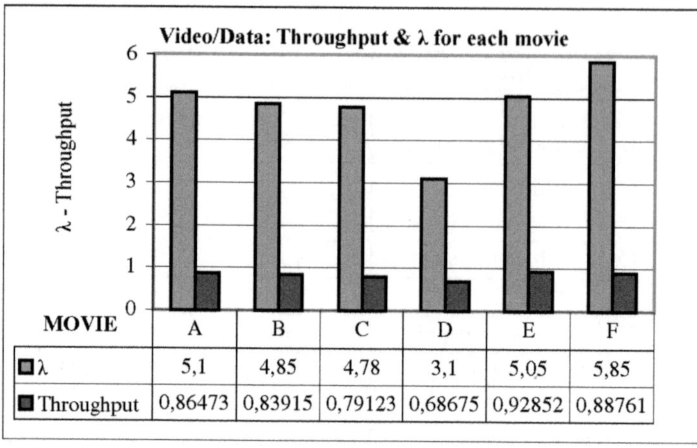

Fig. 1.

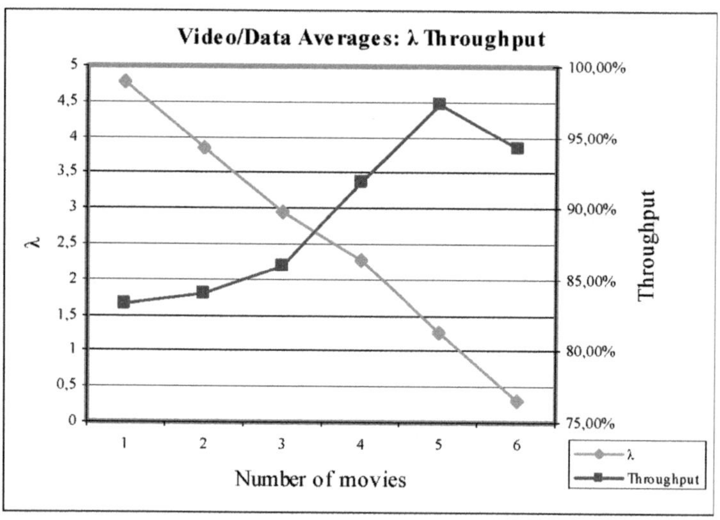

Fig. 2.

Table 3.

System Load	Average λ	Average Throughput %
1 Movie	4.788333	83.29986
2 Movies	3.8366	83.97877
3 Movies	2.96	85.90973
4 Movies	2.2625	91.91385
5 Movies	1.266667	97.34089
6 Movies	0.31	94.18723

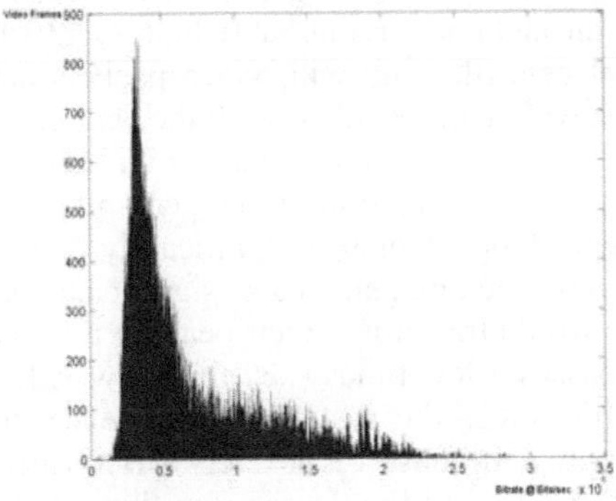

Fig. 3. News Stream

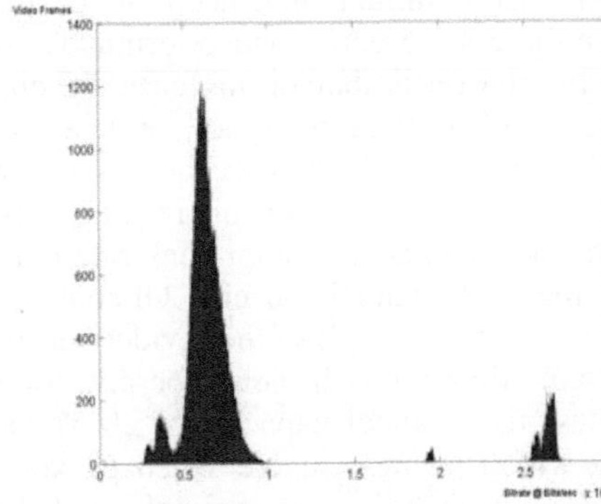

Fig. 4. Parking Lot Camera Stream

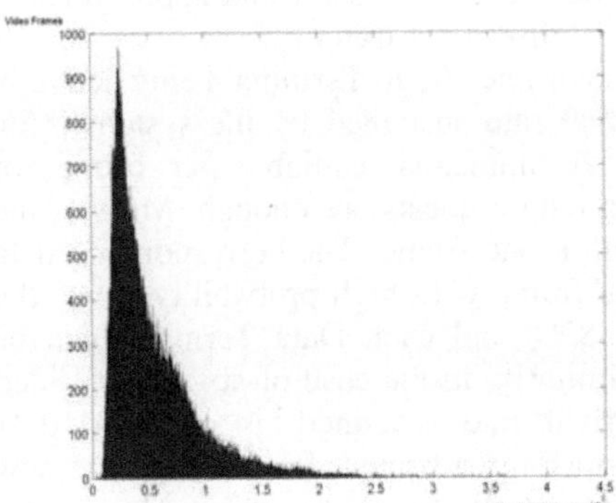

Fig. 5. Silence of the Lambs Stream

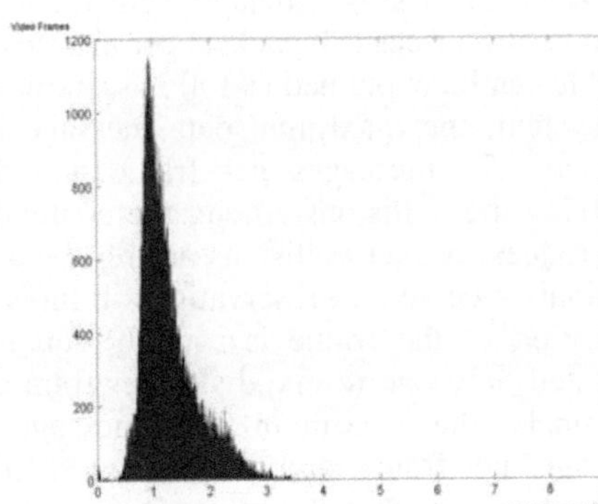

Fig. 6. The Simpsons Stream

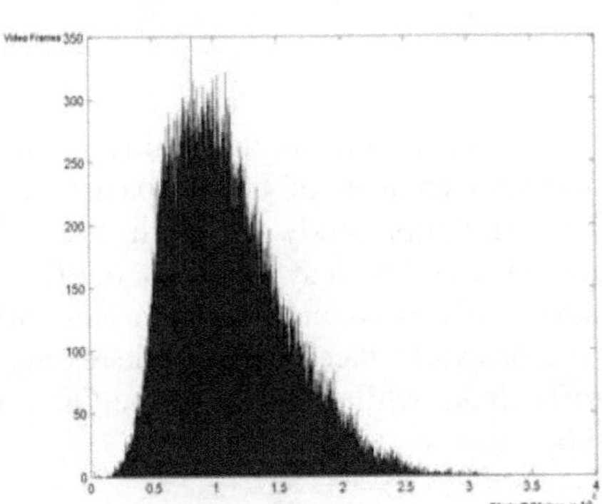

Fig. 7. Soccer Stream

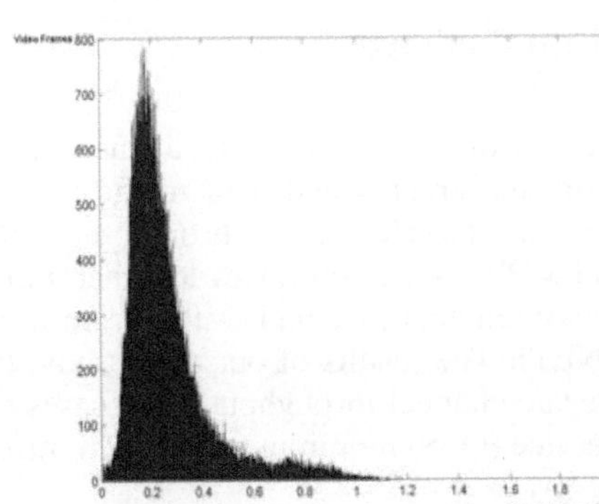

Fig. 8. Star Wars Stream

Another comment that needs to be made is about the throughput of the system when there are 5 active video terminals. This throughput is remarkably high (97,3%), and the reason is that in this case the channel gets filled up with video packets but there is still some empty space for data packets, which almost ideally fill the channel.

On the contrary, when six movies are present in the system, the channel throughput decreases. The reason for this result can be explained if we sum up the peak bit rates of the six movies. The total peak rate reaches 25 Mbps, which is significantly higher than the total channel capacity. Of course, the multiplexing gain is also significant and it is rather rare that all of the 6 video terminals would transmit at their peak bit rates at the same time but still their superposition surpasses a few times (i.e., for a few video frames) the channel capacity of 20 Mbps. This forces the system to experience, in these video frames, severe video packet dropping, because of the very strict video dropping probability requirements, and therefore leads to the decrease in throughput in comparison with the 5 active video terminals scenario.

A final comment concerns the average data messages delays of our mechanism (the corresponding results are not shown here). In all cases examined in this study, the average data messages delays were found to be roughly constant and approximately equal to 960 msecs (80 packets per message * 12 msecs per frame).

This can be explained as follows. In the case of one Video Terminal being active in the system, the maximum data message arrival rate sustained by the system is on average 4.79 messages per frame and the 20 minislots available per frame for resolving the collisions among the contending data requests are enough. Most of the data request packet collisions are resolved within one frame. The corresponding data terminals receive slot reservations in the same frame with high probability, since the utilization of the frame is roughly equal to 83% and each Data Terminal can be allocated only one reserved slot per frame. Similarly, in the case of six active Video Terminals, the maximum data message arrival rate sustained, is equal to 0.31 messages per frame, and the 20 request minislots are sufficient for handling the new video request packets (at most six in each frame, if all Video terminals undergo a bit rate increase at the same time which happens rarely) and the new data message arrivals, which correspond to one new arrival per three frames on average.

7 Conclusions

In this work we proposed a mechanism for transmitting high quality stored multimedia streams and data traffic, over a wireless channel of high capacity. We evaluated its performance through an extensive simulation study in which we used actual MPEG-4 stream traces together with data generated by a synthetic model fitted to measurements of email traffic from a National University and Research Network (FUNET). The results of our study show that the proposed mechanism achieves high aggregate channel throughput in all cases of traffic load, while preserving the Quality of Service (QoS) requirements of each traffic type.

References

1. C. Papadimitriou, S. Ramanathan, P. V. Rangan, and S.SampathKumar, "Multimedia information caching for personalized video-on-demand", Comput. Commun., vol. 18, pp. 204–216, Mar. 1995.
2. Y.Guo, S.Sen, and D.Towsley "Prefix Caching assisted Periodic Broadcast: Framework and Techniques to Support Streaming for Popular Videos", in Proceedings of the IEEE ICC 2002 Conference.
3. C. Vassilakis, M. Paterakis and P. Triantafillou: "Video Placement and Configuration of Distributed Video Servers on Cable TV Networks", ACM Multimedia Systems Journal, Vol. 8, No. 2, 2000, pp. 92–104.
4. A. Basso, S. Varakliotis, R. Castagno, "Transport of MPEG-4 over IP/RTP", in Proceedings of the PV 2000 Workshop.
5. D.A. Dyson and Z.J. Haas, "A Dynamic Packet Reservation Multiple Access Scheme for Wireless ATM", ACM / Baltzer MONET Journal, 4(2), pp. 87–89, Jan 1999.
6. Q. Pang, A. Bigloo, et al.,"Service Scheduling for General Packet Radio Service Classes", in Proc. of the IEEE Wireless Communications and Networking Conference 1999 (WCNC '99), New Orleans, USA.
7. P. Koutsakis, S. Psychis and M. Paterakis, "On the Integration of MPEG-4 Video Streams with Voice and E-mail Data Packet Traffic over Wireless Picocellular Networks", in Proceedings of the IEEE PIMRC 2001 Conference, San Diego, USA.
8. P. Koutsakis, M. Paterakis, and S. Psychis, "Integrating Voice, Video and Bursty Data Packet Traffic over Burst-Error Wireless TDMA Channels with Adjustable Request Bandwidth", in Proceedings of the IEEE PIMRC 2001 Conference, San Diego, USA.
9. A.C. Cleary and M.Paterakis, "On the Voice – Data Integration in Third Generation Wireless Access Communication Networks", European Transactions on Telecommunications and Related Technologies, Vol. 5, No. 1, Jan.–Feb. 1994, pp. 11–18.
10. Leonardo Chiariglione, "MPEG-4, why use it?", Telecom Italia Lab – Italy, http://leonardo.telecomitalialab.com/paper/mpeg-4/
11. http://peach.eas.asu.edu/index.html
12. N.Passas D.Skyrianoglou and L.Merakos, "Traffic Scheduling in Wireless ATM Networks", in Proceedings of the IEEE ATM'97 Workshop, Lisbon, Portugal, May 1997.
13. P. Koutsakis, "Medium Access Control and Traffic Policing for Multimedia Traffic Integrated Access in High Capacity Wireless Channels", PhD Dissertation ECE Dept, Technical University of Crete, Chania 2002
14. M. Paterakis and P. Papantoni – Kazakos, "A Simple Window Random Access Algorithm With Advantageous Properties", IEEE Trans. on Inform. Theory, Vol. IT–35, No. 5, September 1989, pp. 1124–1130.

Embedding MANETs in the Real World

Christian Tschudin[1], Henrik Lundgren[2], and Erik Nordström[2]

[1] Computer Science Department, University of Basel
Bernoullistrasse 16, CH – 4056 Basel, Switzerland
[2] IT Department, Uppsala University
Box 337, SE – 751 05 Uppsala, Sweden

Abstract. We believe that research on ad hoc networking so far has focused on too broad a range of scenarios and optimization issues and we argue that limiting the settings to bare bones services and small networks will allow for quick MANET deployment. Especially if these solutions interoperate easily with the Internet and offer the possibility to grow the MANET networks from there. Starting at this end will also help us to identify and possibly enlarge the "ad hoc horizon", beyond which ad hoc networking is no longer useful. In this paper we report on our experiences from five years of real world ad hoc network experimentation. Our approach is pragmatic and includes several implementations and a testbed software package. We highlight issues that we have encountered and identify crucial problems and potential dangers that deserves special attention in order to make ad hoc networking usable.

1 Introduction

Since the initial work on an IETF charter for mobile ad hoc networks started around 1995/1996, the MANET working group [13] has been working towards a proposed standard for IP level routing in ad hoc networks. Recently, the charter was narrowed as a result of the increasingly broad scope touching on problems like scalability, power awareness and security. Most of these research issues will now be addressed in the context of the Internet *Research* Task Force [8] . The MANET charter as of 2003 will propose four protocols for experimental RFC status, namely AODV, DSR, OLSR and TBRPF. This would indicate a new level of maturity of multi-hop ad hoc networking. However, we believe that the "race for multi-hop" is far from over and point to the fact that MANET solutions, if they exist, are not embedded in our daily network experience.

In this paper we argue for going simple and real. We believe that solving simple scenarios, i.e. small 10-20 node networks with one to three hops together with Internet connectivity would spur innovation and make ad hoc networking go from being a research-only phenomena to being utilized in real world settings. Our suggestion is to focus on implementation, experimentation and integration.

1.1 Implementation

Unfortunately, the deployment of MANET functionality has been slow. The first and main reason is the lack of quality real world implementations. When

M. Conti et al. (Eds.): PWC 2003, LNCS 2775, pp. 578–589, 2003.

we started building the Ad hoc Protocol Evaluation Testbed in the year 2000 [9], we had a hard time finding (a) any implementations, and (b) implementations that worked properly. Out of three protocol implementations (OLSR, AODV, TORA), only one was able to route packets over multiple hops. One implementation even had problems in a single hop scenario. This contrasts sharply with the number of related Internet Drafts and revisions, which seems to indicate that the field is making good progress.

AODV has seen most maturity in this respect whilst other major MANET protocols either lack updated implementations, have intellectual property issues or suffer from the reluctance among researchers to share work and promote open development. In some cases there are no implementations available at all, as is the case with TBRPF[1]. But even AODV is not a product yet: Self-configuration and seamless interfacing with the fixed Internet is still unsolved.

1.2 Experimentation

Experimentation is the bridge between paper specification and real world deployment. This is where the quirks that the draft authors missed and the invalid assumptions they made are discovered. However, although MANET research has been ongoing for some time, little valuable "hands-on" experience is available. It is alarming that protocols are mainly developed and updated in response to findings from simulations which are run using optimized settings that are not available in real life. For example, in simulations, routing protocols typically implement layer interactions (e.g. feeding link layer failures up the stack) that are not available in real world settings.

Another problem is that simulations are almost never validated. It is very tempting to do research through simulations because they are cheap when compared to setting up real experiments and are nicely repeatable. A first aspect of this is the trust which is put into the protocol logic that comes with the simulation packages: The protocol implementations are seldomly assessed in an independent way. The second aspect is that there is rarely an effort to calibrate or directly compare the simulation with findings from actual measurements.

1.3 Integration

Integration is the next step towards deployment. How do we make sure that the functionality we need in operating systems are exposed through standard APIs? How do we make sure that new functionality is integrated?

A consequence of the narrow implementation basis is that the integration of the protocol logic into actual operating systems is similarly scanty. The manipulation of IP routing tables is a good place to point to: "Default IP routes" make little sense in a MANET and the notion of a subnet to which an interface is attached becomes similarly void. This requires "creative solutions" and kernel

[1] A TBRPF implementation was previously available from SRI, but has since then been removed.

workarounds for mapping transient routing paths to IP routing table entries. (At this point one can literally feel the conceptual stress between MANET routing and classical IP routing concepts.)

The other point that integration demonstrates is that routing is more than just an algorithm. Routing is inherently linked to operational matters (e.g., address assignment) and therefore configuration issues. Unfortunately, such questions, although crucial for the end user's comfort, are beyond routing protocol specifications. It is fair to say that we do not have good answers ready for extending Internet connectivity into the MANET realm.

1.4 Half Time for the Multihop Game

Although respectable progress has been achieved we think that research in MANET multihop routing has only done half of the job. Clearly there are no final results. Simulations have not been conclusive on clearing the field of protocol contenders, and hard reports from the user front are missing because there is virtually no deployment of wireless multihop functionality.

So the second part of the MANET job will be decisive for whether wireless multihop routing manages to find a user basis. It could well be that infrastructure solutions (e.g., managed mode IEEE 802.11), which are actually responsible for the deployment of the radio hardware, will simply occupy the market space and that real MANET scenarios are too insignificant in order to create a shift.

In this paper we discuss our pragmatic approach to MANET research and how we believe that it can help turning the MANET concepts into a deployable technology. In section 2 we start with presenting the experimental research we have done in our group during the past years. In section 3, we revisit some of the underlaying ideas and premises behind MANET thinking and present work that we have done outside MANET borders. Section 4 continues with prospects for the future before we conclude this paper in section 5.

2 Reality Checks

Most wireless testbeds [12,11] [16] [14] could be called "measurement setups"; They targeted a single protocol and do not represent a software product which can be used by others for repeating experiments (see table 1 below). By developing the Ad hoc Protocol Evaluation testbed (APE) we aimed at creating an investigation tool that ultimately would be used to compare all MANET implementations at a 1-to-1 scale.

To this end we adopted a modular architecture allowing an experimentalist to plug in *different* protocols and traffic generators. Currently, the APE testbed is prepared for installation of most available protocol implementations running under Linux. Because of "scaling concerns" when setting up experiments with ten or more human participants, APE is designed for easy installation, deployment and handling. Installation is as simple as "downloading and extracting" a single file. It contains a self-extracting stripped-down Linux system that can

Table 1. Comparison of various ad hoc network testbeds.

	CMU	*GA tech*	*BBN*	*APE*
Real-world	yes	yes	yes	yes
W-less emul.	yes	yes	yes	yes
Hardware	laptops + GPS	laptops	Nokia Rooftop	laptops
OS	FreeBSD	Linux	DAWN/FreeBSD	Linux
Routing Prot.	DSR	ABR	NLS	**Independent**
Visualization	yes	no	no	yes
Analysis tools	yes	yes	yes	yes
Software Distribution	no	no	no	**Yes**
Reproduc.	no	no	no	**Yes**

be easily installed and booted on machines running Windows or Linux, or run using a "LiveCD" on a PC compatible machine. During the testruns the participants only need to follow on-screen movement instructions and afterwards the collected measurement data is uploaded to a collect-node, all with minimal user intervention.

To address repeatability we use strict choreography instructions to the participants in conjunction with a new mobility metric, derived from measured signal quality. This gives us a way to compare the "mobility fingerprint" of each testrun and to assess whether testruns were performed in a sufficiently similar manner, independent of the routing protocol used. We have run experiments where we scaled up to a 37 node setting and successfully used the new mobility metric to assess the reproducibility.

2.1 Implementation Issues

As the first thing a routing protocol is usually exposed to is a simulation environment, the implementations tend to be more or less tailored to the particular simulation environment at hand and neglect the interactions with the operating system and its protocol stack. However, sooner or later the protocol will have to be implemented and integrated within an OS.

Although OS kernel interactions are quite complex and make the implementation time-consuming, the most striking experience during our AODV-UU [2] development was that some protocol design assumptions might not be feasible to implement in a real system. For example, the AODV specification suggests to primarily use link-layer feedback for monitoring the link status of active routes and only if that is not possible to rely on periodic HELLO beacons. In simulation, link-layer feedback is easy and therefore commonly used. In reality, there is to this date still no way for upper layers to acquire such information from IEEE 802.11 interfaces. This observation impacts those protocol simulations studies which used link-layer feedback. Repeating them using the HELLO beacons would give a better view of how the protocol would behave and perform in reality.

2.2 Simulation Model Accuracy

Exposing implementations to the real world will not only highlight practical limitations, but will also help reveal protocol behavior that might not show up in simulations due to inaccurate simulation models. We became subject to this statement when we observed spurious performance degradations during experiments with our AODV-UU implementation. Already in a basic scenario with one mobile node we experienced "communication gray zones" [10]. In these gray zones of several meters the routing protocol reports a valid route to the destination node but virtually no data packets would pass through. This effect does not show up when replicating the setting inside the ns-2 simulator (using the same AODV-UU code as in the OS setting) and thus it was clear that this is a discrepancy between the simulation environment and the real-world because data transmission works well in all other areas.

Further real-world experiments revealed that these gray zones form because of the difference in transmission range for specific packet types. In IEEE 802.11b broadcast packets are sent at 2Mbit/s and unicast can be sent at speeds up to 11Mbit/s. The transmission output power is independent of transmission speed and thus lower transmission speed means more energy per bit, which translates to longer transmission range.

The consequence is that at the transmission range border between two nodes the data packets will not reach the destination while the broadcasted HELLO beacons (used for neighbor sensing) will be delivered and make the routing daemon believe that the other node is reachable, thus it will *not* try to discover a new route. This gray zone effect did not show up in the simulations simply because ns-2 uses a flat 2Mbit/s model.

Based on our implementation we then started to test work-arounds. In [10] we showed that artificially limiting the range of the HELLO packets by filtering on the SNR (Signal to Noise Ratio) value eliminates the gray zone effect. In fact, setting the acceptance SNR level such that HELLO packets have slightly *shorter* transmission range than the data packets will force AODV to pick a more robust link earlier, which decreases the inherent packet loss during route switching.

Accuracy of the physical layer model is equally important. This model includes e.g., SNR calculation, signal reception, fading and path loss. However, the study in [15] is revealing: Its authors examined the interaction between two routing protocols and the physical layer models used in ns-2 and GloMoSim. They observed that the physical model had great influence on the result such that the corresponding choice could even change the relative ranking of the compared protocols.

In conclusion, it is important to realize that the simulation models are just models. Although simulation studies are very useful in the protocol development process they can never totally replace real-world testing. In fact, we need more real-world experience earlier in the process.

2.3 The Ad Hoc Horizon

There are many questions still open when it comes to real world experiments. For example, little has been reported in the research literature about problems of TCP over wireless multi-hop *implementations*, although there are some discomforting results from simulations [4,6]: TCP throughput is inversely proportional to the hop distance for a RTS/CTS style MAC layer (i.e., 25% after 4 hops) and even worse if the TCP window is not artificially limited. Another way of grasping this is to say that there exists an "ad hoc horizon" beyond which it is not worth trying to offer MANET services. Our conservative estimate is that using IEEE 802.11 technology this horizon is as narrow as 3 hops.

How can such a horizon form? Reachability information, on which routing protocols base their decisions, loses its value the further it travels in a mobile environment. Also, a loss rate factor applies for each hop that a data packet has to traverse: Crossing 4 hops with an average delivery ratio of 95% brings us down to almost 80% (0.95^4) end-to-end, probably making TCP based applications unusable. Finally, the longer the path is the more places there are where TCP data packets and backward traveling ACK packets can collide. The overall effect, even if the average drop rate is low, is that TCP becomes an unusable transport protocol because it is tricked into seeing a permanently congested network.

In [5] a measurement study is presented for determining the "scaling law" of (static) wireless networks. It demonstrates a dramatic decrease in (UDP-) throughput as the number of nodes goes from 2 to 10 and compares this to theoretical figures. It seems that the IEEE 802.11 standard is far from using the available capacity in an efficient way. Here, we present a simulation study that shows similarly severe throughput degradation when going from 10 to 100 nodes and when adding mobility and ad hoc routing.

Using ns-2 (v2.26) we compared TORA, DSR and two AODV implementations whereof one is bundled with ns-2 (AODV) and the other is our own AODV-UU implementation. The goal of the simulation is to compare network scalability with respect to the number of nodes ranging from 10 nodes to 100 nodes while keeping the density constant. In our simulation this is done by increasing the network area size accordingly, using a rectangular area with a x:y ratio of 1:4, to promote increasing path lengths as the network grows[2].

[2] Node placement and movement is random, using the random waypoint mobility model with a max speed of 20 m/s and pause time of 0.2 s. Traffic is generated using the CMU traffic generation tools. We use CBR traffic with a packet size of 512B, 0.5 connections per node (which translates to $0.5 \times \#nodes$ connections max), with a rate of 10 pkt/s. We experimentally found a node density of $5e^{-5}$, by aiming for a delivery ratio slightly under 100% for the 10 node setting. This translates to an initial area size of $223.6 \times 894.4m^2$. For the 100 nodes setting, the network size has increased to $707.1 \times 2828.4m^2$. The protocols are run with default settings for TORA and DSR that corresponds to ns-2 version 2.26. The ns-2 AODV implementation also had default settings with the exception that HELLO messages were used instead of link layer feedback. AODV-UU was run using default values for version 0.7.1, but also with HELLO messages.

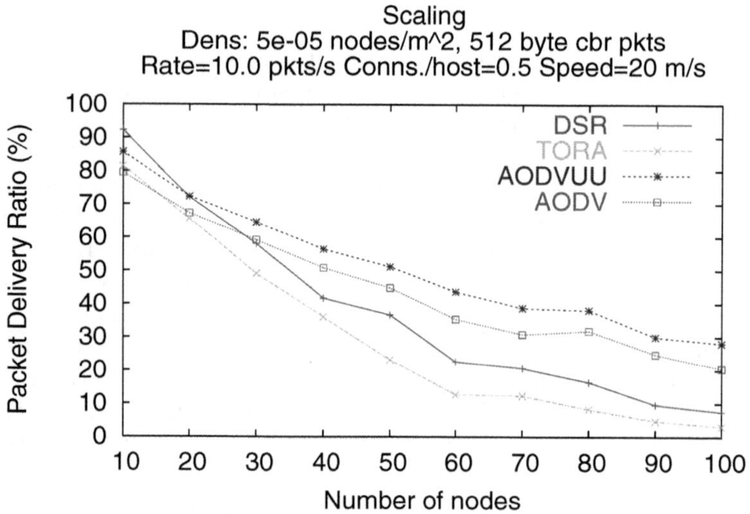

Fig. 1. The effect of scaling where node density is kept fixed.

Figure 1 shows the performance results averaged over 10 simulation runs. Although working with optimistic assumptions (simple UDP traffic, uniformly spread sources and sinks) the delivery ratio is in the best case inversely proportional to the maximal path length: Remember that since density is fixed, increasing the number of nodes means increasing the area size and producing longer routes. An examination of the trace files indicate that the maximum path length increases from 5 to 15 hops when growing the network from 10 to 100 nodes. It is obvious that the performance of *all* protocols decrease rather ungracefully, although at different rates. Our conclusion is that independently of which of the three routing algorithms we chose we will hit the ad hoc horizon well below 100 nodes, that is 15 hops. Adding TCP to the picture we deem it reasonable to suspect the horizon to be 10 nodes and 3 hops – a hypothesis to be further studied with experiments and simulations.

3 Out of the MANET Box Thinking

While we see the APE testbed and the AODV-UU implementations as our contributions to the MANET effort, we also wanted to investigate opportunities outside these borders. One direction uses an active networking approach where protocol logic resides in control packets instead of the nodes (hence one obtains a multi-protocol environment). The other direction is on sidestepping some of the IP interfacing problems and on simplifications of the routing logic. In this section we give a gentle introduction to these approaches and highlight some of our results.

3.1 Active Ad Hoc Routing

The difficulty to agree on "the" set of scenarios and border conditions for a MANET protocol relates to the assumption that a single ad hoc routing protocol could satisfy all (or most) requirements. Our approach relies on mobile code for ad hoc routing implementations. Such an *active networking* approach permits to deploy the routing logic at run time either in the scope of a network or within a single session. This extreme customization also comes in handy for the debugging and testing of protocol variants, as there is no need to visit nodes for installing a new software version.

We gained several insights from this "implementation experiment", which we have documented in [18]. First, routing logic does not need to be executed very fast. The transmission time in the range of milliseconds and the low accuracy of HELLO beacons (multiple seconds) leaves enough room to start up complete command shells for processing routing logic. We used simple Bash scripts which could forward themselves through the network for implementing the routing protocol. Of course, handling the data packets in the same manner would not work well. To this end we created a separate forwarding layer below IP where data packets are forwarded in the kernel without IP processing. This separation of pure data forwarding (without any intervening routing or packet salvation logic) and active routing control packets worked well enough to demonstrate the streaming of MP3 audio packets in a wireless ad hoc network!

Second, we used this setup to experiment with proactive as well as reactive routing schemes i.e., both routing styles can be expressed equally well with active packets. We "scripted" a proactive protocol loosely inspired by Cellular IP [3]: Three different populations of active packets were responsible for a) periodically gathering neighborhood information, b) for building a common delivery tree and c) for establishing delivery paths along this tree. Later on, we also scripted a reactive protocol called AORTA with the classical flood and reverse path pattern. This was a mere matter of using another script, under the full control of the initiating node. Each routing path then operates independently and could use its own variant of a reactive protocol, if desirable.

Third, when looking back we think that current ad hoc routing logic is not complex in its core: Most difficulties stem from optimization that one applies on top of the basic functionality. One example is the sharing of forwarding state where one has to decide which control packet is allowed to modify state created by other (end) nodes' packets. Another example is the salvation of data packets which considerably complicates the simple forwarding mechanism by adding timeouts and new dynamic elements.

3.2 LUNAR – Simple Routing Below IP

Even plain ad hoc routing can be simplified, as we tried out with LUNAR, the "Lightweight Underlay Network Ad hoc Routing protocol" [17]. The main question we stumbled over was: Once we have set up a routing path, why should we bother to maintain it? Protocols using HELLO messages will typically need

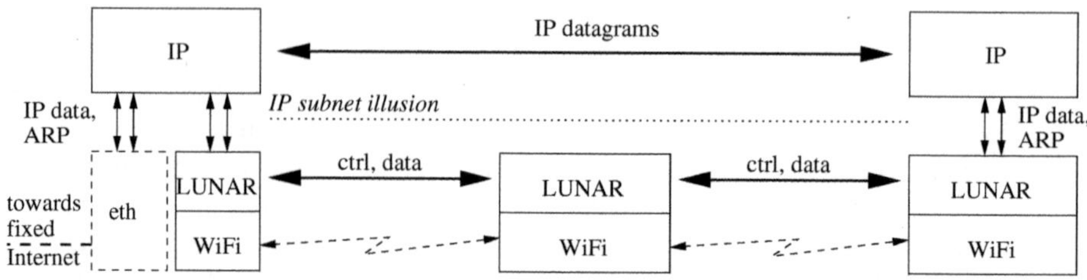

Fig. 2. The LUNAR ad hoc routing underlay for IP.

two HELLO rounds to decide that a link is broken before the protocol starts reacting. So we decided to do a plain discovery every 3 seconds and to drop HELLO messages altogether.

To our surprise this simple approach performed better than OLSR and the original AODV implementation (before applying the "grey zone patch" described above). Admittedly, LUNAR's "periodic-full-rediscovery" will not scale well. However, the crucial question is whether more complex approaches will ever be operated in large networks, as inherent scaling problems as reported in figure 1 degrade their usefulness.

Another useful simplification of LUNAR is the underlaying of IP (see figure 2). LUNAR works by trapping ARP requests which are then forwarded through the wireless network. Once the destination is found a delivery path is set up end-to-end and mapped to a virtual ethernet number. This creates a subnet illusion to IP which makes integration with the IP stack much simpler, as only a single entry has to be added to the IP routing table.

A fun project turned out to be the demonstration of leveraging LUNAR's simplicity. We rewrote LUNAR for embedded devices and ported it to the LEGO mindstorm microcontrollers with 32K of RAM, wherein we managed to fit the LegOS, a TCP/IP stack, a tiny web server as well as the LUNAR protocol and demonstrated it at Mobicom'2002. All code is GPL and available at http://www.docs.uu.se/selnet/lunar/.

4 Ad Hoc Prospects

4.1 Danger of Point Solutions and Too Slow Progress

The basic assumption of research in ad hoc routing is that conventional (IP) routing algorithms behave poorly in wireless environments. This means that the layering approach is flawed because one cannot hide the link layer characteristics from the networking layer: the network layer has to choose different routing strategies depending on the fixed or wireless link layers used. Ironically, this observation will fall back on the ad hoc research community as IEEE 802.11a, b, g and Bluetooth have already sufficiently many differences that conclusions drawn for the "802.11+ad hoc" combination will be invalid for the "Bluetooth+ad hoc" pair.

One possible (worst case) outcome is that new developments and progress at the link and physical layer will outpace the slow progress at the routing level, first by creating more variants which the routing layer has to cope with, second by producing technology which performs well in an infrastructure setting but badly in ad hoc mode, and finally by raising the level of expectations so that the "imperfect" ad hoc solutions never will have no chance to catch up. So users should have a chance to experience the advantages of ad hoc functionality early on: The quicker even limited (but robust) MANET solutions become available, the better.

4.2 Shooting for the Moon Instead of Mars

One impressive goal of MANET work is to aim at networks of 100s and more nodes. Many simulations work with such scenarios and see a high node number as a goal of its own. Recently a special interest group (ANS [1]) was formed to look into scaling issues of ad hoc networks. The problem with such large numbers is that they represent a fake goal. First, such a point of operation is rather arbitrary on the spectrum from a single-site spontaneous network with a handful of mobiles to the Terminodes vision [7] that aims at operating at the global level. Second, such a number is again dependent on link layer technology as well as the scenarios. The HELLO beacons of 100 mobiles in the cocktail party room of a networking conference would kill AODV. Are the protocols at fault? The problem is rather with the scenario as well as the "one-protocol-should-fit-all" assumption. Thus it is wise to aim at a more modest goal.

4.3 Constantly Raising the Barrier

Progress in the lower layers is one front where MANET has to fight, progress at the problem front is the other. In the past couple of years new problems have become fashionable while working on core ad hoc problems has lost attractivity. This pattern certainly applies to the topic of security – the prime knock out candidate – and also extends to QoS and power awareness. The net effect is that the innovation hurdle has been raised dramatically and that the solution to problem ratio is getting smaller and smaller. Maybe we should learn from the sensor net community and declare some network attributes as being currently "out of reach" because of self-chosen constraints. Probably more balanced would be to accept these new research topics, of course, and to make sure that we do not loose focus on hammering out pragmatic solutions.

4.4 What is the Value of Ad Hoc, Really?

The reaction to the growing "grand complication" should be to concentrate on simple but convincing cases. Better have a working solution for a majority of wireless interfaces that might be limited in scaling and that does not address all possible concerns, but which gives an added value right now. As Metcalfe's law tells us it is a matter of getting full coverage which gives value to the network users (and much less the perfection by which this happens). The case today is that although everybody has and uses IEEE 802.11 hardware, virtually nobody is able to run it in multihop ad hoc mode, as kernels are not shipped with ad hoc routing logic and configuration is a nightmare. The real value of ad hoc networking would be that every device is ad hoc enabled, wherefore resolving penetration hurdles should be the prime target for the ad hoc community.

4.5 Ad Hoc for the Lazy

One crucial element in raising the penetration rate of ad hoc functionality is that management must become really simple. We all are lazy and even nerds will become bored after having configured their WiFi cards for the 42nd time. Clearly, there is a management issue here that has to be solved. One could establish a list of the other ten related problems to be looked into too... Again: Cutting a thin slice through the problem space and going for it till the end seems to us the more promising way of getting us lazy people to use multihop ad hoc.

5 Outlook

This papers tries to debunk some of the glory targets of wireless ad hoc networking research. Our main concern is that research on wireless multihop ad hoc networking could look into so many new and important problem areas and technologies that when "acceptable" solutions will eventually emerge nobody is interested anymore. Also, we are concerned about the lack of confrontation of simulation studies and conceptual contributions with actual implementations. A first point to work on is in our opinion the selection of a modest but realistic target e.g., a dozen nodes operating inside the ad horizon. Second, the integration of ad hoc routing with the Internet has to become a no-brainer. And only then, as a third step, we have a chance to push ad hoc functionality into the operating system main stream and the real world.

Acknowledgment. We wish to thank Richard Gold for his careful reading and editing help.

References

1. Ad hoc Network Scaling Research (IRTF group). `https://www1.ietf.org/mailman/listinfo/ans-research`.
2. AODV-UU implementation, 2003. `http://www.docs.uu.se/scanet/aodv`.
3. Andrew T. Campbell, Javier Gomez, and András G. Valkó. An overview of cellular IP. *IEEE Wireless Communications and Networks Conferance 1999 (WCNC'99)*, 2:606–611, 1999.

4. M. Gerla, K. Tang, and R. Bagrodia. TCP performance in wireless multi-hop networks. In *Proceedings of IEEE WMCSA'99, (New Orleans, LA).*, Feb 1999.
5. Piyush Gupta, Robert Gray, and P. R. Kumar. An experimental scaling law for ad hoc networks. Unpublished submission, May 2001, `http://citeseer.nj.nec.com/gupta01experimental.html`.
6. Gavin Holland and Nitin H. Vaidya. Analysis of TCP performance over mobile ad hoc networks. In *Proceedings of IEEE/ACM MOBICOM '99*, pages 219–230, Aug 1999.
7. J. Hubaux, J. Le Boudec, M. Vojnović, S. Giordano, M. Hamdi, L. Blazević, and L. Buttyán. Toward mobile ad-hoc WANs: Terminodes. Technical Report DSC/2000/006, Swiss Federal Institute of Technology, Lausanne, Jan 2000.
8. Internet research task force. `http://www.irtf.org`.
9. H. Lundgren, D. Lundberg, J. Nielsen, E. Nordström, and C. Tschudin. A large-scale testbed for reproducible ad hoc protocol evaluations. In *3rd annual IEEE Wireless Communications and Networking Conference (WCNC 2002)*, pages 412–418, Mar 2002.
10. H. Lundgren, E. Nordström, and C. Tschudin. Coping with communication gray zones in IEEE 802.11b based ad hoc networks. In *5th ACM international workshop on Wireless mobile multimedia (WoWMoM 2002)*, pages 49–55, 2002.
11. D. Maltz, J. Broch, and D. Johnson. Experiences designing and building a multi-hop wireless ad hoc network testbed, 1999.
12. D. Maltz, J. Broch, and D. Johnson. Quantitative lessons from a full-scale multi-hop wireless ad hoc network testbed. In *2nd annual IEEE Wireless Communications and Networking Conference (WCNC 2000)*, Sep 2000.
13. The Official IETF working group MANET webpage. `http://www.ietf.org/html.charters/manet-charter.html`.
14. R. Ramanathan and R. Hain. An ad hoc wireless testbed for scalable, adaptive qos support. In *2nd annual IEEE Wireless Communications and Networking Conference (WCNC 2000)*, Sep 2000.
15. M. Takai, J. Martin, and R. Bagrodia. Effects of wireless physical layer modeling in mobile ad hoc networks. In *ACM International Symposium on Mobile Ad Hoc Networking and Computing (MobiHoc 2001)*, Oct 2001.
16. C-K. Toh, M. Delwar, and D. Allen. Evaluating the communication performance of an ad hoc wireless network. In *Transaction on Wireless Communications*, Jul 2002.
17. C. Tschudin and R. Gold. LUNAR: Lightweight Underlay Network Ad-hoc Routing. *Uppsala University Technical Report 2003-021*, Jan 2002.
18. C. Tschudin, H. Gulbrandsen, and H. Lundgren. Active routing for ad-hoc networks. *IEEE Communications Magazine, Special issue on Active and Programmable Networks*, Apr 2000.

Impact of Directional Antennas on Ad Hoc Routing*

Romit Roy Choudhury[1] and Nitin H. Vaidya[2]

[1] Dept. of Computer Science
[2] Dept. of Electrical and Computer Engineering and
Coordinated Science Laboratory
University of Illinois at Urbana-Champaign, USA

Abstract. Previous research on directional antennas has been confined mostly to medium access control. However, it is necessary to evaluate the impact of directional antennas on the performance of routing protocols as well. In this paper, we identify the issues and evaluate the performance of an omnidirectional routing protocol, DSR, when executed over directional antennas. Using insights gained from simulations, we propose routing strategies suitable for directional communication. Our analysis shows that by using directional antennas, ad hoc networks may achieve better performance. However, scenarios exist in which omnidirectional antennas may be suitable.

1 Introduction

The impact of directional communication on the performance of an ad hoc network is often counter-intuitive. While fewer-hop-routes may be discovered due to higher transmission ranges of directional antennas, performing a simple neighborhood broadcast may now require the antenna system to sweep its transmitting beam sequentially over multiple directions. As a result, neighbors of a node receive broadcast packets at different points of time (unlike with omnidirectional antennas). Also, sweeping incurs greater delay and higher control overhead, partially negating the potential advantages. More subtle tradeoffs also arise.

A routing protocol that remains unaware of such changes in the radio hardware can degrade system performance. To study the impact, we evaluate an omnidirectional routing protocol called DSR (Dynamic Source Routing), over directional antennas, and propose modifications. In [4], [7], [9], [11] authors addressed the problem of medium access control using beamforming antennas. However, work on directional routing protocols is limited [5],[7],[1]. In [5] directional antennas are utilized for the purpose of on-demand routing. In [1], the authors present the notion of *maximally disjoint routes*, in which routes that do not share the same geographical area are chosen. Our contribution in this paper includes identifying issues in directional ad hoc routing, optimizations to address

* This work is supported in part by National Science Foundation (NSF) under grant ANI 01-96410 and 01-25859.

M. Conti et al. (Eds.): PWC 2003, LNCS 2775, pp. 590–600, 2003.

some of the issues, and a new metric to evaluate the control overhead of routing protocols.

2 Antenna Model

The antenna model we use is comprised of N beam patterns. The main lobe of each beam has a conical radiation pattern, spanning an angle of $\frac{2\pi}{N}$ radians. In our implementation of the antenna, motivated by the model in [7], we have approximated all the side lobes into a single sphere, with the node at its center. The antenna system offers two modes of operation: *Omni* and *Directional*. We assume that a node can operate in any one mode at a given time, but can toggle modes with negligible latency. In *Omni* mode, after a signal is detected, the antenna determines the beam on which the received signal power is maximum. Rest of the packet is then received by using this beam. We assume that in omni mode, signals are received with a gain G^o. An idle node stays in the *Omni* mode. In *Directional* mode, a node can select *only one* of its beams and beamform with a gain of G^d, $G^d >= G^o$. Note that the geographical distance over which two nodes may be able to communicate is proportional to the product of the transmission and the reception gain. As a result, the link-length between directional transmitters and omnidirectional receivers can be longer than that between omnidirectional transmitters and omnidirectional receivers. This motivates the notion of directional neighborhood – *Direction-Omni (DO) Neighbors* and *Omni-Omni (OO) Neighbors*. A node B is a DO-neighbor of a node A, if node B can receive a directional transmission from A even if B is in the *Omni* mode. A node B is a OO-neighbor of a node A, if node B can receive an omnidirectional transmission from A even if B is in the *Omni* mode. Observe that all OO-neighbors are also DO-neighbors, but not necessarily the vice versa.

The notion of *broadcast* changes when directional antennas are used. For example, if a node wishes to broadcast a packet to all its surrounding neighbors, it must transmit the same packet N times, once with each conical beam. We call this entire operation, *sweeping*. Observe that a single *sweep* reaches all the DO-neighbors of a node, while an omnidirectional broadcast reaches only the OO-neighbors.

3 DiMAC Protocol

We have designed DiMAC, a directional MAC protocol, for the purpose of evaluating routing over directional antennas. The design of DiMAC is based on the notion of reserving the wireless channel before actual data is transmitted. Channel reservation is performed using a RTS/CTS handshake (between the sender S and receiver R), both being transmitted directionally. DiMAC at S consults a look-up table to determine the antenna beam that must be used to initiate RTS to R (discussed later). Once the beam is determined, S proceeds through the steps of waiting for a DIFS period and backing off for a random interval (similar to the steps in IEEE 802.11[2]) before it transmits the RTS.

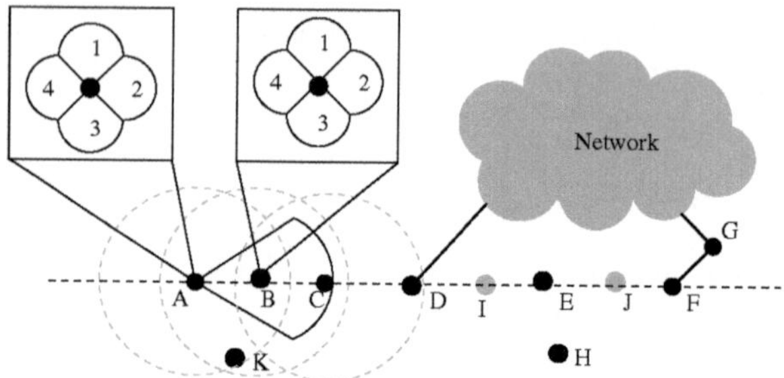

Fig. 1. An example scenario. A 4 beam antenna system is shown in the inset.

Node R, while idle, listens to the channel in omni mode and receives the RTS meant for it. The RTS is received with omnidirectional gain G^o. Node R now determines the antenna beam on which the received signal power of the RTS was maximum, and uses that same beam to send back a CTS directionally. S in the meantime remains beamformed towards R and receives the CTS directionally, with directional gain G^d. Once the RTS/CTS handshake is accomplished, node S sends the data packet directionally to R, to which R replies with a directional ACK. Nodes in the neighborhood of nodes S and R, that overhear RTS, CTS or both, defer transmission for the proposed duration of transfer. DiMAC maintains a directional NAV table (DiNAV), similar to [9],[11], that remembers the beams on which RTSs or CTSs were received. Transmission is deferred for the proposed duration, only on these beams.

A conspicuous problem with a directional MAC protocol such as DiMAC is *"deafness"*, identified and discussed in [9]. Although a MAC layer issue, it affects the performance of routing protocols. Briefly, deafness is caused when a node C attempts to initiate dialogue with a node A, while A is engaged in communication with another node, B. Node A fails to receive signals from C since it remains beamformed towards B over the duration of communication. Node C interprets the absence of a reply from A as indicative of a collision, and retransmits the packet. This can repeat multiple times, until node C is forced to give up and drop the packet.

DSR over DiMAC

We begin our evaluation of ad hoc routing using DSR and suggest improvements later. A node attempting to send a data packet to another node initiates the route discovery process by broadcasting a RREQ to all its neighbors. When using DiMAC, an omnidirectional broadcast is emulated through *sweeping*. Recall that sweeping incurs additional delay (almost N times greater), but can reach the DO-neighbors of a node, unlike omnidirectional broadcasts. For example, in Figure 1, if an RREQ is transmitted directionally by A, link A-C can be on the route. However, if the RREQ is transmitted omnidirectionally, then node C would not receive A's transmission. Now, to unicast packets to a particular neighbor, a node must use the appropriate beam. This information is cached at each node in a look-up table. In mobile scenarios, it is possible that information cached in the look-up table gets stale. We incorporate a *scanning* mechanism to address this problem. A *scan* is essentially "hello" packets, transmitted sequentially over all antenna beams (i.e., swept), whenever necessary. If node B receives a "hello" from node A, B replies using the same antenna beam with which it received the "hello". Node A records the antenna beam with which it receives the reply, and initiates future communication with B using the same beam. In summary, *neighbor discovery* becomes more complex using directional antennas since neighbors are now associated to specific antenna beams at a given instance of time. (In the next section, we suggest a partial scanning optimization to reduce the overhead of scanning.)

4 Performance Evaluation

We use the Qualnet simulator [6] version 2.6.1, for modeling our antenna system and simulating the protocols. We simulate our scenarios in a bounded region of 1500 X 1500 meters, with *Constant Bit Rate* (CBR) traffic for data communication. The tradeoffs we discuss in this section arise from the specific characteristics of our protocol pair - Directional DSR (DDSR) and DiMAC. However, we believe that our broad observations will apply to other protocols as well.

An interesting tradeoff arises from the counteracting effects of large sweeping delay and higher transmission range (thus fewer hops on routes). On one hand, RREQ propagation gets delayed due to sweeping, while on the other, shorter routes could be discovered due to a larger transmission range of directional antennas. Figure 1 may be used to illustrate this. Observe that a RREQ from A can reach D in two hops via node C, if directional antennas are used. In contrast, by using omnidirectional antennas the number of intermediate hops will be greater (through nodes B and C). In addition to this, spatial reuse affects delay in performing route discovery. Referring to Figure 1, note that when node K forwards the RREQ using omnidirectional antennas, node B must defer transmission of RREQ. This is not necessary for directional antennas. To evaluate these tradeoffs, we simulate several static and mobile scenarios, with different flow patterns. We discuss the insights gained from simulation results and propose optimizations that improve the performance of the DDSR protocol.

The simulated topology is a rectangular grid, with nodes placed 200 meters apart. To evaluate network behavior, we measured *route discovery latency* (RDL) and *throughput* for DSR and DDSR. Route discovery latency is the time duration calculated from the point a RREQ is transmitted by the sender till the point a RREP is received by the sender.

Evaluating Route Discovery Latency
Figure 2(b) shows the variation of *route discovery latency* (RDL) versus the physical distance separating source and destination nodes. For a given source node, the destination node is chosen according to the desired distance of separation. Figure 2(b) plots RDL for different values of N, N being the number of beams.

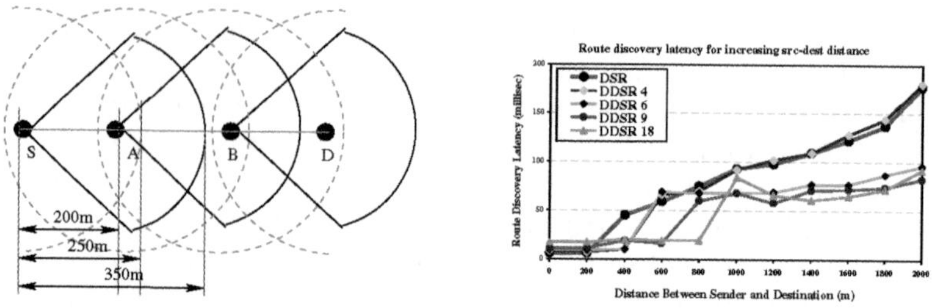

Fig. 2. (a)Left: Section of a grid showing DSR and DDSR4 discover equal-hop routes. (b)Right: RDL at 200m grid-distance.

Hereafter, we would denote Directional DSR with N antenna beams, as "DDSR N". For example, DDSR6 refers to DSR simulated over DiMAC with 6 beams. The beamwidth with N beams is $(2\pi/N)$. By "DSR", we would indicate DSR executed over omnidirectional antennas, using IEEE 802.11 (this is equivalent to DSR1). From Figure 2(b), we observe that the directional and omnidirectional curves cross over frequently when the distance of separation between the source and destination is small. At larger separation, directional antennas (except for DDSR4) exhibit lower route discovery latency in comparison to omnidirectional antennas.

Behavior of *route discovery latency* (RDL) using directional antennas may be intuitively explained as follows. Several factors influence RDL: *fewer hop routes* and *higher spatial reuse* reduce RDL, while *sweeping delay*, *deafness* and *higher directional interference* increase RDL (discussed in detail in [8]). When the separation between the source and destination node is small, the gain due to higher transmission range is only marginal. This is because both directional and omnidirectional antennas are capable of reaching the destination over a few hops. As an example, reaching node D from node S in Figure 2(a) requires three hops for both DDSR4 and DSR (DDSR6 or DDSR9 would require at least two hops). However, in such scenarios, the *sweeping* delay of DDSR is high, easily offsetting the slight gain (if any) of a shorter directional route.

When source-destinations are separated by larger distances, the advantage of higher transmission range begins to dominate. The gains due to a shorter route in DDSR, now offsets the additional delay incurred in sweeping. However, we observe that the RDL for DDSR4 is almost identical to DSR, even at higher source-destination separation. This happens because the transmission range of DDSR4 is 350 meters and the grid distance is 200 meters. Observe that DDSR4 can at best communicate directly to an adjacent node in the grid (Figure 2(a)). DSR, with a transmission range of 250 meters, can also do the same. Thus DDSR4 does not show any benefit over DSR. In fact, due to sweeping delay, the performance of DDSR4 is somewhat worse than DSR.

The behavior of route discovery latency depends on node density as well. To illustrate this, consider a source destination pair separated by 800 meters. Both

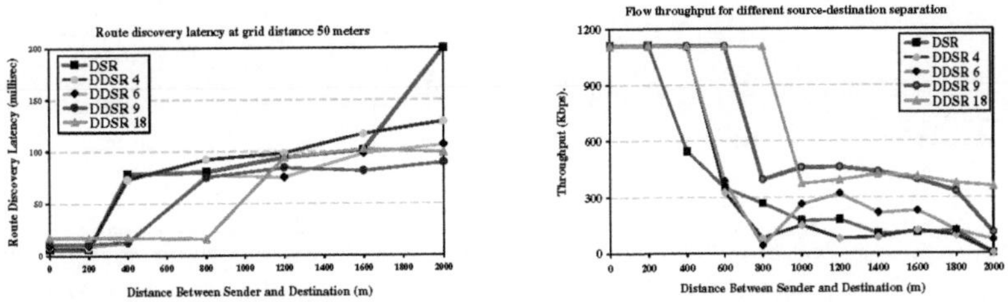

Fig. 3. (a)Left: RDL at 50m node-spacing. (b)Right: Throughput versus source-destination separation, at 200m grid-distance.

DSR and DDSR4 discover 4 hop routes, leaving DDSR4 with no advantage for higher transmission range. However, if the grid distance was 50 meters, DDSR4 is expected to discover a 3 hop route, while DSR must still require 4 hops. This motivated us to evaluate the performance of DDSR in a high density network with grid distance of 50 meters. Figure 3(a) shows the results of simulation. Counter to intuition, the performance of DDSR degrades in denser networks, even at high transmission ranges and large source-destination separation.

The degraded performance of DDSR in Figure 3(a) is a result of interference. Due to closely packed nodes, signals received from unwanted directions using antenna side-lobes, increase the probability of collisions. As a result, RREQs take longer to reach the destination. The net effect is that DDSR performs only marginally better than DSR. The advantage of high transmission range becomes conspicuous only when the source-destination separation is extremely large.

Evaluating Throughput

Figure 3(b) shows the throughput of a single CBR flow for different source-destination separations. The traffic generated is large enough to keep the source backlogged at all times. We observe that DSR performance is comparable to DDSR4's performance. For smaller beamwidths (i.e., higher transmission range), the throughput is not much greater, although the expected hop-count for DDSR can be far lesser than DSR. The 2 key reasons are discussed below.

1. In the simulation of a single flow in the grid topology, we observed that DDSR often chooses sub-optimal routes. This happens because *sweeping* causes DO-neighbors of a node to get the same RREQ at different points of time. Consequently, neighbors that receive the RREQs earlier, have a higher probability of delivering the RREQs earlier to the destination. Observe that earlier delivery in this case may not indicate a shorter-hop route (omnidirectional routing protocols assume that the earliest arriving RREQ traversed the shortest-hop route). If the routing protocol requires the destination to reply only to the earliest arriving RREQ, a potential suboptimal route can be established. Prolonged use of such a suboptimal route may result in serious performance degradation. Interestingly, replying to all RREQs (as optionally performed by DSR) is also not suitable

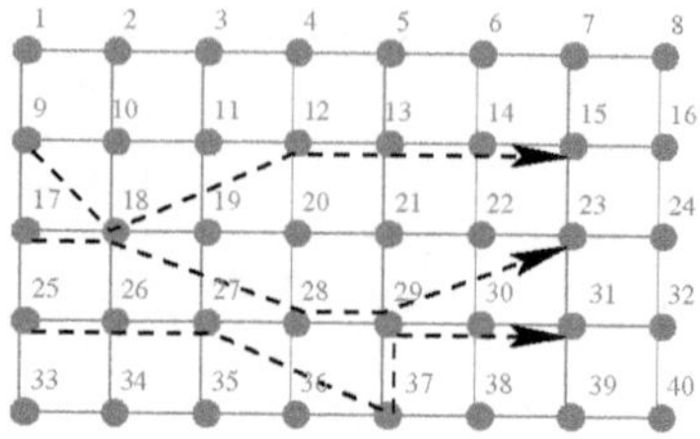

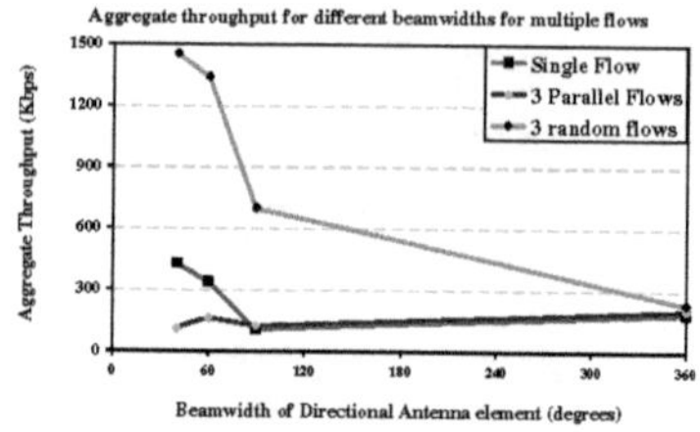

Fig. 4. (a)Left: Grid topology with 3 flows. (b)Right: Performance of single and multiple flows in linear and random topologies.

when using DDSR. In the next subsection, we discuss the reasons for this and evaluate a *"Delayed Route Reply"* optimization, to handle this problem.
2. Furthermore, MAC layer studies indicate that directional antennas perform poorly in "linear" topologies [9][7]. Performance in random scenarios may be significantly different, as observed later.

Delayed Route Reply Optimization: The optimization is motivated by the observation that the earliest RREQ received by the destination may not have traversed the optimal path when *sweeping* is used. This optimization requires the destination node to delay sending the route reply (RREP) by a time duration T, calculated from the time it received the first RREQ. This allows the destination node to choose the best among all the routes that arrive within this time T. We specify T as $T = \rho \times T_{sweep}$, where T_{sweep} is the time taken to complete one full sweep, and ρ is a configuration parameter. We argue that replying to all the RREQs that arrive at the destination, as performed optionally by DSR, is not equivalent to the optimization we propose. This is because, if the destination replies to the first RREQ, it would need to beamform towards one of its neighbors, to whom it must unicast the RREP. Now, while it is beamformed towards one of its neighbors, other RREQs may be transmitted by its neighbors. Due to the effect of deafness, the destination node may fail to receive these RREQs of which one could possibly have traversed the optimal path. Waiting for a sufficiently long time duration T minimizes this possibility. We have observed that with this simple optimization, routes chosen by DDSR are often shorter.

For a single flow, directional antennas achieve higher throughput than omnidirectional antennas. However, in presence of multiple flows, one may expect DDSR to further outperform DSR, due to spatial reuse of the channel. To verify this, we simulated a rectangular grid topology with 3 flows as shown in Figure 4(a). The dashed lines show an example of routes discovered by DDSR. Adjacent nodes in the grid are separated by 200 meters. Surprisingly, Figure 4(b) shows that for multiple flows the performance of directional antennas almost falls below the omnidirectional performance (note that the omnidirectional performance corresponds to the beamwidth of 360 degrees). This happens because, in the presence of multiple flows, several MAC layer issues start affecting performance – specifically, the problems of topological linearities and deafness

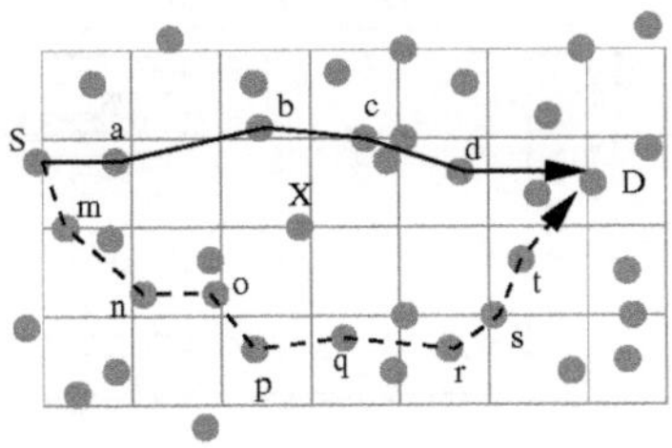

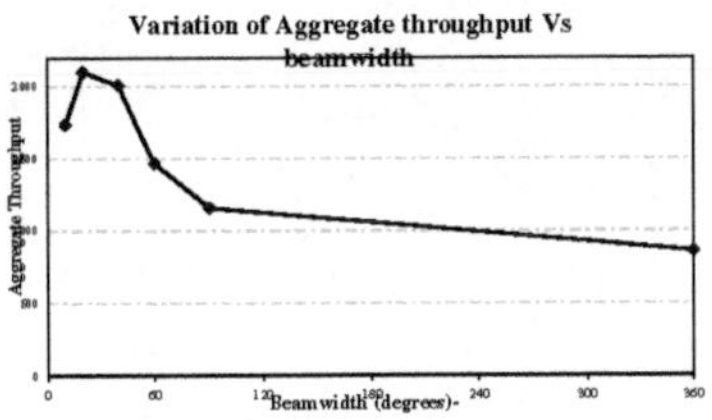

Fig. 5. (a)Left: A random scenario. (b)Right: Throughput variation with beamwidth.

become dominating. A detailed discussion on these issues is available in [8], [9]. To confirm that topological linearities and parallel flow patterns are the cause of degradation in performance, we displaced each node in the grid by a distance randomly selected from the interval [0, 200] meters. We simulated 5 such random topologies, with 3 identical flows used previously for the grid topology. Figure 5(a) shows one such random topology and outlines the routes taken by DDSR and DSR for one of the three flows. Figure 5(b) shows the average throughput of DSR and DDSR, simulated in random topologies (recall that DSR beamwidth equals 360 degrees). Clearly, DDSR performs better than DSR. Closer examination reveals that higher transmission ranges of directional antennas may be effective in bridging "voids" in the topology. Consider Figure 5(a). The solid line shows the DDSR route, while the dashed line indicates the DSR route. Clearly, the DSR route is much longer than DDSR. This happens because the DSR transmission range is not sufficiently large to form a link between nodes a and b. This forces DSR to choose a longer route through nodes m, n, o, etc. On the other hand, higher transmission range of DDSR allows a link between a and b, bridging, what may be called a "void region" for omnidirectional transmissions.

The advantage of smaller hop-count, combined with spatial reuse, can lead to higher aggregate throughput with DDSR as compared to DSR. This is evident in Figure 5(b). The graph shows the average of aggregate throughput over 25 random topologies for increasing beamwidth, hence decreasing transmission ranges. While aggregate throughput decreases with decrease in transmission range, at very high transmission ranges the trend is reversed. This is because at higher transmission ranges (i.e., at narrower beamwidths), the number of beams required for a single sweep increases proportionally. This not only increases *sweeping delay* but exacerbates *deafness* as well. Nodes that are capable of forming the shortest route often do not receive the RREQ. Sub-optimal routes are discovered, and performance can degrade with extremely narrow beamwidths.

Routing Overhead

Sweeping requires N sequential transmission of the same packet in N different directions, increasing control overhead. We compare control overhead (of DSR and DDSR) using the following overhead metric α.

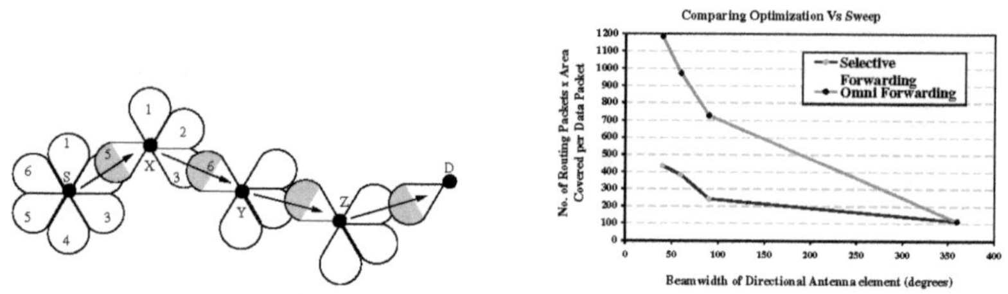

Fig. 6. (a)Left:Propagation of RREQ when using *selective forwarding* optimization. (b)Right: Comparing routing overhead α for DSR and DDSR

$$\alpha = \frac{\Sigma NumberofControlPackets \times AreaBlockedbyEachPacket}{\Sigma No.ofDataPackets}$$

where *Area Blocked by Each Packet* is the approximately conical area covered by the radiation pattern of a single beam. Intuitively, network capacity consumed due to the transmission of each control packet is proportional to the interference footprint of that transmission. The total control overhead is thus equivalent to *Number of Control Packets* $\times$ *Area Blocked by each packet.* This overhead is in exchange of the total *Number of Useful Data Packets* transmitted in the network. We use the ratio of these two quantities as the metric α.

The *sweeping* overhead of directional antennas is clearly larger than omnidirectional antennas. To address this in DDSR, we propose the following optimization.

Selective Forwarding Optimization: In Figure 6(a), when node Y receives a RREQ from node X, it may not be necessary to forward the RREQ back in the direction of node X. This is because, nodes lying in the coverage of node X's beam 3, can receive the packet from X and need not receive a copy of the same packet from Y again. In omnidirectional scenarios, the problem of redundant reception of RREQs has been referred to as the *broadcast storm* problem [10], since control packets are unnecessarily broadcast in the channel. Using directional antennas, we consider an optimization whereby a node forwards a control packet with only n $(n <= N)$ beams. A node initiating the control packet, however, forwards it using all beams. We define a configuration parameter $\gamma = \lceil \frac{n}{N} \rceil$, $\gamma = \frac{1}{2}$ for our simulations. Also, in our simulations, the n beams used to forward control packets are the ones that are diagonally opposite to the beam with which the control packet was received. Figure 6(a) shows how a RREQ may propagate when $\gamma = \frac{1}{2}$ and N = 6. Observe that node X receives the RREQ using beam 5 and forwards it using beams 1, 2 and 3.

Figure 6(b) plots routing overhead α against beamwidth of directional antennas. α was calculated from 25 simulations of random, static topologies. This optimization reduces the routing overhead of DDSR, although it remains higher than DSR.

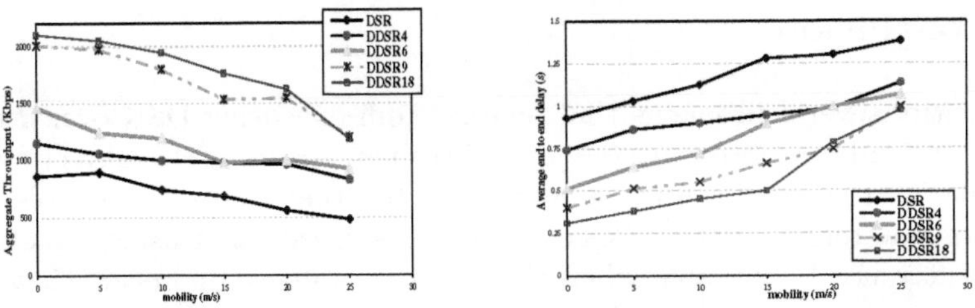

Fig. 7. Aggregate throughput and and Average end to end delay for mobile scenarios.

Mobility and Route Error

Our observations until this point are under the absence of node mobility. In the presence of node mobility, link failures may happen when nodes move too far away, or when a node moves out of the radiation pattern of one beam to that of the other – called *handoff*. Look-up tables thus get stale. In contrast, due to higher transmission range of directional antennas, links may persist for a longer duration when the end nodes of the link are moving away from each other. This reduces the frequency of route errors.

We have handled the problem of handoff at the MAC layer and DDSR remains unaware of it. If DiMAC does not receive a reply from a neighboring node after a small number of attempts, it assumes handoff, and attempts to find the appropriate beam that could re-establish communication. The node performs a partial *scan*, whereby the transmitter node sends "hello" packets using K beams, adjacent to the beam that was in use previously (detailed in [8]). Figure 7(a) shows a graph that measures aggregate throughput for several mobile scenarios and different beamwidths. A *random waypoint* mobility model [6],[3] is assumed. 10 source-destination pairs are chosen, and the time of simulation is proportionally reduced when mobility is increased. From Figure 7(a), aggregate throughput of DDSR degrades with mobility. This happens because the time for route discovery is almost a constant. Consider an example where a simulation is performed for 100 seconds using $mobility = 10m/s$. Also assume that over the entire simulation, exactly one route error occurs at time = 20 seconds, and data transmission is disrupted for 5 seconds. Now consider the scenario using $mobility = 20m/s$ and corresponding simulation time = 50 seconds. Observe that the route error in this scenario occurs at 10 seconds through the simulation and ends at 15 seconds. Consequently, the throughput at higher mobility is expected to decrease since a smaller fraction of the time (45/50) is now utilized for useful data communication (observe that at $mobility = 10m/s$, the fraction of time used for transmitting useful data is 95/100). However, DDSR achieves higher aggregate throughput, even at higher mobility. This is because, due to the higher transmission range of DDSR, links break less frequently, which in turn reduces number of route rediscoveries. Average end-to-end delay of DDSR is also observed to be lower than DSR, although between the variations of DDSR the differences are small.

5 Conclusion

We identify several problems that emerge from executing DSR over directional antennas, and propose suitable modifications - leading to Directional DSR (DDSR). Our strategies may be applicable to other routing protocols that use directional antennas. Performance evaluation suggests that using directional antennas may not be suitable when the network is dense or linear. However, the improvement in performance is encouraging for networks with sparse and random topologies. Under mobile conditions, DDSR outperforms DSR and handoff does not pose a serious problem. Our evaluation and optimizations have produced results that are encouraging, suggesting the need to utilize directional antennas more efficiently in ad hoc networking.

References

1. S. Bandyopadhyay, D. Saha, S. Roy and T. Ueda *"A Network Aware MAC and Routing Protocol for Effective Load Balancing in Ad Hoc Networks with Directional Antennas"*, Proc. of ACM/SIGMOBILE MobiHoc 2003.
2. IEEE, *"Wireless LAN Medium Access Control(MAC) and Physical Layer(PHY) Specifications"*, IEEE Standard 802.11, June 1999.
3. D. Johnson, D. Maltz and J. Broch, *"Dynamic Source Routing for Mobile Ad Hoc Networks"*, Mobile Ad Hoc Network Working Group, IETF, March 98.
4. Y. B. Ko, V. Shankarkumar and N. H. Vaidya, *"Medium access control protocols using directional antennas in ad hoc networks"*, Proc. of the IEEE INFOCOM 2000.
5. A. Nasipuri et al. *"On Demand Routing Using Directional Antennas in Mobile Ad Hoc Networks"*, Proc. of the IEEE WCNC 2000.
6. *"Qualnet 2.6.1"*, Scalable Network Technologies, www.scalable-networks.com.
7. Ram Ramanathan, *"On the Performance of Beamforming Antennas in Ad Hoc Networks"*, Proc. of the ACM/SIGMOBILE MobiHoc 2001, October 2001.
8. R. Roy Choudhury and N. Vaidya, *"Ad Hoc Routing Using Directional Antennas"*, Technical Report, http://www.crhc.uiuc.edu/ nhv/publications.html
9. R. Roy Choudhury, X. Yang, R. Ramanathan and N. Vaidya, *"Using Directional Antennas for Medium Access Control for Ad Hoc Networks"*, Proc. of the ACM MOBICOM 2002, September 2002.
10. Sze-Yao Ni, Yu-Chee Tseng, Yuh-Shyan Chen and Jang-Ping Sheu, *"The Broadcast Storm Problem in a Mobile Ad Hoc Network"*, Proc. of the Fifth Annual ACM/IEEE International Conference on Mobile Computing and Networking (MOBICOM 99), WA, USA.
11. Mineo Takai, Jay Martin, Rajiv Bagrodia and Aifeng Ren, *"Directional Virtual Carrier Sensing for Directional Antennas in Mobile Ad Hoc Networks"*, Proc. of the ACM/SIGMOBILE MobiHoc June 2002.

A Rotational Sector-Based, Receiver-Oriented Mechanism for Location Tracking and Medium Access Control in Ad Hoc Networks Using Directional Antenna

Tetsuro Ueda[1], Shinsuke Tanaka[1], Dola Saha[2], Siuli Roy[2], and Somprakash Bandyopadhyay[2]

[1] ATR Adaptive Communication Research Laboratories
2-2-2 Hikaridai, Seika-cho Soraku-gun, Kyoto 619-0288 Japan
{teueda,shinsuke}@atr.co.jp
[2] Indian Institute of Management Calcutta
Diamond Harbour Road, Joka Calcutta 700104 India
{dola, siuli,somprakash}@iimcal.ac.in

Abstract. The use of directional antenna in wireless ad hoc networks potentially increases simultaneous communication by directing the transmitting and receiving beams towards the receiver and transmitter node as compared to omni-directional antenna, where nodes in the vicinity of a communication are kept silent. However, in order to implement effective directional MAC protocol using directional antenna, a node should know how to set its transmission direction to transmit a packet to its neighbors and to avoid transmission in other directions where data communications are already in progress. So, it becomes imperative to have a mechanism at each node to track the locations of its neighbors and to know the communication status of neighboring nodes. In this paper, we propose a receiver-centric approach for location tracking and MAC protocol. The performance evaluation on QualNet network simulator indicates that our protocol is highly efficient with increasing number of communications and increasing data rate.

1 Introduction

The recent progress in wireless communication and personal computing leads to the research of ad hoc wireless networks, which are envisioned as rapidly deployable, infrastructure-less networks with each node acting as a mobile router, equipped with a wireless transceiver. Usually, in ad hoc networks, all nodes are equipped with omni-directional antenna. However, ad hoc networks with omni-directional antenna uses RTS/CTS based floor reservation scheme that wastes a large portion of the network capacity by reserving the wireless media over a large area. Consequently, lot of nodes in the neighborhood of transmitter and receiver has to sit idle, waiting for the data communication between transmitter and receiver to finish. To alleviate this problem, researchers have proposed to use directional (fixed or adaptive) antennas that direct the transmitting and receiving beams toward the receiver and transmitter node only. This would largely reduce radio interference, thereby improving the utilization of wireless medium and consequently the network throughput [1-11].

M. Conti et al. (Eds.): PWC 2003, LNCS 2775, pp. 601–610, 2003.

In order to fully exploit the capability of directional antenna, it is necessary for each node to know the information of the neighboring nodes (such as node-ID, direction, link quality, etc.) beforehand. Thus, in order to implement effective MAC and routing protocol in this context, a node should know how to set its transmission direction to transmit a packet to its neighbors. So, it becomes imperative to have a mechanism at each node to track the locations of its neighbors. However, this location tracking mechanism in the context of wireless ad hoc networks with directional antenna is a serious problem, since it incurs a lot of control overhead. In this paper, we are proposing a receiver-centric approach for location tracking and MAC protocol. In order to track the location of its neighbor, each node n periodically *collects* its neighborhood information and forms an Angle- Signal Table (AST). Based on AST, a node n knows the direction of node m and controls the *medium access* during transmission-reception. The performance evaluation on QualNet network simulator [12] indicates that our protocol is highly efficient with increasing number of communications and with increase in data rate. The one-hop MAC throughput of our protocol is 1.8 times as compared to that of IEEE 802.11. Moreover, average number of data packet retransmission due to loss of data packets is substantially less (less than 5% of that with IEEE 802.11).

2 Related Work

In spite of the advantages of directional antennas, work on developing efficient MAC protocol using directional antennas in the context of ad hoc networks is limited because of the inherent difficulty to cope up with mobility and de-centralized control in ad hoc networks. Some researchers in the past have tried to address this challenge in several ways [1,2]. In recent years, several MAC protocols that rely on RTS-CTS type handshaking as in IEEE 802.11 have been suggested with directional antennas [3-11]. In [3], a set of D-MAC (Directional MAC) schemes has been proposed where combination of directional/ omni-directional RTS / CTS are used to block nodes from transmitting in directions that would interfere with existing data transmission while allowing them to transmit on other directions. In [4], a MAC protocol to achieve multihop efficiency has been proposed with multihop-RTS-singlehop-CTS using directional antenna. In this mechanism, using larger range of directional beam, a destination is reachable in less number of hops as compared to that using omni directional antenna. In both the schemes [3-4], the mobile nodes are assumed to know the physical locations of themselves and their neighbors using GPS. In [5], the proposed MAC protocol need not know the location information; the source and destination nodes identify each other's direction during omni-directional RTS-CTS exchange in an on-demand basis. In [6], Ramanathan studied the performance of ad hoc networks using beamforming antennas with changing antenna patterns and beam control, channel access mechanisms, link power control and neighbor discovery. The authors assume prior knowledge of location information. In [7], concept of Directional NAV (the network allocation vector) has been used, one for each sector, allowing immediate transmission of control packets on those sectors which are clear instead of having to defer the transmission until it is safe to transmit on all sectors at the same time. In [8] also, Takai et al. proposed a directional NAV with a direction and a width, which is set depending on the signal strength in that direction.

Developing a suitable MAC protocol in ad hoc network to exploit the advantages of directional antenna for overall performance improvement requires proper location tracking and neighborhood knowledge. In [5], the source and destination nodes identify each other's direction during omni-directional RTS-CTS exchange. However, in this mechanism, a node is not aware of its complete neighborhood information. In [3-4], the use of GPS is proposed to track the location of each node but the exact mechanism of information exchange and the consequent overhead has not been discussed. In our earlier work, we have developed a MAC protocol [9], where each node keeps certain neighborhood information dynamically through the maintenance of an Angle-SINR Table. In this method, in order to form AST, each node periodically sends a directional beacon in the form of a directional broadcast, sequentially in all direction at 30 degree interval, covering the entire 360 degree space. The nodes, which receive these signals at different angles, determine the best received signal strength and transmit the information back to the source node as data packet with RTS/CTS handshake. However, the overhead due to control packets is very high in this method [9] of location tracking.

In this paper, we will illustrate a receiver-oriented location tracking mechanism to reduce the control overhead, which is described in Section 4 and a simple MAC protocol for efficient medium utilization, which is depicted in Section 5. We have done extensive performance evaluation using QualNet to demonstrate its effectiveness and is illustrated in Section 6.

3 System Description

3.1 Antenna Model

There are basically two types of smart antennas used in the context of wireless networks: switched-beam or fixed beam antennas and steerable adaptive array antennas [12,13,14]. A switched-beam antenna generates multiple pre-defined fixed non-overlapping directional beam-patterns and applies one at a time when receiving a signal. Signals will be sensed in all sectors and the antenna is capable of recognizing the sector with the maximum gain. When receiving, exactly one sector, which usually is the one chosen by the sensing process, will collect the signals.

In a steerable adaptive array antenna, the beam structure adapts to Radio Frequency (RF) signal environment and directs beams towards the signal of interest to maximize the antenna gain, simultaneously depressing the antenna pattern (by setting nulls) in the direction of the interferers [14]. In adaptive array antennas, an algorithm is needed to control the output, i.e. to maximize the Signal to Interference and Noise Ratio (SINR).

We have developed a wireless ad hoc network testbed using smart antenna [15] where each user terminal uses a small, low-cost smart antenna, known as ESPAR (Electronically Steerable Passive Array Radiator) antenna [16,17]. The adaptive array antennas are normally digital beamforming antennas. On the other hand, ESPAR antenna relies on RF beamforming, which drastically reduces the circuit complexity. Since ESPAR antenna requires only one receiver chain, it is possible to provide drastic improvement in both dc power dissipation and fabrication costs, by eliminating the need for frequency converters and analog-digital converters by the number of array branches [16]. The features of ESPAR are: controlling beam direction, multiple beams (with same frequency) formation, steerable beam (360 degree sweeping) and controlling null steering. For receiver application, the null should be steered in the direction from which an interfering signal is coming.

Developing suitable MAC protocols with adaptive antenna in ad hoc networks is a challenging task. That is why, most of the works in the context of ad hoc networks assume to use simpler switched beam antenna. In this work also, we are using smart ESPAR antenna as a switched beam antenna. ESPAR antenna can also be used as a generalized switched beam antenna or quasi-switched beam antenna, by selecting the value of reactance for one specific directional beam among multiple directional beam patterns, without using multiple receiver chains (frequency converters and analog-digital converters). By including some mechanism to detect *direction of arrival* (DoA) for the signal received from the user (as will be illustrated shortly), continuous tracking can be achieved and it can be viewed as a generalization of the switched beam concept [14]. In this case also, the received power is maximized. The advantage of using ESPAR antenna as generalized switched beam antenna is that, with only one receiver chain, continuous tracking is possible and we can have variable number of beam-pattern. Since ESPAR antenna would be a low-cost, low-power, small-sized antenna, it would help to reduce the power consumption of the user terminals in WACNet and would be able to deliver all the advantages of switched beam antenna.

3.2 A Few Assumptions and the Rationales

- When the antenna of a node operating in omni-directional mode, it is capable of transmitting and receiving signal from all direction with a gain, say, G^{omni}. While idle, a node operates in omni-directional receive mode.
- When the antenna of a node operating in directional mode, a node can points its beam (main lobe) towards a specified direction with beam width *w* and with a gain, say G^{dir} ($G^{dir} >> G^{omni}$). Beam width is around 60 degree in our simulation.
- Consequently, for a given amount of input power, the transmission range R^{dir} with directional antenna will be much larger than that with corresponding omni-directional antenna (R^{omni}).
- We define neighbors of a node n as a set of nodes within the omni-directional transmission range of n. From the perspective of directional data communication, it implies that a neighbor, say, m of a node, say, n is always a *strong* neighbor. Hence, the received signal strength at m from n is always high to ensure proper capture even in presence of other interference. Thus, the chance of m getting disconnected or weakly connected during a data packet transfer from n due to an outward mobility of either m or n is far less.
- This will alleviate the problem of hidden terminal in this context as indicated in [4]. Let us consider Fig. 1 where node n is communicating with node m with directional beam. Node p now wants to communicate with node q. If node p is within the neighborhood of n, this communication will not be initiated, since p is not allowed to form directional beam towards n and/or m. However, if node p is outside the neighborhood of n, node p forms a directional beam towards node q and starts communication. This may interfere with node m's reception. However, since the distance p and m is larger than n and m by at least R^{omni} (the omni-

directional range), the received signal at m from n will predominate and chance of data packets being lost due to this interference will be insignificant. This is reflected in our simulation results, which shows very low packet retransmission under different conditions.

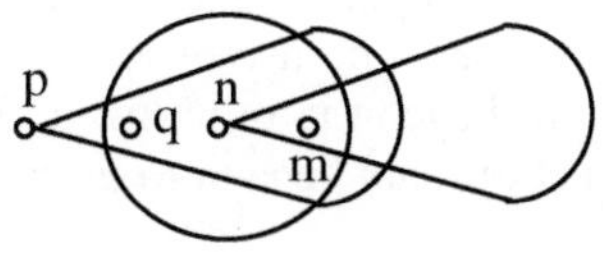

Fig. 1. The capture of receiver m by transmitter n is strong enough to tolerate interference from another transmitter p

However, as a consequence of this assumption, we are sacrificing multihop efficiency which could have been achieved using directional antenna, since using larger range of directional beam, a destination is reachable in fewer number of hops as compared to that using omni directional antenna. However, what we are gaining is SDMA efficiency, as will be demonstrated in the performance evaluation.

4 Location Tracking Mechanism

In this study, each node waits in omni-directional-receive-mode while idle. Whenever it senses some signal above a threshold, it enters into *rotational-sector-receive-mode*. In rotational-sector-receive mode, node n rotates its directional antenna sequentially in all direction at 30 degree interval, covering the entire 360 degree space in the form of the sequential directional receiving in each direction and senses the received signal at each direction. After one full rotation, it decides the best possible direction of receiving the signal with maximum received signal strength. Then it sets its beam to that direction and receives the signal.

However, in order to enable the receiver decoding the received signal, each control packet is transmitted with a preceding tone with a duration such that the time to rotate a receiver's rotational receive beam through 360 degree is less than the duration of the tone. The purpose of this transmitted tone before any control packet is to enable the receiver to track the best possible direction of receiving the signal. Once it sets its beam to that direction, the purpose of tone signal is over and subsequently the control packet is transmitted.

In this proposed framework, we have used three types of control packets: beacon or "hello" packet used to track the location of neighboring nodes), RTS (Request to send) and CTS (clear to send) for medium access control. Beacon is a periodic signal, transmitted from each node at a pre-defined interval. At each periodic interval, each node, say, m, sends an omni-directional beacon to its neighbors, if the medium is free. As indicated earlier, each beacon is transmitted with a preceding tone signal that helps the receivers to detect the best possible direction of receiving the beacon. Then each receiver sets its beam to that direction and receives and decodes the beacon. Thus, the node n which is, say, a neighbor of m forms the Angle-Signal information for node m, and similarly, for other neighbors. An entry in AST of node n for its neighbor m is $SIGNAL^{\alpha}_{n,m}(t)$, which is the maximum strength of received signal at node n from node m at an angle α with respect to n and as perceived by n at any point of time t. Based on AST, a node n knows the direction of node m and controls the medium access during transmission-reception.

Since RTS is a broadcast packet and contains source address, nodes can decode that RTS also to form the Angle-Signal Table. So, we have used RTS as beacon. If an RTS is sent, beacon timer is reset. The use of RTS as beacon is advantageous at high traffic where overhead due to beacon is minimized. This is because, the transmitting nodes don't have to send an additional beacon to inform its neighbors of its presence.

5 Directional Medium Access Control Protocol

In IEEE 802.11 MAC protocol standard, RTS-CTS-DATA-ACK exchange mechanism is used to ensure reliable data communication. In our scheme, initially, when node n wants to communicate with m, it senses the medium and if it is free, sends *omni-directional RTS*. The back-off mechanism is same as in IEEE 802.11. The purpose of RTS is to inform all the neighbors of n, including m, that a communication from n to m has been requested. It also specifies the approximate duration of communication. All the neighboring nodes of n keep track of this request from node n, whose direction is known to the each of them from the received RTS signal. The mechanism for receiving RTS is same as that for beacon.

The target node m sends *omni-directional* CTS to grant the request and to inform the neighbors of m that m is receiving data from n. It also specifies the approx. duration of communication. All the neighboring nodes of m keep track of the receiving node m, whose direction is known to the each of them from the received CTS signal. Once again, the mechanism for receiving CTS is same as that for beacon. It is to be noted that the objective of RTS/CTS here is not to inhibit the neighbors of n from transmitting or receiving (as is the case with omni-directional antenna) but to inform the neighbors of n that m is receiving data from n.

After transmission of omni-directional CTS, the receiving node waits in directional receive mode until Data is transmitted or timeouts and returns to omni-directional receive mode. Also, once the CTS is received, the transmitter transmits Data directionally and waits for Acknowledgement directionally until Acknowledgement is received or timeouts and returns to omni-directional receive mode. The receiver on receiving Data, transmits Acknowledgement directionally and returns to omni-directional receive mode. The directional reception mode ensures proper reception of signal from the required direction and minimization of interference from other direction.

Other nodes in the neighborhood of n and m, who overheard the RTS/CTS exchange, set their Directional Network Allocation Vector (DNAV) in the direction which they detected as the direction of arrival of the RTS or CTS respectively. Now, if they have a packet to send to a node, whose direction as known from AST, is not in the direction of blocked DNAV, then they can issue both RTS and CTS omni-directionally *without disturbing the communication between n and m.* Omni-directional RTS/CTS will not create interference at the receiver end of the on-going communication because receiver directs its receiving beam pattern towards the transmitter. If the direction of receiving node is blocked by DNAV and RTS is issued, it is most probable that CTS will not be issued or their may be RTS collision. As a result, the node will increase its contention window and enter into backoff. This may happen repeatedly and as a result, the node will get less chance to transmit. So, we do not allow transmission of RTS in this case. Here, the node waits for DNAV time and then tries to start communication, which is similar to waiting for NAV as explained in standard IEEE 802.11.

6 Performance Evaluation

6.1 Simulation Environment

The simulations are conducted using QualNet 3.1 [12]. We have simulated ESPAR antenna in the form of a *quasi-switched beam antenna*, which is steered discretely at an angle of 30 degree, covering a span of 360 degree. We have simulated our MAC protocol with (i) Simulated ESPAR Antenna Pattern (ESPAR), (ii) QualNet's default directional antenna pattern (DEFAULT) and (iii) an Ideal directional antenna pattern without sidelobes (IDEAL) as described in Section 3.1. We have done the necessary changes in QualNet simulator to implement Directional Virtual Carrier Sensing in MAC Layer and directional transmission in Physical Layer of QualNet simulator.

ESPAR antenna requires 16 microseconds to measure the level of received signal in each sector at 2Mbps data rate and 10 nanoseconds to change the beam pattern from one sector to another. So, to rotate in 12 patterns and detect the gain in each antenna pattern, ESPAR antenna requires nearly [(12×(16+0.01)) microseconds]= 192.12 microseconds. Hence, in our simulation, we have chosen the duration of preceding tone in control packets to be 200 microseconds.

We have used simple one-hop randomly chosen communication in order to avoid the effects of routing protocols to clearly illustrate the difference between 802.11 and our proposed MAC. Also, we have used static routes to stop all the control packets generated by any routing protocol, whether it is proactive or reactive. In our simulation, we studied the performance of the proposed MAC protocol in comparison with the existing omnidirectional 802.11 MAC protocol by varying the data rate and number of simultaneous communications. In studying our MAC protocol, we have used different antenna patterns as described above to ensure the robustness of our proposed MAC protocol. In doing this, we have used ESPAR antenna as one of the antenna patterns, to evaluate the performance of the ESPAR antenna as well. The set of parameters used are listed in Table 1.

Table 1. Parameters used in Simulation

Parameters	**Value**
Area	1000 x 1000 m
Number of nodes	40
Transmission Power	15 dBm
Receiving Threshold	-81.0 dBm
Sensing Threshold	-91.0 dBm
Data Rate	2Mbps
Packet Size	512 bytes
Duration of Preceding Tone	200 microseconds
CBR Packet Arrival Interval	2 ms to 50 ms
Number of simultaneous communication	2 to 16
Simulation Time	5 minutes

6.2 Results and Discussions

We have used the existing IEEE 802.11 MAC, which we caption as "802.11", as a benchmark to compare and evaluate the performance of our proposed MAC protocol with ESPAR antenna, QualNet's default antenna and an ideal antenna, which we caption as "ESPAR", "DEFAULT", and "IDEAL" respectively. Our evaluation is based on two criteria: ***Average Throughput,*** and ***One Hop Average End-to-End Delay.*** The results are shown in Fig. 2 and 3 respectively. Each result reported is an average of ten executions with different seeds. So, to complete our results, we had to simulate over 400 scenarios, each of which was executed in the simulator for 5 minutes to get an overall average result.

MAC Performance depends much on antenna pattern also. So, we have simulated for 3 different types of antenna patterns. In showing the results with three beam patterns, we had two notions in our mind: (a) E-MAC performs nicely with directional antenna and maximum gain that can be obtained from E-MAC is nearly thrice than that of IEEE 802.11, if beam-width of directional antenna is 45 degree and sectors are 45 degree each, and (b) In context of an ideal antenna pattern with 45 degree beamwidth, which gives maximum gain, performance of ESPAR antenna with 60 degree beamwidth is quite reasonable and it even performs better than a standard antenna pattern with 45 degree beamwidth provided by QualNet Simulator.

In Fig. 2, it is observed that with increasing data rate, average throughput of our proposed MAC protocol with any directional antenna pattern is much better than that of IEEE 802.11 and one hop average end-to-end delay of E-MAC is nearly half of that obtained with IEEE 802.11 protocol. In omnidirectional 802.11, nodes have to sit idle in the omni-directional range of both transmitter and receiver to facilitate the on-going communication. But, with the use of directional antenna, and the implementation of Directional Virtual Carrier Sensing, E-MAC creates an environment of lower contention which "802.11" cannot create with an omnidirectional antenna. Hence, with increasing data rate, average throughput increases sharply in E-MAC as evident from Fig. 2. Also, we can see that due to SDMA efficiency in E-MAC, less contension in E-MAC leads to low one hop average end-to-end delay. With ESPAR antenna, this delay is nearly half that of IEEE 802.11 protocol.

In Fig. 3, it is observed that with increasing number of simultaneous communication, average throughput decreases in both E-MAC and 802.11, but E-MAC shows significant gain in Average Throughput. This is because E-MAC does not inhibit neighboring nodes to transmit, but just informs neighbors of the ongoing communication and its direction, so that they can start communication in other directions. But 802.11 with omnidirectional antenna, keeps all neighboring nodes silent by issuing RTS/CTS. Also, with increasing number of simultaneous communication, one hop average end-to-end delay increases in both IEEE 802.11 and E-MAC, but the increase is much prominent in "802.11" than in E-MAC, irrespective of the directional antenna pattern used. With increasing number of simultaneous communication, interference to each communication increases due to interference of added number of simultaneous communication. But, E-MAC not only informs other nodes in its vicinity of the on-going communication, but also transmits and receives directionally, which minimizes interference from other directions also. Thus E-MAC exploits SDMA efficiency for which more number of simultaneous communication is possible, which leads to lesser queuing delay and lesser one hop average end-to-end delay as observed in Fig. 3.

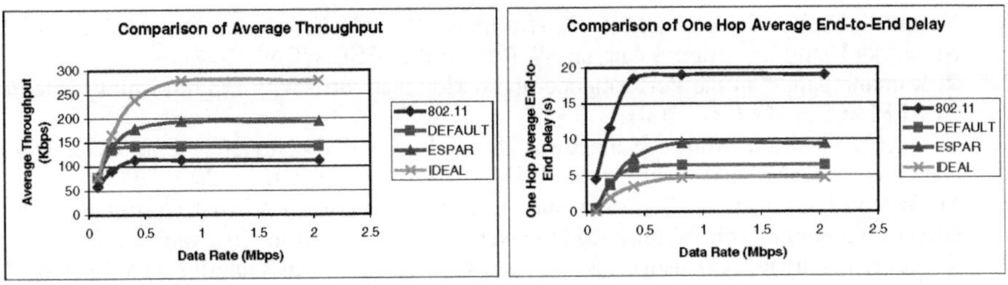

Fig. 2. Performance Evaluation of the proposed MAC protocol with directional antenna with increasing data rate

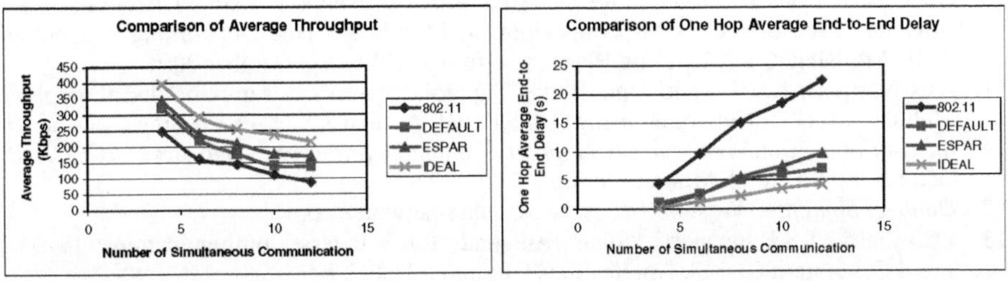

Fig. 3. Performance Evaluation of the proposed MAC protocol with directional antenna with increasing number of simultaneous communication

7 Conclusion

Use of directional antenna in ad hoc wireless network can drastically improve system performance, if proper MAC protocol can be designed. With directional setting of Virtual Carrier Sensing, medium can be utilized to its maximum with directional antenna. Even with the overhead of location tracking, gain obtained in MAC is really significant. Presently, we are working on efficient controlling of transmission power to improve the proposed MAC performance.

References

1. J. Zander, "Slotted ALOHA multihop packet radio networks with directional antennas", Electronic Letters, vol.26, no.25, 1990
2. T.S. Yum and K.W. Hung, "Design algorithms for multihop packet radio networks with multiple directional antennas stations," IEEE Transactions on communications, vol. 40, no. 11, pp. 1716–1724, 1992.
3. Y.B. Ko, V. Shankarkumar and N. H. Vaidya, "Medium access control protocols using directional antennas in ad hoc networks," Proc. of the INFOCOM 2000, March 2000.
4. Romit Roy Choudhury , Xue Yang , Nitin H. Vaidya , Ram Ramanathan, "Using directional antennas for medium access control in ad hoc networks" Proc. of the eighth annual international conference on Mobile computing and networking September 2002

5. Nasipuri, S. Ye, J. You and R.E. Hiromoto, "A MAC Protocol for Mobile Ad Hoc Networks Using Directional Antennas", Proc of the IEEE WCNC 2000.
6. R. Ramanathan, "On the Performance of Ad Hoc Networks with Beamforming Antennas", ACM MobiHoc, October 2001.
7. Kou Kobayashi and Masao Nakagawa, "Spatially divided channel scheme using sectored antennas for CSMA/CA – directional CSMA/CA", *Proc.of PIMRC'2000*, 2000.
8. M. Takai, J. Martin, R. Bagrodia and A. Ren, "Directional Virtual Carrier Sensing for Directional Antennas in Mobile Ad Hoc Networks", ACM MobiHoc, June 2002.
9. S. Bandyopadhyay, K. Hasuike, S. Horisawa, S. Tawara, "An Adaptive MAC Protocol for Wireless Ad Hoc Community Network (WACNet) Using Electronically Steerable Passive Array Radiator Antenna", Proc of the GLOBECOM 2001, November 25–29, 2001, San Antonio, Texas, USA
10. S. Bandyopadhyay, K. Hasuike, S. Horisawa, S. Tawara, "An Adaptive MAC and Directional Routing Protocol for Ad Hoc Wireless Network Using Directional ESPAR Antenna" Proc of the ACM Symposium on Mobile Ad Hoc Networking & Computing 2001 (MOBIHOC 2001), Long Beach, California, USA, 4–5 October 2001
11. Asis Nasipuri, Kai Li, and Uma Reddy Sappidi, "Power Consumption and throughput in Mobile Ad Hoc Networks using Directional Antennas" in Proceedings of the *IEEE International Conference on Computer Communication and Networks (ICCCN2002),* October 14–16, 2002, Miami, Florida.
12. *QualNet Simulator Version 3.1*, www.scalable-networks.com
13. J.C.Liberti, T.S.Rappaport, "Smart Antennas for Wireless Communications: IS–95 and Third Generation CDMA Applications", Prentice-Hall, 1999.
14. Per H. Lehne and Magne Pettersen, An Overview of Smart Antenna Technology for Mobile Communications Systems, IEEE Communications Surveys, http://www.comsoc.org/pubs/surveys, Fourth Quarter 1999, vol. 2 no. 4.
15. T. Ueda, K. Masayama, S. Horisawa, M. Kosuga, K. Hasuike, "Evaluating the Performance of Wireless Ad Hoc Network Testbed With Smart Antenna", Fourth IEEE Conference on Mobile and Wireless Communication Networks (MWCN), September 2002
16. T. Ohira, "Adaptive array antenna beamforming architectures as viewed by a microwave circuit designer", 2000 Asia-Pasific Microwave Conf., Sydney, Dec. 2000.
17. K. Gyoda and T. Ohira, "Beam and Null Steering Capability of ESPAR Antennas, Proc of the IEEE AP-S International Symposium, July 2000.

Alleviating Effects of Mobility on TCP Performance in Ad Hoc Networks Using Signal Strength Based Link Management*

Fabius Klemm[1**], Srikanth V. Krishnamurthy[2], and Satish K. Tripathi[2]

[1] Department of Computer Science,
EPFL, Lasaunne, Switzerland
Fabius.Klemm@epfl.ch

[2] Department of Computer Science and Engineering,
University of California, Riverside, CA, 92521
{krish,tripathi}@cs.ucr.edu

Abstract. Mobility in ad hoc networks causes link failures, which in turn result in packet losses. TCP attributes these losses to congestion. This results in frequent TCP retransmission timeouts and degradation in TCP performance even at light loads. We propose mechanisms that are based on signal strength measurements to alleviate such packet losses due to mobility at *light loads*. Our key ideas are (a) if the signal strength measurements indicate that a link failure is most likely due to a neighbor moving out of range, *in reaction*, facilitate the use of temporary high power to re-establish the link and, (b) if the signal strength measurements indicate that a link is likely to fail, initiate a route re-discovery *proactively* before the link actually fails. We make changes at the MAC and the routing layers to predict link failures and estimate if a link failure is due to mobility. We also propose a simple mechanism that can help alleviate *false link failures* that occur due to congestion when the IEEE 802.11 MAC protocol is used. We compare the above proactive and reactive schemes and also demonstrate the benefits of using them together. We show that, in high mobility, the performance of a TCP session can increase by as much as 45 percent when our methods are incorporated.

Keywords: Power Management, Ad Hoc Networks, TCP, Signal Strength, IEEE 802.11

1 Introduction

TCP performs poorly in wireless ad hoc networks as demonstrated in [8, 12]. The main reason for this poor performance is a high level of packet losses and a resulting high number of TCP retransmission timeouts. First, a node drops a packet if it cannot forward the packet to the next hop of the route as the next hop node has moved out of transmission range. A second reason for packet loss is congestion in the shared

* This work was partially funded by the DARPA Fault Tolerant Networks (FTN) contract no: F30602-01-2-0535.

** This work was done when the author was with University of California, Riverside.

M. Conti et al. (Eds.): PWC 2003, LNCS 2775, pp. 611–624, 2003.

medium. In this case, a node cannot reach the next hop node because there are too many nodes trying to access the channel at the same time. This might even result in a node *capturing* the medium of access if the IEEE 802.11 MAC protocol is used [12]. While congestion can degrade the observed performance of TCP even in wire-line networks, mobility causes a degradation of performance of TCP in ad hoc networks even at very light loads. Our objective in this paper is to mainly stem the degradation of TCP performance due to mobility.

Towards this goal, we propose mechanisms to reduce the number of packet losses. These mechanisms are based on signal strength measurements at the physical layer. Based on these signal strength measurements, when a node *fails* to communicate with a neighbor, the MAC layer at the node guesstimates if the failure is due to congestion or due to the neighbor moving out of range. If the MAC layer predicts that the neighbor has just moved out of range, then it stimulates the physical layer to increase its transmission power and attempts to re-establish the link to the neighbor temporarily. It also prompts the routing layer to search for a new route. The signal strength measurements can also be used to predict possible link failures to a neighbor that is about to move out of range. Thus, if the measurements indicate that the signal strength is going down and the link is likely to break, a search for a new route can be proactively initiated before the link actually fails. While searching for the new route, the routing layer should take care to avoid either the temporary high power link or the weak link (as the case may be). We have made modifications to the ad hoc on-demand distance vector (AODV) routing protocol [11] such that it precludes the use of such links during the computation of a new route. In order to cope with failures that are not due to mobility, we have included a simple mechanism by which, the MAC layer, upon guesstimating that the neighbor is within range, *persists* in its attempt to reach that neighbor for a longer period of time. We re-iterate that our goals are mainly to cope with the effects of mobility on TCP. At high loads, it is more likely that congestion dominates packet losses. In the simulation experiments that we perform to evaluate our schemes, we therefore restrict ourselves to conditions of light load (one or two TCP connections). In such scenarios we show that the performance of a TCP session can improve by as much as 45%.

The use of signal strength and a count of the transmitted packets in the local neighborhood (nodes can overhear other packet transmissions) can provide an estimate of whether or not there is congestion in the local vicinity of a node. A node should only increase its transmission power if the network is lightly loaded. If there is congestion, temporary increases in power levels can actually increase the number of collisions and increase the congestion. This could degrade the performance further. However, congestion estimation mechanisms are a focus of further study and are beyond the scope of this paper.

Several researchers have been working on improving TCP for wireless ad hoc networks. Holland and Vaidya [8] demonstrated that node mobility causes TCP throughput to drop considerably due to "TCP's inability to recognize the difference between link failure and congestion". They introduced the *Explicit Link Failure Notifications (ELFN)* to allow TCP to react appropriately to link failures. Similar failure notification schemes are presented in [4] and [5]. In [13], the authors propose to split long TCP sessions into multiple segments. By doing so, they argue that, even if a link failure occurs on one of these segments, data flow can be sustained on other segments. The previous schemes however, are unable to salvage TCP packets that are in transit. Our efforts are to salvage packets in transit if the a link failure is due to mobility as opposed to congestion.

To the best of our knowledge, ours is the first work to use lower physical layer features such as signal strength and adaptive transmission power levels to improve TCP performance in ad hoc networks. The methods that we propose can be used with the User Datagram protocol (UDP) as well. However, since the effects are unlikely to be as profound we do not consider UDP in this paper.

The rest of this paper is organized as follows: In section 2, we explain the causes of high packet loss. Section 3 will introduce methods to decrease packet loss in ad hoc networks. Section 4 will present the simulation setup and the simulation results. We shall present our conclusions in section 5.

2 Packet Losses in Ad Hoc Networks

In wireless ad hoc networks, TCP performance is affected by packet losses. *Node mobility* and *link layer congestion* are the main reasons for packet loss. A link failure on an active path due to mobility causes the MAC protocol to report a link failure to the routing layer. The routing layer will then have to re-compute routes to the appropriate destinations. When the IEEE MAC protocol [9], which is popular for ad hoc networks, is used, when congestion occurs *false link failures* may be induced. Since our methods should be invoked only when there is a true link failure due to mobility, it is important to correctly identify such failures.

A false link failure occurs when the MAC protocol at a node, say, N_0 declares that the link to a neighbor N_1 is broken, even though N_1 is within its transmission range [1]. The MAC protocol at N_0 fails to establish an RTS-CTS handshake because N_1 cannot respond to its RTS message because it senses another transmission in its vicinity. This is a direct result of the following: In models used, it is often assumed that each node has a transmission range of 250 meters and an interference range of 550 meters[1]. Nodes within the *transmission range* of a node N_0 can receive packets from N_0. Nodes that are not within the transmission range but are within *interference range* can sense a transmission from N_0 but cannot successfully receive packets from it. These nodes are also precluded from performing transmissions if N_0 is in the process of transmitting. Thus, upon receiving RTS messages, they will have to ignore these messages.

At the network layer, the routing protocol has to react appropriately to route failures. We discuss how AODV behaves in light of a link failure report from the MAC layer as we use AODV [11] in our simulations later. AODV would simply drop the packets that are to be routed on the failed link. It brings down the routes to the destinations using the failed link and generates *route error* messages and sends such a message to the source of each connection that uses that link.

---

[1] The network simulator ns-2 uses these values for the transmission and interference range.

3 Reducing Link Failures to Improve TCP Performance

We propose mechanisms that help alleviate packet losses due to link failures due to mobility. Our mechanisms are based on measuring the signal strength at the physical layer. As pointed out in Section 2, it is important to first estimate if those failures that are due to mobility. False link failures, discussed earlier, cannot be overcome by tuning power levels. We propose a simple way to identify and cope with false link failures. The methods we propose however only work at light loads and will have to be complemented by other techniques that can estimate the congestion/load in the network.

3.1 Reducing False Link Failures

The IEEE 802.11 MAC reports a link failure if it cannot establish an RTS-CTS handshake with a neighbor within seven RTS retransmissions [9]. Our idea is to send out more than eight RTSs if the probability that the neighbor is still within transmission range is high. We call our version of the MAC protocol the *Persistent MAC*.

In order to determine if a node is still within range, a node keeps a record of the received signal strengths of neighboring nodes. Received signal strength measurements are taken at the physical layer. When a node receives a packet from a neighbor, it measures the received signal strength P_r. The node then observes how P_r changes over time. This would provide an indication as to whether the node is still within range or has probably moved out of range.

For our implementation with the network simulator ns-2, we used the received signal strength P_r to calculate the distance to the transmitter of the packet. Ns-2 uses the *free-space propagation model* as described in [7]. Using this propagation model, the distance d to the transmitter of a packet can be calculated as follows:

$$d = \sqrt[4]{\frac{P_t \cdot G_t \cdot G_r \cdot h_t^2 \cdot h_r^2}{P_r \cdot L}}, \qquad (1)$$

where, P_t is the default transmission power and P_r the received signal power; G_t and G_r are the antenna gains of the transmitter and the receiver respectively; h_t and h_r are the heights of the antennas, and L is the system loss, which is set to 1 by default. We assumed that the network was homogenous, i.e. all nodes use the same parameters P_t, G_t, G_r, h_t, h_r, and L. If a node transmits with a different signal power P_t, it must include the value of P_t in the options field of the MAC protocol header.

The MAC protocol keeps a record of the distances to neighboring nodes in a *neighbor table.* A table entry consists of five fields: a neighbor ID, a distance d_1 to the neighbor (estimated using **(1)**), the time at which this distance was estimated t_1, and a distance d_2 to the same neighbor estimated at time t_2. When a node receives a packet from a neighbor N_Y, it replaces the older entries of the table, corresponding to that neighbor, with the more recent ones. For simplicity, in our models, we have used only two timestamps and have assumed linear node movement. Thus, at any given time t, we estimate the current distance, d_{est} , as follows:

$$d_{est} = d_2 + \frac{d_2 - d_1}{t_2 - t_1} \cdot (t - t_2), \quad \text{for } t_1 < t_2 < t. \tag{2}$$

If a node N_X cannot establish an RTS-CTS handshake with a neighbor N_Y, it uses the neighbor table to estimate the current distance to N_Y. If d_{est} is smaller than the transmission range of N_X, Persistent MAC will send out up to eight[2] additional RTSs to establish a handshake with N_Y. If d_{est} is greater than the transmission range, or if the information in the neighbor table about N_Y is too old, the Persistent MAC will report a link failure to the routing protocol. The persistent MAC will also report a link failure if the additional attempts to establish a handshake with N_Y were to fail. Note here that the increase in RTS is not likely to increase the actual congestion since the RTS messages are sent only if the channel is sensed idle. Furthermore, for each RTS failure, the node still continues its back-off process, which in turn, would give ample time for the congestion to abate.

We re-iterate that this is a simple methodology to reduce the number of false link failures. During periods of high load, the link failures may mainly be due to congestion. In such cases, a more sophisticated methodology may be required to estimate congestion. Our main objective is to use power management to improve TCP performance in MANETs when the network is lightly loaded. Towards this goal, the above mechanism is useful in differentiating between true and false link failures.

We also note that the linear model to estimate distances is a simple method used to evaluate our mechanisms. We expect that the absolute value and the gradient of the received signal strength might be indicative of whether or not a node is moving out of range and may even be more realistic in practice. However, one might expect similar results with such methods.

3.2 Signal Strength Based Link Management Methods

We propose two mechanisms for alleviating the effects of mobility on TCP. We call these the *Proactive* and the *Reactive Link Management (LM)* schemes. These schemes are implemented at the MAC layer. We also provide a *modification of AODV* at the network layer that can exploit the presence of the link management schemes. Proactive LM tries to predict link breakages, whereas Reactive LM temporarily reestablishes a broken link with higher transmission power to salvage packets. The modified AODV allows the forwarding of packets in transit on a route that is going down while simultaneously initiating a search for a new route.

3.2.1 Proactive Link Management

The idea of Proactive LM is to inform the routing protocol that a link is going to break *before* the link actually breaks. The link break prediction mechanism uses the information from the neighbor table, which we described in section 2.1. Proactive LM estimates the projected distance to a neighbor in the immediate future. For example, if the current time is t, the distance of a particular neighbor, $d_{0.1}$, at $(t + 0.1)$ seconds is:

---

[2] This number will typically be a system parameter. The choice will depend on the density of the network and the congestion levels.

$$d_{0.1} = d_2 + \frac{d_2 - d_1}{t_2 - t_1} \cdot \left(t + 0.1 - t_2\right), \text{ for } t_1 < t_2 < t. \quad (3)$$

Proactive LM informs the routing protocol as soon as $d_{0.1}$ becomes greater than the transmission range. The routing protocol then informs the packet source, which stops sending and initiates a route discovery[3]. In this example, packets in transit have 0.1 seconds to traverse the *weak* link. If the warning comes too late, the weak link breaks before all packets in transit can be salvaged. On the other hand, the warning should not come too early as we want to use the link as long as possible.

In [2], Goff, Abu-Ghazaleh, Kahvecioglu, and Phatak implemented a similar mechanism to predict link breakages. This mechanism also measures the signal strength of received packets. A node informs the packet source when the link to the next hop is close to breaking. The packet source initiates a route request but does not stop sending. If the route discovery is successful, the source switches to the new route.

3.2.2 Reactive Link Management

Reactive LM temporarily increases the transmission range of a node to *reestablish* a broken link. Packets in transit can then traverse the reestablished *high power link.* A node N_X tries to set up a high power link if the RTS-CTS handshake (to a neighbor N_Y) with *default* transmission power is not successful. N_X therefore sends RTSs with high transmission power. N_Y must also switch to high transmission power to send the CTS. Otherwise, the CTS would not reach N_X. The RTS frame must therefore contain the value of the transmission power P_t. N_X and N_Y also send the DATA and the ACK packets with high power. When Reactive LM at N_X establishes the temporary high power link to N_Y, it stimulates the routing protocol to begin a new route discovery.

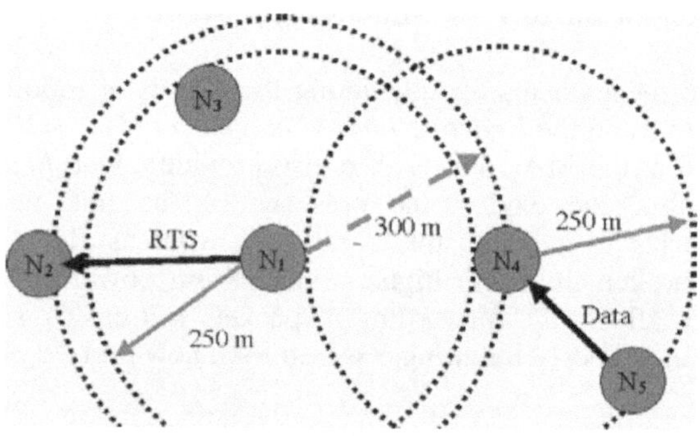

Fig. 1. Effects of Power Asymmetry

Reactive LM maintains a table to record the default and high power links. A node should be able to change its transmission power quickly, because it should not use high transmission power to communicate with neighbors that are within default transmission range. Furthermore, nodes must not broadcast *Route Request* messages

[3] We shall describe in detail how the modified AODV reacts to the notifications from Proactive LM in a later sub-section.

from AODV with high transmission power since new routes should consist of default power links only. As Reactive LM does not use signal strength to estimate the distance to a neighbor, it also raises the transmission power in case of false link failures. However, it can be combined with Persistent MAC, which helps avoid this effect.

The IEEE 802.11 MAC Collision Avoidance (CA) mechanism does not work well if nodes have different transmission ranges [6]. Figure 1 shows an example, in which N_1 increases its transmission range from 250 to 300 meters to reestablish the link to N_2. N_1 does not know about the DATA transmission from N_5 to N_4, because it could not receive N_4's CTS. N_1 therefore disturbs the DATA transfer from N_5 to N_4. Due to this effect, as congestion or the load in the network increases, increasing transmit power can in fact degrade performance. Thus, this method should be incorporated only at light loads.

3.2.3 Modifications to AODV

Proactive and Reactive LM inform the routing protocol of either weak or high power links. In this subsection, we explain how our modified version of *AODV* reacts to these MAC layer notifications. The routing protocol does not necessarily have to distinguish between *weak* and *high power links.* In both cases, the objective is to inform the packet source of the link failure, initiate a new route discovery and to salvage the packets in transit. In AODV, a route to a destination N_D in the routing table of a node N_X can be in either of two states: The route can be UP, which means N_X forwards packets to N_D. If N_X receives a *Route Request (RREQ)* for N_D, it will respond with a *Route Reply (RREP)* because it knows a route to N_D. The second state is DOWN; in this state, N_X does not have a routing entry for N_D. If N_X wants to send packets to N_D, it will initiate a route discovery. If N_X receives an RREQ for N_D, it will broadcast the RREQ. If N_X receives a packet for N_D, it will drop the packet and respond with a *Route Error* (RERR).

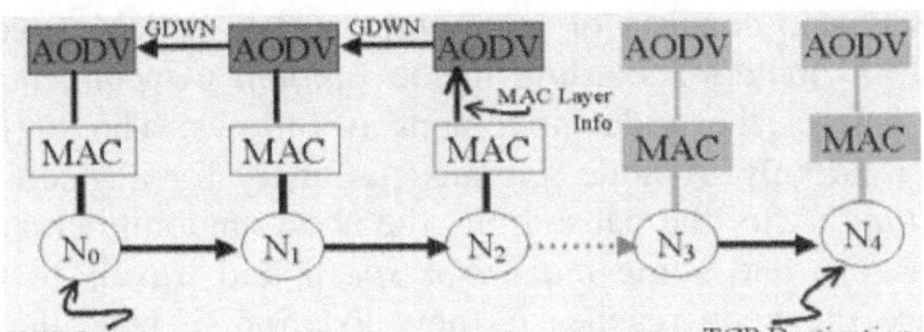

Fig. 2. Modifications to AODV

We have added an *additional route state* to allow nodes to salvage packets in transit. This state has the following characteristics: If N_X receives a packet for N_D, it will *forward* the packet. If it receives a RREQ for N_D, it will not reply with a RREP but *broadcasts the RREQ.* If an application at N_X wants to send packets to N_D, the modified AODV will initiate a route discovery. We call this third route state the *Going Down (GDWN)* state. Figure 2 gives an example, in which the MAC protocol at N_2 informs the modified AODV that the link to N_3 is getting weak (or has become a temporary high power link). The modified AODV then, sends a GDWN packet to its active neighbor N_1, which also sends a GDWN packet to N_0. All three nodes N_0, N_1, and N_2 change the route state for destination N_4 from UP to GDWN. N_1 and N_2 keep forwarding TCP transit packets towards N_4, but N_0 stops sending packets and initiates a route discovery. The old route to N_D via N_1, N_2, and N_3 is then no longer used and finally times out, i.e. the route state is set to DOWN. If the MAC protocol reports a *link breakage,* the modified AODV behaves like the original AODV, i.e. it brings down the route to destination N_D, sends RERR messages to its active neighbors, and drops all packets in transit to N_D.

3.2.4 Transport Layer

The methods we presented in the previous subsections are aimed at reducing the number of packet drops. TCP Tahoe, Reno, and New Reno grow the congestion window until packets are dropped. In wireless ad hoc networks, congestion does not lead to buffer overflow as in wired networks, but rather to false link failures, which cause the routing protocol to bring down the route. Therefore, even in static networks, where we would expect stable routes, the excessive growth of the TCP congestion window and false link failures cause repeated route changes. This behavior was shown by Saadawi and Xu [12], who quantified the performance of several versions of TCP, in ad hoc networks. They showed that TCP Tahoe, Reno, and New Reno suffer from the "instability problem" due to the excessive growth of the congestion window. They suggested restricting the maximum window size. They also showed that *TCP Vegas* does not suffer from this instability problem, because it uses a more conservative mechanism with *Round Trip Time (RTT)* estimations to control the size of the congestion window. TCP Vegas does not need packet losses to stop the growth of the congestion window. Since our goal is to study the effects of mobility as opposed to congestion, we used TCP Vegas for our simulations with ns-2. [3],[10].

4 Simulations and Discussion

The simulation scenario consists of 50 mobile nodes, which move in a rectangular area of 300 by 1500 meters according to the *random waypoint model.* The random waypoint model defines the node movement as follows: The start position of each node is assigned randomly. A node remains stationary for a specified period of time called the *pause time.* After the pause time, the node randomly chooses a location and a speed between zero and some *maximum speed* and travels to that location in a straight line. Once the node reaches its new location, it rests there for *pause time* seconds and then randomly chooses a new location and travel speed. With a pause time of zero seconds, all nodes are constantly in motion.

Figure 3 shows the simulation scenario for two setups that we consider. We call these setup I and setup II. In setup I, there is *one* TCP Vegas connection between two static nodes. Setup II has *two* crossing TCP Vegas connections between four static nodes. The static nodes are placed at the edges of the rectangular area. As mentioned earlier, these scenarios are typical when the network is lightly loaded and are appropriate for studying the effects of mobility. When the network is heavily loaded increase in power or proactively searching for new routes might not be appropriate and might increase congestion levels. We re-iterate that our schemes will have to be supplemented by other schemes that can estimate the congestion levels in the network.

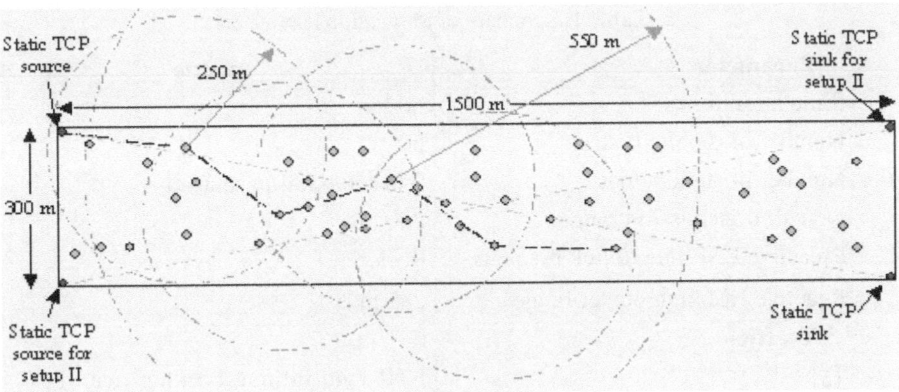

Fig. 3. Simulation Scenario

The traffic over each TCP connection is a file transfer of infinite length, i.e. a TCP source will send TCP data packets for the entire duration of the simulation. The default transmission range of each node is 250 meters and the interference range is 550 meters. A TCP packet travels an average of about 8 hops to get from a source to the corresponding sink. The pause time is zero seconds and the maximum speed of the mobile nodes is set to 0, 4, 8, 12, 16, and 20 m/s for different simulation runs. The simulation time is 600 seconds. These parameters are typically used and are appropriate to study the effects of our protocols with varying levels of mobility.

Proactive LM notifies the routing protocol when the distance estimation to a neighbor for *current time + 0.1 seconds* becomes greater than the default transmission range. This time seems appropriate since we do not want it to be too long (routes are left unused even when the link is fairly stable). The high power transmission range for Reactive LM is 285 meters. This is a system parameter and should be set depending upon the traffic conditions. We choose a reasonable value in this study (approximately a 10% increase from the original range). Table 1 summarizes the simulation parameters.

We used the following metrics to qualify the performance of TCP:

- *Packet loss:* Ratio of the number of dropped TCP packets to the total number of sent TCP packets.
- *TCP goodput:* Number of TCP data packets received by the *application layer* at the TCP sink. Goodput does not count retransmitted packets.
- The number of TCP retransmission timeouts per delivered packet.

In the following two subsections, we shall evaluate our protocols by comparing the performance of five versions: (1) The *Original scheme,* which is the unchanged version with the IEEE 802.11 MAC, (2) Including the *Persistent MAC,* (3) Including the *Proactive LM,* (4) Including the *Reactive LM,* and (5) A *Combined scheme,* which includes all of the above methods, i.e., the Persistent MAC, the Proactive LM, and the Reactive LM. For the Original scheme, we used the original AODV and for versions (2) to (5), the modified AODV was used. Each point in a graph represents an average of thirty simulation runs with different random movement patterns described earlier.

Table 1. Summary of Simulation Parameters

Parameter	**Value**
Simulation time	600 s
Number of mobile nodes	50
Number of static nodes	2 in setup I, 4 in setup II
Default transmission range	250 m
Proactive LM time to link breakage	0.1 s
Reactive LM high power range	285 m
Traffic	
Type	FTP with infinite backlog over TCP
Packet size	1460 bytes
Number of connections	1 in setup I, 2 in setup II
Movement	
Pause time	0 s
Maximum speed	0, 4, 8, 12, 16, and 20 m/s

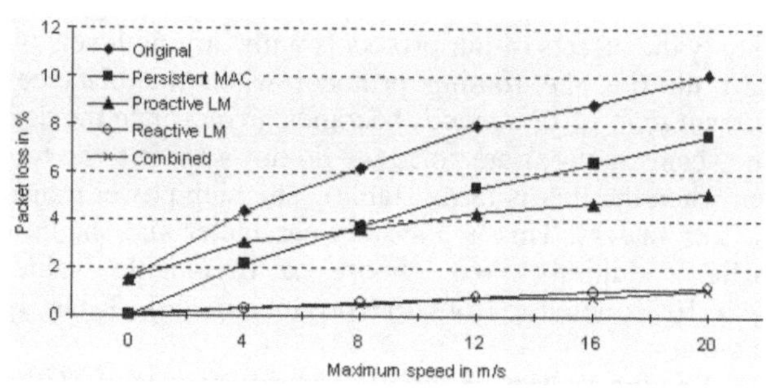

Fig. 4. Performance of our schemes with one TCP session

Figure 4 shows the packet loss as function of maximum speed with set up I. The original scheme drops between 1.5 and 10 percent of the packets. Persistent MAC can significantly reduce the loss in scenarios with low mobility, in which contention-induced link failures dominate. There is no improvement with the Proactive LM in static scenarios as all packet losses are caused by false link failures. When node mobility increases, Proactive LM can approximately reduce the number of losses by half. Reactive LM and Combined MAC can reduce the packet loss to less than 1 percent in all mobility scenarios. The reason why the reactive scheme outperforms the proactive scheme is because, in the proactive scheme, although a route failure is initiated prior to the actual failure, the source does not stop sending in the meantime. Hence, packets in transit are still lost when the link actually fails. In the reactive scheme, when the link fails the packets in transit are salvaged.

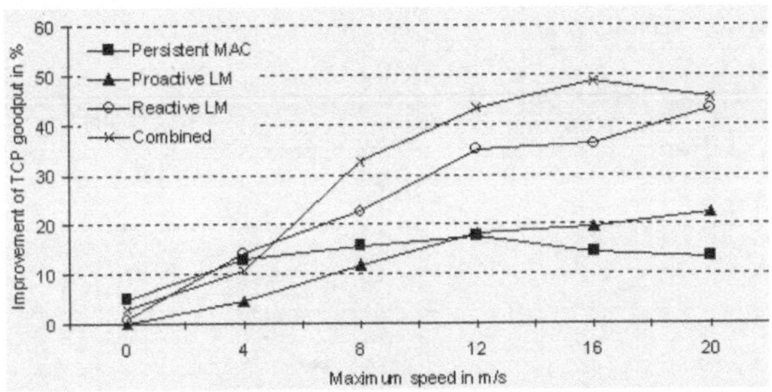

Fig. 5. Improvement of TCP Goodput versus maximum speed for one connection.

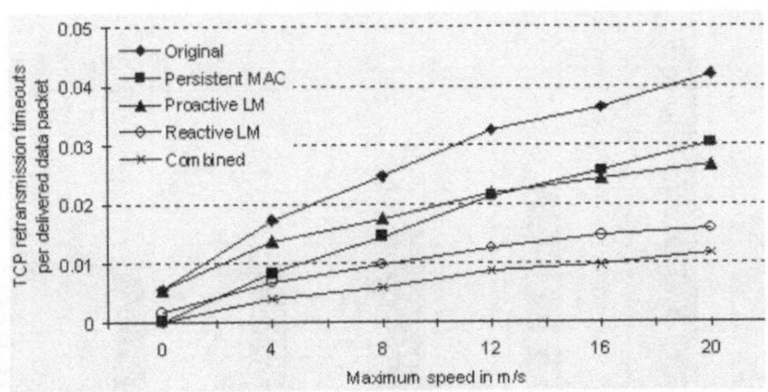

Fig. 6. TCP retransmission time-outs per delivered packet as a function of maximum speed.

Figure 5 shows the effects of decreased packet loss on TCP goodput with set up I. Persistent MAC, Proactive LM, Reactive LM, and the Combined scheme can increase the TCP goodput, especially in high mobility scenarios. The reason for higher goodput is the decreased number of TCP retransmission timeouts. Figure 6 shows the number of TCP retransmission timeouts per delivered data packet. With the Original scheme, TCP times out over 3 times per 100 delivered data packets in high mobility scenarios, whereas with the Combined scheme, there is only about 1 timeout per packet.

Figure 7 compares the Original scheme and the Combined scheme for one and two TCP connections. With two TCP connections, the percentage of dropped packets is higher than with one connection. The reason for this increase in packet losses is increased link layer contention, which leads to a higher percentage of false link failures (Figure 8). At higher levels of congestion, one would notice that the percentage of dropped packets increases. Thus, one would expect that the proposed schemes will be beneficial only at light loads wherein mobility is primarily responsible for link failures.

Figure 9 shows the goodput improvement the Combined scheme for one and two TCP connections. The total goodput improvement for two TCP connections is lower than for one, except with low node mobility. With increased network contention, it is more difficult for the Proactive and Reactive LM to salvage packets in transit as it takes longer for these packets to traverse the weak or high power link. With low node mobility, Persistent MAC is more effective in reducing false link failures in the scenario with two connections.

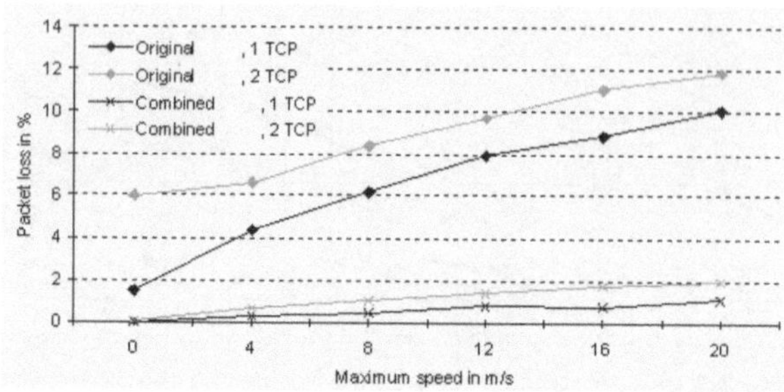

Fig. 7. Comparing the performance in the presence of one and two TCP connections.

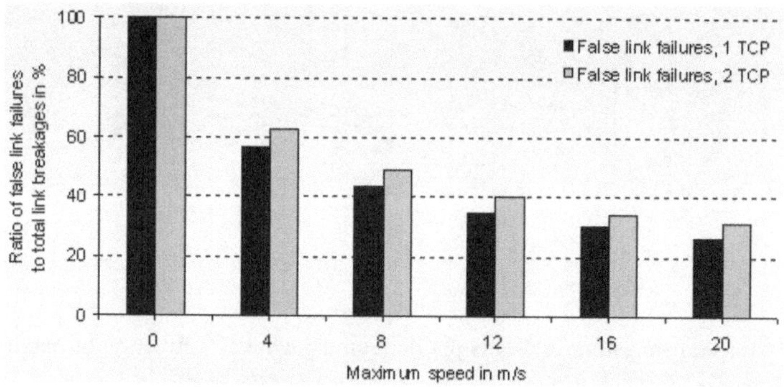

Fig. 8. Fraction of False Link Failures

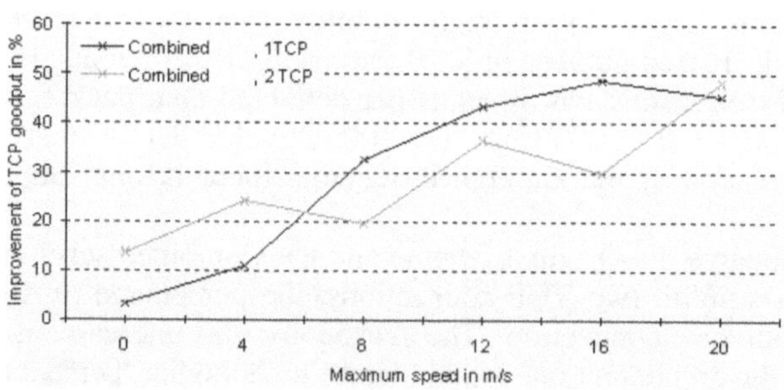

Fig. 9. Improvement in TCP goodput versus maximum speed with one and two TCP connections

In summary, the reduction of packet loss results in fewer TCP retransmission timeouts and therefore in higher TCP goodput. The higher the packet loss, the greater is the improvement by the proposed mechanisms. In our simulation scenario, the Original scheme drops up to 10 percent of all packets on a long multi-hop route with high node mobility. Combined MAC can improve the TCP goodput by up to 45 percent. These simulation results are for 8-hop TCP connections. With shorter connections, there is less packet loss and therefore less improvement of TCP goodput.

5 Conclusions and Future Work

In this paper our objective is to reduce the packet losses due to mobility in ad hoc networks and thereby improve the performance of TCP. We propose a temporary increase in transmit power level when a node moves out of range due to mobility to temporarily re-establish the failed link. This would enable the TCP packets that are already in flight to traverse the link.

The use of the IEEE 802.11 MAC protocol causes *false link failures* due to congestion. We propose a methodology that allows us to distinguish between true link failures due to mobility and false link failures. The methodology is based on the measurement of signal strength at the physical layer and using it to determine if a node is still within range at the MAC layer. We then increase power levels to temporarily re-establish a failed link only if it is determined to be mobility induced. We also investigate a proactive scheme, in which weak links are identified based on these signal strength measurements and routes are proactively found prior to failure. This in turn helps in switching to the new route even before failure occurs and thus can stem packet losses.

The simulation results with ns-2 showed that these mechanisms could considerably reduce the number of packet losses. Consequently, the number of TCP retransmission timeouts was reduced and the TCP sources sent more packets, thereby increasing the goodput by up to 45 percent.

However, our methods are applicable when there is little congestion in the network since they are primarily geared towards coping with mobility. Additional mechanisms are required to correctly determine the levels of congestion in the network. These schemes are to be used only at light loads. At higher loads, mobility induced failures are likely to be much lower as compared to congestion induced losses. Other mechanisms for coping with such losses are needed.

References

1. S. Xu and T. Saadawi, "Does the IEEE 802.11 MAC protocol work well in multi-hop wireless ad hoc networks," *IEEE Communications Magazine*, vol.39, Issue 6., June 2001.
2. T.Goff, *et al*, "Pre-emptive Routing in Ad Hoc Networks", *Proceedings of ACM MOBICOM*, Italy 2001
3. L.S. Brakmo, and L..Peterson, "TCP Vegas: End to End Congestion Avoidance on a Global Internet",. *IEEE Journal On Selected Areas In Communications*, Vol. 13, No. 8, October 1995
4. K. Chandran, K et al, "A Feedback based Scheme for Improving TCP Performance in Ad hoc Wireless Networks", *Proc. ICDCS*, Amsterdam, May 1998.
5. D.Kim, *et al*, "TCP-BuS: Improving TCP Performance in Wireless Ad Hoc Networks", *Journal Of Communications And Networks*, June 2001.
6. N. Poojary, *et al*, "Medium Access Control in a Network of Ad Hoc Mobile Nodes with Heterogeneous Power Capabilities", *Proceedings of IEEE ICC*, Helsinki 2001.
7. K. Fall and K. Varadhan, *The ns Manual.* 2001.
8. G. Holland and N. Vaidya, "Analysis of TCP Performance over Mobile Ad Hoc Networks", *Proceedings of ACM MOBICOM* , p. 219–230, Seattle 1999
9. http://grouper.ieee.org/groups/802/11/main.html.
10. Information Sciences Institute (ISI). *The Network Simulator – ns-2.* http://www.isi.edu/nsnam/ns/

11. C.E. Perkins, and E. M. Royer, "Ad-hoc On-Demand Distance Vector Routing", *Proceedings of the IEEE Workshop on Mobile Computing Systems and Applications*, February 1999.
12. T.Saadawi and S.Xu, "Performance Evaluation of TCP Algorithms in Multi-hop Wireless Packet Networks", *Journal of Wireless Communications and Mobile Computing* 2002.
13. Kopparty, S. et al, "Split-TCP for Ad Hoc Networks", *Proceedings of IEEE GLOBECOM* 2002.

Energy Efficient Multicast Routing in Mobile Ad Hoc Networks

Chor Ping Low and Chau Yuen

School of Electrical and Electronic Engineering
Nanyang Technological University
Singapore 639798

Abstract. In this paper, we address the problem of multicast routing in mobile ad-hoc networks from the viewpoint of energy efficiency. In an ad-hoc wireless network, each node runs on a local energy source which has a limited energy life span. Thus, energy conservation is a critical issue in mobile ad-hoc networks. One approach for energy conservation is to establish routes that require lowest total energy consumption. This optimization problem is referred as the minimum-energy multicast routing problem (MEMRP). In this paper, propose new efficient heuristic algorithms for the construction of energy-efficient multicast routes in mobile ad-hoc networks. These algorithms exploit the broadcast nature of the wireless channel, and address the need for energy-efficient operations. Empirical studies show that our algorithms are able to achieve better performance than algorithms that have been developed for MEMRP.

1 Introduction

Mobile ad-hoc networks are multi-hop wireless networks where nodes cooperatively maintain network connectivity. Unlike wired networks or cellular networks, no wired backbone infrastructure is installed in ad-hoc wireless networks. Such networks have been receiving increasing attention in recent years due to their potential applications in the establishment of dynamic communications for emergency/rescue operations, disaster relief efforts, military networks, etc [1].

In an ad-hoc wireless network, a link between any pair of nodes is considered to exist, if each has enough resources (such as the number of transceivers) available for this purpose and that the signal-to-noise ratio at the receiving node is sufficiently high. Thus, unlike the case of wired networks, the set of wireless network links and their capacities are not determined a priori, but depend on factors such as distance between nodes, transmitted power, users' interference, etc.

Each node in an ad-hoc wireless network is battery operated and thus runs on a local energy source. When a node runs out of energy it must be turned off to replace or recharge its battery. During that time the node cannot be used in the network. Thus, it is important to establish routes with low energy consumption to maximize the energy life of each node in the network.

Consequently, if the destination nodes are in the immediate communication vicinity of a transmitting node, a single transmission suffices for reaching all these

M. Conti et al. (Eds.): PWC 2003, LNCS 2775, pp. 625–639, 2003.

receivers. We assume that the power level of a transmission can be chosen within a given range of values. Hence, there is an incentive to route using maximum power (and thus maximum communication range). However, doing so may result in increased energy usage. Due to the nonlinear attenuation of the received signal power with distance, a transmission over multiple short hops may require less power than a transmission over a single direct link.

In this paper, we address the problem of constructing minimum energy routes for source-initiated multicast (one-to-many) traffic. The objective here is to construct a minimum energy tree, rooted at the source that reaches all of the desired destinations. This problem is referred to as the *minimum energy multicast routing problem (MEMRP)* and it has been addressed in the pioneering work by Wieselthier et. al. in [1,2]. This problem has also been shown to be NP-hard [3,4].

MEMRP involves the designation of which nodes are to transmit and the power levels at which they are to do so. The construction of minimum energy multicast trees will depend on various factors which include node mobility, network resources (such as number of transceivers and number of frequency bands), transmitted power level, etc. In order to assess these complex trade-offs one at a time, we assume that there is no mobility (as in [1,2]). Nevertheless, as argued in [1,2], the impact of mobility can be incorporated into this static model because transmitter power can be adjusted to accommodate the new locations of the nodes, as necessary. In addition, we also assume that there are sufficient network resources. Under these assumptions, we will propose new efficient heuristic algorithms for MEMRP. Extensive empirical studies are carried out to evaluate the performance of our proposed algorithms and to compare them with existing heuristic algorithms.

2 Wireless Communication Model

We consider a network consisting of a set of nodes randomly deployed over a given area. Connectivity of the network depends on the Euclidean distance between nodes, the maximum transmission power level (or maximum transmission range) and the minimum signal power that is required at the receiving node. The received signal power varies as $r^{-\alpha}$, where r is the Euclidean distance and α is a parameter that typically takes on a value between 2 and 4, depending on the characteristics of the communication medium. Based on this model the transmitted power required to support a link between two nodes separated by distance r is proportional to r^{α}. Without loss of generality, we set the normalizing constant equal to 1, resulting in:

P_{ij} = power needed to support connection between node i and node j

$= r_{ij}^{\alpha}$ where r_{ij} is the Euclidean distance between node i and node j.

We assume that omni directional antennas are used by all nodes in a mobile ad-hoc network to transmit and receive signals. Due to the *broadcast nature* of these antennas, all nodes within the communication range of a transmitting node can receive its transmission. Consider the example shown in Figure 2.1, in which node S is the source node (transmitting node) and nodes $D1$ and $D2$ are the destination nodes. The power required to reach node $D1$ from node S is $P1$; the power required to reach node $D2$ from node $D1$ is $P2$; the power required to reach node $D2$ from node S is $P3$.

A single transmission at power $P = \max\{P1, P3\}$ by node *S* is sufficient to reach both nodes *D1* and *D2*, based on the assumption of omni directional antennas. So the total cost (energy consumption) for wireless applications for the above example is equal to *P*. This situation is fundamentally different from wired applications, in which the cost of node *S*'s transmission to nodes *D1* and *D2* would be the sum of the costs to individual nodes (sum of the link cost). The ability to exploit this property of wireless communication, which is referred to as the *"wireless multicast advantage"* in [1,2], makes multicasting an excellent setting to study the potential benefits of energy-efficient protocols.

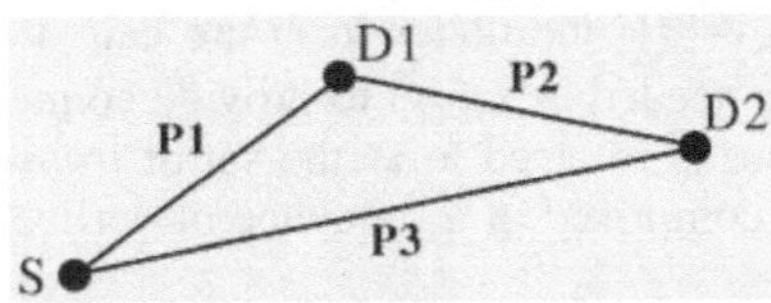

Fig. 2.1. Wireless Multicast Advantage

In wired models, any two nodes can communicate as long as there is a link connecting them. The cost of transmission from a source node to a set of nodes D_r that are within distance *r* from the source node would be the sum of the cost of individual links that are used in the routes from the source to this set of nodes. The situation in wireless network is different, however, due to its wireless broadcast property. Let *v* be a node in D_r that is furthest away from the source node and let the distance between them be denoted by r_{max}. In this case, a transmission at power $r_{\max}^{\alpha}$ by the source node will be received by all nodes in the set D_r without additional expenditure of power from this set of nodes. Thus wired networks can be viewed as *link-based* in contrast to the case of wireless networks which are viewed as *node-based.*

We next consider the example as shown in Figure 2.1 again to determine an appropriate transmission strategy that will result in lower energy consumption. Without loss of generality, we assume node *D2* is further away from node *S* than node *D1*; so *P3* is greater than *P1*. Two alternative strategies can be used to transmit data from node *S* to nodes *D1* and *D2*:

(a) Node *S* transmits at *P1* to reach node *D1*, then node *D1* transmits at *P2* to reach node *D2;* total power required is *P1+P2*
(b) Node S transmits at *P3* to reach both nodes *D1* and *D2;* total power required is *P3*

We would like to choose the alternative that results in lower total energy consumption. So if *P3* is greater than *P1+P2*, strategy (a) should be used; otherwise choose strategy (b) (illustrated in Figure 2.2).

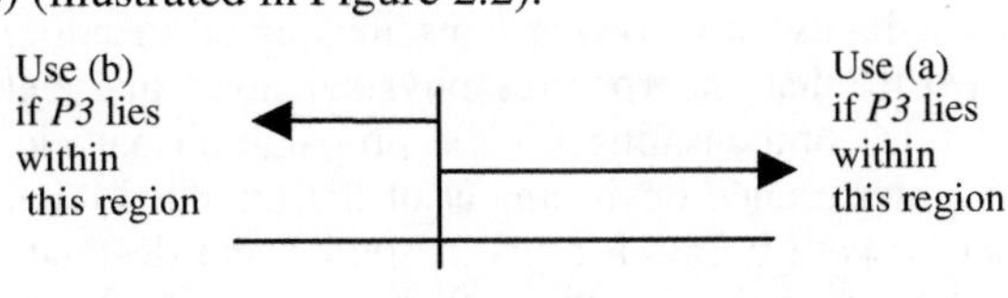

Fig. 2.2. Choices for transmission strategy.

3 Problem Formulation

We consider source-initiated, circuit-switched multicast sessions. We assume that the network nodes are given as a finite point set V in a two-dimensional plane. Any node in the network is permitted to initiate a multicast session. For each node $u \in V$, let P_u denotes the power level at which node u is transmitting. We assume that each node u may transmit at any power level $P_u \leq P_{max}$. Thus, given the value of P_{max}, we can determine the communication range of all nodes and the connectivity of the network. In each multicast session, the transmission from a source node $s \in V$ is to reach all nodes in D, where $D \subseteq V$-$\{s\}$; the nodes in D are called *destination nodes*. A set of nodes $U \subseteq V$-$\{s\}$ may be needed as *relays* to provide connectivity to other nodes in the session. This set of nodes is referred to as the set of *transmitting nodes*. A *path* from node v_0 to node v_{k+1} is comprised of a sequence of vertices $v_0, v_1, v_2, \ldots, v_k, v_{k+1}$ such that

(i) a transmission from node v_0 will reach v_1 but not v_j where $2 \leq j \leq k+1$
(ii) a transmission from v_i will reach v_{i+1} but not v_j where $i \in \{1,2,\ldots,k\text{-}1\}$ and $i+2 \leq j \leq k+1$
(iii) a transmission from v_k will reach node v_{k+1} .

A *multicast (broadcast) tree* T is comprised the source node s, the set of transmitting nodes U, and the set of non-transmitting nodes W. The source node s is the *root* of the multicast(broadcast) tree T. The set of non-transmitting nodes W form the *leaf nodes* of the multicast (broadcast) tree. Note that all leaf nodes are destination nodes, i.e. $W \subseteq D$; otherwise these nodes can be deleted from the tree resulting in lower energy consumption.

The total energy required to maintain the multicast (broadcast) tree is simply the sum of the energy expended by the source node and each of the transmitting nodes in the tree; leaf nodes (which do not transmit) do not contribute to this quantity. The problem of building a minimum energy multicast (broadcast) tree for a given network is called the *Minimum Energy Multicast (Broadcast) Routing Problem.* Due to the wireless broadcast advantage property of mobile ad-hoc networks (as mentioned in section 2), MEBRP is different from the broadcast routing problem in wired network which can be formulated as the minimum cost spanning tree (MST) problem. While the MST problem is solvable in polynomial time, MEBRP is known to be NP-hard. Thus heuristic algorithms are likely to be the only viable approach for this problem.

4 Previous Work

Subbarao [5] proposed a dynamic power-conscious unicast routing scheme (minimum power routing – MPR) that incorporates physical layer and link layer statistics to conserve power, while compensating for the propagation path loses, shadowing and fading effects, and interference environment at the intended receiver. The main idea of MPR is to select a path between a given source and destination (using Bellman-Ford algorithm - shortest path algorithm) that will require the least amount of total power expended, while still maintaining an acceptable signal-to-noise ratio at each receiver.

Chang et. al. [6] proposed an energy conserving routing algorithm for wireless ad-hoc networks by maximizing the lifetime of the system. In particular, the algorithm select routes and the corresponding power levels such that the time until a node's energy source (battery) drains out, is maximized. However, this algorithm does not attempt to minimize the overall energy required to maintain the routing tree. Other existing work in [7], [8], [9] addressed the problem of energy efficient unicast routing (rather than multicasting) for wireless ad-hoc network.

Both MEBRP and MEMRP have been addressed by Wieselthier et. al. in [1,2]. In particular, the works in [1,2] jointly address the issues of transmitting power levels (and hence network connectivity, a Physical layer function) and multicast (broadcast) tree formation (a routing function, associated with the Network layer) in order to achieve energy-efficient communication. Heuristic algorithms that have been proposed to address the issue of constructing energy-efficient broadcast and multicast routing trees include the *Broadcast Incremental Power* (*BIP*) heuristic, *Multicast Incremental Power* (*MIP*) heuristic [1,2] and the *Node-Based Spanning Tree* (*NST*) heuristic [1].

4.1 Broadcast Incremental Power (BIP) Algorithm

BIP is similar to the Prim's Algorithm for constructing minimum spanning tree (MST); new nodes are added to the tree one at a time until all nodes are included into the tree. The cost of adding a new node to the current partial tree is equal to the additional amount of energy that needs to be consumed when the new node is added to the current partial tree.

The fundamental difference between BIP and Prim's algorithm is as follows. The inputs to Prim's algorithm are the link costs, which remain unchanged throughout the execution of the algorithm. However, BIP dynamically update the costs at each step (after the addition of a new node to the partial tree) to reflect the fact that the cost of adding a new node v (which is adjacent to some node u in the partial tree) to T. In particular, the dynamic computation of cost is carried out as follows. Let P_{uv} denotes the power level that node u has to transmit to reach node v and let $P(u)$ denotes the power level at which node u is transmitting (prior to the addition of node v); if node u is currently a leaf node, $P(u) = 0$. The incremental cost (energy consumption) P'_{uv} incurred in the addition of node v into the partial tree is computed as follows: $P'_{uv} = P_{uv} - P(u)$. BIP has a complexity of $O(|V|^3)$.

4.2 Multicast Incremental Power (MIP) Algorithm

A broadcast tree is initially constructed using the BIP algorithm. To obtain a multicast tree, the broadcast tree is pruned by eliminating all transmissions that are not needed to reach the members of the multicast group. In particular, nodes with no downstream destinations will not need to transmit, and some nodes will be able to reduce their transmitted power if their distant downstream neighbors are pruned from the tree.

4.3 Node-Based Spanning Tree Algorithm (NBST)

This heuristic is based on the notion of the metric *n/p*, where *n* is number of "new" nodes (i.e. nodes not in the partial tree) that can be added to a current partial tree by using an additional *p* units of power:

$$\frac{n}{p} = \frac{\textit{number of new nodes reached}}{\textit{total power required to reach them}}$$

Hence, the physical meaning of *n/p* can be interpreted as the number of "new" nodes that can be reached per unit of additional power. The idea of this heuristic is to try to maximize the value of *n/p* in each iteration of the algorithm. Given a partial tree *T*, let V_T denotes the set of nodes in *T* and $\bar{V_T}$ denotes the set of nodes that are not in *T*, i.e. $\bar{V_T} = V - V_T$. Construction of the spanning tree begins at the source node, i.e. $V_T = \{s\}$ and $\bar{V_T} = V - \{s\}$. The transmission power of the source node is chosen to maximize the *n/p* metric. Nodes within its communication range will be included into the partial tree, V_T and $\bar{V_T}$ are both updated to reflect this fact. In the next iteration, each node that is currently in the partial tree (i.e. V_T) will evaluate their respective *n/p* metric. The node which has the maximum *n/p* value will then transmit at the power level which maximizes the *n/p* metric and all nodes within its communication range will next be included into the partial tree. This process continues until all nodes are in the tree. Next, the broadcast tree is pruned to eliminate all unnecessary transmissions. This heuristic has a time complexity of $O(|V|^3)$.

5 New Proposed Algorithms

In this section, we propose two new heuristic algorithms, namely the *Look-ahead Node-to-Power Ratio* (*LANPR*) *Algorithm* and the *Least Incremental-Power-Path* (*LIP*) *Algorithm,* for MEMRP. LANPR is also based on the notion of the *n/p* metric. However, it differs from the NBST algorithm (described in section 4.3) in that it is designed to address a drawback of NBST, which will be described next.

5.1 A Drawback of the NBST Heuristic

Consider the network in Figure 2.1, where *P3* > *P1*. The power required to reach node *D1* from node *S* is *P1*; the power required to reach node *D2* from node *D1* is *P2*; the power required to reach node *D2* from node *S* is *P3*. The two alternative strategies for transmitting data from the source node *S* to nodes *D1* and *D2* has been discussed in section 2 and is illustrated in Figure 2.2.

Using NBST, we will use strategy (a) if (*1/P1*) > (2/*P3*) and if (1/*P2*) > (1/(*P3*-*P1*)); in this case the source node will first transmit to node *D1* using power level *P1* and next node *D1* will transmit to node *D2* using power level *P2*; the total power consumed is thus *P1* + *P2*. Strategy (b) will be applied if (*1/P1*) < (*2/P3*); in this case the source node will transmit to both nodes *D1* and *D2* using power level *P3*. We note that

(i) $$\frac{1}{P1}-\frac{2}{P3}=\frac{P3-2P1}{P1\times P3}\Rightarrow\begin{cases}\frac{1}{P1}>\frac{2}{P3} & \text{if } P3>2P1\\ \frac{1}{P1}<\frac{2}{P3} & \text{if } P3<2P1\end{cases}$$

(ii) $$\frac{1}{P2}-\frac{1}{P3-P1}=\frac{(P3-P1)-P2}{P2(P3-P1)}$$

If $P3 > 2P1$ and $P1 > P2$, then $(P3-P1)-P2 > P3-2P1 > 0 \Rightarrow \frac{1}{P2}-\frac{1}{P3-P1} > 0$

Hence under the assumption of $P3 > P1 > P2$, $\begin{cases}\text{use strategy } (a) \text{ if } P3 > 2P1\\ \text{use strategy } (b) \text{ if } P3 < 2P1\end{cases}$

The choice of transmission strategies under the above assumption is illustrated in Figure 5.1.

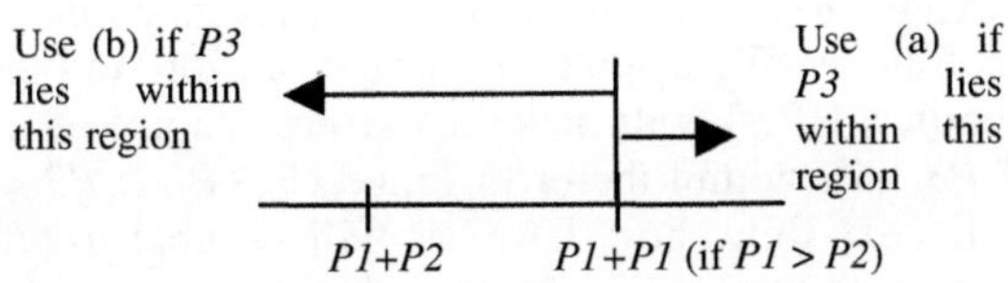

Fig. 5.1. Strategy Selection Range for NBST Algorithm

Based on the selection strategies illustrated in Figure 5.1, a non-optimal decision will be made by NBST if the value of $P3$ lies between the value of $(P1+P2)$ and $(P1+P1)$. In particular, it is easy to see that strategy (a) should be used in this case, as it will result in lower power consumption (since $P3 > P1+P2$). But NBST will opt to use strategy (b) instead, which will result in higher power consumption. This scenario is illustrated in Figure 5.2.

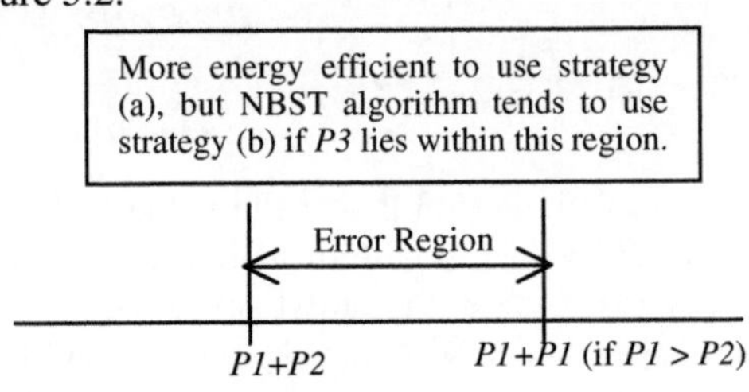

Fig. 5.2. Error Range in Strategy Selection for NBST Algorithm

5.2 Look-Ahead Node-to-Power Ratio (LANPR) Algorithm

In this section, we propose a new heuristic algorithm, called the *Look-ahead Node-to-Power Ratio (LANPR) Algorithm,* that is capable of addressing a deficiency (mentioned in section 5.1) of NBST. Construction of the spanning tree begins at the source node. Given a partial tree T, let V_T denotes the set of nodes in T and $\bar{V}_T$ denotes the set of nodes that are not in T. In each iteration of LANPR, we determine a node v such that this node has the maximum n/p value among all nodes that are currently in the partial tree. Having found the node v, let $(n/p)_v$ denote its maximum n/p value, S_v denote the set of nodes in $\bar{V}_T$ that are within the communication range of node v when it transmits at the power level, say P_v, which is dictated by $(n/p)_v$. If $|S_v| = 1$, then node v will transmit at power level P_v to reach the node in S_v. If $|S_v| > 1$, we will attempt to **look-ahead** (predict) to determine if it would be more efficient (in terms of energy consumption) to first transmit to only one of the nodes in S_v, say w. The node w is selected to be the nearest node to node v. Let the maximum n/p value that corresponds to the transmission of node w to other nodes that are not in the tree be denoted by $(n/p)_w$. If $(n/p)_w > (n/p)_v$ then, the algorithm will choose to transmit from node v to node w. Otherwise, node v will transmit all nodes in S_v using power level P_v. This process continues until all nodes are included into the tree. Next, the broadcast tree is pruned to eliminate all unnecessary transmissions.

The ability of LANRP to address the drawback (described in section 5.1) of NBST can be explained as follows. We again consider the network in Figure 2.1, where $P3 > P1$. As discussed earlier, NBST will make a "wrong" choice of transmission strategy when the value of $P3$ falls within the error range ($P1+P2 < P3 < 2P1$) as shown in Figure 5.2. We will next show how LANRP will be able to address this problem. LANRP will first compute the maximum n/p value that node s should transmit. Since $2/P3 > 1/P1$, node s may choose to transmit to both nodes, $D1$ and $D2$, simultaneously at power level $P3$ (this corresponds to strategy (b)). However, prior to doing so, it will first attempt to "look-ahead" to determine if it would be more energy efficient to first transmit to its nearest neighbor, namely $D1$, and subsequently may use the latter to transmit to $D2$ (strategy (a)). The decision to use either of these strategies by LANRP is determined by the values of $1/P2$ and $2/P3$. In particular, strategy (a) will be used if $(1/P2) > (2/P3)$; otherwise strategy (b) will be chosen.

We note that
$$\frac{1}{P2}-\frac{2}{P3}=\frac{P3-2P2}{P2\times P3}\Rightarrow\begin{cases}>0 \ \ if \ \ P3>2P2\\ <0 \ \ if \ \ P3<2P2\end{cases}$$

Hence, we can deduce from Figure 5.3 that if the value of $P3$ lies between $P1+P2$ and $2P1$, then $P3 > 2P2$ (as $P1 > P2$); this in turns implies that $1/P2 > 2/P3$ and thus LANPR will adopt strategy (a). Hence, the drawback of NBST can be circumvented in this case (illustrated in Figure 5.3). The pseudocode of LANPR is given in Figure 5.4.

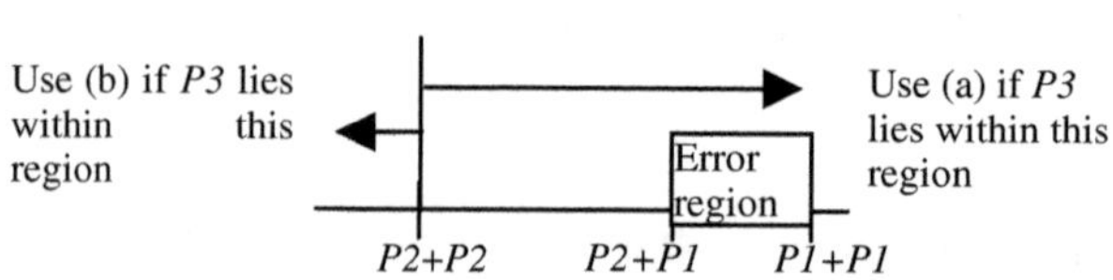

Fig. 5.3. Strategy Selection Range in LANPR Algorithm

Lemma 1. The time complexity of LANPR is $O(|V|^3)$.

Proof. The **while** ***loop*** in step 1 has at most $|V|$ iterations. The same is true for each of the **for** loops in steps 2 and 3. Hence steps 1 to 3 can be done in $O(|V|^3)$. Step 4 can be completed in $O(|V|)$. Pruning of the tree in Step 5 can be done in $O(|V|)$. Thus, the worst case time complexity of LANPR is $O(|V|^3)$.

While LANPR has the same worst case time complexity as the MIP algorithm, it potentially requires fewer steps to build a multicast tree than MIP algorithm in general. In the worst case, it adds in one node at one time, and takes $|V|$-1 steps to build a tree for a network with $|V|$ nodes, which is the same as the number of steps required by MIP algorithm. We also note that LANPR is based on similar paradigm as MIP in the sense that a broadcast tree is first constructed using these algorithms and the tree is subsequently pruned to eliminate all unnecessary transmissions. We will next propose another algorithm, called the *Least Incremental-Power Path Algorithm (LIP)*, that is based on the paradigm to constructing a multicast tree directly (thus eliminating the need to prune the resultant tree).

5.3 Least Incremental-Power Path Algorithm (LIP)

Given a pair of nodes $u,v \in V$, we define the *shortest path* SP_{uv} from u to v to be a path that requires the least total energy consumption for a transmission from node u to reach node v. The energy consumption of path SP_{uv}, say E_{Spuv}, is equal to the sum of the power level at which each node in SP_{uv} has to transmit.

Given a source node $s \in V$ and a set of destination nodes $D \subseteq V - \{s\}$, let $P_{su} = r_{su}^{\alpha}$, where r_{su} denotes the Euclidean distance between s and u, for each $u \in D$. Let w be a node in D such that

$$P_{sw} = \max_{u \in D} \{P_{su}\} .$$

It is clear that if the source node transmits at power level P_{sw}, each node in D will be able to receive the transmission from s. Hence the total energy required to maintain the multicast tree is bounded from above by P_{sw}. Next, we note that the total energy consumption of shortest path from s to w (which may include some other destination nodes) is also no more than P_{sw}, i.e. $E_{Spuw} \leq P_{sw}$.

Procedure LANPR

$V_T = \{s\}$; $\bar{V}_T = V - \{s\}$;

while ($|V_T|$ < total number of nodes) { /* step 1 */
  for (every node $i \in V_T$) { /* step 2 */
    for (every node $j \in \bar{V}_T$) { /* step 3 */
      compute n_{ij} / p_{ij}
    }

$$(n/p)_{(i)} = \max_{j \in \overline{V_T}} \frac{n_{ij}}{p_{ij}}$$

}

Let $(n/p)_v = \max_{\forall i \in V_T} \{(n/p)_i\}$

Let S_v denotes the set of nodes from $\overline{V_T}$ that are able to receive the transmission from node v.

if $|S_v| = 1$ {
$S'_v = S_v$
} **else** {
Let node $w \in S_v$, which is nearest to node v

$\overline{V'_T} = \overline{V_T} - \{w\}$

for (every node $y \in \overline{V'_T}$) { /* step 4 */
compute n_{wy}/p_{wy}
}

$$(n/p)_{(w)} = \max_{y \in \overline{V'_T}} \frac{n_{wy}}{p_{wy}}$$

if ($(n/p)_w > (n/p)_v$) {
Let $S'_v = \{w\}$
} **else** {
$S'_v = S_v$
}
}
}

$\overline{V_T} = \overline{V_T} - S'_v$
$V_T = V_T + S'_v$
}

eliminate all unnecessary transmissions from set V_T; /* step 5 */

Fig. 5.4. LANPR Algorithm

The basic idea of the Least Incremental-Power Path Algorithm (LIP) is as follows. The algorithm first compute a shortest path from the source node s to a node w in D, where $P_{sw} = \max_{u \in D}\{P_{su}\}$. All nodes in this path are included in the set V'. Next, the algorithm builds a multicast tree by adding nodes from the multicast group D-V' to V' one at a time, and stops when V' contains all elements from D. In each iteration, a node from D-V' is selected for inclusion into the partial tree as follows. Given a pair of nodes $a \in V'$ and $b \in D$-V', let IP_{ab} denotes a path from a to b, where each node in this path, except for *a,* belongs to *V-V'*. Let *P(a)* denotes that power level at which node *a* is currently transmitting in the partial tree. Let *q* be the node that is adjacent to node *a* in the path IP_{ab}. The incremental power P'_{aq} that node *a* has to utilize for its transmission to reach node *q* is equal to P_{aq}*-P(a)*. Hence the energy consumption of path IP_{ab}, say E_{IPab}, is equal to the sum of P_{aq}*-P(a)* and the power levels at which each of the other nodes in IP_{ab} has to transmit. Let LIP_{uv} denotes a path from a node $u \in V'$ and $v \in D-V'$ such that $E_{LIPuv} = \min_{a \in V', b \in D-V'} \{E_{IP_{ab}}\}$. We call LIP_{uv} a *least incremental-power path* from *V'* to *D-V'*. Hence, in each iteration, the algorithm attempts to find a least incremental-power path from *V'* to *D-V'*. The nodes on this path will be included into the partial tree if the total energy requirement of the resultant tree is less P_{sw}; otherwise the source will choose to transmit at power level P_{sw} to reach all destination nodes and the algorithm will terminate.

Lemma 2. The time complexity of LIP is $O(|V|^3)$.

Proof. The shortest path in step 1 can be found in $O(|V|^2)$. Each iteration of step 2 is of order $O(|V|^2)$. Since there are at most $|V|$ iterations of step 2, this step can be completed in $O(|V|^3)$. Thus, the worst case time complexity of LIP is $O(|V|^3)$.

Procedure LIP;
Input: $V, s \in V, D \subseteq V$;
Output: multicast tree $T=(s, U \cup W)$;
begin
set $V'=\phi$; $D'=\phi$, $A=\phi$, $dest=1$; $E=0$;
find shortest path SP_{sw} from *s* to *w*; /* **step 1**; $w \in D$ & $P_{sw} = \max_{u \in D}\{P_{su}\}$ */

$V'=V' \cup$ {all nodes in SP_{sw} };
$D'=D' \cup$ {all destination nodes in SP_{sw} };
$E= E_{Spuw}$;

while ($|D'| < |D|$) /* **step 2** */
find a path LIP_{uv} where $u \in V'$, $v \in D-V'$, such that
$E_{LIPuv} = \min_{a \in V', b \in D-V'} \{E_{IP_{ab}}\}$; /* **step 3** */

if $E+ E_{LIPuv} < P_{sw}$ **then**
$V'=V' \cup$ {all nodes in LIP_{uv} except *u* };
$D'=D' \cup$ {all destination nodes in LIP_{uv} except *u* };
else
$V' = \{s,w\}$; /* *s* will transmit using power level P_{sw} to all destination nodes*/

$D' = D;$

end while;

$U \cup W = V' - \{s\}$

end; (Procedure LIP)

Fig. 5.5. Algorithm LIP

6 Simulation Results

Let ALG = {MIP, NBST, LANPR, LIP} denotes the set of algorithms to be evaluated. In our empirical study, the performance of each algorithm from the set ALG is evaluated for various networks. Each network with a specified number of nodes is randomly generated within a 10*10 square region. One of the nodes is randomly chosen to be the source and the some of the nodes are chosen to be the destinations. The value of α (characteristics of the attenuation of communication medium) is chosen to be 2. The performance metric is the total power required to maintain a multicast tree by the respective algorithms. For each algorithm q from the set ALG, let

$Pow_q(m)$ = total power of the multicast tree for network m generated by algorithm q.

$Pow_{opt}(m)$ = total power of the **minimum energy** multicast tree for network m generated using exhaustive search

We define the normalized power associated with algorithm q (against exhaustive search) for network m to be $Pow_q^{opt}(m) = \frac{Pow_q(m)}{Pow_{opt}(m)}$.

Table 6.1 summarizes the performance of all the algorithms whose solutions are normalized against the corresponding optimal solutions that are found using exhaustive search. Only small network sizes (ranging from 10 to 20) are used in this set of results, as it is computationally expensive to apply exhaustive search to large networks. For each network size, 100 networks are randomly generated, and the following performance indicators are computed for each algorithm: (i) mean value of the normalized power (ii) variance of the normalized power and (iii) 90% confidence interval of the normalized power obtained by each algorithm.

As the number of nodes increases, the performance of all algorithms deviates further away from the optimal solution. We observe that NBST has the worst performance among all the algorithms; one possible reason for this phenomenon is given in section 5.1. Both LANPR and LIP consistently perform better than MIP (in terms of mean power) but we observe that the gain in performance of LANPR and LIP over MIP reduces as the number of destination nodes increase. Similarly, LIP performs better than LANPR for networks with no more than 50% of nodes being destination nodes. The results also indicate that LANPR will potentially perform better than LIP for networks with large number of destination nodes (as could be seen from Table 6.1(b)).

Table 6.1. Total Power Required for Multicasting (Normalized against Optimal Solution)

|V|	MIP		NBST		LANPR		LIP	
	Mean	Var	Mean	Var	Mean	Var	Mean	Var
10	1.1754	0.1615	1.2751	0.1455	1.1729	0.1087	1.0343	0.0085
15	1.2303	0.0924	1.3633	0.0974	1.1970	0.0370	1.0794	0.0172
20	1.3050	0.2660	1.4372	0.1146	1.2757	0.0892	1.0942	0.0161

(a) 25% of nodes are destination

|V|	MIP		NBST		LANPR		LIP	
	Mean	Var	Mean	Var	Mean	Var	Mean	Var
10	1.0991	0.0202	1.1493	0.0267	1.0745	0.0141	1.0753	0.0126
15	1.1443	0.0214	1.1887	0.0247	1.0845	0.0088	1.0904	0.0098
20	1.1506	0.0151	1.2429	0.0338	1.1326	0.0141	1.1373	0.0140

(b) 75% of nodes are destination

Table 6.2. Total Power Required for Multicasting (Normalized against MIP)

|V|	MIP		NBST		LANPR		LIP	
	Mean	Var	Mean	Var	Mean	Var	Mean	Var
25	1.0000	0.0000	1.1425	0.0768	1.0131	0.0405	0.9129	0.0211
50	1.0000	0.0000	1.1508	0.0495	1.0021	0.0174	0.9369	0.0127
75	1.0000	0.0000	1.1567	0.0302	1.0052	0.0140	0.9491	0.0109
100	1.0000	0.0000	1.1654	0.0297	1.0042	0.0111	0.9652	0.0082

(a) 25% of nodes are destination

|V|	MIP		NBST		LANPR		LIP	
	Mean	Var	Mean	Var	Mean	Var	Mean	Var
25	1.0000	0.0000	1.0583	0.0370	0.9639	0.0112	0.9862	0.0095
50	1.0000	0.0000	1.0635	0.0247	0.9713	0.0084	1.0060	0.0050
75	1.0000	0.0000	1.0670	0.0170	0.9725	0.0046	1.0175	0.0036
100	1.0000	0.0000	1.0648	0.0135	0.9693	0.0037	1.0262	0.0029

(b) 75% of nodes are destination

Table 6.2 summarizes performance for all algorithms whose solutions are normalized against those obtained using the MIP algorithm. Network sizes ranging from 25 to 100 are used in this study and 1000 networks are randomly generated for each of them. As could be seen from the results, LIP performs the best for networks with small number of destination nodes (Table 6.2(a)). In contrast, LANPR performs better than the other algorithms for networks with a larger number of destination nodes (Table 6.2(b)). Again, we observe that NBST does not perform as well as the other algorithms in all cases.

7 Conclusion

In this paper, we proposed new heuristic algorithms to address some of the fundamental issues associated with energy-efficient multicasting in wireless ad-hoc network. Empirical study shows that our proposed algorithms are able to achieve better performance (in terms of energy consumption) than existing multicast algorithms for ad-hoc wireless networks. In particular, our proposed algorithms (LANPR and LIP) not only exploit the wireless multicast advantage of ad-hoc networks, but also take a broad view of the network's node location before making routing decision. This in turn gives our proposed algorithm an edge over existing algorithms in generating low energy multicast trees.

Our empirical results show that LIP performs best for networks with small number of destination nodes. This implies that pruning a broadcast tree to form a multicast tree is not an effective way of building multicast tree for such networks; i.e. building multicast tree directly would lead to a more energy efficient multicast tree for these networks.

We note however that our proposed algorithms assume complete knowledge of the network topology which in turns limit their scalability. Hence further research is needed to develop distributed algorithms that provide the benefits that have been demonstrated in this paper. Furthermore, it is important to study the impact of limited bandwidth and transceiver resources on MEMRP, as well as to develop mechanisms to cope with node mobility.

References

[1] Wieselthier, J.E.; Nguyen, G.D.; Ephremides, A, "Algorithms for Energy-Efficient Mulcasting in Static Ad Hoc Wireless Networks", *Mobile Networks and Applications (MONET), Vol. 6, June 2001,* 251–263.

[2] Wieselthier, J.E.; Nguyen, G.D.; Ephremides, A, "Energy-Efficient Broadcast and Multicast Trees in Wireless Networks", *Mobile Networks and Applications (MONET), Vol 7, Dec 2002,* 481–492.

[3] Weifa Liang, "Constructing Minimum-Energy Broadcast Trees in Wireless Ad Hoc Networks", *MobiHoc 2002.*

[4] M. Cagalj, J. P. Hubaux, and C. Enz, "Minimum-Energy Broadcast in All-Wireless Networks :NP-completeness and Distribution Issues", *Mobicom 2002.*

[5] Subbarao, M.W., "Dynamic power-conscious routing for MANETs: an initial approach", *Vehicular Technology Conference, 1999. VTC 1999 – Fall. IEEE VTS 50th, Volume: 2, 1999,* 1232–1237.

[6] Jae-Hwan Chang; Tassiulas, L., "Energy conserving routing in wireless ad-hoc networks", *INFOCOM 2000. Nineteenth Annual Joint Conference of the IEEE Computer and Communications Societies. Proceedings. IEEE, Volume: 1, 2000,* 22–31.

[7] Michail, A.; Ephremides, A, "Energy efficient routing for connection-oriented traffic in ad-hoc wireless networks", *Personal, Indoor and Mobile Radio Communications, 2000. PIMRC 2000. The 11th IEEE International Symposium, Vol. 2, 2000,* 762–766.
[8] Stojmenovic, I.; Xu Lin, "Power-aware localized routing in wireless networks", *Parallel and Distributed Processing Symposium, 2000. IPDPS 2000. Proceedings. 14th International, 2000,* 371 –376.
[9] Rodoplu, V.; Meng, T.H., "Minimum energy mobile wireless networks", *IEEE Journal on Selected Areas in Communications, Volume: 17 Issue: 8 , Aug. 1999,* 1333–1344.

Providing Internet Access to IPv6 Mobile Personal Area Networks through UMTS

Nikolaos Alexiou, Georgios Tsiouris, and Efstathios Sykas

National Technical University of Athens
Department of Electrical and Computer Engineering
Communications, Electronics and Information Engineering Division
9 Heroon Polytechniou St., 157 73 Zografou Athens, Greece
Tel: +30 210 772 1493, Fax: +30 210 772 2534
{alen,gtsiouris,sykas}@telecom.ntua.gr

Abstract. This paper describes a mechanism and suggests a protocol enhancement to provide global Internet access to Mobile Personal Area Networks and support mobility under different scenarios. For this purpose the concept of the user's Mobile Station acting as a Mobile Multilink Subnet Router is introduced, which provides the necessary IPv6 connectivity and mobility service to the nodes connected to the Personal Area Network. The Mobile Station can distinguish between the nodes that need mobility support and the ones that do not in order to provide the appropriate type of IPv6 connectivity service. The main issues discussed are UMTS IPv6 connectivity, routing and mobility for the Mobile Station and the other nodes of the Mobile Personal Area Network.

1 Introduction

IPv6 [1] is the next generation protocol developed to replace the current version of the Internet Protocol, IP Version 4 (IPv4). One of the main issues concerning the use of IPv4 is the growing shortage of addresses. Mobile devices with IP capabilities become more popular and provide new services while their cost decreases. Meanwhile, solutions like private addresses and implementation of Network Address Translators cannot provide efficient support for new applications and services like peer-to-peer and Internet telephony. The allocation of public IP addresses to mobile terminals will be only feasible by the introduction of IPv6 since their number is quite high and is set to grow in the near future. In addition to the addressing issue, IPv6 also provides support for mobility, security and automatic configuration for every IPv6 enabled node.

The future user requirements of a mobile access network will not only include seamless internet connectivity for the mobile handset but for all the various devices with IP functionality that a user may carry. Hence the future user should be treated as a Personal Area Network (PAN) which in many cases may provide internet access to some visitor nodes. Future user requirements may also include mobility support of the PAN, which changes frequently its point of attachment.

M. Conti et al. (Eds.): PWC 2003, LNCS 2775, pp. 640–650, 2003.

This paper proposes a way to connect a PAN through a UMTS Mobile Station by using the concept of the Multi-Link Subnet Router [3], which connects different links belonging to the same subnet. This is based on the large number of available IPv6 addresses to the Mobile Station since it is assigned a unique 64-bit IPv6 prefix [9]. The IPv6 interface identifier can be chosen by the MS and can be changed at any time without any disruption to the IPv6 connectivity service since there is no UMTS network involvement.

The concept of connecting a PAN through a UMTS MS is extended by providing mobility support to the nodes comprising the PAN. This can be achieved by conjugating the concept of the Multilink Subnet Router (MSR) and the Mobile Router (MR) [4], which uses a bidirectional tunnel with its home network. However a distinction has to be made between the nodes that support mobility and the nodes that are mobility unaware for issues such as optimal routing, reduced traffic on mobile router's home network etc. For these reasons, a protocol enhancement is proposed, in order for the MS to be able to distinguish between the nodes that need mobility support and the ones that do not, in order to provide the appropriate type of IPv6 connectivity service.

2 IPv6 PAN Connectivity in UMTS

In order to connect a Personal Area Network (PAN), a subnet prefix must be provided. This subnet prefix has to be globally routable in order to provide end-to-end transparency for applications, protocols, mobility and security. In visited networks, the mobile router should acquire a whole IPv6 subnet prefix from a foreign access network.

A typical wireless IP connectivity scenario in the near future will be through UMTS. The main core network elements of the UMTS packet switched domain (General Packet Radio Services, GPRS) [9] are the Serving GPRS Support Node (SGSN) and the Gateway GPRS Support Node (GGSN). The GGSN is a specialized router that functions as the gateway between the GPRS network and the external networks. From the IP point of view, the GGSN can be seen as the first-hop router between the user and the Internet [13]. The SGSN's main functions include authentication, authorization, mobility management, and collection of billing information.

The main concept in UMTS systems regarding IP connectivity is the PDP context which is the connection between the user equipment and the Gateway GPRS Support Node (GGSN), over which the packets are transferred. In order to acquire IP connectivity, the user must initialize the PDP activation process.

In UMTS systems, IPv6 support is possible through both statefull and stateless autoconfiguration procedures [2] [9] [10]. Statefull address allocation mechanism requires a DHCPv6 server [17] while stateless autoconfiguration [6] involves mainly the IPv6 node in the allocation of addresses and does not require any external entity in the address autoconfiguration procedure. In cellular networks like UMTS, some Neighbor discovery messages can cause unnecessary traffic as

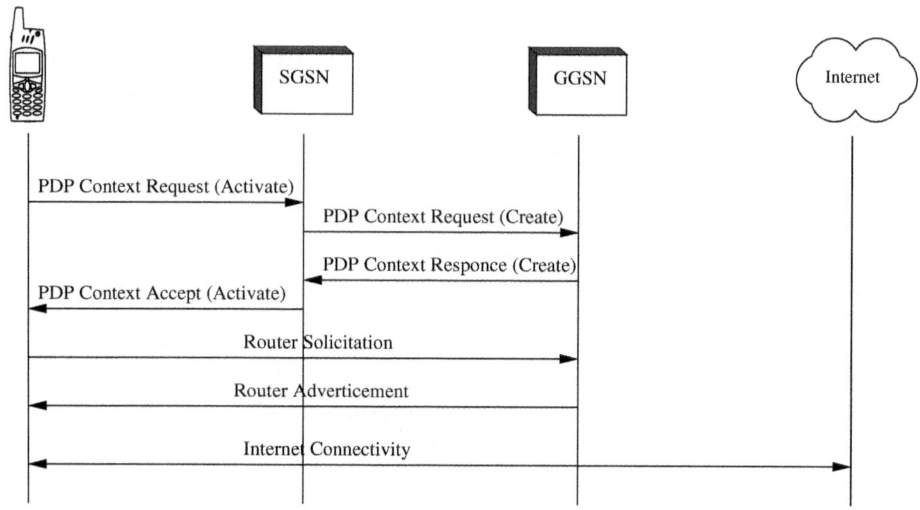

Fig. 1. IPv6 address acquisition in UMTS networks through stateless address autoconfiguration

the established link is a point-to-point link and the host's only neighbor is the default router (the GGSN).

During the autoconfiguration process (see Fig. 1) the Mobile Station uses the interface identifier received from the PDP context activation procedure and the prefix received in the Router advertisement [5] message from the GGSN, in order to create a globally routable IPv6 address. The interface identifier is provided to the mobile station by the GGSN, in order to avoid collisions with the link-local address of the GGSN. The prefix that GGSN provides to mobile station is unique and has a length of 64 bits. A Mobile Station (MS) may be comprised of a standalone IPv6 enabled phone or a Mobile Terminal (MT) and the user's Terminal Equipment (TE), e.g. a laptop or PDA. In the second case, a PPPv6 link [11] exists between MT and TE. The MT performs the PDP context activation on a request from the TE, and provides to the TE the interface identifier suggested by the GGSN.

The mobile station can at any time change the interface identifier used to generate global IPv6 address (e.g. for privacy reasons [7]) without updating the PDP context in the SGSN and the GGSN. Moreover the interface identifier does not need to be unique across all PDP contexts since the MS is considered to be alone on its link toward the GGSN. This practically means that the MS can choose any interface identifier without any network involvement. Thus a full prefix is available to the mobile station.

3 Mobile Station Acting as Multi-link Subnet Router

In this section the concept of a general Multi-link Subnet Router is discussed and the usage of the Mobile Station (MS) as a Multi-link Subnet Router (MSR) is introduced.

3.1 The Concept of a Multi-link Subnet Router

A Multi-link Subnet is defined [3] as a collection of independent links, connected by routers, but sharing a common subnet prefix. A single subnet prefix is sufficient to support multiple physical links.

During start-up the MSR starts as a normal host, discovering routers (if any) in each one of its interfaces. Then it switches to router-mode in all these interfaces where no routers were discovered. In case a router is found in one or more of its interfaces, the MSR chooses one and acts as a proxy mode on that interface. On all the remaining interfaces the MSR advertises itself as the default router and includes copies of the prefix information options that it learned on its proxy-mode interface. In a simple scenario where only one MSR exists, it will have one interface on which will act as a router, and one interface on which will act as a "host", proxying for all nodes on its router interface.

An example of an MSR with a proxy mode interface is depicted in Fig. 2 below. Two links, (1) and (2) are on a common subnet with global prefix G and are connected by an MSR (node B). The top level router of the subnet (node C) is connected on link 1. The MSR discovers that there is a router on link 1 and switches to proxy-mode on that interface. The MSR (node B) is in router-mode on link 2 (since no router exists), where it has link-layer address b2, and IPv6 address Gb2. On link 1, where it acts as a proxy it has link-layer address b1 and IPv6 address Gb1. Node A has link-layer address a on link 2, and has acquired global IPv6 address Ga. Node C has link-layer address c on link 1, and IPv6 address Gc. Node D has link-layer address d on link 2, and IPv6 address Gd.

The MSR depending on its configuration can broadcast router advertisements on the interfaces it acts as a router or respond to router solicitation messages from a specific node by sending a router advertisement message to this node.

3.2 Neighbor Discovery in a Multi-link Subnet

During Neighbor Discovery (ND) there are two possibilities for how an MSR can influence the ND procedure used. This is determined by the value of the flag L at the Prefix Information option of the Routing Advertisement message. It is assumed throughout this document that the hosts perform Autonomous Address Configuration (ADDRCONF) which depends on the flag A be set at the Prefix Information option [5].

Off-link Model. If the MSR sets the L flag all the hosts on the same link will not treat the prefix as being on-link.As a result ND is effectively disabled and

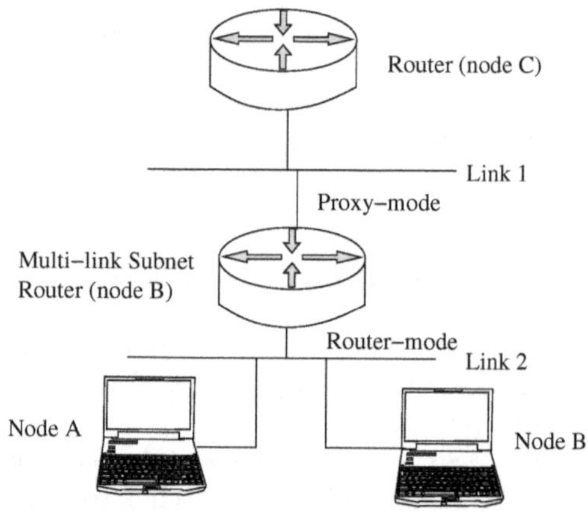

Fig. 2. A simple example of a Multi-link Subnet Router (MSR) connecting two separate physical links on a common prefix. It acts as a proxy on Link 1 where the default router is node C and as the default router on Link 2 where no router exists

packets to new destinations always go to the MSR first, which will then either forward them or redirect them depending on the destination node being on the same or different link. This case is referred as "Off-link Model".

As an example (see Fig. 2), when node A wants to start communication with C, it finds that the destination address matches no on-link prefix, and sends the packet directly to its default router B. B knows that C is on-link to link 1, with link-layer address c, and it forwards the packet to C. When node A wants to communicate with D, it finds that the destination address matches no on-link prefix, and sends the packet directly to its default router B. B knows that D is on-link to the same link as A, and responds with a Redirect message.

On-link Model. When the MSR does not set L the hosts on the link will perform ND by issuing Neighbor Solicitation Messages. The MSR should learn or know a-priori the location of the destination node. Neighbor Advertisements destined for nodes on another link should receive the Link-local Address of the MSR from the MSR. The MSR should refrain from answering Neighbor Solicitation Messages when the nodes are on the same physical link. This case is referred as "On-link Model".

As an example (see Fig. 2), when node A wants to start communication with C, it finds that the destination address matches an on-link prefix, and so sends an Neighbor Solicitation to the solicited-node multicast address. The NS message is received by node B, which listens on all multicast groups. Node B knows that C is on-link to link 1, and responds to A with an Neighbor Advertisement containing its own link-layer address b2 as the Target Link-Layer Address. After this, A can send packets to the address Gc. The packets will be sent to the link address b2 and they will be received by B, which will apply its validation rules (including decrementing the Hop Count in the IPv6 header) and forward them

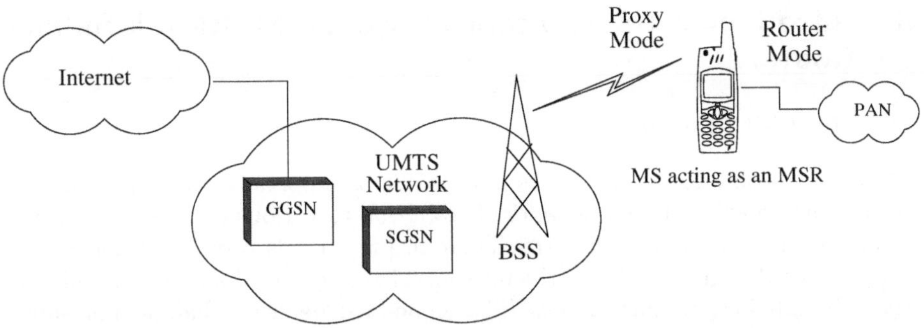

Fig. 3. The use of the mobile terminal as a multi-link subnet router. The MS is in proxy mode from the GGSNs perspective and acts as a router for the PAN

to the address c on link 1. When A wants to communicate with D, it again finds that the destination address matches an on-link prefix, and so sends an NS to its solicited-node multicast address. D receives the NS and responds. B also receives the NS, but knows that D is on the same link as A, and so does not respond.

3.3 Mobile Station Acting as a Multi-link Subnet Router

By conjugating the concept of the Multi-link Subnet Router and the former conclusion drawn, i.e. a full prefix is available to the mobile station, can lead to the concept of using the mobile station as a Multi-link subnet router, in order to make the prefix available to the nodes reside in the PAN.

Figure 3 depicts the use of the Mobile Station as an MSR. It can be seen that the GGSN is the top level router of the subnet and the Mobile Station acts as a Multi-link Subnet Router to the nodes attached to the link below. Assuming the most simple scenario (ie the MS is the only MSR) then a similar situation occurs as in the example mentioned previously. In this case the MS is in proxy mode from the GGSNs perspective and acts as a router for all the connected nodes to it. The point-to-point nature of the MS-GGSN link must be taken into account and the mobile router should not forward any local-scope packets (like Neighbor Solicitation messages) towards the GGSN.

Mobile router must get involved in duplicate address detection for link-local addresses to ensure that all addresses are unique across the personal area network and that the interface identifier provided by the GGSN to the mobile station will not be re-used. The MS can either operate in on-link or off-link mode. The MS acting as a MSR should not forward any local-scope packets towards the interface connecting the MS and the GGSN. The MS should only forward packets with global-scope addresses in the source and destination fields of the IPv6 header.

4 Introducing the Concept of Mobile Multi-link Subnet Router

4.1 Mobility Issues

Mobility support for IP networks is considered important, as a single user can be seen as a personal area network (PAN), due to various devices with IP functionality that the user may carry. There are two kinds of nodes when mobility support is taken into account, the fully enabled nodes that support mobility and the minimal functionality nodes (like embedded devices) that do not support mobility at all.

When a node is mobility-aware, it is desirable to use it's own mobility mechanisms [8] by custom preferences (home address and home agent). The basic requirement for a mobility-aware node is the acquirement of a globally routable IPv6 address. When a node completes the configuration of a global IPv6 care-of address, it exchanges binding update messages with its home agent in order to be reachable through its home address. In order for the mobility-aware node to use route optimization (for example when running real-time applications), it should initiate binding updates to its correspondent nodes so as to use its current care-of address.

There are cases that mobility support for mobility-unaware nodes is desired. One way of providing such support is described in the Mobile Router Tunneling Protocol proposed specification [4] where the concept of the Mobile Router (MR) is introduced. A Mobile Router is responsible for routing and mobility of an IP subnet network which moves with the mobile router. This subnet is attached to one of Mobile Router's network interfaces, while the other interface is connected to Mobile Router's home network. The subnet network (e.g. the PAN) has a routable prefix, which is called Mobile Router's subnet prefix. In general case, the Mobile Router's subnet prefix is different from the prefix of the Mobile Router's home network. The home network forwards every packet with destination an address belonging to the Mobile Router's subnet prefix towards the Mobile Router.

When a Mobile Router is attached to a foreign network, it installs an encapsulation interface towards its home agent, which is comprised of a bi-directional IP tunnel. Through this interface, the MR forwards (reverse-tunnels) all packets not originated from itself towards its home agent. The MR behaves as a normal Mobile Node for packets originating from itself. Hence, when the MR arrives to a visited link it injects a routing path to the reverse tunnel pointing to its home agent for all its depending nodes, which continue to use the same prefix as when the MR is located at home (Mobile Router's subnet prefix). Moreover there exists a default route to its default router on the visited link.

Also, the home agent of the MR injects a routing entry towards the encapsulation interface with ending point the care-of address of MR, for the Mobile Router's subnet prefix. The home agent forwards the packets based on the prefix information and does not need to know what nodes are behind the Mobile Router. The main drawback of this solution is the absence of route optimization

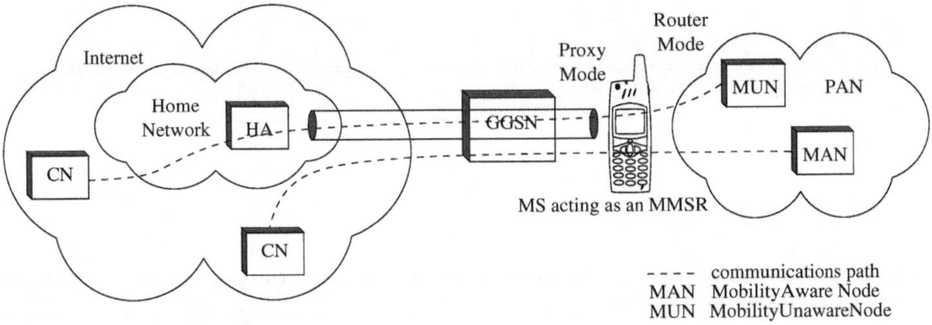

Fig. 4. The IPv6 mobility scenarios. A Mobility Unaware Node (MUN) which has acquired a home prefix address is connected through the bi-directional tunnel between the MS and its Home Agent. A Mobility Aware Node (MAN) which has acquired a foreign prefix address is connected directly

support as the packets get forwarded via the Mobile Router's Home Agent, which could be located far away from the exact access point of the Mobile Personal Area Network.

4.2 Mobile Station Acting as a Mobile Multi-link Subnet Router

By conjugating the concept the Mobile Station being used as a Multi-link subnet router and the mobility issues discussed above, a new concept of using the Mobile Station as a Mobile Multi-link Subnet Router (MMSR) is introduced.

In the case of a mobility-aware node, the IPv6 address that it acquired from the MMSR should belong to the visited network's subnet and the MMSR must advertise the subnet prefix obtained from the foreign network to that particular node. As seen in Fig. 4, the mobility-aware node is connected directly to its home agent and the correspondent nodes. This approach reduces traffic and processing load on mobile router's home network and its home agent.

In the case of a mobility-unaware node, the IPv6 address that it acquired from the MMSR should belong to the Mobile Router's subnet prefix. The MMSR must advertise the Mobile Router's subnet prefix to that particular node, in order to create a global IPv6 address that can be routed through the encapsulation forwarding mechanism mentioned above. If the advertised prefix was different, the packets could not be forwarded through Mobile Router's home network. As seen in Fig. 4, the mobility-unaware node is connected through the bi-directional tunnel between the MS and its Home Agent.

4.3 Introduction of the Mobility Support Bit

Hosts send Router Solicitations messages in order to prompt routers to generate Router Advertisements quickly. In response to a valid solicitation message, a

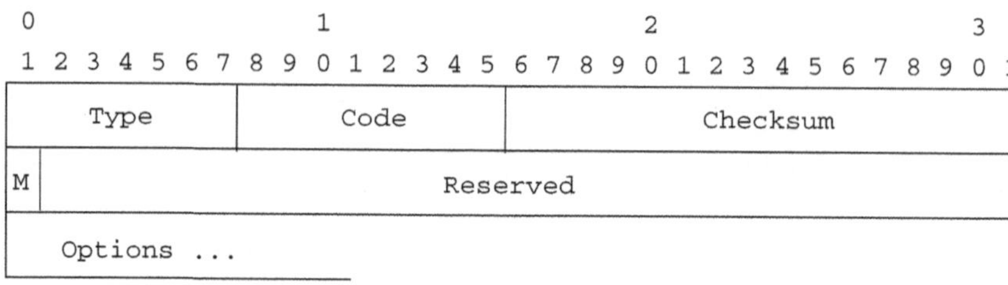

Fig. 5. The enhanced Router Solicitation message showing the position of the Mobility Support bit (M)

router may choose to unicast the response directly to the soliciting host's address or to multicast the response to the all-nodes group [5] [6].

As far as the MMSR is concerned, only the first case should be considered since the MMSR should be able to choose the appropriate network prefix to advertise to a node depending on its mobility support capabilities. For the above reason, a new flag situated at the Reserved Header space of the routing solicitation message is introduced

This flag, which we call Mobility Support Bit (M-bit) should be set by the node if it requests mobility support from the MMSR. In this case, the MMSR responds by a Router Advertisement message containing the Mobile Router's subnet prefix as the prefix to be used in the construction of the IPv6 address. In any other case (the M-bit is not set), the MMSR provides the subnet prefix obtained from the foreign network (e.g. UMTS). The enhanced Router Solicitation message format can be seen in Figure 5

This enhanced routing solicitation message will not affect the functionality of routers that do not support the M-bit, as the particular bit belongs to reserved header space and will be ignored by a normal router.

Based on the above discussion, a node that wishes to have mobility support from the MMSR has to send a router solicitation message with the M-bit on. In such case, the node constructs an IPv6 address with its subnet prefix being the same to the Mobile Router's subnet prefix and has mobility support if the point of attachment of the MMSR changes. Otherwise, a node that does not require mobility support from the MMSR should send a normal router solicitation message (M-bit off) and will receive a subnet prefix belonging to the foreign visited network. Another possibility exists where a mobility-aware node may need to use the mobility support of the MMSR, for example when the communication with its home agent is not possible.

When MMSR is attached to its home network, it advertises the Mobile Router's subnet prefix to be used by all the nodes connected to the PAN, regardless of the M-bit value. However the MMSR must remember the nodes that did not request mobility. This is due to the fact that every time the MMSR changes its point of attachment must sent an unicast Router Advertisement message to the above nodes in order for them to change their network prefix.

In case that the MMSR can not acquire a full prefix from the visited network (e.g. Wireless LAN) and only a global IPv6 address is provided, all the nodes belonging to the Mobile PAN must have addresses with the Mobile Router's subnet prefix in order to use the encapsulation forwarding mechanism, regardless of the value of the M-bit.

5 Security Issues

It is assumed that the user which owns the Mobile Station acting as a Mobile Multi-link Subnet Router and the associated Personal Area Network, will provide access to its network on a restricted basis and not on a freely manner. This means that the link layer access is somehow granted securely, for example by providing cryptographic keys to any friendly nodes that wish to connect to the PAN.

The communication between the MMSR and its Home Agent must be secured when the MMSR is not on its home link. A possible solution would be the use of IPSec [14]. This can be done by the existence of a security association between the mobile router and its home agent [8] [12]. IPSec makes possible the realization of secure connections over insecure networks by using two protocols to provide traffic security, Authentication Header (AH) [15] and Encapsulating Security Payload (ESP) [16]. These protocols may be applied alone or in combination to provide a desired set of security services and access control. The MMSR and its home agent should use ESP to protect payload packets tunnelled between themselves.

6 Conclusions

This paper has presented an approach on providing the necessary connectivity and mobility support to a Personal Area Network through UMTS IPv6 connectivity. UMTS connectivity issues regarding IPv6 mechanisms were examined and the concept of using the Mobile Station as a Multi-link Subnet Router was introduced.

This was extended to include the provision of mobility support to all the nodes comprising the PAN. In this context the concept of the UMTS Mobile Station acting as a Mobile Multi-link Subnet Router was introduced.

A Personal Area Network may contain nodes that support mobile IPv6 as well as nodes that are mobility unaware. In order to provide efficient mobility support for all the nodes of the PAN, the MS needs to treat them differently depending on their mobility capabilities. For the above reason an enhancement to the Router Solicitation mechanism is proposed with the addition of a flag (M-bit) in the reserved header space of the Router Solicitation Message. This should be set by the node that wishes to have mobility support from the PAN's Mobile Multi-link Subnet Router.

Future work should include the statefull autoconfiguration scenario by using a DHCPv6 server [17]. Moreover, security issues as well as the protection of Routing Solicitation and Advertisement Messages should be further investigated. Furthermore multi-homing environment scenarios should be considered where the Mobile Multi-link Subnet Router is connected to more than one access providers. Also the use of more than one Mobile Multi-link Subnet Router should be investigated where a nested architecture is present e.g. a Mobile PAN is connected to another Mobile PAN.

References

1. Deering S., Hinden R.: Internet protocol version 6 (IPv6) specification, Dec. 1998, RFC 2460
2. Loughney J.: IPv6 node requirements, Internet draft, Oct. 2002. Work in progress
3. Thaler D., Huitema C.: Multi-link Subnet Support in IPv6, Internet draft, June 2002. Work in progress
4. Kniveton T., Malinen J., Devarapalli V., Perkins C.: Mobile Router Tunneling Protocol, Internet draft, Nov. 2002. Work in progress
5. Narten T., Nordmark E., Simpson W.: Neighbor discovery for IPv6, Dec. 1998, RFC 2461
6. Thomson S., Narten T.: IPv6 Stateless address autoconfiguration, Dec. 1998, RFC 2462
7. Narten T., Draves R.: Privacy extensions for stateless address autoconfiguration in IPv6, Jan. 2001, RFC 3041
8. Johnson D., Perkins C., Arkko J.: Mobility support in IPv6, Internet draft, Jan. 2003, Work in progress
9. Universal Mobile Telecommunications System (UMTS); General Packet Radio Service (GPRS) Service description; Stage 2, Mar. 2002. 3GPP TS 23.060 version 5.1.0 for Releases 4, 5, 99
10. Wasserman M.: Recommendations for IPv6 in Third Generation Partnership Project (3GPP) Standards, Sep. 2002. RFC 3314
11. Haskin D., Allen E.: IP Version 6 over PPP, Dec. 1998. RFC 2472
12. Arko J., Devarapalli V., Dupont F.: Using IPsec to Protect Mobile IPv6 Signalling between Mobile Nodes and Home Agents, Internet draft, Jan. 2003. Work in progress
13. Universal Mobile Telecommunications System (UMTS); Interworking between the Public Land Mobile Network (PLMN) supporting Packet Based services and Packet Data Networks (PDN), Mar. 2002, 3GPP TS 29.061 version 5.1.0 for releases 99, 4, 5
14. Kent S., Atkinson R.: Security Architecture for the Internet Protocol, Nov. 1998. RFC 2401
15. Kent S., Atkinson R.: IP Authentication Header, Nov. 1998. RFC 2402
16. Kent S., Atkinson R.: IP Encapsulating Security Payload, Nov. 1998. RFC 2406
17. Droms R., Bound J., Volz B., Lemon T., Perkins C., Carney M.: Dynamic Host Configuration Protocol for IPv6 (DHCPv6), Internet draft, Nov. 2002. Work in progress

Smooth Handover and Optimized Uplink Routing in Cellular IPv6 Networks

Rola Naja and Samir Tohmé

Ecole Nationale Supérieure des Télécommunications
46, Barrault Street, 75634 Paris
{rola.naja,samir.tohme}@enst.fr

Abstract. Cellular IPv6 is an efficient IP micro-mobility protocol that provides important features such as easy global migration, cheap passive connectivity, efficient location management and routing, and flexible handover. However, Cellular IPv6 presents some problems when dealing with intra-domain traffic. In this paper, we propose an enhanced uplink routing mechanism coupled with a study of a smooth handover for intra-domain traffic. Simulation is carried out in order to show the performance results.

Keywords: Micro-mobility, Macro-mobility, Cellular IP, Handover.

1 Introduction

Provision of real time multimedia services to mobile users is the main objective of the next generation wireless networks, which will be IP-based and are expected to inter-work with the Internet backbone seamlessly. Two major challenges exist in wireless mobile networks, namely the fast handoff and the provision of Quality of Service (QoS) over IP-based wireless access networks.

Research efforts are oriented towards the design of IP micro-mobility protocols that can handle the IP mobility seamlessly. A suitable IP mobility protocol must resolve the following contradiction: it must keep the mobile's IP address unchanged in order to identify an end-system having a TCP session and at the same time it must deal with an IP address that changes with the mobile location. The Mobile IP architecture [2] was conceived in order to resolve the above-mentioned contradiction by using two IP addresses for a mobile node:

- The Home Address: is a permanent address used to uniquely identify the mobile node on the Internet.
- The Care-of Address (CoA): is a temporary address used to route the datagrams destined to the mobile node to the current point of attachment of the mobile.

As a result, Mobile IP allows the mobile nodes to change their point of attachment in an IP network. Unfortunately, Mobile IP has some limitations when applied to wide-area wireless networks with high mobility users that may require

M. Conti et al. (Eds.): PWC 2003, LNCS 2775, pp. 651–664, 2003.

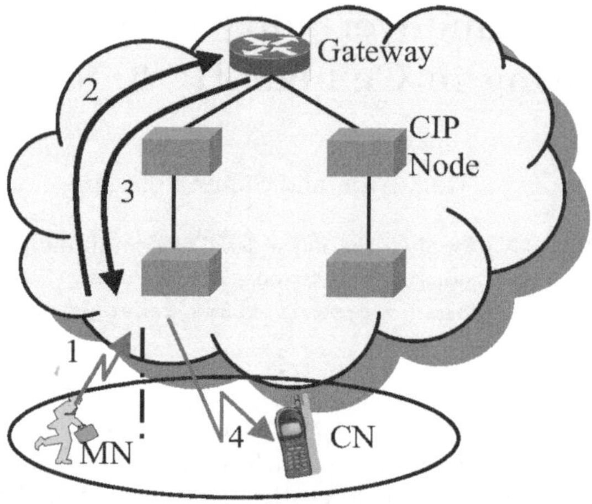

Fig. 1. Non optimization of Uplink Routing Mechanism for Intra-network traffic in Cellular IPv6 Network

quality of service. Thus, Mobile IP needs to be enhanced to cope with micro-mobility, i.e., movement across multiple subnetworks with frequent handovers and within a single network or domain.

IP Micro-mobility protocols were introduced in order to limit the disruption to user traffic during handoff: they complement Mobile IP by providing fast and also seamless handoff control. One promising micro-mobility protocol is Cellular IPv6 (CIP). Cellular IPv6 inherits cellular technology principles for mobility management, passive connectivity and handoff support, but implements these around the IP paradigm. As defined in [3], Cellular IPv6 suffers from a "non-optimization" of the uplink routing in the case of intra-network traffic. In fact, all the traffic coming from the mobile node (MN) must pass through the gateway before being delivered to the corresponding node (CN), even if the MN and the CN are connected to the same base station. Far from an optimal path, this kind of routing increases unduly the delay and the jitter of the packets. It can also imply the waste of bandwidth which constitutes a problem in case of high traffic (figure 1).

D. Gatzounas et al. [9] proposed an uplink routing optimization for Cellular IP networks that improves the protocol performance. In their proposition, even if the corresponding node and the mobile node are not in the same Cellular IP domain, all the Route caches in the route towards the gateway will be checked in order to find if CN has an entry with a "Tear down optimized flag" unset. This would cause an unnecessary delay for the cache processing. On the other hand, the overall performance is made at the expense of an increase of the signaling load.

In our study, we achieve two objectives. First, we enhance the uplink routing mechanism by adopting the mechanism proposed in [9] and by improving it in order to have faster lookup, better scalability and less frequent signaling messages. Second, we propose to study the performance of a smooth and anticipated handover in the Cellular IPv6 network. More precisely, we evaluate the performance of a buffering mechanism while taking into consideration the optimized uplink routing mechanism.

To this end, we start our study by briefly exposing the protocol Cellular IPv6 in section 2. Our proposition is introduced in section 3 discussing the extensions made to the Cellular IPv6 protocol. A simulation exposed in section 4 finalizes the study before concluding the paper.

2 Cellular IPv6 Protocol

The Cellular IPv6 proposal [3] uses specialized domain routers with host-based entries for local mobility and Mobile IP for inter-domain mobility. Thus, updates can be localized, enhancing the scalability of update mechanisms and limiting disruption.

Cellular IPv6 operates on mobile hosts, base stations and Internet gateways. It combines the cellular network with the IP network. Through paging and other handoff techniques, low-latency handoff and efficient management can be achieved.

2.1 Cellular IPv6 Routing

In Cellular IPv6, routing is based on host routes for the mobile host, through all the nodes on the path of the mobile host to the gateway. Cellular IPv6 provides two parallel cache systems that store the host-based routing entries. These caches are the *Route cache* and the *Paging cache*.

Each Cellular IPv6 node has an uplink neighbor and a downlink neighbor. To minimize control messaging, regular data packets transmitted by mobile hosts are used to refresh host-based routing entries stored in the caches. Cellular IPv6 nodes monitor mobile originated packets and maintain a distributed, hop-by-hop location database that is used to route packets to mobile hosts.

2.2 Cellular IPv6 Handover

Handoff is required when there is a change of access point during active transmission. Cellular IPv6 supports two types of handoff schemes. Cellular IPv6 hard handoff uses an algorithm that trades off some packet loss in exchange for minimizing handoff signaling. Cellular IPv6 semi-soft handoff tries to proactively notify the new access point before actual handoff. Semi-soft handoff minimizes packet loss and provides improved TCP and UDP performance over hard handoff. For the wireless technologies that do not provide the simultaneous connections, the indirect semi-soft handoff is proposed.

3 Proposition and Enhancements

The Cellular IPv6 protocol uses two parallel cache systems in order to have faster lookup and better scalability. In fact, since only a portion of the hosts will be in active state at any given time, it is better to separate the caches for active and idle mobile hosts. Following the same reasoning, we propose to apply two parallel uplink routing mechanisms, namely the non-optimized uplink routing mechanism specified in [3] and the optimized uplink routing mechanism that we will detail in subsection 3.2.

When the MN wants to send packets, it checks the subnet prefix of the CN's address.

- If the IPv6 subnet prefix of the MN's CoA is equal to the IPv6 subnet prefix of the CN's CoA, then the MN sets the flag OR, as defined in section 3.1, to 1. As a result, the CIP nodes will apply the optimized uplink routing mechanism.
- Otherwise, the MN sends the packets without setting the flag OR. Then, the CIP nodes will route the packets with the non-optimized uplink routing mechanism.

This routing mechanism distinction will prevent the CIP nodes from scanning the caches mappings at each packet reception. Consequently, this will decrease the processing load in the nodes.

The optimized uplink routing proposed in [9] was made at the expense of an increase of the signaling load. In fact, in order to retain the routing cache consistency, the optimizing node must send a "proxy route-update" packet at a rate just faster than the route-timeout [1]. In our mechanism, we propose to reduce this signaling load.

The loss of packets during the transition between networks should be minimal. It is shown in some research studies that buffering packets improves the global performance of Mobile IP. This paper defines a buffering mechanism that attempts to meet this goal for Cellular IPv6.

3.1 Protocol Extensions

Before detailing the optimized uplink routing mechanism, let us distinguish between the two following types of CIP nodes (figure 2):

1. *Optimizing Node.* This node is the intersection of the path from the MN's access node to the gateway router with the path from the CN's access node to the gateway router. Only one optimizing node at any time performs route optimization for a single pair of communicating mobile nodes in the same CIP network. The optimizing node must route data through the optimal path to the destination address.

[1] The route-timeout is the validity time of mappings in the Route caches.

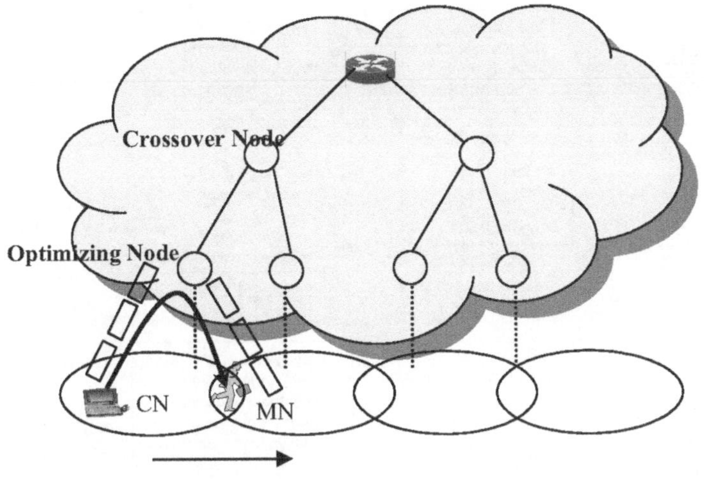

Fig. 2. Optimizing and Crossover Nodes in Cellular IPv6 Networks

2. *Crossover Node.* This node is the intersection of the old path from the gateway to the previous access node (base station) with the path from the gateway to the new access node (base station). The crossover node has two mappings for the mobile during the handover.

Let us define the following messages and flags added to the Cellular IPv6 protocol:

1. *Proxy route-update* message: it is sent by the crossover node towards the gateway [9]. It is an IPv6 packet carrying a Hop-by-Hop Options extension header.
2. *Remove Mapping* message: it is sent by a mobile upon handoff. It is an IPv6 packet carrying a Hop-by-Hop Options extension header.
3. *Optimize Route (OR)* flag : it is carried in the Hop-by-Hop Options extension header of the data packets. When set, the route optimization must be performed by the nodes receiving the packets (default=0).
4. *Duplicate (DUP)* flag: it is in the IP header of the data packets indicating, when set, that the current packets are duplicated (default =0).
5. *Optimize (ON)* flag: it is in the Route caches' mappings (default=3). If a node has, in its Route cache, a mapping for a mobile node MN with ON equal to:

- 0, then the node "prepares" itself to be an optimizing node for the mobile node and its corresponding node. The ON is set to 0 during the handover establishment. The node in question does not send data through the optimal path as long as ON is equal to 0, in order to prevent the packets routing to the new cell during handover.
- 1, then the node can be an optimizing node for the MN and its corresponding mobile node. The optimizing node must send data through the optimal path.

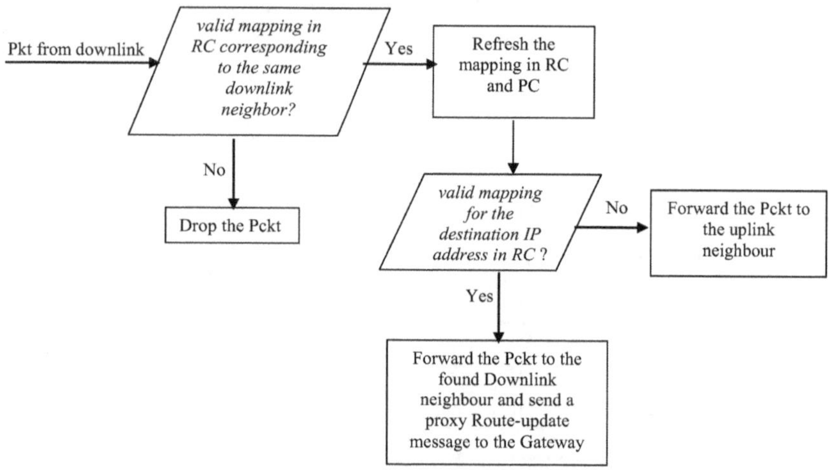

Fig. 3. Optimized Uplink Routing Mechanism in Cellular IPv6 Network (RC and PC are the Route and Paging caches)

– 2, then the data packets sent by the mobile node MN must be duplicated by the node if it is an optimizing node.

3.2 Optimized Uplink Routing Mechanism

When a data packet arrives to a Cellular IPv6 node, the following events occur (figure 3):

If the data packet arrives from a downlink neighbor, the Route cache entry of the source IP address is searched first in the current node's Route cache. If the data packet arrives from a different neighbor than the one that is in its mapping or no mapping exists for the IP address, then the packet is dropped.

Otherwise, if the data packet is coming from the same neighbor, the mapping is refreshed in the Route and Paging caches.

The current node checks if the destination IP address has a valid mapping in the Route cache. There are two cases:

1. If such a mapping exists, the packet is forwarded to the downlink neighbor found in the mapping. In this case, the current node becomes an optimizing node for the two communicating mobile nodes.
2. Otherwise, if the Route Cache contains no mapping for the destination IP address then the packet is forwarded to the uplink neighbor.

In the first case, the optimizing node generates a proxy route-update message towards the gateway. The proxy route-update contains the IP address of the MN in the control field of the Hop-by-Hop Options extension header. The sending rate of the proxy route-update in [9] is controlled by a time interval shorter than the route-timeout interval (9s). Thus, more signaling load is generated than in the non-optimized uplink routing. We propose that the optimizing node sends the proxy route-update message once: this message is sent when the node receives the first data packet that causes the node to become an optimizing node.

The proxy route-update sets the ON, in the mapping corresponding to the MN, to 1. If it finds ON equal to 2, no change in the mapping will be done.

Moreover, the proxy route-update will set the expiration time in the MN's mapping to the sum of the current time and an estimated time. The latter will be estimated according to the dwell time of the mobile in the CIP domain (according to the speed of the mobile) and to the call duration (according to the traffic type). In this case, the CIP nodes will operate in a hard state. This will help to reduce the signaling load and will not affect the route cache consistency. With our mechanism, the caches will always be updated according to the mobiles' movement, and the packets will arrive to the correct destination to the mobiles movement. In fact, a change in the route cache occurs:

- When one of the two communicating mobiles leaves the CIP domain. If the sending mobile node leaves the cell, then it sends a "paging-tear down"[2] The latter removes the mappings corresponding to the sending mobile node in the caches. Whenever the receiver leaves the cell, then it also sends a paging-tear down. The latter removes the mappings corresponding to the receiver in the caches. The packets sent to the receiver, in this case, will pass through the gateway in the uplink direction. And these packets will refresh the caches.
- When the sending mobile node's traffic is forwarded uplink. In this case, the packets will refresh the caches of the optimizing node and will be sent to a new optimizing node.
- At the end of the session and after the expiration of an "idle time", the sending mobile node becomes idle. The mapping of the sending mobile node stored in the Route caches will expire. As a result, the optimizing node will be a regular Cellular IPv6 node. The sending mobile node sends frequent paging-update packets. These packets will refresh the Paging caches of all the nodes in the path leading to the gateway.

As a result, the caches are refreshed according to the mobility of the flows, to the session duration, ... Therefore, the proxy route-update can be sent less frequently without affecting the routing mechanism.

In order to retain the routing cache consistency, the optimizing node of a communicating hosts pair, when receiving the update messages coming from the sending mobile node, must refresh its caches. Nevertheless, it must prevent sending these messages upwards in order to prevent the refreshing of the caches belonging to the branch leading to the gateway. Otherwise, the expiration time in the caches' mappings will change from the estimated time value. In this case,

[2] A paging-teardown packet is an IPv6 packet with a Hop-by-Hop Options extension header where the source address is the IP address of the sending mobile node, the destination address is the Gateway and the Hop-by-Hop option is of Paging-teardown type [3].

this expiration time will be set to the sum of the current time and the Route-update time. This will cause the caches' entries to expire because the caches do not receive frequent proxy route-update messages.

3.3 Handover Handling

Not all wireless technologies have simultaneous connection capability, i.e. they cannot listen to the current BS while sending a route-update packet to the new BS. For this situation an indirect semi-soft handoff is used [3]. We propose to enhance the indirect semi-soft handover handling while taking into account the optimized uplink routing mechanism.

Our study defines a buffering mechanism for the indirect semi-soft handoff. This buffering mechanism reduces the loss and the packet delay during the handover. The key idea is that the optimizing node duplicates the packets destined to the moving receiver. The original packets will be routed via the optimal path. As for the duplicated packets, they are sent to the crossover node where they are stored. These duplicated packets will be routed to the new mobile location after handover. In this way, the delay and the packets loss will be optimized.

We assume that a call is set-up between a mobile node and a corresponding node. We suppose also that a mobile node can not send and receive packets at the same time. This assumption is made for the ease of the handover handling description.

When a mobile node performs handover, the following sequence of events occurs:

1. The mobile node sends a route-update packet to the current BS. This packet has the IP address of the new BS as destination IP address. The route-update packet contains the address of the corresponding mobile in the control field of Hop-by-Hop Options extension header. The I flag is set to indicate indirect semi-soft handoff.
   We distinguish between two cases:
   If the mobile node is a sending mobile node, then it sets the OR flag of the route-update packet to 1.
   If the mobile node is a receiver, then it sets the OR flag of the route-update packet to 0. This implies that the packet will reach the gateway. In this way, the route-update packet will reach the crossover node. This will not happen when the optimized uplink routing mechanism is used (OR=1) and when the optimizing node is hierarchically under the crossover node.
   The current BS forwards the route-update packet to the Gateway. The latter uses then normal IP routing to deliver the packet to the new BS. The route-update packet sets the flag ON to 2 in the Route caches mappings that correspond to the sending mobile node. This is done to all the nodes belonging to the branch starting from the current base station up to the gateway.
   In the following steps, we consider that the mobile node that is performing handover is the receiver. This is an important case, since the packets sent to

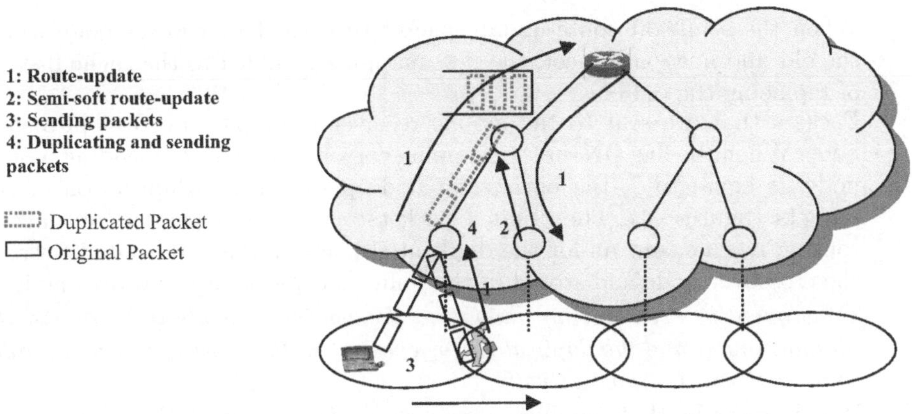

Fig. 4. Packets duplication during handover

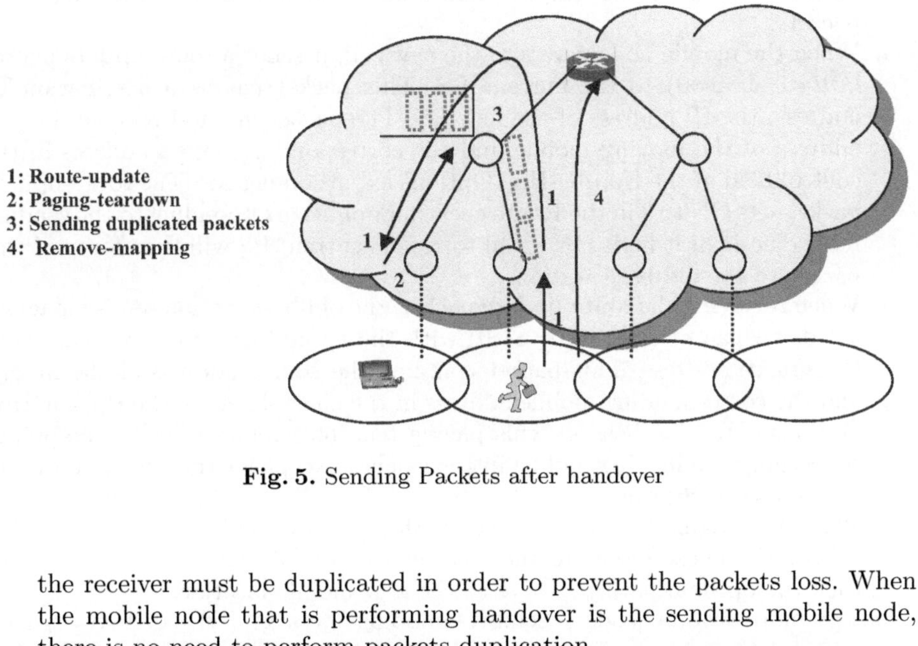

Fig. 5. Sending Packets after handover

the receiver must be duplicated in order to prevent the packets loss. When the mobile node that is performing handover is the sending mobile node, there is no need to perform packets duplication.

2. When the new BS receives the indirect semi-soft handoff packet, a semi-soft route update packet is created (I=0, S=1) with the IP address of the mobile host as the source address. It is then forwarded upstream. The semi-soft route-update packet creates new mappings in the Route and Paging Cache similarly to regular route-update packets. However, it sets the flag ON in the Route cache mapping that corresponds to the sending mobile node to 0. This is done to prevent the routing of packets to the new cell before the handover takes place. Recall that we are proposing to study the indirect semi-soft handoff.

When the semi-soft route-update packet reaches the crossover node where the old and new path meet, the new mapping is added to the cache instead of replacing the old one.

3. Packets that are sent to the mobile receiver must pass by the optimizing node. When the flag ON, in the mapping corresponding to the sending mobile node, is equal to 2, the optimizing node performs the duplication of the packets (figures 4). The original packets are routed using the optimizing routing mechanism. As for the duplicated packets, they are routed towards the crossover node and stored in the buffer located in the crossover node.

   Consequently, even during handover, the packets are always sent via the optimal route, and the duplicated packets wait in the crossover node in order to be sent via the shortest path.

   It is noteworthy that the duplicated packets do not refresh the caches in our proposition. Otherwise, the caches must be refreshed each route-update time: this would incur more signaling load. Thus, the flag DUP of the duplicated packets will be set to 1: the CIP nodes will be able to identify the duplicated packets.
4. When the mobile node moves to the new cell, it sends a route-update packet (OR=1, I=S=0) to the current BS. This packet has as a destination IP address, the IP address of the new BS. The packet in question contains the address of the moving mobile and the corresponding node's address in the control field of the Hop-by-Hop Options extension header. The route-update packet sets ON to 3 in the Route cache mapping corresponding to the sending mobile node, if it finds ON equal to 2. The current BS will then forward this packet to the old base station.
5. When receiving the route-update packet, the old base station sends a paging-teardown packet (OR=0, I=S=0) with the IP address of the receiver as in the source address. This packet contains the source address of the mobile and the corresponding mobile address in the control field of the Hop-by-Hop Options extension header. This paging-teardown removes all the mappings concerning the moving mobile in the Caches except for the ones pointing to the new Base Station.

   When the paging-teardown arrives to the crossover node, it forces the buffer to free the packets sent to the receiver (figure 5). The freed packets take then the optimal path to arrive at the new mobile location.
6. The mobile node sends a remove-mapping message (OR=0, I=S=0) that contains the address of the corresponding mobile. This message will set ON in the mapping corresponding to the sending mobile node to 3, if it finds it equal to 2. This is done in all the caches of the CIP nodes belonging to the branch leading to the gateway.

   The handover is then complete.

4 Simulation Results and Performance Analysis

In order to study the performance of the proposed mechanism, simulations was carried out using OMNeT++ [11]. The CIP network illustrated in figure 2 was

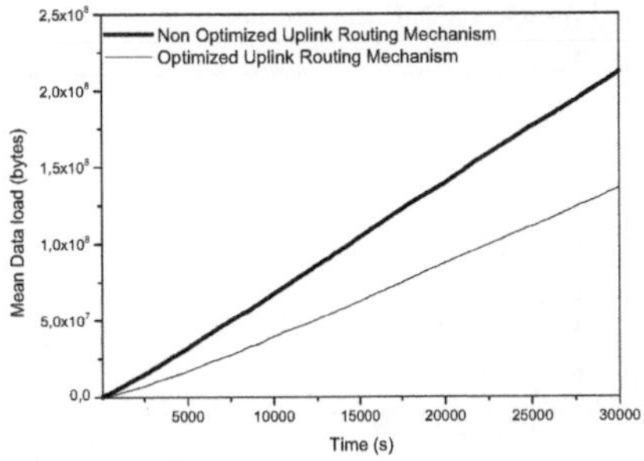

Fig. 6. Mean Data Load

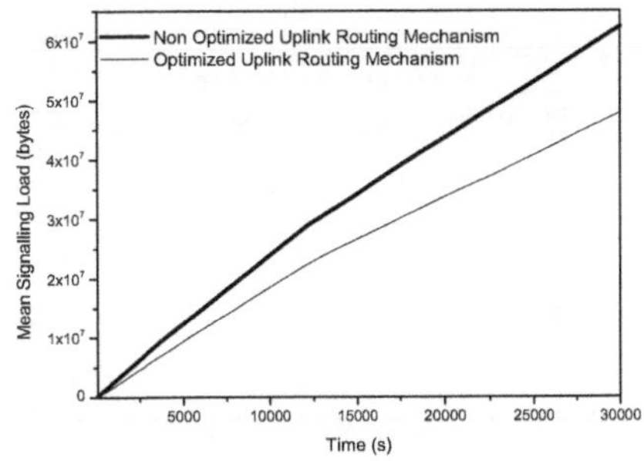

Fig. 7. Mean Signaling Load

simulated. We suppose that the wireless bandwidth is equal to $1Mb/s$, the wired link capacity dedicated for data is $1,92Mb/s$, that for the signaling is $128Kb/s$. The data traffic considered represents a typical WWW session ($64Kb/s$) that consists of a sequence of packet calls. During a packet call, several packets may be generated. A time interval called reading time separates two consecutive packet calls. The parameters and laws that model the data traffic are specified in [10]. Mobile users are considered as pedestrians with mean speed of $1.8Km/h$ moving within the cells of radius $0,1Km$. Moreover, we consider that each CIP node has 3 buffers:

- the first buffer is dedicated for the signaling packets Bs.
- the second buffer is allocated to the data packets Bd.
- the third buffer is for the duplicated data packets during the handover Bh.

Since the signaling packets are important in CIP network, the signaling buffer Bs is allocated a percentage of the link capacity. Thus, Bs does not suffer from the resource contention. One better alternative is to apply the Round-Robin mechanism between the signaling buffer and the data packets. This would improve the bandwidth use but it would not change our results nor the conclusions of our study.

We apply the Head Of the Line (HOL) discipline with no-preemption in order to schedule the packets in Bd and Bh. The HOL serves the packets, stored in Bh first, after the paging-teardown reception. Note that the freed packets are the ones sent to the moving receiver which address is the source address of the paging-teardown.

4.1 Numerical Results

Figure 6 shows the data load measured on the gateway interfaces. One can see that the optimizing routing mechanism lowers the load on the gateway which is considered as a bottleneck in the CIP network. This result has also been obtained by [9]. However and unlike [9], the signaling load on the gateway interfaces is less than the one obtained with the non-optimized uplink routing

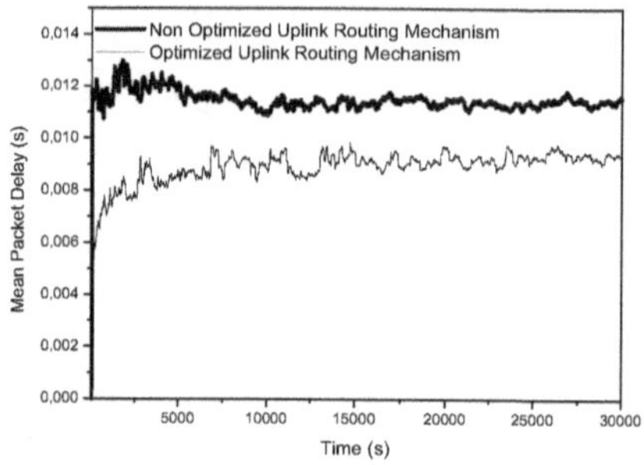

Fig. 8. Mean Data Packet Delay

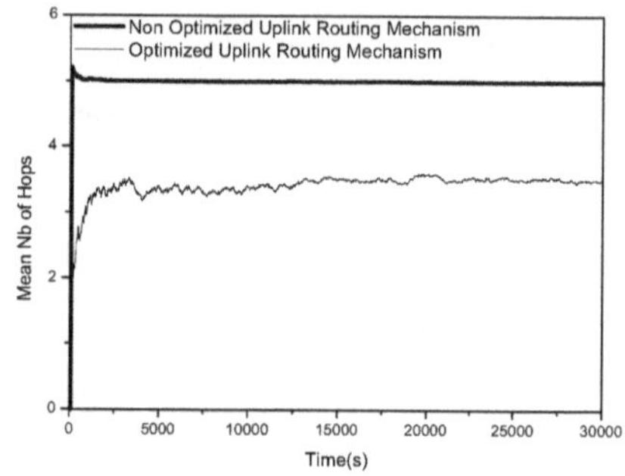

Fig. 9. Mean Number of Hops

mechanism (figure 7). In fact, the optimizing nodes retain all the update packets sent by the sending mobile node. This can alleviate the signaling load on the gateway. On the other hand, we argue that the proxy route-update packet must be sent once and that the expiration time, related to this packet, in the caches mappings must be well-chosen. Moreover, most of the signaling messages needed to establish the handover are sent through the optimal path. As a result, the decrease of the signaling load on the gateway interfaces is a logical consequence of our mechanism.

Figure 9 shows the mean number of hops crossed by the packets before arriving at destination. As we can see, this number of hops is constant with the non-optimized uplink routing mechanism. In fact, the packets always pass by the gateway before being routed to destination. Thus, the number of hops depend on the network topology. With the optimized uplink routing mechanism, the number of crossed hops is reduced. This is because the packets take the optimal path before arriving at destination.

As for the delay experienced by the packets, the curves depicted in figure 8 show better results than with the non-optimized uplink routing mechanism, due to the reduced number of hops crossed by the packets. This delay is also due to the duplicated packets that are received by the mobile upon sending the paging-teardown. As we can see, the localization of the storing buffer on the crossover node and not on the gateway helps to reduce the delay encountered by the packets.

Figure 10 depicts the delay of the packet sent during the handover. It can be seen that this delay is much higher with the non-optimized uplink routing mechanism. In fact, with our enhanced mechanism, the packets are sent through the optimal during and after handover.

We also measured the delay of establishment the handover in figure 11. We found that this delay is higher with the non-optimized uplink routing mechanism. In fact, with the optimized uplink routing mechanism and when the mobile moves to the neighboring cell, the path taken by the route-update packet is shorter than that with the non optimized uplink routing mechanism.

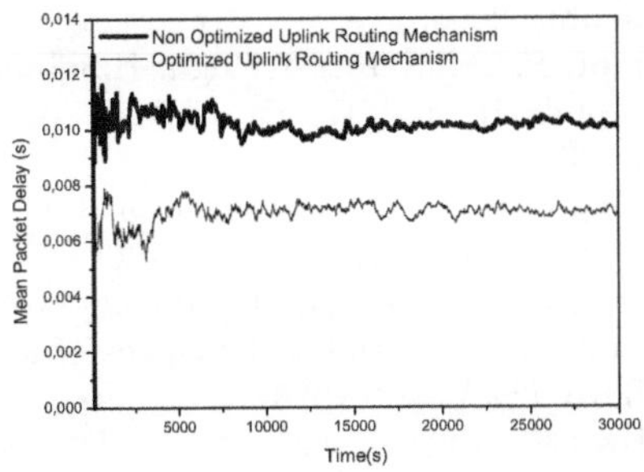

Fig. 10. Mean Data Packet Delay During Handover

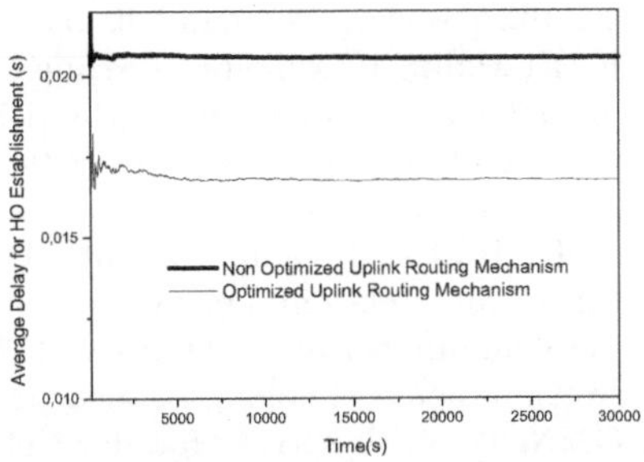

Fig. 11. Mean Handover Establishment Delay

The retrieved numerical results show the benefits of the enhanced optimized uplink routing mechanism. These better results were obtained at the expense of some complexity added to the CIP nodes. One must make the trade-off between better performance and complexity.

5 Conclusion

This paper presents an enhanced uplink routing mechanism coupled with a smooth handoff study. Our proposal aims to minimize the delay and the loss experienced by the data packets during communications and especially during handover. Simulation results show the good performance obtained at the expense of some complexity added to the Cellular IPv6 nodes.

Some open issues still remain. The resource allocation and the study of an efficient call admission control providing quality of service to different classes of mobile users are important issues to be addressed and need further investigation.

References

1. A.Campbell, J.Gomez, S.Kim, C.Wan, Z.Turanyi, A.Valko: Comparison of IP Micromobility Protocols. IEEE Wireless Commun. Mag., vol.9, no.1, February 2002.
2. D.B. Rice, C. Perkins: Mobility Support in IPv6. Internet draft, draft-ietf-mobileip-ipv6-15.txt, July 2001.
3. Z.D.Shelby, D.Gatzounas, A.Campbell, C.Wan: Cellular IPv6. Internet draft, draft-shelby-seamoby-cellularipv6-00.txt, November 2000.
4. H. Soliman, C. Castelluccia, K. El-Malki, L.Bellevier: Hierarchical MIPv6 mobility management (HMIPv6). Internet draft, draft-ietf-mobileip-hmipv6-05.txt, July 2001.
5. R. Ramjee et al.: HAWAII: A Domain-Based Approach for Supporting Mobility in Wide-area Wireless Networks. Proc.IEEE Int'l.Conf.Network Protocols, 1999.
6. R. Koodli, C.Perkins: A framework for smooth handovers with Mobile Ipv6. Internet draft, draft-koodli-mobileip-smoothv6-00.txt, July 2000.

7. K.EL-Malki, H.Soliman: Fast Handoffs in Mobile IPv4. Internet draft, draft-elmalki-mobileip-fast-handoffs-03.txt, September 2000.
8. K. El-Malki, P.Calhoun, T. Hiller, J. Kempf, P.J.McCann, A.Singh, H.Soliman, S.Thalanany: Low Latency Handoffs in Mobile IPv4. Internet draft, draft-ietf-mobileip-lowlatency-handoffs-v4-04.txt,July 2002.
9. P.Mähönen, D.Gatzounas, A. Inzerilli: Cellular IP Route Optimization. Internet draft, draft-shelby-cip-routeoptimization-00.txt, June 2001.
10. European Telecommunications Standards Institute ETSI, Universal Mobile Telecommunications Systems UMTS: Selection procedures for the choice of radio transmission technologies of the UMTS.TR101 112,V3.2.0,1998.
11. OMNeT++:Objective Modular Network Testbed in C++. http://www.hit.bme.hu/phd/vargaa/omnetpp.htm.

IPv6 Stateless Address Autoconfiguration in Ad Hoc Networks

Zhong Fan

Toshiba Research Europe Ltd.
Telecommunications Research Laboratory
32, Queen Square, Bristol BS1 4ND, UK
zhong.fan@toshiba-trel.com

Abstract. A mobile ad hoc network is an infrastructure-free wireless network that is built on the fly. Since central administration or configuration by the users is impractical in such networks, autoconfiguration of network nodes is highly desirable. In this paper, we propose an approach to IPv6 address autoconfiguration in ad hoc networks, where we apply the IPv6 Stateless Address Autoconfiguration Protocol and Neighbor Discovery Protocol to the context of ad hoc networks. It overcomes some of the limitations of existing protocols. In particular, we consider the scenarios of network partitioning and merging. A distributed scheme for duplicate address detection is also discussed.

1 Introduction

Generally there are two variations of mobile wireless networks. The first is known as infrastructured networks, i.e., those networks with fixed and wired gateways. Examples of this type of network include office wireless local area networks (WLANs). The second type of mobile wireless network is the infrastructure-less network, commonly known as a mobile ad hoc network (Manet) [1]. Infrastructure-less networks have no fixed routers. All nodes are capable of movement and can be connected dynamically in an arbitrary manner. Network nodes function as routers which discover and maintain routes to other nodes in the network.

In future ubiquitous computing environments (such as the networked home of the future with various IP-enabled appliances), the large number of network-enabled nodes as well as the need to establish dynamic connections between such nodes, make the manual configuration of individual nodes impractical. A robust and fast plug-and-play solution is therefore needed to provide autoconfiguration capabilities.

In this paper we consider the problem of automatic IPv6 address configuration in ad hoc networks. We have chosen IPv6 because IPv6 has a number of advantages compared to IPv4 [2]:

M. Conti et al. (Eds.): PWC 2003, LNCS 2775, pp. 665–678, 2003.

- Larger address space.
- Autoconfiguration. Hosts can automatically construct link-local addresses on their own and subsequently acquire additional network prefixes from routers [3].
- Mandatory security. IPSec [4] provides authentication, integrity and encryption services to the two points of communication by making sure (by encryption and signature) that nothing can be changed in a packet from the IP layer and above by other entities along the communication route. A variety of security levels are available to meet the needs of different users.
- Mobility: mobile IPv6 [5].
- Automatic device discovery using the Neighbor Discovery Protocol [6] and service discovery using the Service Location Protocol [7].
- Future-proof: applications using IPv6 can completely avoid the problems associated with the use of private IPv4 addressing and network address translations.

In this paper, we first provide an overview of autoconfiguration mechanisms for ad hoc networks, highlighting their features, differences and limitations. In general, the purpose of address autoconfiguration is to assign an address to an interface, which is unique and routable in the network. In ad hoc networks, such a mechanism has to cope with the highly dynamic environment. An approach to IPv6 address autoconfiguration in ad hoc networks is proposed, where the IPv6 Stateless Address Autoconfiguration Protocol and Neighbor Discovery Protocol are applied to the context of ad hoc networks. In particular, a method to support network partitioning and merging is described. A distributed scheme for duplicate address detection (DAD) is also discussed.

2 Address Autoconfiguration Overview

In the following, we provide an overview of address autoconfiguration schemes. The original focus of the first three methods is mainly on fixed Internet, while proposals tailored for autoconfiguration in ad hoc networks are discussed in section 2.4.

2.1 DHCP

The Dynamic Host Configuration Protocol (DHCP) [8][9] has been deployed widely to alleviate administrative requirements for the installation and initial configuration of network devices. Generally speaking, DHCP is used by clients to obtain necessary information like their IP addresses, DNS (Domain Name System) server addresses, domain names, subnet prefixes, and default routers.

DHCP is specified in a general way to allow a very flexible relationship between DHCP servers and DHCP clients. DHCP clients and servers interact through a series of client-initiated request-response transactions. When a client starts up and needs to get an IP address, it first broadcasts a request on the network to which it is attached. If there is a DHCP server on the same network, the server replies to the client using the client's hardware address, which is included in the client's initial message. If, on the other hand, there is only a DHCP relay agent on the network, the relay then proceeds to rebroadcast the request to other networks, or send the client's request to DHCP servers the relay has been configured to contact. After successful exchange of messages between the server and client, the server commits the allocation of the IP address and other information to the client, finally acknowledges that fact to the client, and the whole process is finished. DHCP messages are all transported via UDP (User Datagram Protocol).

2.2 Zero Configuration (Zeroconf) Networking

The IETF Zeroconf working group's goal is to enable direct communications between two or more computing devices via IP, and their focus is mainly on wired networks.

As pointed out in [10], the typical zero configuration networking protocols imply changes to only the lower layers of IP-enabled devices, and hence are transparent to end users. It has been envisaged that four functions will benefit from zero configuration protocols: name-to-address translation at the application level, IP interface configuration at the network level, service discovery at the application layer and multicast address allocation at the network layer.

Address autoconfiguration requirements include allowing a host to configure its interfaces with unique addresses, determine which subnet mask to use, detect duplicate address assignment, and cope with collisions.

Name-to-address translation requirements include obtaining the IP address associated with a name and determining the name associated with an IP address.

2.3 IPv6 Neighbor Discovery and Stateless Address Autoconfiguration

The protocol proposed in this paper is based on the IPv6 Neighbor Discovery and Stateless Address Autoconfiguration Protocols, which will be discussed in detail below. IPv6 supports plug-and-play, whether the connection is to an isolated stand-alone network or to a large corporate network. There are two flavors of address autoconfiguration: stateless and stateful. In stateless autoconfiguration, a node forms addresses by determining the subnet prefixes on the links to which it attaches and then forming an address on that subnet. In stateful autoconfiguration, on the other hand, a node can use DHCPv6 [11] to obtain addresses and other configuration information.

IPv6 supports multiple address scopes. Global-scope addresses are globally unique and can be used anywhere in the Internet. Link-local addresses are unique only on a specific link, such as a LAN. Site-local addresses are analogous to IPv4's private addresses. They can be used only within a site, and routers do not forward packets containing site-local addresses beyond the site.

The IPv6 Stateless Address Autoconfiguration (SAA) protocol [3] provides a useful way to assign IP addresses to nodes in a network with no configuration servers. It is based on the Neighbor Discovery Protocol (NDP) [6] which is specified for links that support a native form of multicast or broadcast. NDP extends and improves on IPv4's ARP (Address Resolution Protocol). NDP defines two main pairs of messages:

- Neighbor Solicitation (NS) and Neighbor Advertisement (NA) messages are used to determine the link-layer addresses of neighbors, as well as to verify that a neighbor is reachable.
- Router Solicitation (RS) and Router Advertisement (RA) messages are used to locate and obtain information from routers.

When a node doesn't know the link-layer address corresponding to the IP address of a neighbor, it resolves the address by sending out a multicast NS message. The neighbor with the requested target IP address responds with an NA containing its link-layer address.

The IPv6 SAA process begins with the construction of a link-local address that is based on a unique interface identifier and a well-known link-local prefix (FE80::/64). IEEE defines a 64-bit Extended Universal Identifier (EUI-64), which is to be converted to the interface identifier. It is derived from the MAC address of an IEEE 802 interface.

The DAD process is needed to ensure that the newly formed address (tentative address) is not already in use by another node on the attached link. A node issues an NS message containing the tentative address as the target address. If the address is already in use by another node, this node responds with an NA message carrying the all-nodes multicast address as the destination IP address. An address conflict is recognized, if the sender receives an NA message in reply to the NS message or if an NS message with the same solicitation target address is received, indicating that another node with the same tentative address is currently performing DAD. Following DAD, if a node ascertains that its tentative link-local address is unique, it assigns it to the interface and the node hence has IP-level connectivity with neighboring nodes.

Once the address is determined to be unique, the node sends out solicitations (RS messages) to locate routers and obtain additional configuration. If RAs containing a subnet ID are received, hosts construct a site-local address using the link-local address, a well-known site-local prefix and the announced subnet ID.

2.4 Autoconfiguration in Ad Hoc Networks

In general, autoconfiguration mechanisms in ad hoc networks can be classified into three categories: those based on IPv6 stateless address autoconfiguration (e.g. the schemes in [12][13]), those based on specific distributed system algorithms (e.g. the scheme in [14]), and those based on DHCP (e.g. the scheme in [15]).

Perkins et al.'s Proposal. A simple solution for address autoconfiguration in ad hoc networks has been proposed by Perkins et al. in [12]. Addresses are randomly chosen on network 169.254/16 in case of IPv4, or on prefix `MANET_PREFIX` in case of IPv6. A Manet node performing autoconfiguration chooses two addresses: a temporary address and the actual address to use. The former is used only once in the uniqueness check to minimize the possibility for it to be non-unique. The uniqueness check is based on sending an Address Request (`AREQ`) and expecting an Address Reply (`AREP`) back in case the address is not unique. If no `AREP` is received, the uniqueness check is passed. For IPv4, the Address Request/Reply messages are ICMP (Internet Control Message Protocol) packets. For IPv6, the `AREQ` is a modified Neighbor Solicitation and the `AREP` is a modified Neighbor Advertisement, as specified in the Neighbor Discovery Protocol [6]. The autoconfiguration mechanism is designed to be independent of the routing protocol.

Duplicate address detection is performed only once by each node. Therefore this approach does not guarantee address uniqueness in partitioned networks that merge later on. If a network is disconnected, the DAD process has to be performed again when the network partition heals. The draft [12] does not specify any method for detecting when the network partition heals, nor any procedure by which such detection would cause new attempts at DAD.

Weniger's Proposal. In [16][13], Weniger describes how the IPv6 Stateless Address Autoconfiguration [3] and Neighbor Discovery Protocols [6] can be applied to hierarchical mobile ad hoc networks. A hierarchical address space is built up to limit the protocol overhead needed for DAD and to enable route aggregation for hierarchical routing protocols.

In their proposal, a node first generates a link-local address as described in [3]. Subsequently, DAD is performed. The node broadcasts a modified NS message extended by the so-called Manet option. In order to distinguish NS messages of different senders, which potentially have the same IP address, a random source ID is introduced. This ID is not changed if the message is forwarded only. The NS message will be flooded within a limited area, the so-called *scope*. A node which has the same address replies with an NA message. Then, the sender of the NS message chooses a new address and repeats the process. This guarantees the uniqueness of the addresses within each node's scope.

In the hierarchical structure, some nodes known as leader nodes are elected which are responsible for part of the address configuration of other nodes. The Manet option contains a weight that implies how well a node qualifies to be a leader node. This should include the number of neighbors, the degree of association with neighboring nodes and the remaining battery power of the node. The node with the highest weight within a scope becomes the leader node. This node sends an RA message containing a randomly chosen subnet ID. All nodes within the scope of the leader node construct a site-local address based on the received subnet ID. In order to guarantee the uniqueness of subnet IDs, duplicate subnet ID detection (DSD) is performed between all leader nodes. As a result, the site-local addresses are guaranteed to be unique within the entire ad hoc network.

Nesargi's Proposal. A distributed, dynamic host configuration protocol for nodes in a Manet is presented in [14]. Specifically, the problem of assigning unique IP addresses to Manet nodes in the absence of a DHCP server is addressed using the Ricart-Agrawala mutual exclusion algorithm. The proposed solution can tolerate message losses, node crashes, network partitioning and mergers.

A new node (requester) entering the network chooses a reachable Manet node as the initiator which performs address allocation on its behalf. All other nodes know a route to the initiator and can forward their responses to it. Ultimately, the initiator conveys the result of the address allocation operations to the requester. Even if the requester moves, except for the initiator none of the Manet nodes has to track the requester. Thus, the initiator acts a proxy for the requester until an IP address is assigned and packets can be routed to the requester. Some of the salient features of this protocol are [14]: use of a two-phase address allocation mechanism, return of released IP addresses to the pool of available addresses, soft state maintenance, concurrent IP address allocation for multiple requesters, and prioritization among concurrent initiations to avoid deadlocks.

McAuley et al.'s Proposal. Another approach [15], called the Dynamic Configuration Distribution Protocol (DCDP), tries to extend DHCP to a stateless autoconfiguration protocol for wired and wireless networks. DCDP evolved from the Dynamic Address Allocation Protocol (DAAP) [17], which was a mechanism to automate the distribution of IP address pools to a hierarchy of DHCP servers. DCDP also provides autoconfiguration of additional IP-related services, such as the location of DNS servers.

DCDP uses a transactional model whereby nodes are either requesters of or responders to individual configuration requests. A requester asks for configuration information from a DCDP entity. The DCDP responder subleases part of the available address pool and gives other configuration information to the requesting node. To distribute the available pool to another DCDP requester, DCDP uses a very simple binary splitting approach: it splits the currently available pool into two equal halves. By recursively splitting the address pool down the distribution hierarchy, DCDP can automatically distribute address pools to each link. This simple partitioning rule simplifies routing and significantly reduces the length of DCDP packets. However, it may lead to scalability problems as a result of many unassigned addresses in the already scarce IPv4 private network address space.

2.5 Discussion

DHCP requires the presence of a centralized DHCP server which maintains the configuration information of all nodes in the network. Obviously, this would be impractical in a Manet. Zeroconf is designed to assign link-local unique IP addresses to network nodes that are reachable through link-level broadcast, which again may not always be the case in ad hoc networks.

The proposal of Perkins et al. [12] does not support dynamic network partitioning and merging. When a node originates an `AREQ`, it sets a timer for `ADDRESS_DISCOVERY` milliseconds. If no `AREP` is returned for the selected address within a timeout period, the node retries the `AREQ` up to `AREQ_RETRIES` times. If, after all retries, no `AREP` is still received, the node assumes that there is no other node with the same address. This timeout mechanism could result in considerable delay when the network size is large. However, in address autoconfiguration there always exists the tradeoff between latency and reliability. In Nesargi's protocol [14], every node has to maintain some data structures keeping track of the address allocation status in the network, which incurs extra overhead. Furthermore, special support (e.g. various timeout schemes) is needed to tackle the relative movement between the requester and initiator as well as crash failure of the initiator. Due to its hierarchical nature, the scheme proposed by Weniger [13] seems to be more suitable for large scale ad hoc networks. Unfortunately, election of leader nodes is not a trivial problem in large networks, given a number of factors that need to be considered. Another problem associated with their hierarchical approach is that the address changes if a node changes its subnet. This situation can lead to interruption of TCP sessions [16].

There are security issues (e.g. denial of service attacks) with nearly all of the above approaches. IP authentication can be used to improve security of autoconfiguration protocols.

3 Stateless Address Autoconfiguration in Ad Hoc Networks

Considering a small or medium scale stand-alone ad hoc network (e.g., a home network), we believe a hierarchical approach may not be necessary, where the selection of a central node, such as a leader node [13][18], or an initiator [14], adds extra complexity to the protocol. On the other hand, the autoconfiguration scheme proposed in [12] is simpler, but does not specify any method for detecting network partitioning and merging, hence duplicate address allocation is possible in such circumstances. Here we propose an address autoconfiguration method that takes this important factor into consideration. Our scheme can thus be viewed as an enhancement to the protocol in [12].

Since an ad hoc network is essentially a peer-to-peer system, address configuration should be performed in a distributed fashion, i.e., without any central server. Our scheme is based on the IPv6 Stateless Address Autoconfiguration Protocol [3] and Neighbor Discovery Protocol [6]. As in [12], IPv6 Address Request and Reply messages are based on modified Neighbor Solicitation and Neighbor Advertisement messages respectively. Since an ad hoc network is a multi-hop environment, it should be considered as a site rather than a link. Essentially, the address scope of these messages has been changed from link-local

```
 0                   1                   2                   3
 0 1 2 3 4 5 6 7 8 9 0 1 2 3 4 5 6 7 8 9 0 1 2 3 4 5 6 7 8 9 0 1
+-+-+-+-+-+-+-+-+-+-+-+-+-+-+-+-+-+-+-+-+-+-+-+-+-+-+-+-+-+-+-+-+
|Version| Prio. |                   Flow Label                  |
+-+-+-+-+-+-+-+-+-+-+-+-+-+-+-+-+-+-+-+-+-+-+-+-+-+-+-+-+-+-+-+-+
|         Payload Length        |  Next Header  |Hop Limit: N.D.|
+-+-+-+-+-+-+-+-+-+-+-+-+-+-+-+-+-+-+-+-+-+-+-+-+-+-+-+-+-+-+-+-+
|                                                               |
+                                                               +
|                                                               |
+                        Source Address:                        +
|                temporary site-local address                   |
+                                                               +
|                                                               |
+-+-+-+-+-+-+-+-+-+-+-+-+-+-+-+-+-+-+-+-+-+-+-+-+-+-+-+-+-+-+-+-+
|                                                               |
+                                                               +
|                                                               |
+                     Destination Address:                      +
|                all-nodes multicast address                    |
+                                                               +
|                                                               |
+-+-+-+-+-+-+-+-+-+-+-+-+-+-+-+-+-+-+-+-+-+-+-+-+-+-+-+-+-+-+-+-+
|     Type      |     Code      |          Checksum             |
+-+-+-+-+-+-+-+-+-+-+-+-+-+-+-+-+-+-+-+-+-+-+-+-+-+-+-+-+-+-+-+-+
|M|N|                       Reserved                            |
+-+-+-+-+-+-+-+-+-+-+-+-+-+-+-+-+-+-+-+-+-+-+-+-+-+-+-+-+-+-+-+-+
|                                                               |
+                                                               +
|                                                               |
+                     Requested IPv6 Address:                   +
|                       site-local address                      |
+                                                               +
|                                                               |
+-+-+-+-+-+-+-+-+-+-+-+-+-+-+-+-+-+-+-+-+-+-+-+-+-+-+-+-+-+-+-+-+
```

Fig. 1. The modified NS message (`AREQ`)

to site-local. For example, in an `AREQ` message (as shown in Figure 1), the hop limit field in the IP header is changed from 255 to a parameter related to the diameter of the ad hoc network to enable multi-hop connectivity. So, unlike what is specified in [6], NDP messages with a hop limit field other than 255 must not be discarded. The destination address is the all-nodes multicast address instead of the solicited-node multicast address. The source address is a site-local, randomly chosen temporary address from a site-local prefix `MANET_INITIAL_PREFIX` (FEC0:0:0:FFFF::/96) [12]. It is used only once in the address uniqueness check messages [1]. The "M" flag indicates that the packet should be sent over a multi-hop network [12]. The "N" (`NetMerge`) flag is used for network merging DAD as discussed later in this paper. In an `AREP` message, the destination address is the temporary, site-local address chosen by the sender of the `AREQ`. An `AREP` is sent back to the originator of the `AREQ` via unicast if an address conflict is detected.

The requested IPv6 address is a site-local address that consists of four fields: a 10-bit site-local format prefix (FEC0::/10), a 38-bit all zeros field, a 16-bit subnet ID and a 64-bit interface ID [19]. The interface ID is generated from

[1] Alternatively, hosts can communicate with the unconfigured node using its link-layer MAC address. In this case, the `AREQ` source address is the unspecified address.

the node's link-layer address, e.g., EUI-64 as described in [3]. The subnet ID is selected randomly from a permissible range: for example, any 16-bit subnet ID other than FFFF so that it will not overlap with MANET_INITIAL_PREFIX. This results in a flat addressing scheme. In [12], the requested IPv6 address is chosen by selecting at random a host number from a site-local prefix MANET_PREFIX (FEC0:0:0:FFFF::/64). Although the address space is quite large, the possibility of address conflict increases with the number of devices in the network. Since MAC addresses are guaranteed to be unique (to some extent), our approach minimizes the possibility of duplicate address assignment.

As an example, consider a device with a unique EUI-64 value of 1:21FF:FE63:7135. Then the interface ID is formed as 201:21FF:FE63:7135. Assuming the randomly generated subnet ID is FFFE, the site-local IPv6 address is FEC0::FFFE:201:21FF:
FE63:7135.

3.1 Network Initialization

When the very first node joins (forms) the network, it follows the procedure detailed in [12][3] by obtaining a non-link-local prefix and a host number from which to form an IPv6 address and sending out AREQ messages. It sets a timer for ADDRESS_DISCOVERY milliseconds during which it waits for a response. Since it is the first and only node in the network, it will not receive any AREP. After it tries AREQ_RETRIES times, it assumes that there is no network in the area and it is the only node. It will then configure itself with the address chosen and the network is initialized. This first node also chooses a unique identifier for this network (for the purpose of detection of partitioning and merging) and includes it in hello messages (or "beacons") it sends out periodically to (future) neighbors. The unique identifier could be the first node's MAC address.

3.2 Joining Nodes

When a node joins the network and requests an address, it chooses an IPv6 address as described earlier and then follows the DAD steps in [12] using AREQ/AREP messages. In case DAD fails, another tentative address may be chosen randomly or manual configuration is needed. One of the neighboring nodes can send out a hello message embedding the unique identifier of the Manet after the new node has configured itself with an IP address. The new node will store this identifier which is to be used to detect network partitioning and merging as elaborated later.

3.3 Leaving Nodes

One of the requirements for zeroconf networks is the timely reclamation of any resources they allocate [20]. A node could either depart gracefully (informing other nodes before it leaves) or abruptly due to failure. In either case, there will be no response from this node to any further address requests. In the current protocol [12], there is not any explicit mechanism for the network to reclaim the address of the leaving node. Generally, address reuse requires the maintenance of state information as well as the cooperation of the departing node (notifying other nodes). A finite-time lease mechanism similar to that in [8] can be used as well, where a lease timer is set when a new address is allocated to a node. If a node's timer expires and other nodes have not received any updating messages (e.g. hello) from that node, it can be assumed that the node is down or has moved out of the network range.

If a node wishes to rejoin the network some time after it has left and use the same address as before, there could be a risk of duplicate address allocation. This situation can be treated as a special case of network merging.

3.4 Network Partitioning and Merging

A Manet may split into two or more partitions. On the other hand, partitions may be connected together, creating a single merged network. An example could be: in a home network two separate clusters of A/V devices (one in the lounge downstairs and the other in one of the bedrooms upstairs) merge into a big network with the devices upstairs moved to the lounge downstairs. Prior to the merge, each autoconfiguration network has independently allocated addresses. After merging, two hosts in the merged network may end up using the same address, thus potentially creating conflicts. To minimize the risk of duplicate address allocation when two or more networks merge, we make use of hello messages in ad hoc routing protocols [21][22].

In ad hoc networks, network connectivity is determined through the reception of broadcast control messages. Any broadcast control message also serves as a hello message, indicating the presence of a neighbor. Hello messages are exchanged periodically among all neighboring nodes. Each node in an ad hoc network maintains a unique *partition identifier* which, for example, could be the MAC address of the first node in the same partition. Here we assume that MAC addresses of interfaces are unique. We propose to include this partition identifier in hello messages to detect network merging. When a node receives a hello message with an identifier equal to its own as well as a hello message which has a different identifier, it will detect that two partitions are merging. This node then triggers the DAD process. It follows the normal DAD procedure as described before, with the exception that the `AREQ` message includes a `NetMerge` flag indicating that this message is used for *network merging* DAD. Therefore other nodes in the same partition (which may be further away from the partition boundary) also learn of the merging of two networks. Upon receiving this `AREQ`, these nodes will start the DAD process for their own addresses as well. However, a node launches DAD for this purpose only once within a reasonable time period (a timer can be used here), in case that it receives multiple `AREQ` messages with the `NetMerge` flag set. Obviously, it will respond to an `AREQ` by sending out an `AREP` as usual if a requested address is already utilized. In addition, to avoid possible congestion caused by DAD messages, a random jitter is introduced between the commencements of DAD among network nodes. Later on, all the nodes in the merged network agree on a new *common* partition identifier. For instance, the new network ID could be something like a combination of the IDs of the two merged networks, say, $ID_{NEW} = ID_A + ID_B$. The first node detecting the merger of two networks include this new ID in the `AREQ` messages (as an extension) it sends out. Other nodes receiving this `AREQ` (with the `NetMerge` flag set) will change their network IDs to the new ID accordingly. The above process therefore guarantees the uniqueness of addresses even when two independent networks merge.

Figure 2 shows an example of two partitions merging. In Figure 2(a), there are two partitions, A and B. In partition A, node 1 has chosen IP address x, while in partition B, node 8 has chosen IP address x too. In Figure 2(b), the two partitions have merged to form a single network. Node 1 then receives hello messages from nodes of both (old) partitions. These hello messages have different partition identifiers so that node 1 detects network merging. In Figure 2(c) node 1 starts the ad hoc DAD procedure by broadcasting `AREQ` messages and eventually detects the existence of duplicate address allocation (node 8).

Inevitably, duplicate address detection uses flooding throughout the merged network, and this could cause a broadcast storm of DAD messages. A caching and counter-based method can be used to lessen this problem [13]. Each node maintains a cache entry that keeps the source addresses from which `AREQ/AREP` messages have been sent. Redundant messages will not be forwarded. Another issue involved in address conflict concerns site renumbering. If an end address changes, existing TCP connections will break. To enable the graceful renumbering of a site's devices, "preferred" and "deprecated" addresses have been introduced [3].

Nodes can detect network partitioning by periodically exchanging hello messages. Failure to receive any hello messages from some nodes for several time intervals would indicate potential partitioning.

Since network merging increases the potential of network conflicts, it may be prudent to ensure that addresses associated with hosts are not immediately reclaimed for reuse after partitioning. When a new node tries to join a partition, the autoconfiguration protocol therefore should choose an address that was not in use prior to partitioning. This will increase the chances of a particular host being allocated the same address should it leave and rejoin the network.

3.5 Related Work

Compared to [13], our protocol has minimal changes to IPv6 header and does not add any options to NDP messages. In [13], leader node election and DAD have to be performed periodically to cope with network partitioning and merging. Our protocol is more efficient because it can automatically detect the event of network merging and take actions accordingly. Although the protocols in [18][14] also use unique identifiers to detect merging partitions, their solutions do not make use of the built-in autoconfiguration capability of IPv6 and involve the selection of leader nodes and maintenance of a large amount of state information

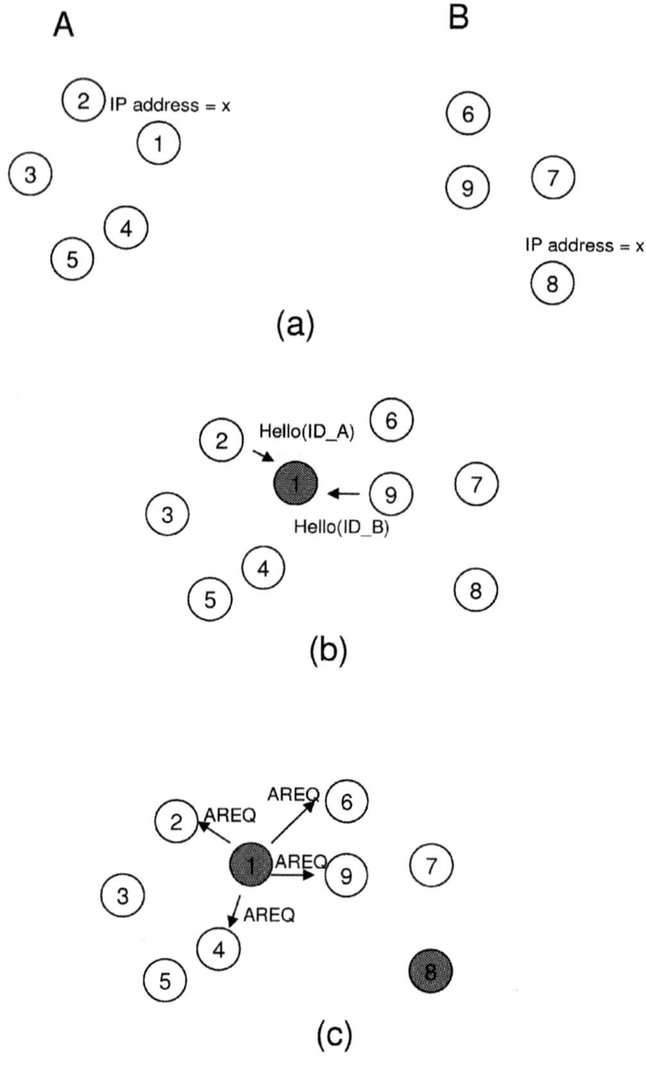

Fig. 2. An example of network merging

at each node. To this end our protocol is much simpler and yet equally effective. The DAD triggering process after network merging in our protocol is significantly different from previous approaches. Furthermore, another difference between our approach and that of [18] is that, in [18] every node sends hello messages to its *logical* neighbors, which may be multiple hops away. In comparison to [12], our approach has the capability of detecting duplicate addresses when networks coalesce. In addition, the address generation in our protocol is different from that in [12]: our method makes use of devices' link-layer addresses, which makes duplicate address allocation even less likely.

One of the limitations of our scheme and other solutions based on hardware addressing (e.g. [3]) is their dependence on unique MAC addresses. Manets can be built on different data link technologies; not all Manet nodes use network interface cards with IEEE-assigned unique MAC addresses. Since MAC addresses may sometimes be duplicated on multiple devices, DAD is required to guarantee the uniqueness of an address.

4 Conclusion

Devices in an ad hoc network need to be assigned a unique address before they can communicate to each other. Ideally, this process should be automatic, fast, and free of error. In this paper, we have examined various autoconfiguration mechanisms for ad hoc networks, highlighting their features, differences and limitations. We have proposed an approach to IPv6 address autoconfiguration in ad hoc networks, where we leverage the IPv6 Stateless Address Autoconfiguration Protocol and Neighbor Discovery Protocol. When a new device joins an ad hoc network, it assigns itself a site-local IPv6 address, which is formed automatically from its link-layer MAC address and a randomly generated subnet ID. Subsequently, duplicate address detection is performed to check the uniqueness of this address using modified Neighbor Discovery Protocol messages. Our proposal enhances previous protocols by taking into consideration the issue of network partitioning and merging. A unique partition identifier is maintained by each node in the network and exchanged via hello messages between neighbors. This identifier is used to detect network merging, after which a DAD procedure is launched to resolve any possible address conflicts. Compared to other existing approaches, the proposed protocol is simple and efficient.

Ideally a prototype should be constructed to experiment with the ideas reported in the paper. However, to obtain any meaningful results for scenarios such as duplicate address detection and network merging/partitioning, it requires a fairly large number of participating nodes in the network. Therefore, our work in the near future aims at the simulation of this scheme (e.g. using ns-2) to test its efficiency/overhead (e.g. number of message exchanges needed, latency) and determine optimal values of operational parameters such as timeout duration and number of retries. As communication links in wireless ad hoc networks are not 100% reliable, the impact of packet losses on address assignment and methods to increase protocol robustness also need investigation. Another key research issue is provisioning of security measures to combat various risks (e.g. denial of service attacks, eavesdropping) present in autoconfiguration ad hoc networks.

Acknowledgements. I would like to thank Y. Zhao for carefully proofreading the manuscript. The constructive comments from the anonymous reviewers are also gratefully acknowledged.

References

1. M. S. Corson, J. Macker, and G. Cirincione. Internet-based mobile ad hoc networking. *IEEE Internet Computing*, pages 63–70, July 1999.
2. A. Williams, B. Haberman, and R. Kermode. IPv6 in the home makes sense. *ISOC Member Briefing Number 7*, 2002.
3. S. Thomson and T. Narten. IPv6 stateless address autoconfiguration. *RFC 2462*, 1998.
4. S. Kent and R. Atkinson. Security architecture for the internet protocol. *RFC 2401*, 1998.
5. D. Johnson and C. Perkins. Mobility support in IPv6. *IETF Internet Draft*, 2000.
6. T. Narten, E. Nordmark, and W. Simpson. Neighbor discovery for IPv6. *RFC 2461*, 1998.
7. E. Guttman. Service location protocol modifications for IPv6. *Work in Progress*, 2000.
8. R. Droms. Dynamic host configuration protocol. *RFC 2131*, 1997.
9. R. Droms. Automated configuration of TCP/IP with DHCP. *IEEE Internet Computing*, pages 45–53, July 1999.
10. E. Guttman. Autoconfiguration for IP networking: enabling local communication. *IEEE Internet Computing*, pages 81–86, May 2001.
11. C. Perkins and J. Bound. DHCP for IPv6. In *IEEE ISCC*, 1998.
12. C. Perkins, T. Malinen, R. Wakikawa, E. Belding-Royer, and Y. Sun. IP address autoconfiguration for ad hoc networks. *IETF Internet Draft*, 2001.
13. K. Weniger and M. Zitterbart. IPv6 autoconfiguration in large scale mobile ad-hoc networks. In *Proceedings of European Wireless*, 2002.
14. S. Nesargi and R. Prakash. MANETconf: configuration of hosts in a mobile ad hoc network. In *IEEE INFOCOM*, 2002.
15. A. Misra, S. Das, A. McAuley, and S. K. Das. Autoconfiguration, registration, and mobility management for pervasive computing. *IEEE Personal Communications*, pages 24–31, August 2001.
16. K. Weniger and M. Zitterbart. IPv6 stateless address autoconfiguration for hierarchical mobile ad hoc networks. *IETF Internet Draft*, 2002.
17. A. J. McAuley and K. Manousakis. Self-configuring networks. In *IEEE Milcom*, 2000.
18. P. Patchipulusu. Dynamic address allocation protocols for mobile ad hoc networks. Technical report, CS Department, Texas A&M University, 2001.
19. R. Hinden and S. Deering. IP version 6 addressing architecture. *RFC 2373*, 1998.
20. A. Williams. Requirements for automatic configuration of IP hosts. *IETF Internet Draft*, 2002.
21. C. Perkins, E. Belding-Royer, and S. Das. The ad hoc on-demand distance vector (AODV) routing protocol. *IETF Internet Draft*, 2002.
22. I. Chakeres and E. Belding-Royer. The utility of hello messages for determining link connectivity. In *IEEE WPMC*, 2002.

Performance Analysis of Fast Handover in Mobile IPv6 Networks*

Sangheon Pack and Yanghee Choi

School of Computer Science&Engineering
Seoul National University, Seoul, Korea
Tel: +82-2-880-1832, Fax: +82-2-872-2045
shpack@mmlab.snu.ac.kr
yhchoi@snu.ac.kr

Abstract. The Fast Handover protocol [1] provides seamless handover in wireless IP networks by minimizing handover latency. To reduce handover latency and to provide faster handover, Fast Handover uses anticipation based on layer 2 (L2) trigger information. Therefore, it incurs higher signaling costs compared with the basic Mobile IP protocol. Furthermore, since the L2 trigger is based on fluctuating wireless channel states, the handover anticipation using the L2 trigger may sometimes be incorrect. In the case of incorrect anticipation, unnecessary buffer space may be used for the purpose of providing a smooth handover. Therefore, it is essential to analyze these overhead costs, in order to evaluate and compare the performance of Fast Handover with that of the basic Mobile IP protocol. In this paper, we analyzed the overhead associated with Fast Handover including the signaling cost and the packet delivery cost. We formulated these costs based on a timing diagram and compared Fast Handover with basic Mobile IPv6 in terms of their packet loss rates and buffer requirements. Also, we studied the impact of the L2 triggering time on the total overhead cost. As a result, we found that the L2 triggering time is an important factor to consider in the optimization of handover performance.

Keywords: Fast Handover, Mobile IP, IP-based wireless/mobile networks, L2 trigger, Performance analysis.

1 Introduction

In IP-based mobile/wireless networks, minimizing handover latency is one of the most important issues. The Mobile IPv6 (MIPv6) specification defines how mobile host (MH) can maintain connectivity to the Internet when its attachment point is changed from one access router (AR) to another. During the handover procedure, there is a time period during which the MH is unable to send or receive

* This work was supported in part by the Brain Korea 21 project of the Ministry of Education and in part by the National Research Laboratory project of the Ministry of Science and Technology, 2003, Korea.

M. Conti et al. (Eds.): PWC 2003, LNCS 2775, pp. 679–691, 2003.

any packets, due to both link switching delay and IP protocol operation considerations. This time period is referred to as *handover latency*. In many instances, handover latency resulting from standard MIPv6 handover procedures could be greater than what is acceptable to support real-time applications. Therefore, an enhanced MIPv6 specification, called *Fast Handover* [1], has been proposed by the Internet Engineering Task Force (IETF), in order to reduce handover latency due to IP protocol operations, for it to be as small as possible in comparison with the inevitable link switching latency. Fast Handover provides seamless handover in wireless IP networks using a layer 2 (L2) trigger. To evaluate the performance of Fast Handover in different network environments, several studies have been done [4] [5] [6].

Koodli et al. presented recent works enabling fast handovers and context transfer between two ARs [4]. Also, Koodli et al. proposed a framework for general context transfer solutions. In terms of performance evaluation, handover latency was studied in two different cases, depending on whether the Fast Binding Acknowledgment message is received on an old link or a new link.

Costa et al. compared the handover latency of different IP mobility management schemes currently being discussed within the IETF. They include basic MIPv6, Fast Handover (FMIPv6), and Hierarchical MIPv6 (HMIPv6). FMIPv6 supports a faster handover procedure compared with basic MIPv6, while HMIPv6 provides an approach allowing for different hierarchies of mobility agents [5]. In this work, they studied handover latency for each protocol and concluded that the best option in order to get the better performance is to implement both HMIPv6 and FMIPv6.

Fikouras et al. studied the performance of MIP handover [6]. According to the results of their study, handover latency is largely dependent on the efficiency of the various movement detection methods such as Lazy Cell Switching, Eager Cell Switching, and Prefix Matching. In their report, two generic formulas were derived for determining the average handover latencies of the Lazy Cell Switching and Eager Cell Switching algorithms.

Blondia et al. investigated the performance of two low latency handover protocols called pre- and post-registration [7]. They proposed a simple analytical model to access the packet loss and the delay characteristics of these protocols. Besides, their scalability properties are investigated by means of an OPNET simulation model implementing IEEE 802.11 as a link layer protocol.

Although the Fast Handover scheme can provide low latency handover, this is only achieved at a certain cost in terms of additional overhead. First, an increased number of signaling messages, between the AR and the MH, are required to provide for the anticipated handover procedure. Also, since Fast Handover is based on the anticipation of future events made using link layer information, there may be some overhead resulting from incorrect predictions. For example, when packet forwarding is supported for smooth handover, the forwarded packets are stored in buffer space. However, if the handover anticipation is wrong, so that the handover doesn't really occur, the allocated buffer space constitutes useless overhead. Therefore, to evaluate the performance of Fast Handover, we need to

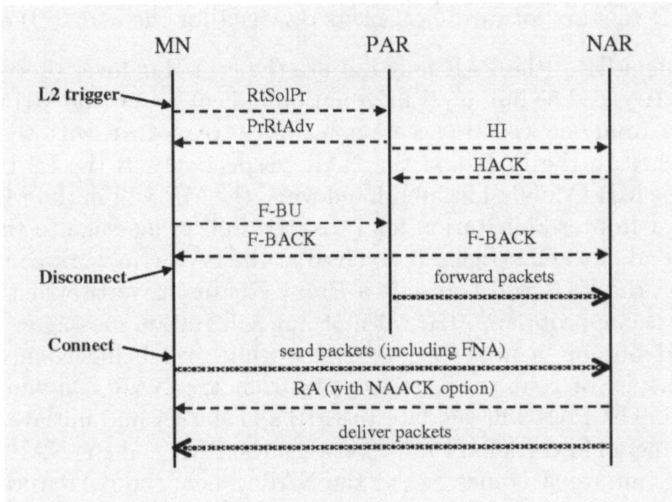

Fig. 1. Fast Mobile IPv6 Handover Protocol

take these different sources of overhead into consideration. However, all of the previous studies [4] [5] [6] [7] on this issue focused on handover latency and didn't consider the effects of overhead, when Fast Handover is employed in wireless IP networks.

In this paper, we analyzed the issue of overhead in Fast Handover. Based on our analysis, we concluded that the total amount of overhead is largely dependent on the question of when the L2 trigger is notified to layer 3 (L3). Therefore, we studied the optimal point in time for the L2 trigger to occur, in order to minimize the total overhead. Also, we compared Fast Handover with basic Mobile IP in terms of the packet loss rate and buffer requirements.

The rest of this article is organized as follows. Section II describes the mechanism of the Fast Handover protocol proposed by the IETF. In Section III, we analyze the overhead of associated with Fast Handover. Section III shows the numerical results of our study based on various parameters. Section IV concludes this paper.

2 Fast Handover

Figure 1 shows the sequence of message flows used in the Fast Handover protocol.

While the MH is connected to its previous access router (PAR) and is about to move to a new access router (NAR), Fast Handover in Mobile IPv6 requires:

- the MH to obtain a new care-of address(CoA) at the NAR while still being connected to the PAR
- the MH to send a Binding Update message to its PAR to update its binding cache with the MH's new CoA

– the PAR to start forwarding packets destined for the MH to the NAR

Either the MH or the PAR may initiate the Fast Handover procedure by using the L2 trigger. The link-layer information indicates that the MH will soon be handed over from one wireless access point (AP) to another, with these two APs being attached to the PAR and the NAR, respectively. If the L2 trigger is received at the MH (Mobile-initiated handover), the MH will initiate L3 handover by sending a Router Solicitation for Proxy (RtSolPr) message to the PAR. On the other hand, if the L2 trigger is received at the PAR (Network-controlled handover), then the PAR will transmit a Proxy Router Advertisement (PrRtAdv) message to the appropriate MH, without any solicitation messages.

The MH obtains a new CoA (NCoA), while still being connected to the PAR, by means of router advertisements from the NAR containing network information. The PAR will validate the MH's new CoA and initiate the process of establishing a bidirectional tunnel between the PAR and the NAR, by sending a Handover Initiate (HI) message to the NAR. Then, the NAR verifies that its new CoA can be used on the NAR's link. Also, in response to the HI message, the NAR sets up a host route for the MH's previous CoA (PCoA) and responds with a Handover Acknowledge (HACK) message.

When the MH receives a PrRtAdv message, it should send a Fast Binding Update (F-BU) message, preferably prior to disconnecting its link. When the PAR receives an FBU message, it must verify that the requested handover is accepted by the NAR as indicated in the HACK message status code. Then, it begins forwarding packets intended for PCoA to the NAR and sends a Fast Binding Acknowledgement (F-BACK) message to the MH.

In basic Mobile IP, there exists a service disruption period during the handover procedure. The period of service disruption usually occurs due to the time required for an MH to inform the Home Agent (HA) of its current location, after it moves into the area of the NAR. During this period, the MH cannot resume or continue communication. On the other hand, Fast Handover [1] involves the use of an L2 trigger which allows L3 handover to be anticipated rather than being performed after the completion of the L2 handover procedure. Thus, Fast Handover ensures that the L3 handover delay is minimized, and also that the period of service disruption, which normally occurs when an MH moves between two ARs, is eliminated. Figures 2 and 3 are cited from [4]. They show timing diagrams corresponding to basic Mobile IPv6 handover and Fast Handover. In basic Mobile IPv6 handover, $t_L + t_I + t_P$ is the service disruption period. On the other hand, in Fast Handover, packet reception latency is equal to IP connectivity latency ($t_P = t_I$). Namely, $t_L + t_I$ is the service disruption time in Fast Handover. However, since packet forwarding from PAR to NAR is supported in Fast Handover, no packets arriving during this period ($t_L + t_I$) are lost.

3 Performance Analysis

In this section, we present analytic cost functions to evaluate Fast Handover and basic MIPv6. We classify the total cost into two types: signaling cost (C_{signal})

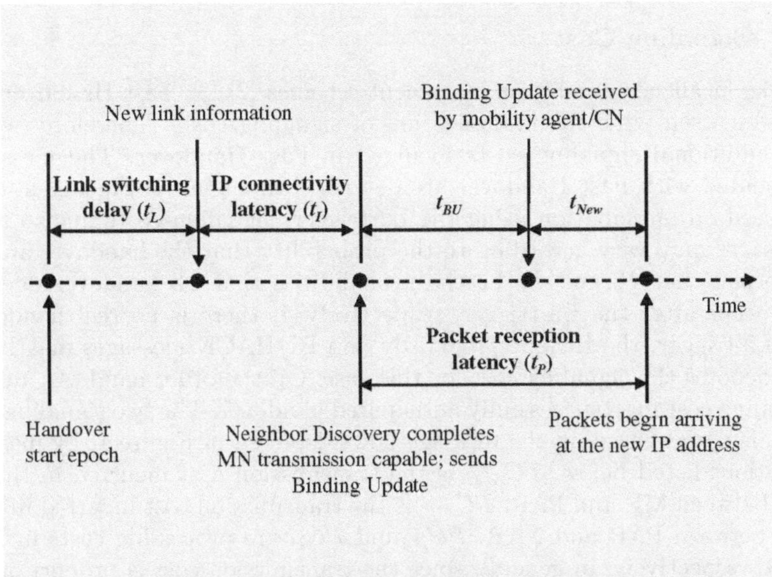

Fig. 2. Timing Diagram in Basic Mobile IPv6

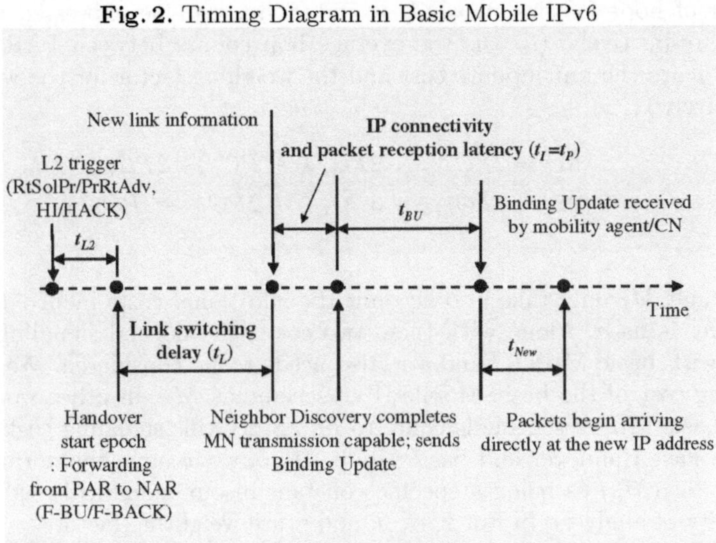

Fig. 3. Timing Diagram in Fast Handover

and packet delivery cost (C_{packet}). The signaling cost is incurred by the signaling messages used for a handover procedure. On the other hand, the packet delivery cost is related to the packet loss rate associated with a handover procedure. Then, the total cost (C_{total}) can be expressed as follows.

$$C_{total} = C_{signal} + C_{packet} \tag{1}$$

3.1 Signaling Cost

Unlike localized mobility management schemes [2] [3], Fast Handover does not concern itself with the minimization of signaling costs. Therefore, we analyze the additional signaling costs incurred in Fast Handover. The message flows associated with Fast Handover are described in Figure 1. Since Fast Handover is based on anticipation using the L2 trigger, signaling costs due to additional messages may vary according to the probability that the handover anticipation is correct. Let P_s and P_f be the probabilities that L3 handover does or does not occur after the L2 trigger, respectively. If there is no real handover after the L2 trigger, the RtSolPr/PrRtAdv and HI/HACK messages may be useless. M_f denotes the signaling cost for this case. On the other hand, M_s denotes the signaling cost for a successfully anticipated handover. The two signaling costs can be calculated based on the message flows depicted in Figure 1, by means of the equations listed below. TC_{MP} is the transmission cost incurred in the wireless link between MH and PAR. TC_{PN} is the transmission cost incurred in the wired link between PAR and NAR. PC_N and PC_P are processing costs in NAR and PAR, respectively. In general, since the transmission cost is proportional to the number of hops in the link, TC_{MP} and TC_{PN} can be expressed as $\rho\theta_U$ and $l_{PN}\theta_U$, respectively. (l_{PN} is the average hop counts between PAR and NAR. θ_U and ρ are the unit update cost and the weighing factor for the wireless link, respectively.)

$$\begin{aligned} M_f &= 2TC_{MP} + 2TC_{PN} + PC_N + 2PC_P \\ M_s &= 4TC_{MP} + 3TC_{PN} + 2PC_N + 3PC_P \end{aligned}$$

M_s and M_f only take into account the additional costs incurred when Fast Handover is used. Along with these two costs, the normal signaling cost associated with basic MIPv6 handover also needs to be considered. We denote the signaling cost of the basic Mobile IPv6 scheme as M_n. In other words, C_{signal} in the basic MIPv6 scheme is equal to M_n. Since this signaling cost is common to both Fast Handover and basic Mobile IPv6, we assume the normal case signaling cost (M_n) as being a specific constant in our numerical analysis for the simplicity of analysis. In Eq. 2, α, β, and γ are weighting factors.

$$C_{signal} = \alpha M_s P_s + \beta M_f P_f + \gamma M_n \tag{2}$$

Definition of Success Probability (P_s). To calculate the signaling cost referred to in the above section, we need to define the success probability (P_s). According to Figure 3, t_{L2} is the time taken from the occurrence of the L2 trigger event to the start of the real L2 switching process. Therefore, a small t_{L2} value means that the L2 trigger occurred at a time which was close to the start of the real link switching process. In this case, the value of P_s, which indicates the probability that L3 handover really occurs after the L2 trigger, will be larger. Therefore, we assume the success probability to be dependent on the timing of

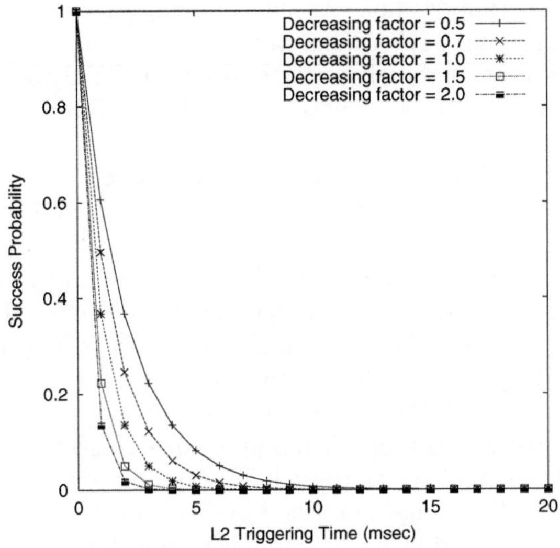

Fig. 4. The determination of P_s as t_{L2}

the L2 trigger. Taking this relationship between t_{L2} and P_s into consideration, we used the following equation for the determination of the success probability. In Eq. 3, τ is a decreasing factor, which is introduced to consider a variety of decreasing patterns.

$$P_s = \frac{1}{e^{\tau t_{L2}}} \tag{3}$$

According to Eq. 3, the success probability is inversely proportional to $e^{\tau t_{L2}}$. In the ideal case, where t_{L2} is 0, the success probability will be 1. On the other hand, if the value of t_{L2} approaches ∞, the success probability will be 0. This means that handover anticipation is done too early without any exact information as to the link condition. Figure 4 shows the relationship between P_s and L_2 as τ is changed.

3.2 Packet Delivery Cost

In terms of the packet delivery cost, we consider the costs associated with both the forwarded packets (*forwarding cost*) and the lost packets (*loss cost*). Eq. 4 shows the packet delivery cost consisted of the forwarding cost and the loss cost, where δ and ϵ are weighting factors.

$$C_{packet} = \delta C_{forwarding} + \epsilon C_{loss} \tag{4}$$

In Fast Handover, packet forwarding from PAR to NAR is supported to avoid packet losses and to support smooth handover. Using the tunnel established after RtSolPr/PrRtAdv and HI/HACK message exchanges, PAR starts the packet forwarding to NAR, after receiving the F-BU message sent by the MH just prior to its disconnecting its link. As shown in Figure 3, there is a delay before the MH begins receiving packets directly at the NCoA in the NAR ($t_L + t_I + t_{BU} + t_{New}$). We assume that if an MH sends an F-BU message to the PAR, then it will definitely start L3 handover to the NAR with no exceptions. In this paper, we consider the forwarding cost as the additional buffer space used by the forwarded packets during this period. Therefore, the forwarding cost ($C_{forwarding}$) is proportional to the packet arrival rate (λ_p) and the forwarding time ($t_L + t_I + t_{BU} + t_{New}$).

In general, since Fast Handover buffers all incoming packets after the L2 trigger, no packets are lost during the handover procedure. However, if an MH moves to another AP's area before establishing a forwarding tunnel, some packets may be lost. Specifically, in Fast Handover, a tunnel is established after the RtSolPr/PrRtAdv and HI/HACK message exchanges. t_{L2} denotes the time period from the L2 trigger to the starting point of link switching and t_R denotes the time required to establish the tunnel. We can obtain the value of t_R using an analytic method similar to that described in [5]. In most cases, since t_{L2} is larger than t_R, packets received during the handover procedure are forwarded to NAR using the already established tunnel. However, in the case of fast moving MHs, t_{L2} may be so small. If t_{L2} is less than t_R, packets arriving at PAR during the ($t_R - t_{L2}$) period may be lost, because the tunnel is not yet established. Therefore, the cost of packet losses (C_{loss}) can be expressed as $\lambda_p max\{(t_R - t_{L2}), 0\}$.

In short, the packet delivery cost (C_{packet}) in Fast Handover can be expressed as Eq. 5.

$$C_{packet} = \delta\lambda_p(t_L + t_I + t_{BU} + t_{New}) + \epsilon\lambda_p max\{(t_R - t_{L2}), 0\} \qquad (5)$$

On the other hand, packet delivery cost in the basic Mobile IP model can be obtained as Eq. 6. In basic Mobile IP, we assume that packet forwarding begins after the Neighbor Discovery procedure. Therefore, the forwarding time is $t_{BU} + t_{New}$. Also, since no handover anticipation is supported in the basic Mobile IP model, all packets arriving during $t_L + t_I$ will be lost.

$$C_{packet} = \delta\lambda_p(t_{BU} + t_{New}) + \epsilon\lambda_p(t_L + t_I) \qquad (6)$$

4 Numerical Results

As mentioned above, Fast Handover is initiated by the L2 trigger. The L2 trigger signals an layer 2 event to layer 3. In the handover procedure, an L2 event can occur due to a change in the signal strength of one or more APs. Therefore, the time when layer 3 is notified of the L2 event is determined by the threshold value of the signal strength used in the protocol implementation. In this paper, we focus on the optimal L2 trigger timing, required to minimize total overhead cost. This value is important in the implementation of Fast Handover, because the total cost is largely dependent on this value.

Table 1. System Parameters for Numerical Analysis

α	β	γ	δ	ϵ	θ_U	ρ	l_{PN}	PC_P	PC_N	λ_p	M_n
0.2	0.8	1.0	0.2	0.8	1	10	2	5	5	1	100

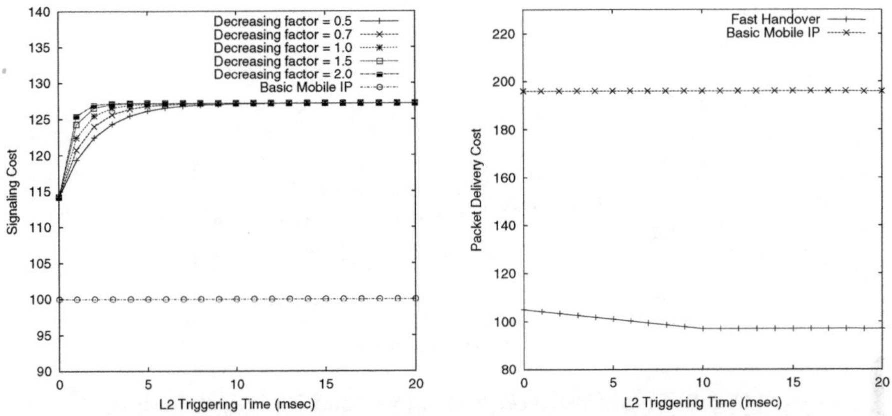

(a) Signaling Cost (b) Packet Deliver Cost

Fig. 5. Cost Comparison: Fast Handover vs. Basic Mobile IP

4.1 Impact of L2 Triggering Timing

In the first analysis, we consider only the influences of t_{L2} and assume that other latency values are determined in advance by the corresponding message procedures. Table 1 shows the parameter values used in the numerical analysis. In this analysis, t_R is set to 10ms. $t_L + t_I$ and $t_{BU} + t_{New}$ are set to 165ms and 320ms using reference values defined in [4].

Figure 5 shows the signaling cost and packet delivery cost as t_{L2} is changed. In terms of the signaling cost, the basic Mobile IP exhibits a constant signaling cost as a function of the L2 triggering time. On the other hand, the signaling cost in Fast Handover increases as the L2 triggering time increases. The rate of increase of the signaling cost is determined by the value of the decreasing factor which is used. However, the signaling cost no loger increases once the L2 triggering time attains a certain critical value. These values are different from the values of decreasing factors which are used.

In contrast to the signaling cost, the packet delivery cost is not dependent on the value of the success probability (P_s). In terms of the packet delivery cost, the basic Mobile IP imposes a higher packet delivery cost than Fast Handover. This is because the handover latency in the basic Mobile IP is larger than that of Fast Handover. Although additional packet loss can occur due to the hasty handover anticipation which sometimes takes place in Fast Handover, the amount of packet loss is negligible compared to that caused by handover latency in basic Mobile IP.

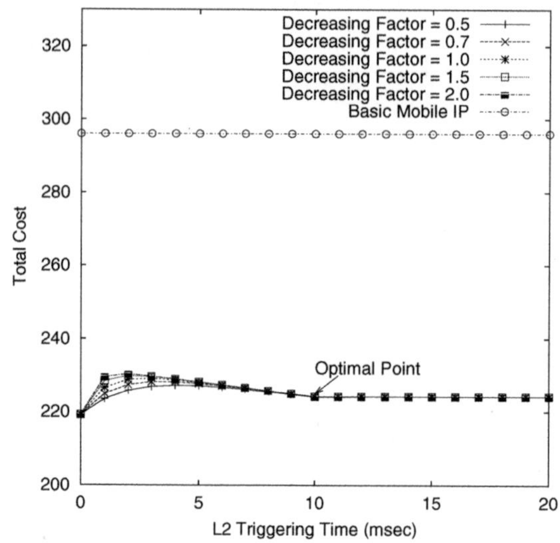

Fig. 6. Total Cost in Fast Handover and Basic Mobile IP

Figure 6 shows the total costs incurred in Fast Handover and basic Mobile IP. In this figure, when the value of t_{L2} is about 10ms, the total cost is minimized. Namely, the total cost is minimized when t_{L2} is equal to t_R. This fact can be proved theoretically using the defined cost functions (refer to Appendix). Therefore, it is necessary to adjust t_{L2} to be close to the value of t_R for exact handover anticipation and optimized handover performance.

4.2 Buffer Space Analysis

Fast Handover anticipates whether or not an MH is likely to move to another cell area using L2 information. To avoid packet losses during the anticipation period, Fast Handover should support packet forwarding during an extended time period compared with basic Mobile IP. In this section, we analyze the buffer requirement of Fast Handover and basic Mobile IP. In this analysis, we assume the average packet size is 200 bytes and consider the buffer space required for one mobile host's handover procedures. Figure 7 shows the required buffer space in Fast Handover and basic Mobile IP. As mentioned above, since Fast Handover begins packet forwarding at an earlier time than basic Mobile IP, it requires more buffer spaces. The buffer space required increases in proportion to the packet arrival rate. Also, the buffer space required increases according to the number of mobile hosts performing handover procedures. In addition, the buffering time may affect the on-time delivery of data packets, especially in real-time applications. For example, if some packets are stored in buffer during a longer time period than the admissible end-to-end delay, they may become useless. Therefore, it is essential to manage the forwarding buffer efficiently to support Fast Handover with minimum overhead and provide more scalable services.

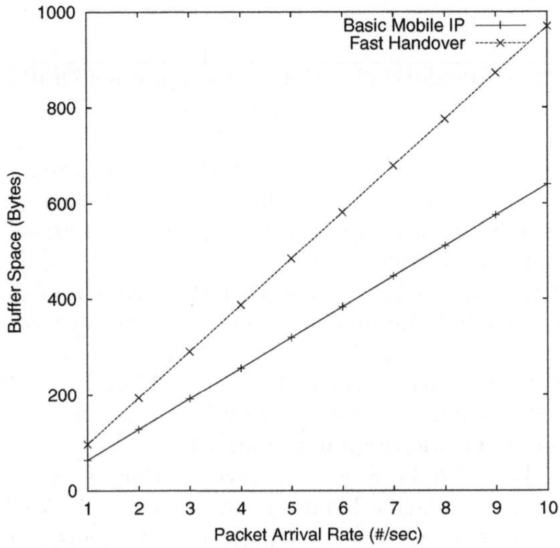

Fig. 7. Buffer Requirement: Basic Mobile IP vs. Fast Handover

5 Conclusion

In IP-based wireless/mobile networks, minimizing handover latency is one of the most important issues. Compared with the existing protocols, Fast Handover has several advantages in terms of its ability to reduce handover latency and support smooth handover. However, this is only achieved at a certain cost in terms of additional overhead.

In contrast to the results published in previous reports, in this paper, we presented several analysis results of the overhead costs incurred in Fast Handover. First, we classified overhead costs into signaling cost and packet delivery cost. Then, we compared and evaluated these two costs both in Fast Handover and in basic Mobile IPv6. In the first analysis, we studied the change in the overhead costs as a function of the L2 triggering time. Based on the results of this study, we found that the cost is minimized when t_{L2} is equal to t_R. For this condition to be met, a more exact notification of the L2 trigger is required. In other words, the support of the lower layers is essential for optimized handover performance. In terms of the buffer space required for packet forwarding, efficient buffer management and optimized forwarding timing are needed to minimize buffer management overhead and to support real-time applications.

The analysis proposed in this paper can be utilized in the determination of the signal threshold value for the L2 trigger, and as a reference for the design of buffer management schemes, which can be used in access routers designed for smooth handover.

References

1. Rajeev Koodli: Fast Handovers for Mobile IPv6. Internet Draft, draft-ietf-mobileip-fast-mipv6-06.txt, Work in Progress, March 2003.
2. E. Gustafsson, A. Jonsson, and C. Perkins: Mobile IP Regional Registration. Internet Draft, draft-ietf-mobileip-reg-tunnel-02, Work in Progress, March 2000.
3. H. Soliman, C. Castelluccia, K. E. Malki, and L. Bellier: Hierarchical Mobile IPv6 mobility management (HMIPv6). Internet Draft, draft-ietf-mobileip-hmipv6-07.txt, Work in Progress, July 2000.
4. Rajeev Koodli and Chales E. Perkins: Fast Handovers and Context Transfers in Mobile Networks. ACM Computer Communication Review, Vol. 31, No. 5, October 2001.
5. X. P. Costa, R. Schmitz, H. Hartenstein, and M. Liebsch: A MIPv6, FMIPv6 and HMIPv6 Handover Latency Study: Ananlytic Approach. Proc. of IST Mobile & Wireless Telecommunications Submit, June 2002.
6. N. A. Fikouras, K. El Malki, S. R. Cvetkovic: Performance Analysis of Mobile IP Handoffs. Proc. of Asia Pacific Microwave Conference (APMC), December 1999.
7. C. Blondia, O. Casals, Ll. Cerda, N. Wijngaert, G. Willems, P. Cleyn: Performance Comparison of Low Latency Mobile IP Schemes. Proc. of Modeling and Optimization in Mobile Ad Hoc and Wireless Networks (WiOpt 2003), March 2003.

Appendix: Proof of the Optimal L2 Triggering Time

To obtain the optimal L2 triggering time minimizing the total cost, we formulated the total cost shown in Eq. 1 as the function of t_{L2} as follows:

$$C_{total}(t_{L2}) = \alpha M_s P_s + \beta M_f P_f + \gamma M_n + \delta\lambda_p(t_L + t_I + t_{BU} + t_{New}) + + \varepsilon\lambda_p max\{(t_R - t_{L2}), 0\} \quad (7)$$

A. $0 < t_{L2} < t_R$

Differentiating $C_{total}(t_{L2})$ with respect to t_{L2}, and set it to zero, we obtain the stationary equation

$$C'_{total}(t_{L2}) = -\tau\alpha M_s e^{-\tau t_{L2}} + \tau\beta M_f e^{-\tau t_{L2}} - \varepsilon\lambda_p = 0 \quad (8)$$

Then, the root (t^*_{L2}) of the equation shown in Eq. 8 is

$$t^*_{L2} = \frac{\log(\tau\beta M_f - \tau\alpha M_s) - \log(\varepsilon\lambda_p)}{\tau} \quad (9)$$

If t_{L2} is smaller than t^*_{L2}, $C'_{total}(t_{L2})$ is larger than zero. On the other hand, if t_{L2} is larger than t^*_{L2}, $C'_{total}(t_{L2})$ is smaller than zero. In other words, the total cost is maximized when t_{L2} is equal to t^*_{L2} in the period of $[0, t_R)$. In addition, the total cost is minimized when t_{L2} approaches to t_R, because $C_{total}(0)$ is larger than $\lim_{t_{L2} \to t_R} C_{total}(t_{L2})$.

B. $t_{L2} > t_R$

Differentiating $C_{total}(t_{L2})$ with respect to t_{L2}, we obtain the following equation.

$$C'_{total}(t_{L2}) = -\tau\alpha M_s e^{-\tau t_{L2}} + \tau\beta M_f e^{-\tau t_{L2}} \tag{10}$$

In the period of $[t_R, \infty)$, $C'_{total}(t_{L2})$ is larger than zero. Namely, $C_{total}(t_{L2})$ is a monotone increasing function in that period. Therefore, the total cost is minimized when when t_{L2} is equal to t_R.

Conclusively, the total cost is minimized when t_{L2} is equal to t_R (note that $C_{total}(t_{L2})$ is a continuous function), which is the required time for establishing a bidirectional tunnel between NAR and PAR.

A Novel Energy Efficient Communication Architecture for Bluetooth Ad Hoc Networks*

Carlos Cordeiro, Sachin Abhyankar, and Dharma P. Agrawal

OBR Center for Distributed and Mobile Computing
Department of ECECS, University of Cincinnati
Cincinnati, OH 45221-0030
{cordeicm,sabhyank,dpa}@ececs.uc.edu

Abstract. Bluetooth is a promising wireless technology aiming at supporting electronic devices to be instantly interconnected into short-range ad hoc networks. The Bluetooth medium access control protocol is based on the Master/Slave concept wherein any communication between slave devices has to go through the Master. While this model provides for simplicity, it incurs a longer delay between any two slave devices due to far from optimal packet forwarding, the use of double the bandwidth, and also additional energy wastage at the Master. Moreover, if more than two devices want to communicate as a group, this can only be achieved by either multiple unicast transmissions or a piconet-wide broadcast, clearly resulting in inefficiency. In this paper, we propose a novel Dynamic Slot Assignment (DSA) scheme whereby the Master device dynamically assigns slots to Slaves so as to allow them to communicate directly with each other without any Master intervention. This proposed communication architecture also provides for Quality of Service (QoS) requests, admission control, and multi-device conversation by which a multicast-like communication is implemented within a piconet. Through extensive simulation, we observe that DSA drastically enhances Bluetooth performance in terms of delay and throughput, while significantly reducing power consumption at the master and also of the overall piconet.

1 Introduction

Bluetooth [1, 22, 24] is a wireless communication technology that provides short-range, semi-autonomous radio network connections in the 2.4GHz ISM band, and can establish ad hoc networks, called piconets. It was also chosen to serve as the baseline of the IEEE 802.15.1 standard for Wireless Personal Area Networks (WPANs) [2], which can support both synchronous traffic such as voice, and asynchronous data communications. In Bluetooth, two or more devices using the same channel form a piconet, where one unit operates as a master and the others (at most seven active at a time) act as slaves. Bluetooth operates on a Master/Slave concept wherein the Master periodically polls the Slave devices and only after receiving such a poll is a Slave

* This work has been supported by the Ohio Board of Regents Doctoral Enhancement Funds and the National Science Foundation under grant CCR-113361.

M. Conti et al. (Eds.): PWC 2003, LNCS 2775, pp. 692–710, 2003.

allowed to transmit. A Master device can directly control up to seven active Slave devices in what is defined as a piconet, with the Master transmitting in even and the Slaves in odd numbered slots. Multiple piconets can be linked together through bridge devices to form what is called a scatternet.

So far, we can envision three waves of Bluetooth-based applications. Initially, Bluetooth was designed to enable a wide range of devices such as laptops, PDAs, mobile phones, and headsets, to form ad hoc networks in a semi-autonomous fashion [1]. The second wave of applications was the development of access points (with functionality similar to IEEE 802.11 access points) enabling hundreds of Bluetooth units to access the wired network in places such as theaters, stadiums, conferences, pavilions, and so on [28, 29]. However, in the third wave of applications, the low cost, effortless and instant connection provided by Bluetooth technology has also become attractive for automatically forming an ad hoc network of a large number of low-power sensor nodes [30, 31]. These sensor network applications are characterized by thousands of nodes embedded in the physical world, and a Bluetooth RF link to enable them to form a network by only bridging these sensor nodes within radio range [22, 32, 33]. These call for solutions to be developed that are applicable to both small scale and large scale Bluetooth networks, and that aim at keeping interference at minimum levels [21, 23] due to the envisioned large number of piconets. As we shall see, DSA satisfies such scalability requirements as opposed to existing solutions.

By virtue of the Master/Slave communication model, the Bluetooth medium access provides for simplicity, low power (as compared to other standards), and low-cost, these being the major forces driving the usefulness of the technology. However, this design choice also brings in major shortcomings, such as the inability for slaves to communicate directly with each other since their packets must be forwarded through the master device [1, 22, 24]. Moreover, in Bluetooth there is no built-in support for the many applications which require group communication [17, 27], while this can only be achieved by either multiple unicast packets or by a piconet-wide broadcast. As a result of these limitations, the packet forwarding among slaves in Bluetooth becomes sub-optimal, bandwidth is wasted by forwarding through the master, end-to-end packet delay increases, and power consumption is significantly increased at the master unit due to its frequent medium access for both transmission and reception. Therefore, the adoption of the Master/Slave paradigm in its present form does not seem to be the most adequate solution.

To overcome these issues and address the shortcomings of the current Bluetooth Master/Slave communication model, we propose a novel Dynamic Slot Assignment (DSA) scheme to be coordinated by the master of a piconet. Based on the piconet traffic patterns, the master device dynamically allocates slots for direct communication between slaves. This way, packets do not need to be forwarded by the master while it periodically re-evaluates slot assignments and changes it accordingly. This novel communication architecture enables for not only direct slave-to-slave communication but also serves as multi-slave communication, hence emulating a group (i.e., multicast-like) communication within the piconet. We have carried out extensive simulations of DSA and observe a drastic enhancement in current Bluetooth performance. In unicast scenarios, piconet throughput increases by up to 300%, delay is reduced to one-third, and overhead is halved, whereas in multicast-like communication throughput boosts up to 500%, delay decreases to approximately one-thirtieth, and overhead is merely one-seventh of existing Bluetooth implementations. Additionally, we have also shown that power consumption at the master is dramatically reduced due to the lesser number of transmissions/receptions, and in certain scenarios the reduction achieved is up to 80%.

The rest of this paper is organized as follows. Section 2 gives an overview of the Bluetooth technology and also provides the motivation for our work, while Section 3 describes our proposed Dynamic Slot Assignment scheme. Next, Sections 4 and 5 present the simulation methodology and results of extensive runs, along with comparisons of our scheme with existing Bluetooth. Related work is then given in Section 6. Finally, the paper is concluded in Section 7.

2 Bluetooth Overview and Motivation

The details of the Bluetooth system, architecture and protocols are defined in [1]. A brief overview is provided here for completeness. Bluetooth is a short-range (up to 10m) wireless link technology aimed at replacing cables that connect phones, laptops, PDAs, and other portable devices. Bluetooth operates in the ISM frequency band starting at 2.402 GHz and ending at 2.483 GHz in the USA and most European countries. A total of 79 RF channels of 1 MHz width are defined, where the raw data rate is 1 Mbit/s. A Time Division Duplex (TDD) technique divides the channel into 625μs slots and, with a 1Mbit/s symbol rate; a slot can carry up to 625 bits. Transmission occurs in packets, where DHx (without forward error correction) and DMx (with forward error correction) packets are usually employed for data traffic, while HVx packets are used for voice applications. Here, *x* represents the number of slots and can be equal to 1, 3 or 5.

The Bluetooth specification [1] defines two different types of links for data and voice applications. The first link type Synchronous Connection Oriented (SCO) is treated as a circuit-switched, point-to-point traffic, whereas the second link type Asynchronous Connectionless Link (ACL) acts as a packet-switched, point-to-point data traffic. Usually, SCO links are used for audio applications with strict quality of service requirements where packets are transmitted at predefined regular intervals, while ACL links are often used in data applications where there is no such strict requirement on end-to-end delay. At most three SCO connections can be supported within a piconet and the polling cycle varies when connections of such a type are present, ranging from 6 for HV3 packets to 3 for HV1 packets. Stating that SCO packets from a given connection have a polling cycle of 6 slots means that one SCO packet has to be sent in every 6 slots, as to achieve the 64 Kbps bandwidth required for voice applications. On the other hand, ACL connections do not have such requirements and the polling cycle can be expanded or shrunk according to the number of slaves in a piconet and their traffic demands. For simplicity, we assume that every SCO connection uses HV3 packets which have a polling cycle of 6 slots, even though our scheme applies to any SCO packet type. Fig. 1 illustrates a full polling cycle highlighting the use of 1, 3 and 5-slot packets.

Fig. 2 depicts a Bluetooth piconet comprised of five devices. Fig. 2(a) illustrates the case where the slave device S_1 communicates with another slave S_3, and all packets have to be forwarded through the master device M. Here, the packet forwarding between S_1 and S_3 is clearly sub-optimal, bandwidth is wasted by forwarding through the master M, end-to-end packet delay increases, and, as shown later, power consumption is significantly increased at the master unit due to its frequent medium access for both transmission and reception. Therefore, we believe that for existing wireless and mobile devices which possess limited battery power and for application efficiency, the Master/Slave paradigm of Bluetooth should not be adopted in its current form.

Moreover, in case slave S_1 wants to send packets (e.g., a business card) to both S_3 and S_4, it has to send two unicast packets as illustrated in Fig. 2(b). On top of that, these multi-slave packets may have to be kept in a device's buffer for as many polling cycles as is the number of recipients, since a device can only address one other unit in a given polling cycle. Note that a piconet-wide broadcast may be a feasible option, however as the master is the only device capable of sending a broadcast, packets will still have to be forwarded by it. Additionally, the trade-off between when to employ multiple unicasts or a broadcast is still unclear since slaves within a piconet are stateless. Hence, given the importance and wide applicability of multicasting, a basic support for group communication should also be provided.

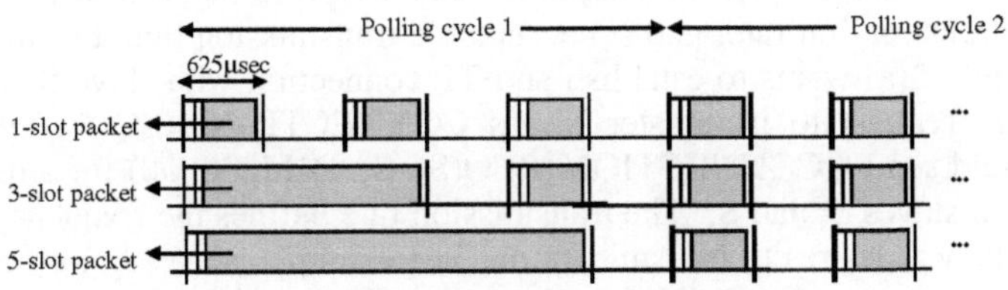

Fig. 1. Packet transmission in Bluetooth

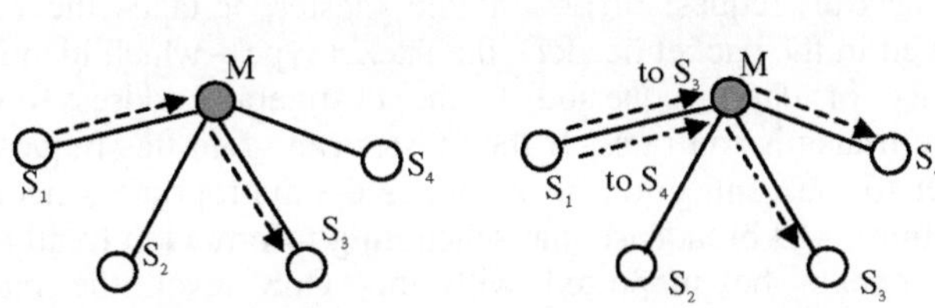

(a) – Forwarding is done through the master

(b) – Communication with more than one device

Fig. 2. Master/Slave communication in Bluetooth

3 The Dynamic Slot Assignment (DSA) Scheme

To address these major shortcomings in the Bluetooth design while keeping the simplicity of the Master/Slave paradigm, we propose a novel Dynamic Slot Assignment (DSA) scheme. The basic idea behind DSA is to appropriately manage the polling cycle conducted by the master of the piconet. As devices initiate or terminate communication with each other within the piconet, we restructure the piconet polling cycle (expanding it with a new connection or shrinking it upon termination), build the transmission schedule for each unit, and then propagate the resulting schedule to the members of the piconet. This way, slaves know exactly in which slot to transmit and/or listen. Therefore, not only is direct communication between slave devices supported, but also a multicast-like communication by having destination slaves listen to the same slot is achieved. Mathematically speaking, if we assume a uniform distribution of connection requests within a piconet, we can conclude that slave-to-slave communication will be present in approximately 75% of all connections, thereby stressing the need for supporting and optimizing such cases.

3.1 Connection Request, Release, and Admission Control

In order for the piconet master to optimally assign and reserve slots for piconet devices, it should know the Quality of Service (QoS) requirements for each connection. In DSA, whenever a slave device wishes to establish a connection with another device (whether another slave or the master itself), it sends a CONNECTION_REQ message to the piconet master, specifying in its payload: i) the destination (if unicast) or destinations (if multicast) address to which the device wants to establish a connection; ii) the baseband packet type to be used in its transmission; iii) desired transmission rate; and iv) acceptable transmission rate. For instance, when slave S_1 in Fig. 2(a) wants to establish an FTP connection with slave S_3, it could send a connection request to its master M, as CONNECTION_REQ(S_3, DH5, 30, 50), while it would send a CONNECTION_REQ(S_3, S_4, DH5, 30, 50) for a multicast FTP connection to slaves S_3 and S_4. The transmission rate defines the frequency (in number of slots) a device desires to transmit. In our last example, slave S_1 notifies the master that it desires to transmit one DH5 packet every 30 slots, but it accepts to transmit in every 50 slots if the desired transmission rate cannot be supported.

When a connection request arrives at the master, it takes the requesting slave address (contained in the packet header), the packet type – which identifies the type of connection (voice or data) requested –, the destination address contained in the payload, and the transmission rates. If the master can grant this request, it allocates a unique identifier to the connection, recalculates the appropriate schedule (detailed in the next subsection), and broadcasts the scheduling information to all active slaves. If the connection cannot be supported with this QoS level, the master returns a CONNECTION_REJ message back to the source. In our earlier example, having requested a DH5 packet does not guarantee the slave that its request will be granted. As explained later, based on the current traffic and schedule, the master decides which packet is best and propagates this information to all slaves with a broadcast. Also, we shall see that the traffic type influences the slot assignment, as SCO packets have to be scheduled periodically due to their time constraints, while no such restriction is present for ACL connections.

An important issue arises in case of bi-directional flows (e.g., TCP traffic) as slots have to be reserved for the reverse traffic. In this case, it is up to the destination device to detect the presence of reverse traffic to the source and similarly make a slot reservation through a CONNECTION_REQ message. However, as this message is sent in response to a bi-directional flow, it is handled differently by the master who immediately allocates the required slots specified within the request (e.g., DH3 for TCP ACKs) to be used in the reverse direction. If the master cannot satisfy this reverse connection request given the absence of enough resources, it returns a CONNECTION_REJ message back to the source and also drops the associated forward connection, along with its reserved slots, which generated this bi-directional flow.

Similar to connection requests, slaves also send connection termination messages to the master device. A slave transmits a CONNECTION_REL message, specifying the connection identifier as originally allocated by the master. Returning to the previous example of Fig. 2(a) where we assume $C_{1,3}$ as being the connection identifier, upon termination of the communication between S_1 and S_3, the slave S_1 would simply send a CONNECTION_REL($C_{1,3}$) to the master device M. The master M would free the slot allocated previously to this connection, calculate the new schedule for the remaining slaves and redistribute the new schedule within the piconet. Note that the master keeps track of all connections within the piconet in order to assign slots to the devices, however it is not generally believed [21] that piconets, which are capable of having at most eight devices [1], will have a large number of connections. Given that, in our design, we assume that at most $CONN_{THRES}$ = 16 connections can be simultaneously present in a piconet, which turns out to be a very reasonable number.

3.2 Slot Scheduling and Assignment

Every time the master of a piconet receives either a connection request or termination, it computes a new schedule for devices. This schedule contains information about which slot(s) belong to which device, and in which slot(s) a given slave needs to listen. With this mechanism, devices are able to directly talk to each other in either a unicast or multicast-like communication.

The schedule has to be transmitted to all the slave devices so that each one of them can determine when to transmit and when to listen. For that, we have defined a new broadcast message called SLOT_SCHED, which has a format depicted in Fig. 3 and is described in detail in section 3.3. For simplicity, we assume that broadcast messages are reliable. Several approaches could be used to achieve reliability such as broadcasting the same message more than once. In our implementation, we assume reliability can be achieved by retransmission of the same broadcast message three times.

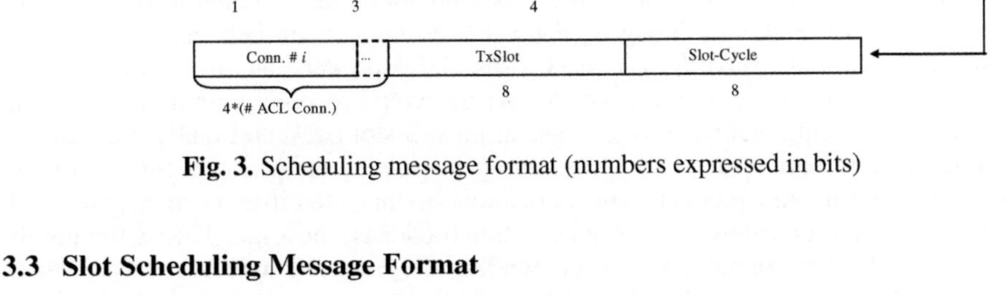

Fig. 3. Scheduling message format (numbers expressed in bits)

3.3 Slot Scheduling Message Format

Fig. 3 illustrates a proposed format of the scheduling message employed in the DSA, with the size of each field in bits as indicated. With this single message format, we can provide both unicast and multicast-like communication within a piconet. As we can see from Fig. 3, the scheduling message consists of three parts: the header, the information about the new connection request which caused the scheduling of this new message, and information about the slot assignment in the new and existing ACL connections. As we shall see, SCO connections are treated differently since they have a fixed scheduling.The header part of the message is composed of three fields. The first field indicates whether the connection request, that caused the transmission of this new scheduling broadcast, was ACL or SCO connection. Alternatively, this field determines whether the information about the new connection contained in the second part of the message is related to an SCO or ACL connection. This makes a difference in the slot scheduling as ACL connections can be scheduled at any time, while SCO connections have to appear at every polling cycle, where a polling cycle is equal to 6 in Bluetooth (see section 2). The second field of the header indicates the number of destinations addressed by this new connection request. In case of a slave-to-slave communication, this field would be equal to one, whereas it would be equal to the group size in a multi-device communication. If this broadcast message is the result of a connection termination, this field contains zero. The third and final field in the header indicates the number of ACL connections contained in the third part of the scheduling message (to be soon described). In other words, this field is used to determine the boundaries of the message.

The second part of the scheduling message provides information about the connection request that has caused the generation of this scheduling broadcast, and absence of this part implies a connection termination. As mentioned earlier, both unicast and multicast-like communications are supported by this scheduling message. With this in mind, the first field indicates the source of data for the connection (i.e., the device which generated the connection request), followed by as many destination devices as specified in the number of destinations field contained in the header. Since each Bluetooth active member address is 3 bits [1], the size of this field is a multiple of 3. Next comes the field which indicates how many slots have been allocated by the master for transmission by the source of the connection. Recall from section 3.1 that together with the connection request message, the requesting device also specifies the type of packet it wishes to use in its transmissions. Depending upon the traffic pattern and the presence or absence of SCO connections, the master may or may not meet the device's request. For example, assume that a device S_1 is currently engaged in a SCO connection with device S_2. In the mean time, another slave S_3 requests to establish a connection with slave S_4 for a file transfer by employing DH5 packets. It is well known that SCO connections are periodically scheduled in every polling cycle due to their strict QoS requirements. Since the Bluetooth slot cycle is equal to 6 and two out of these 6 slots are already being used for a SCO connection between S_1 and S_2, the master cannot satisfy slave S_3's request to use DH5 packets, as only 4 slots out of 6 are available. Therefore, in this case, the master would respond with a value 3 in this field, thus indicating that slave S_3 can use at most 3-slot packets. Finally, the last field of this message part is devoted to assigning a unique connection identifier to each connection within the piconet. The allocation of this identifier is managed by the master and communicated to the slaves in this broadcast message. This is the number that is used by the source slave when sending a connection release message to the master, and is also employed in the third part of the message scheduling.

The third and last part of the scheduling message contains the information about the schedule itself, that is, when each device is supposed to transmit and/or receive.

The first field(s) contains lists of all the connection identifiers currently present in the piconet, and the order in which a connection identifier appears in this field determines the order in which devices associated with this identifier (either as source or destinations) have their slots assigned. To indicate to the slaves the starting slot for transmission (in case of the source of a connection) or reception (in case of the destination(s)), we use the field called *TxSlot* (transmit slot). The values permitted for *TxSlot* are 1, 3, or 5, depending upon the type of packet a connection uses. Lastly, as we have already mentioned, in DSA we employ a scheme where we expand or shrink the slot-cycle according to the number of ACL connections in the piconet. Therefore, conveying the slot cycle information to all slaves is the objective of the last field, called *Slot-Cycle*, as shown in Fig. 3. This field is always a multiple of 6, as this value is the required periodicity of SCO connections. Based on that, a device determines the slot - $slots = (polling\_cycle \times index\_in\_list) + TxSlot$ - in which it is supposed to transmit and/or receive. Here, *polling_cycle* depends upon the SCO packet in use and is equal to 6 in case of HV3 packets, *index_in_list* is the index where the device's associated connection identifier appears in the first field of the third message part, and *TxSlot* is as described earlier.

3.4 Example Scenario

To better illustrate the dynamics of DSA, consider again Fig. 2(a) where we have a piconet composed of 5 devices, one master M and four slaves labeled S_i, for $0 < i < 5$. For simplicity purposes, we do not include transmission rate in our discussion here, but note that QoS admission control is currently incorporated into DSA implementation, and consider unidirectional traffic only while bi-directional traffic is also presently supported by DSA. Assume that no traffic currently exists in the piconet, at which time slave S_1 sends a connection request to the master M in order to establish an ACL connection with slave S_2, specifying DH5 as the desired packet for its transmissions. Upon receipt of the connection request, the master M allocates a connection identifier, say $C_{1,2}$, to the connection between S_1 and S_2 and calculates the scheduling message to be broadcast as follows.

Initially, when there is no traffic, the slot cycle in the piconet is equal to 6 as polling is the only activity and is carried out by the master M. Fig. 4(a) illustrates this scenario, where slots are numbered from 1 to 6. Upon receipt of the aforementioned ACL connection request, the master M increases slot cycle by 6 – making it 12 – and assigns the first slot cycle for direct communication between the requesting device S_1 and destination S_2, while it always keeps the last slot cycle for polling (see Fig. 4(b)). Therefore, we see that the slot cycle is directly proportional to the number of ACL connections existing within a piconet, being always a multiple of 6. During the last slot cycle used by the master for standard polling, priority is given to those devices currently without open connections in order to achieve fairness, while those devices engaged in traffic connections are polled last.

The master also has to indicate the source and destination slaves (S_1 and S_2 in our example) the index within the assigned slot cycle where source (S_1) is supposed to transmit and the destination (S_2) is supposed to listen. This is accomplished by the field *TxSlot* where, in our example, the master M would set *TxSlot* to 1 as to allow slave S_1 to use the entire slot cycle for the transmission of its DH5 packets. The scheduling message is then broadcast to all slaves. If, for instance, another ACL connection request to employ DH3 packets from slave S_3 to S_4 arrives at the master M, it proceeds in a similar manner, by assigning a connection identifier, say $C_{3,4}$, increasing the slot cycle from 12 to 18 to accommodate the new connection, building the new schedule, and broadcasting it throughout the piconet. This new schedule is illustrated in Fig. 4(c).

Assume now that a SCO connection request from device S_1 to S_3 arrives at the master M. Given the periodicity of SCO traffic which has to be present in each and every slot cycle, the master now has to reorganize the slot assignments. In our example, the connection $C_{1,2}$ will no longer be allowed to use DH5 packets as this would prevent SCO packets from being sent during this slot cycle, while the connection $C_{3,4}$ could continue employing the same packet type. To incorporate the SCO traffic within the schedule, the master changes the value of *TxSlot* from 1 to 3 indicating to all ACL connections that they will, from now on, transmit starting at slot 3, while the SCO connection is assigned the first two slots of every slot cycle. As a matter of fact, in DSA, SCO connections always use the first slots of a cycle while ACL connections follow next. Furthermore, note that contrary to ACL connections which have their polling cycle expanded according to the number of existing connections, the slot cycle of SCO connections is always 6, which means that SCO packets have to be transmitted in every slot cycle and hence need not be specified in the scheduling message. This way, we adhere to the requirements of the Bluetooth specification [1]. This new schedule is illustrated in Fig. 4(d).

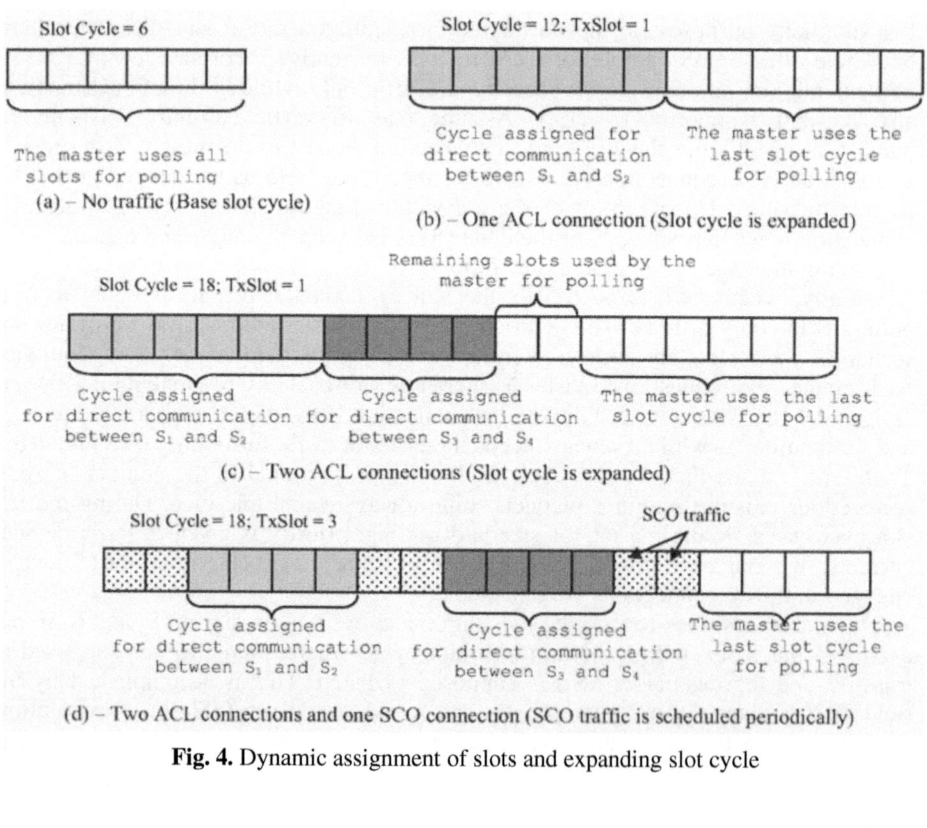

Fig. 4. Dynamic assignment of slots and expanding slot cycle

3.5 Multicast Address Allocation

One very important issue remains to be discussed, namely, how the multicast addressing is performed in DSA given that all active member addresses (AM_ADDR) are allocated with no unused address. In the design of DSA, we considered two options to tackle this issue: we could increase the AM_ADDR field from three to four bits and use one out of these four bit addresses for multicasting, or we could simply temporarily allocate one of the three bit addresses for the purpose of multicasting. The first option would make DSA incompatible with the current Bluetooth specifications. Therefore, we have selected the second option, where we temporarily allocate the broadcast address for multicasting. However, note that by allocating the broadcast address for multicasting does not imply that all devices will receive such a packet since the packet is discarded once the device realizes it is not supposed to listen at that particular slot as specified in the slot scheduling message currently in effect. Hence, only the actual multicast members (specified in the slot scheduling message) will accept the corresponding packet. On the other hand if the packet is actually to be broadcast throughout the piconet, then the given slot carrying the broadcast packet will not be present in the slot scheduling message, resulting in all piconet devices accepting such packet.

Regarding multicast support in DSA, we have chosen a sender-initiated approach, as it is optimal for small groups [18]. It contrasts with receiver-oriented approaches [17, 27] which are best applicable for medium-to-large groups.

4 Simulation Environment and Methodology

To evaluate our proposed DSA scheme, we have implemented it in the Network Simulator (ns-2) [19] and BlueHoc [20], an open-source Bluetooth simulator provided by IBM. Since BlueHoc only provides the basic functionality of Bluetooth, we have made considerable extensions to this simulator in order to conduct our experiments including the support for direct slave-to-slave and multicast-like communication.

The simulation experiments share the same network topology. It uses a single piconet comprised of 8 devices placed randomly within a 7m x 7m region. Radio propagation range is 10m, and nominal channel capacity is set to 1Mbps.

We have thoroughly compared our DSA scheme with the existing Bluetooth under the following metrics:

- Aggregate Throughput – This is the average number of data bytes correctly received by all piconet devices per unit time;
- Delay – This measures the delay per bit in transferring data from a given source to a given destination;
- Overhead – This measures the efficiency of the scheme. It relates the total number of data and control bytes sent to the total number of data bytes received by all devices;
- Power Consumption – With this metric, expressed in Megabytes/Joule (Mbytes/J), we evaluate the power efficiency of DSA as compared to Bluetooth. Bluetooth radios operate at 2.7 V and 30 mA, resulting in a 115 nJ/bit for transmission [1, 25]. Moreover, measurements have shown [26]

that receive:send ratio is usually 1:1.4, while power consumption in idle times is negligible. In our simulations, we have used these relations to compute power consumption for the Bluetooth devices.

Both slave-to-slave and multi-slave communications have been studied, where we vary the number and type of traffic. In order to cover a broad range of scenarios, we have run our simulations with FTP, Telnet, and SCO (voice) traffic, where FTP and Telnet connections use DHx and DMx packets while SCO traffic employs HV3 packets. Each traffic connection is pair-wise distinct nodes, and we consider up to three connections of each type. To achieve reliability in the broadcast of the DSA slot scheduling message and at the same time determine an upper bound on the overhead of DSA, we perform the broadcast of a slot-scheduling message three times in all simulation runs. We believe it to be a reasonable number of broadcasts to compare our DSA with existing Bluetooth, and to serve as a worst-case analysis.

5 Simulation Results

Results for unicast communication and multicast-like communications are given separately.

5.1 Unicast Communication

Here, we study the behavior of our four selected metrics under three different traffic scenarios:

- Scenario A – In this scenario, a total of 3 FTP connections are considered where each connection is initiated consecutively in different points in time;
- Scenario B – This scenario is devoted to analyze voice traffic performance. It employs a total of 2 SCO connections also initiating consecutively at different points in time;
- Scenario C – This scenario considers a mixed case of connections. Total of 3 connections are initiated consecutively where the first one is a SCO connection, the second one a Telnet, and the third one an FTP.

Aggregate Throughput: Figs. 5(a), 5(b), and 5(c) depict the aggregate throughput for DSA and existing Bluetooth for the scenarios under consideration. In general, we can observe that DSA always achieves higher throughput than existing Bluetooth. More specifically, at some points DSA experiences up to 300% improvement. The reason is simple: while in Bluetooth all traffic has to go through the master, path becomes non-optimal and bandwidth is wasted, hence compromising the number of data bytes received per unit time. On the other hand, DSA employs direct slave-to-slave communication, thus avoiding additional delay due to packet transmission, propagation time, and queuing, and hence boosting the number of data packets received per unit time.

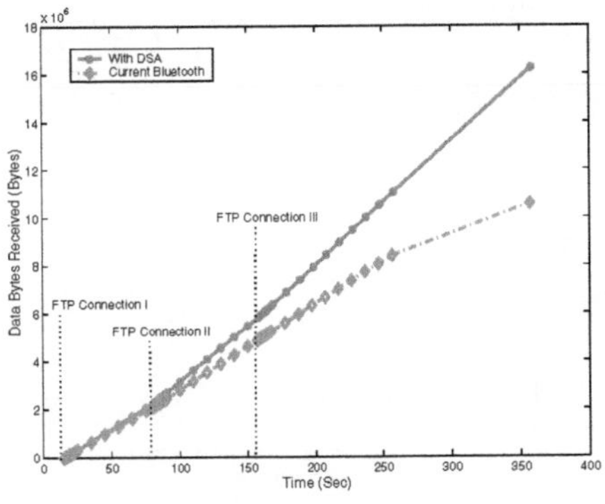

(a) – Scenario A

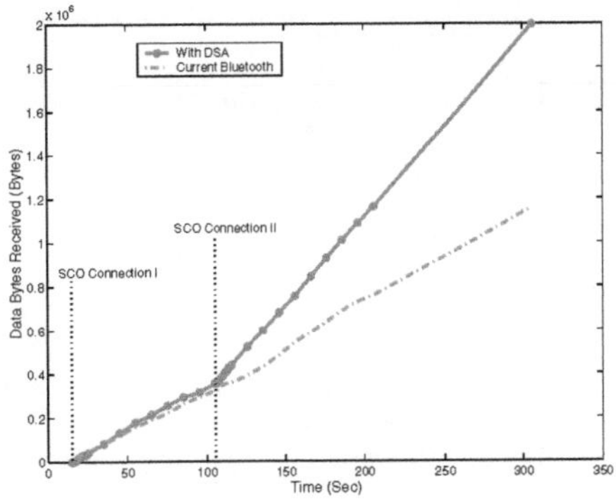

(b) – Scenario B

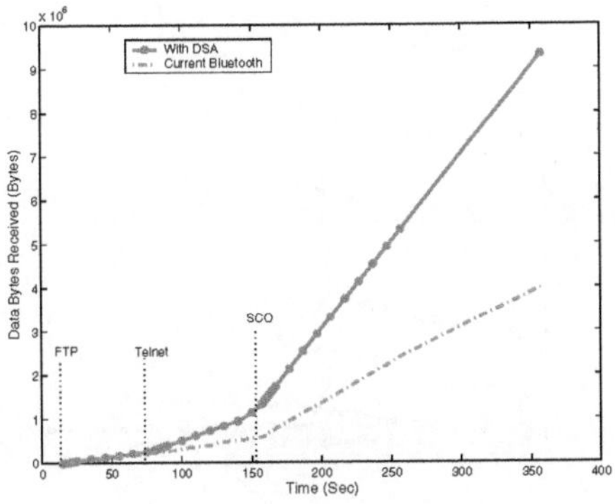

(c) – Scenario C

Fig. 5. Aggregate throughput for the three scenarios considered

In Fig. 5(a), we see that with one FTP connection Bluetooth can still provide a performance comparable with that of DSA as the delay effects are minimized. However, when the number of connections increases to two and three, Bluetooth can no longer sustain the same performance whereas DSA scales approximately linearly.

Delay: The delay experienced in each of the three scenarios are illustrated in Figs. 6(a), 6(b), and 6(c) respectively. Due to the direct communication between slaves, and hence optimal communication path, DSA is shown to provide reduced delay in all scenarios, with delays being as low as one-third of the current delay in Bluetooth. Note that in scenarios A and C where we have traffic connections involving TCP, the delay increases as we increase the number of connections. This is due to the burstiness of TCP traffic which increases the average queue length in every burst.

On the other hand, this is not the case in scenario B where only SCO connections are considered and the traffic pattern follows a constant bit rate (CBR) stream. Here, the delay for DSA is constant due to our direct slave-to-slave communication, while in Bluetooth the delay initially increases rapidly and then tends to stabilize. Note that in all scenarios, the delay experienced by the DSA scheme is dramatically lower than that of Bluetooth.

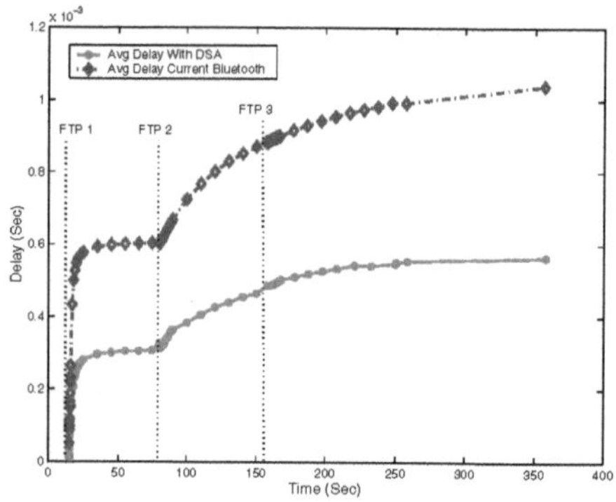

(a) – Scenario A

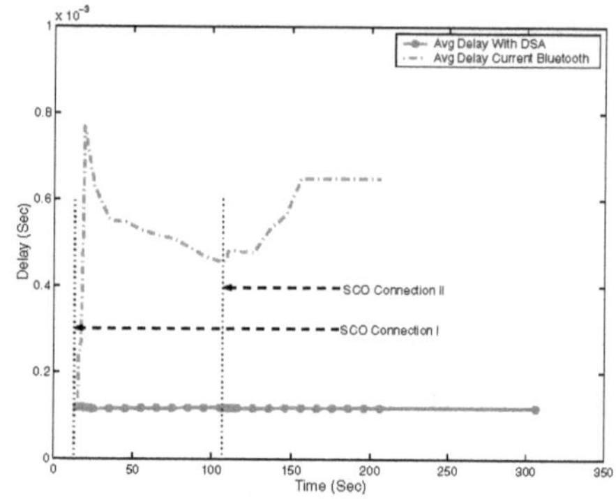

(b) – Scenario B

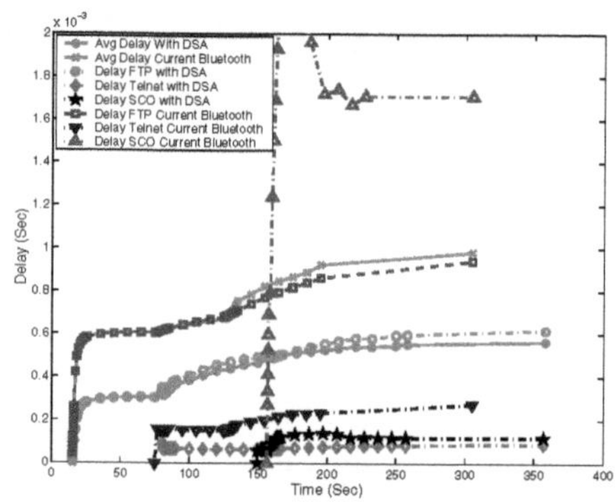

(c) – Scenario C

Fig. 6. Delay for the three scenarios considered

Overhead: Figs. 7(a), 7(b), and 7(c) compare the overhead of Bluetooth and DSA for scenarios A, B, and C respectively. Here, we see that although DSA transmits three broadcasts for each message scheduling, the associated overhead is always approximately half that of Bluetooth for the three scenarios considered. The reason for this is that Bluetooth's overhead is present for every packet since all of them have to be forwarded by and routed through the master hence incurring an additional transmission. In DSA, the overhead is only tied to the broadcast message while all data packets are directly sent to the intended destination.

In Fig. 7(b), we can see that the overhead of Bluetooth for scenario B remains nearly the same as in DSA while only one SCO connection is present in the piconet, whereas it increases rapidly when the second SCO connection arrives while DSA maintains a proportional grade of service. There is an important reason behind that. Current Bluetooth can support at most three simplex SCO connections, that is, having the master as either a source or a destination of SCO traffic. Therefore, when it comes to duplex connections, that is, from slave-to-slave, Bluetooth can support only one SCO connection. With our DSA scheme employing direct slave-to-slave communication, we are able to remove this restriction and enable up to three duplex SCO connections within a piconet.

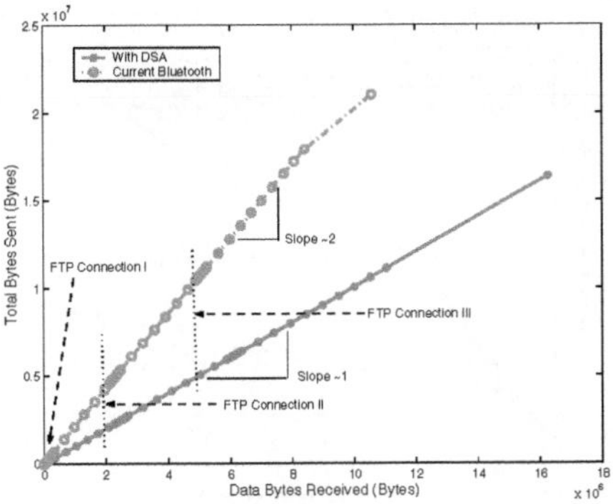

(a) – Scenario A

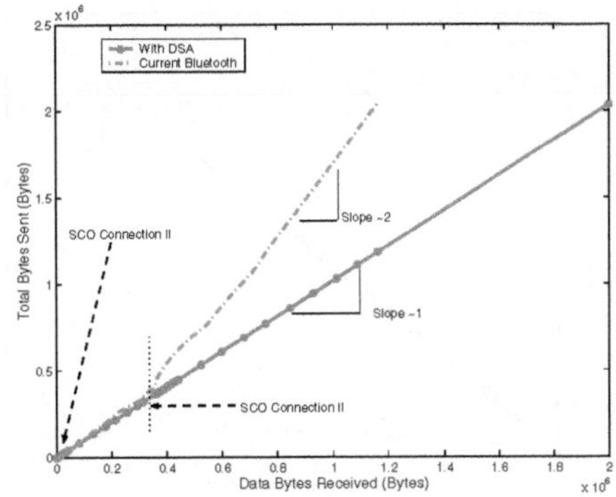

(b) – Scenario B

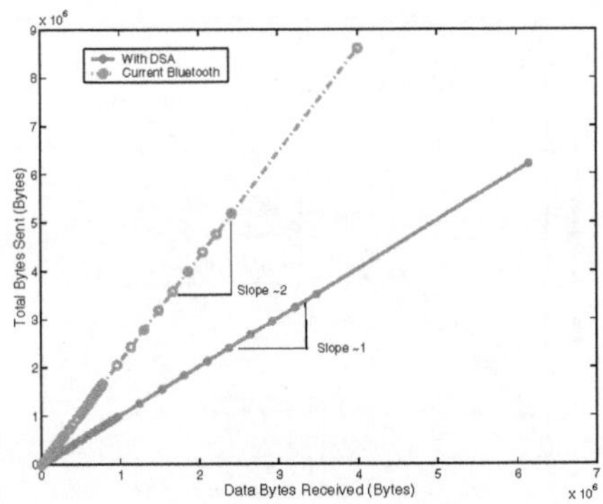

(c) – Scenario C

Fig. 7. Overhead for the three scenarios considered

Power Consumption: Fig. 8 shows the power consumption comparison of DSA and Bluetooth for the three scenarios under evaluation. In all curves, we can clearly notice the dramatic reduction in power consumption at the master device as a result of DSA scheme. As we can see from Fig. 8, as with DSA the master does not get involved in the communication between slaves, this results in a significant increase in the energy efficiency (Mbytes/Joule). Table 1 illustrates the average power consumption improvement of DSA both at the master as well as in all the slaves.

As can be seen from Table 1, the average energy saved at the slaves is mainly due to the reduction in the number of polls by the master device, as these slaves are involved in the data communication. Hence, power consumption at the slaves is much less as compared to the master device where energy conservation is immense as a result of DSA (see Table 1). Finally, it is important to note in Fig. 8 the energy-per-byte improvement ratio of DSA over Bluetooth. In other words, with the adoption of DSA, a significantly larger amount of data can be transmitted with the same energy. A consequence of this fact is that the piconet has its lifetime substantially increased.

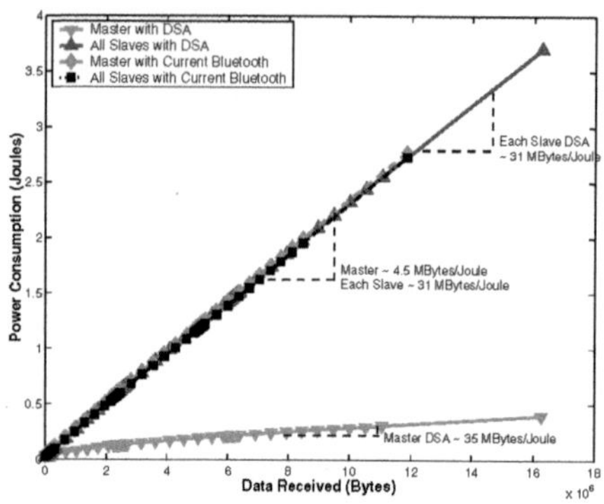

(a) – Scenario A

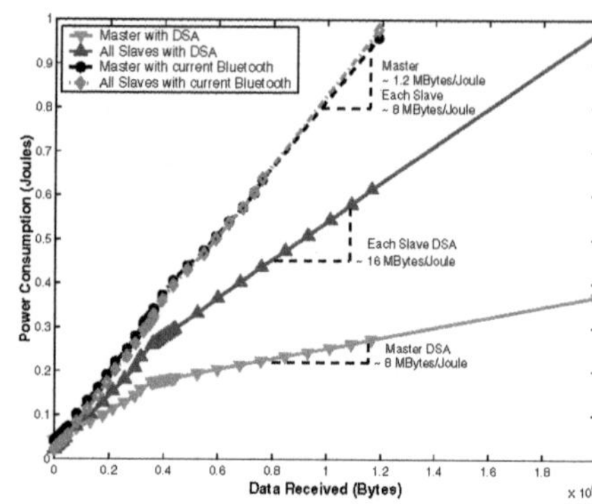

(b) – Scenario B

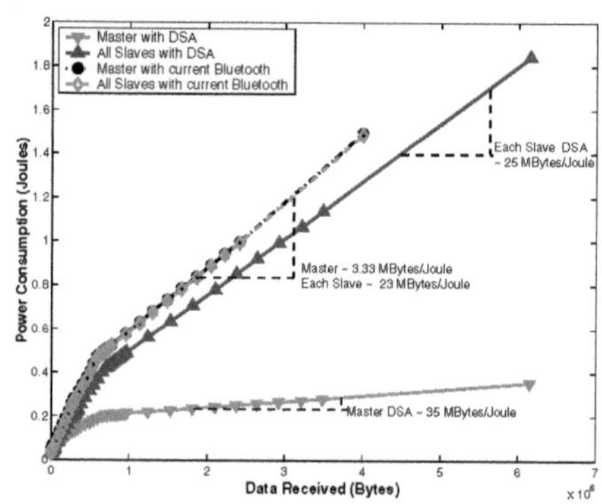

(c) – Scenario C

Fig. 8. Power consumption comparison of DSA and Bluetooth

Table 1. DSA power improvement over Bluetooth

	Master	Slaves
Scenario A	83.06%	2.12%
Scenario B	43.62%	8.95%
Scenario C	54.36%	6.60%

5.2 Multicast Communication

In this section, we study the multicast support in DSA and compare it with the native way of implementing group communication in Bluetooth by multiple unicasts. For this experiment, we consider two multicast sessions wherein each of the two multicast sources has exactly three other devices as group members. Here, the first multicast session is initiated at simulation startup, while the second is initiated 100 seconds later. Exponential traffic sources are considered.

Aggregate Throughput: Fig. 9 depicts the aggregate throughput of both DSA and Bluetooth. As expected, DSA greatly outperforms Bluetooth by employing a single transmission for all group members, while Bluetooth transmits a packet per destination. As a matter of fact, most of the well-known benefits of multicasting [17, 27] over multiple unicasts are embodied in the DSA, making it more efficient than Bluetooth. DSA improves the average throughput by up to 500% (see Fig. 9).

Delay: Delay is another aspect in which the multicasting mechanism built in DSA, brings it to a minimum level. As we can see from Fig. 10, the delay in DSA is practically constant due to its direct one-to-many communication, whereas it is approximately 30 times higher in Bluetooth.

Overhead: As Bluetooth employs multiple unicasts in order to provide a group-like communication mechanism, it is expected to have a much higher overhead than DSA. Fig. 11 confirms this assertion by revealing that DSA experiences around one-seventh the overhead of Bluetooth, despite the control messages broadcasted in DSA.

Power Consumption: Finally, Fig. 12 presents the power consumption comparison of Bluetooth and DSA. Similar to the unicast case, multicasting with DSA also significantly reduces the power consumed at the master device as well as at all slaves. On average, the master experiences a 54.17% reduction in energy consumption, while the slaves have a 33.05% reduction. As a consequence, we can see from Fig. 12 that the energy efficiency (Mbyte/Joule) ratio of DSA overwhelms that of current Bluetooth, hence prolonging the lifetime of the overall network.

6 Related Work

In general, recent research studies have tackled the issue of Bluetooth piconet performance improvement from different angles. From the scheduling perspective, [3, 4, 5, 6] pointed out the drawbacks of existing scheduling techniques and several modifications are suggested to enhance the polling mechanism from the master device. However, the issue of packet forwarding through the master device is not addressed, leading to the problems already mentioned in this paper.

[7, 8, 9, 10, 11, 12] proposes algorithms to generate the scatternet topology with properties such as limiting the maximum roles and degrees of any node, distributing topology generation, limiting the number of piconets and the maximum number of hops between any pair of devices, and so on. However, the major shortcoming of these approaches is the disregard of traffic characteristics when building the topology. That is, even if any two slaves need to communicate frequently, there is no guarantee that they will be within the same piconet in the resulting scatternet.

As we have discussed, another solution would be piconet partitioning (e.g., NPPC), wherein a pair of slaves form a piconet by themselves if frequent communication is taking place between them [14]. However, as we have seen, this approach incurs additional problems such as scatternet scheduling, and, more importantly, this cannot be done indefinitely as interference levels may become unacceptable [16, 21, 23]. Additionally, this solution makes a very unlikely built-in assumption that all connections are pair-wise distinct nodes in order to piconet partitioning to be successful, and also it cannot support multicast communication.

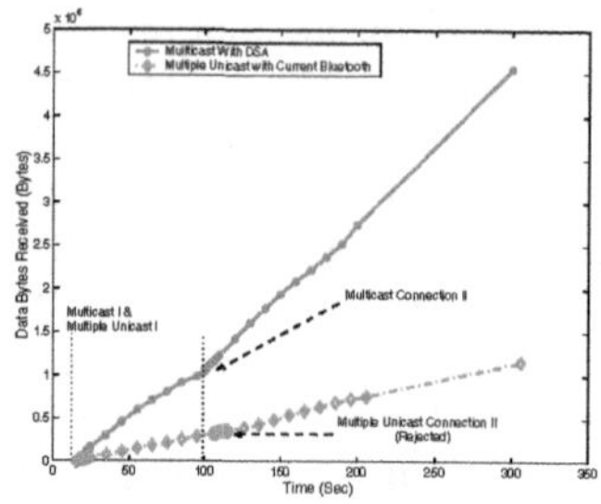

Fig. 9. Aggregate throughput comparison of DSA and Bluetooth

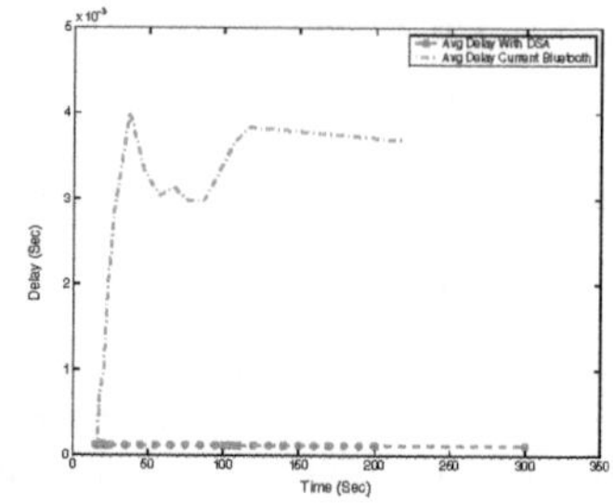

Fig. 10. Delay comparison of DSA and Bluetooth

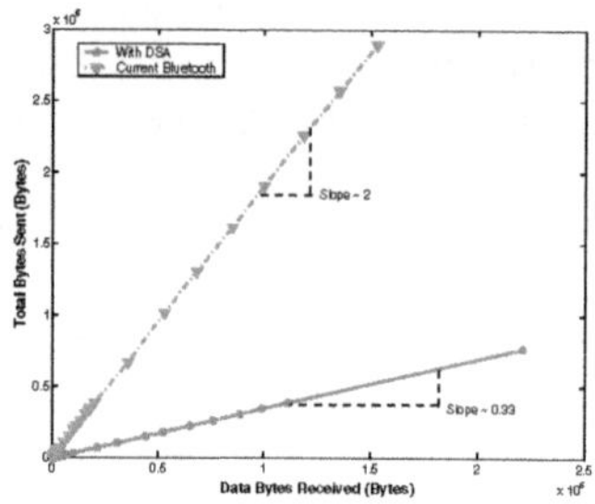

Fig. 11. Overhead comparison of DSA and Bluetooth

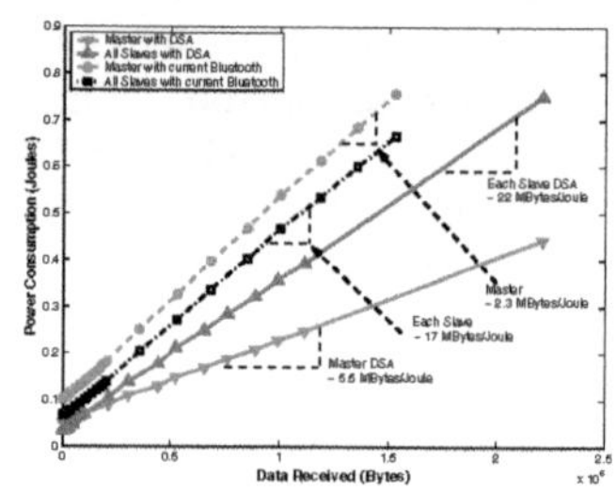

Fig. 12. Power consumption of DSA and Bluetooth

Finally, [15] proposes a time-slot leasing approach where the master device allocates slots to slaves for direct communication. While this scheme has a few points in common with our proposal, there are fundamental differences. Firstly, SCO connections are not evaluated. While this greatly simplifies design, as periodic scheduling of SCO packets has not to be taken into account, it is unrealistic to assume that SCO traffic will not be present [2]. Second, the mechanism in [15] does not provide for multicast-like communication, while this has been a major concern of our DSA scheme. Third, this scheme does not provide any mechanism to adjust bandwidth allocation as traffic demand increases or decreases, while our proposed DSA mechanism dynamically adjusts the slot cycle so as to meet traffic needs.

7 Conclusions and Future Work

By virtue of the Master/Slave communication model, Bluetooth medium access provides for simplicity, low-power, and low-cost, these being the major forces driving the technology. However, this medium access scheme brings with it several problems such as inefficient packet forwarding and wastage of bandwidth by forwarding through the master, increased end-to-end packet delay, and additional power consumption at the master unit due to frequent medium access. Moreover, multicast-like communication is not supported which could be employed by many applications envisioned for WPAN environments.

With that in mind, in this paper we proposed a dynamic scheduling scheme with two novel approaches, namely, the implementation of direct communication between any two slaves within a piconet, and the support for multicast-like transmissions for group communication. Through extensive simulation, we have shown DSA to greatly outperform existing Bluetooth implementation by providing optimal bandwidth usage, lower delay, and low overhead. We believe this work will give rise to new research in the area. Future work includes the evaluation of DSA over scatternets by synchronizing slot assignments of different piconets.

References

1. Bluetooth SIG, "Bluetooth Specification", http://www.bluetooth.com.
2. C. Bisdikian, "An Overview of the Bluetooth Wireless Technology", *IEEE Communications Magazine*, December 2001, pp. 86–94.
3. M. Kalia, S. Garg, and R. Shorey, "Efficient Policies for Increasing Capacity in Bluetooth: An Indoor Pico-Cellular Wireless System," in *Proceedings of IEEE VTC*, May 2000.
4. Capone, M. Gerla, and R. Kapoor, "Efficient Polling Schemes for Bluetooth Piconets," in *Proceedings of IEEE ICC*, June 2001.
5. M. Kalia, D. Bansal, and R. Shorey, "Data Scheduling and SAR for Bluetooth MAC," in *Proceedings of IEEE VTC*, May 2000.
6. Kalia, M. Bansal, and D. Shorey, "MAC scheduling and SAR policies for bluetooth: a master driven TDD Pico-cellular wireless system," in *MoMuC*, May 1999, pp. 384–388.
7. L. Ramachandran, M. Kapoor, A. Sarkar, and A. Aggarwal, "Clustering Algorithms for Wireless Ad Hoc Networks," in *Proceedings of the 4th international workshop on Discrete Algorithms and Methods for Mobile Computing and Communications*, 2000.
8. T. Salonidis, P. Bhagwat, and L. Tassiulas, "Proximity awareness and fast connection establishment in Bluetooth," in *Proceedings of MobiHoc*, 2000.
9. T. Salonidis, P. Bhagwat, L. Tassiulas, and R. LaMaire, "Distributed Topology Construction of Bluetooth Personal Area Networks," in *Proc. of IEEE Infocom, 2001.*
10. C. Law, A. Mehta, and K. Siu, "Performance of a new Bluetooth scatternet formation protocol," in *Proceedings of the ACM Symposium on Mobile Ad Hoc Networking and Computing*, October 2001.
11. C. Law and K. Siu, "A Bluetooth scatternet formation algorithm," in *Proceedings of the IEEE Symposium on Ad Hoc Wireless Networks*, November 2001.
12. G. Zaruba, I. Chlamtac, and S. Basagni, "Bluetrees – scatternet formation to enable Bluetooth-based ad hoc networks," in *Proceedings of IEEE ICC*, June 2001.
13. Z. Wang, R. Thomas, and Z. Haas, "Bluenet – a New Scatternet Formation Scheme," in *Proceedings of the 35th HICSS*, Hawaii, Jan. 2002.
14. W. Zhang, H. Zhu, and G. Cao, "On Improving the Performance of Bluetooth Networks Through Dynamic Role Management," *Technical Report*, http://www.cse.psu.edu/~gcao/paper/bluetooth.ps, May 2001.
15. W. Zhang, H. Zhu, and G. Cao, "Improving Bluetooth Network Performance Through A Time-Slot Leasing Approach," in *Proc. of WCNC*, 2002.
16. C. Cordeiro and D. Agrawal, "Mitigating the Effects of Intermittent Interference on Bluetooth Ad Hoc Networks," *in the 13th IEEE PIMRC*, Lisbon, Portugal, September 2002.
17. H. Gossain, C. Cordeiro, and D. Agrawal, "Multicast: Wired to Wireless," in *IEEE Communications Magazine*, June 2002, pp. 116–123.
18. L. Ji, and M. S. Corson, "Differential Destination Multicast – A MANET Multicast Routing Protocol for Small Groups," in *Infocom 2001*, pp. 1192–1202.
19. The Network Sumulator (ns-2), *http://www.isi.edu/nsnam/ns/.*

20. BlueHoc, IBM Bluetooth Simulator, http://oss.software.ibm.com/developerworks/opensource/bluehoc/.
21. C. Cordeiro, D. Agrawal, and D. Sadok, "Piconet Interference Modeling and Performance Evaluation of Bluetooth MAC Protocol," in *IEEE Transactions on Wireless Comm.*, accepted for publication.
22. D. Agrawal and Q-A. Zeng, "Introduction to Wireless and Mobile Systems," Brooks/Cole Publishing, 438 pages, ISBN 0534-40851-6, August 2002.
23. C. Cordeiro and D. Agrawal, "Employing Dynamic Segmentation for Effective Co-located Coexistence between Bluetooth and IEEE 802.11 WLANs," in *IEEE GLOBECOM*, Taiwan, November 2002.
24. Y. Bin Lin and I. Chlamtac, "Wireless and Mobile Network Architectures," John Wiley & Sons, ISBN: 0471394920, October 2000.
25. W. Heinzelman, A. Chandrakasan, and H. Balakrishnan, "Energy-Efficient Communication Protocol for Wireless Microsensor Networks," in *Proc. of HICSS*, January 2000.
26. W. Ye, J. Heidemann, and D. Estrin, "An Energy-Efficient MAC Protocol for Wireless Sensor Networks," in *Infocom 2002*, June 2002.
27. C. Cordeiro, H. Gossain, and D. Agrawal, "Multicast over Wireless Mobile Ad Hoc Networks: Present and Future Directions," in *IEEE Network, Special Issue on Multicasting: An Enabling Technology*, January/February 2003.
28. Y. Lim, S. Min, and J. Ma, "Performance Evaluation of the Bluetooth-based Public Internet Access Point," in *Proc. of the 15th ICOIN*, pages 643–648, 2001.
29. N. Rouhana and E. Horlait, "BWIG: Bluetooth Web Internet Gateway," In *Proc. of IEEE Simposyum on Computer and Comm.*, July 2002.
30. O. Kasten and M. Langheinrich, "First Experience with Bluetooth in the smart-its Distributed Sensor Network," In *Proc. of the Workshop in Ubiquitous Computing and Communications*, October 2001.
31. F. Siegemund and M. Rohs, "Rendezvous Layer Protocols for Bluetooth-enabled Smart Devices," in *Proc. of International Conference on Architecture of Computing Systems*, April 2002.
32. D. Estrin, R. Govindan, and J. Heidmanm, "New Century Challenges: Scalable Cordination in Sensor Networks," in *ACM Mobicom*, pages 263–270, 1999.
33. J. Kahn, R. Katz, K. Pister, "New Century Challenges: Mobile Networking for Smart Dust," in *ACM Mobicom*, pages 271–278, 1999.
34. C. Cordeiro, S. Abhyankar, R. Toshiwal, and D. Agrawal, "A Novel Architecture and Coexistence Method to Provide Global Access to/from Bluetooth WPANs by IEEE 802.11 WLANs," *in IEEE IPCCC*, Phoenix, Arizona, April 2003.

Design and Analysis of a Bluetooth-Based Indoor Localization System*

Raffaele Bruno and Franca Delmastro

IIT Institute CNR
Via G. Moruzzi, 1 - 56100 Pisa, Italy
{raffaele.bruno,franca.delmastro}@iit.cnr.it

Abstract. The emerging context-aware applications relying on the wireless networking, require the availability of network services that provide even simple notions of context as position and identity. In the last few years several localization and tracking systems for indoor environment have been proposed. However, these systems are often very expensive since they require the installation of dedicated infrastructures of sensors based on radio, ultrasound or infrared technologies, to determine the user position. This paper aims to propose and analyze a novel indoor localization and tracking system based on the Bluetooth technology. This is a short-range, low complexity and low cost wireless technology that supports explicit device discovery services. The proposed localization system fully integrates these device discovery procedures into the infrastructure-based network of Bluetooth access points that provides the wireless connectivity inside a building. This design approach guarantees a cost-effective solution for the positioning service implementation. In this paper we have extensively investigated through simulative analysis the performances of the proposed localization system, showing that it has a negligible overhead, still providing a timely update of user location.

1 Introduction

Currently, the research interest in the design of technologies, systems and protocols that enables the deployment of new context aware applications within distributed and mobile computing systems, is rapidly growing. The term *context* is used to indicate the set of information that describes both the identity, activity and state characterizing a user, or a group of users, and the state of the objects that interact with them. The use of simple forms of context, such as time and identity are already an integral part of mobile computing. The proliferation of wireless networking and sensor technologies are enabling portable computers and handheld devices to be used in more places and to have a greater awareness of the dynamic context. In the last few years a wide class of new context-aware applications that use elementary notions of context (e.g. location, proximity and activity) have been developed [1]. All these proposals and prototypes demonstrated that one of the basic descriptions of the context is the physical location

* This work has been funded by MIUR in the framework of the VICOM Project.

M. Conti et al. (Eds.): PWC 2003, LNCS 2775, pp. 711–725, 2003.

of objects and users. The localization systems proposed so far differ in many parameters: techniques and physical phenomena exploited, technologies which are based on, the precision of the position information and so on. The interested reader can find an exhaustive taxonomy with a brief description of main localization systems for mobile computing applications in [2]. The aim of this paper is to propose and analyze a novel system that provides position information of mobile users inside a building based on the Bluetooth wireless technology [3]. Hereafter it will be indicated as *BIPS* (*Bluetooth Indoor Positioning System*). The Bluetooth technology is one of the most promising and cost-effective candidates to provide short-range wireless connectivity for mobile users. Key features of this technology are the robustness, the low complexity, the low power consumption and the short range (below 10 meters).The choice of the Bluetooth technology as base for the BIPS system is motivated by the use of this technology as a *de facto* standard to construct Wireless Personal Area Networks. The core of the BIPS system is constituted of a network of Bluetooth access points, acting as masters, distributed inside the building. The Bluetooth access points are interconnected through a dedicated wired network (e.g. an Ethernet LAN) and their operations are coordinated by a central server machine. A single room of the building is the granule of the position information that BIPS considers. We define a single room as the space that can fit into a circle of 10 meter radius since this is the dimension of the greatest coverage area of a Bluetooth piconet. The BIPS system is devoted to track the mobile users that walk (or stand) inside the building at a low speed. The user localization and tracking services are fully integrated within the same network infrastructure that provides the wireless connectivity for the data transfer, guaranteeing a cost-effective implementation of the localization system. The other proposal that fully integrates the tracking system with the same infrastructure that provides the wireless networking is the RADAR system developed by the Microsoft Research group [4] and based on the IEEE 802.11 Wireless LAN technology [5]. The RADAR system achieves a better accuracy than the BIPS system (about$3-4$meters of precision), paying the cost of higher complexity, since it requires either scene analysis (i.e., an *a priori* measured map of the radio signal strength inside the building) or multi-lateration (i.e., the triangulation of distances between at least three base stations). The paper is organized as follows. In section 2, the Bluetooth technology is briefly outlined. Particular attention is dedicated to describe the support for device discovery through the inquiry and page phase. Section 3 details the BIPS-system organization and an accurate analysis of the device discovery performances. The performance evaluation and comparison between the solutions based on different discovery approaches is showed in section 4. Concluding remarks and considerations on the ongoing work are presented in section 5.

2 Bluetooth Basics

The building block of a Bluetooth network is the *piconet*, a small cluster of devices that share a common physical channel, i.e., they are synchronized to the same clock and adopt the same frequency hopping sequence. Within the piconet, one device assumes the role of *master*, and all the other devices associated to the piconet play the role of *slaves*. The slaves derive the channel-hopping sequence as a function of the master's clock and address. The transmission scheme adopted in the piconet is the Time Division Duplex (TDD) scheme. The channel is divided into time slots, each $625\,\mu s$ long. To establish a connection it is required that the devices agree on the Bluetooth role to assume, either master or slave, and exchange their 48-bit address (BD_ADDR) and 28-bit native clock (CLKN). To achieve this the Bluetooth specification standardized a link establishment process constituted of two different phases. The first initial phase is named *inquiry*, and it allows the *inquirer* to discover the identity of possible slaves. The second final phase is named *page*, and it corresponds to initial connection setup, in which the *pager* informs the *paged unit* about its identity and imposes its clock as the piconet clock. In both phases the master and the slaves perform different actions that we will briefly summarize in the following subsections. At the end of these phases, the devices are synchronized on the clock of the inquirer/pager that becomes the master of the piconet. Then the connection is established and the devices can start exchanging data.

2.1 Inquiry Phase

A master begins the device discovery by entering into the inquiry state. In this state, the master broadcasts messages using a pool of 32 frequencies called inquiry hopping sequence, splitted into two 16-hop parts (train A and B). On every even slot, the master sends two ID packets switching between two frequencies of the same train every $312.5\,\mu s$. Thus $10\,ms$ are required to scan a single train, and it must be repeated for at least $N_{inquiry} = 256$ times before a new train is used. In order to collect all responses in an error-free environment, at least three train's switches must take place. A slave that wants to be discovered enters in the inquiry scan state, during which it listens for ID packets on the same 32 dedicated frequencies used from the master. The slave changes listening frequency every $1.28\,s$ and it keeps listening on the same frequency for the $T_{w\_inquiry\_scan}$ time, that must last enough to completely scan one train. $T_{inquiry\_scan}$ is the time between the beginning of two consecutive inquiry scan cycles and it is not longer than $2.56\,s$. If $T_{inquiry\_scan} = T_{w\_inquiry\_scan}$ then the inquiry scan activity is continuous. Both interval values can be changed and they are set by default as $T_{inquiry\_scan} = 1.28\,s$ and $T_{w\_inquiry\_scan} = 11.25\,ms$.

2.2 Page and Connection Phases

During the inquiry phase, the inquirer collects the BD_ADDRs and CLKNs of all the units that acknoledge the inquiry messages. However, the inquiry messages sent by the inquirer don't carry any information about the sender. Therefore, a further phase is needed by the master to establish a connection with a slave. This is the page phase, in which the master tries to capture a specific slave sending page messages, containing its own synchronization data. As in the inquiry phase, the master uses 32 frequencies belonging to the page hopping sequence, splitted into two 16-hop parts (train A and B). Since the master does not know when the slave will enter into the page scan substate, each train must be repeated at least N_{page} times. N_{page} is defined depending on the T_{page_scan} value, which is the interval between the beginning of two consecutive page scans. Default values for page scan parameters are equal to the inquiry scan default values: $T_{page\_scan} = 1.28\,s$, $T_{w\_page\_scan} = 11.25\,ms$. At the end of the page phase, the master enters into the connection state and sends a connection request to the slave. If this agrees and acknowledges the request, the connection is established and the devices can start exchanging data packets.

3 BIPS System Design Principles

BIPS is an indoor positioning service designed to track mobile users that may be in motion inside a corporate building. Each mobile user is represented by a handheld device equipped with a Bluetooth interface in order to interact with the static infrastructure of the BIPS system. This is constituted of a set of Bluetooth access points, placed in fixed positions inside a building and interconnected through a dedicated network (for instance a wired Ethernet LAN). They always assume the master role in order to discover and enroll mobile devices in their coverage area. The intelligence of the BIPS system is located in a central machine, hereafter denoted as BIPS-server, that performs the coordination of the masters to locate users and track their movements[1]. Each access point executes two main tasks: one is the discovery of the units entering into the radio coverage area of its Bluetooth interface; the other one is the data transfer towards/from users that are associated to its piconet. The BIPS system should guarantee to achieve the timely update of the user's position without negatively affecting the throughput of data connections. It is straightforward to observe that these two goals are complementary since a prompt user-movement tracking can be obtained performing frequently the device discovery procedures. In subsections 3.1 and 3.2 we will analyze the overhead introduced by the device discovery task inside an isolated piconet. Then in section 4 the impact of the device discovery on the data transfer will be quantified. It is worth pointing out that the main objective of this paper is to design an effective discovery scheme that minimizes the localization system overheads preserving the promptness of the tracking service.

3.1 Device Discovery

The BIPS system implements the localization and tracking functionalities through the coordination of the device discovery phases executed by the master units. The performance evaluation of the overhead introduced by the execution

[1] The BIPS-server may implement further functionalities in addition to the user position update, as user authentication, mobile users lookup and user profiling. The reader is referred to [6] for a detailed discussion about the services that can be implemented in the BIPS system.

of the device discovery task will be very usefull in the design of the BIPS system. Because of the users' mobility, some links may disappear since the devices exit from the radio visibility of the master. In the same way new links can be established when two nodes are close enough to allow the receivers to listen to the transmitted radio signals. The localization of the users can be realized by tracking the links that disappears and the new links that are established. An efficient design of the coordination among masters' operations to promptly identify disconnections and, at the same time, to open new connections, cannot be done without precisely identifying the limitations of the device discovery procedures standardized by the Bluetooth technology. The link establishment process requires that each device becomes aware of the other unknown devices that enter into its radio coverage area. The connection establishment procedure described in section 2 are designed to enable a device to discover which other devices are in range, and their addresses and clocks. In previous studies the Bluetooth device discovery procedures have been analyzed only when the link establishment is accomplished in an ad hoc fashion without any coordination, namely, without any explicit master or slave role pre-assignment. In [7] a symmetric mechanism that guarantees an ad hoc point-to-point connection between two devices by forcing the nodes to independently alternates between Inquiry mode and Inquiry Scan mode, was proposed. However, the effectiveness of the proposed protocol was evaluated only in a simplified scenario where all the nodes are in radio visibility of each other, i.e., the nodes form a single-hop network. In [8] the protocol defined in [7] was generalized to the case of a multi-hop networks and the performance of the Bluetooth device discovery procedures where evaluated in terms of the percentage of neighbors discovered within a given discovery-time interval. These previous works clearly show that the device discovery procedures standardized in [3] are robust against interference due to the concurrent operations of nodes in highly dense networks, but they introduce significant delays in the neighboring discovery process. In this section we investigate the Bluetooth discovery procedures from a completely different perspective. Specifically, we will consider a scenario where the devices' operations are coordinated and there is a pre-assignment of the master and slave roles. The designated master will be the *active inquirer*, namely the node in charge on scanning its coverage area to discover the presence of other units willing to connect to it, whereas the designated slaves will be the *passive listeners*, waiting for being discovered from the master closest to them. There are many cases in which is reasonable and advantageous to construct the Bluetooth network in an infrastructure-based fashion, and to select a-priori the nodes that will act as masters. For instance, in a Bluetooth-based indoor tracking system, we can assume that there is a network of *inquirers* that actively search the users and track their movements. The simulations we have conducted are aimed to investigate the amount of resources that an inquirer should spend in gathering an exact knowledge of the units present in its coverage area, at least with a very high probability. Specifically, a first set of simulations were conducted to evaluate the setup connection delay in realistic conditions, without the biasing factor of periodic alternation between Inquiry and Inquiry

Scan mode. Moreover, these simulations allow us to better understand the weight of the different phases in the link establishment process. A second set of simulations has been carried out to determine the amount of time the inquirer should dedicate to the device discovery task to localize a given number of neighbors. These simulations provide an estimate of the efficiency of the Bluetooth device discovery procedures, and allow us to better determine how many mobile units can be reasonably tracked by a single inquirer.

3.2 Numerical Results

In all the simulations carried out the master is completely dedicated to execute the discovery procedure, i.e., it performs the inquiry phase until all the slaves have been discovered. The units to be discovered act as slaves and they start their operations in inquiry scan mode. To analyze the most favorable case from the perspective of device discovery, the slave executes a continuous scanning. As soon as the master receives a FHS packet from the slaves it goes to page mode to enroll the new discovered unit in its piconet. In the same way, as soon as the slave sends an FHS reply it goes to Page Scan mode to open a connection with the master. The Inquiry and Page mode parameters were selected according to the default values. In particular $N_{inquiry} = 256$ and $N_{page} = 128$. As explained in the previous section, the first set of simulations is aimed to evaluate the link setup delay between a couple of Bluetooth units. Fig. 1 shows the time distribution of the link setup delay. The shape of the distribution clearly highlights that the link setup delay is uniformly distributed into two separate ranges, the lower one is the range around $[0.1s, \ldots, 0.7s]$, instead the upper one is the range around $[2.6s, \ldots, 3.4s]$. The explanation of this particular distribution is straightforward. Let us consider the case in which both master and slaves are tuned to a frequency picked from the same train, and calculate the maximum time required to setup the link. The slave receives the first inquiry message that triggers the back-off timer at most within the interval needed to the inquirer to completely scan the train (i.e., $10\,ms$). The maximum back-off length is 1023 time slots (i.e., $639.375\,ms$). After the back-off timer expiration, another scanning interval of $10\,ms$is needed to the inquirer to receive the FHS packet. The page phase that immediately follows the inquiry phase has a significant lower duration. Considering that both the master and the slave starts the Page mode and Page Scan mode respectively picking up a frequency from the same train, a 16-frequencies scan interval is needed to achieve the coarse synchronization between the master and the slave. When this coarse synchronization is obtained the remaining part of the connection setup requires a few couples of time slots to be completed[2]. The sum of these contributes is equal to $673.425\,ms$. The second lobe of delay distribution is obtained when the master and slaves pick a frequency from opposite trains and a further startup delay of $2.56\,s$ is added to the link establishment process. The numerical results showed

[2] The interested reader is reminded to the Bluetooth specification [3] for the details of message exchange during the Page Response mode and at the beginning of the connection establishment.

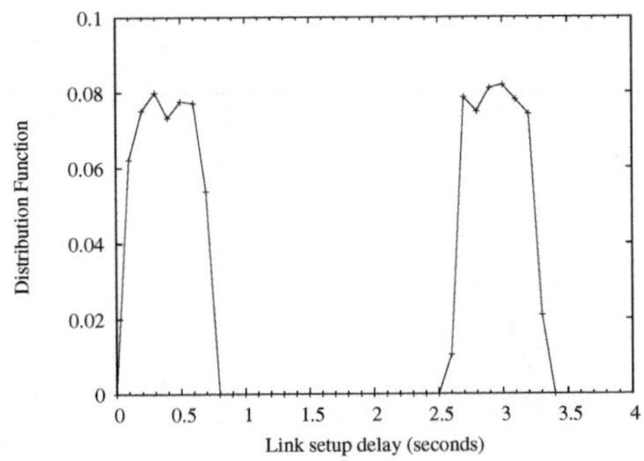

Fig. 1. Link setup delay.

in Fig. 1 quantify the time spent by a master in opening a single connection. It is necessary to investigate how the device discovery procedures scale up with the number of slaves that should be discovered (i.e., the number of connections that should be opened). To this end, we performed a second set of simulations where master's and slave's behavior is the same of the first set of simulations, but the number of slaves ranges between 2 and 15. The results obtained are showed in Fig. 2. The figure shows that less than $1.5\,s$ is always sufficient to establish a connection towards half of the slaves that are close to the master. This time reduces to $0.8\,s$ for piconets smaller than 7 slaves. The remaining slaves have to wait $2.56\,s$ to be discovered because the master has to change its frequency train. All the N slaves are connected to the master within $4\,s$. The percentage of discovered slaves shows a step-function shape due to the random selection of the train used during the inquiry phase. Specifically, the 50% of slaves start to listen on a frequency picked from a train different from the one used by the master. In this case the link setup delay is due to the second part of the Fig. 1. This implies that at least $2.56\,s$ are needed before the slaves are reachable by the master, therefore on average half of the slaves are reachable before $2.56\,s$, whereas the remaining half is reachable only after the first $2.56\,s$. Two useful observations can be taken comparing the numerical results showed in Fig. 1 and Fig. 2. The first one is that the back-off part of the link establishment process is the main cause of delay in forming the Bluetooth connection. The second one is that the back-off range introduces an excessive spread of the slave replies since, even in dense piconets, the collision probability is negligible because of the not-correlated nature of the slaves' native clocks. To summarize, the design of a localization system, relying upon the Bluetooth device discovery procedures, has to deal with three main intrinsic technological limitations: *i*) the use of two separated frequency trains that can introduce a delay of up to $2.56\,s$ during the link establishment; *ii*) the use of a long random back-off delay that increases the average link setup delay of about $300\,ms$; and *iii*) the use of a broadcast inquiry message that doesn't carry any information about the inquirer. In the following section, we will show that, by a limited tuning of the Inquiry mode parameters,

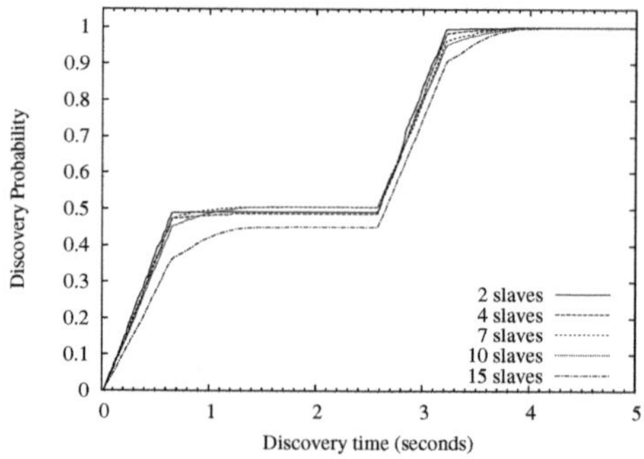

Fig. 2. Device discovery efficiency for a piconet of N slaves.

and designing a localization scheme relying only upon the paging phase, these drawbacks can be significantly overcome.

3.3 Page versus Inquiry

Observing the results obtained from previous simulations, we can notice that the overhead introduced by the device discovery task based on the inquiry phase negatively influences the user tracking performances. For this reason we propose and analyze a scheme for implementing the user positioning service relying on the page phase, in order to reduce the impact of the device discovery on other services like data transfer. Hereafter, we refer to the proposed localization system as $\mathrm{BIPS_P}$ system. Specifically, the $\mathrm{BIPS_P}$ system assumes that the BIPS-server commands the execution of a new page phase when a user disconnection is discovered by the network access points. This solution requires that masters and BIPS-server exchange information about the current status of their piconets, e.g. user disconnections and associations, allowing only the discovery of known devices. For this reason it is always necessary the execution of a periodic inquiry phase with a period much greater than page's one. The $\mathrm{BIPS_P}$ system will be compared with a reference architecture in which the localization task is performed by the network access points without exploiting masters' communication and coordination. Specifically, each master performs the discovery procedures independently from the other masters. Information about user disconnections are not exchanged. Since masters don't know the identity of the devices to track, and when it is necessary to execute a new discovery phase, this scheme has to rely on the inquiry phase (outlined in section 3.1). For the sake of brevity this second solution will be denoted as $\mathrm{BIPS_I}$ system. The remaining of this section details the $\mathrm{BIPS_I}$ and $\mathrm{BIPS_P}$ system operations, considering only the overhead introduced by the discovery of known devices. A critical design choice for the localization system is to define the rules adopted by the master and slaves to become aware of a link disconnection due to a slave departed from the radio coverage area of the piconet to which it is associated. The Bluetooth specification introduces a supervision timeout, named *supervisionTO*, used by both masters and slaves to monitor link losses. Specifically, if a device doesn't receive any correct packet which is destined to itself before the supervisionTO expiration, then the connection is automatically closed. The default value is 20 s, but it is an exceedingly high value to define any meaningful localization and tracking service. Therefore the supervisionTO should be modified to better deal with the users' dynamics. We have considered an ideal scenario in which no packet losses are allowed during the Connection state, hence the master can assume a slave as disconnected when it doesn't answer immediately to its polls. As far as the slaves' behavior, we have considered only active slaves that are involved in high loaded data connections served with a Round Robin scheduler. Therefore, each slave is polled at least once per polling cycle[3]. Hence the supervisionTO needs to take into account only the maximum size of time interval in which the master suspends serving data connections to execute the device discovery phases. In section 4, the proper value for the slaves' supervisionTO will be evaluated. In the following subsections we analyze the parameters values used for simulations in both systems.

BIPS$_\mathbf{I}$ System. Each master periodically performs an active scanning of its coverage area, broadcasting inquiry messages. The period between two successive inquiry phases is referred to as T_{rep}, while the lenght of each of them is indicated as W. The ratio between W and T_{rep} represents the percentage of channel resources dedicated to the discovery task. This overhead is almost independent of the user mobility since the discovery phase is periodically performed by the masters. To guarantee that a Bluetooth device, in Inquiry Scan mode, has the chance of listening to at least one inquiry message , the W period is equally divided into two parts dedicated to scan both train A and B. This requires the modification of the $N_{inquiry}$ default value (i.e., 256). On the other hand, when a user realizes to be disconnect from the piconet (thanks to the supervisionTO expiration), it immediately goes to Inquiry Scan mode. As soon as the master receives the reply to its inquiry messages from a new device entered into its coverage area, it tries to connect to it. To this end, it performs immediately a page procedure. The duration of this phase should be limited to reduce the time the master is not serving data connections. We select it equal to W. Now we are able to estimate a conservative value for the supervisionTO, which should be used by the slaves[4]. In fact, the maximum delay between two consecutive master's polls, under the ideal conditions of no packet losses during the Connection state, is lower bounded by $2 \times W + C^7_{MAX}$, and the supervisionTO should be selected greater than that value.

---

[3] The maximum length of a polling cycle for a Round Robin scheduler used in a piconet with N slaves (C^N_{MAX}) when only DH_5 packets occupy the channel (i.e., the maximum length baseband packet) is $C^N_{MAX} = 2 \times N \times DH_5$. In a piconet with 7 slaves we have $C^7_{MAX} = 8.75\,ms$.

[4] It is worth reminding that the slaves' promptness to discover the disconnection from its master depends on the supervisionTO.

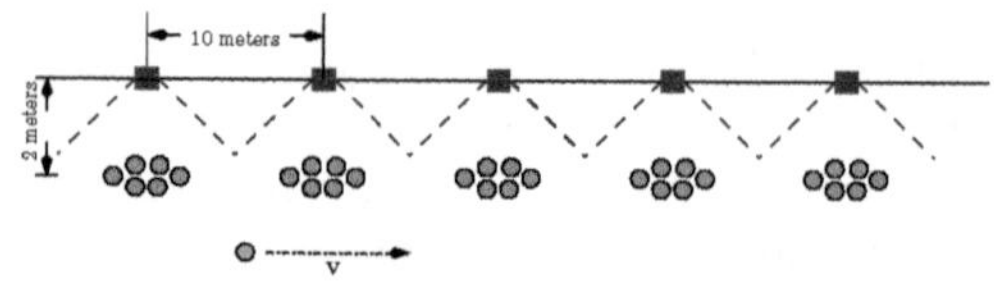

Fig. 3. Simulations scenario.

BIPS$_P$ System. The masters perform only page procedures to track the movement of the users. Specifically, as soon as a master identifies a link disconnection, it informs the BIPS-server via a specific "alarm" message that carries the identity of the lost user, say U_L, and the up to date clock information available at the master. These information are delivered to all the masters "close" to the one that originated the alarm. The notion of proximity can be implemented in several ways. The simplest is the physical proximity, i.e., the masters nearest to the one that originated the alarm are designated to search U_L. However, more complex policies can be thought if the user profile is available at the BIPS-server, e.g. the most visited access points, the final destination etc.. Then, these designated masters begin the execution of a page phase, trying to open a connection towards U_L. This phase lasts for W and it is periodically repeated with period T_{rep}. On the other hand, when a user realizes to be disconnected from the piconet (thanks to the supervisionTO expiration), it immediately goes to Page Scan mode. As soon as a master succeeds in finding the user, it notifies the BIPS-server about the detection, so that it stops the other designated masters to seek for U_L. The slaves' supervisionTO is chosen as $2 \times W + C^7_{MAX}$, to be as much possible close to the BIPS$_I$ system operations.

4 Performance Evaluation

To compare BIPS$_I$ with BIPS$_P$ system we have identified a realistic scenario, that is showed in fig3. Specifically, we consider a corridor, where 5 Bluetooth access points are located at a distance of 2 meters from the ceiling, spaced by ten meters. Each access point has 6 static slaves associated to its piconet. Highly loaded data connections are active towards each slave (e.g. FTP data transfers), served according to a Round Robin polling scheme. A single user moves with a constant speed v, and no active data connections are active towards the mobile user. The length of the TCP payload has been selected equal to 1000 bytes that allows us to have a maximum TCP throughput per piconet around 620 Kbps [9]. During all the simulation we have assumed $W = 100\,ms$, a time long enough to scan five times both the train A and B[5]. The performance indexes we investigated, are aimed to quantify both the systems promptness in tracking the user movement and the impact of the discovery task on the data connections performances (e.g., connections throughput, MAC delay and End-to-End delay). We present a comparison between the two systems about the

[5] This choice implies that the supervisionTO cannot be lower that $210\,ms$. In all the simulations we have assumed the supervisionTO equal to $250\,ms$.

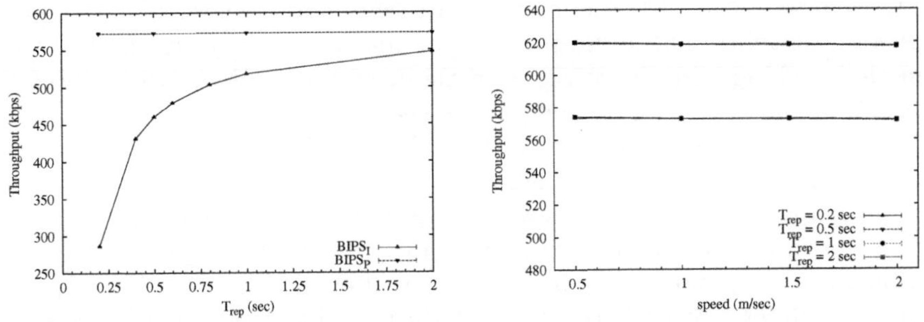

(a) TCP throughputs for the user speed equal to $1\,m/s$.

(b) $BIPS_P$ throughputs (TCP and L2CAP).

Fig. 4. Throughputs comparison.

TCP throughput per piconet versus the T_{rep} parameter, the link setup delay after a link disconnection and the MAC and End-to-End delays. The MAC delay represents the transmission time required by a packet at the head of the queue, and the End-to-End delay is equal to the MAC delay plus the queueing time. The $BIPS_I$ and $BIPS_P$ system operations were implemented in the *ns2* simulator [10], extended with the IBM module *BlueHoc* [11]. All the simulative results showed hereafter are presented with a 95% confidence interval.

4.1 Simulative Results

In this section we describe the simulative results related to the performances of the $BIPS_I$ and $BIPS_P$ systems using the scenario showed in Fig. 3. Fig. 4a) shows the throughput per piconet achieved by the TCP connections versus the T_{rep} parameter, when the user speed is $1\,m/s$. For the $BIPS_I$ system the same results are obtained with different user speed values. This is explained observing that it executes the device discovery procedures periodically, independently of the user behavior. We vary the T_{rep} parameter to evaluate how the frequency, used by the system to perform the localization task, affects the data connections throughput. The buffer size at masters and slaves queues has been selected large enough to avoid packet losses due to buffer overflow. It is straightforward to notice that the throughput scales up rapidly with the T_{rep} value, since the time dedicated to the discovery phase decreases. For instance, the choice $T_{rep} = 1\,s$ implies that 10% of the channel is occupied by the execution of inquiry phases, with the remaining 90% dedicated by the master to serve data connections. Instead $T_{rep} = 0.2\,s$ implies that only the 50% of the channel time is dedicated to serve data connections. In the $BIPS_P$ system the masters page the mobile user only when it is disconnected, and they continue to periodically page it until it is not enrolled again in a piconet (with period T_{rep}). Therefore, we expect that the T_{rep} has a negligible impact on the throughput performance if a few pages are enough to establish the connection. The numerical results clearly confirm that the throughput dependency on the T_{rep} parameter is negligible. On the other hand, because of the user sojourn time into the radio coverage area of a piconet is inversely proportional to its speed, the frequency, used by the masters to perform the tracking task, increases with the user speed. Hence, we expect that, differently from theBIPS$_\mathrm{I}$ system, the speed value may affect the throughput performance. However, Fig. 4b) clearly highlights that the throughput decrease is negligible, also when the user mobility is high (e.g. $2\,m/s$). Figure 4b) shows the throughput per piconet achieved by the L2CAP and TCP connections versus the speed of the mobile user. We investigate the throughput performance varying the T_{rep} parameter. The TCP throughput is represented by the lower curve. The difference with the L2CAP throughput is due to the presence of the TCP acknowledgment traffic that flows from the slaves towards the master in the active L2CAP connections. It is worth pointing out the BIPS$_\mathrm{P}$ system adapts its behavior to the user mobility, minimizing the overhead introduced by the tracking service. Therefore, even in low mobility scenarios (e.g. $0.5\,m/s$) the piconet throughput is always higher in BIPS$_\mathrm{P}$ than in BIPS$_\mathrm{I}$ (580 Kbps against the 550 Kbps achieved when $T_{rep} = 2\,s$). To complete the performance analysis of data traffic, Fig. 5(a) and 5(b) show respectively the MAC and End-to-End delays for the L2CAP packets versus the T_{rep} parameter for the user speed equal to $1\,m/s$. In [12] it was proved that the aggregate MAC delay of L2CAP packets measured in asymptotic conditions (i.e., when there is always a queued packet in the master's and in the slaves' local queues) for a Round Robin scheduler, depends only on the number of slaves and on the average number of time slots needed to transmit the L2CAP packet. Instead, the End-to-End delay mainly depends on the buffer size. However, the BIPS$_\mathrm{I}$ system introduces a further W delay to fragmented L2CAP packets, which fragments are not all delivered within the channel time between two successive inquiry phases. It is straightforward to observe that the probability to introduce this fixed delay grows by decreasing the T_{rep} value. This implies that both MAC and End-to-End delays significantly decreases as the T_{rep} value increases. At the opposite, the BIPS$_\mathrm{P}$ system shows a significant improvement compared with the BIPS$_\mathrm{I}$ system. In fact the adaptive policy used to decide when to page a mobile user, together with the exploitation of the page's properties, guarantees to keep the packet delays almost constant. In this case, both MAC and End-to-End delays are almost independent of the T_{rep} value, and further results have demonstrated that they are slightly affected by the increase of the user speed. Even in low mobility scenarios (e.g. $0.5\,m/s$) the measured MAC and End-to-End delays are always lower in the BIPS$_\mathrm{P}$ system than in the BIPS$_\mathrm{I}$ system. As final result we have to investigate the effectiveness of each scheme in providing a prompt update of the user position. To quantify this performance we have identified two indexes: the first is the average time the system requires to track the new user position, say T^M_{blind}; the latter is the average time the user is not associated to any piconet, say T^S_{blind}. Specifically, T^M_{blind} is the average time between two events: the first is a master becoming aware of the mobile user disconnection, and the second is a master opening a connection towards the mobile user. In the same way, T^S_{blind} is the average time between two events: the first is the mobile user becoming aware of not being associated

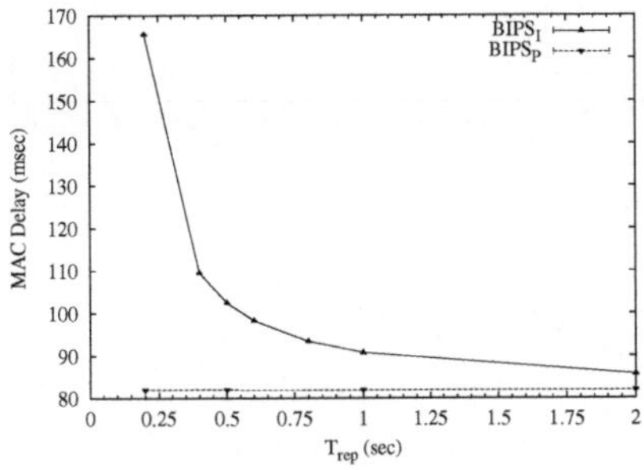

(a) Comparison between MAC delays

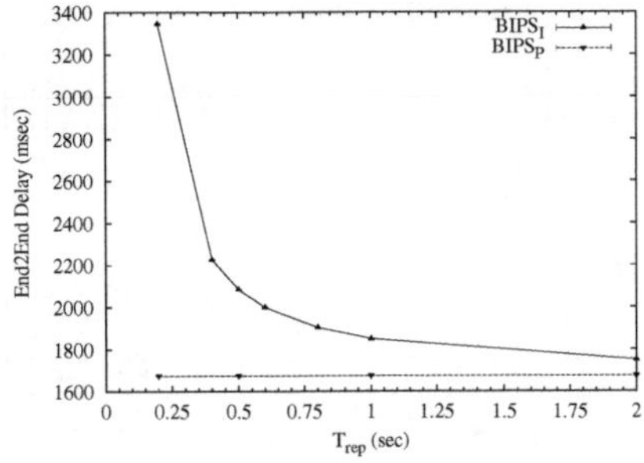

(b) Comparison between end-to-end delays

Fig. 5. MAC and end-to-end delays.

anymore with the master, and the second is the mobile user enrollment into another piconet. The T_{blind}^{M} value assesses the system promptness in discovering the mobile user location, whereas the T_{blind}^{S} value measures the user perception of the network failure. Fig. 6 shows the comparison between T_{blind}^{M} in the two systems. It is straightforward to notice that, for the $BIPS_I$ system, as the T_{rep} increases, also the delay between tracking updates increases. In fact, at least two successive inquiry phases are required to open a connection: the first one to trigger the back-off selection, the second one to trigger the page phase. Hence, T_{blind}^{M} and T_{blind}^{S} cannot be lower than the T_{rep} duration. Nevertheless, the inquiry and page procedures, thanks to the very fast hopping scheme adopted, are very robust against the interference due to the contemporary data transmissions coming from overlapping piconets. Therefore, the probability of connection failures after the back-off expiration is very low. This implies that usually two inquiry phases are enough to establish a new connection. Hence, the $BIPS_I$ system introduces an overhead independent of the mobile user speed, but depending only on the T_{rep} value. All the performance indexes related to data connections worse as the T_{rep} value reduces, since the time dedicated to locate the mobile user increases. For the $BIPS_P$ system, to better explain the numerical results, we have to detail the timing of alarms generation in the masters. Specifically, when a master realizes that a user departed from its piconet, it sends an alarm to the BIPS-server only after a supervisionTO. This guarantees that the masters, designated by the BIPS-server to discover the new user position, begin the page procedure when the user is already seeking for a new network access point. However, this choice also implies that the first page phase is executed by the designated masters with a starting delay of the supervisionTO. Hence, the T_{blind}^{M} value cannot be lower than the suspervionTO. The numerical results plotted in Fig. 6a) clearly show that in both systems, the T_{blind}^{M} value increases as the T_{rep} increases. However, since the page procedure is less time-consuming than the inquiry (there isn't a back-off delay before the slave responses), the T_{blind}^{M} value is always lower in $BIPS_P$ than in $BIPS_I$. As we can see in Fig. 6b), the T_{blind}^{S} value is always about

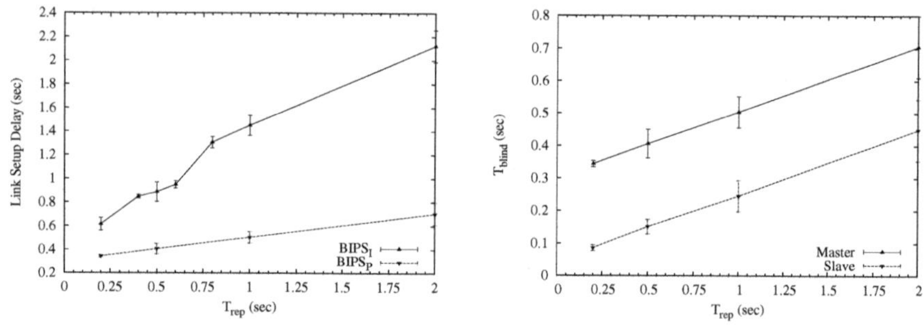

(a) BIPS_{I} and BIPS_{P} T^{M}_{blind} (b) BIPS_{P} localization delays

Fig. 6. Localization delays.

$200\,ms$ lower than the T^{M}_{blind} since the slaves became aware of a link disconnection after the supervisionTO delay. Simulations with different speed values have provided the same results, and are not reported here. To conclude, adapting the discovery phases lenght to the user mobility, we can guarantee to keep the overhead due to the execution of the localization service very low. Hence, there is a negligible impact on the performances perceived by the users data connections. On the other hand the use of the page procedure guarantees the reduction of the setup delay needed to open new connections towards the mobile user.

5 Conclusions and Future Works

In this section we have proposed a solution to offer an indoor localization service that is fully integrated in the infrastructure that guarantees the wireless networking inside a building. Our proposal is based on the exploitation of an emerging technology for short range and ad hoc wireless connectivity: the Bluetooth technology. We have proposes a scheme that adapts, to the user mobility profile, the frequency used by network access points to perform user tracking. The localization service has been implemented using the paging procedures defined in the Bluetooth standard. This design approach guarantees a cost-effective solution for the implementation of the localization service. Our simulative investigations have shown that throughput and packet delays are negligible affected by the load generated by localization service. In addition, our scheme is able to guarantee a prompt update of the user localization, significantly limiting the handoff delays between different piconets. During the analysis we assumed a symmetric traffic distribution for the slaves associated to the network access points. The master serves the data connection using a simple Round Robin scheduler. Further studies are required to design novel scheduling algorithms that can efficiently integrate the device discovery procedures inside the polling scheme. The final goal should be the development of an architecture that provides the wireless networking adapting the access point operations to the user mobility and traffic demands.

References

1. R. Malladi and D.P. Agrawal. Current and future applications of mobile and wireless networks. *Communications of ACM*, 45(10):144–146, October 2002.
2. G. Hightower and G. Borriello. Location systems for ubiquitous computing. *IEEE Computer*, pages 57–66, August 2001.
3. Bluetooth SIG. *Specification of the Bluetooth System, Volume 1–Core.* Version 1.1, February 2001.
4. P. Bahl and V. Padmanabhan. RADAR: An in-building RF-based user location and tracking system. In *Proceedings of the IEEE Infocom 200*, pages 775–784, Tel-Aviv, Israel, March 2000.
5. IEEE Standard for Wireless LAN. *Medium Access Control and Physical Layer Specification.* P802.11, November 1999.
6. G. Anastasi, R. Bandelloni, M. Conti, F. Delmastro, E. Gregori, and G. Mainetto. Experimenting an indoor Bluetooth-based positioning service. In *Proceedings of the International Workshop on Mobile Distributed Computing*, Providence, RI, May 2003.
7. T. Salonidis, P. Bhagwat, L. Tassiulas, and R. LaMaire. Distributed topology construction of Bluetooth personal area networks. In *Proceedings of the IEEE Infocom 2001*, pages 1577–1586, Anchorage, AK, April 2001.
8. S. Basagni, R. Bruno, and C. Petrioli. Device discovery in Bluetooth networks: A scatternet perspective. In *Proceedings of the Second IFIP-TC6 Networking Conference (Networking 2002)*, volume 2345 of *LNCS*, Pisa, Italy, May 2002. Springer.
9. R. Bruno, M. Conti, and E. Gregori. Wireless access to internet via bluetooth: Performance evaluation of the edc scheduling algorithm. In *First ACM International Workshop on Wireless Mobile Internet (WMI 2001)*, pages 43–49, Rome, Italy, July 2001.
10. The VINT Project. *The ns Manual.* http://www.isi.edu/nsnam/ns/, 2002.
11. IBM. *BlueHoc: Bluetooth Ad Hoc Network Simulator.* Version 2.0, June 2001.
12. R. Bruno, M. Conti, and E. Gregori. Bluetooth: Architecture, protocols and scheduling algorithms. *Kluwer Cluster Computing*, 5(2):117–131, April 2002.

On Efficient Topologies for Bluetooth Scatternets*

Daniele Miorandi, Arianna Trainito, and Andrea Zanella

Department of Information Engineering
University of Padova, Via Gradenigo 6/B
35131 Padova, Italy
phone [fax]: +39 049 827 7656 [7699]
{daniele.miorandi,arianna.trainito,andrea.zanella}@dei.unipd.it

Abstract. In this paper we investigate the relationship between network capacity and topology for Bluetooth scatternets. We start by considering the intrinsic capacity limits of a scatternet structure, and we show that limiting capacity may be achieved for very local traffic and under specific conditions on the scatternet structure. Then, we provide a description of the performance achievable with two basic scatternet configurations, namely star and closed–loop topologies, and show the role played by inter–piconet interference in the choice of efficient configurations. Finally, we present some efficient topologies, based on Platonic solids structure.

Keywords: Bluetooth, scatternet, efficient configuration, performance evaluation, topology.

1 Introduction

Originally born as a wireless replacement for cables connecting electronic devices, Bluetooth [1] has been gaining a lot of consideration and attention by the scientific community in the last few years. The development of this technology is now focused on the area of the so–called Wireless Personal Area Networks (WPANs), where Bluetooth is expected to play a major role in the short and mid-term future. The commercial success of WPANs is intimately linked to their ability to support advanced digital services, like audio and video streaming, web browsing, etc. [2]. In such a scenario, the performance aspects of the radio technologies involved appear of primary importance.

Bluetooth has been designed to work in a scattered ad–hoc environment, where multiple independent overlapping networks, called piconets, may coexist and be inter–connected to form a multi-hop network, called *scatternet*. Recently, much attention has been devoted, by both academic and industrial world, to the issues concerning scatternets formation and management. In particular, scatternet formation algorithms [3],[4],[5],[6] have been widely investigated, and many solutions have been proposed to build up a scatternet starting from disconnected units. The focus is now moving to the characterization and design of efficient scatternet topologies [7], [8], [9], [10], since it is clear that the way piconets are interconnected to form scatternets may dramatically impact

* This work was partially supported by MIUR within the framework of the PRIMO project FIRB RBNE018RFY

M. Conti et al. (Eds.): PWC 2003, LNCS 2775, pp. 726–740, 2003.

network performance. To the best of authors knowledge, however, the literature still lacks in thorough investigation of the optimal scatternet topologies.

Clearly, the optimality of a scatternet topology does depend on the performance indexes considered. Typically, scatternet performance metrics are: number of piconets, number of bridge nodes, number of roles per node etc. When traffic connections are considered, average traffic delay, throughput and maximum traffic delay are shown as metrics of interest. Nevertheless, packet delay may lead to misleading conclusions whether network capacity is not taken into account. In particular, results obtained by simulating an over–loaded network may be affected by the setting of the simulator parameters, like buffers capacity, simulations run–time, traffic generation pattern, and so on. For this reason, we prefer to consider a different performance metric, namely the *network capacity*. This metric represents the maximum aggregated average traffic that nodes can inject into the network without overflowing. In general, we say that a network configuration is *stable* if the total traffic offered to the network does not exceed the network capacity, otherwise the network is said to be *unstable*. It may be worth underlining that, for unstable configurations, some traffic connections will either experiment always increasing average packet delays or packet losses due to overflows. Conversely, in stable configurations the average packet delay is limited and, provided that the dimensions of the buffers are adequately chosen, packets are never dropped because of overflows.

In this paper, we investigate the relation between the scatternet topology, i.e., the way piconets are interconnected to form the scatternet, and the network capacity. We first discuss some conditions to achieve efficient piconets interconnection. Then, we investigate the performance achieved by some specific scatternet topologies, both "planar" and "solid", in case of uniform traffic matrix, that is, assuming that every node in the network generates an equal amount of traffic towards any other node.

This work is a first attempt to provide a mathematical insight into the relation between scatternet topology and performance. The analysis we provide allows us to determine some optimality criteria and to evaluate the performance achieved by some scatternet topologies. Such an analysis may be worth, for instance, for designing networks of static sensors or domestic appliances, where the end—to—end traffic matrix among the nodes may be known a priori. At this phase of the work, however, many aspects related to the network creation and management issues have still to be considered and the discussion presented in this paper is far to be complete. Nevertheless, it may provide a starting point for further, more practical–oriented, investigations in the field.

The paper is organized as follows. Section 2 outlines the primary characteristics of the Bluetooth technology. Section 3 describes the system model used for our analysis. Section 4 provides the analysis of the limiting performance, in terms of capacity, achievable by Bluetooth scatternets. In Section 5, system performance is evaluated under the assumption of a uniform traffic matrix. Section 6 concludes the paper with some remarks and directions for future work.

2 Background: The Bluetooth Technology

Bluetooth operates in the 2.4 GHz ISM unlicensed band, providing a raw bit rate of 1Mb/s by using a binary Gaussian-shaped FSK modulation. In order to reduce interference with other devices operating in the ISM band, Bluetooth makes use of a frequency hopping (FH) spread spectrum technique, spanning 79 RF carriers, 1-MHz wide each. Two up to eight Bluetooth units may be connected in a small network, called *piconet*. In each piconet, a unit acts as *master*, controlling the channel access by means of a simple polling scheme, while the other units act as *slaves*. Time is divided into consecutive slots of $625\mu s$ each, which are used for downlink (master–to–slave) and uplink (slave–to–master) transmissions, alternatively, in a time division duplex (TDD) fashion. Namely, each time–slot is associated to a hop in the hopping sequence, resulting in a nominal hop rate of 1600 hop/s.

Different piconets are associated to independent FH channels. This allows more piconets to share the same physical space and spectrum without excessively increasing the mutual interference. Piconets may communicate by sharing a device on a time division basis, forming what is commonly referred to as a *scatternet*. The shared device, usually named *gateway*, spends a given amount of time in each piconet it belongs to, forwarding the inter–piconet traffic. The scatternet configuration, in turn, is determined by:

- the assignment of the master and slave roles in each piconet composing the scatternet;
- the identification of the shared units for each piconets.

3 System Model

Before presenting our analysis, we need to introduce some simplifying assumptions for the network behavior.

At this phase of the work, we assume all nodes are in range, neglecting the spatial constraints that, however, may have a dramatic impact on the scatternet topology. For the sake of simplicity, we assume that only one–slot long packets are used (namely, $DH1$ packets). Furthermore, we limit our analysis to the simplest polling scheme, i.e., Pure Round Robin (PRR). With PRR, all slaves get polled cyclically by the master, whether they have data to transmit or not. Although many efficient polling schemes have been proposed in the last few years [11,12,13], at this time available devices basically support only PRR. This is essentially due to the necessity of keeping the complexity of the firmware as low as possible, in order to reduce manufacturing costs and power consumption. PRR does not require complex logic to be embedded on the chip and, thus, it represents the most attractive choice for low–cost, energy–conserving products.

Finally, we neglect the *switchover time*, i.e., the time a gateway takes to switch from one piconet to another. Such a time may determine a capacity loss to the system, since during the switchover time the gateway is not synchronized to any piconet and, thus, it cannot communicate. Nevertheless, the capacity loss may be reduced by increasing the *persistence time*, i.e., the period the gateway spends in each piconet before switching to the next one. Hence, we assume that the switchover time is a small fraction of the persistence time, so that the capacity loss caused by the inter–piconet switchover time may be neglected.[1] Furthermore, we suppose gateways are equally shared among the

[1] Note that, the persistence time cannot be increased indefinitely, since the longer the persistence time, the greater the forwarding delay introduced by the gateway and the buffer size required to hold the packets that have to be forwarded to the other piconets.

piconets they belong to, so that a gateway connected to γ_i piconets spends $\frac{1}{\gamma_i}\%$ of the time in each piconet it belongs to.

Let us denote by N the total number of nodes (which will be considered static) in the scatternet and by M the number of resulting piconets.[2] For the generic piconet $i \in \{1, 2, \ldots, M\}$, let n_i be the total number of nodes (master, slaves and gateways) the piconet consists of. Moreover, let ϑ_i denote the number of units that act as gateways in the i-th piconet. Clearly, it has to be $\vartheta_i \leq n_i$ and $n_i \leq 8$, since a piconet cannot have more than 8 active units. It may be worth noting that each gateway is counted in the ϑ_i of all the piconets it belongs to. The total number of gateways in the scatternet is denoted by Θ. Thus, the following relation holds:

$$N = \sum_{i=1}^{M} (n_i - \vartheta_i) + \Theta \,. \tag{1}$$

We denote by $\Delta = [\{\delta_{i,j}\}]$ the $N \times N$ *end–to–end traffic matrix*, so that the (i, j)–th entry $\delta_{i,j}$, with $i, j \in \{1, 2, \ldots, N\}$, is the average user traffic (expressed in packet/slot) generated by node i to node j. Note that, the end–to–end traffic matrix Δ refers to the traffic produced at the transport layer, and, hence, it does not depend on the specific network topology.

Given the end–to–end traffic matrix, network topology and routing tables,[3] we can univocally determine the *effective traffic matrix* $\Lambda = [\{\lambda_{i,j}\}]$, which gives the actual data traffic flowing in the physical network links. Since, in Bluetooth, direct communication can occur between master and slave (or gateway) units only, the (i, j)–th entry$\lambda_{i,j}$ of the effective matrix Λ can be greater than zero only if either i or j is a master node.

4 Network Capacity

We are interested in determining the ability of a given scatternet topology to carry information flows. Such an ability will be measured in terms of *network capacity*, defined as the maximum aggregate traffic that can be injected into the network preserving the network stability. In this context, a network is said to be stable if, for any of its nodes, the number of buffered packets does not grow indefinitely.

As a first step, network capacity will be evaluated without considering the inter–piconet interference [14,15]. The effect of interference, however, will be considered in the successive sections.

Given the scatternet topology, we can define the set Ω of all the end–to–end traffic matrices Δ that preserve the network stability.[4] Then, for a traffic matrix $\Delta = [\{\delta_{i,j}\}]$, we define the *offered load*, $G(\Delta)$, as:

[2] We have as many piconets as nodes playing the role of master.

[3] In the following we will consider balanced routing algorithms, in the sense that, if node A wants to communicate with node B and there are P shortest paths, the traffic $A \to B$ is equally divided among the P routes.

[4] Note that, Ω is strictly related to the network topology.

$$G(\Delta) = \sum_{i=1}^{N} \sum_{j=1}^{N} \delta_{i,j} \, . \tag{2}$$

It may be worth stressing that G gives the effective aggregate user traffic offered to the network, since the traffic carried over multiple hops is counted just once. Given these preliminary definitions, we can finally define the ***network capacity*** C as follows:

$$C = \max_{\Delta \in \Omega} \{G(\Delta)\} \, . \tag{3}$$

Analogously, for a given network topology, we can define the set Ω^* of all stable effective matrices $\Lambda = [\{\lambda_{i,j}\}]$. Hence, we define the *aggregated throughput* $S(\Lambda)$ of the network as:

$$S(\Lambda) = \sum_{i=1}^{N} \sum_{j=1}^{N} \lambda_{i,j} \, ; \tag{4}$$

where also the traffic forwarded in a multi–hop fashion is taken into account. The parameter S gives the total traffic flowing in the physical links of the network.
Finally, we define the ***transport capacity*** T as:

$$T = \max_{\Lambda \in \Omega^*} \{S(\Lambda)\}. \tag{5}$$

Trivially, under stability conditions, the offered load G cannot exceed the aggregated throughput S and the two quantities are equal if all the traffic goes through a single hop. Taking the maxima of G and S and using the notation defined above, we get the following lemma.

Lemma 1. *The network capacity C equals the transport capacity T: $C = T$.*

Proof. First we note that, for a given scatternet topology, the set of stable end–to–end matrices Ω includes the set of stable effective matrices Ω^*. Thus, if $\Lambda \in \Omega^*$ achieves the transport capacity T, then Λ belongs to $\Omega^* \subset \Omega$ and, consequently, by taking $\Delta = \Lambda$, we get an offered traffic $G(\Delta) = T$. Therefore, $C \geq T$.
Now, let $\Delta \in \Omega$ achieve the network capacity C. Thus, the corresponding effective matrix Λ, which is univocally associated to Δ (the mapping is surjective but not one–to–one), is stable: $\Lambda \in \Omega^*$. Since $S(\Lambda) \geq G(\Delta) = C$, then $T \geq C$ and the lemma is proved.

In short, Lemma 1 trivially says that the maximum aggregate traffic that can be injected into the network, maintaining stability, is equal to the aggregate link capacity of the network. Consequently, the optimal network utilization is reached when communication occurs between nodes that are directly connected. In other words, the system capacity is reached when no forwarding over multiple hops is needed (local traffic) and, thus, all traffic gets through a single hop only.

4.1 Capacity of Isolated Piconets

As far as the capacity of a single piconet is concerned, it is clear that, under the collision–free MAC scheme provided by the Bluetooth system, the resulting capacity is $C = 1$ packet/slot. Hence, we have the following lemma:

Lemma 2. *Given M isolated piconets,*[5] *the network capacity is equal to $C = M$ packet/slot.*

This is clearly an upper bound on the capacity achievable by M piconets, since the effect of inter–piconet interference is not taken into account.

4.2 Capacity of Connected Piconets

In our terminology, a piconet is said to be *connected* if it shares at least one unit with some other piconet.

Let us focus on a piconet with a single gateway. Let us assume the gateway is shared among β piconets in such a way it is present in each piconet for a fraction $\alpha = 1/\beta$ of the time. We may distinguish two cases.

a) The shared unit is the master of the piconet. Hence, the piconet is active only for a fraction α of the time and the piconet capacity is $C = \alpha$ packet/slot.
b) The shared unit is a slave of the piconet. Therefore, the piconet capacity is $C = 1$ packet/slot when the piconet consists of more than two nodes, $C = \alpha$ packet/slot otherwise.

Two considerations arise from this simple analysis: to approach the bound given by Lemma 2 all interpiconet units ought to be slaves in all the piconets they belong to. Furthermore, to fully exploit the transport capacity of the network, each piconet has never to be empty, i.e., at least a slave must be present in each piconet at any time. Summing up, we get the following proposition:

Proposition 1. *A scatternet consisting of M piconets may achieve an aggregated throughput of at most M packet/slot. This happens if the following conditions hold:*

1. *the shared units are slaves in all the piconets they belong to;*
2. *in each piconet, there must be at least one slave present at any time.*

It is not difficult to find structures that satisfy the above conditions. An example is given by the simple chain configuration depicted in Fig. 1, where it suffices to synchronize the gateways in an odd–even fashion, in such a way that there is not more than one gateway in each piconet at any time.

Under the conditions of Proposition 1, the capacity turns out to be linearly increasing with the number of piconets.[6] Therefore, it may be interesting to determine the maximum capacity a scatternet of N nodes may achieve. The answer is given by the following proposition.

[5] A piconet is said to be *isolated* when the nodes in the piconet do not belong to any other piconet.
[6] We recall that, in this section, interpiconet–interference is not considered.

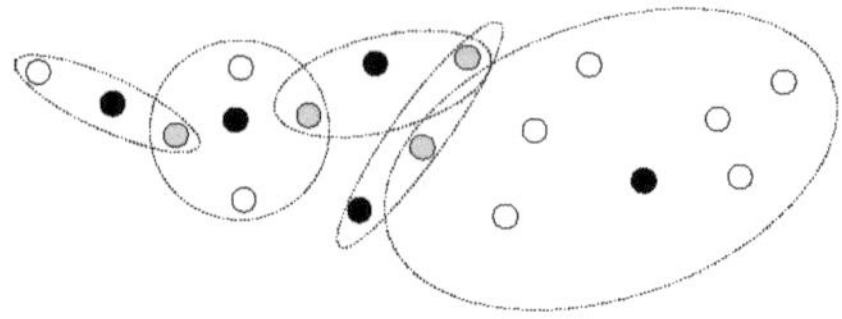

Fig. 1. Chain configuration.

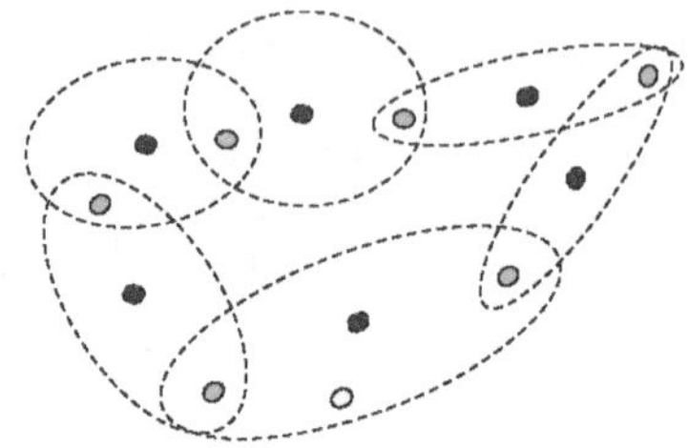

Fig. 2. Closed loop configuration.

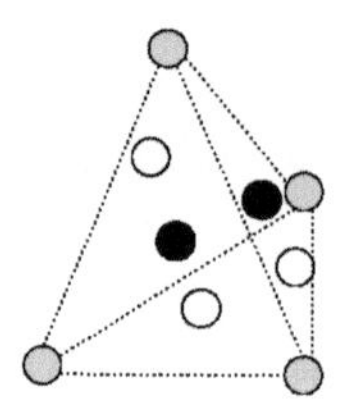

Fig. 3. Tetrahedral configuration (only the nodes of two piconets are shown).

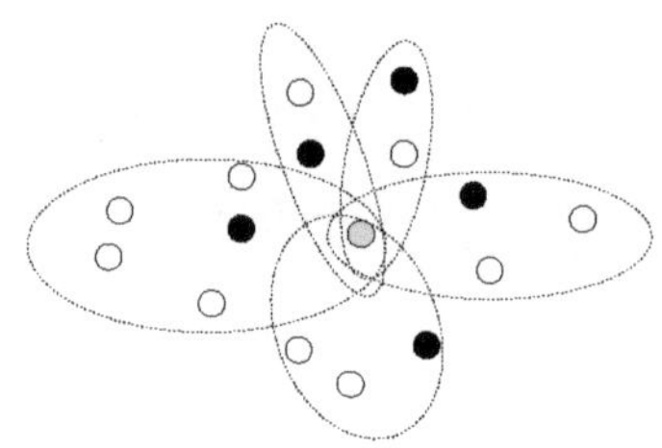

Fig. 4. Single gateway configuration.

Proposition 2. *A scatternet consisting of N interconnected nodes may achieve an aggregated throughput of at most $\lfloor \frac{N}{2} \rfloor$ packet/slot.*

Proof. The proof is constructive. We first find a configuration which achieves the bound. To this purpose, we organize the nodes in a closed–loop topology, as depicted in Fig. 2, so that each piconet consists of $n = 3$ nodes (except for at most one piconet, with $n_M = 4$ if N is odd) and $\vartheta = 2$ gateways. The total number of piconets and gateways are, then, equal to $M = \lfloor \frac{N}{2} \rfloor$ and $\Theta = \lfloor \frac{N}{2} \rfloor$. The regularity of the structure enables us to find a scheduling of the gateways such that there is always at least one slave (or gateway) in each piconet. Thus, conditions of Proposition 1 are satisfied and, consequently, the limit capacity of the network is equal to M packet/slot.

To prove the proposition, it remains to verify that the capacity of a scatternet containing N nodes cannot exceed $\lfloor \frac{N}{2} \rfloor$ packet/slot. Let us assume, by contradiction, that a scatternet topology of N nodes achieves a network capacity C greater than $\lfloor \frac{N}{2} \rfloor$ packet/slot. This requires that, at each time, C master–slave pairs are actively exchanging data–packets. Since we cannot extract more than $\lfloor \frac{N}{2} \rfloor$ distinguished pairs of nodes out of a total of N nodes, the assert is clearly a contradiction. Hence, the scatternet capacity cannot exceed the limit $\lfloor \frac{N}{2} \rfloor$ packet/slot and the proposition is proved.

Note that, the above results may provide useful insight into efficient configuration design. For instance, we can observe that the closed–loop configuration is asymptotically optimal, since it achieves limiting capacity for any N. For a fixed N, however, it is easy to find different configurations which achieve the bound. For example, when $N = 8$, the limiting capacity may be achieved by a tetrahedral structure, as that shown in Fig. 3.

Another interesting observation concerns the ring architecture considered in [16], where all nodes act as master/slave relays. It is easy to prove that also this configuration (which can be, in a certain sense, considered a degeneration of the closed–loop architecture) is asymptotically optimal. (In fact, it seems that these are the only two asymptotically optimal architectures). The ring topology, however, presents scalability problems and performance rapidly worsens when traffic has to be forwarded in a multi–hop fashion. Furthermore, it can be observed from the above propositions that also tree structures [4] are inefficient from the point of view of network capacity (and, hence, spectrum utilization).

In the above analysis, the inter–piconet interference has been neglected. However, the impact of inter–piconet interference will be considered within the following sections, where some possible scatternet configurations are analyzed in case of homogeneous end–to–end traffic matrices.

5 Efficient Configurations for Bluetooth Scatternets

From the analysis presented in the previous section, it turns out that network capacity may be achieved with very local traffic only. However, a scatternet topology ought to be efficient under a wide range of traffic patterns.

In this section we provide a sort of worst case analysis, considering uniform end–to–end traffic matrices, where each node generates, on average, the same amount of traffic to every other node in the network. It is clear that, in this situation, any reordering of the nodes would not increase the traffic locality. Consequently, network performance for uniform traffic matrices will dependent on the scatternet topology only, and not on single node position.

Therefore, we consider a uniform end–to–end traffic matrix Δ, having the following structure:

$$\Delta = \begin{pmatrix} 0 & \delta & \dots & \delta & \delta \\ \delta & 0 & \dots & \delta & \delta \\ \vdots & \ddots & \ddots & \ddots & \vdots \\ \delta & \delta & \dots & 0 & \delta \\ \delta & \delta & \dots & \delta & 0 \end{pmatrix}; \tag{6}$$

where the (i, j)-th entry, $i, j = 1, 2, \dots, N$, represents the average traffic generated from node i to node j (in packet/slot). In this situation, which models a cooperative computing scenario, the scatternet configuration we look for should present three desirable properties:

1. **efficiency**: the scatternet ought to support maximum traffic;
2. **fairness**: all the piconets should experience the same traffic load;
3. **scalability**: the resulting structure should scale well with respect to the number of nodes.

The simplest configuration that meets the above requirements is the single–gateway topology. In this case, the resulting scatternet is star–shaped, as depicted in Fig. 4.

Next, we will consider the closed–loop topology, to end up with a discussion on some efficient configurations based upon platonic solids.

We first introduce some stability issues, which are at the basis of the successive capacity evaluation. Then, we introduce a model for inter–piconet interference, derived from [14]. Finally, we show some numerical results.

5.1 Stability Conditions

Let us consider again the effective traffic matrix Λ, introduced in Section 3. Each element $\lambda_{i,j}$ of this matrix represents the average number of packets per slot sent along the link from node i to node j. We remark that effective links are present only between master and slave or gateway units. Thus, if $\lambda_{i,j} > 0$ either i or j has to be a master unit, while the other has to be a slave or gateway unit. Applying basic queueing theory, it can be shown that for an admissible link (i, j), the stability condition, under the assumption of Pure Round Robin (PRR) polling scheme, is given by [17]:

$$\lambda_{i,j} < \frac{1}{2(n_{i,j} - 1)},$$

where $n_{i,j}$ is the number of nodes in the piconet that i and j belong to. The above inequality trivially means that for a piconet to be stable each link must not receive, on average, more than a packet per polling cycle.

When connected piconets are considered, however, we have to take into account the presence of gateways, which will be active only for a fraction of the time. Hence, adding dependence on the time t, we may write:

$$\lambda_{i,j}(t) < \frac{1}{2(n_{i,j}(t) - 1)} \cdot b_{i,j}(t), \tag{7}$$

where $n_{i,j}(t)$ represents the number of nodes in the considered piconet at a given time instant t, and $b_{i,j}(t)$ is either 1 or 0 according to the active or idle state of link (i, j) at time t. Clearly, both $n_{i,j}(t)$ and $b_{i,j}(t)$ depend on the presence of gateways in the piconet of interest. For the regular configurations we are interested in, the network shows a periodical behavior, in the sense that there exists a cycle period of duration T_C,[7] such that $n_{i,j}(t) = n_{i,j}(t + kT_C)$ and $b_{i,j}(t) = b_{i,j}(t + kT_C)$ for each integer k. Furthermore, we may assume that the cycle time T_C may be partitioned in intervals in which the scatternet configuration does not change. Denoting with Q the cardinality of the partition, we have:

$$(0, T_C] = (0, T_1] \cup (T_1, T_2] \cup \ldots \cup (T_{Q-1}, T_Q]. \tag{8}$$

Let n_m and b_m be the values assumed by $n_{i,j}(t)$ and $b_{i,j}$ in $(T_{m-1}, T_m]$, respectively. Furthermore, let $\overline{\lambda}_{i,j}$ be the average traffic offered over a period T_C, given by

$$\overline{\lambda}_{i,j} = \frac{1}{T_C} \int_0^{T_C} \lambda_{i,j}(t) dt \ .$$

[7] It is worth stressing that this cycle time has not to be confused with the cycle time of the polling algorithm.

Table 1. Relationship between N, n and M for different scatternet topologies.

Topology	Number of piconets
single–gateway	$M = \frac{N-1}{n-1}$
closed–loop	$M = \frac{N}{n-1}$
tetrahedron	$M = 4$

Averaging both sides of inequality (7) over the cycle period, we get:

$$\overline{\lambda}_{i,j} < \frac{1}{2T_C} \sum_{m=1}^{Q} \frac{b_m}{n_m - 1}(T_m - T_{m-1}) ; \tag{9}$$

where T_0 is assumed equal to zero. Thus, inequalities (9) for each pair (i, j) of network nodes give the overall network stability conditions.

5.2 Performance Evaluation for Uniform Traffic

Given the conditions on the offered traffic that assure network stability, we can derive the maximum throughput that can be achieved by the network for different scatternet topologies. Once more, we neglect the effect of inter–piconet interference, thus obtaining upper bounds for network performance. Such bounds will be, then, used in the following subsection to derive performance estimation in the presence of interference.

Let N be the total number of units in the scatternet; M the number of piconets and $n_i,\quad i = 1, \ldots, M$ the number of hosts in the i-th piconet. For the sake of simplicity, in the following we assume that all the piconets have the same number of units, i.e. $n_i = n$, $i = 1, \ldots, M$. Tab. 1 shows the relationship that ties N, n and M for some scatternet topologies that will be considered in the following.

The basis for the evaluation of network capacity is the stability condition given by (9). Applying equation (9) to the end–to–end traffic matrix Δ given by (6), we find a condition of the form $\delta < \zeta$, where ζ depends on the scatternet topology. Thus, the offered traffic $G(\Delta)$ will satisfy the relation $G(\Delta) < N(N-1)\zeta$. Considering the set $\mathcal{U}$ of all the uniform end–to–end traffic matrices Δ which preserve the network stability, we define the *uniform capacity* $\tilde{C}$ as the maximum offered traffic that can be reached within $\mathcal{U}$, i.e.,

$$\tilde{C} = \max_{\Delta \in \mathcal{U}} G(\Delta) = N(N-1)\zeta \ . \tag{10}$$

Due to our assumptions on routing and scatternet structure, the computation of $\tilde{C}$ for the different topologies we consider may be reduced to the evaluation of the maximum traffic rate achievable in a generic master–gateway link. Indeed, it is easy to realize that, with uniform end–to–end traffic connections, the stability conditions (9) are satisfied for each link (i, j) as long as they are verified by all the master–gateway links.

Planar Topologies. Let us, first, evaluate the uniform capacity for two simple planar structures, namely the star–shaped and the close–loop topologies.

The star topology relies on a single gateway to forward traffic among the M piconets. The average aggregate traffic offered to the gateway by the nodes of each piconet is equal to $(n-1)(N-n+1)\zeta$ packets per slot. Since the gateway spends a fraction $1/M$ of its time in each piconet, getting a fraction $1/(2(n-1))$ of the piconet capacity during its permanence time, the limiting capacity (expressed in packet/slot) turns out to be

$$\tilde{C} = \frac{N(N-1)}{2M(n-1)^2(N-n+1)} = \frac{N}{2(n-1)(N-n+1)} . \tag{11}$$

It is clear that, in such a situation, the gateway represents a bottleneck that limits the performance of the network. In order to overcome this weakness, a closed–loop structure, as depicted in Fig. 2, can be considered. In this case, routing should be taken into account. We assume that the routing algorithm preserves somehow the symmetry of the structure (i.e., each gateway handles equal load), and that gateways are synchronized, so that a single gateway at a time is active in each piconet. Under these assumptions, each gateway node gets a traffic rate of approximately $\zeta \cdot N^2/8$ packet/slot from each one of the two piconets it belongs to. Since the gateway gets, on average, a fraction $1/(2(n-2))$ of the piconet capacity, the uniform capacity (in packet/slot) is approximately equal to:[8]

$$\tilde{C} = \frac{8N(N-1)}{4(n-2)N^2} = \frac{2(N-1)}{N(n-2)} . \tag{12}$$

Solid Topologies. In order to achieve better performance, we might add another dimension to the network, so that the resulting structure may be represented by a *solid* structure. In this case, each piconet can be represented as a regular polygon having the gateways as vertices. To have a "fair" structure, the resulting polyhedron has to be regular. To avoid problems with the routing schemes employed, the resulting polyhedron has also to be convex. However, it is a well–known result (which goes back to the ancient Greeks) that there are just five regular convex polyhedra, namely tetrahedron, cube, octahedron, dodecahedron and icosahedron. These are usually referred to as the Platonic solids.

Denoting by F and V the number of facets and vertices of the polyhedron, respectively, and by E the number of edges of the building polygon, we have that the minimal and maximal number of Bluetooth nodes such a structure may handle are equal to:

$$N_{min} = F + V; \tag{13}$$
$$N_{max} = (8-E) \cdot F + V. \tag{14}$$

We focus on the tetrahedron structure, the analysis for the other cases being similar. We assume perfect synchronization among gateways, so that at each time instant there will be only one gateway present in each piconet. Such a scheduling of the gateways is made possible by the regular structure of the tetrahedron topology. Thus, a gateway will spend one third of its time in each piconet it belongs to. During the permanence in a given piconet, the gateway gets a fraction $1/(n-3)$ of the piconet capacity. Let us, now, introduce some notation. Let GW_1, GW_2, GW_3, GW_4 denote the four gateways

[8] For the sake of conciseness, we omit the derivation of the exact expression of the uniform capacity that, however, can be easily obtained by distinguishing the cases of M even and odd.

at the vertices of the tetrahedron. Furthermore, let P_i denote the only piconet that GW_i does not belong to, i.e., the piconet on the facet opposite to vertex i. Finally, we will use the notation $\mathcal{X}_i \rightarrow \mathcal{Y}_j$ to indicate the aggregate traffic produced by the node(s) in $\mathcal{X}_i$ to the node(s) in $\mathcal{Y}_j$, where $\mathcal{X}_i$ and $\mathcal{Y}_j$ can be piconets or gateways.

Given the symmetry of the network, we can focus on a specific link, for instance the link between the master of P_4 and the gateway GW_1. The traffic along this link is given by the sum of the following terms:
a) $P_4 \rightarrow GW_1$, equal to $(n-3)\zeta$;
b) one half of $\{P_2, P_3\} \rightarrow P_4$, equal to $(n-3)^2\zeta$;
c) one third of $P_4 \rightarrow GW_4$, equal to $(n-3)\zeta/3$;
d) one third of $P_1 \rightarrow GW_1$, equal to $(n-3)\zeta/3$;
e) one sixth of $GW_2 \rightarrow P_2$ and $GW_3 \rightarrow P_3$, equal to $2(n-3)\zeta/6$;
f) one half of $\{GW_2, GW_3\} \rightarrow GW_1$, equal to ζ.
Thus, the uniform capacity turns out to be:

$$\tilde{C} = \frac{N(N-1)}{6(n-3)(n-2)^2} \,. \tag{15}$$

5.3 Performance with Inter–piconet Interference

The mutual interference among overlapping piconets may have a strong effect on scatternet performance. Let $P_s(M)$ be the probability that two stations in piconet A successfully exchange a packet when $M-1$ other piconets are collocated in the scatternet. Therefore, the effective uniform network capacity can be expressed as:

$$C' = \tilde{C} \cdot P_s(M) \, ; \tag{16}$$

where $\tilde{C}$ is given by (11), (12) and (15). We evaluate inter–piconet interference in the worst case, i.e., with always–transmitting unsynchronized piconets. Thus, according to [14], we have $P_s(M) = a^{M-1}$, where the parameter a is equal to $a = 0.9852$.

Fig. 5 depicts the uniform capacity C' versus the number of nodes. We have plotted two sets of curves, obtained for $n = 8$ and $n = 3$, respectively. (Clearly, the parameter n has been varied only for the single–gateway and closed–loop configurations, since for the tetrahedral configuration, the number of nodes per piconet is univocally determined by N.) As expected, the single–gateway topology always performs worse than the others, due to the bottleneck introduced by the gateway. The tetrahedron topology achieves the highest capacity, but only for a very small number of nodes N. On the contrary, the closed–loop topology achieves good performance for a wide range of N values, provided that the piconets are "slim", i.e., the number n of nodes per piconet is small.

In Fig. 6, the uniform capacity for single–gateway and closed–loop topologies is depicted for a high number of nodes N. As expected, the overall capacity tends towards zero as N grows, because of the increasing interference.

Given the number of nodes N and the scatternet topology, it may be interesting to evaluate the uniform capacity for different number of piconets M. The left–hand side of Fig. 7 shows the number of piconets M that maximizes the uniform network capacity at the varying of the total number N of nodes in the scatternet, for the single–gateway

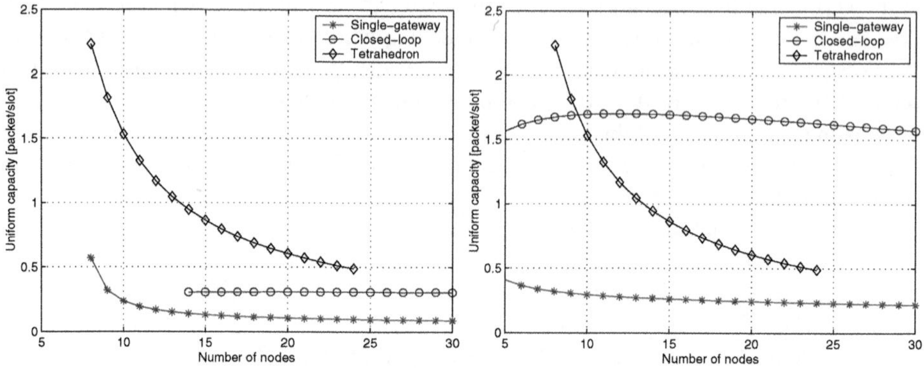

Fig. 5. Uniform capacity C' versus number of nodes N for $n = 8$ (left–hand side) and $n = 3$ (right–hand side).

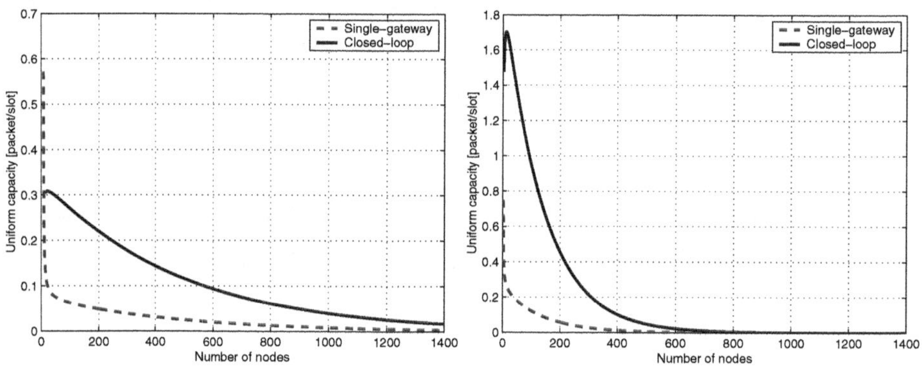

Fig. 6. Uniform capacity C' for high number of nodes N, for $n = 8$ (left–hand side) and $n = 3$ (right–hand side).

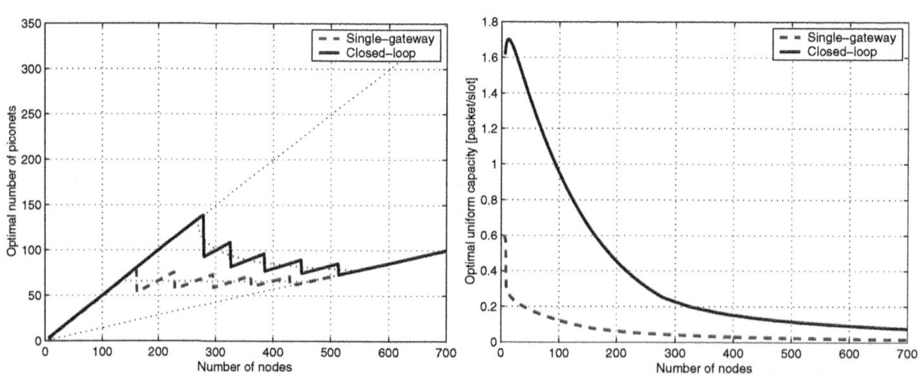

Fig. 7. Optimal number of piconets M and maximum capacity versus the number of nodes N, for the closed–loop and single–gateway topologies.

and closed–loop topologies. The right–hand side of Fig. 7 shows the maximum capacity achieved by choosing the optimal M for each N. We can note that building up "slim" piconets seems to be preferable, with respect to the overall throughput, for N smaller than ≈ 250. Then, the behavior is rather variable until N is bigger than ≈ 550, after which "fat" piconets, i.e., piconets consisting of $n = 8$ nodes, represent the best choice because of the strong influence of interference. Note that, the "comb"–like trend that can be observed in the left–hand side plot of Fig. 7 for values of N between 250 and 550 is due to the assumption that all the piconets contain the same integer number n of nodes.

6 Conclusions

In this paper we presented an analysis on the impact of the nodes configuration in a Bluetooth scatternet. We discussed how Bluetooth nodes should be organized to build up a scatternet, where the efficiency of the resulting configuration is measured in terms of network capacity. We presented a theoretical study of intrinsic capacity limits of a scatternet, where the maximum throughput may be achieved under local traffic. Hence, we performed a sort of worst–case analysis by considering uniform end–to–end traffic matrices. We showed the impact of mutual interference term on network capacity, and studied how it affects the optimal configuration choice. We thus introduced some regular configurations based upon Platonic solids, which achieve high capacity, even if just for a low number of nodes. As an overall result, closed–loop structure seems to represent a valuable choice, since it scales well with respect to the number of nodes. Some promising directions for future work seem to be the definition of scatternet formation algorithms which connect nodes building up an efficient network and the incorporation into such a framework of the constraints due to the spatial dimension of the network, which has not been treated in the present paper. Moreover, the analysis may be extended to more realistic traffic matrices as, for instance, sparse matrices, for which not only the scatternet topology but also the nodes disposition in the piconets may determine substantial performance variations.

References

1. Bluetooth special interest group. [Online]. http://www.bluetooth.com
2. J. Karaoguz. High-rate wireless personal area networks. *IEEE Communications Magazine*, pages 96–102, Dec. 2001.
3. C. Law, A. K. Mehta, and K.-Y. Siu. Performance of a new bluetooth scatternet formation protocol. In *Proc. ACM MobiHoc*, Long Beach, CA, 2001.
4. G. V. Zaruba, S. Basagni, and I. Chlamtac. Bluetrees - scatternet formation to enable bluetooth-based ad hoc networks. In *Proc. IEEE ICC*, Helsinki, 2001.
5. S. Basagni and C. Petrioli. Multihop scatternet formation for bluetooth networks. In *Proc. IEEE VTC (Spring)*, Birmingham, AL, 2002.
6. G. Tan, A. Miu, J. Guttag, and H. Balakrishnan. Forming scatternets from bluetooth personal area networks. Technical Report TR-826, MIT Laboratory for Computer Science, 2001.
7. M. Kalia, S. Garg, and R. Shorey. Scatternet structure and inter-piconet communication in the bluetooth system. In *Proc. IEEE National Conference on Communications*, New Delhi, 2000.

8. S. Baatz, C. Bieschke, M. Frank, P. Martini, C. Scholz, and C. Kuehl. Building efficient bluetooth scatternet topologies from 1- factor. In *Proc. IASTED WOC*, Banff, Alberta, 2002.
9. S. Baatz, M. Frank, C. uhl, P. Martini, and C. Scholz. Bluetooth scatternets: An enhanced adaptive scheduling scheme. In *Proceedings 21st Annual Joint Conference of the IEEE Computer and Communications Societies, INFOCOM 2002.*, 2002.
10. P. Bhagwat and S. P. Rao. On the characterization of bluetooth scatternet topologies, 2001. Submitted for publication.
11. A. Capone, M. Gerla, and R. Kapoor. Efficient polling schemes for bluetooth picocells. In *Proc. IEEE ICC*, Helsinki, 2001.
12. M. Kalia, D. Bansal, and R. Shorey. Data scheduling and sar for bluetooth mac. In *Proc. IEEE VTC (Spring)*, Tokyo, 2000.
13. R. Kapoor, A. Zanella, and M. Gerla. A fair and traffic dependent polling scheme for bluetooth. In *Proc. of the joint IEEE conferences ICWLHN 2002 and ICN 2002*, Atlanta, USA, 26–29 Aug. 2002.
14. Amre El-Hoiydi. Interference between bluetooth networks – upper bound on the packet error rate. *IEEE Comm. Lett.*, 5(6):245–247, Jun 2001.
15. A. Zanella, A. M. Tonello, and S. Pupolin. On the impact of fading and inter-piconet interference on bluetooth performance. In *Proc. of IEEE WPMC 2002*, Honolulu, Hawaii, Oct. 27–30, 2002.
16. C. C. Foo and K.C. Chua. Bluerings - bluetooth scatternets with ring structures. In *Proc. IASTED WOC*, Banff, Canada, 2002.
17. L. Kleinrock. *Queueing Systems*. John Wiley & Sons, New York, 1975.

Enhancing Scatternets Performance via Scheduling Algorithm Parametrization

Werner Priess[1], José Ferreira de Rezende[2], and Luci Pirmez[1]

[1] NCE – Universidade Federal do Rio de Janeiro
Caixa Postal 2324,
20001-970 Rio de Janeiro, RJ, Brazil
priess@posgrad.nce.ufrj.br
luci@nce.ufrj.br

[2] COPPE/PEE – Universidade Federal do Rio de Janeiro
Caixa Postal 68504,
21945-970 Rio de Janeiro, RJ, Brazil
rezende@gta.ufrj.br

Abstract. Bluetooth specification still has open issues, including the intra and interpiconet scheduling topics. This paper proposes an interpiconet scheduling algorithm, referred to as AISA (Adaptive Interpiconet Scheduling Algorithm). AISA is characterized by: its adaptability to varying network traffic conditions; and its ability to optimize specific performance metrics via parametrization. Both features enable it to be employed in a variety of scenarios with improved performance showed by the simulation results.

Keywords: *ad hoc* networks, Bluetooth, scatternet, scheduling, QoS.

1 Introduction

Bluetooth is a promising radio technology for *ad hoc* networking. Its interfaces are small in size and are energy efficient at an increasingly lower cost. The Bluetooth network topology, or *piconet*, has a maximum of eight devices. In order to extend the network range and the number of devices, the scatternet concept was created. A scatternet is a network formed of two or more piconets interconnected by shared nodes that will be referred to as *bridges* throughout the paper.

Research in scatternets is recent and a number of open issues remains. Interpiconet scheduling is one of them and it defines the mode by which the bridges participate in the piconets that they are connected to. Since a bridge is usually attached to only one Bluetooth interface, it must alternate in time its participation in multiple piconets. Most existing interpiconet scheduling proposals evaluate aggregate throughput and packet delay metrics, but few proposals handle the power consumption performance metric. Moreover, their application is usually limited to specific scenarios.

M. Conti et al. (Eds.): PWC 2003, LNCS 2775, pp. 741–755, 2003.

This paper presents an algorithm for interpiconet scheduling, referred to as AISA (Adaptive Interpiconet Scheduling Algorithm). Parametrization is the key concept behind AISA, enabling the bridges to be configured so as to optimize one chosen performance metric like, for example, throughput, packet delay or power consumption. The fact that AISA is applied solely to the bridges minimizes the need for adaptations to the Bluetooth standard. In this study AISA performance was evaluated by way of three simulated scenarios whose focused metrics differed. The results showed that the algorithm performed well in all three situations.

This paper is organized as follows: Section 2 presents a brief overview of Bluetooth technology; Section 3 summarizes the related work on interpiconet scheduling; Section 4 describes AISA; Section 5 presents the simulated scenarios and their results; and the conclusions are presented in Section 6.

2 Bluetooth Specification

Bluetooth [1,2] is a short range and low power radio technology intended to replace cable connections between electronic devices. It may also be used to create small wireless *ad hoc* networks. The Bluetooth Special Interest Group (SIG) [1] released an open specification with two parts: the Core and the Profiles. The Core Specification defines radio characteristics and the protocol stack. The Profiles define which protocols of the stack should be implemented for each application. In this section we provide some information about the Core Specification.

2.1 Physical and Link Layers

The Bluetooth radio operates in the unlicensed Industrial, Scientific and Medical (ISM) band at 2.4 GHz and it uses a fast frequency hopping scheme. It hops over 79 channels (23 in some countries) displaced by 1 MHz at the rate of 1600 hops per second, corresponding to a 625 μs time slot.

The baseband is responsible for creating piconets and links. The piconet is a network with at most eight active devices that share the same frequency hopping scheme. One device becomes the piconet master and the others behave as slaves. The master dictates the hopping sequence and its phase.

A time division duplex (TDD) scheme is used where master and slaves alternatively transmit. A slave is allowed to transmit only if it has been addressed by the master in the prior slot. In every new slot, the devices of a piconet hop to the next frequency of the hopping sequence.

The baseband handles two types of links: Synchronous Connection-Oriented (SCO) and Asynchronous Connectionless (ACL). SCO is a symmetric point-to-point link between the master and a single slave, maintained with slot reservation at regular intervals by the master. ACL is a point-to-multipoint link between the master and all slaves participating on the piconet. ACL traffic may only occupy slots not reserved for SCO links.

A major Bluetooth concern is power consumption. There are three low power operation modes: Sniff, Hold and Park. In Sniff mode, a slave listens to the piconet only at periodic time slots, called sniff slots. In Hold mode, a slave goes into sleep for a specified time period, after which it returns to Active mode. In Park mode, a slave releases its active member address, but remains synchronized to the piconet for future activation.

The Link Manager Protocol (LMP) and the Logical Link Control and Adaptation Protocol (L2CAP) form the Bluetooth link layer. LMP is responsible for configuring and managing baseband connections. For example, when a bridge wants to enter Hold power saving mode, it communicates this fact to the piconet master through LMP signaling packets. L2CAP provides connection-oriented and connectionless services to upper layer protocols with protocol multiplexing capability and segmentation and reassembly operation (SAR).

2.2 Scatternets

If devices on different piconets want to communicate, these piconets may be interconnected, creating a scatternet. The scatternet is formed when at least one device, referred to as a bridge, participates of two or more piconets. Bridges may be attached to only one Bluetooth interface, preventing them from being active in more than one piconet at the same time. Therefore, a bridge alternates in time its participation in multiple piconets. This task is called interpiconet scheduling. A bridge can be a slave in more than one piconet, but a master in only one.

The scatternet concept brought along new issues regarding its implementation. These issues have been addressed by several research studies. Research related to interpiconet scheduling is discussed in the next section.

3 Related Work

Research on scatternets is concentrated in three main topics: topology formation [3,4], packet forwarding [5,6] and interpiconet scheduling [7,8,9,10,11]. The first two topics are out of scope of this paper.

According to our point of view, the interpiconet scheduling algorithms may be divided into two categories, depending on the device that coordinates the scheduling process [12]: mechanisms with isolated decision and with distributed decision. In the first case, the bridge itself decides about its presence in the piconets it belongs to and it may communicate (or not) this decision to the masters of these piconets (if acting as a slave). These algorithms are usually simpler to implement, requering few modifications in Bluetooth Specification. In mechanisms with distributed decision, decisions about future meeting points between the bridge and each master from the piconets in which it participates as a slave result from agreements between both devices. These agreements enable information exchange about the scatternet topology, making better throughput results possible. However, the distributed decision approach requires new LMP messages, which increases the implementation complexity.

Initial studies dealt with scatternet scheduling in a generic way. Their major concern was to study traffic behavior and not specifically scheduling algorithms [13]. Gerla et al. [14] introduced the *rendezvous point* concept, meaning the slot in which a bridge and a piconet master decide to communicate. They also presented the *rendezvous window* which is basically the rendezvous time period.

Among the distributed decision proposals, Johansson et al. [7] presented a scatternet scheduling framework, based on their proposed JUMP mode. A device in JUMP mode on a piconet is, by default, absent of that piconet. A bridge is able to alternate between piconets without explicitly notifying it. However, the inclusion of a new mode to the Bluetooth link controller may not be possible.

Kapoor et al. [8] presented another distributed decision mechanism based on the *rendezvous point* (RP) concept. Each master maintains a list containing its RPs with the bridges, and a list containing its bridges RPs with other piconets. This information enables the master to optimize further RP allocation. It is not completely clear in the paper the way information is passed to the master. The main drawback of this algorithm is that it demands a lot of messages to be exchanged to keep masters up-to-date.

The Tree Scatternet Scheduling (TSS) scheme [9] was designed to work with the Tree Scatternet Formation mechanism from the same authors. A tree-based network topology simplifies the scheduling task, and it enables a global coordination among all piconets to be achieved. On the other hand, the algorithm applicability is restricted to some specific topologies without loops.

As for the research of isolated decision mechanisms, Racz et al. proposed the Pseudo-Random Coordinated Scatternet Scheduling (PCSS) algorithm [10]. Devices assign meeting points with their peers through a pseudo-random process. These meeting points will be different for each pair of nodes. An advantage of PCSS is the coordination among devices with no explicit signaling needed. However, as the number of devices increases, the meeting points are likely to collide.

Har-Shai et al. proposed the Load Adaptive Algorithm (LAA) [11] that operates only on bridges. In this algorithm, each bridge adapts to traffic variations by observing its queues and receiving information about the other end node queue. Probably a new field will be necessary in data packets to piggyback this queue information. Currently, LAA is applicable to small scatternets, in which bridges connect only two piconets.

Summarizing, most proposals present the following limitations. A bridge may belong to only two piconets, always working as a slave. They do not scale in terms of the number of scatternet nodes. Proposals evaluate aggregate throughput and packet delay metrics, but only PCSS [10] handles the power consumption performance metric.

4 Proposed Interpiconet Scheduling Algorithm: AISA

This section presents the Adaptive Interpiconet Scheduling Algorithm (AISA). AISA differs from other interpiconet scheduling mechanisms in that it enables the

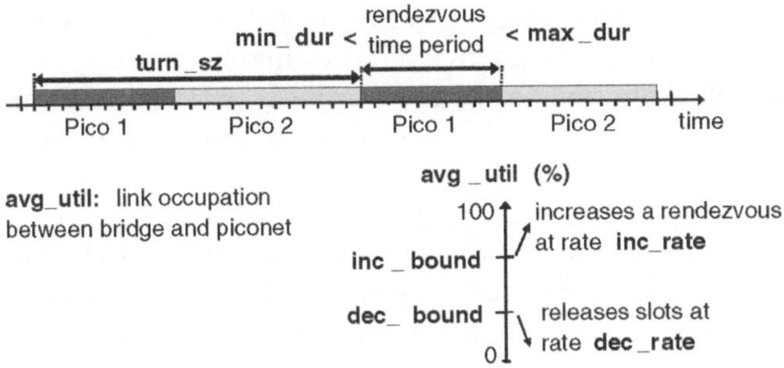

Fig. 1. AISA parameters

choice of a performance metric to be optimized by configuring parameters that work as performance metric tuning knobs. More specifically, depending on the algorithm parametrization it is possible to prioritize one of three metrics: traffic aggregate throughput, packet delay or power consumption. Moreover AISA adapts to varying traffic conditions and provides fairness among flows that cross a bridge.

AISA operates only on bridges. Therefore, it is up to a bridge to decide how long it will remain in each piconet (the corresponding of a RP window). AISA was developed to behave in this way to avoid having to create new signaling packets specific to the scheduling task. As a result, it is possible to minimize modifications in the Bluetooth Specification. According to the classification described in the previous section, AISA fits into the isolated decision mechanism category. The remaining of this section explains how AISA works in terms of its parameters.

A bridge [1] schedules its piconets in a Weighted Round Robin (WRR) [15] fashion. Time is divided up into turns all with the same time period (*turn_size* in Figure 1). During each turn the bridge will be connected to each one of the piconets for a certain length of time (rendezvous time period in Figure 1). The lower limit of the rendezvous time period is referred to as *min_dur* and the upper limit as *max_dur*. When leaving a piconet, the bridge calculates the rendezvous time period with this piconet for the next turn, based on the percentage of slots occupied by data packets in the current rendezvous, i.e., the average link occupation (*avg_util* in Figure 1). As the link occupation increases or decreases, the bridge updates the next rendezvous time period.

If *avg_util* goes below the *dec_bound* limit, the bridge will try to reduce the rendezvous time period by releasing slots at the *dec_rate* rate (see Figure 1). Conversely, if *avg_util* exceeds the *inc_bound* limit, the bridge will try to extend the next rendezvous time period by acquiring slots at the *inc_rate* rate. Note that the *turn_size* is kept constant during all bridge operation.

[1] All explanations consider bridges as slaves in all piconets. When a bridge works as a master in one piconet, it controls this piconet and doesn't need to signal its presence.

It may happen that the bridge requires an extension of a rendezvous time period but there are not enough free slots available. In this case, the bridge tries to remove slots from the longest rendezvous it participates in. However, the rendezvous from which slots are removed cannot have a shorter time period than the one that is being extended. This rule aims to ensure fairness for all piconets.

Before leaving a piconet, the bridge sends a LMP_Hold_Req packet to the piconet master, informing the exact moment that it will enter Hold mode and the moment it will return to Active mode. Thus, the piconet master knows about the bridge absence and removes it from the intrapiconet scheduling scheme during the agreed period.

Bridges try to save power by entering Hold mode, what may happen in two ways. In the first case, as long as the traffic between a bridge and a piconet reduces, slots are set free. These slots are accumulated, and in the remaining slots at the end of each turn the bridge may enter Hold mode until the beginning of the next turn. The second case takes place when a rendezvous time period is already at minimum (*min_dur*) and even so the link utilization (*avg_util*) between the bridge and the piconet is low. Then, the bridge may not schedule this piconet on the next turn, returning to schedule it on the following turn. A piconet may be skipped only if the boolean parameter *skip_pico* is enabled.

Summarizing, the following parameters were introduced: *turn_size*, *min_dur*, *max_dur*, *inc_bound*, *dec_bound*, *inc_rate*, *dec_rate*, and *skip_pico*. Depending on their configuration, a metric performance may be improved or not.

5 Simulations and Results

This section presents three simulated scenarios and their results. Each scenario focuses on one specific performance metric, which are: aggregate throughput, packet delay and power consumption. Depending on the chosen metric, some parameters remain unchanged, while others vary in order to obtain the best metric configuration.

We developed a Bluetooth extension to the Network Simulator (ns-2) [16]. It is referred to as BlueNetS (Bluetooth Network Simulator) and was introduced and validated in [12]. BlueNetS tool models physical and link layer Bluetooth characteristics necessary to simulate traffic communication. The connection establishment procedures are not simulated, and we employ static scatternet configurations. Regarding upper layer protocols (TCP, UDP, IP) and applications, the available ns-2 modules were used.

5.1 Scenario 1 – Throughput Metric

In Scenario 1, AISA parameters were configured to maximize the interpiconet traffic throughput. A scatternet, composed of three piconets interconnected by way of one bridge, was used to demonstrate this bridge capability via parametrization (Figure 2).

Table 1. Fixed AISA parameters in Scenario 1

min_dur	max_dur	inc_rate	dec_rate	skip_pico
20 *slots*	turn_size *slots*	20%	20%	0 (bool)

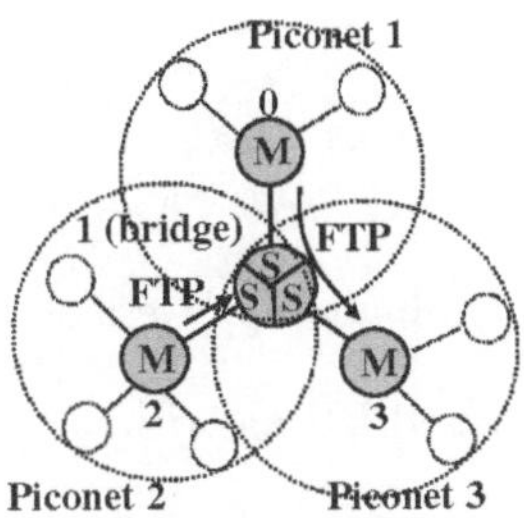

Fig. 2. Scenario 1

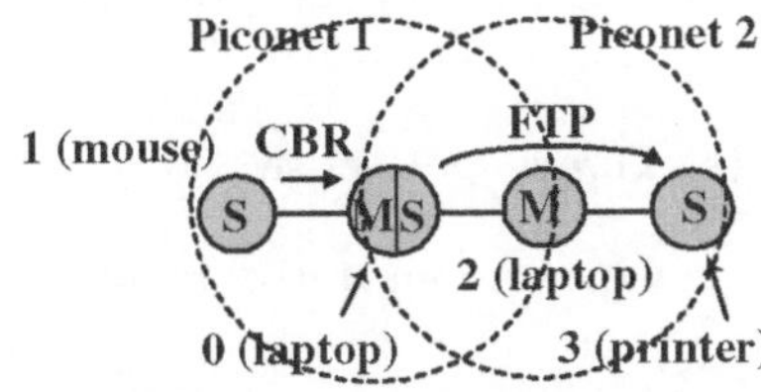

Fig. 3. Scenario 2 (M=master; S=slave)

File transfer traffic (FTP) was configured from 0 to 3, and from 2 to 1. After each file transfer, the FTP sources wait an interval before they initiate the next file transfer. Both file size and time interval between files follow exponential distributions. In traffic between 0 and 3, these distributions have mean values of 30 Kbytes and 1 sec, respectively, and between 2 and 1, 40 Kbytes and 1 sec, respectively. As an example, this traffic model may represent photo file transmissions from a digital camera to a laptop or small printing jobs.

Table 1 shows the parameters that remained unchanged throughout Scenario 1 simulations. The parameters were chosen to enable fast bridge adaptation to traffic changes. In order to maximize aggregate throughput the minimum rendezvous time period between the bridge and a piconet (*min_dur*) should be as small as possible, which allows the bridge to grant slots to piconets with more interpiconet traffic load. As for the increase rate (*inc_rate*) and decrease rate (*dec_rate*), their effect is limited by the number of free and busy slots, respectively, and it has minimum influence on the scenario (so, the chosen values were an average of previously tested values). Finally, as the goal of *skip_pico* is power saving, it was disabled.

5.2 Simulation Results from Scenario 1

The simulation provides comparative results between AISA and Round Robin (RR). AISA was tested with two different configurations: AISA 1, with the limits *inc_bound* (above which a bridge tries to increase the rendezvous time period) and *dec_bound* (below which a bridge tries to release slots) equal to 80% and 60%, respectively; and AISA 2, with *inc_bound* and *dec_bound* equal to 90% e 50%, respectively. The results are averages of 10 simulation runs, each of them with 120 seconds of simulation time. The uncertainty is expressed as 95% confidence intervals (CI).

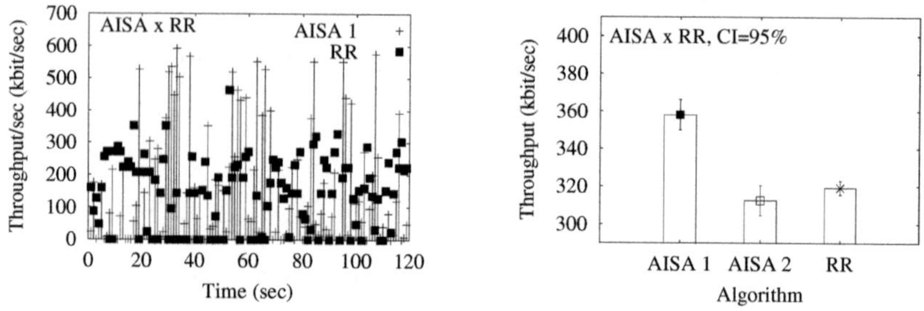

(a) Throughput per second

(b) Simulation average

Fig. 4. Comparative results between AISA and RR algorithms

Since different piconets are not synchronized in time, a bridge loses up to two slots while switching from one piconet to the other. Therefore, as the turn gets smaller, the percentage of lost slots caused by piconet switches grows. In this simulation, the turn duration (*turn_size*) was set to 240 slots [2] (equivalent to 80 slots/piconet in the beginning of each run).

Figure 4 (a) shows the FTP aggregate throughput, measured in each subsequent second interval, to configurations AISA 1 and Round Robin (RR). The curve oscillation is caused by the silence and transmission periods in the modeled traffic. AISA enables traffic peaks above 500 kbit/sec, represented by vertical bars. These peaks are caused by the dynamic slot redistribution of AISA. In RR case, almost all points are under 400 kbps.

Figure 4 (b) shows the average aggregate throughput of FTP flows in the entire simulation. AISA 1 obtained a 15% gain over RR. This difference is even larger with bigger files since peak periods are longer.

AISA 1 provided a better scatternet performance than AISA 2. As *inc_bound* approaches to 100% (the link occupation *avg_util* needs to become larger to reach it), it is more difficult to increase a rendezvous time period. Similarly, as *dec_bound* decreases, it is more difficult to release slots. AISA 2 configuration makes the rendezvous time period more stable, reducing AISA adaptability.

5.3 Scenario 2 – Delay Metric

This scenario is formed by two piconets. At a certain moment, the laptop (master of Piconet 1) establishes a connection to the computer (master of Piconet 2) in order to print a file. In Piconet 2, the laptop is a slave, defining a master/slave bridge. Figure 3 illustrates the scenario.

Since the mouse (in Piconet 1) is an interactive device, it may be considered a delay-sensitive traffic source. It generates one 16-byte packet every 65 ms (similar to the one used by Racz et al. [10]). Background traffic is characterized by the printing traffic flow, which is modeled as a long-sized file transfer (FTP).

[2] *turn_size* values above 240 slots give close results in terms of throughput(see [12]).

Table 2. AISA fixed parameters in Scenario 2

inc_bound	dec_bound	inc_rate	dec_rate	max_dur	skip_pico
80%	60%	20%	20%	turn_size *slots*	0 (bool)

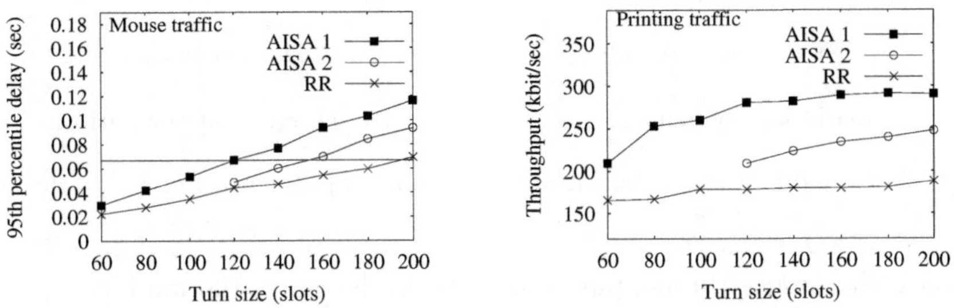

(a) Delay comparison (b) Throughput comparison

Fig. 5. Comparative results between AISA and RR, with two piconets

Table 2 presents AISA parameters that remained unchanged throughout simulations of Scenario 2. Since the mouse traffic rate is low, parameters that influence on how fast a bridge increases or decreases a rendezvous time period (including *inc_bound*, *dec_bound*, *inc_rate* and *dec_rate*) have no effect over mouse generated traffic. However, these parameters affect background traffic throughput. So, the same values tested in AISA 1 from Scenario 1 were adopted in this scenario.

5.4 Simulation Results from Scenario 2

The mouse traffic was tested in the presence of the printing job. The results are averages of 10 simulation runs, each of them with 120 seconds of simulation time. Two AISA configurations and Round Robin (RR) were compared. The AISA configurations are AISA 1, with *min_dur* (minimum rendezvous period) equal to 20 slots, and AISA 2, with *min_dur* equal to 50 slots. AISA 1 and RR were tested with the turn size (*turn_size*) varying from 60 to 200 slots. In AISA 2, simulations started at 120 slots (equivalent to 60 slots/piconet).

Figure 5 (a) shows the 95th percentile delay for the three configurations. Since the mouse traffic was modeled with one packet per 65 ms, we decided to use this value as the maximum acceptable delay.

AISA 1 reaches the maximum acceptable delay close to 120 slots. By increasing the minimum rendezvous period (in AISA 2), this delay limit is reached close to 150 slots. This difference is due to the fact that increasing a minimum rendezvous time period (*min_dur*) reduces the number of slots left to other piconets.

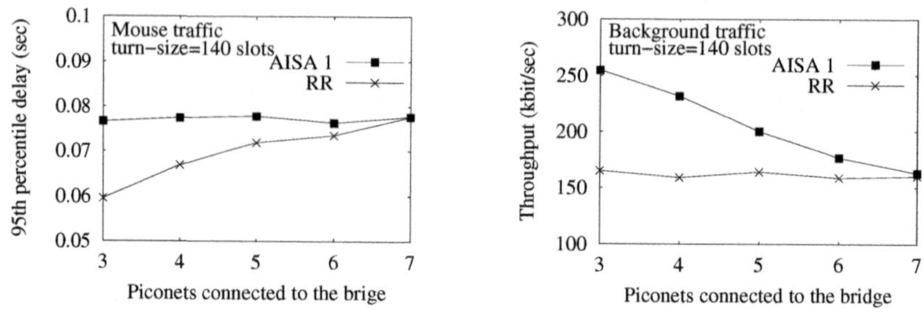

(a) Delay comparison

(b) Throughput comparison

Fig. 6. AISA and RR results, increasing the number of piconets connected to the bridge

Since the mouse throughput is low, the bridge stays the minimum possible period (*min_dur*) in its piconet. When *min_dur* is 20 slots and *turn_size* is 120 slots, 100 slots (62.5 ms) are left to the printer piconet, which is close to the maximum acceptable delay.

Round Robin (RR) guarantees smaller delay values than AISA, but this reduction comes with smaller values of background traffic throughput. This effect is presented in Figure 5 (b). For a *turn_size* of 120 slots, the AISA throughput result (280 kbit/sec) is 60% better than the best RR result (180 kbit/sec).

We have also observed the mouse packet delays when the bridge participates in more than two piconets. The number of piconets that the bridge participates in was varied from three to seven. The simulation was run with a 140 slot turn size, ensuring that even with seven piconets, the 20-slot *min_dur* is honored for all piconets. The 95th percentile mouse packet delay and background traffic aggregate throughput are shown in Figure 6.

Regardless of the number of piconets connected to the bridge, the AISA 1 delay curve is stable, since the bridge always stays the minimum rendezvous period in Piconet 1. The bridge distributes the rest of the turn among other piconets. On the other hand, RR distributes the turn equally among all piconets. Consequently, the delay increases as the number of piconets increases.

Regarding the throughput, RR curve is almost constant, but AISA curve decreases as the number of piconets increases. Each new piconet connected to the bridge reduces the printing job throughput. Besides, more slots are lost because of piconet switching. Even so, in the configuration with six piconets AISA outperforms RR in about 10%.

5.5 Scenario 3 – Power Consumption Metric

The goal of Scenario 3 is to show that a correct AISA parametrization may reduce bridge power consumption and, consequently, overall consumption. This metric evaluation is important to *ad hoc* networks, and, more specifically, it may be useful to sensor network implementations.

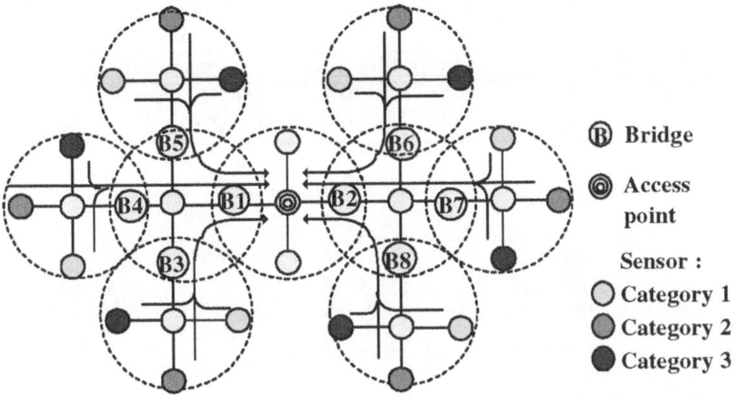

Fig. 7. Scenario 3

The topology is a scatternet formed by nine piconets as illustrated in Figure 7. Only the border piconets have data sources. Sources are divided in three categories, representing different sensoring data values. Each piconet has one source from each category. The central node is an access point, which is the sink for all sources. Sources generate packets at a 3 kbit/sec constant rate. Categories 1, 2 and 3 use 300-, 100- and 20-byte packet length, respectively, resulting in different packet intervals.

The AISA parameters that remained unchanged throughout this simulation set are: *min_dur* (20 slots), *dec_rate* (20%) and *skip_pico* (1). The mininum rendezvous period (*min_dur*) should be small so that a bridge stays less time in piconets with low link occupation. The parameter *skip_pico* enables power saving.

In order to measure power consumption we defined an energy unit (e.u.). One e.u. is the transmission power of a one-slot packet [3]. Note that in Bluetooth a packet transmission dissipates almost the same power of a packet reception ([17, 18]).

5.6 Simulation Results from Scenario 3

Two AISA configurations and Round Robin (RR) were simulated. RR does not save power but it was simulated for packet delay comparison. Although the delay metric is not focused in this scenario, power saving should not cause excessive packet delay. AISA 1 and AISA 2 parameter configurations are presented in Table 3.

Power consumption and packet delay were measured with AISA 1, AISA 2 and RR. The results are averages of 10 simulation runs, each of them with 120 seconds of simulation time. The uncertainty is expressed as 95% confidence intervals (CI). In each run, the turn size was varied from 60 to 140 slots.

[3] More details about power levels used in this paper may be obtained in [12].

Table 3. AISA 1 and AISA 2 configurations

	inc_bound	dec_bound	inc_rate	max_dur
AISA 1	80%	60%	20%	turn_size slots
AISA 2	90%	70%	10%	(60% of turn_size) slots

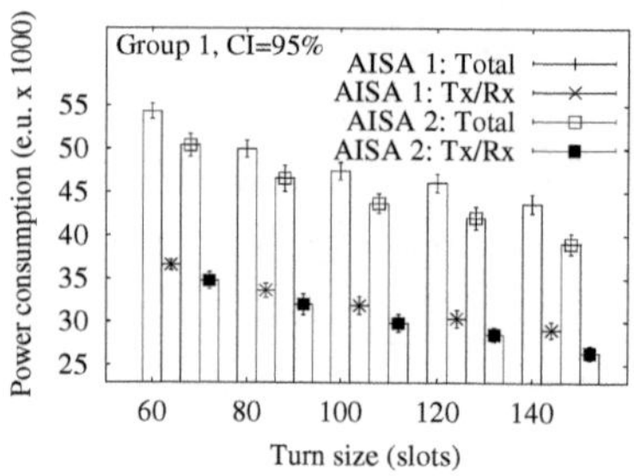

(a) Group 1 power consumption

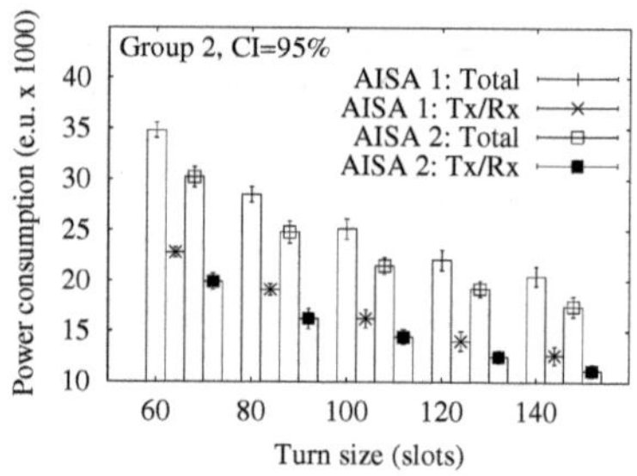

(b) Group 2 power consumption

Fig. 8. Power consumption of bridge Groups 1 and 2, with configurations 1 and 2

There are eight bridges in this scenario. The bridges are divided in two groups, regarding its position in the network topology: Group 1, with Bridges 1 and 2, and Group 2, with Bridges 3 to 8. Due to the traffic source positions in the scatternet, the bridges from Group 1 will dissipate almost the same power. Analogously, results within Group 2 will be close to each other. Therefore, all results are presented in terms of averages obtained for each group. Figure 8 presents both groups dissipated power in this simulation [4].

As expected, the results show that, for all turn sizes, the bridges from Group 1 consume more power than those from Group 2, because all packets addressed to the access point are routed by Group 1. Certainly, the lifetime of Group 1 will be smaller than the lifetime of Group 2. A solution to this problem will be discussed later.

In all cases, as the turn size increases bridge power consumption is reduced. This result is explained as follows. The bridges normally remain the mininum rendezvous time period (*min_dur*) in each piconet. Increasing the turn size will enable bridges to accumulate free slots at the end of each turn. During these free slots bridges enter Hold mode.

Comparing AISA 1 and AISA 2 results, one may note that AISA 2 causes less power consumption than AISA 1 for both groups (Figure 8). In AISA 2, it is easier for a bridge to release slots (because $dec\_bound_{AISA2} > dec\_bound_{AISA1}$ in Table 3), and more difficult to increase a rendezvous time period (because

[4] RR does not save power, since bridges are always active in a piconet. As an example, for Group 1 with turn of 60 slots, the average power consumption was 70461 ± 1023 e.u., representing about 30% more power than with AISA 1.

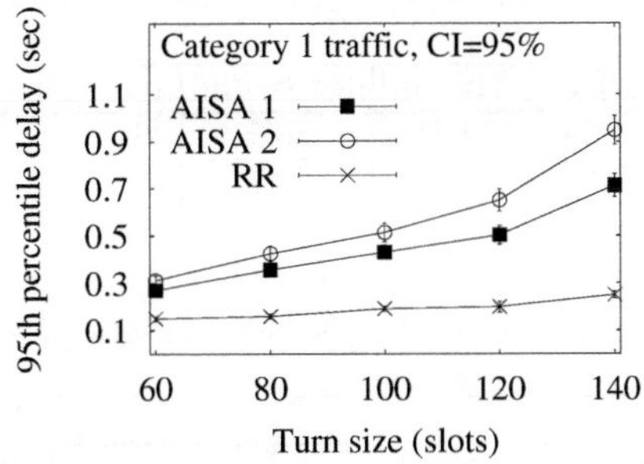

Fig. 9. Packet delay for category 1 traffic

$inc\_bound_{AISA2} > inc\_bound_{AISA1}$). When the link utilization (*avg_util*) goes beyond *inc_bound*, AISA 2 increasing rate is smaller than that of AISA 1. Finally, AISA 1 lets a bridge occupy all free slots in a rendezvous, while AISA 2 has an upper limit for a rendezvous time period (*max_dur*).

RR was included in the study of packet delay. Figure 9 shows the average packet delay for category 1 sources. The other two categories presented similar curves.

RR algorithm causes the smallest delay values, but without power saving. AISA 2 presents delay values larger than AISA 1. So, there is a trade off between power saving and packet delay.

Due to the bridges position in the scatternet, Bridges 1 and 2 consume more power than those belonging to Group 2. If all bridges start with the same battery power, Bridges 1 and 2 will cease to function before the others (Group 2), and there will be no route to the access point. So, we should find a specific configuration for each bridge group in order to equalize power consumption. Also, new configurations should not increase packet delay.

Scenario 3 was tested with one different parametrization for each bridge group. Table 4 summarizes both configurations. Note that Group 1 was configured similarly to AISA 2 above in order to minimize power consumption by this group. However, delay increases with this configuration. To counterbalance the delay problem, although increasing consumption, Group 2 was configured similarly to AISA 1.

The average power consumption was calculated for each bridge group, with uncertainty expressed as 95% confidence intervals (CI). Table 5 contains this result. One important conclusion is that the difference between the dissipated power by bridges from Groups 1 and 2 was less than 3%.

The average packet delay for the three traffic categories was also calculated. The results for categories 1, 2 and 3 were, respectively, in seconds: 0,627 ± 0,015; 57 ± 0,032; and 0,58 ± 0,024. From the power consumption and packet delay results, we conclude that Bridges 1 and 2 consumed 20% less power than in case of AISA 2 for the same delay results (AISA 2 with 120 slot turn in the previous simulation). This confirms that it is possible to configure AISA to extend network lifetime.

Table 4. AISA parametrization to equalize power consumption

	turn_size	inc_bound	dec_bound	inc_rate	max_dur
Group 1	140 *slots*	90%	70%	10%	80 *slots*
Group 2	60 *slots*	80%	60%	20%	turn_size *slots*

Table 5. Power consumption, with different configurations for Groups 1 and 2

	Group 1	Group 2
Total consumption (e.u.)	35134 ± 775	34436 ± 591
Consumption of transmited and received packets (e.u.)	23136 ± 534	22952 ± 467

Table 6. AISA parametrization guidelines

Parameter	Performance metric		
	Throughput	**Delay**	**Consumption**
turn_size	↑*	↓*	↑*
inc_bound	↓*	↓	↑*
dec_bound	↑*	↑	↑*
inc_rate	↑	↑	↓
dec_rate	↓	↓	↑
max_dur	↑*	↑	↓*
min_dur	↓*	↑*	↓
skip_pico (boolean)	0	0*	1*

'*' indicates the most relevant parameters for each metric.
'↑' means that increasing the parameter improves the metric performance.
'↓' means that reducing the parameter improves the metric performance.

5.7 Parametrization Summary

After several simulations, AISA parameters that significantly affect each metric performance were identified. Although absolute values are topology dependent, Table 6 presents general guidelines about parameter configuration.

6 Conclusions and Future Work

This paper presented an interpiconet scheduling algorithm herein referred to as Adaptive Interpiconet Scheduling Algorithm (AISA). AISA enables bridges to adapt to varying traffic conditions. Moreover, AISA parameters can be configured so that one chosen performance metric will be optimized. Also, the creation of explicit signaling packets was avoided by applying AISA solely to the bridges, thereby simplifying its implementation and facilitating adherence to the standard.

Three scenarios with different performance metric constraints were created to test AISA. The three chosen metrics were: traffic aggregate throughput, packet delay and power consumption. Simulations were performed in BlueNetS, an adaptation of the ns-2 simulator. Various attempts were made to configure the parameters in order to optimize the performance of each scenario metric with minimum degradation to the other two. This goal was achieved and a set of guidelines to optimize the performance of each metric was established. Encouraged by these results, plans are underway to apply AISA to additional traffic models that would include video and web traffic for the purpose of studying performance metrics like, for example, jitter and response time.

References

1. Bluetooth Special Interest Group. *http://www.bluetooth.com.* visited nov. 2002.
2. J. Haartsen. The bluetooth Radio System. *IEEE Personal Communications*, 7(1):28–36, February 2000.
3. T. Salonidis, P. Bhagwat, L. Tassiulas, and R. LaMaire. Distributed Topology Construction of Bluetooth Personal Area Networks. *IEEE Infocom*, April 2001.
4. C. Law, A. Mehta, and K.-Y. Siu. Performance of a new Bluetooth Scatternet Formation Protocol. *Proc. the 2001 ACM MobiHoc*, October 2001.
5. P. Bhagwat and A. Segall. A Routing Vector Method (rvm) for Routing in Bluetooth Scatternets. *The 6th IEEE MOMUC*, November 1999.
6. M. Sun, C.K. Chang, and T.H. Lai. A Self-routing Topology for Bluetooth Scatternets. *Proc. I-SPAN 2002*, May 2002.
7. N. Johansson, F. Alriksson, and U. Jönsson. Jump Mode - A Dynamic Window-based Scheuling Framework for Bluetooth Scatternets. *Proc. 2001 ACM MobiHoc*, October 2001.
8. P. Johansson, R. Kapoor, M. Kazantzidis, and M. Gerla. Rendezvous Scheduling for Bluetooth Scatternets. *Proc. ICC 2002*, April 2002.
9. G. Tan. Self-organizing Bluetooth Scatternets. *Master Thesis*, January 2002.
10. A. Racz, G. Miklos, F. Kubinszky, and A. Valko. A Pseudo Random Coordinated Scheduling Algorithm for Bluetooth Scatternets. *Proc. 2001 ACM MobiHoc*, October 2001.
11. L. Har-Shai, R. Kofman, G. Zussman, and A. Segall. Interpiconet Scheduling in Bluetooth Scatternets. *Proc. OPNETWORK 2002 Conference*, August 2002.
12. W. Priess. Scheduling Mechanisms with Quality of Service for Bluetooth Networks. *Master Science Thesis, NCE/UFRJ*, January 2003.
13. P. Johansson, N. Johansson, U. Körner, J. Elgg, and G. Svennarp. Short Range Radio Based Ad Hoc Networking: Performance and Properties. *Proc. ICC'99*, 1999.
14. P. Johansson, R. Kapoor, M. Gerla, and M. Kazantzidis. Bluetooth: an Enabler of Personal Area Networking. *IEEE Network, Special Issue in Personal Area Networks*, October 2001.
15. E. L. Hahne. Round Robin Scheduling for Fair Flow Control in Data Communication Networks. *Phd Thesis*, December 1986.
16. The Network Simulator (ns-2). http://www.isi.edu/nsnam/ns, visited jun. 2002.
17. Philips Semiconductors. UAA3558 Bluetooth RF Transceiver. www.semiconductors.philips.com/technologies/bluetooth, visited sep. 2002.
18. Ericsson Microeletronics. PBA 313 01/3 Bluetooth Radio. www.ericsson.com/microe/products/bluetooth_solutions, visited sep. 2002.

Security in Ad Hoc Networks*

Refik Molva and Pietro Michiardi

Institut Eurecom
2229 Route des Crêtes
06904 Sophia-Antipolis, France

Abstract. In wirelesss ad hoc networks basic network operations are carried out through the cooperation of all available nodes. Due to the inherent lack of a managed infrastructure the nodes of an ad hoc network cannot be considered as trustworthy as in a dedicated infrastructure. Wireless ad hoc networks are thus vulnerable to various exposures threatening the basic network operations like routing and packet forwarding. This paper presents a survey of current research activities dealing with routing security, cooperation enforcement and key management in wireless ad hoc networks. Existing solutions seem to only partially address the threats and fall short of providing a comprehensive answer. Wireless security mechanisms in layer 2 that are often considered as part of the solution domain do not meet the specific requirements of wireless ad hoc networks.

1 Introduction

A mobile *ad hoc* network (MANET) consists of a set of mobile hosts that carry out basic networking functions like packet forwarding, routing, and service discovery without the help of an established infrastructure. Nodes of an ad hoc network rely on one another in forwarding a packet to its destination, due to the limited range of each mobile host's wireless transmissions. Security in MANET is an essential component for basic network functions like packet forwarding and routing: network operation can be easily jeopardized if countermeasures are not embedded into basic network functions at the early stages of their design. Unlike networks using dedicated nodes to support basic functions like packet forwarding, routing, and network management, in ad hoc networks those functions are carried out by all available nodes. This very difference is at the core of the security problems that are specific to ad hoc networks. As opposed to dedicated nodes of a classical network, the nodes of an ad hoc network cannot be trusted for the correct execution of critical network functions. These security problems call on the other hand for different solutions based on the organizational links between the nodes of a MANET:

- in managed environments, the nodes are controlled by an organization (or a structured set of organizations) and an *a priori trust* relationship between the nodes can be derived from the existing trust relationship of the organization;

---

* This research was partially supported by the Information Society Technologies program of the European Commission, Future and Emerging Technologies under the IST-2001-38113 MOBILEMAN project and by the Institut Eurecom.

M. Conti et al. (Eds.): PWC 2003, LNCS 2775, pp. 756–775, 2003.

– in open environments whereby nodes and their owners aren't linked by any organizational relationship, network security mechanisms cannot rely on any existing trust relationship among the nodes.

In managed environments, entity authentication can be sufficient to verify the trust level of each node in the organization and correct execution of critical network functions is assured based on the organizational trust. Such a priori trust can only exist in a few special scenarios like military networks and corporate networks, where a common, trusted authority manage the network, and requires tamper-proof hardware for the implementation of critical functions. Entity authentication in a large network on the other hand raises key management requirements.

In managed environments without tamper-proof hardware and strong authentication infrastructure, or in open environments where a common authority that regulates the network does not exist, any node of an ad hoc network can endanger the reliability of basic functions like routing. The correct operation of the network requires not only the correct execution of critical network functions by each participating node but it also requires that each node performs a fair share of the functions. The latter requirement seems to be a strong limitation for wireless mobile nodes whereby power saving is a major concern. The threats considered in the MANET scenario are thus not limited to maliciousness and a new type of misbehavior called selfishness should also be taken into account to prevent nodes that simply do not cooperate.

With lack of a priori trust, classical network security mechanisms based on authentication and access control cannot cope with selfishness and cooperative security schemes seem to offer the only reasonable solution. In a cooperative security scheme, node misbehavior can be detected through the collaboration between a number of nodes assuming that a majority of nodes do not misbehave.

The remainder of this article presents security threats and major solutions from the literature along routing, cooperation enforcement and key management in MANET.

2 Routing Security in MANET

Unlike traditional networks whereby routing functions are performed by dedicated nodes or routers, in MANET, routing functions are carried out by all available nodes. Likewise, common routing security mechanisms consist of node and message authentication referring to an priori trust model in which legitimate routers are believed to perform correct operations. Authentication of a node or its messages does not guarantee the correct execution of routing functions in open environments with lack of a priori trust like MANET.

Security exposures of ad hoc routing protocols are due to two different types of attacks: *active* attacks through which the misbehaving node has to bear some energy costs in order to perform some harmful operation and *passive attacks* that mainly consist of lack of cooperation with the purpose of energy saving. Nodes that perform active attacks with the aim of damaging other nodes by causing network outage are considered to be *malicious* while nodes that perform passive attacks with the aim of saving battery life for their own communications are considered to be *selfish*.

Malicious nodes can disrupt the correct functioning of a routing protocol by *modifying* routing information, by *fabricating* false routing information and by *impersonating* other nodes. Recent research studies [10] brought up also a new type of attack that goes under the name of *wormhole* attack. On the other side, selfish nodes can severely degrade network performances and eventually partition the network [11] by simply not participating to the network operation.

In the existing ad hoc routing protocols nodes are trusted in that they do not maliciously tamper with the content of protocol messages transferred among nodes. Malicious nodes can easily perpetrate *integrity attacks* by simply altering protocol fields in order to subvert traffic, deny communication to legitimate nodes (denial of service) and compromise the *integrity* of routing computations in general. As a result the attacker can cause network traffic to be dropped, redirected to a different destination or to take a longer route to the destination increasing communication delays. A special case of integrity attacks is *spoofing* whereby a malicious node impersonates a legitimate node due to the lack of authentication in the current ad hoc routing protocols. The main result of spoofing attacks is the misrepresentation of the network topology that possibly causes network loops or partitioning. Lack of integrity and authentication in routing protocols can further be exploited through "fabrication" referring to the generation of bogus routing messages. Fabrication attacks cannot be detected without strong authentication means and can cause severe problems ranging from denial of service to route subversion.

A more subtle type of active attack is the creation of a tunnel (or wormhole) in the network between two colluding malicious nodes linked through a private connection by-passing the network. This exploit allows a node to short-circuit the normal flow of routing messages creating a virtual vertex cut in the network that is controlled by the two colluding attackers.

Another exposure of current ad hoc routing protocols is due node selfishness that result in lack of cooperation among ad hoc nodes. A selfish node that wants to save battery life for its own communication can endanger the correct network operation by simply not participating in the routing protocol or by not forwarding packets as in the so called black hole attack. Current ad hoc routing protocols do not address the selfishness problem.

3 Secure Routing Proposals

Current efforts towards the design of secure routing protocols are mainly oriented to reactive (on-demand) routing protocols such as DSR [12] or AODV [13], where a node attempts to discover a route to some destination only when it has a packet to send to that destination. On-demand routing protocols have been demonstrated to perform better with significantly lower overheads than proactive routing protocols in many scenarios since they are able to react quickly to topology changes while keeping routing overhead low in periods or areas of the network in which changes are less frequent. It is possible to find, however, interesting security solutions for proactive routing protocols which are worthwhile to mention.

Current secure routing protocols proposed in the literature take into account *active attacks* performed by malicious nodes that aim at intentionally tampering with the execution of routing protocols whereas *passive attacks* and the selfishness problem are not addressed. Furthermore the prerequisite for all the available solutions is a *managed* environment characterized by some security infrastructure established prior to the secure routing protocol execution. The most significant proposals for secure routing in ad hoc networks are outlined in the sequel of this section.

3.1 Secure Routing Protocol

The Secure Routing Protocol (SRP) [1] is designed as an extension compatible with a variety of existing *reactive* routing protocols. SRP combats attacks that disrupt the route discovery process and guarantees the acquisition of correct topological information: SRP allows the initiator of a route discovery to detect and discard bogus replies. SRP relies on the availability of a *security association* (SA) between the source node (S) and the destination node (T). The SA could be established using a hybrid key distribution based on the public keys of the communicating parties. S and T can exchange a secret symmetric key ($K_{S,T}$) using the public keys of one another to establish a secure channel. S and T can then further proceed to mutual authentication of one another and the authentication of routing messages.

SRP copes with non-colluding *malicious* nodes that are able to modify (corrupt), replay and fabricate routing packets. In case of the Dynamic Source Routing (DSR) protocol [12], SRP requires including a 6-word header containing unique identifiers that tag the discovery process and a message authentication code (MAC) computed using a keyed hash algorithm. In order to initiate a route request (RREQ) the source node has to generate the MAC of the entire IP header, the basic protocol RREQ packet and the shared key $K_{S,T}$.

The intermediate nodes that relay the RREQ towards the destination measure the frequencies of queries received from their neighbors in order to regulate the query propagation process: each node maintains a priority ranking that is inversely proportional to the query rate. A node that maliciously pollutes network traffic with unsolicited RREQ will be served last (or ignored) because of its low priority ranking.

Upon reception of a RREQ, the destination node verifies the *integrity* and *authenticity* of the RREQ by calculating the keyed hash of the request fields and comparing them with the MAC contained in the SRP header. If the RREQ is valid, the destination initiates a route replay (RREP) using the SRP header the same way the source did when initiating the request. The source node discards replays that do not match with pending query identifiers and checks the integrity using the MAC generated by the destination.

The basic version of SRP suffers from the route cache poisoning attack: routing information gathered by nodes that operate in promiscuous mode in order to improve the efficiency of the DSR protocol could be invalid, because of potential fabrication by malicious nodes. The authors propose two alternative designs of SRP that use an Intermediate Node Reply Token (INRT). INRT allows intermediate nodes that belong to the same group that share a common key (K_G) to validate RREQ and provide valid RREP messages.

SRP suffers also from the lack of a validation mechanism for route maintenance messages: route error packets are not verified. However, in order to minimize the effects of fabricated error messages, SRP source-routes error packets along the prefix of the route reported as broken: the source node can thus verify that each route error feedback refers to the actual route and that it was originated at the a node that is part of the route. A malicious node can harm only routes it actually belongs to.

Assuming that the neighbor discovery mechanism maintains information on the binding of the medium access control and the IP addresses of nodes, SRP is proven to be essentially immune to IP spoofing [1].

SRP is, however, not immune to the wormhole attack: two colluding malicious nodes can misroute the routing packets on a private network connection and alter the perception of the network topology by legitimate nodes.

3.2 ARIADNE

Hu, Perrig and Johnson present an *on-demand* secure ad hoc routing protocol based on DSR that withstands node compromise and relies only on highly efficient *symmetric* cryptography. ARIADNE guarantees that the target node of a route discovery process can authenticate the initiator, that the initiator can authenticate each intermediate node on the path to the destination present in the RREP message and that no intermediate node can remove a previous node in the node list in the RREQ or RREP messages.

As for the SRP protocol, ARIADNE needs some mechanism to bootstrap authentic keys required by the protocol. In particular, each node needs a shared secret key ($K_{S,D}$, is the shared key between a source S and a destination D) with each node it communicates with at a higher layer, an authentic TESLA [3, 4] key for each node in the network and an authentic "Route Discovery chain" element for each node for which this node will forward RREQ messages.

ARIADNE provides point-to-point *authentication* of a routing message using a message authentication code (MAC) and a shared key between the two parties. However, for authentication of a broadcast packet such as RREQ, ARIADNE uses the TESLA broadcast authentication protocol. ARIADNE copes with attacks performed by *malicious* nodes that modify and fabricate routing information, with attacks using impersonation and, in an advanced version, with the wormhole attack. Selfish nodes are not taken into account.

In ARIADNE, the basic RREQ mechanism is enhanced by eight additional fields used to provide authentication and integrity to the routing protocol as follows:

<ROUTE REQUEST, initiator, target, id, time interval, hash chain, node list, MAC list>

The initiator and target are set to the address of the initiator and target nodes, respectively. As in DSR, the initiator sets the id to an identifier that it has not recently used in initiating a Route Discovery. The time interval is the TESLA time interval at the pessimistic expected arrival time of the request at the target, accounting for clock skew. The initiator of the request then initializes the hash chain to $MAC_{KS,D}$ (initiator, target, id, time interval) and the node list and MAC list to empty lists.

When any node *A* receives a RREQ for which it is not the target, the node checks its local table of <initiator, id> values from recent requests it has received, to determine if it has already seen a request from this same Route Discovery. If it has, the node discards the packet, as in DSR. The node also checks whether the time interval in the request is valid: that time interval must not be too far in the future, and the key corresponding to it must not have been disclosed yet. If the time interval is not valid, the node discards the packet. Otherwise, the node modifies the request by appending its own address (*A*) to the node list in the request, replacing the hash chain field with H [A, *hash chain*], and appending a MAC of the entire REQUEST to the

MAC list. The node uses the TESLA key K_{Ai} to compute the MAC, where i is the index for the time interval specified in the request. Finally, the node rebroadcasts the modified RREQ, as in DSR.

When the target node receives the RREQ, it checks the validity of the request by determining that the keys from the time interval specified have not been disclosed yet, and that the hash chain field is equal to:

$$\mathbf{H}\,[\eta_n, \mathbf{H}\,[\eta_{n-1}, \mathbf{H}\,[\ldots, \mathbf{H}\,[\eta_1, \mathbf{MAC}_{KSD}\,(\text{initiator, target, id, time interval})\,]\ldots]\,]\,]$$

where η_i is the node address at position i of the node list in the request, and where n is the number of nodes in the node list. If the target node determines that the request is valid, it returns a RREP to the initiator, containing eight fields:

<ROUTE REPLY, target, initiator, time interval, node list, MAC list, target MAC, key list>

The target, initiator, time interval, node list, and MAC list fields are set to the corresponding values from the RREQ, the target MAC is set to a MAC computed on the preceding fields in the reply with the key KDS , and the key list is initialized to the empty list. The RREP is then returned to the initiator of the request along the source route obtained by reversing the sequence of hops in the node list of the request.

A node forwarding a RREP waits until it is able to disclose its key from the time interval specified, then it appends its key from that time interval to the key list field in the reply and forwards the packet according to the source route indicated in the packet. Waiting delays the return of the RREP but does not consume extra computational power.

When the initiator receives a RREP, it verifies that each key in the key list is valid, that the target MAC is valid, and that each MAC in the MAC list is valid. If all of these tests succeed, the node accepts the RREP; otherwise, it discards it.

In order to prevent the injection of invalid route errors into the network fabricated by any node other than the one on the sending end of the link specified in the error message, each node that encounters a broken link adds TESLA authentication information to the route error message, such that all nodes on the return path can authenticate the error. However, TESLA authentication is delayed, so all the nodes on the return path buffer the error but do not consider it until it is authenticated. Later, the node that encountered the broken link discloses the key and sends it over the return path, which enables nodes on that path to authenticate the buffered error messages.

ARIADNE is protected also from a flood of RREQ packets that could lead to the cache poisoning attack. Benign nodes can filter out forged or excessive RREQ packets using *Route Discovery chains*, a mechanism for authenticating route discovery, allowing each node to rate-limit discoveries initiated by any other node. The authors present two different approaches that can be found in [2].

ARIADNE is immune to the wormhole attack only in its advanced version: using an extension called TIK (TESLA with Instant Key disclosure) that requires tight clock synchronization between the nodes, it is possible to detect anomalies caused by a wormhole based on timing discrepancies.

3.3 ARAN

The ARAN secure routing protocol proposed by Dahill, Levine, Royer and Shields is conceived as an on-demand routing protocol that detects and protects against malicious actions carried out by third parties and peers in the ad hoc environment. ARAN introduces *authentication*, message *integrity* and *non-repudiation* as part of a minimal security policy for the ad hoc environment and consists of a preliminary certification process, a mandatory end-to-end authentication stage and an optional second stage that provides secure shortest paths.

ARAN requires the use of a trusted certificate server (T): before entering in the ad hoc network, each node has to request a certificate signed by T. The certificate contains the IP address of the node, its public key, a timestamp of when the certificate was created and a time at which the certificate expires along with the signature by T. All nodes are supposed to maintain fresh certificates with the trusted server and must know T's public key.

The goal of the first stage of the ARAN protocol is for the source to verify that the intended destination was reached. In this stage, the source trusts the destination to choose the return path. A source node, *A*, initiates the route discovery process to reach the destination *X* by broadcasting to its neighbors a route discovery packet called RDP:

$$[\mathbf{RDP};\ \mathbf{IP_X}\ ;\ \mathit{cert_A}\ ;\ \mathbf{N_A}\ ;\ \mathbf{t}]\mathbf{K_{A-}}$$

The RDP includes a packet type identifier ("RDP"), the IP address of the destination (IP_X), *A*'s certificate ($cert_A$), a nonce N_A, and the current time t, all signed with A's private key. Each time A performs route discovery, it monotonically increases the nonce.

Each node records the neighbor from which it received the message. It then forwards the message to each of its neighbors, signing the contents of the message. This signature prevents spoofing attacks that may alter the route or form loops. Let *A*'s neighbor be *B*. It will broadcast the following message:

$$[[\mathbf{RDP};\ \mathbf{IP_X}\ ;\ \mathit{cert_A}\ ;\ \mathbf{N_A}\ ;\ \mathbf{t}]\mathbf{K_{A-}}\ ]\mathbf{K_{B-}}\ ;\ \mathit{cert_B}$$

Nodes do not forward messages for which they have already seen the (N_A ; IP_A) tuple. The IP address of A is contained in the certificate, and the monotonically increasing nonce facilitates easy storage of recently-received nonces.

Upon receiving the broadcast, *B*'s neighbor *C* validates the signature with the given certificate. *C* then rebroadcasts the RDP to its neighbors, first removing B's signature:

$$[[\mathbf{RDP};\ \mathbf{IP_X}\ ;\ \mathit{cert_A}\ ;\ \mathbf{N_A}\ ;\ \mathbf{t}]\mathbf{K_{A-}}\ ]\mathbf{K_{C-}}\ ;\ \mathit{cert_C}$$

Eventually, the message is received by the destination, *X*, who replies to the first RDP that it receives for a source and a given nonce. There is no guarantee that the first RDP received traveled along the shortest path from the source. The destination unicasts a Reply (REP) packet back along the reverse path to the source. Let the first node that receives the RDP sent by X be node *D*. *X* will send to *D* the following message:

$$[\mathbf{REP};\ \mathbf{IP_A}\ ;\ \mathit{cert_X}\ ;\ \mathbf{N_A}\ ;\ \mathbf{t}]\mathbf{K_{X-}}$$

The REP includes a packet type identifier ("REP"), the IP address of A, the certificate belonging to *X*, the nonce and associated timestamp sent by *A*. Nodes that receive the REP forward the packet back to the predecessor from which they received the original RDP. All REPs are signed by the sender. Let *D*'s next hop to the source be node *C*. *D* will send to *C* the following message:

$$[[\mathbf{REP}; \mathbf{IP_A}; \mathbf{\mathit{cert}_X}; \mathbf{N_A}; \mathbf{t}]\mathbf{K_{X-}}]\mathbf{K_{D-}}; \mathbf{\mathit{cert}_D}$$

C validates *D*'s signature, removes the signature, and then signs the contents of the message before unicasting the following RDP message to *B*:

$$[[\mathbf{REP}; \mathbf{IP_A}; \mathbf{\mathit{cert}_X}; \mathbf{N_A}; \mathbf{t}]\mathbf{K_{X-}}]\mathbf{K_{C-}}; \mathbf{\mathit{cert}_C}$$

A node checks the signature of the previous hop as the REP is returned to the source. This avoids attacks where malicious nodes instantiate routes by impersonation and re-play of X's message. When the source receives the REP, it verifies that the correct nonce was returned by the destination as well as the destination's signature. Only the destination can answer an RDP packet. Other nodes that already have paths to the destination cannot reply for the destination. While other protocols allow this networking optimization, ARAN removes several possible exploits and cuts down on the reply traffic received by the source by disabling this option.

The second stage of the ARAN protocol guarantees in a secure way that the path received by a source initiating a route discovery process is the shortest. Similarly to the first stage of the protocol, the source broadcasts a *Shortest Path Confirmation* (SPC) message to its neighbors: the SPC message is different from the RDP message only in two additional fields that provide the destination *X* certificate and the encryption of the entire message with *X*'s public key (which is a costly operation). The onion-like signing of messages combined with the encryption of the data prevents nodes in the middle from changing the path length because doing so would break the integrity of SPC the packet.

Also the route maintenance phase of the ARAN protocol is secured by digitally signing the route error packets. However it is extremely difficult to detect when error messages are *fabricated* for links that are truly active and not broken. Nevertheless, because messages are signed, malicious nodes cannot generate error messages for other nodes. The non-repudiation provided by the signed error message allows a node to be verified as the source of each error message that it sends.

As with any secure system based on cryptographic certificates, the key revocation issue has to be addressed in order to make sure that expired or revoked certificates do not allow the holder to access the network. In ARAN, when a certificate needs to be revoked, the trusted certificate server T sends a broadcast message to the ad hoc group that announces the revocation. Any node receiving this message re-broadcast it to its neighbors. Revocation notices need to be stored until the revoked certificate would have expired normally. Any neighbor of the node with the revoked certificate needs to reform routing as necessary to avoid transmission through the now un-trusted node. This method is not failsafe. In some cases, the un-trusted node that is having its certificate revoked may be the sole connection between two parts of the ad hoc network. In this case, the un-trusted node may not forward the notice of revocation for its certificate, resulting in a partition of the network, as nodes that have received the revocation notice will no longer forward messages through the un-trusted node, while all other nodes depend on it to reach the rest of the network. This only lasts as long as the un-trusted node's certificate would have otherwise been valid, or until the un-trusted node is no longer the sole connection between the two partitions. At the time that the revoked certificate should have expired, the un-trusted node is unable to renew the certificate, and routing across that node ceases. Additionally, to detect this situation and to hasten the propagation of revocation notices, when a node meets a new neighbor, it can exchange a summary of its revocation notices with that neighbor; if these summaries do not match, the actual signed notices can be forwarded and re-broadcasted to restart propagation of the notice.

The ARAN protocol protects against exploits using *modification*, *fabrication* and *impersonation* but the use of asymmetric cryptography makes it a very costly protocol to use in terms of CPU and energy usage. Furthermore, ARAN is not immune to the *wormhole* attack

3.4 SEAD

Hu, Perrig and Johnson present a *proactive* secure routing protocol based on the Destination-Sequenced Distance Vector protocol (DSDV). In a proactive (or periodic) routing protocol nodes periodically exchange routing information with other nodes in attempt to have each node always know a current route to all destinations [7]. Specifically, SEAD is inspired by the DSDV-SQ version of the DSDV protocol. The DSDV-SQ version of the DSDV protocol has been shown to outperform other DSDV versions in previous ad hoc networks simulations [8, 9].

SEAD deals with attackers that *modify* routing information broadcasted during the update phase of the DSDV-SQ protocol: in particular, routing can be disrupted if the attacker modifies the sequence number and the metric field of a routing table update message. *Replay attacks* are also taken into account.

In order to secure the DSDV-SQ routing protocol, SEAD makes use of efficient *one-way hash chains* rather than relaying on expensive asymmetric cryptography operations. However, like the other secure protocols presented in this chapter, SEAD assumes some mechanism for a node to distribute an authentic element of the hash chain that can be used to authenticate all the other elements of the chain. As a traditional approach, the authors suggest to ensure the key distribution relaying on a trusted entity that signs public key certificates for each node; each node can then use its public key to sign a hash chain element and distribute it.

The basic idea of SEAD is to authenticate the sequence number and metric of a routing table update message using hash chains elements. In addition, the receiver of SEAD routing information also authenticates the sender, ensuring that the routing information originates form the correct node.

To create a one-way hash chain, a node chooses a random initial value $x \in \{0,1\}^{\rho}$, where ρ is the length in bits of the output of the hash function, and computes the list of values $h_0, h_1, h_2, h_3, \ldots, h_n$, where $h_0 = x$, and $h_i = H(h_{i-1})$ for $0 < i \leq n$, for some n. As an example, given an authenticated h_i value, a node can authenticate h_{i-3} by computing $H(H(H(h_{i-3})))$ and verifying that the resulting value equals *hi*.

Each node uses a specific authentic (i.e. signed) element from its hash chain in each routing update that it sends about itself (metric 0). Based on this initial element, the one-way hash chain provides authentication for the lower bound on the metric in other routing updates for that node. The use of a hash value corresponding to the sequence number and metric in a routing update entry prevents any node from advertising a route to some destination claiming a greater sequence number than that destination's own current sequence number. Likewise, a node can not advertise a route better than those for which it has received an advertisement, since the metric in an existing route cannot be decreased due to the on-way nature of the hash chain.

When a node receives a routing update, it checks the authenticity of the information for each entry in the update using the destination address, the sequence number and the metric of the received entry, together with the latest prior *authentic* hash value received from that destination's hash chain. Hashing the received elements the correct number of times (according to the prior authentic hash value) assures the authenticity of the received information if the calculated hash value and the authentic hash value match.

The source of each routing update message in SEAD must also be authenticated, since otherwise, an attacker may be able to create routing loops through the *impersonation* attack. The authors propose two different approaches to provide node authentication: the first is based on a broadcast authentication mechanism such as TESLA, the second is based on the use of Message Authentication Codes, assuming a shared secret key between each couple of nodes in the network.

SEAD does not cope with *wormhole* attacks though the authors propose, as in the ARIADNE protocol, to use the TIK protocol to detect the threat.

3.5 Notes on the Wormhole Attack

The wormhole attack is a severe threat against ad hoc routing protocols that is particularly challenging to detect and prevent. In a wormhole attack a malicious node can record packets (or bits) at one location in the network and tunnel them to another location through a private network shared with a colluding malicious node. Most existing ad hoc routing protocols, without some mechanism to defend them against the wormhole attack, would be unable to find consistent routes to any destination, severely disrupting communication.

A dangerous threat can be perpetrated if a wormhole attacker tunnels all packets through the wormhole honestly and reliably since no harm seems to be done: the attacker actually seems to provide a useful service in connecting the network more efficiently. However, when an attacker forwards only routing control messages and not data packets, communication may be severely damaged. As an example, when used against an on demand routing protocol such as DSR, a powerful application of the wormhole attack can be mounted by tunneling each RREQ message directly to the destination target node of the request. This attack prevents routes more than two hops long from being discovered because RREP messages would arrive to the source faster than any other replies or, worse, RREQ messages arriving from other nodes next to the destination than the attacker would be discarded since already seen.

Hu, Perrig and Johnson propose an approach to detect a wormhole based on *packet leashes* [10]. The key intuition is that by authenticating either an extremely precise timestamp or location information combined with a loose timestamp, a receiver can determine if the packet has traversed a distance that is unrealistic for the specific network technology used.

Temporal leashes rely on extremely precise time synchronization and extremely precise timestamps in each packet. The travel time of a packet can be approximated as the difference between the receive time and the timestamp. Given the precise time synchronization required by temporal leashes, the authors propose efficient broadcast authenticators based on symmetric primitives. In particular they extend the TESLA broadcast authentication protocol to allow the disclosure of the authentication key within the packet that is authenticated.

Geographical leashes are based on location information and loosely synchronized clocks. If the clocks of the sender and the receiver are synchronized within a certain threshold and the velocity of any node is bounded, the receiver can compute an upper bound on the distance between the sender and itself and use it to detect anomalies in the traffic flow. In certain circumstances however, bounding the distance between the sender and the receiver cannot prevent wormhole attacks: when obstacles prevent communication between two nodes that would otherwise be in transmission range, a distance-based scheme would still allow wormholes between the sender and the receiver. To overcome this problem, in a variation of the geographical leashes the receiver verifies that every possible location of the sender can reach every possible location of the receiver based on a radio propagation model implemented in every node.

In some special cases, wormholes can also be detected through techniques that don't require precise time synchronization nor location information. As an example, it would be sufficient to modify the routing protocol used to discover the path to a destination so that it could handle multiple routes: a verification mechanism would then detect anomalies when comparing the metric (e.g. number of hops) associated to each route. Any node advertising a path to a destination with a metric considerably lower than all the others could raise the suspect of a wormhole.

Furthermore, if the wormhole attack is performed only on routing information while dropping data packets, other mechanisms can be used to detect this misbehavior. When a node doesn't correctly participate to the network operation by not executing a particular function (e.g. packet forwarding) a collaborative monitoring technique can detect and gradually isolate misbehaving nodes. Lack of cooperation and security mechanism used to enforce node cooperation to the network operation is the subject of the next section.

4 Selfishness and Cooperation Enforcement

Selfishness is a new type of misbehavior that is inherent to ad hoc networks and cooperation enforcement is the countermeasure against selfishness. A selfish node does not directly intend to damage other nodes with active attacks (mainly because performing active attacks can be very expensive in terms of energy consumption) but it simply does not contribute in the network operation, saving battery life for its own communications. Selfishness can cause serious damage in terms of global network throughput and delay as shown by a simulation study on the impact of selfish behavior on the DSR routing protocol [11]. The node selfishness problem has only recently been addressed by the research community, and still very few cooperation enforcement mechanisms are proposed to combat such misbehavior. Current cooperation enforcement proposals for MANET fall in two categories: currency-based solutions whereby some form of digital cash is used as an incentive for cooperation and monitoring solutions based on the principle that misbehaving nodes will be detected through the shared observations of a majority of legitimate nodes. The most significant proposals in each category are outlined in the sequel of this section.

4.1 Nuglets

In [14], Buttyan and Hubaux present two important issues targeted specifically at the ad hoc networking environment: first, end-users must be given some incentive to contribute in the network operation (especially to relay packets belonging to other nodes); second, end-users must be discouraged from overloading the network. The solution consists of a virtual currency call Nuglet used in every transaction. Two different models are described: the Packet Purse Model and the Packet Trade Model. In the Packet Purse Model each packet is loaded with nuglets by the source and each forwarding host takes out nuglets for its forwarding service. The advantage of this approach is that it discourages users from flooding the network but the drawback is that the source needs to know exactly how many nuglets it has to include in the packet it sends. In the Packet Trade Model each packet is traded for nuglets by the intermediate nodes: each intermediate node buys the packet from the previous node on the path. Thus, the destination has to pay for the packet. The direct advantage of this approach is that the source does not need to know how many nuglets need to be loaded into the packet. On the other hand, since the packet generation is not charged, malicious flooding of the network cannot be prevented. There are some further issues that have to be solved: concerning the Packet Purse Model, the intermediate nodes are able to take out more nuglets than they are supposed to; concerning the Packet Trade Model, the intermediate nodes are able to deny the forwarding service after taking out nuglets from a packet.

4.2 CONFIDANT

Buchegger and Le Boudec proposed a technique called CONFIDANT (Cooperation Of Nodes, Fairness In Dynamic Ad-hoc NeTworks) [15,16] aiming at detecting malicious nodes by means of combined monitoring and reporting and establishing routes by avoiding misbehaving nodes. CONFIDANT is designed as an extension to a routing protocol such as DSR. CONFIDANT components in each node include a network monitor, reputation records for first-hand and trusted second-hand observations about routing and forwarding behavior of other nodes, trust records to control trust given to received warnings, and a path manager to adapt the behavior of the local node according to reputation and to take action against malicious nodes. The term reputation is used to evaluate routing and forwarding behavior according to the network protocol, whereas the term trust is used to evaluate participation in the CONFIDANT meta-protocol.

The dynamic behavior of CONFIDANT is as follows. Nodes monitor their neighbors and change the reputation accordingly. If they have a reason to believe that a node misbehaves, they can take action in terms of their own routing and forwarding and they can decide to inform other nodes by sending an ALARM message. When a node receives such an ALARM either directly or by promiscuously listening to the network, it evaluates how trustworthy the ALARM is based on the source of the ALARM and the accumulated ALARM messages about the node in question. It can then decide whether to take action against the misbehaved node in the form of excluding routes containing the misbehaved node, re-ranking paths in the path cache, reciprocating by non-cooperation, and forwarding an ALARM about the node.

The first version of CONFIDANT was, despite the filtering of ALARM messages in the trust manager, vulnerable to concerted efforts of spreading wrong accusations. In a recent enhancement of the protocol, this problem has been addressed by the use of Bayesian statistics for classification and the exclusion of liars.

Simulations with nodes that do not participate in the forwarding function have shown that CONFIDANT can cope well, even if half of the network population acts maliciously. Further simulations concerning the effect of second-hand information and slander have shown that slander can effectively be prevented while still retaining a significant detection speed-up over using merely first-hand information.

The limitations of CONFIDANT lie in the assumptions for detection-based reputation systems. Events have to be observable and classifiable for detection, and reputation can only be meaningful if the identity of each node is persistent, otherwise it is vulnerable to spoofing attacks.

4.3 CORE

The security scheme proposed by Michiardi and Molva [18, 19], stimulates node cooperation by a collaborative monitoring technique and a reputation mechanism. Each node of the network monitors the behavior of its neighbors with respect to a requested function and collects observations about the execution of that function: as an example, when a node initiates a Route Request (e.g., using the DSR routing protocol) it monitors that its neighbors process the request, whether with a Route Reply or by relaying the Route Request. If the observed result and the expected result coincide, then the observation will take a positive value, otherwise it will take a negative value.

Based on the collected observations, each node computes a reputation value for every neighbor using a sophisticated reputation mechanism that differentiates between subjective reputation (observations), indirect reputation (positive reports by others), and functional reputation (task-specific behavior), which are weighted for a combined reputation value. The formula used to evaluate the reputation value avoids false detections (caused for example by link breaks) by using an aging factor that gives more relevance to past observations: frequent variations on a node behavior are filtered. Furthermore, if the function that is being monitored provides an acknowledgement message (e.g., the Route Reply message of the DSR protocol), reputation information can also be gathered about nodes that are not within the radio range of the monitoring node. In this case, only positive ratings are assigned to the nodes that participated to the execution of the function in its totality.

The CORE mechanism resists to attacks performed using the security mechanism itself: no negative ratings are spread between the nodes, so that it is impossible for a node to maliciously decrease another node's reputation. The reputation mechanism allows the nodes of the MANET to gradually isolate selfish nodes: when the reputation assigned to a neighboring node decreases below a pre-defined threshold, service provision to the misbehaving node will be interrupted. Misbehaving nodes can, however, be reintegrated in the network if they increase their reputation by cooperating to the network operation.

As for the other security mechanism based on reputation the CORE mechanism suffers from the spoofing attack: misbehaving nodes are not prevented from changing their network identity allowing the attacker to elude the reputation system. Furthermore, no simulation results prove the robustness of the protocol even if the authors propose an original approach based on game theory in order to come up with a formal assessment of the security properties of CORE.

4.4 Token-Based Cooperation Enforcement

In [20] Yang, Meng, Lu suggest a mechanism whereby each node of the ad hoc network is required to hold a token in order to participate in the network operations. Tokens are granted to a node collaboratively by its neighbors based on the monitoring of the node's contribution to packet forwarding and routing operations. Upon expiration of the token, each node renews its token through a token renewal exchange with its neighbors: the duration of a token's validity is based on the duration of the node's correct behavior as monitored by the neighbors granting/renewing the token. This mechanism typically allows a well-behaved node to accumulate credit and to renew its token less frequently as time evolves.

The token-based cooperation enforcement mechanism includes four interacting components: *neighbor verification* through which the local node verifies whether neighboring nodes are legitimate, *neighbor monitoring* that allows the local node to monitor the behavior of each node in the network and to detect attacks from malicious nodes, *intrusion reaction* that assures the generation of network alerts and the isolation of attackers, and *security enhanced routing protocol* that consists of the ad hoc routing protocol including security extensions.

A valid token is constructed using a group signature whereby a mechanism based on polynomial secret sharing [25] assures that at least k neighbors agree to issue or renew the token. The key setup complexity of polynomial secret sharing and the requirement for at least k nodes to sign each token both are incompatible with high mobility and call for a rather large and dense ad hoc network. Furthermore the duration of a token's validity increases proportionally with the duration of the node's correct behavior as monitored by its neighbors; this feature again calls for low mobility. The token-based cooperation enforcement mechanism is thus suitable for ad hoc networks where node mobility is low. Spoofing attacks through which a node can request more than one token claiming different identities, are not taken into account by the proposal even if the authors suggest that MAC addresses can be sufficient for node authentication purposes.

5 Authentication and Key Management

Authentication of peer entities involved in ad hoc routing and the integrity verification of routing exchanges are the two essential building blocks of secure routing. Both entity authentication and message integrity call on the other hand for a key management mechanism to provide parties involved in authentication and integrity verification with proper keying material. Key management approaches suggested by current secure routing proposals fall in two categories:

- manual configuration of symmetric (secret) keys: the pair-wise secret keys can serve as key encryption keys in a point-to-point key exchange protocol to establish session keys used for authentication and message integrity between communicating nodes. If some dedicated infrastructure including a key server can be afforded, automatic distribution of session keys with a key distribution protocol like Kerberos can also be envisioned.
- public-key based scheme: each node possesses a pair of public and private keys based on an asymmetric algorithm like RSA. Based on this keypair each node can perform authentication and message integrity operations or further exchange pair-wise symmetric keys used for efficient authentication and encryption operations.

Secure routing proposals like SRP assume manual configuration of secure associations based on shared secret keys. Most of other proposals such as Ariadne rely on a public-key based scheme whereby a well known trusted third party (TTP) issues public key certificates used for authentication. The requirement for such a public-key infrastructure does not necessarily imply a managed ad hoc network environment and an open environment can be targeted as well. Indeed, it is not necessary for the mobile nodes that form the ad hoc network to be managed by the public-key certification authority. However, the bootstrap phase requires an external infrastructure, which has to be available also during the lifetime of the ad hoc network to provide revocation services for certificates that have expired or that have been explicitly revoked.

Two interesting proposals presented in the next section tackle the complexity of public-key infrastructures in the ad hoc network environment through self-organization: public-key management based on the concept of web of trust akin to Pretty Good Privacy (PGP) and a public-key certification mechanism based on polynomial secret sharing.

5.1 Self-Organized Public-Key Management Based on PGP

Capkun, Buttyan and Hubaux propose a fully self-organized public key management system that can be used to support security of ad hoc network routing protocols [21]. The suggested approach is similar to PGP [22] in the sense that users issue certificates for each other based on their personal acquaintances. However, in the proposed system, certificates are stored and distributed by the users themselves, unlike in PGP, where this task is performed by on-line servers (called certificate directories). In the proposed self-organizing public-key management system, each user maintains a *local certificate repository*. When two users want to verify the public keys of each other, they merge their local certificate repositories and try to find appropriate certificate chains within the merged repository.

The success of this approach very much depends on the construction of the local certificate repositories and on the characteristics of the certificate graphs. The vertices of a certificate graph represent public-keys of the users and the edges represent public-key certificates issued by the users. The authors investigate several repository construction algorithms and study their performance. The proposed algorithms take into account the characteristics of the certificate graphs in a sense that the choice of the certificates that are stored by each mobile node depends on the connectivity of the node and its neighbors in the certificate graph.

More precisely, each node stores in its local repository several directed and mutually disjoint paths of certificates. Each path begins at the node itself, and the certificates are added to the path such that a new certificate is chosen among the certificates connected to the last node on the path (initially the node that stores the certificates), such that the new certificate leads to the node that has the highest number of certificates connected to it (i.e., the highest vertex degree). The authors call this algorithm the *Maximum Degree Algorithm*, as the local repository construction criterion is the degree of the vertices in a certificate graph.

In a more sophisticated extension called the *Shortcut Hunter Algorithm*, certificates are stored into the local repositories based on the number of the shortcut certificates connected to the users. The shortcut certificate is a certificate that, when removed from the graph makes the shortest path between two users previously connected by this certificate strictly larger than two.

When verifying a certificate chain, the node must trust the issuer of the certificates in the chain for correctly checking that the public key in the certificate indeed belongs to the node identification (ID) named in the certificate. When certificates are issued by the mobile nodes of an ad hoc network instead of trusted authorities, this assumption becomes unrealistic. In addition, there may be malicious nodes who issue false certificates. In order to alleviate these problems, the authors propose the use of authentication metrics [23]: it is not enough to verify a node ID key binding via a single chain of certificates. The authentication metric is a function that accepts two keys (the verifier and the verified node) and a certificate graph and returns a numeric value corresponding to the degree of authenticity of the key that has to be verified: one example of authentication metric is the number of disjoint chains of certificates between two nodes in a certificate graph.

The authors emphasize that before being able to perform key authentication, each node must first build its local certificate repository, which is a complex operation. However this initialization phase must be performed rarely and once the certificate repositories have been built, then any node can perform key authentication using only local information and the information provided by the targeted node. It should also be noted that local repositories become obsolete if a large number of certificate are revoked, as then the certificate chains are no longer valid; the same comment applies in the case when the certificate graph changes significantly. Furthermore, PGP-like schemes are more suitable for small communities because that the authenticity of a key can be assured with a higher degree of trustiness. The authors propose the use of authentication metrics to alleviate this problem: this approach however provides only probabilistic guarantees and is dependent on the characteristics of the certificate graph on which it operates. The authors also carried out a simulation study showing that for the certificate graphs that are likely to emerge in self-organized systems, the proposed approach yields good performances both in terms of the size of the local repository stored in each node and scalability.

5.2 Authentication Based on Polynomial Secret Sharing

In [24] Luo and Lu present an authentication service whereby the public-key certificate of each node is cooperatively generated by a set of neighbors based on the behavior of the node as monitored by the neighbors. Using a group signature mechanism based on polynomial secret sharing, the secret digital signature key used to generate public-key certificates is distributed among several nodes. Certification services like issuing, renewal and revocation of certificates thus are distributed among the nodes: a single node holds just a share of the complete certificate signature key. The authors propose a *localized trust model* to characterize the localized nature of security concerns in large ad hoc wireless networks. When applying such trust model, an entity is trusted if any k trusted entities claim so: these k trusted entities are typically the neighboring nodes of the entity. A locally trusted entity is globally accepted and a locally distrusted entity is regarded untrustworthy all over the network.

In the suggested security architecture, each node carries a certificate signed by the shared certificate signing key SK, while the corresponding public key PK is assumed to be well-known by all the nodes of the network, so that certificates are globally verifiable. Nodes without valid certificates will be isolated, that is, their packets will not be forwarded by the network. Essentially, any node without a valid certificate is considered a potential intruder. When a mobile node moves to a new location, it exchanges certificates with its new neighbors and goes through mutual authentication process to build trust relationships. Neighboring nodes with such trust relationship help each other to forward and route packets. They also monitor each other to detect possible attacks and break-ins. Specific monitoring algorithms and mechanisms are left to each individual node's choice. When a node requests a signed certificate from a coalition of k nodes, each of the latter checks its records about the requesting node. If the requestor is recorded as a legitimate node, a partial certificate is computed by applying the local node's share of SK and returned to the requestor. Upon collecting k partial certificates, the requesting node combines them to generate the complete certificate of its public-key as if issued by a centralized certification authority.

The multiple signature scheme used to build the certificate is based on a k-threshold polynomial secret sharing mechanism. This technique requires a bootstrapping phase where a "dealer" has to privately send each node its share of the secret signature key SK. The authors propose a scalable initialization mechanism called "self-initialization" whereby the dealer only has to initialize the very first k nodes, regardless of the global network span. The initialized nodes collaboratively initialize other nodes: repeating this procedure, the network progressively self-initializes itself. The same mechanism is applied when new nodes join the network.

Certificate revocation is also handled by the proposed architecture and an original approach to handle roaming adversaries is presented in order to prevent a misbehaving node that moves to a new location from getting a valid certificate. Roaming nodes are defeated with the flooding of "accusation" messages that travel in the network and inform distant nodes about the behavior of a suspect node.

The main drawback of the proposed architecture is the requirement for a trusted dealer that initializes the very first k nodes of a coalition to the choice of the system-wide parameter k. To cope with the first problem, the authors propose to use a distributed RSA key pair generation [25] for the very first k nodes. The other major limitation of the scheme is the strong assumption that every node of the network has at least k trusted neighbors. Moreover, the authors assume that any new node that joins the system already has an initial certificate issued by an offline authority or by a coalition of k neighbors.

6 MANET and Data Link Layer Security

Various security mechanisms have been proposed as part of 802.11[26] and Bluetooth [27] specifications. While the robustness of these mechanisms has often been argued [29], the main question is the relevance of security mechanisms implemented in the data link layer with respect to the requirements of MANET. This question deserves careful analysis in the light of requirements raised by the two different environments in which these mechanisms can potentially be deployed:

1. wireless extension of a wired infrastructure as the original target of 802.11 and Bluetooth security mechanisms,
2. wireless ad hoc networks with no infrastructure.

In case of 1 the main requirement for data link layer security mechanisms is the need to cope with the lack of physical security on the wireless segments of the communication infrastructure. Data link layer security is then perfectly justified as a means of building a "wired equivalent" security as stated by the objectives of Wired Equivalent Privacy (WEP) of 802.11. Data link layer mechanisms like the ones provided by 802.11 and Bluetooth basically serve for access control and privacy enhancements to cope with the vulnerabilities of radio communication links. However, data link layer security performed at each hop cannot meet the end-to-end security requirements of applications neither on wireless links protected by 802.11 or Bluetooth nor on physically protected wired links.

In case of wireless ad hoc networks as defined in 2 there are two possible scenarios:

- managed environments whereby the nodes of the ad hoc network are controlled by an organization and can thus be trusted based on authentication,
- open environments with no a priori organization among network nodes.

The managed environment raises requirements similar to the ones of 1. Data link layer security is justified in this case by the need to establish a trusted infrastructure based on logical security means. If the integrity of higher layer functions implemented by the nodes of a managed environment can be assured (i.e. using tamper-proof hardware) then data link layer security can even meet the security requirements raised by higher layers including the routing protocol and the applications.

Open environments on the other hand offer no trust among the nodes and across communication layers. In this case trust in higher layers like routing or application protocols cannot be based on data link layer security mechanisms. The only relevant use of the latter appears to be ad hoc routing security proposals whereby the data link layer security can provide node-to-node authentication and data integrity as required by the routing layer. Moreover the main impediment to the deployment of existing data link layer security solutions (802.11 and Bluetooth) would be the lack of support for automated key management which is mandatory in open environments whereby manual key installation is not suitable.

7 Conclusion

Security of ad hoc networks has recently gained momentum in the research community. Due to the open nature of ad hoc networks and their inherent lack of infrastructure, security exposures can be an impediment to basic network operation and the countermeasures should be included in network functions from the early stages of their design. Security solutions for MANET have to cope with a challenging environment including scarce energy and computational resources and lack of persistent structure to rely on for building trust.

The solutions presented in this article only cover a subset of all threats and are far from providing a comprehensive answer to the security problem in ad hoc networks. They often address isolated issues away from a global approach to security: for instance, secure routing proposals do not take into account lack of cooperation and do not include cooperation enforcement mechanisms. Most routing security solutions also make unrealistic assumptions about the availability of key management infrastructures that are in contrast with the very nature of ad hoc networks. As the technology for ad hoc wireless networks gains maturity, comprehensive security solutions based on realistic trust models and addressing all prevalent issues like routing, key management and cooperation enforcement are expected to appear.

References

[1] P. Papadimitratos, Z. Haas, Secure Routing for Mobile Ad Hoc Networks, in proceedings of CNDS 2002.

[2] Y-C Hu, A. Perrig, D. B. Johnson, Ariadne : A secure On-Demand Routing Protocol for Ad Hoc Networks, in proceedings of MOBICOM 2002.

[3] A. Perrig, R. Canetti, D. Song, J.D. Tygar, Efficient and secure source authentication for multicast, in proceedings of NDSS 2001.

[4] A. Perrig, R. Canetti, J.D. Tygar, D. Song, Efficient authentication and signing of multicast streams over lossy channels, in IEEE Symposium on Security and Privacy, 2000.

[5] B. Dahill, B. N. Levine, E. Royer, C. Shields, ARAN: A secure Routing Protocol for Ad Hoc Networks, UMass Tech Report 02-32, 2002.

[6] Y-C Hu, D. B. Johnson, A. Perrig, SEAD: Secure Efficient Distance Vector Routing for Mobile Wireless Ad Hoc Networks, in the Fourth IEEE Workshop on Mobile Computing Systems and Applications.

[7] C. E. Perkins, P. Bhagwat, Highly Dynamic Destination-Sequenced Distance-Vector Routing (DSDV) for Mobile Computers, in proceedings of SIGCOMM 1994.

[8] J. Broch, D. A. Maltz, D. B. Johnson, Y-C Hu, J. G. Jetcheva, A performance Comparison of Multi-Hop Wireless Ad Hoc Network Routing Protocols, in proceedings of MOBICOM 1998.

[9] P. Johansson, T. Larsson, N. Hedman, B. Mielczarek, M. Degermark, Scenario-based Performance Analysis of Routing Protocols for Mobile Ad Hoc Networks, in proceedings of MOBICOM 1999.

[10] A. Perrig, Y-C Hu, D. B. Johnson, Wormhole Protection in Wireless Ad Hoc Networks, Technical Report TR01-384, Dep. Of Computer Science, Rice University.

[11] P. Michiardi, R. Molva, Simulation-based Analysis of Security Exposures in Mobile Ad Hoc Networks, in proceedings of European Wireless Conference, 2002.

[12] D. B. Johnson, D. A. Maltz, Dynamic Source Routing in Ad Hoc Wireless Networks, Mobile Computing, edited by Tomasz Imielinski and Hank Korth, Chapter 5, pages 153–181, Kluwer Academic Publishers, 1996.

[13] Charles Perkins, Ad hoc On Demand Distance Vector (AODV) Routing, Internet draft, draft-ietf-manet-aodv-00.txt.

[14] L. Buttyan, J.-P. Hubaux, Nuglets: a virtual currency to stimulate cooperation in self-organized ad hoc networks, Technical Report DSC/2001/001, Swiss Federal Institute of Technology – Lausanne, 2001.
[15] S. Buchegger, J.-Y. Le Boudec, Nodes Bearing Grudges: Towards Routing Security, Fairness, and Robustness in Mobile Ad Hoc Networks, in proceedings of the 10th Euromicro Workshop on Parallel, Distributed and Network-based Processing.
[16] S. Buchegger, J.-Y. Le Boudec, Performance Analysis of the CONFIDANT Protocol, in proceedings of MobiHoc 2002.
[17] S. Marti, T. Giuli, K. Lai, and M. Baker, Mitigating routing misbehavior in mobile ad hoc networks, in proceedings of MOBICOM 2000.
[18] P. Michiardi, R. Molva, Core: A COllaborative REputation mechanism to enforce node cooperation in Mobile Ad Hoc Networks, IFIP – Communication and Multimedia Security Conference 2002.
[19] P. Michiardi, R. Molva, Game Theoretic Analysis of Security in Mobile Ad Hoc Networks, Institut Eurecom Research Report RR-02-070 – April 2002.
[20] H. Yang, X. Meng, S. Lu, Self-Organized Network-Layer Security in Mobile Ad Hoc Networks.
[21] S. Capkun, L. Buttyan and J-P Hubaux, Self-Organized Public-Key Management for Mobile Ad Hoc Networks, in ACM International Workshop on Wireless Security, WiSe 2002.
[22] P. Zimmermann, The Official PGP User's Guide. MIT Press, 1995.
[23] M. Reiter, S. Stybblebine, Authentication metric analysis and design, ACM Transactions on Information and System Security, 1999.
[24] H. Luo, S. Lu, Ubiquitous and Robust Authenticaion Services for Ad Hoc Wireless Networks, UCLA-CSD-TR-200030.
[25] A. Shamir, How to share a secret, Communications of ACM 1979.
[26] IEEE 802.11b-1999 Supplement to 802.11-1999,Wireless LAN MAC and PHY specifications: Higher speed Physical Layer (PHY) extension in the 2.4 GHz band
[27] "Specification of the Bluetooth System", Bluetooth Special Interest Group, Version 1.1, February 22, 2001, http://www.bluetooth.com/pdf/Bluetooth_11_Specifications_Book.pdf
[28] "Applied Cryptography", Bruce Schneier, John Wiley & Sons, 1996
[29] "Using the Fluhrer, Mantin, and Shamir Attack to Break WEP", Stubblefield, Loannidis, and Rubin, AT&T Labs Technical Report 2001
[30] "Providing robust and ubiquitous security support for manet", J. Kong, P. Zerfos, H. Luo, S. Lu, and L. Zhang, In Proc. IEEE ICNP, 2001.

Equilibrium Analysis of Packet Forwarding Strategies in Wireless Ad Hoc Networks – The Static Case*

M. Félegyházi, L. Buttyán, and J.-P. Hubaux

[1] Laboratory of Computer Communications and Applications,
Swiss Federal Institute of Technology – Lausanne
[2] Department of Telecommunications,
Budapest University of Technology and Economics

Abstract. In multi-hop wireless networks, every node is expected to forward packets for the benefit of other nodes. Yet, if each node is its own authority, then it may selfishly deny packet forwarding in order to save its own resources. Some researchers have proposed to introduce an incentive mechanism in the network that motivates the nodes to cooperate. In this paper, we address the question of whether such an incentive mechanism is necessary or cooperation between the nodes exists in the absence of it. We define a model in a game theoretic framework and identify the conditions under which cooperative strategies can form an equilibrium. As the problem is somewhat involved, we deliberately restrict ourselves to a static configuration.

1 Introduction

By definition, in multi-hop wireless networks, every node is expected to forward packets for the benefit of other nodes. This is probably the most important difference of this family of networks with respect to those that are more conventional, such as the Internet or cellular networks.

As long as the whole network is under the control of a single authority, as is usually the case in military networks or for rescue operations, this situation is not problematic because the interest of the mission by far exceeds the vested interest of each participant. However, if each node is its own authority, the situation changes dramatically: In this case, the most reasonable assumption is to consider that each node will try to maximize its benefits by exploiting the network, even if this means adopting a selfish behavior. This selfishness can mean not participating in the unfolding of mechanisms of common interest (e.g., route setup, packet forwarding, or mobility management), notably to spare resources, including batteries.

Over the last few years, several researchers have proposed incentive techniques to encourage nodes to collaborate, be it by circumventing misbehaving nodes [MGLB00], by making use of a reputation system [BL02], or by relating the right to benefit from the

* The work presented in this paper was supported (in part) by the National Competence Center in Research on Mobile Information and Communication Systems (NCCR-MICS), a center supported by the Swiss National Science Foundation under grant number 5005-67322 (http://www.terminodes.org)

M. Conti et al. (Eds.): PWC 2003, LNCS 2775, pp. 776–789, 2003.

network to the sense of the common interest demonstrated so far [BH03]. To the best of our knowledge the only paper that focuses on the conditions of cooperation without any incentive mechanism is [SNCR03], where the authors build up a theoretical framework based on different energy classes.

In this paper, we study the problem of cooperation for the most basic mechanism, namely packet forwarding. In order to do so, we define a model in a game theoretic framework and identify the conditions under which an equilibrium based on cooperation exists. As the problem is somewhat involved, we deliberately restrict ourselves to a static configuration.

The rest of the paper is organized in the following way. In Section 2, we show how packet forwarding can be modelled in a game theoretic framework. In Section 3, we analyze the model introduced in Section 2. We provide our main results, stated as theorems in Section 4 along with our simulation results. In Section 5, we present the related work. Finally, we conclude the paper in Section 6.

2 Modelling Packet Forwarding as a Game

Connectivity graph. Let us consider an ad hoc network of n nodes. Let us denote the set of all nodes by N. Each node has a given power range and two nodes are said to be neighbors if they reside within the power range of each other. We represent the neighbor relationship between the nodes with an undirected graph, which we call the connectivity graph. Each vertex of the connectivity graph corresponds to a node in the network, and two vertices are connected with an edge if the corresponding nodes are neighbors.

Routes. Communication between two non-neighboring nodes is based on multi-hop relaying. This means that packets from the source to the destination are forwarded by intermediate nodes. Such a chain of nodes (including the source and the destination) is represented by a path in the connectivity graph, which we call a route. We assume that one of the shortest paths is chosen randomly as a route for each source-destination pair.

Time. We use a discrete model of time where time is divided into slots. We assume that both the connectivity graph and the set of existing routes remain unchanged during a time slot, whereas changes may happen at the end of each time slot. We assume that the duration of the time slot is much longer than the time of a packet relaying from the source to the destination. This means that a node is able to send several packets within one time slot.

Forwarding game. We model the packet forwarding operation as a game, which we call the forwarding game. The players of the forwarding game are the nodes. In each time slot t, each node i chooses a cooperation level $p_i(t) \in [0, 1]$, where 0 and 1 represent full defection and full cooperation, respectively. Here, defection means that the node does not forward traffic for the benefit of other nodes and cooperation means that it does. Thus, $p_i(t)$ represents the fraction of the traffic routed through i in t that i actually forwards. Note that i has a single cooperation level $p_i(t)$ that it applies to every route in which it is involved as a forwarder. We preferred not to require the nodes to be able to distinguish the packets that belong to different routes, because this would require identifying the source-destination pairs and applying a different cooperation level to each of them. This would probably increase the computation significantly.

Let us assume that in time slot t there exists a route r with source node s and ℓ intermediate nodes $f_1, f_2, \ldots, f_\ell$. The normalized value of the throughput $\tau(r,t)$ experienced by the source s on r in t is defined as the fraction of the traffic sent by s on r in t that is delivered to the destination. Since we are studying cooperation in packet forwarding, we assume that the main reason for dropping packets in the network is the non-cooperative behavior of the nodes. In other words, we assume that the network is not congested and that the number of packets dropped because of the limited capacity of the nodes and the links is negligible. Hence, $\tau(r,t)$ can be computed as the product of the cooperation levels of all intermediate nodes:

$$\tau(r,t) = \prod_{k=1}^{\ell} p_{f_k}(t) \tag{1}$$

The payoff $\xi_s(r,t)$ of s on r in t depends on the normalized throughput $\tau(r,t)$. In general, $\xi_s(r,t) = u(\tau(r,t))$, where u is some non-decreasing function. In this paper, we assume that u is linear and has the following form: $u(x) = (G+L)\cdot x - L = G\cdot x - L\cdot(1-x)$, where $G = u(1) > 0$ represents the gain that s obtains if all its traffic sent on r in t is delivered to the destination, and $-L = u(0) < 0$ represents the loss that s suffers if nothing is delivered in t to the destination. Thus, we get that

$$\xi_s(r,t) = (G+L)\cdot\tau(r,t) - L \tag{2}$$

For simplicity, we assume that every node in the network uses the same function u and the same parameters G and L, which we assume to be independent of r and t.

The payoff $\eta_{f_j}(r,t)$ of the j-th intermediate node f_j on r in t is non-positive and represents the cost for node f_j to forward packets on route r during time slot t. It depends on the fraction $\tau_j(r,t)$ of the traffic sent by the source on r in t that is forwarded by f_j. The value $\tau_j(r,t)$ is the normalized throughput on r in t leaving node j, and it is computed as the product of the cooperation levels of the intermediate nodes from f_1 up to and including f_j:

$$\tau_j(r,t) = \prod_{k=1}^{j} p_{f_k}(t) \tag{3}$$

Let $-C < 0$ represent the forwarding cost for the first forwarder if it forwards all the traffic sent by the source on r in t. The payoff $\eta_{f_j}(r,t)$ of the j-th intermediate node is then defined as follows:

$$\eta_{f_j}(r,t) = -C\cdot\tau_j(r,t) \tag{4}$$

Again, for simplicity, we assume that forwarding one unit of traffic costs the same for every node in the network and that the parameter C is independent from r and t.

By definition, the payoff of the destination is 0. In other words, we assume that only the source has a benefit if the traffic reaches the destination (information pushing). We note however, that our model can be applied and all of our results hold in the reverse case when only the destination benefits from receiving traffic; an example for this scenario would be a file download.

Table 1. Three special strategies. Here, we assume that σ_i takes as input a one dimensional vector (i.e., a scalar)

Strategy	*Initial move*	*Function*
AllD (always defect)	0	$\sigma_i(x) = 0$
AllC (always cooperate)	1	$\sigma_i(x) = 1$
TFT (Tit-For-Tat)	1	$\sigma_i(x) = x$

The total payoff $\pi_i(t)$ of node i in time slot t is then computed as

$$\pi_i(t) = \sum_{r \in S_i(t)} \xi_i(r,t) + \sum_{r \in F_i(t)} \eta_i(r,t) \tag{5}$$

where $S_i(t)$ is the set of routes in t where i is source, and $F_i(t)$ is the set of routes in t where i is an intermediate node.

Strategy space. In every time slot, each node i updates its cooperation level using a strategy function σ_i. In general, i could choose a cooperation level to be used in time slot t based on the information it obtained in *all* preceding time slots. In order to make the analysis feasible, we assume that i uses only information that it obtained in the previous time slot. More specifically, we assume that i chooses its cooperation level in time slot t based on the normalized throughput it experienced in time slot $t-1$ on the routes where it was a source. Formally:

$$p_i(t) = \sigma_i([\tau(r, t-1)]_{r \in S_i(t-1)}) \tag{6}$$

where $[\tau(r, t-1)]_{r \in S_i(t-1)}$ represents the normalized throughput vector for node i in time slot $t-1$, each element of which being the normalized throughput experienced by i on a route where it was source in $t-1$. The strategy of a node i is then defined by its strategy function σ_i and its initial cooperation level $p_i(0)$.

Note that σ_i takes as input the normalized throughput and not the total payoff received by i in the previous time slot. The rationale is that i should react to the behavior of the rest of the network, which is represented by the normalized throughput in our model.

There are an infinite number of strategy functions; we highlight three special cases (in Table 1) that we will encounter later in the analysis [Axe84]. If the output of the strategy function is independent of the input, then the strategy is called a *non-reactive strategy* (e.g., always defecting and always cooperating in Table 1). If the output depends on the input, then the strategy is *reactive* (e.g., Tit-For-Tat in Table 1).

Our model requires that each source be able to observe the throughput in a given time slot on every route. We assume that this is made possible with sufficiently high precision by using some higher level control protocol above the network layer.

3 Meta-model

In this section, we introduce a meta-model in order to formalize the properties of the packet forwarding game defined in the previous section. Thus, we study the behavior of the model through the analysis of the meta-model (i.e., a model of the model). Unlike

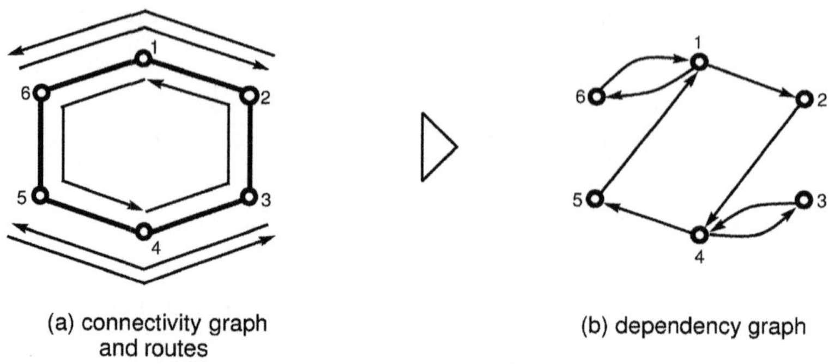

Fig. 1. Example of dependency graph. In (a), the arrows represent the routes (e.g., there is a route between nodes 1 and 4 passing through nodes 6 and 5). The corresponding dependency graph is shown in (b)

in the model presented in the previous section, in the meta-model and in the rest of the paper, we will assume that routes remain unchanged during the lifetime of the network. In addition, we also assume that each node is a source on only one route.

Dependency graph. Let us consider a route r. The normalized throughput experienced (hence the payoff received) by the source on r depends on the cooperation levels of the intermediate nodes on r. We represent this dependency relation between the nodes with a directed graph, which we call the dependency graph. Each vertex of the dependency graph corresponds to a network node. There is a directed edge from vertex i to vertex j, denoted by the ordered pair (i, j), if there exists a route where i is an intermediate node and j is the source. Intuitively, an edge (i, j) means that the behavior (cooperation level) of i has an effect on j. The concept of dependency graph is illustrated in Fig. 1.

Game automaton. Now we define the automaton A that will model the unfolding of the forwarding game. The automaton is built on the dependency graph. We assign a machine M_i to every vertex i of the dependency graph and interpret the edges of the dependency graph as links that connect the machines assigned to the vertices. Each machine M_i thus has some input and some (possibly 0) output links.

The internal structure of the machine is illustrated in Fig. 2. Each machine M_i consists of a multiplication gate $\prod$ followed by a gate that implements the strategy function σ_i of node i. The multiplication gate $\prod$ takes the values on the input[1] links and passes their product to the strategy function gate[2]. Finally, the output of the strategy function gate is passed to each output link of M_i.

The automaton A works in discrete steps. Initially, in step 0, each machine M_i outputs some initial value $x_i(0)$. Then, in step $t > 0$, each machine computes its output $x_i(t)$ by taking the values that appear on its input links in step $t - 1$.

[1] The multiplication comes from the fact that the experienced throughput for the source is the product of the cooperation levels of the forwarders on its route.

[2] Note that here σ_i takes a single real number as input, instead of a vector of real numbers as we defined earlier, because we assume that each node is source on only one route.

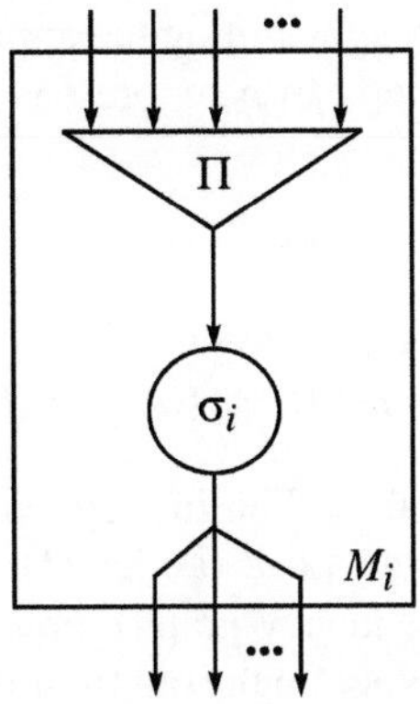

Fig. 2. Internal structure of machine M_i

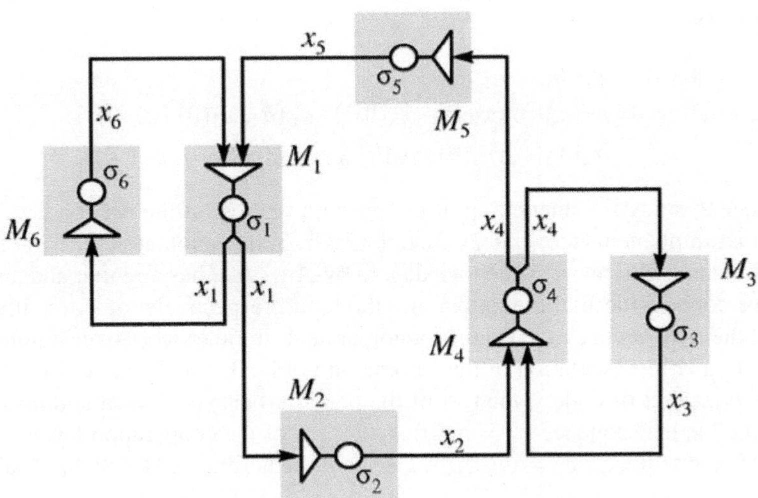

Fig. 3. The automaton that corresponds to the dependency graph of Fig. 1

Formally, the operation of A can be described by a set of equations, where each equation corresponds to a machine M_i. As an example, let us consider the automaton in Fig. 3, which is the automaton that corresponds to the dependency graph of Fig. 1. Its operation is described by the following set of equations:

$$
\begin{aligned}
x_1(t) &= \sigma_1(x_5(t-1) \cdot x_6(t-1)) \\
x_2(t) &= \sigma_2(x_1(t-1)) \\
x_3(t) &= \sigma_3(x_4(t-1)) \\
x_4(t) &= \sigma_4(x_2(t-1) \cdot x_3(t-1)) \\
x_5(t) &= \sigma_5(x_4(t-1)) \\
x_6(t) &= \sigma_6(x_1(t-1))
\end{aligned}
$$

It is easy to see that by iteratively substituting the appropriate equations into each other, the value of any output in any step can be expressed as a function of the initial values $x_i(0)$. For instance:

$$\begin{aligned} x_1(3) &= \sigma_1(x_5(2) \cdot x_6(2)) \\ &= \sigma_1(\sigma_5(x_4(1)) \cdot \sigma_6(x_1(1))) \\ &= \sigma_1(\sigma_5(\sigma_4(x_2(0) \cdot x_3(0))) \cdot \sigma_6(\sigma_1(x_5(0) \cdot x_6(0)))) \end{aligned}$$

Note that if $x_i(0) = p_i(0)$ for all i, then in step t, each machine M_i will output the cooperation level of node i in time slot t (i.e., $x_i(t) = p_i(t)$), as we assumed that the set of routes (and hence the dependency graph) remains unchanged in every time slot. Therefore, the evolution of the values on the output links of the machines (which, in fact, represent the state of the automaton) models the evolution of the cooperation levels of the nodes in the network. It also follows that the cooperation level $p_i(t)$ of any node i in any time slot t can be expressed as a function of the initial cooperation levels of the nodes. For instance:

$$\begin{aligned} p_1(3) &= x_1(3) \\ &= \sigma_1(\sigma_5(\sigma_4(x_2(0) \cdot x_3(0))) \cdot \sigma_6(\sigma_1(x_5(0) \cdot x_6(0)))) \\ &= \sigma_1(\sigma_5(\sigma_4(p_2(0) \cdot p_3(0))) \cdot \sigma_6(\sigma_1(p_5(0) \cdot p_6(0)))) \end{aligned}$$

In order to study the interaction of node i with the rest of the network, we extract the gate that implements the strategy function σ_i from the automaton A. What remains is the automaton without σ_i, which we denote by A_{-i}. A_{-i} has an input and an output link; if we connect these to the output and the input, respectively, of σ_i (as illustrated in Fig. 4) then we get back the original automaton A. In other words, the automaton in Fig. 4 is another representation of the automaton in Fig. 3, which captures the fact that from the viewpoint of node i, the rest of the network behaves like an automaton: The input of A_{-i} is the sequence $\overline{x}_i = x_i(0), x_i(1), \ldots$ of the cooperation levels of i, and its output is the sequence $\overline{y}_i = y_i(0), y_i(1), \ldots$ of the normalized throughput values for i.

By using the system of equations that describe the operation of A, one can easily express any element $y_i(t)$ of sequence $\overline{y}_i$ as some function of the preceding elements $x_i(t-1), x_i(t-2), \ldots, x_i(0)$ of sequence $\overline{x}_i$ and the initial values $x_j(0)$ ($j \neq i$) of the machines within A_{-i}. We call such an expression of $y_i(t)$ the t-th *input/output formula* or shortly the t-th *i/o formula* of A_{-i}. It is important to note that the i/o formulae of A_{-i} may involve any strategy function σ_j where $j \neq i$, but they never involve σ_i. Taking the automaton of Fig. 3 again, and extracting, for instance, σ_6, we can determine the first few i/o formulae of A_{-6} as follows:

$$\begin{aligned} y_6(0) &= x_1(0) \\ y_6(1) &= \sigma_1(x_5(0) \cdot x_6(0)) \\ y_6(2) &= \sigma_1(\sigma_5(x_4(0)) \cdot x_6(1)) \\ y_6(3) &= \sigma_1(\sigma_5(\sigma_4(x_2(0) \cdot x_3(0))) \cdot x_6(2)) \\ \ldots &\quad \ldots \end{aligned}$$

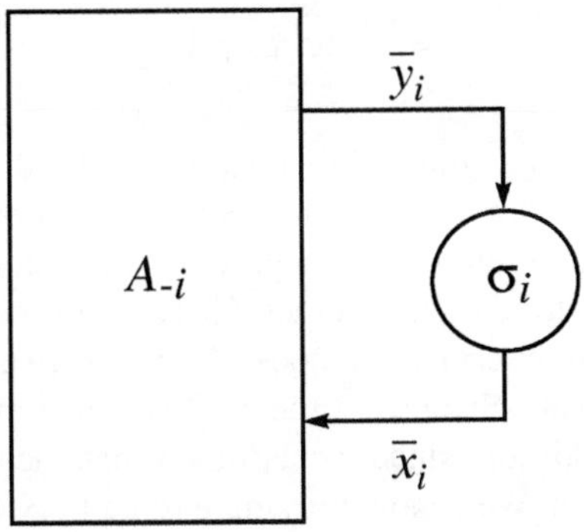

Fig. 4. Model of interaction between node i and the rest of the network represented by the automaton A_{-i}

Dependency loops. Node i has a dependency loop if there exists a sequence $(i, v_1), (v_1, v_2), \ldots, (v_{\ell-1}, v_\ell), (v_\ell, i)$ of edges in the dependency graph. The existence of dependency loops is important: If node i has no dependency loops, then the cooperation level chosen by i in a given time slot has no effect on the normalized throughput experienced by i in future time slots.

We define two types of dependency loops depending on the strategies played by the nodes on the dependency loop. If all nodes j (where $j \neq i$) on the dependency loop of i play reactive strategies, then we talk about a *reactive dependency loop* of i. If, on the other hand, there exists at least one node j (where $j \neq i$) on the dependency loop of i that plays a non-reactive strategy, then the loop will be a *non-reactive dependency loop* of i.

4 Results

Our goal, in this section, is to find possible Nash equilibria [Nash50] and investigate the conditions for them. The existence of a Nash equilibrium based on cooperation would mean that there are cases in which cooperation is "naturally" encouraged, meaning without using incentive mechanisms. In the following, we use the model presented in Fig. 4. In Section 4.1, we present analytical results that define the best strategies for a node under specific network conditions. In Section 4.2, we present the results of simulations that we performed to estimate the probability that these conditions are satisfied by the network.

4.1 Analytical Results

The aim of the nodes is to maximize the payoff that they accumulate over time. We compute this cumulative payoff by applying the standard technique used in the theory of iterative games [Axe84]: The cumulative payoff $\overline{\pi}_i$ of a node i is computed as the weighted sum of the payoffs $\pi_i(t)$ that i obtains in each time slot t:

$$\overline{\pi}_i = \sum_{t=0}^{\infty} [\pi_i(t) \cdot \omega^t] \tag{7}$$

where $0 < \omega < 1$, and hence, the weights exponentially decrease with t. The value ω is often called the *discounting factor*, as it represents the degree to which the payoff of each time slot is discounted relative to the previous time slot.

Recall that $S_i(t)$ denotes the set of routes where i is the source, and $F_i(t)$ denotes the set of routes where i is an intermediate node. Since we assumed that the routes remain static, $S_i(t)$ and $F_i(t)$ do not change in time, and we will simply write S_i and F_i instead of $S_i(t)$ and $F_i(t)$. In addition, since we assumed that each node is source on exactly one route, S_i is a singleton. We denote the single route in S_i by r_i. The cardinality of F_i will be denoted by $|F_i|$. Let us assume that node i is an intermediate node on route r. We denote the set of intermediate nodes on r upstream from node i including node i in time slot t by $\Phi(r, i)$. If node i has a reactive dependency loop with some of its sources, then we consider the shortest loop for each source. We denote the length of the longest of such reactive dependency loops minus one by Δ_i.

Theorem 1. *If a node i does not have any dependency loops, then its best strategy is AllD (i.e., to choose cooperation level 0 in every time slot).*

Proof. Node i wants to maximize its cumulative payoff $\overline{\pi}_i$ defined in (7). In our case, $\pi_i(t)$ can be written as:

$$\begin{aligned} \pi_i(t) &= \xi_i(r_i, t) + \sum_{r \in F_i} \eta_i(r, t) \\ &= [(G + L) \cdot y_i(t) - L] - C \cdot \sum_{r \in F_i} \prod_{k \in \Phi(r,i)} x_k(t) \end{aligned}$$

Given that i has no dependency loops, $y_i(t)$ is independent of all the previous cooperation levels of node i. Thus, $\overline{\pi}_i$ is maximized if $x_i(t) = 0$ for all $t \geq 0$. □

Theorem 2. *If a node i has only non-reactive dependency loops, then its best strategy is AllD.*

Proof. The proof is similar to the proof of Theorem 1. Since all dependency loops of i are non-reactive, its experienced normalized throughput $\overline{y}_i$ is independent of its own behavior $\overline{x}_i$. This implies that its best strategy is not to forward. □

Corollary 1. *If every node j ($j \neq i$) plays AllD, then the best response of i to this is AllD. Hence, every node playing AllD is a Nash equilibrium.*

In order to illustrate the case in which some nodes have no dependency loops, we modify the network of Fig. 1. In the modified network, shown in Fig. 5, nodes 2 and 5 have no dependency loops.

In the case when the conditions of Theorems 1 and 2 do not hold, we could not determine the best strategy of a node i in general, because it very much depends on the particular scenario (dependency graph) in question and the strategies played by the other nodes. However, in certain cases, the best strategy can be determined, as shown by the following theorem:

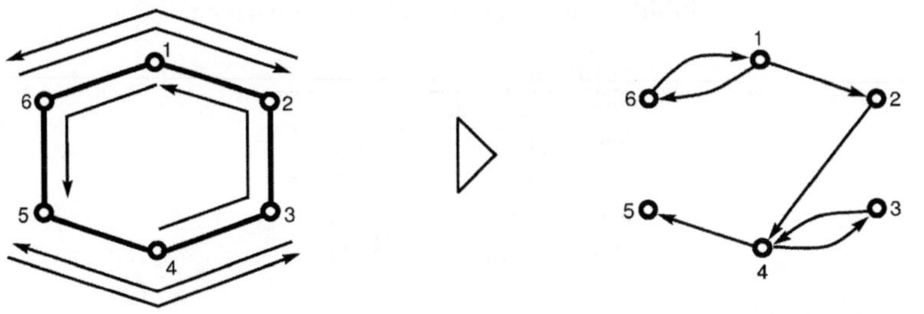

Fig. 5. A network with routes where nodes 2 and 5 have no dependency loops

Theorem 3. *The best strategy for node i is TFT, if:*

1. *Node i has a reactive dependency loop with at least one of its sources. In addition, for each of its sources s, either node i has a reactive dependency loop with s or i is the first forwarder for s,*
2. *every other node j ($j \neq i$) plays TFT, and*
3. $(G + L) \cdot \omega^{\Delta_i} > |F_i| \cdot C$.

The expression $(G + L) \cdot \omega^{\Delta_i} > |F_i| \cdot C$ means that the possible benefit for node i must be greater than its maximal forwarding cost considering every source node with which node i has a reactive dependency loop.

Because of space limitations we only provide the idea of the proof: We give an upper bound for the total payoff of any node using the concept of dependency graph. Then, we show that the node playing TFT reaches the maximum payoff defined by the upper bound. The complete proof is provided in [FBH03].

Corollary 2. *If conditions 1 and 3 of Theorem 3 holds for every node, then all nodes playing TFT is a Nash equilibrium.*

As an example, let us consider Fig. 1 again. If the third condition of Theorem 3 holds for each node, then all nodes playing TFT is a Nash equilibrium for this network.

The first two theorems state that if the behavior of node i has no effect on its experienced normalized throughput, then defection is the best choice for i. In addition, Corollary 1 says that if every node defects constantly, it is a Nash equilibrium. The condition of Theorem 2 implies that it would be a naïve approach to program all devices with the strategy that always forwards (AllC), because this would make the network exploitable by selfish participants. Instead, manufacturers should program a reactive strategy into the devices. Theorem 3 and its corollary mentions the existence of a cooperative equilibrium (each node playing TFT). Yet, the conditions for this existence are extremely stringent.

Table 2. Parameter values for the simulation

Parameter	*Value*
Number of nodes	100
Area type	Torus
Area size	1500 m^2
Radio range	250 m
Route length	4 hops

4.2 Simulation Results

In this subsection our goal is to determine the likelihood that a cooperative equilibrium exists (i.e., the conditions of Theorem 3 hold). To this end, we performed simulations in randomly generated scenarios. We summarize the parameters in Table 2.

In our simulations, we randomly placed nodes in a toroid[3] area. Then, for each node, we randomly chose a destination at a fixed distance (in hops) from the node and we determined a route to this destination using any shortest path algorithm. If several routes existed, then we randomly chose a single one. From the routes, we built up the dependency graph of the network. On this dependency graph, we performed a search for reactive dependency loops while varying the fraction of nodes that played non-reactive strategies. All the results presented in the following subsections are the mean values of 100 simulations with a 95% confidence interval.

Figure 6 shows the proportion of nodes that (a) can play reactive strategies (i.e. are not programmed to play non-reactive strategies), (b) are forwarders at least on one route, (c) have at least one reactive dependency loop and (d) fulfill the first condition of Theorem 3.

The nodes that cannot play reactive strategies are irrelevant to our investigations because they follow a fixed packet forwarding strategy, thus they do not make a decision about packet forwarding at all. Additionally, there is a fraction of nodes that are not forwarders in any of the routes, thus they have no effect on other nodes. The nodes in this situation are also irrelevant to our study. In Fig. 7 we show the same results as in Fig. 6 excluding these two types of nodes. We can see that if initially every node can play reactive strategy, the proportion of nodes that fulfill the first condition of Theorem 3 is about 50 percent. This proportion decreases as we increase the proportion of nodes that play constantly non-reactive strategies.

If there exists at least one node that needs an incentive mechanism to cooperate, then we say that an incentive mechanism must be applied for the whole network. Since the proportion of such nodes is always around or above 50 percent, we can conclude that in general an incentive mechanism is almost always needed under the assumption we have considered.

5 Related Work

Energy-efficient cooperation in ad hoc networks. Srinivasan *et al.* [SNCR03] provide a mathematical framework for cooperation in ad hoc networks, which focuses on

[3] We used this area type to avoid border effects. In a realistic scenario, the toroid area can be considered as an inner part of a large network.

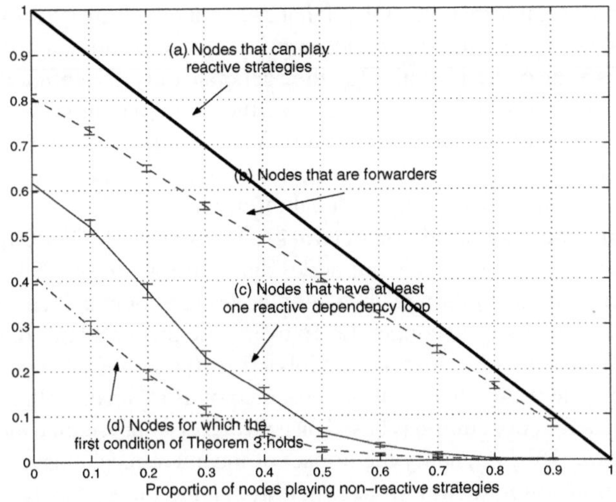

Fig. 6. The proportion of nodes that (a) can play reactive strategies (i.e. are not programmed to play non-reactive strategies), (b) are forwarders at least on one route, (c) have at least one reactive dependency loop and (d) fulfill the first condition of Theorem 3

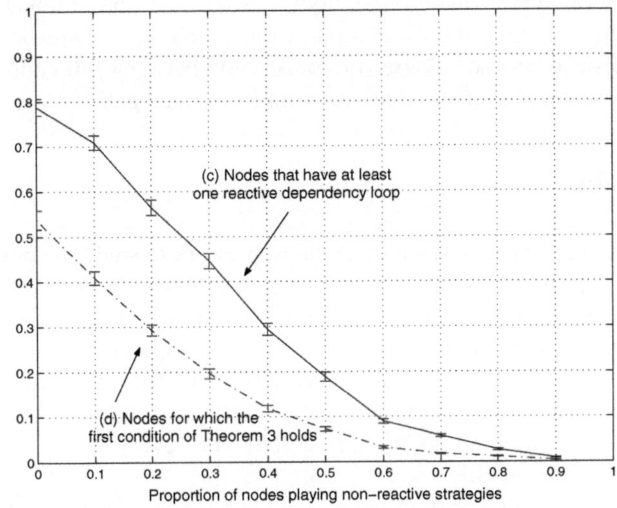

Fig. 7. The proportion of nodes (c) have at least one reactive dependency loop and (d) fulfill the first condition of Theorem 3

the energy-efficient aspects of cooperation. In their solution, the nodes are classified in different energy classes. The nodes differentiate between the connections based on the energy classes of the participants and apply different behavior according to the type of the connection. This framework relies on an ideal mechanism that distributes class information. We introduce a game theoretic model that does not rely on an additional mechanism, thus our investigations are more generic. Similarly to our framework, Srinivasan *et al.* define time slots as an operating unit. But, they generate only one communication session in the network in each time slot. They choose the participating nodes for this session uniformly. In our model we take the constitution of the network also into account, which is in our opinion an important factor. Finally, we develop a model for the nodes where they can make decisions based only on local information.

Incentive mechanism in ad hoc networks. Marti *et al.* [MGLB00] consider an ad hoc network where some misbehaving nodes agree to forward packets but then fail to do so. However, in their solution misbehaving nodes are not punished, and thus there is no motivation for the nodes to cooperate. To overcome this problem, Buchegger and Le Boudec [BL02] as well as Michiardi and Molva [MM02] define protocols that are based on a reputation system. Another solution presented by Buttyan and Hubaux [BH03] that gives an incentive for cooperation is based on a virtual currency, called *nuglets or credits*. Plaggemeier *et al.* [PWL03] study the impact of incentive mechanisms in general.

Application of game theory to networking. Game theory has been used to solve problems both in fixed and cellular networks. Researchers addressed problems like power control (e.g. Ji and Huang [JH98]), resource allocation (e.g. Korilis *et al.* [KLO95]) and pricing (e.g. Yaïche *et al.* [YMR00]).

Cooperation studies in other areas of science. The emergence of cooperation has also been previously studied in an economic context using the *Iterated Prisoner's Dilemma (IPD)* game as an underlying model(see e.g. Axelrod [Axe84,CRA99]). Wahl and Nowak [WN99] study the *Continuous valued Prisoner's Dilemma (CPD)* game, where the nodes can choose a degree of cooperation between full cooperation and full defection. Our model is also based on a continuous valued game.

6 Conclusion

In this paper, we have proposed a theoretical framework to study the possible existence of cooperation in packet forwarding in multi-hop wireless networks. We have introduced the concept of dependency graphs, based on which we were able to prove several theorems. As a main result, we proved the existence of a cooperative equilibrium of packet forwarding strategies. But, our simulation results show that, in general, the conditions for such a cooperative equilibrium are satisfied for at most half of the nodes. Thus, the likelihood that the condition for Corollary 2 holds (i.e., every node has TFT as the best strategy) is extremely small.

It is important to notice that our approach does not require a node to keep track of the individual behavior of other nodes: indeed, the considered node is served simply in a way that reflects the way it served others. As we have shown, this concept is captured by the game in which a given node perceives itself as playing with the rest of the network. In this way, the node does not even need to know the identity of the nodes forwarding (or deciding to not forward) its packets.

In this paper, we have assumed the routes to be static; this drastic decision was motivated by the complexity of the problem and by our willingness to provide formal results. In terms of future work, we intend to study the impact of mobility; we expect mobility to have beneficial effects, as it will increase the dependency between the nodes. In this work, we focused on the possible *existence* of equilibria; in the future, we will also investigate the possible *emergence* of such equilibria. Finally, we intend to relax the assumption of linearity of the utility function.

References

[Axe84] Robert Axelrod, *The Evolution of Cooperation*, Basic Books, New York, 1984.

[BL02] S. Buchegger, J-Y. Le Boudec, "Performance Analysis of the CONFIDANT Protocol (Cooperation Of Nodes–Fairness In Dynamic Ad-hoc NeTworks)," *In Proc. 3rd ACM International Symposium on Mobile Ad Hoc Networking and Computing (MobiHoc'02)*, Lausanne, Switzerland, pp. 80–91, June 9–11, 2002.

[BH03] L. Buttyán and J.-P. Hubaux, "Stimulating Cooperation in Self-Organizing Mobile Ad Hoc Networks," to appear in *ACM/Kluwer Mobile Networks and Applications (MONET) Special Issue on Mobile Ad Hoc Networks*, Vol. 8 No. 5, October 2003.

[CRA99] M. C. Cohen, R. L. Riolo, R.Axelrod, "The Emergence of Social Organization in the Prisoner's Dilemma: How Context-Preservation and Other Factors Promote Cooperation," Santa Fe Institute *Technical report Nr. 99-01-002*, submitted to *Nature*

[FBH03] M. Félegyházi, L. Buttyán, J.-P. Hubaux, "Equilibrium Analysis of Packet Forwarding Strategies in Wireless Ad Hoc Networks – the Static Case (extended version)," Technical Report Nr. IC/2003/33

[JH98] H. Ji, C-Y. Huang, "Non-Cooperative Uplink Power Control in Cellular Radio Systems," *Wireless Networks (WINET)*, vol. 4, pp 233–240, Apr. 1998

[KLO95] Y. Korilis, A. Lazar, A. Orda, "Architecting Noncooperative Networks," *IEEE Journal on Selecte Areas in Communication*, vol. 13, No. 8, 1995

[MGLB00] S. Marti, T.J. Giuli, K. Lai, M. Baker, "Mitigating Routing Misbehavior in Movile Ad Hoc Networks," *Proceedings of ACM/IEEE International Conference on Mobile Computing and Networking (Mobicom'00)*, pp. 255–265, 2000.

[MM02] P. Michiardi, R. Molva, "Core: A COllaborative REputation mechanism to enforce node cooperation in Mobile Ad Hoc Networks," *Communication and Multimedia Security 2002*, Portoroz, Slovenia, September 26–27, 2002.

[Nash50] J. Nash, "Equilibrium Points in N-person Games," *Proceedings of the National Academy of Sciences*, 36:48–49, 1950.

[PWL03] M. Plaggemeier, D. Westhoff, B. Lamparter, "About the Impact of Co-operation Approaches for Ad Hoc Networks," Poster presentation at *ACM International Symposium on Mobile Ad Hoc Networking and Computing (MobiHoc'03)*, Annapolis, MD, USA, June 1–3, 2003.

[SNCR03] V. Srinivasan, P. Nuggehalli, C. F. Chiasserini, R. R. Rao, "Cooperation in Wireless Ad Hoc Networks," *In Proceedings of IEEE INFOCOM'03*, San Francisco, Mar 30 – Apr 3, 2003

[WN99] L. M. Wahl, M. A. Nowak, "The Continuous Prisoner's Dilemma: I. Linear Reactive Strategies," *Journal of Theoretical Biology* vol. 200, 307–321., 1999.

[YMR00] H. Yaïche, R. R. Mazumdar, C. Rosenberg, "A Game Theoretical Framework for Bandwidth Allocation and Pricing in Broadband Networks," *IEEE/ACM Transactions on Networking*, vol. 8, No. 5, October 2000.

Towards Reliable Forwarding for Ad Hoc Networks*

Marco Conti, Enrico Gregori, and Gaia Maselli

IIT Institute - CNR,
Via G. Moruzzi, 1 - 56124 Pisa, Italy
{conti.marco,gregori.enrico,maselli.gaia}@iit.cnr.it

Abstract. Ad hoc networking is a new paradigm of wireless communications for mobile nodes. Mobile ad hoc networks work properly only if the partecipating nodes cooperate to network protocols. Cooperative algorithms make the system vulnerable to user misbehavior as well as to malicious and selfish misbehavior. Nodes act selfishly to save battery power, by not cooperating to routing-forwarding functions. Lack of cooperation may severely degrade the performance of the ad hoc system. This paper presents a new approach to cope with cooperation misbehavior, focusing on the forwarding function. We present a general framework, based on reliability indices taking into account not only selfish/malicious misbehavior, but also situations of congestion and jammed links. We aim at avoiding unreliable routes and enforcing cooperation, thus increasing network "performability" (performance and reliability).

1 Introduction

A mobile ad hoc network is composed by a group of wireless nodes that cooperatively form the network without the support of any fixed infrastructure. All nodes are capable of movement and can be connected dynamically in an arbitrary manner. Nodes of these networks function as routers which discover and maintain routes to other nodes in the network. Essentially, the mobile ad hoc networking technology enables an autonomous system of mobile nodes, and introduces the notion of a *spontaneous network* [1], created when a group of people come together for some collaborative activity. Such networks have been proposed for several goals: data collection in sensor arrays, providing a communication mean in hostile environment (battlefield) or rescue operations, providing connectivity to people attending conferences, meetings or lectures with their laptops. Another emerging application of ad hoc networks concerns vehicular networks, where intervehicle communications and vehicle to road communications are considered to have extensive potential for the development of efficient safety systems installed in vehicles [2].

* This work was partially funded by the Information Society Technologies programme of the European Commission, Future and Emerging Technologies under the IST-2001-38113 MOBILEMAN project

M. Conti et al. (Eds.): PWC 2003, LNCS 2775, pp. 790–804, 2003.

The lack of centralized points leads to the necessity of distributing basic functions like packet routing and forwarding to all available nodes in the network that must cooperate, and provide services to each other. The lack of a fixed infrastructure, and consequently of a centralized authority, leads adversaries to exploit this vulnerability for new types of attacks designed to break the cooperative paradigm. For example, routing mechanism is vulnerable in ad hoc networks because each device acts as a router. Forwarding mechanism is cooperative as well: communications between nodes more than 1-hop away are performed by intermediate nodes that act as relays.

A node that does not cooperate is called a misbehaving node, and routing-forwarding misbehavior can be caused by nodes that are broken, overloaded, malicious or selfish [3]. A broken node is not able to cooperate because of a software/hardware fault. An oveloaded node does not cooperate because it lacks CPU cycles, buffer space, or network bandwidth. A malicious node does not cooperate because it wants to intentionally damage network functioning by dropping packets. A selfish node is unwilling to spend battery life, CPU cycles, or available network bandwidth to forward packets not of direct interest to it, even though it expects others to forward packets on its behalf. It uses the network but does not cooperate, saving battery life for its own communications: it does not intend to directly damage other nodes. While the first two cases (broken and overloaded) define misbehavior due to uncontrollable events, in case of a malicious or selfish node, cooperation misbehavior is motivated by an intentional action of the node, even if with different aim: malicious or not. In any case, a cooperation misbehavior can have severe effects on the network functioning. Cooperation misbehavior due to an intentional action is a new problem that arises in the context of ad hoc networks, so new mechanisms are needed to face the problem of service availability.

Cooperation among nodes has been previously addressed in [4], [5], [6]. The starting step is given in [4]. They present a solution aimed at detecting and avoiding misbehaving nodes through a mechanism based on a *watchdog* and a *reputation system*. The watchdog identifies misbehaving nodes by performing a neighborhood monitoring: it observes the behavior of neighbors by promiscuously listening to communications of nodes in the same transmission range. According to collected information, the reputation system maintains a value for each observed node that represents a reputation of its behavior. The reputation mechanism allows to avoid sending packets through misbehaving nodes. In this way, malicious nodes are rewarded and strengthened, while cooperation enforcing is totally absent. The following works, CONFIDANT [5] and CORE [6], extend such a scheme with a punishment mechanism that isolates misbehaving nodes by not serving their requests. When a neighbor's reputation falls down a predefined threshold, service provision to the misbehaving node is interrupted. In such a way, there is no advantage for a node to misbehave because any resource utilization will be forbidden. These solutions present some limitations. First, the watchdog's weaknesses are not negligible: in presence of collisions, dishomogeneous transmission ranges, or directional antennas, the watchdog is not able to properly monitoring the neighborhood, and misbehaving nodes detection can fail. Another important aspect that must be considered is the employing of cooperation in security mechanisms. This approach may have severe drawbacks in term of traffic overhead and wrong accusation spreading. The CONFIDANT protocol generates some additional traffic for reputation propagation. The produced overhead may result heavy, and malicious nodes may perform a new attack by sending false alarms about other nodes. Furthermore, it is worth noticing that both CONFIDANT and CORE do not take into account the network utilization: by totally avoiding all routes containing misbehaving nodes, they risk deviating all the traffic on well behaving nodes, with the result of overloading them and links between them. Finally, both mechanisms work as extensions to the Dynamic Source Routing (DSR) [7] protocol. This can be a big constraint as there is not yet a standard routing protocol for ad hoc networks [8].

2 Estimating Routes Reliability

We address forwarding misbehavior due to intentional actions, malicious or selfish, as well as uncontrollable events, such as congestion or jammed links. The basic idea is to control nodes' in/out traffic to optimize network utilization, and enforce cooperation. The system is based on the principle that we can trust only ourselves and we cannot solve the cooperation misbehavior problem by using cooperation. So, the proposed mechanism is distributed, but not cooperative, and based on nodes internal knowledge. Every node acts independently, without sharing any information with other nodes, and trusts only information coming from the other communication peer (communication between peers can be encrypted to avoid forged acknowledgements by intermediate malicious nodes). According to our approach a node is responsible not only for forwarding a packet, but it shall forward it on the route that maximizes its success probability.

The framework we define is based on reliability indices. Every node has a dynamically updated reliability table containing a value for every outgoing link to a neighbor. Such a value represents a reliability index for paths rooted at that neighbor. Every time the node sends a packet on a path, it updates the reliability value associated to the neighbor through which the packet has passed: the updating is positive whenever source node receives an acknowledgement from destination, negative otherwise. The reliability value is unique for all paths rooted on that neighbor (see Fig. 1(a)). If source node observes that the reliability index of that subtree decreases, then it should immediately reduce the traffic sent through that neighbor, by preferring routes passing through a neighbor with a higher reliability index. Figure 1(b) shows an example: source node S has three possible routes to send a packet to destination node D. Each route passes through a different neighbor (I, J, K), and each link to a neighbor has a reliability index[1]. By comparing such values, source S finds out that path through K is the better

[1] Node S does not have the knowledge of the reliability of all the links, but it has a reliability index for each subtree rooted on its neighbors. This index summarizes the reliability of all the links crossed by the S-D path.

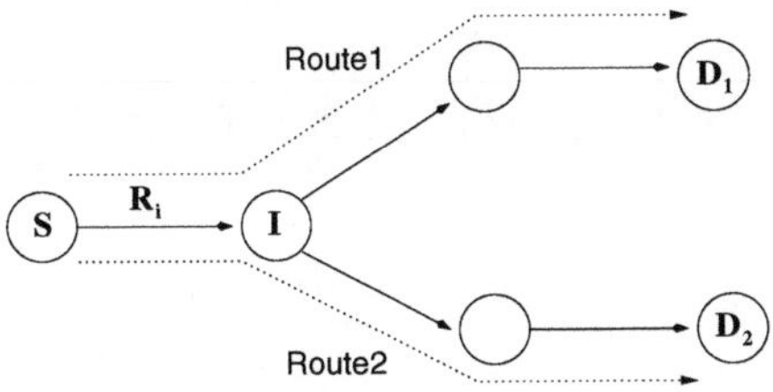

(a) Every time node S sends a packet on Route 1 or Route 2, it updates index R_i, that is associated to neighbor I. Thus R_i indicates the reliability level of the network subtree rooted at I.

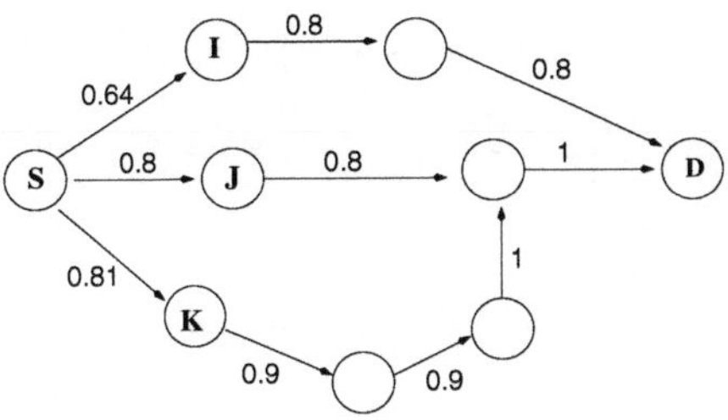

(b) The graph shows for each node, the reliability indices associated to its neighbors. Node S has three possible routes to reach destination D and can choose one of them according to the reliability index associated to its neighbors (route through K seems the most reliable)

Fig. 1. Reliability index

one, even if the longer in terms of hops number. Source S may decide to take that path to maxime the success probability of the packet forwarding. This procedure of choosing the best next hop can be executed by each intermediate node between source and destination, if the routing protocol adopted does not provide that the packet shall follow a predefined route (i.e., DSR protocol specifies the whole route a packet has to follow to reach destination, so intermediate nodes cannot choose the next hop to which forward packets). The only difference between source node and intermediate nodes is the possibility to update reliability values. While source node expects an acknowledgement from destination, intermediate nodes do not have any way to understand if the packet has reached destination. Thus, intermediate nodes can use reliability values (calculated when sending their own traffic) to take decisions about available routes, but they cannot modify them, because in such a case they do not get any acknowledgement back.

The reliability index associated to each neighbor reflects the behavior of all paths passing through it. In this way, we do not point to single misbehaving nodes, but to zones (potentially) containing one or more of them. In reality, when communication between nodes increases (most of nodes communicate with each other), there is a strong relation between a reliability value and the node pointed out by it: closer is the misbehaving node, lower is the reliability value in its direction. Figure 2 shows an example. M is a selfish node dropping packets with probability p. At first all reliability values are initialized to 1. Node S starts sending packets to Node D through K. On such a route there is a selfish node (M) that drops some packets. Thus, every time Node M drops a packet, Node S updates J's reliability that decreases, for example, to value 0.72 (Fig. 2(a)). Than Node S starts sending all packets on the other available route (through neighbor K) that keeps a good reliability value. When communication between nodes increases (e.g., Node S starts communicating with Y, J with T, and S with

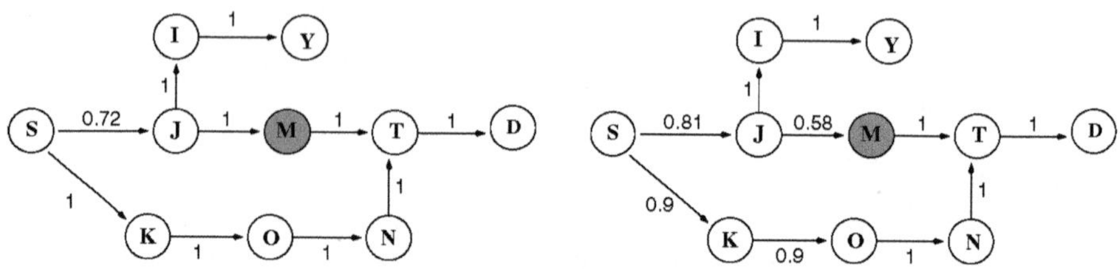

(a) M is a selfish node that drops packets with a probability p. When S starts sending packets to D through J, reliability associated to J decreases because M's behavior affects it. Thus S starts deviating its traffic through J.

(b) At full capacity, J's reliability increases because of communication from S to Y, while reliability on link (J,M) continues decreasing every time M drops a packet sent by J to T or D.

Fig. 2. Relation between node's behavior and reliability calculated by its neighbors.

D), J's reliability calculated by S increases, thanks to successful communication between S and Y, while M's reliability calculated by J decreases deeply, because of the selfish node M (Fig. 2(b)). This example highlights how reliabilities of all nodes sending packets on a path with a selfish node are affected by that misbehavior. Closer is the misbehaving node, lower is the reliability calculated in that direction.

To enforce cooperation we propose an approach analogous to the outgoing traffic control. The basic idea is to accept an incoming packet according to the reliability associated with the incoming link: if the value is low, then the receiving node may neglet the packet. This reciprocal approach to in/out traffic allows to use links in a way that agrees with their functioning level. A link that has a low reliability value is used to send messages with a frequency proportional to its reliability value, as well as messages received from that link are accepted according to that value. The correspondence between the reliability index and the accepting level of incoming traffic must still be defined.

The formula used to choose a route, in case of multiple available choices, takes into account the network utilization problem: besides looking at the probability value, the formula considers also traffic balance.

Every node builds its own reliability table by only referring to its internal actions, without any cooperation with other nodes. In this manner, there is no way for a malicious node to deceive other nodes because there is no information exchange. This absence of cooperation in the security mechanism has two important advantages. First, the system is resistant to attacks performed using the security mechanism itself: it is impossible for a node to maliciously decrease another node's probability. Second, there is an advantage in terms of traffic overhead, because no additional traffic is generated.

Another advantage of the proposed scheme is the overcoming of watchdog's weaknesses: probabilities are calculated by means of local traffic observations instead of neighborhood monitoring. Furthermore, the mechanism is flexible because it is independent on the routing protocol adopted.

3 The Proposed Scheme

3.1 Model and Assumptions

We model a network as a graph G = (V, L), where V is a set of mobile nodes and L is a set of direct links. Each node i ∈ V has a unique node identifier (ID). A link (i, j) ∈ L represents a connection between the two nodes i and j, meaning that j is in the transmitting range of i, and viceversa. In that case, nodes i, j are said adjacent (or neighbors), and we call N(i), neighbor set of i, the set of nodes adjacent to a given node i:

$$N(i) = \{j | j \in V \wedge (i, j) \in L\}$$

In this model we assume the following:

1. links allow two-way communication (bidirectional links), so that connected nodes can communicate with each other in either direction;
2. an end-to-end acknoledgement notifies packets delivery between peer nodes[2];
3. we know multiple routes to a destination, and for each route the source node knows the next hop to reach destination.

Regarding to hypothesis 3, it can be easily satisfied in a sensor network where each node has view of the network topology that is static. In a mobile ad hoc network, a slight modification of one of the existing routing protocols may allow the executing node to store the different routes it identified.

Given any node i ∈ V, for each j ∈ N(i) we have a probability value R_j that represents the reliability level of link (i, j). Probability R_j is dinamically updated every time node i sends a packet on link (i, j) and represents a reliability measure for paths rooted at neighbor j: it increases if the sent packet reaches the destination, it decreases otherwise. We suppose to have an end-to-end notification acknowledgement on packet delivery: if node s is the source node, node d the destination, with an arbitrary number n of hops between s and d ($n > 0$), when destination node d receives the packet, it sends back to s an ack message. If s does not receive any acknowledgement before a specified timeout, then we assume the packet did not reach destination, and some node on the path did not forward it. For each packet sent, we denote with M the result of the packet delivery and we estimate a smoothed reliability value R_j using a low-pass filter, with the same approach used in the TCP protocol for Round-Trip Time measurement [9]:

$$R_j \leftarrow \alpha R_j + (1 - \alpha) M \tag{1}$$

[2] Communication between source and destination nodes can be encrypted to avoid forged acknowledgements

where α, $0 \leq \alpha \leq 1$, is a smoothing factor and represents the percentage of the previous estimate considered on each new estimate. If $\alpha = 0.9$, then ninety percent of the new estimate is from the previous estimate, and 10% is from the new measurement.

M is the result of a packet delivery process from s to d, and it can assume the following values:

$$M = \begin{cases} 0 \text{ if } s \text{ does not receive ack from } d \\ 1 \text{ if } s \text{ receives ack from } d \end{cases}$$

If packet does not reach destination, then the reliabilty on outgoing link of source node decreases by a α factor. If packet reaches destination then nodes are cooperating and reliability on outgoing link of source node is smoothed by a α factor and increased by $(1 - \alpha)$.

In the following, we show how reliability indices are used to control outgoing and incoming traffic, achieving reliable forwarding and enforcing cooperation.

3.2 Outgoing Traffic Control

In case of multiple routes available for packet forwarding to a destination node, source node can choose one of them according to a certain principle. Routing protocols for ad hoc networks usually choose the shorter one, in terms of hops number, or the fresher one in terms of discovering time. Such criteria do not take into account links reliability. Hereafter, we propose two route selection policies dealing with reliability values associated to outgoing links, and we investigate their effectiveness.

Policy-1. Source node takes always the most reliable route. In such a case, source node compares reliability values for available routes and forwards packets on the link with the greatest value. This policy assures source node of taking always the most reliable route. The main drawback of such a choice is the deviation of all traffic on most reliable links which, in case of high traffic load, can quickly get congested.

Policy-2. This policy relates reliability values of available routes to build a probabilistic scheme. Let us suppose we have several possible routes to a destination through different source's neighbor nodes, i_1, i_2,..., i_n. Each neighbor has its respective reliability value R_{i_1}, R_{i_2},...,R_{i_n}. We associate a probabilistic value to each of such neighbors, p_{i_j}, $1 \leq j \leq n$, defined in the following way:

$$p_{i_j} = \frac{R_{i_j}}{\sum_{k=1}^{n} R_{i_k}} \tag{2}$$

where $\sum_{j=1}^{n} p_{i_j} = 1$. Equation (2) relates reliability values so that the resulting probabilistic value reflects the link reliability level. Routes are chosen according to the probabilistic value associated to the first node on the path: the greater the probability, the higher the route selection frequency. This probabilistic policy allows nodes to take even less reliable routes: traffic forwarding function is better distributed on all available routes and links congestion becomes rarer.

3.3 Incoming Traffic Control

Reliability indices can be used also to enforce cooperation. The basic idea is to reverse the service we obtain from each neighbor. For example, an incoming packet is accepted according to the reliability index associated to the incoming link. Thus we can define the probability to accept a packet from a neighbor I as:

$$P\{accept\ from\ I\} = R_i \tag{3}$$

If a node accepts an incoming packet according to such a probability, then the sender will see its packets accepted in a way that agrees with its reliability level. In fact, as we said in section 2, there is a strong relation between a reliability value and the behavior of the node pointed out by such a value. If a node has a very low reliability value for a neighbor, it is quite likely that the pointed node is misbehaving. This mechanism stimulates nodes cooperation: a good reliability value will allow them to see their traffic forwarded through the network.

The example above represents a first possible approach coping with cooperation enforcing. Further work will investigate an efficient utilization of reliability indices. In particular, we aim at finding out refined policies able to control incoming traffic, without penalizing too much sender nodes.

So far we considered only forwarding misbehavior as it is the topic we address in this paper. Let us see what happens in case of a routing misbehavior, caused by a malicious or a selfish node. A typical example for the former case is the Black Hole attack [10]. A node uses the routing protocol to advertise itself as having the shorter path to the node whose packets it wants to intercept. Once it has created the forged route, it can drop packets passing by it. As the malicious node drops packets, destination node will never receive such packets. Consequently, source node will not receive any acknoledgement from destination, and will decrease reliability in that direction. In the future, the outgoing traffic will avoid it and packets incoming from that zone will be refused with a high probability. On the other hand, in case of a selfish disruption of the routing protocol, the node does not partecipate to the routing function to save energy. In this situation, the selfish node will not appear in any route. This will not affect our system, that will continue working correctly, but it will not be able to detect the routing misbehavior. We intend to approach routing misbehavior in the future work.

4 Evaluation

The objective of this evaluation is to test the effectiveness of the reliability formula, given by (1), and of the policies for route selection, defined in Section 3.2. Specifically, we want to check if the defined reliability correctly reflects links behavior. In addition we want to investigate the throughput achieved by the different policies.

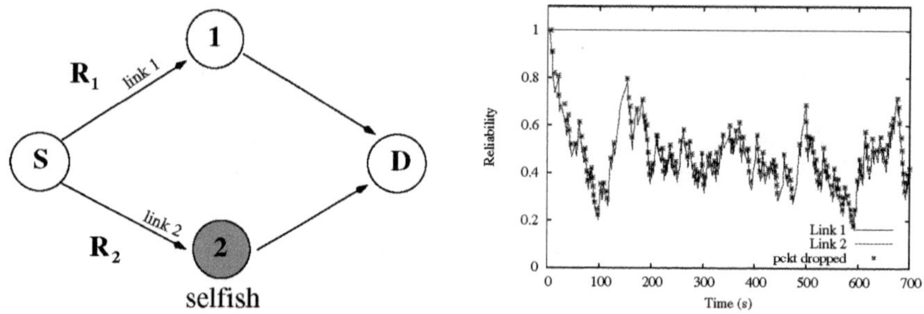

(a) (b)

Fig. 3. Test of reliability formula

4.1 Simulation Setup

To test the reliability formula we simulated a double path between two source and destination nodes (see Figure 3(a)). On the first path, Node 1 is cooperative while, on the second path, there is a selfish node discarding packets with probability $p = 0.5$. Every time the node sends a packet on one of two paths, it updates the reliability value associated to the neighbor through which the packet was forwarded[3]. Figure 3(b) shows the reliability function for both neighbors. Both reliability values, R_1 and R_2 are inizialized to 1. On Link 1 reliability keeps constant the value 1 because node on that path never discards packets. On Link 2 reliability decreases every time a packet is dropped (a cross on the plotting indicates a packet dropping on Link 2), assuming a mean value of 0.5.

To evaluate network performance for the forwarding policies we considered a similar context (see Figure 4). Source node S and destination node D are more than one hop away (we denote intermediate nodes between source and destination as *zone*). Let us call Node 1 and Node 2, the neighbors of source node respectively on Route 1 and Route 2. We associate a reliability value R_1 to link towards Node 1 and, R_2 to link towards Node 2. We also suppose that all nodes in Zone 1 are cooperative, while in Zone 2 there is a selfish node discarding packets with a probability p.

We investigated throughput for different route-selection policies, by varying the traffic loads and the transmission speeds. We defined four different scenarios and conducted experiments applying the different policies to such scenarios.

Scenario 1. All links have the same transmission speed (1 Mbit). Nodes on *zone 1* are all cooperative, so that there is no packets dropping on Route 1, while there is a selfish node in *zone 2*, that provokes packets' losses with probability $p = 0.5$.

[3] For the sake of semplicity, simulations implement immediate acknowledgments of delivered packets and no loss of ACKs. This choise does not affect the meaning of obtained results and effects of ACK delay on reliability estimates will be studied in further work.

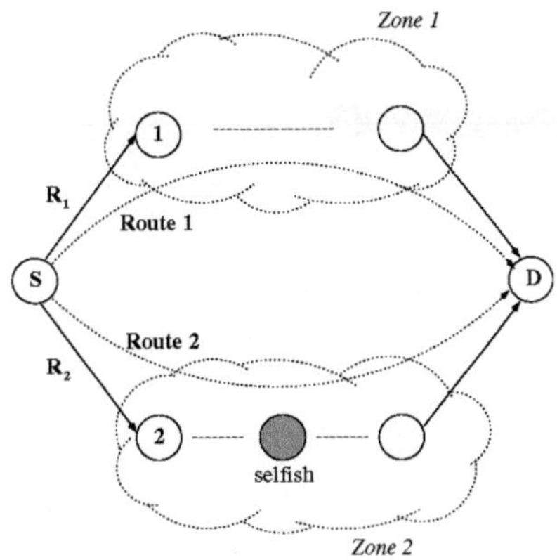

Fig. 4. Network simulated

Scenario 2. We inserted in *zone 1* a node with an outgoing link with transmission speed of 0.75 Mbit. Node N has a buffer capacity of 10 packets. Consequently, N can get congested, and drops packets arriving while the buffer is full.

Scenario 3. We considered the same parameters of the second scenario, by increasing buffer capacity to 100 for Node N.

Scenario 4. We increase the transmission speed of sender link to 2 Mbit, keeping the other parameters the same as Scenario 3.

4.2 Simulation Results

Experiments have been performed in each scenario previously defined, by applying the different route selection policies. We simulated also the conventional case in which the forwarding node does not consider reliability values, and distribute packets equally on both routes. Choosing both routes with the same probability represents the best compromise when no selection criterion is used, because it minimizes congestion events. In such a case the forwarding node would probably choose always the same route (e.g., the shorter one), by overloading nodes on that route. In the following, we call such a criterion of traffic distribution *load-balancing*, to indicate that reliability values are not considered, but traffic is equally spread on both routes.

We evaluated the model with the same traffic load. We observed the reliability function and network throughput in time[4]. In particular, we studied situations of congestion. In experiments concerning Scenario 2 and 3, Zone 1 contains a node that may get congested in case of high traffic load, as it has limited buffer capacity and link speed passes from 1 Mbit (incoming link) to 0.75 Mbit (outgoing link). Figures 5 and 7 show the reliability function obtained by applying policy-1 and policy-2 to scenario 2 (Figures 5(a) and 7(a)) and to scenario 3 (Figures 5(b) and 7(b)).

[4] Time is always measured in seconds.

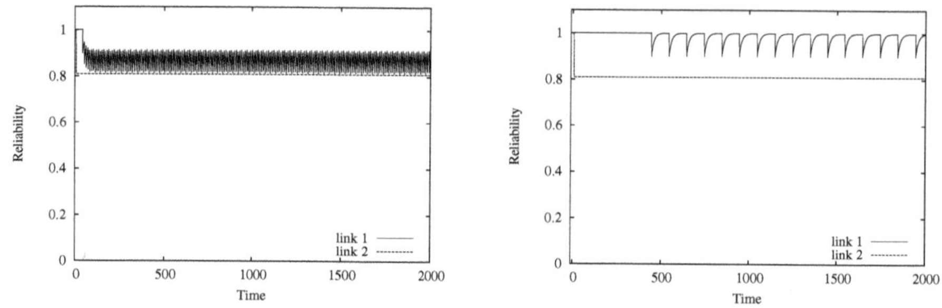

(a) Scenario 2 (b) Scenario 3

Fig. 5. Reliability function for link 1 and link 2, by applying policy-1

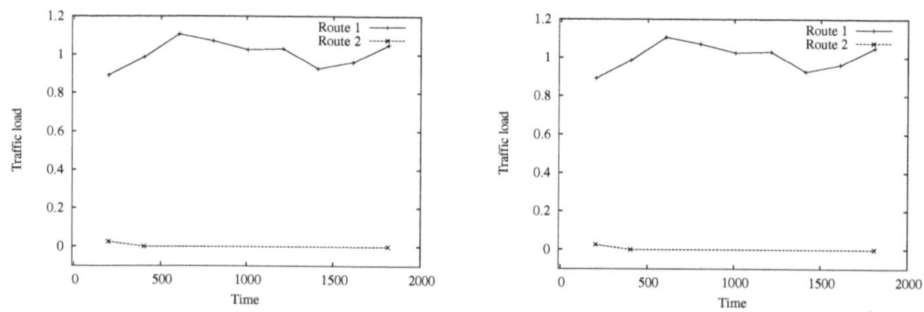

(a) Scenario 2 (b) Scenario 3

Fig. 6. Traffic sent on both routes by policy-1

When we apply policy-1, we choose always the most reliable route. Figure 5(a) shows the reliability function for Scenario 2. As reliability R_2 on Link 2 descreases because of the selfish node in Zone 2, source node S starts forwarding packets always on Route 1 (Fig. 6(a)), as soon as R_2 slopes down the value 0.81. In fact, R_1 is always greater than such a value. On Link 1 reliability decreases whenever Node N get congested. Such events are very frequent because Node N cannot buffer more that 10 packets. By increasing the buffer capacity (figure 5(b)), congestion events become rarer. This phenomenon is due to burstiness in the traffic. A longer buffer space allows to absorb temporary overload conditions. Alike the previous case, all traffic is sent on Route 1 as soon as R_2 slopes down the value 0.81 (Fig. 6(b)).

While applying policy-2, packets are distributed on both links, even if most of traffic is sent on Route 1 (Fig. 8(a)). In fact, when reliability value for Link 2 decreases greatly, probability to choose Route 1 increases, leading to some packets dropping due to congestion. Figure 7(a) shows that when the reliability value on Link 2 drops, the consequent deviation of most of traffic on Route 1 causes several events of congestion. Again, increasing buffer capacity for Node

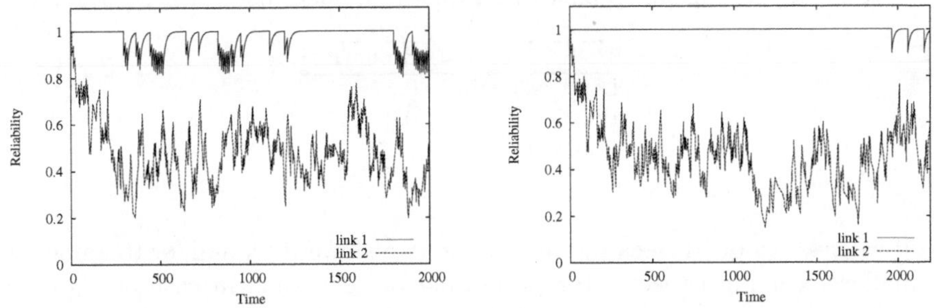

(a) Scenario 2 (b) Scenario 3

Fig. 7. Reliability function for link 1 and link 2, by applying policy-2

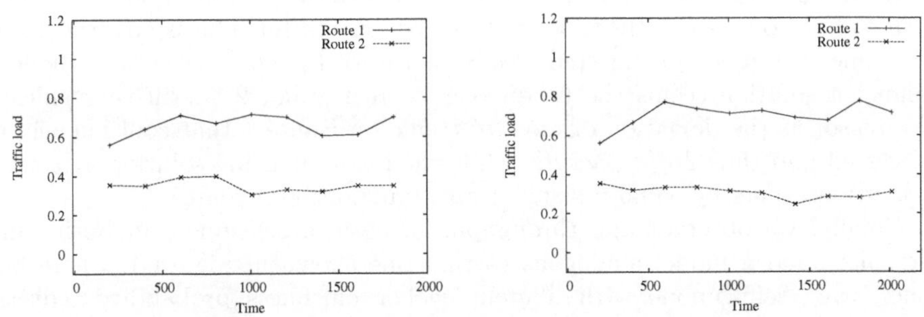

(a) Scenario 2 (b) Scenario 3

Fig. 8. Traffic sent on both routes by policy-2

Table 1. Throughput comparison

	Scenario 1	Scenario 2	Scenario 3	Scenario 4
Load-balancing	0.77	0.76	0.77	1.16
Policy-1	1.00	0.66	0.66	0.66
Policy-2	0.86	0.82	0.79	0.98

N, congestion becomes rarer (Figure 7(b)). Traffic sent on both routes is shown in Figure 8(b).

We want now to investigate the efficiency of policies previously defined by comparing their throughputs. Table 1 compares policies based on reliability (policy-1 and -2) with the load-balancing case, and shows the throughputs observed in each scenario. It is quite obvious that in Scenario 1 policies based on reliability values prevail over load-balancing. In fact, as Zone 1 never fails in forwarding packets, Route 1 is always chosen in case of policy-1, and very often chosen when policy-2 is applied.

Table 2. Throughput comparison with selfish nodes both in Zone 1 and 2

	Scenario 1
Load-balancing	0.67
Policy-1	0.81
Policy-2	0.70

If we insert a bottleneck in Zone 1, due to a slower link and/or to a node with a smaller receiving capacity (or processing power), then in case of high traffic load, Zone 1 can get congested, and drops packets. In such cases (Scenario 2 and 3), policy-1 is not very successful as it deviates all the traffic on Route 1, by causing frequent congestion events. Policy-2 appears better as it allows a more balanced traffic distribution on both routes, selecting more frequently the most reliable one.

Both reliability-based policies increase network throughput: policy-1 is better in case of no congestion, there policy-2 is more successful. This is not true if Zone 1 becomes completely congested (see Scenario 4). In such a situation, the load-balancing solution seems the better one, even if policy-2 is still very efficient. The reason is the deviation of most of traffic on Route 1 that continuously get congested and thus drops packets, while the load-balancing solution avoids any congestion events by equally distributing traffic on both routes.

Finally, we observed the throughput in case of selfishness in both zones. We considered a model analogous to that one shown in Figure 4, where both zones have a selfish node, with different level of selfishness: probability to discard packets is 0.2 in Zone 1 and 0.5 in Zone 2. Even in this situation reliability-based policies improve network performance in packet forwarding. Values reported in Table 2 show the efficiency of such policies, highlighting again the success of policy-1 in case of no congestion.

5 Policy Refinement

Results obtained by applying our policies to scenario 4 led us to a new possible route-selection policy, based on congestion control. In fact, in case of high traffic load and frequent congestion, reliability-based policies become ineffective. Policy-1 especially aims at choosing always the most reliable route, without taking into account the amount of traffic sent on it. To improve network performance, the forwarding node should stop sending packets on a route that is becoming overloaded, even if it has the bigger reliability value. Thus a good policy would be:

Choose the most reliable link until it does not get congested.

We repeated Scenario 4 experiments, inserting such a restriction on packet forwarding. For the sake of semplicity, we simulated the ability to understand if a node has reached the maximum of its receiving capacity by checking the occupancy of its buffer before sending a packet to it. Results obtained through

Table 3. Throughput comparison for congestion-control based policies.

	Scenario 4
Load-balancing	1.14
Policy-1	1.19
Policy-2	1.16

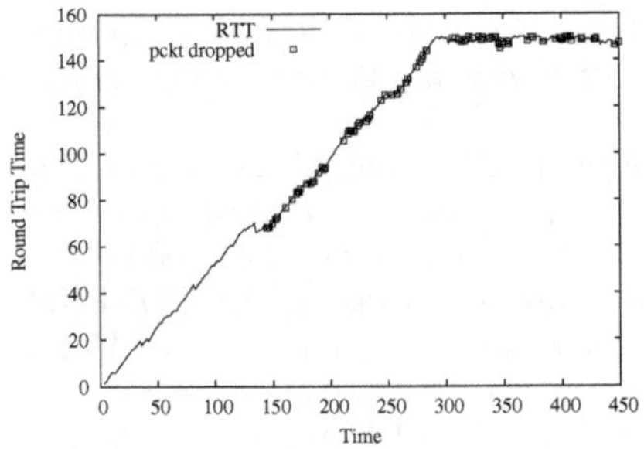

(a) Route 1

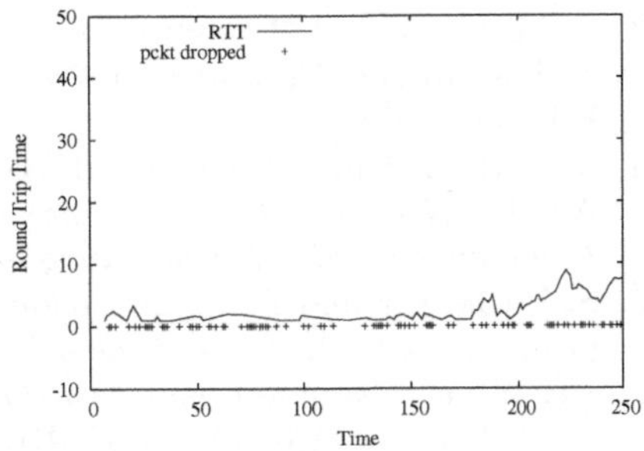

(b) Route 2

Fig. 9. Round Trip Time

such a simulation are shown in Table 3. Both policy-1 and policy-2 are better than load-balancing and improve network throughput in comparison with results obtained without taking congestion into account.

In reality, the simulated method is not applicable because a node cannot know other nodes' capacity. Instead, it is possible to understand if a zone on a path is becoming congested by observing the Round Trip Time (RTT). A RTT increase indicates a possible congestion condition. In fact, if we relate forwarding failures and RTT values measured in the experiment performed for policy-2 in Scenario 4, we note that, on Route 1, nodes start dropping packets when RTT reaches value 68 (see fig. 9(a)). After that value we have frequent packets dropping due to the high level of congestion. On the other route, there is no relation between RTT and packets dropping, as it is caused by a selfish node. Figure 9(b) shows how packets dropping on Route 2 happens even if RTT value keeps very low. We intend to face in a further study the identification of congestion conditions by observing the RTT value, its derivative and variation.

This led us to observe that the reliability index provides a good link characterization for selfish nodes that discard packets intentionally. On the other hand, to better cope with congestion situations, reliability can be characterized by a two states index R, to distinguish between congestion events and intentional misbehavior. We will address this new kind of reliability index in the future work.

References

1. L.M. Feeney, B. Ahlgren and A. Westerlund, Spontaneous Networking: An Application-oriented Approach to Ad Hoc Networking. In *IEEE Communications Magazine*, June 2001.
2. I. Chisalita and N. Shahmehri, A novel architecture for supporting vehicular communication. In *IEEE 56th Vehicular Technology Conference* pp. 1002–1006, September 2002.
3. P. Michiardi and R. Molva, Simulation-based Analysis of Security Exposures in Mobile Ad Hoc Networks. In *Proceedings of European Wireless 2002 Conference*, February 2002.
4. S. Marti, T. Giuli, K. Lai and M. Baker, Mitigating Routing Misbehavior in Mobile Ad Hoc Networks, In *Proceedings of the Sixth annual ACM/IEEE International Conference on Mobile Computing and Networking*, pages 255–265, 2000.
5. S. Buchegger and J. Y. Le Boudec, Performance analysis of the CONFIDANT protocol. In *Proceedings of the* 3rd *ACM International Sysmposium on Mobile Ad Hoc Networking and Computing (MobiHoc)*, pages 226–236, June 2002.
6. P. Michiardi and R. Molva, CORE: A COllaborative REputation mechanism to enforce node cooperation in Mobile Ad Hoc Networks. In *Proceedings of Communication and Multimedia Security 2002 Conference*, September 2002.
7. D. B. Johnson and D. A. Maltz, The Dynamic Source Routing for mobile Ad Hoc Networks. Internet Draft, Mobile Ad Hoc Network (MANET) Working Group, IETF, October 1999.
8. E.M. Royer and C-K Toh, A Review of Current Routing Protocols for Ad-Hoc Mobile Wireless Networks. *IEEE Personal Communications*, Apr. 1999.
9. V. Jacobson, Congestion Avoidance and Control. In *Proceedings of SIGCOMM '88* Stanford, CA, Aug. 1988, ACM.
10. Y.-C. Hu, A. Perrig and D.B. Johnson, Ariadne: A secure on-demand routing protocol for ad hoc networks. In *The 8th ACM International Conference on Mobile Computing and Networking*, September 2002.

Towards a Novel Transport Protocol for Ad Hoc Networks*

G. Anastasi and A. Passarella

University of Pisa, Dept. of Information Engineering
Via Diotisalvi 2 - 56122 Pisa, Italy
{g.anastasi,a.passarella}@iet.unipi.it

Abstract. The TCP protocol exhibits poor performance in multi-hop Mobile Ad Hoc Networks (MANETs). The ultimate reason for this is that MANETs behave in a significantly different way from traditional wired networks (like the Internet) for which the TCP protocol was originally designed. In particular, route failures and route changes due to node mobility may be frequent events in MANETs. Furthermore, congestion phenomena in MANETs are essentially different from traditional wired networks. In this paper we propose a novel transport protocol for MANETs. Unlike other proposals, our protocol is not a modification of the TCP but is specifically tailored to the characteristics of the MANET environment. It is able to manage efficiently route changes and route failures. Furthermore, it includes a completely re-designed congestion control mechanism. Finally, it is designed in such a way to reduce as much as possible the number of useless retransmissions. This is extremely important since retransmissions consume energy.

Keywords: Ad Hoc Networks, Mobility, Transport Protocols, TCP.

1 Introduction

The proliferation of portable computers (notebooks, palm tops, PDAs, smart phones) and the development of wireless technologies has spurred the interest towards Mobile Ad Hoc Networks (MANETs). Thanks to their self-organizing nature, MANETs are suitable for applications that must be deployed in high dynamic scenarios (e.g., peer to peer communications within a campus, support in rescue operations, etc.).

In the last years the research activities in the field of MANETs have mainly focused on routing protocols [1,2]. In addition, several papers have pointed out that the TCP behavior in a multi-hop ad hoc network is far from ideal. Many aspects contribute to this non-ideal behavior, some of which are discussed in Sect. 2. To improve the performance of the TCP protocol in multi-hop ad hoc networks several proposals have been presented [3,4,5,6,7,8]. To the best of our

* This work was partially funded by the Information Society Technologies programme of the European Commission, Future and Emerging Technologies under the IST-2001-38113 MOBILEMAN project.

M. Conti et al. (Eds.): PWC 2003, LNCS 2775, pp. 805–810, 2003.

knowledge, all these proposals are modified versions of the legacy TCP protocol. However, as shown in Sect. 2, MANETs behave in a completely different way from wired networks, (e.g., Internet), for which the TCP protocol was originally conceived. Therefore, we think that it is more fruitful to think in terms of a new transport protocol optimized for MANETs rather than trying to adapt the TCP protocol to the ad hoc environment. Moreover, the compatibility with host connected to the fixed Internet (i.e., running the TCP protocol) can be achieved by exploiting the Indirect-TCP model [14,11].

In this paper we propose a novel transport protocol, named TPA (Transport Protocol for Ad hoc networks), specifically tailored to the characteristics of the MANET environment. It provides a reliable, connection-oriented type of service and includes several innovations with respect to the legacy TCP protocol. In particular, the TPA is able to manage situations that may arise due to nodes' mobility (e.g., route failures and route changes). Furthermore, the congestion control mechanism is completely re-designed with respect to the legacy TCP. Finally, the TPA implements a novel retransmission policy aimed at reducing the number of useless retransmissions and, hence, energy consumption.

2 Motivations

The TCP protocol was originally conceived for wired networks, like the Internet, where nodes are static. However, nodes' movements and failures in MANETs are very frequent, and cause phenomena like *link failures, route failures,* and *route changes.* The TCP is not able to manage such phenomena efficiently. In particular, the sender TCP misinterprets duplicated ACKs and timeouts caused by to route failures or route changes as congestion and activates the congestion control mechanism. This leads to both unnecessary retransmissions and throughput degradation [3,4,10].

Even assuming that nodes in the MANET are static, the MANET behavior is significantly different from that of a traditional wired network. In traditional wired networks, like the Internet, packet losses are almost totally due to congestions causing buffer overflows at intermediate routers. This is not true in MANETs where buffer overflows at intermediate nodes are rare events, while packet losses due to link-layer contentions are largely predominant. The TCP protocol reacts to such packet losses by activating the legacy congestion-control mechanism. A severe drawback of using this mechanism is that the TCP window size is allowed to grow beyond its optimal value, which is typically very small (i.e., 1 - 4 TCP packets [9,10]). This behavior exacerbates the problem, since it produces new link-layer congestions at intermediate nodes.

3 TPA Protocol Description

The TPA protocol provides a reliable, connection-oriented type of service. The set up and tear down phases are similar to the TCP protocol and are thus omitted for the sake of space. In the following we only describe the data transfer phase.

The TPA protocol is based on a sliding-window scheme where the window size varies dynamically according to the flow control and the congestion control algorithms (like the TCP protocol [15]). The congestion control mechanism is described in Sect. 3.3, while the flow control mechanism is similar to the corresponding TCP mechanism [15] and is thus omitted. The TPA tries to minimize the number of (re)transmissions in order to save energy. To this end, packets to be transmitted are managed in blocks, with a block consisting of K packets. Specifically, the source TPA grabs a number of bytes – corresponding to K TPA packets – from the transmit buffer[1], encapsulates these bytes into TPA packets, and tries to transmit them reliably to the destination. Only when all packets belonging to a block have been acknowledged the TPA takes care to manage the next block. Each packet header includes a *sequence number* field that identifies the block to which the packet belongs, and a *data_bitmap* field consisting of K bits to identify the position of the packet within the block. The TPA header also includes two fields for piggybacking ACKs into data packets: *acknowledgement number* and *ack_bitmap*. The *acknowledgement number* identifies the block containing the packet(s) to be acknowledged, while a bit set in the *ack_bitmap* indicates that the corresponding packet within the block has been received correctly by the destination. Please note that it is possible to acknowledge more than one packet by setting the corresponding bits in the bitmap.

Packet transmissions are handled as follows. Whenever sending a packet, the source TPA sets a timer and waits for the related ACK from the destination. Upon receiving an ACK for an outstanding packet the source TPA performs the following steps: i) evaluates the new window size according to the congestion and flow control algorithms; ii) shifts the window forward, so that it starts with the packet next to the last acknowledged one; and iii) sends packets included in the current window (see Fig. 1-a). On the other hand, if all timeouts related to packets in the current window expire, the source TPA still executes steps i)-iii) above, just as in the case the last outstanding packet has been acknowledged (see Fig. 1-b). In other words, the TPA performs a transmission round during which it tries to send all packets within the block, without retransmitting missed packets.

After the first one, the sender performs a second round for retransmitting packets in the block not yet acknowledged, which are said to form a "retransmission stream" (see Fig.1-c). Again, this stream is managed according to steps i)–iii) above. If a packet within the retransmission stream is acknowledged before being retransmitted, it is dropped from the stream. This procedure is repeated until all packets within the original block have been acknowledged by the destination.

The proposed scheme has several advantages with respect to the retransmission scheme used in the TCP. First, the probability of useless retransmissions

[1] A block may include less than K packets if the buffer does contain a sufficient number of bytes.

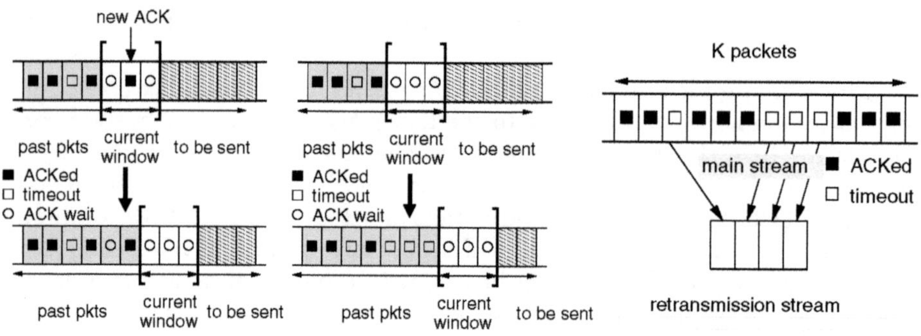

(a) ACK received (b) all timeouts (c) retr. stream

Fig. 1. Management of the sender sliding window (a,b) and management of the retransmission stream (c)

is reduced since packets for which the ACK is not received before the timeout expiration are not retransmitted immediately (as in the TCP protocol). This is particularly important in MANETs where nodes are highly mobile and, thus, the timeout value might not reflect the current RTT of the connection (see also Sect. 3.2). It should also be observed that the longer waiting time in the TPA protocol does not result in a throughput degradation since during this time interval the sender transmits other packets. Second, the TPA is resilient against ACK losses because a single ACK is sufficient to notify the sender about all missed packets in the current block. Third, the sender does not suffer from the out-of-order arrivals of packets. This implies that the TPA can operate efficiently also in MANETs using multi-path forwarding [16], where, on the contrary, the TCP performs very poorly [5].

The TPA protocol also includes some mechanisms to dynamically adapt to the network conditions. Specifically, it is able to detect and manage three kinds of events: *route failures*, *route changes* and *congestions.*

3.1 Route Failure Management

Like many other solutions [3,4,5,6], the TPA protocol relies on the network-layer support to detect route failures [1,2]. Specifically, if an intermediate node realizes that a packet cannot be forwarded to the next node because of a link failure, and no alternative route to the destination is available, it sends an *Explicit Link Failure Notification* (ELFN) back to the sender node. At the sender node the ELFN is notified by the network layer to the transport layer. Upon receiving an ELFN, the source TPA enters into a *freeze* state where it refrains from transmitting new packets (there is no available path to the destination).

We assume that the network layer does not provide route re-establishment notifications. Therefore, while in the freeze state, the source TPA sends a probe packet every t_{pr} seconds in order to look for a new route. Upon receiving an ACK from the receiver it realizes that the route has been re-established. Therefore, it i) leaves the freeze state; ii) sets the congestion window to the maximum value $cwnd_{max}$; and iii) sends the packet that has originated the ELFN.

3.2 Route Change Management

Similarly to the TCP, the TPA protocol estimates the RTT of the connection and, then, uses these estimate to set the retransmission timeout. Both parameters are derived in the same way as in the TCP protocol, i.e.,

$$\begin{aligned} \mu_{RTT}(n) &= g \cdot RTT(n) + (1-g) \cdot \mu_{RTT}(n-1) \\ \sigma_{RTT}(n) &= h \cdot |RTT(n) - \mu_{RTT}(n)| + (1-h) \cdot \sigma_{RTT}(n-1) \\ Timeout(n) &= \mu_{RTT}(n) + 4 \cdot \sigma_{RTT}(n) \end{aligned} ,$$

where $\mu_{RTT}(n)$ and $\sigma_{RTT}(n)$ are, respectively, the average value and standard deviation of the RTT estimated at the *n-th* step, $RTT(n)$ is the *n-th* RTT sample, $Timeout(n)$ is the retransmission timeout computed at the *n-th* step and, finally, g and h $(0 < g, h < 1)$ are real parameters (see [15] for details).

When a route change occurs, packets typically experience a variation in the RTT and the retransmission timeout might be no longer appropriate for the new path. To avoid possible useless retransmissions the TPA protocol must detect route changes as soon as they occur, and modify the RTT estimation method accordingly. In practice, the TPA detects that a route change has occurred either i) when a new route becomes available after an ELFN; or ii) when th_{RC} consecutive samples of the RTT are found to be external to the interval $[\mu_{RTT} - \sigma_{RTT}, \mu_{RTT} + \sigma_{RTT}]$. Upon detecting a route change, the TPA replaces the g and h values in the μ_{RTT} and σ_{RTT} estimators to greater values g_1 and h_1 so that the new RTT estimates is heavily influenced by the new RTT sample. This allows to achieve a reliable estimate of the new RTT immediately after the route change has been detected. Finally, after n_{RC} updates of the estimated RTT, the parameter values are restored to the normal values g and h.

3.3 Congestion Control Mechanism

Congestions due to link-layer contentions manifest themselves at the transport layer in two different ways. An intermediate node may fail in relaying *data* packets to its neighboring nodes and, thus, it sends an ELFN back to the sender node. This case, throughout referred to as *data inhibition*, cannot be distinguished by the sender TPA from a real route failure. On the other hand, an intermediate node may fail in relaying *ACK* packets. In this case, throughout referred to as *ACK inhibition*, the ELFN is received by the destination TPA, while the source TPA experiences consecutive timeouts without receiving any ELFN. As soon as the source TPA detects th_{CONG} consecutive timeout expirations it assumes that an *ACK inhibition* has occurred, and reacts by entering the *congested* state. This state will be exited when the source TPA receives th_{ACK} consecutive ACKs from the destination.

The TPA congestion control mechanism is window-based as in the TCP protocol. However, as anticipated, in the TPA the maximum congestion window size $cwnd_{max}$ is very small (in the order of 3 or 4 TPA packets). Therefore, the TPA congestion control algorithm is very simple. When the TPA is not in the congested state, the congestion window is set to the maximum value, $cwnd_{max}$. On the other hand, during the congested state, the congestion window is reduced to 1 to allow congestion to disappear.

References

1. J. Broch, D.B. Johnson and D.A. Maltz, "The dynamic source routing protocol for mobile ad hoc networks", Internet Draft of the IETF MANET Working Group, December 1998.
2. C. Perkins, E. Royer and S. Das, "Ad-hoc on demand distance vector (aodv) routing", in IETF Internet Draft, November 2000.
3. K. Chandran, S. Raghunathan, S. Venkatesan, R. Prakash, "A Feedback Based Scheme for Improving TCP Performance in Ad-Hoc Wireless Networks", Proceedings of ICDCS '98, pp. 472–479.
4. G. Holland and N. Vaidya, "Analysis of TCP performance over mobile ad hoc networks", Wireless Networks, Vol. 8, pp. 275–288, 2002.
5. J. Liu and S. Singh, "ATCP: TCP for Mobile Ad Hoc Networks", IEEE J-SAC, Vol. 10, No. 7, July 2001.
6. D. Sun and H. Man, "ENIC – An Improved Reliable Transport Scheme for Mobile Ad Hoc Networks", Proceedings of the IEEE Globecom Conference, 2001.
7. D. Kim, C. Toh and Y. Choi, "TCP-Bus: Improving TCP Performance in Wireless Ad-Hoc Networks", ICC, 2000.
8. F. Wang and Y. Zhang, "Improving TCP Performance over Mobile Ad-Hoc Networks with Out-of-Order Detection and Response", MobiHoc 2002.
9. Z. Fu, P. Zerfos, H. Luo, S. Lu, L. Zhang and M. Gerla, "The Impact of Multihop Wireless Channel on TCP Throughput and Loss", Proceedings of IEEE INFOCOM 2003, San Francisco (CA), March 30–April 3, 2003.
10. S. Xu, T. Saadawi, "Revealing the problems with 802.11 medium access control protocol in multi-hop wireless ad hoc networks", Computer Networks 38 (2002), pp. 531–548.
11. K. Xu, S. Bae, S. Lee and M.Gerla, "TCP Behavior accross Multihop Wireless Networks and the Wired Internet", The Fifth International Workshop on Wireless Mobile Multimedia (WoWMoM 2002), Atlanta (GA), September 28, 2002.
12. P. Gupta and P.R. Kumar, "The Capacity of Wireless Networks", IEEE Transactions on Information Theory, Vol. 46, No. 2, pp. 388–404, March 2000.
13. P. Gupta, R. Gray, and P.R. Kumar, "An Experimental Scaling Law for Ad Hoc Networks", http://black.csl.uiuc.edu/~prkumar/postscript_files.html, 2001.
14. A.Bakre, B.R.Badrinath, "Implementation and Performance Evaluation of Indirect TCP", IEEE Transactions on Computers, Vol.46, No.3, March 1997.
15. W.R. Stevens, "TCP/IP Illustrated", Vol. 1, Addison Wesley, 1994.
16. V.D. Park and M.S. Corson, "A Highly Adaptive Distributed Routing Algorithm for Mobile Wireless Networks", Proceedings of IEEE INFOCOM '97, Kobe, Japan, 1997.

An Enhanced MAC Architecture for Multi-hop Wireless Networks

R. Bernasconi, I. Defilippis, S. Giordano, and A. Puiatti

University of Applied Sciences, (SUPSI),
Computer Science and Electrical Engineering Department,
6928 Manno, Switzerland
{bernascr,ivan.defilippis,silvia.giordano,puiatti}
@die.supsi.ch
http://www.die.supsi.ch

Abstract. The IEEE 802.11 MAC family of standards has been designed for communication in traditional wireless LANs. The enhancement of the IEEE 802.11 family of standards is mainly seen in terms of channel speeds [1][2]. This fact, combined with advances in portable/handable computing and wireless technologies are expected to further increase possibilities and significant interest in the design and development of instantly deployable, wireless networks. Its usability in multi-hop wireless networks has been discussed in several papers. Starting from several known problems and issues, and analyzing the cause of these problems, we design the architecture of an enhanced IEEE 802.11 MAC card for multi-hop wireless networks[1].

1 Introduction

Instantly deployable, multi-hop wireless networks can be both fully self-organized (ad-hoc networks) and connected to the wired backbone via multiple wireless hops (multi-hop wireless LANs). Potential examples of this include rescue, citizen and guerrilla networks, where users self-organize to communicate or to access the Internet; multi-hop Hot Spots, wireless networks in public areas where traditional WLAN technology is augmented with ad-hoc wireless communications. However, several recent works [3], [4], [5], expressed serious doubt about the workability of IEEE 802.11 in real multi-hop scenarios. In this paper, we propose the architecture for an enhanced IEEE 802.11 MAC card to seamlessly work in a multi-hop wireless network, but, at the same time, it is still fully compatible with traditional IEEE 802.11 implementations. This architecture is designed to address: 1) a backoff scheme more adequate to multi-hop mode; 2) channel access and switching; 3) routing performed at MAC layer; 4) cross-layering.

---

[1] This work was partially funded by the Information Society Technologies programme of the European Commission, Future and Emerging Technologies under the IST-2001-38113 MOBILEMAN project.

M. Conti et al. (Eds.): PWC 2003, LNCS 2775, pp. 811–816, 2003.

2 Problems Definition

The 802.11 MAC defines two different shared access methods to the wireless medium, the Distributed Coordination Function (DCF), described as support for ad hoc architectures - where each node acts a peer to all nodes within its transmission range, and the Point Coordination Function (PCF). The DCF method, which involves a purely distributed mechanism for contention resolution, is indeed employed in ad hoc and multi-hop wireless networks, even if does not present any real forwarding capability. DCF is mainly a carrier sensing with collision avoidance mechanism: station N1, in order to send to station N2, uses a 4-way handshake mechanism (request-to-send - RTS, clear-to-send - CTS, Data, Ack) to both avoid collisions and verify reliable packet forwarding. If the medium is free for a specified time (Distributed Inter Frame Space - DIFS), then the station is allowed to transmit. The exponential backoff mechanism is further used after collision, to solve contention situations. If N1 senses the medium busy then it defers for a random backoff time in the range (0, Congestion Window). Upon failure of the RTS packet, the congestion window is doubled. The backoff timer is decreased whenever the channel is free; the node makes a fresh attempt at sending an RTS packet upon the expiration of the timer. These mechanisms, designed for WLAN environments, do not work well with multi-hop mode.

Backoff: In the 802.11 MAC, the channel access scheme is regulated by the exponential backoff: nodes failing to obtain the channel have to backoff a random time before trying again. It has been shown in [3], [4] that, depending on the network configuration, the standard IEEE 802.11 protocol can be not fair and operate very far from the theoretical limit of the wireless network. Both unfairness and low channel utilization are reflected higher, especially at transport layer, if TCP is used. This was shown in simulations [3], [4], and resulted even worst when tested in real experiments [5]. On the opposite, the capacity of an IEEE 802.11 protocol with a backoff window tuned on the optimal value is fairer and close to the theoretical limit.

Channels: Traditionally, IEEE 802.11 implementations allow the use of a small set of channels (usually channels 10 and 11). There are also serious difficulties to switch from one channel to another at a single node (it is necessary to tear down the network interface, modify the operating channel in the configuration file and then restart it). In order to fully exploit the capacity of multi-hop mode (see, for example [6]) and reduce contention under network congestion status, it is important to work with multiple operating channels and easily switch from one to another.

Routing: The main motivation for integrating MAC and forwarding functionality is highlighted in [7] and arises out of the fact that wireless nodes have conventionally just one interface, which is involved twice during the forwarding process. That is: a packet received at the wireless interface, it must be passed up to the routing layer (in order to discover the next hop), and further down to the same wireless interface for transferring it to the next hop. This adds undesirable delay and overhead at both MAC and routing layer.

Cross-layering: In several recent works [8], [9], [10]; the authors point out the need of a cross-layering design. More than routing, research has shown that several mechanisms can profit by the knowledge of some parameters that are typically confined at the MAC: transport, power management, cooperation, etc…

3 Features of the Enhanced Architecture

The main innovative characteristics of our implementation, as illustrated in Figure 1, are: P-persistent IEEE 802.11, Multi-channels communication mode, MAC packet forwarding, Exportable parameters module.

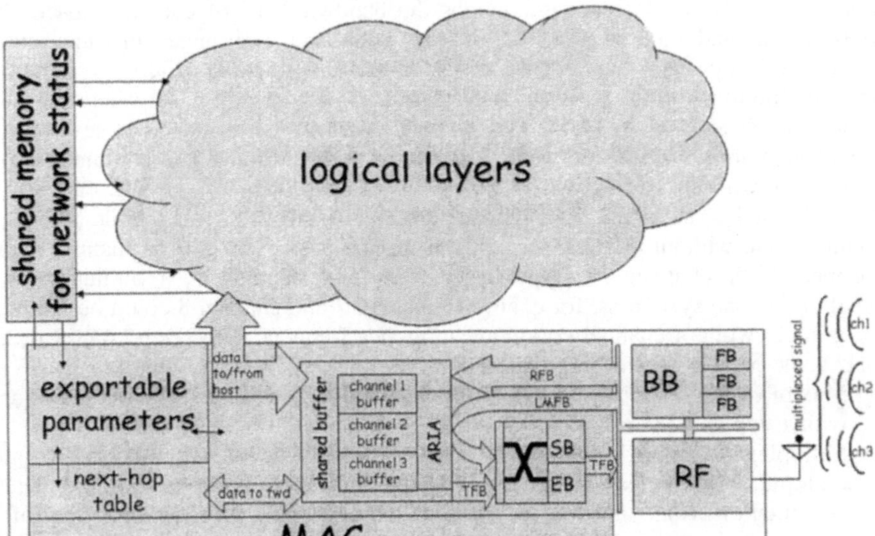

Fig. 1. Enhanced MAC card architecture. Multiple transmissions (up to 3 non-interfering channels) can be received as multiplexed signal at the antenna. The BaseBand (BB) receives it from the Radio Frequency (RF), and demultiplexes it towards up to 3 Frame Buffers (FB). The content of each FB is checked as usual (with CRC) and sent it to MAC on the Received Frames Bus (RFB) or on the Local Managed Frames Bus (LMFB). For a frame received on LMFB, the Additional Routing Information Analyzer (ARIA) checks if it contains some information for further forwarding. If so, it contacts the Next Hop Table to discover the next hop. The MAC has to manage the contention by assigning different priorities and handling the communication on the different channels (bufferized in separate memory spaces). The transmission is sent to the BB over the Transmitted Frames Bus (TFB) following the active backoff algorithm: Standard Backoff (SB), which implements the standard IEEE 802.11, or Enhanced Backoff (EB)

P-persistent IEEE 802.11: Several works have shown that, with an opportune tuning of the backoff algorithm, the protocol capacity increases [12] [13]. Our card will have the possibility to switch from standard IEEE 802.11 backoff to a backoff more adapt for the multi-hop mode. Our design is based on algorithms for approaching the performance of the p-persistent protocol [4]. The p-persistent IEEE 802.11 protocol differs from the standard protocol only in the selection of the backoff interval. Instead of the binary exponential backoff used in the standard, the backoff interval of the p-persistent IEEE 802.11 protocol is sampled from a geometric distribution with parameter p. In [4] it was shown that the p-persistent IEEE 802.11 protocol closely approximates the standard protocol (at least from the protocol capacity standpoint) if the average backoff interval is the same.

Multi-channels communication mode: our card can use all the available standard channels and switch from one channel to another in order to increase the node performances and the bandwidth utilization. In the Direct Sequence (DS) transmission mode the IEEE 802.11 standard divides the bandwidth in 14 channels spaced by 5 MHz with a bandwidth of about 22 MHz for each channel. It means that in a multiple cell network topology, overlapping and/or adjacent cells using different channels can operate simultaneously without interference if the distance between the center frequencies is at least 30 MHz. This already limits the performance of several nodes in the same area. Suppose we have 3 groups of nodes wishing to communicate in the same area without interfering: if group1 is using channel 1, group2 can use any channel equal to or larger than channel 7 (i.e. channel 7, 8, …) , while group3 can communicate without interferences only if group2 uses channel 7 or channel 8, using channel 13 or channel 14 respectively. This can be seen as a limitation to the flexibility of the system as, for example, channel 7 and channel 8 could be not usable for group2, while a channel communicating at a frequency between 2.448 and 2.451 GHz could be. The new design includes:

- Maximize the flexibility of use of the bandwidth by setting each carrier frequency with a step of 1 MHz;
- Multiplexed channel sensing to receive transmissions on multiple channels; multiple buffers baseband, to manage multiple frames; multiple backoff computation (that allows to choose the channel with smaller backoff in homogeneous enhanced MAC-based scenarios);
- Priority mechanism for traffic on different channels. This includes the introduction of new broadcast messages for alerting the other possible channels in case of scenarios that includes nodes without enhanced card;
- MAC buffer must be shared among different channels in use, allowing for transmission on different channels, transfer between two channels at the same node (switching), as well as the traditional transfer of data.

MAC packet forwarding: our implementation, which executes routing operations at MAC, will be based on a next-hop address lookup in conjunction with a path strategy as, for example, the fixed length labels architecture as defined in [7]. Basically, the packet forwarding protocol builds on the IEEE 802.11 DCF MAC using RTS/CTS and uses some additional information in the control packets (RTS/CTS) to allow the forwarding node to determine the next hop node while contending for the channel. The nature of this additional information is matter of current research, as strongly related with the distribution of information among neighbors. Moreover, we intend to add channel information (as defined in previous point) for routing decisions.

Exportable parameters module: the architecture design foresees the access to MAC parameters for a full integration of mechanisms traditionally working at different layers. This is enabled by a cross-layering architecture as the one proposed for the MobileMAN project [10]. In this architecture the shared memory component acts as exchange area of networking information (parameters, status, etc…) for all the layers. As drown in Figure1, this allows the MAC layer to distribute "physical" information up to the higher levels, as well as to profit from some higher layers elaborations too complex to be performed at MAC. A typical example is the interaction between

MAC, routing and transport information for congestion and network utilization purposes. If the transport is aware of the link status, it can distinguish between congestion due to physical failure and congestion due to the amount of traffic, and acts consequently. As well as the routing can decide different routing paths or strategies, and the MAC can modify the distribution of some information as consequence.

4 Some Trials Scenarios

We will implement the above described mechanisms and perform several levels of validation tests. Firstly, during the development phase, we will continuously test any new module inside the card in a stand-alone fashion and in communication with the other modules. Secondly, we will perform some 802.11 performance tests, similarly to [5]. Finally, we will put together some multi-hop wireless networks and test the workability of this solution. Some of the scenarios we plan to use are:

Ad hoc heterogeneous scenario - some of the ad hoc network nodes have not the enhanced MAC card: a) N is the only node with the enhanced card, it will select the standard backoff and use only one channel; b) N is not the only one with the enhanced card, N will use multiple channels: it will receive a multiplexed signal; agree with the peer(s) about the channel; select the enhanced backoff on the channels where this is possible. If the transmission goes through the LMFB for further forwarding, the MAC has to access the additional information about routing channels.

Ad hoc homogeneous scenario - all nodes in the ad hoc network use the enhanced MAC card: node N will use multiple channels and enhanced backoff: it will receive a multiplexed signal; select the channel with better backoff for transmitting.

Multi-hop WLAN heterogeneous scenario - some of the network nodes have not the enhanced MAC card: a) N is the only node with the enhanced card, it will select the standard backoff and, in the best case, use only channel 10 in ad hoc mode and 11 in WLAN mode; b) N is not the only one with the enhanced card, N will use multiple channels, (with a fixed one for the WLAN): it will receive a multiplexed signal; agree with the peer(s) about the channel; select the enhanced backoff on the channels where this is possible. If the transmission goes through the LMFB for further forwarding, the MAC has to access the additional information about routing channels.

Multi-hop WLAN homogeneous scenario - all nodes use the enhanced MAC card: node N will use multiple channels, (with a fixed one for the Access Point (AP)), and enhanced backoff: it will receive a multiplexed signal; select the channel with better backoff for transmitting.

5 Conclusions

In this paper, we presented the architecture for an enhanced IEEE 802.11 MAC card for improved performance in multi-hop wireless network, which is still fully compatible with current implementations. The main novel elements of this card are:

- the access and management of operating channels, which allows for a more timely and efficient use of available capacity in several types of scenarios;
- a dynamically tuned backoff algorithm that allows to obtaining performance close to the optimum;
- a modular and exportable set of parameters for interaction with higher layers;
- a packet forwarding strategy integrated at MAC layer.
- This work has several open issues – among them:
- discovery mechanism for other nodes strategies (channels, backoff, routing, etc…);
- priority mechanisms;
- best additional information for performing MAC packet forwarding;

Future work includes the completion of the implementation (see [14] for more details), as well as a validation and a trial phases that will involve the higher layers for feedbacks.

References

1. IEEE Computer Society. 802.11: Wireless LAN Medium Access Control (MAC) and Physical Layer (PHY) Specifications, June 1997.
2. B. Crow, I. Widjaja, J. G. Kim and P. T. Sakai. IEEE 802.11: Wireless Local Area Networks. IEEE Communications Magazine, Sept 1999.
3. Xu Shugong and T. Saadawi: Does the IEEE 802.11 MAC Protocol Work Well in Multihop Wireless Ad Hoc Networks?, IEEE Communications Magazine, Jun 2001.
4. R. Bruno, M. Conti, and E. Gregori: Optimization of Efficiency and Energy Consumption in p-Persistent CSMA-Based Wireless LANs, IEEE Transaction on Mobile Computing, Jan 2002.
5. G. Anastasi, E. Borgia, M. Conti, and E. Gregori: IEEE 802.11 Ad Hoc Networks: Performance Measurements, Proceedings of MWN 2003.
6. R. Battiti, M. Conti, E. Gregori: Price-based Congestion-Control in Wi-Fi Hot Spots, Proceedings of Wiopt 2003.
7. A.Acharya, A. Minsra, S. Bansal: A label-switching packet forwarding architecture for multi-hop wireless LANs, Proceedings of Wowmom 2002.
8. S. Toumpis and A. J. Goldsmith: Performance, Optimization, and Cross-Layer Design of Media Access Protocols for Wireless Ad Hoc Networks, Proceedings of ICC 2003.
9. P. Stanforth and V. Hasty: Meshing Together: Advantages and Challenges of Deploying Ad Hoc Wireless, Networks http://grouper.ieee.org/groups/802/meeting/meeting_files/Ad%20Hoc%20And%208022.pdf
10. M. Conti, S. Giordano, G. Maselli, G. Turi: MobileMAN Mobile Metropolitan ad hoc Network, pap, to appear in Proceedings of PWC 2003.
11. G. Bianchi, L. Fratta and M. Olivieri: Performance evaluation and enhancement of the CSMA/CA MAC Protocol for 802.11 Wireless LANs, Proceedings of PIMRC 1996.
12. F. Cali', M. Conti and E. Gregori: Dynamic IEEE 802.11: design, modeling and performance evaluation, IEEE Journal on Selected Areas in Communications, Sept. 2000.
13. R. Bernasconi, I. Defilippis, A. Puiatti: Issues and implementation plan of an enhanced MAC, draft version available at http://b.die.supsi.ch/projects/MobileMAN

On Demand Network Level Service Deployment in Ad Hoc Networks

Daniel Prince, Andrew Scott, and W.D. Shepherd

DMRG, Computing Department Lancaster University,
Lancaster, LA1 4YR, UK
{d.prince,acs,doug}@comp.lancs.ac.uk

Abstract. Many network level services such as routing, auto-addressing and, security have been proposed for ad hoc networks. Each has characteristics that may cause it to outperform similar services under different network conditions; therefore the choice of optimal services depends on the network deployment. Furthermore, since ad hoc network groups may be mobile it is probable that networks operating different or varying service implementations will interact. This paper presents work to provide dynamic, on demand network service selection and deployment, with minimal user interaction, in ad hoc networks.

1 Introduction

The demand for ad hoc network technology is growing as computing becomes increasingly pervasive and mobile. The IETF Mobile Ad hoc NETwork (MANET) Working Group [4] and wider research community have produced designs for various network level services such as routing protocols [9, 11], auto-addressing [7], and gateway protocols [2]. Simulations [3, 8] show that routing protocols perform optimally for particular network characteristics (node mobility/density). We expect all classes of network level service to display this property. With heterogeneous network level services, it is difficult to predict the services present at ad hoc network formation and inefficient to provide them on all devices. Additionally, some may be suboptimal for current networking conditions making it important to select the most appropriate.

As with Packet Radio Networks [10] before them, ad hoc networks require predetermination of network services in order to interact. However, technologies such as PCMCIA/compact flash or Software Defined Radio (SDR) allow wireless interfaces to be dynamically selected and deployed "on the fly" to suit conditions. We believe that similar capabilities for network level services would greatly enhance the adaptability and self organisational properties of ad hoc networks and that Programmable Networking provides a suitable basis for an appropriate solution.

In this paper, we discuss our motivating scenario and a real world example. We then present a new node mobility model, used in the design of a solution, which captures the scenarios behaviour. We go on to present our system design and finally conclude with a discussion on further and related work.

M. Conti et al. (Eds.): PWC 2003, LNCS 2775, pp. 817–822, 2003.

2 Scenario

The *'squad and base'* model is the name given to our generic scenario. It comprises two main elements, large *base* and small *squad* ad hoc networks. Nodes in both networks are equipped with a common type of wireless interface. *Base* ad hoc networks are formed by large groups of users, typically around 100+, who ordinarily have very low mobility levels. Therefore, the resultant ad hoc network of mobile nodes has a fairly stable network topology. *Squad* ad hoc networks are much smaller, 5 to 10 nodes, and display increased dynamicity over base networks. The typical mobile-temporal relationships of these two types of networks are shown in Figure 1.

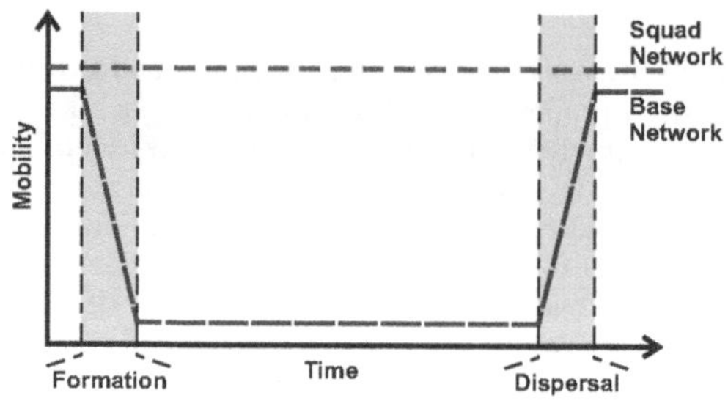

Fig. 1. Mobile Temporal Relationships of a Squad and Base Network.

2.1 Real World Example

Consider a scenario in which base networks are lecture groups using real-time multimedia applications and squad networks are student groups around a campus sharing files and information. Here, service selection is not only based on network characteristics but also application constraints. To reduce latency the real-time application requires timely route creation, suggesting a proactive routing protocol such as OLSR [1]. In contrast, file sharing has minimal latency constraints suggesting an on-demand protocol such as DSR [9] or AODV [11].

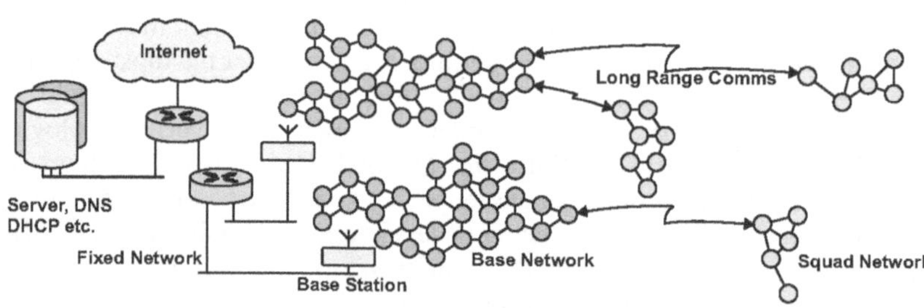

Fig. 2. Complete Scenario Diagram

The convergence of a student and lecture group raises an initial problem as the potential lack of common network level services will make network interaction and inter-network node migration difficult. Network separation, through lecture group dispersal, presents additional problems as previously common lecture group network level services may have characteristics that cause sub-optimal performance or in extreme cases network failure, in the newly formed student groups. So far only autonomous networks have been discussed. However, connectivity to high speed fixed network infrastructures need to be considered. Here, mobile nodes that provide interconnectivity between LAN and ad hoc networks may need additional services such as gateway protocols and auto-addressing. **Fig. 2** shows the overall scenario.

3 Dynamic Group Mobility

In developing the scenario depicted in Figure 2, the authors concluded that the scenario presented complex network interactions which existing node mobility models [6,9] do not fully capture. As a result, suitable system design parameters and constraints could not be taken from simulations based upon these models. This led to the development of the Dynamic Group Mobility (DGM) model.

DGM is based on the Group Mobility (GM) model presented in [6]. Both GM and DGM groups have a central point which moves through a way point set and an area in which nodes move. However, inside this area DGM nodes move with a Random Waypoint behaviour [9] where as GM nodes move around randomly distributed fixed points. Additionally, DGM nodes are free to change group membership.

On initialisation each node is given random *associativity* and *fickle* values between 0 and 100. Group changes occur when a node detects a neighbour which is in a different group. The nodes then examine their associativity values to determine whether they should stay together. If they should, then the fickle factor is used to decide which node should change groups. Figure 3 gives the algorithm. To compensate for changes in group size, the group radius changes proportionally to the number of nodes in the group.

```
Node currentBest = null;
int bestAssociativity = 0;
For each Node n in the current neighbour set
{
      tempAssociativity = n.associativity - this.associativity

      if ((tempAssociativity >= Threashold) && (tempAssociativity > bestAssociativity))
      {
            bestAssociativity = tempAssociativity;
            currentBest = n;
       }
}

if (currentBest != null)
{
      if (this.Fickle >= currentBest.Fickle)
            Migrate this to n.Group
      else  Migrate n to this.Group
}
```

Fig. 3. Migration algorithm pseudo code.

A simulator has been implemented in C# to evaluate node behaviour with the various models and is available for download. Statistical analysis of interaction type and duration are being used as design parameters for algorithms and protocols used in the ANOM service described in the following section.

4 System Design and Implementation

Many ad hoc network level protocols and services must manipulate packets at various points in the processing path. For example, DSR inserts and processes source route data while MobileIP based gateway protocols [2] manipulate header options. In many Active Network platforms, installed services access packets using dynamic packet filtering. The system under development is based on the Lancaster Active Router Architecture (LARA++) [12]. In LARA++ received data traverses a dynamic Classification Graph. At each classification node services install filters in order to receive packets. LARA++ also allows services to be dynamically instantiated.

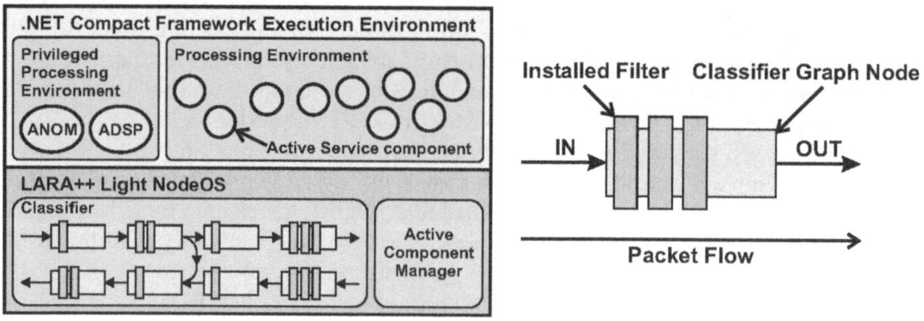

Fig. 4. System design with a classification node diagram

The system design, shown in Figure 5, consists of three main elements:

1. **LARA++ Light** – A lightweight version of LARA++, based on WinCE and the .NET compact framework, capable of running on limited hardware such as PDAs.
2. **Active Service Discovery Protocol (ASDP)** [13] – Allows clients to query and publish services on active nodes.
3. **Ad hoc Network Observer and Manager (ANOM)** - ANOM is responsible for monitoring network characteristics, distributing/evaluating service characteristics and then deciding with rest of the network the best service to deploy.

Network level services are then implemented as active components.

To demonstrate node bootstrapping and continued operation, consider a mobile device with LARA++ light and ASDP attempting to access an ad hoc network. Network nodes running the ANOM component can detect whether an ANOM component is instantiated on the new node. If not, a network neighbour can use ADSP to request the node install the ANOM component. Subsequently the new node can retrieve other services from its neighbours to bootstrap itself into the network. Additionally, if a node arrives with new services, the ANOM components would distribute information about those services, evaluate their characteristics against various constraints and potentially elect to run a new service deemed more optimal for the given environment.

5 Further Work

The following three main areas have yet to be fully addressed in this work.

5.1 Security

Existing security mechanisms for active networks, such as code signing, static code checks and sandboxing, software fault isolation can protect an active node from malicious or buggy services attempting to disrupt its operation. However, these mechanisms cannot prevent misbehaviour within allowed system constraints such as providing incorrect results, i.e. routes. Such misbehaviour may cause extreme network degradation, making recovery exceedingly difficult. To this end a node must be capable of assessing service performance to determine intentional misbehaviour or whether current network characteristics are outside its nominal operational limits.

5.2 Service Characteristics

To be able to characterise services, benchmark parameters for each class of service must be identified. Simulation work [3,8] and the MANET WG have already defined benchmark parameters such as Packet Delivery Ratio for routing protocols. Similar parameters for other services are yet to be defined. An extensible metadata language must be defined so that service benchmark parameters and other variables may be formed into service descriptions similar to the one in Figure 6. Note that .NET framework assemblies include an augmentable XML metadata code description.

```
<Protocol Name="AODV" Author="Lancaster University">
    <Type>on_demand</Type>
    <Description>
        Lancaster University implementation of AODV Draft 10.
    </Description>
    <Network_characteristics>
        <mobility_rate>1.7</mobility_rate>
        <net_diameter>50</net_diameter>
    </Network_characteristics>
    <Performance At="Peak">
        <Normalised_Routing_Load>10</Normalised_Routing_load>
        <Ave_Throughput>5</Ave_Throughput>
        <Ave_Delay>10</Ave_Delay>
        <Delivery_Ratio>98</Delivery_Ratio>
    </Performance>
    <Performance At="Highest_Mobility"> … </Performance>
    <Performance At="Lowest_Mobility"> … </Performance>
    <Signature>1A 2B 3C 4D 5E 6F 78 9A</Signature>
</Protocol>
```

Fig. 6. An example XML service description

5.3 Distributed Service Evaluation

As described previously the ANOM component is responsible for evaluating service characteristics against the current network state and selecting the most suitable service. This must be a distributed operation as every node must migrate to the new service in order to maintain network connectivity. This means that the ANOM components must be capable of conversing, evaluating and voting on different service metadata descriptions in a distributed, stable and scalable manner.

6 Related Work and Conclusions

Using programmable networking for routing in ad hoc networks has been, and still is being researched. SoftNET [5] was an early active node developed for routing in PRNs. More recently underlay networks [15] and modular versions of existing routing protocols [14] have been implemented using programmable networking.

This work examines the wider issue of interoperability in heterogeneous ad hoc environments and aims to provide a platform to enable nodes to flexibly and optimally deploy network level services on demand. Additionally, the requirements of network level services form a strong argument for an active network based solution.

This paper has presented the squad and base scenario which we believe provides a strong argument for adaptive, on demand network level service selection and deployment in ad hoc networks. The Dynamic Group Mobility model is introduced as a mechanism to inform the development of a suitable solution. Finally, a work in progress system has been presented based on an existing active network platform.

References

1. Adjih C. et al.: Optimized Link State Routing Protocol (OLSR), IETF Draft 2003, draft-ietf-manet-olsr-09.txt, (April 2003), (Work in Progress).
2. Belding-Royer, E. et al.: Global Connectivity for IPv4 Mobile Ad hoc Networks. IETF Draft, draft-royer-manet-globalv4-00.txt, (Nov, 2001), (Work in Progress).
3. Broch, J. et al.: A Performance Comparison of Multihop Wireless Ad Hoc Network Routing Protocols. Proc. IEEE/ACM MOBICOM '98, (Oct, 1998).
4. Corson, S. and, Macker, J.: Mobile Ad Hoc Networking (MANET): Routing Protocol Performance Issues and Evaluation Considerations, RFC 2501 (1999).
5. Forchheimer R. and Zander J.: Softnet – Packet Radio in Sweden. In Proceedings of AMRAD Conference, (1981).
6. Hong X. et al.: A Group Mobility Model for Ad Hoc Wireless Networks, In Proceedings of ACM/IEEE MSWiM'99, Seattle, WA, (Aug, 1999).
7. Jeong, J et al.: Ad Hoc IP Address Autoconfiguration, IETF Draft, draft-jeong-adhoc-ip-addr-autoconf-00.txt, (May 2003), (Work in Progress).
8. Johansson, P. et al.: Scenario-Based Performance Analysis of Routing Protocols for Mobile Ad-Hoc Networks. Proceedings of MOBICOM'99, (Aug, 1999).
9. Johnson, D. and Maltz, D.: Dynamic source routing in ad hoc wireless networks, in Mobile Computing (ed. T. Imielinski and H. Korth), Kluwer Academic (1996).
10. Jubin J. and Tornow J.: The DARPA Packet Radio Network Protocols, The proceedings of the IEEE, Vol. 75, No. 1, (Jan, 1987).
11. Perkins, C. et al.: Ad Hoc on Demand Distance Vector (AODV) Routing. draft-ietfmanet-aodv-10.txt, IETF Draft, (2002), (Work In Progress).
12. Schmid, S. et al.: Component-Based Active Network Architecture. Proceedings of 6th IEEE Symposium on Computers and Communications, (2001).
13. Sifalakis M. et al.: A Generic Active Service Deployment Protocol. To appear in the proceedings of ANTA2003, (May 2003).
14. Tornquist Modular and Adaptive routing in ad hoc networks, Masters thesis Department of Computer Science, University of Colorado, (2001).
15. Tschudin and R. Gold, "LUNAR: Lightweight Underlay Network Ad-Hoc Routing," tech. rep., University of Basel, Switzerland, (Jan, 2002).

A Context-Aware Location Service for Metropolitan Ad Hoc Networks

Giovanni Turi

IIT Institute, CNR,
Via G. Moruzzi 1, 56124 Pisa, Italy
Giovanni.Turi@iit.cnr.it

Abstract. This paper presents a node location service for metropolitan Ad Hoc networks. The new system is innovative because it exploits context information in its internal polices. Context information is referred as knowledge about metropolitan user mobility patterns. In urban areas people do not uniformly distribute. Specific zones, defined as *hot spots*, work like assembly points, where people show predictive mobility. This zones are office buildings, shopping malls and similar operational contexts. The paper shows how a *hot spots aware* node location service could better find nodes where to publish and retrieve updated position. This proposal paper mainly focuses on protocol aspects, showing mechanisms to let the system adaptable to the inherent dynamic conditions of an Ad Hoc network.

1 Introduction

Infrastructureless mobile networks, also called Ad Hoc networks, are systems where mobile and resource-constrained nodes organize spontaneously, forming a network with dynamic topology. With an increasing use, in urban areas, of ubiquitous technologies, like PDA's, smart phones and wearable computers, users will hopefully form Ad Hoc networks on a metropolitan scale. In this scenario, tracking and delivering the current node positions is an important task that a node location service should carry out.

A node location service is a *database* of nodes position information. The system implements two main operations: *position update* to submit a node's current position, and *position lookup* to retrieve it. In Figure 1(a) node N updates its current position *(x,y)* submitting it to the service along with its unique identifier. Node M in turn, queries the service looking for the position of node N, and receives back the correct answer.

On a Metropolitan Ad Hoc network, the absence of an infrastructure implies that nodes have to implement the location service themselves. First of all, the position database has to be distributed. Many distribuition schemes have been already proposed [1]. Essentially *all* or *some* of the nodes maintains position information, these are the location servers, and each of them maintains a copy of the *whole* or only *a part* of the database. A fair database distribution scheme involves all the nodes as potential location servers, making each responsible only

M. Conti et al. (Eds.): PWC 2003, LNCS 2775, pp. 823–828, 2003.

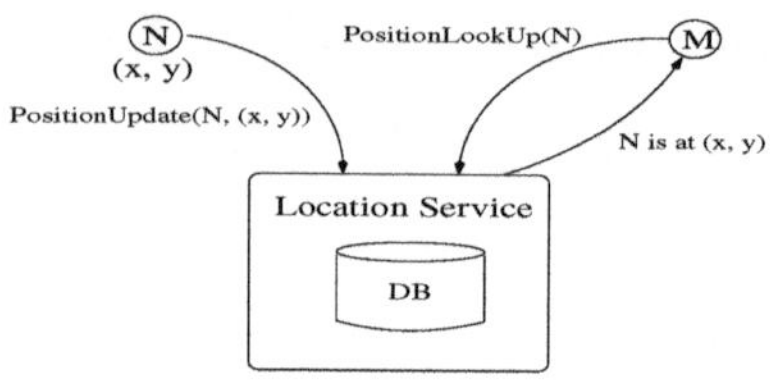

(a) Classic architecture

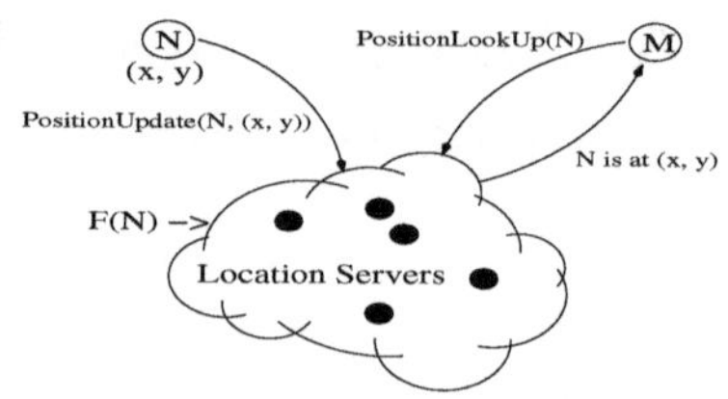

(b) Ad Hoc variation

Fig. 1. Location services.

for a part of the database. The location servers of a given node N are identified through a function F on the unique identifier (see Figure 1(b)).

In the Virtual Home Region [2], a consistent hash function *H()* maps each node unique identifier to the centre of a circular area called the *home region.* Those nodes located inside the home region of a given node N are its potential location servers. When N updates its position, it geographically forwards a request packet to its home region, where some nodes will eventually cache it. A position lookup proceeds similarly: when node M needs to retrieve N's current position, it geographically forwards a position lookup request packet to N's home region, identified through the same hash function *H()*. Eventually, one of the location servers of N will answer to the request. The hash function uniformly maps node identifiers over the network area, independently from the geographic nodes distribuition.

For the same node *N*, location servers in the Grid Location Service [3] are spread all over the network area. The density of location servers decreases logarithmically moving away from *N*. On one hand, position lookups are guaranteed to find closer location server. On the other hand, position updates have higher costs as they cover the whole network area. The system virtually partitions the network area in squares of increasing size, and tries to guarantee for N the existence of a location server in each square. Inside a square, the location server is again identified through the node identifier, as the node that has the least greater identifier compared with *N*.

This paper presents a location service where the association between nodes and location servers is context aware. On a metropolitan area, nodes tend to assemble around well defined contexts, instead of uniformly distribute. As detailed in Section 2, urban user mobility patterns can be analysed and exploited to improve the location service perfomance together with its adaptability to the Ad Hoc dynamics. Section 3 briefly outlines the conclusions.

2 Exploiting Metropolitan User Mobility Patterns

On a metropolitan area, people do not uniformly distribute, but tend to assemble around *hot spots.* These are for example offices, shopping malls, theathers or similar places where people is plunged in a well defined context. Beside that, a user tends to frequent the same contexts, where often the same people can be met. Of daily habits change, but over long periods of time.

If thousends of PDA-equipped users are connected on a metropolitan Ad Hoc network, the observations above will result in users possibly aware of places where connectivity and node density are usually good. People can understand and easily remember hot spots locations. Moreover, this process could be directly supported using mobile hosts. For example each mobile host could maintain a ranking of positions where its user passed over and where connectivity and node density was good. Once a user identifies a hot spot context, the system can exploit it to colocate its location servers. The user will then compose its address with the unique identifier and the hot spot's coordinates (X_{hs}, Y_{hs}), resulting in $UID@(X_{hs}, Y_{hs})$. This address format is not far from present e-mail addresses like *my_id@my_domain*, where *my_id* identifies a mailbox in the domain *my_domain*. Finally, users distribute context aware addresses as present e-mail addresses are exchanged today. This give seekers nodes a way to find the location servers of sought-after nodes.

2.1 Assumptions and Definitions

Each node is able to sense its current position through an on-board GPS device. A standard GPS device samples the current position one time per second, $f_{GPS} = 1$. Besides, routing is assumed to implement a geographic forwarding procedure, like those described in [1].

Consider a node N and a hot spot with coordinates X_{hs}, Y_{hs} specifyed by its user. The system potentially identifies the location servers of N, as those nodes currently located around the hot spot. More formally, the location servers of N belong to its *Friends* set F_N, defined as the set of nodes located at distance less than or equal to a value r from the point X_{hs}, Y_{hs}. Note that not all the F_N nodes are location servers for N. Some of them might be unreachable, while others might have just entered the area, without having yet received a position update. Infact, node N has no control over its F_N area, and no knowledge about its location servers. The node periodically sends a position update request towards the F_N area, simply asking to cache the position information for a limited amount of time. Infact, the context aware location service is a *soft state* system. Positions are only temporarily cached on location servers. This eliminates the problem of handling stale entries.

In order to be locatable, node N must also deliver its F_N references to its peers. As outlined before, the address of N is composed by the unique identifier and the hot spot coordinates, $Addr_N = [UID_N, (X_{hs}, Y_{hs})]$. The hot spot coordinates are the centre of F_N, so that far away peers holding the address of N, will eventually retrieve its current position querying nodes towards this point. Consider at first the distance parameter r as a constant value. Later on it is shown how this can vary dynamically according to the density of nodes in the area.

(a) Flooding (b) Leaving

Fig. 2. Handling dynamics inside F.

2.2 Position Update

Node N periodically updates its current position to its location servers, as the system is soft state. The caching time, and respectively the update frequency, vary accordingly to the node's mobility pattern (see Section 2.3). A postion update consists in the following steps:

1. An update request packet is geographically forwarded towards F_N. By definition, the forwarding procedure will deliver the packet to the reachable node L that closest to the F_N coordinates.
2. If L is at distance greater than r, then a failure will return back to N. Otherwise, L floods the position update request bounding to the F_N area. Each node receiving the position update request caches N's position and broadcasts again the packet only if it belongs to F_N, otherwise it simply drops it.
3. Node L collects acknoledgements from caching nodes and sends back a cumulative packet to N.

Nodes distribution inside F_N can be far from uniform. In this case, the packet flooding described above, does not guarantee N to have its position cached in a satisfying number of nodes spanning the F_N area. To minimize this condition, node L could drive the flooding to start also from other three points located inside F_N as shown in Figure 2(a). Having the position (x, y) of node L and the F_N coordinates (X_{hs}, Y_{hs}), points P_1, P_2 and P_3 are simple orthogonal projections. The coordinates for the points are $P_1 = (X_{hs} - y + Y_{hs}, Y_{hs} + x - X_{hs})$, $P_2 = (2X_{hs} - x, 2Y_{hs} - y)$, $P_3 = (X_{hs} + y - Y_{hs}, Y_{hs} - x + X_{hs})$.

2.3 Position Update Frequency

Each node updates its current position periodically, with a frequency depending from its linear speed (or displacement). The faster the node moves linearly, the more frequent it updates its position.

Each node is assumed to move not faster than a maximum speed S_{MAX}. Moreover, a motionless node updates at a given base frequency f_b to refresh its location servers. Given that, the position update frequency should range between $[f_b, K * f_b]$. The upper bound is K times the base frequency, and has to be reached when the node moves at the maximun linear speed S_{MAX}. If S_n

indicates the current linear speed, given by the distance between the position at step n and the position ar step *n-1*, then, assuming $f_0 = f_b$ and $S_0 = 0$, the formula

$$f_n = f_{n-1} + \frac{S_n - S_{n-1}}{S_{MAX}} * (K-1) * f_b = f_b * \left(1 + (K-1) * \frac{S_n}{S_{MAX}}\right)$$

has the right features, as S_n ranges in $[0, S_{MAX}]$. Finally, considering f_{GPS} and the cost of a position update operation, fixing $K * f_b = \frac{f_{GPS}}{H}$, where $H = 10^h$ and h is a natural number, results in having $f_n \in \left[\frac{f_{GPS}}{K*H}, \frac{f_{GPS}}{H}\right]$.

2.4 Position Lookup

A node M willing to communicate with node N has to have its address $Addr_N = [UID_N, (X_{hs}, Y_{hs})]$. There are two cases:

1. N is sufficiently close to M, so that they directly see each other. In this case there is no need to query the location service;
2. N is far away from M, the two nodes do not directly see each other, and M needs to query the location service to obtain N's position and furtherly use geographic forwarding to exchange packets with it.

A position lookup is needed in the second case. Node M geographically forwards a position lookup request packet towards (X_{hs}, Y_{hs}). Eventually a node selected by the forwarding procedure, caching N's position, will send back a lookup answer packet. If none exists, an error is reported to M. It is worth of note, that in order to lookup the current position of a node, the size r of the F_N area is not needed.

2.5 Varying the Size of the *Friends* Set

For a node N, the F_N area has been defined with a fixed distance parameter r. A direct improvement is to let r be variable in order to guarantee a percentage of successfull position lookups coming between subsequent position updates. Infact, it is possible that in the time frame between two position updates, enough caching nodes leave the F_N, determining the failure of incoming position lookups. The goal is to minimize the probability of this event.

As described in Section 2.2, the leader L of the position update request flooding, returns a cumulative acknowledgement to node N. This contains at least the number of caching nodes in F_N. Let this number be n. As shown in Figure 2(b), consider one of these nodes moving out of F_N at a given discrete speed $s \in [1, .., S_{MAX}]$, choosing one of the directions $\overrightarrow{d_1}$, $\overrightarrow{d_2}$, $\overrightarrow{d_3}$, $\overrightarrow{d_4}$.

Considering only these four directions is of course an approximation. Suppose, without loss of generality, to have the node positioned along the radius at point (x, Y_{hs}) with x variable, and the position update frequency to be f. Approximating with a uniform nodes distribution inside F_N, the probability $P\{out_f\}$ that the node leaves the area before the next position update, is given by:

$$\int_0^r \sum_{s=1}^{S_{MAX}} \sum_{i=1}^{4} P\{out_f | S = s, D = \overrightarrow{d_i}, x = dx\} P(S = s) P(D = \overrightarrow{d_i}) P(x = dx) dx$$

where $P\{out_f | S = s, D = \overrightarrow{d_i}, x = dx\}$ is 0 or 1 depending on the frquency f, the position X, the speed s and the direction $\overrightarrow{d}$. The condition $P\{out_f\}^k \leq 1\%$ gives the minimum number of nodes k that the F_N should contain to guarantee the success of incoming position lookups. By comparing k with the returned number of caching nodes n, a policy to vary the size of F_N can be implemented. The policy could be continuos, as $P\{out_f\}$ depends from r, or discrete, for example advancing at multiples of T_x, equal to the MAC layer maximum transmission range. This is currently under evaluation.

3 Conclusions and Future Work

This short paper describes the key polices of a context aware location service. The role of a location service is essential at the networking layer when the size of an Ad Hoc network grows up to a metropolitan scale, but nevertheless it could also be used by location aware middlewares and applications. None of the previously proposed location services are context aware. This system uses context to realize policies maximizing the percentage of successful operations, while keeping costs comparable to other approaches like VHR and GLS. The results of a preliminary performance evaluation could be found in [4].

References

1. M. Mauve, J. Widmer, and H. Hartenstein. A Survey on Position-Based Routing in Mobile Ad-Hoc Networks. *IEEE Network*, Vol. 15 (2001), Nr. 6, pg. 30–39, 2001.
2. S. Giordano, and M. Hamdi. Mobility Management: The virtual home region. *Technical report*, EPFL-(ICA), Lausanne, Switzerland, October 1999.
3. J. Li, J. Jannotti, D. S. J. De Couto, D. R. Karger, and R. Morris. A scalable location service for geographic ad hoc routing. *Proc. of the 6th Annual ACM/IEEE Int. Conf. on Mobile Computing and Networking (MOBICOM) 2000*, pg. 120–130, Boston, MA, USA, 2000.
4. M. Conti, E. Gregori, and G. Turi. Desing and Analysis of a Context Aware Node Location Service for Ad Hoc Networks. *Submitted for publication*, 2003.

A Novel Approach for Distributed Simulation of Wireless Mobile Systems*

Luciano Bononi and Gabriele D'Angelo

Dipartimento di Scienze dell'Informazione,
Università degli Studi di Bologna,
Mura Anteo Zamboni 7, 40126, Bologna, Italy
{bononi,gdangelo}@cs.unibo.it

Abstract. This position paper introduces the motivation and preliminary implementation issues of a distributed simulation middleware designed to increase the performance and speed-up in the distributed simulation of wireless systems characterized by mobile hosts. Topology changes due to simulated hosts' mobility map on dynamic causality effects in the "areas of influence" of each mobile device. We analyze the preliminary definition of a new dynamic mechanism for the runtime management and distributed allocation of model-components executed over a cluster of Physical Execution Units (PEUs). A migration mechanism dynamically adapts the topology changes in the wireless network to a reallocation of model components over the PEUs. The aim is the reduction of communication overheads, between the PEUs, required to distribute the event-messages between model components. The distributed simulation framework is based on HLA-compliant runtime infrastructure and preliminary, adaptive load-balancing and migration heuristics.

1 Introduction

A wide research work has been done in recent years in the field of tools and methodologies for modeling and efficient simulation of wireless systems [1,2,11, 12,13,15,18,20,22]. Among the relevant scenarios considered Cellular systems, PCS networks and the Mobile Ad Hoc Networks (MANETs) are gaining an increasing relevance. Wireless networks currently considered appealing for the analysis may include a potentially high number of simulated hosts. The simulation of every host may require a relevant computation time (e.g. due to simulation of protocol-stacks and applications running on top). This is often unpractical or impossible to simulate on a classical Von Neumann (mono-processor) architecture [17,18]. The simulation is unlikely to have success because of huge memory requirements and large amount of time required to complete.

Many practical experiences have demonstrated that a speed-up in the simulation of network systems is achievable using parallel and distributed models and architectures, i.e. a Parallel Discrete Event Simulation (PDES) approach [7,8],

* This work is supported by MIUR FIRB funds, under the project: "Performance Evaluation of Complex Systems: Techniques, Methodologies and Tools"

M. Conti et al. (Eds.): PWC 2003, LNCS 2775, pp. 829–834, 2003.

e.g. Glomosim [22] based on PARSEC [1], Maisie [18], parallel and distributed implementations based on Network Simulator (ns-2) [10,15,16,20] based on RTI-KIT [15], on ANSE/WARPED [14], on TeD [13], USSF over Warped/Notime [Rao99], Wippet [11], SWiMNET [2], and many others [10,12,18]. In order to exploit the maximum level of computation parallelism, many research activities dealt with dynamic balancing of logical processes' executions (both cpu-loads and virtual time-advancing speeds) by trading-off communication, synchronization and speed-up, both in optimistic and conservative approaches [4,5,9,19,21].

High Level Architecture (HLA) is a recently approved standard (IEEE 1516) dealing with component-oriented distributed simulation [3,6]. It defines rules and interfaces allowing for heterogeneous components' interoperability in distributed simulation. The definition of model components (formally known as federates) with standard management APIs brings to a high degree of model re-usability. The HLA standard defines APIs for the communication tasks and synchronization between federates. The simulation is supported by a runtime middleware (RTI). The RTI is mainly responsible for providing support for time management, distributed objects' interaction, attributes' ownership and many other optimistic and conservative management policies [3,6].

2 Work Motivation

Many approaches have been investigated in order to reduce the overhead effects of distributed synchronization and communication in both optimistic and conservative distributed simulations. The motivation for this communication-reduction approach is the frequent adoption of networked cluster of PCs, in the place of shared-memory or tightly-coupled multiprocessors, as the execution units of the distributed simulation, primarily for cost reasons. The high network latency in these clusters could play a fundamental role in determining the weight of communication and synchronization between the distributed model components. Solutions have been proposed, relying on the reduction of communication obtained when the update of a event- or state-information (e.g. event and/or anti-message) does not need to be flooded to the whole system, but is simply propagated to all the causally-dependent components. This is the basis of publishing/subscribing mechanisms for sharing state-information and event-notifications between causally dependent components [3,6,15].

Simulation models for wireless systems have to deal with at least two innovative concepts with respect to wired networks' models: *i)* the user-mobility and *ii)* the open-broadcast nature of the wireless medium.

Communication is commonly defined between neighbor hosts and determines a causal effect in the simulated system. The causal effect of communication may be extended to all the neighbor-hosts (even if their communication-links are partitioned over different logical channels) due to the broadcast nature of the wireless medium. A multi-hop communication between non-adjacent hosts can be considered as a communication based on local interaction between a chain of neighbor hosts. Thus, a high degree of causality in the simulation of the wireless hosts' communication is driven by the local-topology interaction (i.e. messages) between neighbor hosts. This interaction can be considered the most relevant causality effect to be modeled in many wireless system simulations (e.g. when evaluating MAC protocols, routing protocols, or physical interference effects).

Topology changes (due to simulated hosts' mobility) map on causality effects in the "areas of influence" of each mobile device, resulting in dynamically shaped causality-domains. Accurate simulation results would require accurate details to be modeled, and many fine-grained, low-level causal effects (i.e. events) to be kept into account in the simulation process. We define a dynamic system as a system where the interactions (i.e. the causal effects of events) are dynamically subject to fast changes driven by the system (and model) evolution over time. Given this definition, a wireless network (e.g. a MANET) can be considered a highly dynamic system. A static definition of publishing/subscribing lists, groups and causal domains could not be more convenient, in presence of a dynamic system, than implementing a complete state-sharing. A dynamic approach for the distribution of events and state-information (e.g. dynamic lists and groups updates) would lead to network communication overheads. In some scenarios, the cost of list-updates or fine-grained events' communication between a dynamically variable set of components could be traded-off with the migration cost needed to cluster the whole interacting components on a single Physical Execution Unit (PEU). This would be more attractive if the object migration could be implemented incrementally as a simple data-structure (i.e. state) transfer, and if the object interaction would be maintained for a significant time (time-locality). A certain degree of time-locality of local communication can be considered an acceptable assumption in many mobile wireless systems, depending on the motion model characteristics.

Our proposal is to define and investigate a simple mechanism allowing for Simulated Mobile Hosts' (SMHs) migrations in HLA-based distributed simulations. We realized a prototype migration framework, adopting it together with heuristic load-balancing and migration policies, whose aim is to dynamically partition and cluster the interacting SMHs among federates executed over a set of PEUs. SMHs in our approach have a common definition and can be migrated simply by serializing and transferring their state-information. This would realize an example of a prototype framework and an adaptive, tunable mechanism able to react to dynamic systems' behavior (like the mobile and wireless systems') under the communication-reduction viewpoint. In this position paper we sketch the prototype implementation of this mechanism and future work.

3 The Distributed Simulation Framework

We implement a parallel discrete event simulation of model components (federates) by using a set of physical execution units (PEUs) connected by a physical network (e.g. a networked cluster of PCs). The PDES simulator built to obtain an experimental evidence of our proposal is based on a distributed architecture made by a set of federates glued together by HLA middleware. Our approach at this level is mainly focused on the communication reduction between the PEUs where federates are executed. We implement a federate as a single logical process, managing and updating the state information (data structures) of a set of (at least one) Simulated Mobile Hosts (SMHs). We follow a time-stepped, conservative approach for ease of implementation, and given the high number of unpredictable, simulated entities with a high ratio of interactions performed. This means a conservative, fixed-lookahead time-management based on a HLA RTI implementation.

On top of the RTI we built a middleware extension called Generic Adaptive Interaction Architecture (GAIA). GAIA provides the interaction to the simulation core, the location and distribution data management, the random number generator, tracefile-logging and other simulation facilities. The HLA definition and components to handle a distributed simulation from the federate viewpoint is left untouched. The target of GAIA is to provide migration and service APIs to the simulation developer. We implement SMH models as code (they have a common code definition in the federates) with data structures to define and maintain the SMH state information. We choose to migrate the data structure, i.e. the state information of SMHs between federates. This requires to design and implement a supporting middleware for the data structure management and distribution of SMH entities between HLA federates. Our models can be simply executed as HLA federates in the static approach. If it is required to exploit migration, then the models should be defined to deal with migration APIs provided by GAIA on the top of HLA middleware. The equivalent of data distribution management and object ownership in the HLA meaning has been re-implemented in the GAIA framework. This allow us to deal with controlled overheads and ad hoc implementation for our purposes. In the current release, GAIA includes only prototype data-location and distribution-management services whose overheads have been preliminarily considered as an implementation issue. Currently, the GAIA framework includes also simple heuristic functions defined to evaluate the migration of SMHs between PEUs. The migration is performed to the PEU which is the target of the majority of interactions required. The steady state behavior of the proposed heuristic in isolation would lead to a long-term concentration of the SMHs over a restricted set of the available execution units, because the adaptive effect is focused on the "external" communication overhead. For this reason, we introduced the migration heuristic on the top of a simple load-balancing policy implemented by the GAIA middleware supporting the set of federates.

4 Preliminary Results

In our preliminary tests, we assume a high number of simulated mobile hosts (SMHs), each one following a mixed variation of the Natural Random and Random Waypoint (RWP) motion models. This motion model gives unpredictable and uncorrelated mobility pattern of SMHs under the motion viewpoint. In this way, any heuristic definition governing the mechanism cannot rely on any assumption about the motion correlation and predictability of SMHs. Space is modeled as a torus-shaped 2-D grid-topology, populated by a constant number of SMHs. In this way the clustering of SMHs is not trivially determined by high concentration in small areas. The modeled communication between SMHs is a constant flow of ping messages (i.e. constant bit rate), transmitted by every SMH to its nearest device. We ran our simulation experiments over a variable set of PEUs equipped by Dual-Pentium III 600Mhz, 256MB RAM, connected by a FastEthernet (100Mb/s) LAN. Preliminary results shown a speed-up up to 23% given by adaptive distributed simulation with respect to a static distributed simulation, for 900 SMHs simulated over 3 PEUs.

5 Conclusions and Future Work

We propose an adaptive framework, named Generic Adaptive Interaction Architecture (GAIA), for the dynamic allocation of model entities (ME) in a HLA-based framework for distributed simulations. GAIA is based on runtime migration and load-balancing policies, to reduce in adaptive way the amount of external communication between the PEUs. We tested our mechanism for simple, sub-optimal migration and load-balancing heuristics in the testbed simulation of a prototype mobile wireless system, characterized by Simulated Mobile Hosts (SMHs). The runtime mechanism adapts the MEs' allocation over the PEUs to the dynamic interactions of SMHs. Preliminary results demonstrated the effectiveness of the proposed mechanism and performance enhancements, with controlled overhead for the worst-case scenario defined. We expect increasing performance to be obtained by optimizing the framework code, by tuning the mechanism's heuristics on the hypothesis and assumptions related to real system models.

Our future work will extend our analysis from a qualitative to a quantitative one. Additional efforts will be done in the code-optimization for the migration mechanism implemented. We will study new Data Distribution Management (DDM) implementations (both centralized and distributed) to enhance the message reduction and filtering. We will also design new HLA-based interaction management and filtering, development of detailed heuristics based on analysis of multiple metrics and parameters, many different hardware and network architectures. The ad hoc network scenario will be extended to deal with protocols and complex SMHs' behaviors. Specifically, the migration mechanism will be evaluated with respect to many dynamic factors to be modeled, in addition to the host mobility (e.g. dynamic communication-session establishment). The migration-based approach of this work could be extended to a wide set of simulations where the sequential approach gives low performance (e.g. multi-agent systems, genetic and molecular systems, P2P models).

References

1. Bagrodia, R., Meyer, R., Takai, M., Chen, Y., Zeng, X., Martin, J., Song, H.Y.: PARSEC: a parallel simulation environment for complex systems, IEEE Computer, 31(10):77–85, October 1998

2. Boukerche, A., Fabbri, A.: Partitioning Parallel Simulation of Wireless Networks, Proc. of the 2000 Winter Simulation Conference (WSC), 2000
3. Dahmann, J., Fujimoto, R. M., Weatherly, R. M.:, High Level Architecture for Simulation: an update, Winter Simulation Conference, December 1998.
4. Das, S. R.: Adaptive protocols for Parallel Discrete Event Simulation, Proc. of Winter Simulation Conference, 1996.
5. Deelman, E., Szymanski, B.K.: Dynamic load balancing in parallel discrete event simulation for spatially explicit problems, Proc. of the twelfth workshop on Parallel and distributed simulation PADS'98, July 1998
6. Defence Modeling and Simulation Office (DMSO): High Level Architecture RTI Interface Specification, Version 1.3, 1998
7. Ferscha, A.: Parallel and Distributed Simulation of Discrete Event Systems, In Handbook of Parallel and Distributed Computing, McGraw-Hill, 1995
8. Fujimoto, R.M: Parallel and Distributed Simulation Systems, Wiley & Sons, 2000
9. Gan, B.P., Low, Y.H., Jain, S., Turner, S.J., Cai, W., Hsu, W.J., Huang, S.Y.: Load balancing for conservative simulation on shared memory multiprocessor systems, Proc. of 14-th Workshop PADS'00, May 2000, Bologna, Italy
10. Jones, K.G., Das, S.R.: Parallel Execution of a sequential network simulator, Proc. of the 2000 Winter Simulation Conference, 2000
11. Kelly, O. E., Lai, J., Mandayam, N. B., Ogielski, A. T., Panchal, J., Yates, R. D.: Scalable parallel simulations of wireless networks with WiPPET: modeling of radio propagation, mobility and protocols, MONET, v.5, n.3, Sep. 2000.
12. Liu, W.W., Chiang, C.C., Wu, H.K., Jha, V., Gerla, M., Bagrodia, R.: Parallel simulation environment for mobile wireless networks, Proc. of Winter Simulation Conference 1996.
13. Perumalla, K., Fujimoto, R., Ogielsky, A.: TeD – A language for modeling telecommunications networks, Performance Evaluation Review 25(4), 1998
14. Rao, D.M., Wilsey, P.A.: Parallel Co-simulation of Conventional and Active Networks, Proc. of MASCOTS'00, August 2000
15. Riley, G.F., Fujimoto, R.M., Ammar, M.H.: A generic framework for parallelization of network simulations, Proc. of MASCOTS'99, College Park, MD, October 1999
16. Riley, G.F., Ammar, M.F., Fujimoto, R., Perumalla, K., Xu, D.: Distributed Network Simulations using the Dynamic Simulation Backplane, Proc. of MASCOTS'01.
17. Riley, G.F., Ammar, M.H.: Simulating Large Networks How Big is Big Enough? Proc. of Intern'l Conf. on Grand Challenges for Modeling and Simulation, Jan 2002.
18. Short, J., Bagrodia, R., Kleinrock, L.: Mobile wireless network system simulation, Wireless Networks 1, August 1995, (pp. 451–467).
19. Som, T.K., Sargent, R.G.: Model structure and load balancing in optimistic parallel discrete event simulation, Proc. of 14-th workshop PADS'00, Bologna, Italy
20. UCB/LNBL/VINT The NS2 network simulator, http://www.isi.edu/nsnam/ns/
21. Vee, V-Y, Hsu, W-J: Locality-preserving load-balancing mechanisms for synchronous simulations on shared-memory multiprocessors, Proc. of 14-th workshop PADS'00, May 2000, Bologna, Italy
22. Zeng, X., Bagrodia, R., Gerla, M.: GloMoSim: A library for parallel simulation of large-scale wireless networks, Proc. of 12-th workshop PADS'98, May 1998, Alberta, Canada.

Enhanced Power Saving Scheme for IEEE 802.11 DCF Based Wireless Networks*

Jong-Mu Choi, Young-Bae Ko, and Jai-Hoon Kim

Graduate School of Information and Communication
Ajou University, Republic of Korea
jmc@dmc.ajou.ac.kr
{youngko,jaikim}@ajou.ac.kr

Abstract. Providing energy efficiency in MAC (Medium Access Control) layer, while achieving desirable throughput, is an important research issue in the area of wireless networking. A wireless LAN such as IEEE 802.11 using the Distributed Coordination Function (DCF) also provides a mechanism for power conservation which allows each node to "sleep" for some amount of periods, but also requires the nodes to wake up periodically and stay "awake" for a certain duration called the *ATIMWindow*. In this paper, we propose a new energy-efficient MAC protocol that allows the nodes to go to sleep early, without the need to be "on" for the whole ATIM interval, in the case they are acknowledged that no data is buffered within an ad hoc network and therefore no data transmission will be taken place.

1 Introduction

A key characteristic of wireless hosts is their operation with a limited energy resource, like batteries in general. Because battery capacity is quite restricted, the issue of reducing power consumption in all aspects of wireless host design has been an important challenge to lengthen battery lifetime. Along with many approaches addressing energy conserving issue at various layers of the network protocols stack, this paper studies on the design of an energy-efficient medium access control (MAC) protocol for wireless networks, especially for IEEE 802.11-based wireless LANs.

Past work on energy-efficient MAC protocols can generally be categorized into two: *power-control* approaches and *power-saving* approaches. The first approach, so called power-control mechanism, is to adjust the transmission power level for the purpose of energy saving[1]. The main objective of these power control schemes is to find an optimal power level for successfully transmitting data with the minimum transmission power consumption. Another approach for energy-efficient MAC is often named as power-saving schemes because they try

* This work is supported by grant No. R05-2003-000-10607-0 from Korea Science & Engineering Foundation, and grant for upcoming research program of the Korea Research Foundation(KRF-2002-003-D00262).

M. Conti et al. (Eds.): PWC 2003, LNCS 2775, pp. 835–840, 2003.

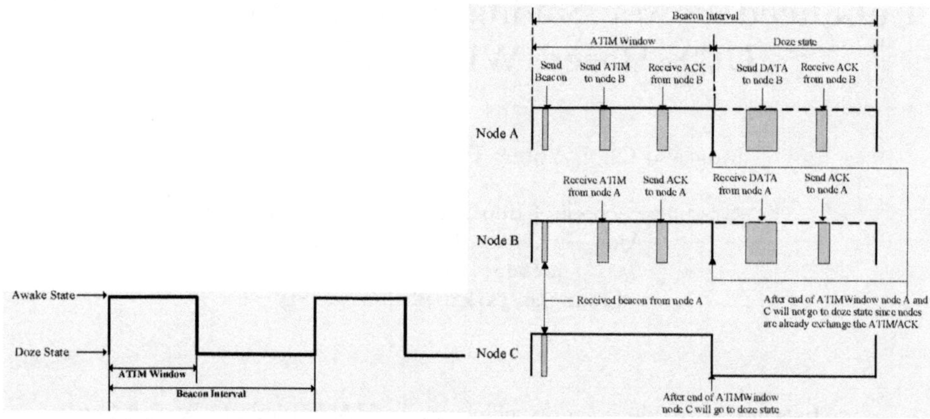

(a) Two operation states (b) Data exchange

Fig. 1. Operation in IEEE 802.11 DCF (PSM) mode

to achieve energy efficiency by powering off each node's wireless network interface in a certain amount of time[2,3] whenever needed. Thus, in the power saving mechanism, the wireless hosts are allowed to enter a "sleep" mode by putting their wireless interfaces in a doze state. The IEEE 802.11, standard for WLANs (Wireless Local Area Networks)[4], also provides a similar power saving mechanism using the doze-state and results in some energy conservation by allowing a node to be awake only for certain amount of time – this time is so called ATIMWindow and its value is known to affect the network throughput[5]. Clearly, the larger value of ATIMWindow a node has, the less energy saving it can achieve. Since the node has to be powered on for longer period of time with a shorter sleeping time in a doze state.

This paper proposes enhanced power saving scheme for IEEE 802.11 DCF (Distributed Coordination Function) mode. The new scheme, presented in this paper, guarantees the energy efficiency without degradation of throughput by utilizing the traffic information. According to our proposed scheme, the node can vary the duration of awake-state to adapt the traffic on the network.

2 Background and Motivation

In this section, we summarize the PSM (Power Saving Mode) in IEEE 802.11 DCF standard and present a motivation for our work by pointing out some limitation of the standard.

2.1 Power Saving Mechanism in IEEE 802.11 DCF

The basic approach of the power saving mechanism in IEEE 802.11 DCF is to periodically change power state between awake and doze, as shown in Fig. 1(a). The figure also shows two terms of "Beacon interval" and "ATIMWindow (Ad hoc Traffic Indication Message Window)". The beacon interval means a fixed interval for changing state to awake (i.e., all nodes must be awake at the beginning of every beacon interval) and the ATIMWindow means a fixed awake-period during each beacon interval. Note that, each node should be powered on during an ATIMWindow interval for any possible packet communication (beacon or ATIM). To achieve the change between awake and doze state periodically, all nodes need to be synchronized with their neighbors. For the synchronization, the IEEE 802.11 provides TSF (Timing Synchronization Function), which uses a beacon to notify the time stamp. Every node in DCF is responsible for generating a beacon at the beginning of every beacon interval, and each beacon contains the node's local timestamp. When beacon interval begins, all nodes compete for transmission of their beacon using the standard back-off algorithm. As an example lets assume that there are three nodes competing for sending their own beacons (see Fig. 1(b)). Here node A wins the competition so transmits its beacon. Upon receiving the beacon sent by node A, all other receiving nodes (node B and C) should cancel their beacon transmission schedule and adjust their local timer to the timestamp of the received beacon (i.e., node A's time value).

Fig. 1(b) also shows how data packets can be exchanged in IEEE 802.11 DCF power saving mode. In the figure node A is assumed to have data packets to unicast for node B and therefore transmits its ATIM packet to node B in ATIMWindow. Once the ATIM packet is successfully received, node B may acknowledge to node A with its ACK packet. Successful exchange of ATIM-ACK packets between two nodes implies that they can now exchange any buffered data packets and thus both nodes should stay awake until their data communication completes (i.e., until the next beacon interval). However, all other nodes that are not involved in the exchange phase of ATIM-ACK packets (Node C in our example) are allowed to enter a doze mode to sleep at the end of ATIMWindow so they can save their energy during the remaining beacon interval.

2.2 Motivation

As explained earlier the current IEEE 802.11 standard has a *fixed value* of ATIMWindow (and a beacon interval), and thus every node must stay awake for the same duration no matter how long to be actually involved in real data communication. Remind that, in the previous Fig. 1(b), node C had to stay awake for the whole ATIMWindow unnecessarily because it is not allowed to turn its radio off until this fixed awake duration. It is not difficult to imagine that such inflexibility can cause all nodes to awake and just waste their energies in the worst case when no nodes have data packets to send. We believe that one of the reasons causing this problem is not to consider the actual traffic information in the network. This observation leads us to propose a new energy-efficient MAC protocol that allows the nodes to go to sleep early in the case they are acknowledged that no data is buffered within an ad hoc network and therefore no data transmission will be taken place.

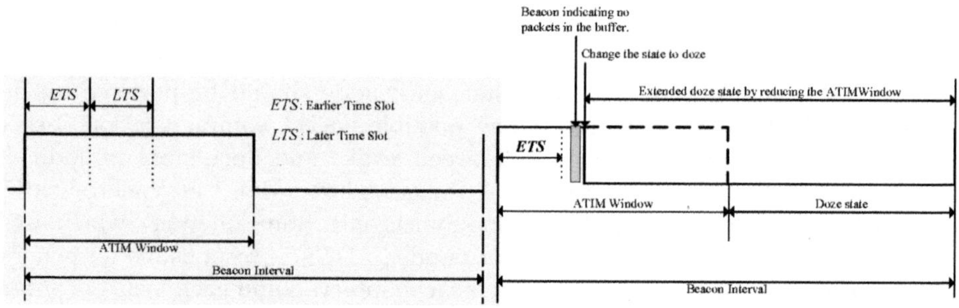

(a) Two time slots for beacon transmission (b) ATIMWindow size reduction

Fig. 2. Operation of TIPS

3 Proposed Power Saving Scheme

In this section, we present a basic idea of our proposed power saving scheme, TIPS (Traffic Indication-based Power Saving). The TIPS protocol is identical to PSM of IEEE 802.11 DCF, with some modification. We will explain these modifications by using the following example in Fig. 2.

In contrast to the original IEEE 802.11 DCF (where a beacon is used for a synchronization purpose only), the basic idea of TIPS is to give one more functionality on a beacon as a future data transmission indicator. To achieve this idea, we made one modification of varying the time point of a beacon transmission based on the beacon transmitter's buffer status. Thus, if the transmitter issuing a beacon has any data packet to send in its buffer, it tries to transmit the beacon in the earlier time slot(ETS in Fig. 2(a)) , otherwise, it tries to delay the transmission time in the later time slot (LTS in Fig. 2(a)). If no nodes transmit any beacon in the ETS, all nodes can assume that there are no nodes currently buffering data to send within a present beacon interval. This means that all nodes are allowed to change their state into a doze mode, right after receiving a beacon in the LTS (see Fig. 2(b)). Therefore, in the TIPS, any node buffering data packet must try to send its beacon in the ETS duration.

One of the important issues is how to decide an optimal size of ETS and LTS. There may be several approaches, but in this paper, we propose to use $aCWmax$ value, which is used in the standard random backs off algorithm of IEEE 802.11 as the maximum delay before issuing a beacon. Thus, each node having data packets backoff for a random time chosen from the interval $[0, aCWmax]$ before transmitting its beacon just like the IEEE 802.11 standard. However, this interval becomes $[aCWmax, 2 \times aCWmax]$ for some nodes having no data packet in their buffer – note that ATIMWindow size is larger than doubled $aCWmax$ value.

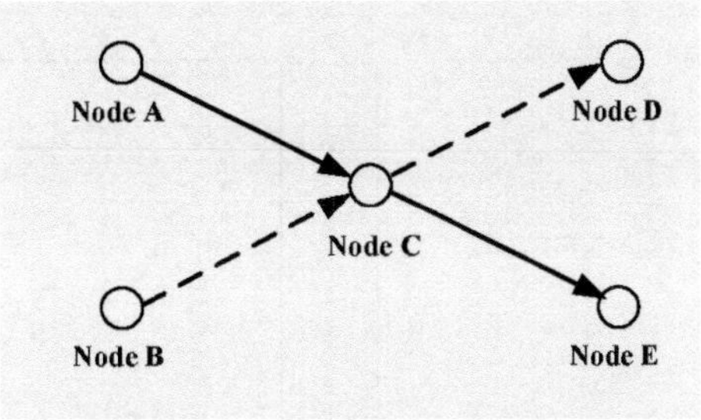

Fig. 3. Topology used in our simulation

4 Performance Evaluation

In this section we evaluate the performance of our TIPS, compared to the IEEE 802.11 PSM. For our simulations we have modified the CMU wireless extended version of ns-2.

4.1 Simulation Model

The objective of our proposed algorithm is to reduce the energy consumption during the idle time. Thus we use percentage of sleep mode as performance metric.

Fig. 3 shows a simple topology used for our preliminary simulation study. It is a two-hop network with two sources (node A and B) and two destinations (node D and E). Packets from node A and B are delivered through node C to node D and E, respectively. Each nodes are separated by 200 meters and transmission range of each node is 250 meters. We assume that the maximum bit rate of channel is 2Mbps and packet size is 512 Bytes. The value of beacon interval is 0.1 seconds. For measuring energy consumption, we use 1.4W, 0.95W, 0.805W, and 0.06W as value of power consumed by the MAC layer in transmit, receive, idle, and doze state respectively. In our simulation, after delivering 10 messages, we measure energy consumption and percentage of sleep time of two schemes (TIPS and PSM).

We vary the traffic load by having different inter-arrival period of the message issued from upper layer (from 0.1 to 1seconds). We also vary the value of ATIMWindow to observe the effect of its size (from 40 to 80ms).

4.2 Preliminary Simulation Result

Fig. 4(a) shows the percentage of sleep time when the ATIM window is 40ms. When there is little traffic, each node has more chance to make a longer sleep. However, by adjusting sleep time according to traffic information, TIPS always has more sleep time than PSM. This is because, when the traffic is sparse, the TIPS can reduce the ATIMWindow more often. Fig. 4(b) shows the energy consumption ratio between TIPS and PSM when varying the ATIMWindows and message inter-arrival. This value can be measured by dividing the energy

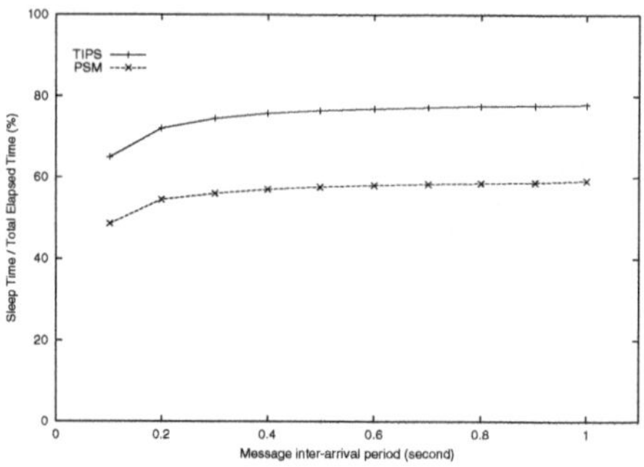

(a) Percentage of sleep time

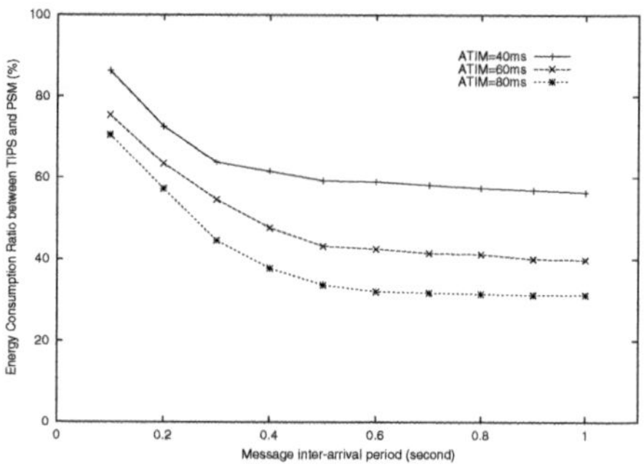

(b) Energy consumption ratio (TIPS/PSM)

Fig. 4. Simulation results

consumption of TIPS to that of PSM. According to the figure, when the network traffic becomes sparse and ATIMWindow is increased, TIPS is more efficient on the aspect of energy consumption. For instance, when the message-interarrival period is equal to 1 the energy improvement of TIPS against PSM is at least 40% (when ATIMWindow = 40ms).

5 Conclusions

In this paper, we propose an enhanced power saving scheme for IEEE 802.11 DCF by utilizing traffic information. To verify the excellence of proposed scheme, we have done some preliminary simulation study, and made very positive results of producing over 70% energy save in our TIPS compared to the IEEE 802.11 DCF method. Future works include further simulation studies with more realistic environments such as multihop scenarios.

References

1. E.-S. Jung and N. H. Vaidya, "A Power Control MAC Protocol for Ad Hoc Networks," *ACM Mobicom 2002,* Oct. 2002.
2. E.-S. Jung and N. H. Vaidya, "An Energy Efficient MAC Protocol for Wireless LANs," *IEEE INFOCOM 2002,* Jun. 2002.
3. Y.-C. Tseng, C.-S. Hsu, and T.-Y. Heieh, "Power-saving Protocols for IEEE 802.11-based Multihop Ad Hoc Networks," *IEEE INFOCOM 2002,* Jun. 2002.
4. IEEE Computer Society LAN MAN Standard Committee, "Wireless LAN medium access control and physical specification," *IEEE 802.11 Standard,* Aug. 1998.
5. H. Woesner, J.-P. Ebert, M. Schlager, and A. Wolisz, "Power-Saving Mechanisms in Emerging Standards for Wireless LANs: The MAC Level Perspective," *IEEE Personal Communication,* Jun. 1998.

Are Classes of Nodes with Different Power Amplifiers Good for Wireless Multi-hop Networks?*

Martin Kubisch, Holger Karl, and Adam Wolisz

Telecommunication Networks Group
Technische Universität Berlin
Sekr. FT 5-2, Einsteinufer 25
10587 Berlin, Germany
{kubisch,karl,wolisz}@ee.tu-berlin.de

Abstract. It has been shown that optimization of wireless network operations can be achieved by transmission power control of wireless nodes. Controlling the transmission power is mostly performed as a regulation of the final power amplifier, which is the major power sink. Depending on the desired transmission power level, currently popular amplifiers work at different levels of efficiency: highest in case of maximum emitted power, lower in case of reduced emitted power.
In this paper we present a novel approach for improving the liftime of mobile devices in multi-hop wireless networks by using nodes with *heterogeneous* efficiency characteristics of their power amplifiers.
To assess the potential of such amplifier-heterogeneous networks, we have performed multi-hop network simulations under the idealistic assumption of optimal routing based on global knowledge. The obtained results demonstrate an appealing potential of a reduction in energy consumption to less than 40 % compared to classical networks with uniform amplifiers.

Keywords: Multi-hop wireless networks, power control, sensor network.

1 Introduction

Today's popular power amplifiers used in wireless network cards are designed to have the highest power efficiency at the maximum output power. When the output power is reduced, the power efficiency of the amplifier decreases, i.e., the power consumed by an amplifier does not reduce with the same ratio as the output power decreases. As an example: the RF2155 power amplifier [1], which is designed for cordless phones and other application in the 915 MHz ISM band, has four different output power levels. The power efficiency of these power level range from 54 % efficiency for the highest level of output power to 1 % efficiency for the lowest level of output power.

Power amplifiers are usually chosen according to the maximum range one wants to overcome. Assuming a desired PER and other parameters of the receiver characteristic, this leads to a requirement on the output power necessary to overcome the maximal

* This work is funded by the German Ministry of Education and Research (BMBF) under the project IBMS[2].

M. Conti et al. (Eds.): PWC 2003, LNCS 2775, pp. 841–846, 2003.

distance. In realistic setups however, the distances between communicating nodes are variable. Thus, the usage of the highest output power is rarely necessary, even when direct communication between the sender and receiver is possible. Further options for using lower transmission power appear when (as in IEEE 802.11) the coding/modulation might be dynamically adjusted (slower data rates require lower transmission power to achieve the same PER). In addition, the usage of multi-hop schemes instead of direct communication offers even more freedom for using lower output power. It is well known that reducing the output power is beneficial because of both reduction of the energy of the transmitter as well as reduction of the potential interference with other nodes operating in the same frequency band.

The desired and beneficial reduction of the output power does, unfortunately enough, not lead to a proportional reduction of power used to drive the amplifier, as the amplifier is moving into a less efficient operation range. But, in fact, there is no physical rule that mandates that power amplifiers have the highest efficiency at the highest output power. A practical example for a different amplifier designs is CRIPPS [2], which presents amplifiers which are developed with *efficiency enhancement techniques* in mind. The presented "Doherty" amplifier is an example of an adaptation of power efficiency. It has the highest power efficiency at a power level 6 dB less than the maximum one. Obviously, using such amplifiers with shifted power efficiency would pay off if — mostly — lower output power is applied, even if un-proportionally high power consumption for the highest power level would be used.

As this could be extended with the use of dynamically adjustable characteristics of power amplifiers to result in an optimal selection for a specific network topology, we constrain ourselves only to nodes with two types of amplifiers, further referred to as short- or long-range specialists (note that even the short-range specialists in our model can transmit at the highest power if necessary, but the costs are comparably high). Intuitively, such a mixture of nodes might improve energy efficiency. In this paper, we want to assess whether this intuition actually holds: we study a varying percentage of long- and short-range specialist nodes at various node densities and study the resulting energy efficiency in a wireless network.

Our considerations are structured as follows: first we give a short overview of related work proving that a differentiation of output power level is an attractive way to optimize wireless networks. Further we describe the methodology of this study, details of our simulation set up and some encouraging results. As this paper pertains to work in progress, we complete our considerations with comments on ongoing investigations.

2 Related Work

There exist numerous possibilities for reduction of the communication energy in multi-hop wireless networks. We will not discuss the papers which put their nodes into sleep, even if this is an interesting functionality of some MAC designs (including IEEE 802.11); it can be combined with the approach discussed in this paper.

Relevant to this paper are approaches pertaining to the selection of the transmission power level. This might be globally unified for the whole network, e.g. with respect to desired connectivity [3] [4] or individually for each pair of nodes [5] [6] [7] [8] [9]. [10]

As shown in [5] and [3], the use of lower power level, which shorten the distance and require more intermediate hops, can increase the energy efficiency as well as the capacity of a network. Especially the consideration of the remaining battery capacity can lead to an extension in network lifetime. Another approach is the COMPOW protocol [4] which exchanges life messages on separate, discrete power levels to find a common power level throughout the network. Unlike these networks based on IEEE 802.11, is the search for the minimum power level of a node [6] a more general approach, which can be applied to low data rate networks as well, e.g., sensor networks. They determine the power level while relying on measurements of the received signal power strength. Additionally to this power level calculation, [7] uses a path recalculation in every node to determine if it could provide a more energy-efficient path.

A proven way of finding a minimum transmission power is done in [9] and [10]. While these algorithms rely either on location information or on angle of arrival, the algorithms in [8] do not. There, locally available information is used to adapt the power level in an sensor network and it is shown that this increases the network lifetime.

Out of this partial list one can see that setting different transmission power levels is beneficial for the energy consumption and other optimization criteria. When optimizing the power setting it is important to understand that mostly the emitted power (assuring a certain SNR at the receiver, defined by a desired PER) is used as the optimization criterion.

In all of these papers it has been assumed that the parameters of the amplifier of all nodes are *identical*, and only individual settings are possible. To our knowledge, this is the first paper considering a mixture of nodes with *different* types of amplifier characteristics.

3 Evaluation Based on Global Knowledge

The system scenario under consideration is a wireless multi-hop network, e.g., a mobile ad hoc network. Nodes want to communicate with other nodes over longer distances than is possible even with the highest transmission power, necessitating multi-hop communication. For such networks we need an ad hoc routing protocol which chooses energy-efficient routes and thus the nodes can use transmission power level small enough to safely communicate with their next hop neighbor.

The advantage of such energy-efficient routing schemes is that the reduction of distance, useful for multi-hop communication, implies a reduction of necessary transmission power, but as the efficiency of the final amplifier does decrease as well, the gain is not as high as it could be. Our hypothesis is that this shortcoming can be overcome by using different nodes in the network, equipped with different power amplifiers, optimized for different distances. We claim that such "specialists" for long- and short-range communication will increase the energy efficiency.

Ideally, an ad hoc routing protocol should be modified to exploit the different characteristics of such nodes, routing short-range communication over nodes with amplifiers that operate efficiently at low transmission powers and bridging long distances with other nodes. Developing such an extension to an ad hoc protocol is actually a non-trivial activity. Hence, to first obtain an understanding of the principle feasibility and benefits of this approach, we used a simplified approach and examine it.

In our approach we have a fixed number of uniform randomly distributed nodes, where each node has only two output power levels (the output power levels are the same for short- and long-range specialists), and we change the percentage of nodes which are short-range specialists. For each node and every power level in such a network the maximum transmission range is calculated. Using the global knowledge of all node's location, we determine which node can be reached and annotate the edges between the nodes with the energy consumed per packet as the cost function.

Then we use Dijkstra's algorithm to calculate the optimal path between any two nodes in the network, resulting in a forwarding table for every node which also contains the necessary transmission power level. Having these paths, we used the following traffic pattern to determine the total energy used: Every node randomly selects one destination node in the network and transfers one packet to it. At the end, the total consumed energy of all nodes serves as a figure of merit for different network densities and ratio of long-/ short-range specialists.

4 Model Assumptions

We have considered various model assumptions here. In all scenarios, 100 nodes are uniform randomly distributed in an area between 5 km $*$ 5 km and 18.25 km $*$ 18.25 km. This is equal to network densities between 4 (all nodes can communicate directly) and 0.3 nodes per km^2 (lower values result in partitioned networks). The efficiency model for the transmission power is based on values for the RF 2155 power amplifier: Each node has two transmission power levels of 70 mW and 447 mW. The long-range specialists have an efficiency of 54 % (consumed power of 826 mW) for the high power level and 20 % (consumed power of 337.5 mW) for the low power level. The short-range specialists have an efficiency of 20 % (consumed power of 2235 mW) for the high power level and 54 % (consumed power of 129.6 mW) for the low power level. The other values of -85 dBm receiver sensitivity (implying a PER of 1 %), 200 mW reception power and 200 mW computation power while transmitting are based on the "SieMo S50037 Bluetooth Module"[11]. The amplifier characteristic used in a particular node in a particular scenario depends on the percentage necessary for this scenario, e.g., out of this 100 nodes 10 are selected to be short-range specialists the others are long-range. For the traffic we used a packet size of 1500 byte as well as a immediate acknowledgement of 30 bytes at a data rate of 1 Mbit/s — taking into account the acknowledgment as well is important because of the heterogeneous energy costs of different devices!

5 Simulation Results

Figure 1(a) displays the total energy consumed, each averaged over 40 different random placements of nodes. A path loss coefficient of 3 was used. On the x-axis, the percentage of short-range specialists is displayed and on the y-axis the energy necessary for the used traffic pattern which results in a transfer of 100 packets. The lower line is the energy

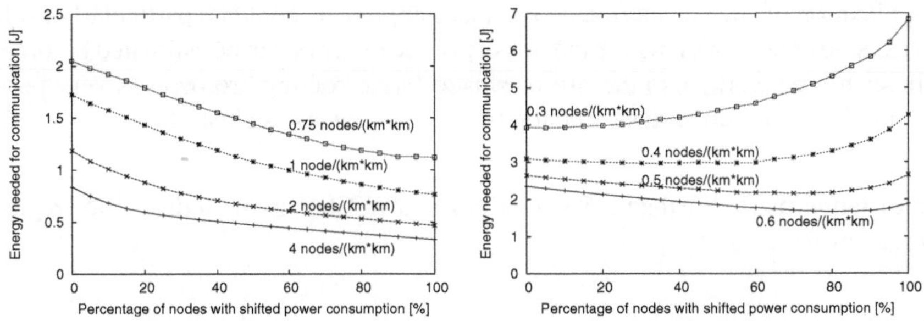

(a) High density (b) Low density

Fig. 1. Total consumed energy over percentage of short range specialists

average for a density of 4 nodes per km^2, the upper one for 0.75 nodes per km^2. The lines between are intermediate densities.

For the high-density networks, using only short-range specialists is beneficial and the energy needed is less then 38.4 % compared to a network having only long-range ones.

Figure 1(b) also displays the total energy consumed over the percentage of short-range specialists, but the networks are less dense. The lower line is the energy average for a density of 0.6 nodes per km^2, the upper one for 0.3 nodes per km^2.

The three lower lines are the energy curves where the network density is sparse and using only short-range specialists is not most beneficial. Instead, there is an optimal point depending on the density. For a density of 0.6 nodes per km^2 this ratio is 80 %, for 0.5 nodes per km^2 the ratio is around 70 % and for 0.4 nodes per km^2 it is around 30 %. When the density is 0.3 nodes per km^2 or smaller, it is not beneficial to use short-range specialists at all. Additionally, further reduction in density leads to another problem: the connectivity. When a much lower density is used, the probability of having a disconnected network becomes higher, thus further curves are left out.

6 Conclusions and Further Work

As the results show, heterogeneity of nodes with differently optimized power amplifiers are beneficial for energy efficiency. The best ratio between long- and short-range specialists depends on the density of the network, the node layout (one could imagine a network where particular nodes are of better use when they adapt their amplifier characteristic) and also, implicitly, on the characteristics of the different amplifiers. Thus, it is a complex problem, but we intend to tackle it step by step.

Although this idea was demonstrated for WLAN networks, we believe it could also be applied to other wireless networks, e.g., sensor networks. In these networks the problem is quite similar apart from the fact that other packet sizes and traffic patterns are used.

As a next step we intend to integrate the characteristics of such different short- and long-range specialists into ad hoc routing protocols. The particular challenge is to handle the different energy costs for both directions in a simple and efficient manner. Moreover, the application of such a heterogeneous nodes approach should be particularly useful in wireless sensor networks, where the density of the network can be estimated beforehand.

In such a network, also the use of cluster-based routing protocols is very popular, and it might be attractive to use the long-range specialists as clusterheads.

Acknowledgement. We thank Prof. G. Böck and W. Chen for their discussion regarding power amplifiers.

References

1. RF Micro Devices: RF 2155 3V Programmable Gain Power Amplifier. Designer's Handbook (2003) 2–173–2–180
2. Cripps, S.C.: RF Power Amplifiers for Wireless Communications – Efficiency Enhancement Techniques. Artech House Publishers, Boston, MA, USA (2000)
3. Monks, J.P., Ebert, J.P., Wolisz, A., Hwu, W.: A study of the energy saving and capacity improvement potential of power control in multi-hop wireless networks. In: Workshop on Wireless Local Networks, Tampa, Florida, USA, also Conf. of Local Computer Networks (LCN). (2001)
4. Narayanaswamy, S., Kawadia, V., Sreenivas, R.S., Kumar, P.R.: Power control in ad-hoc networks: Theory, architecture, algorithm and implementation of the COMPOW protocol. In: European Wireless Conference, Florence, Italy, EUREL, VDE (2002) 156–162
5. Kubisch, M., Mengesha, S., Hollos, D., Karl, H., Wolisz, A.: Applying ad-hoc relaying to improve capacity, energy efficiency, and immission in infrastructure-based WLANs. In Irmscher, K., ed.: Kommunikation in Verteilten Systemen (KiVS), 13. ITG/GI-Fachtagung, Leipzig, Germany, Informatik Aktuell, Springer (2003)
6. Bergamo, P., Maniezzo, D., Giovanardi, A., Mazzini, G., Zorzi, M.: Distributed power control for power-aware energy-efficient routing in ad hoc networks. In: European Wireless Conference, Florence, Italy, EUREL, VDE (2002) 237–243
7. Doshi, S., Bhandare, S., Brown, T.X.: An on-demand minimum energy routing protocol for a wireless ad hoc network. Mobile Computing and Communications Review **6** (2002) 50–66
8. Kubisch, M., Karl, H., Wolisz, A., Zhong, L.C., Rabaey, J.: Distributed algorithms for transmission power control in wireless sensor networks. In: Wireless Communications and Networking Conference (WCNC), New Orleans, LA, IEEE (2003)
9. Rodoplu, V., Meng, T.H.Y.: Minimum energy mobile wireless networks. IEEE Journal on Selected Areas in Communications **17** (1999) 1333–1344
10. Wattenhofer, R., Li, L., Bahl, P., Wang, Y.M.: Distributed topology control for power efficient operation in multihop wireless ad hoc networks. In: INFOCOM – The Conference on Computer Communications, Anchorage, Ak, IEEE (2001)
11. Siemens AG Austria: Siemens S50037 Bluetooth Module (2001)

Solar-Aware Routing in Wireless Sensor Networks

Thiemo Voigt, Hartmut Ritter, and Jochen Schiller

Institut für Informatik,
Freie Universität Berlin,
Takustr. 9, D-14195 Berlin, Germany
{voigt,hritter,schiller}@inf.fu-berlin.de

Abstract. In wireless sensor networks energy is a valuable but not necessary scarce resource. While it is often assumed that nodes in sensor networks are powered by batteries, other energy sources such as solar power may provide unlimited energy resources to a changing subset of the nodes. Since these nodes can receive and transmit packets without consuming battery power, routing via these nodes is appealing. In this paper, we present solar-aware routing, a routing protocol for wireless sensor networks that preferably routes traffic via nodes powered by solar energy. Simulations show that solar-aware routing can provide significant energy savings in many scenarios.

1 Introduction

Routing in wireless sensor networks has to take into account the very limited resources of the nodes. While many researchers assume that all nodes in a sensor network are battery-driven [1], [5], [6], nodes can also be powered by other energy sources such as gravitation or solar power. Nodes powered by such a source can receive and transmit packets without consuming battery energy. Therefore, routing packets via such nodes is appealing. It is, however, complicated by the fact, that the energy source is not permanent. In our lab, we have developed sensor boards which we will soon equip with solar cells. This motivated us to investigate if preferably routing via solar-powered nodes is feasible and can provide energy savings. We present simulation results using a protocol we call solar-aware routing. As other routing protocols for sensor networks this protocol is mainly based on local interactions between adjacent nodes. Our results indicate that solar-aware routing can provide significant energy savings in certain scenarios.

Our routing protocol has similarities to directed diffusion [2] and reactive protocols for ad-hoc routing such as AODV. Willig et al. have developed a routing protocol that considers nodes with permanent power supply [4] while our protocols assumes that the set of nodes having "unlimited" solar energy resources changes over time.

This paper presents our hardware (Section 2), the routing protocol (Section 3) as well as simulation results (Section 4) and concludes with conclusions and future work.

M. Conti et al. (Eds.): PWC 2003, LNCS 2775, pp. 847–852, 2003.

Fig. 1. Sensor board with attached solar cell.

2 Sensor Board Hardware

The hardware used in our labs consists of a Texas Instruments MSP430 controller as core and a set of associated sensor hardware. These sensors are:

- A light sensor for the detection of visible light
- A passive infrared sensor for detection of movement
- A temperature sensor
- A gravitation sensor for the detection of movement of the sensor board
- A microphone for determination of the ambient noise level

The sensors communicate via an RF module in the 868 MHz band. The sensor board can be operated in three different power modes. The energy needed even in the most power-intensive mode (ca. 40 mA) can be provided by a solar cell. Fig. 1 shows the sensor board and an attached solar cell. Note that much smaller solar cells could be used as long as they provide enough power. The first generation of the sensor board has the size of three AAA batteries.

Thus the hardware demonstrates that an operation mode we further call *solar-powered* is possible. We are currently building large quantities of the sensor boards (see `www.scatterweb.net`).

3 Protocol Description

As in directed diffusion, data propagation in this version of our protocol is based on localized interactions, i.e. interactions with neighbouring nodes. In the protocol description, we use terminology known from directed diffusion. A *sink* is a node that is interested in some information a sensor in the network might be able to deliver, i.e. the node has a certain *interest*. A sensor node being able to deliver the desired information is called a *source*. Sinks express their interest by sending an *Interest message* into the network. The requested data is named as an attribute-value pair.

While interest messages are propagated towards a potential source, nodes set up *gradients*. A gradient denotes state information about the sensor data the sink desires and about the neighbours, which we call other nodes in transmission range. The gradient also contains information about the *solar state* of the neighbours that denotes if a node is battery-driven or if it is running on solar power. A source delivers its information back to the sink using *Data messages.*

As in directed diffusion, our protocol assumes that at least adjacent nodes can be distinguished by identifiers.

3.1 Interest Propagation

Interest messages are generated by a sink and are simply flooded through the network. Note that other methods such as directional flooding based on location or previously cached data are more efficient and could be used as well [2]. Interest messages contain the attribute value pair, source and destination identifier, an interval that states the rate at which the sink wants to receive sensor data as well as a hop counter and the solar state of the sender. Nodes establish gradients to all other nodes from which they receive interest messages. A gradient to a neighbour is updated when an interest message for the same interest is received with a smaller hop count.

For every interest, nodes rebroadcast the first interest message they receive as well as following interests with a lower hop count. In most of the cases, however, the first interest message should have the lowest hop count.

3.2 Data Propagation

When a source receives an interest, it senses the environment at the specified rate and sends a data message towards the sink. A data message contains the attribute-value pair that allows the intermediate nodes to identify the gradients and the next hop towards the sink. When choosing the next hop, care must be taken to avoid loops. Propagating data along the shortest path is one method to avoid a loop. However, if there is exactly one shortest path we will not be able to preferably route via solar-powered nodes. Therefore, we allow the source and exactly one intermediate node to choose a next hop different from the one closest to the sink. The latter nodes are those neighbours from which a node has received the smallest hop counter in the interest message. A flag in the data message indicates if a node already has utilized the option to route via an intermediate node not on the shortest path to the sink. If this is not the case, an intermediate node may choose to forward the data to another node, that is solar-powered rather than a node on the shortest path. Once the flag is set an intermediate nodes must forward the message to a neighbour that is on the shortest path. If one of its neighbours is both on the shortest path and solar-powered, intermediate nodes forward the message to such a node.

3.3 Status Updates

When nodes change their solar status, they should inform the nodes closer to the source about their new status to make them route via them if they are solar-powered or to give them the possibility to route via a solar-powered node instead.

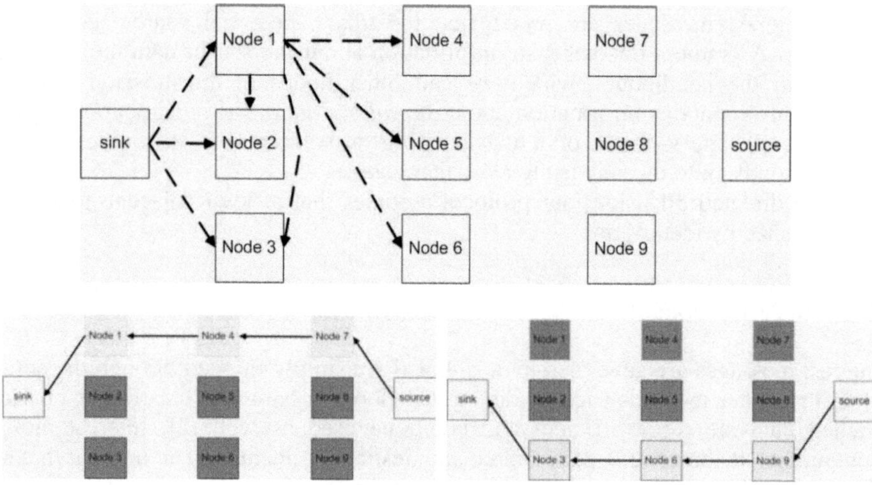

Fig. 2. Experiment with simple topology

There are two design options here. One is that a node simply broadcasts its new status; the other option is that it waits until it hears some communication going on or itself receives a data packet. The advantage of the first option is that a more energy-efficient route can be found at once while using the second option some data messages might still be routed via a less energy efficient route. The advantage of the second option is that it avoids unnecessary broadcasts, meaning broadcasts that do not lead to more energy-efficient routes.

4 Experiments

We have conducted simulation studies using the OMNet++ discrete event simulator [3]. The aim of the study was to investigate if solar-aware routing provides energy savings by preferably routing via solar-powered nodes.

4.1 Experiment with a Simple Topology

The setup for the first experiment is shown in Figure 2. In this experiment, each node is in the transmission range of the nodes in the same column and the neighbouring columns. For example, the sink is in the transmission range of the nodes 1 to 3 and vice versa. Node 7 is in the transmission range of the nodes 4-6, 8-9 and the source. In the experiment the sun is moving from North to Source, i.e. in the beginning node 1, 4 and 7 are solar-powered. When the interest messages are flooded from the sink to the source as depicted in the top part of the figure, each node learns which adjacent nodes are solar-powered. The same is true for the source that therefore transmits the first data message to node 7. Since node 8 and 9 are also in the transmission range, they receive this data message as well, but do not forward it. We have not yet designed a sleeping protocol which would enable nodes 8 and 9 to sleep. It is obvious that

Table 1. Improvement with solar-aware routing

Number of nodes	20	50
Improvement (%)	26,8	6,9

assuming there is at least one solar-powered node in each column, solar-aware routing is able to find a path using solar-powered nodes only. The bottom left of Figure 2 shows the data flow when nodes 1, 4 and 7 are solar-powered, the bottom right of the figure shows the data flow when nodes 3, 6 and 9 are solar-powered.

4.2 Experiments with Random Topologies

We generated random topologies with the nodes distributed randomly across a certain area *a* for a given number of nodes. The sink and source are placed in the middle of the eastern and western boundary of this area *a*. The transmission range of the nodes is fixed. In our scenarios we let the sun move across the area from North to South, covering an area which is about a third to a fifth of the area *a*. We compare solar-aware routing with shortest-path routing, i.e. a route that would be found by directed diffusion. For comparison we use a simple metric, namely the number of data messages that must be transmitted using battery power. We do not include the interest messages in our comparison since they do not differ between solar-routing and shortest path routing. However, we include the extra status messages generated by solar-aware routing in our comparison. The sensor networks contain 20 and 50 intermediate nodes.

As shown in Table 1, with 20 nodes, the average improvement of solar-aware routing over shortest-path routing is 26.8 %. For several scenarios, the improvement was more than 40%, while for other scenarios the improvements were only marginal. The latter scenarios often contained one bottleneck "link" each packet had to pass and thus solar-aware routing was not able to deviate from the shortest path for more than a few hops.

With 50 nodes, solar-aware routing was only 6.9 % better than shortest-path routing. Also here, the improvement depended on the random scenario. We saw improvements of 15 % to 20% but also only marginal improvements. However, solar-aware routing was only in very rare cases (and in these negligibly) worse than shortest path routing. The improvement with 50 nodes is less than with 20 nodes, because solar-aware routing based on localized interactions cannot take advantage of paths containing solar-driven nodes in the northern or southern areas.

5 Conclusions and Future Work

We have presented solar-aware routing, a routing protocol for wireless sensor networks that preferably routes traffic via nodes powered by solar energy. Our simulations show that this can provide significant energy savings in many scenarios.

Localized interactions are an important feature that enables the energy-efficiency of directed diffusion [5]. While for solar-aware routing this is true for small networks, for larger networks other methods need to be explored. Routing based solely on local interactions in a large sensor network makes it, for example, almost impossible to route via the nodes in the North if only these are solar powered and the shortest path is more towards the South. We are currently designing and exploring protocols with more global knowledge suitable for larger sensor networks. Among other tasks of future work, we are considering multiple sinks and sources as well as the definition of a suitable protocol that enables nodes to decide when to go into sleep mode.

References

1. Akyildiz, I., Su, W., Sankarasubramaniam, Y., Cayirci, E.: Wireless Sensor Networks: a survey. Computer Networks, 38 (2002) 393–422
2. Intanagonwiwat, C., Govindan, R., Estrin, D., Heidemann, J., Silva, F.: Directed Diffusion for Wireless Sensor Networking. Transactions on Networking, Feb. 2003
3. Varga, A.: The OMNeT++ Discrete Event Simulation System. European Simulation Multiconference, Prague, Czech Republic, June 2001
4. Willig A., Shah R., Rabaey, J., Wolisz, A.: Altruists in the PicoRadio Sensor Network. Factory Communication Systems, Västerås, Sweden, August 2002
5. Zhao, Y., Govidan R., Estrin, D.: Residual Energy Scan for Monitoring Sensor Networks. Wireless Communications and Networking Conference, Orlando, FL, USA, March, 2002
6. Heinzelman, W., Chandrakasan, A., Balakrishnan, H.: Energy-Efficient Communication Protocol for Wireless Microsensor Networks, International Conference on System Sciences, Hawaii, January 2000

Author Index

Lecture Notes in Computer Science

For information about Vols. 1–2741
please contact your bookseller or Springer-Verlag

Vol. 2742: R. N. Wright (Ed.), Financial Cryptography. Proceedings, 2003. VIII, 321 pages. 2003.

Vol. 2743: L. Cardelli (Ed.), ECOOP 2003 – Object-Oriented Programming. Proceedings, 2003. X, 501 pages. 2003.

Vol. 2744: V. Mařík, D. McFarlane, P. Valckenaers (Eds.), Holonic and Multi-Agent Systems for Manufacturing. Proceedings, 2003. XI, 322 pages. 2003. (Subseries LNAI).

Vol. 2745: M. Guo, L.T. Yang (Eds.), Parallel and Distributed Processing and Applications. Proceedings, 2003. XII, 450 pages. 2003.

Vol. 2746: A. de Moor, W. Lex, B. Ganter (Eds.), Conceptual Structures for Knowledge Creation and Communication. Proceedings, 2003. XI, 405 pages. 2003. (Subseries LNAI).

Vol. 2747: B. Rovan, P. Vojtáš (Eds.), Mathematical Foundations of Computer Science 2003. Proceedings, 2003. XIII, 692 pages. 2003.

Vol. 2748: F. Dehne, J.-R. Sack, M. Smid (Eds.), Algorithms and Data Structures. Proceedings, 2003. XII, 522 pages. 2003.

Vol. 2749: J. Bigun, T. Gustavsson (Eds.), Image Analysis. Proceedings, 2003. XXII, 1174 pages. 2003.

Vol. 2750: T. Hadzilacos, Y. Manolopoulos, J.F. Roddick, Y. Theodoridis (Eds.), Advances in Spatial and Temporal Databases. Proceedings, 2003. XIII, 525 pages. 2003.

Vol. 2751: A. Lingas, B.J. Nilsson (Eds.), Fundamentals of Computation Theory. Proceedings, 2003. XII, 433 pages. 2003.

Vol. 2752: G.A. Kaminka, P.U. Lima, R. Rojas (Eds.), RoboCup 2002: Robot Soccer World Cup VI. XVI, 498 pages. 2003. (Subseries LNAI).

Vol. 2753: F. Maurer, D. Wells (Eds.), Extreme Programming and Agile Methods – XP/Agile Universe 2003. Proceedings, 2003. XI, 215 pages. 2003.

Vol. 2754: M. Schumacher, Security Engineering with Patterns. XIV, 208 pages. 2003.

Vol. 2756: N. Petkov, M.A. Westenberg (Eds.), Computer Analysis of Images and Patterns. Proceedings, 2003. XVIII, 781 pages. 2003.

Vol. 2758: D. Basin, B. Wolff (Eds.), Theorem Proving in Higher Order Logics. Proceedings, 2003. X, 367 pages. 2003.

Vol. 2759: O.H. Ibarra, Z. Dang (Eds.), Implementation and Application of Automata. Proceedings, 2003. XI, 312 pages. 2003.

Vol. 2761: R. Amadio, D. Lugiez (Eds.), CONCUR 2003 - Concurrency Theory. Proceedings, 2003. XI, 524 pages. 2003.

Vol. 2762: G. Dong, C. Tang, W. Wang (Eds.), Advances in Web-Age Information Management. Proceedings, 2003. XIII, 512 pages. 2003.

Vol. 2763: V. Malyshkin (Ed.), Parallel Computing Technologies. Proceedings, 2003. XIII, 570 pages. 2003.

Vol. 2764: S. Arora, K. Jansen, J.D.P. Rolim, A. Sahai (Eds.), Approximation, Randomization, and Combinatorial Optimization. Proceedings, 2003. IX, 409 pages. 2003.

Vol. 2765: R. Conradi, A.I. Wang (Eds.), Empirical Methods and Studies in Software Engineering. VIII, 279 pages. 2003.

Vol. 2766: S. Behnke, Hierarchical Neural Networks for Image Interpretation. XII, 224 pages. 2003.

Vol. 2768: M.J. Wilson, R.R. Martin (Eds.), Mathematics of Surfaces. Proceedings, 2003. VIII, 393 pages. 2003.

Vol. 2769: T. Koch, I. T. Sølvberg (Eds.), Research and Advanced Technology for Digital Libraries. Proceedings, 2003. XV, 536 pages. 2003.

Vol. 2773: V. Palade, R.J. Howlett, L. Jain (Eds.), Knowledge-Based Intelligent Information and Engineering Systems. Proceedings, Part I, 2003. LI, 1473 pages. 2003. (Subseries LNAI).

Vol. 2774: V. Palade, R.J. Howlett, L. Jain (Eds.), Knowledge-Based Intelligent Information and Engineering Systems. Proceedings, Part II, 2003. LI, 1443 pages. 2003. (Subseries LNAI).

Vol. 2775: M. Conti, S. Giordano, E. Gregori, S. Olariu (Eds.), Personal Wireless Communications. Proceedings, 2003. XVI, 855 pages. 2003.

Vol. 2776: V. Gorodetsky, L. Popyack, V. Skormin (Eds.), Computer Network Security. Proceedings, 2003. XIV, 470 pages. 2003.

Vol. 2777: B. Schölkopf, M.K. Warmuth (Eds.), Learning Theory and Kernel Machines. Proceedings, 2003. XIV, 746 pages. 2003. (Subseries LNAI).

Vol. 2778: P.Y.K. Cheung, G.A. Constantinides, J.T. de Sousa (Eds.), Field-Programmable Logic and Applications. Proceedings, 2003. XXVI, 1179 pages. 2003.

Vol. 2779: C.D. Walter, Ç.K. Koç, C. Paar (Eds.), Cryptographic Hardware and Embedded Systems – CHES 2003. Proceedings, 2003. XIII, 441 pages. 2003.

Vol. 2780: M. Dojat, E. Keravnou, P. Barahona (Eds.), Artificial Intelligence in Medicine. Proceedings, 2003. XIII, 388 pages. 2003. (Subseries LNAI).

Vol. 2781: B. Michaelis, G. Krell (Eds.), Pattern Recognition. Proceedings, 2003. XVII, 621 pages. 2003.

Vol. 2782: M. Klusch, A. Omicini, S. Ossowski, H. Laamanen (Eds.), Cooperative Information Agents VII. Proceedings, 2003. XI, 345 pages. 2003. (Subseries LNAI).

Vol. 2783: W. Zhou, P. Nicholson, B. Corbitt, J. Fong (Eds.), Advances in Web-Based Learning – ICWL 2003. Proceedings, 2003. XV, 552 pages. 2003.

Vol. 2786: F. Oquendo (Ed.), Software Process Technology. Proceedings, 2003. X, 173 pages. 2003.

Vol. 2787: J. Timmis, P. Bentley, E. Hart (Eds.), Artificial Immune Systems. Proceedings, 2003. XI, 299 pages. 2003.

Vol. 2788: S. Anderson, M. Felici, B. Littlewood (Eds.), Computer Safety, Reliability, and Security. Proceedings, 2003. XIX, 426 pages. 2003.

Vol. 2789: L. Böszörményi, P. Schojer (Eds.), Modular Programming Languages. Proceedings, 2003. XIII, 271 pages. 2003.

Vol. 2790: H. Kosch, L. Böszörményi, H. Hellwagner (Eds.), Euro-Par 2003 Parallel Processing. Proceedings, 2003. XXXV, 1320 pages. 2003.

Vol. 2792: T. Rist, R. Aylett, D. Ballin, J. Rickel (Eds.), Intelligent Virtual Agents. Proceedings, 2003. XV, 364 pages. 2003. (Subseries LNAI).

Vol. 2794: P. Kemper, W. H. Sanders (Eds.), Computer Performance Evaluation. Proceedings, 2003. X, 309 pages. 2003.

Vol. 2795: L. Chittaro (Ed.), Human-Computer Interaction with Mobile Devices and Services. Proceedings, 2003. XV, 494 pages. 2003.

Vol. 2796: M. Cialdea Mayer, F. Pirri (Eds.), Automated Reasoning with Analytic Tableaux and Related Methods. Proceedings, 2003. X, 271 pages. 2003. (Subseries LNAI).

Vol. 2798: L. Kalinichenko, R. Manthey, B. Thalheim, U. Wloka (Eds.), Advances in Databases and Information Systems. Proceedings, 2003. XIII, 431 pages. 2003.

Vol. 2799: J.J. Chico, E. Macii (Eds.), Integrated Circuit and System Design. Proceedings, 2003. XVII, 631 pages. 2003.

Vol. 2801: W. Banzhaf, T. Christaller, P. Dittrich, J.T. Kim, J. Ziegler (Eds.), Advances in Artificial Life. Proceedings, 2003. XVI, 905 pages. 2003. (Subseries LNAI).

Vol. 2803: M. Baaz, J.A. Makowsky (Eds.), Computer Science Logic. Proceedings, 2003. XII, 589 pages. 2003.

Vol. 2804: M. Bernardo, P. Inverardi (Eds.), Formal Methods for Software Architectures. Proceedings, 2003. VII, 287 pages. 2003.

Vol. 2805: K. Araki, S. Gnesi, D. Mandrioli (Eds.), FME 2003: Formal Methods. Proceedings, 2003. XVII, 942 pages. 2003.

Vol. 2806: J. Favela, D. Decouchant (Eds.), Groupware: Design, Implementation, and Use. Proceedings, 2003. XII, 382 pages. 2003.

Vol. 2807: V. Matoušek, P. Mautner (Eds.), Text, Speech and Dialogue. Proceedings, 2003. XIII, 426 pages. 2003. (Subseries LNAI).

Vol. 2810: M.R. Berthold, H.-J. Lenz, E. Bradley, R. Kruse, C. Borgelt (Eds.), Advances in Intelligent Data Analysis V. Proceedings, 2003. XV, 624 pages. 2003.

Vol. 2811: G. Karlsson, M.I. Smirnov (Eds.), Quality for All. Proceedings, 2003. XII, 295 pages. 2003.

Vol. 2812: G. Benson, R. Page (Eds.), Algorithms in Bioinformatics. Proceedings, 2003. X, 528 pages. 2003. (Subseries LNBI).

Vol. 2815: Y. Lindell, Composition of Secure Multi-Party Protocols. XVI, 192 pages. 2003.

Vol. 2816: B. Stiller, G. Carle, M. Karsten, P. Reichl (Eds.), Group Communications and Charges. Proceedings, 2003. XIII, 354 pages. 2003.

Vol. 2817: D. Konstantas, M. Leonard, Y. Pigneur, S. Patel (Eds.), Object-Oriented Information Systems. Proceedings, 2003. XII, 426 pages. 2003.

Vol. 2818: H. Blanken, T. Grabs, H.-J. Schek, R. Schenkel, G. Weikum (Eds.), Intelligent Search on XML Data. XVII, 319 pages. 2003.

Vol. 2819: B. Benatallah, M.-C. Shan (Eds.), Technologies for E-Services. Proceedings, 2003. X, 203 pages. 2003.

Vol. 2820: G. Vigna, E. Jonsson, C. Kruegel (Eds.), Recent Advances in Intrusion Detection. Proceedings, 2003. X, 239 pages. 2003.

Vol. 2821: A. Günter, R. Kruse, B. Neumann (Eds.), KI 2003: Advances in Artificial Intelligence. Proceedings, 2003. XII, 662 pages. 2003. (Subseries LNAI).

Vol. 2822: N. Bianchi-Berthouze (Ed.), Databases in Networked Information Systems. Proceedings, 2003. X, 271 pages. 2003.

Vol. 2823: A. Omondi, S. Sedukhin (Eds.), Advances in Computer Systems Architecture. Proceedings, 2003. XIII, 409 pages. 2003.

Vol. 2824: Z. Bellahsène, A.B. Chaudhri, E. Rahm, M. Rys, R. Unland (Eds.), Database and XML Technologies. Proceedings, 2003. X, 283 pages. 2003.

Vol. 2825: W. Kuhn, M. Worboys, S. Timpf (Eds.), Spatial Information Theory. Proceedings, 2003. XI, 399 pages. 2003.

Vol. 2827: A. Albrecht, K. Steinhöfel (Eds.), Stochastic Algorithms: Foundations and Applications. Proceedings, 2003. VIII, 167 pages. 2003.

Vol. 2830: F. Pfenning, Y. Smaragdakis (Eds.), Generative Programming and Component Engineering. Proceedings, 2003. IX, 397 pages. 2003.

Vol. 2831: M. Schillo, M. Klusch, J. Müller, H. Tianfield (Eds.), Multiagent System Technologies. Proceedings, 2003. X, 229 pages. 2003. (Subseries LNAI).

Vol. 2832: G. Di Battista, U. Zwick (Eds.), Algorithms – ESA 2003. Proceedings, 2003. XIV, 790 pages. 2003.

Vol. 2834: X. Zhou, S. Jähnichen, M. Xu, J. Cao (Eds.), Advanced Parallel Processing Technologies. Proceedings, 2003. XIV, 679 pages. 2003.

Vol. 2836: S. Qing, D. Gollmann, J. Zhou (Eds.), Information and Communications Security. Proceedings, 2003. XI, 416 pages. 2003.

Vol. 2837: N. Lavrač, D. Gamberger, H. Blockeel, L. Todorovski (Eds.), Machine Learning: ECML 2003. Proceedings, 2003. XVI, 504 pages. 2003. (Subseries LNAI).

Vol. 2838: N. Lavrač, D. Gamberger, L. Todorovski, H. Blockeel (Eds.), Knowledge Discovery in Databases: PKDD 2003. Proceedings, 2003. XVI, 508 pages. 2003. (Subseries LNAI).

Vol. 2839: A. Marshall, N. Agoulmine (Eds.), Management of Multimedia Networks and Services. Proceedings, 2003. XIV, 532 pages. 2003.

Vol. 2840: J. Dongarra, D. Laforenza, S. Orlando (Eds.), Recent Advances in Parallel Virtual Machine and Message Passing Interface. Proceedings, 2003. XVIII, 693 pages. 2003.

Vol. 2849: N. García, J.M. Martínez, L. Salgado (Eds.), Visual Content Processing and Representation. Proceedings, 2003. XII, 352 pages. 2003.

Vol. 2850: M.Y. Vardi, A. Voronkov (Eds.), Logic for Programming, Artificial Intelligence, and Reasoning. Proceedings, 2003. XIII, 437 pages. 2003. (Subseries LNAI)

Vol. 2853: M. Jeckle, L.-J. Zhang (Eds.), Web Services – ICWS-Europe 2003. Proceedings, 2003. VIII, 227 pages. 2003.

GPSR Compliance

The European Union's (EU) General Product Safety Regulation (GPSR) is a set of rules that requires consumer products to be safe and our obligations to ensure this.

If you have any concerns about our products, you can contact us on ProductSafety@springernature.com

In case Publisher is established outside the EU, the EU authorized representative is:

Springer Nature Customer Service Center GmbH
Europaplatz 3
69115 Heidelberg, Germany

Batch number: 10682269

Printed by Printforce, the Netherlands